DYNAMICS

Engineering Mechanics

Second Edition

Benson H. Tongue

University of California at Berkeley

SI Version

WILEY John Wiley & Sons, Inc.

ISBN: 978-0-470-55304-6

Printed in Asia

10 9 8 7 6 5 4 3 2 1

Mechanics courses have historically confronted engineering students as precise, mathematical, and, all too frequently, dry and pretty boring. This approach has appeal in that it presents mechanics as a relatively uncluttered "science," but the material often comes across as a mysterious body of facts and "tricks" that allow the student to attack idealized problems having little obvious relevance to the real world. When confronted with more realistic systems, students are often at a loss as to how to proceed. What is lacking is an appreciation for and understanding of the material that will empower students to tackle meaningful problems at an early stage in their undergraduate education.

In this book I've tried to present the best of both worlds—combining rigor with "user friendliness." Dynamics is one of the toughest subjects to "get" and it can be a major frustration to people who don't immediately relate to the logic behind the material (and this includes most everybody!) Thus the presentation in this text is a very personalized one; one in which you'll feel that you're having a one-on-one discussion with me. This approach minimizes the air of mystery that a more austere presentation can create, and I believe very strongly that it will aid you immensely in your ability to actually learn the material, which is, after all, the whole point of the exercise.

Be reassured that I'm not going to skimp on rigor but at the same time I'll be working tirelessly to make the material as accessible as possible and, as far as I can, fun to learn. People learn best when they're interested, and I'll definitely try to keep it interesting.

Features

The goals outlined above are supported by a number of unique features in this text:

Emphasis on sketching: The importance of communicating solutions through graphics is continuously emphasized. Most engineering students, indeed most people, are visual learners.[1] This fact, coupled with the importance of graphical information and communication in engineering practice (e.g., the use of sketching during conceptual design), make graphical representation of information a key element of the book. To reinforce the importance of drawing, this volume includes:

a. An innovative illustration format that uses engineering graph paper background and a ***hand-sketched*** look that reinforces in students the notion of how they themselves should be documenting their solutions. An ideal response from a reader regarding a graphical element of the book would be, "the sketch in Figures 7.14 and 7.15 made the concept more understandable AND I think that I could create a similar drawing to illustrate the concept to someone else."

Figure 7.13 Author riding up Marin Avenue

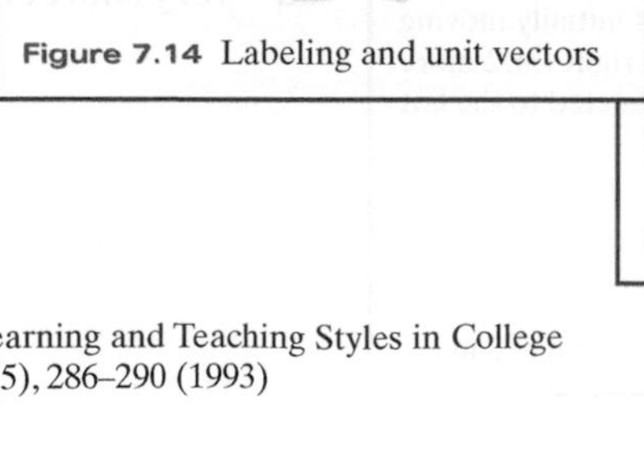

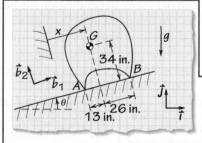

Figure 7.14 Labeling and unit vectors

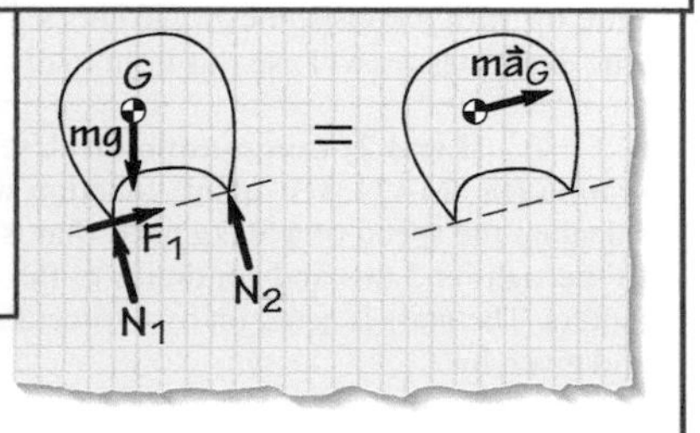

Figure 7.15 FBD=IRD

[1] Felder, Richard, "Reaching the Second Tier: Learning and Teaching Styles in College Science Education." *J. College Science Teaching*, 23(5), 286–290 (1993)

b. A **Draw** step is included in every worked example, oftentimes with commentary on how to create the most useful sketch.

To reinforce the drawing concept, vectors in the "hand-drawn" figures appear with an "arrow over top" notation, mirroring how they would be drawn in a hand-written solution.

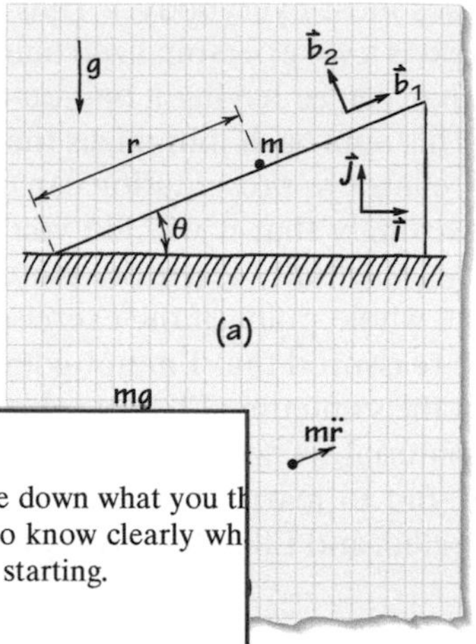

c and FBD=IRD

Draw **Figure 3.8*a*** shows a schematic of our system. The bowl is represented by a mass particle m. Two sets of unit vectors are shown: $\boldsymbol{i}, \boldsymbol{j}$ for right-left/up-down motion and $\boldsymbol{b}_1, \boldsymbol{b}_2$ for motion along the slope and orthogonal to it. The coordinate r is introduced to track motion along the slope.

Our FBD and IRD are shown in **Figure 3.8*b***. The gravity force (mg) is oriented vertically, but the two reaction forces created by the bowl/slope interface (N_1 and N_2) are aligned with the $\boldsymbol{b}_1$ and $\boldsymbol{b}_2$ directions. (Recall from Statics that when you break a system into a free body you will have, in general, three orthogonal forces at the connection point or, in the planar case, two.)

Formulate Equations Because the motion will be along the slope (in the $\boldsymbol{b}_1$ direction), it makes sense to express our equation of motion in the $\boldsymbol{b}_1, \boldsymbol{b}_2$ coordinate set. Applying (3.1) gives us

$$m\ddot{r}\boldsymbol{b}_1 = -mg\boldsymbol{j} + N_1\boldsymbol{b}_1 + N_2\boldsymbol{b}_2$$

To re-express the gravity term, we use a coordinate transformation array

Goal

- This is where you write down what you t find. It always helps to know clearly wh before jumping in and starting.

Given

- Here's where you remind yourself of wha given with which to solve the problem.

Draw

- A picture is worth a kilo-word, or so I've ing them here. Simplified diagrams, free- seems helpful.

Assume

- For many problems you will need to mak in order to get very far. This is where y regard to kinematic constraints, slip vers so on.

Formulate Equations

- Having laid the groundwork, here's whe erning equations with which you'll actuall

Solve

- Solve the problem in closed form or num

Check

- Come up with a consistency/logic check in the examples we will use this item to dis

4.1.14. **[Level 2] Computational** Use numerical integration (the MATLAB "quad" function will do nicely) to calculate the speed of a 50 kg mass that is initially moving to the right at 3 m/s and continues to the right for 2 more meters. The mass is acted on by a force directed to the left and given by

$$f = -a - b\frac{e^{-cx}}{d + x}$$

where $a = 100$ N, $b = 50$ N·m, $c = 1.1\ \text{m}^{-1}$, and $d = 2$ m.

Development of structured problem-solving procedures: A consistent analysis procedure is introduced early in the text and used consistently throughout all worked examples. Several key steps are emphasized here more so than in most other texts, including explicitly listing **Assumptions** made, and the importance of **Draw** and **Check** as part of the solution.

Analyzing dynamic motion with computers: Throughout the book, examples and exercises make use of computational approaches. Those exercises that require the computer are labeled **Computational**, making it easy for the instructor to seek out or, conversely, avoid them, depending on his/her preference.

At this point in their studies students should be as familiar and comfortable with computers and computer programs as with hand-held personal calculators. We'll be using the computer as a tool, nothing more. Any program that contains integration and plotting can be used, so we won't explicitly be teaching any of them. We've included an appendix that contains a brief introduction to MATLAB® for those students who may want a quick "bootstrap" to get them started—very short and to-the-point.

Application of principles to engineering systems: End-of-chapter **System Analysis (SA) Exercises** offer students the opportunity to apply mechanics principles to broader systems. These exercises are more open-ended than those in other parts of the text, and sometimes have more than one "correct" answer. We hope these exercises will provide opportunities for group work, exploration of similar systems near the students' own campus, and in general show how the principles in the text apply to analysis of real artifacts.

SYSTEM ANALYSIS (SA) EXERCISES

SA3.1 Escape from Colditz

Colditz Castle (**Figure SA3.1.1*a***) in Germany was probably the most infamous of all German prisoner-of-war camps during World War II. The castle, with its dark granite walls, barred windows, and medieval towers, sat high on a cliff overlooking the small town of Colditz in Saxony, Germany. Over 300 escape attempts were made during the $5\frac{1}{2}$ year war. However, one escape plan stands apart, the Colditz Glider.

Flight Lieutenants Bill Goldfinch, Antony Rolt, and Jack Best conceived the idea of launching two prisoners in a glider from the castle roof across the Mulde River to freedom. The glider was built, along with a launching mechanism that included dropping a bathtub full of concrete 18 m to catapult the glider off the roof; see **Figure SA3.1.1*b***. However, the Allied prisoners of war were liberated in April 1945 before the glider could be launched.

a. A model of the simplified launching system is shown in **Figure SA3.1.1*b***. Assume that the glider with two prisoners weighs 230 kg and that the concrete-filled tub weighs 460 kg. In addition, assume that the glider rests on a freely rolling cart and that the pulley is frictionless and weightless. Using a force balance, calculate the initial acceleration of the glider when the concrete-filled tub is released.

b. Why is the glider's acceleration less than the acceleration of gravity (9.8 m/s^2)?

c. Is the glider's acceleration constant while the concrete-filled tub falls 18 m?

d. Assuming the glider has 18 m of runway, how long does it take the glider to reach the edge after the concrete-filled tub falls 18 m?

e. How fast is the glider going at the end of the runway in **d**?

Reuters/New Media Inc./Corbis Images

(*a*)

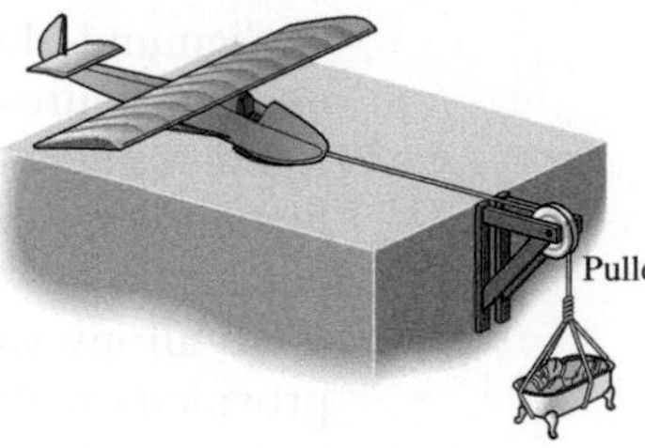

(*b*)

Figure SA3.1.1 (*a*) Colditz Castle; (*b*) escape glider

Inclusion of useful study tools: Most students will read and review the text to find key information as quickly as possible. To facilitate speedy access to key content, we have included review and study tools, such as **Chapter Objectives** at the start of each chapter, and a **Just the Facts** section at the end of each chapter that summarizes key terms, key equations, and key concepts from the chapter. To the greatest extent possible, all in-text figures include ***descriptive figure captions*** that show at a glance what is being illustrated. ***Key equations*** are highlighted in yellow, and ***key terms*** in bold blue type when they first appear.

Instructor Resources

The following resources are available to faculty using this text in their courses:

Solutions Manual: Fully worked solutions to all exercises in the text, using the same solution procedure as the worked examples.

Electronic figures: All figures from the text are available electronically, for use in creating your own lectures.

Student Resources

The following resources are available to students:

Answers to selected exercises: The text website, www.wiley.com/go/global/tongue, includes answers to selected exercises from the text, to help students check that they have solved the exercises correctly. MATLAB M-files used in the text are found there as well.

Author Acknowledgments

No book can come together without help from a whole lotta people.

John Rogosich and Techsetters showed once again how powerful TeX is when it's in the hands of experts.

A very special shout out to Carol Sawyer. She is a proofing/editorial goddess and I am in awe.

Editors Joe Hayton and then Mike McDonald were there to pull the strings when strings needed pulling to keep the project moving along smartly.

Sigmund Malinowski and J.C. Morgan worked mightily to keep those exercise figures looking good.

Sheri Shepard and Thalia Anagnos began the long journey toward the first edition with me and hopefully we'll rejoin our projects in the not too distant future.

To all my students who learned from early versions of this text and provided excellent feedback to me—thanks and thanks again.

Two students need special acknowledgement. Bayram Orazov lent his considerable skills to the task of exercise checking and in preparing much of the Wiley+ material. Huge job. Many thanks Bayram. Daniel Kawano provided absolutely invaluable assistance in problem creation, proofing, and organizing. Daniel has an inborn talent for teaching and if he doesn't watch out he's going to end up being a truly world-class educator.

Finally, there's my wife Claire. Only she knows the true facts of what went on in my cramped office, and she's not telling.

READ THIS! PLEASE!!!

⇓⇓⇓⇓⇓⇓⇓⇓⇓⇓⇓⇓⇓⇓⇓⇓⇓⇓⇓⇓⇓⇓⇓⇓⇓⇓⇓⇓⇓⇓

Hi and welcome to Dynamics! With this text you've got something new—a virtual professor in addition to your actual professor. I've written this book in just the same way as I teach in my classes and in office hours. So you now have a double resource. A real professor during the day and me when you're studying in the evening. Or whenever you study.

I won't take up too much time, but it's imperative that you understand a few things about the course you're starting in on. If you were in my class, I'd be jumping up and down on the first day telling you this, but likely you're not and so this written welcome will have to do. At the start, many of you will think, "Hey, I learned all this in high school." You didn't. Others will think, "This isn't new—I did all this last year in Physics." Once again—wrong.

The problem is that the very early stuff in this book *is* familiar. In high school you looked at some simple force problems, likely without calculus and only with regard to point masses. In first-year college Physics you had a quarter or semester in which large parts of Newtonian dynamics were shot at you with a fire hose. If your course was like most of them, there was very little tie-in to real-world applications, and the problems were very simplified.

So here's the deal. Note the car model. It's got an engine (with an applied torque), gears, wheels, and a body. All of these extended bodies have mass. If you can find the acceleration of this model when it's placed on a rough surface (so there's friction between the tires and ground) and all of the associated interbody forces, then you should go to your instructor and politely ask to be excused from the course on the grounds that you already know all that you need. If you're like everyone I've ever taught, however, all of whom already had a couple of Physics courses, you don't know how to solve this problem. By the end of this course you will.

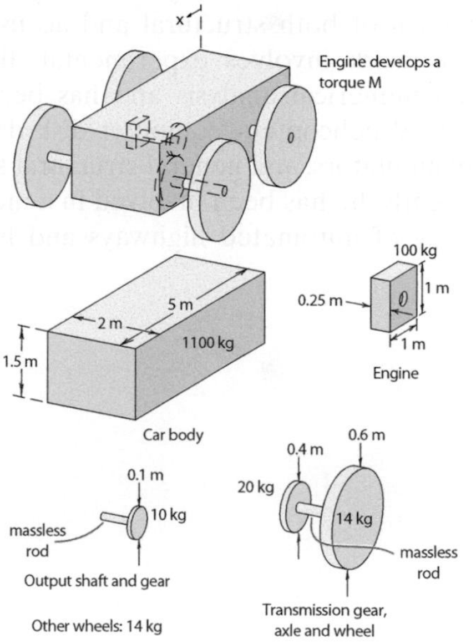

The danger, and the reason for my yelling at the top of the page, is that it's very easy for a student to be lulled into a false sense of security. The material at the start of this course is familiar, and the complexity is low. It's easy to not read the text and even, although this is really getting extreme, not go to lecture and still be able to handle the homework. At first. The fact is, though, that the material has a sneaky way of getting more difficult in a gradual way until one day you wake up and realize you're completely and totally lost.

Believe me; I've seen this happen. And it's not pretty. There's absolutely no doubt in my mind that every one of you reading this has the ability to learn the material and do well in the course. As long as you pay attention, which means listening at lecture and reading the text, you *will* learn how to do dynamics. Real dynamics, not make-believe problems of particles on a string, but cars and engines and mechanisms and more. The key is to stay focused and make sure you're solid on the concepts as they're introduced. Don't get complacent or think you can put studying the material off and learn everything the night before the exam. This might be an okay strategy for some courses but not for dynamics. Pay attention and it'll be easy. Ignore it and you're doomed. Your choice.

Benson H. Tongue

Ph.D., is a Professor of Mechanical Engineering at University of California—Berkeley. He received his Ph.D. from Princeton University in 1988, and currently teaches graduate and undergraduate courses in dynamics, vibrations, and control theory. His research concentrates on the modeling and analysis of nonlinear dynamical systems and the control of both structural and acoustic systems. This work involves experimental, theoretical, and numerical analysis and has been directed toward helicopters, computer disk drives, robotic manipulators, and general structural systems. Most recently, he has been involved in a multidisciplinary study of automated highways and has directed research aimed at understanding the nonlinear behavior of vehicles traveling in platoons and in devising controllers that optimize the platoon's behavior in the face of non-nominal operating conditions. His most recent research has involved the active control of loudspeakers, a biomechanical analysis of human fall dynamics, and green engineering.

Benson is the author of *Principles of Vibration*, a senior/first-year graduate-level vibrations textbook. He has served as Associate Technical Editor of the *ASME Journal of Vibration and Acoustics* as a member of the ASME Committee on Dynamics of Structures and Systems. He is the recipient of the NSF Presidential Young Investigator Award, the Sigma Xi Junior Faculty award, and of multiple Pi Tau Sigma Excellence in Teaching awards. He serves as a reviewer for numerous journals and funding agencies and is the author of more than eighty publications.

In his spare time Benson races his bikes up and down mountains, draws and paints, birdwatches, and creates espresso art.

CONTENTS

◆ CHAPTER 1

BACKGROUND AND ROADMAP 1

1.1 Newton's Laws 2

1.2 How You'll Be Approaching Dynamics 2

1.3 Units and Symbols 6

1.4 Gravitation 10

1.5 The Pieces of the Puzzle 11

◆ CHAPTER 2

MOTION OF TRANSLATING BODIES 16

2.1 Straight-Line Motion 17
EXAMPLES 23
EXERCISES 2.1 28

2.2 Cartesian Coordinates 33
EXAMPLES 38
EXERCISES 2.2 43

2.3 Polar and Cylindrical Coordinates 47
EXAMPLES 54
EXERCISES 2.3 58

2.4 Path Coordinates 64
EXAMPLES 67
EXERCISES 2.4 71

2.5 Relative Motion and Constraints 76
EXAMPLES 82
EXERCISES 2.5 87

2.6 Just the Facts 93
SYSTEM ANALYSIS 97

◆ CHAPTER 3

INERTIAL RESPONSE OF TRANSLATING BODIES 99

3.1 Cartesian Coordinates 100
EXAMPLES 102
EXERCISES 3.1 108

3.2 Polar Coordinates 119
EXAMPLES 120
EXERCISES 3.2 127

3.3 Path Coordinates 133
EXAMPLES 134
EXERCISES 3.3 139

3.4 Linear Momentum and Linear Impulse 143
EXAMPLES 145
EXERCISES 3.4 147

3.5 Angular Momentum and Angular Impulse 155
EXAMPLES 158
EXERCISES 3.5 161

3.6 Orbital Mechanics 163
EXAMPLES 176
EXERCISES 3.6 178

3.7 Impact 183
EXAMPLES 189
EXERCISES 3.7 191

3.8 Oblique Impact 193
EXAMPLES 196
EXERCISES 3.8 200

3.9 Just the Facts 203
SYSTEM ANALYSIS 206

◆ CHAPTER 4

ENERGETICS OF TRANSLATING BODIES 209

4.1 Kinetic Energy 210
EXAMPLES 212
EXERCISES 4.1 215

4.2 Potential Energies and Conservative Forces 220
EXAMPLES 225
EXERCISES 4.2 231

4.3 Power and Efficiency 243
EXAMPLES 247
EXERCISES 4.3 250

4.4 Just the Facts 255
SYSTEM ANALYSIS 257

◆ CHAPTER 5

MULTIBODY SYSTEMS 258

5.1 Force Balance and Linear Momentum 259
EXAMPLES 263
EXERCISES 5.1 268

5.2 Angular Momentum 273
EXAMPLES 277
EXERCISES 5.2 279

5.3 Work and Energy **282**
EXAMPLES **284**
EXERCISES 5.3 **287**

5.4 Stationary Enclosures with Mass Inflow and Outflow **288**
EXAMPLES **291**
EXERCISES 5.4 **293**

5.5 Nonconstant Mass Systems **299**
EXAMPLES **303**
EXERCISES 5.5 **305**

5.6 Just the Facts **310**
SYSTEM ANALYSIS **313**

◆ **CHAPTER 6**

KINEMATICS OF RIGID BODIES UNDERGOING PLANAR MOTION **314**

6.1 Relative Velocities on a Rigid Body **315**
EXAMPLES **320**
EXERCISES 6.1 **325**

6.2 Instantaneous Center of Rotation (ICR) **333**
EXAMPLES **335**
EXERCISES 6.2 **341**

6.3 Rotating Reference Frames and Rigid-Body Accelerations **346**
EXAMPLES **350**
EXERCISES 6.3 **356**

6.4 Relative Motion on a Rigid Body **361**
EXAMPLES **365**
EXERCISES 6.4 **371**

6.5 Just the Facts **378**
SYSTEM ANALYSIS **380**

◆ **CHAPTER 7**

KINETICS OF RIGID BODIES UNDERGOING TWO-DIMENSIONAL MOTION **382**

7.1 Curvilinear Translation **384**
EXAMPLES **385**
EXERCISES 7.1 **392**

7.2 Rotation about a Fixed Point **396**
EXAMPLES **401**
EXERCISES 7.2 **411**

7.3 General Motion **422**
EXAMPLES **425**
EXERCISES 7.3 **444**

7.4 Linear/Angular Momentum of Two-Dimensional Rigid Bodies **457**
EXAMPLES **460**
EXERCISES 7.4 **462**

7.5 Work/Energy of Two-Dimensional Rigid Bodies **468**
EXAMPLES **471**
EXERCISES 7.5 **475**

7.6 Just the Facts **482**
SYSTEM ANALYSIS **484**

◆ **CHAPTER 8**

KINEMATICS AND KINETICS OF RIGID BODIES IN THREE-DIMENSIONAL MOTION **487**

8.1 Spherical Coordinates **488**

8.2 Angular Velocity of Rigid Bodies in Three-Dimensional Motion **489**
EXAMPLES **493**

8.3 Angular Acceleration of Rigid Bodies in Three-Dimensional Motion **495**
EXAMPLES **496**

8.4 General Motion of and on Three-Dimensional Bodies **497**
EXAMPLES **498**
EXERCISES 8.4 **502**

8.5 Moments and Products of Inertia for a Three-Dimensional Body **506**

8.6 Parallel Axis Expressions for Inertias **508**
EXAMPLES **510**
EXERCISES 8.6 **511**

8.7 Angular Momentum **513**
EXAMPLES **517**
EXERCISES 8.7 **520**

8.8 Equations of Motion for a Three-Dimensional Body **521**
EXAMPLES **524**
EXERCISES 8.8 **526**

8.9 Energy of Three-Dimensional Bodies **532**
EXAMPLES **534**
EXERCISES 8.9 **536**

8.10 Just the Facts **537**
SYSTEM ANALYSIS **541**

◆ **CHAPTER 9**

VIBRATORY MOTIONS 542

9.1 Undamped, Free Response for Single-Degree-of-Freedom Systems 543
EXAMPLES 546
EXERCISES 9.1 549

9.2 Undamped, Sinusoidally Forced Response for Single-Degree-of- Freedom Systems 555
EXAMPLES 558
EXERCISES 9.2 560

9.3 Damped, Free Response for Single-Degree-of-Freedom Systems 563
EXAMPLES 567
EXERCISES 9.3 568

9.4 Damped, Sinusoidally Forced Response for Single-Degree-of- Freedom Systems 569
EXAMPLES 572
EXERCISES 9.4 575

9.5 Just the Facts 576
SYSTEM ANALYSIS 579

◆ **APPENDIX A**

NUMERICAL INTEGRATION LIGHT 580

◆ **APPENDIX B**

PROPERTIES OF PLANE AND SOLID BODIES 588

◆ **APPENDIX C**

SOME USEFUL MATHEMATICAL FACTS 592

◆ **APPENDIX D**

MATERIAL DENSITIES 595

◆ **BIBLIOGRAPHY** 597

◆ **INDEX** 598

OBJECTIVES

Upon completion of this chapter you'll have been introduced to:

- Newton's Laws of Motion
- A helpful problem-solving methodology
- Necessary symbols and units
- Vector conventions
- The ultimate goal of the course

CHAPTER 1

BACKGROUND AND ROADMAP

The man most responsible for what you'll be learning in this book is Sir Isaac Newton. Even among geniuses, Newton stands out.

He invented calculus because he needed a new mathematical approach to handle his investigations. The same year, he revolutionized optics by discovering that white light is made up of a spectrum of colors. To top it all off, he laid down his three laws of motion. What is even more amazing is that he did all of these when he was in his early twenties while taking a short break from London to avoid the plague.

(cont.)

Benson Tongue

(cont.)

He was not only one of the supreme scientists the world has seen, but he was also a card-carrying alchemist. He spent a lot of time doing things like searching for the philosopher's stone—a useful item (if you could make it), which supposedly turns base metals into gold. He was quite a fascinating guy.

Interestingly, his laws of motion were not overly complicated and you've undoubtedly run across them in physics class. The early material in this book will be somewhat familiar, and that's a good thing. You will be better prepared for what comes later by reacquainting yourself with the material in the early sections. We will certainly go quite a bit beyond what you've seen before, and the material has a sneaky way of getting difficult when you're not looking. So don't get bored as we start off; I promise, within a very short time, you'll be solving problems that would have stumped the best minds in Europe a few hundred years ago.

1.1 NEWTON'S LAWS

You have seen them previously in physics class and have dealt with at least one of them in Statics, but they're always worth repeating, and so, without further ado, here are Newton's Laws of Motion:

Law 1: *A body at rest (or moving in a straight line at a particular speed) will remain at rest (or continue moving in a straight line at that speed) unless acted on by a nonzero net force.* The phrase "nonzero net force" indicates that several forces might act on the body. If they all cancel one another, then the net force is zero. But if they don't cancel, then there is a nonzero net force that will change the body's motion. It's pretty intuitive. If no forces are acting on the body and it's not already moving, it will stay put. And if it is moving, it will keep moving in the same direction with the same speed.

The obvious question is—What happens when a nonzero net force *does* act on the body? The answer is **Law 2**: *When a body is acted on by a nonzero net force, the net force is equal to the time rate of change of the body's linear momentum.* Using this relationship, we can determine what happens, in a dynamic sense, when a force acts on a body. When a net force F acts on a body having a constant mass m, the relationship described in Law 2 simplifies to the familiar $\boldsymbol{F} = m\boldsymbol{a}$.

Lastly, we have **Law 3**: *For every action, there is an equal and opposite reaction.* This means that if we act on a mass by pushing it, the mass reacts against our hand with a force that is equal in magnitude and opposite in direction. This principle is violated all the time in the movies. Whenever a gun is fired, you should see the actor's hand move backward as it's acted on by the reaction force exerted by the gun. But does this happen on the screen? Rarely. Action/reaction is probably the most commonly ignored physical reality in the movies.

These three little laws were stunning when first published and haven't lost any of their punch since then. As long as you're dealing with human-scale problems, you're going to find them absolutely invaluable. By human-scale I mean we're dealing with speeds substantially slower than lightspeed and dimensions substantially larger than atomic scale. Near the speed of light, Newtonian dynamics breaks down and we have to resort to relativity. And a hot new area of research, nano-engineering, is concerned with structures that are defined on the atomic level. When you get this small, you also have some problems with a traditional Newtonian approach. But for everything in between (and that's most of what you as an engineer will be dealing with), Newton's laws work just fine.

1.2 HOW YOU'LL BE APPROACHING DYNAMICS

Historically, dynamics texts have tended to concentrate on a single approach to problem solving, an approach that could be termed *semi-static*. A typical problem would ask you to use the relationship $\boldsymbol{F} = m\boldsymbol{a}$ to calculate the value of some dynamical quantity *at a particular instant in time.* For instance, you might be told that the force acting on a body is 3 N at

$t = 0$ and then be asked to determine the acceleration of the body's center of mass at $t = 0$. I call this type of problem semi-static because you're asked for the result at only a single instant of time. True, you have to use dynamics to get the answer through Newton's Laws, but there's nothing particularly dynamic about the result—it's simply what the body is doing at one instant in time.

Because the let's-find-the-acceleration-at-this-instant approach is the one that's been used for decades, you might well assume it's a good one. After all, what's changed in the world that would make us want to alter this viewpoint? Well, what's changed is the presence of computers. In the past, texts limited themselves to semi-static problems because this was the only viable option before computers became widely available. Now that almost everybody has access to high-quality computation, there's no reason to shy away from it. By the time you're done, you should be as comfortable with numerically determining a rocket's trajectory as you would be if you were asked to take a square root with your calculator. Computation will simply become an accepted part of your arsenal of tools as an engineer.

When you finish this book you will have the ability to write the equations of motion for a system—the equations that govern the system's motion over time. You will be able to put these equations into a solvable form, and you will even spend time integrating them to compute the system's long-term trajectories. This means you will be able to calculate the trajectory of a spacecraft that's being injected into orbit, predict where billiard balls will end up after multiple collisions, calculate the motions of a robotic arm under commanded torque inputs, and on and on. If you later take a course in controls, you will be able to provide the equations of motion for complex systems such as aircraft, disk drives, and robotic manipulators. Having accurate equations of motion is a necessary first step in the design of good control strategies.

We're going to follow the same approach in all the problems we'll be encountering. It's a multistep program to success that can be used in all fields of human endeavor, but it is particularly effective in a dynamics class. The litany goes like this:

Goal

- This is where you write down what you think you're being asked to find. It always helps to know clearly what it is you're trying to do before jumping in and starting.

Given

- Here's where you remind yourself of what information you've been given with which to solve the problem.

Assume

- For many problems you will need to make appropriate assumptions in order to get very far. This is where you decide what to do with regard to kinematic constraints, slip versus no-slip conditions, and so on.

Draw

- A picture is worth a kilo-word, or so I've been told. So start drawing them here. Simplified diagrams, free-body diagrams—whatever seems helpful.

Formulate Equations

- Having laid the groundwork, here's where you formulate the governing equations with which you'll actually solve the given problem.

Solve

- Solve the problem in closed form or numerically, as required.

Check

- Come up with a consistency/logic check on the results. Sometimes in the examples, we will use this item to discuss related problems and interesting results.

*Optional

- Celebrate with the beverage of your choice.

Let's take these one at a time. The first two, **Goal** and **Given**, are to make sure you really understand what the problem is asking for. Often, students taking tests will jump into the problem, thinking they understand what's being asked, and either waste a lot of energy or get the wrong result because they didn't really understand what was being asked. Taking a second to decide what you're trying to accomplish and how you're going to do so is time well spent. Putting down the given facts can help clarify exactly what's known at the start and thereby make what's needed more obvious.

The **Assume** step shows up because you may be looking at a system that can exhibit different behaviors depending on the particulars of the situation. For instance, the wheels of a car might be rolling along without slip or they might be slipping on the surface as well, as they would if you were trying to accelerate on ice. In order to proceed, you will need to assume that one of these is true. Perhaps you decide that slip isn't occurring. That assumption will introduce constraints that allow you to solve the problem. Of course, once you've solved it you've got to go back and verify that your assumption was correct. If it wasn't, you start over with a different assumption.

Another example of where you would need some assumptions would be if you were analyzing the motion of a car that was driving over a rise in the road. Go fast enough and the car will launch itself into the air and become an unrestrained body. This will involve a particular set of equations. But if the car is moving slowly enough, it will follow the road surface. This will imply a different equation set.

Sometimes the problem will be defined well enough that no assumptions are needed. In this case we'll just say do and move on.

Drawing the system on paper (the **Draw** step) is a good way to organize everything you need to proceed. First, you will likely draw a coordinate system. This will be a convenient way to orient the system in space and give you a sense of how the object you're examining can move in that system.

Next, you should add any necessary labels that identify displacement, angle of inclination, length of links, and so forth. One or more sets of unit vectors will generally be needed.

Of course, you're already familiar with a **Free-body diagram (FBD)** from your Statics course. For Dynamics, this is only half the story. To make progress in Dynamics you'll also need the complementary **Inertial-response diagram (IRD)**, the thing that puts the "*ma*" in your $F = ma$. To do a kinetic analyses, you'll put the two together and draw a **Free-body diagram=Inertial-response diagram**, or **FBD=IRD** for short, something like the one shown in **Figure 1.1**.

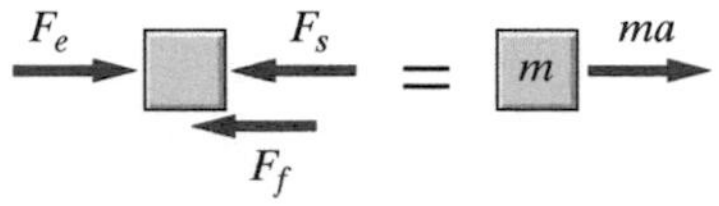

Figure 1.1 FBD=IRD example

This is the big, big deal of the course. Burn it into your mind. Repeat it as a mantra before going to sleep: FBD=IRD ... FBD=IRD... FBD=IRD.... I can't possibly emphasize this enough. Failing to draw an FBD=IRD is the first step into darkness and despair; drawing one brings contentment and peace. You have been warned.

Don't worry about being too rigid with the order of the **Assume** and **Draw** steps. In some problems it'll make more sense to sketch out the system and from the sketch come up with your assumptions, whereas at other times the assumptions will come easily at the start and allow the sketch to follow naturally from them.

Formulate Equations means just what it sounds like. You go through the book (or your memory) and pick out the equations you will need in order to reach a solution.

Solving the equations in the **Solve** step can be quick or not-so-quick, depending on what you're asked for. If all that's requested is the acceleration associated with a given force (or conversely, the force needed to produce a given acceleration), then you just solve for this quantity using your equation of motion. If, on the other hand, what's desired is a complete time history of the displacement and velocity of the object from some initial time to some final time, then generally you will have to input your equation of motion into a computer program that integrates system equations with respect to time and outputs the desired time histories.

Finally, it's not a bad idea to verify your result in some way, and this is the purpose of the **Check** step. The best scenario is having a second, independent, way of reaching the solution. If the answers from both approaches match, then you can be confident that it's correct. You can also apply your engineering intuition to determine if the result "makes sense." This doesn't work for all cases because dynamic problems can sometimes fool you: what seems counterintuitive at first may actually be correct. But if you're solving for the acceleration of a mass on a slope and your answer says it goes *up* the slope, even though gravity is pulling it down, that's a strong hint that you have done something wrong in your analysis. Similarly, applying a force to the right and seeing the object move to the left isn't a very reasonable answer. You can also check the general magnitude of your answer. If you're examining a 2-kg mass under the influence of gravity and your answer says the acceleration is 128,900 m/s^2, you can be fairly confident that there's an error somewhere. The answer 128,900 is so out of line with the magnitudes of the mass and gravity that it should immediately raise a red flag in your mind. This would be the time to pause, realize that your answer is absurdly high, and go back to find the error.

1.3 UNITS AND SYMBOLS

As you're probably aware of by now, two major systems of units are used in the world today: the **International System** (the logical one that's used by many, many people worldwide) and the **U.S. Customary System** (still used in the United States but that's pretty much it).

In the International System (SI), the fundamental quantity of interest is mass. A kilogram of aluminum is always a kilogram of aluminum, even though its weight will vary with variations in gravity. In SI units, the relationship between weight and mass is

$$1\ \mathrm{N} = (1\ \mathrm{kg})(1\ \mathrm{m/s^2})$$

where N is the abbreviation for newton, the basic SI force unit; kg is the abbreviation for kilogram, the basic SI mass unit; and m is the abbreviation for meter, the basic SI length unit. In the International System, force is defined in terms of mass and acceleration.

When using SI units, we express everything metrically, which means they involve factors of 10. You simply need to learn a few Greek prefixes—1000 **milli**meters in a meter, 1000 meters in a **kilo**meter, and so on. All the prefixes necessary for this course are listed in **Table 1.1**.

What we need in SI system are the variables that allow us to apply Newton's second law ($\boldsymbol{F} = m\boldsymbol{a}$), namely, time, length, mass, and force. **Table 1.2** shows these quantities.

The basic unit of time is the second, and there's no metric simplicity available; even in Europe you have 60 seconds in a minute, 24 hours in a day, and so forth. Strange but true. The second entry is length.

Mass is the interesting one. As already mentioned, the SI basic mass unit is the kilogram. This unit is in very wide use. If you're shopping

Table 1.1 SI Prefixes

Multiple	Prefix	Symbol	Example
10^{12}	tera	T	terabyte
10^{9}	giga	G	gigahertz
10^{6}	mega	M	megaton
10^{3}	kilo	k	kilogram
10^{2}	hecto	h	hectoliter
10^{1}	deca	da	dekagram
10^{-1}	deci	d	deciliter
10^{-2}	centi	c	centimeter
10^{-3}	milli	m	milliampere
10^{-6}	micro	μ	micrometer
10^{-9}	nano	n	nanosecond
10^{-12}	pico	p	picofarad

Table 1.2 SI Units

Dimension	SI units
Time	second (s)
Length	meter (m)
Mass	kilogram (kg)
Force	newton (N)

anywhere in Europe, you will be asked how many kilograms of cheese you want.

Units of measure (newton, meter, second), will *always* use a roman font—N, m, s. Points located either on a body or somewhere on a reference frame will always be represented by italicized capital letters—A, C, and so on. When we are dealing with a system variable, like mass and a spring constant, a lower-case italic font—m, k, and so on—is used. Thus, the m in "1 m" stands for meter, and $m = 1$ kg tells you that the mass m is equal to 1 kg.

You're already familiar with the two common ways to describe angles: degrees and radians. Although either can be used when we're concerned with simply an angle, we will soon be looking at rotating bodies, which necessarily involve angular rotation rates. When dealing with an angular speed or acceleration, you will want to be sure to use rad/s or rad/s^2, not °/s or °/s^2.

Following the usual convention, use of the word *speed* signifies a scalar quantity (The bicycle's speed is 5 m/s), whereas the word *velocity* refers to a vector quantity, one that contains an object's speed as well as the direction in which it's moving.

Speaking of velocity, a dot will often be used to indicate differentiation with respect to time:

$$\dot{x} \equiv \frac{dx}{dt}$$

and two dots will indicate acceleration:

$$\ddot{x} \equiv \frac{d^2x}{dt^2}$$

Two kinds of arrows will be used as a shorthand: $\rightarrow$, which means "substituted into," and $\Rightarrow$, which means "yields." Here's an example:

$$(1.3.4) \rightarrow (1.3.3) \quad \Rightarrow \quad c = 4.0\,\text{m}$$

In words, this expression reads "equation (1.3.4) substituted into equation (1.3.3) yields $c = 4.0$ m." It's a quick way to let you know which equations are being used and the results of the operations.

Occasionally, "notes" will be added to some of the equations to remind you of what the quantities represent or where they came from. For example,

$$F_1\boldsymbol{i} + F_2\boldsymbol{j} + F_3\boldsymbol{k} = \overbrace{(1.5 \times 10^5\,\text{kg})}^{m_A}\,\overbrace{[(1.0\boldsymbol{i} + 1.1\boldsymbol{j} + 0.050\boldsymbol{k})\,\text{m/s}^2]}^{\boldsymbol{a}_A}$$

lets you see that the numerical values on the right side of the equation correspond to the variables m_A and $\boldsymbol{a}_A$.

Now that we have used some vectors in the preceding discussion, it's definitely time to talk more precisely about how vectors are represented. When printed in the text they'll be presented as bold italic letters, like $\boldsymbol{F}$ and $\boldsymbol{r}_m$. This choice was made to keep the equations as uncluttered as possible. As a student, however, since you have no easy way to write the variables down in boldface, you'll need another approach. What you'll do instead is put an arrow over the variable: $\vec{F}$, $\vec{r}_m$. When the figures in the text are supposed to represent your work, the artwork will include arrows. Otherwise, boldface will be used.

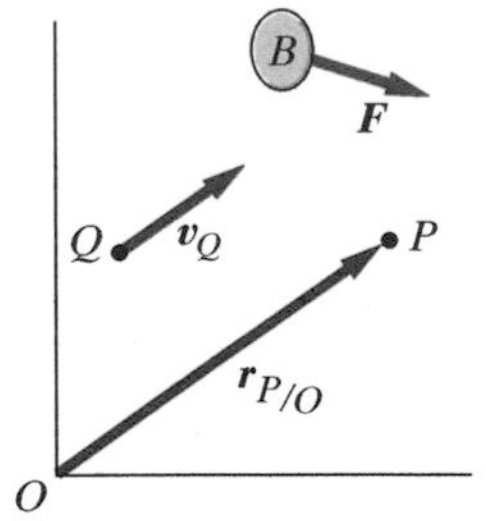

Figure 1.2 Position, velocity, and force vectors

Position vectors, ones that tell you where something is with respect to something else, will be shown as $\boldsymbol{r}_{P/O}$ when we're being completely precise. This means it's the vector that points from O to P, as shown in **Figure 1.2**. Note that this vector is colored blue. Blue vectors will always indicate position vectors.

Figure 1.2 also shows the velocity vector $\boldsymbol{v}_Q$ of the particle Q. There's no "with respect to" element (like $_{Q/O}$) because velocities don't depend on any particular fixed origin. Whether you're measuring the velocity from O or from some other fixed point O', the velocity of P will be the same. Note that this vector is green. All vectors in this text that involve velocity or acceleration will be green.

The final color we'll be using for a vector is red, as shown in Figure 1.2 by the force vector $\boldsymbol{F}$ acting on the body B. Red will always be used in this text when we are representing force vectors.

Often we will want the **magnitude** of the vector, which is simply its "length." Also called the **Euclidean norm**, it's defined as

$$\|\boldsymbol{F}\| \equiv \sqrt{\boldsymbol{F} \cdot \boldsymbol{F}}$$

that is, the square root of the dot product of the vector with itself.

Taking the arrow off the variable ($\vec{F} \rightarrow F$) or, equivalently, changing from boldface to a regular font ($\boldsymbol{F} \rightarrow F$) gives us the vector's **signed magnitude**. The signed magnitude differs from the magnitude in that the signed magnitude can be positive or negative, whereas the magnitude is always positive.

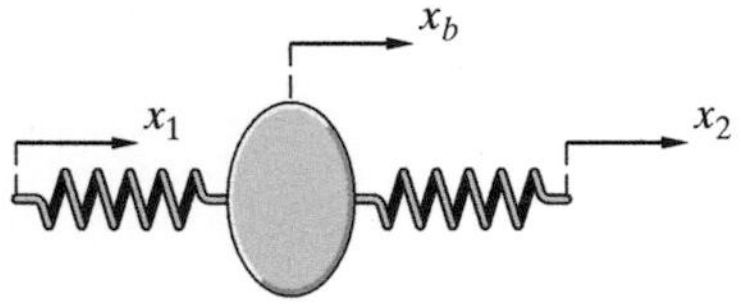

Figure 1.3 Spring-restrained mass

Here's how such a situation will show up. Say we're considering the mass with two attached springs that's shown in **Figure 1.3**.

As part of our analysis, we want to solve for the force exerted on the mass by the right spring. To do this, we represent the force as shown on the left of **Figure 1.4**. The direction of the arrow is to the right, but this is purely an arbitrary choice at this point. As shown on the right of **Figure 1.4**, the final result might be a positive F, meaning that the force indeed does point right, or a negative F, in which case the arrow's actual direction is to the left.

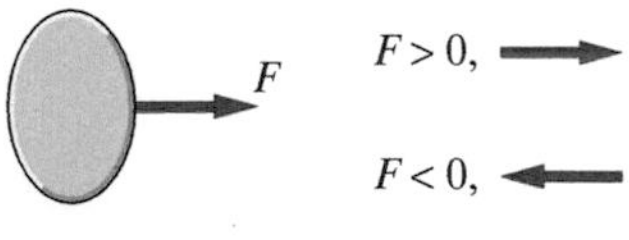

Figure 1.4 Force from right spring

Thus, we have the situation in which we know the direction but not the sign of an interacting force. Using the signed magnitude allows us to capture the orientation but leave the final detail of positive/negative determination until later in the problem solution. If our solution is positive, it means the direction we chose initially is the right one. If negative, we know that the actual direction of action is opposite to our initial choice. We'll be encountering this sort of thing when dealing with frictional force. Often we'll know that the force points along a certain line but we won't know the precise direction (forward or backward). Our approach will be to guess a direction, solve the problem, and then see if we guessed correctly.

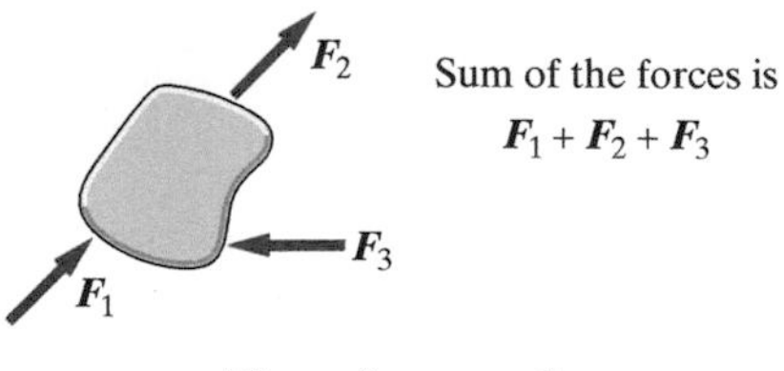

Figure 1.5 Three forces acting on a body

What if we *don't* know the directions of the forces but just want to indicate that the forces exist? In this case we'd label as shown in **Figure 1.5**. The use of boldface for the vectors indicates that the particular directions shown in the figure are arbitrary. If nonbold variables had been used, it would have indicated that the line of action is correctly shown and only the sign remains unknown.

Mathematically we'd say that the sum of the forces shown in Figure 1.5 is simply $\boldsymbol{F}_1 + \boldsymbol{F}_2 + \boldsymbol{F}_3$, which is most easily represented as

$$\sum_{i=1}^{3} \boldsymbol{F}_i$$

in which the $\sum_{i=1}^{3}$ indicates summation from $i = 1$ to $i = 3$. No particular knowledge of the force orientations is presumed. But then let's say we start analyzing the system and decompose the forces along particular directions. **Figure 1.6** shows such a case for $\boldsymbol{F}_1$. The horizontal component F_{1_x} and the vertical component F_{1_y} are shown. In this case we'd use the nonbold variables F_{1_x} and F_{1_y} because the only thing unknown about the vector is its magnitude and sign; the orientation has been assigned and can't vary. The signed magnitudes F_{1_x} and F_{1_y} are therefore the variables that we will solve for, and eventually we will come up with some particular answers, perhaps $F_{1_x} = -1.3$ N and $F_{1_y} = 11.4$ N, for example.

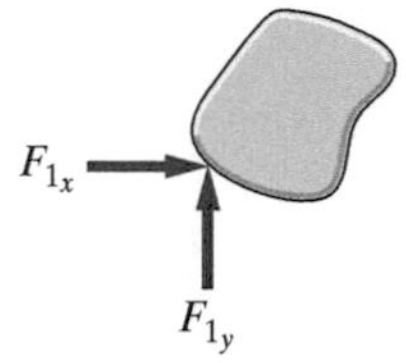

Figure 1.6 Directed magnitudes of $\boldsymbol{F}_1$

As a final example, consider **Figure 1.7**. The system is a box that's held in position by two taut ropes. The problem might be to determine the power supplied by the motors in order to move the box at a given rate. Because the orientation of the forces is known, we can use the directed magnitudes F_1 and F_2 to represent the unknown forces. We can go further in this case because we know the forces have to be positive in the direction shown in as much as the ropes can only support tensile loads; under compression they would go limp. Thus, if we solved the problem and found that either F_1 or F_2 was negative, we'd know that we'd made an error because a negative force implies compression, given the direction arrows shown on the figure, and such a solution wouldn't be physically meaningful.

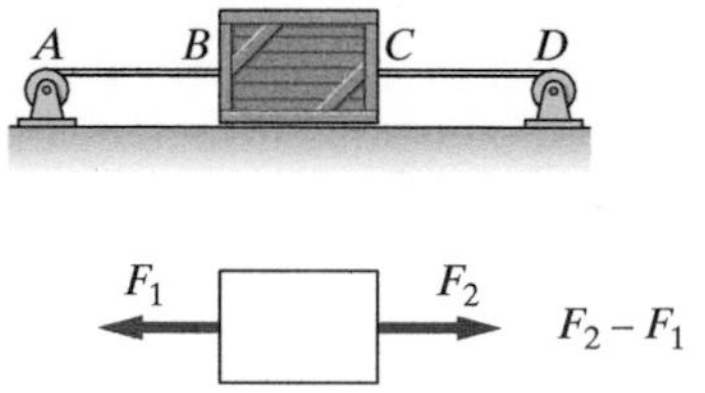

Figure 1.7 Force on box

Seeing as we're talking about forces, this would be a good place to mention a particular kind of force, that which is due to friction. As you will recall from physics, there are two kinds of frictional coefficients—a coefficient of static friction μ_s and a coefficient of dynamic friction μ_d. Both are illustrated in **Figure 1.8**. The body in **Figure 1.8*a*** is sliding across the ground with speed v, and the normal force acting on the body is given

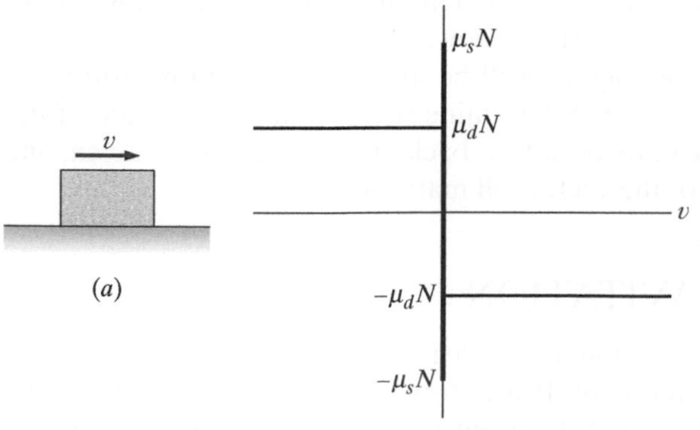

Figure 1.8 Sliding block

by N. The force parallel to the ground which resists the body's motion, the friction force, is given by $\mu_d N$. It's negative for positive v, as shown in **Figure 1.8*b***. When the block slides to the left, the force resisting the motion is positive (pointing to the right). If the body isn't moving at all, a range of forces can develop at the body/ground interface, up to a maximum of $\mu_s N$ and a minimum of $-\mu_s N$.

Often we will consider the simplified case for which $\mu_s = \mu_d$, in which instance we will just refer to a single coefficient of friction and denote it with μ.

This is a good example of how signed magnitudes come into play. A free-body diagram of the system just discussed is shown in **Figure 1.9**. The frictional force $\boldsymbol{F}$ is indicated as pointing to the left. However, that's just our initial guess. If the subsequent analysis determines that the block is actually moving left, then the frictional force will point to the right. Mathematically, this means that when F is solved for, it will turn out to be negative.

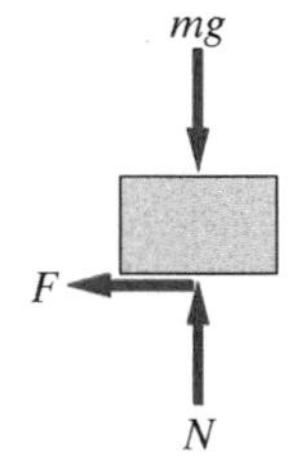

Figure 1.9 FBD of a sliding block

Before leaving our vector discussion, let's look for just a moment at the right side of our force balance, the $\boldsymbol{ma}$ part of $\boldsymbol{F} = \boldsymbol{ma}$. The equals sign indicates that the overall vector on the left has to be the same as that on the right. However, you needn't worry about making sure your sketch shows this. Actually, it's pretty much impossible to do so early in the problem because that's going to be the whole point of our exercises—finding the answer of how much a body accelerates or what forces act on it. If we knew the complete solution ahead of time, then we could certainly draw all the vectors appropriately so that they "matched" on each side of the equals sign. Here's an example. Say we had a block that was being pushed to the right by means of a rocket (thrust equal to F_L) and also pushed to the left because of a strong headwind (with force F_R). The force due to the headwind might depend on the block's speed, and thus we won't immediately know its magnitude until we calculate it later in the problem. A reasonable person would draw the forces as shown in **Figure 1.10** and not worry about the relative lengths of the two vectors at this point.

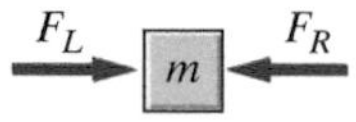

Figure 1.10 Forces acting on block

What do we do about the $\boldsymbol{ma}$ part? Without knowing the actual magnitudes of the applied forces, we don't know if the acceleration will be to the left or right. What I'll do in such a case is draw the acceleration in the positive direction (to the right for a typical x coordinate axis) and not worry about whether that's ultimately the correct direction or not. I also won't worry about the length of the vector and will simply draw a "typical" vector length, as shown in **Figure 1.11**.

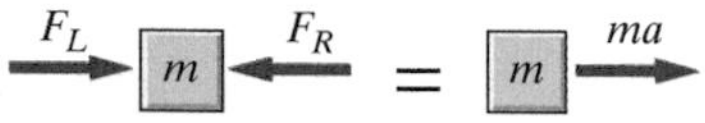

Figure 1.11 FBD=IRD

This sort of diagram will be all I'll need to move forward and solve the problem. After I determine the sign and magnitude of $\boldsymbol{a}$, I could, if I were sufficiently bored, go back and create a new sketch, one in which the lengths of the vectors all matched up properly.

1.4 GRAVITATION

Gravity played a dominant role in Newton's work, and it's worth reviewing a few key points here. If you stop to think about it, gravity is pretty weird. It's a force that exists between any two masses, and it operates whether or not there's anything between the masses. That by itself is kind of odd. A thin sheet of paper will block light, metal shielding will block radio waves, but gravity seems to penetrate anything without any trouble.

Experimentation has shown that the force between two bodies depends on their individual masses and on the distance between them. The formula is

$$F = \frac{Gm_1m_2}{r^2}$$

where m_1 and m_2 are the masses of the two bodies, r is the distance between them, and G is the gravitational constant. Expressed in SI units, G is approximately $6.67 \times 10^{-11}\ \frac{\text{m}^3}{\text{kg}\cdot\text{s}^2}$. This is the value we will be using in the examples and homework assignments.

We'll be using this relationship when we start looking at orbital mechanics, which involves such systems as the space shuttle as it orbits the earth and the earth as it orbits the sun. We won't be using it when we're looking at the force of gravity as applied to an object near the earth's surface because in this case r is the distance between the object and the earth's center, a distance that is pretty much constant. This means that three of the four factors in our formula for gravitational force ($G, m_1 = m_{\text{earth}}$ and r^2) are constant and can therefore be combined into a single new constant, the gravitational acceleration g. Unless otherwise stated, g will be equal to 9.81 m/s^2 in this book's examples and homework problems. The quantity $\frac{Gm_{\text{earth}}}{r^2}$ can therefore be considered to be a fixed quantity.

A word about numbers. All numerical results will be displayed with three significant figures. When doing the calculations, however, we will crunch the numbers on a calculator or computer (as you'll likely be doing), and double precision accuracy will be retained. So don't be surprised if you get a result that's slightly different from the one in the text if you jump into a calculation near the end. One of the calculated variables might be displayed midway through the problem as 1.51 but stored on the computer as 1.514735. If you start from the beginning of the example and follow through, you should get the same answer that the book shows. Also, all calculations will be started just by keying in the initial information in the problem statement, just as you would. Thus, if we were trying to find the speed of a car that had traveled 10 m in 64 seconds, we'd enter 10, divide by 64, and get a displayed result of 0.15625. All the example would show, however, would be 0.156.

1.5 THE PIECES OF THE PUZZLE

It's always nice to know *why* you're going to be learning something before starting out; it helps to keep everything in perspective. So let's look at where you'll end up if you're taking a normal one-quarter or one-semester course and see why we can't just jump to that point.

Figure 1.12 shows the sort of system you'll be able to analyze by the end of the course. Once you know the engine's torque output, transmission gearing, and the frictional tire/road interface conditions, you'll be able to formulate the equations of motion and solve them numerically. The difficulty in this task is the fact that the objects that interact with the

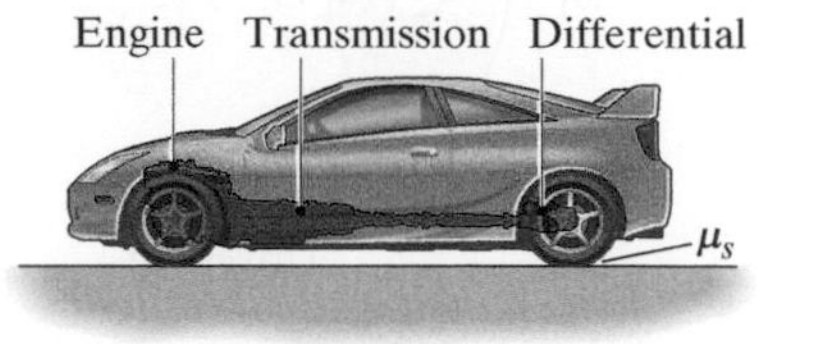

Figure 1.12 Simplified car model

ground (and make the car "go") are the tires, and they're far removed from the engine. Somehow you'll have to be able to link the torque-producing engine to the tires by means of the drivetrain. Here's some of what's needed.

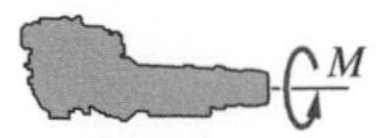

Figure 1.13 Simplified engine

Consider the engine indicated in **Figure 1.13**. If we really want a detailed analysis, we might want to consider the operation of the valves, pistons, and so on. For now, we'll just assume that it can produce torque and spin an output shaft. But how do we handle the interactions of the engine with the body? The engine is sitting on the chassis, and as it rotates, forces will be generated at the attachment points. Do we have to consider them? The answer is "sometimes." But for now let's ignore that issue and just focus on what happens to the engine's torque.

As you probably know, the engines of typical cars idle at about 1000 rpm and reach a top rotational speed of between 5000 and 6000 rpm. If the wheels actually turned at 6000 rpm, the car would be traveling at over 644 km/h. Obviously, this doesn't happen, so what's providing the necessary speed reduction? The answer is that there's a gearbox, a simple two-speed, manual version of which is shown in **Figure 1.14**. The problem is that when gears of differing radii interact, both the rotational speeds and the transmitted torque change. To determine how this happens, you will have to be conversant with angular velocity and angular acceleration of rigid bodies (the gears) and be able to solve for their accelerations due to the applied moment and forces.

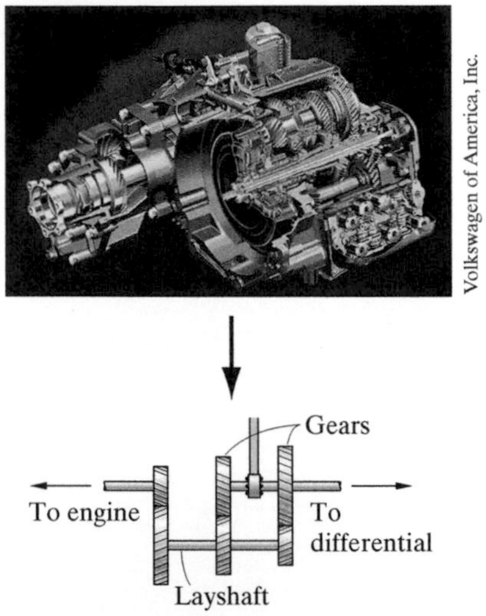

Figure 1.14 Two-speed gearbox

Once through the gearbox, the driveshaft terminates in a differential, a mechanism that shifts the spin of the driveshaft into a spin of the rear axle, reorienting the rotational velocity by 90 degrees (**Figure 1.15**).

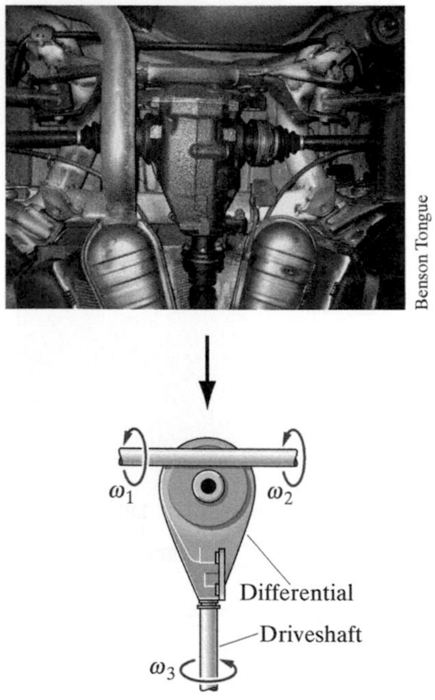

Figure 1.15 Simplified differential

Finally, and nontrivially, we have to determine how the torque, applied to the wheel, causes a force that can move the car forward (**Figure 1.16**). The details of that force will depend on the interface between the tire and the ground and on whether the tire is undergoing pure rolling or rolling while slipping at the same time. This involves a force and moment balance on the wheel, something you likely don't know how to do yet.

Although you can't just jump in and solve such a problem right now, in three or four months you'll be able to do so. Of course, you'll be able to do far more than just analyze a car's motion. You'll be able to determine the equations that govern the motion of a read/write head on a computer hard drive, and you'll be able to analyze the forces and moments driving a robotic arm as it moves from place to place. Analyzing biomechanical motions is certainly in the realm of what you'll be able to do, and we'll be looking at a few examples before we're done.

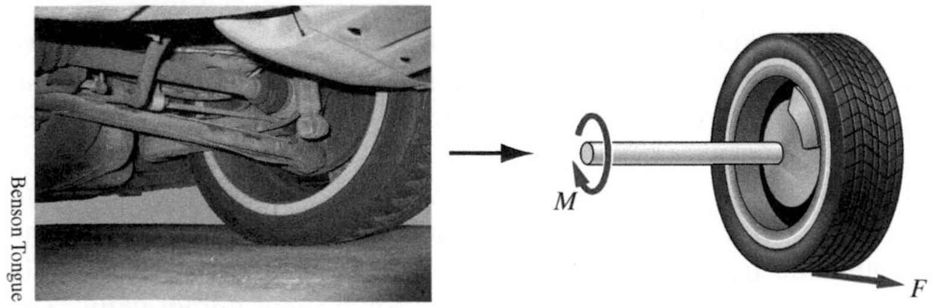

Figure 1.16 Wheel/tire

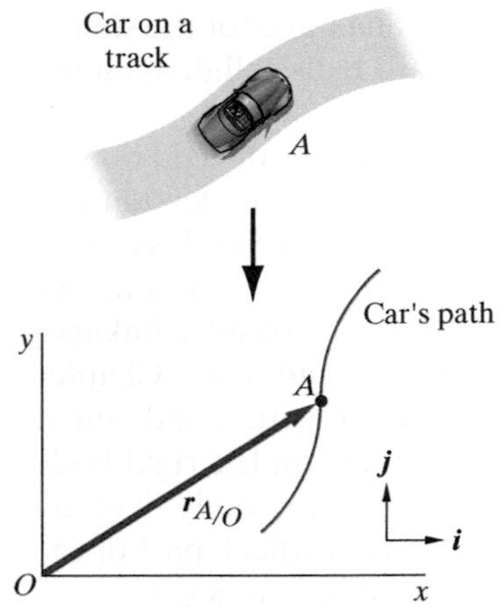

Figure 1.17 Chapter 2: Tracking the motion of a car

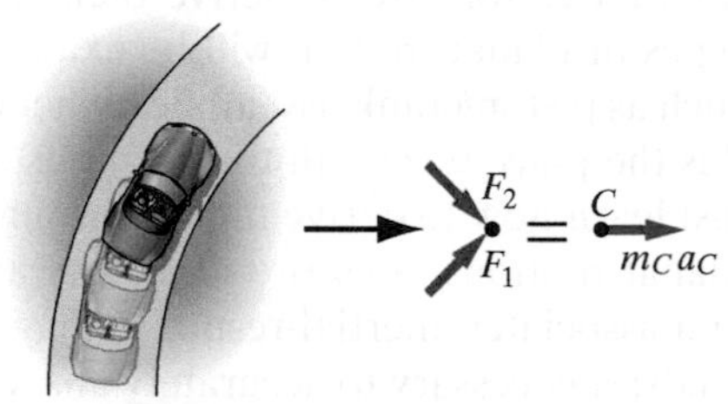

Figure 1.18 Chapter 3: Newton's second law applied to a lumped-mass car model

Now that you know what our ultimate goal is, let's see how the upcoming chapters work to get us there.

Chapter 2 is concerned with the study of motion, known as kinematics. **Figure 1.17** shows both an aerial view of a car traveling along a racetrack and what our schematic representation might look like. By deciding how we can describe the car's position, velocity, and acceleration, we will be giving ourselves the key to half of Newton's second law—the *ma* part.

Chapter 3 marks the start of actual dynamics. We take the descriptive kinematics of Chapter 2 and put it together with the forces acting on our system to derive the governing equations of motion. The car just mentioned works well as an example. A variety of forces act on the car, forces from the road/tire interface and aerodynamic drag being the primary ones. **Figure 1.18** shows how we might set up the free-body diagram and the associated inertial-response diagram. Taken together these diagrams allow us to determine the equations of motion, which we can then integrate to find the car's position and velocity as a function of time.

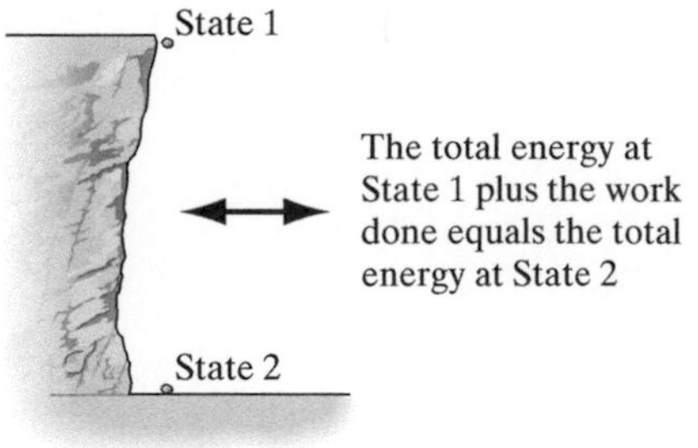

Figure 1.19 Chapter 4: Rock falling off a cliff

In **Figure 1.19** we see a rock dropping from a cliff. At the start it has a certain energy, and at the bottom its energy makeup has changed. The kinetic energy (initially zero) has increased to a maximum just before the car strikes the ground. This increase occurs because gravity did work on the mass as it fell. The concept of work and energy is a powerful one, and in this chapter you will see some of its uses.

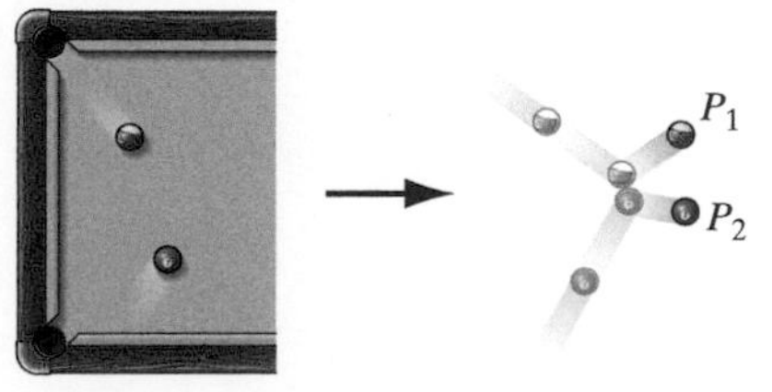

Figure 1.20 Chapter 5: Colliding billiard balls

Figure 1.20 shows a representative problem from Chapter 5—two particles colliding. Chapter 5 is a bridge chapter between the single-particle dynamics you have already learned and the rigid-body dynamics of Chapters 6 and 7, which deal with systems of particles. We will learn that the center of mass of a collection of objects has some nice features that can aid in the analysis of multimass systems. After seeing how large collections of particles behave, we'll let the number of particles go toward infinity as the individual masses go toward zero. By requiring that they remain in a fixed orientation with respect to each other, we'll arrive at a model for a rigid body (which is, after all, a huge number of very small atoms that remain locked in a fixed orientation). A physical analogue for the simple two-mass collision is, as shown, a couple of billiard balls colliding on the billiard table.

Chapter 6 introduces you to the kinematical behavior of rigid bodies, such as the rotating wheel shown in **Figure 1.21**. Only by understanding how to express the velocity and acceleration of bodies such as this can we then move forward to derive their governing equations of motion. All types of planar motion will be examined, and we will consider linkages such as piston/crank assemblies, wheels, robotic arms, and so on. Chapter 7 is the place where all the previous chapters come together and you at last learn how to derive the governing equations for complex, rigid-body planar motions. **Figure 1.22** shows an example of a free-body diagram and associated inertial-response diagram for a driven wheel, part of the analysis necessary to accurately analyze the full response of a car.

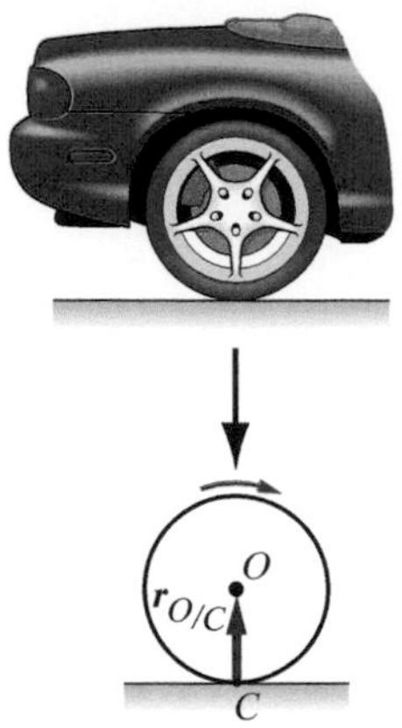

Figure 1.21 Chapter 6: Kinematics of a rolling wheel

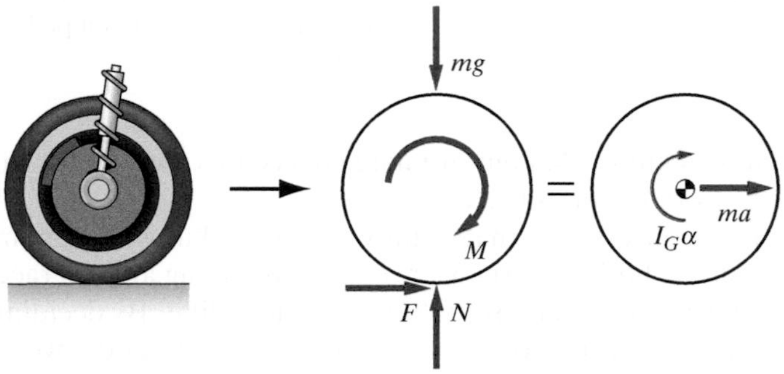

Figure 1.22 Chapter 7: Forces and moments affecting a driven wheel

Most courses don't get this far, but if you get to Chapter 8 you will learn how to extend the two-dimensional analyses of the previous chapters to fully three-dimensional rigid-body motion. This will allow you to analyze such three-dimensional problems as spinning tops, gyroscopes, aircraft and spacecraft, submarines, and so on. **Figure 1.23** shows a robotic arm, a typical three-dimensional problem, and a schematic representation that we would use for analysis purposes.

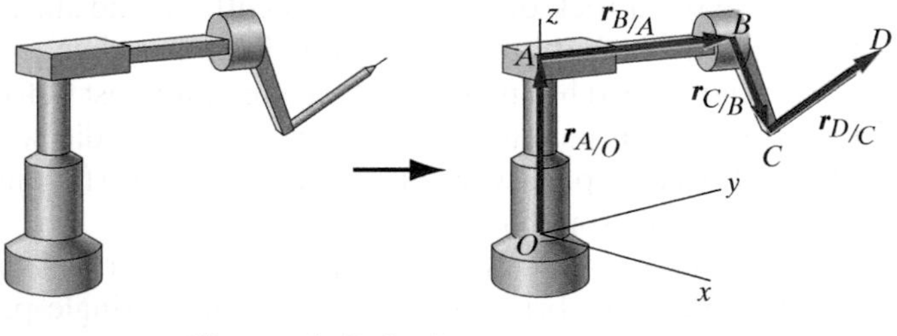

Figure 1.23 Chapter 8: Robotic arm

Figure 1.24 illustrates a typical system from Chapter 9. On the right is a side view of a car. If we were doing a first-cut analysis of the car's vibratory motion that ignored any lateral motions, we would have a simplified model such as the one shown on the left: two springs supporting a rigid bar. The bar models the car body, and the springs model the front and rear suspension elements. Vibrations is a huge field, and just about any

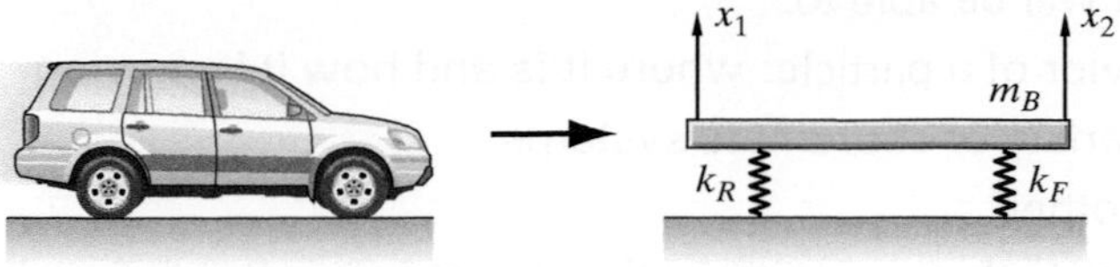

Figure 1.24 Chapter 9: Half-car model

system you can name experiences vibrations of one sort or another, from skyscrapers to nano-devices. Although just an introduction, this chapter will equip you with the tools necessary to analyze a surprisingly large variety of problems.

OBJECTIVES

Upon completion of this chapter you will be able to:

- Analyze the time-dependent behavior of a particle: where it is and how it is moving
- Describe a particle's motion in a variety of coordinate systems
- Switch from one description to another
- Deal with relative motion (one particle that's moving with respect to another) as well as constrained motion

CHAPTER 2
MOTION OF TRANSLATING BODIES

Before we start in on any of the heavy-duty material, we have to nail down the basics and review what's meant by such terms as *velocity* and *acceleration*. You have certainly come across these concepts in previous classes, but we are going to be extending the concepts in the upcoming chapters and so a little review isn't a bad idea. We'll start by examining purely in-line motion, such as you would have if you were analyzing the position, velocity, and acceleration of a bullet as it travels down a rifle barrel. When we're done, you should be very comfortable with finding a body's position if you're given its acceleration and starting conditions or, conversely, with determining its acceleration if you're given its position as a function of time. Then we'll complicate matters by adding a second dimension and considering several different ways of expressing our motion variables.

Benson Tongue

2.1 STRAIGHT-LINE MOTION

Before we start, I have to ask if you've read the "TO THE STUDENTS" note at the front of the book. No? I didn't think so. Well, DO IT NOW!! If you were in my class, I'd be going over this on the first day, but since you're likely at a different school, the only way you're going to hear it is if you read it. It's important—no kidding. Maybe the difference between doing well and failing. It's not long, I promise. Just read it, okay?

Right. With that out of the way, let's start. To motivate our discussion, let's consider the system shown in **Figure 2.1**, a car (an M coupe, to be precise) being tested on a dragstrip. For those not familiar with the sport, a large vertical set of lights is set up within sight of the driver. The lights come on in sequence, and once the green light is lit, the driver drives as fast as possible to the end of the track.

Figure 2.1 M coupe waiting on dragstrip

What can we say about this situation? Well, at any time after the run starts, the car will be somewhere along the track. Its position along the track is shown diagramatically in **Figure 2.2** as C. The start of the race is labeled O, and the distance from O to C is given by x.

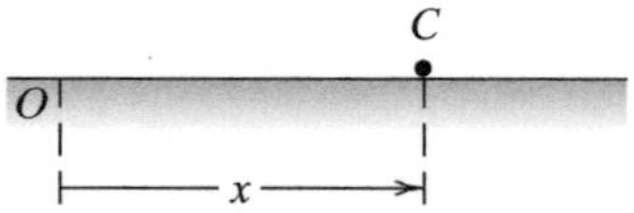

Figure 2.2 Schematic of the M coupe's position along the track

IMPORTANT NOTE! Note how C doesn't look much like a car. Instead, we use a dot because our plan is to treat the car as a *single particle*. All parts of the body will have the same velocity. This is a key point to understand. The title of this chapter is "Motion of Translating Bodies." The word *translating* in that title tells you that the material we'll be covering applies to any body that isn't rotating. In fact, even if there is some actual rotation, as long as we can neglect it as being unimportant for the problem at hand, we can still stick with the approaches presented in this chapter. Thus, we might look at a race car traveling along a curve and ask what the translational acceleration of the car is. We'll treat the car like the M coupe above and model it as a single particle. This will be absolutely fine for telling us, for instance, what the acceleration of the car's mass center is. Later (in Chapter 6), we'll worry about the real acceleration of all points on the car's body and will need to extend our analysis.

Now let's return to our car. If the driver glances at his speedometer, he will see something like what's shown in **Figure 2.3*a***, a needle indicating his speed, in this case 161 km/h. He might then glance over to a digital readout (**Figure 2.3*b***) that displays his acceleration, shown as 0.82 (which stands for $0.82g$, or 0.82 times the acceleration due to gravity).

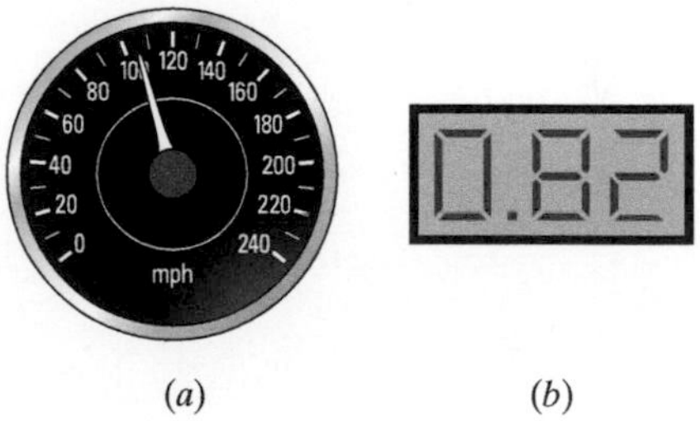

Figure 2.3 (*a*) Speed readout; (*b*) acceleration readout

What the driver is seeing represents some of the key questions of this section. How do we determine

- Where something is?
- How fast is it moving?
- What's its acceleration?

With that as background, let's begin. Determining the location of something isn't too hard if it can move only along a straight line. You start by defining an origin O and then marking the line in the positive and negative directions with some evenly spaced indicators of position. In **Figure 2.4** the scale is in meters, and positions from -1.8 m to 1.8 m are labeled. A body C moves along this line and is currently 1.2 m from the origin. Let's define the current time as t_1, with time given in seconds. At some time t_2 in the future ($t_2 = t_1 + \Delta t$), the body has moved 0.6 m

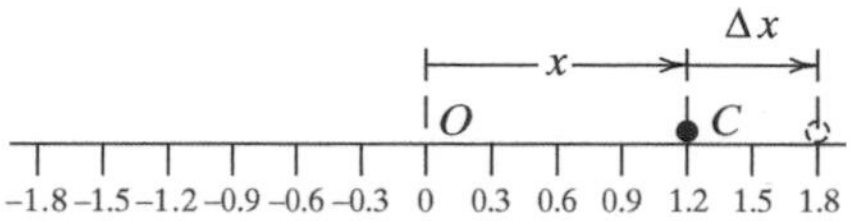

Figure 2.4 Position of a body

farther from the origin. Mathematically, we say that $x(t_1) = 1.2$ m and $x(t_2) = 1.8$ m and the difference between these two positions is indicated by Δx. Note that **position** is used to indicate where something is and **displacement** indicates the difference between that something's final and initial positions. The body has been displaced from its initial position at $x = 1.2$ m to its new position at $x = 1.8$ m.

The location of the body relative to the origin can be presented as a vector because we need two pieces of information to pin it down: its distance from the origin and its orientation with respect to the origin (in front or behind). For our current situation, the orientation is represented by the sign of the answer: + means to the right and – indicates to the left of the origin.

IMPORTANT NOTE! We really *need* vectors only when dealing with two- and three-dimensional motion. Although we could include the formalism of boldface symbols to represent our vectors in the one-dimensional case, it's not really necessary. We already know the orientation of all the vectors we'll be dealing with because they have to point along the x axis. The only unknown is whether they're pointing in the positive or the negative direction, and this is easily handled by using a minus sign when needed. For simplicity, therefore, I'll use the signed magnitude: $x_{c/o}$. To keep the notation even simpler, I'll drop the $_{c/o}$ when there's no possibility of confusion.

Once again, I'm not using boldface for vectors in this section because x itself is sufficient. As soon as we get to Section 2.2, though, I'll begin using boldface and will continue to do so throughout the rest of the text.

Finding the body's **average velocity** from t_1 to t_2 is simply a matter of dividing the distance traveled by the elapsed time:

$$v_{avg} = \frac{\Delta x}{\Delta t} \tag{2.1}$$

Thus if we know t_1 is zero and t_2 is 4 s, we have

$$v_{avg} = \frac{0.6 \text{ m}}{4 \text{ s}} = 0.15 \text{ m/s}$$

Note that when we talk of the body's **speed** we are referring to how fast it's going—that is, the magnitude of the velocity. Thus if car A is going to the right at 16 km/h and car B is going to the left at 16 km/h, they have different velocities but the same speed.

Now that we know how to find the average velocity, how do we find the **instantaneous velocity**, the velocity at a particular instant of time t_1? All we have to do is take the limit as Δt goes to zero:

$$\begin{aligned} v(t_1) &= \lim_{\Delta t \to 0} \frac{x(t_2) - x(t_1)}{t_2 - t_1} = \lim_{\Delta t \to 0} \frac{x(t_1 + \Delta t) - x(t_1)}{(t_1 + \Delta t) - t_1} \\ &= \lim_{\Delta t \to 0} \frac{\Delta x}{\Delta t} = \frac{dx}{dt} = \dot{x} \end{aligned} \tag{2.2}$$

Finding the acceleration a is straightforward at this point. It is simply the time rate of change of velocity:

$$a(t_1) = \lim_{\Delta t \to 0} \frac{v(t_1 + \Delta t) - v(t_1)}{\Delta t} = \lim_{\Delta t \to 0} \frac{\Delta \dot{x}}{\Delta t} = \frac{d^2 x}{dt^2} = \ddot{x} \qquad (2.3)$$

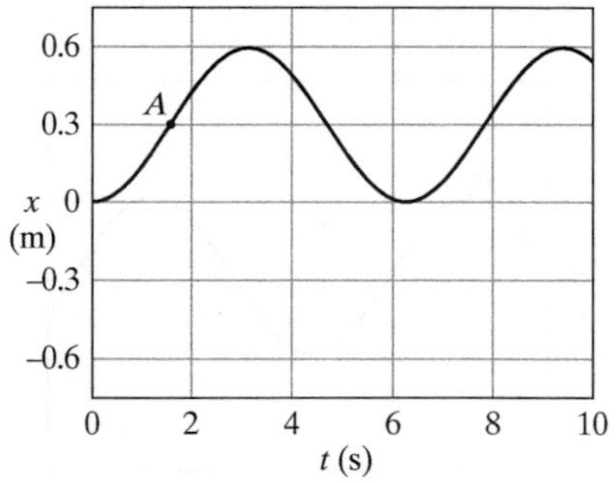

Figure 2.5 x versus t

Note that a negative acceleration is commonly referred to as a deceleration. **Figures 2.5, 2.6,** and **2.7** illustrate what $x, \dot{x}$, and $\ddot{x}$ look like when the body's motion is given by

$$x(t) = 1 - \cos t$$

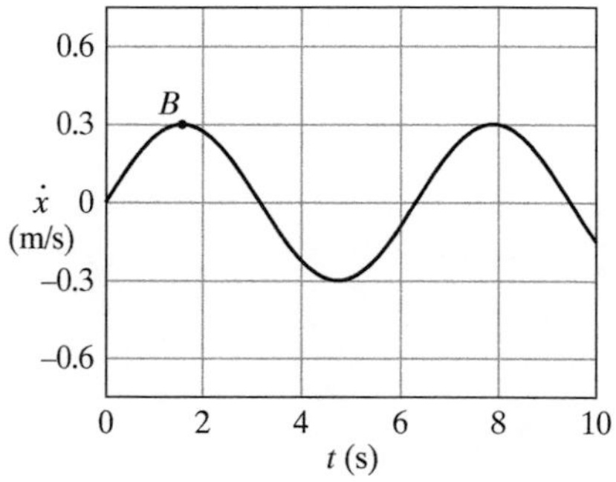

Figure 2.6 $\dot{x}$ versus t

(for which $\dot{x} = \sin t$ and $\ddot{x} = \cos t$). Note that you can easily see how the derivatives used in the preceding derivations come into play. In **Figure 2.5** the x position at $t = 1.6$ s is marked with the letter A. As you can determine from the plot, the slope at this point appears to be equal to 0.3. Going to **Figure 2.6** shows that the value of $\dot{x}$ at $t = 1.6$ s is indeed 0.3. The slope at B seems to be zero (the curve is flat at that point), and looking at **Figure 2.7** verifies that observation; $\ddot{x}$ is zero for $t = 1.6$ s.

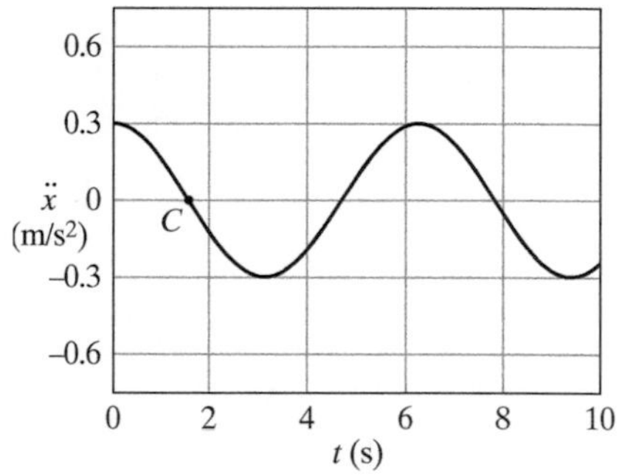

Figure 2.7 $\ddot{x}$ versus t

There's nothing to stop us from reversing this process and progressing from $\ddot{x}$ to $\dot{x}$ and then from there to x. In fact, this is the most common scenario for a dynamics analysis. Typically, we will have solved for the acceleration of some body and will then want to determine its velocity and position over time. Doing so, whether analytically or with the computer, will involve integration, namely, finding the area beneath the appropriate curves. **Figure 2.8** shows $\ddot{x}$ versus t, and the shaded area beneath the $\ddot{x}$ curve and between 5 s and 6 s has an area of $\Delta\dot{x}$. What this means is that if you know $\dot{x}$ at $t = 5$ s, all you need to do is add $\Delta\dot{x}$ to it in order to find $\dot{x}(6)$. Mathematically, for any two values of time t_1 and t_2, we have

$$\dot{x}(t_2) = \dot{x}(t_1) + \int_{t_1}^{t_2} \ddot{x}\, dt \qquad (2.4)$$

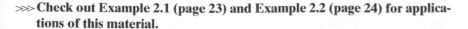

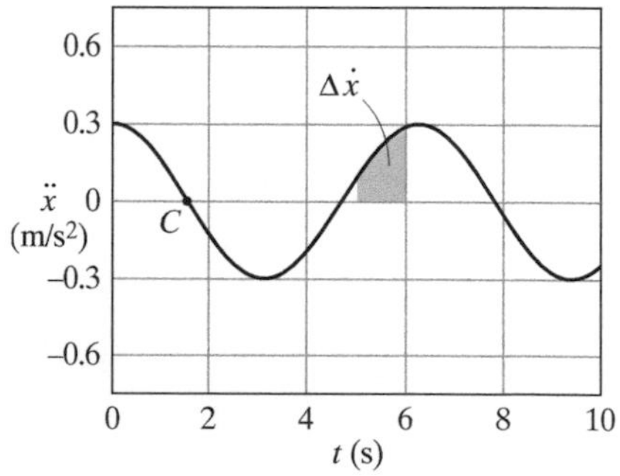

Figure 2.8 $\ddot{x}$ versus t

In an exactly analogous manner, the area under the $\dot{x}$ curve shown in **Figure 2.9** represents the change in position between $t = 5$ and $t = 6$ s. The general mathematical representation is given by

$$x(t_2) = x(t_1) + \int_{t_1}^{t_2} \dot{x}\, dt \qquad (2.5)$$

Notice that in this case the area lies beneath the $\dot{x} = 0$ line and thus represents a negative change in position; that is, the body moves in the $-x$ direction, as can be seen in **Figure 2.5**.

>>> **Check out Example 2.1 (page 23) and Example 2.2 (page 24) for applications of this material.**

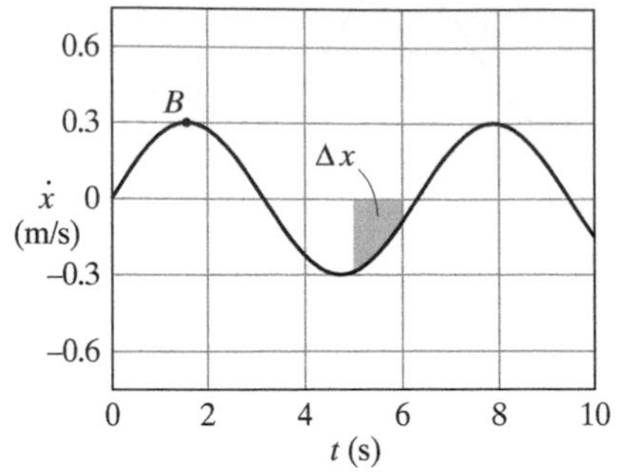

Figure 2.9 $\dot{x}$ versus t

When Acceleration Is Constant

If the problem at hand isn't too complex, we'll be able to solve it analytically. In the next few pages we'll do just that, both to help you brush up on your integration skills and to illustrate some nice dynamics results.

We have already seen the integral expressions for velocity and position based on a general acceleration ((2.4) and (2.5)). What we will now do is assume that the acceleration $\ddot{x}$ is constant and equal to $\overline{a}$. In this case, the velocity change will be linear with time and our general expression for $\dot{x}$ can be written as

$$\dot{x}(t_2) = \dot{x}(t_1) + \int_{t_1}^{t_2} \overline{a}\, dt = \dot{x}(t_1) + \overline{a} \int_{t_1}^{t_2} dt = \dot{x}(t_1) + \overline{a}(t_2 - t_1)$$

Letting t_1 be zero, denoting $\dot{x}(0)$ as v_0, and replacing t_2 by the general t give us

$$\overbrace{v(t)}^{\dot{x}(t_2)} = \overbrace{v_0}^{\dot{x}(t_1)} + \overbrace{\overline{a}t}^{\overline{a}(t_2-t_1)} \tag{2.6}$$

where $v \equiv \dot{x}$.

Integrating once again will yield the body's position as a function of time:

$$x(t) = x_0 + v_0 t + \overline{a}t^2/2 \tag{2.7}$$

where x_0 indicates the position at $t = 0$.

What (2.7) says is that the position of a body at some time t depends on where the body was initially (x_0), on what its initial velocity was ($v_0 t$), and on what its acceleration is ($\overline{a}t^2/2$). In fact, this result is very physical. If the body had no acceleration and no initial velocity, then it would always be at the initial position x_0. If there was an initial velocity but still no acceleration, the velocity would stay constant; thus, the body's distance from the starting position would increase linearly with time: $x_0 + v_0 t$.

Remember: These results are valid *ONLY if the acceleration is CONSTANT!* I can't tell you how often I've had students forget this and say that the displacement from some original position x_0 is equal to $a(t)t^2/2$ even though the acceleration $a(t)$ was time-varying. Don't let this happen to you!

If the acceleration isn't constant but depends explicitly on t, you can still integrate by hand to find velocity and position (as long as the time dependence is relatively simple). But if you have something annoyingly complex such as

$$\ddot{x} = \frac{\dot{x}^2}{x} + \dot{x}\sqrt{\frac{g}{x}}$$

then you're going to have to use a computer because there's no simple way to solve this problem with paper and pencil.

We can go even further by looking more closely at how position, velocity, and acceleration are related. The acceleration is just the time rate of change of the velocity:

$$a = \frac{dv}{dt}$$

Next, consider what velocity is, namely, the time rate of change of position,

$$v = \frac{dx}{dt}$$

Solving both equations for dt gives

$$a = \frac{dv}{dt} \quad \Rightarrow \quad dt = \frac{dv}{a}$$

$$v = \frac{dx}{dt} \quad \Rightarrow \quad dt = \frac{dx}{v}$$

Equate these two expressions for dt and you get

$$\frac{dv}{a} = \frac{dx}{v}$$

which, when rearranged, becomes

$$v\,dv = a\,dx \tag{2.8}$$

This relationship is a reasonably big deal. As in BIG deal. You'll be using it a lot. You'll note that time doesn't show up explicitly in (2.8). That's one of the interesting things about this relationship. It shows another way of looking at position, velocity, and acceleration and how they relate to one another.

Now let's see what we can get by letting the acceleration be constant. If you think back to your calculus class, you will recognize that both the left- and right-hand sides of (2.8) are exact differentials, meaning that we can immediately integrate them:

$$\int_{v_1}^{v_2} v\,dv = \bar{a} \int_{x_1}^{x_2} dx$$

Because $\bar{a}$ is constant, it can safely be pulled out of the integral on the right-hand side, as I did. Carrying through the integration gives us

$$\frac{v_2^2}{2} - \frac{v_1^2}{2} = \bar{a}(x_2 - x_1)$$

which can be expressed more simply as

$$v_2^2 = v_1^2 + 2\bar{a}(x_2 - x_1) \tag{2.9}$$

>>> **Check out Example 2.3 (page 25) for an application of this material.**

When Acceleration Depends on Position

Another case that comes up *very* frequently is when a body's acceleration depends on its position. This situation occurs, for instance, when the body is attached to a spring. In this case, the force acting on the body, and consequently the body's acceleration, depend on the body's position because the spring force is position-dependent (as the spring stretches, the force it exerts changes).

We can rewrite (2.8) to show this dependency:

$$v\,dv = a(x)\,dx$$

What we have now are two exact differentials, which means we can integrate both sides without having to find any sort of integrating factor:

$$\int_{v_1}^{v_2} v\,dv = \int_{x_1}^{x_2} a(x)\,dx$$

Integrating the left-hand side is quick. The integral of $v\,dv$ is $v^2/2$, and so we have

$$\frac{1}{2}(v_2^2 - v_1^2) = \int_{x_1}^{x_2} a(x)\,dx$$

$$v_2^2 = v_1^2 + 2\int_{x_1}^{x_2} a(x)\,dx \qquad (2.10)$$

What this tells us is that, because of the acceleration, the velocity of the body is going to change in some easily computable way.

>>> **Check out Example 2.4 (page 26) for an application of this material.**

When Acceleration Depends on Velocity

So far we have looked at constant acceleration and position-dependent acceleration. The last acceleration relationship that lends itself to hand calculation is when a body's acceleration depends on its velocity. This is the case, for example, when a body is experiencing a drag force that depends on velocity, as most do. Again we need to use our acceleration/velocity relationships. We have

$$a(v) = \frac{dv}{dt} \quad \Rightarrow \quad dt = \frac{dv}{a(v)}$$

$$v = \frac{dx}{dt} \quad \Rightarrow \quad dt = \frac{dx}{v}$$

Taken together, these relationships imply that

$$\frac{dx}{v} = \frac{dv}{a(v)} \quad \Rightarrow \quad dx = \frac{v\,dv}{a(v)}$$

Both sides can be integrated to yield

$$x_2 = x_1 + \int_{v_1}^{v_2} \frac{v}{a(v)}\, dv \tag{2.11}$$

This result shows that if you have an acceleration that depends in a reasonably simple fashion on velocity, you can determine the body's new position by carrying out the indicated integration.

>>> **Check out Example 2.5 (page 27) and Example 2.6 (page 28) for applications of this material.**

EXAMPLE 2.1 VELOCITY DETERMINATION VIA INTEGRATION (Theory on page 19)

A piston P with an initial speed of $v_0 = 3.20\,\text{cm/s}$ experiences an acceleration defined by $\ddot{x} = (a_1 + a_2 t)$ for 1.15 s ($a_1 = 2.50\,\text{cm/s}^2$ and $a_2 = -2.50\,\text{cm/s}^2$). What is the piston's final velocity?

Goal Determine the velocity of a piston.

Given Piston's initial speed, acceleration profile, and elapsed time.

Assume No assumptions are needed.

Draw **Figure 2.10** shows our system.

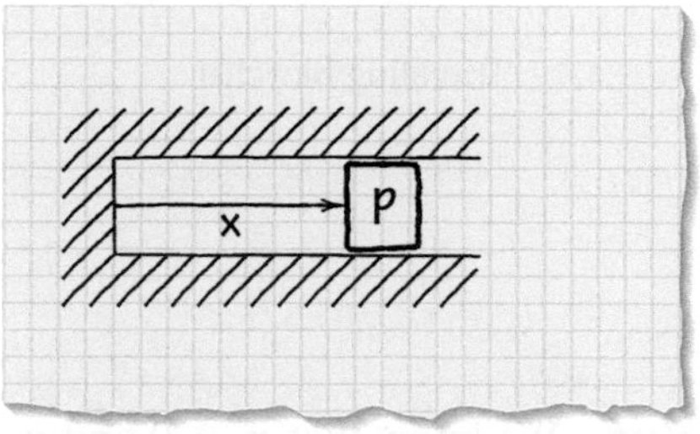

Figure 2.10 Moving piston

Formulate Equations

(2.4) ⇒

$$v(1.15) = v(0) + \int_0^{1.15\text{ s}} (a_1 + a_2 t)\, dt$$

Solve Carrying out the integration gets us

$$\begin{aligned} v(1.15) &= v(0) + \left(a_1 t + \frac{a_2}{2} t^2\right)\Big|_0^{1.15\text{ s}} \\ &= 3.20\,\text{cm/s} + (2.50\,\text{cm/s}^2)(1.15\,\text{s}) + (-2.50\,\text{cm/s}^2)(1.15\,\text{s})^2/2 \\ &= \boxed{4.42\,\text{cm/s}} \end{aligned}$$

Check We can do a geometric integration to validate our analytical one. The acceleration at $t = 0$ is $2.5\,\text{cm/s}^2$ and it changes linearly to $-0.375\,\text{cm/s}^2$. The average acceleration is therefore $1.06\,\text{cm/s}^2$. The net change in speed is therefore $(1.06\,\text{cm/s}^2)(1.15\,\text{s}) = 1.22\,\text{cm/s}$. Hence the final speed is the initial speed plus the net change: $3.20\,\text{cm/s} + 1.22\,\text{cm/s} = 4.42\,\text{cm/s}$.

EXAMPLE 2.2 DECELERATION LIMIT DETERMINATION (Theory on page 19)

Figure 2.11 Bicyclist and mammoth

Starting in the early days of the twenty-first century, Japanese scientists began trying to create a modern-day mammoth from frozen prehistoric mammoths. Assume that they succeeded, and, several years later, you're riding your bike in a trans-Siberian race when, after rounding a curve, you see a mammoth standing directly in front of you (**Figure 2.11**). You're traveling at 48 km/h and have only 6.7 m in which to stop. Hitting the mammoth is out of the question—mammoths are very irritable animals and you wouldn't want to annoy him. What constant negative acceleration is needed to just avoid a collision?

Goal Determine the constant negative acceleration that will reduce your speed from 48 km to zero in 6.7 m.

Given Initial and final speed as well as distance traveled.

Assume All 6.7 m will be used when coming to a stop.

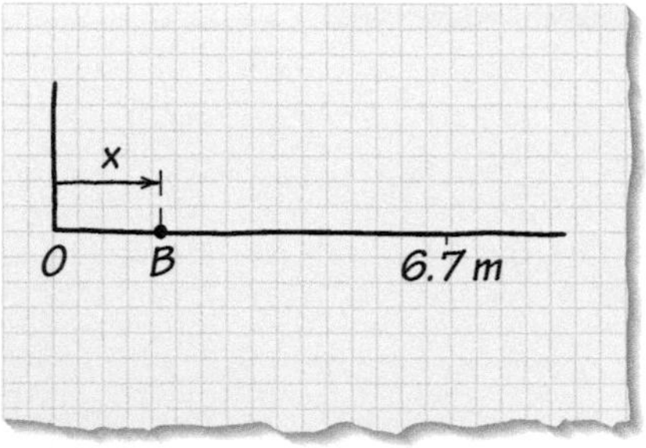

Figure 2.12 Traveling bicyclist

Draw **Figure 2.12** shows our one-dimensional situation. x gives the position of the bicyclist B, who starts at $x(0) = 0$ ($t_1 = 0$) and comes to a complete stop at $x(t_2) = 6.7$ m.

Formulate Equations We'll use $\overline{a}$ to represent the constant acceleration. Let $x_0 \equiv x(0)$ and $v_0 \equiv \dot{x}(0)$.

$$(2.4) \Rightarrow \qquad \dot{x}(t_2) = v_0 + \int_0^{t_2} \overline{a}\, dt = v_0 + \overline{a}t_2 \tag{2.12}$$

More generally, for any final time t, we have

$$\dot{x}(t) = \overbrace{v_0 + \overline{a}t}^{(2.6)} \quad \Rightarrow \quad \overbrace{x(t) = x_0 + v_0 t + \overline{a}t^2/2}^{(2.7)} \tag{2.13}$$

Solve We know that $x_0 = 0$ and $v_0 = 48\,\text{km/h} = 13.3$ m/s. Using these data in (2.13), along with the known distance to the mammoth of 6.7 m, yields

$$6.7\,\text{m} = (13.3\text{ m/s})t_2 + \overline{a}t_2^2/2 \tag{2.14}$$

where t_2 is the time at which the cyclist reaches $x = 6.7$ m. This equation has too many unknowns ($\overline{a}$ and t_2), and so we'll invoke (2.12), evaluated at $t = t_2$:

$$(2.13) \Rightarrow \qquad 0 = 13.3\text{ m/s} + \overline{a}t_2 \quad \Rightarrow \quad \overline{a} = -(13.3\text{ m/s})/t_2 \tag{2.15}$$

$$(2.15) \rightarrow (2.14) \Rightarrow \qquad 6.7\,\text{m} = (13.3\,\text{m/s})t_2 - (13.3\text{ m/s})t_2/2 \quad \Rightarrow \quad t_2 = 1.00\,\text{s} \tag{2.16}$$

$$(2.16) \rightarrow (2.15) \Rightarrow \qquad \boxed{\overline{a} = -13.3\text{ m/s}^2 = -1.36g}$$

Check We can work backwards and ask, starting from an initial speed of zero, how long it takes to reach a speed of 13.3 m/s if we accelerate at 13.3 m/s^2. Clearly, the answer is 1 s, matching the result just found. In 1 s, starting from rest and assuming a constant acceleration of 13.3 m/s^2, we would travel a distance $x = \frac{(13.3\text{ m/s}^2)(1\text{ s})^2}{2} = 6.7$ m.

EXAMPLE 2.3 **CONSTANT ACCELERATION/SPEED/DISTANCE RELATIONSHIP** (Theory on page 21)

Imagine that you're a spy journalist for a major car magazine, secretly observing preproduction tests on a new sports car. From your vantage point high atop a tree overlooking the track, you can see the radar gun being used to track the car. Your purpose in spying is to determine the magnitude of the car's average acceleration. (Later, back in the safety of your cubicle, you can use the acceleration data to make some good guesses about the engine's horsepower and torque.) In one time trial, you see that the car, starting from rest, has moved exactly 100 m at the point where it hits 100 km/h. Unfortunately for you, the battery of your digital watch ran out just 3 min before testing began, thus giving you no way to time the trial. But you need to get the acceleration information. How do you do it?

Goal Find the magnitude of the car's average acceleration.

Given Initial and final speed and distance traveled.

Assume Acceleration is constant.

Draw **Figure 2.13** shows our system.

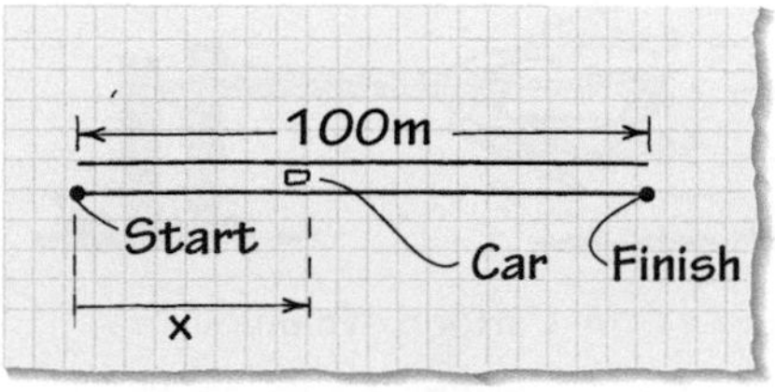

Figure 2.13 Car and test track

Formulate Equations Because you're looking for the average acceleration, which is constant by definition, all you need to find is the magnitude of the constant acceleration that, over 100 m, brings the car to 100 km/h. Equation (2.9) is ideally suited for the task.

First, convert km/h to m/s:

$$(100\,\text{km/h})\left(\frac{1000\,\text{m/km}}{3600\,\text{s/h}}\right) = 27.8\,\text{m/s}$$

Using (2.9) you will have

$$(27.8\,\text{m/s})^2 = (0)^2 + 2\bar{a}(100\,\text{m} - 0)$$

Solve

$$\bar{a} = \frac{(27.8\,\text{m/s})^2}{200\,\text{m}} = 3.86\,\text{m/s}^2 = 0.393g \qquad (2.17)$$

Take a moment to be a little impressed with this. In the bad old days, you would have required speed/time data if you were trying to find acceleration. Now you can do it with speed/position data. Your bag of tricks just got larger.

Check For this example we will just apply a "Does it make sense?" check. Our result is an acceleration of $0.393g$. That's a reasonable answer. A typical acceleration during normal driving is around $0.1g$, and $0.393g$ is about what you would expect from a midrange sports car undergoing hard acceleration.

EXAMPLE 2.4 POSITION-DEPENDENT ACCELERATION (Theory on page 22)

Figure 2.14 Cartoon dynamics

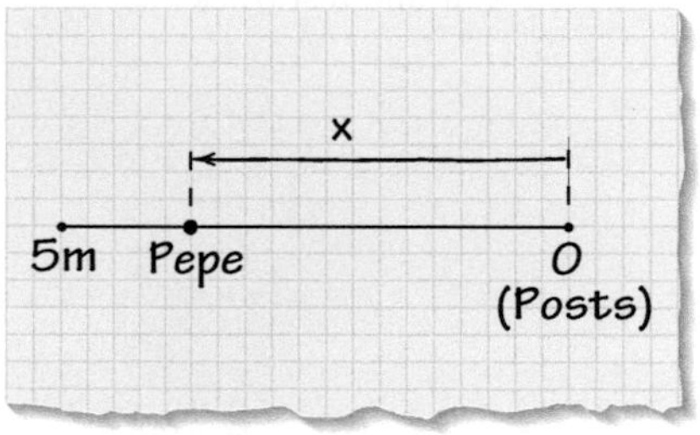

Figure 2.15 Cartoon schematic

Here's a cartoon dynamics problem. Pepe Coyote is standing within a stretched rubber band, and Rodney Rabbit is about to snip the rope that's keeping the rubber band stretched (**Figure 2.14**). After Rodney snips the restraining rope, Pepe zips backward, pulled along by the retracting rubber band. Determine Pepe's speed at the instant he reaches the rubber band's support posts. Initially he's 5 m from the posts and experiences an acceleration of $a_1 x$ toward them (x, given in meters, is measured from the posts to Pepe, and $a_1 = 2\,\text{s}^{-2}$).

Goal Find Pepe's speed at the instant he reaches the support posts.

Given Initial distance from posts and acceleration as a function of position.

Assume Pepe moves horizontally.

Draw **Figure 2.15** shows our simplified setup.

Formulate Equations Pepe's acceleration is given as $a_{\text{Pepe}} = -a_1 x$. We've defined the positive x direction to the left, and Pepe's acceleration is to the right, thus requiring the minus sign in the preceding equation. Unlike the previous examples, here the acceleration depends explicitly on displacement rather than being constant. The formula to use is (2.10):

$$v_2^2 = v_1^2 + 2\int_{x_1}^{x_2} a(x)\,dx$$

Solve Let v_1 be Pepe's initial speed (0) and v_2 his speed at the posts. We also know that $x_2 = 0, x_1 = 5$ m, and $a_1 = 2\,\text{s}^{-2}$. Using these values in (2.10) gives

$$v_2^2 = -2\int_5^0 a_1 x\,dx = -a_1 x^2\Big|_5^0 = a_1(25\,\text{m}^2) = 50\,(\text{m/s})^2$$

$$\boxed{v_2 = 5\sqrt{2}\ \text{m/s}}$$

Note that there's no mathematical reason to choose the positive square root over the negative one. Right now we have to rely on our engineering insight, knowing that Pepe is going to be moving to the right when he reaches the posts because that's the direction in which he's been accelerating.

Check There's not much to check mathematically. We can ask if the result seems reasonable. (Of course, asking cartoons to be reasonable is going a bit far, but let's check anyway.) The speed $5\sqrt{2}$ m/s is about 7 m/s. That's not an unreasonable answer—it's the speed I cruise at when I'm on my bicycle. Well within believable limits for a cartoon coyote propelled by a rubber band.

EXAMPLE 2.5 VELOCITY-DEPENDENT ACCELERATION (A) (Theory on page 23)

A sandbag is dropped from a hot air balloon (**Figure 2.16**). The net effect of both aerodynamic drag and gravity is to produce an acceleration $\ddot{y} = g - cv^2$, where c (equal to $6.00 \times 10^{-4}\ \mathrm{m}^{-1}$) is a drag coefficient with units of m^{-1} when v is expressed in m/s. The positive y direction is downward. Determine the sandbag's speed after it has fallen 400 m.

Figure 2.16 Balloon drop

Goal Find a released sandbag's speed as a function of position.

Given Sandbag drops 400 m. $\ddot{y} = g - cv^2$

Assume No assumptions are needed.

Draw **Figure 2.17** shows our system.

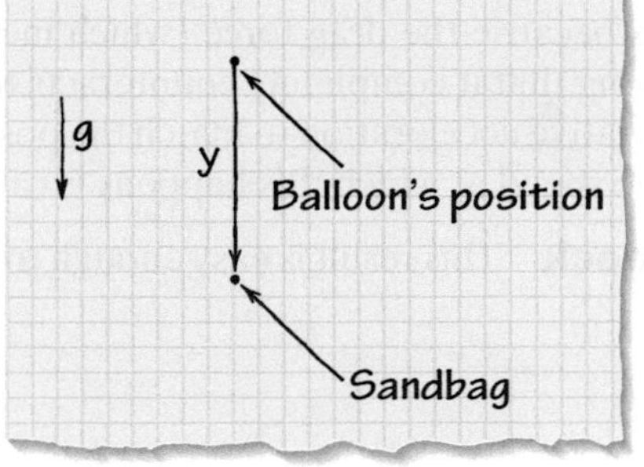

Figure 2.17 Coordinates for problem

Formulate Equations The equation we need to solve is $\ddot{y} = g - c\dot{y}^2$, which is nonlinear and therefore a tough nut to crack. However, thanks to (2.11), we can handle it. Using $\ddot{y} = g - cv^2$ and (2.11) yields

$$y_2 = \int_{v_1}^{v_2} \frac{v}{g - cv^2}\, dv \tag{2.18}$$

where $y_1 = y(0) = 0$.

Solve If you managed to stay awake in calculus (and even if you didn't, I'm going to tell you anyway), you'll recall that $d(\ln y)/dy = 1/y$ and, by using the chain rule, leads to

$$\frac{d}{dy}\ln(f(y)) = \frac{f'(y)}{f(y)}$$

If we identify the function f as $g - cv^2$, with v the independent variable, then we can apply this differential relationship to our system and obtain

$$y_2 = -\frac{1}{2c}\ln\left(g - cv^2\right)\Big|_0^{v_2} = \frac{1}{2c}\left[\ln\left(g - cv_2^2\right) - \ln(g)\right]$$

Using the given values for g, c, and y_2 lets us solve for v_2:

$$v_2 = 79.0\ \mathrm{m/s}$$

Check 79 m/s is around 284 km/h, about the top speed of a very high-performance sports car. This seems like a believable speed, given that the sandbag has fallen for about a quarter of a mile. Another check is to go numerical and use the computer. We know that the acceleration is given by $\ddot{y} = g - c\dot{y}^2$. All we need to do to solve this is use a numerical integrator. Appendix A, an introduction to MATLAB, provides one approach: ode45.m. This m-file contains a Runge-Kutta integrator, a general-purpose numerical integrator that will easily handle our equation. Numerically integrating for 9.4 s produces a displacement of 400 m and a speed of 79.0 m/s.

EXAMPLE 2.6 VELOCITY-DEPENDENT ACCELERATION (B) (Theory on page 23)

Increase your insight into falling bodies by resolving the previous example for the speed corresponding to a drop of 40,000 m. Show that the solution you obtain can be deduced from the original acceleration formula.

Goals Find the sandbag's speed when it has traveled 4.00×10^4 m. Show that the result can be found from $\ddot{y} = g - cv^2$.

Given Sandbag falls 4.00×10^4 m.

Assume Assume that the acceleration due to gravity stays constant at 1 g over the 40,000 m.

Draw See **Figure 2.17**.

Formulate Equations From Example 2.5 we have $a = g - cv^2$ and

$$y_2 = -\frac{1}{2c}\left[\ln(g - cv_2^2) - \ln(g)\right]$$

Solve Using $y_2 = 4.00 \times 10^4$ m and solving give us $v_2 = 128$ m/s. This is faster than the speed found in Example 2.5, which makes sense because the sandbag has now fallen farther. But actually, this new speed is not much faster. After all, the sandbag has traveled 100 times as far, but the speed hasn't even doubled. Why is this? Well, it's because the drag force (which increases with the square of the speed) gets large pretty quickly and soon is so big that it completely counteracts the acceleration due to gravity. So, in the limit, we expect the drag-induced negative acceleration to match the positive gravitational acceleration and thus produce a net acceleration of zero. The speed at which this happens is called the body's **terminal speed**.

Check This result is easy enough to verify. Just set the expression for acceleration equal to zero and solve for v:

$$g - cv^2 = a = 0$$

$$v = \sqrt{\frac{g}{c}} = \sqrt{\frac{9.81 \text{ m/s}^2}{6.00 \times 10^{-4} \text{ m}^{-1}}} = 128 \text{ m/s}$$

This matches our first solution, showing that by the time the sandbag has dropped 40,000 m, it has already reached its terminal speed.

EXERCISES 2.1

2.1.1. [Level 1] The 2007 BMW Z4 Coupe 3.0si can accelerate from 0 to 96 km/h in 5.6 s, and its maximum speed is $\dot{x}_{max} = 249$ km/h. Assuming the car accelerates from 0 to 96 km/h at a constant rate and that it can continue to accelerate at this rate without redlining, find how far the car travels and how long it takes to reach its maximum speed.

E2.1.1

2.1.2. [Level 1] A map tells you to go 3 paces east, 2 paces south, 5 paces east, 7 paces north, and 4 paces west. Plot the complete path as well as the overall vector from the initial to final position. Then plot the same thing but in reverse order, that is, 4 paces west, 7 paces north, and so on. Do you end up at the same position?

2.1.3. [Level 1] A particle moves as shown. What is the average speed over the illustrated time interval? Estimate the maximum and minimum speeds from the graph.

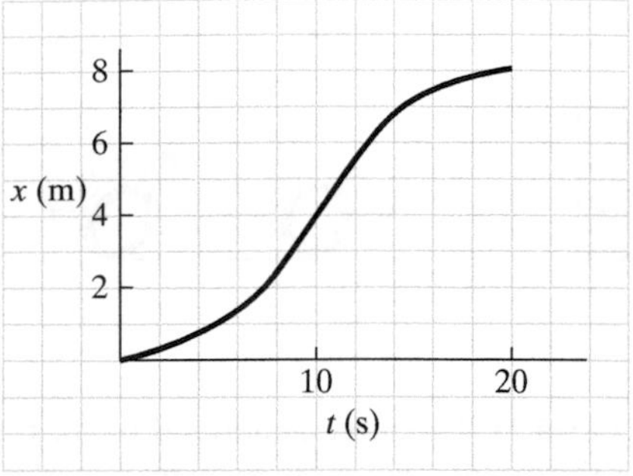

E2.1.3

2.1.4. [Level 1] Find $\dot{x}(0)$ such that $\dot{x}(2.5) = 0$ m/s for a particle whose acceleration is given in the graph.

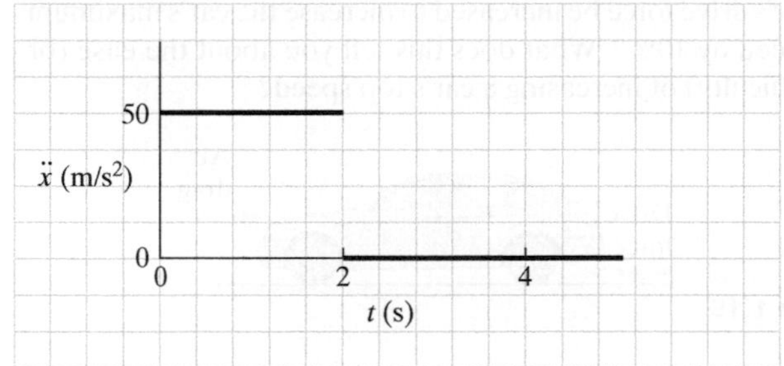

E2.1.4

2.1.5. [Level 1] A ball bearing is accelerated from rest by means of an electromagnet according to $\ddot{s} = as + be^{cs}$, where $a = 3\ \text{s}^{-2}$, $b = 0.3\ \text{m/s}^2$, and $c = 0.06\ \text{m}^{-1}$. How fast is the ball bearing moving after it has traveled $d = 3$ m?

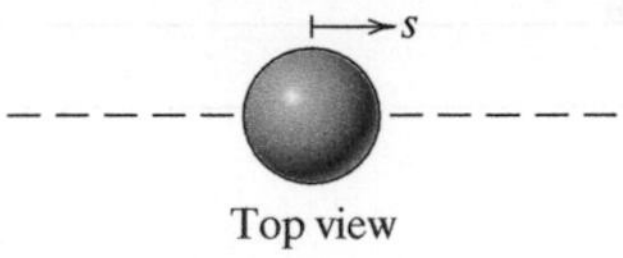

E2.1.5

2.1.6. [Level 1] One of the severe dynamics realities of aircraft operations on an aircraft carrier is the need to stop the fighter plane from airspeed to zero in a very short distance. To do this the jet actually trails behind it a hook which catches onto a cable that very decisively brings the jet to a stop.

Assume that a TOPGUN jet touches down at 274 km/h and its speed is reduced to zero in 73 m. Assuming a constant deceleration, calculate the time elapsed and the acceleration acting on the jet (expressed in both m/s^2 and g's).

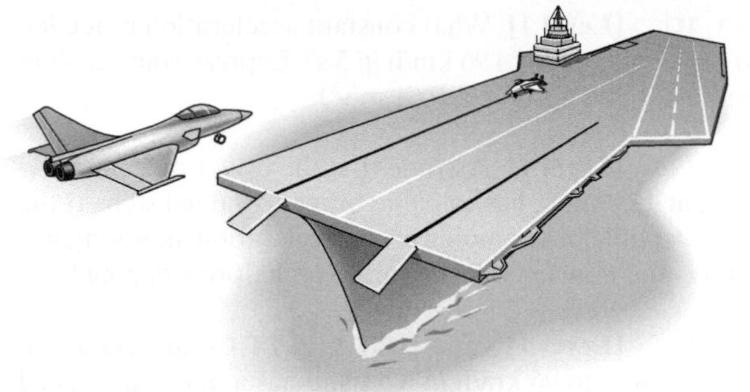

E2.1.6

2.1.7. [Level 1] Taking off from an aircraft carrier is a nontrivial operation. Basically, there's not enough room for the jet to taxi normally, and thus an additional means of getting the jet up to takeoff speed is needed. What's commonly used is a steam-driven catapult. Large pistons move within their cylinders, driven by steam pressure at one end, and tow the jet forward. Assume that the jet is brought from a stationary state to 265 km/h in 2.23 s and that the acceleration was constant. How much deck does this imply must be in front of the jet?

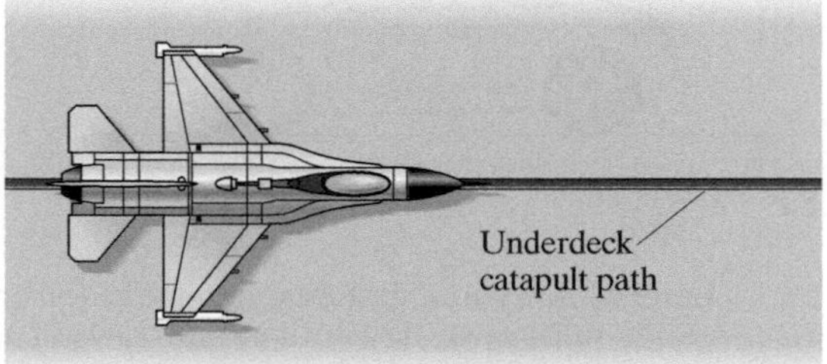

E2.1.7

2.1.8. [Level 1] The displacement of a particle is given by

$$x(t) = a_1 + a_2 t + a_3 t^3$$

where $a_1 = 10$ m, $a_2 = -16$ m/s, and $a_3 = \frac{1}{3}$ m/s^3. Plot the speed and acceleration for $0 \le t < 6$ s. Find the value of t for which $\dot{x} = 0$ m and indicate it on your plot of $\dot{x}$ and $\ddot{x}$.

2.1.9. [Level 1] The displacement of a car is given by

$$x(t) = a_1 + a_2 t + a_3 t^2$$

where $a_1 = 30$ m, $a_2 = -27$ m/s, and $a_3 = 3$ m/s^2.

a. What is the car's speed at $t = 2$ s?
b. What is the car's acceleration at $t = 10$ s?

2.1.10. **[Level 1]** What constant acceleration is needed to get a car from 0 to 96 km/h in 5 s? Express your result in both m/s^2 and g's ($1\ g = 9.81$ m/s^2).

2.1.11. **[Level 1]** Assume that the driver of a car, traveling at 113 km/h, has to panic-brake to a dead stop. If the car is capable of sustaining a 1 g deceleration, how long will it take and how far will the car travel before stopping?

2.1.12. **[Level 1]** The 2008 Audi TT Coupe can accelerate from 0 to 96 km/h in 6.5 s and has a maximum speed of $\dot{x}_{max} = 237$ km/h. Assume that the car accelerates from 0 to 96 km/h at a constant rate and that it can continue to accelerate at this rate without redlining.

a. How far does the car travel before reaching its maximum speed?

b. Suppose a competitor's car needs 0.50 km to accelerate to its maximum speed (which is slightly higher) under the same assumptions. How does the TT compare?

c. What change in the 0-to-60 time is needed so that both vehicles reach their respective maximum speeds in the same length of road?

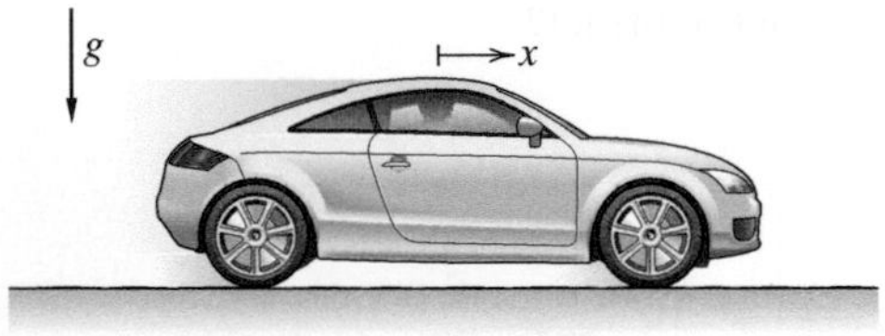

E2.1.12

2.1.13. **[Level 1]** A favorite sports car of mine has an engine for which the bore/stroke is 3.40/8.97 cm. This means that the piston moves 8.97 cm (and then back again) during one cycle of operation. Let's assume that it does this in a sinusoidal fashion:

$$x(t) = \frac{8.97 \text{ cm}}{2} \sin(\omega t)$$

The engine operates at a maximum of 7000 rpm. What is the maximum speed of the piston, and what is its maximum acceleration (in g's)?

2.1.14. **[Level 1]** Here's a nice exercise that has some important real-world applications. Safety experts are always telling us that it's important to buckle up, but how important is it really? I mean, come on—moving at city speeds can't be all that dangerous, can it? You can get a physical feel for the answer to this question by asking how far you would have to fall vertically in order for your contact speed with the ground to equal 56 km/h (a "typical" speed in an automotive collision). You should picture yourself falling from that height, face first, onto a hard floor because that's essentially what will be happening if you get into a crash without a fastened seatbelt. The car will smash into whatever it's colliding with and essentially stop immediately. Your body, meanwhile, having nothing to restrain it, will continue moving at 56 km/h until you smash into the dash. Find the height h from which you will contact the floor at 56 km/h.

E2.1.14

2.1.15. **[Level 1]** The force generated by air resistance that opposes the motion of a car is given by $F_a = Av^2$, where A is a constant that depends on the density of the air and the shape of the car and v is the car's speed. When the car is traveling as fast as its engine can drive it, there exists a balance between the drag force and the drive force due to the car's engine: $F_d = F_a$. By what percentage must the car's drive force be increased to increase the car's maximum speed by 10%? What does this tell you about the ease (or difficulty) of increasing a car's top speed?

E2.1.15

2.1.16. **[Level 1]** Two bicycle racers are racing along a straight course. Bicyclist A has a lead of 229 m and has a mile to go until he reaches the finish line. He maintains a constant speed of 29 km/h. At what constant speed will bicyclist B need to travel to catch bicyclist A at the finish line?

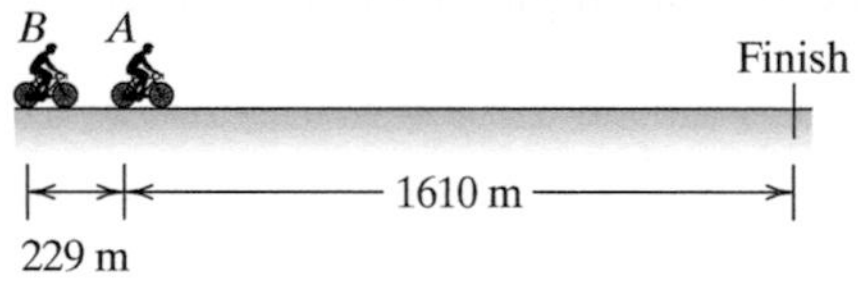

E2.1.16

2.1.17. **[Level 1]** **Computational** A body traveling in a straight line starts from rest and experiences an acceleration equal to $a_0 - a_1 v^3$ where v is the body's speed. $a_0 = 24{,}384$ m/s^2 and $a_1 = 0.108$ s/m^2. Determine the body's terminal speed v_{term} (speed as time approaches infinity). How long does it take for the body to reach 95% of v_{term}?

2.1.18. **[Level 2]** A plot of acceleration versus time for a particle is shown. Estimate the position of the particle at $t = 3$ s if it starts at $x = 1$ m with a speed of $\dot{x} = 0$ m/s.

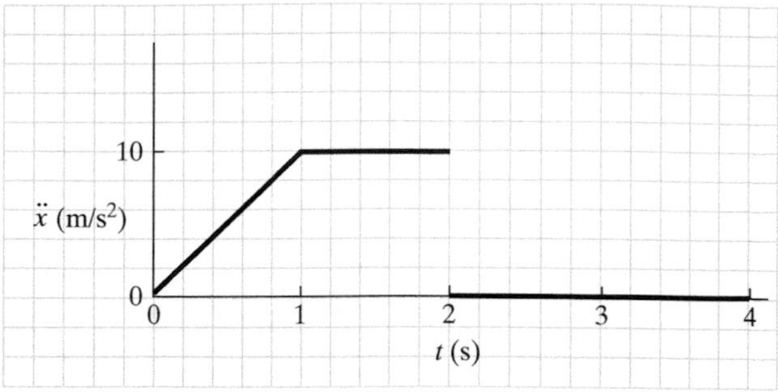

E2.1.18

2.1.19. **[Level 2]** A plot of acceleration versus time for a particle is shown. What's the difference between its position at $t = 4$ s and $t = 0$ s if $\dot{x}(0) = -4$ m/s?

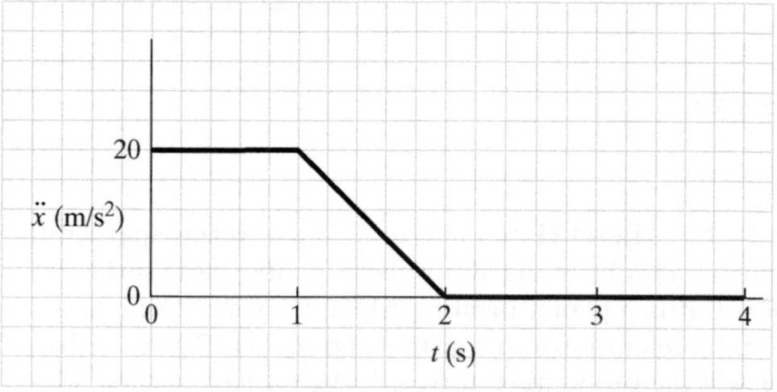

E2.1.19

2.1.20. **[Level 2]** A ball is launched vertically with an initial speed of $\dot{y}_0 = 50$ m/s, and its acceleration is governed by $\ddot{y} = -g - c_D\dot{y}^2$, where the air drag coefficient c_D is given by $c_D = 0.001\ \text{m}^{-1}$. What is the maximum height that the ball reaches? Compare this to the maximum height achieved when air drag is neglected.

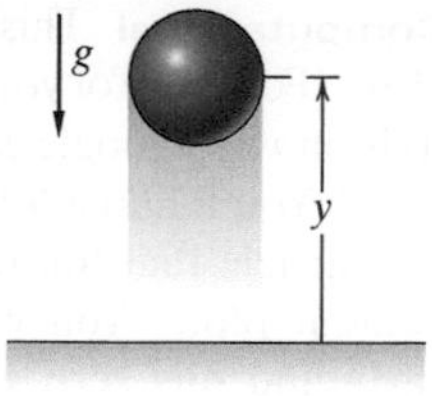

E2.1.20

2.1.21. **[Level 2]** An important area in controls research is to determine minimum-time solutions to a problem. Consider the following case. You're asked to start from rest, go a distance $L = 105$ m, and end up at rest once again, all in minimum time. Your vehicle has a maximum acceleration capability of a_1 and a maximum braking capability of a_2. The minimum-time solution is to use maximum braking and acceleration. Determine t_1, t_2, and the car's maximum speed. $a_1 = 7.5\ \text{m/s}^2$ and $a_2 = -10\ \text{m/s}^2$.

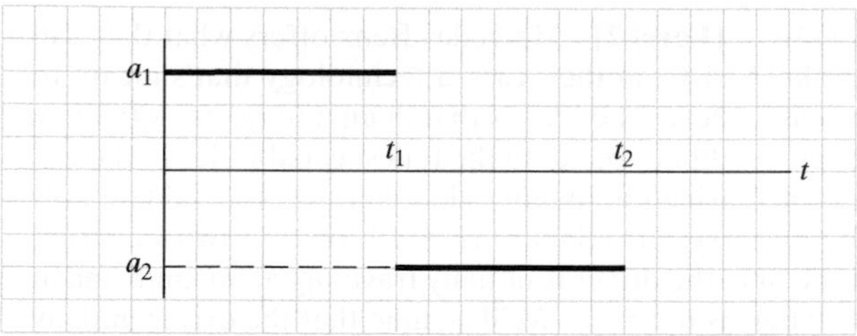

E2.1.21

2.1.22. **[Level 2]** In this problem we'll analyze a car collision. Assume that both cars A and B are initially traveling to the right at 30 m/s. Car A is tailgating, leaving very little room between the front (F) of his car and the rear (R) of car B. In spite of this, the driver in car A starts fiddling with the radio and stops paying any attention to car B. At this instant, the driver in car B notices an accident ahead of him and starts to brake, causing a deceleration of 6 m/s². By the time the driver of car A looks up, the gap between the two cars has narrowed to 6.9 m and the speed of car B has dropped to 24 m/s.

Assume that the driver of car A immediately reacts and tries to apply maximum braking. The driver of car A has good reflexes and it therefore takes "only" 0.5 s for him to hit the brakes. Assume that his car has better brakes than the one in front and that his car begins to decelerate at 9 m/s² upon contact with the brake pedal (at $t = 0.5$ s). Calculate when the collision occurs and the relative speed of the collision.

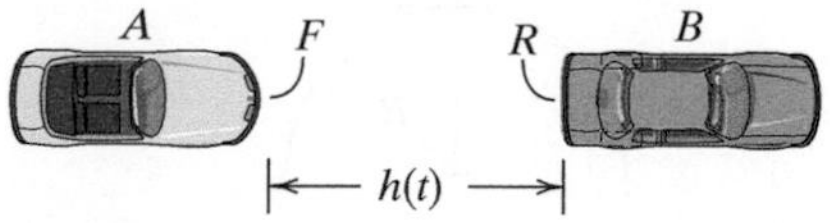

E2.1.22

2.1.23. **[Level 2]** In this problem we'll examine a two-car system similar to that of **Exercise 2.1.22**. Again, both cars A and B are initially traveling to the right at 30 m/s. In this case the following car has worse brakes than the car in front, thus making it more difficult to avoid a collision when the lead car starts braking. At $t = 0$ the driver in car B sees an accident taking place ahead of him and jams on his brakes, causing a deceleration of 10.7 m/s². It takes 1 s for the driver of car A to react to this and begin braking with a deceleration of 9 m/s². The initial separation between the cars is 9 m. Determine whether or not car A collides with car B. If it does, what is the relative collision speed?

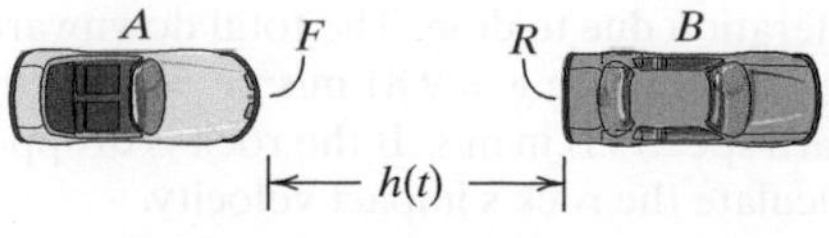

E2.1.23

2.1.24. **[Level 2]** Mercedes Benz offers what they call "Brake Assist" on their cars, a technology that's meant to increase safety. The idea is that in an emergency situation, people might not brake to the fullest capabilities of the car, thus increasing the likelihood of a crash. To evaluate how well such a system might work, we'll consider two scenarios. In the first, the driver is initially traveling at 96 km/h and at $t = 0$ begins to brake. We'll assume that the driver initially uses only part of the car's full braking, causing a deceleration of 6 m/s^2. After 1 s, we'll assume that the driver then jams the brakes on fully, achieving a braking deceleration of 9 m/s^2.

Compare the absolute and percentage difference between this scenario and one in which the Brake Assist system ensured maximum braking (9 m/s^2) for the entire braking interval.

2.1.25. **[Level 2]** A Separatist Droid Army battlestation is damaged during an intense battle over Coruscant, and it begins to fall toward the planet with an acceleration of

$$\ddot{s} = -\frac{g_0 R^2}{s^2}$$

where s is the battlestation's distance from Coruscant's center and R is Coruscant's radius. $g_0 = 9$ m/s^2, $R = 6440$ km, and the initial height above the planet is $H = 644$ km. What is the battlestation's impact speed?

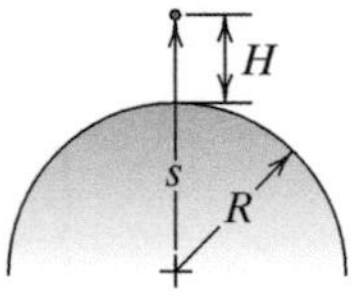

E2.1.25

2.1.26. **[Level 2]** A particle is initially at $x_0 \equiv x(0) = 20$ m, and its time history after $t = 0$ is given by

$$x(t) = [x_0 + c_1 t + c_2 t^3]$$

where $c_1 = 180$ m/s, $c_2 = -30$ m/s^3, and t is expressed in seconds. If you differentiate this to solve for the time at which the speed equals zero, you will obtain two results, one positive and the other negative. What's the meaning behind the negative result?

2.1.27. **[Level 2]** A rock that's dropped from the top of a cliff will experience an acceleration due to gravity, along with a deceleration due to drag. The total downward acceleration is $g - c_d \dot{s}^2$, where $g = 9.81$ m/s^2, $c_d = 0.01$ m^{-1}, and the downward speed $\dot{s}$ is in m/s. If the rock is dropped from 30 m up, calculate the rock's impact velocity.

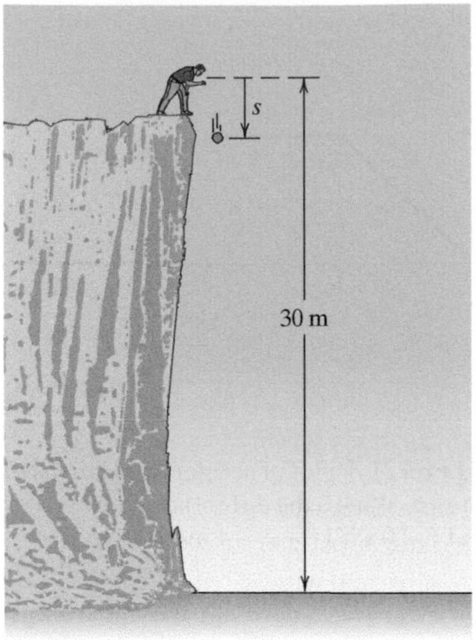

E2.1.27

2.1.28. **[Level 2]** In the year 2137 a hovering spacecraft releases a probe in the atmosphere far above the surface of New Earth, a planet orbiting Alpha Proximi. The gravitational acceleration g_1 is constant and equal to 7.5 m/s^2. How far will the probe have traveled when it has reached 98% of its terminal speed? The probe's speed is governed by

$$\ddot{y} = -g_1 + c\dot{y}^2$$

where y is the probe's position above the planet's surface (expressed in m). $c = 1.2 \times 10^{-4}$ m^{-1}.

2.1.29. **[Level 3] Computational** This exercise is a variation on Zeno's Paradox. (For those of you who haven't run across this, run a search on www.google.com with the keywords "Zeno's paradox." You'll find a lot of information.) In our case we have a particle that follows the illustrated speed versus displacement plot. You will note that the speed drops off linearly and hits zero when the particle reaches $s = 8$ m.

a. Use a numerical integrator (see Appendix A) to determine how fast it is going and how far it has traveled if it starts from $\dot{x} = 10$ m/s and travels for 5 s. How about after 10 s?

b. Determine how long it takes to get to 7.99 m, 7.999 m, and 7.9999 m. (Note that each time we're getting 10 times closer to 8.0000 meters than the time before.)

c. What's the relationship between these successive jumps toward $x = 8.0000$ m? Can you analytically predict the change in time associated with one of these 90% jumps?

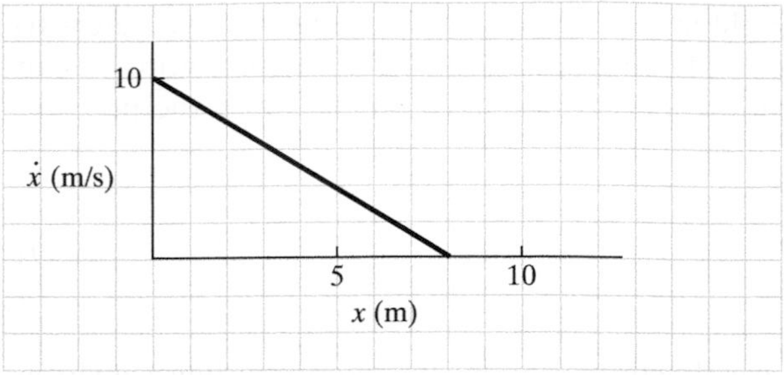

E2.1.29

2.2 CARTESIAN COORDINATES

Coordinate systems—you can't live with 'em, and you can't live without 'em. Well, okay, most people probably can live without them, but we're going to have to deal with them. So, what are coordinate systems anyway? Simply put, they're tools that allow you to define where you are in space now and how to get to where you're going next. We were able to sidestep the issue in the previous section because the body we were looking at was always constrained to be somewhere on a straight line. But what happens when a body is free to wander on a flat surface? For instance, think about a pirate searching for some buried treasure. He's got a map (**Figure 2.18**) that says, "Start at the flat rock. Walk 3 paces north. Walk 4 paces east. Dig." This is a valid coordinate system. The units of measure are paces, and the system is oriented with respect to a reference frame (north/south, east/west) fixed to the ground.

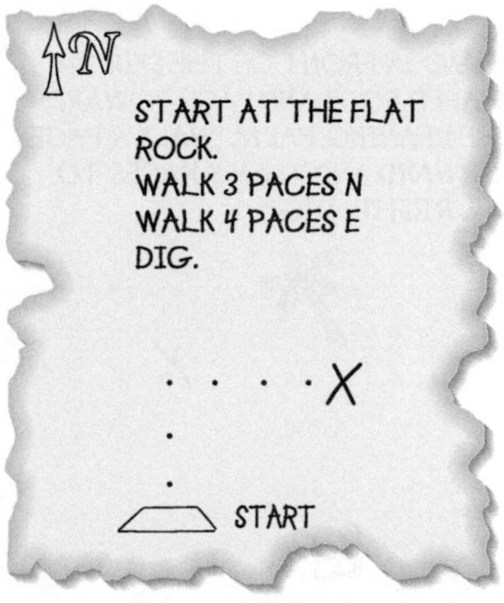

Figure 2.18 Ground-fixed coordinate system

We need to examine coordinate systems because we want to study the motion of bodies, and to do so, we need a way of determining where the bodies are and where they're going. This falls under the heading of **kinematics**, which is the study of the geometry of motion without regard to forces. It is kinematics that lets us represent the $\boldsymbol{a}$ part of $\boldsymbol{F} = m\boldsymbol{a}$. If someone designs a linkage as part of a multilink rear suspension for a car, that person uses kinematics. No attention is paid to forces or moments in the system—all that's looked at is how the system moves. Of course, before the suspension can be used on a car, an analysis that looks at the forces of the system must be done, but initially only the kinematics is analyzed.

Once you have learned how to describe the motion of a system, it will be a relatively simple matter to add forces to the problem and then solve for the system's equations of motion. The next few sections introduce the coordinate systems we will be using to describe the motion of systems.

The first set of coordinates we will look at is the simplest to deal with—**Cartesian coordinates**. These coordinates are named after the seventeenth-century philosopher and mathematician René Descartes and are also known as **rectangular coordinates**. They're the ones used by the pirate just mentioned.

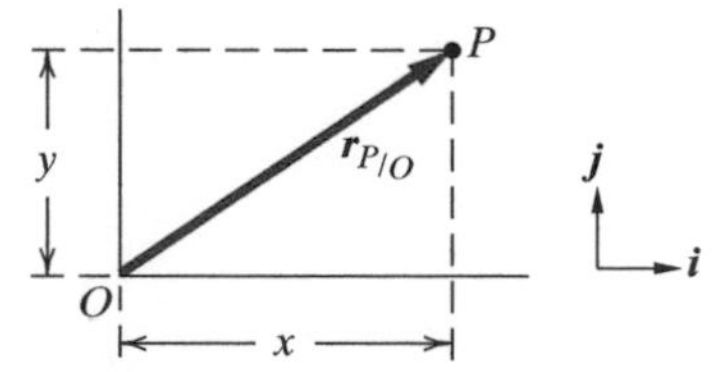

Figure 2.19 Cartesian coordinates

Figure 2.19 shows a typical Cartesian setup. There are two axes (one horizontal and one vertical) and a position P. The vectors $\boldsymbol{i}$ and $\boldsymbol{j}$ are called **unit vectors** and serve the same purpose as writing "go east" or "go north" on the treasure map. They allow us to get from the origin O to the location P. These two vectors constitute an **orthogonal** set of vectors; that is, they're oriented at right angles to each other. The word *unit* refers to the fact that each vector is of unit length: $\|\boldsymbol{i}\| = \|\boldsymbol{j}\| = 1$. Therefore we can express P's position with respect to the fixed point O with the position vector $\boldsymbol{r}_{P/O}$:

$$\boldsymbol{r}_{P/O} = x\boldsymbol{i} + y\boldsymbol{j} \tag{2.19}$$

To use our pirate map to illustrate the concept of representing a position vector with unit vectors, we would say that in **Figure 2.18** the location of the X relative to the flat rock is given by

$$\boldsymbol{r}_{X/\text{rock}} = [4\boldsymbol{i} + 3\boldsymbol{j}]\text{ paces}$$

where $\boldsymbol{i}$ points east and $\boldsymbol{j}$ points north.

This is a good place to introduce the concept of a **moving reference frame**. For each of the preceding examples, the reference frame associated with the coordinate system is fixed in space; that is, it's a **fixed reference frame**. Going north or east means walking in a direction that's referenced to the ground, something that's not moving about (unless you live in California). Although this is fine, it isn't the only way to give directions. Consider **Figure 2.20**, for instance. In this map the instructions are to first stand in front of the skull-shaped rock and face the leaning palm. The skull-shaped rock provides an initial starting point, and facing the leaning palm provides an initial orientation. The instructions are then based purely on the person doing the walking: "Walk 5 paces forward and then 4 paces to the right." If you happened to start off facing north, then walking 5 paces "forward" means you're going north. But if after walking 4 paces to the "right" you're asked to go "forward" again, then you'll be going east. "Forward" and "right" don't have a fixed meaning with respect to the ground, although they're certainly well defined with respect to the person doing the walking. A reference frame that's fixed to the ground is called, not surprisingly, **ground-fixed** and a reference frame attached to a moving body is called, equally unsurprisingly, **body-fixed**.

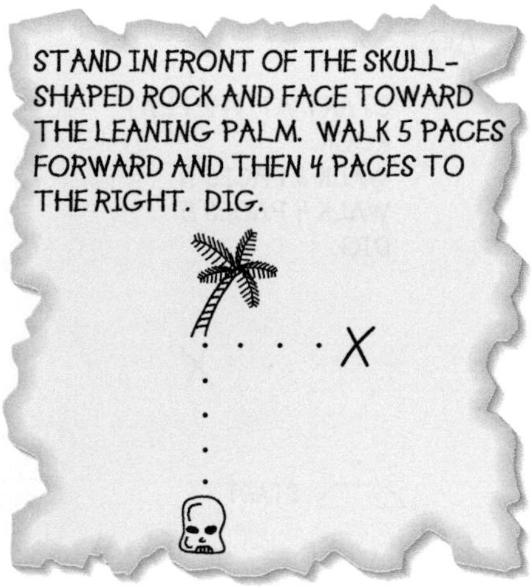

Figure 2.20 Body-based coordinate system

What we need in our study of dynamics is an effective way to relate these frames of reference to each other so that we can use whichever is more convenient. An example of where we would encounter multiple frames of this sort is an air traffic control system. A ground-fixed viewpoint used by flight control might be to chart an airplane's position with respect to the airport in terms of the plane's x, y, z coordinates (16 km east, 32 km north, and 3.2 km up). The airplane's crew, however, would use a reference frame based on the airplane (body-fixed) because that's the frame they work with when altering their course (bank left, pitch up, and so on).

Figure 2.21*a* shows a ground-fixed reference frame (the X and Y axes) with unit vectors $\boldsymbol{i}$ and $\boldsymbol{j}$ "attached" to it. These unit vectors are the equivalent of east/north for the pirate example. In this text, $\boldsymbol{i}$ and $\boldsymbol{j}$ will *always* be ground-fixed, nonrotating, nontranslating unit vectors.

The unit vectors $\boldsymbol{b}_1$ and $\boldsymbol{b}_2$ shown in the drawing are "attached" to the body B, which has its own personal set of axes x, y attached to it. Thus, if B rotates, so do $\boldsymbol{b}_1$ and $\boldsymbol{b}_2$ and so do the x and y axes, as illustrated

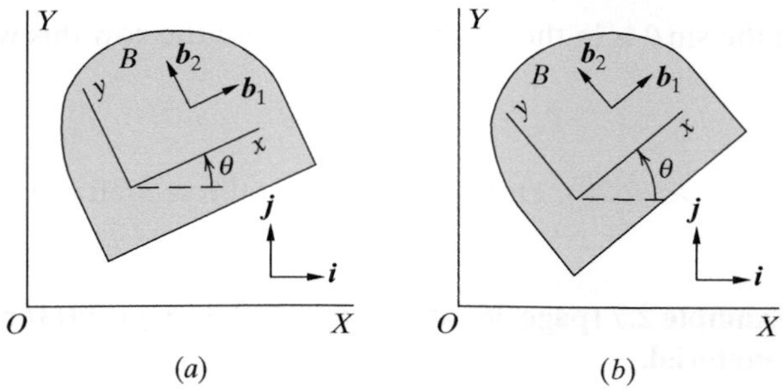

Figure 2.21 Coordinate transformation

in **Figure 2.21*b***. The unit vectors $\boldsymbol{b}_1$ and $\boldsymbol{b}_2$ are the equivalent not of east/north but rather of right/forward directions that make sense only with respect to the body that's moving.

Okay, the problem is to relate these two sets of unit vectors to each other so that someone using the $\boldsymbol{i}, \boldsymbol{j}$ ground-fixed frame of reference can figure out what's going on if told what's happening only in terms of the $\boldsymbol{b}_1, \boldsymbol{b}_2$ body-fixed reference frame.

In **Figure 2.22** all four unit vectors are redrawn so that their tails are touching. This is fine to do because the only function of these vectors is to supply a direction—where I draw them doesn't matter. I have also indicated the angle between the two sets. You can see that the $\boldsymbol{b}_1, \boldsymbol{b}_2$ frame is rotated counterclockwise from the $\boldsymbol{i}, \boldsymbol{j}$ frame by the angle θ.

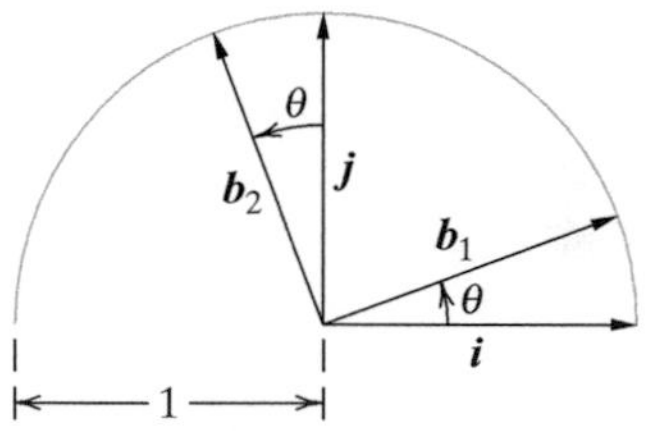

Figure 2.22 Overlay of two unit vector sets

Figure 2.23*a* indicates how we go from one reference frame to the other. Notice that because all the vectors are of unit length, their tips all touch the indicated circle of radius 1. Look at $\boldsymbol{b}_1$. How can we get from the origin O to the tip of $\boldsymbol{b}_1$ by using $\boldsymbol{i}, \boldsymbol{j}$? We can form a right triangle by dropping a line from the tip of $\boldsymbol{b}_1$ to $\boldsymbol{i}$. The legs of the triangle are oriented in the $\boldsymbol{i}, \boldsymbol{j}$ directions, and their lengths are found from trigonometry to be $\cos\theta$ for $\boldsymbol{i}$ and $\sin\theta$ for $\boldsymbol{j}$. Thus, to get from O to the tip of $\boldsymbol{b}_1$ using the $\boldsymbol{i}, \boldsymbol{j}$ vectors, we need to go to the right (in the $\boldsymbol{i}$ direction) a distance $\cos\theta$ and then go up (in the $\boldsymbol{j}$ direction) a distance $\sin\theta$, as illustrated in **Figure 2.23*b***.

In a similar manner, getting to the tip of $\boldsymbol{b}_2$ via $\boldsymbol{i}, \boldsymbol{j}$ means going along $\boldsymbol{j}$ an amount $\cos\theta$ and then traveling in the direction opposite the direction of $\boldsymbol{i}$ by an amount $\sin\theta$. In equation form, we have

$$\boldsymbol{b}_1 = \cos\theta\boldsymbol{i} + \sin\theta\boldsymbol{j} \tag{2.20}$$

$$\boldsymbol{b}_2 = -\sin\theta\boldsymbol{i} + \cos\theta\boldsymbol{j} \tag{2.21}$$

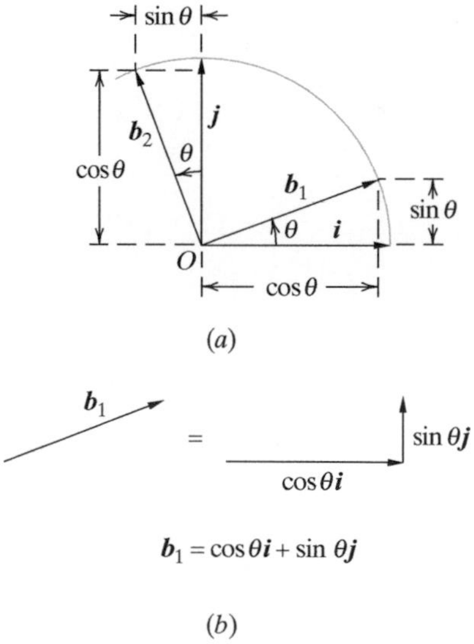

Figure 2.23 Detailed overlay of two unit vector sets

These equations can easily be represented in an array called a **coordinate transformation array**:

	$\boldsymbol{i}$	$\boldsymbol{j}$
$\boldsymbol{b}_1$	$\cos\theta$	$\sin\theta$
$\boldsymbol{b}_2$	$-\sin\theta$	$\cos\theta$

(2.22)

You can read the array in two ways. In the first approach, you read left to right. The first row contains $\boldsymbol{b}_1$, $\cos\theta$, and $\sin\theta$. The $\cos\theta$ is in the $\boldsymbol{i}$

column, and the $\sin\theta$ is in the $\boldsymbol{j}$ column. You read the row this way:

$$\boldsymbol{b}_1 = \cos\theta\boldsymbol{i} + \sin\theta\boldsymbol{j}$$

which is the same as (2.20). Doing the same for the second row gives you (2.21).

⋙ **Check out Example 2.7 (page 38) and Example 2.8 (page 39) for applications of this material.**

The nifty thing is what happens when you read down rather than across. Take the first column: $\boldsymbol{i}$, $\cos\theta$, $-\sin\theta$. Repeating what you did when reading across, you read this as

$$\boldsymbol{i} = \cos\theta\boldsymbol{b}_1 - \sin\theta\boldsymbol{b}_2$$

And this is precisely the correct formula for expressing $\boldsymbol{i}$ in terms of $\boldsymbol{b}_1$ and $\boldsymbol{b}_2$! You can see this for yourself by looking at **Figure 2.23**.

Of course, you can read the second column in the same way to get

$$\boldsymbol{j} = \sin\theta\boldsymbol{b}_1 + \cos\theta\boldsymbol{b}_2$$

Thus you see that setting up an array for finding the coordinate transformation from $\boldsymbol{i}$, $\boldsymbol{j}$ to $\boldsymbol{b}_1, \boldsymbol{b}_2$ gives you the inverse transformation from $\boldsymbol{b}_1$, $\boldsymbol{b}_2$ to $\boldsymbol{i}$, $\boldsymbol{j}$) for free.

Finding the velocity of the particle in **Figure 2.19** involves differentiating its position vector with respect to time:

$$\begin{aligned} \boldsymbol{v}_P &= \frac{d}{dt}\left(\boldsymbol{r}_{P/O}\right) \underbrace{=}_{(2.19)} \frac{d}{dt}(x\boldsymbol{i} + y\boldsymbol{j}) \\ &= \frac{dx}{dt}\boldsymbol{i} + x\frac{d\boldsymbol{i}}{dt} + \frac{dy}{dt}\boldsymbol{j} + y\frac{d\boldsymbol{j}}{dt} \quad \text{(product rule)} \end{aligned} \tag{2.23}$$

As you can see, we have four terms that need differentiation. The coordinates x and y are certainly time-dependent, for as P moves around, these values will necessarily change. Therefore $\dot{x}$ and $\dot{y}$ will, in general, exist. But what about $\boldsymbol{i}$ and $\boldsymbol{j}$? Are these unit vectors also time-dependent? Well, the answer is no, they're not. I've already mentioned that $\boldsymbol{i}$ always points to the right and $\boldsymbol{j}$ always points directly up. And, because they're unit vectors, their magnitude is constant (and equal to 1.0) for all time. The conclusion is that both $d\boldsymbol{i}/dt$ and $d\boldsymbol{j}/dt$ are identically zero because nothing about $\boldsymbol{i}$ and $\boldsymbol{j}$ changes with time—neither their magnitude nor their orientation.

The $d\boldsymbol{i}/dt$ and $d\boldsymbol{j}/dt$ terms are shown in the above equation in order to highlight the fact that for *this* case the unit vectors' derivatives are zero. This is decidedly *not* the case when we're using polar or path coordinates (which are covered later in this chapter). So please be careful when differentiating a term involving a unit vector so that you don't mistakenly ignore the differentiation when it's needed.

Letting the unit vector derivatives go to zero means that (2.23) becomes

$$\boldsymbol{v}_P = \dot{x}\boldsymbol{i} + \dot{y}\boldsymbol{j} \tag{2.24}$$

To find the particle's acceleration, all we need do is differentiate one more time (remembering that $\boldsymbol{i}$ and $\boldsymbol{j}$ have no time dependence) to obtain

$$\boldsymbol{a}_P = \ddot{x}\boldsymbol{i} + \ddot{y}\boldsymbol{j} \tag{2.25}$$

This is, by far, the easiest representation for acceleration. If you want to know the acceleration in the x direction, you just differentiate the expression for the x coordinate of P twice. The same holds true for acceleration in the y direction.

>>> **Check out Example 2.9 (page 40) and Example 2.10 (page 42) for applications of this material.**

Extending this analysis to three dimensions requires only one more thing: the addition of a z term:

$$\begin{aligned}\boldsymbol{r}_{P/O} &= x\boldsymbol{i} + y\boldsymbol{j} + z\boldsymbol{k}\\ \boldsymbol{v}_P &= \dot{x}\boldsymbol{i} + \dot{y}\boldsymbol{j} + \dot{z}\boldsymbol{k}\\ \boldsymbol{a}_P &= \ddot{x}\boldsymbol{i} + \ddot{y}\boldsymbol{j} + \ddot{z}\boldsymbol{k}\end{aligned}$$

where $\boldsymbol{k}$ is a unit vector pointing out of the x, y plane.

Integrating in order to go from acceleration to velocity and from velocity to position proceeds just as we have already seen in the one-dimensional analyses. Of course, it's now twice (or even three times) as much work because we have to integrate the response in the y (and perhaps z) direction(s) as well as that in the x direction.

EXAMPLE 2.7 COORDINATE TRANSFORMATION (A) (Theory on page 35)

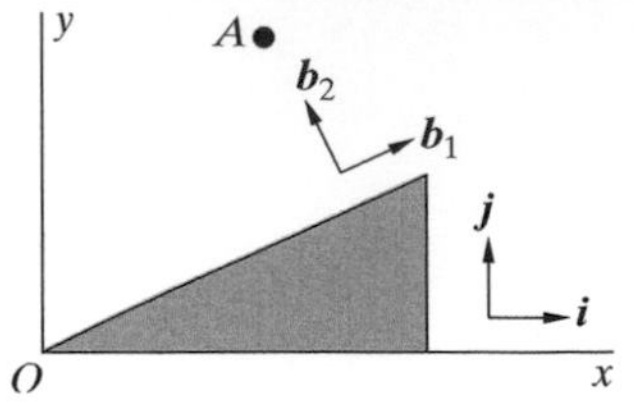

Figure 2.24 Two sets of unit vectors

The position vector $\boldsymbol{r}_{A/O}$ of a point A in **Figure 2.24** is measured with respect to the unit vectors $\boldsymbol{b}_1$ and $\boldsymbol{b}_2$ aligned with the slanted face of a wedge-shaped object:

$$\boldsymbol{r}_{A/O} = [4\boldsymbol{b}_1 + 3\boldsymbol{b}_2] \text{ cm}$$

Express this vector in terms of the $\boldsymbol{i}$, $\boldsymbol{j}$ unit vectors.

Goal Express $\boldsymbol{r}_{A/O}$ in terms of $\boldsymbol{i}$, $\boldsymbol{j}$.

Given Position of A with respect to O in terms of $\boldsymbol{b}_1, \boldsymbol{b}_2$ unit vectors.

Assume No assumptions are needed.

Draw **Figure 2.25** shows our system.

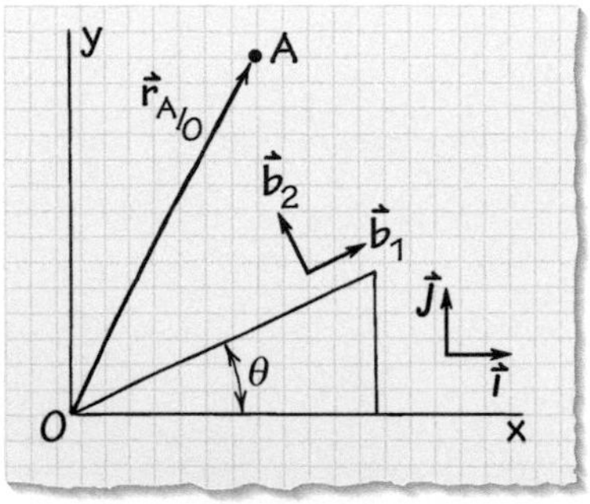

Figure 2.25 Fully detailed parameters

Formulate Equations All we need to do is construct a transformation array and use it to change from $\boldsymbol{b}_1, \boldsymbol{b}_2$ to $\boldsymbol{i}$, $\boldsymbol{j}$:

	$\boldsymbol{i}$	$\boldsymbol{j}$
$\boldsymbol{b}_1$	$\cos\theta$	$\sin\theta$
$\boldsymbol{b}_2$	$-\sin\theta$	$\cos\theta$

Solve

$$\begin{aligned}\boldsymbol{r}_{A/O} &= [4\boldsymbol{b}_1 + 3\boldsymbol{b}_2] \text{ cm}\\ &= [4(\cos\theta\boldsymbol{i} + \sin\theta\boldsymbol{j}) + 3(-\sin\theta\boldsymbol{i} + \cos\theta\boldsymbol{j})] \text{ cm}\\ &= [(4\cos\theta - 3\sin\theta)\boldsymbol{i} + (4\sin\theta + 3\cos\theta)\boldsymbol{j}] \text{ cm}\end{aligned}$$

Check As a check, we can look at the overall magnitude of the position vector, something that can't change just because our particular reference frame has. Our original representation $\boldsymbol{r}_{A/O} = [4\boldsymbol{b}_1 + 3\boldsymbol{b}_2]$ cm has a magnitude of $\sqrt{(4 \text{ cm})^2 + (3 \text{ cm})^2} = 5$ cm. The new representation has a magnitude of

$$\begin{aligned}\sqrt{(4\cos\theta - 3\sin\theta)^2 + (4\sin\theta + 3\cos\theta)^2} &= \sqrt{(\sin^2\theta + \cos^2\theta)(4^2 + 3^2)}\\ &= 5 \text{ cm}\end{aligned}$$

The two magnitudes match, and so we can have confidence that the analysis is correct. Of course, it's possible that the magnitudes match but the vectors are pointing in different directions. A further check would therefore be to determine the orientation of the two vectors and see that they're the same, something that's left as an exercise for you.

EXAMPLE 2.8 COORDINATE TRANSFORMATION (B) (Theory on page 35)

Sometimes you're given a problem in which the angles and unit vectors (**Figure 2.26**) are a bit awkward to deal with because the included angle is large. In this problem we will see how easy it is to create the correct transformation array if you follow the right procedure.

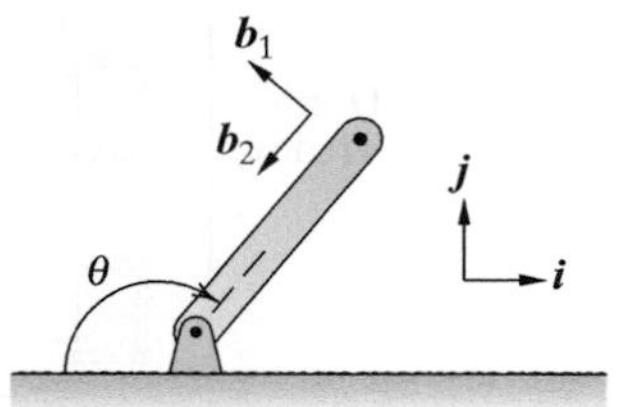

Figure 2.26 Two different sets of unit vectors

Goal Determine the coordinate transformation array from $\boldsymbol{b}_1, \boldsymbol{b}_2$ to $\boldsymbol{i}$, $\boldsymbol{j}$ for the illustrated system.

Given Two sets of unit vectors and an angle θ.

Assume No assumptions are needed.

Draw **Figure 2.27** shows our system with θ reduced to a small value.

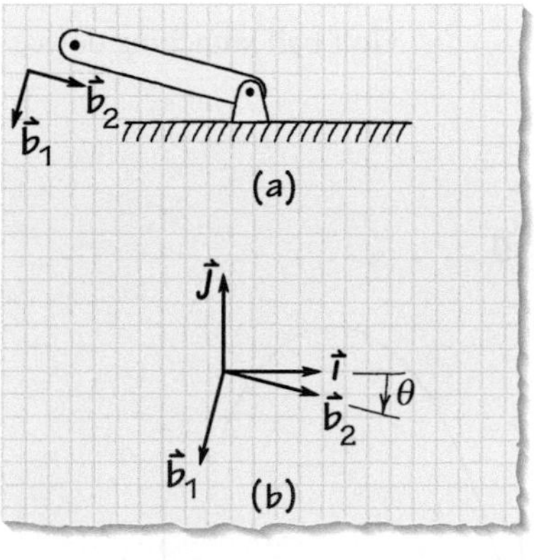

Figure 2.27 Reoriented system

Formulate Equations Once again we will use a coordinate transformation array to go from the $\boldsymbol{b}_1, \boldsymbol{b}_2$ unit vectors to $\boldsymbol{i}$, $\boldsymbol{j}$.

Solve The trick to getting the transformation array is to refuse to look at the problem the way it's presented and instead to make the relevant angle between the two unit vector sets very small. For the problem at hand, that means rotating the arm counterclockwise until it's nearly horizontal, as shown in **Figure 2.27*a***. **Figure 2.27*b*** shows the relationship between the two unit vector sets once we've rotated the arm: $\boldsymbol{i}$ and $\boldsymbol{b}_2$ are almost aligned (because θ is small) and $\boldsymbol{j}$ and $\boldsymbol{b}_1$ point in almost completely opposite directions. As we have seen, two unit vectors that almost line up are related by $\cos\theta$, and if they are almost aligned but point in different directions, they are related by $-\cos\theta$.

I've superimposed the two unit vector sets in **Figure 2.27*b***. From this you can see that

$$\boldsymbol{i} = \cos\theta \boldsymbol{b}_2 - \sin\theta \boldsymbol{b}_1$$

and

$$\boldsymbol{j} = -\cos\theta \boldsymbol{b}_1 - \sin\theta \boldsymbol{b}_2$$

and our coordinate transformation array is

	$\boldsymbol{i}$	$\boldsymbol{j}$
$\boldsymbol{b}_1$	$-\sin\theta$	$-\cos\theta$
$\boldsymbol{b}_2$	$\cos\theta$	$-\sin\theta$

Check An easy check is to try some particular values of θ in **Figure 2.26** for which the correct answer is clear. For instance, if $\theta = 90°$, then $\boldsymbol{b}_1$ points in the direction opposite to the direction of $\boldsymbol{i}$. Looking at our coordinate transformation array shows this to be the case: $\sin(90°) = 1$ and $\cos(90°) = 0$, giving $\boldsymbol{b}_1 = -\boldsymbol{i}$, as expected.

EXAMPLE 2.9 RECTILINEAR TRAJECTORY DETERMINATION (A) (Theory on page 36)

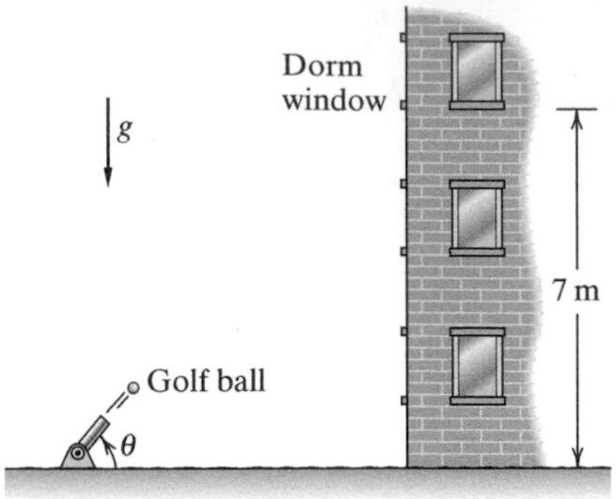

Figure 2.28 Golf ball launching device

Let's consider a classic example of motion in a Cartesian reference frame. A student wants to launch a golf ball up and into her friend's third-floor window (**Figure 2.28**) located 7 m above ground level. The ball is to be launched out of a homemade pressure gun, a device that will use compressed air to project the ball at whatever angle θ the student chooses. The muzzle speed v_0 of the ball is determined by the pressure the student can pump into the gun. Because pumping air is hard work and because, being a good engineer, she always seeks to minimize her work, she wants to figure out the smallest value of v_0 that will allow the ball to reach the window.

Goal Find the minimum launch speed v_0 that allows the ball to reach the third-floor window.

Given During its ball's trajectory, gravity continuously accelerates it toward the ground.

Assume To simplify the analysis we will neglect air drag.

Draw **Figure 2.29** shows our coordinate system and a typical trajectory.

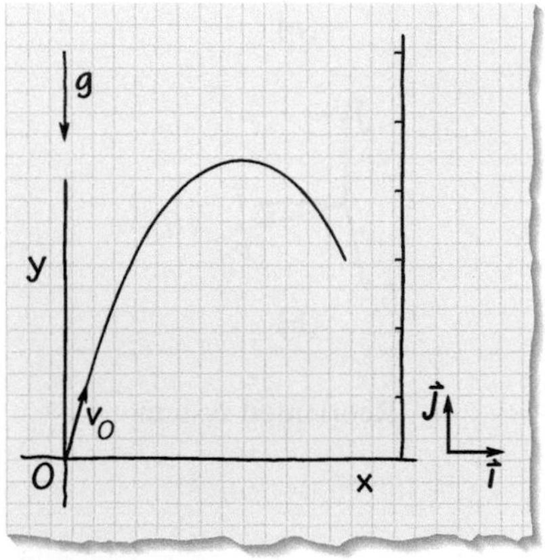

Figure 2.29 Coordinate system and trajectory for launched golf ball

Formulate Equations Our task is to find the trajectory that maximizes the height for a given muzzle speed v_0. Because the muzzle speed is fixed, our constraint is

$$v_0^2 = v_{0_x}^2 + v_{0_y}^2 = \text{constant} \tag{2.26}$$

The ball's initial velocity components in the x and y directions depend on the barrel's orientation. Once it leaves the barrel, the ball will be acted on only by gravity and will feel a downward acceleration g. Thus

$$\boldsymbol{a}_{\text{ball}} = -g\boldsymbol{j} \tag{2.27}$$

Note that the acceleration is expressed as a vector, having components in the $\boldsymbol{i}$ and $\boldsymbol{j}$ directions (although in this case the $\boldsymbol{i}$ component is zero). This means that when we integrate with respect to time, we will still have two components (one in the $\boldsymbol{i}$ direction and one in the $\boldsymbol{j}$ direction) and thus two constants of integration.

Integrating (2.27) with respect to time and matching initial conditions give us, from (2.6),

$\boldsymbol{i}$: $$v_x(t) = v_{0_x} \tag{2.28}$$

$\boldsymbol{j}$: $$v_y(t) = -gt + v_{0_y} \tag{2.29}$$

or, in vector form, $\boldsymbol{v}_{\text{ball}}(t) = v_{0_x}\boldsymbol{i} + (-gt + v_{0_y})\boldsymbol{j}$. Thus we can view two-dimensional motion as either a single-vector equation or two scalar equations.

Integrating (2.28) and (2.29) once more gives us the ball's position as a function of time:

$\boldsymbol{i}$: $$x = v_{0_x}t + x_0 \tag{2.30}$$

$\boldsymbol{j}$: $$y = -\frac{gt^2}{2} + v_{0_y}t + y_0 \tag{2.31}$$

$$\boldsymbol{r}_{\text{ball}/O}(t) = (v_{0_x}t + x_0)\boldsymbol{i} + \left(-\frac{gt^2}{2} + v_{0_y}t + y_0\right)\boldsymbol{j}$$

If, as shown in **Figure 2.29**, we let the origin of our coordinate system be at the base of the gun (and neglect the length of the barrel), then x_0 and y_0 are equal to zero, giving us the $\boldsymbol{i}$ and $\boldsymbol{j}$ components of our final trajectory:

$$\boldsymbol{i}: \qquad x = v_{0_x}t \tag{2.32}$$

$$\boldsymbol{j}: \qquad y = -\frac{gt^2}{2} + v_{0_y}t \tag{2.33}$$

You may remember that free projectile motion follows a parabolic trajectory. We can show this in a couple of lines.

$$(2.32) \Rightarrow \qquad t = \frac{x}{v_{0_x}} \tag{2.34}$$

$$(2.34) \rightarrow (2.33) \Rightarrow \qquad y = -\left(\frac{g}{2v_{0_x}^2}\right)x^2 + \left(\frac{v_{0_y}}{v_{0_x}}\right)x \tag{2.35}$$

Solve You should also remember that at the top of the trajectory, the slope $\frac{dy}{dx}$ is zero. Differentiating (2.35) with respect to x and setting $\frac{dy}{dx}$ equal to 0 give

$$\left.\frac{dy}{dx}\right|_{x=x_{\text{max}}} = \frac{v_{0_y}}{v_{0_x}} - \frac{2gx_{\text{max}}}{2v_{0_x}^2} = 0 \tag{2.36}$$

$$(2.36) \Rightarrow \qquad x_{\text{max}} = \frac{v_{0_x}v_{0_y}}{g} \tag{2.37}$$

$$(2.37) \rightarrow (2.35) \Rightarrow \qquad y_{\text{max}} = \frac{v_{0_y}^2}{2g} \tag{2.38}$$

What this means physically is that the more v_{0_y} you have, the higher the ball will go. From (2.26) we realize that, because $v_{0_x}^2 + v_{0_y}^2$ is constant, the way to maximize $v_{0_y}^2$ is by minimizing $v_{0_x}^2$. The minimum that v_{0_x} could equal is zero, which would mean that the ball is shot straight up—no horizontal velocity at all. We can now solve for the ball's minimum muzzle speed $v_0 = v_{0_y}$ using (2.38):

$$v_{0_y} = \sqrt{2gy_{\text{max}}} = \sqrt{2(9.81\ \text{m/s}^2)(7\ \text{m})} = 11.7\ \text{m/s} \tag{2.39}$$

Check So, what does this all mean in terms of our problem? It means that we haven't formulated it correctly! The student's goal was to get the golf ball *into* her friend's window. If she shoots it straight up, it might get up to the window's level, but it surely won't go in—it will just drop straight down again. Thus we need to reconsider the problem, which we do in the next example.

EXAMPLE 2.10 RECTILINEAR TRAJECTORY DETERMINATION (B) (Theory on page 36)

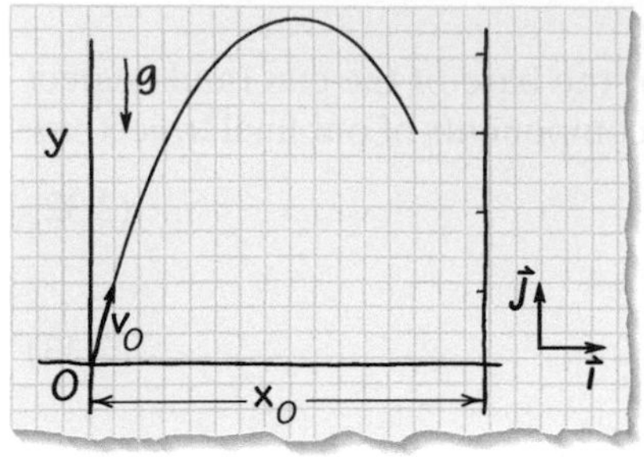

Figure 2.30 Trajectory for golf ball with gun located $x_0 = 2$ m to left of building

Because our goal in Example 2.9 was to get the ball into a third-floor window, our result isn't a workable solution. What we will do now is add another constraint—that the student will position the gun a fixed distance x_0 from the building and aim it toward the building, thus ensuring a horizontal velocity component. We will let $x_0 = 2$ m.

Goal Find the minimum muzzle speed that gets the ball to the third-floor window with a positive horizontal component of velocity.

Given Distance to building is 2 m.

Assume Air drag is neglected.

Draw **Figure 2.30** shows the distance x_0 between the gun and the building.

Formulate Equations We know from (2.37) that

$$x_{\text{max}} = \frac{v_{0_x} v_{0_y}}{g}$$

Solve We will use (2.37) to relate v_{0_x} and v_{0_y} by realizing that we want the height to be a maximum when the ball is just reaching the building—in other words, we want the distance x_0 between the gun and the building to be

$$x_{\text{max}} = x_0 = \frac{v_{0_x} v_{0_y}}{g} \quad \Rightarrow \quad v_{0_x} = \frac{x_0 g}{v_{0_y}} \tag{2.40}$$

Because v_{0_x} depends on v_{0_y}, we won't end up with an unusable solution the way we did in Example 2.9.

Getting the ball up to the window means, from (2.39), that $v_{0_y} = \sqrt{2gy_{\text{max}}}$, and substituting this result into (2.40) tells us that $v_{0_x} = x_0\sqrt{\frac{g}{2y_{\text{max}}}}$. The necessary muzzle speed is therefore

$$v_0 = \sqrt{v_{0_x}^2 + v_{0_y}^2} = \sqrt{2gy_{\text{max}}\left(1 + \left(\frac{x_0}{2y_{\text{max}}}\right)^2\right)} = \sqrt{2(9.81\ \text{m/s}^2)(7\ \text{m})\left(1 + \left(\frac{2\ \text{m}}{2(7\ \text{m})}\right)^2\right)}$$

$$= \boxed{11.8\ \text{m/s}}$$

Thus, $v_0 = 11.8$ m/s, a 1% increase over the speed needed to just reach the window but not go inside.

Check One check on consistency is to find v_{0_x} and v_{0_y} and take the inverse tangent of v_{0_y}/v_{0_x}, which gives us a muzzle inclination of $\theta = 81.9°$. Then we can evaluate $v_0 \sin\theta$ to find the vertical velocity component at launch. Doing this produces an exact match to the result of Example 2.9: 11.7 m/s. This validates our analysis because we know that speed in the y direction at launch has to be the same for both cases. The window is 7 m high in each case, and thus the initial speed in the y direction at launch has to be the same if the ball is to reach the window.

EXERCISES 2.2

2.2.1. [Level 1] Determine the coordinate transformation array between $\boldsymbol{c}_1, \boldsymbol{c}_2$ and $\boldsymbol{b}_1, \boldsymbol{b}_2$.

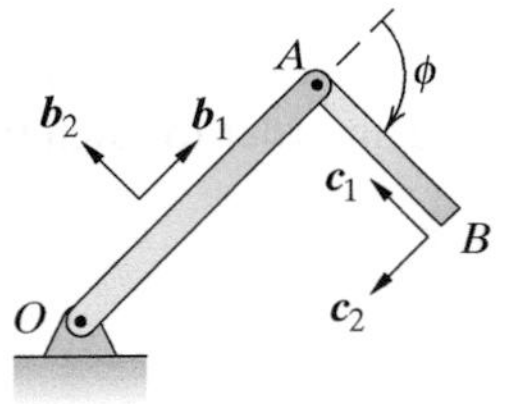

E2.2.1

2.2.2. [Level 1] The position of a particle P moving in the horizontal plane is described by $\boldsymbol{r}(t) = (5t^2\boldsymbol{i} + 2te^{0.1t}\boldsymbol{j})$ m. Find the magnitude of the particle's position, velocity, and acceleration at time $t = 2$ s.

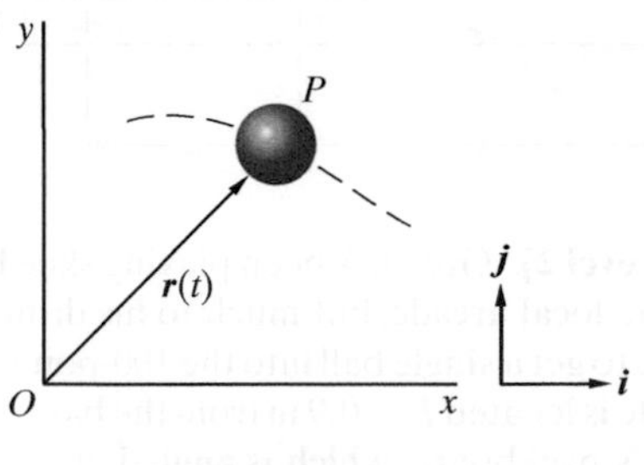

E2.2.2

2.2.3. [Level 1] The figure contains three sets of unit vectors. $\boldsymbol{i}, \boldsymbol{j}$ are fixed with respect to the ground. $\boldsymbol{b}_1, \boldsymbol{b}_2$ are fixed with respect to a gauge, which is itself fixed to an airplane which has begun a roll maneuver. The final set of unit vectors $(\boldsymbol{c}_1, \boldsymbol{c}_2)$ is fixed with respect to the indicator needle within the gauge.

Determine the coordinate transformation array between $\boldsymbol{b}_1, \boldsymbol{b}_2$ and $\boldsymbol{c}_1, \boldsymbol{c}_2$ and between $\boldsymbol{i}, \boldsymbol{j}$ and $\boldsymbol{c}_1, \boldsymbol{c}_2$.

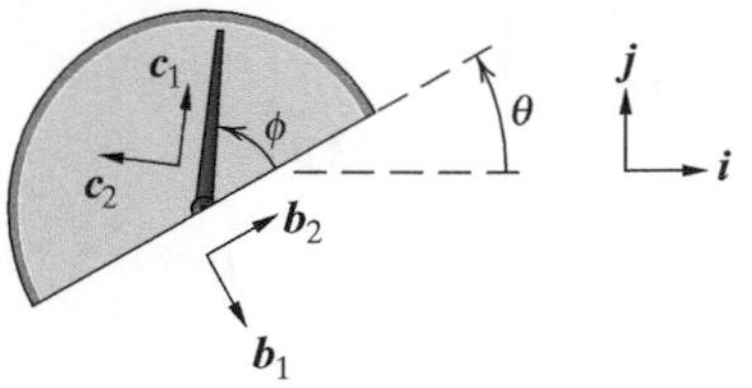

E2.2.3

2.2.4. [Level 1] A sailor is stuck on a tropical island and emergency supplies are airlifted to him before a ship can come to his rescue. The supply plane (at an altitude of $H = 46$ m) is moving horizontally and approaching the island at a speed of $v = 322$ km/h. How far before the island must the supply package be released if it is to just reach the island?

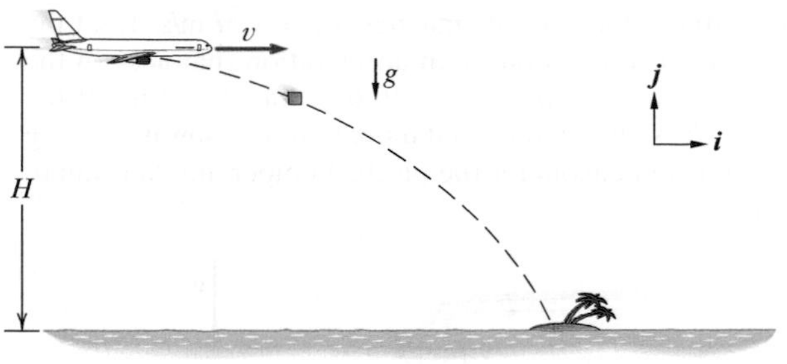

E2.2.4

2.2.5. [Level 1] A particle P is moving in the horizontal plane according to $\boldsymbol{r}(t) = (0.09t^3\boldsymbol{i} + 0.12t^2 \sin 2t\boldsymbol{j})$ m. Find the magnitude of the particle's position, velocity, and acceleration at time $t = 1$ s.

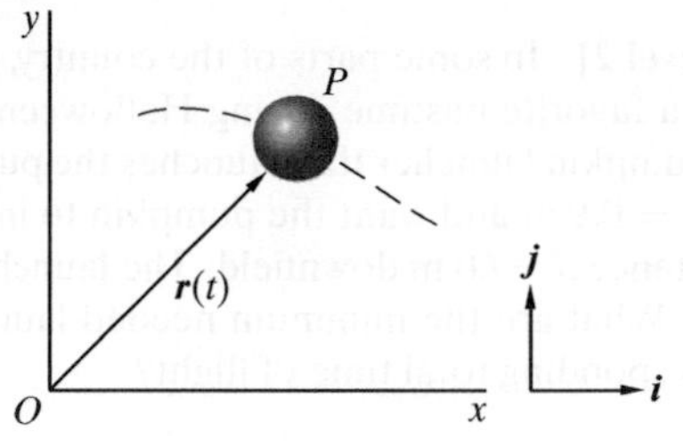

E2.2.5

2.2.6. [Level 1] Construct a coordinate transformation array from $\boldsymbol{i}, \boldsymbol{j}$ to $\boldsymbol{b}_1, \boldsymbol{b}_2$ and express the vector $\boldsymbol{p} = 4\boldsymbol{i} - 8\boldsymbol{j}$ in terms of $\boldsymbol{b}_1, \boldsymbol{b}_2$ for $\theta = 130°$. The $\boldsymbol{b}_1, \boldsymbol{b}_2$ unit vectors are attached to the illustrated link $\overline{AB}$.

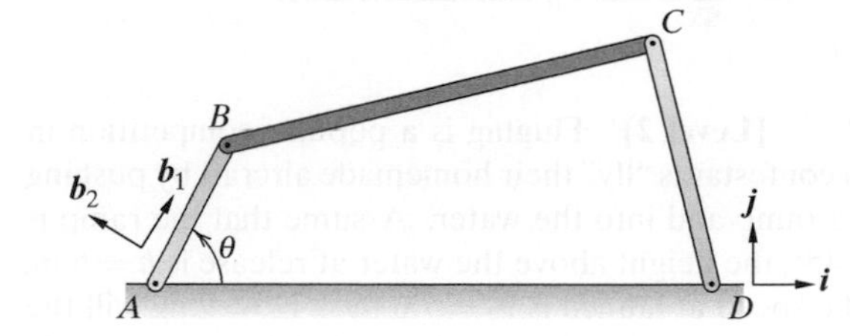

E2.2.6

2.2.7. [Level 1] Construct a coordinate transformation array from $\boldsymbol{i}, \boldsymbol{j}$ to $\boldsymbol{b}_1, \boldsymbol{b}_2$ and express the vector $\boldsymbol{p} = 3\boldsymbol{i} + 4\boldsymbol{j}$ in terms of $\boldsymbol{b}_1, \boldsymbol{b}_2$ for $\theta = 53°$. The $\boldsymbol{b}_1, \boldsymbol{b}_2$ unit vectors are attached to the pivoting bar.

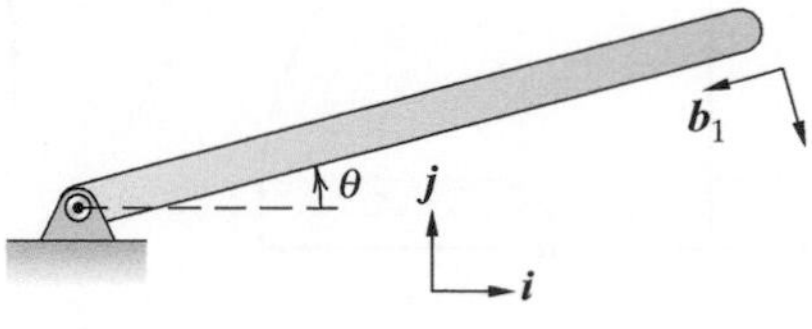

E2.2.7

2.2.8. [Level 1] A child tosses out the back of a station wagon a very strong magnet m, which is subsequently attracted by the sheet metal of the car. Will the magnet return and stick to the back of the car or fall to the ground? The initial velocity of the magnet is $\boldsymbol{v} = \sqrt{5}\boldsymbol{i}$ m/s. Treat the magnet/car interaction as an acceleration that acts on the magnet equal to $-a_m\boldsymbol{i}$. $a_m = 10$ m/s^2 and $L = 1$ m. If the magnet hits the car at a point more than L below its starting height, it will encounter the plastic bumper and thus fail to stick.

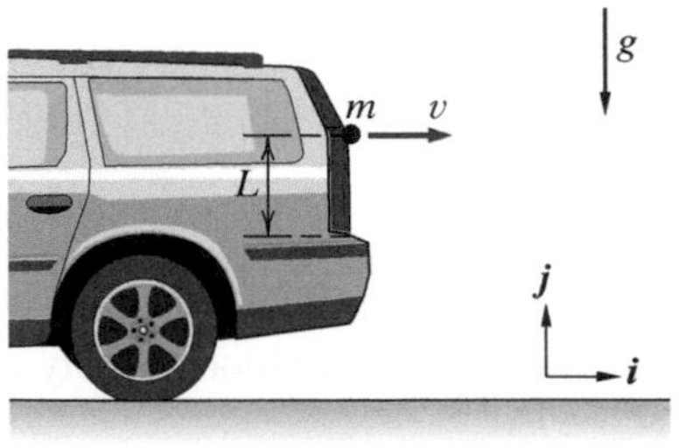

E2.2.8

2.2.9. [Level 2] In some parts of the country, pumpkin launching is a favorite pastime during Halloween. You've designed a pumpkin launcher that launches the pumpkin at a height of $h = 0.9$ m and want the pumpkin to impact the ground a distance $d = 60$ m downfield. The launch speed is $v_0 = 30$ m/s. What are the minimum needed launch angle and the corresponding total time of flight?

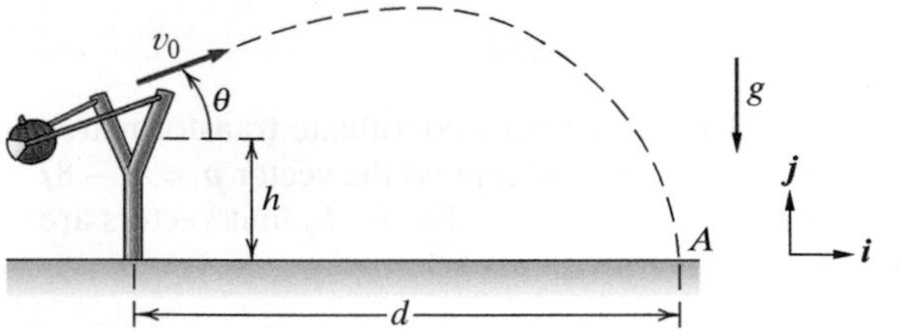

E2.2.9

2.2.10. [Level 2] Flugtag is a popular competition in which contestants "fly" their homemade aircraft by pushing it off a ramp and into the water. Assume that the ramp is set at 45°, the height above the water at release is $h = 6$ m, and the speed at launch is $v_0 = 7.6$ m/s. How long will the aircraft be in the air, what maximum height does it reach, and how far does it travel horizontally?

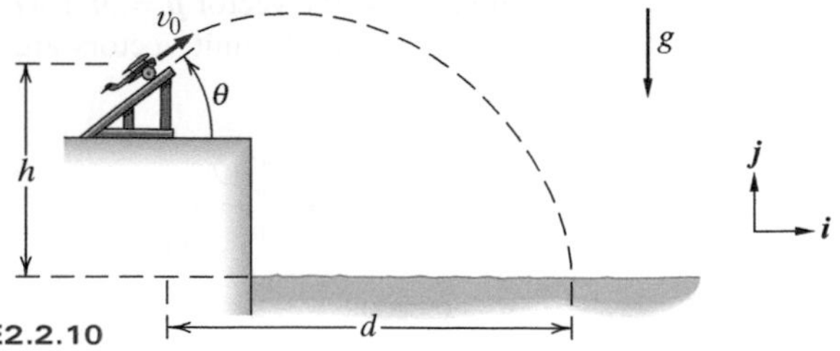

E2.2.10

2.2.11. [Level 2] Kyle (A) throws a ball from the left bank of a ravine ($H = 9$ m) to his buddy Clem (B) who's on the (lower) right bank ($h = 3$ m). Kyle throws the ball up at a 30° angle with a release speed of $v = 6$ m/s. Ignore the height of the two people.

a. What's the needed horizontal separation s between Kyle and Clem for Clem to catch the ball without moving?

b. If Clem is actually at point C, a distance $d = 6$ m farther than the distance calculated in part **a**, how fast must he run, at a constant speed, in order to catch the ball?

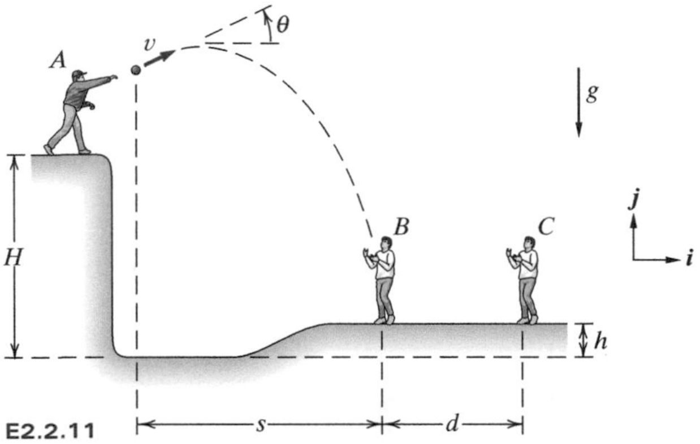

E2.2.11

2.2.12. [Level 2] Greg has been playing skee ball all afternoon at the local arcade, but much to his dismay, he has not been able to get a single ball into the 100-point hole. The 100-point hole is located $L = 0.9$ m from the base of the skee ball machine's backboard, which is angled at $\beta = 20°$ with respect to the horizontal. If the end of the launch ramp is $h = 0.6$ m above the backboard's base and oriented at $\theta = 45°$ to ground, how fast should Greg project a ball up the launch ramp so that it lands in the 100-point hole? How long does it take for the ball to reach the hole?

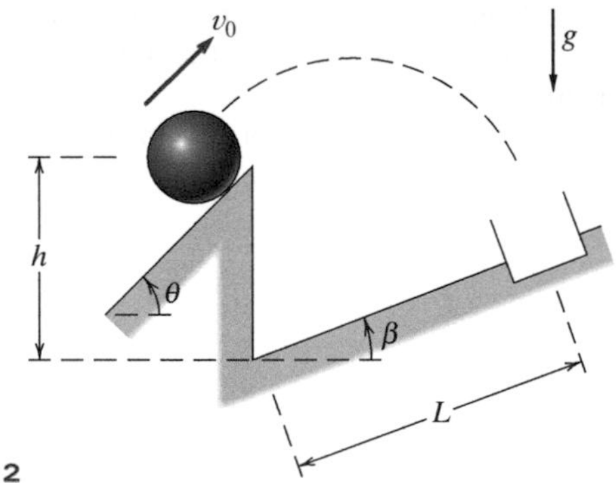

E2.2.12

2.2.13. [Level 2] A football is kicked from the ground at an angle of $\theta = 50°$ and speed $v_0 = 12$ m/s. A player downfield (Karl) can jump up with initial speed $v_j = 2.7$ m/s. Karl can reach a height of 2.1 m without jumping and catches the football at the top of a 0.3 m jump.

a. How long after kickoff did Karl initiate his jump in order to catch the football?

b. How far away was Karl from the kickoff position?

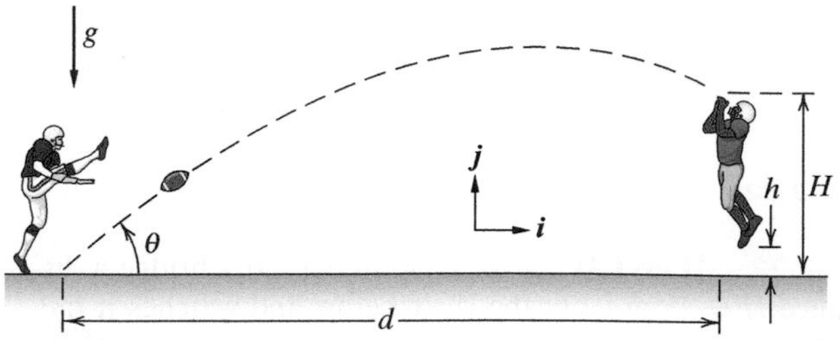

E2.2.13

2.2.14. **[Level 2]** A ball is launched with a speed v_0 at an angle $\beta = 50°$ with respect to the horizontal toward point A, which is located $d = 5$ m along a surface inclined at $\theta = 15°$ to ground. If the launch site is $s = 4$ m from the base of the incline, at what speed v_0 should the ball be launched so that it lands at A? How long is the ball airborne?

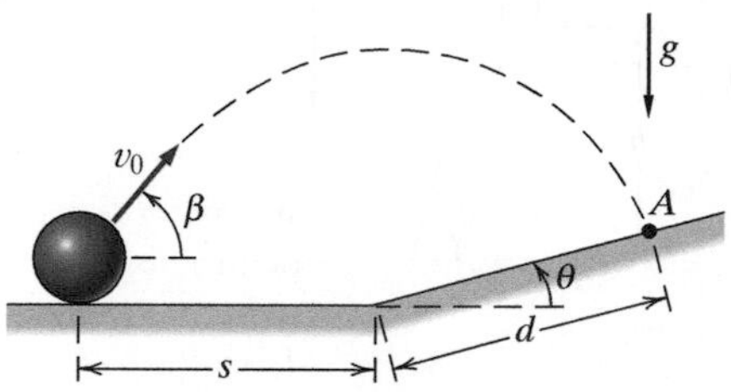

E2.2.14

2.2.15. **[Level 2]** Construct coordinate transformation arrays from $\boldsymbol{i}$, $\boldsymbol{j}$ to $\boldsymbol{b}_1, \boldsymbol{b}_2$ and from $\boldsymbol{i}$, $\boldsymbol{j}$ to $\boldsymbol{c}_1, \boldsymbol{c}_2$ for the illustrated mechanism. Use these to express $\boldsymbol{b}_1$ and $\boldsymbol{b}_2$ in terms of $\boldsymbol{c}_1, \boldsymbol{c}_2$. Show that the overall transformation array between $\boldsymbol{c}_1, \boldsymbol{c}_2$ and $\boldsymbol{b}_1, \boldsymbol{b}_2$ can be expressed in terms of $\sin(\theta - \phi)$ and $\cos(\theta - \phi)$.

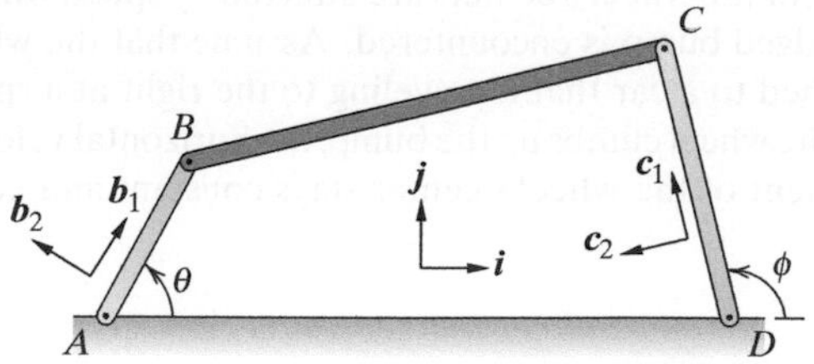

E2.2.15

2.2.16. **[Level 2]** Ronald, a clever kid, has designed a snowball launcher to get his snowballs over the neighbors' fence and into their yard. The fence is 17.58 m high (these neighbors are very unsociable), the launcher is positioned 67.5 m from the wall, and the neighbor's son, Rupert, is sitting on the ground 79.5 m from Ronald (12 m from the wall). What is the only allowable launch angle that, with a sufficient launch velocity, will cause the snowball to just miss hitting the top of the wall and then land on Rupert?

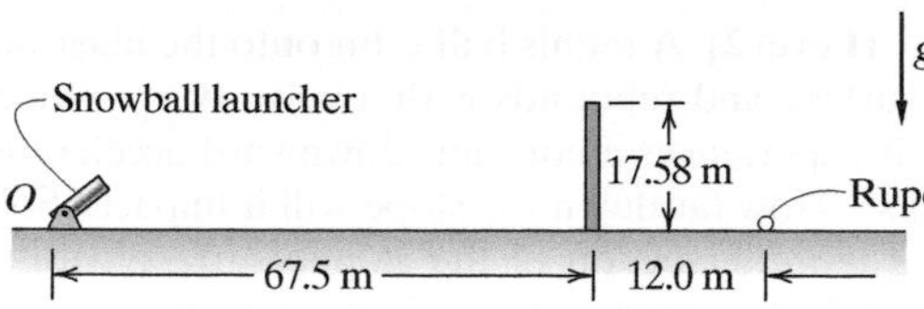

E2.2.16

2.2.17. **[Level 2]** A dried pea is projected upward at an angle of η degrees with speed v. Derive the formula that will let you solve for h (the distance up the slope at which the pea contacts the surface) as a function of η, g, and v.

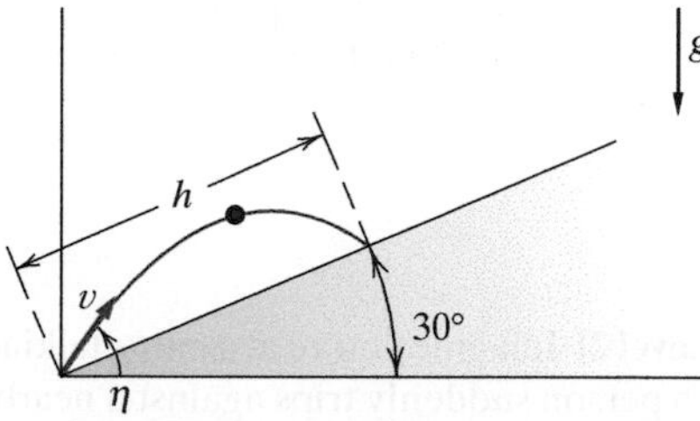

E2.2.17

2.2.18. **[Level 2]** A cricket wants to leap onto a spot that's 0.30 m higher than the spot it's currently on. It's currently resting on a 30° slope. Assuming that it can launch itself at an angle η with speed $2.44\sqrt{2}$ m/s, determine the necessary value of η for it to attain its goal.

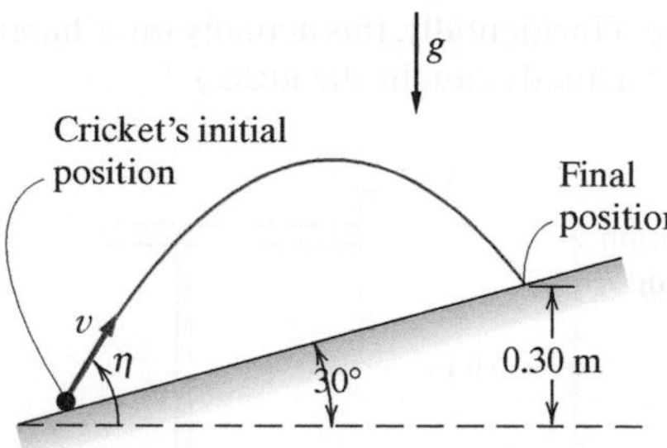

E2.2.18

2.2.19. **[Level 2]** A watermelon launcher is designed to launch watermelons short distances so as to gauge the watermelon's ability to withstand rough handling during shipment. If the melon is launched at a 45° angle and has to land 3 meters along a 35° slope, how fast must the watermelon be launched?

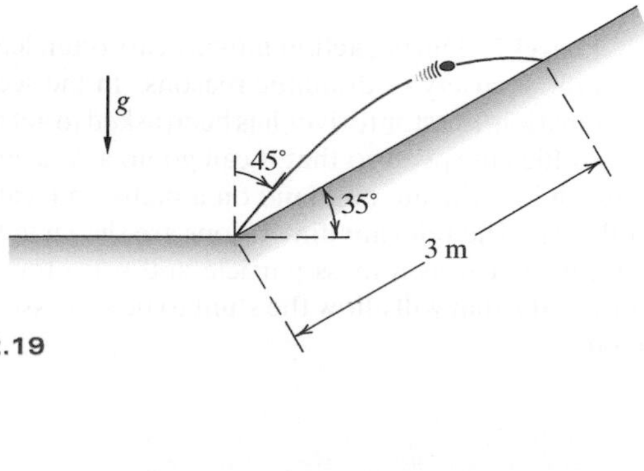

E2.2.19

2.2.20. [Level 2] A tennis ball is hit onto the illustrated inclined surface and rebounds with a velocity $\boldsymbol{v}_1$. Once it bounces, it experiences a constant downward acceleration of 9.81 m/s^2. How far down the slope will it impact? Solve by finding the intersection of the sloped surface with the ball's parabolic trajectory. $\|\boldsymbol{v}_1\| = 25$ m/s and $\beta = 30°$.

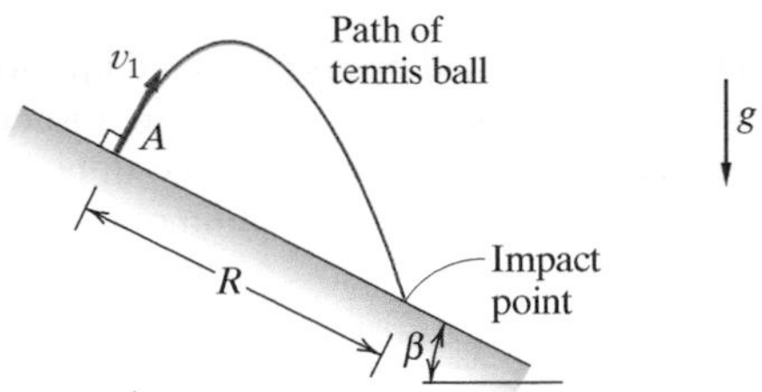

E2.2.20

2.2.21. [Level 2] Imagine you're at a party, talking with a friend, when a person suddenly trips against a nearby table, causing a glass to fall off the edge. An excellent human reaction time is 0.25 s. In that time, how far will the glass fall? Assume that you immediately (after the 0.25 s has elapsed) begin to accelerate your hand so that you grab the glass when it is 15 cm from the floor. What constant acceleration was necessary, and how fast was your hand traveling when you contacted the glass? Assume that your hand moved in a straight line. (Incidentally, this actually once happened to me and, yes, I actually caught the glass.)

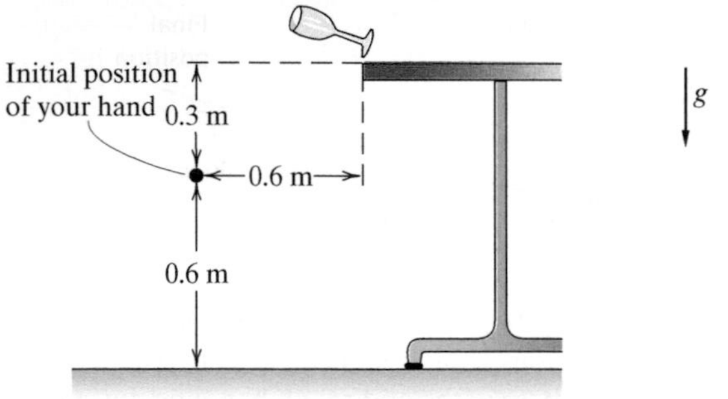

E2.2.21

2.2.22. [Level 2] During action movies, cars often leave the ground for a variety of dramatic reasons. In the scene under examination, the stuntdriver has been asked to get the car up to a sufficient speed so that it can go up a 20 degree incline, launch into the air, and land on a platform located ahead of the car. The relevant dimensions are shown in the sketch. Treat the car as a mass particle and solve for the minimum speed v that will allow the stunt to be successfully undertaken.

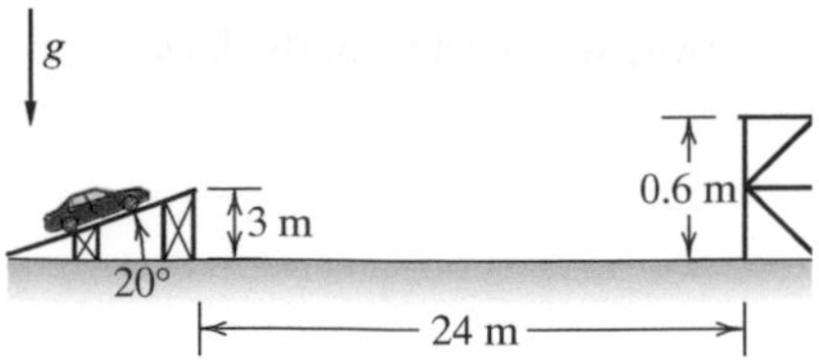

E2.2.22

2.2.23. [Level 3] Ernst drops a ball off a bridge a distance $H = 21$ m from the river below and watches it fall. When it reaches the point A (15 m above the river) a strong gust of wind pushes it to the right with a speed of $v_w = 24$ m/s.

a. How long does the ball take to hit the ground?
b. Where does it hit?
c. What are the impact angle and speed?

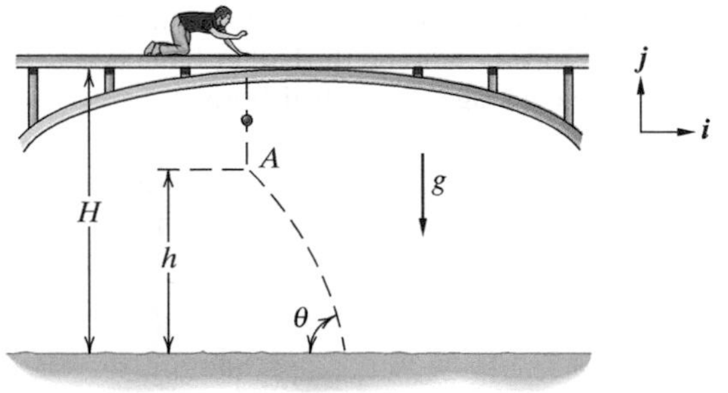

E2.2.23

2.2.24. [Level 3] You already know that acceleration is the time derivative of velocity, but did you know that the time derivative of acceleration is known as "jerk"? Yep, that's the real word for it. Imagine what it feels like to be in a car in which the driver alternately mashes the accelerator and then jams on the brakes. That's what jerk measures. Calculate how the acceleration and jerk in the vertical direction (of the wheel's center) are affected by speed when a sharp-edged bump is encountered. Assume that the wheel is attached to a car that is traveling to the right at a speed v_0. As the wheel climbs up the bump, the horizontal velocity component of the wheel's center stays constant and equal to v_0.

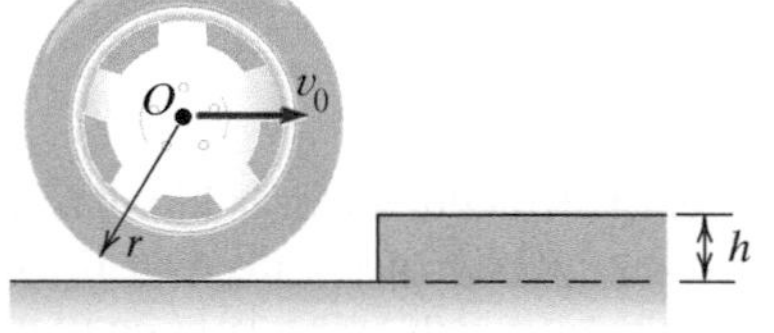

E2.2.24

2.2.25. [Level 3] A paintball is fired from a starting position $x = y = 0$ with speed v_0 and inclination θ, the intention being to hit a car when the car has traveled a distance L up a slope. The car travels at a constant speed v_C and starts from (x_1, y_1) at the same instant that the paintball is fired. Find an expression for the necessary launch angle θ in terms of $\phi, d, h, L, g,$ and v_C.

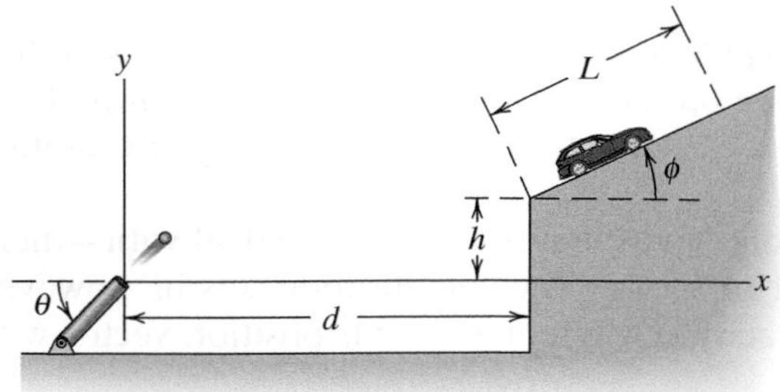

E2.2.25

2.2.26. [Level 3] A simple batting game allows the user to fire a mass m from a tube when a button is pressed. Initially, the bat is horizontal. The bat has a length L, and θ varies as $\ddot{\theta} = \alpha t, \alpha$ a constant. At what speed must the mass be fired so that it strikes the middle of the bat when $\theta = \frac{\pi}{2}$ rad? The launch angle ϕ is fixed. Let the time of impact be denoted by t^*.

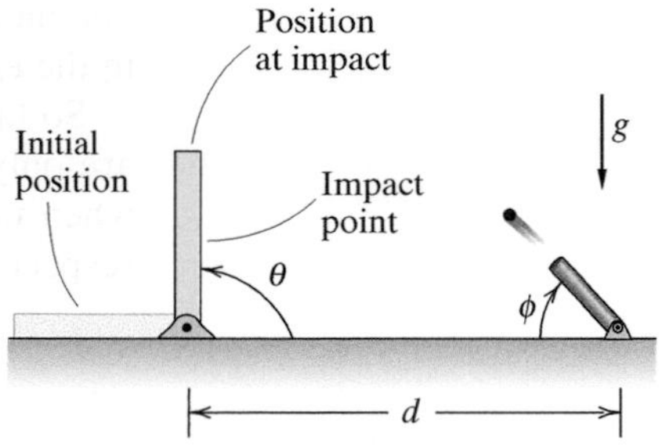

E2.2.26

2.3 POLAR AND CYLINDRICAL COORDINATES

The next set of coordinates we will examine are called **polar coordinates** because they are radially based, like a map of the world would be if it were centered on the North or South Pole, showing lines of latitude and longitude. A typical polar plot is shown in **Figure 2.31**.

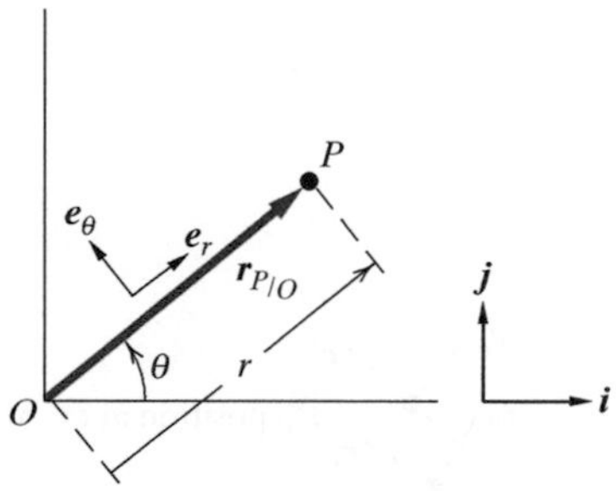

Figure 2.31 Polar plot

Rather than determining the position of a body P by going forward some amount and then right some amount, as we did in Figure 2.19, the polar approach determines P's position by the distance (r) it is from the origin and the angle θ that P's position vector makes with the positive horizontal axis. The **radial unit vector** $\boldsymbol{e}_r$ always points from the origin to the body. The total position vector is given by

$$\boldsymbol{r}_{P/O} = r\boldsymbol{e}_r \tag{2.41}$$

which says that to get to P, you must go a distance r in the $\boldsymbol{e}_r$ direction. Unlike the Cartesian case, in which we needed two terms to define position, in the polar case we need just one.

You would use this viewpoint, for example, if you were an airport traffic controller. Because you were using radar, the information you would have about an airplane would be its straight-line distance from you (r) and its vertical inclination (θ).

As just mentioned, $\boldsymbol{e}_r$ is not oriented permanently vertical or permanently horizontal as a Cartesian unit vector would be. Rather, it is determined by where P is with respect to the origin O. The unit vector $\boldsymbol{e}_r$ always points from the origin to P, and therefore if P moves around O, the orientation of $\boldsymbol{e}_r$ will change. This is going to make life difficult when we try to find velocities and accelerations. But before we do that, let's introduce $\boldsymbol{e}_\theta$, the other unit vector we'll be using (and which we'll call the **angular unit vector**). This vector is found by imagining that the length r is fixed and then increasing θ. Our normal convention will be that θ

increases in a positive direction when $r_{P/O}$ rotates in a counterclockwise direction. The angular unit vector $\boldsymbol{e}_\theta$ points in the direction that P moves as θ increases.

Another way to think of polar coordinates is to imagine you're standing at the origin O and looking out at P. Motion *toward* you or *away* from you means motion along the $\boldsymbol{e}_r$ direction. Motion *around* you is motion in the $\boldsymbol{e}_\theta$ direction.

So far, it seems that polar coordinates are easier to deal with—there are only half the number of terms. The complication sets in, however, when we move on to velocity. Differentiating our position vector with respect to time gives

$$\boldsymbol{v}_P = \frac{d}{dt}(r\boldsymbol{e}_r) = \frac{dr}{dt}\boldsymbol{e}_r + r\frac{d\boldsymbol{e}_r}{dt} \tag{2.42}$$

The $\frac{dr}{dt}$ term isn't too bad. It's just the time rate of change of r, namely, how fast the body P is moving either away from or toward the origin. The $d\boldsymbol{e}_r/dt$ is a bit trickier because here we're being asked to calculate the time rate of change of the radial unit vector. You'll recall that in the Cartesian coordinate system there are no time rates of change associated with $\boldsymbol{i}$ or $\boldsymbol{j}$ because these vectors neither rotate nor change in magnitude. Here, however, the situation is quite different. The magnitude of the polar unit vectors $\boldsymbol{e}_r$ and $\boldsymbol{e}_\theta$ can't ever change (it's always unity by definition), but the direction of these vectors can definitely change as the body rotates about the origin. Because they're completely determined by P's position with respect to O, if P moves around O, then the vectors are going to rotate as well.

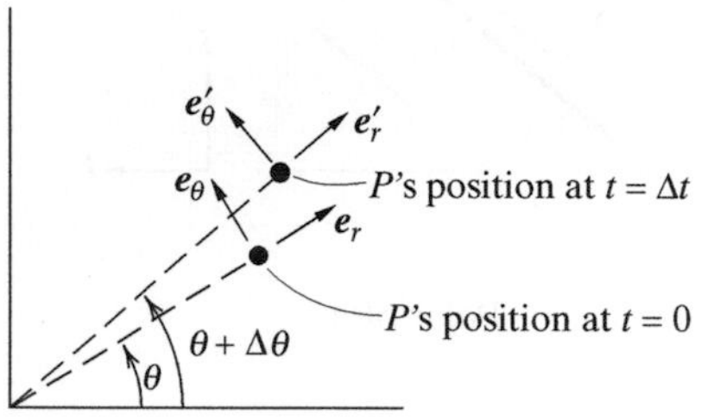

Figure 2.32 Polar representation of a particle at two points

We'll sneak up on the answer to what $\dot{\boldsymbol{r}}_P$ is equal to by applying the fundamental definition of a derivative. Recall that a point's velocity is found by dividing its displacement (the change in its position over some time Δt) by Δt and taking the limit as Δt goes to zero.

Figure 2.32 shows P at $t = 0$ and at $t = \Delta t$. Both r and θ increased when P moved. To keep track of the two states, $\boldsymbol{e}_r$, $\boldsymbol{e}_\theta$ are used for P's initial position and $\boldsymbol{e}'_r$, $\boldsymbol{e}'_\theta$ for its final position. Because the unit vectors can't change their length, we need worry only about their change in direction. As you can see, $\boldsymbol{e}'_r$ and $\boldsymbol{e}'_\theta$ are rotated counterclockwise from $\boldsymbol{e}_r$ and $\boldsymbol{e}_\theta$ by the amount $\Delta\theta$.

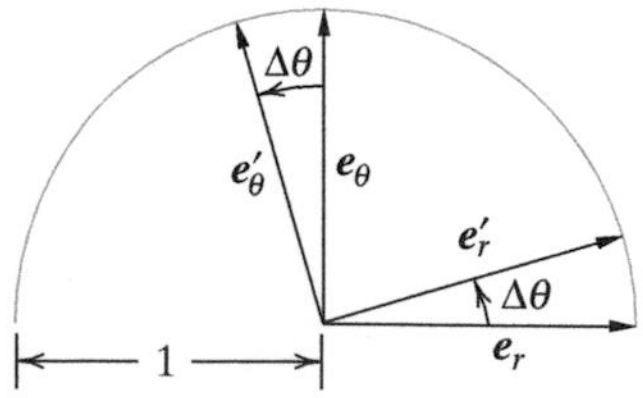

Figure 2.33 Overlay of two unit vector sets

Figure 2.33 shows both sets of unit vectors with their tails aligned. This looks just like **Figure 2.22** (with $\boldsymbol{i}$, $\boldsymbol{j}$, $\boldsymbol{b}_1$, $\boldsymbol{b}_2$ replaced by $\boldsymbol{e}_r$, $\boldsymbol{e}_\theta$, $\boldsymbol{e}'_r$, $\boldsymbol{e}'_\theta$), which means that we can express $\boldsymbol{e}_r$, $\boldsymbol{e}_\theta$ in terms of $\boldsymbol{e}'_r$, $\boldsymbol{e}'_\theta$, and vice versa, something that we'll do in just a second. But first, we'll write the limiting expression for $\dot{\boldsymbol{e}}_r$:

$$\dot{\boldsymbol{e}}_r = \lim_{\Delta t \to 0} \frac{\boldsymbol{e}'_r - \boldsymbol{e}_r}{\Delta t}$$

In order to do this operation, we have to express $\boldsymbol{e}'_r$ in terms of $\boldsymbol{e}_r$ and $\boldsymbol{e}_\theta$. We can construct a coordinate transformation array just as we did in

(2.22) and obtain

	$\boldsymbol{e}_r$	$\boldsymbol{e}_\theta$
$\boldsymbol{e}'_r$	$\cos(\Delta\theta)$	$\sin(\Delta\theta)$
$\boldsymbol{e}'_\theta$	$-\sin(\Delta\theta)$	$\cos(\Delta\theta)$

(2.43)

Using this array to re-express $\boldsymbol{e}'_r$ gives us

$$\dot{\boldsymbol{e}}_r = \lim_{\Delta t \to 0} \frac{\boldsymbol{e}_r \cos(\Delta\theta) + \boldsymbol{e}_\theta \sin(\Delta\theta) - \boldsymbol{e}_r}{\Delta t} \tag{2.44}$$

Recall that for small x, $\sin x \approx x$ and $\cos x \approx 1$. As we're taking the limit $\Delta t \to 0$, $\Delta\theta$ is going to zero as well. Using these limiting values in (2.44) gives us

$$\dot{\boldsymbol{e}}_r = \lim_{\Delta t \to 0} \frac{\boldsymbol{e}_r + (\Delta\theta)\boldsymbol{e}_\theta - \boldsymbol{e}_r}{\Delta t} = \lim_{\Delta t \to 0} \left(\frac{\Delta\theta}{\Delta t}\right) \boldsymbol{e}_\theta$$

In this limit, as Δt goes to zero, $\frac{\Delta\theta}{\Delta t}$ becomes $\frac{d\theta}{dt} = \dot{\theta}$, the angular speed (often written as ω). Thus

$$\dot{\boldsymbol{e}}_r = \dot{\theta}\boldsymbol{e}_\theta = \omega\boldsymbol{e}_\theta \tag{2.45}$$

With this knowledge in hand, we can write (2.42) as

$$\boldsymbol{v}_P = \dot{r}\boldsymbol{e}_r + r\dot{\theta}\boldsymbol{e}_\theta \tag{2.46}$$

This expression tells us that there are two fundamental components to a velocity in a polar coordinate system: a linear (radial) component representing the motion along a radius originating at O (the radial velocity $\dot{r}\boldsymbol{e}_r$) and a component representing the circular motion about O (the transverse velocity $r\dot{\theta}\boldsymbol{e}_\theta$). Picture it this way: Imagine swinging a baseball bat that, unlike the usual kind, gets longer as you're swinging it (maybe it's made of telescoping pieces). You can definitely see what the two terms we've just derived correspond to. The derivative $\dot{r}$ is the speed at which the tip of the bat is moving radially away from you as the bat lengthens, and the product $\dot{r}\boldsymbol{e}_r$, being a vector quantity, is the *velocity* with which the tip is moving away from you. The scalar quantity $r\dot{\theta}$ is the speed at which the tip of the bat is going in the direction of the swing, and the vector quantity $r\dot{\theta}\boldsymbol{e}_\theta$ is the velocity of this circular motion. When we add together the velocity in the swing direction and the velocity in the extensional (radial) direction, we get the overall velocity vector.

IMPORTANT NOTE! Always make sure that you use radians rather than degrees when calculating $\dot{\theta}$ (and later on, $\ddot{\theta}$). Degrees are fine for determining geometries in a problem, but once you start looking at time rates of change you have to switch to radians. An extremely common mistake is to leave your calculator in degree mode rather than switching to radian mode, thereby giving you very wrong results from your calculations.

>>> **Check out Example 2.11 (page 54) for an application of this material.**

Having successfully dealt with velocity, now let's look at acceleration. Differentiating (2.46) leads to

$$\begin{aligned} \boldsymbol{a}_P &= \frac{d}{dt}(\dot{r}\boldsymbol{e}_r) + \frac{d}{dt}(r\dot{\theta}\boldsymbol{e}_\theta) \\ &= \ddot{r}\boldsymbol{e}_r + \dot{r}\dot{\boldsymbol{e}}_r + \dot{r}\dot{\theta}\boldsymbol{e}_\theta + r\ddot{\theta}\boldsymbol{e}_\theta + r\dot{\theta}\dot{\boldsymbol{e}}_\theta \end{aligned} \tag{2.47}$$

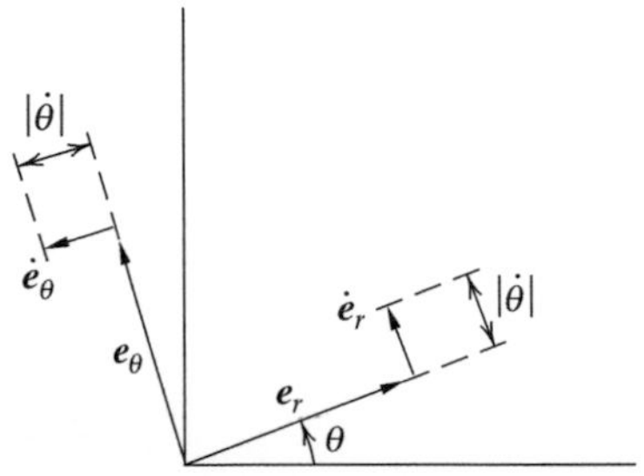

Figure 2.34 Velocities of the unit vector's tips

As you can see, we have again got to deal with the time derivative of unit vectors. We already know what $\dot{\boldsymbol{e}}_r$ is from (2.45): $\dot{\theta}\boldsymbol{e}_\theta$. To figure out $\dot{\boldsymbol{e}}_\theta$, we could go through the procedure we just suffered through, substituting $\boldsymbol{e}_\theta$ for $\boldsymbol{e}_r$. Luckily, though, there's an easier way. Physically, what $\dot{\boldsymbol{e}}_\theta$ represents is the velocity of the tip of the vector $\boldsymbol{e}_\theta$ as the tip rotates about its base. (Remember that the vector can't get any longer, being a unit vector, so the only thing left to it is rotation.) **Figure 2.34** shows both $\boldsymbol{e}_\theta$ and $\boldsymbol{e}_r$. If you imagine θ increasing, you'll see that the tips of the two vectors rotate counterclockwise around a circle of radius 1. The speed of $\boldsymbol{e}_r$'s tip is given by the rotation rate: $\dot{\theta}$. Now, is there any reason for $\boldsymbol{e}_\theta$'s tip to move at a speed different from $\boldsymbol{e}_r$'s? Probably not, seeing as they're both unit vectors and are rotating at the same rate. Thus, we can deduce that the magnitude of $\boldsymbol{e}_\theta$'s velocity is $\dot{\theta}$, and as for direction, you can see from **Figure 2.34** that the tip of $\boldsymbol{e}_\theta$ is moving in the negative $\boldsymbol{e}_r$ direction. Thus

$$\dot{\boldsymbol{e}}_r = \dot{\theta}\boldsymbol{e}_\theta \quad \text{and} \quad \dot{\boldsymbol{e}}_\theta = -\dot{\theta}\boldsymbol{e}_r \tag{2.48}$$

Substituting these values for $\dot{\boldsymbol{e}}_r$ and $\dot{\boldsymbol{e}}_\theta$ into (2.47) gets us

$$\boldsymbol{a}_P = \ddot{r}\boldsymbol{e}_r + \dot{r}\dot{\theta}\boldsymbol{e}_\theta + \dot{r}\dot{\theta}\boldsymbol{e}_\theta + r\ddot{\theta}\boldsymbol{e}_\theta - r\dot{\theta}^2\boldsymbol{e}_r \tag{2.49}$$

Note that the term $\dot{r}\dot{\theta}\boldsymbol{e}_\theta$ shows up twice in this expression. If you look at the derivation, you will see that these two terms come from two different places. The first stems from the $\frac{d}{dt}(\dot{r}\boldsymbol{e}_r)$ term of (2.47), and the other comes from the $\frac{d}{dt}(r\dot{\theta}\boldsymbol{e}_\theta)$ term. Two seemingly different differentiations produce the same term. These two terms are fascinating, partly because without them we wouldn't have those huge cyclonic storm systems that show up on the Weather Channel. Taken together, the two $\dot{r}\dot{\theta}\boldsymbol{e}_\theta$ terms make up what's called the **Coriolis acceleration**. If you want to understand a little more about this acceleration, read on for a few more paragraphs. This material isn't absolutely necessary to your understanding of polar coordinates, but I think it's pretty nifty, and so I'm going to talk about it.

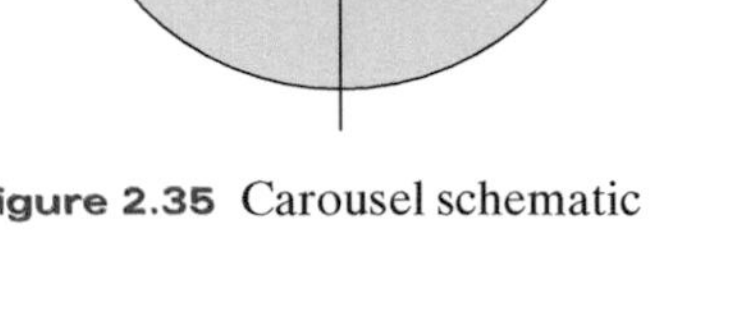

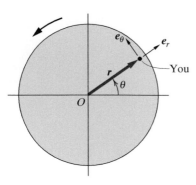

Figure 2.35 Carousel schematic

Imagine you're on a large rotating platform, like the carousel shown on the opening page of this chapter. **Figure 2.35** shows a top view, with a polar coordinate system set up with respect to you standing still on the platform. As the carousel rotates, you move along a circle centered at O. Now start walking radially outward, as if you're trying to walk off the carousel. (*Note:* I actually did this at Disneyland, and it was a lot of fun. Of course, the Mousepolice won't let me in anymore, but that's what I'm willing to sacrifice for science.) If you try this, you'll find that a force seems to appear that knocks you sideways (to the right in **Figure 2.35**). The faster you walk radially, the stronger the force. Very weird. What's happening is that the floor of the carousel is reacting against you as you try to go straight while it's trying to go around in a circle. The force involved is called the **Coriolis force** and is related to the Coriolis acceleration through the relationship $\boldsymbol{F} = m\boldsymbol{a}$.

Can we see why this force and acceleration come about from a physical argument? Absolutely. First, think about what happens when you take a step radially outward. Just before taking the step, you were at a particular place on the platform and had a certain velocity resulting from the platform's rotation. When you move radially outward, you're stepping onto a spot on the platform that has a *higher* speed than you currently have. (Although I shouldn't be talking about rigid-body motion effects yet, I'm going to assume you remember running into rotating bodies in your physics class and will do so anyway. If the explanation doesn't make sense, you can always come back to it after you have read the rigid-body material later in the book.) It takes a small time interval Δt to move from your original position to your new position, and after you get there, you're moving faster in the $\boldsymbol{e}_\theta$ direction by $(\dot{r}\Delta t)\dot{\theta}$ (**Figure 2.36*a***). So what does that mean? It means you accelerated (speed increased over time). If we divide this change in speed by the time it took (Δt), we get an acceleration $\dot{r}\dot{\theta}$. That's the half of the Coriolis effect stemming from the $\frac{d}{dt}(r\dot{\theta}\boldsymbol{e}_\theta)$ term of (2.47).

The $\frac{d}{dt}(\dot{r}\boldsymbol{e}_r)$ half comes about because you have a radial velocity vector that is itself rotating (because it's attached to you and you're on a rotating platform). **Figure 2.36*b*** shows the radial component of your velocity vector both initially and after you've stepped radially outward and the platform has rotated a little. Note that the magnitude of your radial velocity hasn't changed but the direction has. **Figure 2.37** shows the two radial velocity vectors with their tails touching so that you can see the effect of the rotation more clearly. For small rotations, the difference between the two vectors is approximately $\dot{r}\dot{\theta}\Delta t$ in the $\boldsymbol{e}_\theta$ direction. Divide by Δt and we have $\dot{r}\dot{\theta}\boldsymbol{e}_\theta$, the half of the Coriolis acceleration from the $\frac{d}{dt}(\dot{r}\boldsymbol{e}_r)$ term of (2.47).

Okay, now you know everything you need to know about Coriolis acceleration. So let's look at all the polar acceleration components. Grouping terms in (2.49) gives us

$$\boldsymbol{a}_P = (\ddot{r} - r\dot{\theta}^2)\boldsymbol{e}_r + (2\dot{r}\dot{\theta} + r\ddot{\theta})\boldsymbol{e}_\theta \tag{2.50}$$

The derivative $\ddot{r}$ is the radial acceleration component that involves only r. If r is fixed, then this term is zero. If θ isn't changing, then we have motion in only a single direction (r) and $\ddot{r}$ represents acceleration toward or away from the origin. The term $-r\dot{\theta}^2$ is the centripetal acceleration. This is the only acceleration if r is constant as a particle rotates about some

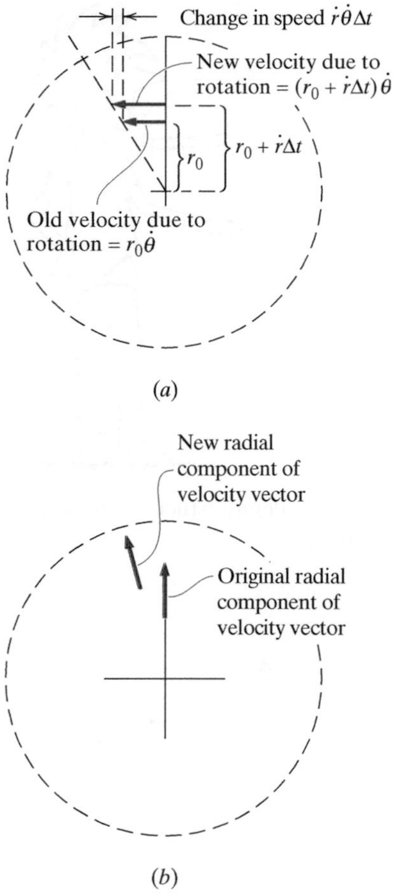

Figure 2.36 Velocity variations on a rotating body

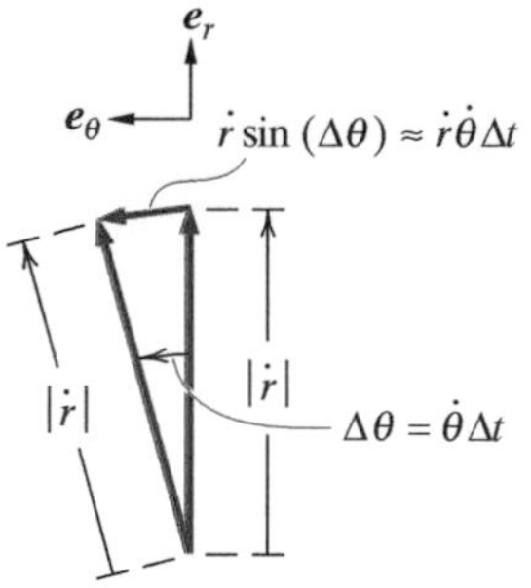

Figure 2.37 Detail of velocity difference

Figure 2.38 Penny "stuck" to hand due to rotation

origin. This centripetal acceleration is the one that comes into play when we think about the orbit of the earth around the sun. It's the term that lets us deduce that a penny won't fall off the surface of your palm if you whirl your hand around as shown in **Figure 2.38** (at a sufficient speed), even though there's nothing physically stopping the penny from falling when your hand is at the top of the swing.

As already noted, $2\dot{r}\dot{\theta}$ is the Coriolis acceleration—the acceleration that occurs when something is both rotating *and* moving radially outward or inward at the same time. And finally, there's $r\ddot{\theta}$, the component of transverse acceleration that doesn't involve $\dot{r}$. This is the dominant term when you're swinging at a baseball. When you swing a bat, the distance from your hand to the bat's tip doesn't change (constant r), and the tip is accelerating along its direction of travel (as you go from swing initiation to home run speed).

>>> **Check out Example 2.12 (page 55) and Example 2.13 (page 56) for applications of this material.**

Now that you've mastered polar coordinates, getting comfortable with **cylindrical coordinates** is going to be easy. If you take a look at **Figure 2.39** you'll see that cylindrical coordinates are just polar coordinates with the addition of a z axis and the unit vector k associated with this axis. That's where the name *cylindrical* comes from—a cylinder is just a circle projected upward from its original plane. Thus polar coordinates can be thought of as a special case of cylindrical coordinates, one for which the motion in the z direction is set to zero.

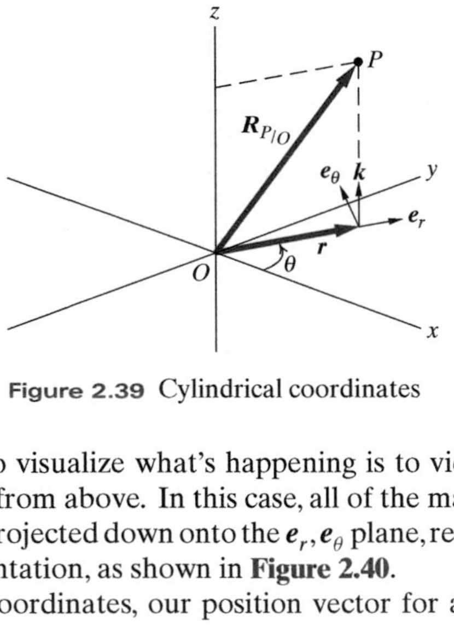

Figure 2.39 Cylindrical coordinates

Another way to visualize what's happening is to view the cylindrical coordinate frame from above. In this case, all of the material lying along the $\boldsymbol{k}$ direction is projected down onto the $\boldsymbol{e}_r, \boldsymbol{e}_\theta$ plane, reducing everything to a polar representation, as shown in **Figure 2.40**.

In cylindrical coordinates, our position vector for a particle P is expressed

$$\boldsymbol{R}_{P/O} = r\boldsymbol{e}_r + z\boldsymbol{k} \tag{2.51}$$

which is simply (2.41), our polar representation, with a $z\boldsymbol{k}$ term added on. Just as in the Cartesian case, the $\boldsymbol{k}$ unit vector doesn't change in length

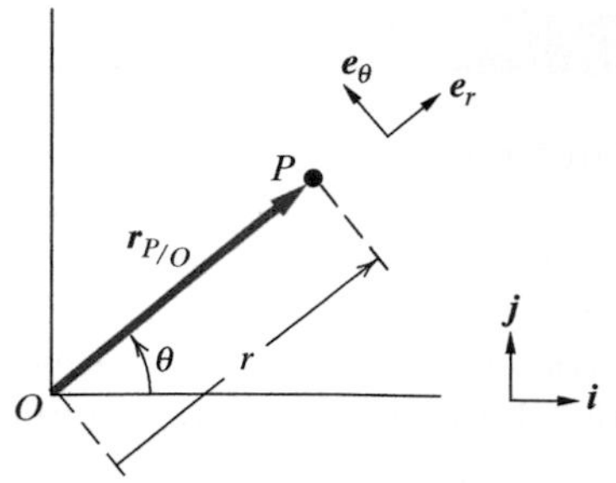

Figure 2.40 Projection onto polar coordinates

or orientation and therefore has a time derivative of zero. Our velocity expression will therefore be (2.46), our polar result, with the addition of a $\dot{z}\boldsymbol{k}$ term:

$$\boldsymbol{v}_P = \dot{r}\boldsymbol{e}_r + r\dot{\theta}\boldsymbol{e}_\theta + \dot{z}\boldsymbol{k} \tag{2.52}$$

Acceleration requires one more differentiation and, as in the velocity result, simply tacks on one additional term to the polar result, (2.50):

$$\boldsymbol{a}_P = (\ddot{r} - r\dot{\theta}^2)\boldsymbol{e}_r + (r\ddot{\theta} + 2\dot{r}\dot{\theta})\boldsymbol{e}_\theta + \ddot{z}\boldsymbol{k} \tag{2.53}$$

Cylindrical coordinates will come in handy when we have an obvious polar component along with some sort of vertical translation. Modeling the path of someone walking up a spiral staircase is a good example. Seen from above, the person would appear to be walking in a circle, but viewed from the side she or he would exhibit an obvious upward motion as well.

>>> **Check out Example 2.14 (page 57) for an application of this material.**

EXAMPLE 2.11 VELOCITY—POLAR COORDINATES (Theory on page 49)

Figure 2.41 Dog pulling leash

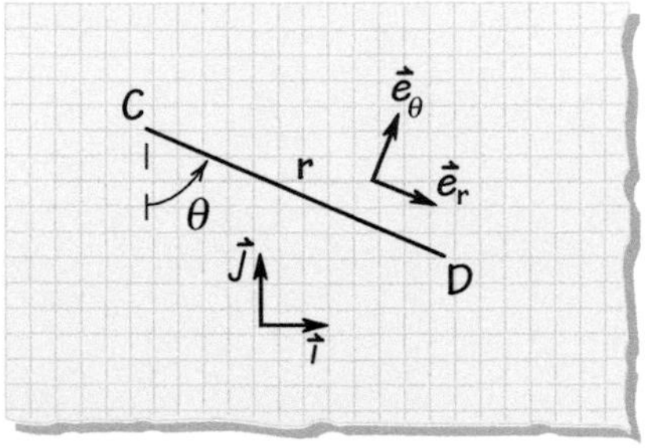

Figure 2.42 System labeling

Crystal is standing, chatting on her cell phone, when her dog Derek suddenly takes off after a cat. Because she hadn't locked the retractable leash, the cord begins to reel out of the housing. The cord is taut, and at the instant illustrated in **Figure 2.41**, the attachment to Derek's collar is 3 m to the right and 1 m down from the leash housing. If Derek is running at 10 m/s in the $\boldsymbol{i}$ direction, how fast is the cord coming out of the housing?

Goal Determine how fast the dog leash is unreeling.

Given Orientation of dog to housing. $\boldsymbol{v}_D = 10\boldsymbol{i}$ m/s.

Assume Our constraint is that the dog is running in a horizontal direction. Thus we can assume that its velocity is given by

$$\boldsymbol{v}_D = v_D\boldsymbol{i} \tag{2.54}$$

Draw The geometry of our system is shown schematically in **Figure 2.42**.

Formulate Equations The dog is running in the $\boldsymbol{i}$ direction (Cartesian coordinates), and the leash is best described with a polar representation. Thus we'll need a coordinate transformation array between the two:

	$\boldsymbol{i}$	$\boldsymbol{j}$
$\boldsymbol{e}_r$	$\sin\theta$	$-\cos\theta$
$\boldsymbol{e}_\theta$	$\cos\theta$	$\sin\theta$

as well as the formula for velocity in terms of polar coordinates:

$$\boldsymbol{v}_D = \dot{r}\boldsymbol{e}_r + r\dot{\theta}\boldsymbol{e}_\theta \tag{2.55}$$

Solve We have Derek's velocity in terms of two sets of coordinates: the actual velocity in terms of Cartesian coordinates and a general velocity form in terms of polar coordinates. By equating the two, we'll be able to translate the known horizontal speed v_D into the unknown unreeling speed ($\dot{r}$):

(2.54), (2.55) $\Rightarrow$
$$\dot{r}\boldsymbol{e}_r + r\dot{\theta}\boldsymbol{e}_\theta = v_D\boldsymbol{i} = v_D(\sin\theta\boldsymbol{e}_r + \cos\theta\boldsymbol{e}_\theta) \tag{2.56}$$

matching $\boldsymbol{e}_r$ components:
$$\dot{r} = v_D\sin\theta$$

The angle θ is found from $\theta = \tan^{-1}(3\,\text{m}/1\,\text{m}) = 71.6°$. Using $v_D = 10$ m/s and this value of θ in our expression for $\dot{r}$ gives us

$$\dot{r} = (10\,\text{m/s})\sin(71.57°) = 9.49\,\text{m/s}$$

Check A consistency check would be to calculate $\dot{\theta}$ and then use $\dot{r}$ and $\dot{\theta}$ in our polar expression for velocity and see if we obtain a purely horizontal result (matching the problem statement). The $\boldsymbol{e}_\theta$ component of (2.56) gives us $r\dot{\theta} = v_D(\cos\theta) \Rightarrow \dot{\theta} = 1.0$ rad/s. From (2.55) we have

$$\begin{aligned}\boldsymbol{v}_D &= \dot{r}\boldsymbol{e}_r + r\dot{\theta}\boldsymbol{e}_\theta \\ &= \dot{r}(\sin\theta\boldsymbol{i} - \cos\theta\boldsymbol{j}) + r\dot{\theta}(\cos\theta\boldsymbol{i} + \sin\theta\boldsymbol{j}) \\ &= (\dot{r}\sin\theta + r\dot{\theta}\cos\theta)\boldsymbol{i} + (-\dot{r}\cos\theta + r\dot{\theta}\sin\theta)\boldsymbol{j} \\ &= (10\,\text{m/s})\boldsymbol{i} + 0\boldsymbol{j}\end{aligned}$$

EXAMPLE 2.12 ACCELERATION—POLAR COORDINATES (A) (Theory on page 51)

An extensible robotic arm with a small payload is shown in **Figure 2.43**. At the current instant, $\theta = 30°, \dot{\theta} = 1.2$ rad/s, $\ddot{\theta} = -5$ rad/s^2, $r = 1$ m, $\dot{r} = 5$ m/s, and $\ddot{r} = 0.06$ m/s^2. What are the current magnitudes of the payload's velocity and acceleration?

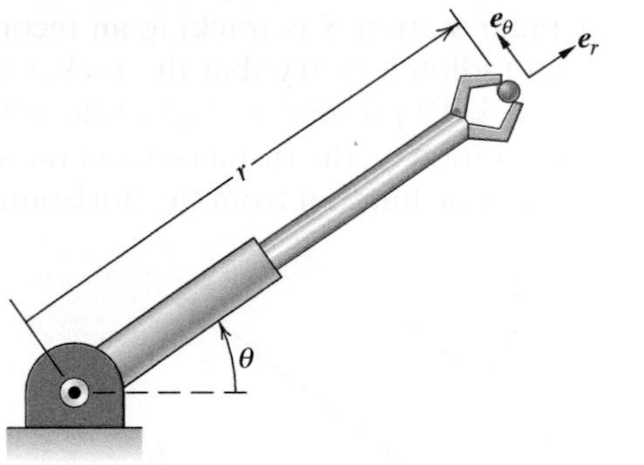

Figure 2.43 Robotic arm

Goal Obtain the velocity and acceleration of the payload at the end of a robotic arm.

Given $\theta = 30°, \dot{\theta} = 1.2\,\text{rad/s}, \ddot{\theta} = -5\,\text{rad/s}^2, r = 1\,\text{m}, \dot{r} = 5\,\text{m/s}$, $\ddot{r} = 0.06\,\text{m/s}^2$.

Assume This problem doesn't require any assumptions.

Draw The geometry of our system is shown schematically in **Figure 2.44.**

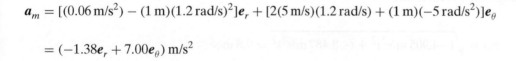

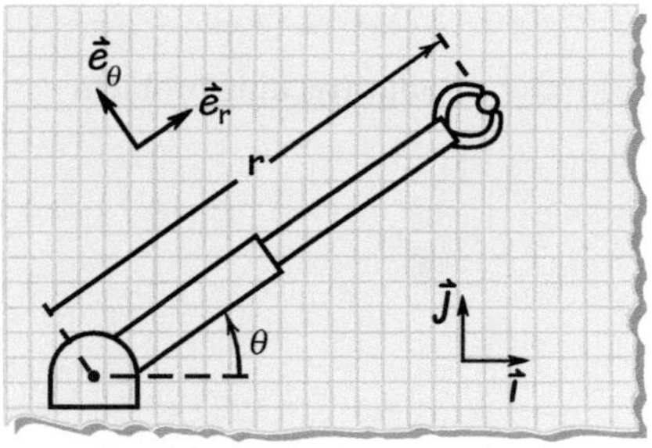

Figure 2.44 Schematic of robotic arm

Formulate Equations The equations for velocity and acceleration in polar coordinates are (2.46) and (2.50):

$$\boldsymbol{v}_m = \dot{r}\boldsymbol{e}_r + r\dot{\theta}\boldsymbol{e}_\theta$$

$$\boldsymbol{a}_m = \left(\ddot{r} - r\dot{\theta}^2\right)\boldsymbol{e}_r + \left(2\dot{r}\dot{\theta} + r\ddot{\theta}\right)\boldsymbol{e}_\theta$$

Solve Using the provided parameters in (2.46) and (2.50) gives

$$\boldsymbol{v}_m = [(5\,\text{m/s})\boldsymbol{e}_r + (1\,\text{m})(1.2\,\text{rad/s})\boldsymbol{e}_\theta]\ \text{m/s} = \left(5\boldsymbol{e}_r + 1.2\boldsymbol{e}_\theta\right)\ \text{m/s}$$

$$\|\boldsymbol{v}_m\| = \sqrt{(5\,\text{m/s})^2 + (1.2\,\text{m/s})^2} = 5.14\,\text{m/s}$$

$$\boldsymbol{a}_m = [(0.06\,\text{m/s}^2) - (1\,\text{m})(1.2\,\text{rad/s})^2]\boldsymbol{e}_r + [2(5\,\text{m/s})(1.2\,\text{rad/s}) + (1\,\text{m})(-5\,\text{rad/s}^2)]\boldsymbol{e}_\theta$$

$$= (-1.38\boldsymbol{e}_r + 7.00\boldsymbol{e}_\theta)\ \text{m/s}^2$$

$$\boldsymbol{a}_m = \sqrt{\left(-1.38\,\text{m/s}^2\right)^2 + \left(7.00\,\text{m/s}^2\right)^2} = 7.13\,\text{m/s}^2$$

Check There's not much that we can check for this problem, beyond making sure that the magnitudes of the velocity and acceleration are greater than or equal to their respective individual components (which they are).

EXAMPLE 2.13 ACCELERATION—POLAR COORDINATES (B) (Theory on page 51)

A radar station S is tracking an incoming rocket O and produces the data listed in this example. It's known from radiotelemetry that the rocket's engine ceased firing several seconds before the radar station picked up the rocket's presence. Neglect the effect of air drag and determine the magnitude of the rocket's velocity and acceleration. r, the distance from the station to the rocket, is 3000 m, $\dot{r} = -483.0\,\text{m/s}$, and $\ddot{r} = 0.6782\,\text{m/s}^2$. θ, the rocket's inclination from the horizontal, is 30°, $\dot{\theta} = -0.04314\,\text{rad/s}$, and $\ddot{\theta} = -0.01672\,\text{rad/s}^2$.

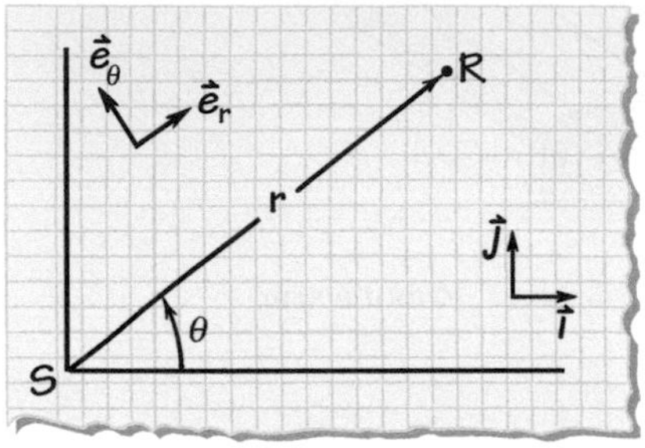

Figure 2.45 Radar tracking problem

Goal Determine the rocket's velocity and acceleration.

Given Rocket's position, velocity, and acceleration.

Assume No assumptions are needed.

Draw **Figure 2.45** shows our system.

Formulate Equations We can find the velocity from our polar representation, (2.46):

$$\begin{aligned} \boldsymbol{v}_O = \dot{r}\boldsymbol{e}_r + r\dot{\theta}\boldsymbol{e}_\theta &= -483.0\boldsymbol{e}_r\,\text{m/s} + (3000\,\text{m})(-0.04314\,\text{rad/s})\boldsymbol{e}_\theta \\ &= -(483.0\boldsymbol{e}_r + 129.4\boldsymbol{e}_\theta)\,\text{m/s} \end{aligned}$$

The velocity magnitude is therefore

$$v_O = \sqrt{(-483.0\,\text{m/s})^2 + (-129.4\,\text{m/s})^2} = 500\,\text{m/s}$$

The acceleration is, from (2.50),

$$\begin{aligned} \boldsymbol{a}_O &= (\ddot{r} - r\dot{\theta}^2)\boldsymbol{e}_r + (r\ddot{\theta} + 2\dot{r}\dot{\theta})\boldsymbol{e}_\theta \\ &= [0.6782\,\text{m/s}^2 - (3000\,\text{m})(-0.04314\,\text{rad/s})^2]\boldsymbol{e}_r + [(3000\,\text{m})(-0.01672\,\text{rad/s}^2) + 2(-483.0\,\text{m/s})(-0.04314\,\text{rad/s})]\boldsymbol{e}_\theta \\ &= -(4.905\boldsymbol{e}_r + 8.487\boldsymbol{e}_\theta)\,\text{m/s}^2 \end{aligned}$$

Solve The acceleration magnitude is therefore

$$a_O = \sqrt{(-4.905\,\text{m/s}^2)^2 + (-8.487\,\text{m/s}^2)^2} = 9.8\,\text{m/s}^2$$

Check We're given the fact that the rocket's engine isn't working and that we can neglect air drag. This means that the only acceleration we should see is that produced by gravity. Let's check by expressing the acceleration in $\boldsymbol{i}$, $\boldsymbol{j}$ components:

	$\boldsymbol{i}$	$\boldsymbol{j}$
$\boldsymbol{e}_r$	cos(30°)	sin(30°)
$\boldsymbol{e}_\theta$	−sin(30°)	cos(30°)

$$\begin{aligned} \boldsymbol{a}_O &= -\{4.905[\cos(30°)\boldsymbol{i} + \sin(30°)\boldsymbol{j}] + 8.487[-\sin(30°)\boldsymbol{i} + \cos(30°)\boldsymbol{j}]\}\,\text{m/s}^2 \\ &= -9.80\boldsymbol{j}\,\text{m/s}^2 \end{aligned}$$

The whole of the acceleration vector is simply gravitational acceleration (9.8 m/s^2 pointing straight down), implying that the rocket doesn't carry an active source of propulsion and that it's therefore moving along a **ballistic trajectory**, the trajectory followed by objects moving purely under the influence of gravity.

EXAMPLE 2.14 VELOCITY AND ACCELERATION—CYLINDRICAL COORDINATES (Theory on page 53)

Figure 2.46 shows a slotted screw partially screwed into a wooden board (screwdriver removed for clarity). It's being unscrewed at a constant rate of 20 rad/s, and the pitch of the threads causes it to move up 3.0 mm for every rotation. The distance from the screw's longitudinal centerline to the point A (on the outside edge of the head) is 3.5 mm. Find A's velocity and acceleration.

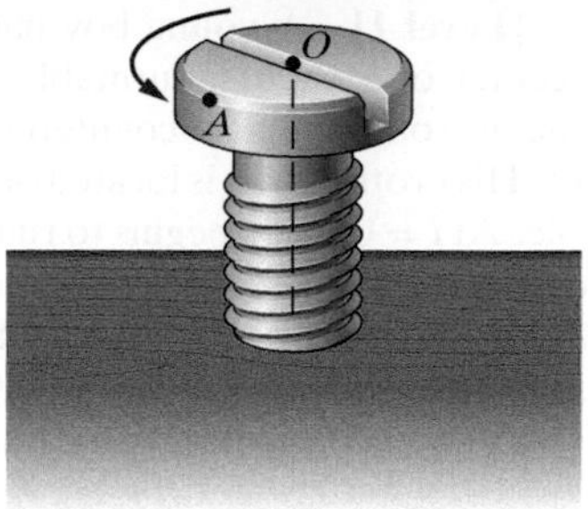

Figure 2.46 Screw being unscrewed

Goal Find the velocity and acceleration of a point on the outer edge of a moving screw.

Given Dimensions of screw and its rotational speed.

Assume Counterclockwise screw rotation is taken as positive.

Draw **Figure 2.47** shows a simplified model of our system. The position vector $\boldsymbol{r}_{A/O}$ goes from the centerline of the screw to A and defines the $\boldsymbol{e}_r$ unit vector. The $\boldsymbol{e}_\theta$ and $\boldsymbol{k}$ unit vectors follow immediately, $\boldsymbol{e}_\theta$ indicating the direction of rotation. Because we're concerned with a point that's a fixed distance from the rotation axis, the extensional speed and acceleration, $\dot{r}_{A/O}$ and $\ddot{r}_{A/O}$, are zero. Also, the vertical speed and rotation rate are constant, and therefore the vertical acceleration and $\ddot{\theta}$ are both zero.

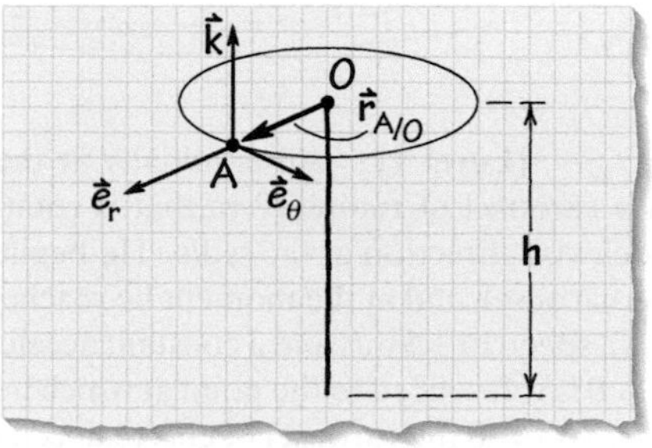

Figure 2.47 Simplified screw model

Formulate Equations The screw moves up a distance of 3.0 mm for every rotation and rotates at $\omega = 20$ rad/s (counterclockwise rotation taken to be positive, as usual). The vertical translation rate is therefore related to the rotation rate by

$$\dot{h} = 3.0\text{ mm}\left(\frac{20\text{ rad/s}}{2\pi\text{ rad}}\right) = 9.55\text{ mm/s}$$

The governing velocity and acceleration equations are given by (2.52) and (2.53):

$$\boldsymbol{v}_A = \dot{r}\boldsymbol{e}_r + r\dot{\theta}\boldsymbol{e}_\theta + \dot{z}\boldsymbol{k}$$

$$\boldsymbol{a}_A = (\ddot{r} - r\dot{\theta}^2)\boldsymbol{e}_r + (r\ddot{\theta} + 2\dot{r}\dot{\theta})\boldsymbol{e}_\theta + \ddot{z}\boldsymbol{k}$$

Solve Using the given constraints in the governing equations yields

$$\boldsymbol{v}_A = 0\boldsymbol{e}_r + (3.5\text{ mm})(20\text{ rad/s})\boldsymbol{e}_\theta + 9.55\boldsymbol{k}\text{ mm/s}$$

$$= (70\boldsymbol{e}_\theta + 9.55\boldsymbol{e}_z)\text{ mm/s}$$

$$\boldsymbol{a}_A = [0 - (3.5\text{ mm})(20\text{ rad/s})^2]\boldsymbol{e}_r + (0+0)\boldsymbol{e}_\theta + 0\boldsymbol{k}$$

$$= -1.40\boldsymbol{e}_r\text{ m/s}^2$$

Check There's nothing in particular to check beyond double checking the calculations.

EXERCISES 2.3

2.3.1. **[Level 1]** A young boy named Andy is located at the center of a circular turntable of radius 10 m. The turntable is rotating in a counterclockwise direction at 0.4 rad/s. His brother Ben is located on the outer edge of the turntable. At $t = 0$ Andy begins to run directly toward Ben, moving with a constant acceleration of $3\ \text{m/s}^2$. Ultimately he collides with Ben and they both fall off the turntable. At the instant of collision, what was the absolute value of Andy's velocity and acceleration with respect to ground?

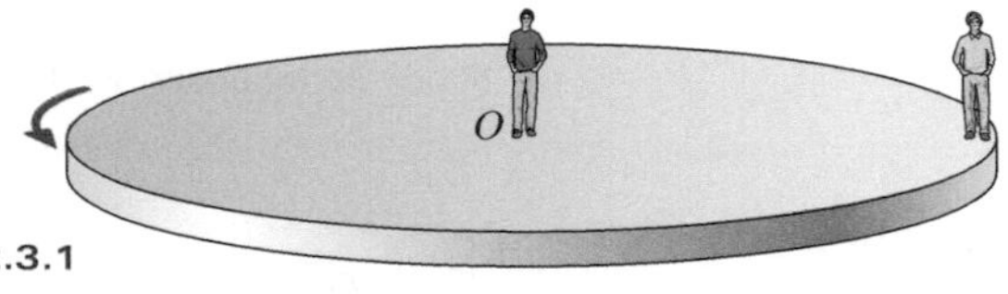

E2.3.1

2.3.2. **[Level 1]** Bill is initially at the outer edge (A) of a carousel of radius 12 m that is rotating in a counterclockwise direction at 0.6 rad/s. He begins to move across the carousel, and at the moment he reaches the center O he has a speed of 1.5 m/s and a constant acceleration of $0.6\ \text{m/s}^2$ ($t = 0$ is referenced to the point at which he reaches the center). At $t = 0$ the carousel begins to decelerate at a constant rate of $0.1\ \text{rad/s}^2$. What are Bill's velocity and acceleration with respect to the ground when he reaches point B?

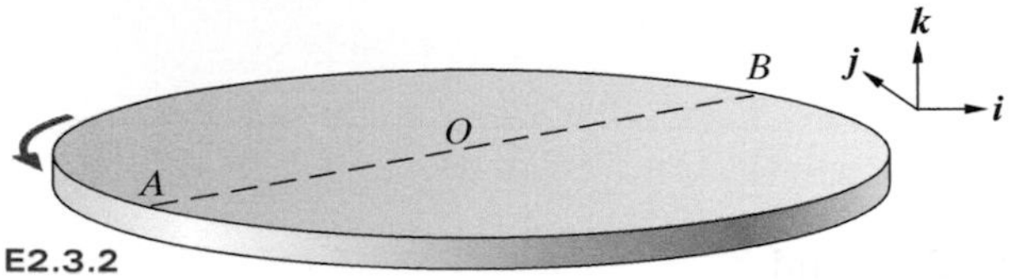

E2.3.2

2.3.3. **[Level 1]** A ceiling fan is moving at a constant rotation rate and makes one revolution in 1.3 s. The distance from the fan's center to the outermost tip of a blade is 1 m. A fly F is resting right at the tip. What is the fly's acceleration?

E2.3.3

2.3.4. **[Level 1]** Give an example of a motion of point A for which

$$\frac{d\|\boldsymbol{r}_A\|}{dt} = 0$$

and

$$\|\boldsymbol{v}_A\| = 1$$

2.3.5. **[Level 1]** The lower portion of a fire ladder ($\overline{OA}$) rotates about its hinge O at $\dot{\theta} = 0.05$ rad/s and $\ddot{\theta} = 0.04\ \text{rad/s}^2$. The upper portion extends out from the lower portion such that $\dot{c} = 0.4$ m/s and $\ddot{c} = 0.1\ \text{m/s}^2$. Determine the velocity and acceleration of B with respect to $\boldsymbol{e}_r, \boldsymbol{e}_\theta$ as well as $\boldsymbol{i}, \boldsymbol{j}$ for $\theta = \frac{\pi}{6}$ rad and $c = 2$ m.

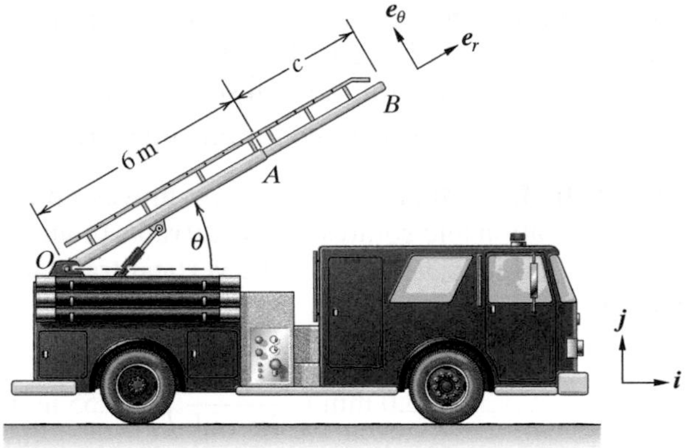

E2.3.5

2.3.6. **[Level 1]** A hydraulic piston controls the opening of a car's trunk lid. What is the magnitude of the velocity and acceleration of the end B if $\dot{\theta} = -0.683$ rad/s, $\ddot{\theta} = -0.1585\ \text{rad/s}^2$, $\|\boldsymbol{r}_{B/O}\| = 0.3$ m, $\frac{d}{dt}\|\boldsymbol{r}_{B/O}\| = -0.0549$ m/s, and $\frac{d^2}{dt^2}\|\boldsymbol{r}_{B/O}\| = 0.0375\ \text{m/s}^2$?

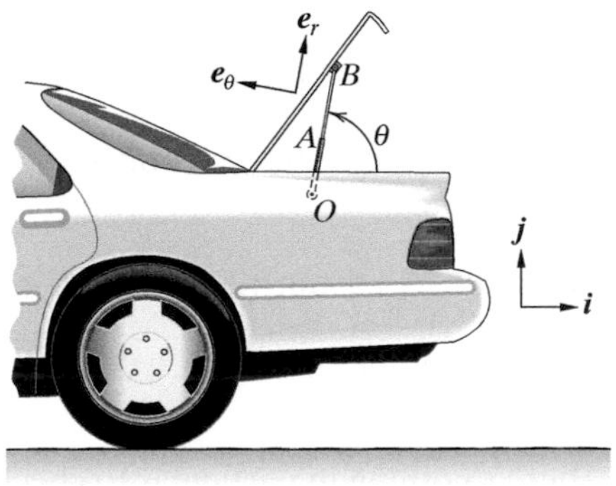

E2.3.6

2.3.7. **[Level 1]** You're tracking an airplane from the ground. The aircraft is at a constant height h from the ground, at a distance r_0 from you at the illustrated instant, and at an inclination θ. The aircraft's speed is constant at 1200 km/h. Find the rate at which your tracking dish must rotate if $r_0 = 3$ km and $\theta = 30°$. Does the acceleration of gravity make any contribution to your answer?

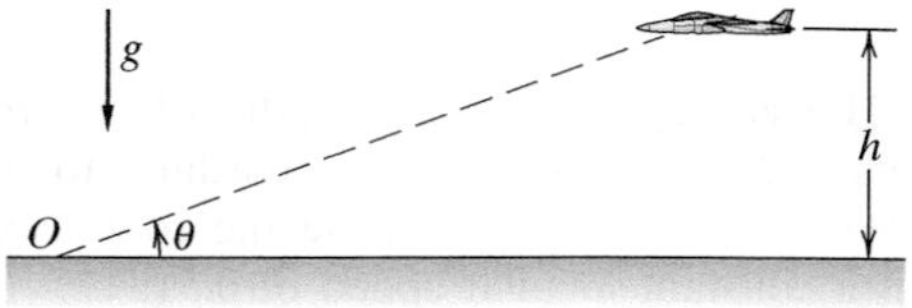

E2.3.7

2.3.8. **[Level 1]** Car B is driving straight toward the point O at a constant speed v. An observer, located at A, tracks the car with a radar gun. What is the speed $|\dot{r}_{B/A}|$ that the observer at A records?

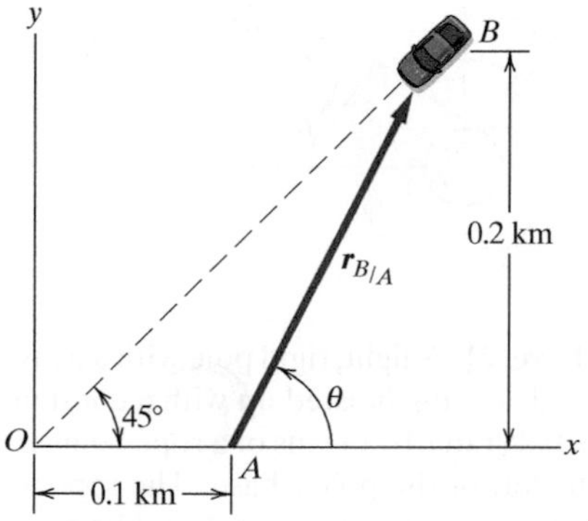

E2.3.8

2.3.9. **[Level 1]** Explain why $\left\| \frac{d\boldsymbol{r}(t)}{dt} \right\|$ is not in general equal to $\frac{d}{dt}\|\boldsymbol{r}(t)\|$.

2.3.10. **[Level 1]** When is $\frac{d}{dt}\|\boldsymbol{r}(t)\|$ equal to $\left\| \frac{d}{dt}(\boldsymbol{r}(t)) \right\|$?

2.3.11. **[Level 1]** If $\dot{r} = 0.90$ m/s and $r\dot{\theta} = -0.90$ m/s, will $\frac{d}{dt}(\boldsymbol{r}) = 0$?

2.3.12. **[Level 2]** In this exercise we'll see how path curvature can affect the acceleration felt by an object. In case (a) the path traveled by a particle A is described by a sinusoid: $y = r_0 - r_1 \cos(2\omega t)$. In case ($b$) we have the same variation of height, but in this case we vary the radial variable r rather then the height measure y. The particle travels around at a constant rate ($\theta = 2\omega t$, with ω constant) and $r = r_0 - r_1 \cos(\theta)$. Calculate and compare the acceleration in each case, $\ddot{y}$ for case (a) and a_r for case (b).

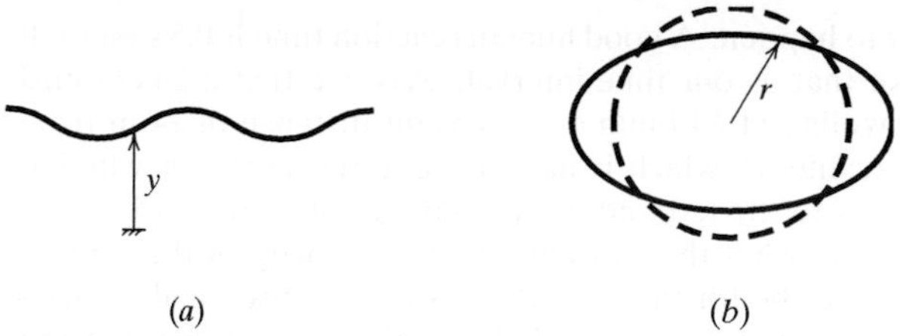

E2.3.12

2.3.13. **[Level 2]** When I'm out photographing birds in flight, I often have a problem keeping the bird in view. Let's analyze this problem. Assume that I've spotted a Violet-Green Swallow that's moving in a straight line at a constant speed, as shown. My camera's telephoto lens captures 102 cm of target image when the object is 24 m away. Thus a bird 102 cm long would stretch from end to end in the picture. Luckily, swallows are smaller than this, being only 20 cm long. Assume that the bird is at its closest position to me ($\theta = 90°, h = 24$ m) and traveling at 64 km/h.

a. At what angular speed will I have to rotate my camera to keep the bird centered?

b. What is the acceleration of the far end of the lens? Assume a distance of 43 cm from the far end to the center of my head (the assumed center of rotation).

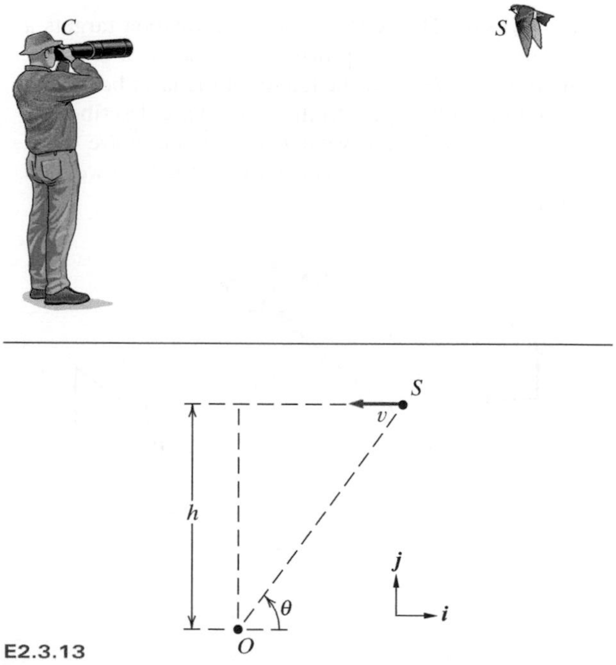

E2.3.13

2.3.14. **[Level 2]** In this exercise we'll look more deeply into the problems of photographing moving objects. Consider **Exercise 2.3.13**, in which we're trying to photograph a moving swallow. One of the key problems is that a slight change in speed (on the part of either the swallow or the photographer) will cause the bird to disappear from view. Let's determine the accuracy that one needs in order for this *not* to happen. A good human reaction time is 0.5 s, so we'll take that as our time interval. Assume that a 20 cm bird (traveling at 64 km/h at a constant distance of 24 m from the camera), which is initially centered in the viewfinder, moves to the left and completely out of view in 0.5 s. Determine what the constant angular velocity of the camera needs to be for this to occur. Compare this angular speed to the angular speed needed to perfectly track the bird and determine the percent variation from the ideal case.

2.3.15. [Level 2] In this exercise we'll continue to examine the problems of photographing moving objects. Consider **Exercise 2.3.13**, in which we're trying to photograph a moving swallow. One of the problems with tracking a moving bird is that the rate at which one has to rotate the camera isn't constant; there exists a significant angular acceleration. Assume that a swallow is moving at a constant 64 km/h in a straight line, as shown. The swallow starts off far to the right ($\theta \approx 0$) and then proceeds to move to the left. Calculate at what angle θ the camera's angular acceleration is a maximum. What is its angular acceleration at $\theta = 0$ and $\theta = 90°$?

2.3.16. [Level 2] A Republic laser weapon targets a Trade Federation battleship fleeing to space at a constant velocity of $\boldsymbol{v} = 25\boldsymbol{j}$ m/s. If the length of the laser beam and its orientation with respect to the ground are described by r and θ, respectively, find what r, $\dot{r}$, $\ddot{r}$, $\dot{\theta}$, and $\ddot{\theta}$ are when $\theta = 50°$. The horizontal distance between the laser weapon and battleship is 300 m.

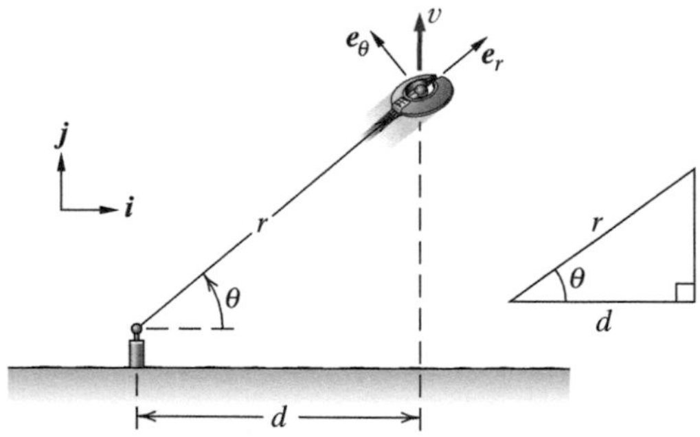

E2.3.16

2.3.17. [Level 2] Sarah has attached a resistance cable to her foot to give her leg a workout. Let r be the length of cable from the spool to Sarah's foot and θ be the angle the cable makes with the ground. At a certain instant, the cable is 0.6 m long and is angled at 15°, and Sarah's foot has a velocity and acceleration of $\boldsymbol{v} = (1.32\boldsymbol{i} + 0.76\boldsymbol{j})$ m/s and $\boldsymbol{a} = (0.27\boldsymbol{i} + 0.15\boldsymbol{j})$ m/s^2, respectively. Calculate $\dot{r}$, $\ddot{r}$, $\dot{\theta}$, and $\ddot{\theta}$ at this instant.

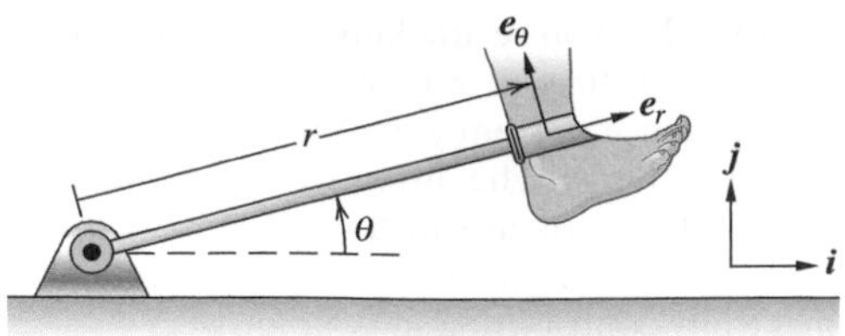

E2.3.17

2.3.18. [Level 2] The length of a robotic arm to its end effector varies according to $r(\theta) = (0.5 + 0.25[1 - \cos(2\theta)])$ m. Suppose the robotic arm is rotating at a constant angular speed of 0.5 rad/s. Determine the magnitude of the velocity and acceleration for the robotic arm's end effector when $\theta = 45°$.

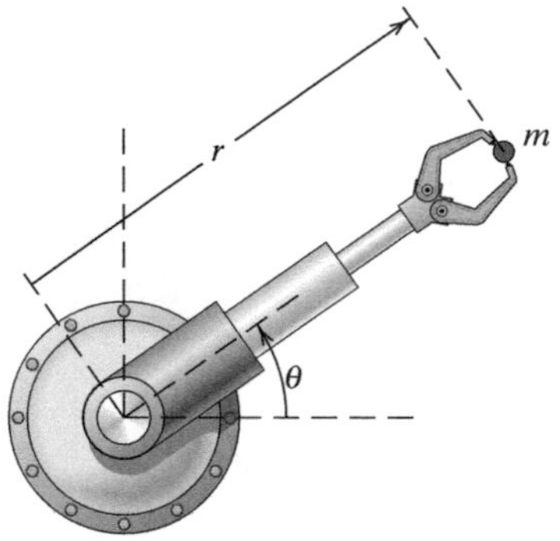

E2.3.18

2.3.19. [Level 2] A light, rigid pole with a heavy lumped mass on its end is being hoisted up with a constant angular speed of $\omega = 0.25$ rad/s by means of a rope being reeled into a spool to the left of the pole's base. The rope is attached to the lumped mass on the pole's end, and let r describe the length of rope from the spool to the lumped mass. At a certain instant the rope makes an angle of 40° with the ground, and the pole is vertical. If the pole is 6 m long, find what r, $\dot{r}$, $\ddot{r}$, $\dot{\theta}$, and $\ddot{\theta}$ are at this instant.

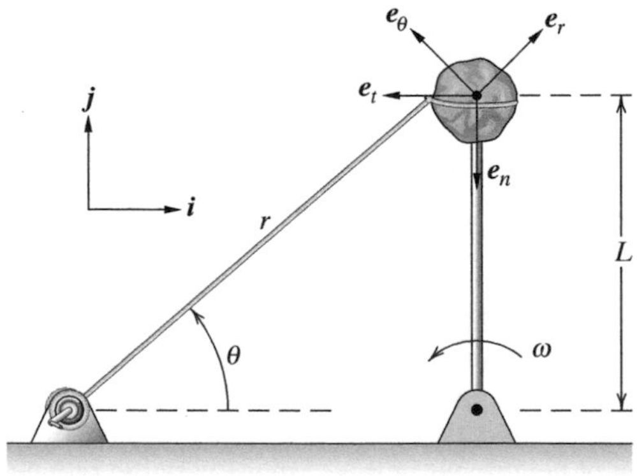

E2.3.19

2.3.20. [Level 2] A little kid places his RC toy car at the center of a merry-go-round at the local playground. He then spins the merry-go-round to a constant angular speed of 3 rad/s and proceeds to drive the RC car directly outward to A on the merry-go-round's edge with an acceleration of 0.9 m/s^2 relative to the merry-go-round. What is the magnitude of the toy car's velocity and acceleration at A? The radius of the merry-go-round is 1.5 m.

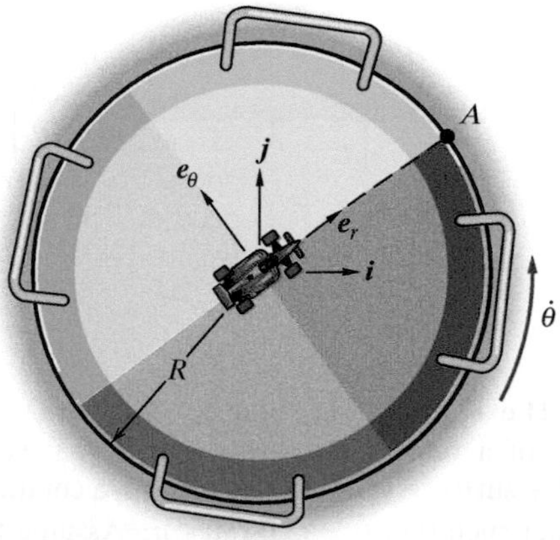

E2.3.20

2.3.21. **[Level 2]** A dolphin, being tracked by sonar, produces readings of $r = 300$ m, $v_r = 10$ m/s, $v_\theta = 17.3$ m/s, $a_r = 0$ m/s^2, and $a_\theta = 0$ m/s^2. What are the corresponding values of $\dot{r}, \ddot{r}, \dot{\theta}$, and $\ddot{\theta}$?

2.3.22. **[Level 2]** Imagine that you're a poor graduate student who has been given the job of determining how fast a gecko (a small lizard) can climb walls. Since your research budget is small, all you have available is a spool of thread and a lightweight spinometer. The spinometer records the rate at which it spins and from that calculates the rate at which the thread leaves the spool. You put the spool of thread in the spinometer and attach the thread's free end to the gecko's neck, as shown. When the gecko is 2 m up the wall, your spinometer reads 0.78 m/s. What is the gecko's speed?

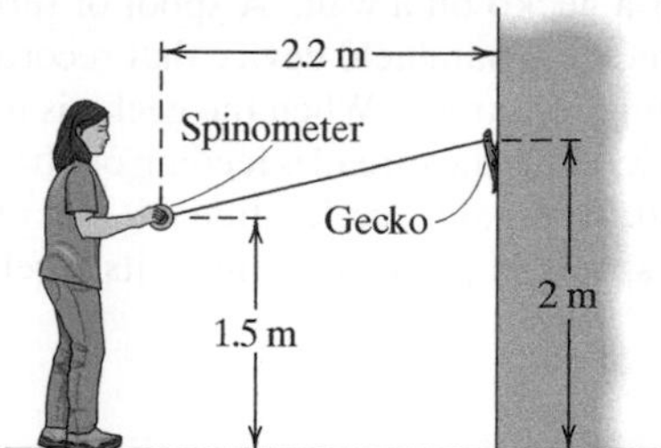

E2.3.22

2.3.23. **[Level 2]** The position of a charged particle is given by $r = at - bt^3$ and $\theta = ce^{-t}$. Derive expressions for the magnitudes of the particle's velocity and acceleration.

2.3.24. **[Level 2]** A stunt plane is at the bottom of a circular loop. The loop's radius is 305 m, its center is directly above O, and the plane is traveling at a constant speed. At the illustrated instant it is 30.5 m above an observer at O and is traveling at 485 km/h. The acceleration of the plane is oriented upward (toward the center of the loop) and has a magnitude of 58.8 m/s^2 (the centripetal acceleration of the plane moving in a circle about C). Is $\ddot{r}$ equal to 58.8 m/s^2 as well? $\theta = \frac{\pi}{2}$ rad.

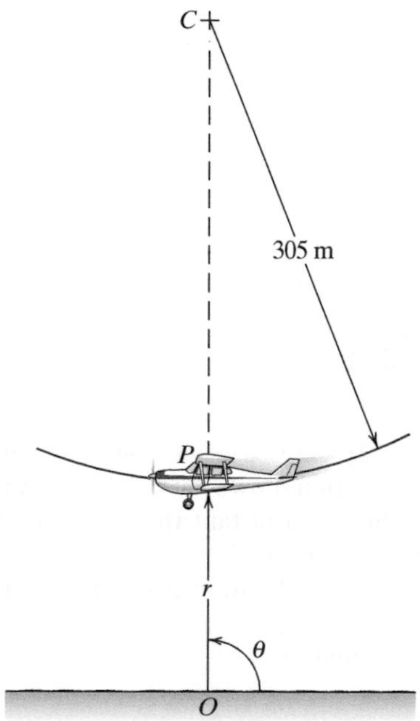

E2.3.24

2.3.25. **[Level 2]** A rope runs from a reel at O to an eyelet A attached to a vertically operating door. Ignore the dimensions of the reel. Assume that rope is being reeled in at a constant rate of 2 m/s and at the illustrated instant $r = 2$ m and $\theta = 25°$. What are the velocity and acceleration of A at this instant?

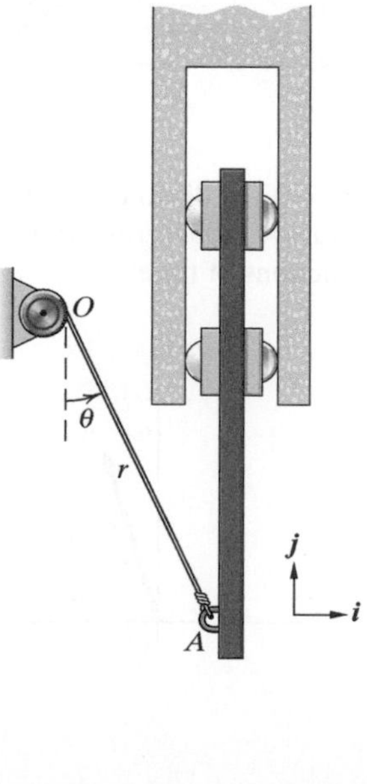

E2.3.25

2.3.26. **[Level 2]** A dual-axis accelerometer at the end of an extensional robotic arm gives a reading of $\boldsymbol{a} = (4\boldsymbol{e}_r + 5\boldsymbol{e}_\theta)$ m/s^2. At this instant $r = 1.5$ m, $\ddot{r} = 10$ m/s^2, $\theta = 45°$, and $\ddot{\theta} = -0.5$ rad/s^2.

a. Express the acceleration in the form $a_1\boldsymbol{i} + a_2\boldsymbol{j}$.

b. Determine $\dot{\theta}$ and $\dot{r}$.

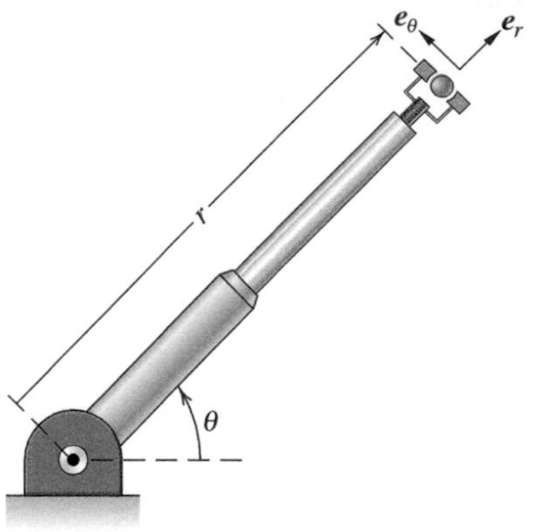

E2.3.26

2.3.27. **[Level 2]** Car C is driving on a circular section of track and at the position shown has a velocity of $\boldsymbol{v} = -80.5\boldsymbol{j}$ km/h. At the illustrated instant the car is accelerating at $0.3\,g$ in the direction of travel.

a. What are the acceleration components of the car in the $\boldsymbol{i}$ and $\boldsymbol{j}$ directions?

b. What is $\ddot{L}$ equal to?

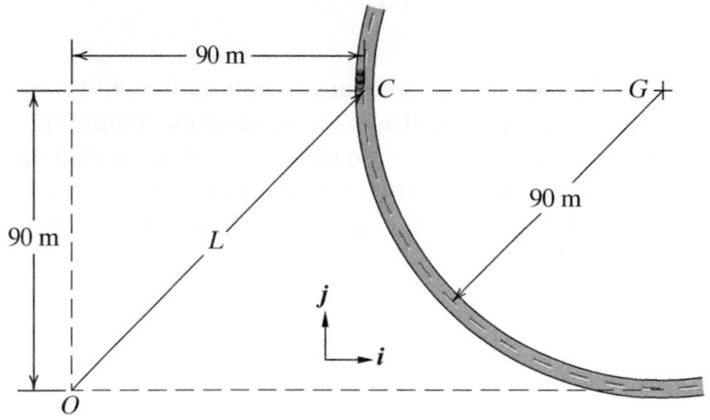

E2.3.27

2.3.28. **[Level 2]** Let the distance from O to A be given by $r = a\theta$, with a equal to 3.05 m/rad. If $\dot{\theta} = 10$ rad/s, what are $\boldsymbol{v}_A$ and $\boldsymbol{a}_A$ as functions of time?

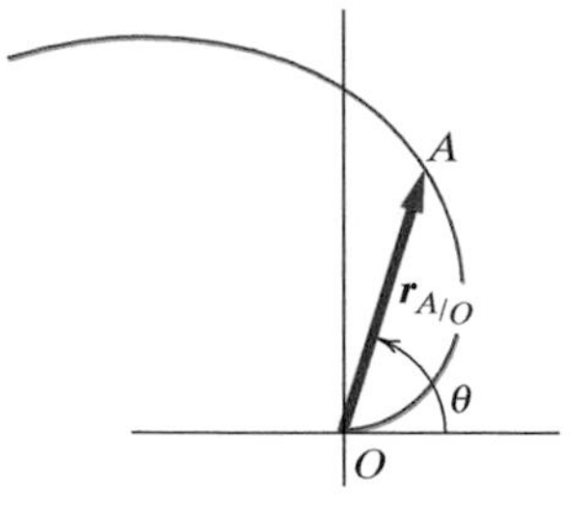

E2.3.28

2.3.29. **[Level 2]** What is the acceleration $\boldsymbol{a}_A$ of point A on the periphery of a drill bit that's rotating at a constant 5000 rpm? $d = 6$ mm. What would the drill bit's constant, negative angular acceleration need to be for the magnitude of the $\boldsymbol{e}_\theta$ component of A's total acceleration to equal the value of $\|\boldsymbol{a}_A\|$ you just calculated? How long would it take for the drill to spin down to rest at this value of acceleration?

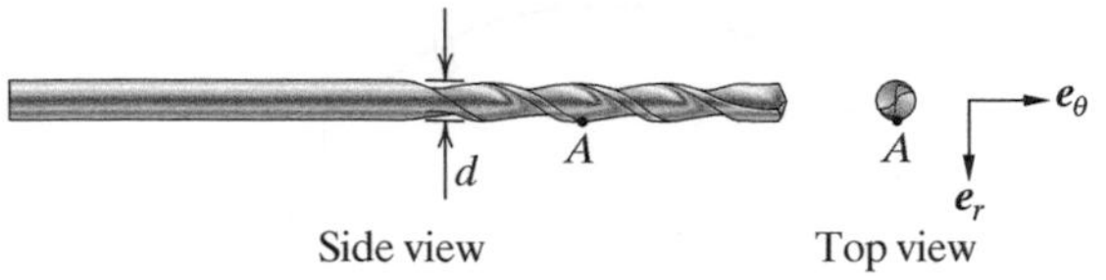

E2.3.29

2.3.30. **[Level 2]** A laser pointer, located at O, is aimed at the top of a tunnel, the shape of which is such that $r_{A/O} = L(1 + \sin\theta)$. The pointer is swung in a counterclockwise manner such that $\dot{\theta} = (2.0 \text{ rad/s}^2)t$. Assume that the swing began at $\theta = 0$ when $t = 0$ s. What are $\boldsymbol{v}_A$ and $\boldsymbol{a}_A$ when $\theta = \frac{\pi}{2}$ rad, where A is the point on the tunnel where the laser beam hits?

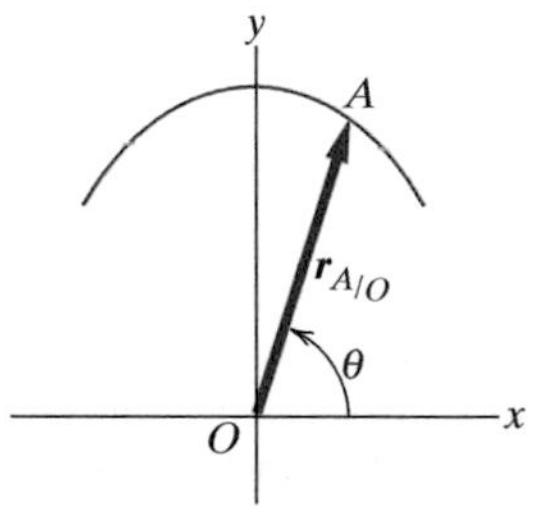

E2.3.30

2.3.31. **[Level 2]** This exercise is an extension of **Exercise 2.3.22.** As in **Exercise 2.3.22**, you're trying to analyze the motion of a gecko on a wall. A spool of thread is fed from a spinometer, a handheld device that records the rate at which the thread unreels. When the gecko is in the position shown in **E2.3.22**, the thread is feeding out from the reel at a speed of 0.78 m/s and acceleration of -1.2 m/s^2. What is the gecko's speed? Can you determine its acceleration as well?

2.3.32. **[Level 2]** A Confederation battlestation has captured a derelict Xiyalian transport vessel in its tractor beam, causing the distance from the battlestation to the transport to decrease as $r = a(t_0 - t)$. The Xiyalian vessel's speed in the $\boldsymbol{e}_\theta$ direction is constant at b. What are the velocity and acceleration of the Xiyalian transport (before it hits the battlestation) as a function of time?

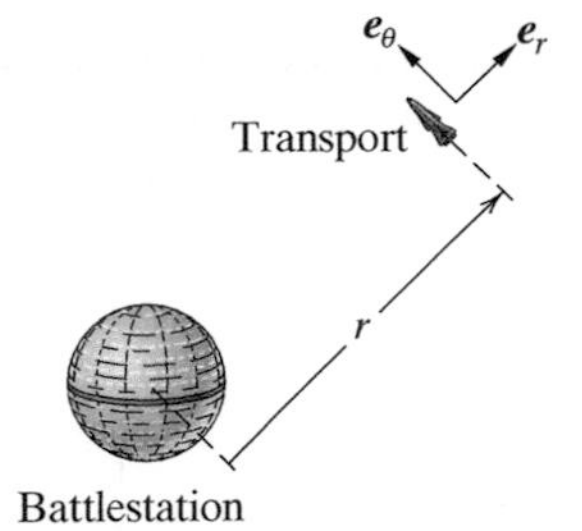

E2.3.32

2.3.33. **[Level 3]** A car B is traveling as illustrated, and follows a path given by

$$\boldsymbol{r}_{B/O} = bt^2\boldsymbol{e}_r, \quad \theta = at^2$$

What are $\boldsymbol{r}_{B/O}$, $\boldsymbol{v}_B$, and $\boldsymbol{a}_B$ when $\theta = \frac{\pi}{2}$ rad?

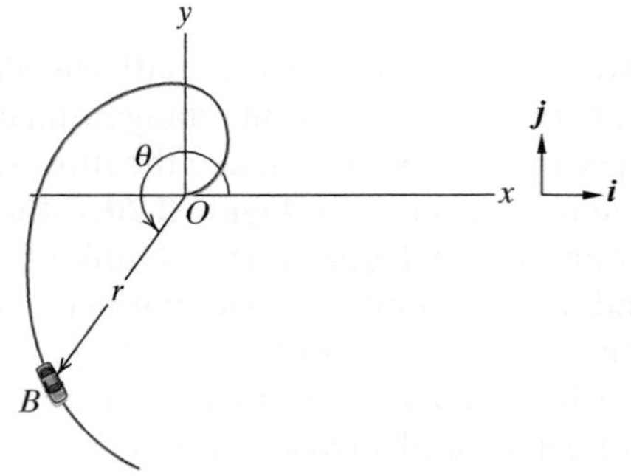

E2.3.33

2.3.34. **[Level 3]** Two tracking stations are observing a satellite S. Station B is directly under the satellite, and Station A lies under the satellite's path at an angle of 45 degrees from the satellite's current position. Assume that stations A and B are positioned as shown, both on the same level. This neglects the earth's curvature, a reasonable assumption for these distances. Station B records the values $\beta = 90°$, $\dot{\beta} = -4.88 \times 10^{-2}$ rad/s, $r_2 = 161$ km, and $\dot{r}_2 = 0$ km/h. The total acceleration of the satellite is $\boldsymbol{a} = -9.36\boldsymbol{e}_{r_2}$ m/s^2.

a. Find $\ddot{r}_2$ and $\ddot{\beta}$.

b. Find $r_1, \dot{r}_1, \ddot{r}_1, \dot{\theta}$, and $\ddot{\theta}$.

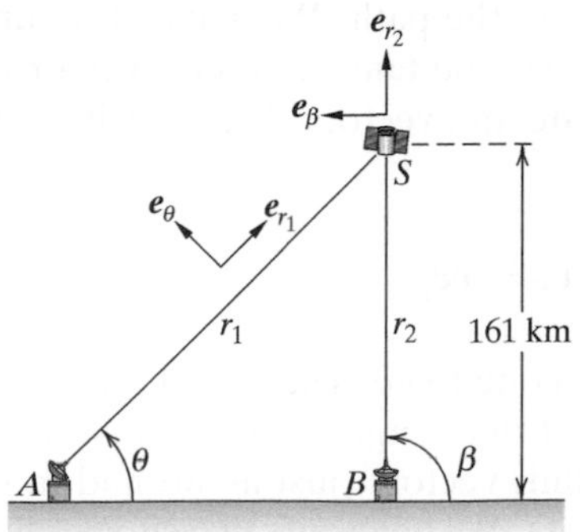

E2.3.34

2.3.35. **[Level 1]** The illustrated device is a robotic probe that tests the surface integrity of curved panels. The arm's length L is constant, but the height h of the probe tip and the orientation θ of the arm are variable and given by

$$\theta(t) = a[1 - \cos(\omega_1 t)]$$

$$h(t) = b[1 - \cos(\omega_2 t)]$$

What is the total acceleration felt by the probe tip?

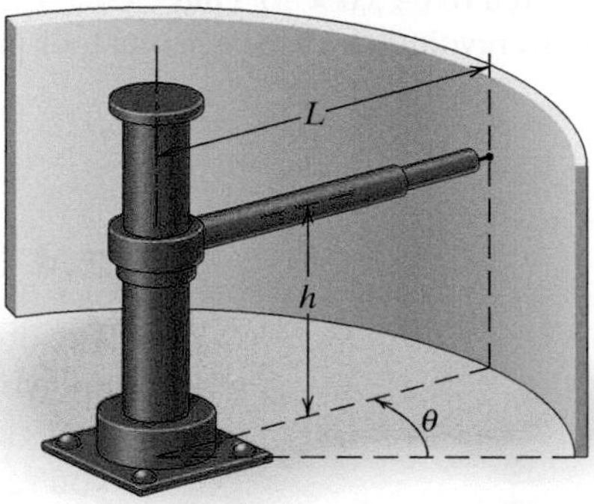

E2.3.35

2.3.36. **[Level 2]** Consider a carousel ride. Each carved horse is attached to a fixed vertical post that's fixed to the carousel's floor. As the carousel rotates, the horse moves up and down the post. We will analyze the motion as the carousel ride is slowing to a stop. Let the vertical motion of the horse be given by $z = z_0 + z_1 \sin(\omega_2\theta)$, where θ is the angular rotation of the floor. The distance from the carousel's axis of rotation and the fixed post is r_h, and the carousel's rotational speed is given by $\omega_1 e^{-bt}$. What are the velocity and acceleration of the horse?

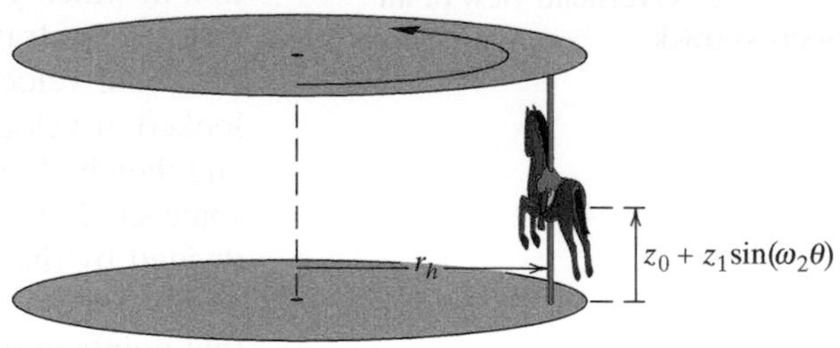

E2.3.36

2.3.37. **[Level 2]** The Guggenheim Museum in New York is designed so that the floor is a continuously rising spiral, letting people view the artwork as they slowly climb the ramp. Assume that an art thief has nabbed a painting and is running down toward the main door. Given the illustrated dimensions and assuming a running speed of 20 km/h, determine the magnitude and direction of the thief's velocity and acceleration.

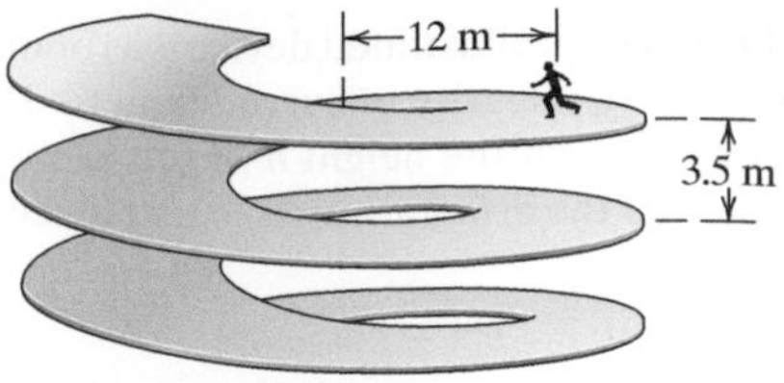

E2.3.37

2.3.38. **[Level 2]** A wingnut is shown, attached to a fixed, threaded shaft. The wingnut is spun rapidly (assume a constant rotation rate) and moves down 2.3 cm in 1.2 s. The acceleration of point A, located 2 cm from the shaft's centerline, is measured to be 5.6×10^3 cm/s^2.

a. How many revolutions of the wingnut took place during this time?

b. By what percentage does point A's speed increase due to the wingnut moving down, as compared to the case in which it turns at the same rate but also stays at the same height?

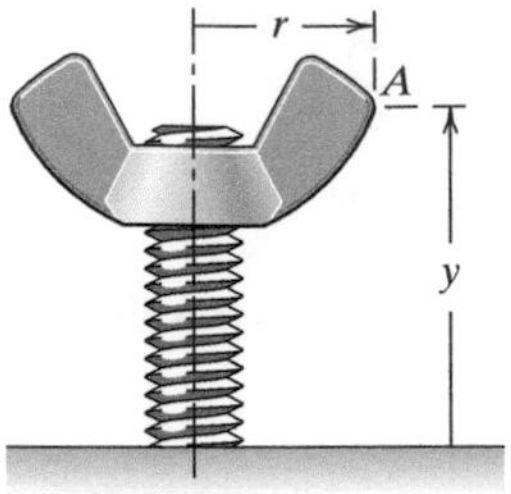

E2.3.38

2.4 PATH COORDINATES

The third type of coordinates we will examine are called **path coordinates**, and you will sometimes see them referred to as normal/tangential coordinates as well. These coordinates are the ones that make the most sense if you're referencing your motion to yourself, as in **Figure 2.20**, rather than to some external reference frame (as in **Figure 2.18**). Unlike the case with Cartesian coordinates and polar coordinates, with path coordinates we define our unit vectors with respect not to position but to velocity.

Imagine you're driving a car in an autocross (a race against the clock in which you have to pilot your car around a twisty path, such as the one shown in **Figure 2.48**). (By the way, the width of the track isn't drawn badly in **Figure 2.48**—the path width is always variable in an autocross track so that you can slide your car around using throttle-induced oversteer. Great fun, but tough on your tires.) What would you say if a passenger asked you to describe the most basic thing about your motion? You would probably say, "I'm going that way," as you pointed straight ahead.

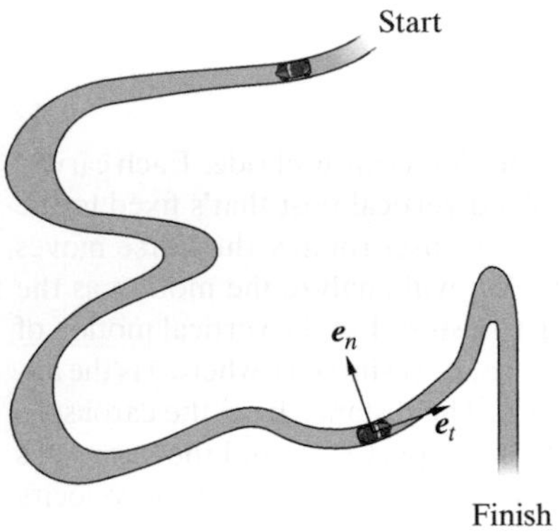

Figure 2.48 Overhead view of an autocross track

That's the most immediate thing you would be able to see—the direction in which you're traveling. And what is always pointing along your direction of travel? Your velocity vector! This isn't in any way mysterious. Your velocity vector always defines your path. In fact, when we first looked at velocity, we found it by saying a particle moves along a path and then by looking at two positions along the path separated in time by some small interval Δt. We then defined velocity as the position change divided by the time change as the time change went to zero. Thus the velocity vector is always tangent to the path. We will callthe unit vector that points in the direction of travel the **tangential unit vector e_t**. This is the first of our two path coordinate unit vectors. The velocity of a moving particle P is therefore given by

$$\boldsymbol{v}_P = v\boldsymbol{e}_t$$

That's one direction down and one to go. The most logical thing to do is introduce another unit vector at right angles to $\boldsymbol{e}_t$ because we know we will want an orthogonal set of unit vectors (just as we had in Cartesian coordinates and polar coordinates). We have only two choices here: point to the left of the path or to the right. The way we decide is to realize that the section of path you're on is, in general, curving; in other words, it looks like part of a circle. And circles have centers. So what we will do is always point the **normal unit vector** $\boldsymbol{e}_n$ (the second unit vector we need to round out our path coordinate set) toward that center. This unit vector is called normal because it's always oriented in a direction normal to the path.

Though not proved here, it's a fact of kinematic analysis that, for short distances, the path of a particle moving on a plane can be approximated by a circular arc. The center of the arc is called the **center of curvature**, and the distance from this center to the particle is the **radius of curvature**. There isn't a single center of curvature for the entire path unless the path is a perfect circle. Rather, there is a different center for each position along the path. **Figure 2.49** shows a few of the centers of curvature and their associated radii of curvature.

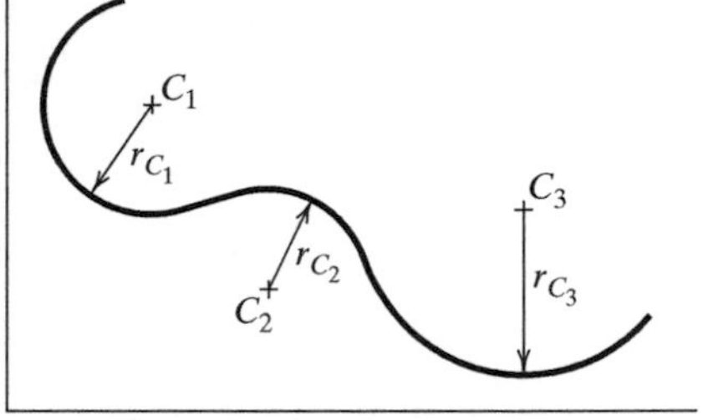

Figure 2.49 Various centers of curvature

Thus any curved path, such as the racetrack of **Figure 2.48**, has any number of different centers of curvature. So, how can we describe the behavior of a particle moving along such a path if things are always changing? We do this by "freezing" the motion long enough to figure out what we need to know regarding acceleration components. For the center of curvature C_1 of **Figure 2.50*a***, for instance, β is the angle the radius of curvature vector $\boldsymbol{r}_C$ makes with the y axis, and $\dot{\beta}$ is the rate at which a particle P is rotating about C_1.

This seems very much like a polar representation, and we can see that P's speed is related to r_C (the length of $\boldsymbol{r}_C$) and $\dot{\beta}$ by

$$v = r_C\dot{\beta} \tag{2.57}$$

If the path is well approximated by a circular arc, there are no equivalent $\dot{r}$ and $\ddot{r}$ terms in a path analysis because in a circle the radius r doesn't change over time. Thus all we see for our acceleration are the $r\ddot{\theta}$ and $r\dot{\theta}^2$ terms in (2.50). In terms of our path-coordinate parameters, $r\ddot{\theta}$ corresponds to the tangential acceleration $\dot{v}$ (acceleration along the path) and $r\dot{\theta}^2$ corresponds to $r_C\dot{\beta}^2$, the normal acceleration. Alternatively, and more commonly, the normal acceleration $r_C\dot{\beta}^2$ can be written as v^2/r_C, using (2.57). Both the tangential and normal components of acceleration are illustrated in **Figure 2.50*b***, along with the unit vectors $\boldsymbol{e}_n$ and $\boldsymbol{e}_t$. Our complete velocity and acceleration expressions in path coordinates are therefore

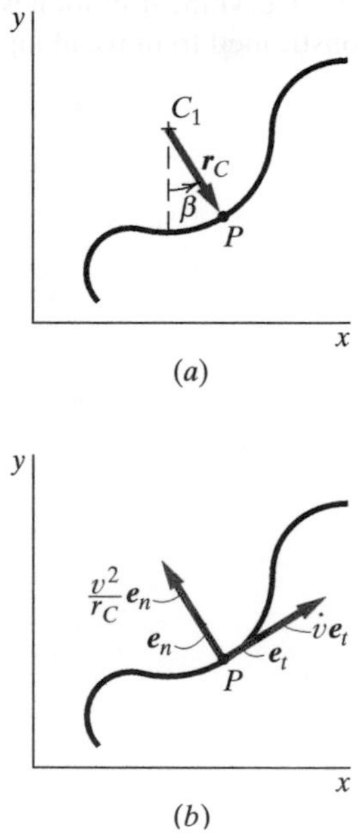

Figure 2.50 Path coordinates

$$\boldsymbol{v}_P = v\boldsymbol{e}_t \tag{2.58}$$

$$\boldsymbol{a}_P = \dot{v}\boldsymbol{e}_t + \frac{v^2}{r_C}\boldsymbol{e}_n \tag{2.59}$$

A good question is how exactly do you figure out the radius of curvature for some curve? This question makes an interesting exercise (and, in fact, is in the Exercises section), but because it's so useful, I thought I'd let you know the result ahead of time. If your curve is $y(x)$, then the radius of curvature is given by

$$r_C = \frac{\left[1 + \left(\frac{dy}{dx}\right)^2\right]^{\frac{3}{2}}}{\left|\frac{d^2y}{dx^2}\right|} \tag{2.60}$$

>>> **Check out Example 2.15 (page 67) and Example 2.16 (page 68) for applications of this material.**

One-Dimensional Motion Along a Curve

In Section 2.1 we derived (2.8), $a\,dx = v\,dv$, and integrated it for some special cases (a is constant, a depends on x, and a depends on v). This formula works just as well for motion along a curve as for motion along a straight line. Consider the point P, constrained to move along the curved path shown in **Figure 2.51**. Let $s(t)$ indicate the position along the curve. For this system the applicable relationship among displacement, speed, and acceleration is $\ddot{s}\,ds = \dot{s}\,d\dot{s}$.

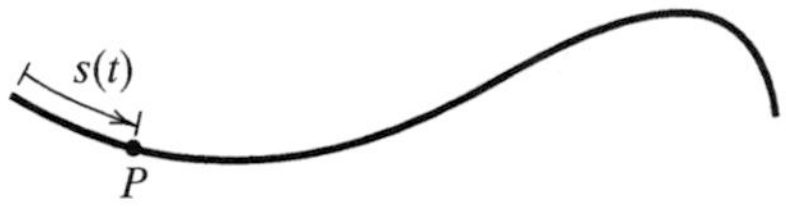

Figure 2.51 Curvilinear motion with a body P constrained to move along the curve

The formulas (2.9), (2.10), and (2.11) for motion along a straight line have exact analogues for motion along a curved line:

$$\dot{s}_2^2 = \dot{s}_1^2 + 2\ddot{s}(s_2 - s_1) \tag{2.61}$$

$$\dot{s}_2^2 = \dot{s}_1^2 + 2\int_{s_1}^{s_2} \ddot{s}(s)\,ds \tag{2.62}$$

$$s_2 = s_1 + \int_{\dot{s}_1}^{\dot{s}_2} \frac{\dot{s}}{\ddot{s}(\dot{s})}\,d\dot{s} \tag{2.63}$$

which apply to the constant $\ddot{s}$ case, the $\ddot{s}(s)$ case, and the $\ddot{s}(\dot{s})$ case, respectively.

>>> **Check out Example 2.17 (page 70) for an application of this material.**

EXAMPLE 2.15 ANALYTICAL DETERMINATION OF RADIUS OF CURVATURE (Theory on page 66)

A radio-controlled model car is driving from left to right along a hilly road. A cross section through the road surface is in the shape of a sine wave, with an amplitude of 6 cm and a wavelength of 8 m (**Figure 2.52**). At point A, the car's speed is 10 m/s. An on-board accelerometer shows that the net acceleration at A is 5 m/s^2. How much of this is due to acceleration along the path? Can you tell if this acceleration is positive or negative?

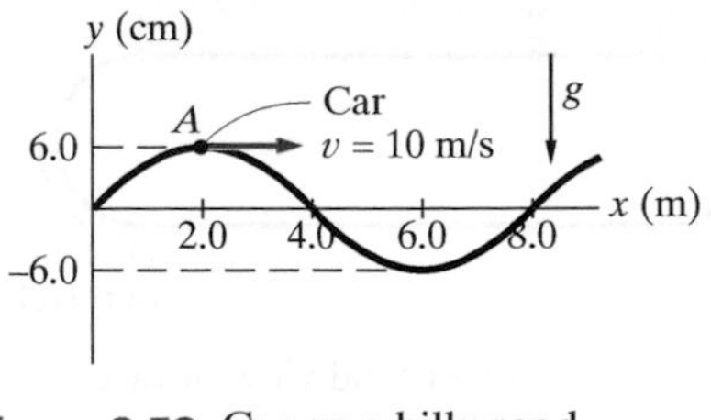

Figure 2.52 Car on a hilly road

Goal Find the tangential component of acceleration and determine the direction of the net acceleration, if possible.

Given Shape of car's path, car's speed, and car's acceleration magnitude.

Assume As you know from watching the movies, a sufficiently high speed will allow a car to become airborne. That's fun, but for this example we'll assume that the car remains in contact with the road.

Draw **Figure 2.53** shows our system.

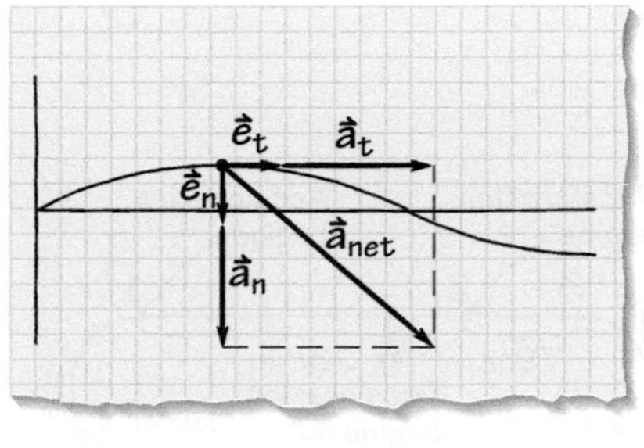

Figure 2.53 Acceleration components

Formulate Equations The normal component of the acceleration at A can be found by using v^2/r_C. Because we know the speed but not the radius of curvature, this is the sort of place where (2.60) comes in handy. The road surface's profile is sinusoidal, and so we can express it as

$$y(x) = (0.06\ \text{m}) \sin\left(\frac{2\pi x}{8\ \text{m}}\right)$$

Solve We can see from **Figure 2.52** that at A the slope $\frac{dy}{dx}$ is zero. The second derivative $\frac{d^2y}{dx^2}$ isn't zero, however, and has to be calculated:

$$\frac{d^2y}{dx^2} = -(6 \times 10^{-2}\ \text{m})\left(\frac{2\pi}{8\ \text{m}}\right)^2 \sin\left(\frac{2\pi x}{8\ \text{m}}\right)$$

which, when evaluated at $x = 2$ m, yields

$$\frac{d^2y}{dx^2} = -0.037\ \text{m}^{-1}$$

Using these values in (2.60) yields $r_C = 27$ m.

This information tells us that $a_n = \frac{v^2}{r_C} = \frac{(10\ \text{m/s})^2}{27\ \text{m}} = 3.7\ \text{m/s}^2$. We know that $|a_{\text{net}}| = \sqrt{a_t^2 + a_n^2}$. Because we know $a_{\text{net}} = 5\ \text{m/s}^2$ and have just found a_n, we can solve this expression for the tangential acceleration:

$$(5\ \text{m/s}^2)^2 = a_t^2 + (3.70\ \text{m/s}^2)^2 \quad \Rightarrow \quad a_t = 3.4\ \text{m/s}^2$$

Based just on this result, there's no way to determine whether this acceleration is positive or negative because the formula we used to determine a_t squares all the component values.

Check Not much to check here beyond unit consistency.

EXAMPLE 2.16 ACCELERATION—PATH COORDINATES (Theory on page 65)

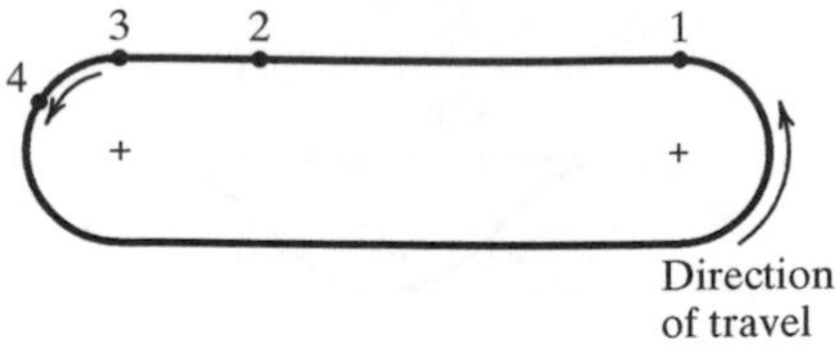

Figure 2.54 Overhead view of race track

You're driving a race car on a test track made up of two straight sections and two circular sections, as shown in **Figure 2.54**, with each curved section having a radius of 53 m. Your speed is 80 km/h while you are in the circular section to the right of point 1. Your speed increases in a linear fashion to 160 km/h on exiting this region, reaching 160 km/h at point 2, when you're 238 m beyond the semicircular portion. You then brake with a linear negative acceleration back down to 80 km/h in another 79 m (point 3). Your speed is 80 km/h as you enter the circular section to the left of point 3 and remains 80 km/h through the curve. Determine your acceleration (a) at point 4, (b) between points 1 and 2, and (c) between points 2 and 3. Draw e_t, e_n unit vectors for each region and determine where the magnitude of acceleration is least and where it is greatest.

Goal Determine the acceleration at three points along the track.

Given Speeds at particular points along the track and track shape.

Assume Acceleration positive from 1 to 2 and negative from 2 to 3.

Draw **Figure 2.55** shows our system.

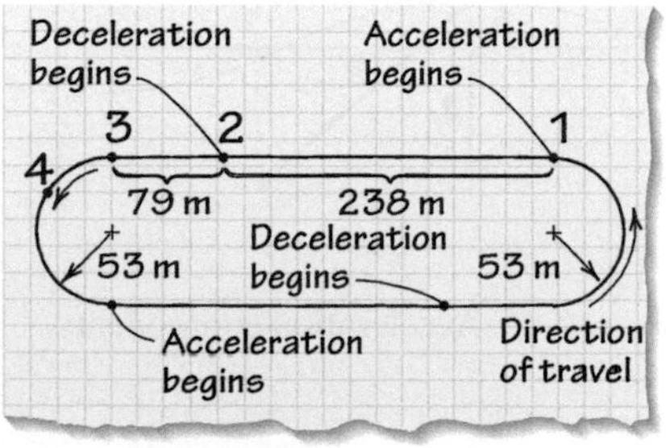

Figure 2.55 Track details

Formulate Equations Because the speed changes are linear, we know that the acceleration is constant. We don't know the time necessary to reach first 160 km/h and then 80 km/h, but we do know the distances needed, telling us that (2.9) is in order. First we will re-express our speeds in m/s: 80 km/h is 22.2 m/s and 160 km/h is 44.4 m/s.

Solve For the positive acceleration phase we have

$$\tfrac{1}{2}\left(v_2^2 - v_1^2\right) = (x_2 - x_1)a_{(1-2)}$$
$$(44.4\ \text{m/s})^2 - (22.2\ \text{m/s})^2 = 2(238\ \text{m})a_{(1-2)}$$

which gives us $a_{(1-2)} = 3.1\ \text{m/s}^2$. For the negative acceleration phase we have

$$\tfrac{1}{2}\left(v_3^2 - v_2^2\right) = (x_3 - x_2)a_{(2-3)}$$
$$(22.2\ \text{m/s})^2 - (44.4\ \text{m/s})^2 = 2(79\ \text{m})a_{(2-3)}$$

or $a_{(2-3)} = -9.36\ \text{m/s}^2$.

Because the straight section is, well, straight, this is the only component of acceleration we will have. Thus we've obtained

$$\boldsymbol{a}_{(1-2)} = 3.1\boldsymbol{e}_t\ \text{m/s}^2 \quad \text{or} \quad a_{(1-2)_t} = 3.1\ \text{m/s}^2, \quad a_{(1-2)_n} = 0\ \text{m/s}^2$$

and

$$\boldsymbol{a}_{(2-3)} = -9.36\boldsymbol{e}_t\ \text{m/s}^2 \quad \text{or} \quad a_{(2-3)_t} = -9.36\ \text{m/s}^2, \quad a_{(2-3)_n} = 0\ \text{m/s}^2$$

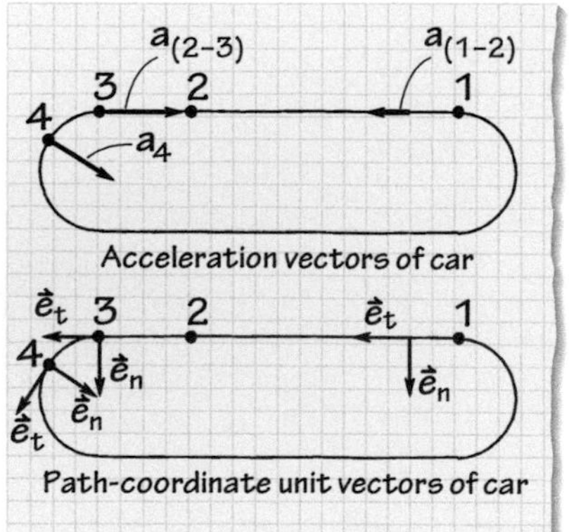

Figure 2.56 Acceleration vectors and unit vectors during car's motion

These accelerations and associated path-coordinate unit vectors are shown in **Figure 2.56**. Although not generally true, in this case the path-coordinate unit vectors are the same for both straight-line accelerations, a consequence of the motion being in a straight line and the velocity being oriented to the left in both cases. (Because the two sections of track we are interested in are straight, and therefore the normal direction is not defined, I have chosen to orient $\boldsymbol{e}_n$ toward the interior of the track to be consistent with the semicircular regions.)

Now on to the circular region containing point 4. The speed in this region is constant at 80 km/h, which means that $\dot{v}$, the tangential acceleration component, is zero. Because the path follows a circular arc, the radius of curvature is the same for all points along the curve and is equal to 53 m. The normal acceleration component, v^2/r_C, is therefore $(22.2\,\mathrm{m/s})^2/53\,\mathrm{m} = 9.3\,\mathrm{m/s^2}$. Thus

$$\boldsymbol{a}_4 = (9.3\boldsymbol{e}_n)\ \mathrm{m/s^2} \quad \text{or} \quad a_{4_t} = 0.0\ \mathrm{m/s^2}, \quad a_{4_n} = 9.3\ \mathrm{m/s^2}$$

Because there's only one component of acceleration for each of these cases, we don't have to take the square root of the sum of the squares to get the acceleration magnitudes. The smallest acceleration occurs during the $1 \rightarrow 2$ phase of the motion, and the (just barely) largest occurs during the straight-line braking from $2 \rightarrow 3$.

Check This all matches physical reality. Unless they're defective, cars always brake much more quickly than they can accelerate. The normal acceleration they can support on a circle is also much higher than their forward acceleration ability. Why do you think this is so?

Waiting ...

The reason is that a car's forward acceleration capability is determined by the engine's torque production. The tires aren't being stressed to their limit, and if a much more powerful engine were used, the tires would be able to transmit the increased load to the ground and therefore give the car greater acceleration. (This assumes a road surface with sufficient grip.) On a circular track (known as a skid pad), there's no tangential acceleration. The car has simply gotten up to speed, and then the acceleration comes about because the car is continuously turning. In this case, the limiting factor is the tires—the car can go around as long as the force necessary to keep it on the track is within the limits the tires can provide. Forward acceleration is generally limited by the size of the car's engine, whereas braking and normal acceleration are limited by the tires' coefficient of friction.

EXAMPLE 2.17 SPEED ALONG A CURVE (Theory on page 66)

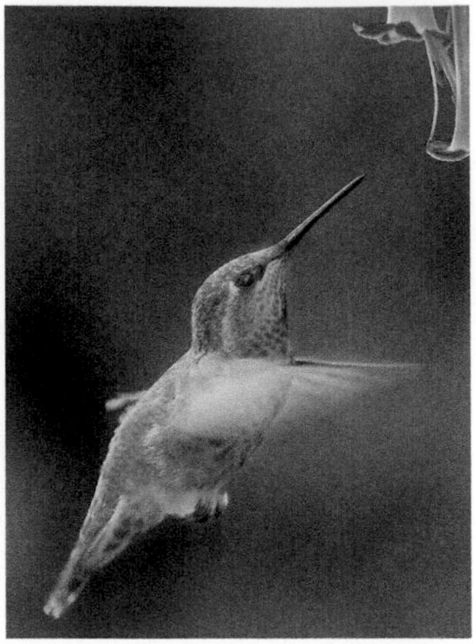

Figure 2.57 Hovering hummingbird

An Anna's hummingbird (**Figure 2.57**) can flap its wings (length 6.35 cm) at the amazing rate of 60 beats per second. We will approximate the wingtip's path as a circular arc and say that s measures distance along this path. At any instant, the tip's position $s(t)$ along the path is given by

$$s(t) = (5 \text{ cm}) \left[1 - \cos\left(\frac{2\pi t}{T}\right)\right]$$

where t is in seconds and T is the time needed for a single wing beat.

(a) What is the wingtip's maximum acceleration along its path? Express your result in cm/s^2 as well as in g's.

(b) Assume that the wing starts from $s(0) = 0$ cm, $\dot{s}(0) = 0$ cm/s, and accelerates at the maximum acceleration found in part (a). How fast will it be moving in kilometers per hour after traveling 1.3 cm?

Goal Determine the maximum tangential acceleration at the tip of a hummingbird's wing as well as its speed after it has covered a given distance (at maximum acceleration).

Given Dimensions of wing, beat rate, and position of wingtip versus time.

Assume Assume the wing is stationary where it attaches to the body.

Draw **Figure 2.58** shows our system.

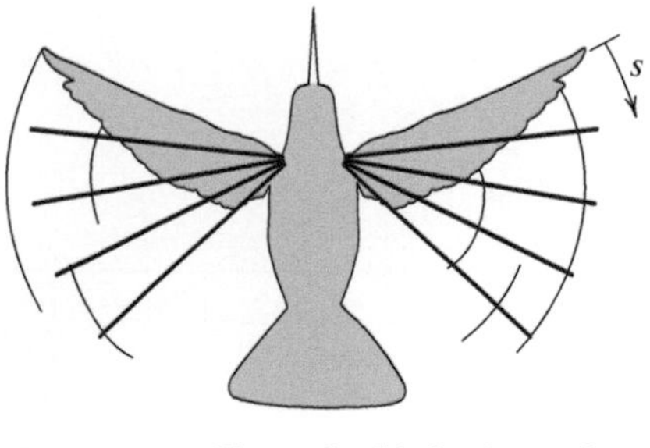

Figure 2.58 Hummingbird schematic

Formulate Equations We're given an expression for $s(t)$ and told that the wingtip moves along a circular arc. To answer (a), we need to find T and then differentiate $s(t)$ twice. Formula (2.61) can then be used to solve (b).

Solve **(a)** The expression for $s(t)$ is periodic, with period T. If we know the period of the motion, then we know T. Luckily for us, that's given in the problem statement. We're told that the wings complete 60 full flaps (down and then back up) in 1 s. Therefore a single flap takes $\frac{1}{60}$ s and so $T = \frac{1}{60}$ s. Our expression for $s(t)$ is thus

$$s(t) = (5 \text{ cm}) [1 - \cos(120\pi t)]$$

Differentiating twice yields

$$\ddot{s}(t) = 5(120\pi)^2 [\cos(120\pi t)] \text{ cm/s}^2$$

which is a maximum when $\cos(120\pi t)$ equals one. Thus

$$\ddot{s}_{max} = 5(120\pi)^2 \text{ cm/s}^2 = 7.11 \times 10^5 \text{ cm/s}^2 = 725g$$

Take a second to think about this result because the numbers used here are accurate. The wingtip feels a tangential acceleration that's 725 times higher than the standard acceleration due to gravity. That's a lot.

(b) Using $\ddot{s} = 7.11 \times 10^5 \text{ cm/s}^2$ in (2.61) gives us

$$\dot{s}_2^2 = 0 + 2(7.11 \times 10^5 \text{ cm/s}^2)(1.3 \text{ cm})$$

or

$$\dot{s}_2 = 1360 \text{ cm/s} = 49 \text{ km/h}$$

From zero to 49 km/h in 1.3 cm—not bad at all.

Check A nice check on the believability of the answers is to do an average calculation. The wings go from $s = 0$ to $s = 5$ cm in $T/4 = 4.17 \times 10^{-3}$ s. This implies an average speed of $5 \text{ cm}/4.17 \times 10^{-3} = 1199 \text{ cm/s} = 43$ km/h. This is about the same as the speed we decided the wingtips would reach after traveling 1.3 cm under maximum acceleration. Thus our prior result seems reasonable. Furthermore, if our average speed is 1199 cm/s, then we can approximate the maximum speed as double this (assuming the average is taken between zero and maximum) and find an associated tangential acceleration of

$$2398 \text{ cm/s}/4.17 \times 10^{-3} \text{ s} = 5.75 \times 10^5 \text{ cm/s}^2 = 586g$$

Our quick estimate of $586g$ is right around our actual answer of $725g$, giving us some confidence in the answer for (a) as well.

EXERCISES 2.4

2.4.1. [Level 1] The acrobatic airplane (P) is currently flying level at a speed of 90 ms/s. What is the normal acceleration felt by the airplane if the pilot pulls into an upward loop with a radius of 396 m?

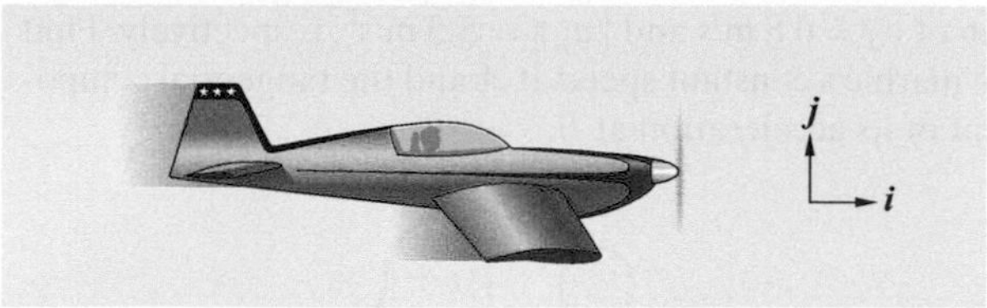

E2.4.1

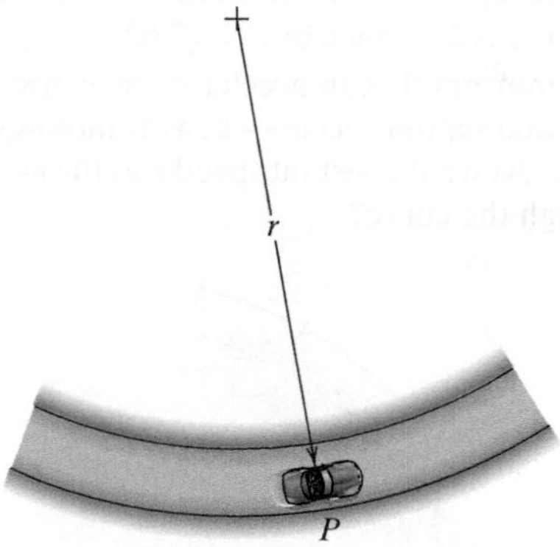

E2.4.2

2.4.2. [Level 1] A car P is moving along the track and the on-board accelerometer indicates an acceleration of $\boldsymbol{a}_P = (-7.0\boldsymbol{e}_t + 5.0\boldsymbol{e}_n)$ m/s^2. At this instant it's midway through a curve having a radius of 100 m. What is the car's speed?

2.4.3. [Level 1] A marble M is constrained to move in the horizontal plane along a wire whose shape is as illustrated. At point A, the marble is traveling at a constant speed $v_A = 5$ m/s, and the radius of curvature at that location is $r_A = 2$ m. When the marble reaches point B, it has a speed of $v_B = 3$ m/s and a total acceleration of $\|\boldsymbol{a}_B\| = 10$ m/s^2. The tangential component of the marble's acceleration at B is $a_{B,t} = 2$ m/s^2. What are the marble's total acceleration at A and its radius of curvature at B?

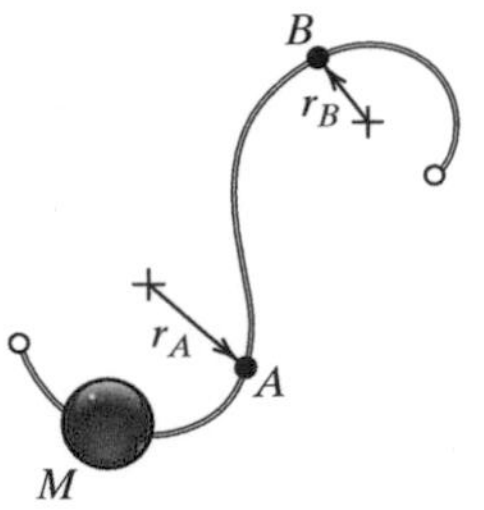

E2.4.3

2.4.4. **[Level 1]** A GPS tracking system determines that the race car P in a curve of radius $r = 90$ m has an acceleration of $\boldsymbol{a}_P = (0\boldsymbol{e}_t + 0\boldsymbol{e}_n)$ m/s^2. What is its speed?

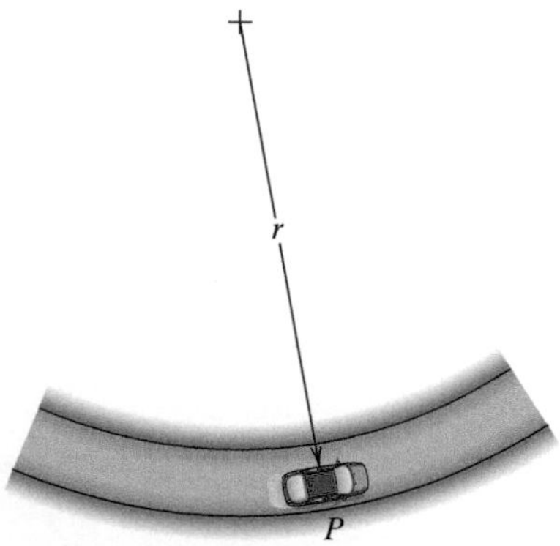

E2.4.4

2.4.5. **[Level 1]** A motorcyclist enters a curve whose radius of curvature is described by $r_c = (7.62 - 6 \times 10^{-4}s^3)$ m. Suppose the motorcycle can accelerate at 9 m/s^2 before it begins to slip, and slipping occurs when the motorcycle is 6 m into the curve. At what constant speed was the motorcyclist driving through the curve?

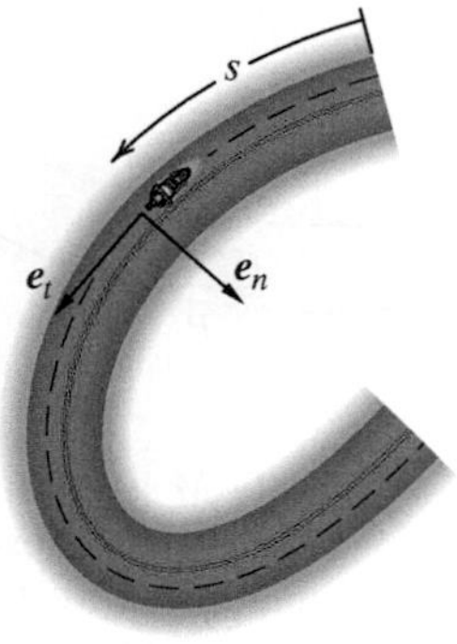

E2.4.5

2.4.6. **[Level 1]** A person P on a Ferris wheel travels in a circle with a 9 m radius and experiences an acceleration of magnitude 0.1 m/s^2. The person's speed is constant during the ride. What is the Ferris wheel's rotational speed?

E2.4.6

2.4.7. **[Level 1]** A car enters a circular offramp (radius of 180 m) at 30 m/s. The car's on-board accelerometers sense a total acceleration magnitude of 7.07 m/s^2. What is the car's acceleration magnitude a_t (the component tangent to its path)?

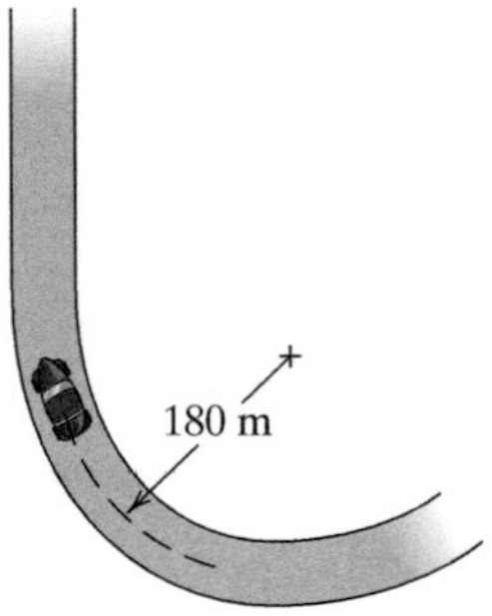

E2.4.7

2.4.8. **[Level 1]** Consider a marble M constrained to move in the horizontal plane along the illustrated wire. When the marble arrives at point A, its total acceleration is given by $\|\boldsymbol{a}_A\| = 4$ m/s^2. The radius of curvature at A is $r_A = 0.2$ m. At point B, the radius of curvature is $r_B = 0.12$ m, and the marble is traveling with a speed and total acceleration of $v_B = 0.8$ m/s and $\|\boldsymbol{a}_B\| = 5.5$ m/s^2, respectively. Find the marble's constant speed at A and the tangential component of its acceleration at B.

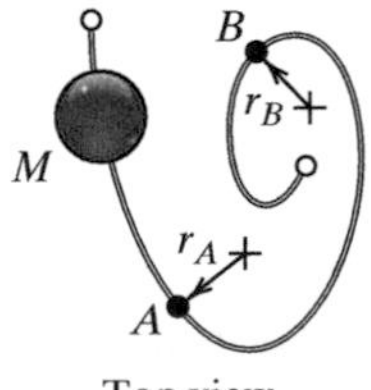

E2.4.8

2.4.9. **[Level 1]** When designing a rollercoaster, the designer wants to ensure that the passengers survive the ride. At the bottom of a circular loop, the designer wants the cars to move at 96.5 km/h, and safety considerations demand that the normal acceleration not exceed $3.5g$. What is the minimum allowable radius r?

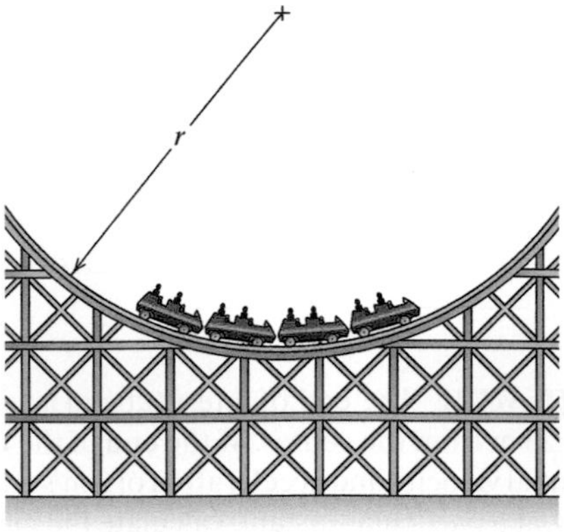

E2.4.9

2.4.10. **[Level 1]** A Ferris wheel with radius 9 m is decelerating such that at $t = 0$ s the tangential speed of a point P on the rim is $v = 3$ m/s and $\frac{dv}{dt} = ct$. $c = -0.12$ m/s^3. What is the acceleration of P at $t = 3$ s?

E2.4.10

2.4.11. **[Level 1]** Suppose two cars are traveling on a wide road at constant speeds when they encounter a U-turn. Car A is in the inside lane, moving at $v_A = 24$ m/s in a semicircular path of radius $r_A = 90$ m. Car B has a speed of $v_B = 28$ m/s in the outside lane, and the radius of its path is $r_B = 95$ m. Assume that the cars can handle a maximum normal acceleration of $a_{n,\max} = 0.75g$ before their tires lose traction and they slide off the road. Will either vehicle make it around the turn without sliding off the road, and if so, how long does it take?

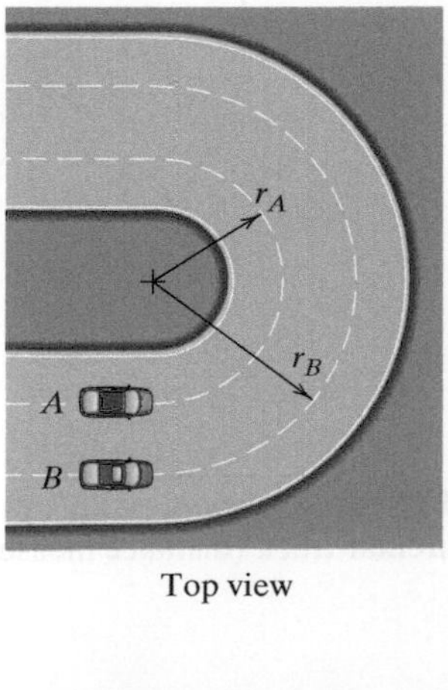

Top view

E2.4.11

2.4.12. **[Level 1]** A skier using the illustrated ski jump starts from rest at A and accelerates at a constant 8.5 m/s^2 to B. Just after B the straight slope changes to a curve having a radius of curvature equal to 67 m. What is the magnitude of the total acceleration felt by the skier at this point? Assume that the tangential acceleration doesn't immediately change. Comment on the magnitude.

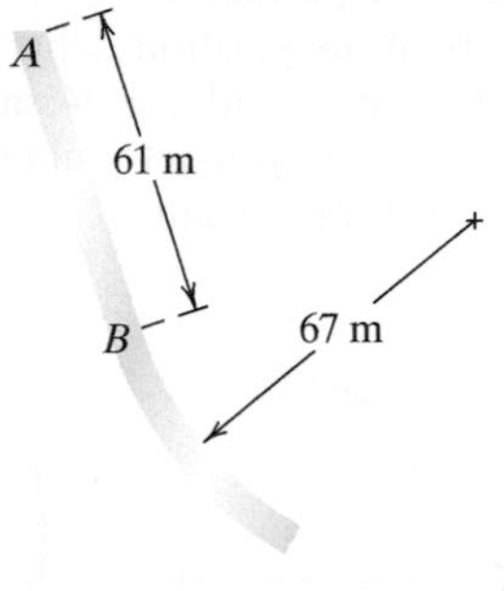

E2.4.12

2.4.13. **[Level 1]** The small particle b slides to the right at a constant speed v_b toward A, the end of the table. When it passes A, it begins to accelerate downward at 9.81 m/s^2. Thus there is a very marked difference in its acceleration just before and just after reaching A. Is the same true for the velocity? What are the direction and magnitude of the velocity vector the instant before the particle reaches A as well as the instant after?

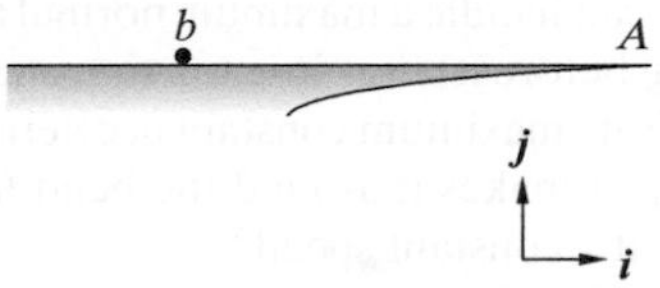

E2.4.13

2.4.14. **[Level 2]** The sports car A starts from rest on a circular track (radius 100 m), and its speed is given by $v = v_0\left(1 - e^{-at}\right)$ where $v_0 = 60$ m/s and $a = 0.3\,\text{s}^{-1}$. How much time must elapse before the sports car's tangential acceleration is 10% of its normal acceleration?

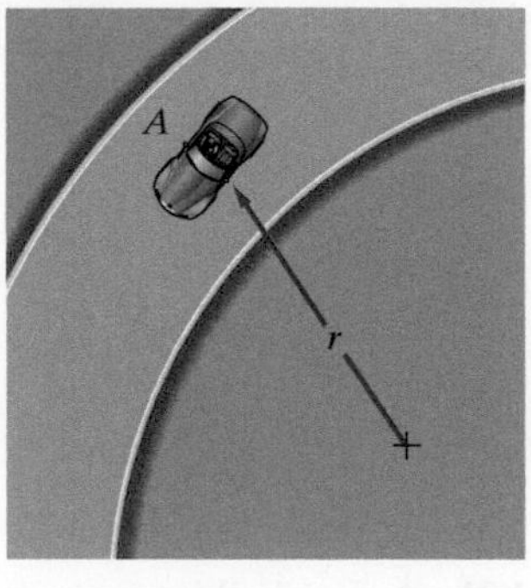

E2.4.14

2.4.15. **[Level 2]** A particle P is moving along a curve defined by $y = ax^2$, where $a = 20\ \text{m}^{-1}$. $\dot{y} = bx$, $b = 12\,\text{s}^{-1}$. Determine where along the curve P's speeds in the $\boldsymbol{i}$ and $\boldsymbol{j}$ directions are equal in magnitude, and calculate the normal and tangential acceleration at that point.

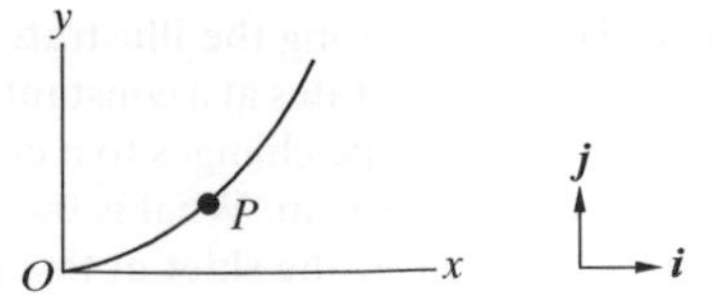

E2.4.15

2.4.16. **[Level 2]** A particle P is moving along a sinusoidally varying path, its position defined by $y = a[1 - \cos(\lambda x)]$ with $\lambda = 0.01\ \text{m}^{-1}$ and $a = 24$ m. $a_P = (2.4\boldsymbol{i} + 0.6\boldsymbol{j})\ \text{m/s}^2$ when $x = \frac{\pi}{2\lambda}$. Can you determine what the particle's speed is equal to at this instant?

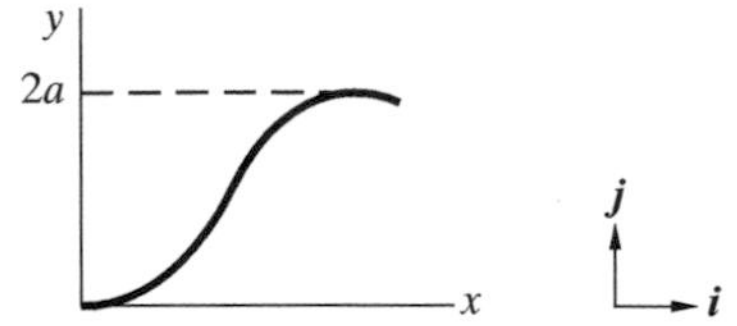

E2.4.16

2.4.17. **[Level 2]** A car is traveling at a constant speed $v_A = 40$ km/h when it encounters a 90° bend to the right at point A, where the path is given by an arc of radius $r_1 = 21$ m. The car exits the bend at point B and then travels a straight-line distance of $d = 12$ m before encountering another 90° bend (this time to the left) at point C, for which the path's radius is $r_2 = 24$ m.

a. If the car can handle a maximum normal acceleration of $a_{n,\text{max}} = 0.75g$ before its tires lose traction and it slides off the road, what is its maximum constant acceleration from B to C so that it just makes it around the bend from C to D while traveling at a constant speed?

b. How long does it take for the car to go from A to D under these conditions?

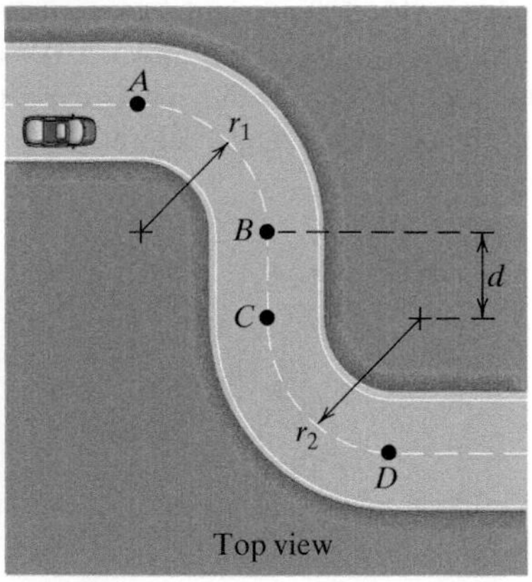

E2.4.17

2.4.18. **[Level 2]** A snowball is careening down a hill whose surface can be approximated by $y = 0.25x^2$ m. At $x = 4$ m, the snowball has a tangential velocity and acceleration of 10 m/s and 2 m/s², respectively. Find the magnitude of the snowball's total acceleration at this position.

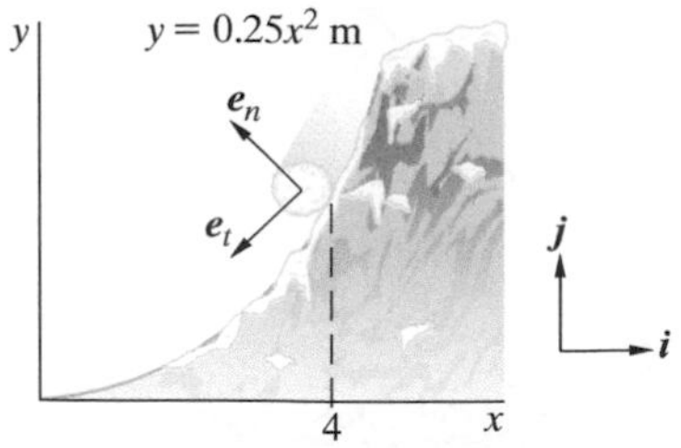

E2.4.18

2.4.19. **[Level 2]** The end of a test tube in a centrifuge is located 0.23 m from the center. If the end of the test tube has a tangential velocity and acceleration of 3 m/s and 3 m/s², respectively, at a particular instant, calculate the magnitude of the total acceleration it experiences in g's. Also, find the angular velocity and acceleration at this instant.

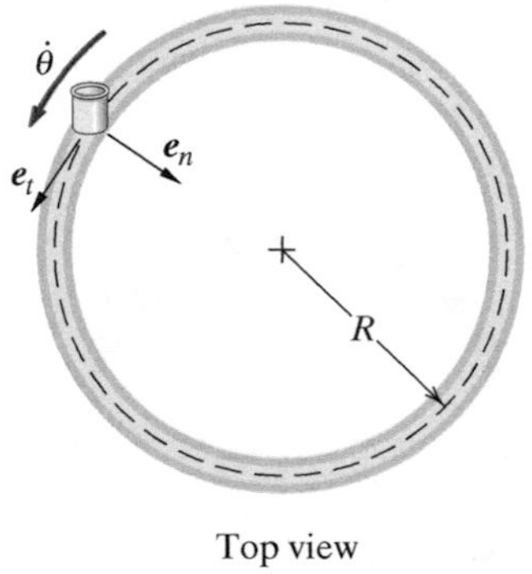

E2.4.19

2.4.20. **[Level 2]** A snowboarder is on the right edge of a halfpipe, just about to propel himself into the air. Immediately before leaving the halfpipe's edge, the snowboarder experiences a tangential acceleration of 3 m/s² and a total acceleration of 6 m/s². What are his tangential speed and normal acceleration at this instant? The halfpipe has a radius of 6 m. How high does the snowboarder get in the air, and how long is he airborne?

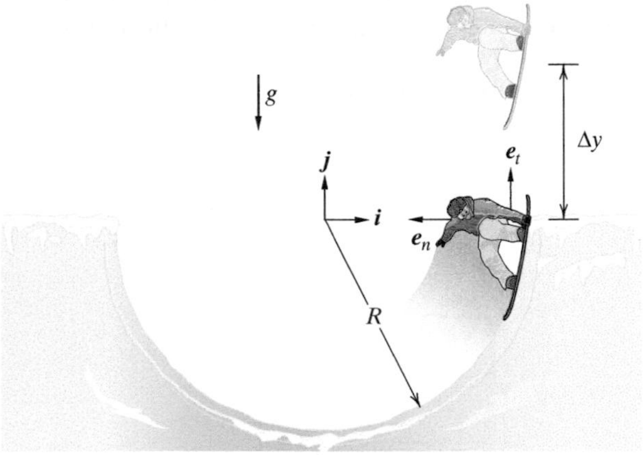

E2.4.20

2.4.21. **[Level 2]** A car (at point A) is 100 m away from point B along a circular track (distance measured along the track's perimeter) and begins to accelerate from rest. What constant speed increase is needed for the total acceleration magnitude to be 8 m/s^2 when the car reaches B? The radius of the track is 150 m.

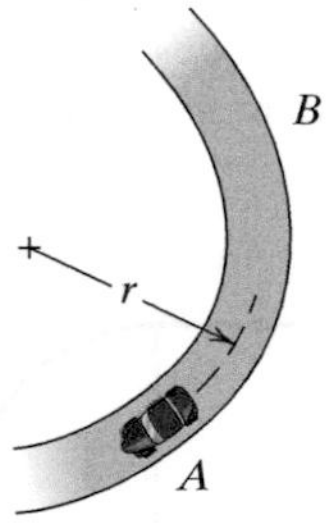

E2.4.21

2.4.22. [Level 2] A bicyclist comes around a decreasing-radius turn and maintains a constant speed of 32 km/h. The radius of curvature varies as $r_C = a + bs^2$, where s indicates motion along the curve expressed in meters (a = 15 m and $b = -8.2 \times 10^{-3}\ \text{m}^{-1}$). How long will it be until the bicycle slips if the maximum acceleration it can sustain without slip is 8.5 m/s^2?

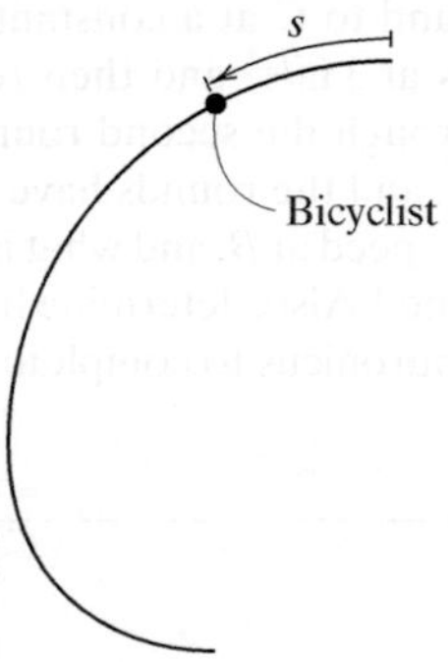

E2.4.22

2.4.23. [Level 2] A Ferris wheel is started up from rest such that its rotational speed is given by $\dot{\theta} = ct$, where θ measures the rotational position of the wheel (c = 0.05 rad/s^2). What are the magnitudes of the velocity and acceleration of a point P on the wheel's rim at t = 5 s? The distance from the center of rotation to the rim is 11 m.

E2.4.23

2.4.24. [Level 2] A car is traveling along a horizontal race track. From A to B the track is straight, and from B to D it is in the form of a circular arc. The car is accelerating from A to C with a constant tangential acceleration of 1.5 m/s^2. Its velocity at A is 24 m/s. The distance from A to B is 90 m and from B to C the distance is 90 m as well. The magnitude of the car's total acceleration (accounting for both normal and tangential components) is 1.5 m/s^2 at A and 4.45 m/s^2 at C. What is the track's radius of curvature at C?

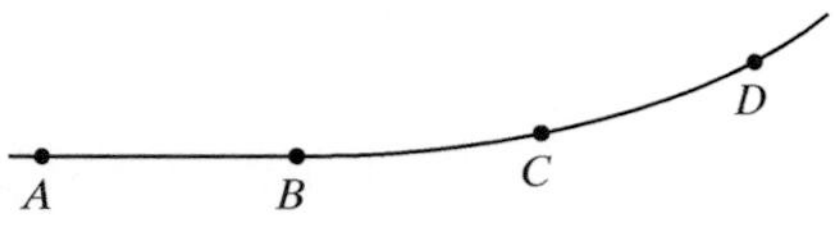

E2.4.24

2.4.25. [Level 2] When a particular meteoroid enters our atmosphere (thus becoming a meteor and soon to be a meteorite), it encounters aerodynamic drag. At the instant it is first noticed by a tracking station, it's experiencing a gravitational acceleration of 9.5 m/s^2 and an aerodynamic deceleration of 5 m/s^2. Its speed is 25,000 kph, and its angular heading is 3 degrees down from horizontal.

What is the radius of curvature of the trajectory at this instant? Note that the deceleration due to aerodynamic drag points in the opposite direction to the meteor's velocity vector.

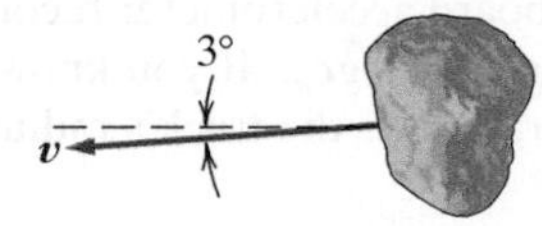

E2.4.25

2.4.26. [Level 2] At the instant illustrated, a rocket is experiencing a thrust from its engine that is producing a forward acceleration and is also being accelerated downward due to gravity. The overall acceleration of the rocket is

$$\boldsymbol{a} = (5.657\boldsymbol{i} - 3.843\boldsymbol{j})\ \text{m/s}^2$$

The acceleration due to gravity at the rocket's position has a magnitude of 9.5 m/s^2, and the rocket's velocity vector is

$$\boldsymbol{v} = (5000\boldsymbol{i} + 2000\boldsymbol{j})\ \text{m/s}$$

a. Determine the acceleration due to the rocket's engine.

b. Determine a_t and a_n.

c. Determine the radius of curvature of the rocket's path at this instant.

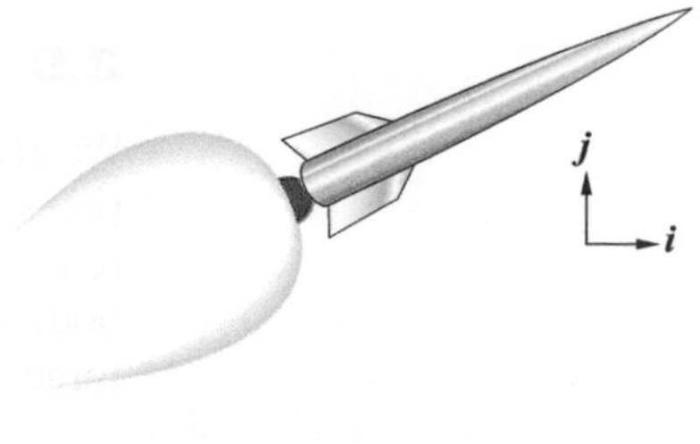

E2.4.26

2.4.27. [Level 2] A particle moves along the path defined by $xy = 36\,\text{m}^2$. At time t_0 its speed is constant at 10 m/s, and its velocity has a positive $\boldsymbol{i}$ component. At the illustrated instant it's at $x = 9$ m. What are the acceleration components of the particle in the $\boldsymbol{i}$ and $\boldsymbol{j}$ directions?

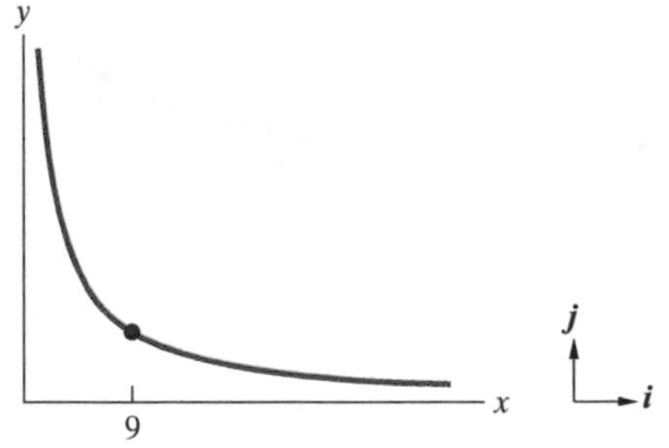

E2.4.27

2.4.28. [Level 2] A car starts from rest on a circular track with a radius of 300 m. It accelerates with a constant tangential acceleration of $a_t = 0.75\,\text{m/s}^2$. Determine the distance traveled and the time elapsed when the magnitude of the car's overall acceleration is equal to $0.9\,\text{m/s}^2$.

2.4.29. [Level 2] A race car C is being tracked via telemetry. The on-board accelerometer records an acceleration level of $-0.7g\boldsymbol{e}_t + 0.5g\boldsymbol{e}_n$. If you know that v at this instant is 200 km/hr, what is the track's radius of curvature at this instant?

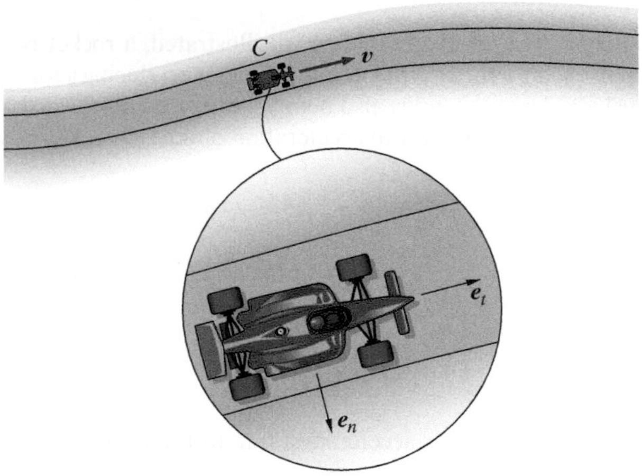

E2.4.29

2.4.30. [Level 2] A sports car C is being driven around a decreasing-radius turn. On entry, the car's speed is 161 km/h, and the driver keeps the speed constant throughout the curve. The radius of curvature at entry is $r_{C_1} = 305$ m, and the radius of curvature at exit is $r_{C_2} = 152.5$ m. Will the car complete the turn without losing traction and sliding? Assume that the tires can maintain rolling contact up to an acceleration of 11 m/s.

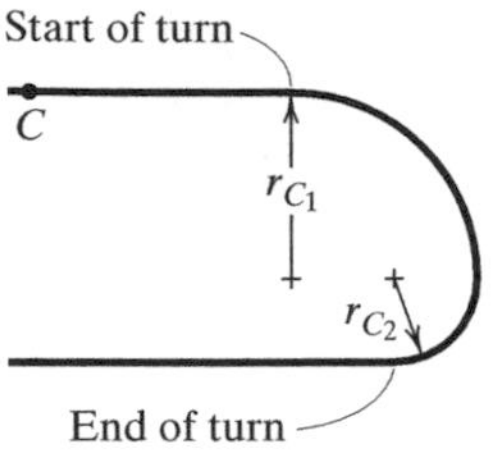

E2.4.30

2.4.31. [Level 3] Roman chariot racer Andronicus is competing in the big race at the Circus Maximus. Starting at A, Andronicus accelerates to B at $2\,\text{m/s}^2$ and then travels through the first round to C at a constant speed. From C to D, he accelerates at $1\,\text{m/s}^2$ and then returns to A with a constant speed through the second round. The straightaways are 50 m long, and the rounds have a radius of 15 m. What is Andronicus' speed at B, and what is his acceleration through the first round? Also, determine his speed at D and how long it takes Andronicus to complete one circuit.

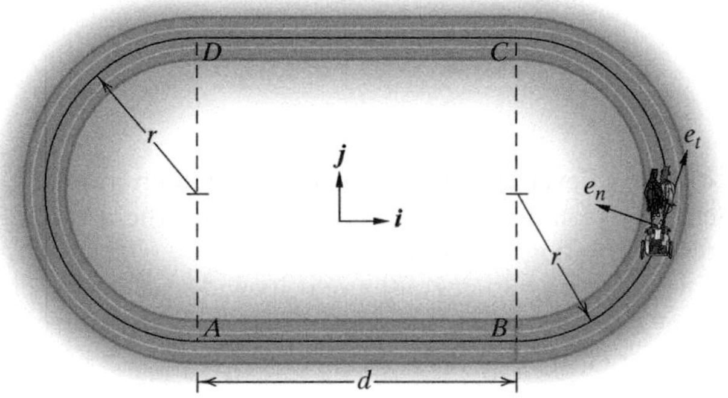

E2.4.31

2.4.32. [Level 3] Derive equation (2.60).

2.5 RELATIVE MOTION AND CONSTRAINTS

We are not always going to restrict our attention to a single body. Sometimes we will be concerned with how two or more bodies are moving relative to one another, or we will want to know what's happening to a body from the viewpoint of something that is itself moving. That's the topic of this section: how **relative motions** are handled.

Figure 2.59 shows two bodies A and B lying in a plane. We want to characterize the motion of each, as well as the relative motion of one with respect to the other. Three position vectors are drawn on the figure: $\boldsymbol{r}_{A/O}$, $\boldsymbol{r}_{B/O}$, and $\boldsymbol{r}_{A/B}$. The vectors $\boldsymbol{r}_{A/O}$ and $\boldsymbol{r}_{B/O}$ represent the positions of A and B, respectively, with respect to the fixed origin O, whereas $\boldsymbol{r}_{A/B}$ indicates the position of A with respect to B. We can relate the vectors to one another by the expression

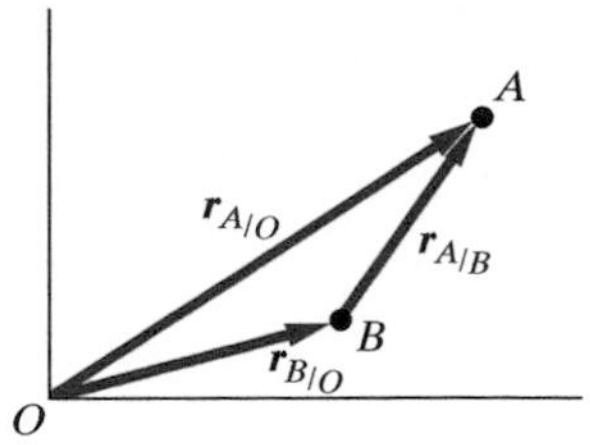

Figure 2.59 Two particles in a plane

$$\boldsymbol{r}_{A/O} = \boldsymbol{r}_{B/O} + \boldsymbol{r}_{A/B} \tag{2.64}$$

This vector relationship says that to get to A, you can go straight there from $O(\boldsymbol{r}_{A/O})$ or, alternatively, you can go first to $B(\boldsymbol{r}_{B/O})$ and then to $A(\boldsymbol{r}_{A/B})$.

This isn't too thrilling—in fact, it probably seems pretty obvious, but what we're doing is slowly coming to the notion of multiple reference frames. To see this more clearly, look at **Figure 2.60**. Here O, A, and B are still present, but another set of axes has been drawn, and both sets are labeled: X–Y for the original set and x–y for the new set. What we want to do now is think of the x–y axes as being attached to B. They stay aligned with the X–Y axes but can move around on the plane as B moves, whereas the X–Y axes are fixed.

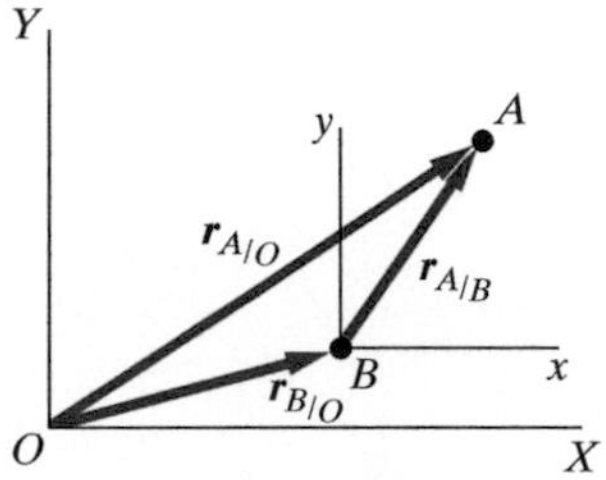

Figure 2.60 Two particles and two coordinate frames

IMPORTANT NOTE! In this section we restrict the x–y axes to translational motion only, which means the axes cannot rotate and so must always stay parallel to X–Y. The reason for this is that the position of body B is represented by a single point, and so there's no way it can rotate. This often confuses people, but think about it for a second. Rotation can occur only when you can define a direction vector within a finite body, which can then change its orientation as the body rotates. But you'll recall that we're currently neglecting all body rotations and are referencing to a point. There's no way to define any directions on a point (you need two points to define a line segment), and so there's no way to define rotation with respect to that point. When we get to rigid bodies, we will define rotations, and the analysis we are presenting here will become substantially more complex and more fun.

Now that we have positions covered, how about velocities? All we need do is differentiate (2.64) with respect to time and we will have them:

$$\frac{d}{dt}\boldsymbol{r}_{A/O} = \frac{d}{dt}\boldsymbol{r}_{B/O} + \frac{d}{dt}\boldsymbol{r}_{A/B}$$

or

$$\boldsymbol{v}_A = \boldsymbol{v}_B + \boldsymbol{v}_{A/B} \tag{2.65}$$

where $\boldsymbol{v}_{A/B}$ represents the relative velocity of A with respect to B. Please remember that $\boldsymbol{v}_{A/B}$ is a **relative** velocity (velocity referenced to some moving point), not an **absolute** velocity like $\boldsymbol{v}_A$ and $\boldsymbol{v}_B$, which are defined with respect to a fixed origin.

Differentiate again and we have acceleration:

$$\frac{d}{dt}\boldsymbol{v}_A = \frac{d}{dt}\boldsymbol{v}_B + \frac{d}{dt}\boldsymbol{v}_{A/B}$$

or

$$\boldsymbol{a}_A = \boldsymbol{a}_B + \boldsymbol{a}_{A/B} \tag{2.66}$$

Just as we saw with velocity, we have both absolute and relative acceleration terms, with $\boldsymbol{a}_{A/B}$ representing the relative acceleration of A with respect to B. When we get into kinetics, the differences between **relative acceleration** (acceleration referenced to some moving point) and **absolute acceleration** (acceleration defined with respect to an inertial reference frame) will become very important.

So, what if you want to know the acceleration of A with respect to B? Just use (2.66) to find

$$\boldsymbol{a}_{A/B} = \boldsymbol{a}_A - \boldsymbol{a}_B \tag{2.67}$$

How about the acceleration of B with respect to A? It's just $-\boldsymbol{a}_{A/B}$. If you don't see why, switch A and B in (2.66).

>>> **Check out Example 2.18 (page 82), Example 2.19 (page 84), and Example 2.20 (page 84) for applications of this material.**

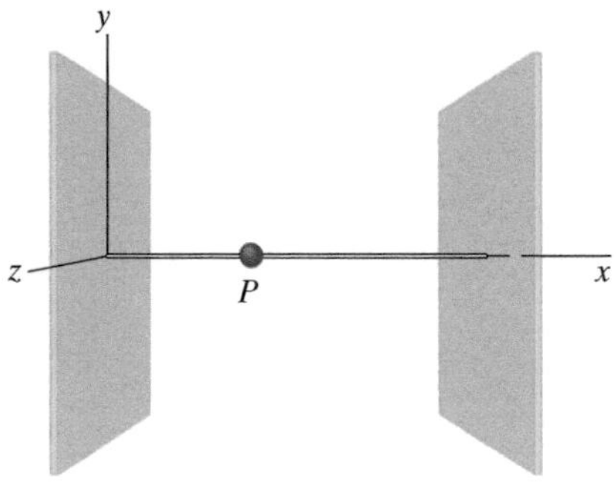

Figure 2.61 Bead constrained by a wire has one degree of freedom

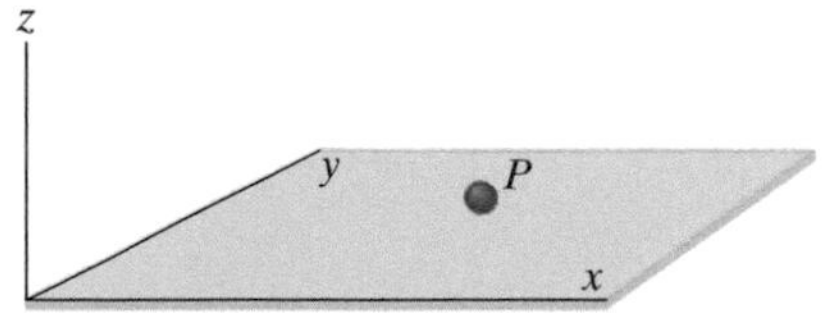

Figure 2.62 Bead constrained to a tabletop has two degrees of freedom

Although we have already been dealing with constraints without thinking explicitly about them, it's now time to take a closer look at the subject. A constraint simply means that a particle isn't completely free to move in all directions. The simplest type of constrained motion is illustrated in **Figure 2.61**: a bead strung on a stiff, straight wire aligned with the x axis. The only motions we need consider are those that involve x because the wire prevents the bead from moving in the y or z direction. The bead has only one **degree of freedom**. The wire has constrained the bead—without the wire, the bead could move anywhere in the three-dimensional space defined by x, y, and z, in which case it would have three degrees of freedom.

The number of degrees of freedom in a system is determined by the number of objects in the system as well as by the number of degrees of freedom each object individually has. If, instead of one bead, we have three beads on a wire, the system has three degrees of freedom. Each *bead* has one degree of freedom, but the *system* has three degrees of freedom. Two beads free to move in three-dimensional space have a total of six degrees of freedom (three for each bead).

Constraints reduce the number of degrees of freedom an object (or system) has. As just discussed, a bead on a wire has one degree of freedom (or 1-DOF for short). This is true even if the wire is bent; the bead is still constrained to move along a one-dimensional "surface." This means that only one variable is needed to define where the bead is along the wire. If you say that the bead has moved 5 cm along the wire, starting from the left end, then there's no confusion about its location. If the bead is taken off the wire, however, and put on a tabletop (**Figure 2.62**), it now has two degrees of freedom because it can be moved independently in two directions (x, y).

Getting back to constraints, we see that the wire of **Figure 2.61** limits the number of degrees of freedom. This kind of constraint—an impenetrable surface—is pretty easy to understand. There's another type of constraint, however, that's a bit less obvious and also more interesting. The simplest (and most boring) example of this constraint is shown in **Figure 2.63**. Here we have two bodies, m_1 and m_2. Now, normally you would say that their positions are determined by x_1 and x_2, specified independently. Note the line joining the two bodies, however. If this line represents a massless rod of length L, then x_1 and x_2 aren't really independent of each other. If you know x_1 and L, then you also know x_2 because

$$x_2 = x_1 + L \tag{2.68}$$

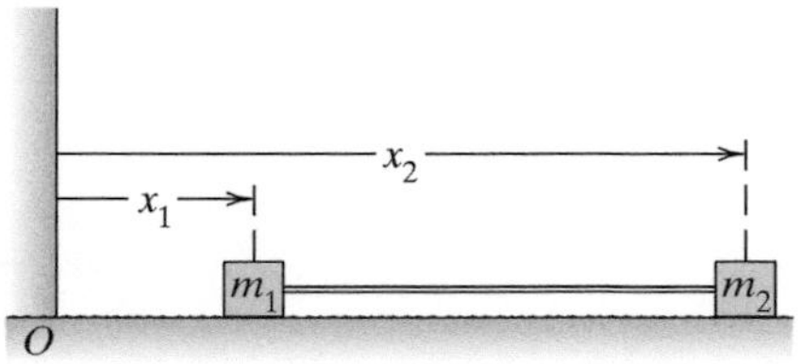

Figure 2.63 Two bodies joined by a massless rod

This means that the system has only one degree of freedom, not the two that you might expect from two individual blocks that can move in a line. So how does this matter? Well, imagine we are a few pages farther along in the book and are dealing with $\boldsymbol{F} = m\boldsymbol{a}$. If you apply a force to m_2, you would like to apply the equation $\boldsymbol{F} = m\boldsymbol{a}$ to it in order to calculate the block's acceleration. In this case, because of the constraint, you have to deal with m_1 as well because it accelerates any time m_2 does. For this case, the relationship is pretty straightforward. Just differentiate (2.68) twice with respect to time to get

$$\ddot{x}_2 = \ddot{x}_1$$

All right, I said this first example wouldn't be too thrilling. So let's move on to the star of this section—the pulley. That's right, after you've finished with this book, you'll be an expert on pulleys, blocks, and tackles. You may not be fully qualified to sail a four-master from New England to Cape Horn, but at least you'll understand how all those complicated pulleys you see on sailing ships actually work. Pulleys involve positional constraints that allow heavy objects to be lifted with small forces (something we'll see when we study kinetics). To understand how that happens, we need to analyze how pulleys constrain the motions of attached objects.

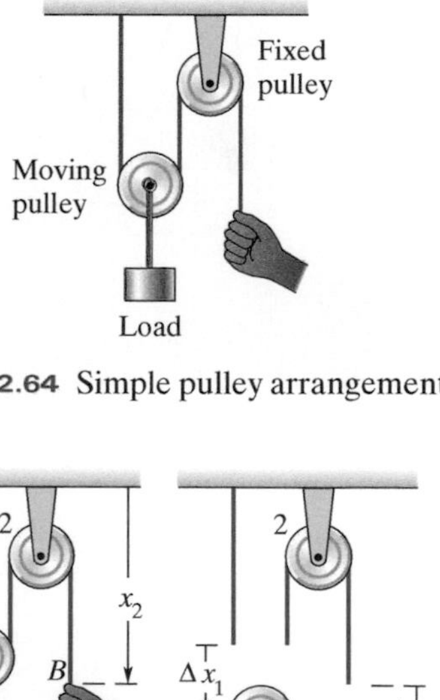

Figure 2.64 Simple pulley arrangement

Figure 2.64 shows a simple pulley setup and consists of one fixed pulley, one moving pulley, a load, some rope, and a disembodied hand that can pull down on the rope's free end. To analyze how this pulley works, we're going to use the concept of **conservation of rope**. This is a very simple concept, namely that the rope used in a pulley doesn't stretch or contract. No matter what configuration the pulley is in, *the length of rope has to always stay constant.* I'm going to go through the process slowly for this first problem so that each step is clear.

Figure 2.65 shows what we'll need for our analysis. All positions are referenced to the ceiling: x_1 is the distance from the ceiling to the movable pulley's axle, and x_2 is the distance from the ceiling to the free end of the rope being pulled. What we need to do is to identify each active rope. For our case there's just one, and **Figure 2.65*a*** shows it in pink. Ropes that aren't active are ones that act as fixed-length connections, such as a fixed length of rope that connects a load to a moving pulley.

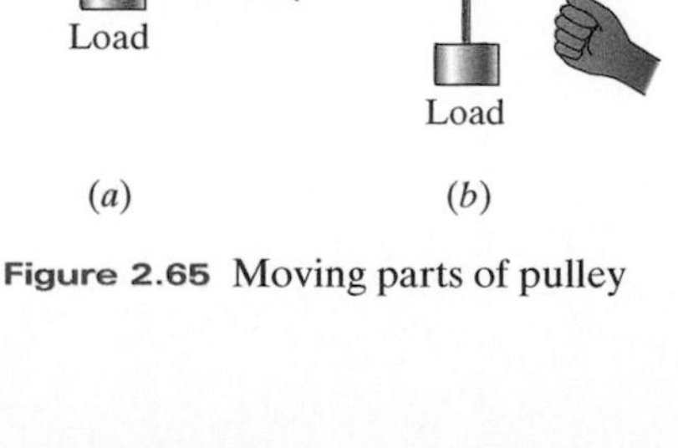

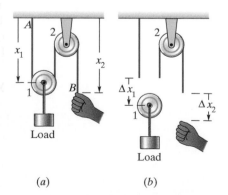

Figure 2.65 Moving parts of pulley

The next step is to ask if any portion of the active rope goes around a movable pulley and also look at each of the rope's ends. If an end is attached to a nonmoving surface, then we don't do anything more with that end. That's the case for the end at point A. The rope end at point B definitely can move; the hand holding it will either pull down or let up on the end, causing its position to change. In addition, pulley 1 can move, as we've already mentioned.

Although it might seem counterintuitive, we *don't* want to try and figure out from the figure which part of the system goes up and which part goes down. In simple systems (like this one) it's clear what happens physically: When the hand pulls down, pulley 1 will go up. But this is a simple pulley system, and the correct motions are not necessarily so clear in more complex examples. Therefore, what we'll do is assume that when the end of the rope in the hand is moved, every movable part of the pulley system moves to a new position. Also, each movable part has its own coordinate axis, and the motion of each part is assigned as being in the *positive* direction of its coordinate axis.

We will *not* draw the rope as if it has stretched or shortened to accommodate this change. Rather, we'll leave the rope exactly as initially drawn and move the associated pulley elements and rope ends. You can imagine that the rope has been snipped with scissors and both pulley 1 and the hand have been moved down by unknown amounts, labeled as Δx_1 and Δx_2 in **Figure 2.65*b***.

This snipping and moving means there are now gaps in the rope. Because pulley 1 has moved down, we need a length of rope equal to Δx_1 for each side of pulley 1 (a total of two Δx_1 lengths). Without this extra rope, there wouldn't be a continuous line of rope going around pulley 1. In addition to this, the hand has moved to a lower position, which means an extra Δx_2 length of rope is needed here as well. Hence at this point we have figured out that to get the system into the positions shown in **Figure 2.65*b*** we need to add a length $2\Delta x_1 + \Delta x_2$ to the rope.

Finally, realizing that conservation of rope tells us the rope can't get any longer or shorter, and thus the net change in length has to be zero, we have

$$2\Delta x_1 + \Delta x_2 = 0 \quad \Rightarrow \quad \Delta x_2 = -2\Delta x_1 \tag{2.69}$$

The minus sign tells us that the hand has to move down twice as far as pulley 1 moves *up*. Conservation of rope has given us both the correct signs and the correct ratio that governs motions of the movable pulley and the rope's free end. This is one of the key observations about pulleys. The amount of rope you have to move is more than the change in height of the pulley. In more complicated pulleys, the ratio is higher than 2:1 because multiple pulleys are involved.

Divide (2.69) by Δt, take the limit as Δt goes to zero, and you have your speed relationship:

$$\dot{x}_2 = -2\dot{x}_1 \tag{2.70}$$

Differentiating again leads to our acceleration relationship

$$\ddot{x}_2 = -2\ddot{x}_1 \tag{2.71}$$

an equation we will come back to once we get into kinetics.

Before moving on to some examples, I'd like to introduce a simpler notation. The pulley problem we just worked on was done in great detail, redrawing pulley 1 in a new position and redrawing the disembodied hand as well. Normally you won't want to take this much trouble. The key point to the problem is that in order to move pulley 1 and the hand down, you'll need to make three cuts in the rope. Rather than drawing all the pieces as was done in **Figure 2.65**, you can simply make a couple of slashes through the relevant ropes to indicate the cuts and label them with the Δx amounts needed at those points, as shown in **Figure 2.66**. You then simply need to follow along the rope, count up all the bits of extra rope needed, and sum them to zero.

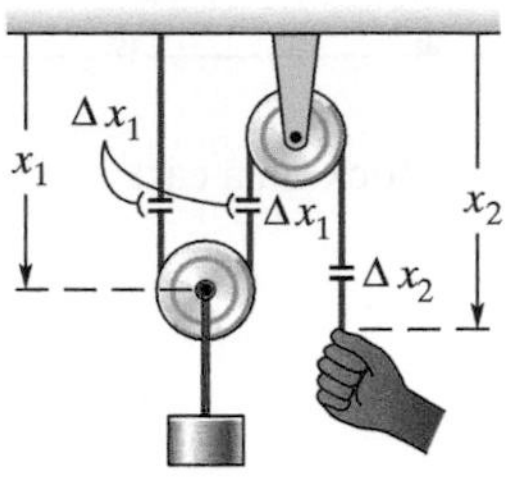

Figure 2.66 Simpler diagram for pulley problem. The three green-highlighted spots are where the rope was cut.

One key point has to be kept in mind, though. The pieces of the pulley system were moved in the positive direction indicated by the coordinates. The axis labeled x_1 is drawn pointing down, indicating that a positive addition to x_1 drives the pulley lower. Hence we'd need to *add* rope, as was done. Counting along the rope we'd have $\Delta x_1 + \Delta x_1 + \Delta x_2 = 0$. However, if for some reason you wanted some of your coordinates to point up, as in **Figure 2.67*a*** for x_1, then an increase in that coordinate indicates that pulley 1 is moving up. For this case we'd have to *subtract* rope. If we didn't, then the rope segments would get all loose and floppy, as shown in **Figure 2.67*b***. Hence the correct application of the concept of conservation of rope for **Figure 2.67** would be $-2\Delta x_1 + \Delta x_2 = 0$, leading to the relationship

$$\Delta x_2 = 2\Delta x_1$$

This all makes good sense. It says that for pulley 1 to rise a distance Δx_1 (rising being associated with an increase in x_1), x_2 needs to increase by twice that amount. The hand goes down, and pulley 1 goes up. Same result as before, just with a different sign on our equation due to the coordinate flip.

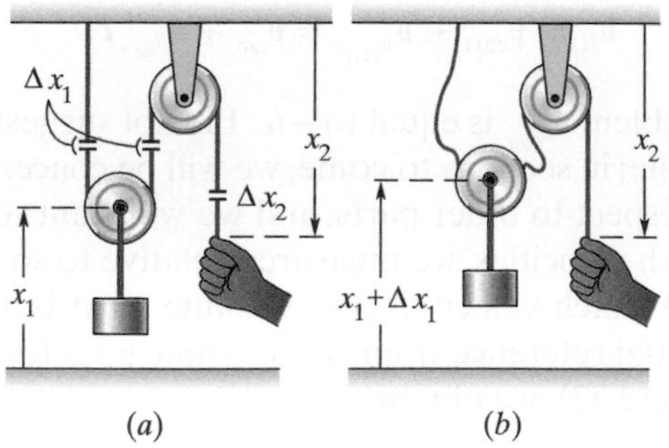

Figure 2.67 Different coordinate directions

>>> **Check out Example 2.21 (page 85) and Example 2.22 (page 86) for applications of this material.**

EXAMPLE 2.18 ONE BODY MOVING ON ANOTHER (Theory on page 77)

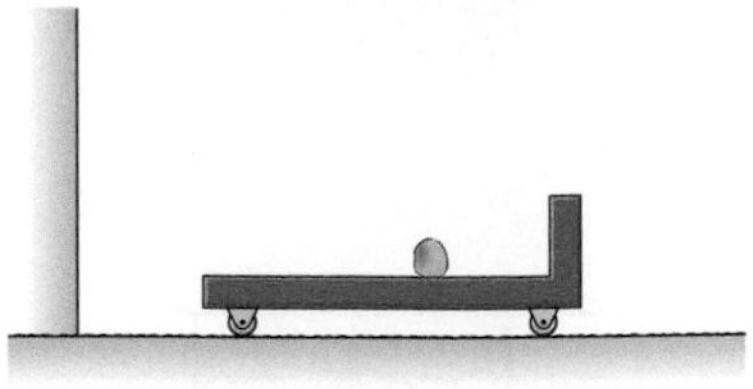

Figure 2.68 Rock on a cart

This example looks easy but can cause quite a bit of confusion once we start dealing with forces and friction. **Figure 2.68** shows a long cart on which a rock R can move. The left edge of the cart is located a horizontal distance x from a wall. A vertical stop on the right of the cart is a distance L from the left edge. We'll reference the position of the rock with respect to the cart as its distance u from the vertical stop. Determine the absolute velocity and acceleration of R with respect to the ground-fixed X–Y frame, as well as its relative velocity and acceleration with respect to the cart.

Goal Find the absolute and relative velocities and accelerations of the rock R in two reference frames.

Given Coordinates defining position of R with respect to cart and position of cart with respect to wall.

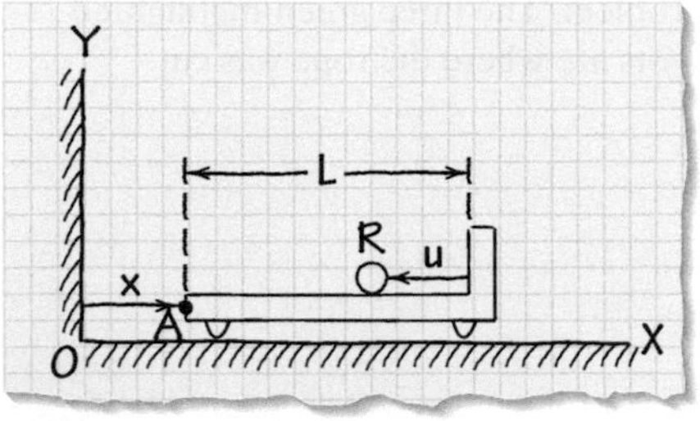

Figure 2.69 Parameters for rock on a cart

Assume: The position vector of the rock is given by

$$\boldsymbol{r}_{R/O} = (x + L - u)\boldsymbol{i}$$

Draw **Figure 2.69** shows our system.

Formulate Equations We will be using (2.65) and (2.66) to find the velocity and acceleration.

Solve We can differentiate $\boldsymbol{r}_{R/O}$ to find the rock's absolute velocity:

$$\boldsymbol{v}_R = \dot{x}\boldsymbol{i} - \dot{u}\boldsymbol{i} \tag{2.72}$$

In terms of (2.65), we would identify $\boldsymbol{v}_R$ as $\boldsymbol{v}_A$, $\dot{x}\boldsymbol{i}$ as $\boldsymbol{v}_B$, and $-\dot{u}\boldsymbol{i}$ as $\boldsymbol{v}_{A/B}$. You might also want to think of $-\dot{u}\boldsymbol{i}$ as the velocity of R *relative* to the cart and express the velocity relationship as

$$\boldsymbol{v}_R = \boldsymbol{v}_{\text{cart}} + \boldsymbol{v}_{R/\text{cart}} = \boldsymbol{v}_{\text{cart}} + v_{R/\text{cart}}\boldsymbol{i}$$

where for this problem $v_{R/\text{cart}}$ is equal to $-\dot{u}$. I'm not suggesting this just to complicate your life; in sections to come, we will be concerned with parts that move with respect to other parts, and we will want to keep straight in our minds which velocities are measured relative to something that is itself moving and which velocities are absolute (that is, measured with respect to an inertial reference frame). The same goes for acceleration.

Differentiating (2.72), we obtain

$$\boldsymbol{a}_R = \ddot{x}\boldsymbol{i} - \ddot{u}\boldsymbol{i}$$

Following the path we took with velocity, we see that this is in the form

$$\boldsymbol{a}_A = \boldsymbol{a}_B + \boldsymbol{a}_{A/B}$$

where $\boldsymbol{a}_{A/B}$ is equal to $-\ddot{u}\boldsymbol{i}$, and if we use the view

$$\boldsymbol{a}_R = \boldsymbol{a}_{\text{cart}} + \boldsymbol{a}_{R/\text{cart}}$$

then

$$\boldsymbol{a}_{R/\text{cart}} = a_{R/\text{cart}}\,\boldsymbol{i} \quad \text{and} \quad a_{R/\text{cart}} = -\ddot{u}$$

The importance of this result is that $-\ddot{u}$ (which is the acceleration of the rock with respect to the cart) IS NOT the rock's absolute acceleration. We have to include the cart's acceleration to get the rock's absolute acceleration. The relative acceleration $a_{B/\text{cart}}$ is the actual acceleration of R measured by an observer standing on the cart. So, in order to determine R's absolute acceleration (in an inertial reference frame), that observer would need to know the cart's acceleration with respect to the inertial frame and add that to the acceleration measured in the cart's frame. This makes a LOT of difference when we start throwing $\boldsymbol{F} = m\boldsymbol{a}$'s around because this formula is based on the assumption that we will be using an absolute acceleration (measured with respect to an inertial frame), not a relative one.

Check An easy physical check is to use our velocity relationship:

$$\boldsymbol{v}_R = \dot{x}\boldsymbol{i} - \dot{u}\boldsymbol{i} = (\dot{x} - \dot{u})\boldsymbol{i}$$

Say that the cart moves to the right at 2 m/s and at the same time the rock moves to the left at 2 m/s *with respect to the cart.* Its net velocity should be zero, and that's precisely what we get from evaluating $(\dot{x} - \dot{u})$ when $\dot{x} = \dot{u}$.

EXAMPLE 2.19 TWO BODIES MOVING INDEPENDENTLY (A) (Theory on page 77)

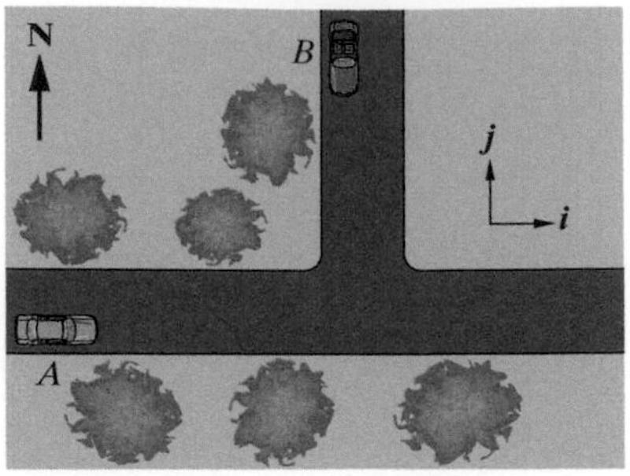

Figure 2.70 Dangerous intersection

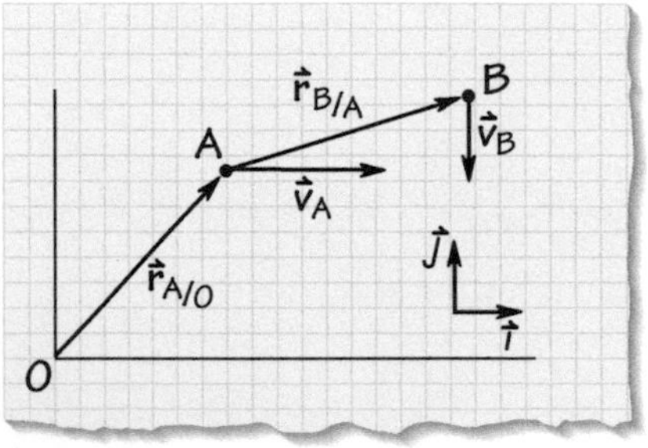

Figure 2.71 Position and velocity vector details

Assume you're in car B in **Figure 2.70**, moving south at 64 km/h, while car A is moving east at 96 km/h at the same time. Find $\boldsymbol{v}_{A/B}$, the velocity of A with respect to you. Does A's position relative to you matter?

Goal Find $\boldsymbol{v}_{A/B}$, the velocity of car A with respect to you riding in car B.

Given Velocity of car A and car B.

Assume No additional assumptions are needed.

Draw **Figure 2.71** shows our system.

Formulate Equations All we will need is (2.65).

Solve

$$\boldsymbol{v}_A = \boldsymbol{v}_B + \boldsymbol{v}_{A/B}$$
$$(96\boldsymbol{i})\ \text{km/h} = (-64\boldsymbol{j})\ \text{km/h} + \boldsymbol{v}_{A/B}$$

$$\boldsymbol{v}_{A/B} = (96\boldsymbol{i} + 64\boldsymbol{j})\ \text{km/h}$$

Note that to you in car B, it seems as if car A is moving north (positive $\boldsymbol{j}$) as well as east (positive $\boldsymbol{i}$).

The particular positions of the cars never came up in the calculations and are thus irrelevant to the problem. This is always the case when dealing with velocities, in the same way that velocities wouldn't be relevant if we were looking at acceleration. The act of differentiating (to go from position to velocity) transforms position information to zero.

Check The most straightforward check is to think physically about the problem and see that it matches the mathematics. If you're going south, then everything to the south of you will seem to be moving toward you—that is, moving north. Thus, even if stationary, car A will seem to be moving north as you're moving south. If car A is moving east as well, then its vector velocity with respect to you (a *relative* velocity) will contain this $\boldsymbol{i}$ component as well as the $\boldsymbol{j}$ component due to your velocity, just what we got in our algebraic result.

EXAMPLE 2.20 TWO BODIES MOVING INDEPENDENTLY (B) (Theory on page 77)

Assume that car A in Example 2.19 collides into the front passenger door of car B while in the intersection. Will car A strike car B directly or obliquely?

Goal Find the orientation of the collision.

Given Velocity of car A and velocity of car B from Example 2.19.

Assume No assumptions are needed.

Draw **Figure 2.72** shows $\boldsymbol{v}_{A/B}$.

Formulate Equations Same as in Example 2.19.

Solve The solution here follows directly from the solution to Example 2.19 but illustrates what can sometimes be a subtle point. We know $\boldsymbol{v}_{A/B} = (96\boldsymbol{i} + 64\boldsymbol{j})$ km/h. The speed (equal to the square root of the sum of the squares of the two components) is 115 km/h. As **Figure 2.72** shows, the velocity vector is oriented at an upward angle of 0.59 rad. The conclusion is that the collision is oblique.

The moral of the story is that relative velocities and accelerations really mean something; they're not just items introduced for no particular reason. Although much of the work we will be doing will involve absolute quantities, the relative information is often of equal importance.

Check There isn't anything in particular to check this time.

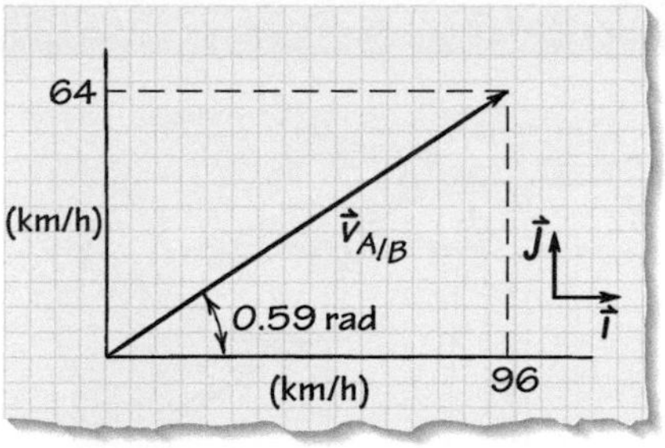

Figure 2.72 Relative velocity of car *A* with respect to car *B*

EXAMPLE 2.21 SIMPLE PULLEY (Theory on page 80)

Figure 2.73 shows a pulley arrangement designed to get you to a seat that you've built in a tree as part of the "contacting my inner child" phase you're currently in. (The pressures of college must finally be getting to you.) What you have is a movable platform with a chain attached at each corner. The free ends of the four chains are joined together and attached to the movable pulley. The free end of the pulley rope lies on the platform. If your assistant, standing off to the side of the platform, grabs the free end and starts pulling it down at 0.90 m/s while you sit on the platform, how long will it take you to reach the seat?

Goal Find the time needed to reach the seat.

Given Free end of the pulley rope moving at 0.90 m/s, distance to seat is 6 m.

Assume No assumptions are needed.

Draw A schematic of our problem is shown in **Figure 2.74**, with both coordinates referenced to the fixed tree branch supporting the pulley system.

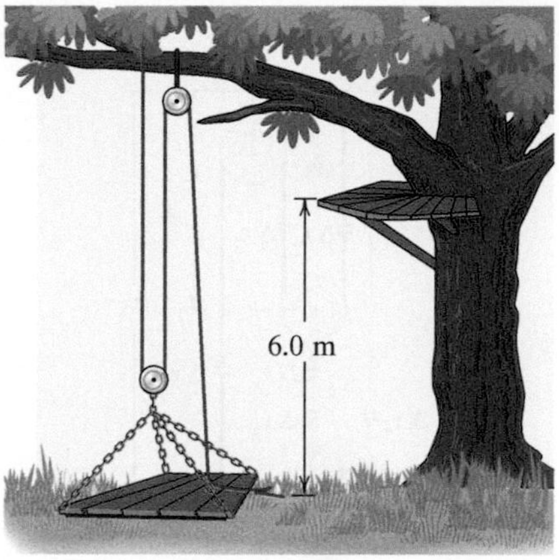

Figure 2.73 Treehouse elevator

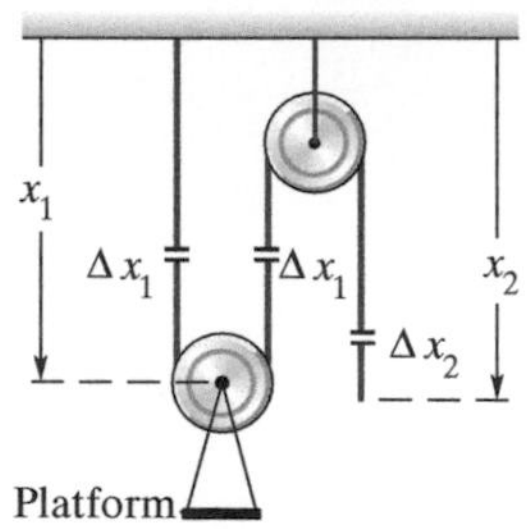

Figure 2.74 Treehouse pulley schematic

Formulate Equations From **Figure 2.74** we have $\Delta x_1 + \Delta x_1 + \Delta x_2 = 0$ because increases in both x_1 and x_2 require the addition of rope.

Solve Our conservation of rope analysis leads to $\dot{x}_1 = -\frac{1}{2}\dot{x}_2$. The platform rises at half the speed at which the free end is pulled. Thus, because the rope is being pulled at 0.90 m/s, the movable pulley and platform will rise at 0.45 m/s, and so the time needed to reach the seat is

$$t = \frac{6\text{ m}}{0.45\text{ m/s}} = 13.3\text{ s}$$

Check This pulley is so simple that there's no problem checking consistency—when the free end of the rope is pulled down, the movable pulley will go up.

EXAMPLE 2.22 DOUBLE PULLEY (Theory on page 80)

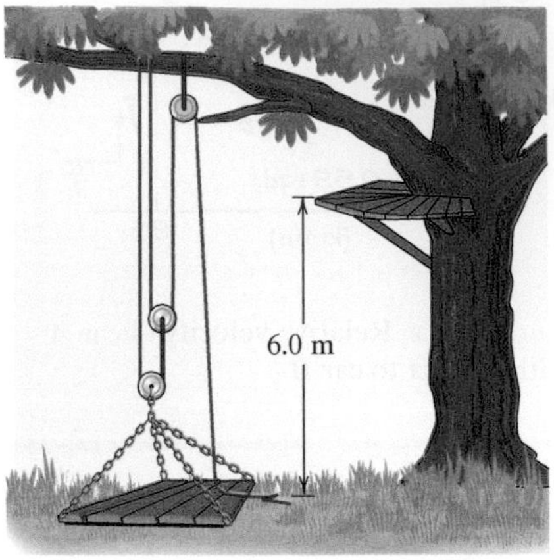

Figure 2.75 Improved treehouse elevator

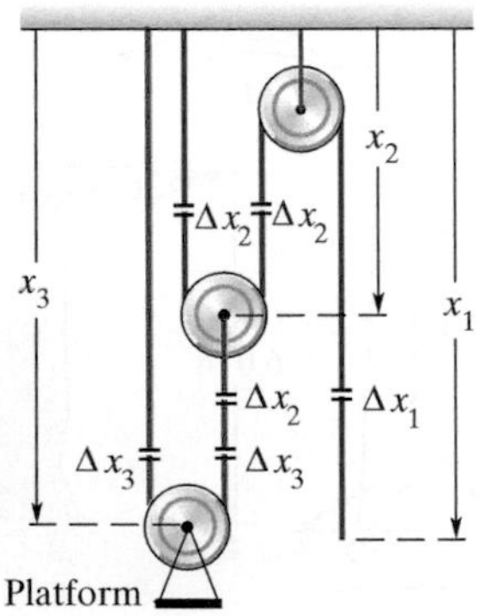

Figure 2.76 Improved elevator schematic

Repeat Example 2.21 for the pulley system shown in **Figure 2.75**.

Goal Find the time needed to reach the seat.

Given Free end is pulled at 0.90 m/s. Distance to seat is 6 m.

Assume No assumptions are needed.

Draw **Figure 2.76** shows our system.

Formulate Equations We will be using (2.70) a couple of times.

Solve A color-coded schematic of this more complicated pulley system is shown in **Figure 2.76**. As you can see, the system can be regarded as being made up of two of the simpler pulley systems introduced in **Figure 2.64**. Each of these systems has a pulley and a rope with one fixed and one moving end. This is our first example of a **compound pulley**, one made up of more than a single simple pulley.

We approach this problem in the same way we did previously: by using the concept of conservation of rope. I've drawn coordinates x_1, x_2, x_3 in **Figure 2.76** to quantify everything that's moving in the system.

We'll first look at the orange rope. An increase in x_3 will require two pieces of rope to be added, each of length Δx_3. An increase in x_2, however, will "compress" the orange rope on the right, meaning that we need to subtract Δx_2 worth of rope. Thus conservation of rope gives us

$$2\Delta x_3 - \Delta x_2 = 0 \tag{2.73}$$

Next we'll examine the blue rope. An increase in x_2 will require two pieces of rope, each of length Δx_2, and an increase in x_1 will require a piece of length Δx_1. For this rope, conservation of rope gives us

$$2\Delta x_2 + \Delta x_1 = 0 \tag{2.74}$$

Solving for Δx_1 and Δx_2 yields

$$\Delta x_2 = -0.5\Delta x_1 \quad \text{and} \quad \Delta x_3 = 0.5\Delta x_2 = -0.25\Delta x_1$$

Therefore, pulling down on the free end of the blue rope at 0.90 m/s will cause the platform to rise at 0.225 m/s. The time needed to reach the seat is 6 m/0.225 m/s = 26.7 m/s.

Check If you pull down the free end of the blue rope with speed v, you can see that pulley 2 will rise at half that rate: $v/2$. Then, because pulley 3 must rise at half the rate at which the right side of the orange rope is moving, it moves up at speed $v/4$, one-quarter the speed of your original pull rate.

EXERCISES 2.5

2.5.1. **[Level 1]** The Red Devils high-performance, radio-controlled model team is performing at a local airshow, and two jets are in the middle of a death-defying maneuver. Jet A is in the bottom of a loop with a radius of 30 m, and at this instant it has a velocity and acceleration of $\boldsymbol{v}_A = 201\boldsymbol{i}$ m/s and $\boldsymbol{a}_{t,A} = 15\boldsymbol{i}$ m/s, respectively. Just a few meters below jet A, jet B is traveling at $\boldsymbol{v}_B = -213\boldsymbol{i}$ m/s with an acceleration of $\boldsymbol{a}_B = -12\boldsymbol{i}$ m/s^2. What are the velocity and acceleration of jet A as seen by jet B at this instant?

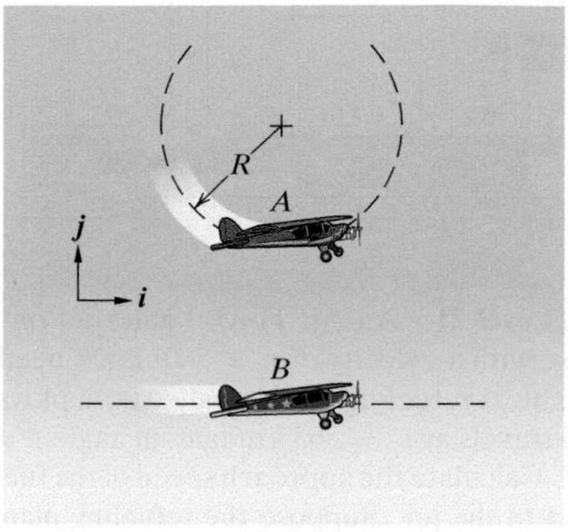

E2.5.1

2.5.2. **[Level 1]** A commuter plane is en route from A to B with a velocity of $\boldsymbol{v}_p = 322\boldsymbol{j}$ km/h in a straight-line path. At a point $d = 161$ km from B, the plane encounters some strong winds of velocity $\boldsymbol{v}_w = 129\boldsymbol{i}$ km/h. If the plane maintains its current speed and direction relative to the wind, how far over from B will the plane be at the anticipated time of arrival (that is, with no wind)? At what speed and angle should the plane fly relative to the wind to actually arrive at B in this time?

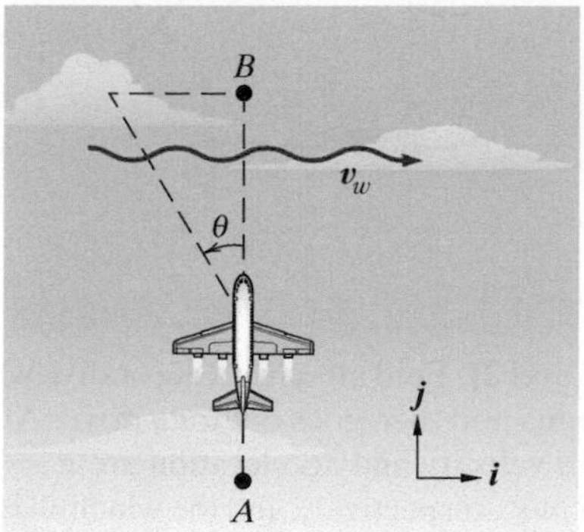

E2.5.2

2.5.3. **[Level 1]** In an old James Bond movie, 007 used an ejector seat to remove a bad guy from his Aston Martin. If 007 was traveling at 30 m/s when he activated the ejector and the launch velocity relative to the Aston Martin was 10 m/s (oriented vertically with respect to the Aston Martin), what was the overall velocity of the bad guy as he left the car?

E2.5.3

2.5.4. **[Level 1]** A disk of radius $R = 18$ cm spins at a constant angular speed of $\omega = 500$ rpm in the clockwise direction. If the disk is released from rest far above the ground, what are the speed and acceleration of the disk's top C after it has fallen $h = 0.9$ m?

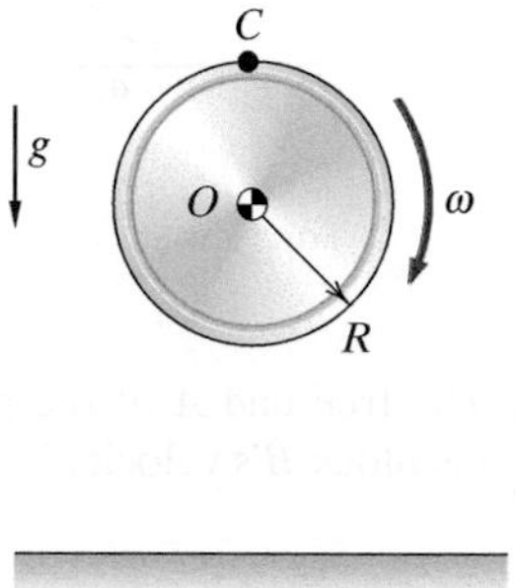

E2.5.4

2.5.5. **[Level 1]** A worker is dragging a large drum of mass m through the use of a pulley arrangement. If she's able to move to the left at 2 m/s, how fast will the mass move to the right?

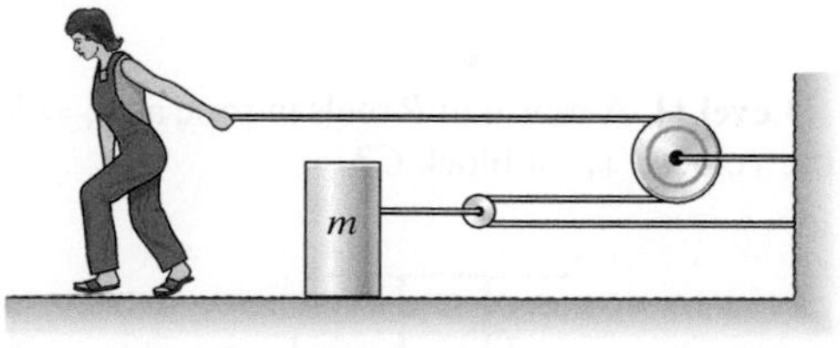

E2.5.5

2.5.6. **[Level 1]** Two masses, A and B, are connected by a rope that goes over a central reel C. The reel is attached to a horizontal bar that rides in two vertical guides. The bar's motion is given by x, and the height of A above the ground is given by y. What is the velocity of B in terms of $\dot{x}$ and $\dot{y}$?

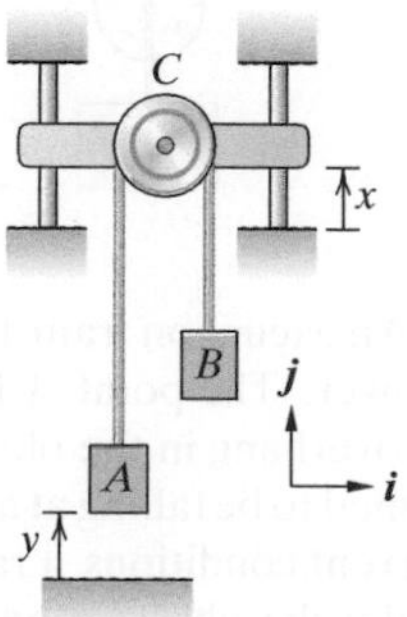

E2.5.6

2.5.7. **[Level 1]** Suppose two ball bearings (A and B) are constrained to move in the horizontal plane along concentric circular paths, where $r_A = 0.4$ m and $r_B = 0.9$ m. At the illustrated instant, the two ball bearings are collinear with respect to the vertical, and it is measured that A is traveling at a constant $\boldsymbol{v}_A = -5\boldsymbol{i}$ m/s and B is moving at $\boldsymbol{v}_B = 8\boldsymbol{i}$ m/s with a tangential acceleration of $\boldsymbol{a}_{B,t} = 0.7\boldsymbol{i}$ m/s^2. What are the velocity and acceleration of B as seen by A?

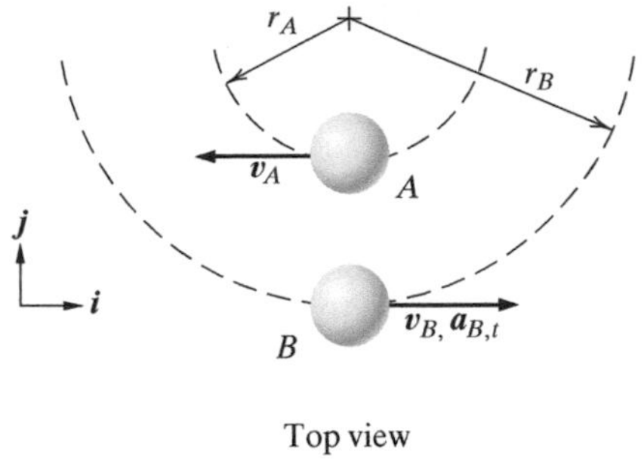

E2.5.7

2.5.8. **[Level 1]** The free end A of the pulley is pulled down at 2 m/s. What is block B's velocity?

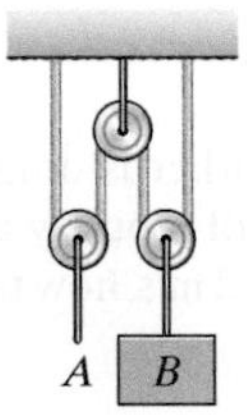

E2.5.8

2.5.9. **[Level 1]** A motor at R reels in rope at $v_R = 3$ m/s. What is the velocity v_C of block C?

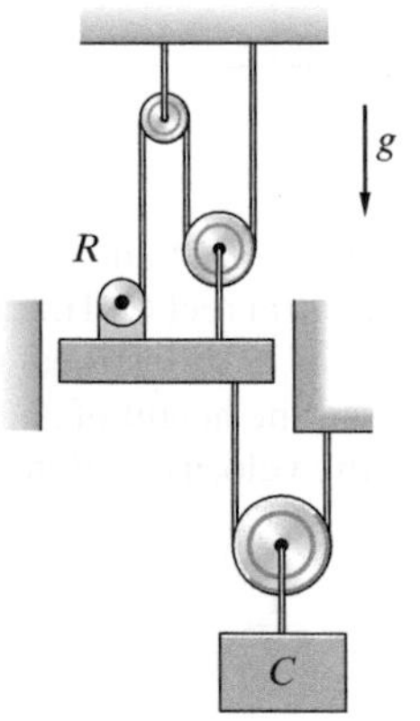

E2.5.9

2.5.10. **[Level 2]** An excursion train, traveling at 24 m/s, encounters a rain shower. The point A lies directly 2.4 m beneath the lip of an overhang in the observation car. The raindrops can be assumed to be falling at a constant speed of 12 m/s. Under the current conditions, a raindrop just clearing the lip would strike the observation car's floor 4.8 m from the point A. At what constant rate would the train need to be decelerating (deceleration starting just as the drop passes the lip) for the drop to hit the floor 2.4 m from A? What angle does the raindrop's velocity vector make with respect to the floor at the time of impact, as seen by an observer on the train?

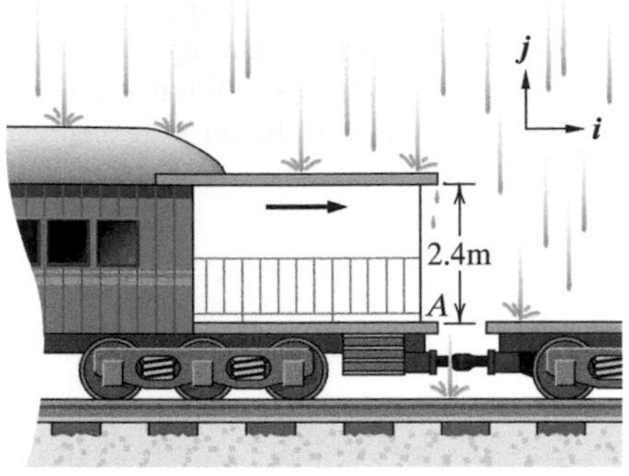

E2.5.10

2.5.11. **[Level 2]** An Air Force fighter jet on its way back to base with a velocity of $\boldsymbol{v}_j = 580\boldsymbol{i}$ km/h needs to refuel mid-flight. A refueling plane is called in, deploys its fuel nozzle, and travels at $v_p = 563$ km/h at an angle $\beta = 85°$ to the vertical. Calculate the approach speed of the fuel nozzle with respect to the jet. Suppose the refueling plane slows down to 547 km/h. What speed should the fighter pilot slow down to if he wants to maintain the same nozzle approach speed?

E2.5.11

2.5.12. **[Level 2]** Fred attaches a decorative windmill to his car's antenna and then goes out for a drive. At a certain instant, Fred's velocity and acceleration are $\boldsymbol{v}_c = 48\boldsymbol{i}$ km/h and $\boldsymbol{a}_c = 1.5\boldsymbol{i}$ m/s^2, respectively, and the windmill is spinning at $\dot{\theta} = 200$ rpm counterclockwise. Take A to be the edge of the windmill directly above the center of rotation. If the windmill has a radius of $r = 10$ cm, find the magnitude of the velocity and acceleration of A at the given instant, for which $\boldsymbol{e}_t = -\boldsymbol{i}$ and $\boldsymbol{e}_n = -\boldsymbol{j}$. Assume that the antenna doesn't flex.

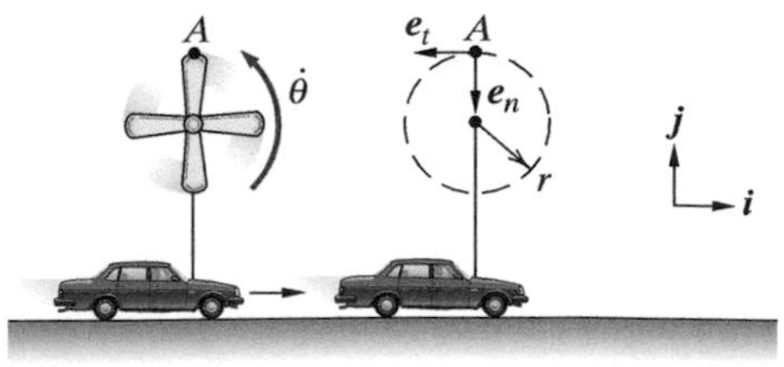

E2.5.12

2.5.13. [Level 2] A secret agent is running toward the back of a moving bus, his intent being to jump off the back before the explosive he planted at the front goes off. At the illustrated instant, the bus has a constant velocity of $\boldsymbol{v}_{\text{bus}} = 16\boldsymbol{i}$ km/h. Currently, our agent is 9 m from the rear of the bus and moving at 0 km/h *with respect to the bus*.

a. What constant acceleration will he need *with respect to the bus* so that he will leave the rear of the bus with zero velocity relative to the ground?

b. Where will he land with respect to point A?

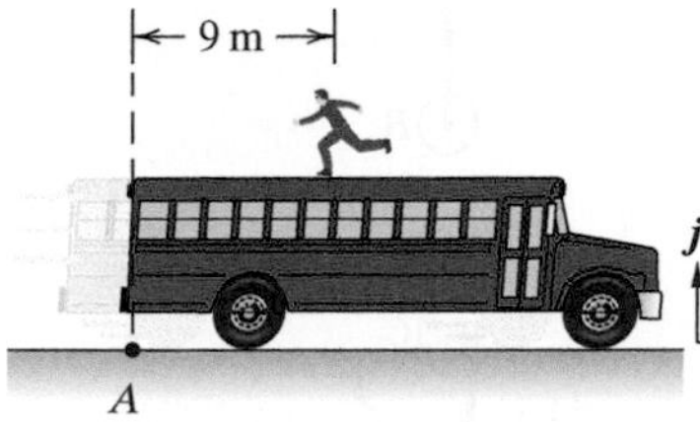

E2.5.13

2.5.14. [Level 2] Two cars are passing by each other, moving in the directions illustrated. Car A is moving at a constant 48 km/h around a circle with radius 30 m, and car B is moving at a constant 96 km/h in the direction $0.5\boldsymbol{i} + \frac{\sqrt{3}}{2}\boldsymbol{j}$. What are the velocity and acceleration of car A with respect to car B? $\boldsymbol{r}_{A/O} = 30\boldsymbol{i}$ m and $\boldsymbol{r}_{B/A} = 60\boldsymbol{i}$ m.

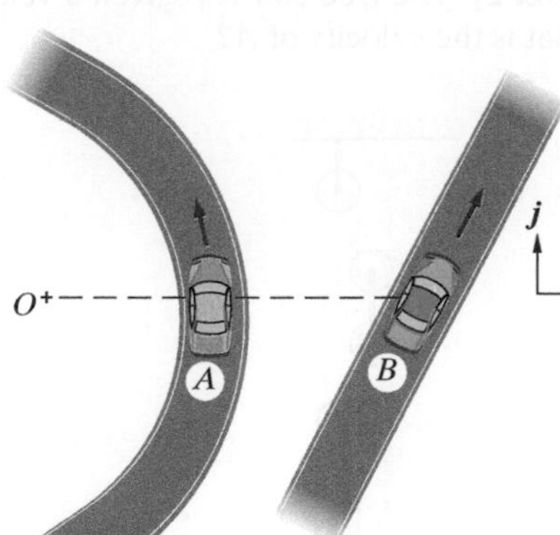

E2.5.14

2.5.15. [Level 2] A swordfish has been hooked by a fisherman who has 500 m of line remaining in his fishing reel. The initial position of his boat (B) and the swordfish (S) are shown. $\boldsymbol{v}_B = -v_B\boldsymbol{j} = -3\boldsymbol{j}$ m/s and $\boldsymbol{v}_S = -v_S\boldsymbol{i} = 10\boldsymbol{i}$ m/s, both constant.

a. How long before he runs out of line?

b. What are $\boldsymbol{v}_{S/B}$ and $\boldsymbol{a}_{S/B}$?

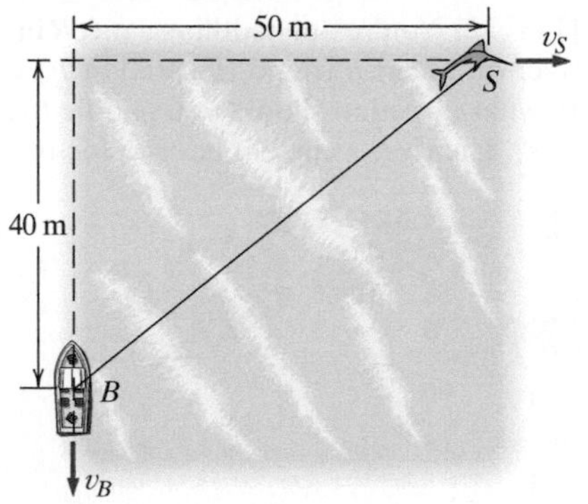

E2.5.15

2.5.16. [Level 2] An excursion train, traveling at 48 km/h from left to right in the figure, encounters an unexpected rain shower. Point A lies directly 2.1 m beneath the lip of an overhang in the observation car. How far toward the rear of the train will the raindrops reach if they're falling at 40 km/h? What is the velocity of a raindrop with respect to an observer in the train?

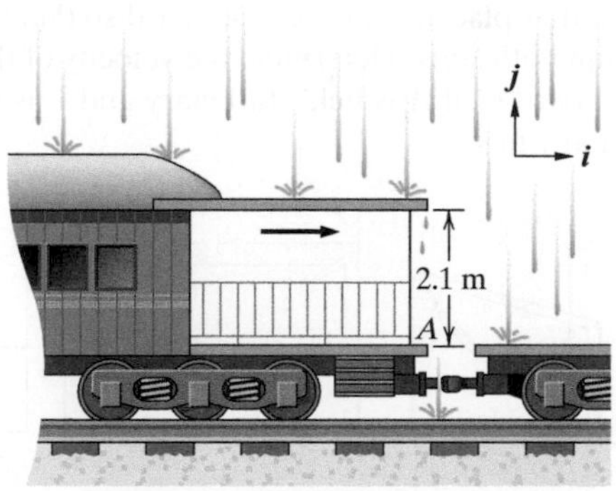

E2.5.16

2.5.17. [Level 2] Assume you're out paddling in a river that has a 3 m/s current. You're midstream, 11 m from either shore. If you paddle toward the left shore at 4 m/s and at a 45° angle ($\boldsymbol{i} + \boldsymbol{j}$ direction), where along the shore will you land?

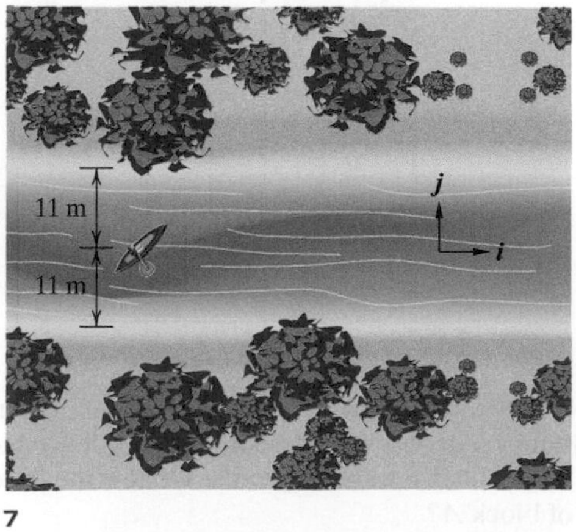

E2.5.17

2.5.18. **[Level 2]** You're out paddling a raft R in the middle of the river and hear a friend (located at F) call from the shore. In what direction should you paddle to arrive at F in 2 minutes? The river's current has a velocity $1\boldsymbol{i}$ m/s.

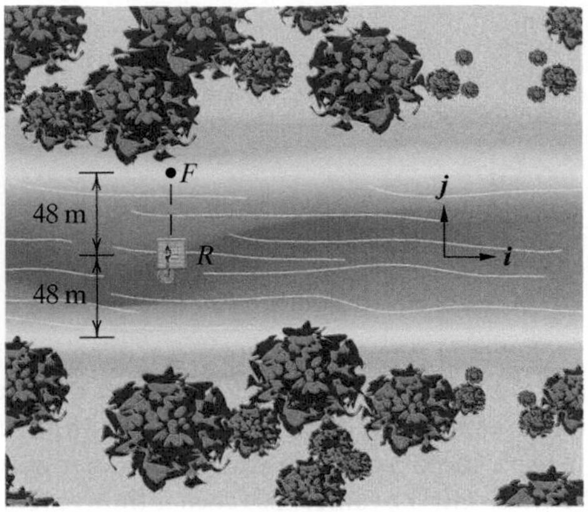

E2.5.18

2.5.19. **[Level 2]** A rod B is free to slide vertically within the slot cut into board A. Another board, C, with an inclined slot cut into it, is placed over the first board so that the rod is held within both slots. Determine the velocity of the rod, relative to board C, if A is held stationary and C is moved to the right at 1 m/s.

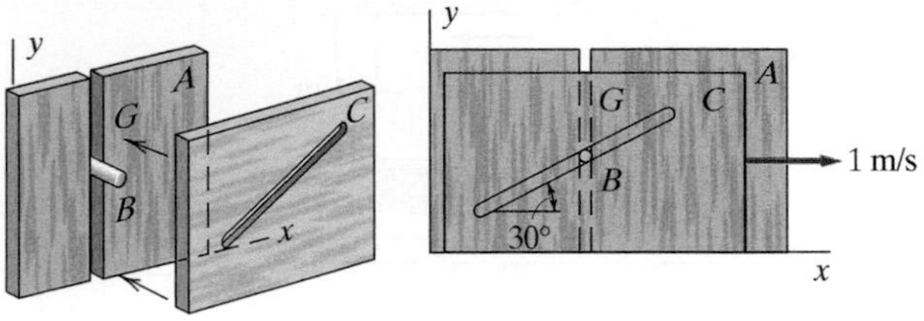

E2.5.19

2.5.20. **[Level 2]** The free end A of the pulley is pulled down at 3 m/s. What is block B's velocity?

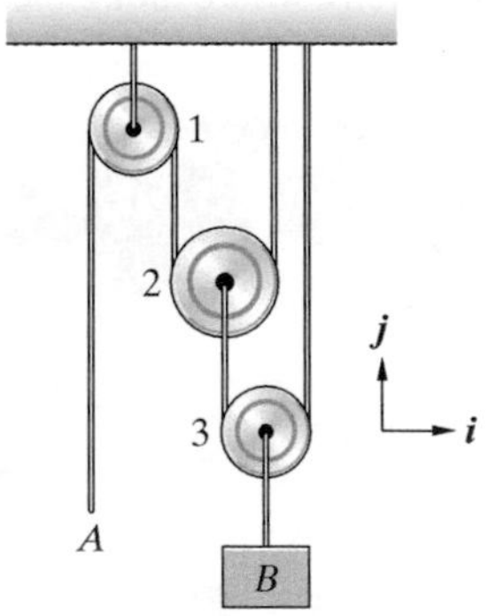

E2.5.20

2.5.21. **[Level 2]** A motor at O reels in the illustrated rope at 36 cm/s. An observer measures the absolute velocity of block B and finds it to be $-5\boldsymbol{j}$ cm/s. What is the absolute velocity of block A?

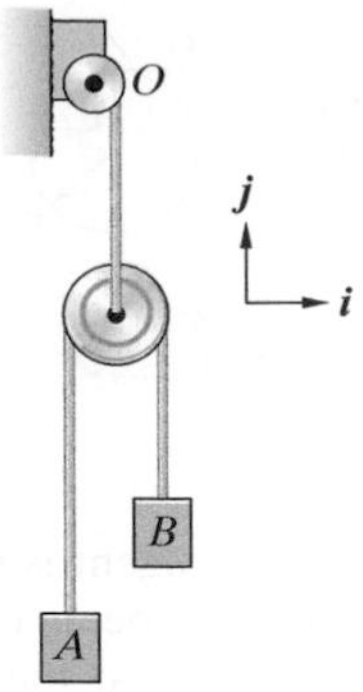

E2.5.21

2.5.22. **[Level 2]** A motor at A reels in the illustrated rope at 0.4 m/s. What is the absolute velocity of block D?

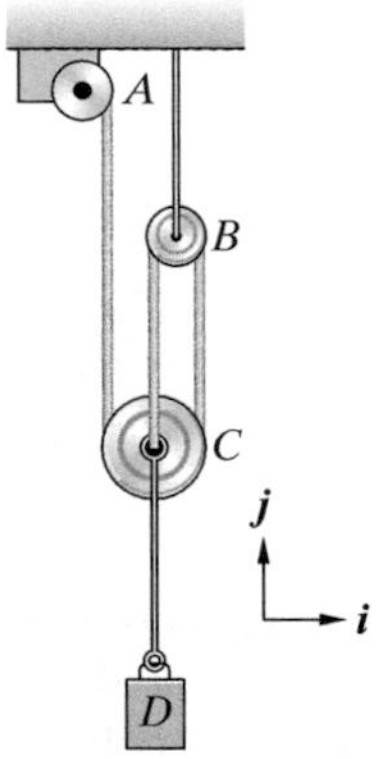

E2.5.22

2.5.23. **[Level 2]** The free end B is given a velocity of $-1.2\boldsymbol{j}$ m/s. What is the velocity of A?

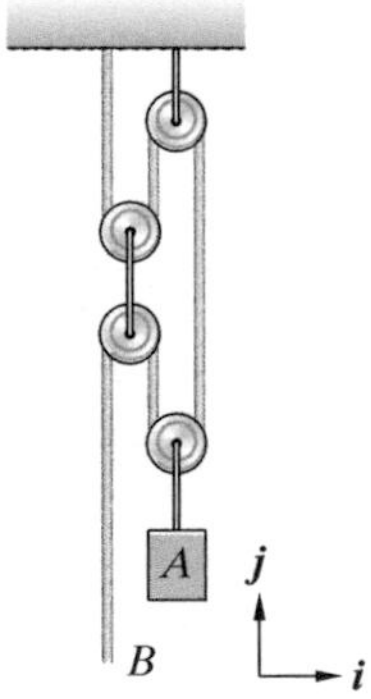

E2.5.23

2.5.24. **[Level 2]** A motorized reel R pulls in rope at a rate of 25 cm/s. The rope goes around pulley B, up and around pulley C, and then terminates at the center axle of pulley B. What is the velocity of D, a point on the rightmost piece of rope connecting C and B?

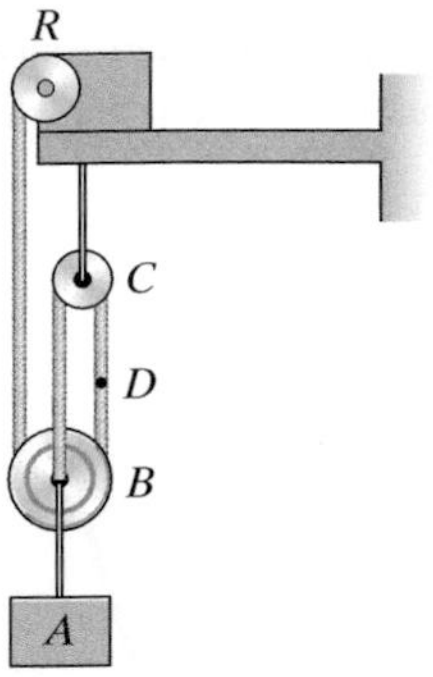

E2.5.24

2.5.25. **[Level 2]** Rope is drawn into the motorized reel at A at a rate of v_0. How fast does the weight B rise above the floor ($\dot{y}$)?

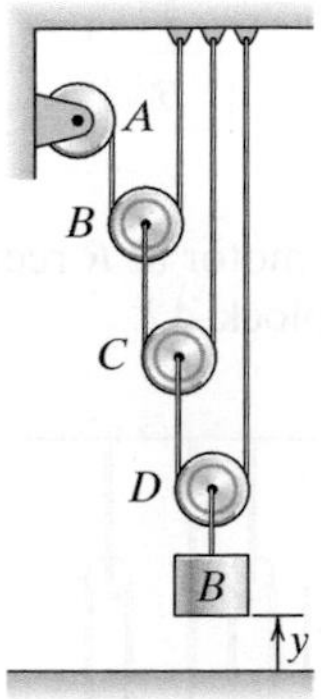

E2.5.25

2.5.26. **[Level 2]** If the free end of the pulley rope is pulled down with an acceleration of 1.2 m/s^2, what will the acceleration of m be? For simplicity, assume that all the straight rope segments are vertical.

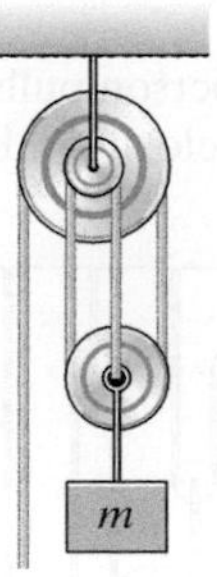

E2.5.26

2.5.27. **[Level 2]** What is $\dot{y}$ equal to if $\dot{x} = 10$ m/s and the rope remains taut?

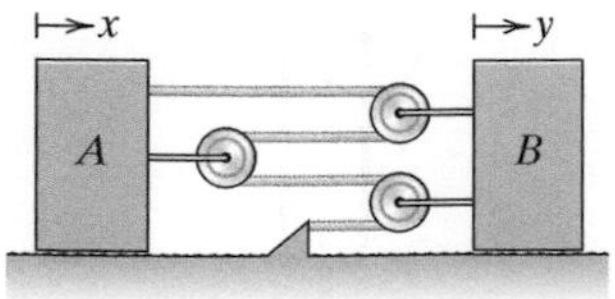

E2.5.27

2.5.28. **[Level 2]** A complicated system of pulleys and motors is shown. At A is a motor that reels in the pulley rope at a rate of 50 cm/s. Another motor, at G, reels in pulley rope at 25 cm/s. What is block B's velocity?

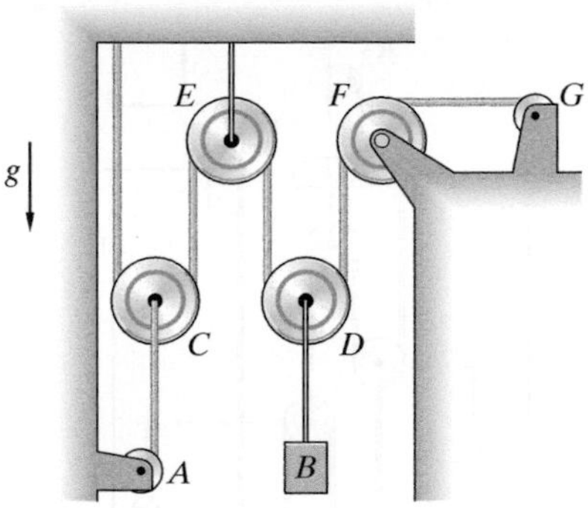

E2.5.28

2.5.29. **[Level 2]** The free end D of the pulley is pulled down with a speed of 25 cm/s. What is block A's velocity? For simplicity, assume that all the straight rope segments are oriented vertically.

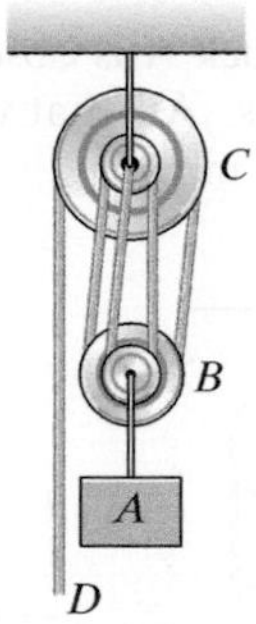

E2.5.29

2.5.30. **[Level 2]** If the free end at A is pulled down at a speed v_A, what is the resultant velocity of block B?

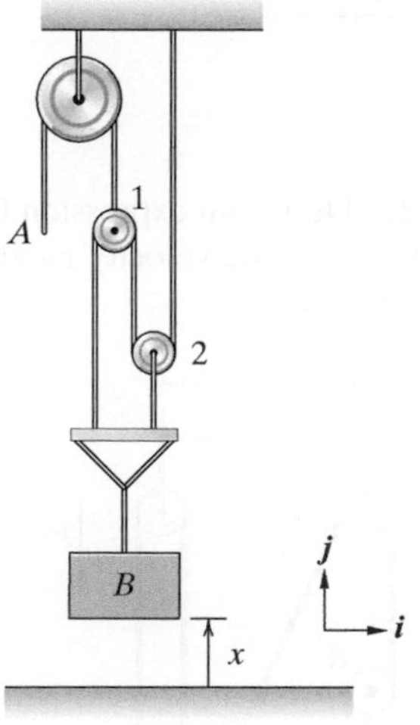

E2.5.30

2.5.31. **[Level 2]** The reels R_1 and R_2 are taking in rope at speeds v_1 and v_2, respectively. Determine $\dot{y}_3$ and $\dot{y}_4$ in terms of v_1 and v_2.

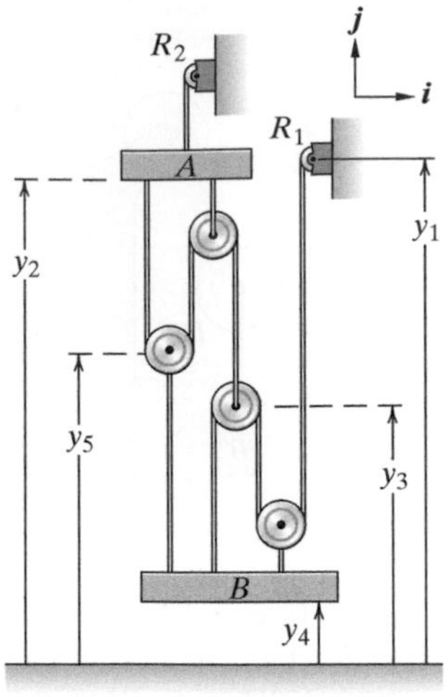

E2.5.31

2.5.32. [Level 2] Block A is observed to be dropping down at a steady 0.27 m/s. At what velocity must the free end of the pulley rope be moving?

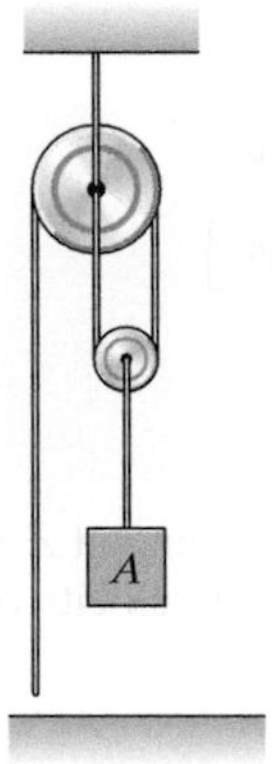

E2.5.32

2.5.33. [Level 2] Derive an expression for the velocity v_A of block A in terms of the velocity v_R at which rope is reeled in by a motor at R.

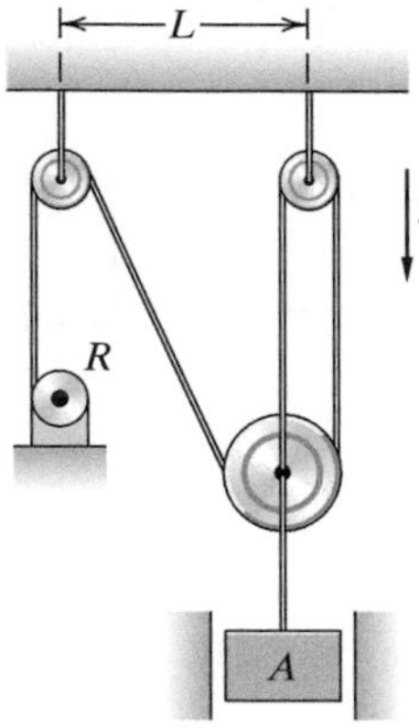

E2.5.33

2.5.34. [Level 2] Derive an expression for the velocity v_B of block B in terms of the velocity v_A at which rope is pulled down at A.

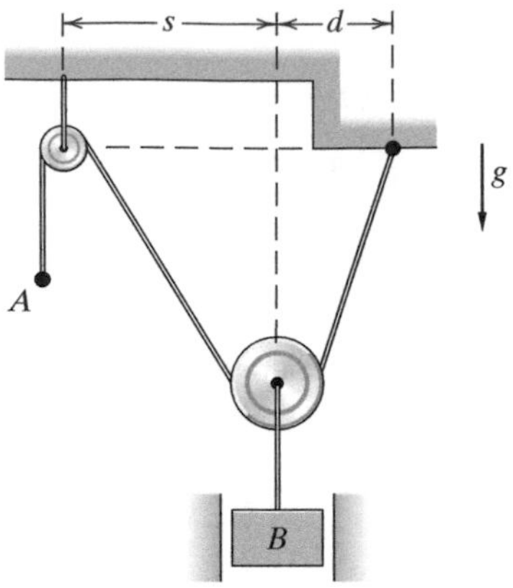

E2.5.34

2.5.35. [Level 2] A motor at R reels in rope at 0.9 m/s. What is the velocity of block A?

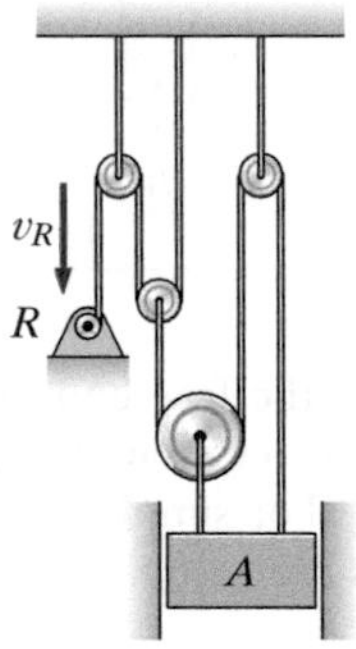

E2.5.35

2.5.36. [Level 2] A person pulls down the rope at A at 0.6 m/s. Determine the velocity of block B.

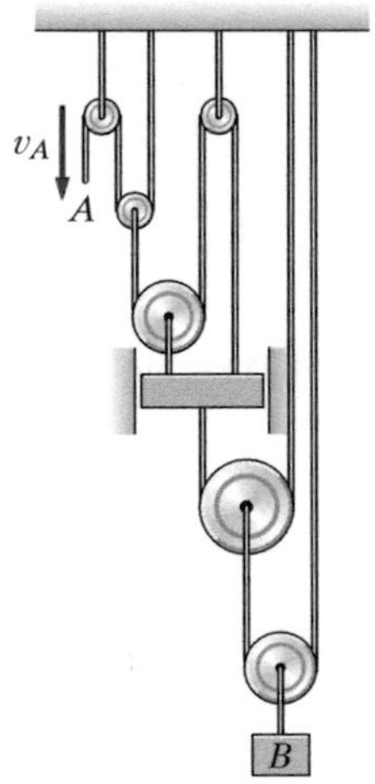

E2.5.36

2.5.37. [Level 3] Today, Judy is kayaking across a 70 m-wide pond from A to B with a velocity of $\boldsymbol{v}_k = 2\boldsymbol{j}$ m/s. In the first 20 m stretch of the pond, the water has a current of velocity $\boldsymbol{v}_w = 3\boldsymbol{i}$ m/s, and she winds up at C. How far over is Judy at C? If the remainder of the pond is still water, at what angle does Judy need to travel to get to B? Suppose Judy was kayaking in the pond yesterday, and the entire pond was still. How much longer did it take her to get from A to B today than it took yesterday?

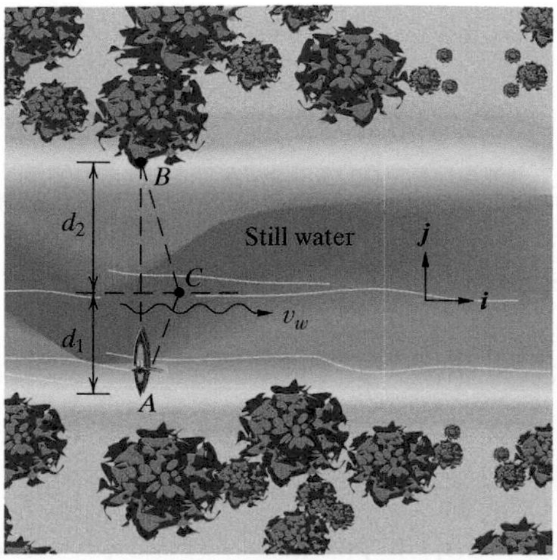

E2.5.37

2.5.38. **[Level 3]** A student is out in her canoe one day and suddenly realizes that she is midstream and heading for a waterfall. It's 10 m to shore and 100 m to the waterfall. The river's current is 4 m/s. If she wants to reach the left (with respect to her) shore 5 m before the waterfall's edge, in what direction should she paddle the boat? Assume that she can paddle at 3 m/s.

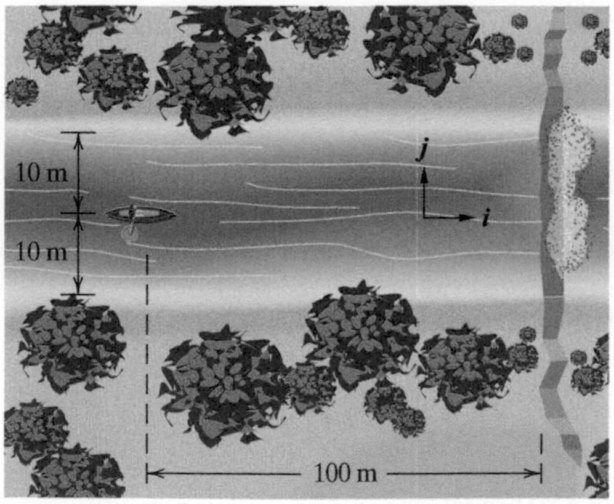

E2.5.38

2.5.39. **[Level 3]** A single length of rope runs from B, up around two pulley wheels, and terminates at C. Derive expressions for the velocity and acceleration of A if the end C is pulled horizontally such that $\boldsymbol{v}_C = \dot{x}\boldsymbol{i}$ and $\boldsymbol{a}_C = \ddot{x}\boldsymbol{i}$.

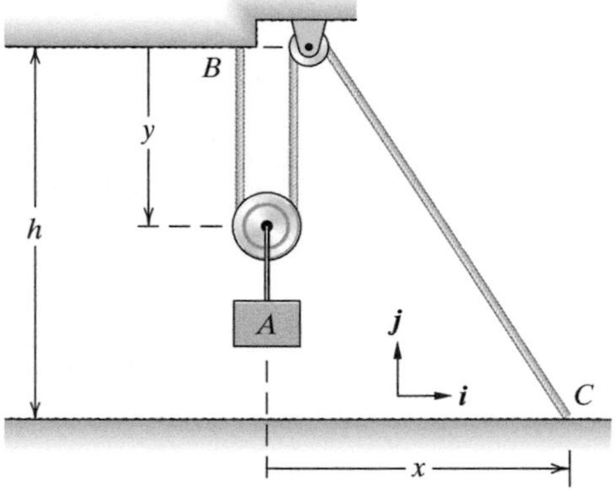

E2.5.39

2.5.40. **[Level 3]** Two cars are traveling as illustrated, with $\boldsymbol{v}_A = v_A\boldsymbol{i}$ (v_A a constant and car starting from O) and car B following a path given by

$$\boldsymbol{r}_{B/O} = bt^2\boldsymbol{e}_r, \quad \theta = at^2$$

a. How long does it take for car B to reach a position of $\frac{\pi}{2}$ rad?

b. What are $\boldsymbol{r}_{B/A}$, $\boldsymbol{v}_{B/A}$, and $\boldsymbol{a}_{B/A}$ at that time?

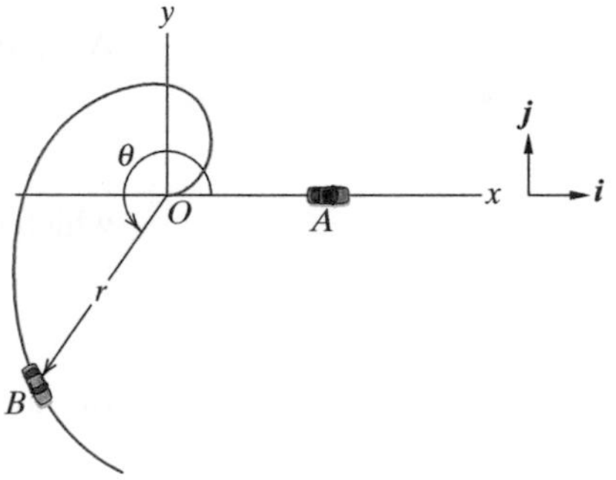

E2.5.40

2.6 JUST THE FACTS

This chapter dealt with **kinematics**—the study of where a body is and how it moves. We got comfortable with the idea of degrees of freedom and saw that we can describe motion with respect to moving reference frames as well as with respect to a fixed inertial frame. Finally, we got some experience with constraints and how to deal with them.

We started off by seeing how to represent the average speed of a body:

$$v_{\text{avg}} = \frac{\Delta x}{\Delta t} \tag{2.1}$$

and then defined **instantaneous velocity**:

$$v_P(t_1) = \lim_{\Delta t \to 0} \frac{x(t_2) - x(t_1)}{t_2 - t_1} \tag{2.2}$$

and **acceleration**:

$$a_P(t_1) = \lim_{\Delta t \to 0} \frac{v_P(t_1 + \Delta t) - v_P(t_1)}{\Delta t} \tag{2.3}$$

The inverse procedure, integration, is

$$\dot{x}(t_2) = \dot{x}(t_1) + \int_{t_1}^{t_2} \ddot{x}\, dt \tag{2.4}$$

and

$$x(t_2) = x(t_1) + \int_{t_1}^{t_2} \dot{x}\, dt \tag{2.5}$$

Allowing the acceleration to be constant gave us the simpler expressions

$$v(t) = v_0 + \bar{a}t \tag{2.6}$$

$$x(t) = x_0 + v_0 t + \frac{\bar{a}t^2}{2} \tag{2.7}$$

A useful general relationship among position, speed, and acceleration is

$$v\, dv = a\, dx \tag{2.8}$$

which helped us find the speed change for a constant acceleration:

$$v_2^2 = v_1^2 + 2\bar{a}(x_2 - x_1) \tag{2.9}$$

the speed change for a position-dependent acceleration:

$$v_2^2 = v_1^2 + 2\int_{x_1}^{x_2} a(x)\, dx \tag{2.10}$$

and the position change for a velocity-dependent acceleration:

$$x_2 = x_1 + \int_{v_1}^{v_2} \frac{v}{a(v)}\, dv \tag{2.11}$$

Adding another direction brought us to two-dimensional **Cartesian coordinates** and an expression for position that utilizes a set of **orthogonal unit vectors**:

$$\boldsymbol{r}_{P/O} = x\boldsymbol{i} + y\boldsymbol{j} \tag{2.19}$$

Once we were in a two-dimensional world, it became useful to have a way to translate between differently oriented sets of coordinates, which brought us to the concept of a **coordinate transformation array**:

	$\boldsymbol{i}$	$\boldsymbol{j}$
$\boldsymbol{b}_1$	$\cos\theta$	$\sin\theta$
$\boldsymbol{b}_2$	$-\sin\theta$	$\cos\theta$

(2.22)

Differentiating gave us the equations for velocity and acceleration in terms of Cartesian unit vectors:

$$\boldsymbol{v}_P = \dot{x}\boldsymbol{i} + \dot{y}\boldsymbol{j} \tag{2.24}$$

$$\boldsymbol{a}_P = \ddot{x}\boldsymbol{i} + \ddot{y}\boldsymbol{j} \tag{2.25}$$

Our second set of coordinates, **polar coordinates**, let us express our motion vectors in terms of distance from an origin and angular orientation and utilized a **radial unit vector** $\boldsymbol{e}_r$ and an **angular unit vector** $\boldsymbol{e}_\theta$. The position, velocity, and acceleration are given by

$$\boldsymbol{r}_{P/O} = r\boldsymbol{e}_r \tag{2.41}$$

$$\boldsymbol{v}_P = \dot{r}\boldsymbol{e}_r + r\dot{\theta}\boldsymbol{e}_\theta \tag{2.46}$$

$$\boldsymbol{a}_P = (\ddot{r} - r\dot{\theta}^2)\boldsymbol{e}_r + (2\dot{r}\dot{\theta} + r\ddot{\theta})\boldsymbol{e}_\theta \tag{2.50}$$

Keep in mind that the $2\dot{r}\dot{\theta}$ term, the **Coriolis acceleration**, comes into play only when a body is both moving away from the origin ($\dot{r}$) and moving around it at the same time ($\dot{\theta}$).

The extension of polar coordinates to **cylindrical coordinates** simply required the addition of an "out-of-plane" axis Z and a unit vector $\boldsymbol{k}$ to point the way to position, velocity, and acceleration:

$$\boldsymbol{R}_{P/O} = r\boldsymbol{e}_r + z\boldsymbol{k} \tag{2.51}$$

$$\boldsymbol{v}_P = \dot{r}\boldsymbol{e}_r + r\dot{\theta}\boldsymbol{e}_\theta + \dot{z}\boldsymbol{k} \tag{2.52}$$

$$\boldsymbol{a}_P = (\ddot{r} - r\dot{\theta}^2)\boldsymbol{e}_r + (r\ddot{\theta} + 2\dot{r}\dot{\theta})\boldsymbol{e}_\theta + \ddot{z}\boldsymbol{k} \tag{2.53}$$

The final set of coordinates we looked at was **path coordinates**, sometimes referred to as normal/tangential coordinates. They involved the **tangent unit vector** $\boldsymbol{e}_t$ and the **normal unit vector** $\boldsymbol{e}_n$. Velocity and acceleration in terms of these unit vectors are given by

$$\boldsymbol{v}_P = v\boldsymbol{e}_t \tag{2.58}$$

$$\boldsymbol{a}_P = \dot{v}\boldsymbol{e}_t + \frac{v^2}{r_C}\boldsymbol{e}_n \tag{2.59}$$

A nice closed-form solution for the radius of curvature was also given, though not derived:

$$r_C = \frac{\left[1 + \left(\frac{dy}{dx}\right)^2\right]^{\frac{3}{2}}}{\left|\frac{d^2y}{dx^2}\right|} \tag{2.60}$$

After finishing with coordinates, we moved on to consider **relative motion**, deriving expressions for position, velocity, and acceleration:

$$\boldsymbol{r}_{A/O} = \boldsymbol{r}_{B/O} + \boldsymbol{r}_{A/B} \tag{2.64}$$

$$\boldsymbol{v}_A = \boldsymbol{v}_B + \boldsymbol{v}_{A/B} \tag{2.65}$$

$$\boldsymbol{a}_A = \boldsymbol{a}_B + \boldsymbol{a}_{A/B} \tag{2.66}$$

SYSTEM ANALYSIS (SA) EXERCISES

SA2.1 Kinematics of Variable Geometry Pulleys

All the analysis we have done has involved pulleys for which the ropes are always oriented in a purely vertical or horizontal fashion. Often, however, pulley ropes are oriented at an angle, and this angle can change as the ropes are pulled. **Figure SA2.1.1** shows two pulley arrangements. In case (*a*) the pulley ropes are always vertical, whereas in (*b*) two of the rope segments are vertical and two are angled. Neglect the dimensions of the pulleys and consider only the effect of the rope orientation. The ropes for both cases are taken up by a motorized reel R at a winding rate of 1 m/s.

a. Plot the speed at which block A is raised as a function of time for $0 < t < 3.5$ s for both cases on the same graph. Initially, $a = 1$ m and $b = 5$ m. Assume that block A remains horizontal at all times.

b. Assume that the dimension a is fixed at 1 m but b is a design variable that you can choose. How would you choose b so that b is minimized and the variation of block A's speed is within ±10% of its starting value over a total distance raised of 2 m?

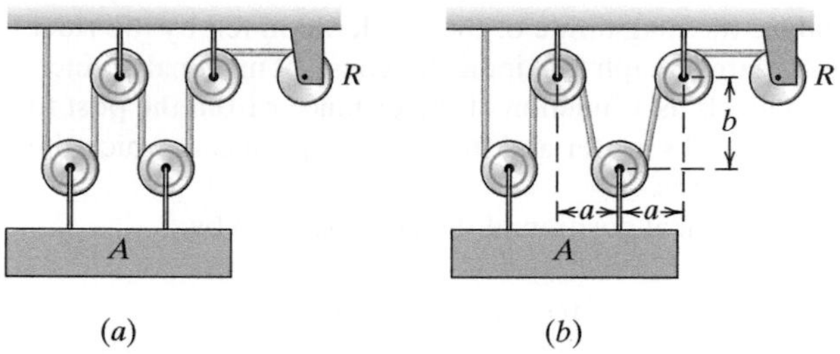

Figure SA2.1.1 Effect of finite dimensions on a pulley system

SA2.2 Multi-Axis Seat Ejection (MASE) Sled

The 846th Test Squadron at Holloman Air Force Base has operated a high-speed test track for over 50 years. The Multi-Axis Seat Ejection sled (**Figure SA2.2.1**) is used on the track to test life-support equipment and ejection seats. The track is nearly 16 km long and has also been used for supersonic testing.

Modern ejection seats are relatively complex systems. They employ telescoping stabilization booms after ejection, activate leg and arm restraints, and deploy a deflector to protect the ejectee from windblast. Thrusters are also used to help stabilize the seat around the pitch axis.

You have been given the task of analyzing a hypothetical ejection taking place under the following conditions:

a. The seat is moving down the track at a speed of 600 KEAS (knots equivalent airspeed: 1 KEAS = 1.85 km/h).

b. The ejection begins at a pitch angle of 20 degrees.

The ejection sequence begins with the catapult phase, which results in an acceleration along the long axis of the seat of $a - bt^2$, where $a = 16g$ and $b = 9g/\text{s}^2$. In addition, the seat accelerates downward at 9.81 m/s² due to gravity. Assume that the ejection rocket fires for 800 ms.

A drogue chute is then released to slow the chair and to orient it with respect to the horizontal airstream. Model this event as occurring immediately after the rocket has ceased firing.

The drogue chute and air drag create a drag force that causes the chair to decelerate with magnitude cv_x^2, where v_x is in meters per second and $c = 9.8 \times 10^{-3}\ \text{m}^{-1}$, until the chair reaches a speed of 100 KEAS. Assume that this slowing occurs solely in the x direction. Once the speed has decreased to 100 KEAS, the main parachute is deployed.

What are the position and the velocity of the seat after rocket burnout? How far has the seat traveled when it reaches 100 KEAS? Plot the x and y coordinates of the seat/chair center of mass as a function of time and let the initial point of ejection be the origin of your coordinate system.

Let c, the drag coefficient of the chute, vary from 0.001 to 0.009. Plot the total distance traveled as a function of c. How does the total distance traveled vary with rocket-fire duration?

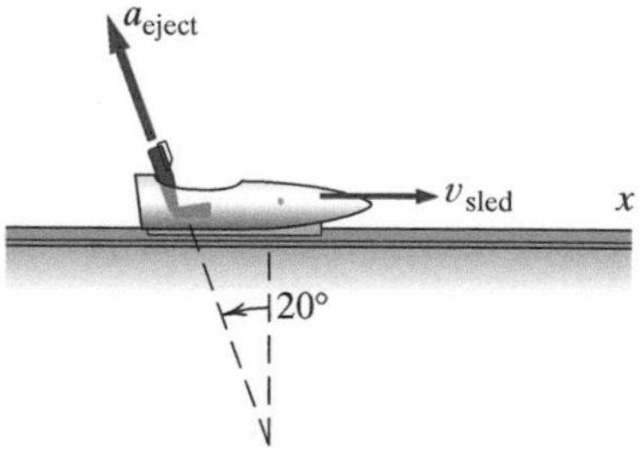

Figure SA2.2.1 Schematic of ejection seat test device

SA2.3 Carousel Ride

Each horse on a carousel (**Figure SA2.3.1**) follows a complex path. The pole to which the horse is attached circles the carousel's center while the horse oscillates up and down along the pole. The ride isn't meant to be overly exciting, and thus the accelerations need to be limited. Your task is to analyze the magnitude of the acceleration felt by the rider and create a graph that indicates constant maximal acceleration levels as a function of the distance r from the post to the carousel's center and the angular speed ω at which the carousel turns.

The vertical position of the horse is given by

$$z(t) = z_0 + z_1 \cos(\omega_h t)$$

where ω_n is the frequency at which the horse oscillates in the z direction. Assume that $\omega_h = 2\omega$. The coefficient $z_1 = 0.39$ m, r can vary from 2.4 to 4.8 m, and the time for the carousel to complete one revolution varies between 3 and 5 s.

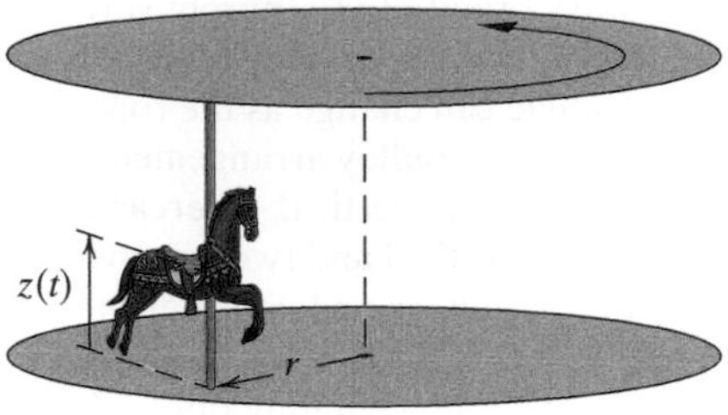

Figure SA2.3.1 Carousel horse

SA2.4 Carnival-Style Golf Game

A progressive golf course has added a twist to its offerings: a carnival-type game (**Figure SA2.4.1**) in which the golfer has to hit a ball through an opening that's a good distance away. The catch is that a door immediately starts to close over the opening as soon as the ball is struck. Assume that the golfer can start the ball off at $v = 40$ m/s. The height of the opening is initially L_2, and the golfer is a distance L_4 from the wall. The door begins to close the opening (moving from bottom to top) at the instant he strikes the ball and moves at a speed v_B. What you want to know is what launch angles θ will allow the ball to pass through the part of the opening that isn't yet covered. $L_1 = 14$ m, $L_2 = 25$ m, $L_3 = 2$ m, $L_4 = 100$ m, and $v_B = 6$ m/s.

a. Construct a coordinate system with its origin at the golfer's feet.

b. Calculate how long it takes for the ball to reach the wall.

c. Calculate the height of the ball when it reaches the wall.

d. Calculate the height of the door when the ball reaches the wall.

e. Write a computer program that plots the distance between the ball and the top of the door, the height of the ball, and the height of the door as a function of θ. From this plot, you should be able to determine the range of launch angles for which the ball will arrive at the wall below the top of the opening (A) and above the door (B).

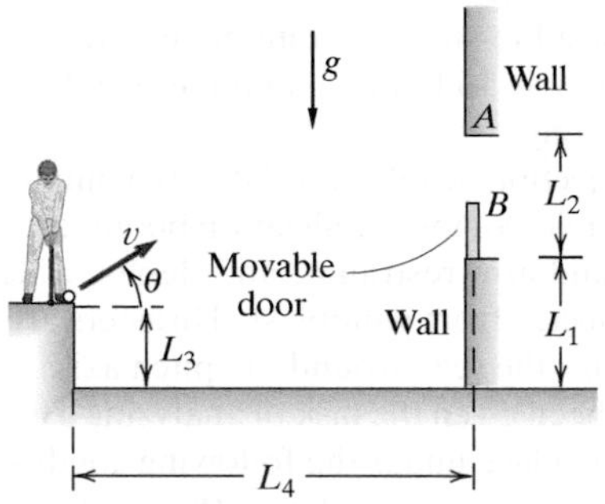

Figure SA2.4.1 Golf game

OBJECTIVES

Upon completion of this chapter you will be able to:

- Apply Newton's Laws of Motion to single-particle problems
- Recognize when a momentum approach is useful
- Apply this knowledge to some planar orbital mechanics problems (satellite orbiting the earth, comet passing by the sun, and so on) as well as to both normal and oblique impact problems

CHAPTER 3

Inertial Response of Translating Bodies

$\boldsymbol{F} = m\boldsymbol{a}$—that's the big news about this chapter. You've seen it in high school and you've seen it in freshman physics. So, I'm sure it will be somewhat familiar to you at this point. What's different now is that we're really going to delve into what's behind this equation. We will see how it can be used to solve a wide variety of useful problems—the motion of the earth around the sun for one. That's easily described with the material we'll be going over in this chapter. We'll go over some basic aspects of how cars move and learn exactly why it's harder to drive up a hill than down one.

But we won't stop with just $\boldsymbol{F} = m\boldsymbol{a}$. We'll integrate this equation with respect to time so that we are looking at momentum rather than at forces. Momentum is a physical quantity that's probably more familiar than force. When you're running fast and want to change direction quickly, it's your momentum that you need to alter. Doing so requires the application of a force over a finite time interval, something that we will look at in some detail. And when the time intervals get very small, we will have reached the subject of impact. Impact problems are fun to work with. Basically, whenever two objects collide with each other, an interchange of momentum takes place, and we will see how a couple of simple laws will let us predict what happens after the impact. This covers everything from the response of billiard balls to the aftermath of a multicar collision.

3.1 CARTESIAN COORDINATES

It's finally time to do some real dynamics. Newton's second law (for a constant mass) tells us that

$$\sum_{i=1}^{n} \boldsymbol{F}_i = m\boldsymbol{a} \tag{3.1}$$

This can be called the **force balance** law. It expresses the fact that there's a relationship between the **total applied forces** on a body (the left-hand side of the equation) and the body's **inertial response** (the right-hand side of the equation). To solve a problem of this sort you'll draw a free-body diagram and an inertial-response diagram and then equate them to produce an overall "free-body diagram equals inertial-response diagram" plot and corresponding equations. Both aspects of the solution are important and we'll be addressing both in just a second.

The correct notation to use is the one just shown, in which explicit mention is made that many individual forces may be acting on the mass m. That being understood, when I'm just talking about the equation in a general way, I'll drop the summation and simply write $\boldsymbol{F}$ for the force side of the equation. In this case, $\boldsymbol{F}$ is understood to be the net result of all the individual forces acting on the mass.

Note that we need to apply this equation in an **inertial reference frame**, one which doesn't introduce accelerations due to its own motion. A Cartesian frame that's fixed to the ground is an inertial reference frame and a Cartesian frame that's translating at a fixed speed is also an inertial reference frame. However, if the frame is rotating or accelerating, then it's not an inertial frame.

We weren't in a position to do much with this expression until we had pinned down what acceleration is and how to specify it. Having done that work in Chapter 2, we can now use (3.1) in different ways. We might be given the force and asked to determine the acceleration:

$$\boldsymbol{a} = \frac{\boldsymbol{F}}{m} \tag{3.2}$$

Or we might be given the acceleration and asked to evaluate the force needed to produce that acceleration:

$$\boldsymbol{F} = m\boldsymbol{a} \tag{3.3}$$

Or, more interesting perhaps, we might be told some quantities of interest (like $\dot{r}$ and F_r) and be asked to determine others (perhaps $\ddot{\theta}$). We'll spend the next few sections doing just this—playing with single-particle force/acceleration relationships and seeing what we can do with them.

Using Cartesian coordinates gives us the most straightforward situation. We've seen that the acceleration terms are decoupled from each other, meaning that motion in the x and y directions can be viewed as individual rectilinear problems. Applying (3.1) to a Cartesian coordinate system gives

$$F = m(\ddot{x}\boldsymbol{i} + \ddot{y}\boldsymbol{j}) \tag{3.4}$$

If we break up the applied force into two components, one in the $\boldsymbol{i}$ direction and one in the $\boldsymbol{j}$ direction ($\boldsymbol{F} = F_x\boldsymbol{i} + F_y\boldsymbol{j}$), we have

$\boldsymbol{i}$:
$$F_x = m\ddot{x} \tag{3.5}$$

$\boldsymbol{j}$:
$$F_y = m\ddot{y} \tag{3.6}$$

It's no trouble to extend this coordinate system to three-dimensional motion—just add a unit vector $\boldsymbol{k}$ as shown in **Figure 3.1**. This vector $\boldsymbol{k}$ adheres to a right-hand rule: when you point the fingers of your right hand in the $\boldsymbol{i}$ direction and then curl your fingers toward $\boldsymbol{j}$, your thumb points in the direction of $\boldsymbol{k}$. In terms of an earth-based reference system, if $\boldsymbol{i}$ points east and $\boldsymbol{j}$ points north, then $\boldsymbol{k}$ points up, away from the earth. **Figure 3.2** shows the same plot but with the individual vectors decomposed into the $\boldsymbol{i}, \boldsymbol{j}, \boldsymbol{k}$ directions. Our equations of motion for the three-dimensional case are

$\boldsymbol{i}$:
$$F_x = m\ddot{x} \tag{3.7}$$

$\boldsymbol{j}$:
$$F_y = m\ddot{y} \tag{3.8}$$

$\boldsymbol{k}$:
$$F_z = m\ddot{z} \tag{3.9}$$

>>> **Check out Example 3.1 (page 102), Example 3.2 (page 103), Example 3.3 (page 104), Example 3.4 (page 106), and Example 3.5 (page 107) for applications of this material.**

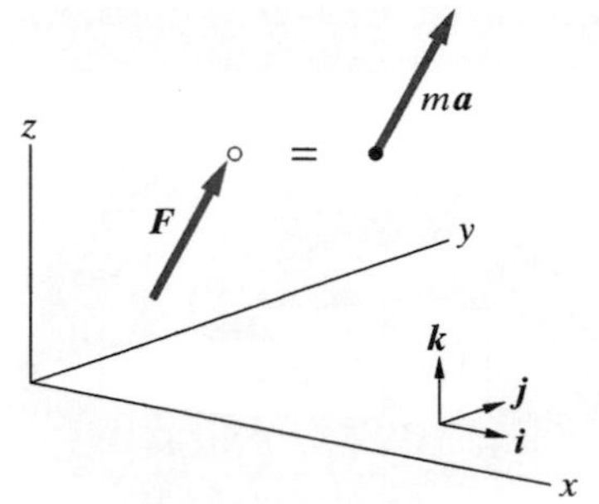

Figure 3.1 Three-dimensional coordinates: FBD=IRD—vectors

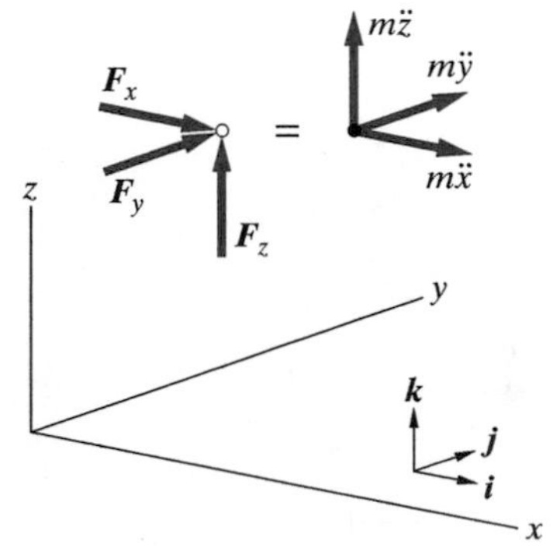

Figure 3.2 Three-dimensional coordinates: FBD=IRD—vector components

EXAMPLE 3.1 ANALYSIS OF A SPACESHIP (Theory on page 101)

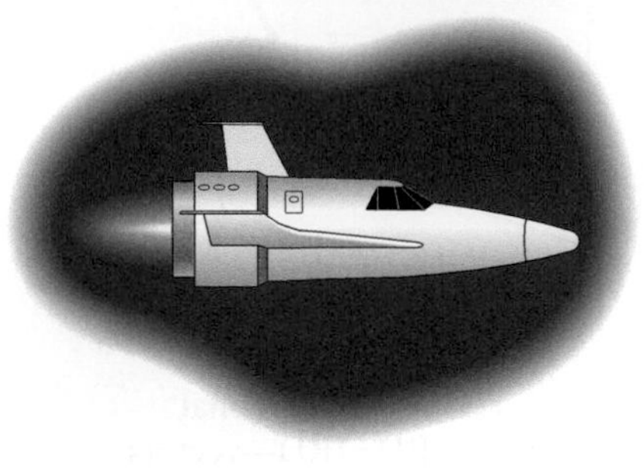

Figure 3.3 Advanced spacecraft

Consider a spacecraft moving through the far reaches of uncharted space, as shown in **Figure 3.3**. The main engine is firing, causing the craft to accelerate in a straight line, and, unlike in *Star Trek*, the silence of space is serenely unbroken because sound can't travel in a vacuum. (Just so you've got the right mental image.) The craft is moving along the $\boldsymbol{i}$ direction, conveniently defined as the direction from the earth to Alpha Proximi. The acceleration magnitude recorded by the ship's navicomputer is 0.2 m/s^2. The mass of the spacecraft is 1×10^6 kg. Determine the magnitude of the force being exerted by the engine on the spacecraft—this is called the engine's **thrust**. Neglect any change in mass due to fuel being burned.

Goal Find the magnitude T of the engine's thrust.

Given Mass and acceleration of the spaceship.

Assume No additional assumptions are needed.

Draw The simplified system is shown in **Figure 3.4**. Hard to get simpler, isn't it? Because the spacecraft is completely unrestrained, the only thing we have to worry about is the thrust—we need not deal with any other force. For convenience, the $\boldsymbol{i}$ unit vector is drawn aligned with the force vector, and $\ddot{x}$ indicates the acceleration along the spaceship's trajectory. The appropriate free-body diagram (FBD) is shown on the left in **Figure 3.4** with its corresponding inertial-response diagram (IRD) on the right. Note that the mass is drawn as an outline on the FBD and filled in on the IRD. This is done to emphasize that in the FBD it's the forces that matter, whereas in the IRD the mass is a central part of the $\boldsymbol{ma}$ term.

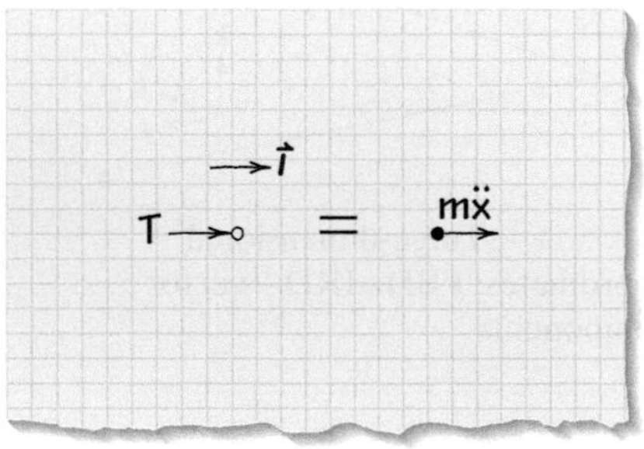

Figure 3.4 FBD=IRD for spacecraft

Formulate Equations Equation (3.5) tells us that

$$T = m\ddot{x}$$

Solve There's not much to do here except plug in the given values:

$$T = m\ddot{x} = (1.0 \times 10^6\,\text{kg})(0.2\,\text{m/s}^2) = 2.0 \times 10^5\,\text{N}$$

To put this result into perspective, 2.0×10^5 N is roughly the force a stack of 140 BMW 325s would exert on the ground.

Check The units in our result work out correctly. Beyond that there's not much to check because there wasn't much going on in the problem.

EXAMPLE 3.2 **FORCES ACTING ON AN AIRPLANE** (Theory on page 101)

Coming closer to home than the vast uncharted reaches of space, you are now aboard a jetliner winging its way over the Atlantic, taking you to a long-deserved holiday in Paris (**Figure 3.5**). For reasons best known to himself, the pilot announces over the intercom that the plane's acceleration vector is currently $(1.0\boldsymbol{i} + 1.1\boldsymbol{j} + 0.050\boldsymbol{k})\ \text{m/s}^2$, where $\boldsymbol{i}$ and $\boldsymbol{j}$ indicate east and north, respectively, and $\boldsymbol{k}$ points up, away from the earth's surface. Knowing that the mass of the jetliner is 1.5×10^5 kg, calculate the individual components of the force acting on it.

Figure 3.5 Airliner

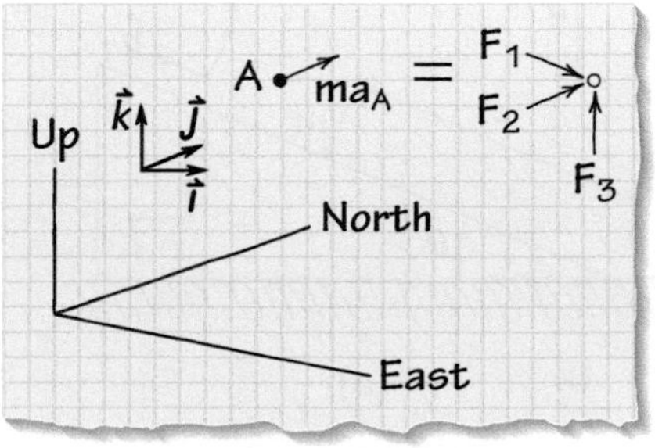

Figure 3.6 FBD=IRD for airliner

Goal Find the force components acting on the plane.

Given Mass and acceleration of the airplane.

Assume No additional assumptions are needed.

Draw The simplified system is shown in **Figure 3.6**, with the jetliner shown as the mass particle A with mass m. The given acceleration vector is illustrated, along with the $\boldsymbol{i}, \boldsymbol{j}, \boldsymbol{k}$ unit vectors. The jetliner has acceleration components in three orthogonal directions, and, as shown in **Figure 3.6**, we expect to have force components along these directions as well.

Formulate Equations This problem, like the preceding one, simply requires an application of $\boldsymbol{F} = m\boldsymbol{a}_A$, although in this case we have to break the acceleration and forces into components.

Using (3.1) gives us

$$F_1\boldsymbol{i} + F_2\boldsymbol{j} + F_3\boldsymbol{k} = (1.5 \times 10^5\ \text{kg})[(1.0\boldsymbol{i} + 1.1\boldsymbol{j} + 0.050\boldsymbol{k})\ \text{m/s}^2]$$

Solve

$$\begin{aligned} F_1\boldsymbol{i} + F_2\boldsymbol{j} + F_3\boldsymbol{k} &= (1.5 \times 10^5\ \text{kg})[(1.0\boldsymbol{i} + 1.1\boldsymbol{j} + 0.050\boldsymbol{k})\ \text{m/s}^2] \\ &= (1.5 \times 10^5\boldsymbol{i} + 1.7 \times 10^5\boldsymbol{j} + 7.5 \times 10^3\boldsymbol{k})\ \text{N} \end{aligned}$$

Matching coefficients tells us that

$\boldsymbol{i}$: $\quad F_1 = 1.5 \times 10^5\ \text{N}$

$\boldsymbol{j}$: $\quad F_2 = 1.7 \times 10^5\ \text{N}$

$\boldsymbol{k}$: $\quad F_3 = 7.5 \times 10^3\ \text{N}$

Check There isn't much to check here beyond making sure the units make sense, which they do.

EXAMPLE 3.3 SLIDING MING BOWL (Theory on page 101)

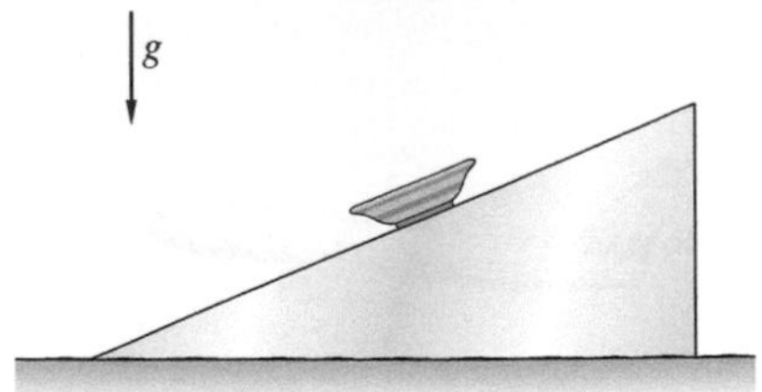

Figure 3.7 Bowl on a slope

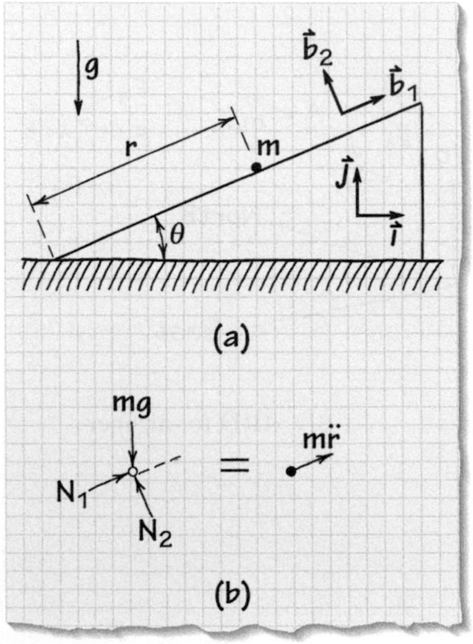

Figure 3.8 Schematic and FBD=IRD for bowl on slope

A precious Ming bowl of mass m and price \$49,999.95, initially held in place on a frictionless slope (**Figure 3.7**), is suddenly released. Treating the vase as a mass particle, find its acceleration at the time of release.

Goal Find the bowl's acceleration.

Given Physical description of sloped surface.

Assume The bowl remains in contact with the sloped surface.

Draw **Figure 3.8*a*** shows a schematic of our system. The bowl is represented by a mass particle m. Two sets of unit vectors are shown: $\boldsymbol{i}, \boldsymbol{j}$ for right-left/up-down motion and $\boldsymbol{b}_1, \boldsymbol{b}_2$ for motion along the slope and orthogonal to it. The coordinate r is introduced to track motion along the slope.

Our FBD and IRD are shown in **Figure 3.8*b***. The gravity force (mg) is oriented vertically, but the two reaction forces created by the bowl/slope interface (N_1 and N_2) are aligned with the $\boldsymbol{b}_1$ and $\boldsymbol{b}_2$ directions. (Recall from Statics that when you break a system into a free body you will have, in general, three orthogonal forces at the connection point or, in the planar case, two.)

Formulate Equations Because the motion will be along the slope (in the $\boldsymbol{b}_1$ direction), it makes sense to express our equation of motion in the $\boldsymbol{b}_1, \boldsymbol{b}_2$ coordinate set. Applying (3.1) gives us

$$m\ddot{r}\boldsymbol{b}_1 = -mg\boldsymbol{j} + N_1\boldsymbol{b}_1 + N_2\boldsymbol{b}_2$$

To re-express the gravity term, we use a coordinate transformation array

	$\boldsymbol{i}$	$\boldsymbol{j}$
$\boldsymbol{b}_1$	$\cos\theta$	$\sin\theta$
$\boldsymbol{b}_2$	$-\sin\theta$	$\cos\theta$

to get

$$m\ddot{r}\boldsymbol{b}_1 = -mg\left[(\sin\theta)\boldsymbol{b}_1 + (\cos\theta)\boldsymbol{b}_2\right] + N_1\boldsymbol{b}_1 + N_2\boldsymbol{b}_2 \tag{3.10}$$

Note that a $\boldsymbol{b}_2$ term is not included in the acceleration because our assumption is that the bowl always remains in contact with the slope. If there were any possibility it might leave the slope, then I would have to include this term.

Combining terms and writing out the two scalar equations arising from (3.10) give us

$\boldsymbol{b}_1$: $$m\ddot{r} = N_1 - mg\sin\theta$$

$\boldsymbol{b}_2$: $$0 = N_2 - mg\cos\theta$$

Solve I've already mentioned one constraint—the bowl remains in contact with the slope. This means no motion in the $\boldsymbol{b}_2$ direction; consequently, our second equation of motion has zero on the left-hand side. Unfortunately, we have two equations with three unknowns: $\ddot{r}$, N_1, and N_2. That's not a good situation. I'm sure you remember that you're not going to get very far trying to solve for three unknowns with only two equations. The conclusion must therefore be that either (1) we need a third equation or (2) one of the "unknowns" is, in fact, known.

The correct choice is the second one—one of our labeled unknowns isn't really unknown. Two reaction forces were included (for full generality)—N_1 and N_2—but we were told that the bowl/slope interface is frictionless. This means that N_1 has to be zero, thus giving us our final equations:

$$\begin{aligned} m\ddot{r} &= -mg\sin\theta \\ 0 &= N_2 - mg\cos\theta \end{aligned}$$

Our first equation gives us the answer we were after, that the acceleration of the bowl is

$$\ddot{r} = -g\sin\theta$$

where the minus sign tells us that the bowl accelerates downslope. This makes sense, for it would be difficult to imagine how gravity could cause the bowl to accelerate upslope.

The second equation gives us something we were not asked to find—the normal force exerted by the slope on the vase:

$$N_2 = mg\cos\theta$$

Check We can do a few checks to assure ourselves that the solution is correct. First, we can check units in our two equations. Our acceleration $\ddot{r}$ is proportional to g, which has the units of acceleration, and the normal force N_2 is proportional to mg, which has the units of force. So the units seem fine. This is a crucial first check—if the units don't work out, there's a real problem.

We can next go beyond a simple unit consistency check to ask ourselves if the answer makes sense. How might it not? Well, we might have gotten an expression for $\ddot{r}$ that said the acceleration was 1,342,782 m/s^2. That wouldn't make a whole lot of sense because gravity is only 9.81 m/s^2.

Our solution, $\ddot{r} = -g\sin\theta$, tells us the acceleration is less than g. That's good—it makes sense that the acceleration is less than what a freely falling object would experience. The direction of the acceleration is downslope; this is yet another good physical check on the solution. We know from everyday experience that objects slide downhill, not up, and that's what we would expect our bowl to do. Finally, we see that the magnitude of the acceleration is proportional to $\sin\theta$. This makes lots of sense. If θ were zero (horizontal slope), the acceleration would be zero, just as intuition suggests. If the slope were completely vertical, the bowl would be in free-fall and we would expect an acceleration equal to g, which is exactly what we get: $g\sin 90° = g$.

The same sort of logic supports our solution for the normal force. If the slope were horizontal, we would expect to get $N_2 = mg$, which is exactly what our equation tells us ($mg\cos 0° = mg$). If the slope were 90°, we would expect $N_2 = 0$ because the bowl would be falling straight down and not interacting with the slope. This result is also given by our equation: $N_2 = mg\cos 90° = 0$.

EXAMPLE 3.4 RESPONSE OF AN UNDERWATER PROBE (Theory on page 101)

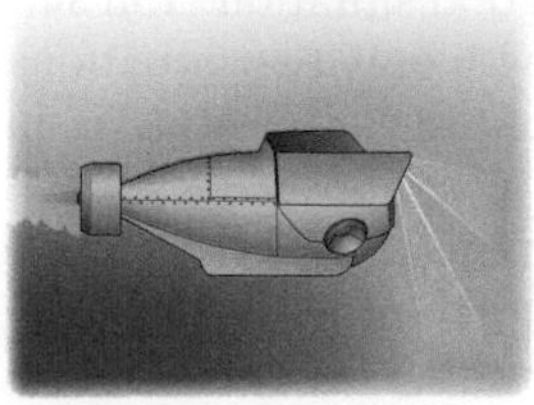

Figure 3.9 Underwater probe

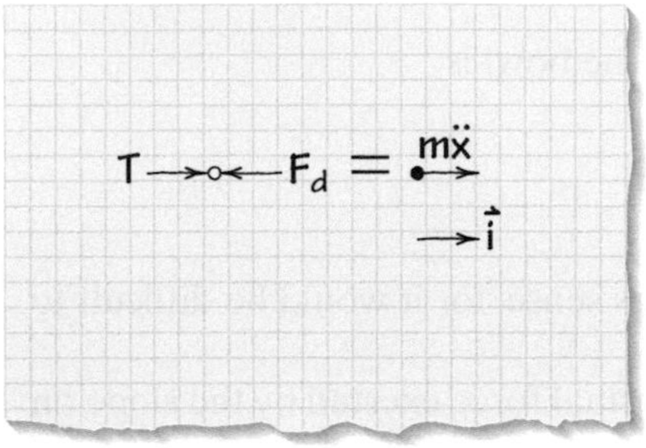

Figure 3.10 FBD=IRD for the probe

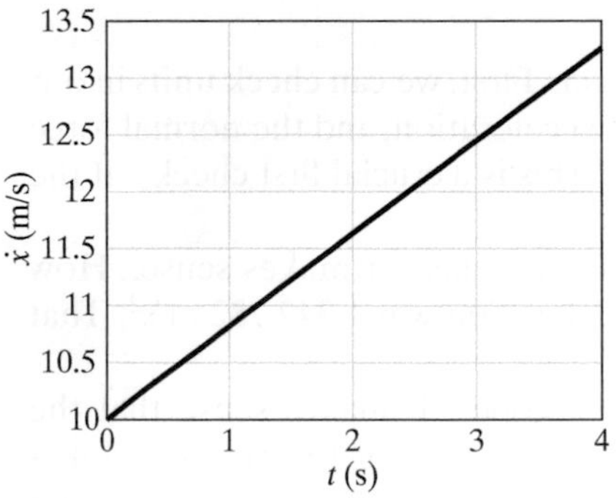

Figure 3.11 Speed versus time

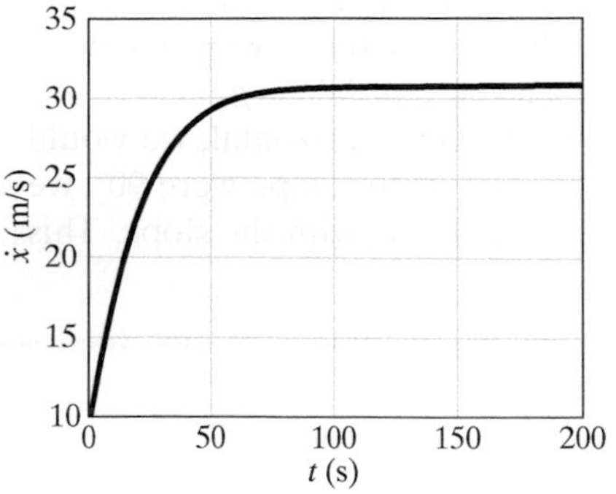

Figure 3.12 Speed versus time

Now let's try a little numerical integration. **Figure 3.9** shows an underwater probe traveling in a straight line. The rear thruster provides a propulsive force T which, without anything else acting on the probe, would result in a speed that goes to infinity. Something else acts on it, however, namely, a drag force $F_d = a + bv^2$, where v is the probe's speed, $a = 500$ N, and $b = 10$ kg/m. The thrust T is equal to a constant 10,000 N, the probe has a mass of 10,000 kg, and it is initially moving to the right (positive x direction) at 10 m/s. How fast will it be moving 4 s later?

Goal Find the probe's speed at $t = 4$ s.

Given Mass of the probe and form of forces acting on it.

Assume No additional assumptions are needed.

Draw The free-body diagram and the associated inertial-response diagram are shown in **Figure 3.10**.

Formulate Equations Applying (3.5) gives us

$$m\ddot{x} = T - F_d \tag{3.11}$$

Solve Dividing by m and using the given drag expression yield

$$\ddot{x} = \frac{1}{m}[T - (a + bv^2)]$$

To determine the change in speed, we'll use the MATLAB© function `ode45`. Check out Appendix A if you'd like to see more details.

The initial conditions are $x(0) = 0$, $\dot{x}(0) = 10$. We'll need to define a column vector `y` such that `y`(1,1) $= x$ and `y`(2,1) $= \dot{x}$. Our initial condition vector `y0` is therefore given by `y0=[0; 10]`. The integration time is 4 s, and thus our integration vector `tspan` is specified by `tspan=[0 4]`.

Typing `[t,y]=ode45('probe',tspan,y0);` will produce the desired position and speed data as the columns of the `y` matrix, after which typing `plot(t,y(:,2));` produces the plot shown in **Figure 3.11** and the result $v(4) = 13.25$ m/s.

Check One way to make sure you have integrated correctly is to look at a limiting case. The drag force grows with speed, and thus we expect that as $t \to \infty$, the craft will stop accelerating because the drag force counterbalances the thrust. We can solve for this by setting $\ddot{x}$ to zero in (3.11):

$$T - F_d = T - a - bv(\infty)^2 = 0 \quad \Rightarrow \quad v(\infty) = \sqrt{\frac{T - a}{b}} = 30.82 \text{ m/s}$$

Letting the integration run for 200 s, we **Figure 3.12**. As you can see, the probe's speed is asymptotically approaching 30.8 m/s, verifying our analytical solution.

EXAMPLE 3.5 PARTICLE IN AN ENCLOSURE (Theory on page 101)

A particle m is suspended in a rigid container by two massless, inextensible strings, $\overline{AB}$ and $\overline{BC}$ (**Figure 3.13**). A force is applied to the container so that it accelerates to the right at $\ddot{x}$. Under this constant acceleration, the tension in string $\overline{BC}$ is four times greater than that in string $\overline{AB}$. Determine both the acceleration $\ddot{x}$ and the tension in $\overline{AB}$.

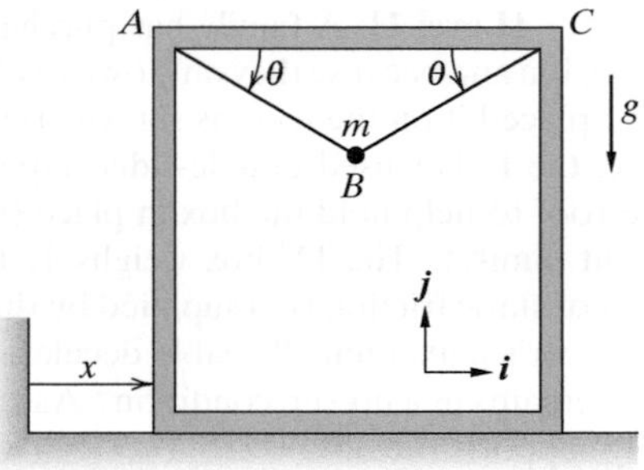

Figure 3.13 Suspended mass in an enclosure

Goal Find the acceleration of the suspended mass and the string tensions.

Given The orientation of the supporting strings, the acceleration of the enclosure, and the relative tension in the strings.

Assume The strings can't stretch and therefore the enclosure's motion is transmitted directly to the mass; knowing $\ddot{x}$ gives us the particle's acceleration.

Draw **Figure 3.14** shows a free-body = inertial-response diagram for the mass.

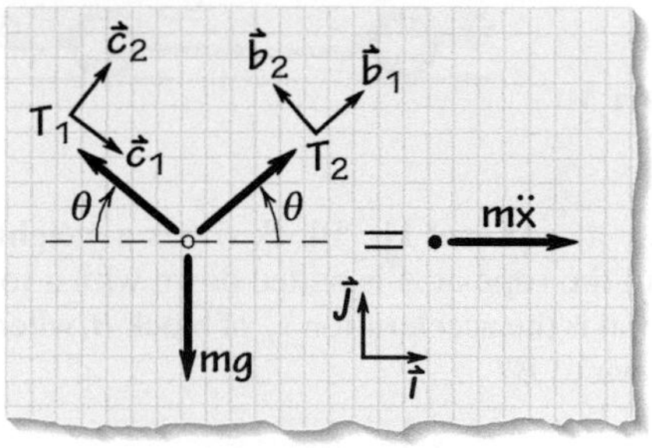

Figure 3.14 FBD=IRD for mass

Formulate Equations We will need to apply (3.1) and realize that for this case we have a sum of three forces: two tension forces and a gravity-induced body force.

$$T_2\boldsymbol{b}_1 - T_1\boldsymbol{c}_1 - mg\boldsymbol{j} = m\ddot{x}\boldsymbol{i} \tag{3.12}$$

To solve this, we will have to create the coordinate transformation arrays that let us express $\boldsymbol{b}_1$ and $\boldsymbol{c}_1$ in terms of $\boldsymbol{i}$ and $\boldsymbol{j}$:

	$\boldsymbol{i}$	$\boldsymbol{j}$
$\boldsymbol{b}_1$	$\cos\theta$	$\sin\theta$
$\boldsymbol{b}_2$	$-\sin\theta$	$\cos\theta$

	$\boldsymbol{i}$	$\boldsymbol{j}$
$\boldsymbol{c}_1$	$\cos\theta$	$-\sin\theta$
$\boldsymbol{c}_2$	$\sin\theta$	$\cos\theta$

Solve Let $T_2 = 4T_1 = 4T$ and decompose (3.12) into the $\boldsymbol{i}$ and $\boldsymbol{j}$ directions:

$$-T\cos\theta + 4T\cos\theta = m\ddot{x} \tag{3.13}$$

$$T\sin\theta + 4T\sin\theta - mg = 0 \tag{3.14}$$

(3.13) (3.14) $\Rightarrow$ $\quad T = \dfrac{mg}{5\sin\theta}, \quad \boxed{\ddot{x} = \dfrac{3g}{5\tan\theta}}$

Check Our final result has the correct units of m/s^2. That's a plus.

Going back and using the derived values of $\ddot{x}$ and T in (3.13) and (3.14) produces $0 = 0$ and $0 = 0$, indicating that, in addition to having the right units, the solutions have the correct magnitudes.

EXERCISES 3.1

3.1.1. **[Level 1]** A family has purchased a new 114 cm television and because they only own a subcompact car, they have placed it on the roof, as shown. Because they had no rope, the father used double-sided tape between the box and roof to help hold the box in place (this isn't an overly bright family). The TV box weighs 41 kg, and the coefficient of static friction (μ_s) supplied by the tape is 0.9. What is the car's maximum allowable deceleration for which the box remains in a no-slip condition? Assume a flat roof that fully supports the box.

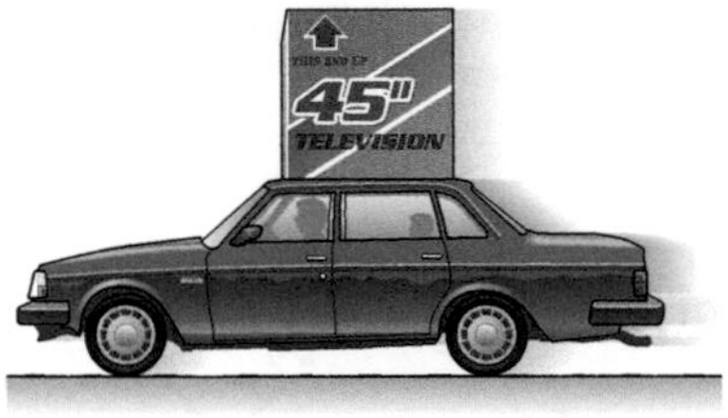

E3.1.1

3.1.2. **[Level 1]** Block B has a weight of $W_B = 18$ kg, and the rope at A is pulled down with a force of $F = 67$ N. What is the acceleration $\ddot{y}_B$ of block B, where y_B is measured down to B?

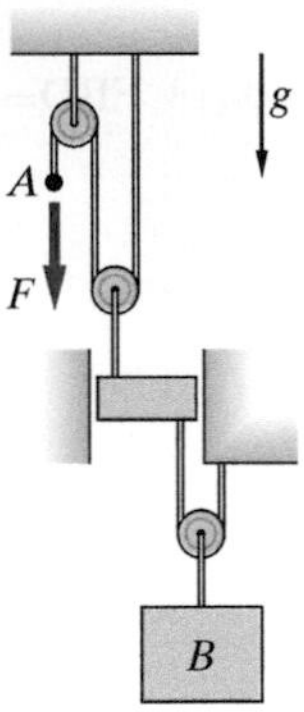

E3.1.2

3.1.3. **[Level 1]** A cyclist is moving along a straight path at a speed of 32 km/h. $\mu_s = \mu_d = 0.50$. Determine the minimum distance h required for her to come to a complete stop, assuming both tires brake and remain in contact with the ground.

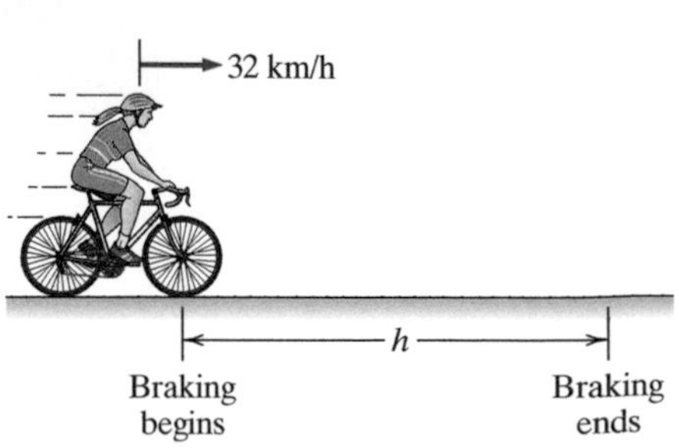

E3.1.3

3.1.4. **[Level 1]** A cyclist is moving down a steep hill that has a grade of 20%. Assuming a steady speed of descent, what must the minimum coefficient of static friction be between the bicycle and the ground? (The coefficient of static friction is the appropriate frictional coefficient to use for bicycles that are rolling without slipping.)

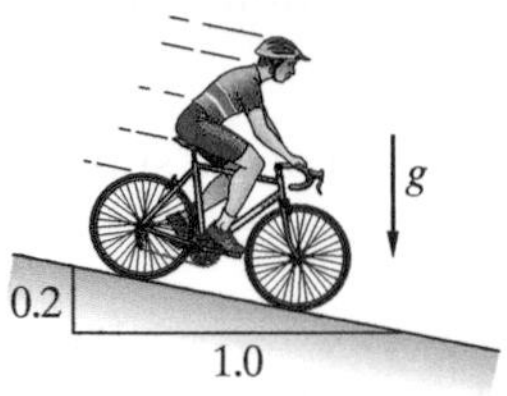

E3.1.4

3.1.5. **[Level 1]** The illustrated framework starts moving from rest and is then given an acceleration of $\ddot{x} = a_1 t$, where $a_1 = 6$ m/s^3. Two strings constrain the motion of a particle of mass m located at B. When will the tension in the strings be such that $T_{\overline{AB}} = T_{\overline{BC}}$?

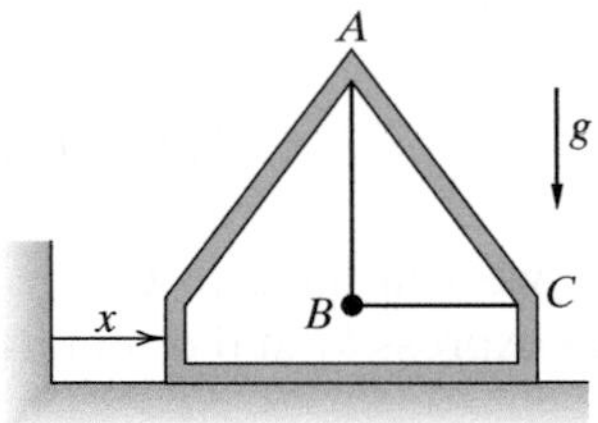

E3.1.5

3.1.6. **[Level 1]** Consider the illustrated block on a truck. The truck starts from rest and accelerates forward at a constant rate. The coefficient of friction between the block and the truck bed is 0.75. What is the maximum acceleration such that the block doesn't slip along the bed? The block has a mass of 100 kg.

E3.1.6

3.1.7. **[Level 1]** A mass is connected via two linear springs to a movable cart. The coefficient of friction is 0.6. $m = 1$ kg and $k = 50$ N/m. The unstretched length of the springs is 0.1 m. What is the steady-state equilibrium position of m if the platform is accelerating at a constant rate of 4 m/s^2?

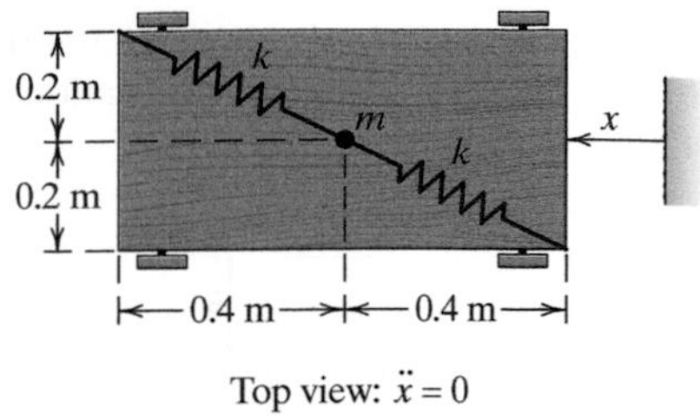

E3.1.7

3.1.8. [Level 1] The simplest model of a car undergoing braking is a lumped mass with a given initial speed that is acted on by a decelerative force due to the brakes. Let the car weigh 1315 kg and go from 96 km/h to zero in 2 s. What is the overall braking force F_b acting on the car? Assume a constant deceleration.

E3.1.8

3.1.9. [Level 1] Your friend always complains that you brake too abruptly, and so you decide to prove him wrong once and for all by making the illustrated device. The mass m is conductive, and when the spring k pushes it against the enclosure, the little bulb is lit. If the enclosure decelerates too quickly, the mass will move away from the wall, breaking the circuit and cutting off the light. $m = 0.03$ kg, $k = 15$ N/m, and the prestress on the mass is initially equal to $F_0 = 0.15$ N. What level of deceleration will cause the circuit to break?

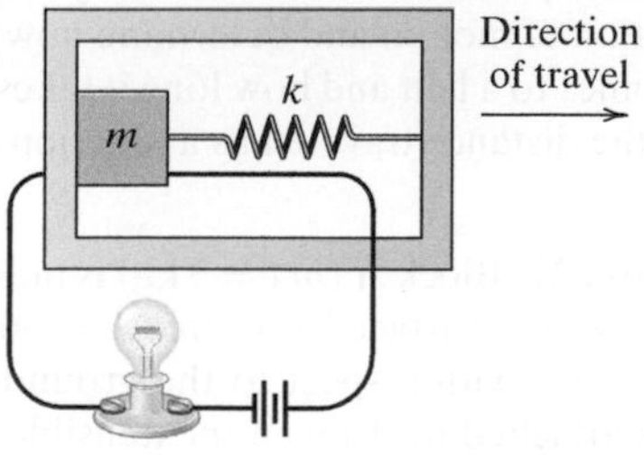

E3.1.9

3.1.10. [Level 1] A downward force of 50 N acts on the free end of the illustrated pulley. $m_A = 10$ kg. What is block A's acceleration? Assume that all straight portions of rope are oriented vertically. The pulleys and ropes are massless.

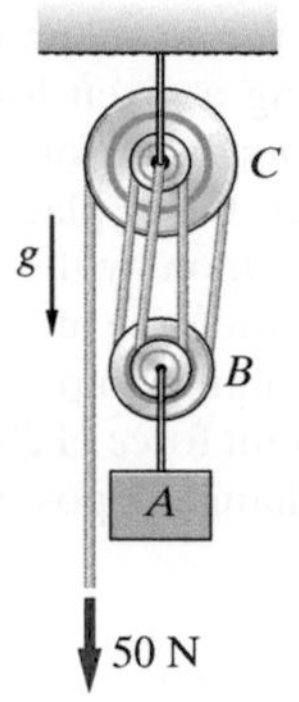

E3.1.10

3.1.11. [Level 1] A fighter jet landing on an aircraft is made to brake very quickly by means of a cable that the plane connects to via a hook. Assume that a 9525 kg jet touches down at 274 km/h and its speed is reduced to zero in 73 m. Assuming a constant deceleration, calculate the force applied to the jet by the braking cable. In order to get a feel for the magnitude of the force, calculate the braking force acting on a 15.6 kN car in order to bring it (at constant deceleration) from 96 km/h to zero in 42.7 m (a realistic number). How many of these "car force" units are needed to obtain the force acting on the jet?

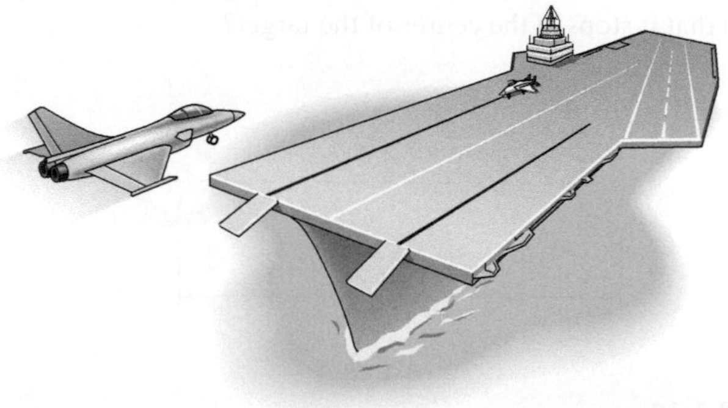

E3.1.11

3.1.12. [Level 1] Illustrated is a conveyor belt inclined at an angle $\theta = 15°$. A 2 kg mass A is placed on the conveyor in the position shown with zero velocity with respect to the ground. The belt is moving at 4.0 m/s. $L_1 = 0.85$ m, $L_2 = 1.05$ m, $\mu = 0.5$. How long will it take for the mass to reach the end of the conveyor belt?

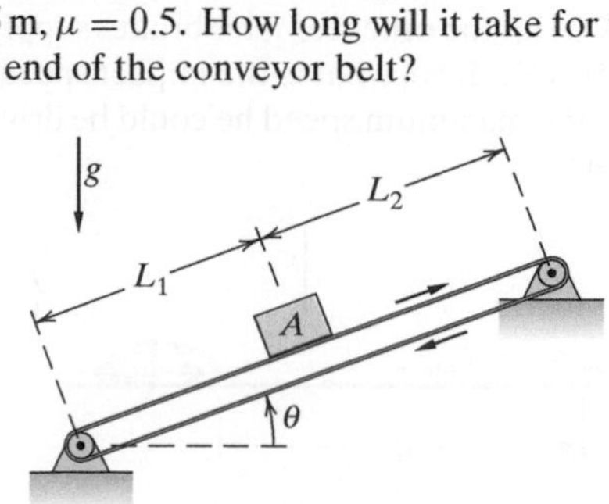

E3.1.12

3.1.13. **[Level 1]** In this problem we'll analyze the difference between applying a given force for a given time versus doubling the applied force for half the time. The key question is how this affects the displacement and final speed of the object being acted upon by the force.

Case 1: Apply a constant force of 100 N to a 100 kg block for 10 s. Determine the change in position and final speed.

Case 2: Apply a constant force of 200 N to a 100 kg block for 5 s. Determine the change in position and final speed.

Compare and discuss.

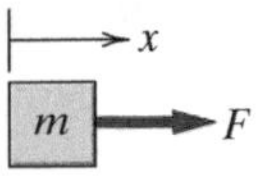

E3.1.13

3.1.14. **[Level 1]** A curling team is competing at the Winter Olympics, and one of the team members launches the curling stone with a speed of $v_0 = 5\,\text{m/s}$ toward the target's center $s_c = 28\,\text{m}$ away. The curling stone has a mass of $m = 18\,\text{kg}$.

a. Assuming that the other players do a good job sweeping and keep the curling stone in a straight-line path, determine where the curling stone comes to a stop if the coefficient of friction between the stone and playing surface is $\mu = 0.05$.

b. How fast does the curling stone need to be launched so that it stops in the center of the target?

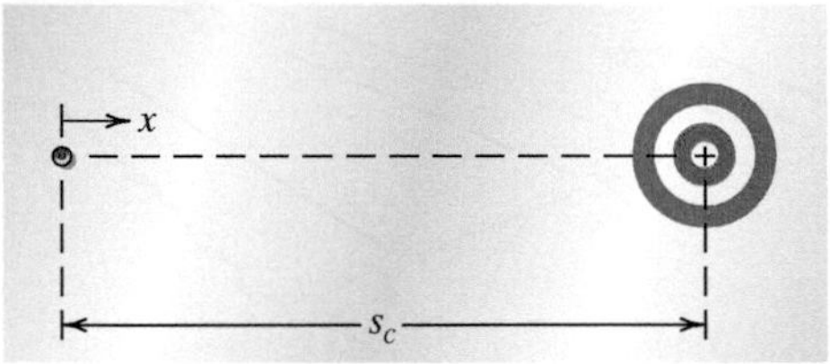

E3.1.14

3.1.15. **[Level 1]** Karl is on his way to work and isn't paying enough attention to the road to notice that the car ahead of him has abruptly stopped. When Karl finally realizes what's going on, he immediately slams on the brakes. He is traveling at $v_0 = 72\,\text{km/h}$ and is $d = 15\,\text{m}$ behind the car ahead of him at this instant. Karl's vehicle weighs 227 kg.

a. Does Karl hit the other car if his brakes supply a force of $F_{\text{brake}} = 7.56\,\text{kN}$? If so, what is the impact speed?

b. What is the maximum speed he could be driving at to avoid collision?

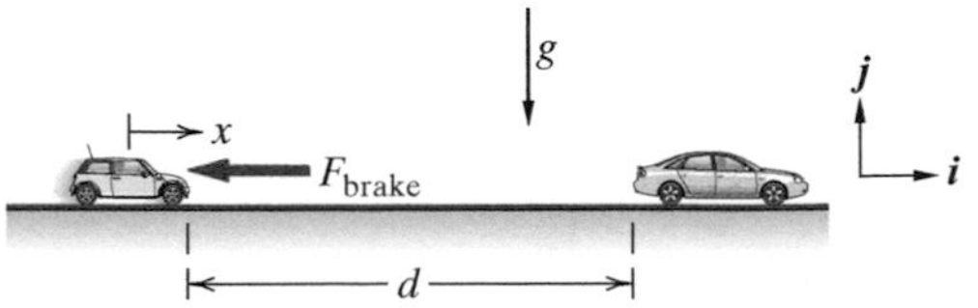

E3.1.15

3.1.16. **[Level 1]** A person pulls down on the rope at B with a force of $F = 44\,\text{N}$. What is the acceleration of block A if it weighs 23 kg?

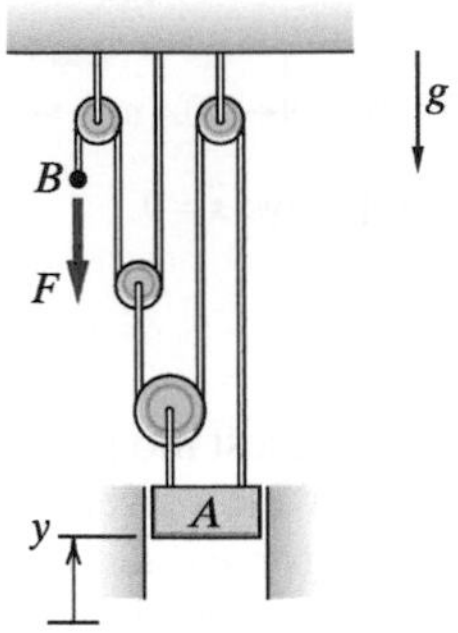

E3.1.16

3.1.17. **[Level 1]** A block is pushed on a frictionless surface with a varying force $F(t)$ at an angle of $\theta = 30°$ with respect to the ground. The force varies according to $F(t) = (9t + 10e^{-0.5t})\,\text{N}$, and the block has a mass of 5 kg. If the block starts from rest, determine its speed at $t = 2\,\text{s}$.

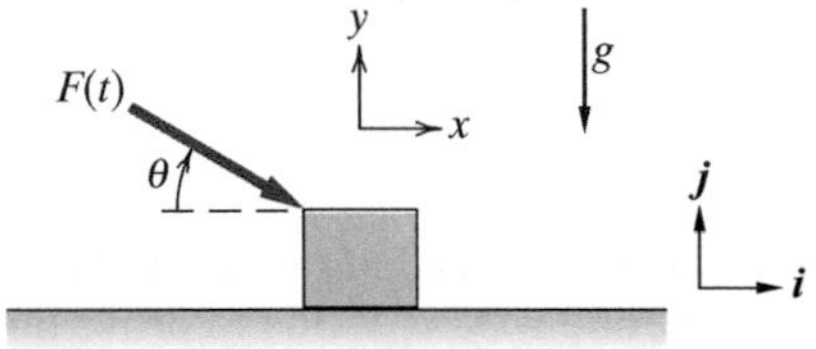

E3.1.17

3.1.18. **[Level 2] Computational** Consider the system discussed in Example 3.3. Let the bowl's mass be 1.5 kg, let the slope have an inclination of 30°, and add friction between the bowl and slope ($\mu_s = \mu_d = 0.7$). The bowl is projected downslope at 4 m/s. Numerically integrate the system's equation of motion and determine how far it travels before it comes to a halt and how long it takes for this to happen. Plot the distance traveled as a function of time.

3.1.19. **[Level 2]** Block A ($m_A = 3\,\text{kg}$) is pulled along a frictionless horizontal surface by a force $F = 50\,\text{N}$ applied at an angle $\theta = 30°$ with respect to the ground. Block B ($m_B = 5\,\text{kg}$) is attached to A by an inextensible cable, and the block slides on top of a rubber mat, nailed onto the surface, with a coefficient of friction $\mu = 0.6$. Find the system's acceleration and the tension in the cable.

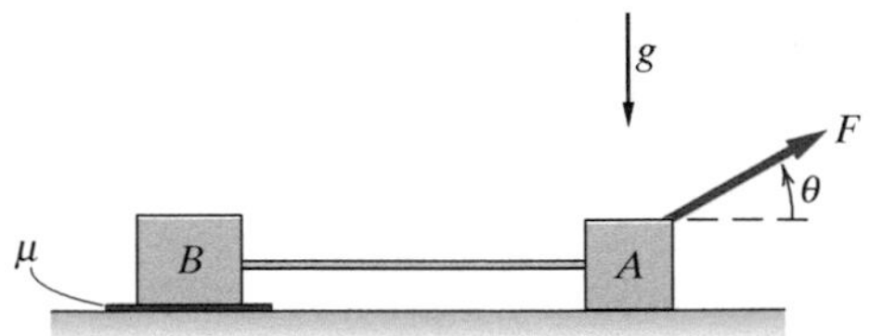

E3.1.19

3.1.20. **[Level 2]** E3.1.20 shows the same system that was considered in Example 3.5—a particle m suspended in a rigid container by two massless, inextensible strings. A force is applied to the container so that it accelerates to the right at a constant rate $\ddot{x}$. Determine the maximum value of $\ddot{x}$ for which both supporting strings remain taut.

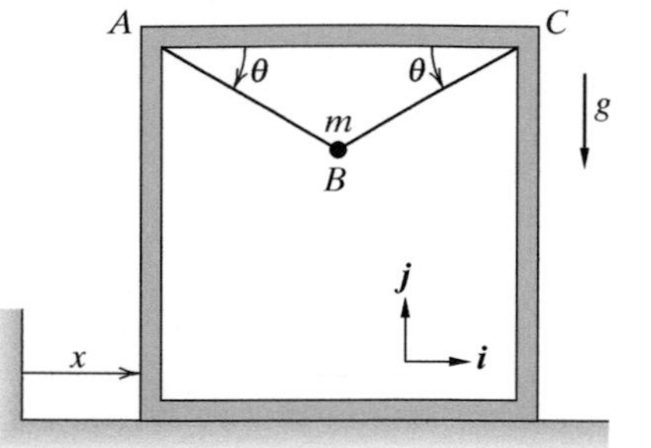

E3.1.20

3.1.21. **[Level 2]** A car is braking at a constant acceleration, causing the fuzzy dice hanging from the mirror to take the position shown. Approximate the fuzzy dice as a lumped mass m and solve for the steady-state value of θ if the car is decelerating at $0.2\,g$.

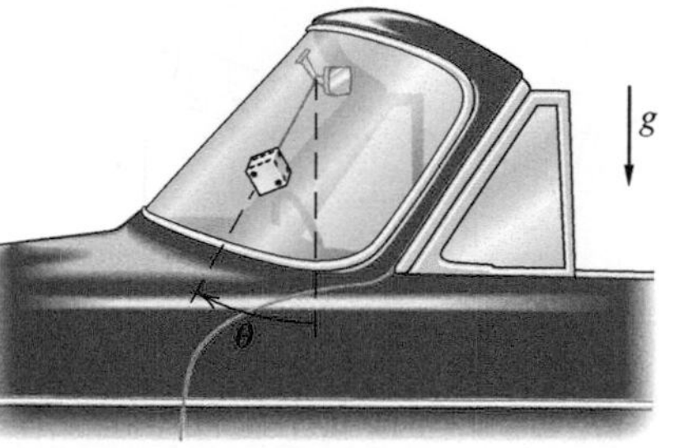

E3.1.21

3.1.22. **[Level 2]** A car is climbing a hill and accelerating at a constant rate of $0.3\,g$ as it does so. $\phi = 10°$. What is the steady-state value of θ, the angle made by a mass hung from a cord that's attached to the rearview mirror of the car?

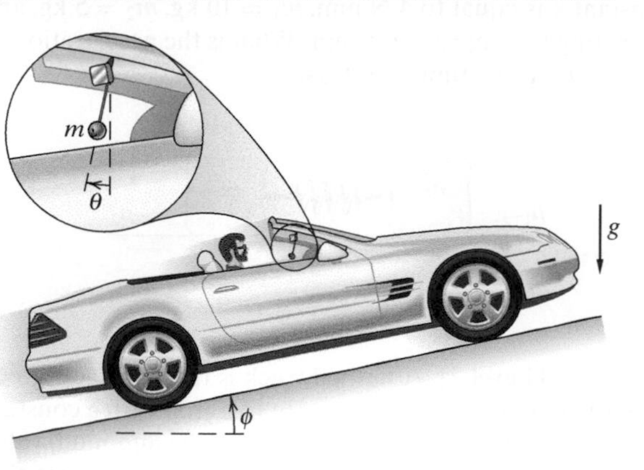

E3.1.22

3.1.23. **[Level 2]** As illustrated, block A ($m_A = 10$ kg) is free to move along a smooth horizontal surface under the action of an applied force of $F = 125$ N. Block B ($m_B = 15$ kg) is attached to A via an inextensible cable, and the block slides on a rough surface angled at $\theta = 20°$ with respect to the horizontal. Find the system acceleration and cable tension if the coefficient of friction between B and the incline is $\mu = 0.4$.

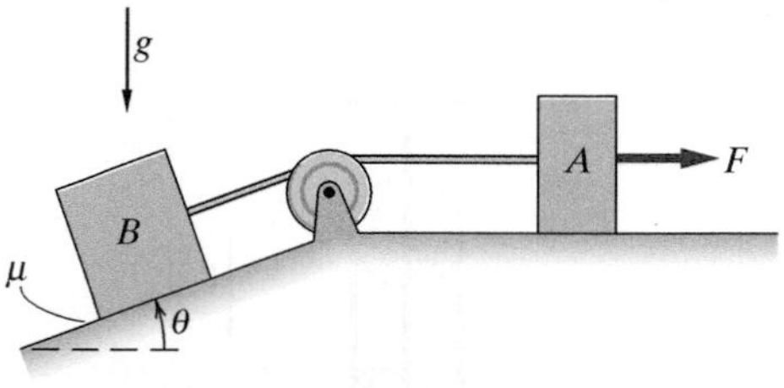

E3.1.23

3.1.24. **[Level 2]** A box is moving in the $\boldsymbol{i}$ direction with a constant acceleration. A mass, suspended from the top of the box by a linear spring (unstretched length of zero), is moving to the right as well and isn't moving with respect to the box (the spring isn't changing in length and θ is constant). The force exerted by the spring on the mass is given by its spring constant k times its extension. For the present case, the spring is extended 5 cm from its rest length, $k = 30$ N/m and $m = 0.15$ kg. What is the angle θ equal to, and what is the acceleration of the box? You can solve for this without using the computer.

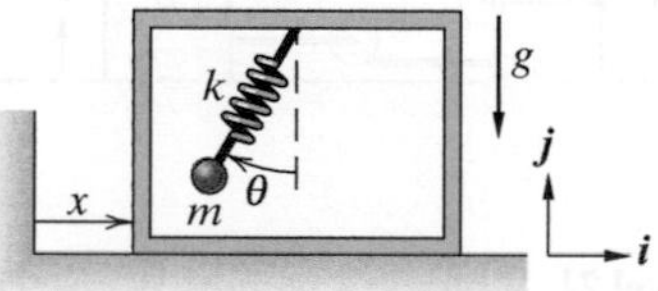

E3.1.24

3.1.25. **[Level 2]** **Computational** Two masses are joined together by a linear spring. The magnitude of the force exerted on each mass by the spring is given by $k(y - L)$, where L is the spring's unstretched length. $L = 0.6$ m, $m_1 = 1.2$ kg, $m_2 = 1.7$ kg, and $k = 100$ N/m.

a. Determine analytically the steady-state value of y if $\ddot{x} = 4.8\,\text{m/s}^2$.

b. Numerically integrate the system's equation of motion and plot y versus t for the system starting from rest and then moving for 10 s.

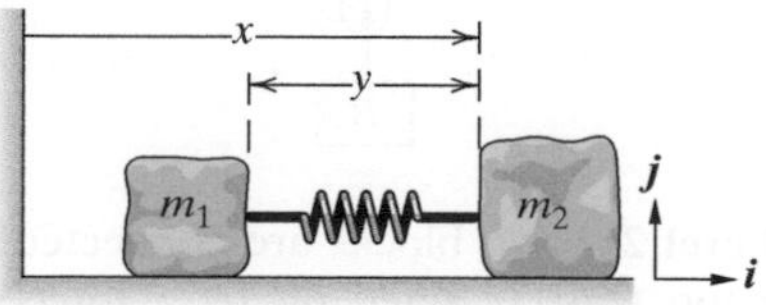

E3.1.25

3.1.26. [Level 2] Computational

a. For what m_1 will the illustrated pulley system be in static equilibrium if $m_2 = 100$ kg? Designate this value as m_1^*.

b. What will the acceleration of m_2 be if m_1 is set equal to $1.1m_1^*$?

c. Numerically integrate the system's equation of motion for case **b.** Let the masses be released from rest. Plot x_1 and x_2 versus t.

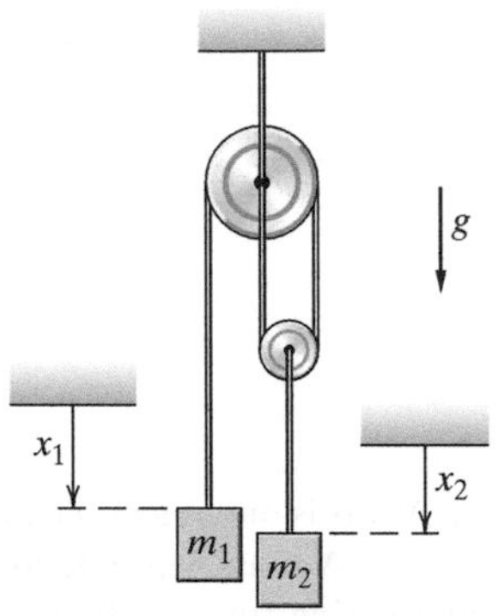

E3.1.26

3.1.27. [Level 2] Consider the illustrated system. A constant force of 10 N acts in the $\boldsymbol{i}$ direction on block B. What are the accelerations of blocks A and B? $m_A = 10$ kg and $m_B = 15$ kg. Assume a frictionless interface between the two blocks and the ground.

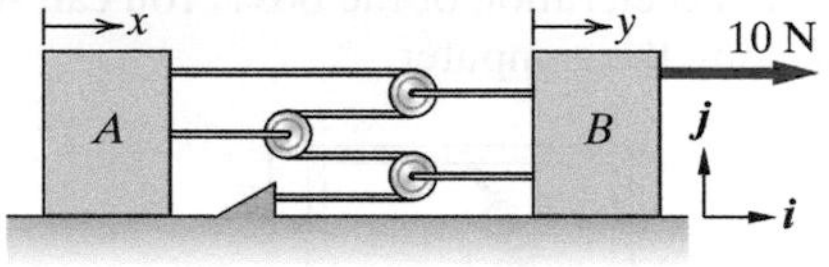

E3.1.27

3.1.28. [Level 2] If block A weighs $W_A = 9$ kg and block B has a weight of $W_B = 23$ kg, find the acceleration $\ddot{y}_B$ of B when A is released from rest, where y_B is measured down to B.

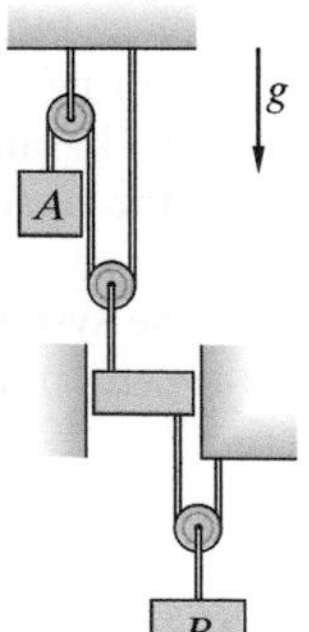

E3.1.28

3.1.29. [Level 2] Two blocks are connected by a pulley, and the top mass is acted on by a force such that $\ddot{x}_A = -7.13\text{ m/s}^2$. $m_A = 2.8$ kg and $m_B = 2.8$ kg. Determine a_B, the magnitude of the force acting on the top mass and the tension in the massless pulley ropes.

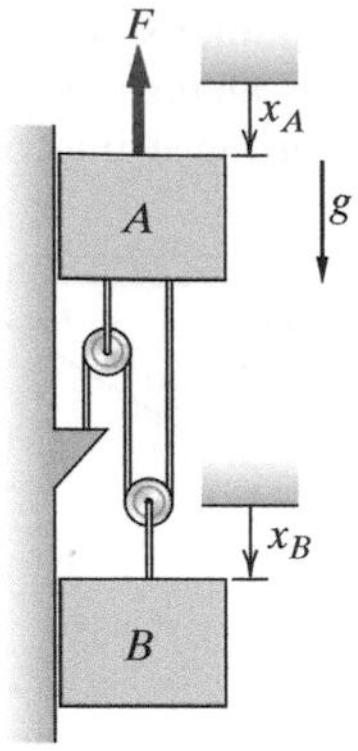

E3.1.29

3.1.30. [Level 2] A mass particle m is free to move within a frictionless tube that's inclined at an angle θ. The entire tube accelerates upward at $\ddot{z}$. What value of $\ddot{z}$ is necessary for the mass particle to accelerate (with respect to the tube) at 9.81 m/s^2 (oriented downslope)? The mass weighs 0.23 kg and $\theta = 28°$.

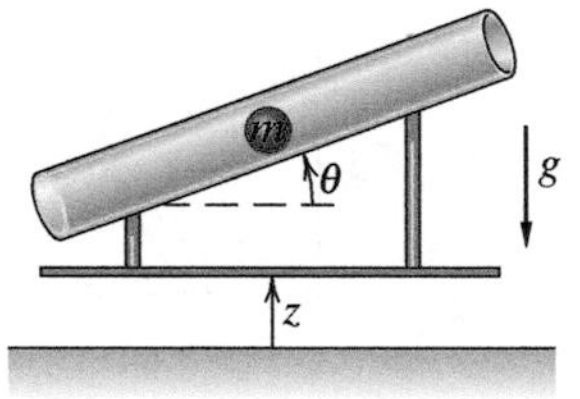

E3.1.30

3.1.31. [Level 2] Two masses are shown, connected by a linear spring. The masses are composed of different materials and have different static and kinetic coefficients of friction. $\mu_{1_s} = \mu_{1_d} = 0.4$ and $\mu_{2_s} = \mu_{2_d} = 0.1$. The spring constant k is equal to 4 N/mm, $m_1 = 10$ kg, $m_2 = 5$ kg, and the spring is compressed 5 mm. What is the acceleration of m_1 and m_2 at the time of release?

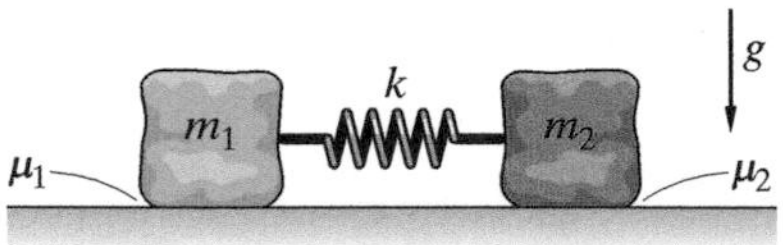

E3.1.31

3.1.32. [Level 2] A flatbed truck is traveling at a speed v_1 in the $\boldsymbol{i}$ direction and is brought to a stop with a constant deceleration over a distance D. What is the minimum value of the coefficient of static friction μ_s that will ensure that the block does not slip with respect to the trailer bed? The mass of the block is m.

E3.1.32

3.1.33. **[Level 2]** Remember how you used to go into an elevator when you were a kid and jump up right when it started moving and then again just when it was finishing its motion? Depending on whether it was going up or down, you would feel lighter or heavier than normal. Let's analyze this problem. We will model you as a lumped mass m. The distance between the starting point and the elevator floor is given by z. The velocity profile is given. $v_0 = 5.5$ m/s, $\Delta_1 = 1.5$ s, $\Delta_2 = 1.5$ s, and $m = 36$ kg.

a. How heavy do you feel during the acceleration phase?

b. How heavy do you feel during the deceleration phase?

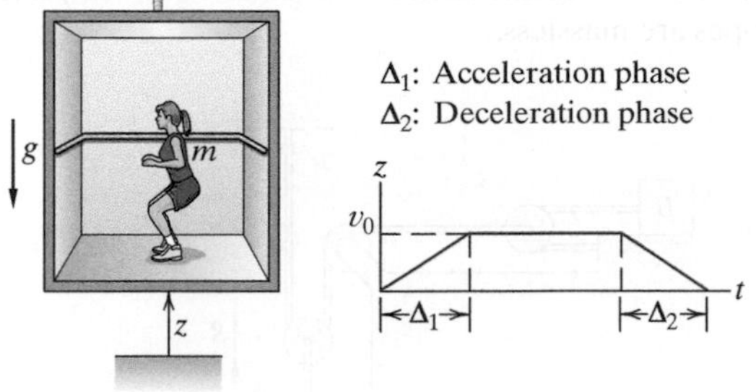

E3.1.33

3.1.34. **[Level 2]** A magician's trick involves pulling a tablecloth out from under a set table, leaving the dishes and cups on the uncovered table. Calculate how far the tableware moves. Assume that the initial situation is as shown. A cup is initially stationary on the tablecloth. At $t = 0$ the cloth is jerked away at 15 m/s. Assume a coefficient of static and dynamic friction equal to 0.6.

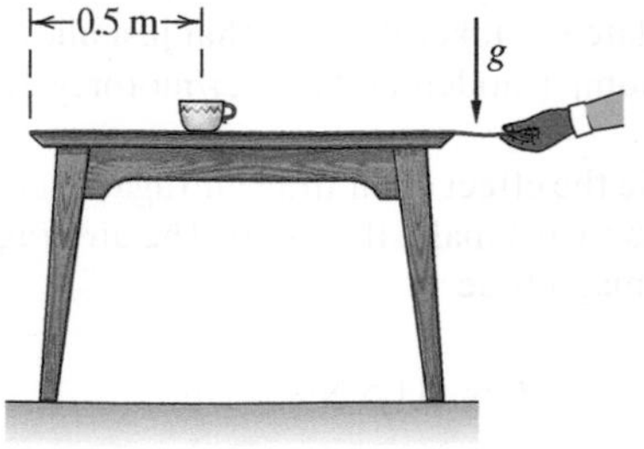

E3.1.34

3.1.35. **[Level 2]** A person is standing outside on a windy day and throws a 0.5 kg ball up at an angle θ and speed $v_0 = 10$ m/s. The wind produces a force on the ball equal to $-2\boldsymbol{i}$ N. What should θ be so that the ball returns to the person's hand?

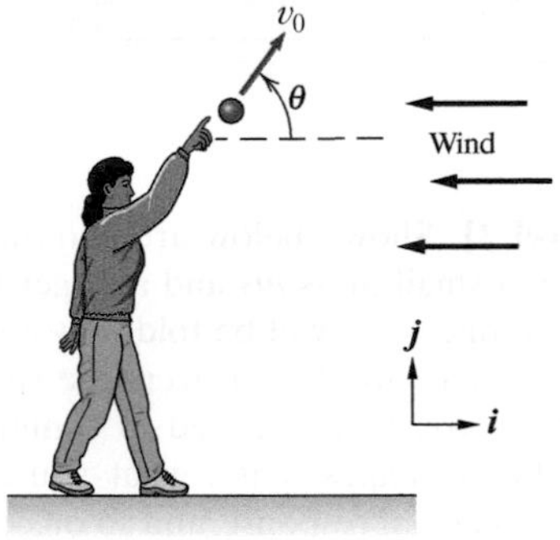

E3.1.35

3.1.36. **[Level 2]** A 18 kg block is acted on by a 133 N force via an attached rope. The coefficient of static and dynamic friction between the block and the floor is 0.6. For what two values of θ will the acceleration of the block be 2.68 m/s^2?

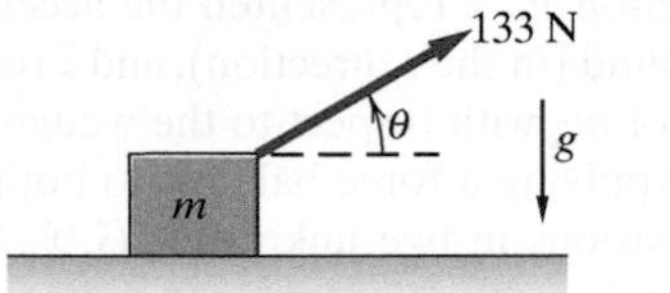

E3.1.36

3.1.37. **[Level 2]** The crush dynamics of a car crashing into a wall can be roughly approximated by modeling the body of the car as a mass m and the front crush zone as a spring k. We will only let the spring compress (no rebound). $m = 1.46 \times 10^3$ kg and $k = 1.09$ MN/m. How long does it take for the spring to bring the car to a speed of zero from an initial impact speed of 56 km/h?

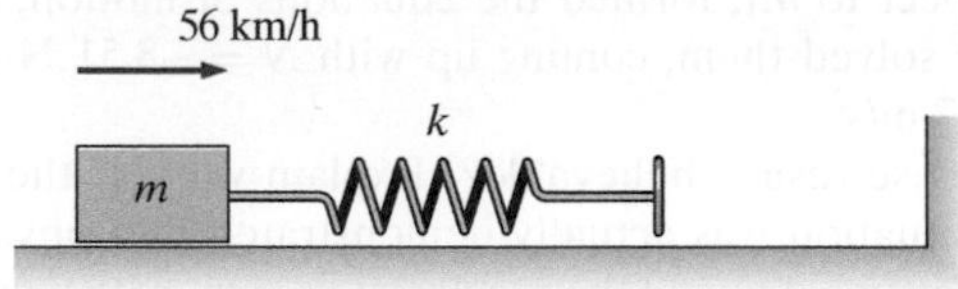

E3.1.37

3.1.38. **[Level 2]** If a car can decelerate at 1.1 g, how many car lengths will it travel if it decelerates from 96 km/h to zero? Assume an average car length of 4.6 m.

3.1.39. **[Level 2]** The illustrated device is a rigid enclosure with an internal, movable mass m (0.5 kg). The spring holding the mass against the wall has been compressed 0.1 m and has a spring constant of $k = 20$ N/m. The enclosure, initially moving at a constant speed, begins to accelerate at a constant rate of 6 m/s^2. What is the mass's acceleration just after the enclosure begins to accelerate?

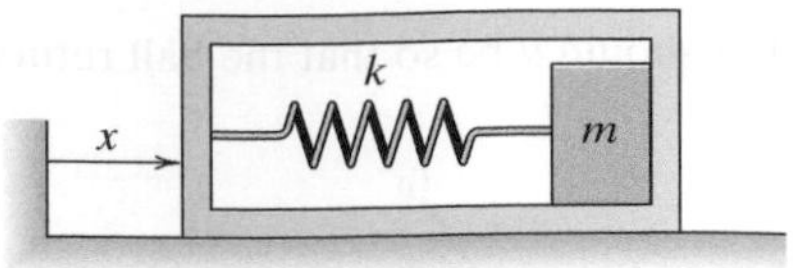

E3.1.39

3.1.40. **[Level 2]** Shown below are two different scenarios involving a small mass m_2 and a larger mass m_1 in the shape of a wedge. You will be told how a student approached the problem and the answers the student came up with, and then you will be asked to comment on the solution—whether it makes sense, what you can deduce about the system's actual response, and so on.

a. Consider the system shown in **E3.1.40*a***. m_2 is released onto m_1 with zero velocity. The coefficients of kinetic and static friction between m_1 and m_2 are the same for this problem. m_1 is free to slide on a frictionless surface. The student assumed sliding contact between m_1 and m_2 and drew the free-body diagram shown. The interaction forces were the normal forces N and P and the force due to sliding friction F. $\ddot{x}$ represented the acceleration of m_1 along the ground (in the $\boldsymbol{i}$ direction), and $\ddot{s}$ represented the acceleration of m_2 with respect to the wedge m_1 (in the $\boldsymbol{b}_1$ direction). Applying a force balance to both masses gave him four equations in five unknowns ($\ddot{s}$, $\ddot{x}$, N, F, and P). Setting F equal to μN (to represent sliding of m_2 on m_1) reduced the number of unknowns to four. The student correctly solved his equations and got the results $\ddot{s} = 0.509\ \text{m/s}^2$, $\ddot{x} = -0.197\ \text{m/s}^2$, $N = 19.05$ N, and $P = 49.31$ N.

Did the student do a correct job in his analysis? If yes, why do you think so? If no, why not? If it's not correct, what should the correct answer be?

b. In **E3.1.40*b*** we have a slightly different problem. A force G is applied to m_1 such that the acceleration of m_1 is equal to 15 m/s². The student assumed slip of m_2 with respect to m_1, formed the equations of motion, and correctly solved them, coming up with $N = -8.51$ N and $\ddot{s} = -21.7\ \text{m/s}^2$.

Are these results believable? Explain why. If the described situation was actually demonstrated in a physical experiment, what would the acceleration of m_2 with respect to a ground-fixed reference frame be equal to?

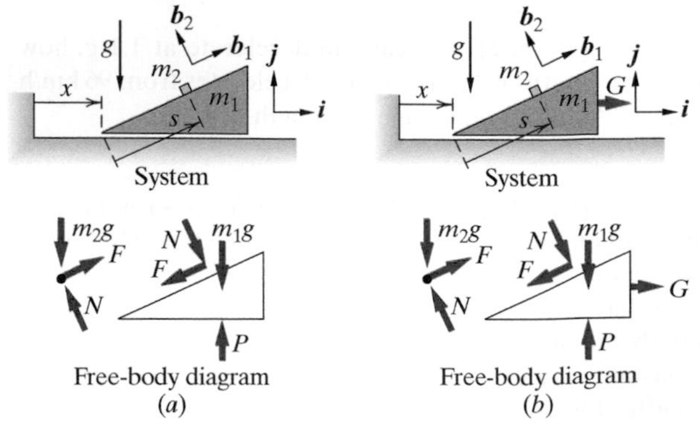

E3.1.40

3.1.41. **[Level 2]** The illustrated pulley system is released from rest. How much time is needed for B to contact the ground? $m_A = 20$ kg and $m_B = 160$ kg. The pulleys and ropes are massless.

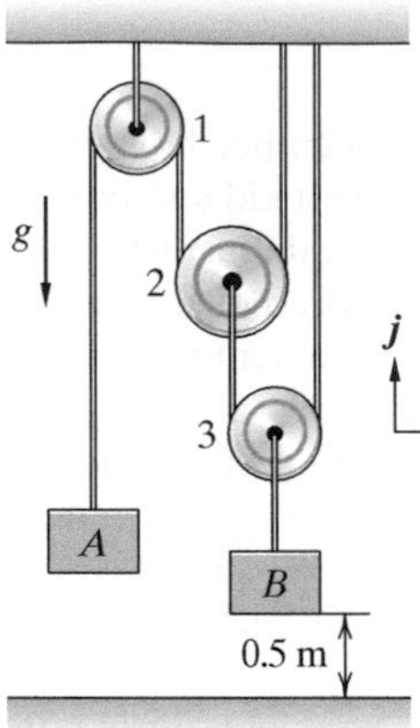

E3.1.41

3.1.42. **[Level 2]** Block A has a mass of 10 kg and is released from rest. How fast is it traveling after 1.5 s? Block B has a mass of 8 kg and slides without friction. The pulleys and ropes are massless.

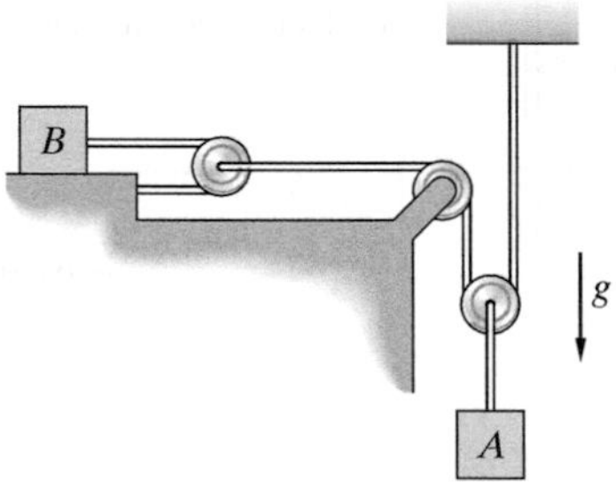

E3.1.42

3.1.43. **[Level 2] Computational** You are given the task of determining the proper launch parameters to allow a motorcycle stunt rider to clear a 9 m tank of water (which presumably contains man-eating piranhas so as to add to the excitement of the jump). The takeoff ramp has a 45° slope. The rider and motorcycle have a weight of 227 kg.

a. Calculate the takeoff speed that just allows the rider to make the jump (modeling the rider/motorcycle as a point mass).

b. Include the effect of air drag and again determine the speed needed to just make the jump. The air drag produces a force with magnitude

$$F = (11.5\,\text{N}\cdot\text{s}^2/\text{m}^2)v^2$$

which always points directly opposite to the rider's velocity vector. v is in the units of m/s. You will need to use numerical integration to solve this part.

c. By what percentage did the speed change when air drag was included?

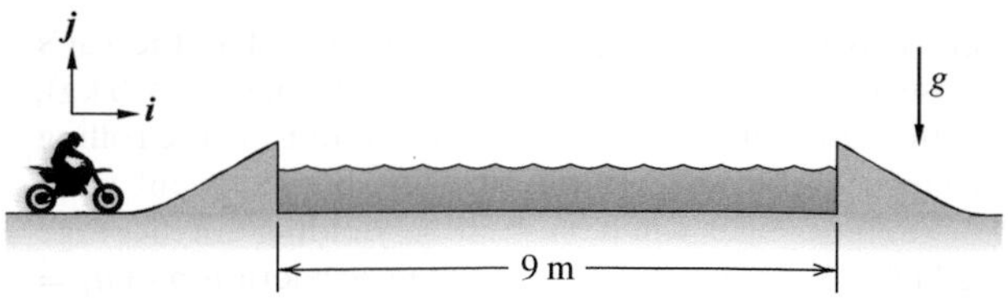

E3.1.43

3.1.44. [Level 2] One means of evaluating the braking of a car is to run two tests. In the first test the car is made to brake on a dry road, and in the second the car is operated such that the left wheels are on dry road ($\mu = 0.9$) and the right are on a low-coefficient-of-friction surface ($\mu = 0.1$) used to mimic icy conditions. Assume that the car can be approximated by a mass particle with mass $m = 1500$ kg. For Case 1 assume that the frictional force simply depends upon the weight of the car and acts opposite the car's direction of travel as it decelerates. For Case 2 approximate the dry/ice conditions by averaging the two coefficients of friction and applying this average coefficient to the weight of the car.

Compare the time required to come to rest, and the distance traveled, for the two cases if maximum braking is applied from a speed of 100 km/hr and the maximum braking force is developed between the tires and the road at all times.

3.1.45. [Level 2] A conveyor belt is shown, inclined at an angle $\theta = 15°$. A 2 kg mass A is placed on the conveyor in the position shown with zero velocity with respect to ground. The belt is moving at 3.0 m/s. $L_1 = 0.85$ m, $L_2 = 1.05$ m, $\mu = 0.8$. What is the acceleration of the mass when it reaches the end of the conveyor belt?

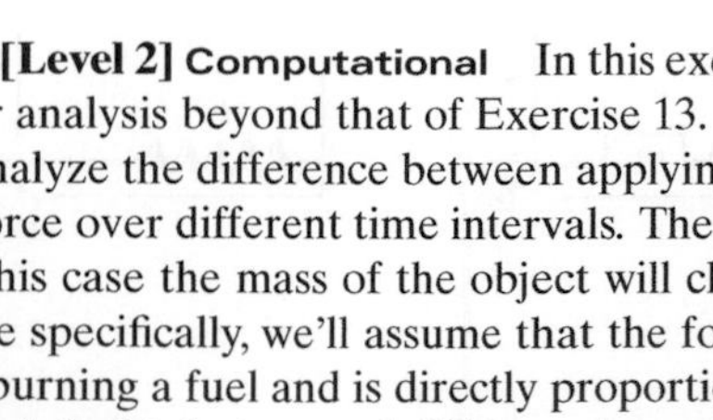

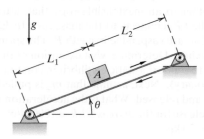

E3.1.45

3.1.46. [Level 2] Computational In this exercise we'll extend our analysis beyond that of Exercise 13. We're still going to analyze the difference between applying different levels of force over different time intervals. The difference is that in this case the mass of the object will change with time. More specifically, we'll assume that the force is generated by burning a fuel and is directly proportional to the rate at which fuel is being used. Ultimately we again want to see how these different scenarios affect the displacement and final speed of the object being acted upon by the force.

Case 1: Assume that the mass of the body decreases at a rate of 1 kg/s and that the force is equal to 100 N. The body starts at 100 kg and after 10 s is down to 90 kg. Determine the change in position and final speed.

Case 2: Assume that the mass of the body decreases at a rate of 2 kg/s and that the force is equal to 200 N. The body starts at 100 kg and after 5 s is down to 90 kg. Determine the change in position and final speed.

Compare and discuss.

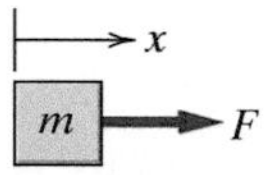

E3.1.46

3.1.47. [Level 2] Tired from spending all day moving into his new apartment, Ken decides to push the last box into his new place instead of carrying it. The box weighs 18 kg, and Ken pushes on it at $\theta = 45°$ with respect to the ground. The coefficients of static and dynamic friction between the box and ground are $\mu_s = 0.7$ and $\mu_d = 0.3$, respectively.

a. If the box is initially at rest, determine how much force F Ken needs to push with to get the box sliding and the resulting initial acceleration.

b. Suppose Ken is pushing the box at a constant speed of $v_0 = 1.2$ m/s and then slips. How far does the box travel before coming to a stop?

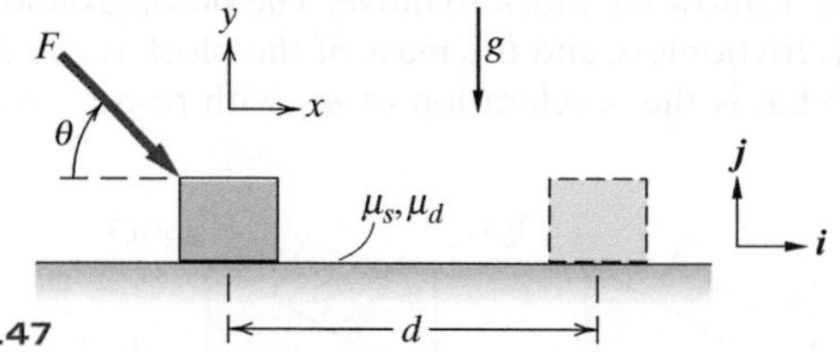

E3.1.47

3.1.48. [Level 2] John is pulling a sled with some fire wood on it back to his winter cabin at the top of a snowy hill with a $\theta = 30°$ incline. The combined weight of the sled and wood is 4.5 kg, and the coefficient of friction between the sled and snow is $\mu = 0.3$. John pulls on a rope attached to the sled at an angle of $\phi = 20°$ with respect to the hill surface and accelerates up the hill at $\ddot{x} = 0.6$ m/s^2.

a. Calculate the tension T in the rope needed to accelerate at this rate.

b. Suppose the sled is 6 m up the hill, moving at $v_0 =$ 1.2 m/s, when John accidently slips and lets go of the rope. How long does it take for the sled to slide to the base of the hill?

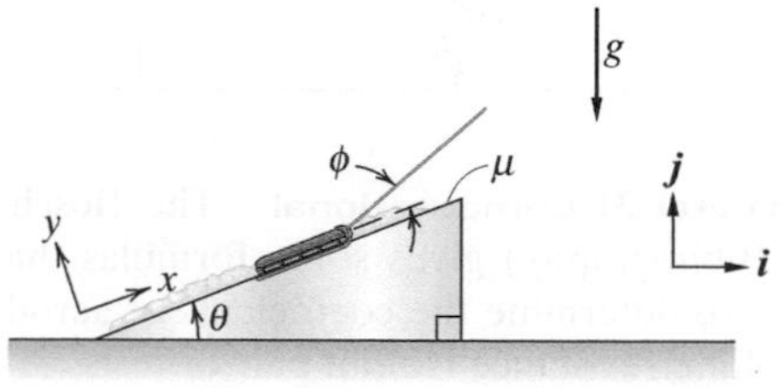

E3.1.48

3.1.49. [Level 2] A heavy box is placed at the top of a rough ramp of length $L = 4$ m, and the coefficients of static and dynamic friction between the box and ramp are $\mu_s = 0.8$ and $\mu_d = 0.4$, respectively.

a. Calculate the ramp angle θ that will just cause the box to slip.

b. How long does it take for the box to slide to the base of the ramp for this angle?

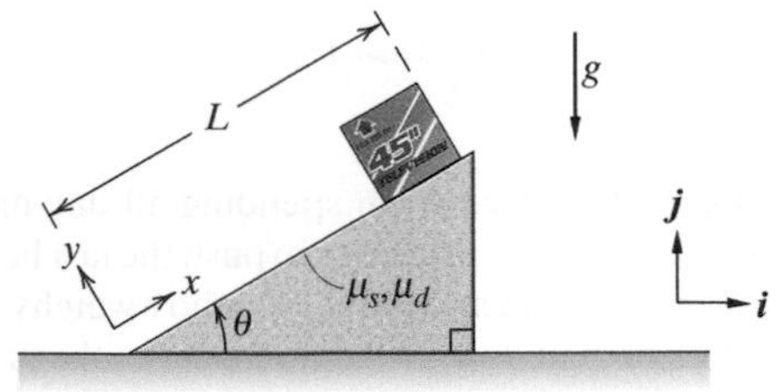

E3.1.49

3.1.50. **[Level 3]** Consider the problem of a 10 kg mass m_1 that's released (with zero initial velocity) on the surface of a sloped block. The coefficient of friction between m_1 and the block is 0.3 ($\mu = 0.3$).

a. Determine the acceleration of m_1 with respect to the block. Assume that the block is fixed in place.

b. Next, allow the block to move. The block/ground interface is frictionless, and the mass of the block is equal to 20 kg. What is the acceleration of m_1 with respect to the block?

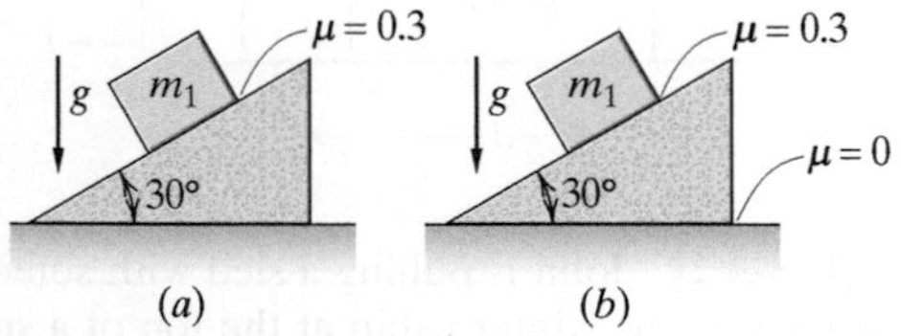

E3.1.50

3.1.51. **[Level 3]** Two masses are connected via a pulley. $\mu_1 = 0.15, \mu_2 = 0, m_1 = 10$ kg, $m_2 = 5$ kg, and $\theta = 45°$. Determine whether the masses slide with respect to one another. If they do, determine their absolute accelerations and discuss the results. They are released from rest.

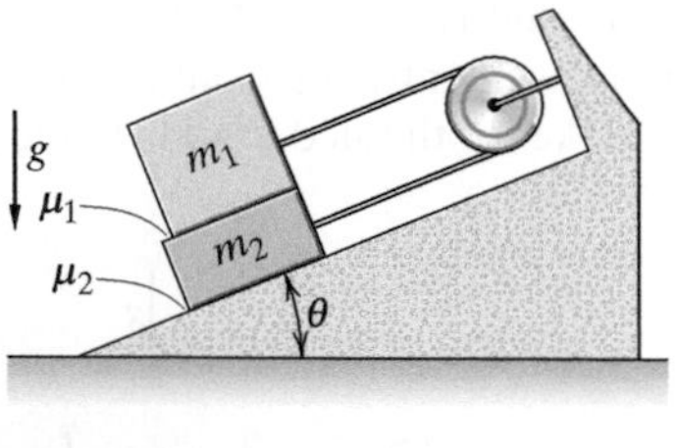

E3.1.51

3.1.52. **[Level 3] Computational** The Bosch Handbook (see Bibliography) gives some formulas that allow a researcher to determine the coefficients of aerodynamic drag and rolling resistance from a pair of tests. A vehicle is allowed to coast in neutral on a level surface (no wind). From initial and final speeds, along with elapsed time, average speed and acceleration are calculated. The governing equation for a car of mass m moving in the x direction is

$$m\ddot{x} = -\frac{1}{2}\rho C_a A\dot{x}^2 - C_r mg$$

where ρ is the density of air (1.20 kg/m^3), A is the car's cross-sectional area (2.20 m^2), m is the car's mass (1450 kg), C_a is the aerodynamic drag coefficient, and C_r is the rolling resistance coefficient. In the first test the car went from 60.0 km/h to 55.0 km/h in 6.50 s, yielding an average speed of $v_1 = 16.0$ m/s and an average acceleration of $a_1 = -0.214$ m/s^2. In the second test the car went from 15.0 km/h to 10.0 km/h in 10.5 s, giving an average speed of $v_2 = 3.47$ m/s and an average acceleration of $a_2 = -0.132$ m/s^2. The formulas given for the two coefficients are

$$C_a = \frac{2m}{\rho A}\frac{a_2 - a_1}{v_1^2 - v_2^2}$$

$$C_r = \frac{a_2 v_1^2 - a_1 v_2^2}{g(v_2^2 - v_1^2)}$$

a. Determine how these formulas were derived from the given equation of motion.

b. What are C_a and C_r equal to?

c. Using the values C_a and C_r you obtained in **b**, numerically integrate the equation of motion and try to replicate the given data. Start at $\dot{x} = 60$ km/h, integrate for 6.5 s, and see if the car's speed is 55 km/h. Likewise, start at $\dot{x} =$ 15 km/h, integrate for 10.5 s, and see if the car's speed is 10 km/h.

3.1.53. **[Level 3]** Two different scenarios are shown. In **E3.1.53a**, a mass m_1 is free to slide on a frictionless surface. One end of the mass is attached to a linear spring k and the other to an inextensible rope that, after going over the pulley P, connects to the mass m_2. In **E3.1.53b**, both m_1 and m_2 are suspended beneath P. The spring is initially unstretched. Determine what extension of the spring is necessary to support a static equilibrium condition (both masses stationary). Next, assume that m_2 is pulled downward 0.01 m and released. What is the acceleration of m_2 at the time of release for the two cases? $k = 1000$ N/m, $m_1 = 10$ kg, and $m_2 = 20$ kg.

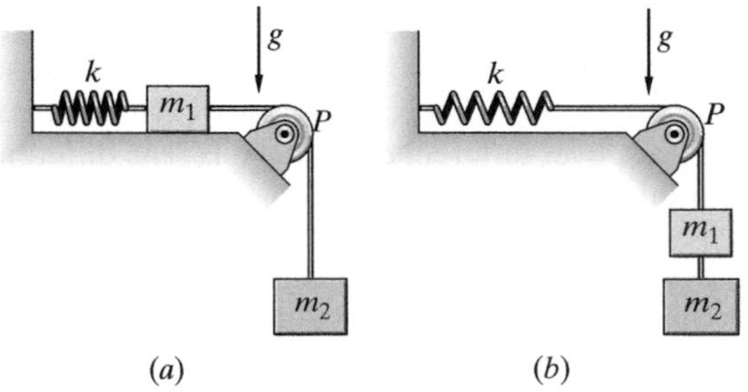

E3.1.53

3.1.54. **[Level 3]** Determine the acceleration of blocks A (mass m_A) and B (mass m_B) in terms of m_A, m_B, θ, and g. Assume a frictionless interface between A and B and between B and the ground. A and B are initially at rest.

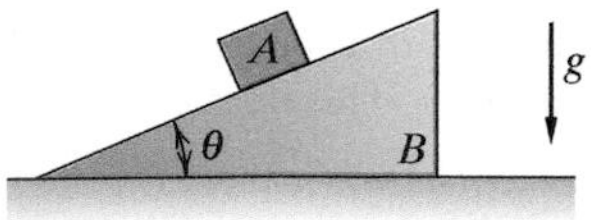

E3.1.54

3.1.55. **[Level 3] Computational** A mass m_2 is suspended by a frictionless pulley. One end of the pulley cord is attached to the ceiling and the other, after passing over a massless roller, attaches to a mass m_1. m_1 lies on a moving surface, and the coefficients of friction between the two are $\mu_s = \mu_d = 0.3$. $m_2 = 1$ kg. Assume that the speed of the moving surface always has a greater magnitude than that of m_1.

a. Initially the system is in static equilibrium and $\theta = 45°$. What is m_1 equal to?

b. Assume that the system is in the static equilibrium just described, and at $t = 0$ a piece of m_2 drops off, decreasing its mass to 0.9 kg. What will the new static equilibrium value of θ be?

c. Numerically simulate and plot the system's response for the scenario described in **b**. Does the system go to the predicted equilibrium position? Why or why not? The initial conditions are $y(0) = 1.2$ m and $\dot{y}(0) = 0$ m/s.

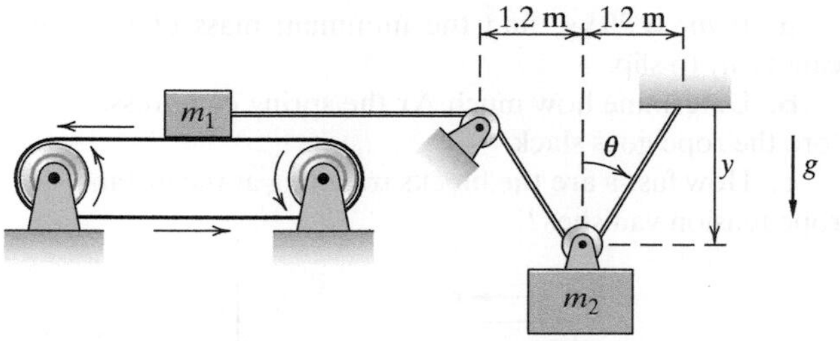

E3.1.55

3.1.56. **[Level 3]** Assume that initially block A is stationary.

a. If m_B is slowly increased, block B will eventually begin to move. Find the minimum m_B (m_B^*) such that m_B will move downward.

b. What are $\boldsymbol{a}_A$ and $\boldsymbol{a}_B$ with $m_B = 2m_B^*$?

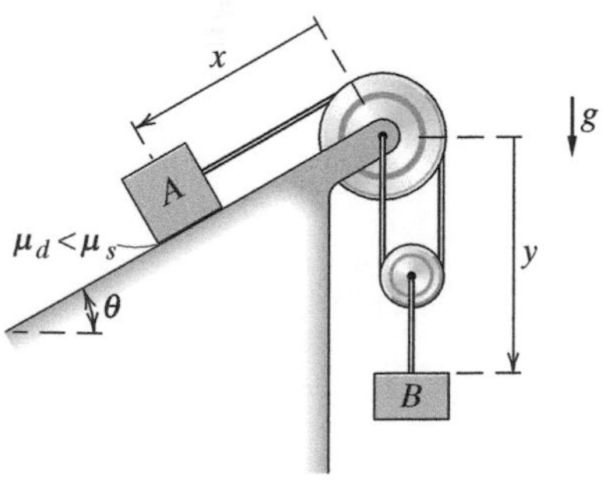

E3.1.56

3.1.57. **[Level 3]** Do the following for the illustrated multiblock system under the assumption that $m_A > m_C$:

a. Find the minimum m_B (m_B^*) needed to permit nonzero accelerations of all masses.

b. Find the tension in the massless cord between mass A and pulley D when $m_B = 1.5m_B^*$.

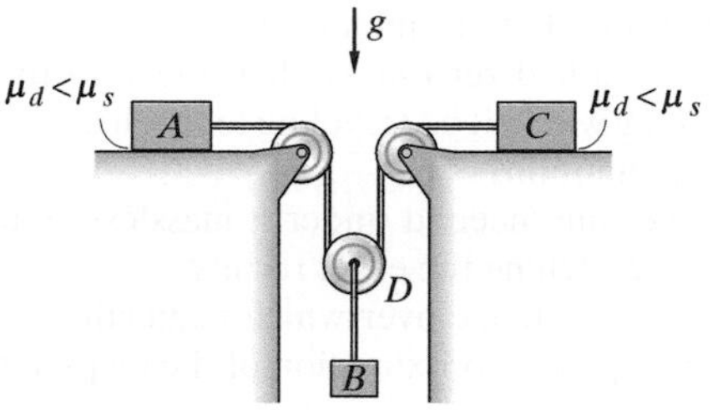

E3.1.57

3.1.58. **[Level 3] Analytical/Computational** A particle m is dropped into a thick liquid with an initial speed $\dot{x}_0$. The drag force due to the mass/liquid interaction is given by $F = -c\dot{x}$, where c is a positive constant and x measures the distance that the mass has traveled into the liquid.

a. Formulate the differential equation governing the mass's speed and show analytically by solving this equation that the mass's speed goes to a constant limiting value as time t goes to infinity.

b. Show how the limiting speed found in **a** can be deduced from an examination of the equation of motion.

c. Numerically integrate the equation of motion with initial conditions $x(0) = 0$, $\dot{x}(0) = 10$ m/s, $m = 10$ kg, and $c = 5$ N·s/m and plot $x, \dot{x}$ versus t.

3.1.59. **[Level 3]** A mass m is connected to two identical springs that are themselves attached to the sides of a movable platform. The unstretched length of the springs is 0.3 m and $k = 1000$ N/m. Where will the mass be with respect to the platform if the platform accelerates at the constant rate of $\ddot{x} = 20$ m/s^2? (Ignore any transient motions and just concentrate on the steady-state equilibrium position.) $m = 10$ kg. The mass/platform interface is friction-free.

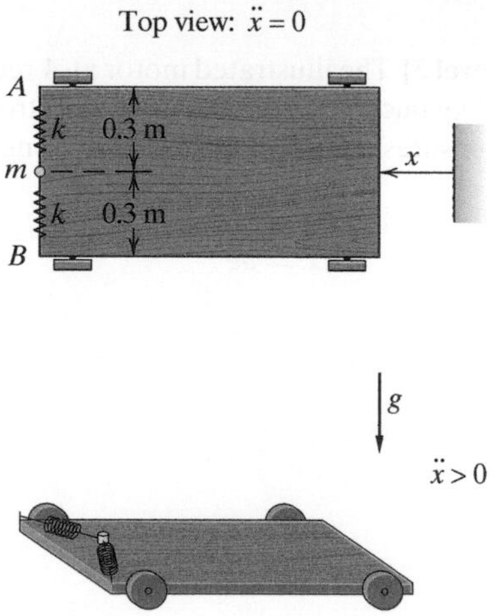

E3.1.59

3.1.60. **[Level 3]** A test sled has a mass of 400 kg. A constant force of 200,000 N acts on it, causing it to accelerate. A viscous drag with magnitude $1600|\dot{x}|$ acts to retard the motion.

a. Assume the sled starts from rest and determine how long the force must act before the sled travels 2000 m.

b. How fast is it traveling when $x = 2000$ m?

c. What is the sled's terminal velocity?

d. If the sled had zero mass, then the velocity is constant. How long would it be traveling to reach 2000 m under a massless assumption?

e. Did the time needed under a massless assumption come close to matching the exact result?

f. Is there a distance over which neglecting the sled's mass leads to a poor approximation of the elapsed time?

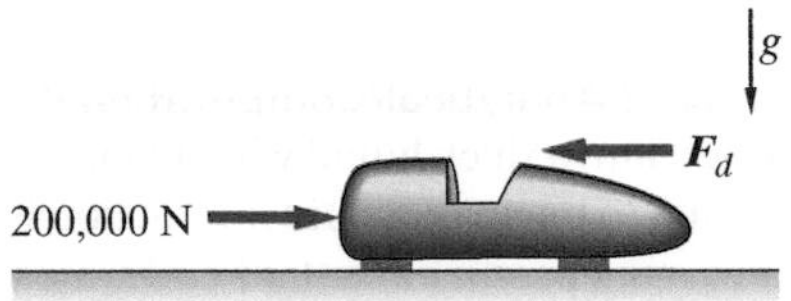

E3.1.60

3.1.61. **[Level 3]** Two masses are coupled by a pulley and restrained by a spring. The spring has an unstretched length of L and is stretched by an amount δ and then attached to block A, as shown. The pulleys and ropes are massless.

a. Determine the minimum mass m_{B^*} needed for block A to move to the right and for block B to move down the inclined surface.

b. What is the maximum coefficient of static friction that makes sense for this problem?

c. Letting m_B be greater than m_{B^*}, find expressions for the accelerations of blocks A and B.

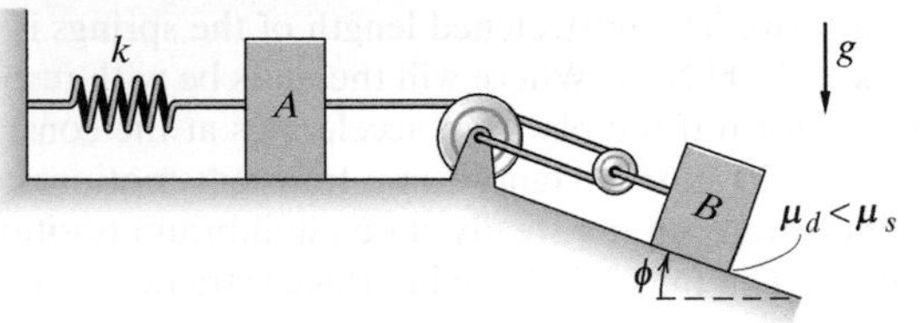

E3.1.61

3.1.62. **[Level 3]** The illustrated motor at A reels rope in to raise the 1.1 kg bucket B from the floor. Control circuitry in the motor ensures a tension in the rope being reeled in of

$$T = ae^t$$

where $a = 1.1$ N and t is in seconds. Consider the pulleys and ropes to be massless. The bucket is initially at rest on the ground. How long does it take to reach a height of 1 m?

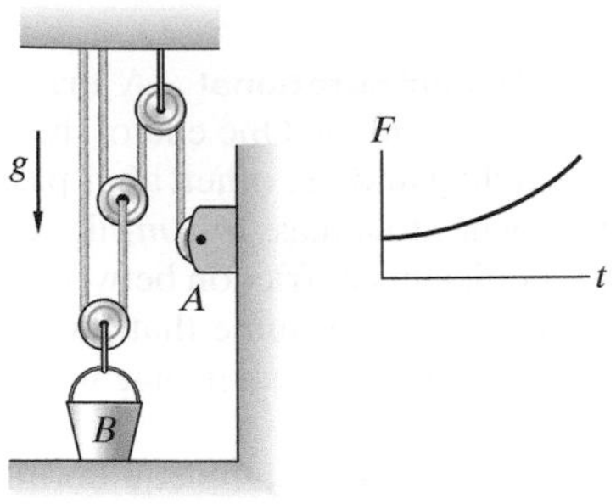

E3.1.62

3.1.63. **[Level 3]** **Computational** Mass m_1 slides on a rough surface whose coefficients of static and dynamic friction are $\mu_s = 0.6$ and $\mu_d = 0.3$, respectively. Mass m_2 is suspended under a pulley and connected to m_1 by an inextensible rope that goes over the pulley. The system starts from rest, and a spring with a stiffness of $k = 1000$ N/m is located $h = 3$ m directly below m_2.

a. If $m_1 = 20$ kg, find the minimum mass of m_2 that causes m_1 to slip.

b. Determine how much Δx the spring compresses before the rope goes slack.

c. How fast v are the blocks traveling at the instant the rope tension vanishes?

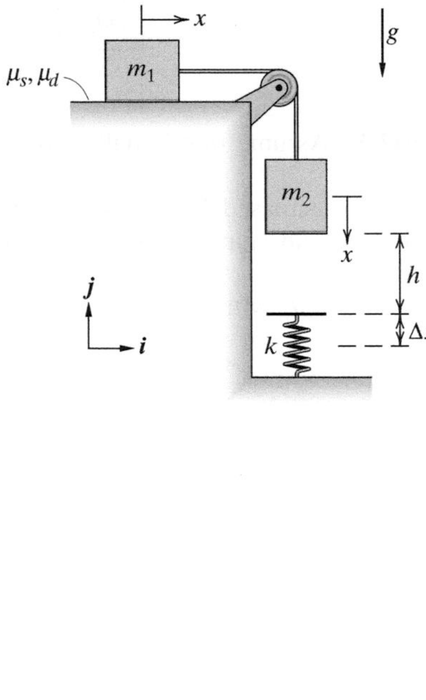

E3.1.63

3.2 POLAR COORDINATES

Life gets more interesting when we move to polar coordinates. We can break $\boldsymbol{F}$ up into components $F_r \boldsymbol{e}_r + F_\theta \boldsymbol{e}_\theta$ and write the two equations

$$F_r = ma_r$$
$$F_\theta = ma_\theta$$

or we can substitute the complete expressions for these acceleration components (2.50) and rearrange into the most commonly used order:

$$m(\ddot{r} - r\dot{\theta}^2) = F_r \qquad (3.15)$$

$$m(r\ddot{\theta} + 2\dot{r}\dot{\theta}) = F_\theta \qquad (3.16)$$

Unlike the Cartesian case, the polar representation is strongly coupled. A force component in the $\boldsymbol{e}_r$ direction affects both $\ddot{r}$ and $\dot{\theta}$, and a force component in the $\boldsymbol{e}_\theta$ direction affects $\dot{r}$, $\dot{\theta}$, and $\ddot{\theta}$. With the Cartesian representation, you could change a particle's acceleration in the x direction by applying a force in that direction:

$$\ddot{x} = \frac{F_x}{m}$$

There's no direct way to alter only $\ddot{r}$ or $\ddot{\theta}$ by applying a force in a particular direction, however. The value of $\ddot{r}$ or $\ddot{\theta}$ depends on the force as well as on the particle's positions and velocities. In the $\boldsymbol{e}_r$ direction, for example, we have

$$\ddot{r} = \frac{F_r}{m} + r\dot{\theta}^2$$

>>> **Check out Example 3.6 (page 120), Example 3.7 (page 121), Example 3.8 (page 122), Example 3.9 (page 124), and Example 3.10 (page 126) for applications of this material.**

EXAMPLE 3.6 MING BOWL ON A MOVING SLOPE (Theory on page 119)

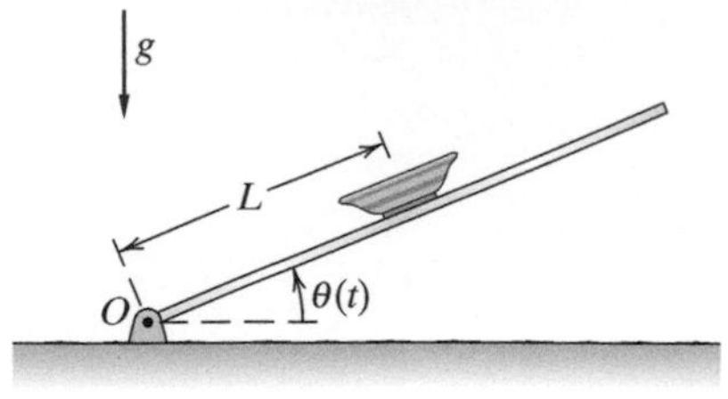

Figure 3.15 Bowl on a rotating support

Let's endanger our Ming bowl even further. Now, instead of lying on a fixed slope, it's sitting on a frictionless, flat board, a distance L away from the left edge (**Figure 3.15**). At $t = 0$, the board is made to rotate counterclockwise about the left edge. As it rotates, the board makes an angle θ with the ground. Find the equation of motion that governs the position of the bowl along the board.

Goal Determine the equation of motion in terms of motion along the board.

Given System geometry.

Assume Our only assumption is that we can treat the bowl as if it's a simple translating body.

Draw **Figure 3.16*a*** shows a schematic with two sets of unit vectors and a distance measure r, where $r(0) = L$. The free-body part of the FBD=IRD (**Figure 3.16*b***) looks much like the one in Example 3.3. The fact that the slope angle is changing with time doesn't affect the configuration of the forces. It does, however, change the expression for the bowl's acceleration, as you will see next.

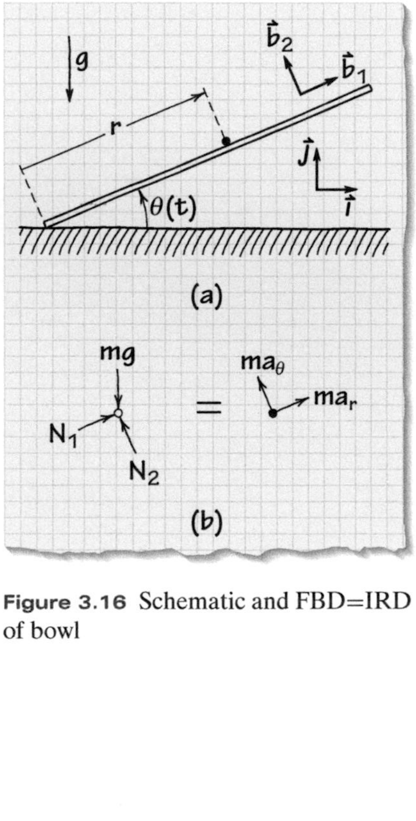

Figure 3.16 Schematic and FBD=IRD of bowl

Formulate Equations We'll start with our position vector, $\boldsymbol{r} = r\boldsymbol{b}_1$, and differentiate to find the particle's acceleration.

Solve Differentiating $\boldsymbol{r}$ gives us $\dot{\boldsymbol{r}} = \dot{r}\boldsymbol{b}_1 + r\dot{\boldsymbol{b}}_1 = \dot{r}\boldsymbol{b}_1 + r\dot{\theta}\boldsymbol{b}_2$. One more differentiation gives us

$$\begin{aligned}\ddot{\boldsymbol{r}} &= \ddot{r}\boldsymbol{b}_1 + \dot{r}\dot{\theta}\boldsymbol{b}_2 + \dot{r}\dot{\theta}\boldsymbol{b}_2 + r\ddot{\theta}\boldsymbol{b}_2 - r\dot{\theta}^2\boldsymbol{b}_1 \\ &= (\ddot{r} - r\dot{\theta}^2)\boldsymbol{b}_1 + (2\dot{r}\dot{\theta} + r\ddot{\theta})\boldsymbol{b}_2\end{aligned}$$

Multiplying by m and equating to the applied forces give us our equation of motion:

$$m(\ddot{r} - r\dot{\theta}^2)\boldsymbol{b}_1 + m(2\dot{r}\dot{\theta} + r\ddot{\theta})\boldsymbol{b}_2 = -mg\left(\cos\theta\boldsymbol{b}_2 + \sin\theta\boldsymbol{b}_1\right) + N_1\boldsymbol{b}_1 + N_2\boldsymbol{b}_2$$

$\boldsymbol{b}_1$: $$m(\ddot{r} - r\dot{\theta}^2) = N_1 - mg\sin\theta$$

$\boldsymbol{b}_2$: $$m(2\dot{r}\dot{\theta} + r\ddot{\theta}) = N_2 - mg\cos\theta$$

Check Comparing our result to the text shows that we've recovered the acceleration components of a particle moving in a polar coordinate system.

EXAMPLE 3.7 MING BOWL IN MOTION (Theory on page 119)

In Example 3.6 we found the equations of motion for a Ming bowl (**Figure 3.17**) on a rotating surface. Assume that the rotation rate $\dot{\theta}$ is constant and find an analytical solution for the bowl's response. Treat the bowl as a simple mass particle.

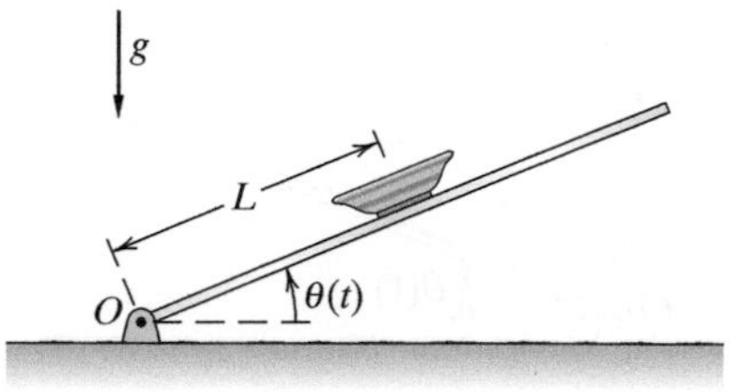

Figure 3.17 Bowl on a rotating support

Goal Determine the response of a bowl on a moving slope.

Given System geometry. $\dot{\theta}$ is constant.

Draw **Figure 3.18*a*** shows a schematic of the system.

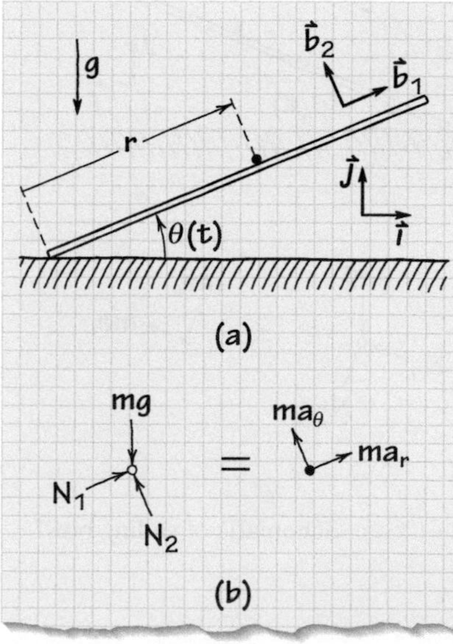

Figure 3.18 Schematic and FBD=IRD of bowl

Formulate Equations From (3.15) and (3.16) we have

$$b_1: \quad m(\ddot{r} - r\dot{\theta}^2) = N_1 - mg\sin\theta$$

$$b_2: \quad m(2\dot{r}\dot{\theta} + r\ddot{\theta}) = N_2 - mg\cos\theta$$

Assume We can now apply the assumption that the angular speed is constant ($\dot{\theta} = \omega_0$), which means $\ddot{\theta} = 0$. Our equations simplify to

$$m(\ddot{r} - r\dot{\theta}^2) = -mg\sin\theta \tag{3.17}$$

$$2m\dot{r}\omega_0 = N_2 - mg\cos\theta \tag{3.18}$$

Solve The one governing the motion up or down the slope is given by (3.17), which can be solved exactly. It can be written as $\ddot{r} - r\dot{\theta}^2 = -g\sin\theta$, and because $\theta = \theta_0 + \omega_0 t$, we have

$$\ddot{r} - \omega_0^2 r = -g\sin(\theta_0 + \omega_0 t) \tag{3.19}$$

Equations like this one are known as forced, ordinary differential equations. In our case the forcing is due to gravity and initial conditions: $-g\sin(\theta_0 + \omega_0 t)$. The general solution for this equation is

$$r(t) = d_1 e^{\omega_0 t} + d_2 e^{-\omega_0 t} + \frac{g}{2\omega_0^2}\sin(\theta_0 + \omega_0 t)$$

The constants d_1 and d_2 are found from matching this solution to the problem's initial conditions ($r(0) = L$, $\dot{r}(0) = 0$). Simplifying the final result in terms of hyberbolic sines and cosines yields

$$r(t) = \left[L - (g/2\omega_0^2)\sin\theta_0\right]\cosh(\omega_0 t)$$
$$- (g/2\omega_0^2)\cos\theta_0\sinh(\omega_0 t) + (g/2\omega_0^2)\sin(\theta_0 + \omega_0 t)$$

Check The units in our answer all check out. What we can also do is differentiate our result and see if (3.19) is satisfied. I'll leave this to you.

EXAMPLE 3.8 MING BOWL ON A MOVING SLOPE WITH FRICTION (Theory on page 119)

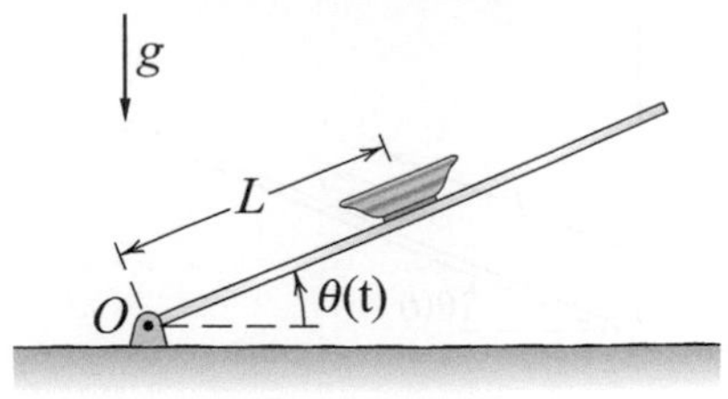

Figure 3.19 Bowl on a rotating support

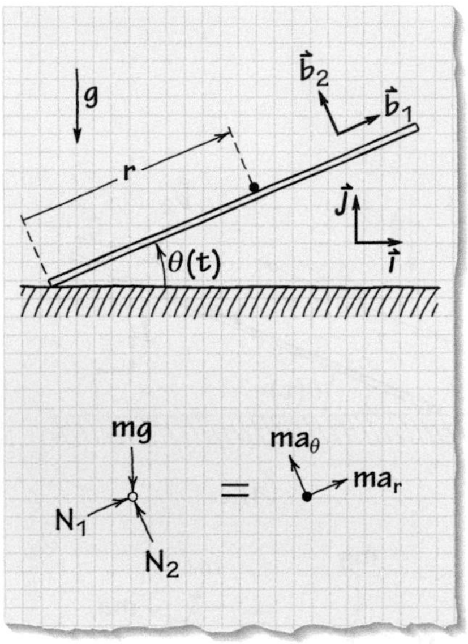

Figure 3.20 Schematic of sliding bowl

I just can't get enough of this Ming bowl. For this example, let's again have the Ming bowl on a board (**Figure 3.19**) and see what happens if there exist both static (μ_s) and dynamic (μ_d) coefficients of friction between the bowl and board. Let $\omega_0 = 0.7$ rad/s, $m = 0.4$ kg, $L = 2.0$ m, $\mu_s = 0.6$, and $\mu_d = 0.3$. At what time t does the bowl begin to slip, and how fast is it moving when it reaches O (the left end of the board)?

Goal

(a) Find the angle of inclination at which slip occurs and from this determine the time of slip.

(b) Find the equation of motion for motion along the board once slip begins, and integrate to find the time to travel 2 m downslope.

Given Angular rotation rate of the supporting board and friction coefficients between the bowl and board.

Draw **Figure 3.20** shows the associated schematic of our problem.

Formulate Equations We need to consider two phases: before and after slip. The governing equations (derived in Example 3.6) are

$$\boldsymbol{b}_1: \qquad m(\ddot{r} - r\dot{\theta}^2) = N_1 - mg\sin\theta \tag{3.20}$$

$$\boldsymbol{b}_2: \qquad m(2\dot{r}\dot{\theta} + r\ddot{\theta}) = N_2 - mg\cos\theta \tag{3.21}$$

Assume Before slip occurs ($\dot{r} = \ddot{r} = 0$), the friction force developed along the board (N_1) is sufficient to hold the bowl in place. The limit occurs when N_1 equals the frictional maximum $\mu_s N_2$. From (3.20) and (3.21), this happens when

$$\ddot{r} - r\dot{\theta}^2 + g\sin\theta = \mu_s(2\dot{r}\dot{\theta} + r\ddot{\theta} + g\cos\theta)$$

Applying the no-slip conditions and the constant rate of rotation gives

$$-L\omega_0^2 + g\sin\theta = \mu_s\, g\cos\theta \tag{3.22}$$

Once this limit is exceeded, the bowl starts sliding. As a result, $\ddot{r}$ becomes an unknown to be dealt with. Luckily, we also lose an unknown, namely N_1, which will equal $\mu_d N_2$. Our governing equations (with constant rotation rate) become

$$\boldsymbol{b}_1: \qquad m(\ddot{r} - r\omega_0^2) = \mu_d N_2 - mg\sin\theta$$

$$\boldsymbol{b}_2: \qquad 2m\dot{r}\omega_0 = N_2 - mg\cos\theta$$

Eliminating N_2 yields the equation of motion

$$\ddot{r} - 2\mu_d\omega_0\dot{r} - \omega_0^2 r = -g\sin\theta + \mu_d g\cos\theta \tag{3.23}$$

Solve Using the given values in (3.22) gives

$$-(2.0\,\text{m})(0.7\,\text{rad/s})^2 + (9.81\,\text{m/s}^2)\sin\theta^* = 0.6(9.81\,\text{m/s}^2)\cos\theta^* \quad (3.24)$$

where θ^* represents the critical value of θ beyond which slip occurs.

Equation (3.24) can be solved in a variety of ways. One way is to cast it in the form

$$f(\theta) \equiv -(2.0\,\text{m})(0.7\,\text{rad/s})^2 + (9.81\,\text{m/s}^2)(\sin\theta^* - 0.6\cos\theta^*) \quad (3.25)$$

and look for the zeros of the equation—that is, where $f(\theta^*) = 0$. Doing so, using MATLAB's `fzero` function, for instance, will give us $\theta^* = 0.626$ rad as the angle at which slip begins. The time at which the board angle reaches this value is found from $\omega_0 t^* = (0.7\,\text{rad/s})t^* = 0.626$:

$$t^* = 0.894\,\text{s}$$

That's one part down; we know when slip begins. Now we need to find the resulting time response of r. The initial conditions are $r(0) = 2$ m, $\dot{r}(0) = 0$ m/s. (For simplicity I will define the new time zero to be the moment slip begins.) The angle of the board at any instant after slipping begins is given by

$$\theta = 0.626\,\text{rad} + \omega_0 t = 0.626\,\text{rad} + (0.7\,\text{rad/s})t \quad (3.26)$$

From (3.23) we have the relationship

$$\ddot{r} = 2\mu_d\omega_0\dot{r} + \omega_0^2 r - g\sin\theta + \mu_d g\cos\theta$$

Using the given parameter values and integrating with MATLAB (**Figure 3.21**) show that r goes to zero at 0.90 s. The corresponding value of $\dot{r}(0.9)$ is -5.69 m/s, showing that the bowl is headed downslope, as expected. The inclination of the board at this time, from (3.26), is 72°. As you can see from **Figure 3.21**, the speed $\dot{r}$ is increasing rapidly as both the slope of the board increases and gravity continues to act on the bowl, an observation that's fully in line with intuition.

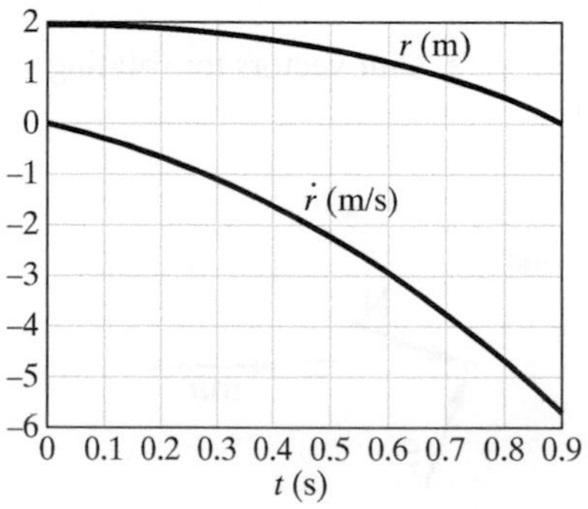

Figure 3.21 $r(t)$ and $\dot{r}(t)$ for bowl

Check The main caution here is to check whether our assumptions were correct. The direction in which N_1 is drawn on the FBD assumes that the bowl will slip downslope. This isn't necessarily true. If the board is rotating very quickly, the bowl can be flung up the slope, away from the left edge. What we need to look at is the pre-slip condition and to determine in what direction the frictional force acts to maintain a no-slip condition. If this direction matches the one chosen for the FBD, then all is well. If not, we must change the sign of N_1 when solving the equation of motion. For this problem, the initial assumption turns out to be a good one. The static force of friction points upslope, indicating that when slip occurs the bowl will move downslope because the frictional force, though trying to hold it in place, isn't strong enough to do so anymore.

EXAMPLE 3.9 NO-SLIP IN A ROTATING ARM (Theory on page 119)

Figure 3.22 shows an arm that rotates in a horizontal plane about O. Attached to the end of the arm is a small enclosure that holds a particle P. The surface farthest from O is flat and angled at 80°, as shown. What is the minimum coefficient of static friction μ_s that would permit P to remain fixed with respect to the angled surface? Assume that $\dot{\theta} = \omega$ and is constant.

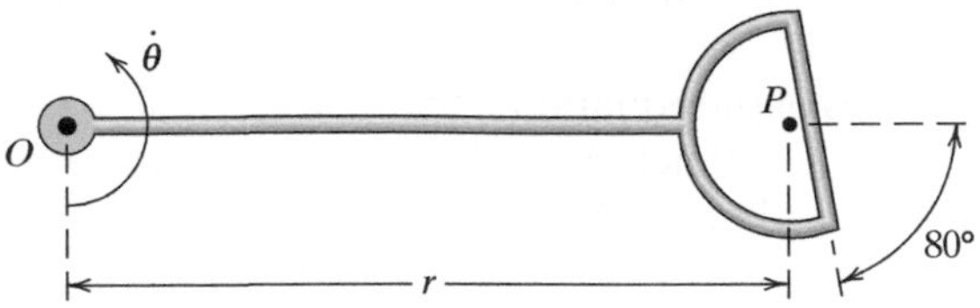

Figure 3.22 Rotating arm with enclosed mass

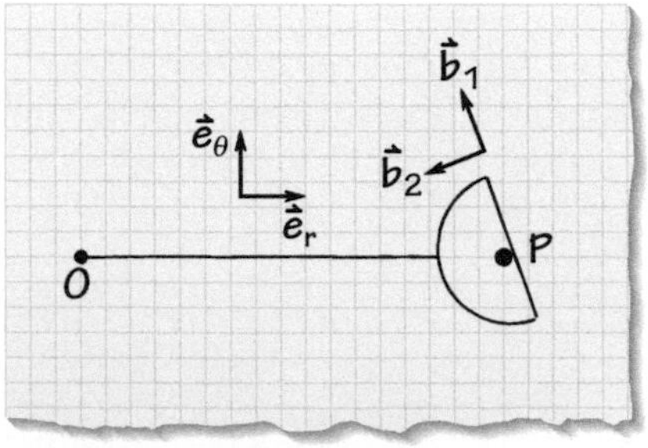

Figure 3.23 Unit vectors for rotating arm

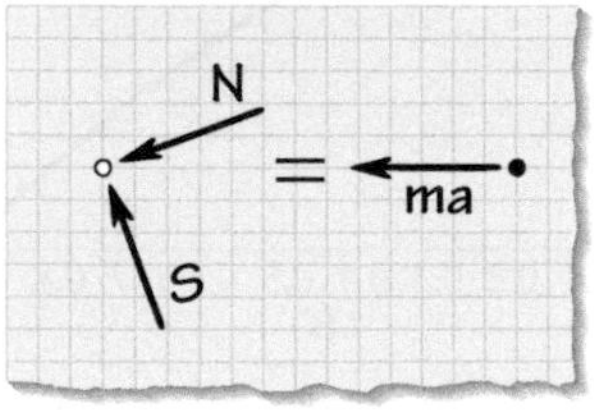

Figure 3.24 FBD=IRD for constrained mass

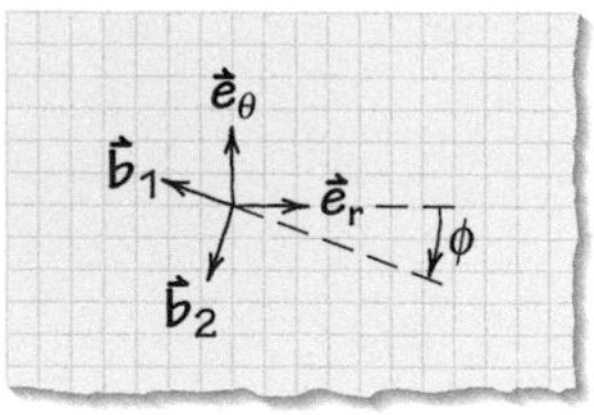

Figure 3.25 Unit vectors aligned to simplify construction of coordinate transformation array

Goal Calculate the minimum value of μ_s such that P won't slide relative to the spinning enclosure.

Given Orientation of the enclosure and distance of P from center of rotation.

Draw **Figure 3.23** shows the system under consideration along with two sets of unit vectors, both fixed to the rotating arm, while **Figure 3.24** shows the relevant free-body/inertial-response diagram.

This problem is an example of when it's convenient to set up the unit vectors in a way that's not so convenient when it comes time to produce the transformation array. Logically, it makes sense to align the $\boldsymbol{b}_1$, $\boldsymbol{b}_2$ vectors as shown in **Figure 3.23**, with $\boldsymbol{b}_1$ directed along the surface that P contacts and $\boldsymbol{b}_2$ normal to that surface. We know that there will be a normal force between the constraining surface and P and a sliding force developed along the surface (shown as N and S, respectively, in **Figure 3.24**). Although this is nice from the standpoint of physics, we derived the coordinate transformation arrays for angles that are small and for which it was then easy to say $\cos\phi \approx 1$ and $\sin\phi \approx \phi$.

So what do we do? Just allow the 80° angle of our problem to become much smaller and rotate the unit vectors $\boldsymbol{b}_1$ and $\boldsymbol{b}_2$ along with it. Doing so will give us **Figure 3.25**. It's now straightforward to create our transformation. $\boldsymbol{b}_1$ and $\boldsymbol{e}_r$ are essentially opposed; hence they are related by $-\cos\phi$. The same holds between $\boldsymbol{b}_2$ and $\boldsymbol{e}_\theta$. Getting to the tip of $\boldsymbol{b}_1$ means going in the opposite direction to $\boldsymbol{e}_r$ and then just a little bit in the $\boldsymbol{e}_\theta$ direction. Hence $\boldsymbol{b}_1 = -\cos\phi\boldsymbol{e}_r + \sin\phi\boldsymbol{e}_\theta$. Similarly, getting to the tip of $\boldsymbol{b}_2$ means going in the opposite direction to $\boldsymbol{e}_\theta$ and then just a bit along the negative $\boldsymbol{e}_r$ direction: $\boldsymbol{b}_2 = -\cos\phi\boldsymbol{e}_\theta - \sin\phi\boldsymbol{e}_r$. The entire transformation array is thus given by

	$\boldsymbol{e}_r$	$\boldsymbol{e}_\theta$
$\boldsymbol{b}_1$	$-\cos\phi$	$\sin\phi$
$\boldsymbol{b}_2$	$-\sin\phi$	$-\cos\phi$

where ϕ, as shown in **Figure 3.25**, should be set equal to 80° to match the example's conditions.

Formulate Equations A force balance gives us

$$m[(\ddot{r} - r\dot{\theta}^2)\boldsymbol{e}_r + (r\ddot{\theta} + 2\dot{r}\dot{\theta})\boldsymbol{e}_\theta] = N\boldsymbol{b}_2 + S\boldsymbol{b}_1$$

Assume Assume that the mass does not move relative to the enclosure.

Solve Using the constraints to simplify our set of unknowns, we will need to apply the given system constraints $r = L, \dot{r} = \ddot{r} = 0,\ \ddot{\theta} = 0$ in our equation of motion. Expressing everything in terms of $\boldsymbol{b}_1$ and $\boldsymbol{b}_2$ gives us

$$-mL\dot{\theta}^2\boldsymbol{e}_r = N\boldsymbol{b}_2 + S\boldsymbol{b}_1$$
$$mL\omega^2 \cos 80°\boldsymbol{b}_1 + mL\omega^2 \sin 80°\boldsymbol{b}_2 = N\boldsymbol{b}_2 + S\boldsymbol{b}_1$$

Equating coefficients leads to

$\boldsymbol{b}_1$: $$S = mL\omega^2 \cos 80° \tag{3.27}$$

$\boldsymbol{b}_2$: $$N = mL\omega^2 \sin 80° \tag{3.28}$$

When S is a maximum, $S = \mu_s N$. Using (3.27) and (3.28) gives us

$$L\omega^2 \cos 80° = \mu_s L\omega^2 \sin 80°$$

$$\boxed{\mu_s = \cot 80° = 0.176}$$

Check Why is the value we just solved for equal to the minimum value of μ? It all has to do with our assumptions. We approached the problem by *assuming* that the mass was stationary with respect to the enclosure. For this to be the case, the coefficient of static friction has to be "large enough"—that is, sufficiently large so that the forces trying to get the mass to slide aren't able to overcome the static friction. If the coefficient were very large, it would mean that $\mu_s N$, the maximum force that can be generated without causing slip, would be much larger than S, the force that's actually required to keep the mass from slipping at the given rotation rate. If you slowly decreased the coefficient of static friction, $\mu_s N$ would decrease as well and would approach S. The critical condition would be when the coefficient had been reduced so far that $\mu_s N$ just equaled S—any smaller and the mass would slip. Thus setting S equal to $\mu_s N$ allows us to solve for the minimum value.

EXAMPLE 3.10 FORCES ACTING ON A PAYLOAD (Theory on page 119)

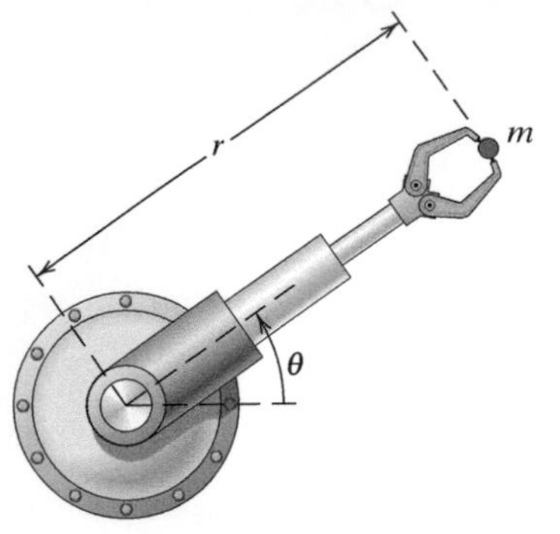

Figure 3.26 Mass in a robotic arm

A robotic arm that operates in a horizontal plane is shown in **Figure 3.26**. The programming for the arm is such that the 4 kg payload with mass m follows a path given by $r(t) = a_1 + a_2 t + a_3 t^2$ where $a_1 = 0.7$ m, $a_2 = 0.1$ m/s, $a_3 = 0.05$ m/s^2, and $\theta(t) = b_1 t + b_2 t^2$ where $b_1 = 0.8$ rad/s, $b_2 = -0.03$ rad/s^2. What are the forces acting on the payload at $t = 1.1$ s? Ignore gravity.

Goal Determine the forces acting on the payload at a particular time.

Given Description in time of the motion variables.

Assume Because r and θ are given as explicit functions of time, they and their derivatives are not unknowns. Thus we have two equations with two unknowns (F_r and F_θ).

Draw **Figure 3.27*a*** shows a system schematic. The arm doesn't come into play except as the means by which the payload is moved. Thus our diagram includes only the payload itself. Because the motion is given in terms of r and θ, we'll use a polar coordinate representation.

Figure 3.27*b* shows that all the applied forces are represented by the two net forces F_r and F_θ.

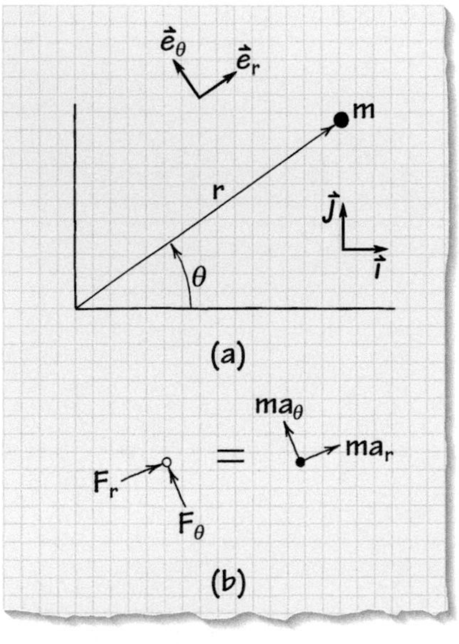

Figure 3.27 Schematic and FBD=IRD of mass

Formulate Equations Applying (3.15) and (3.16) gives us

$$m(\ddot{r} - r\dot{\theta}^2) = F_r \quad \text{and} \quad m(r\ddot{\theta} + 2\dot{r}\dot{\theta}) = F_\theta$$

Solve Differentiating our expressions for r and θ leads to

$$\begin{aligned} r(t) &= a_1 + a_2 t + a_3 t^2 & \theta(t) &= b_1 t + b_2 t^2 \\ \dot{r}(t) &= a_2 + 2a_3 t & \dot{\theta}(t) &= b_1 + 2b_2 t \\ \ddot{r}(t) &= 2a_3 & \ddot{\theta}(t) &= 2b_2 \end{aligned}$$

and evaluating at $t = 1.1$ s, we find

$$\begin{aligned} r(1.1\,\text{s}) &= 0.871\,\text{m} & \theta(1.1\,\text{s}) &= 0.844\,\text{rad} \\ \dot{r}(1.1\,\text{s}) &= 0.210\,\text{m/s} & \dot{\theta}(1.1\,\text{s}) &= 0.734\,\text{rad/s} \\ \ddot{r}(1.1\,\text{s}) &= 0.100\,\text{m/s}^2 & \ddot{\theta}(1.1\,\text{s}) &= -0.060\,\text{rad/s}^2 \end{aligned}$$

Using these values in our equations of motion yields

$$F_r = -1.48\,\text{N} \quad \text{and} \quad F_\theta = 1.02\,\text{N}$$

Check There's not much to check beyond making sure there weren't any calculation errors. By the way, this is a *very* common place to make an error. If your answers don't match the results given for a problem and you're sure your approach was correct, you should re-solve the problem using a different sequence of steps than the ones you used the first time. The human mind tends to run in the same pattern over and over again, and if you just go over the problem in the same way you did originally, you will probably just make the same arithmetic error each time. Often when students come to me and can't figure out what's wrong with their approach, it's just a simple mathematical error that they're making each time they run through the solution.

EXERCISES 3.2

3.2.1. **[Level 1]** A car is driving over a hill at a constant velocity. The hill's surface is well approximated as circular. What is the traction force between the tires and the road at $\theta = 30°$?

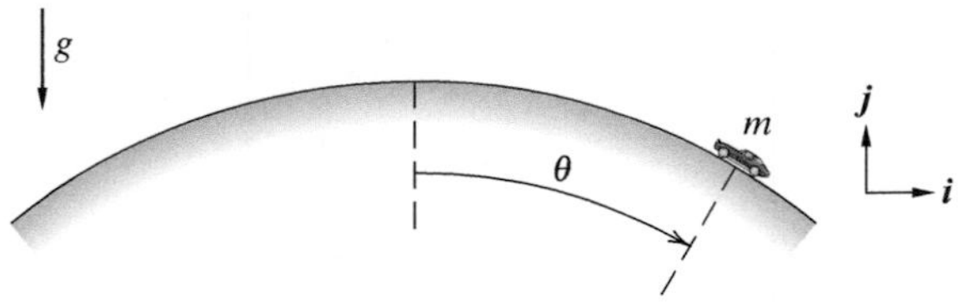

E3.2.1

3.2.2. **[Level 1]** A carnival ride is shown in which people are spun around in a rotating cylinder. When the angular velocity reaches a certain value, the floor drops down but the people remain stuck to the wall. If the coefficient of friction is $\mu = 0.8$ and the radius of the cylinder is 2 m, calculate the critical value of the cylinder's rotational speed ω to ensure that the people don't slip down when the floor moves away from them.

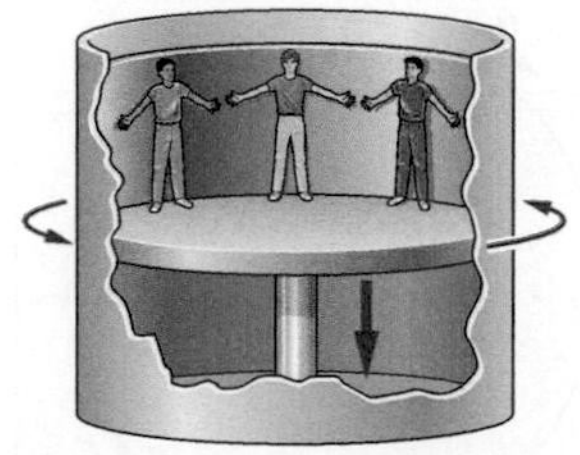

E3.2.2

3.2.3. **[Level 1]** Consider the Ming bowl problem of Example 3.6. Does an ω_0 exist such that the bowl remains fixed as the board rotates (i.e., the centrifugal acceleration due to rotation cancels the acceleration due to gravity)?

3.2.4. **[Level 1]** Assume that a driver (car modeled as a point mass) is negotiating a circular turn with a radius of 49 m. The car and driver have a mass of 1724 kg and the coefficient of friction between the car and road is $\mu_1 = 0.85$. What is the maximum constant speed for which the car can travel at the given radius?

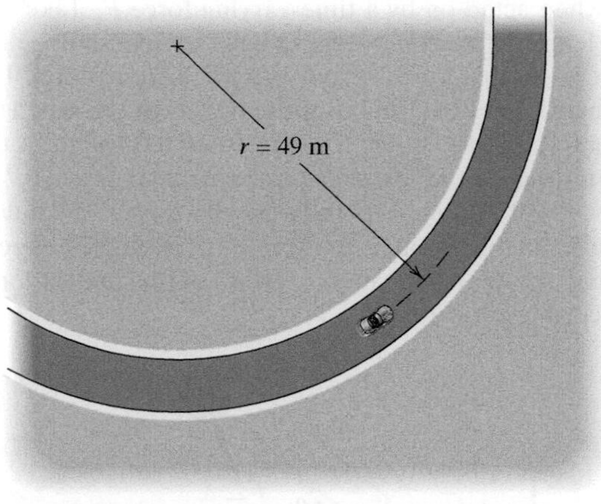

E3.2.4

3.2.5. **[Level 2]** A ball of mass $m = 2$ kg is attached to a fixed point O by an elastic cord and is constrained to rotate about O in the horizontal plane under the action of an applied force F that varies with time. Let r denote the length of cord between O and the mass, and take θ to be the angle that the cord makes with the horizontal. At a particular instant, $r = 0.9$ m, $\theta = 50°$, $\dot{r} = 1.2$ m/s, $\dot{\theta} = 6$ rad/s, $\ddot{r} = 0$, $\ddot{\theta} = 1$ rad/s^2, and F acts on the ball in the $\boldsymbol{j}$ direction. If the surface on which the ball moves has a coefficient of friction of $\mu = 0.7$, find the instantaneous value of the applied force F and the tension T in the cord.

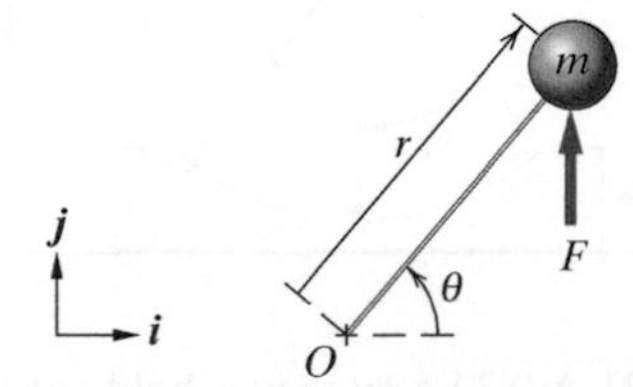

E3.2.5

3.2.6. **[Level 2]** A mass particle is placed against a cylinder and is held in place by small retaining walls. Initially stationary at $\theta = 0$, the cylinder begins to rotate counterclockwise in such a manner that $\theta = t^2$ rad/s^2. The radius of the cylinder is 2 m. Determine if the mass particle will leave the cylinder's surface before it reaches the top.

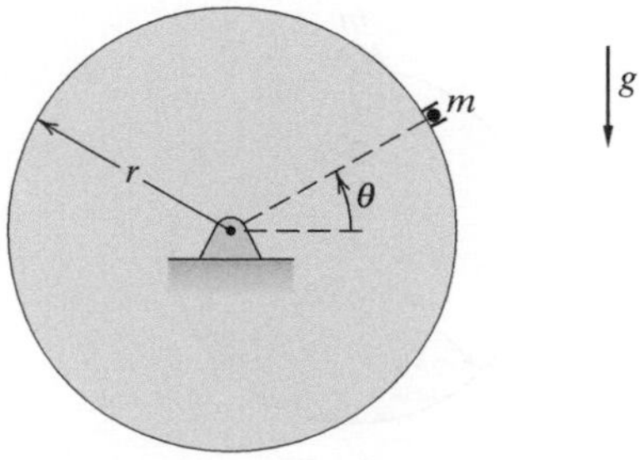

E3.2.6

3.2.7. [Level 2] A ball attached to a fixed point O by an elastic cord is allowed to rotate about O in the vertical plane when acted on by a time-varying force F. Take r to be the length of cord between O and the mass, and define θ as the angle that the cord makes with the horizontal. The ball's mass is $m = 1$ kg, and suppose that it experiences a total drag force of $F_d = b\|\boldsymbol{v}\|^2$, where $b = 0.001\ \text{Ns}^2/\text{m}^2$ is the ball's air drag coefficient. At a particular instant, $r = 0.7$ m, $\theta = 40°$, $\dot{r} = 0.9$ m/s, $\dot{\theta} = 15$ rad/s, $\ddot{r} = 0.2\ \text{m/s}^2$, $\ddot{\theta} = 2\ \text{rad/s}^2$, and F acts on the ball at an angle $\beta = 70°$ from the horizontal. What are the instantaneous values of the applied force F and cord tension T?

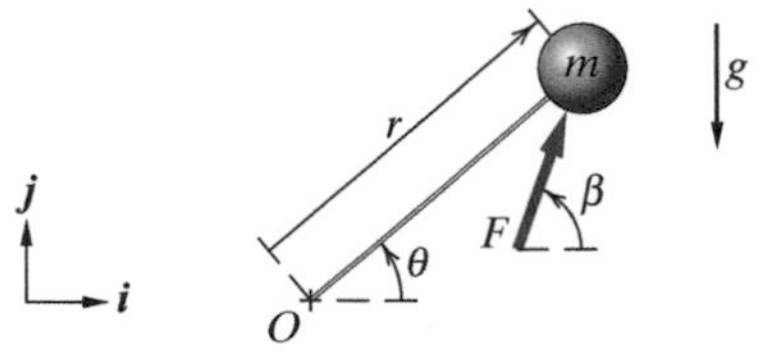

E3.2.7

3.2.8. [Level 2] A friend of yours has designed an acceleration tester. A mass can slide within the friction-free circular bowl. It's clear that if placed in a car, which then accelerates at a constant rate of $\ddot{x} = a_x\ \text{m/s}^2$ to the right, the mass will slide up the left slope. Your friend isn't skilled enough in dynamics to determine the acceleration for which θ_0 will just barely come up to the edge of the bowl. Your task is to answer this question. Obtain a_x as a function of θ_0.

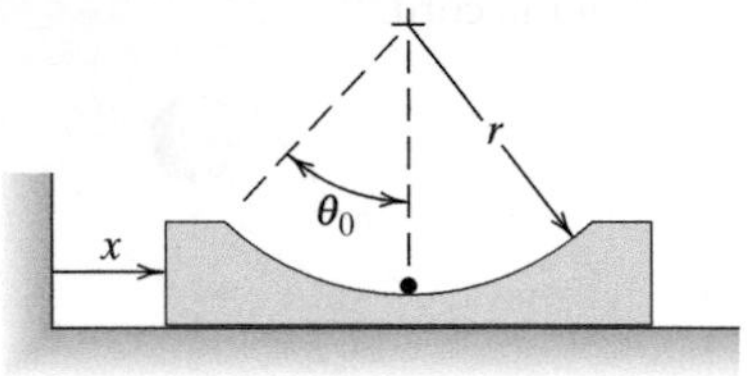

E3.2.8

3.2.9. [Level 2] A 0.3 kg mass m is held onto the surface of a horizontal, rotating disk by a stretched spring k that presses the mass onto the disk's surface with a force of 20 N. Small projections to either side of the mass keep it from sliding along the disk's periphery. The disk is initially at rest and at $t = 0$ begins spinning with a constant angular acceleration of $\ddot{\theta} = 1.0\ \text{rad/s}^2$. When will the mass first lose contact with the disk's surface? $r = 0.5$ m.

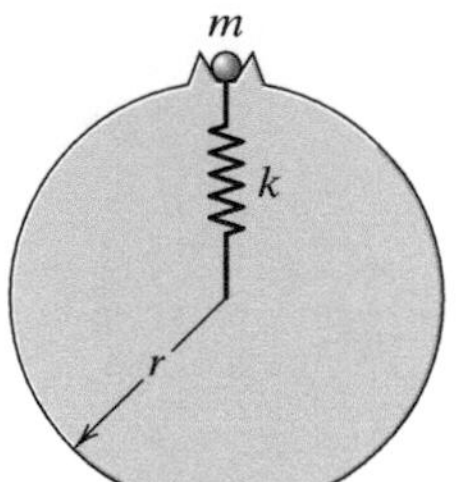

E3.2.9

3.2.10. [Level 2] A mass is resting on a horizontal turntable. The coefficients of friction between the mass and turntable are $\mu_s = \mu_d = 0.5$. At $t = 0$ the initially stationary turntable is accelerated at the constant rate of $\ddot{\theta} = 4\ \text{rad/s}^2$. At what time will the mass begin to move? $L = 1$ m.

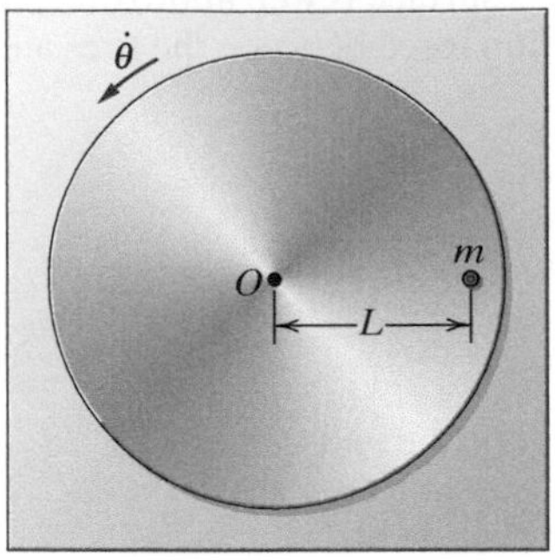

E3.2.10

3.2.11. [Level 2] An engineer sets up an experiment to determine the coefficient of static friction μ_s for an unknown material. She cuts the material into a disk and places a test mass on top, $L = 0.75$ m from the center, and she proceeds to spin the disk with an angular acceleration of $\ddot{\theta}(t) = 40t\ \text{rad/s}^2$ counterclockwise. The engineer notes that the test mass slips at $t = 0.2$ s. What is μ_s?

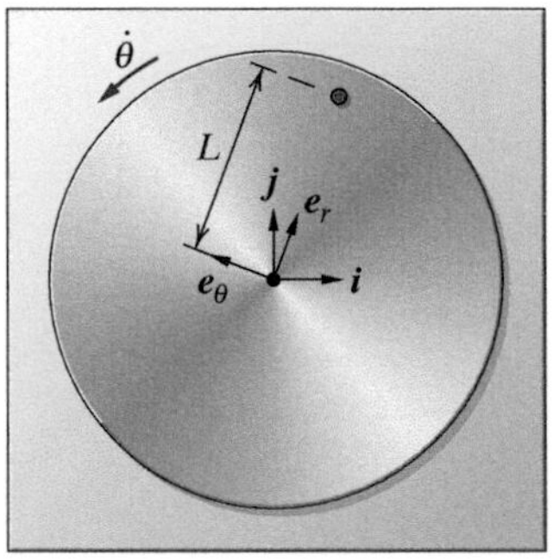

E3.2.11 Top view

3.2.12. [Level 2] The collar with mass m can slide freely along the rod $\overline{AB}$. The rod is rotating at a constant rate of $\dot{\theta} = 6$ rad/s. At $t = 0$ the collar is located a distance r_0 from A and $\dot{r} = 0$. You can neglect gravity. Also, the solution to $\ddot{x} + b^2x = 0$ is $x(t) = a_1 \sin(bt) + a_2 \cos(bt)$, and the solution to $\ddot{x} - b^2x = 0$ is $x(t) = c_1 \sinh(bt) + c_2 \cosh(bt)$.

a. Find the equations of motion for m.

b. Solve for the forces exerted on the mass by the rod as a function of time.

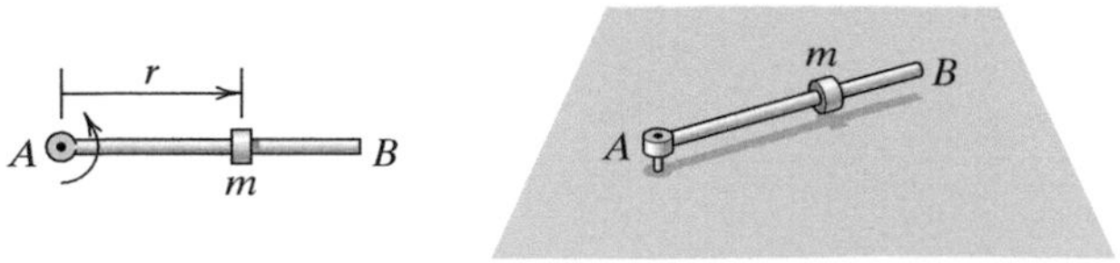

Top view Perspective view

E3.2.12

3.2.13. **[Level 2]** Consider the following idealization of a motorcyclist riding on a banked, circular track. The motorcycle and rider are approximated as a single particle. The rider is in an equilibrium condition—that is, always at a constant distance h from the ground and traveling at a constant velocity v. The coefficient of friction between the motorcycle and track is μ.

a. Find the maximum and minimum velocities allowable as a function of $\mu, \theta, g,$ and R.

b. If the motorcyclist's speed exceeds $v_{\max}$, what will happen?

c. Is there any difference in the analysis if, instead of the motorcycle moving with respect to the track, the motorcycle is at rest with respect to the track and the whole track rotates about $\overline{OO}$ at a constant angular rate w (where $\omega r = v$)?

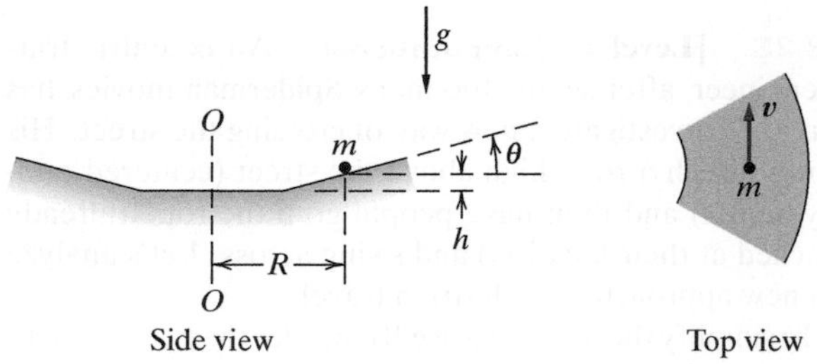

E3.2.13

3.2.14. **[Level 2]** A 2 kg collar C rides along a horizontally rotating arm $\overline{AB}$. A spring k connects C to A. The arm $\overline{AB}$ rotates at a constant rate of 20 rpm. At the illustrated instant $r = 0.9$ m, $\dot{r} = 1.2$ m/s, and $\ddot{r} = 2$ m/s^2. The coefficient of dynamic friction, μ_k, is equal to 0.2. What is the total force exerted by the spring on the collar?

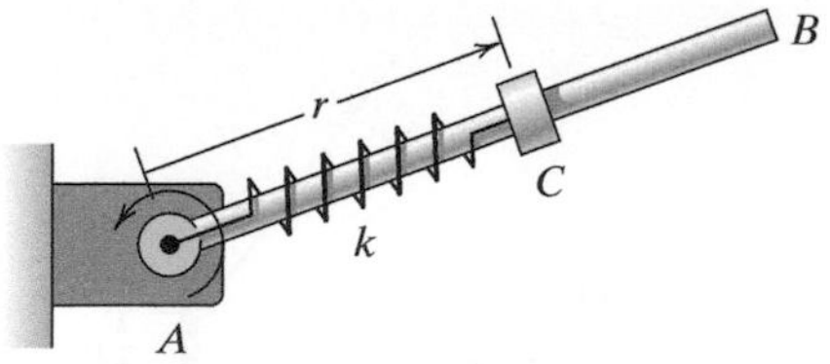

E3.2.14

3.2.15. **[Level 2]** A mass is attached to a rope of length L. Initially held horizontal, it is then released. At what angle θ will the rope break if the rope can withstand a maximum tensile force equal to twice the weight of the mass?

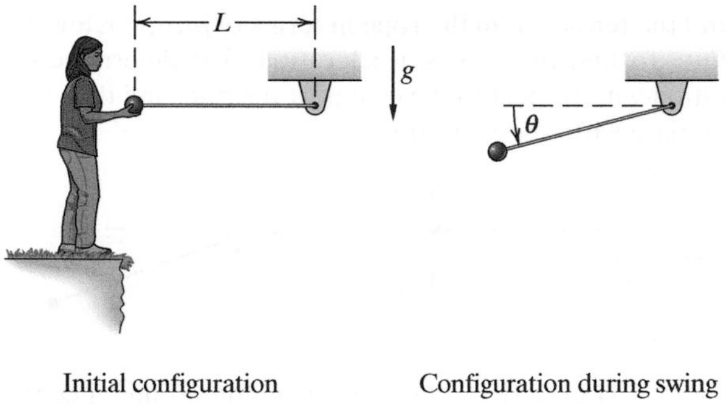

E3.2.15

3.2.16. **[Level 2]** We have all seen movies in which the hero is on the outside of some huge structure and starts rolling off. Let's analyze this from a dynamical point of view. We will assume that it's James Bond on the roof, model him as a point mass, and assume that the structure has a circular cross section. If the interface is frictionless and at $\theta = 0$ rad $\dot{\theta} = 0.5$ rad/s, at what angle will contact be lost? (*Hint:* You will need to find the equation of motion in terms of θ and determine where the normal force goes to zero.) $r = 20$ m.

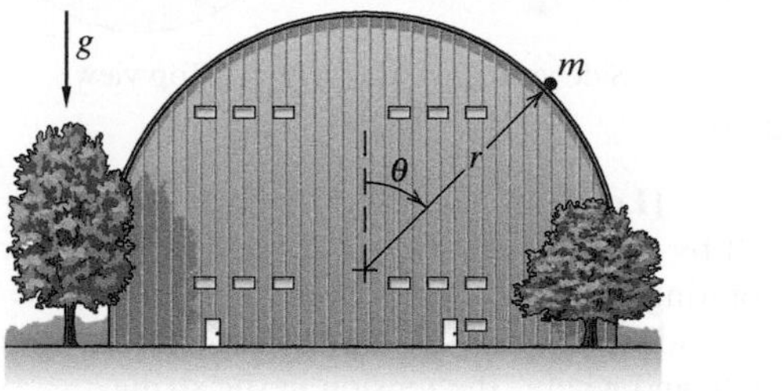

E3.2.16

3.2.17. **[Level 2]** A rotating space station is shown. High-speed cars run in the special connecting tubes to get from the central hub to the outer ring. The illustration shows a car traveling in the north connecting tube from the core toward the outer ring. If the tube wall is too weak and the car breaks through the tube's wall, on which side of the tube will the car emerge, E or W? Assume a constant rate of rotation, $\dot{\theta}$, as shown.

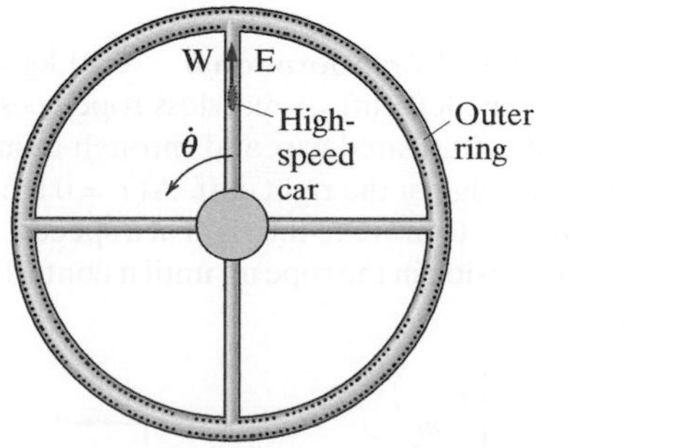

E3.2.17

3.2.18. **[Level 2]** A mass m is tied by a massless rope to a central post (top view shown). The rope's free length is initially L, and the mass has speed v_m. Derive an expression for

$\ddot{\theta}$ and the tension T in the rope in terms of $m, \theta, \dot{\theta}, r,$ and L. (*Hint:* To find the mass's acceleration, first determine its position with respect to the center of the pole and then differentiate with respect to time.)

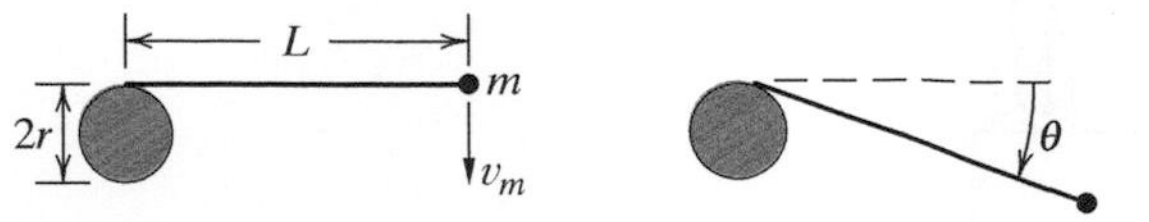

E3.2.18

3.2.19. **[Level 2]** A car (m) is running around a cone-shaped track. Treat the car as a point mass. The track's inclination angle θ is 45°, $r = 180$ m, $\mu_s = \mu_d = 0.7$, and the speed of the car is given by v. What are the minimum and maximum speeds of the car such that it doesn't drift either upslope or downslope?

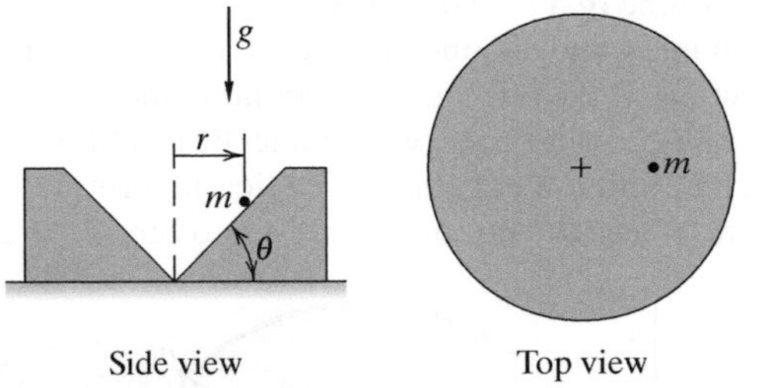

E3.2.19

3.2.20. **[Level 2]** The 2 kg collar B is being drawn to the left by the massless string attached to the collar and to the rotating link $\overline{OA}$. $L = 0.5$ m. Let $\theta = 30°$, $\dot{\theta} = 2$ rad/s, $\ddot{\theta} = 0.5$ rad/s^2, and $\beta = 45°$. What is the acceleration of the collar B, and what is the tension in the string?

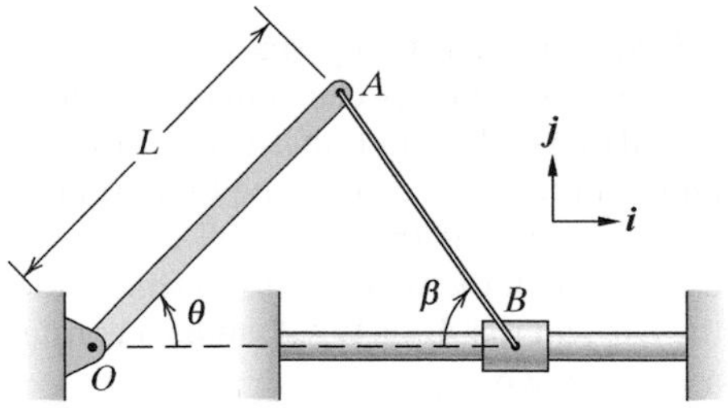

E3.2.20

3.2.21. **[Level 2] Computational** A 10 kg mass m is on the edge of the left cliff. A massless rope goes from the mass, across the illustrated gap, and through a small eyelet attached to the edge of the right cliff. At $t = 0$ the right end of the rope begins to move to the right at a speed $v = 0.2$ m/s. Calculate the tension in the rope up until it contacts the right cliff face.

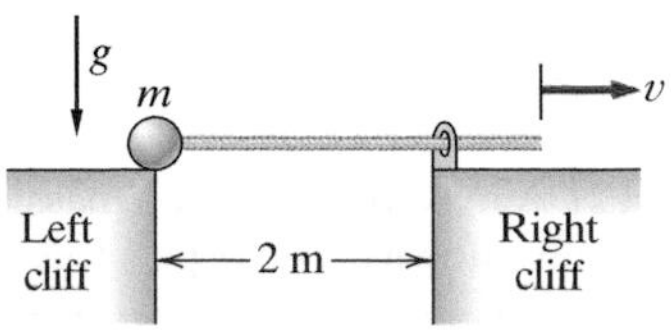

E3.2.21

3.2.22. **[Level 2] Computational** The 8 kg mass m_1 is on the edge of the left cliff. A massless rope goes from the mass, across the illustrated gap, through a small eyelet, and to a 12 kg mass m_2, sitting back from the edge of the right cliff. At $t = 0$, m_1 is just barely nudged off the edge of the left cliff. Calculate the tension in the rope up until it contacts the right cliff face, as well as its position at contact. The mass m_2 can slide without friction.

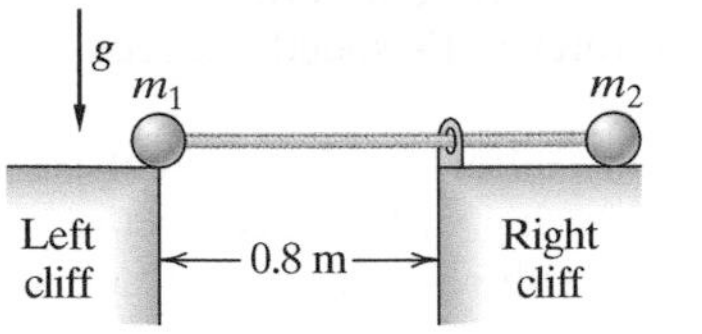

E3.2.22

3.2.23. **[Level 2] Computational** An eccentric traffic engineer, after seeing too many Spiderman movies, has started to investigate a new way of crossing the street. His plan is attach a rope high above the street (centered midway across) and then have people grab the rope (already attached at their location) and swing across. Let's analyze this new approach to pedestrian travel.

To simplify the modeling, we'll consider the pedestrian to be a lumped mass, attached to the rope at a height $h = 1.2$ m above the ground and initially positioned at the rightmost edge of the street. The width of the street is $L = 12$ m.

a. Determine what the rope's length L must be for the person to pass 0.3 m above the ground at the midway point.

b. Using the value of L found in **a**, determine how long the total swing will take to get from one side of the street to the other.

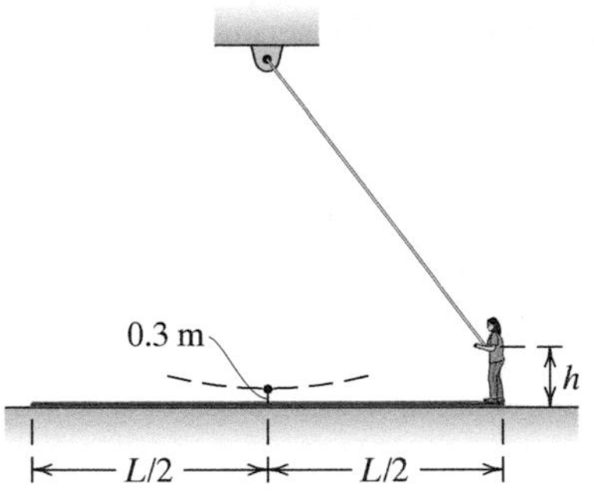

E3.2.23

3.2.24. **[Level 2] Computational** Two mass particles, A and B, each lie on a friction-free surface and are each joined by a massless string to the point O. Both have a mass of 0.4 kg. The distance from O to each particle is given by L $(L = 0.5$ m$)$. Initially the particles are at rest and $\theta = 45°$. At $t = 0$ a horizontally oriented force $F = 2$ N is applied, causing the particles to move to the right and toward each other. Determine when the particles collide and what their absolute speed is equal to at that time.

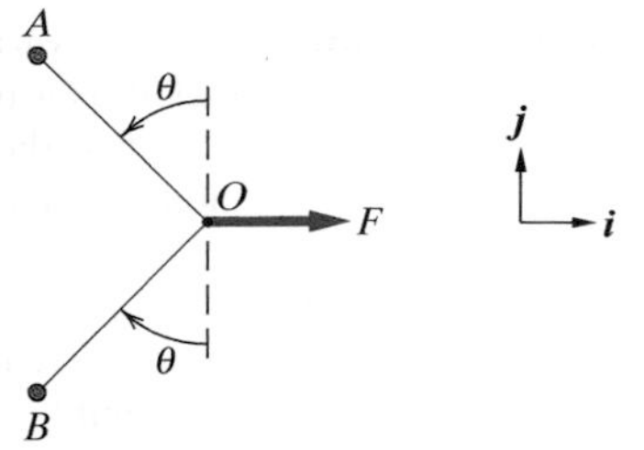

E3.2.24

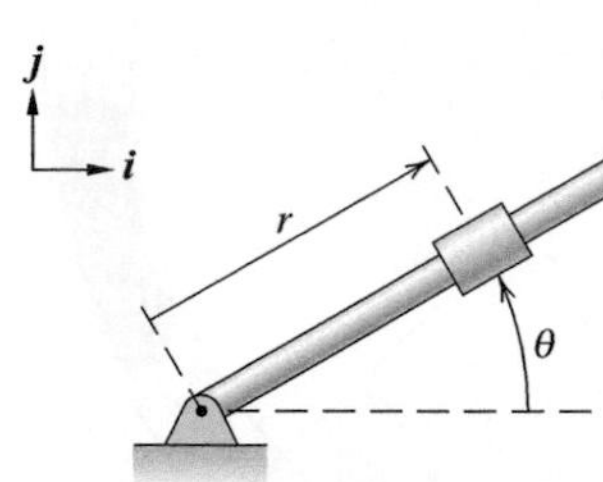

E3.2.27

3.2.25. **[Level 2]** Suppose the engineer from **Exercise 3.2.11** is using the same experimental procedure to determine the coefficient of static friction μ_s for another unknown material. This time, she spins the disk from rest with a constant angular acceleration of $\ddot{\theta} = 0.7\,\text{rad/s}^2$, and she notes that the test mass slips after the disk rotates $R = 1$ rev. What is μ_s for the material?

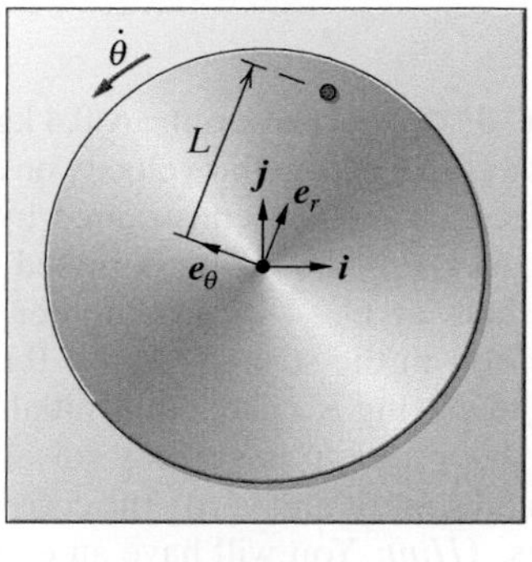

E3.2.25 Top view

3.2.26. **[Level 2]** **Computational** A 70 g marble is free to roll around on the inside of a hollow cylinder with a radius of $L = 0.25$ m which is lying on its side. The inner surface of the cylinder has a coefficient of friction of $\mu = 0.4$. If the marble starts at the bottom ($\theta = 0°$) and is projected with a velocity of $\boldsymbol{v}_0 = 5\boldsymbol{i}$ m/s, find when and where the marble loses contact with the cylinder's inner surface.

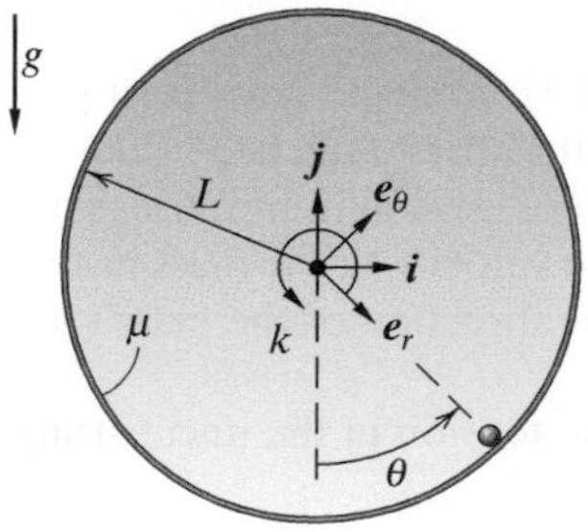

E3.2.26

3.2.27. **[Level 2]** A 2 kg collar is free to slide on a rough rod rotating in the horizontal plane. The collar's radial position along the rod is governed by $r(t) = \left(0.1 + \frac{1}{3}e^t\right)$ m, and its angular position is described by $\theta(t) = \dot{\theta}t$ rad, where $\dot{\theta} = 1.2$ rad/s. What are the in-plane forces acting on the collar at $t = 1$ s?

3.2.28. **[Level 2]** This exercise differs somewhat from **Exercise 3.2.16**. Now, instead of a stationary structure, a mass moves on the outer periphery of a counterclockwise-rotating cylinder. The coefficient of dynamic friction is $\mu_d = 0.2$. Initially, the relative velocity of the mass with respect to the cylinder's periphery is 33 m/s. Determine the value for θ for which the mass loses contact with the cylinder. $m = 50$ kg, $r = 30$ m, $\omega_0 = 0.6$ rad/s, and $\mu_s = 0.2$. Assume that the mass starts from the same initial conditions as in **Exercise 3.2.12**.

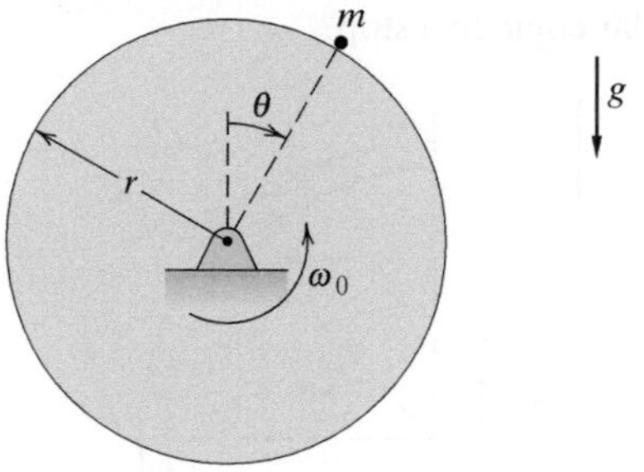

E3.2.28

3.2.29. **[Level 2]** **Computational** In this exercise we'll extend **Exercise 3.2.4** and see what can happen when a driver pushes the limit in a turn and encounters an unexpected road hazard. Assume that the driver is negotiating a turn at a constant speed and a radius of 49 m. He is traveling at the maximum speed allowed by a coefficient of friction of 0.85. At $t = 0$ s the car encounters a section of road covered with oil, reducing the coefficient of friction to $\mu_2 = 0.10$. Numerically integrate the car's equations of motion to determine how long it will take for it to leave the road ($r = 52$ m). Assume during this phase that the car simply slides along the road, experiencing a retarding force due to the frictional drag generated at the car/road interface.

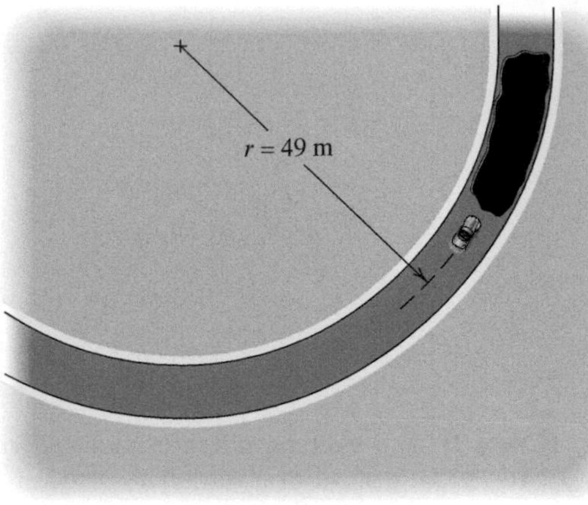

E3.2.29

3.2.30. [Level 3] As illustrated, a collar of mass $m = 1$ kg is constrained to move in the vertical plane along a smooth rod bent into a semicircle of radius $r = 0.5$ m. The collar starts from rest at the rod's top A ($\theta = 90°$) and is acted on by a constant force of $F = 20$ N applied at an angle $\beta = 45°$ with respect to the horizontal. At what angular position does the collar come to a stop?

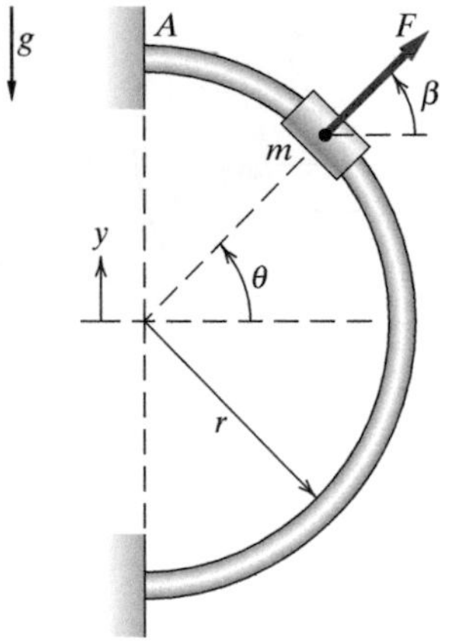

E3.2.30

3.2.31. [Level 3] A top view of a door, hinged at O, is shown. A young boy comes running through the door, causing it to swing counterclockwise at ω_0 rad/s. As the door swings, the mass m, previously resting quietly on the floor, is forced to slide along the door's surface. Calculate when the mass reaches the end of the door A and how far the door has opened at that instant.

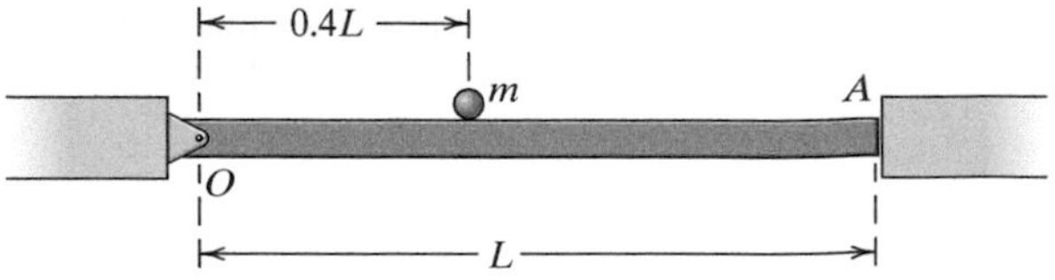

E3.2.31

3.2.32. [Level 3] Computational In a variant of **Exercise 3.2.16**, assume that you (modeled as a point mass) are on the top of a circular structure, such as the dome of a large telescope. At $t = 0$, the clam-shell roof starts to open. The coefficients of friction between you and the dome are $\mu_s = 0.65, \mu_d = 0.4$. $\dot{\beta} = 0.31$ rad/s, $\ddot{\beta} = 0$, and $r = 10$ m.

a. How fast will you be moving when you reach $\theta = 45°$?
b. What is the normal force at this point?
c. When will you lose contact with the surface?

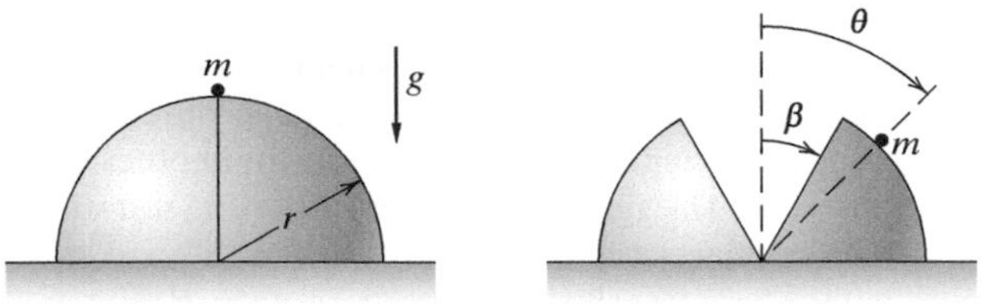

E3.2.32

3.2.33. [Level 3] Computational A 0.4 kg mass is initially rotating at a constant angular velocity on a horizontal, frictionless surface, its motion constrained by a massless, flexible string. At $t = 0$ the string is pulled at a rate $2t$ m/s^2 through a hole at O. Use a computer to calculate and plot the tension in the string from $t = 0$ to 1.2 s. The initial length of the string is 2.5 m. $\dot{\theta}$ is initially equal to 8 rad/s. Because this problem is actually solvable in closed form, go ahead and solve it and verify the correctness of the computer results. (*Hint:* You will have an equation in the form $\frac{d\dot{\theta}}{dt} = f_1(\dot{\theta})f_2(t)$. Separate this into $\frac{d\dot{\theta}}{f_1(\dot{\theta})} = f_2(t)\,dt$ and integrate both sides of the equation.)

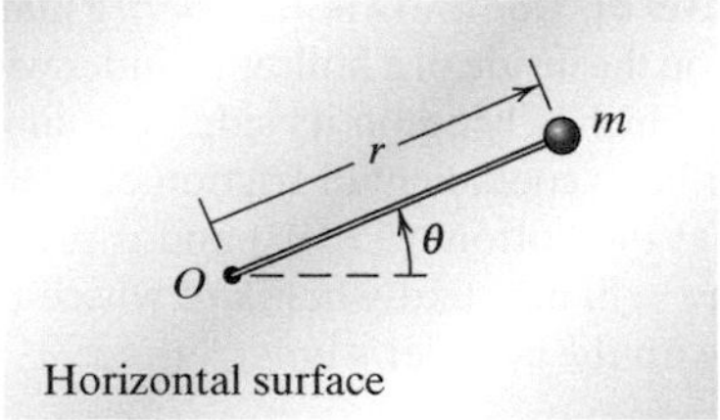

E3.2.33

3.2.34. [Level 3] A single mass is suspended from two strings. At $t = 0$ the string attached at B is cut.

a. What was the tension in the strings before the cut?
b. What is the velocity of the mass just after the cut?
c. What is the direction of the mass's acceleration just after the cut?
d. What is the tension in the uncut string just after the cut? $L = 2$ m.

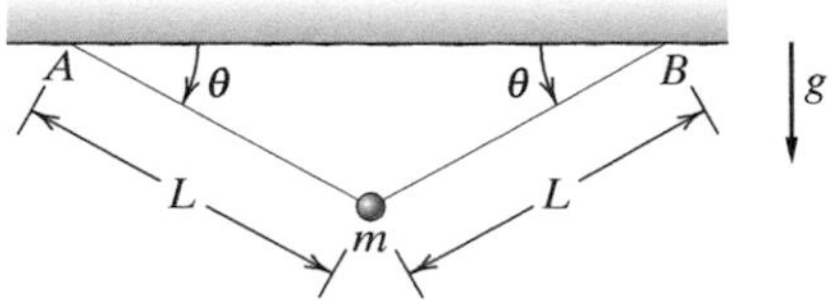

E3.2.34

3.2.35. [Level 3] Computational A 2 kg mass particle B is supported by the massless rope $\overline{AB}$ and is attached to a vertical wall through the massless rope $\overline{BC}$. At $t = 0$ the rope $\overline{AB}$ breaks. Calculate when and where the mass strikes the vertical wall for two cases.

a. Treat the rope $\overline{BC}$ as inextensible.

b. Treat the rope $\overline{BC}$ as elastic. The force exerted by the rope on the mass is equal to kx, where $k = 10$ N/m and x indicates the stretch of the rope beyond its rest length of 2 m.

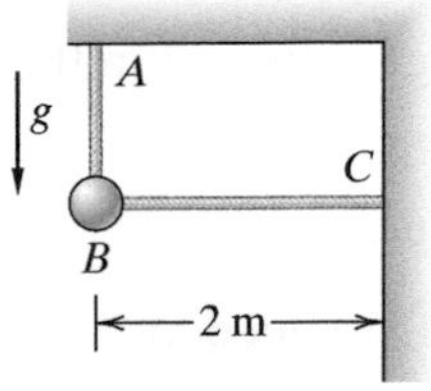

E3.2.35

3.2.36. [Level 3] Computational A curling team is competing in the Continental Cup tournament, and a judge watches from the lower left corner of the curling sheet (playing surface). Let r be the distance from the judge to the curling stone and θ be the angle his line-of-sight makes with the edge of the curling sheet. A player launches the curling stone directly in front of the judge (i.e., $\theta = 90°$), and a little while later the judge notices that the stone begins to deviate from a straight-line path. At this instant, the stone has a speed of $v_0 = 8$ m/s and a deviation of $\phi = 5°$ with respect to the curling sheet's centerline, and $\theta = 60°$. Determine where the curling stone stops. The stone has a mass of 18 kg, the centerline is $L = 2$ m from the curling sheet's edge, and the coefficient of friction is $\mu = 0.2$.

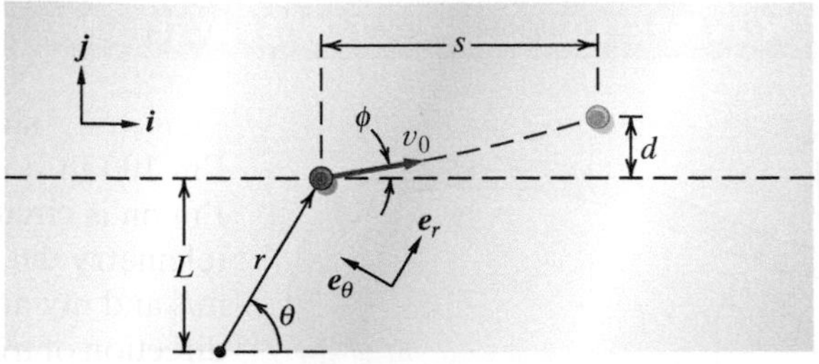

E3.2.36

3.2.37. [Level 3] Computational The 4 kg mass m_1 is on the edge of the left cliff. A massless rope goes from the mass, across the illustrated gap, and through a small eyelet to a linear extensional spring. The spring is initially unstretched. At $t = 0$ the mass is just barely nudged off the edge of the left cliff. Calculate the tension in the rope up until it contacts the right cliff face as well as its position at contact, knowing that the force supplied by the spring is equal to kx, where $k = 2$ N/m and x is equal to the stretch in the spring.

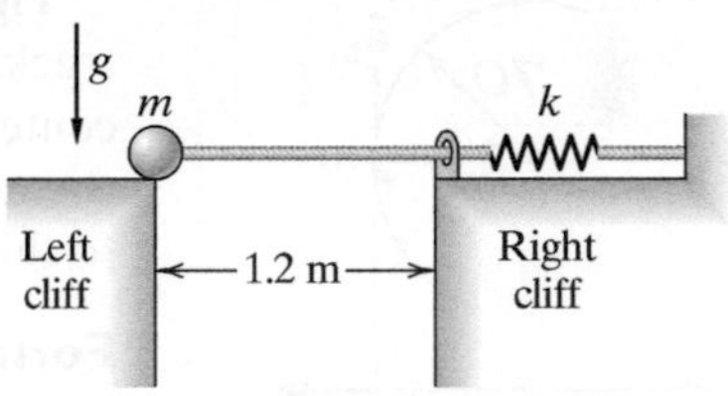

E3.2.37

3.3 PATH COORDINATES

Our final approach to looking at a system's motion over time uses path coordinates. For this case, we break the applied force into $\boldsymbol{e}_t$ and $\boldsymbol{e}_n$ components ($\boldsymbol{F} = F_t\boldsymbol{e}_t + F_n\boldsymbol{e}_n$), and, using (2.59), we obtain

$$F_t = m\dot{v} \tag{3.29}$$

$$F_n = m\frac{v^2}{r_C} \tag{3.30}$$

Recall that r_C is the radius of curvature of the path and v is the particle's speed.

As you can see, the basic procedure is identical to that shown in the previous two sections. You are given a known quantity (perhaps force) and have to find an unknown quantity (such as acceleration).

⋙ **Check out Example 3.11 (page 134), Example 3.12 (page 135), Example 3.13 (page 136), and Example 3.14 (page 138) for applications of this material.**

EXAMPLE 3.11 FORCES ACTING ON MY CAR (Theory on page 133)

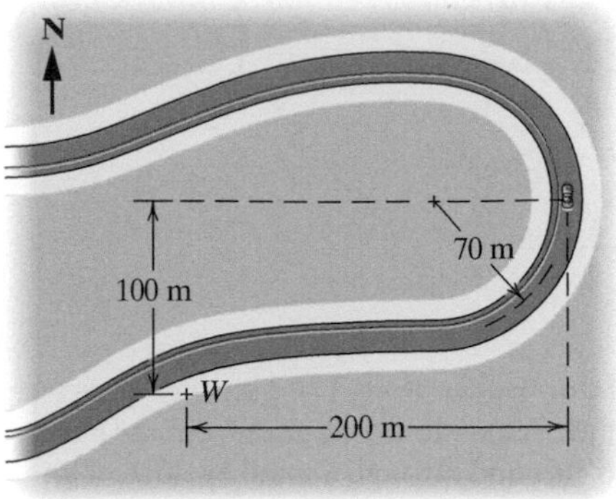

Figure 3.28 Z3 on a race track

Figure 3.28 shows me driving a Z3 at the track. At the illustrated instant, I'm 100 m north and 200 m east of my wife (W). The section of track I'm on is circular with a radius of 70 m, and I'm heading due north. The telemetry data my wife is receiving tells her that my current speed is 23 m/s and my net acceleration is 9.0 m/s^2. What forces are acting along my direction of travel and at right angles to it? The mass of my car plus myself equals 1450 kg, and we can, as usual, be treated as a particle.

Goal Find the forces F_t and F_n acting on the car in the $\boldsymbol{e}_t$ and $\boldsymbol{e}_n$ directions.

Given Speeds and positions of the car.

Assume No additional assumptions are needed.

Draw **Figure 3.29*a*** shows a simplified view of the problem. The unit vector $\boldsymbol{e}_t$ points along the car's direction of travel, and $\boldsymbol{e}_n$ points toward the center of curvature.

Figure 3.29*b* shows that two forces are acting in the plane of the racetrack: F_t pointing in the direction of travel and F_n pointing toward the center of curvature. (There also exist a vertical force due to gravity and the normal force of the ground against the car, but they are not relevant to this problem and so are being ignored.)

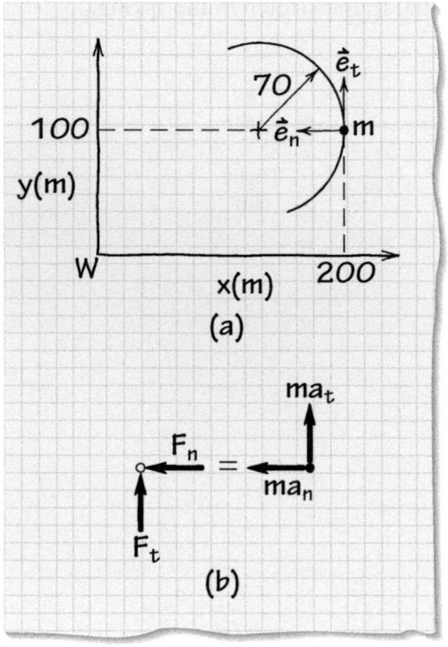

Figure 3.29 Schematic and FBD=IRD for Z3 on a racetrack

Formulate Equations Because the car is running along a track with known curvature, we can deduce the normal acceleration if we know the car's speed. Our equations of motion are (3.29) and (3.30)

$$F_t = m\dot{v} \quad \text{and} \quad F_n = mv^2/r_C$$

Solve In the normal direction, we have, from (2.59),

$$a_n = \frac{v^2}{r_C} = \frac{(23\,\text{m/s})^2}{70\,\text{m}} = 7.56\,\text{m/s}^2$$

We now use the given net acceleration magnitude of 9.0 m/s^2 in the expression $a^2 = a_n^2 + a_t^2$: $(9.0\,\text{m/s}^2)^2 = a_t^2 + (7.56\,\text{m/s}^2)^2$. Solving for a_t gives us $a_t = 4.89\,\text{m/s}^2$. Now we can to use $\boldsymbol{F} = m\boldsymbol{a}$ to find the forces:

$$F_t = (1450\,\text{kg})(4.888\,\text{m/s}^2) = 7.09 \times 10^3\,\text{N}$$

$$F_n = (1450\,\text{kg})(7.557\,\text{m/s}^2) = 1.10 \times 10^4\,\text{N}$$

Check A physical check is to ask if the values seem reasonable. About the highest g level you will see in a production car is around 1.0 (in other words, an acceleration of around 9.8 m/s^2). The acceleration components for this problem are 4.89 m/s^2 and 7.56 m/s^2, both within this limit.

EXAMPLE 3.12 FINDING A ROCKET'S RADIUS OF CURVATURE (Theory on page 133)

Figure 3.30 illustrates a rocket traveling through the atmosphere. At the instant illustrated, the 200 kg rocket has a velocity of $\boldsymbol{v} = 150\boldsymbol{i}$ m/s and is being acted on by gravity and by a thrust $\boldsymbol{T}$ given by $\boldsymbol{T} = 6000(\cos 15°\boldsymbol{i} + \sin 15°\boldsymbol{j})$ N. Determine the trajectory's radius of curvature at this position.

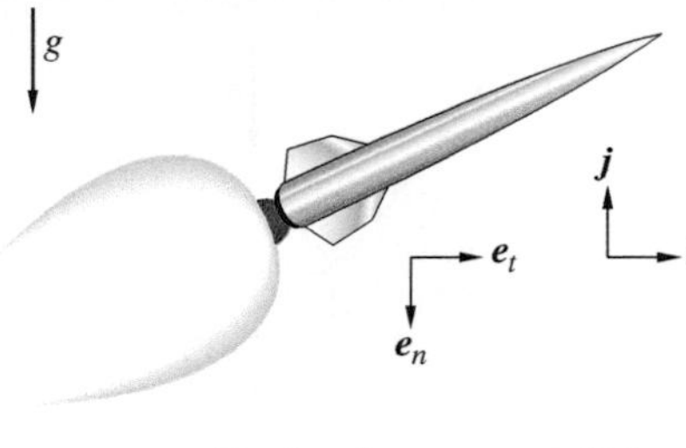

Figure 3.30 Rocket in powered flight

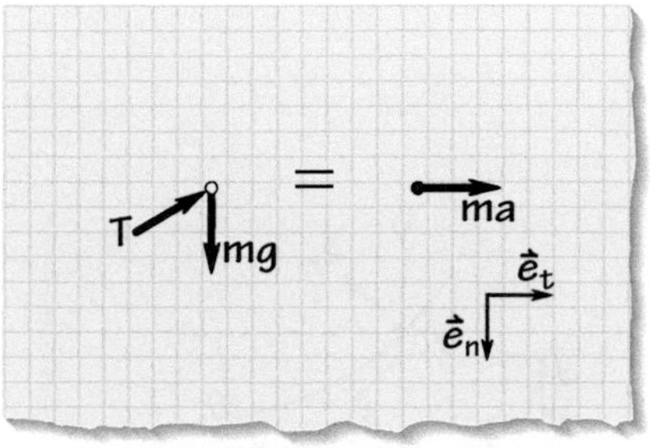

Figure 3.31 FBD=IRD for rocket

Goal Determine r_C.

Given Velocity of rocket and the forces acting on it.

Assume Assume that the center of curvature is below the rocket.

Draw **Figure 3.31** shows the forces acting on the rocket. $\boldsymbol{e}_t$ points along the direction of travel, and $\boldsymbol{e}_n$ points toward the assumed center of curvature.

Formulate Equations A force balance gives us

$$m\boldsymbol{a} = T(\cos 15°\boldsymbol{i} + \sin 15°\boldsymbol{j}) - mg\boldsymbol{j}$$

and so the acceleration is given by

$$\boldsymbol{a} = (1/m)T\cos 15°\boldsymbol{i} + (1/m)(T\sin 15° - mg)\boldsymbol{j} \qquad (3.31)$$

Solve All we need to do now is equate (3.31) with $\boldsymbol{a} = \dot{v}\boldsymbol{e}_t + \frac{v^2}{r_C}\boldsymbol{e}_n$. Our life is made simpler by the fact that for this problem $\boldsymbol{e}_t = \boldsymbol{i}$ and $\boldsymbol{e}_n = -\boldsymbol{j}$. Equating the two expressions yields

$$(T\cos 15°/m)\boldsymbol{e}_t + [g - (T\sin 15°/m)]\boldsymbol{e}_n = \dot{v}\boldsymbol{e}_t + \frac{v^2}{r_C}\boldsymbol{e}_n$$

$\boldsymbol{e}_t$:
$$\dot{v} = \frac{T\cos 15°}{m}$$

$\boldsymbol{e}_n$:
$$\frac{v^2}{r_C} = g - \frac{T\sin 15°}{m} \quad \Rightarrow \quad r_C = \frac{v^2}{g - T\sin 15°/m}$$

Using our known parameter values: $r_C = 11{,}000\ \text{m} = 11\ \text{km}$

Check Thus units check out; that's always important. The radius of curvature is large and that makes sense. The path is horizontal at the instant we are examining, and the accelerations along the path are

$$a_t = T\cos 15°/m = (6000\ \text{N})\cos 15°/(200\ \text{kg}) = 29\ \text{m/s}^2$$

$$a_n = g - T\sin 15°/m = 9.81\ \text{m/s}^2 - (6000\ \text{N})\sin 15°/(200\ \text{kg}) = 2\ \text{m/s}^2$$

This tells us that the normal acceleration is quite a bit smaller than the tangential, implying a close to flat trajectory and therefore a large radius of curvature, just as we found. Finally, we see that the normal component of acceleration is indeed pointed down, matching the assumption we made at the start of the example.

EXAMPLE 3.13 FORCE AND ACCELERATION FOR A SLIDING PEBBLE (Theory on page 133)

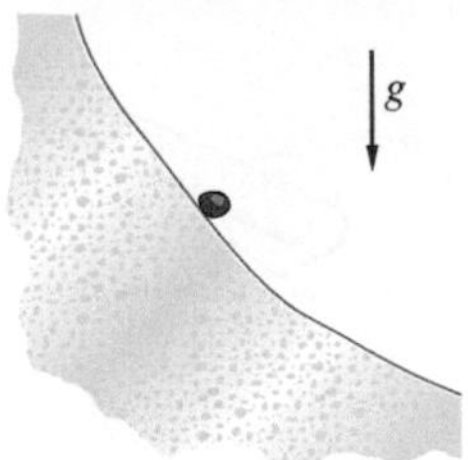

Figure 3.32 Pebble sliding down a mountainside

A small pebble is dislodged from a mountainside in Alaska and slides down an icy surface, as shown in **Figure 3.32**. The position of the pebble is accurately described by

$$y(x) = y_0 - ax + bx^2 + cx^3$$

in the region centered about the pebble, where x is a horizontal axis, y is a vertical axis, and $y(0) = y_0$. a, b, and c are all positive. At $x = 0$ the pebble has speed v. Determine the force being exerted by the icy surface on the pebble and the pebble's acceleration a_P along the surface in terms of the pebble's speed v, mass m, the acceleration of gravity g, a, and b at $x = 0$.

Goal Find the total force being exerted on the pebble by the surface that's supporting it and its acceleration along the surface.

Given The surface's profile as a function of x.

Draw **Figure 3.33** shows a schematic of the sliding pebble, and **Figure 3.34** shows the relevant free-body diagram and inertial-response diagram. In **Figure 3.34**, F is the frictional force between the surface and pebble, and as usual, N is the normal force exerted by the surface.

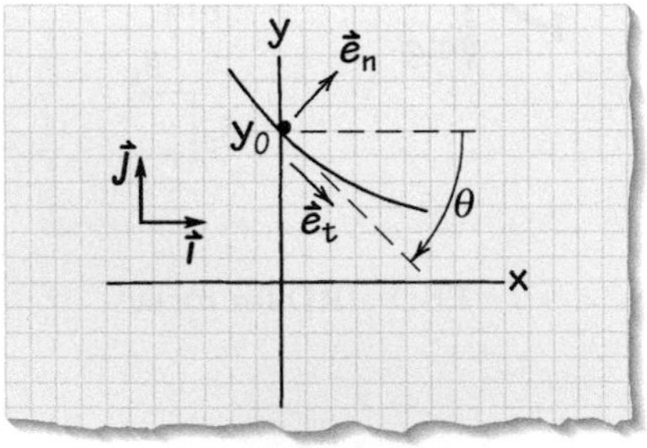

Figure 3.33 Unit vectors and relevant angle

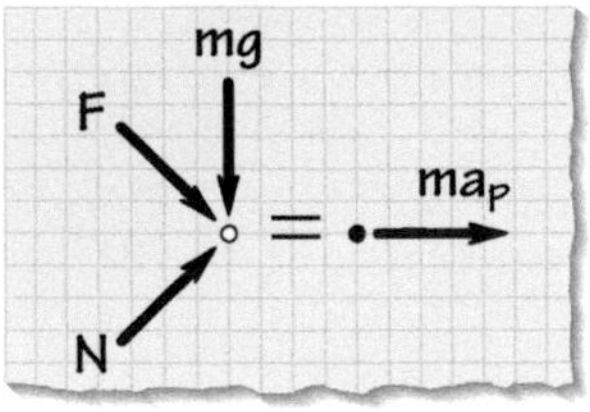

Figure 3.34 FBD=IRD for pebble

Formulate Equations The acceleration of the pebble in path coordinates is given by

$$\boldsymbol{a}_P = a_t \boldsymbol{e}_t + \frac{v^2}{r_C} \boldsymbol{e}_n$$

and a force balance, using the variables shown in **Figure 3.34**, yields

$$m\left(a_t \boldsymbol{e}_t + \frac{v^2}{r_C} \boldsymbol{e}_n\right) = -mg\boldsymbol{j} + N\boldsymbol{e}_n + F\boldsymbol{e}_t$$

We have been given enough to calculate some of the physical parameters of the problem. The angle θ is found from

$$\theta = \tan^{-1}\left(\frac{dy}{dx}\right) = -\tan^{-1}(a)$$

where $\frac{dy}{dx}$ is evaluated at $x = 0$.

Remember (2.60)? That equation, along with our knowledge of $y(x)$, gives us the radius of curvature:

$$r_C = \left[1 + \left(\frac{dy}{dx}\right)^2\right]^{\frac{3}{2}} \Bigg/ \left|\frac{d^2y}{dx^2}\right| = (1 + a^2)^{\frac{3}{2}}/(2b) \qquad (3.32)$$

Assume We are told the pebble is sliding on a surface of ice, and in addition we are not given any coefficient of friction. Thus we can logically assume that the surface is frictionless and therefore $F = 0$.

Solve We need a coordinate transformation array to express $-mg\boldsymbol{j}$ in terms of $\boldsymbol{e}_t$ and $\boldsymbol{e}_n$:

	$\boldsymbol{i}$	$\boldsymbol{j}$
$\boldsymbol{e}_t$	$\cos\theta$	$-\sin\theta$
$\boldsymbol{e}_n$	$\sin\theta$	$\cos\theta$

Remembering that $F = 0$, we find that our equation of motion becomes

$$m\left(a_t\boldsymbol{e}_t + \frac{v^2}{r_C}\boldsymbol{e}_n\right) = mg\sin\theta\boldsymbol{e}_t + (N - mg\cos\theta)\boldsymbol{e}_n$$

Matching coefficients gives us

$\boldsymbol{e}_t$:
$$ma_t = mg\sin\theta \tag{3.33}$$

$\boldsymbol{e}_n$:
$$m\frac{v^2}{r_C} = N - mg\cos\theta \tag{3.34}$$

Equation (3.33) gives us

$$a_t = g\sin\theta$$

and (3.32) used in (3.34) gives us

$$N = m\left[g\cos\theta + \frac{2bv^2}{(1+a^2)^{\frac{3}{2}}}\right]$$

Check Let's think about the normal force N. The $mg\cos\theta$ certainly makes sense. If the slope were horizontal, this would become mg, the weight of the pebble, and if $\theta = 90°$, a vertical slope, $mg\cos\theta$ becomes zero. This makes sense because the pebble would be falling straight down and therefore would not be interacting with the icy surface at all. But does the $\frac{2mbv^2}{(1+a^2)^{\frac{3}{2}}}$ term make sense? Everything in this term is positive, and therefore it is *adding* to the normal force that we would get for a straight slope of θ. A positive b means there is a positive curvature to the curve. In other words, it's concave at this point, as if the pebble were sliding along the inside of a bowl. And we know that for such a case we would have a greater normal force. The additional force is similar to the force the pebble would encounter if it were being whirled around the center of curvature on the end of a string. The tension force would produce a force pointing toward the center of rotation, or in this case the center of curvature. There's no string for this case, but the supporting surface provides the exact same kind of force to keep the pebble moving along the concave upward curve.

EXAMPLE 3.14 DETERMINING SLIP POINT IN A TURN (Theory on page 133)

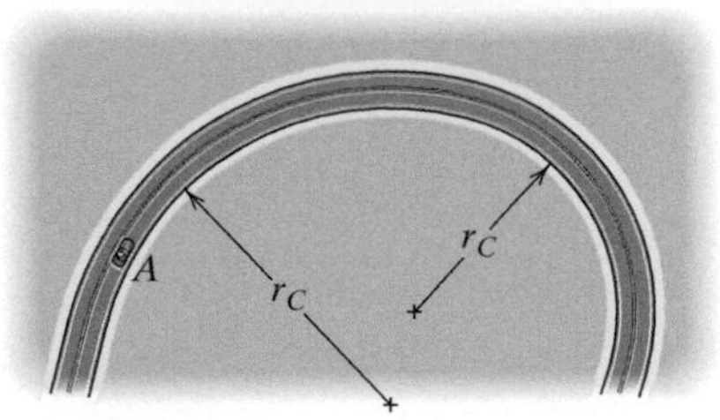

Figure 3.35 Car in a decreasing radius turn

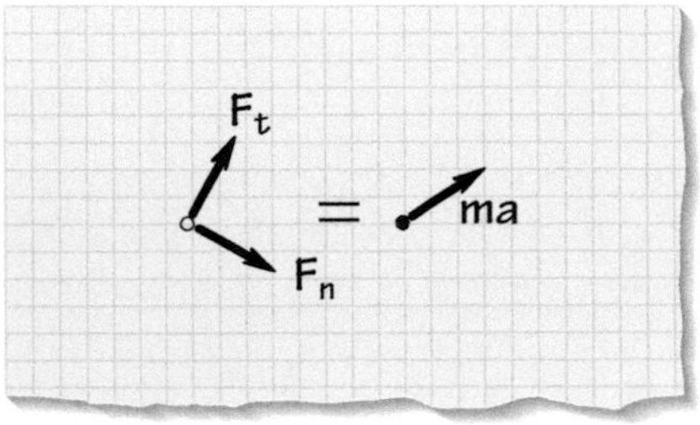

Figure 3.36 FBD=IRD for car

Consider a car that has entered a decreasing-radius turn (**Figure 3.35**). The driver has set the cruise control, and thus the car's speed stays constant at 48 km/h. At the illustrated point along the curve, the radius of curvature is 36 m and the radius of curvature changes as $r_C = (36 - 0.4s)$ m, where s indicates the number of meters the car has moved. When will the car begin to slip? Approximate the car as a mass particle ($m = 1.46 \times 10^3$ kg). $\mu_d = \mu_s = 0.8$.

Goal Determine when the car slips.

Given The radius of curvature as a function of position along the car's path, the car's speed and mass, and the coefficients of friction.

Assume Our primary assumption follows from the fact that the cruise control is set. This means that the car will provide enough driving force to exactly counter any road drag and will cause the overall force in the $\boldsymbol{e}_t$ direction to be zero.

Draw **Figure 3.36** shows the forces acting on the car in both the $\boldsymbol{e}_t$ and $\boldsymbol{e}_n$ directions.

Formulate Equations Using the fact that F_t is zero for this example means that we only need apply (3.30):

$$F_n = m\frac{v^2}{r_C}$$

The maximum F_n that can be generated by the frictional interface is $\mu_s mg$.

Solve First let's find the maximum frictional force:

$$F_{\max} = \mu_s mg = 0.8(1.46 \times 10^3 \text{ N})(9.81 \text{ m/s}^2) = 11.5 \text{ kN}$$

Our limit case will occur when $m\frac{v^2}{r_C} = F_{\max}$. Since 48 km/h corresponds to 13.3 m/s, we have

$$r_C = \frac{mv^2}{F_{\max}} = \frac{(1.46 \times 10^3 \text{ kg})(13.3 \text{ m/s})^2}{11.5 \times 10^3 \text{ N}} = 22.5 \text{ m}$$

From our expression for the radius of curvature we have

$$22.5 \text{ m} = 36 \text{ m} - 0.4s \text{ m} \quad \Rightarrow \quad s = 33.8$$

The car is traveling at 13.3 m/s, and thus the time to travel 33.8 m is

$$t = 33.8 \text{ m}/13.3 \text{ m/s} = 2.54 \text{ s}$$

Check Our answer of 22.5 m implies that the car could circle a 15 m circular track at 48 km/h and experience a lateral acceleration of 0.8g. These numbers match physical expectations.

EXERCISES 3.3

3.3.1. [Level 1] Consider a marble of mass m moving along the inner surface of an inverted cone in a circular path of radius r at a constant height. The cone's inner surface has a coefficient of friction of μ, and take θ to be the angle that the cone's outer surface makes with the horizontal. Neglecting frictional effects along the marble's path, what is the minimum constant angular speed ω at which the marble can go around its path before it starts to slide down the cone?

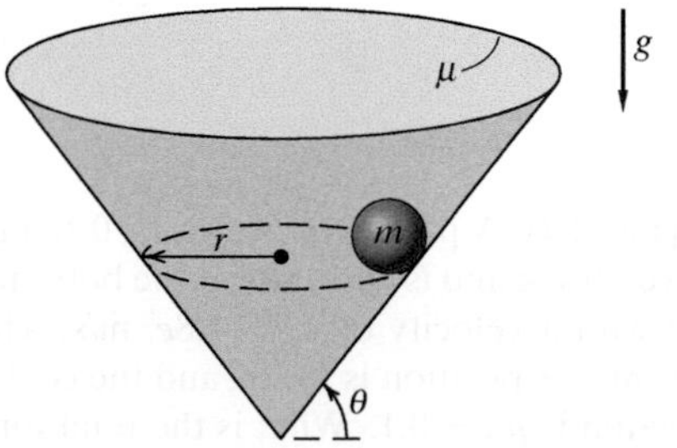

E3.3.1

3.3.2. [Level 1] A cyclist is rounding a curve having a radius of curvature of 15 m. Determine the maximum speed possible without encountering slip. $\mu_s = \mu_d = 0.5$.

3.3.3. [Level 1] Suppose a marble with mass m travels along the rough outer surface of a cone in a circular path of radius r at a fixed height with a constant angular speed of ω. Define θ as the angle that the cone's outer surface makes with the vertical. Taking frictional effects along the marble's path to be negligible, find the minimum coefficient of friction μ for which the marble will not slide down the cone.

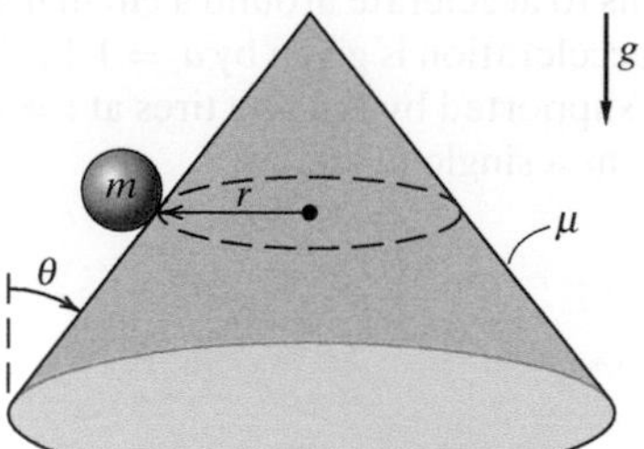

E3.3.3

3.3.4. [Level 1] A car exiting a highway travels along the path shown. At A the road's radius of curvature is 76 m, at B it's 305 m, and at C it's 120 m. Assume a constant speed of 72 km/h and calculate the overall side force generated between the car's tires and the road at each of these points.

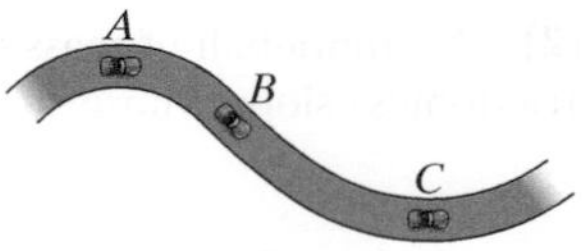

E3.3.4 Top view

3.3.5. [Level 1] The coefficient of friction between a car's tires and the road is such that the car can generate a maximum deceleration of 1.1 g without experiencing slip. What is the minimum radius of curvature for a curve such that the car can negotiate the curve without slipping at a steady speed of 64 km/h?

3.3.6. [Level 1] The pivot of each 120 kg seat on a Ferris wheel travels in a circle with a 10 m radius, and the Ferris wheel rotates with a rotational speed of 0.1 rad/s. What is the interaction force between the Ferris wheel framework and the seat H when $\theta = 45°$, where θ is the angle between the horizontal and the radial line intersecting seat H.

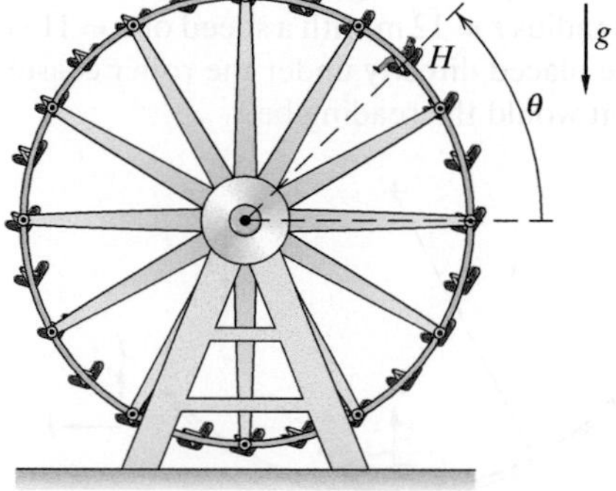

E3.3.6

3.3.7. [Level 1] A pebble P is within a cup at the end of a 0.64 m arm, swinging in a clockwise direction about O. When the arm becomes horizontal it strikes a hard stop, coming instantly to rest as the pebble leaves the cup. Just before striking the stop (at $t = t_1$) the pebble was experiencing a tangential acceleration of 15 m/s^2 and a tangential speed of 12 m/s. Determine the overall magnitude of the acceleration felt by the pebble just before and just after the collision. Express your answer in terms of $\boldsymbol{e}_t$ and $\boldsymbol{e}_n$.

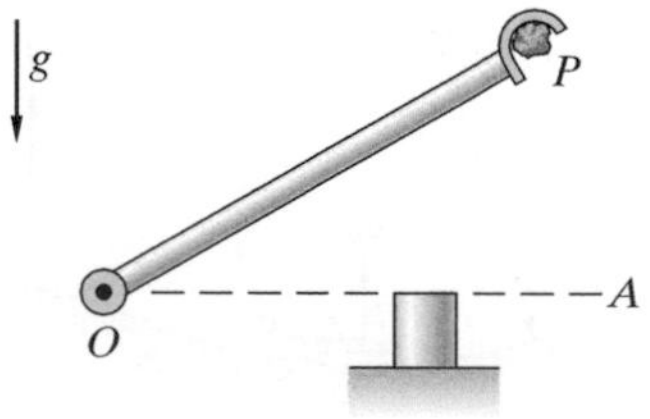

E3.3.7

3.3.8. [Level 1] A cyclist C (cyclist plus bicycle weighing 72.6 N) is riding along a decreasing-radius turn at a constant speed of 32 km/h. The radius of curvature varies as $r_C = a - bs^2$, where s indicates motion along the curve (expressed in meters). $a = 18$ m and $b = 9.8 \times 10^{-3}$ m^{-1}. The tires can support a maximum side force (accounting for both tangential and normal forces) of 534 N. How far will the cyclist travel before the tires slip?

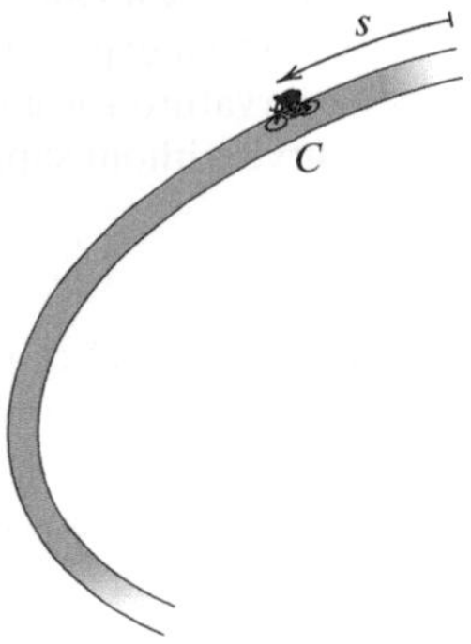

E3.3.8

3.3.9. **[Level 1]** A 160 kg roller coaster is at the bottom of a loop of radius $r = 12$ m with a speed of $v = 113$ km/h. If a scale were placed directly under the roller coaster at this instant, what would the reading be?

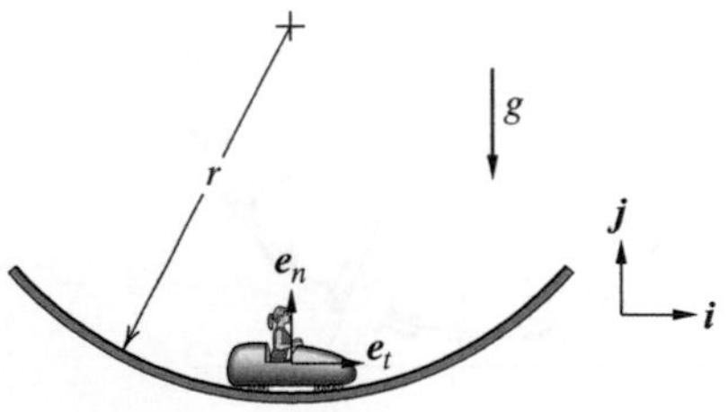

E3.3.9

3.3.10. **[Level 1]** An SUV is at the bottom of a dip, where the radius of curvature is $r_c = 18$ m. The SUV weighs 3.63×10^3 kg, and each of the four springs in its suspension has a stiffness of $k = 1.23$ kN/cm. Find how fast the SUV needs to be going in the bottom of the dip for it to bottom out if the maximum compression of the suspension is $\Delta x = 0.3$ m.

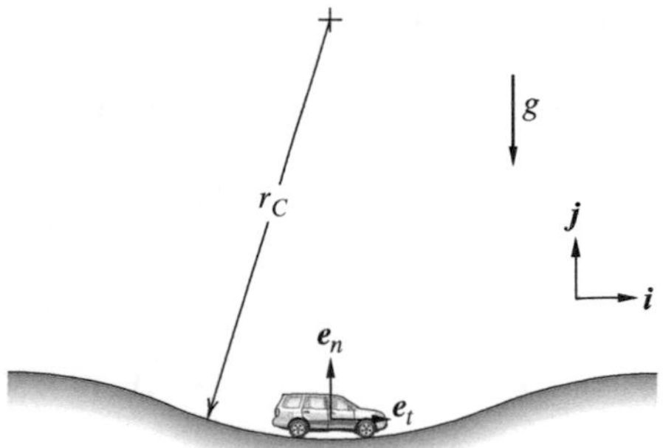

E3.3.10

3.3.11. **[Level 2]** As illustrated, a car starts at point A and travels along a curved path toward point B. The car's mass is $m = 1300$ kg, and it starts from rest at A. When the vehicle reaches B, its speed is $v_B = 60$ kph. Suppose the total traction force acting on the vehicle at B is $F = 4800$ N, and the radius of curvature at that location is $r_B = 80$ m. What is the car's instantaneous acceleration $\dot{v}$ at B? Suppose the car accelerated from A to B at a constant rate of $\dot{v}$. How far s did it travel?

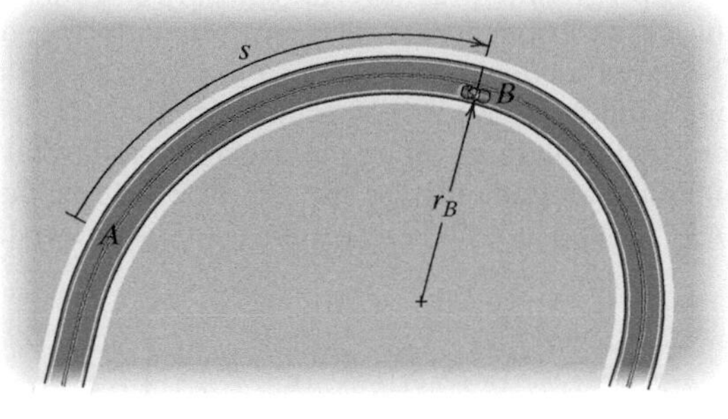

E3.3.11

3.3.12. **[Level 2]** A particle m weighing 0.18 kg has slid down a curved track and is currently at the bottom, moving to the right with a velocity of $\boldsymbol{v}_m = 1.8\boldsymbol{e}_t$ m/s. The radius of curvature at this position is 1.8 m, and the coefficient of dynamic friction is $\mu_d = 0.1$. What is the total force being exerted by the particle on the track? What is the particle's tangential acceleration?

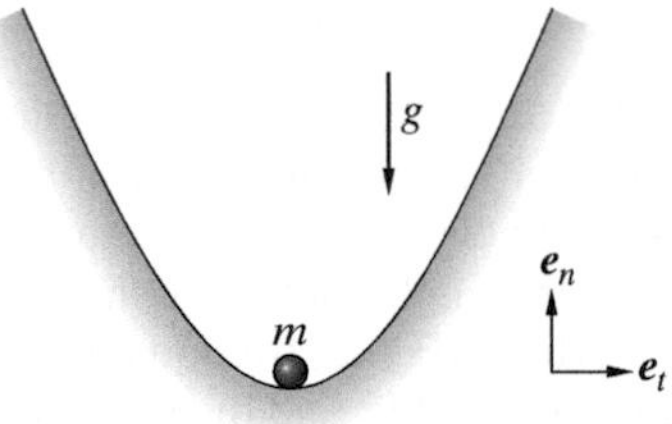

E3.3.12

3.3.13. **[Level 2]** A 300 kg motorcycle and rider, starting from rest, begins to accelerate around a circular track of radius 80 m. Its acceleration is given by $a_t = 1.1$ m/s^2. What is the total force supported by the two tires at $t = 4$ s? (Treat the motorcycle as a single mass.)

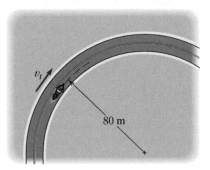

E3.3.13

3.3.14. **[Level 2]** A cannonball of mass m is placed at the ledge A of a frictionless slope whose surface profile is best described by

$$y(x) = a\cos\left(\frac{2\pi x}{L}\right) + a$$

where a is the amplitude, L is the wavelength, and y is measured from the base of the slope. The cannonball is given a slight push, which causes it to roll down the slope into the illustrated dip B, where the cannonball's speed is $v_B = 2\sqrt{ga}$. Find the value of L for which the normal force N_B acting on the cannonball in the dip is limited to twice that at the slope's ledge A.

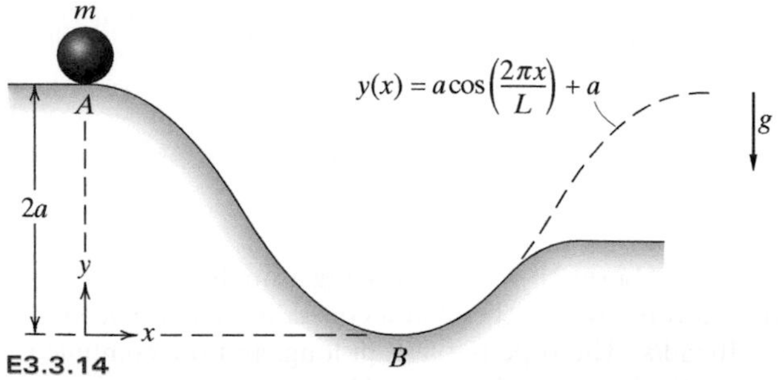

E3.3.14

3.3.15. **[Level 2]** A 1450 kg race car is on a circular portion of track ($r = 90$ m) and is traveling at 161 km/h when it starts to decelerate such that $a_t = -3$ m/s^2. What is the ground force acting on the race car after it has traveled 15 m?

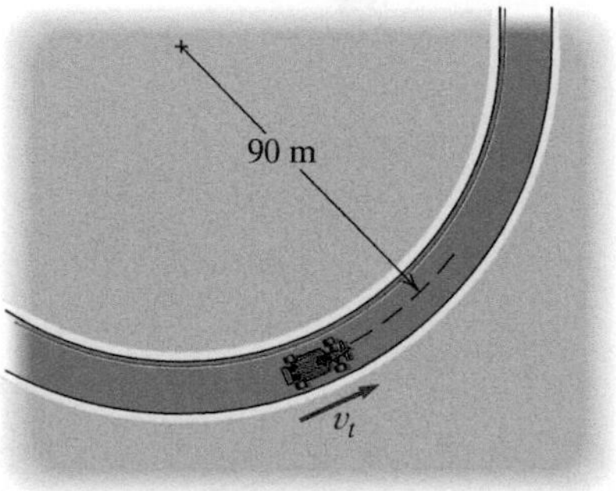

E3.3.15

3.3.16. **[Level 2] Computational** A particle is projected up the side of a parabolically shaped hill, its profile given by

$$y = y_0 - ax^2$$

where $y_0 = 6$ m and $a = 0.066$ m^{-1}. The particle starts at $x = -9$ m with a speed of 6.4 m/s. Assume a friction-free interface and plot both the normal force acting on the particle and the particle's tangential acceleration as a function of x for $-9\,\text{m} \le x \le 9\,\text{m}$.

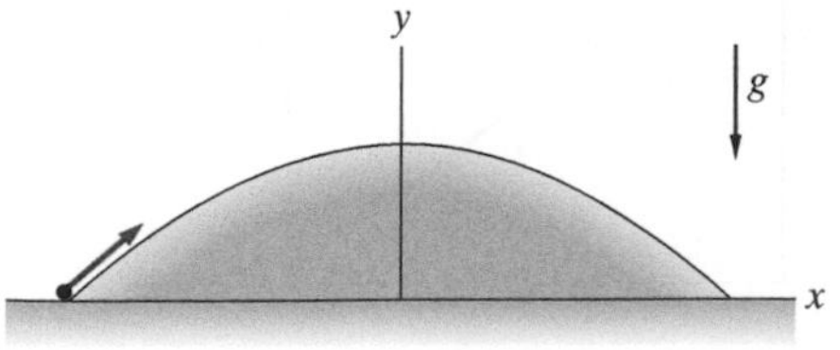

E3.3.16

3.3.17. **[Level 2]** A cyclist is moving around a curved and banked path at a constant speed of 24 km/h. The bicycle and rider have a combined weight of 54 kg. What is the force applied by the path to the bicycle? Treat the bicycle/rider system as a particle.

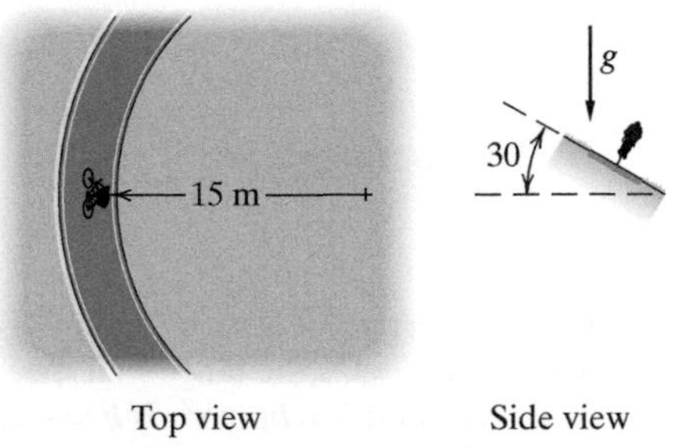

E3.3.17

3.3.18. **[Level 2]** Highways are banked to increase the maximum safe speed at which cars can negotiate a turn. Assume a road with a radius of curvature of 30 m and a banking angle θ. Let the coefficient of friction between the car and road be 0.9.

a. Find the maximum steady speed at which the car can traverse the curve without slipping if $\theta = 0$.

b. Find the maximum steady speed for $\theta = 10°$.

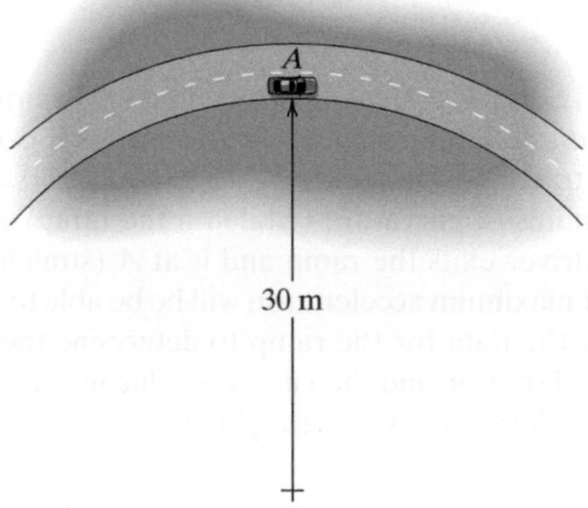

E3.3.18

3.3.19. **[Level 2]** A car, traveling at 32 km/h, encounters a dip in the road. The radius of curvature at the bottom of the dip is 15 m. Each of the car's four springs has a spring constant of 1.42 kN/cm (the spring compresses 2.5 cm for every 3.56 kN applied to it). Determine the deflection of the springs from their unloaded state when the car is at the bottom of the dip. The weight of the car supported by the springs is 1633 kg.

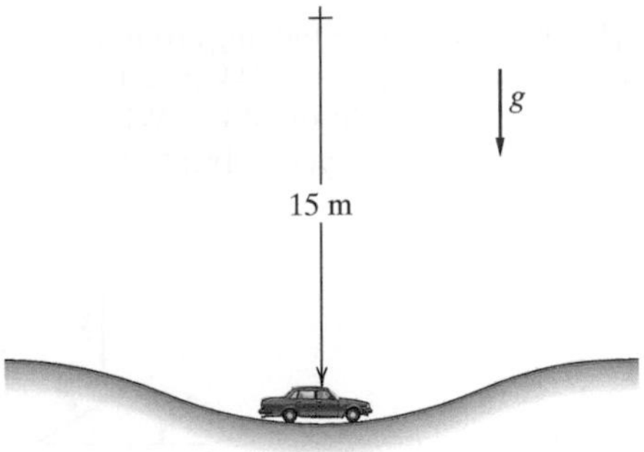

E3.3.19

3.3.20. **[Level 2]** A Ferris wheel with a radius of 8.5 m is decelerating such that at $t = 0$ s the tangential speed of the 109 kg seat S is $v = 3$ m/s and $\dot{v} = bt$ m/s^2. If $b = -0.12$ m/s^3 and $\theta = 0$ at $t = 0$, what is the force acting on the Ferris wheel's frame when $t = 3$ s?

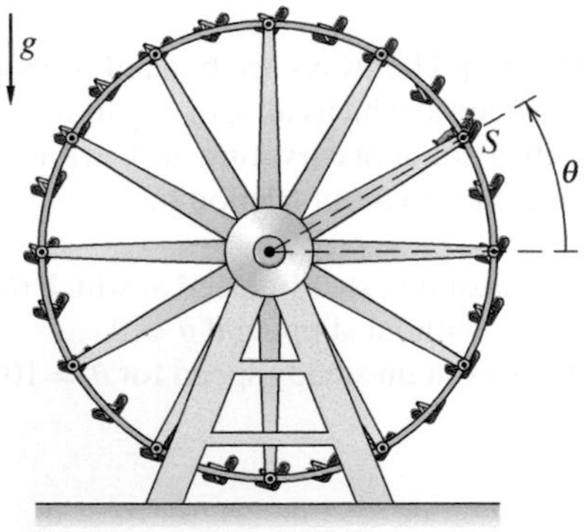

E3.3.20

3.3.21. **[Level 2]** A car C is traveling up a spiral ramp at a constant 48 km/h and traveling as fast as it can without losing traction. The grade of the ramp is 0.10 and the effective radius of curvature (while on the ramp) is 23 m. Once the driver exits the ramp and is at A (straight, level road) what maximum acceleration will he be able to attain? (Hint: Use the data for the ramp to determine the road's coefficient of friction and then use this value in determining the car's acceleration on the straightaway.)

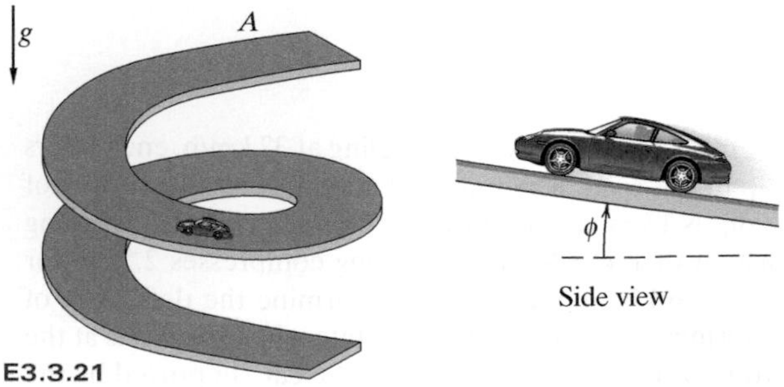

E3.3.21

3.3.22. **[Level 2]** Each day Bob takes the same route to work and tries to maintain a constant speed through a sharp turn. He recently purchased a new sports car and thinks he can negotiate the turn at a constant 100 kph. The coefficient of friction between road and car is 0.80, the mass of the car and driver is 1300 kg, and at the tightest part of the turn the radius of curvature is 90 m. A speed of 100 kph produces 400 N of drag. Determine if Bob can successfully complete the turn at 100 kph.

E3.3.22

3.3.23. **[Level 2]** José is swinging a small bucket of water by a rope over his head at a constant angular speed of $\dot{\theta} = 10$ rad/s. The rope is $L = 1$ m long, and the combined mass of the bucket and water is 4 kg.

a. Calculate the tension in the rope when $\theta = 60°$.

b. What is the magnitude of the bucket's acceleration at this instant?

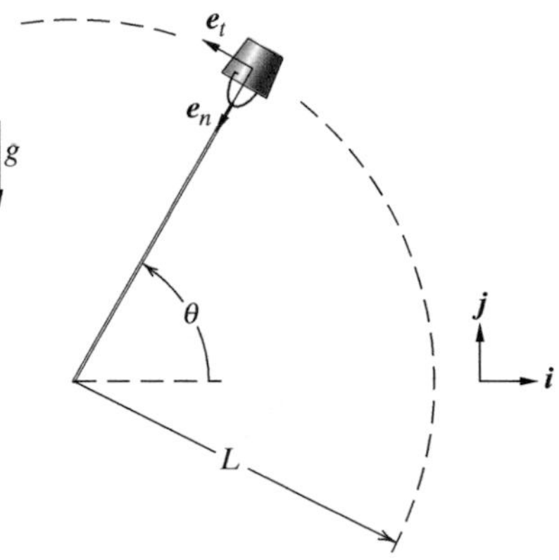

E3.3.23

3.3.24. **[Level 3] Computational** An 80 kg skier travels down a ski jump having a profile given by $y = y_0 - ax + bx^3$ with $y_0 = 50$ m, $a = 1$, and $b = \frac{1}{12{,}500}$ m^{-2}. Plot the normal force exerted by the slope on the skier as a function of time from $x = 0$ to $x = 50$ m.

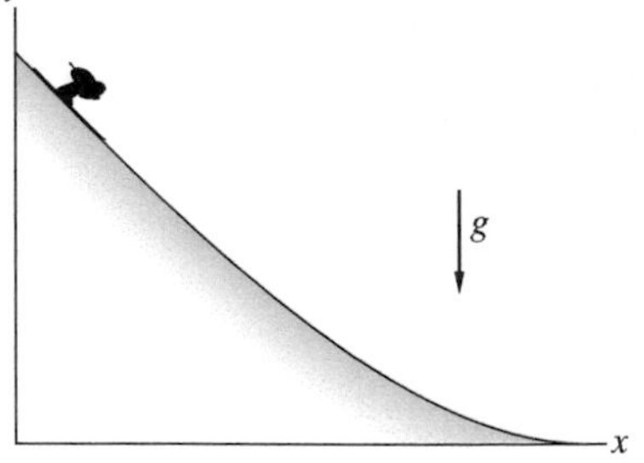

E3.3.24

3.4 LINEAR MOMENTUM AND LINEAR IMPULSE

Remember Newton's first law? Stated roughly, it says, "A body in uniform motion remains that way unless acted on by an unbalanced force." Note the words "body in uniform motion." That means we're thinking about linear momentum: mass times velocity. We've already seen that an unbalanced force acting on a mass particle causes the particle to accelerate. In fact, that's what we have looked at in the last few sections. We can examine the problem in a slightly different way, however—a way that will give us just a bit more insight into dynamics.

Let's begin by integrating $\boldsymbol{F} = m\boldsymbol{a}$ with respect to time:

$$\int_{t_1}^{t_2} \boldsymbol{F}\, dt = m \int_{t_1}^{t_2} \boldsymbol{a}\, dt = m\boldsymbol{v}(t_2) - m\boldsymbol{v}(t_1)$$

which, when slightly rearranged, gives us

$$m\boldsymbol{v}(t_2) = m\boldsymbol{v}(t_1) + \int_{t_1}^{t_2} \boldsymbol{F}\, dt \tag{3.35}$$

This result tells us that the particle's linear momentum at t_2 is equal to its linear momentum at t_1 plus an integral of the applied force for the intermediate times.

To make things easier to write, let's give these terms names. The linear momentum already has a name (*linear momentum*), and so let's give it a one-letter symbol instead:

$$\boldsymbol{L} \equiv m\boldsymbol{v} \tag{3.36}$$

Now that $\boldsymbol{L}$ has been introduced, it should be mentioned that the actual form of Newton's second law is

$$\dot{\boldsymbol{L}} = \boldsymbol{F} \tag{3.37}$$

That is, the time rate of change of a particle's linear momentum is equal to the net force acting on it. Because we're assuming a constant mass, m can come out of the time differentiation of $\boldsymbol{L}$, thereby leaving us with the familiar $\boldsymbol{F} = m\boldsymbol{a}$.

As for the integral term in (3.35), we're going to call it the **linear impulse** ($\mathcal{LI}$)

$$\mathcal{LI} \equiv \int m\boldsymbol{a}\, dt$$

To make it explicit that we're talking about the applied linear impulse from t_1 to t_2 we will use the notation

$$\mathcal{LI}_{1-2} = \int_{t_1}^{t_2} m\boldsymbol{a}\, dt \tag{3.38}$$

Using these new symbols, (3.35) can be rewritten as

$$\boldsymbol{L}(t_2) = \boldsymbol{L}(t_1) + \boldsymbol{\mathcal{LI}}_{1-2} \tag{3.39}$$

This is a useful bit of information. It tells us that if a force acts on a body over some interval of time (the $\mathcal{LI}$ term), the only thing that will change is the body's momentum (and because the mass of the body is constant in our calculations, this really means that only the velocity changes). Left to itself, the body moves along at a constant velocity. Hit it with some force, and the velocity changes in direct proportion to the amount of linear impulse ($\mathcal{LI}$) applied. Why is this cool? As one example, you may recall a scene in the movie *Apollo 13* where the astronauts were doing just this. They were on their way back to earth after the explosion had crippled their ship, and they had to fire their rocket for just the right amount of time so that their spacecraft would enter the earth's atmosphere at precisely the correct angle. If they had gotten it wrong, they would have either skipped off into space or burned up in the atmosphere. It was vitally important to fire the rocket for just the right length of time so that the spacecraft's velocity was altered by the correct amount.

>>> **Check out Example 3.15 (page 145) and Example 3.16 (page 146) for applications of this material.**

A nice way to extend this concept of linear impulse is to ask what happens when two independent bodies interact with each other, a collision between two cars being a good example. Before the collision, the two cars are traveling independently. If the velocities are such that the two cars collide, then there will be a complicated interaction during the collision. What we know is that the forces of interaction are always equal and opposite; that is, if car A is pushing on car B with a force of 1000 N, then car B is pushing on car A with 1000 N as well. This means that the linear impulses between the two cars are identical in magnitude.

If the cars remain together as a result of the collision, we are left with two cars traveling in the same direction with the same speed. What's interesting is that, even without knowing any details of the collision forces, we can calculate the final velocity of the two cars by using our linear impulse relationships.

>>> **Check out Example 3.17 (page 146) for an application of this material.**

EXAMPLE 3.15 CHANGING THE SPACE SHUTTLE'S ORBIT (Theory on page 143)

Imagine you're in the space shuttle, orbiting the earth with speed v_1 (**Figure 3.37**). You want to change your velocity so that you will transition into a different orbit, and you have been told by mission control by how much your speed has to change in order to do so. How do you figure out how long you have to fire your main thruster? Assume that the thruster produces a constant thrust F and that the shuttle has mass m_s.

Figure 3.37 Space shuttle orbiting earth

Goal Find the duration of firing needed to change the shuttle's speed by the required amount.

Given Required change in velocity and available thrust.

Assume There are no major constraints for this problem, though I will mention one—gravity. It's the earth's gravity that keeps the shuttle in its orbit; without gravity, the shuttle would continue in a straight line. For this problem, however, the changes in velocity are occurring over a small enough distance that we can ignore the curvature of the orbit and treat the shuttle as if it were simply experiencing a change in its linear momentum as a result of the thruster. Also, we'll neglect any mass change due to fuel usage.

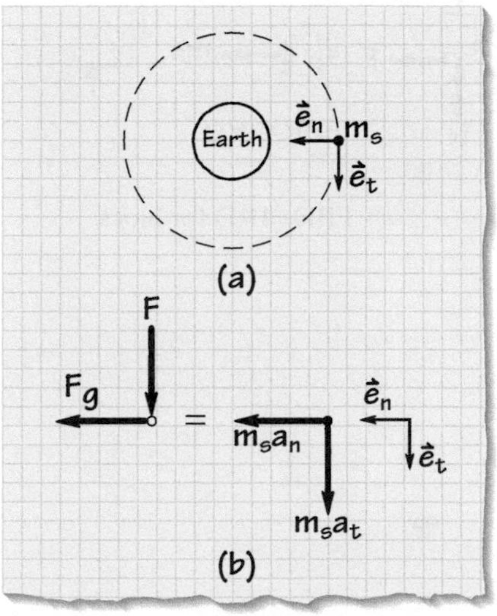

Figure 3.38 FBD=IRD for orbiting shuttle

Draw **Figure 3.38*a*** shows a mass particle m_s (representing our space shuttle) in orbit around the earth. The thruster exerts a force F on the shuttle in its direction of travel, as shown in **Figure 3.38*b***. It's interesting to note that this is the only force needed to alter orbits, as you will see if you check out Section 3.6 on orbital mechanics.

Formulate Equations Using (3.39) in the tangential direction gives us

$$m_s v(t_2) = m_s v(t_1) + F(t_2 - t_1) \tag{3.40}$$

Note that these speeds have to be in the $\mathbf{e}_t$ direction and that the assumption of a constant thrust allows us to replace the linear impulse integral with the simpler force/elapsed-time expression.

Solve Our final speed $v(t_2)$ is simply our original speed $v(t_1)$, plus the prescribed change in speed:

$$v(t_2) = v(t_1) + \Delta v \tag{3.41}$$

The firing duration $t_2 - t_1$ is found from (3.40) and (3.41) to be

$$t_2 - t_1 = \frac{m_s [v(t_2) - v(t_1)]}{F} = \frac{m_s \Delta v}{F}$$

Check Our solution should have the units of seconds. The quantity $\frac{m_s \Delta v}{F}$ has the units of seconds. Thus we have consistent units and therefore some reasonable confidence in the result.

EXAMPLE 3.16 BLOCK ON A SANDING BELT (Theory on page 143)

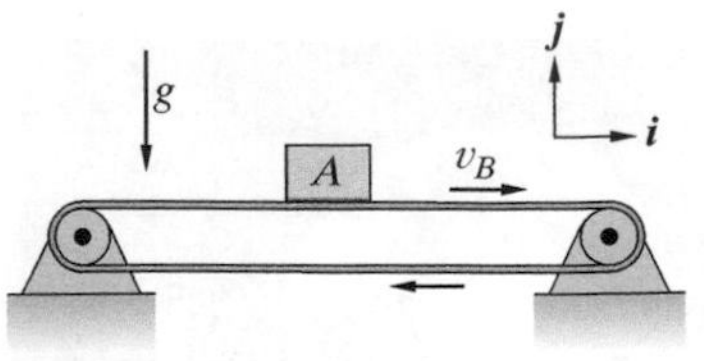

Figure 3.39 Block moving on a sanding belt

A block A is held on the top surface of a moving sanding belt ($v_B = 3\,\text{m/s}$) and then slid to the left with an initial speed $v_0 = 2\,\text{m/s}$ (measured with respect to the fixed ground), as shown in **Figure 3.39**. The coefficient of friction between the block and the belt is $\mu = 0.4$. Determine how much time must elapse until the block is moving at zero speed relative to the belt.

Goal Find the time for the block to match the belt's speed.

Given The coefficient of friction between the block and belt ($\mu = 0.4$) as well as the block's initial velocity ($\boldsymbol{v}_0 = -2\boldsymbol{i}\,\text{m/s}$) and the belt's speed ($v_B = 3\,\text{m/s}$).

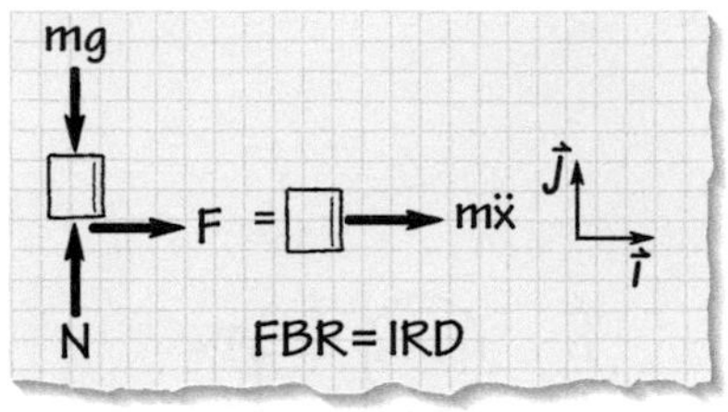

Figure 3.40 FBD=IRD for block

Assume Assume that the block stays in continuous contact with the belt.

Draw **Figure 3.40** shows our FBD=IRD.

Formulate Equations The initial and final linear momenta in the $\boldsymbol{i}$ direction are related through

$$mv = m(-v_0) + Ft \tag{3.42}$$

where t is the elapsed time and v is the block's speed with respect to the ground. Noting that $N = mg$ and utilizing a slip assumption give us

$$F = \mu N = \mu mg \tag{3.43}$$

Solve Use (3.42) and (3.43) with the assumption that the final speed is equal to that of the belt: $t = (v_0 + v_B)/(\mu g) = 1.27\,\text{s}$

Check There's nothing in particular to check for this problem.

EXAMPLE 3.17 TWO-CAR COLLISION (Theory on page 144)

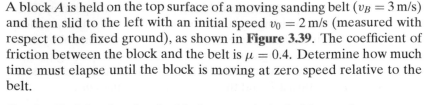

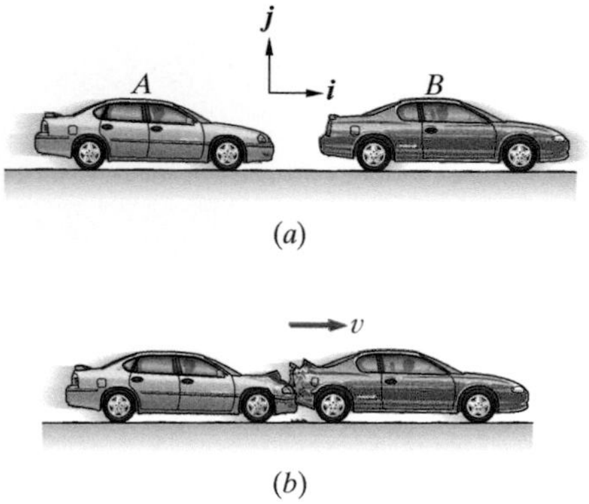

Figure 3.41 Two-car collision

Figure 3.41*a* shows two cars about to have a collision. Car A has mass $m_A = 1500$ kg and car B has mass $m_B = 1600$ kg. The initial velocities are $\boldsymbol{v}_A = 20\boldsymbol{i}$ m/s and $\boldsymbol{v}_B = -10\boldsymbol{i}$ m/s. Determine the velocity of the two vehicles after they collide, assuming that they remain together after the collision.

Goal Find the velocity of the post-collision wreck of cars A and B.

Given Initial speed of the vehicles.

Assume Assume that we can model the deforming cars as simple interacting particles.

Draw **Figure 3.41*b*** shows both vehicles after collision, moving together with velocity $\boldsymbol{v}$.

Formulate Equations Assume that the force acting on car A during the collision is given by $-F_{AB}\boldsymbol{i}$ and the force acting on car B is given by $F_{AB}\boldsymbol{i}$.

The linear impulse relationships are given by

$$m_A \boldsymbol{v}_A(t_2) = m_A \boldsymbol{v}_A(t_1) - \int_{t_1}^{t_2} \boldsymbol{F}_{AB}\,dt$$

$$m_B \boldsymbol{v}_B(t_2) = m_B \boldsymbol{v}_B(t_1) + \int_{t_1}^{t_2} \boldsymbol{F}_{AB}\,dt$$

Add these two equations together and you'll get

$$m_A \boldsymbol{v}_A(t_2) + m_B \boldsymbol{v}_B(t_2) = m_A \boldsymbol{v}_A(t_1) + m_B \boldsymbol{v}_B(t_1) \tag{3.44}$$

Thus you can see that the total linear momentum *before* the collision (at t_1) is equal to the total linear momentum *after* the collision (at t_2). This is what we call **conservation of linear momentum**; we will see it again when we deal with impact and multibody problems.

Solve Putting the given values into (3.44) while calling the after-collision speed v gives

$$(1500\text{ kg} + 1600\text{ kg})v = (1500\text{ kg})(20\text{ m/s}) + (1600\text{ kg})(-10\text{ m/s})$$

$$v = \frac{14{,}000\text{ kg}\cdot\text{m/s}}{3100\text{ kg}} = 4.52\text{ m/s}$$

Thus the final velocity of the two cars is $\boldsymbol{v} = 4.52\boldsymbol{i}$ m/s.

Check Both cars had about the same mass, and car A was traveling to the right at twice the speed that car B was traveling left. Does it seem reasonable that after the collision they would both be going right at 4.52 m/s? Sure it does. That's very close to the average of the two velocities $[\frac{1}{2}(-10\boldsymbol{i}\text{ m/s} + 20\boldsymbol{i}\text{ m/s}) = 5\boldsymbol{i}\text{ m/s}]$, which is what we would logically expect to see for identical-mass vehicles. Thus the answer seems pretty much in line with expectations.

EXERCISES 3.4

3.4.1. **[Level 1]** Two blocks (A and B) are traveling to the right on a smooth horizontal surface, where A is moving at $v_A = 4$ m/s and B has a speed of $v_B = 2$ m/s just ahead of A. Suppose the blocks stick when they collide, and they move together with a speed v immediately after impact. If $m_A = 2$ kg and $m_B = 3$ kg, what are the post-impact speed v and the magnitude of the linear impulse $\mathcal{LI}$ applied to both blocks?

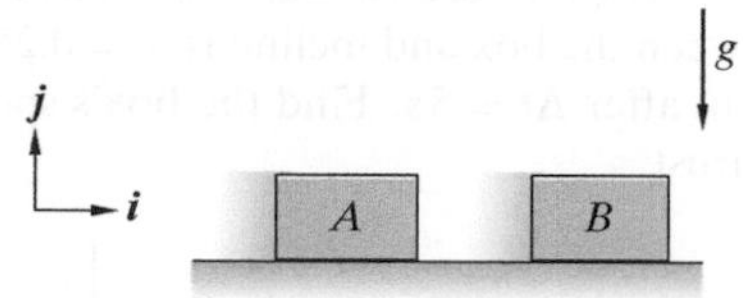

E3.4.1

3.4.2. **[Level 1]** In order to split a log, a wedge is placed in the log and is then struck by a 3 kg splitting maul. Assume that after contact both maul and wedge move together downward through the log. The maul is traveling at 48 km/h when it strikes the wedge and 0.21 s later comes to rest (with the log still not fully split). What is the average force exerted by the wedge against the maul during this log-chopping exercise? (Ignore the linear impulse due to gravity.)

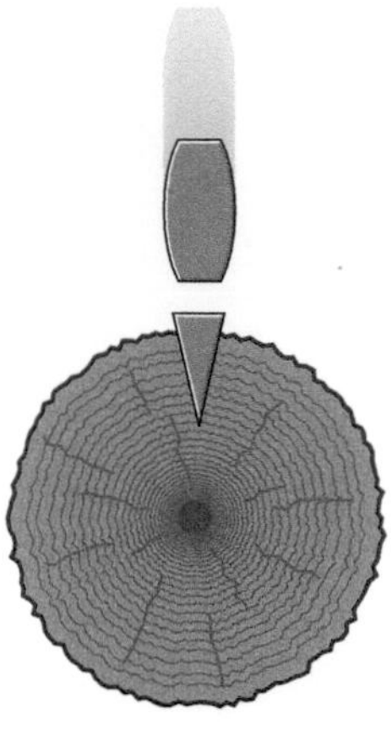

E3.4.2

3.4.3. **[Level 1]** A 200 kg lander is approaching Mars. Assume a gravitational acceleration of 3.7 m/s^2. At the illustrated instant, $h = 15$ m and $v = 2.5$ m/s. What is the average thrust needed to land with a touchdown speed of 0.5 m/s after 10 s?

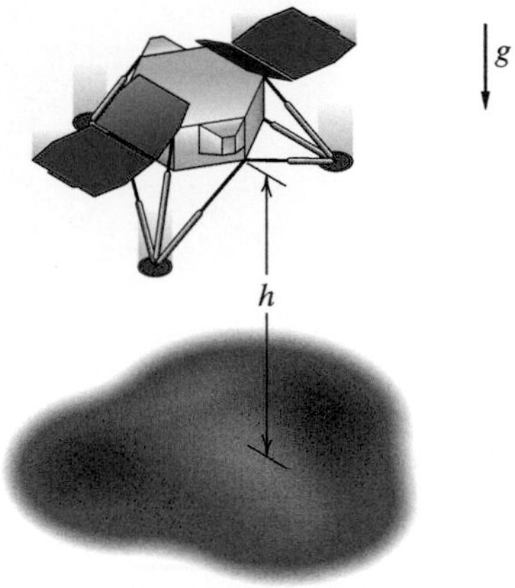

E3.4.3

3.4.4. **[Level 1]** A box of mass $m = 6$ kg is projected with a speed $v_0 = 5$ m/s up a rough incline angled at $\theta = 15°$ with respect to the horizontal. A small booster strapped to the box suddenly ignites, generating a thrust that varies according to $T(t) = T_0\sqrt{t}$, where $T_0 = 25\ \text{N/s}^{\frac{1}{2}}$. The coefficient of friction between the box and incline is $\mu = 0.25$, and the booster cuts out after $\Delta t = 5$ s. Find the box's speed v just after loss of thrust.

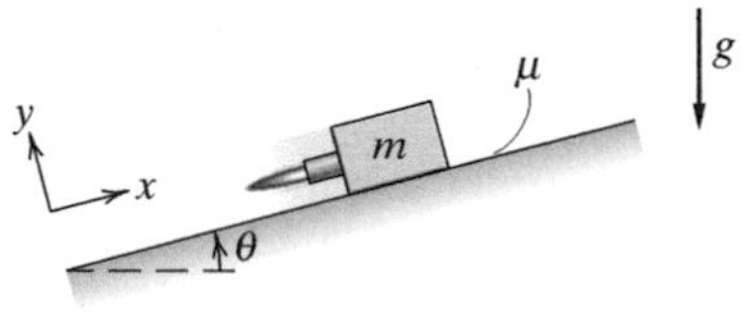

E3.4.4

3.4.5. **[Level 1]** What is the average force exerted on a 1360 kg car that crashes into a rigid wall and goes from 56 km/h to zero in 0.11 s?

3.4.6. **[Level 1]** A 1270 kg car brakes from 96 km/h to zero in 2.7 s. What is the average total force acting between the ground and the car's tires during the braking maneuver? Assume that each tire experiences the same force level.

3.4.7. **[Level 1]** A passenger is playing shuffleboard during a cruise. The shuffleboard disk is sent along the deck at an initial speed of 6 m/s, and it comes to rest after sliding for 2 s. What is the coefficient of dynamic friction between the shuffleboard disk and the deck?

E3.4.7

3.4.8. **[Level 1]** A crate is pushed along a rough floor by a horizontal force $F(t)$ that varies over time as illustrated. The crate starts from rest at time $t = 0$, and its speed $\Delta t = 4$ s later is $v = 2$ m/s. If the crate weighs $W = 18$ kg, what is the coefficient of friction μ between it and the floor?

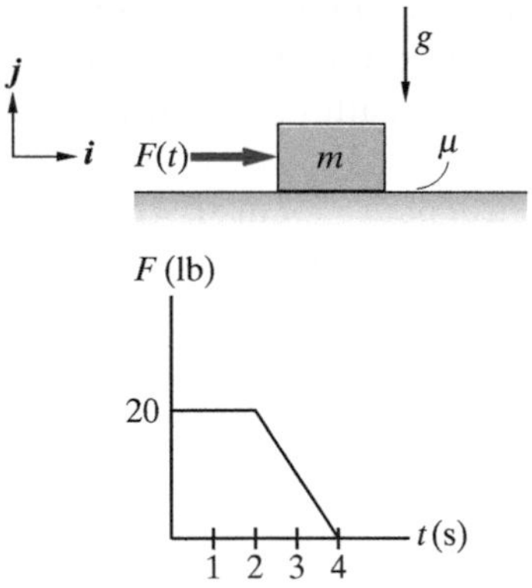

E3.4.8

3.4.9. **[Level 1]** A flat rock is projected along a rough surface (region A) with an initial speed of 10 m/s. After 1.7 s, its speed has dropped to one-third of its initial value. At this point it enters region B, which provides a different coefficient of dynamic friction. After 1.7 s it comes to rest. Knowing that the coefficient of dynamic friction is twice as great in one of the regions as that in the other, determine the particular values of the two coefficients of friction.

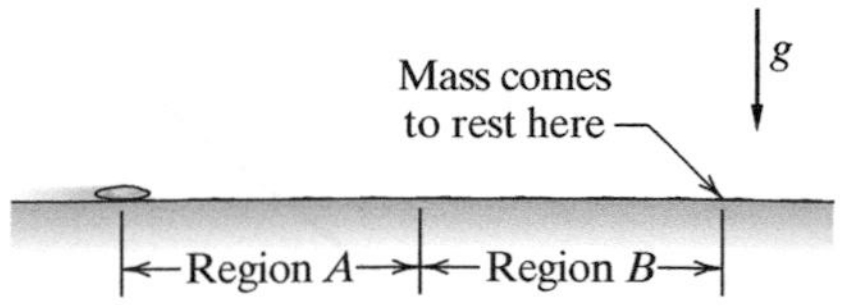

E3.4.9

3.4.10. **[Level 1]** A 2 kg iron disk is projected at 10 m/s across a flat surface, underneath which is an array of electromagnets. The electromagnets are activated sequentially, attracting the disk and thereby increasing the effective coefficient of dynamic friction. The frictional force is given by

$$f = f_0 + f_1 t^2$$

where $f_0 = 1.0$ N and $f_1 = 3$ N/s^2. When will the disk's speed be reduced to 5 m/s?

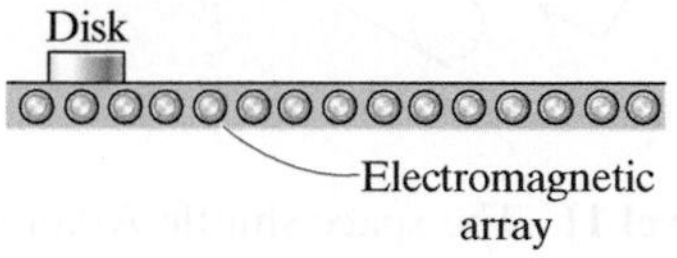

E3.4.10

3.4.11. **[Level 1]** A block of mass $m = 10$ kg is initially at rest on a rough floor with a coefficient of friction of $\mu = 0.2$. The block is acted on by a constant force F applied at an angle $\theta = 30°$ to the horizontal. In $\Delta t = 2$ s, the block reaches a speed of $v = 5$ m/s. How much force is applied to the block?

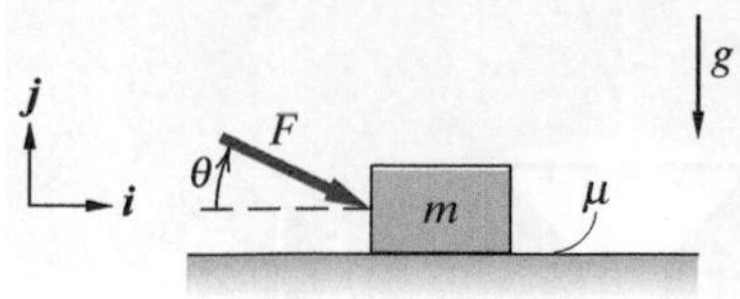

E3.4.11

3.4.12. **[Level 1]** How long would a spacecraft need to keep its thruster running if the ship had a mass of 200,000 kg, the thruster could produce a thrust of 5.0×10^6 N, and the desired final speed was 70,000 km/s? The spacecraft starts from rest. Ignore mass loss due to fuel usage.

3.4.13. **[Level 1]** A standard circus stunt calls for a companion to suddenly drop from a dangling rope onto the back of a cyclist passing beneath her. Assume that the cyclist and bicycle have a mass of 80 kg, they are moving at 5 m/s, and the companion has a mass of 40 kg. Determine the change in speed of the bicycle immediately after she has dropped onto the cyclist's back.

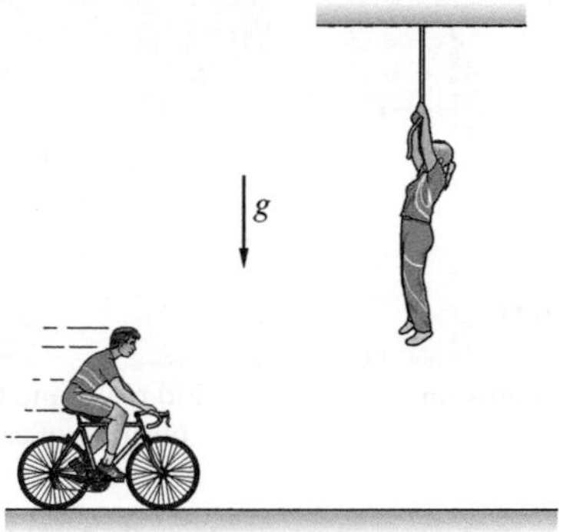

E3.4.13

3.4.14. **[Level 1]** As part of the new Bicycle Corps, a soldier is pedaling down the road at 25 km/h, holding a missile launcher on his shoulder. The combined mass of the bicycle and soldier is 80 kg, and the single missile in the launcher has a mass of 0.5 kg. He launches the missile and, as it leaves the launcher, it has a velocity of 500$\boldsymbol{i}$ m/s *relative to the launcher*. What is the velocity of the soldier/bicycle immediately following launch?

E3.4.14

3.4.15. **[Level 1]** A child puts a 5 g spitball in a straw and blows into the end so that the net internal pressure against the spitball is 4000 N/m^2. The effective cross-sectional area of the spitball is 75 mm^2. After 0.2 s, how fast is the spitball traveling?

3.4.16. **[Level 1]** A bat changes the velocity of a 0.14 kg ball from $-130\boldsymbol{i}$ km/h to $200\boldsymbol{i}$ km/h in 0.17 s. What are the applied linear impulse and the average force acting on the ball during contact?

3.4.17. **[Level 1]** In this problem we'll perform an elementary accident reconstruction. Car A is speeding and is about to run through a red light and collide with car B. At the time of collision $\boldsymbol{v}_A = v_A\boldsymbol{i}$ m/s and $\boldsymbol{v}_B = v_B\boldsymbol{j}$ m/s. $m_A = 1500$ kg, $m_B = 2700$ kg. They move together as a single body after the collision (collision point labeled O) and slide at an angle of $\theta = 31°$ for 9.4 m. The coefficient of friction between the cars and road is 0.85. Determine the speed of each car at the time of collision.

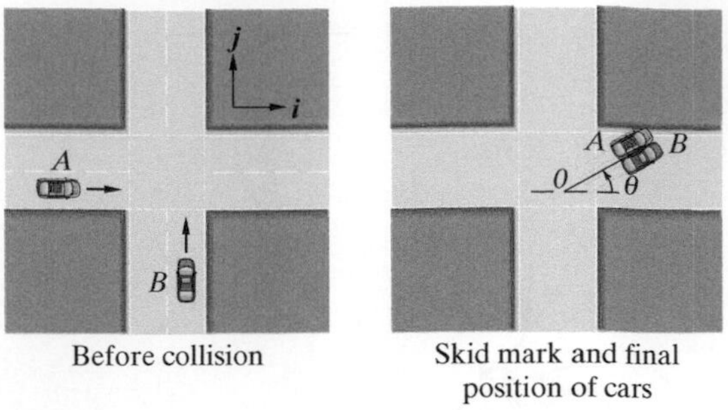

E3.4.17

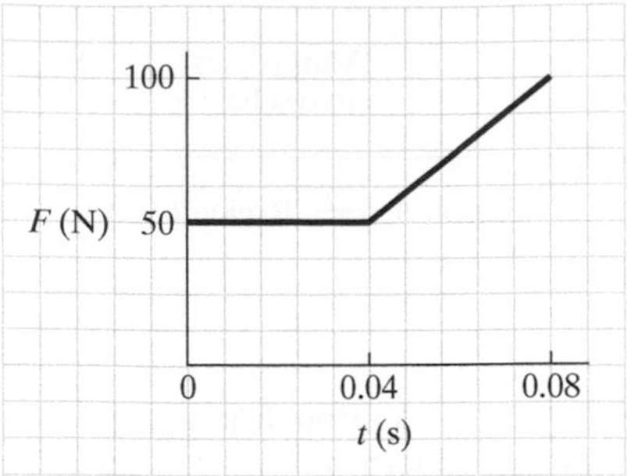

E3.4.20

3.4.18. **[Level 1]** Engineers are testing a new maglev train by timing how quickly it accelerates from rest to $v_f = 161$ km/h. If the train weighs 13,600 kg and takes 10 s to accelerate to v_f, what is the average electromagnetic force F_e that needs to be supplied?

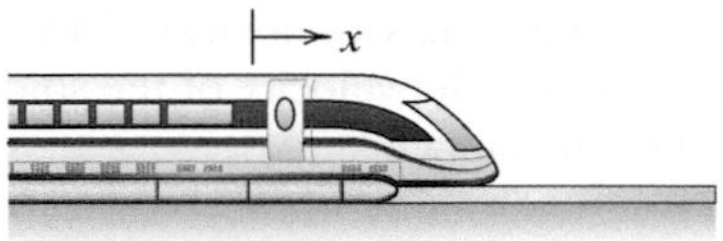

E3.4.18

3.4.19. **[Level 1]** Out of curiosity, a curling team decides to calculate the coefficient of friction μ between the curling stone and the playing surface (one of the players is an engineer). Suppose a team member launches the curling stone at $v_0 = 5.4$ m/s, and it travels in a straight-line path to the center of the target, a distance of $s_c = 28$ m, where it comes to a stop. The travel time is $\Delta t = 10.4$ s, and the curling stone has a mass of 18 kg. What is μ?

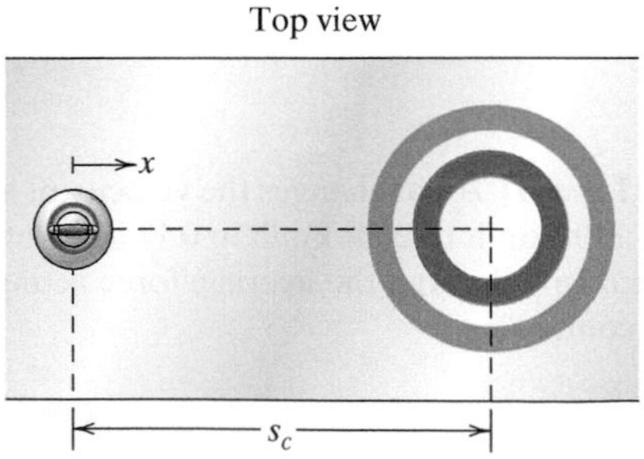

E3.4.19

3.4.20. **[Level 1]** An electromagnetic launcher applies a force to a 0.007 kg projectile that follows the illustrated force/time profile. How fast is the initially stationary mass moving at $t = 0.08$ s?

3.4.21. **[Level 1]** A mass is projected up a rough ramp inclined at $\theta = 30°$ with an initial speed of $v_0 = 10$ m/s. The coefficient of friction between the ramp and mass is $\mu = 0.3$. Find when the mass comes to a stop.

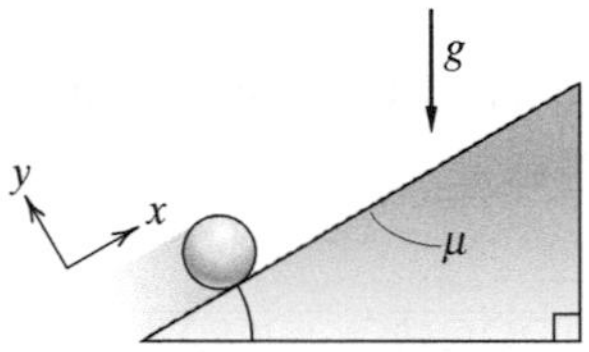

E3.4.21

3.4.22. **[Level 1]** The space shuttle Atlantis is preparing to dock with the International Space Station (ISS), and both maintain an orbit of constant radius. Atlantis has a mass of $m_{ss} = 6 \times 10^4$ kg and is moving at $\boldsymbol{v}_{ss} = 8100\boldsymbol{i}$ m/s. The ISS has a mass of $m_{ISS} = 1.8 \times 10^5$ kg and a velocity of $\boldsymbol{v}_{ISS} = 8000\boldsymbol{i}$ m/s. What is the velocity of the shuttle/ISS system after docking?

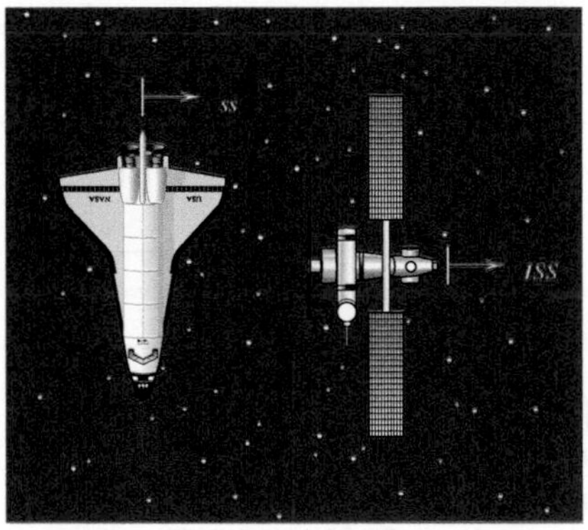

E3.4.22

3.4.23. **[Level 1]** A 0.06 kg tennis ball is traveling with a velocity of $177\boldsymbol{i}$ km/h when struck by a tennis racket. As a result of the ball's being hit, its velocity changes to $\boldsymbol{v} = (-128\boldsymbol{i} + 16\boldsymbol{j})$ km/h. What was the applied linear impulse (in N·s)?

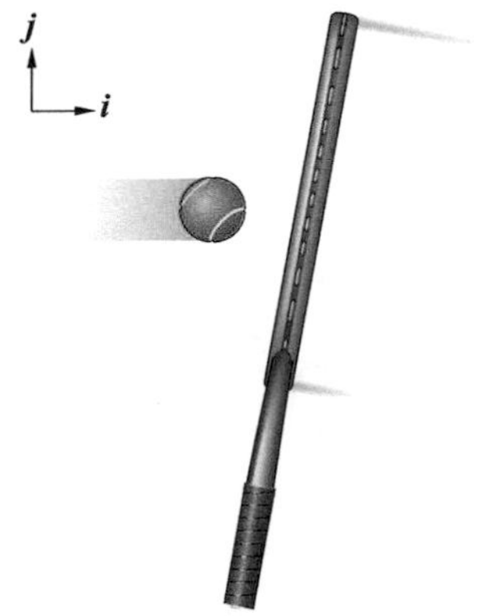

E3.4.23

3.4.24. **[Level 1]** A hockey puck having a mass of 0.3 kg is slid across the ice, starting with a speed of 13 m/s and slowing to 11.3 m/s after traveling for 3 s. What is the coefficient of kinetic friction between the puck and the ice?

3.4.25. **[Level 1]** An astronaut finds himself adrift, 10 m from his spaceship and stationary with respect to it. Lucky for him, he has a 2 kg wrench with him. He throws the wrench as hard as he can *away* from the spaceship, and the reaction causes him to drift toward his ship. He has a mass of 80 kg, and after being released the wrench has a speed of 15 m/s, as measured with respect to the stationary spaceship. How long does it take for the astronaut to reach the safety of his spaceship?

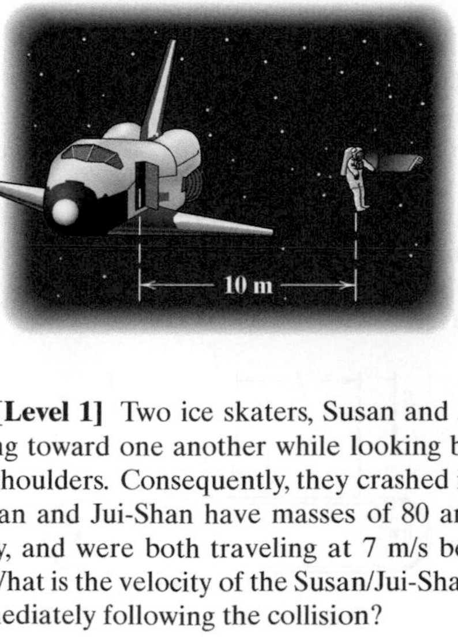

E3.4.25

3.4.26. **[Level 1]** Two ice skaters, Susan and Jui-Shan, were skating toward one another while looking backward over their shoulders. Consequently, they crashed into each other. Susan and Jui-Shan have masses of 80 and 70 kg, respectively, and were both traveling at 7 m/s before the collision. What is the velocity of the Susan/Jui-Shan combination immediately following the collision?

E3.4.26

3.4.27. **[Level 1]** Two Revolutionary War bullets fired from two Revolutionary War rifles (during the Revolutionary War) collided head-on and stuck together. After the collision, the two-bullet glob had a velocity of $-100\boldsymbol{i}$ m/s. Bullet 1 had a mass of 27 g and bullet 2 had a mass of 31 g. Can you determine the velocity of the two bullets before the collision? If not, what additional information would you need?

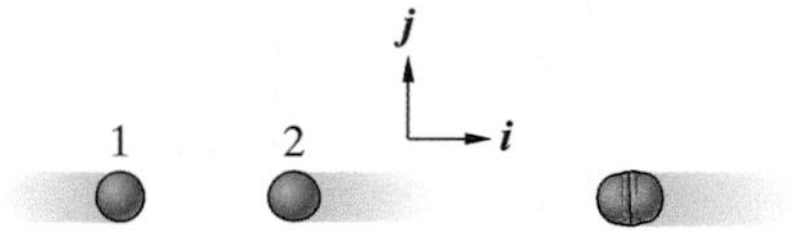

E3.4.27 Before collision After collision

3.4.28. **[Level 1]** Two cars collide with each other, car A going $10\boldsymbol{i}$ m/s and car B $-15\boldsymbol{i}$ m/s. Car A has a mass of 1000 kg and car B has a mass of 1500 kg. Both cars have sophisticated crumple zones in the front which deform in a collision, helping to reduce the severity of the impact. After both vehicles have fully deformed, what is the velocity of the resultant tangle of metal? How does this compare to two point-masses (having the same masses and velocities as the cars) that collide? What's the point of a crumple zone?

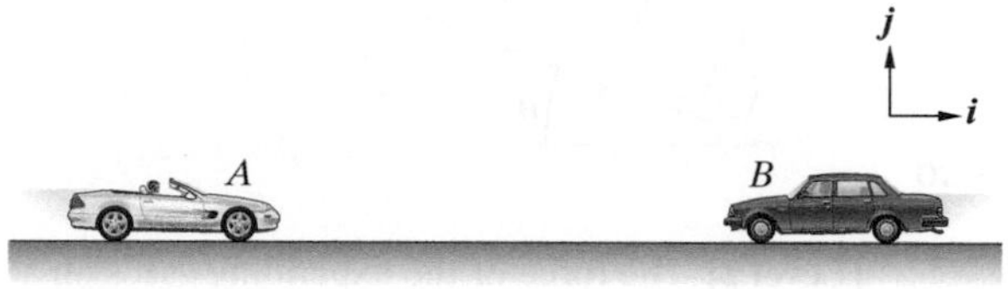

E3.4.28

3.4.29. **[Level 1]** A carpenter uses her hammer to strike a nail. Approximate all the hammer's mass as being concentrated at the head (0.68 kg) and assume that at the instant of impact the head is traveling in the $-\boldsymbol{j}$ direction. If the hammer contacts the nail at 27 m/s and the entire impact phase takes 0.011 s, what is the magnitude of the average force exerted by the nail on the hammer?

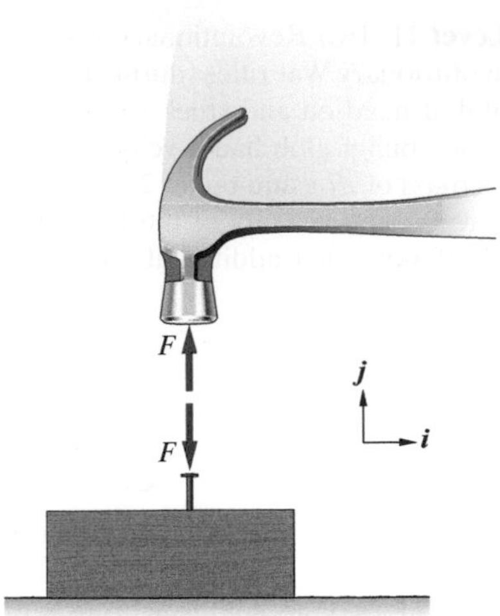

E3.4.29

3.4.30. **[Level 2]** A block A (mass m) is projected up a slope at 16 m/s. The coefficient of dynamic friction is 0.15 and $\theta = 15°$. How long will it take for the mass's speed to drop to 4 m/s?

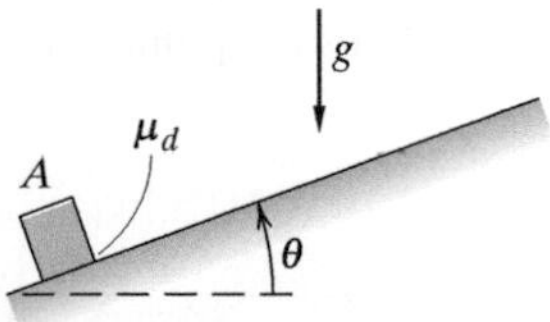

E3.4.30

3.4.31. **[Level 2]** The force of the expanding gases on a 0.02 kg bullet is given by $f = f_0 - f_1 t$. Initially stationary at $x = 0$, the bullet reaches a speed of 1000 kph by the time it exits the gun. $f_0 = 7000$ N and $f_1 = 2.9 \times 10^6$ N/s.

a. What is the total linear impulse acting on the bullet?

b. When does the bullet exit the gun?

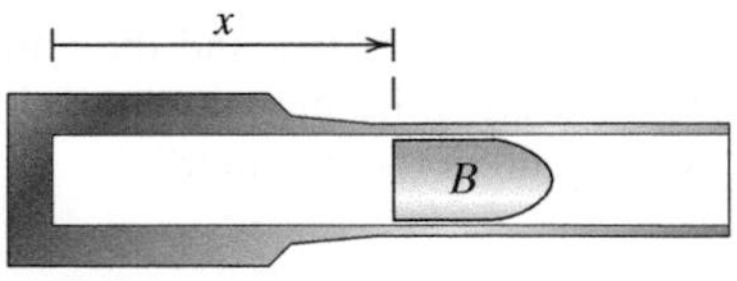

E3.4.31

3.4.32. **[Level 2]** A large rock (R) was on top of a hill with a rope attached to it. It was accidentally knocked off the flat surface and onto the slope. John (J) saw what was happening and grabbed the rope. As John was dragged downhill, his feet dug up an ever-larger pile of dirt, slowing both him and the rock. When John first grabbed the rope, the rock had a speed of 6 m/s. The rock's mass was 50 kg, and John's was 60 kg. The tension in the rope grew linearly:

$$T = T_1 + T_2 t$$

$\theta = 25°$, $\phi = 10°$, $T_1 = 100$ N, and $T_2 = 73$ N/s. How long did it take for John to bring the rock's slide to a halt?

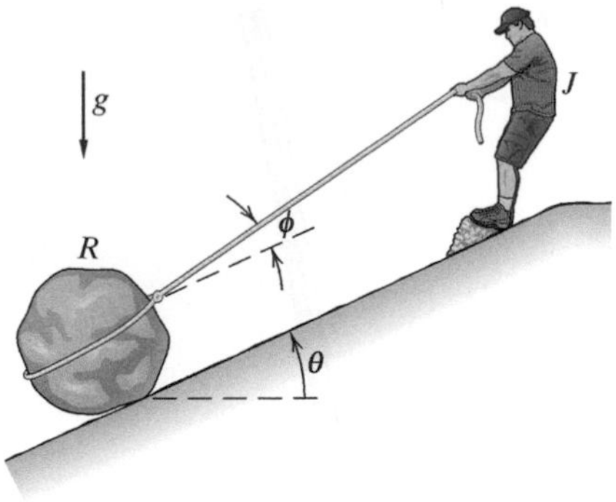

E3.4.32

3.4.33. **[Level 2]** A 0.2 kg bouncing ball is observed to rebound to a height of 1.6 m. What is the linear impulse applied to the ball while it is in contact with the floor?

3.4.34. **[Level 2]** A backpack is dropped while a climber is traversing an ice field having a 20° slope. The coefficient of dynamic friction is 0.20. How fast is the pack moving after 2 s? Assume zero initial speed.

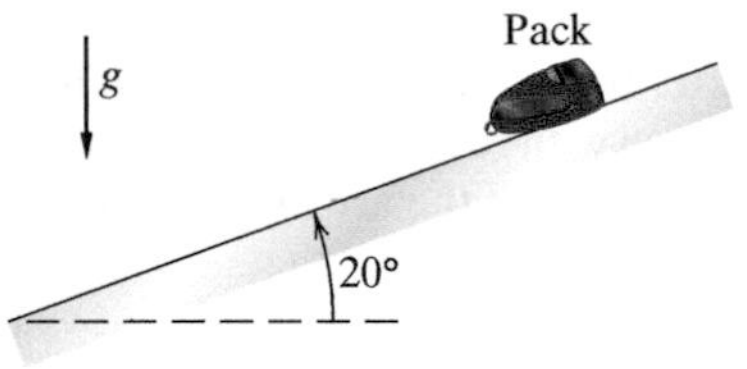

E3.4.34

3.4.35. **[Level 2]** In Example 3.16 a block A was held on the top surface of a moving sanding belt ($v_B = 3$ m/s) and then slid to the left with an initial speed $v_0 = 2$ m/s (measured with respect to the fixed ground). The coefficient of friction between the block and the belt was $\mu = 0.4$. Assume for this exercise that initially the block is a distance L from the left roller ($L = 1.2$ m) and determine if the block reaches the left roller.

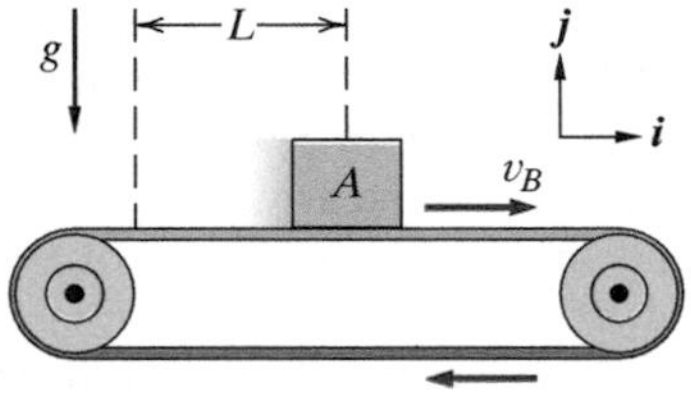

E3.4.35

3.4.36. **[Level 2]** A block A is held on the top surface of a moving sanding belt and then slid to the left with an initial speed $v_0 = 2.0$ m/s (measured with respect to the fixed ground). Eventually it stops moving left and begins to accelerate to the right, reaching a maximum speed equal to the belt's speed. The question to be answered here is what scenario will get the block to a point furthest to the right of the initial point in 4 s? For scenario (a) the coefficient of friction is 0.4 and the belt is moving to the right at 3 m/s.

In scenario (b) the coefficient of friction is 0.2 and the belt is moving to the right at 5 m/s. The answer isn't obvious because scenario (b) will (given enough time) result in a higher speed for the block *but* because of the lowered coefficient of friction it will take relatively longer for the block to slow to a stop and then accelerate up to the belt's speed. Assume in each case that L is long enough that the block stays in contact with the belt during its deceleration phase and also that there's enough belt to the right of the block so that it stays in contact up to $t = 4$ s.

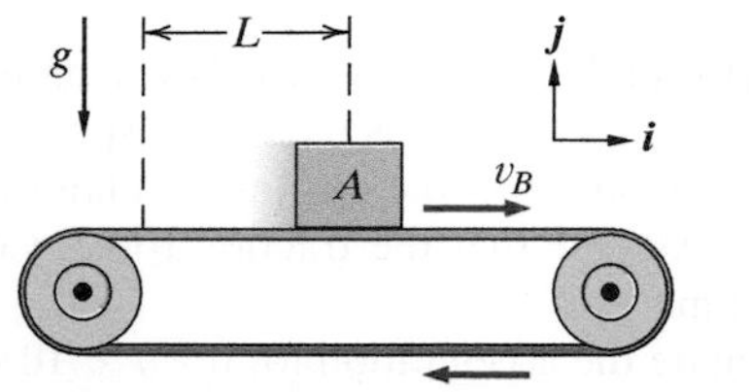

E3.4.36

3.4.37. [Level 2] A pyrotechnician lights a stationary 1 kg firework, and the generated thrust varies according to $T(t) = 200e^{-0.5t}$ N. Determine how long it takes for the firework to reach a speed of $v_f = 100$ m/s if the firework shoots straight up. Neglect the effects of gravity.

E3.4.37

3.4.38. [Level 2] Angela is sledding down a 10° snowy slope. Consider the sled/snow interface as frictionless. She starts at point A with zero velocity. At point B, located 40 m downslope, she plows into her friend, Grace, with the result that Grace ends up on Angela's shoulders, both of them still traveling on the sled. Angela and the sled have a mass of 23 kg, and Grace has a mass of 20 kg. How fast are the two girls traveling immediately following the collision?

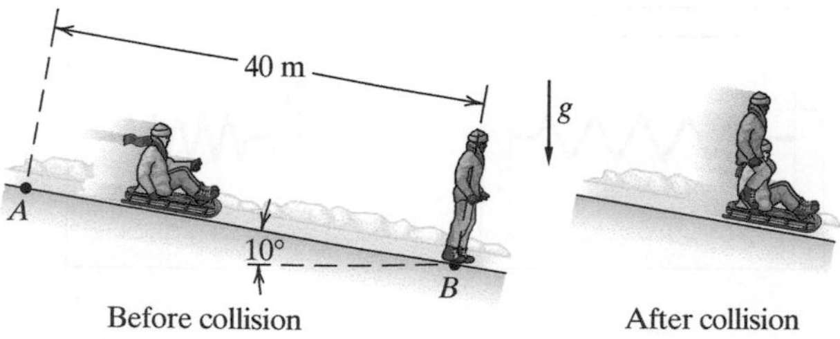

E3.4.38

3.4.39. [Level 2] A 50,000 kg submarine moving forward at 5 m/s launches a 100 kg torpedo. The ejection mechanism acts on the torpedo for 0.25 s, at the end of which the torpedo enters the water with a velocity of 10 m/s *relative to the submarine*.

a. Calculate the velocity of the submarine and torpedo just after ejection is complete.

b. Calculate the average force acting on the torpedo during the launch.

c. Calculate the average force exerted by the torpedo on the submarine during the launch.

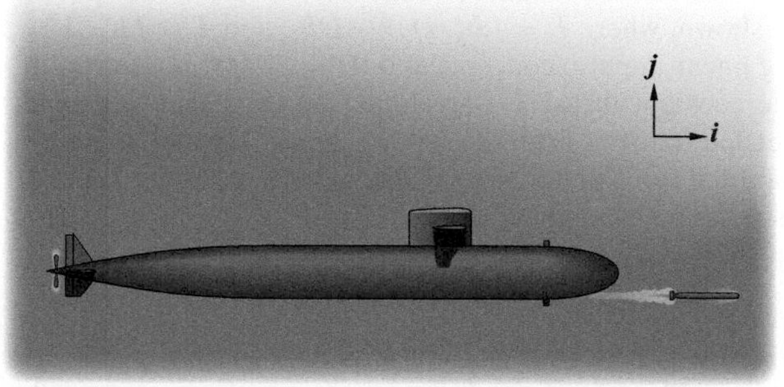

E3.4.39

3.4.40. [Level 2] After dropping from a height of 2 m, a 0.06 kg tennis ball rebounds to a height of 1 m. The impact takes 0.07 s.

a. What is the average force exerted by the ball on the ground during impact?

b. What is the average acceleration experienced by the ball over the impact interval?

3.4.41. [Level 2] In *The Mask of Zorro*, a gigantic soldier grabbed Zorro and threw him about 6 m. Is it likely that someone could actually do this? Let's analyze the problem and see if some conclusions can be drawn. Assume that Zorro weighed 80 kg. Let's give Mr. Soldier the benefit of the doubt and assume he launched Zorro at a 45° angle and from a height of 1.2 m. (The 45° launch angle would produce the greatest range.) Treat Zorro as a point mass.

a. Calculate how fast Zorro had to be moving at launch to hit the ground 6 m in front of the soldier.

b. What is the magnitude of the linear impulse applied to Zorro?

c. If the soldier applied the same linear impulse to a baseball, which weighs 0.14 kg, how fast would the ball be moving as it left his hand?

d. From your answer to **c**, can you conclude that the movie toss is highly improbable?

3.4.42. [Level 2] The SMG II transmission on the BMW M3 requires only 80 ms to shift from first to second gear, all under computer control. That's awfully quick. How does it do it? Well, the transmission linkages are moved by means of 8275 kPa hydraulic fluid, and that's extremely high pressure. Let's work out some numbers to see that those quick times make sense. Assume that the high-pressure fluid acts on a 0.4 cm^2 area. The shift linkage weighs 0.9 kg. Calculate how far the linkage will move if the high pressure is maintained for 80 ms. How fast will it be moving at that time?

3.4.43. **[Level 2]** A 35,000 kg airplane begins to climb at an angle of 3°. Its initial speed is 150 m/s, and after one minute its speed has increased to 180 m/s. The airplane experiences an aerodynamic drag force D that opposes its velocity. The lift, drag, and thrust forces can be decomposed as shown, where $\boldsymbol{L} = L\boldsymbol{b}_2$, $\boldsymbol{D} = -D\boldsymbol{b}_1$, and $\boldsymbol{T} = T\boldsymbol{b}_1$. Using linear impulse concepts and $|D| = 10{,}000$ N, determine the constant thrust T (developed by the engines) that is needed to offset both the aerodynamic drag and the gravitational force and allow the airplane to increase its speed as described above.

E3.4.43

3.4.44. **[Level 2]** In one of the James Bond movies, secret agent 007 leapt onto a dolly and zoomed through a newspaper factory on it. Assume that 007 weighed 83 kg and the dolly weighed 14 kg. The dolly was initially stationary owing to static friction, but as soon as 007 contacted it (traveling at a speed of 7 m/s), it began to roll down the slightly sloped floor of the factory. After he had rolled for 10 s on a sloped floor, his speed had increased to 8.7 m/s. Assuming that no drag force acted on 007, calculate the slope of the floor.

3.4.45. **[Level 2]** **Computational** In **Exercise 3.2.16** Bond is sliding off a circular roof with a frictionless interface. Now I want to make the problem more realistic. Assume that there exists a coefficient of kinetic friction between Bond and roof. Keep all the parameters the same as in **Exercise 3.2.16** (radius of 20 m and initial $\dot{\theta}$ of 0.5 rad/s) and include an additional $\mu_d = 0.1$. Use MATLAB (or whatever code you prefer) to determine the angle at which he leaves the roof or, perhaps, whether he leaves it at all.

3.4.46. **[Level 2]** A grocery cart plus groceries having a combined mass of 25 kg starts rolling downhill at point A. At point B, it strikes a stationary 70 kg postal worker, and both continue down the hill, worker in basket. How fast will the cart/postal worker be traveling when they reach point C? Assume friction-free rolling.

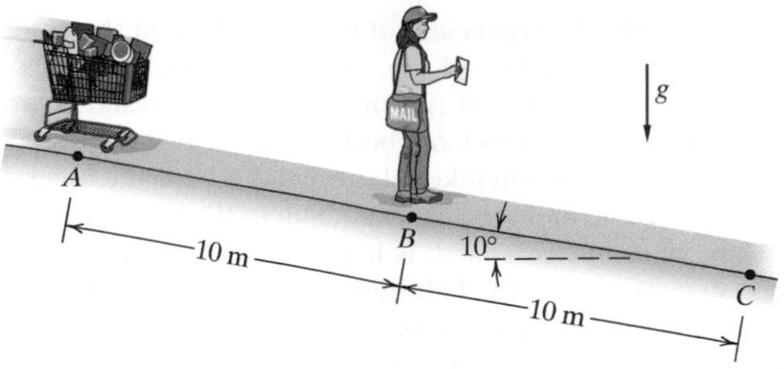

E3.4.46

3.4.47. **[Level 2]** A 1200 kg car is traveling up a 6% grade (tangent of the slope is 0.06) at a speed of 90 kph. At $t = 0$, the driver floors the gas and 10 s later is traveling at 125 kph. Assume that the traction force was constant during this time.

a. Calculate the acceleration for $0 < t < 10$ s in g's.

b. Calculate the traction force acting to drive the car up the hill.

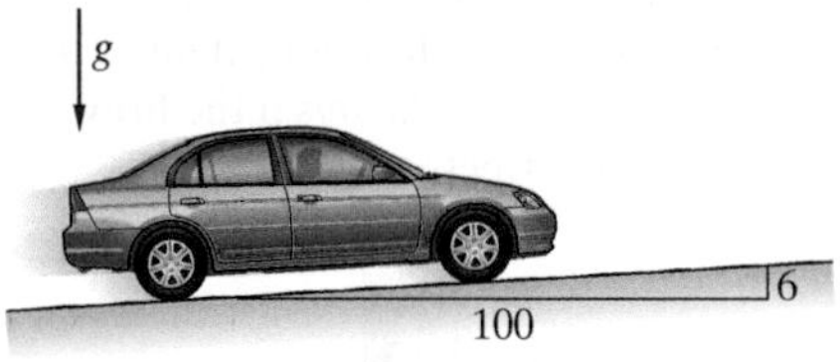

E3.4.47

3.4.48. **[Level 2]** A massless spring originally stretched between two opposing walls of a rectangular enclosure. Because of environmental degradation, the left wall developed cracks that continued to grow until they became so severe that a chunk of the wall pulled free from the left side. When the chunk reached the halfway point across the enclosure ($x = 0.4$ m), it was moving at a speed of $\dot{x} = 5$ m/s. Assume the enclosure moves without friction. x is measured relative to the enclosure. The entire enclosure had a mass of 2.4 kg, and the mass of the broken chunk was 0.21 kg. What is the enclosure's velocity with respect to a stationary observer at the instant the small chunk is at the halfway point?

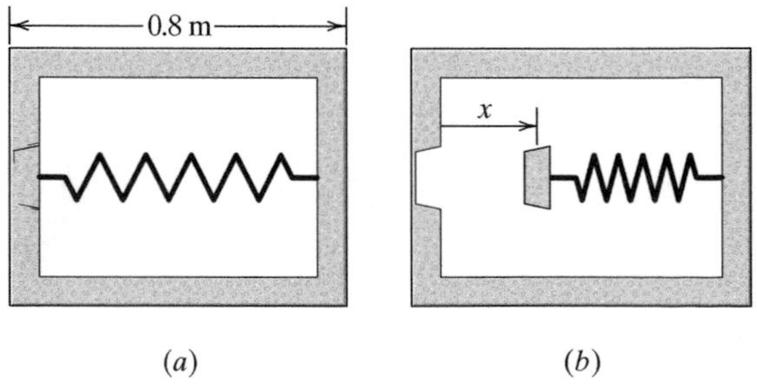

E3.4.48

3.5 ANGULAR MOMENTUM AND ANGULAR IMPULSE

There are rotational analogues to linear momentum and linear impulse called, not surprisingly, **angular momentum** and **angular impulse**. They're very useful, even though they may seem a bit artificial at first. Because we're still concerning ourselves only with mass particles, angular momentum refers only to a particle's motion about some fixed point in space. We will definitely see some interesting behavior (the orbits of the planets are one of the topics we'll be examining), but the really important application will come in Chapter 7, where we will be dealing with rigid bodies (including blocks, wheels, and spacecraft). To fully explain their motion, we have to focus both on where they're going (linear momentum) and on whether or not they're spinning as they go there (angular momentum). So just keep in mind that the applications get better and better as we go along.

Angular momentum, which we will represent by $\boldsymbol{H}_O$, where O is the reference point about which the angular motion takes place, can be thought of as the momentum equivalent of applied moment (also known as torque). Recall that a moment is given by $\boldsymbol{r} \times \boldsymbol{F}$—that is, the moment arm crossed with the force vector. $\boldsymbol{H}_O$, the angular momentum about O, is just $\boldsymbol{r} \times m\boldsymbol{v}$, as shown in **Figure 3.42**:

$$\boldsymbol{H}_O \equiv \boldsymbol{r} \times m\boldsymbol{v} \tag{3.45}$$

Recall that the cross product of two vectors $\boldsymbol{a}$ and $\boldsymbol{b}$ is a third vector $\boldsymbol{c}$, the magnitude of which is given by $ab \sin\theta$, where θ is the angle between $\boldsymbol{a}$ and $\boldsymbol{b}$. Thus, if you know the angle between $\boldsymbol{r}$ and $m\boldsymbol{v}$ (as shown in **Figure 3.43**), you can express the *magnitude* of the angular momentum about O as

$$H_O = mvr \sin\theta \tag{3.46}$$

or, using the illustrated distance d, as

$$H_O = mvd \tag{3.47}$$

because $d = r \sin\theta$.

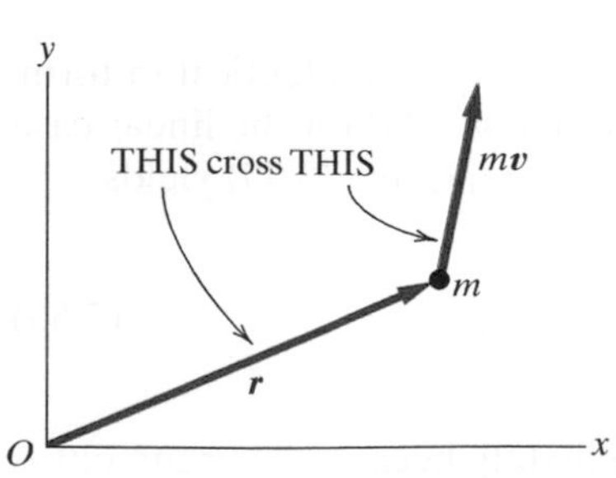

Figure 3.42 Angular momentum of a particle

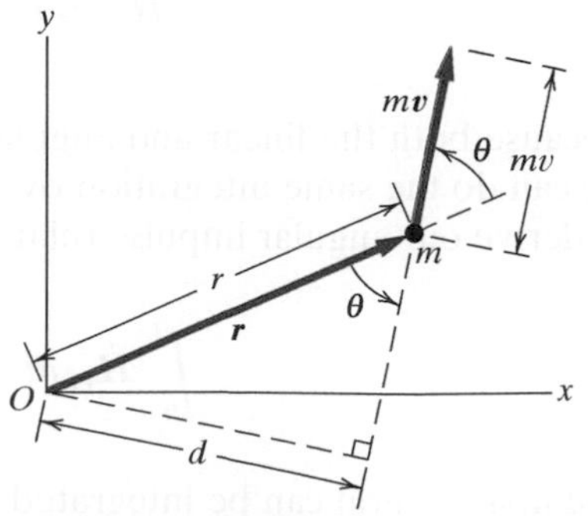

Figure 3.43 More details—angular momentum of a particle

We can also express $\boldsymbol{H}_O$ in terms of the components of $\boldsymbol{r}$ and $m\boldsymbol{v}$. If we let $\boldsymbol{r} = r_1\boldsymbol{i} + r_2\boldsymbol{j}$ and $\boldsymbol{v} = v_1\boldsymbol{i} + v_2\boldsymbol{j}$ then, remembering that

$$
\begin{aligned}
\boldsymbol{i}\times\boldsymbol{i} &= 0\\
\boldsymbol{j}\times\boldsymbol{j} &= 0\\
\boldsymbol{i}\times\boldsymbol{j} &= \boldsymbol{k}\\
\boldsymbol{j}\times\boldsymbol{i} &= -\boldsymbol{k}
\end{aligned}
$$

we have

$$\boldsymbol{H}_O = (r_1\boldsymbol{i} + r_2\boldsymbol{j})\times m(v_1\boldsymbol{i} + v_2\boldsymbol{j}) = m(r_1v_2 - r_2v_1)\boldsymbol{k} \tag{3.48}$$

Okay, so now we have an expression for the angular momentum of a particle about some point. What's this information going to do for us? Well, remember how we just saw that the force applied to a particle and the particle's linear momentum are related? We're going to see the same kind of relationship here, and it's easy to show. To do it, just differentiate (3.45) with respect to time:

$$\dot{\boldsymbol{H}}_O = m(\dot{\boldsymbol{r}} \times \boldsymbol{v} + \boldsymbol{r} \times \dot{\boldsymbol{v}}) = m(\boldsymbol{v} \times \boldsymbol{v} + \boldsymbol{r} \times \boldsymbol{a})$$

Because $\boldsymbol{v} \times \boldsymbol{v}$ is zero, this leaves us with

$$\dot{\boldsymbol{H}}_O = m(\boldsymbol{r} \times \boldsymbol{a}) = \boldsymbol{r} \times (m\boldsymbol{a}) = \boldsymbol{r} \times \boldsymbol{F}$$

What is $\boldsymbol{r} \times \boldsymbol{F}$? It's just the applied moment $\boldsymbol{M}_O$. What we have obtained is an expression that tells us that the time rate of change of the particle's angular momentum $\boldsymbol{H}_O$ is equal to the applied moment about O. This is exactly analogous to what we saw in the previous section, namely, that the time rate of change of a particle's linear momentum $\boldsymbol{L}$ is equal to the net applied force $\boldsymbol{F}$. In this case, however, we have a statement that says the time rate of change of a particle's angular momentum about a fixed point O is equal to the sum of the moments taken about O. We will refer to this equation as a **moment balance**.

Moment Balance:

$$\dot{\boldsymbol{H}}_O = \boldsymbol{r} \times \boldsymbol{F} = \boldsymbol{M}_O \tag{3.49}$$

Because both the linear and angular relationships are identical in form, we can do the same integration over time that we did for the linear case to derive our angular impulse relationship. Integrating (3.49) yields

$$\int_{t_1}^{t_2} \dot{\boldsymbol{H}}_O\, dt = \int_{t_1}^{t_2} \boldsymbol{M}_O\, dt \tag{3.50}$$

The first integral can be integrated immediately because its argument is an exact differential. In keeping with our linear work, we will call the second integral the **angular impulse** ($\mathcal{AI}$):

$$\mathcal{AI}_{O_{1-2}} \equiv \int_{t_1}^{t_2} \boldsymbol{M}_O \, dt \tag{3.51}$$

Equation (3.50) therefore becomes

$$\boldsymbol{H}_O(t_2) - \boldsymbol{H}_O(t_1) = \mathcal{AI}_{O_{1-2}}$$

or

$$\boldsymbol{H}_O(t_2) = \boldsymbol{H}_O(t_1) + \mathcal{AI}_{O_{1-2}} \tag{3.52}$$

Paralleling what we saw in the previous section for linear momentum, the angular momentum of a particle at some time t_2 is equal to the angular momentum at an earlier time t_1 plus the angular impulse that acts on the particle from t_1 to t_2.

⋙ **Check out Example 3.18 (page 158), Example 3.19 (page 159), and Example 3.20 (page 160) for applications of this material.**

EXAMPLE 3.18 CHANGE IN SPEED OF A MODEL PLANE (Theory on page 157)

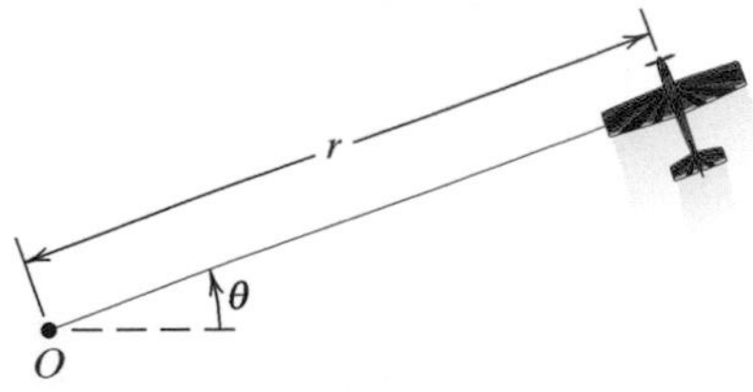

Figure 3.44 Tethered model airplane

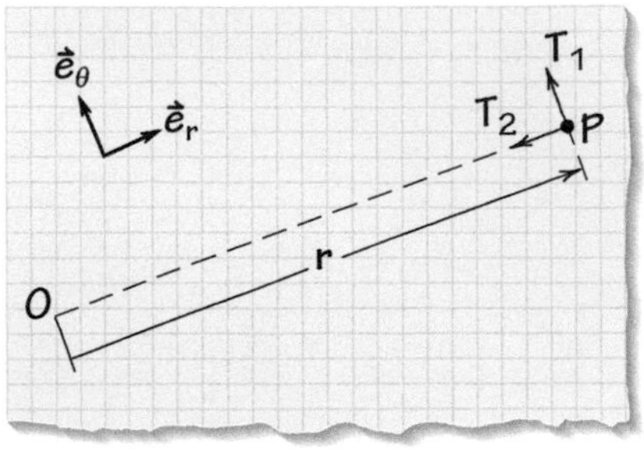

Figure 3.45 Force acting on model airplane

Figure 3.44 shows a 0.1 kg model airplane tethered to an operator at O by a massless control line. At the instant considered $\theta = 0, r = 6\,\text{m}, v_t = 15\,\text{m/s}$, and the thrust generated by the airplane is equal to 0.3 N. Use (3.52) to determine the angular speed of the airplane around O at $t = 3$ s.

Goal Determine the plane's angular speed around O after 3 s of applied thrust.

Given The airplane's mass and thrust and the distance from the airplane to its center of rotation.

Assume We will neglect the effect of gravity, assuming that the lift on the plane's wings counteracts it. Thus the only forces we need concern ourselves with are the thrust and the tension in the control wire.

Draw **Figure 3.45** shows the forces acting on the airplane as well as a convenient set of cylindrical unit vectors.

Formulate Equations Equation (3.52) is directly applicable. The two forces that could produce moments about O are $\boldsymbol{T}_1$ and $\boldsymbol{T}_2$, and their moment contribution is found from

$$\sum \boldsymbol{M}_O = \boldsymbol{r}_{P/O} \times (T_1\boldsymbol{e}_\theta - T_2\boldsymbol{e}_r) = r\boldsymbol{e}_r \times (T_1\boldsymbol{e}_\theta - T_2\boldsymbol{e}_r) = rT_1\boldsymbol{k}$$

Thus
$$H_O(3\,\text{s}) = H_O(0) + \int_0^{3\,\text{s}} rT_1\,dt$$

The airplane's initial speed is 15 m/s, and so

$$H_O = (0.1\,\text{kg})(6\,\text{m})(15\,\text{m/s}) = 9.0\,\text{kg}\cdot\text{m}^2/\text{s}$$

Solve Putting the preceding equations together gives us

$$H_O(3\,\text{s}) = 9.0\,\text{kg}\cdot\text{m}^2\text{s}^{-1} + (6\,\text{m})(0.3\,\text{N})(3\,\text{s}) = 14.4\,\text{kg}\cdot\text{m}^2/\text{s} \qquad (3.53)$$

The magnitude of the angular momentum is equal to rmv_θ, which can be expressed in terms of $\dot{\theta}$ by utilizing $v_\theta = r\dot{\theta}$ to get $H = r^2m\dot{\theta}$. Using (3.53) gives us

$$14.4\,\text{kg}\cdot\text{m}^2/\text{s} = (6\,\text{m})^2(0.1\,\text{kg})\,\dot{\theta} \quad \Rightarrow \quad \dot{\theta} = 4\,\text{rad/s}$$

Check We can actually resort to linear momentum to check our result. The initial speed of the model airplane was 15 m/s, and after applying a force of 0.3 N for 3 s it reached a speed of (6 m)(4 rad/s) = 24 m/s. The change in speed along the path of the airplane divided by the elapsed time is $(24\,\text{m/s} - 15\,\text{m/s})/(3\,\text{s}) = 3\,\text{m/s}^2$, giving us an acceleration of $3\,\text{m/s}^2$. And what should the acceleration be if a force of 0.3 N is applied to a 0.1 kg mass? Why, $3\,\text{m/s}^2$, of course. Thus we've validated our angular momentum result.

EXAMPLE 3.19 ANGULAR MOMENTUM OF A BUMPER (Theory on page 157)

Figure 3.46*a* shows a parked car. The unobservant driver, never noticing that someone has tied a rope to a solid post O and to the car's rear bumper, gets in and drives off. The rope is 15 m long, and initially the bumper is 9 m from the post. **Figure 3.46*b*** is a top view that shows the car's position when the rope has gone taut, just before the bumper is torn off. The car has a speed $v = 14$ m/s at this point. Assuming that the bumper can be approximated as a particle B with mass m, what will the angular speed of the bumper about O be just after parting company with the car? Don't worry about gravity acting down on the bumper; just concern yourself with horizontal motions. Neglect the force between the car and bumper as well.

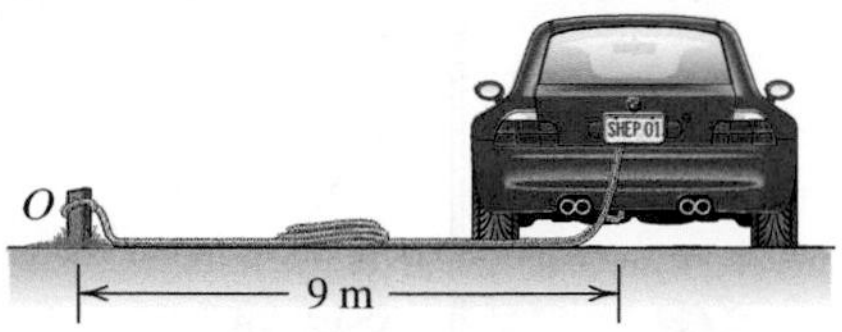

(*a*) Rear view

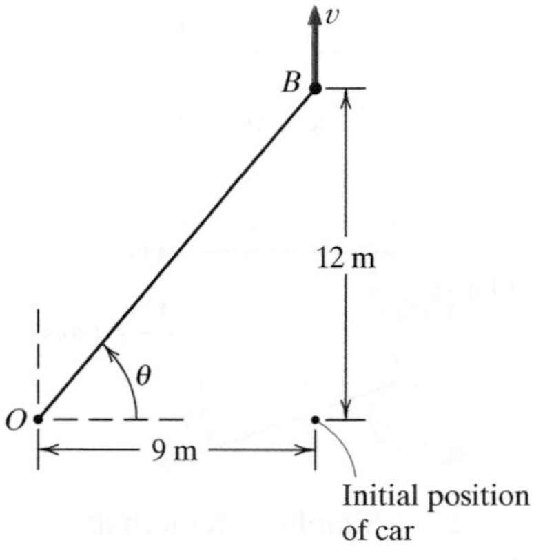

(*b*) Schematic top view

Figure 3.46 Car with rope tied to bumper

Goal Find the angular speed of the bumper about the fixed point O.

Given Speed of car, length of rope, and relative position of the car and O.

Assume We will concern ourselves only with in-plane motions.

Draw **Figure 3.47** shows the relevant force and unit vectors.

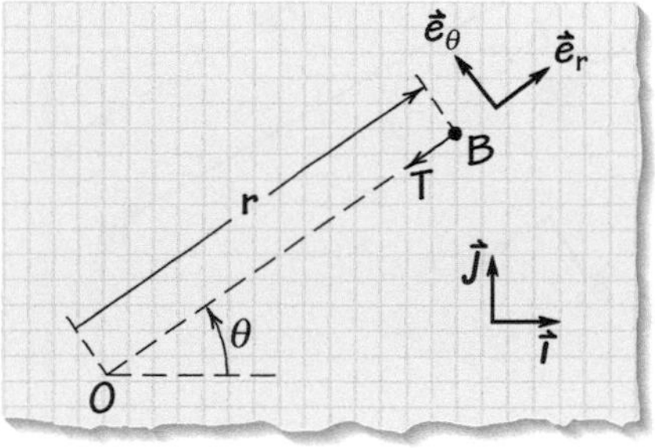

Figure 3.47 Forces when bumper is pulled off

Formulate Equations The relevant equation to use is (3.52):

$$\boldsymbol{H}_O(t_2) = \boldsymbol{H}_O(t_1) + \mathcal{AI}_{O_{1-2}}$$

In this case the times t_1 and t_2 are almost simultaneous—just before the rope goes taut and just after the bumper is off the car.

Solve Just before the rope goes taut, the bumper's angular momentum is

$$\boldsymbol{H}_O(t_1) = \boldsymbol{r}_{B/O} \times mv\boldsymbol{j} = [(9m\boldsymbol{i} + 12\boldsymbol{j})\text{ m}] \times mv\boldsymbol{j} = (9\text{ m})mv\boldsymbol{k}$$

The force applied to the bumper by the rope points directly toward O and therefore doesn't induce any moment. Thus the angular moment just after the bumper pops off is the same as the angular momentum just before that happens. The rope is inextensible, and so there is no velocity component in the $\boldsymbol{e}_r$ direction; the bumper's motion is only in the $\boldsymbol{e}_\theta$ direction.

$$\boldsymbol{H}_O(t_1) = \boldsymbol{H}_O(t_2) \quad \Rightarrow \quad r^2m\dot{\theta} = (9\text{ m})mv \quad \Rightarrow \quad \boxed{\dot{\theta} = 0.56\text{ rad/s}}$$

Check The angle θ at the point the rope goes taut is given by $\theta = \tan^{-1}(12/9) = 53.1°$. The coordinate transformation between $\boldsymbol{i}, \boldsymbol{j}$, and $\boldsymbol{e}_r$, $\boldsymbol{e}_\theta$ is given in **Figure 3.48**. If we decompose the bumper's velocity into the $\boldsymbol{e}_r$, $\boldsymbol{e}_\theta$ directions, we get $v\boldsymbol{j} = 0.8v\boldsymbol{e}_r + 0.6v\boldsymbol{e}_\theta$. The magnitude of the angular momentum about O is equal to rmv_θ, and so we have

$$H_O = 0.6rmv = 0.6(15\text{ m})(14\text{ m/s})m = (126\text{ m}^2/\text{s})m$$

Equating this with $rmv_\theta = r^2m\dot{\theta} = (15\text{ m})^2\dot{\theta}m$ gives us $\dot{\theta} = 0.56$ rad/s, matching our previous result.

	i	j
e_r	0.6	0.8
e_θ	−0.8	0.6

Figure 3.48 Transformation array

EXAMPLE 3.20 ANGULAR MOMENTUM OF A TETHERBALL (Theory on page 157)

Figure 3.49 Tetherball game

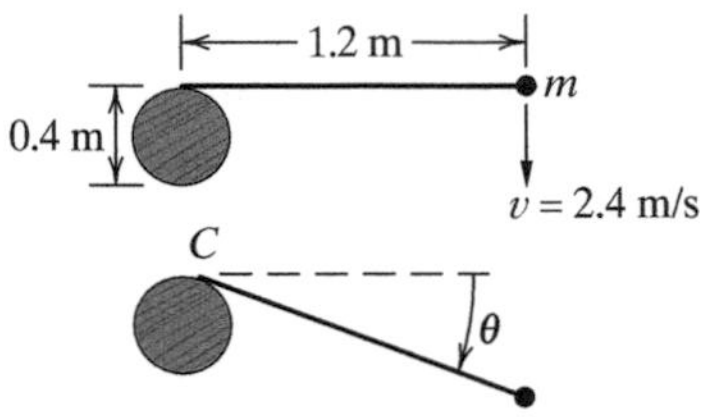

Figure 3.50 Simplified tetherball analysis

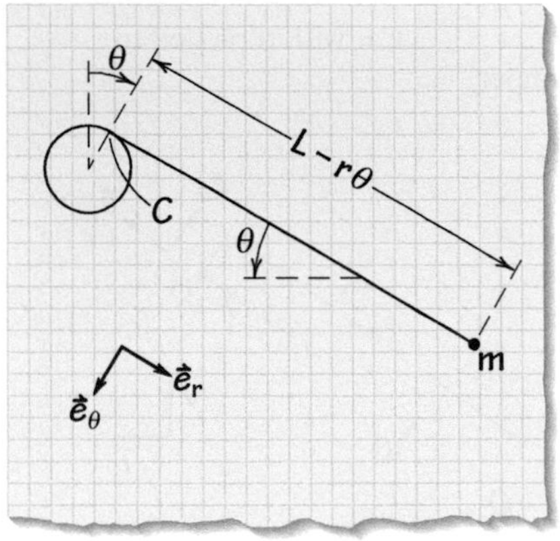

Figure 3.51 Variation of rope with wrap angle

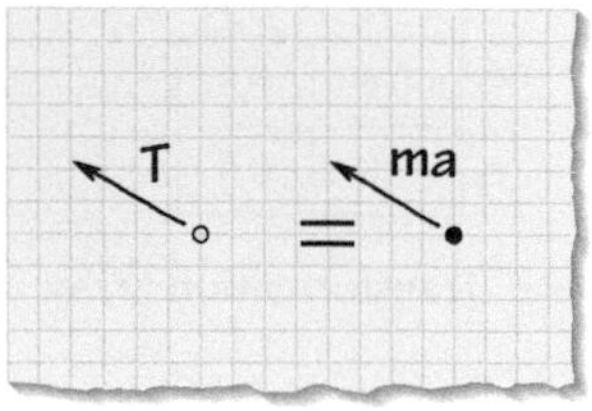

Figure 3.52 FBD=IRD for moving mass

Shown in **Figure 3.49** is a representation of the game tetherball, which consists of a vertical pole with a rope attached to the top and a ball attached to the rope's other end. Two competing players hit the ball, trying to get it to wind around the pole in a particular direction (clockwise for one player, counterclockwise for the other, for example). Projecting onto a flat plane, let the pole be a round disk (radius $r = 0.2\,\mathrm{m}$) and the ball be a point mass m, connected to the disk by means of a massless, inextensible rope (**Figure 3.50**). Once struck, the mass will orbit the disk, being drawn closer as the rope wraps around the disk. Initially, the mass is moving at 2.4 m/s, and the distance between the disk and mass is 1.2 m. What is the angular momentum of the mass about the contact point C? Note that the contact point C is continuously changing position as m moves around the disk. We will consider only the case for which m can still move; that is, the rope hasn't fully wound itself around the disk.

Goal Find the angular momentum of a moving mass about its attachment point to a disk.

Given Initial length of the connecting rope and initial velocity of the mass.

Assume Our assumption is that the rope hasn't fully wrapped around the disk.

Draw From geometry, the length of the rope will decrease linearly with θ as the rope gets wrapped around the pole, as shown in **Figure 3.51**. Thus when the rope is inclined at an angle θ, the amount of rope wrapped on the disk is equal to $r\theta$.

Formulate Equations The position and velocity of the mass m with respect to the disk's center are given by

$$\boldsymbol{r}_{m/O} = -r\boldsymbol{e}_\theta + (L - r\theta)\boldsymbol{e}_r$$

$$\boldsymbol{v}_m = (L - r\theta)\dot{\theta}\boldsymbol{e}_\theta$$

Solve **Figure 3.52** shows a FBD=IRD for the mass. Notice that the tension force acts along the rope and thus the acceleration does as well. The consequence of this is that the speed v_m is constant. The mass is always moving at right angles to the rope with the same speed.

If we apply (3.45) we'll obtain

$$\boldsymbol{H}_C = \boldsymbol{r}_{m/C} \times m\boldsymbol{v}_m = (L - r\theta)\boldsymbol{e}_r \times m(L - r\theta)\dot{\theta}\boldsymbol{e}_\theta = m(L - r\theta)^2\dot{\theta}\boldsymbol{k}$$

Check This isn't going to be a check so much as an observation. This example is interesting because the acceleration is always directed toward the point of attachment. Thus the mass behaves in a manner similar to the way in which the earth behaves as it orbits the sun or a satellite behaves as it orbits the earth. We'll see in Section 3.6 how this plays out for actual orbiting bodies, bodies for which the connection to the central body isn't a physical one, as in this example, but is due to the force of gravity.

EXERCISES 3.5

3.5.1. [Level 1] A rigid massless rod of length $L = 1$ m pivots counterclockwise about its center O in the horizontal plane at a constant angular speed of $\omega = 20$ rad/s. A ball of mass m is attached to both ends of the rod. At a particular instant, the rod is oriented in the $\boldsymbol{i}$ direction, and another ball of mass m strikes the left end of the rod with a velocity of $\boldsymbol{v} = 8\boldsymbol{j}$ m/s. If the ball sticks to the rod at impact, what is the rod's angular speed immediately after collision?

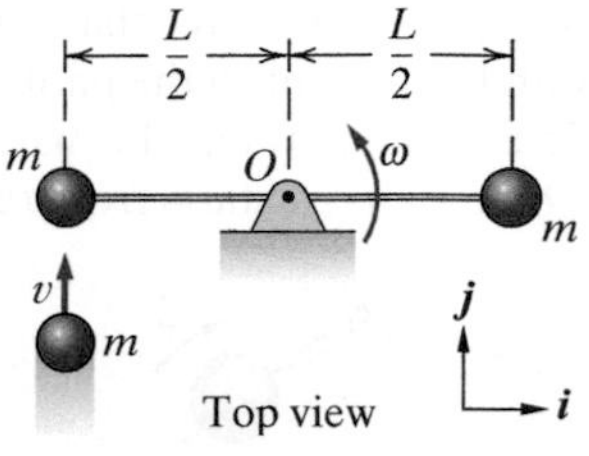

E3.5.1

3.5.2. [Level 1] Find the angular momentum of the 6 kg particle A about O. $\boldsymbol{v}_A = (5.2\boldsymbol{i} - 3.4\boldsymbol{j})$ m/s and $\boldsymbol{r}_{A/O} = (5.6\boldsymbol{i} + 2.5\boldsymbol{j})$ m.

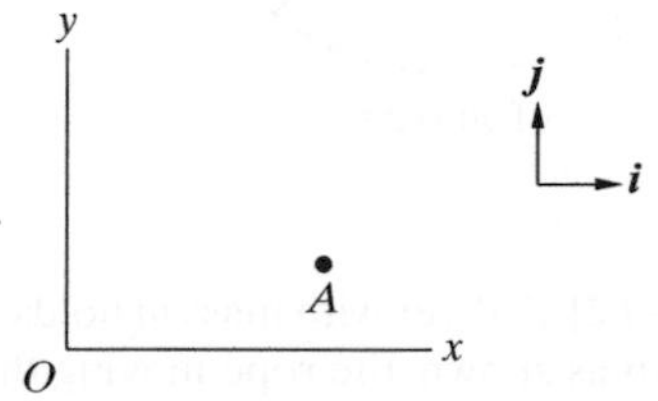

E3.5.2

3.5.3. [Level 1] Find the angular momentum of the 2 kg particle B about both O and A. $\boldsymbol{v}_B = (-3\boldsymbol{i} - 3\boldsymbol{j})$ m/s, $\boldsymbol{r}_{A/O} = 2\boldsymbol{i}$ m, and $\boldsymbol{r}_{B/O} = 2\boldsymbol{j}$ m.

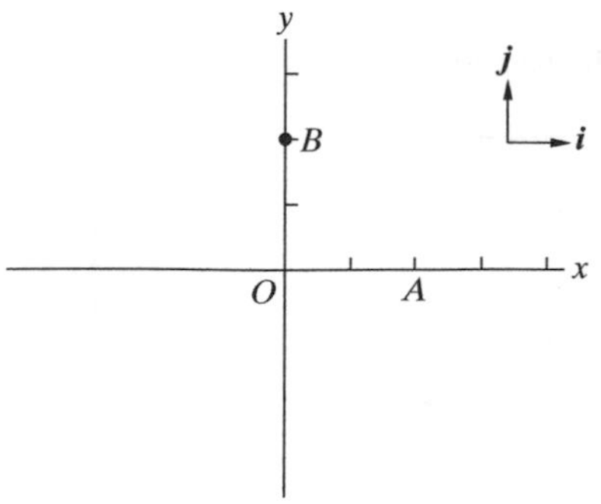

E3.5.3

3.5.4. [Level 1] A ball with mass m is attached to a rigid massless rod of length L that can freely pivot about its end O in the horizontal plane. The ball initially rotates counterclockwise about O with an angular speed of ω_0, and it moves over a rough surface with a coefficient of friction μ. Find how long it takes for the ball to come to a stop.

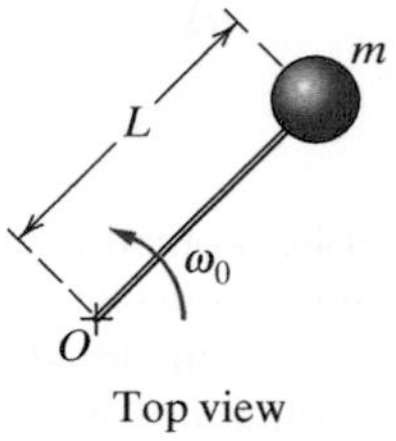

E3.5.4

3.5.5. [Level 1] If $\boldsymbol{r}_{C/O} = (3\boldsymbol{i} - 6\boldsymbol{j})$ m, $m_C = 73$ kg, $H_{C/O} = 47$ kg·m²/s, and C has a velocity component equal to 0.9 m/s in the $\boldsymbol{i}$ direction, what is the velocity of C?

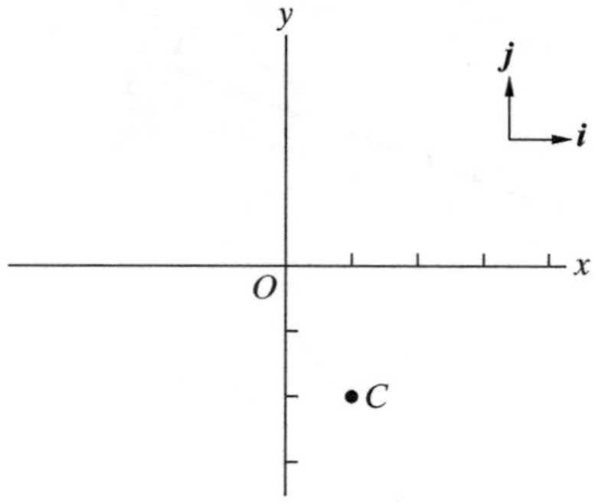

E3.5.5

3.5.6. [Level 1] The 10 kg mass particle A is acted on by the force $\boldsymbol{F}$ as shown. $\boldsymbol{F} = 4\boldsymbol{j}$ N, $\boldsymbol{r}_{A/O} = (4.0\boldsymbol{i} + 1.0\boldsymbol{j})$ m, and $\boldsymbol{v}_A = 10\boldsymbol{j}$ m/s. Find $\boldsymbol{H}_{A/O}$ and $\dot{\boldsymbol{H}}_{A/O}$.

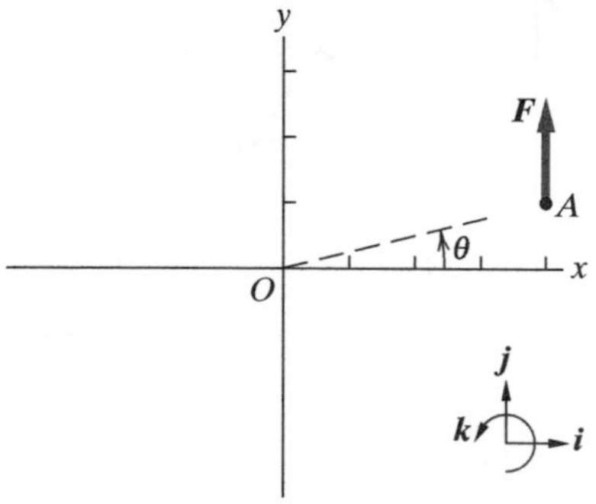

E3.5.6

3.5.7. [Level 1] A mass is spinning on the end of a massless string on a horizontal, frictionless table. The mass is initially rotating at $\dot{\theta} = \omega_0$ and r is constant at r_1. At t_0 the string is pulled into the hole at O, and once r is reduced to r_2 ($r_2 < r_1$) the string length is kept constant at r_2.

What is $\dot{\theta}$ when the mass has moved to r_2 ($r_2 < r_1$)?

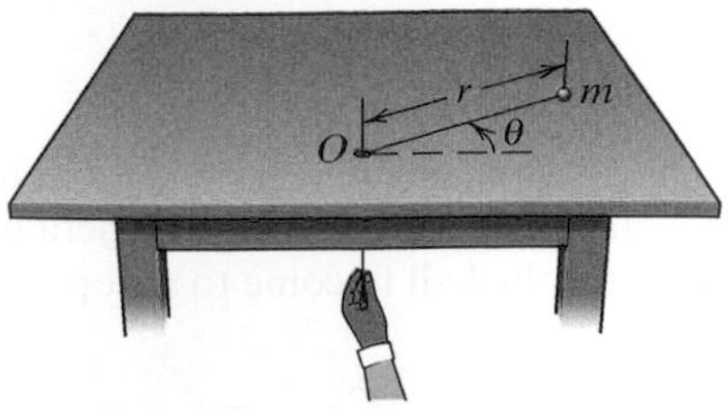

E3.5.7

3.5.8. **[Level 1]** A fisherwoman F has hooked a shark S, and the shark maintains a constant speed of v_0 while the fisherwoman reels it in. She remains stationary during the reeling. She's winding fishing line into her reel at a speed v. What is the angular momentum of the shark about F as a function of time? Assume that the initial value of r is r_0, the shark has a mass m_s, and the shark moves in a counterclockwise direction around the fisherwoman.

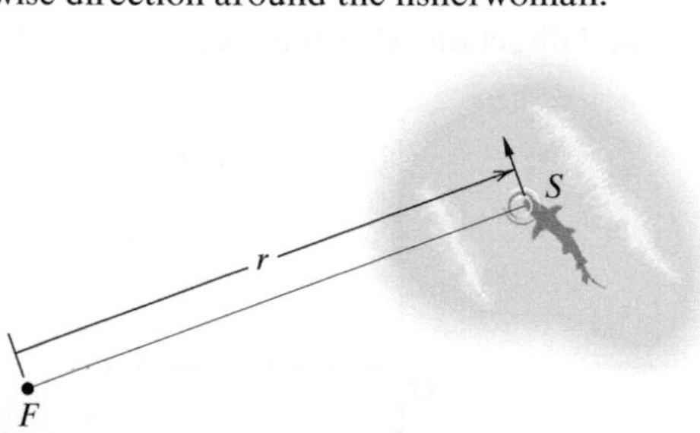

E3.5.8

3.5.9. **[Level 1]** A particle P of mass $m = 6\,\text{kg}$ is traveling at $\boldsymbol{v}_P = -8\boldsymbol{i}$ m/s and has a net force $\boldsymbol{F} = -5\boldsymbol{i}$ N acting on it. At a certain instant, $\theta = 25°$ and the distance from O to P is $L = 3\,\text{m}$. Calculate $\boldsymbol{H}_{P_{/O}}$ and $\dot{\boldsymbol{H}}_{P_{/O}}$ at this instant.

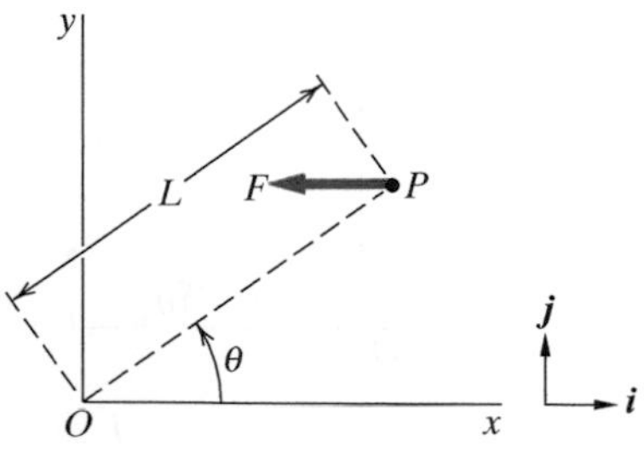

E3.5.9

3.5.10. **[Level 1]** At a particular instant, the location of a particle P of mass $m = 29\,\text{kg}$ relative to O is $\boldsymbol{r}_{P_{/O}} = (2\boldsymbol{i} - 0.9\boldsymbol{j})$ m. The particle has a velocity component in the $\boldsymbol{j}$ direction of $v_{P,y} = 0.3$ m/s and a net force component in the $\boldsymbol{i}$ direction equal to $F_x = 0.45$ kg. If $H_{P_{/O}} = 20\,\text{kg·m}^2\text{/s}$ and $\dot{H}_{P_{/O}} = 7$ N·m, what are the particle's velocity $\boldsymbol{v}_P$ and the net force $\boldsymbol{F}$ acting on P?

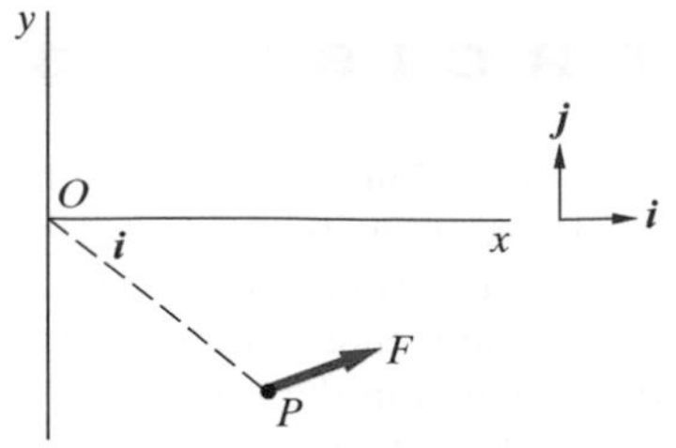

E3.5.10

3.5.11. **[Level 1]** A 100 g marble is constrained to move in a circular path on a frictionless horizontal surface under the action of an electromagnetic force that varies according to $F_e(t) = 2e^{0.1t}$ N and is tangent to the path. The radius of the path is $R = 0.75$ m. Find how fast the marble is spinning about the center O after 2 s if it starts from rest.

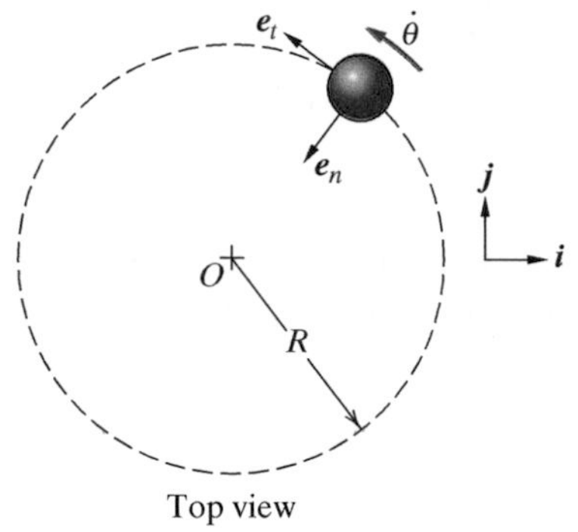

E3.5.11

3.5.12. **[Level 2]** A diver with mass m holds on to a rope and swings down as shown, the rope moving through a 90° arc. For our purposes the diver is modeled as a point mass A. Once at the bottom of the arc, traveling to the right at a speed v, she lets go of the rope. If we define an x, y coordinate system with its origin at the diver's position at release, what will her angular momentum about O be for the rest of her trajectory? Express your answer in terms of m, g, t, and v.

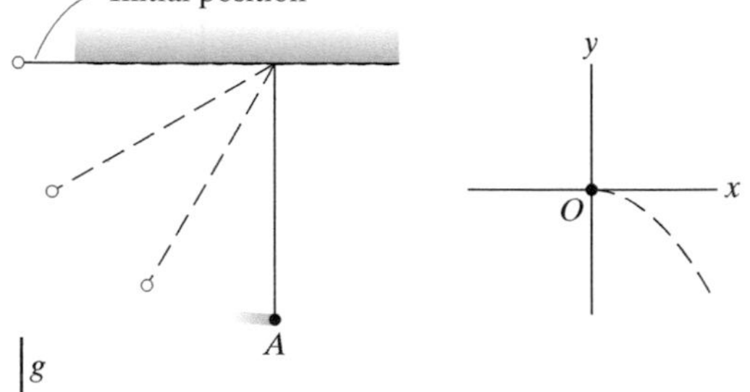

E3.5.12

3.5.13. **[Level 2]** **Computational** You have been called in as a consultant for the North Dakota Coconut Company to evaluate two coconut-opening concepts. The first, (a), consists of a mass m that's dropped from a height h onto a coconut. The second, (b), utilizes a mass m that's attached to the end of a pivoted, massless rod of length h. The rod is raised to a horizontal position and then released. In this case the mass strikes the coconut at the bottom of its swing, smashing the coconut against a vertical wall. Is one device superior to the other in its coconut-cracking ability?

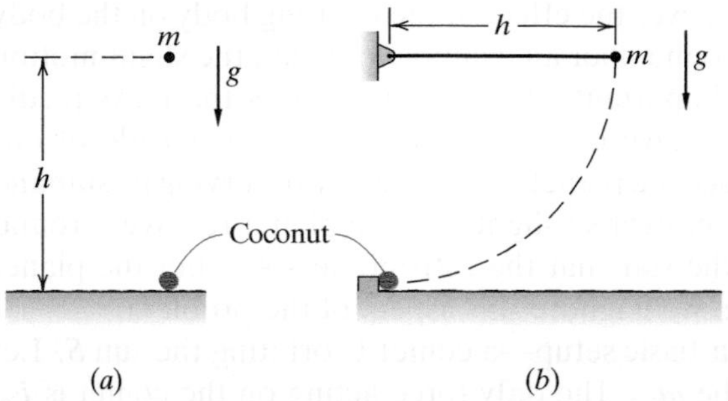

E3.5.13

3.5.14. **[Level 2]** A 0.4 kg mass m is rotating on a horizontal, frictionless surface, its motion constrained by a massless, inextensible string. The distance from the string's attachment point, O, to the mass is a constant. A tiny rocket in the mass exerts a constant thrust in the positive $\boldsymbol{e}_\theta$ direction with a magnitude of 1.2 N. $l = 0.6$ m, and the angular velocity $\dot{\theta}$ is initially equal to 2.5 rad/s. What is the tension in the string from $t = 0$ to $t = 4$ s?

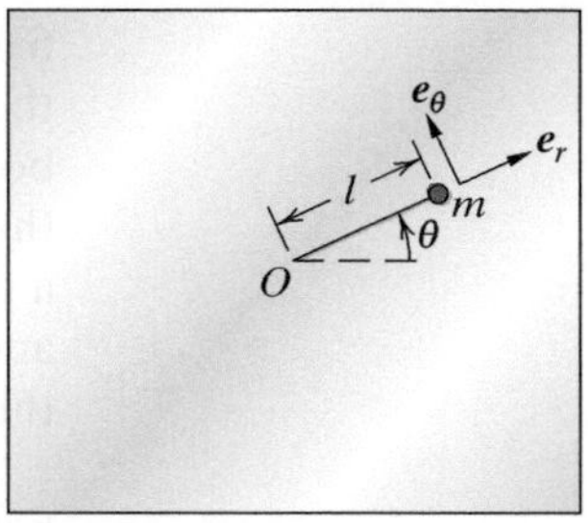

E3.5.14

3.5.15. **[Level 2]** We'll consider two slightly different situations in order to illustrate the implications of having (or not having) an angular momentum about a point. In Case A we have a particle P of mass m moving in the negative $\boldsymbol{j}$ direction with speed v and its position is given by $\boldsymbol{r}_P = -vt\boldsymbol{j}$. In Case B a particle P' (also mass m) is moving in the negative $\boldsymbol{j}$ direction but, unlike Case A, its position is given by $\boldsymbol{r}_{P'} = a\boldsymbol{i} - vt\boldsymbol{j}$. Assume that the massless string has a length of $\sqrt{10}a$. For each case determine the magnitude of the angular speed of the particles about O after the string has gone taut.

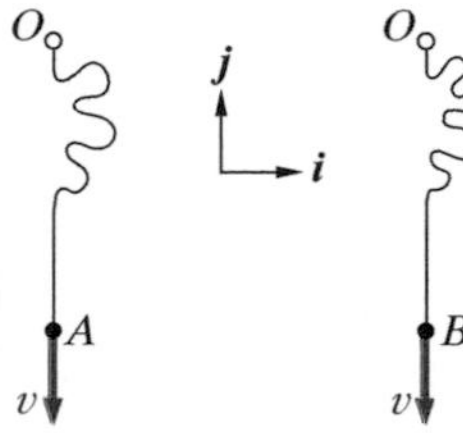

E3.5.15

3.6 ORBITAL MECHANICS

This section is optional. But it's also one of the most fascinating sections, not so much for the problems you can solve (although that's pretty impressive) but rather for the fact that this is material that has confounded scientists since the beginning of recorded history. It's the stuff that almost got Galileo executed by the Inquisition, sucked up years of effort from many, many people, and helped inspire Newton to essentially invent the field of dynamics. And the kicker is that, with what you've already learned, plus some facts you went over in high school, you will be able to understand it all. So what is it that I'm talking about? The orbits of the planets, that's what. Not to mention the trajectories of comets, Apollo 13, space shuttles, and on and on. All of this appears in a single section. If your course is skipping this material because of time constraints, you might want to look at it anyway, just so you can say you've seen it.

The key fact from which everything else will be derived is the equation of motion for a mass particle being acted on by a gravitational force. We will deal with the simplest problem in this section, the "one-body problem." What this name means is that we will see how an orbiting body behaves when attracted by the gravitational force of a large body that is fixed in space. One example is the motion of the space shuttle around the earth. In this case, the orbiting body is much, much smaller than the body it's orbiting, and as a result, we can neglect the gravitational effect that the smaller body has on the larger one. In reality, the shuttle exerts a gravitational attraction that will move the earth slightly (very slightly), and so a more precise analysis would let both bodies move. Yet neglecting this effect doesn't alter the final results in a significant way.

Although we will not cover the effect of the orbiting body on the body being orbited, you may remember an example of when the slight motion of the orbited body is important. Every few months the news media report how astronomers have discovered a new planet outside of our solar system. They deduce the planet's existence by observing its star and seeing how the star moves ever so slightly as the planet revolves around it. It is the motion of the star that the astronomers see, not the planet itself. For now, however, we'll ignore this aspect of the problem.

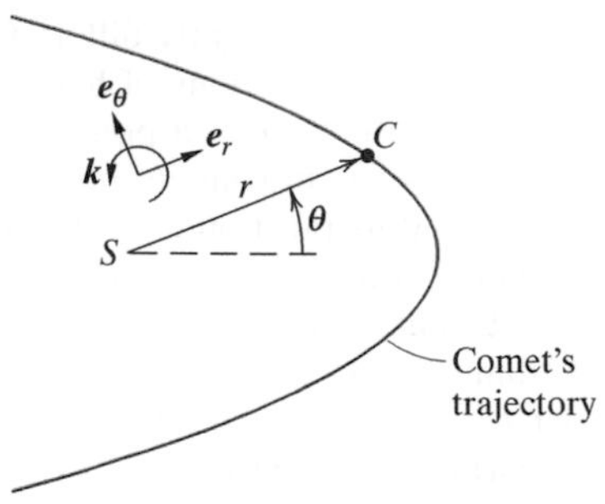

Figure 3.53 Coordinates for orbiting body

Figure 3.53 shows our basic setup—a comet C orbiting the sun S. Let the mass of the comet be m_C. The only force acting on the comet is F_g, the force due to the sun's gravitational pull, which acts in the negative $\boldsymbol{e}_r$ direction.

From (3.15) and (3.16) we have

$$m_C(\ddot{r} - r\dot{\theta}^2) = -F_g \tag{3.54}$$

$$m_C(r\ddot{\theta} + 2\dot{r}\dot{\theta}) = 0 \tag{3.55}$$

We are going to use (3.55) in a second, but first let me explain why. An interesting fact regarding a body that's moving about a fixed point is that if the forces acting on the body act only in the radial direction, the body's angular momentum is conserved. In the case of the earth, this means that because the sun's gravitational attraction always points radially (toward the sun), the earth's angular momentum about the sun is conserved. This is a very good thing and explains why, even after billions of years, we are still orbiting in about the same way as we have always done. It's easy to show this, and I'll use the comet example to do so. The angular momentum of the comet with respect to the sun is

$$\boldsymbol{H}_{C/S} = \boldsymbol{r}_{C/S} \times m_C\boldsymbol{v}_C = r\boldsymbol{e}_r \times m_C(\dot{r}\boldsymbol{e}_r + r\dot{\theta}\boldsymbol{e}_\theta) = m_C r^2\dot{\theta}\boldsymbol{k} \tag{3.56}$$

where $\boldsymbol{k}$ points out of the orbital plane.

As I already mentioned, the angular momentum of a mass orbiting a fixed body is constant, but I haven't proved it yet. So let's demonstrate it in a very straightforward way. We'll start by *assuming* that the angular momentum really is constant. If this is true, then differentiating with respect to time will produce zero. Because $\boldsymbol{k}$ has no time-dependence, the only thing that might be time-dependent in (3.56) is $m_C r^2\dot{\theta}$. Differentiating this with respect to t gives

$$\frac{d}{dt}(m_C r^2\dot{\theta}) = m_C(2r\dot{r}\dot{\theta} + r^2\ddot{\theta}) = rm_C(2\dot{r}\dot{\theta} + r\ddot{\theta}) \tag{3.57}$$

Ah ha! Look at that last term—we've seen it before. Equation (3.55) has already demonstrated that $(2\dot{r}\dot{\theta} + r\ddot{\theta})$ is equal to zero, and therefore we have

$$\frac{d}{dt}(m_C r^2 \dot{\theta}) = \frac{d}{dt} H_C = 0$$

Thus we see that the time derivative of the angular momentum is zero, validating my assertion that the angular momentum is constant. That's fascinating fact number one: the angular momentum of a mass orbiting a much larger body is constant. The mass is constant as well, and thus the really relevant relationship is

$$r^2\dot{\theta} = h \tag{3.58}$$

where h is a constant.

What's amazing about the next fact is the way in which it was discovered. An astronomer, Tycho Brahe (1546–1601), spent his entire life painstakingly plotting the positions of the planets. We're talking years of data here. His hope was to use the data to help shore up the theory, which was then popular, that everything revolves around the earth. Needless to say, it didn't help a great deal. Luckily for science, Johannes Kepler was working for Brahe and, after Brahe died, Kepler managed to get his hands on these data. Convinced that each planet followed a rational and predictable path, Kepler tried to find the description of these paths that lay buried within the numbers. Think about what that meant. What he had were books and books worth of numerical data. He had to look at this overwhelming mass of data to try and make some sense of it. The amazing thing is that he succeeded.

One of Kepler's conclusions was that orbiting bodies sweep out equal areas in equal times. Look at **Figure 3.54** to see what I mean. Say you know a planet is at time position 1 at t_1. After some interval of time Δt has passed, the planet is at position 2. I have shaded the area that's been swept out over this interval and labeled it A_{1-2}. After another time interval Δt has passed, the planet is at position 3, having swept through the shaded area labeled A_{2-3}. The time intervals were the same, and so the swept-out areas are the same.

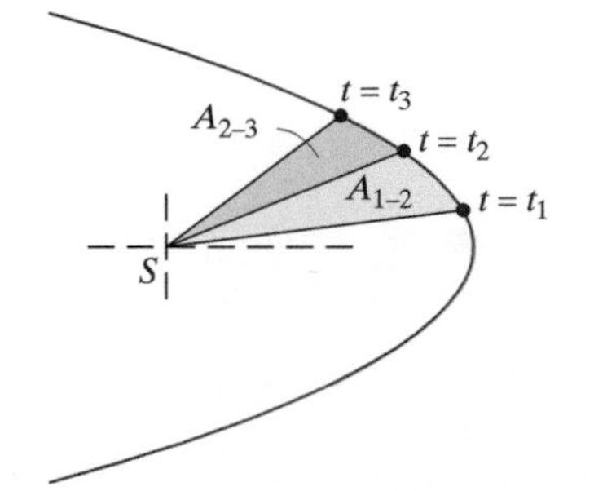

Figure 3.54 An orbiting body sweeps out equal areas in equal time intervals

We can see this very quickly from what we already know. Look at the system at two points in time, t and $t + \Delta t$ (**Figure 3.55**). The swept area ΔA is approximately triangular and has an area approximated by $\frac{(r\Delta\theta)(r)}{2}$. We can divide by the time interval Δt and obtain

$$\frac{\Delta A}{\Delta t} = \frac{r^2}{2}\frac{\Delta\theta}{\Delta t}$$

which, in the limit of $\Delta t \to 0$ gives

$$\frac{dA}{dt} = \frac{1}{2}(r^2\dot{\theta}) \tag{3.59}$$

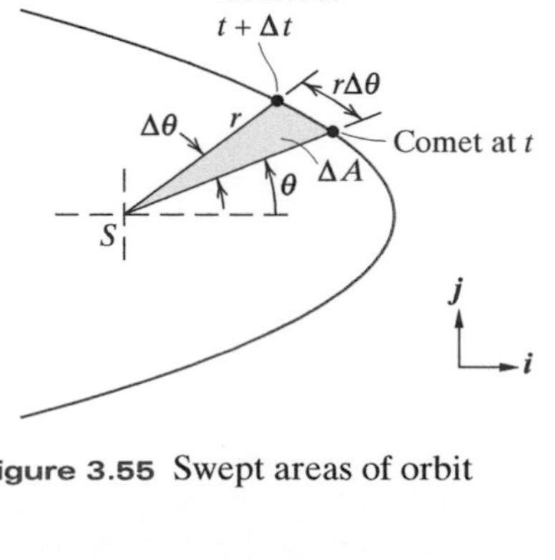

Figure 3.55 Swept areas of orbit

Thus, just from geometry, we can see that the time rate of change of the swept area is proportional to $r^2\dot{\theta}$. We have already seen that $r^2\dot{\theta}$ is constant from (3.58), and so $\frac{dA}{dt}$ must be constant as well—and this is our fascinating fact number two. All this is just an immediate consequence of geometry and Newton's laws. Of course, Newton's laws didn't exist when this observation was first made.

The next observation about orbiting bodies is the most important one, and it ties into conic sections, an area of geometry that deals with regularly shaped solids. A conic section is what you get when you slice a cone with a flat plane. Circles, ellipses, parabolas, and hyperbolas—all are conic sections. Interestingly, all orbits are described by conic sections as well. Given the right initial speed, a satellite can move in a circle around the earth. A bit more initial speed and it will move in an ellipse. More initial speed and the ellipse becomes a parabola. Finally, with enough initial speed, the orbit will become hyperbolic.

Because they involve the three variables t, r, and θ, (3.54) and (3.55) aren't in the right form to let us see that they actually do produce elliptical (or circular, parabolic, or hyperbolic) orbits. What we need to do is get rid of t so that we have a single equation that enables us to find r in terms of θ. Once we've done this, it will be easy to graph the trajectories. All we need do is pick an angle θ, solve for the corresponding r, and put that point down on a piece of graph paper. Increment θ, plot the new r, and so on. So from now on we will consider r to be a function of θ: $r(\theta)$. Rather than differentiating with respect to t, we'll be concerned with differentiation with respect to θ. To distinguish this from time differentiation, I'll use a prime symbol $'$ instead of a dot to indicate differentiation with respect to θ:

$$\frac{d(\,)}{d\theta} \equiv (\,)'$$

This means that differentiation with respect to time will require the chain rule:

$$\dot{r} = r'\dot{\theta} \tag{3.60}$$

Although solving for r directly as a function of θ isn't easy, it turns out that finding its reciprocal as a function of θ isn't too bad. We start by giving r's reciprocal a name: p. Our first equation is therefore

$$r = p^{-1} \tag{3.61}$$

Differentiating p with respect to time (and using the chain rule) gives us

$$\dot{r} = -p^{-2}\dot{p} = -p^{-2}p'\dot{\theta} \tag{3.62}$$

Using $h = r^2\dot{\theta} = p^{-2}\dot{\theta}$ in (3.62) gives us

$$\dot{r} = -hp' \tag{3.63}$$

We can differentiate once more to find

$$\ddot{r} = -hp''\dot{\theta} = -hp''hp^2 = -h^2p^2p'' \tag{3.64}$$

$$(3.58), (3.64) \rightarrow (3.54) \Rightarrow \quad -h^2p^2p'' - p^{-1}h^2p^4 = -Gm_Ep^2 \tag{3.65}$$

which immediately simplifies to

$$p'' + p = \frac{Gm_E}{h^2} \tag{3.66}$$

$$p'' + p = \frac{F_g}{m_C h^2 p^2} \tag{3.67}$$

One more step and we're done. The force of gravity acting on the orbiting body is given by

$$F_g = \frac{Gm_C m_S}{r^2}$$

as was already mentioned in Section 1.4. Substituting this value for F_g into (3.67) gives us

$$p'' + p = \frac{Gm_S}{h^2} \tag{3.68}$$

This equation governs the motion of p, and hence r. Solving it will allow us to plot out the trajectories we are seeking. The form of this equation looks very much like

$$m\ddot{y} + ky = mg \tag{3.69}$$

doesn't it? This is good news because (3.69) is the equation of motion for a mass hung from a spring, just like the one shown in **Figure 3.56**.

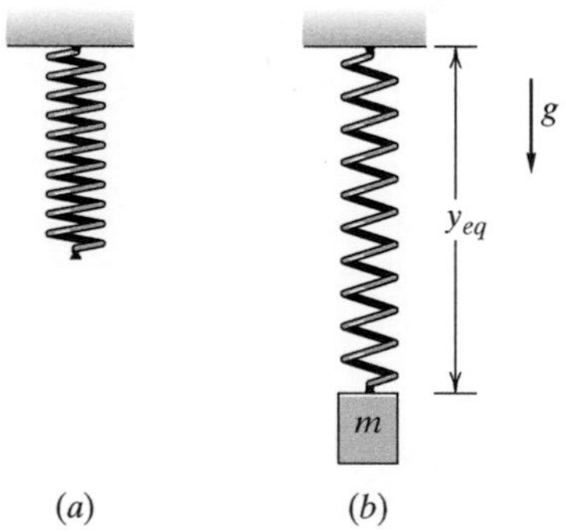

Figure 3.56 Suspended spring-mass system

That's the result of all this work—the observation that the trajectories of a mass around a fixed gravitational source are governed by the same oscillator equation that governs the motion of a spring-mass system. Spring-mass systems are a fascinating topic in themselves; if you have the interest, you can take a look at Chapter 9 to learn more about them.

The solutions to (3.69) are very easy to visualize physically. You know that the spring has an unstretched length, as shown in **Figure 3.56*a***. Once attached, the mass can remain at rest, suspended beneath the spring, if the downward force due to gravity is countered by the upward spring force due to the spring's extension, as shown in **Figure 3.56*b***. y_{eq} indicates how far beneath the attachment point the mass is when it's in equilibrium. Now imagine that you grab hold of the spring, pull it down an additional distance y, and then release it. **Figure 3.57** shows three possible solutions. Multiple positions of the mass are shown in my attempt to illustrate a dynamic motion in a single picture. In **Figure 3.57*a*** the mass bounces up and down considerably. The long-term motion is a large-amplitude oscillation about its steady-state position y_{eq}. In **Figure 3.57*b*** the mass is released closer to its steady-state position, and so the oscillations are smaller in magnitude. In **Figure 3.57*c***, the mass is released right at its steady-state equilibrium position (where the force of gravity is exactly countered by the force in the stretched spring); therefore the mass doesn't oscillate at all but merely sits there.

That's the physical response. This can all be handled mathematically as well, inasmuch as the equation is linear (all dependent variables and their

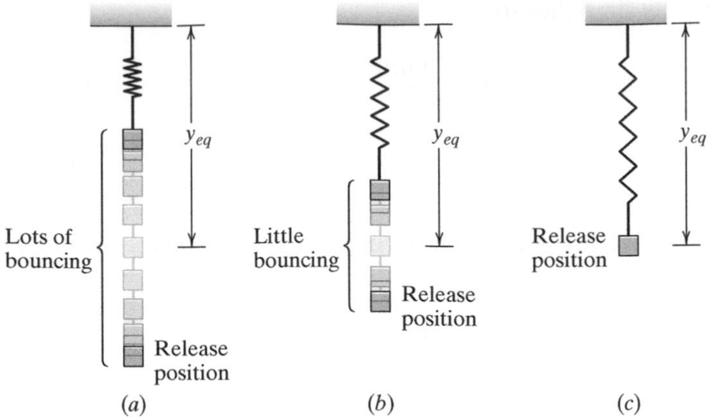

Figure 3.57 Response when released from different heights

derivatives appear only to the first power) and has constant coefficients. It's one of the few equations that has an easily expressed, closed-form solution. And, for the parameters given in (3.68), that solution is

$$p(\theta) = \frac{1}{r(\theta)} = \frac{Gm_S}{h^2} + A\cos\theta \tag{3.70}$$

where A is a constant.

The $\frac{Gm_S}{h^2}$ term is the steady-state solution, also known as the particular solution. It's the part of the solution due to the right-hand side of (3.68) being nonzero. The $A\cos\theta$ part is the unforced (homogeneous) solution, the part that depends on the initial conditions. This is the part that governs whether the oscillation amplitude is large or small in **Figure 3.57.**

Because p is the inverse of r, the radial distance of the comet from the sun varies from a minimum value up to some maximum or even to infinity. **Figure 3.58** shows all possible cases. In **Figure 3.58*a***, the initial conditions are such that the cosine term isn't present. In this case, p is constant and therefore so is r, and the orbit, shown in **Figure 3.58*b***, is circular.

Figure 3.58c shows the case that occurs for all the planets in the solar system, for comets that orbit the sun, and for satellites that orbit planets. As shown in **Figure 3.58*d***, the orbit is elliptical, with a distance between the two bodies that varies from $\left(\frac{Gm_S}{h^2} + A_1\right)^{-1}$ to $\left(\frac{Gm_S}{h^2} - A_1\right)^{-1}$. When looking at an elliptical orbit around the earth, we use the term **perigee** to indicate that part of the orbit that is closest to the earth and the term **apogee** to indicate the furthest point.

If the amplitude of the cosine is large enough, then the $p(\theta)$ curve will osculate with the $p = 0$ line. (Neat word, no? Osculate means, in Latin, "to kiss," and that's what's happening in **Figure 3.58*e***—the p curve just barely "kisses" the horizontal line $p = 0$. And you thought dynamicists were stuffy!) Mathematically, the $p(\theta)$ curve is zero at isolated points, and the slope $\frac{dp}{d\theta}$ is zero at those points. The trajectory that corresponds to this case is parabolic and is shown in **Figure 3.58*f***. The parabolic trajectory is the dividing case between elliptical orbits (in which the orbiting body

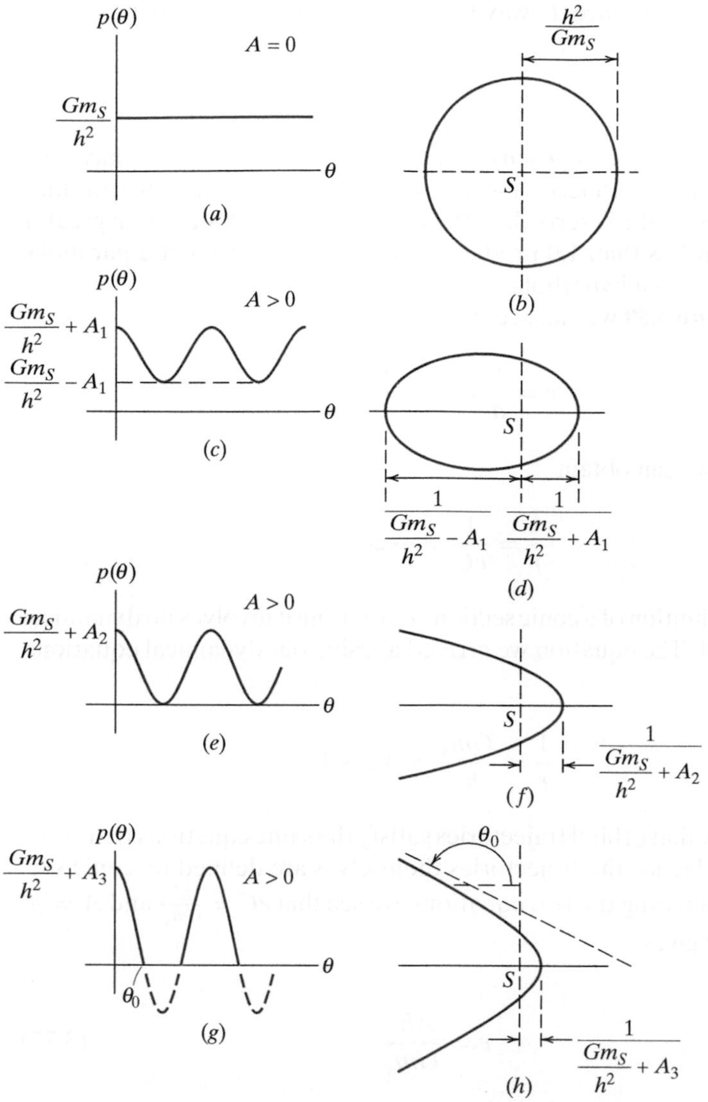

Figure 3.58 Conic solutions and corresponding orbits

always returns to its starting point) and hyperbolic trajectories (in which the moving body goes off to infinity and never returns).

The last two plots show typical hyperbolic solutions. The amplitude of the cosine is so large that the $p(\theta)$ curve hits zero and then goes negative (**Figure 3.58*g***). The actual trajectory would never reach this region and asymptotically goes to θ_0 as r goes to infinity, as shown in **Figure 3.58*h***.

So that is what actually happens. Now I'm going to show you that it can all be predicted from a knowledge of conic sections. The classic conic section diagram is shown in **Figure 3.59**. Point F is the focus of the conic section, and the line located a perpendicular distance C from F is called the **directrix**. The point m is located a distance r away from F and

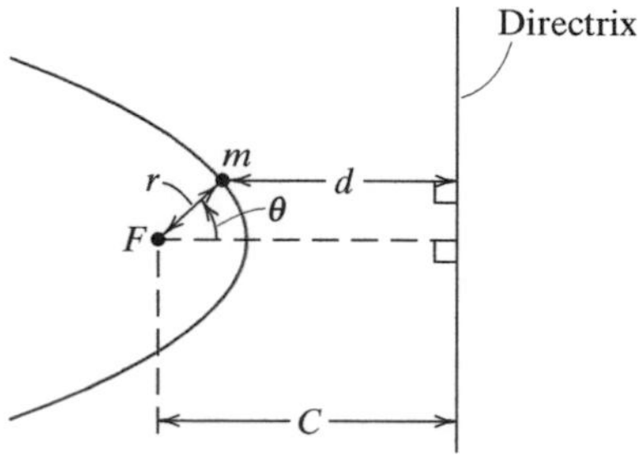

Figure 3.59 Graphical way of describing conic sections

a perpendicular distance d away from the directrix. The ratio e

$$e = \frac{r}{d}$$

is called the curve's **eccentricity**. The loci of all the points that have the same eccentricity e define a conic section, and different values of e produce different curves. If e is zero, then the curve is a circle. Values of e greater than zero but less than 1.0 produce ellipses. If $e = 1$ we get a parabola, and $e > 1$ implies a hyperbola.

From **Figure 3.59** we can see that

$$e = \frac{r}{d} = \frac{r}{C - r\cos\theta}$$

from which we can obtain

$$\frac{1}{r} = \frac{1}{eC} + \frac{1}{C}\cos\theta$$

This is the definition of a conic section, and getting it involves no dynamical analysis at all. The equation we arrived at using our dynamical equations, (3.70),

$$\frac{1}{r} = \frac{Gm_s}{h^2} + A\cos\theta \tag{3.71}$$

shows clearly that orbital trajectories satisfy the same equations that conic sections do. Hence the trajectories themselves are defined by conic sections. By comparing the two equations, we see that $eC = \frac{h^2}{Gm_s}$ and $A = \frac{1}{C}$. Solving for e gives

$$e = \frac{Ah^2}{Gm_s} \tag{3.72}$$

Because by now you're surely starting to space out from all the parameters, let me remind you of what's actually happening. The eccentricity e tells us what sort of trajectory we've got—elliptical, hyperbolic, circular, or parabolic. It's clearly an important thing to know because, if we're putting a satellite into orbit, we wouldn't want it to have a hyperbolic trajectory and end up on Alpha Centauri. Likewise, if we're trying to send a probe to Alpha Centauri, we would not want the trajectory to be elliptical because then the probe would keep looping back around the earth. Equation (3.72) lets us determine the type of trajectory in terms of parameters we know. The gravitational constant G is a known quantity, as presumably is the mass of the body being put into orbit. $h = r^2\dot{\theta}$ is determined by how fast our satellite is moving with respect to the body it is orbiting, and A is determined by the particular initial conditions at the point where the body is injected into orbit.

⋙ **Check out Example 3.21 (page 176) for an application of this material.**

We have seen that if a mass is injected into orbit with no radial velocity, then we can predict the rest of its trajectory. But what if the situation wasn't that well set up? What if we have a mass that's got both a radial and an angular velocity component? Can we figure out the trajectory anyway? Yes, indeed. And here's how.

We already have the formula relating r to θ (3.70):

$$\frac{1}{r} = \frac{Gm_E}{h^2} + A\cos\theta$$

What we will need is an expression that involves both $\dot{r}$ and $\dot{\theta}$, because an orbiting mass's velocity will, in general, be given by

$$\boldsymbol{v} = \dot{r}\boldsymbol{e}_r + r\dot{\theta}\boldsymbol{e}_\theta = v_r\boldsymbol{e}_r + v_\theta\boldsymbol{v}_\theta$$

We can differentiate our expression for $\frac{1}{r}$ with respect to time to obtain

$$-\frac{\dot{r}}{r^2} = -A\dot{\theta}\sin\theta$$

which gives us

$$\dot{r} = Ar^2\dot{\theta}\sin\theta \tag{3.73}$$

If the radial velocity isn't zero, then the trajectory will be starting not at $\theta = 0$ but rather at some other point along the curve. Our job is to figure out where along the curve we are. **Figure 3.60** shows the orbit and the r, θ coordinates we will be using.

Our equations for $\frac{1}{r}$ and $\dot{r}$ allow us to solve for A and θ because we have two equations in two unknowns:

$$\frac{1}{r} = \frac{Gm_E}{h^2} + A\cos\theta \tag{3.74}$$

(3.69) $\Rightarrow$ $$\dot{r} = v_r = Arv_\theta\sin\theta \tag{3.75}$$

where r is the distance of the mass from the focus, v_r is the radial speed relative to the focus, v_θ is the angular speed relative to the focus, and $v_\theta = r\dot{\theta}$ was used to simplify (3.75).

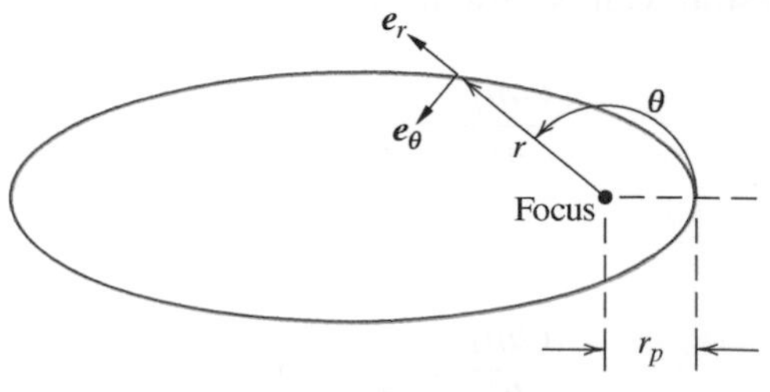

Figure 3.60 Parameters for elliptical orbit

From (3.75) we get

$$A = \frac{v_r}{r v_\theta \sin\theta} \tag{3.76}$$

which, when used in (3.74), gives us

$$\tan\theta = \frac{v_r}{v_\theta r\left(\frac{1}{r} - \frac{Gm_E}{h^2}\right)} = \frac{v_r}{v_\theta - \frac{Gm_E}{h}} \tag{3.77}$$

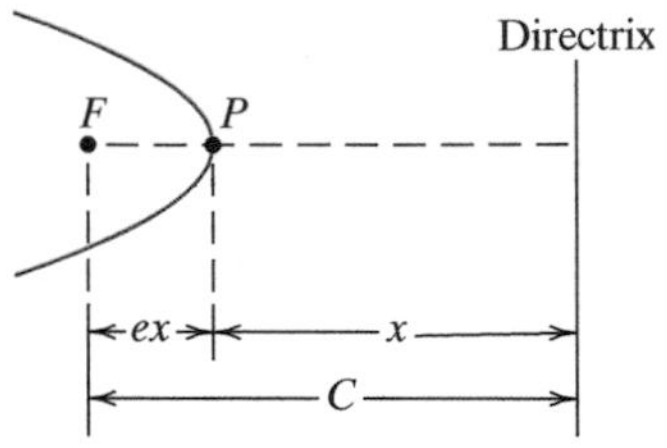

Figure 3.61 Conic section dimensional relations

This equation lets us solve for θ if we know the mass's initial conditions. Knowing θ, we can find A using (3.76). Inverting A gives us C, the distance of the directrix from the focus for our system. At this point it's straightforward to find r_p. Look at **Figure 3.61**. When $\theta = 0$, the curve is a distance x away from the directrix and a distance ex away from the focus.

We can find e from

$$e = \frac{Ah^2}{Gm_E}$$

Because $x + ex = C$, we have

$$x = \frac{C}{1+e} \tag{3.78}$$

Once x, the minimum distance from the point P to the directrix, is known, the distance from the focus to P is easily found by evaluating ex.

>>> **Check out Example 3.22 (page 177) for an application of this material.**

If we know we're dealing with an elliptical orbit, then we can do a bit more with our equations. For an elliptical orbit around the earth, we know that

$$\frac{1}{r_p} = \frac{Gm_E}{h^2} + A$$

$$\frac{1}{r_a} = \frac{Gm_E}{h^2} - A$$

These two expressions can be rewritten as

$$\frac{Gm_E}{h^2} = \frac{1}{r_p} - A$$

and

$$\frac{Gm_E}{h^2} = \frac{1}{r_a} + A$$

Because the left sides of the expressions are equal, so too must the right sides be equal, giving us

$$\frac{1}{r_p} - A = \frac{1}{r_a} + A$$

or

$$A = \frac{1}{2}\left(\frac{1}{r_p} - \frac{1}{r_a}\right) \tag{3.79}$$

Thus, if we are given the maximum and minimum distances for an elliptical orbit, we can immediately solve for A. Or, given A and the minimum distance, we can find the maximum distance.

If we're dealing with an elliptical orbit around the earth, we can derive a nice relationship between the velocity at perigee and the eccentricity of the orbit. Evaluating (3.71) at perigee gives us

$$\frac{1}{r_p} = \frac{Gm_E}{h^2} + A \tag{3.80}$$

We know from (3.72) that

$$e = \frac{Ah^2}{Gm_E} \tag{3.81}$$

Substituting (3.81) into (3.80) gives us

$$\frac{1}{r_p} = A\left(\frac{1+e}{e}\right) \tag{3.82}$$

or

$$A = \frac{e}{r_p(1+e)} \tag{3.83}$$

If we use (3.83) to eliminate A in (3.81), we will be able to solve for h:

$$h = \sqrt{Gm_E r_p(1+e)}$$

Realizing that for this case $h = r_p^2\dot{\theta}$ and $v_p = r_p\dot{\theta}$, let us calculate v_p:

$$v_p = \left[\frac{Gm_E(1+e)}{r_p}\right]^{\frac{1}{2}} \tag{3.84}$$

Thus we can determine the speed needed at perigee to support an orbit of the given eccentricity.

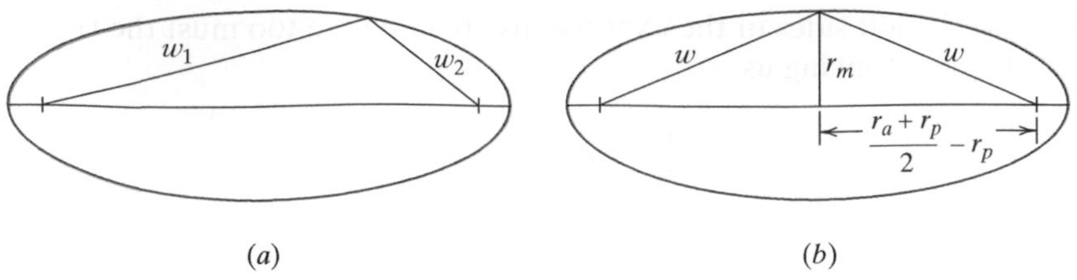

Figure 3.62 Graphical ellipse construction

Using (3.79) and (3.82) gives us an expression for r_a as a function of A and e:

$$\frac{1}{r_a} = A\left(\frac{1-e}{e}\right) \tag{3.85}$$

We now know *almost* everything about an arbitrary elliptical orbit. We know the distance from a focus to the perigee position and to apogee as well. We know how long the ellipse is. But what we don't know yet is the width of the ellipse. That's the final piece. To figure it out, recall that one definition of an ellipse is that (referring to **Figure 3.62*a***) the sum $w_1 + w_2$ is constant for all points on the ellipse. For our case, if we look at the position of the orbit corresponding to r_p, point P in **Figure 3.63**, we have a sum equal to

$$r_a + r_p$$

and therefore w (shown in **Figure 3.62*b***) is given by

$$w = \frac{r_a + r_p}{2}$$

From trigonometry we have

$$w^2 = \left(\frac{r_a + r_p}{2} - r_p\right)^2 + r_m^2$$

which, when expanded out and simplified, gives us

$$r_m = \sqrt{r_a r_p} \tag{3.86}$$

This provides not only an expression for the ellipse's dimensions, but also the key to determining an orbit's period. Recall from your geometry class that S, the area of an ellipse (again referring to **Figure 3.63**), is given by

$$S = \pi ab \tag{3.87}$$

For our case we have $a = \frac{r_a + r_p}{2}$ and $b = \sqrt{r_a r_p}$.

Back at the beginning of the section, we derived an expression for the rate at which area is swept out by an orbiting body (3.59): $\dot{S} = h/2$. Thus,

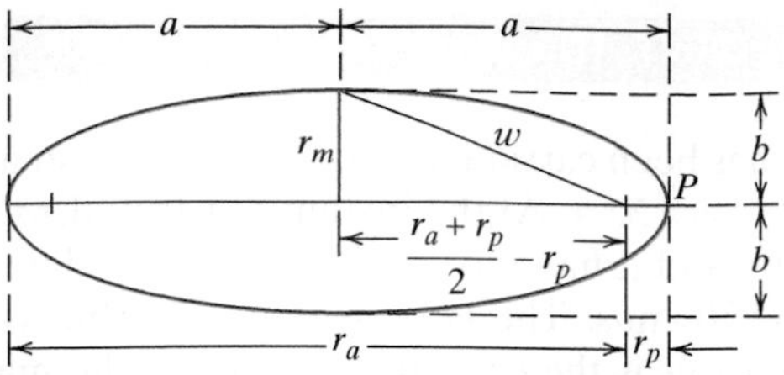

Figure 3.63 Orbital ellipse dimensions

the area swept out in one orbital period is found from

$$S = \frac{1}{2} h T_p \tag{3.88}$$

where T_p is the orbit's period. Equating (3.87) and (3.88) and solving for T_p give

$$T_p = \frac{\pi (r_a + r_p)\sqrt{r_a r_p}}{h} \tag{3.89}$$

As a final observation, note that by eliminating A from (3.83) and (3.85), we obtain a nice expression for the ratio of r_p and r_a in terms of e:

$$\frac{r_p}{r_a} = \frac{1 - e}{1 + e} \tag{3.90}$$

EXAMPLE 3.21 ANALYSIS OF AN ELLIPTICAL ORBIT (Theory on pages 168 and 170)

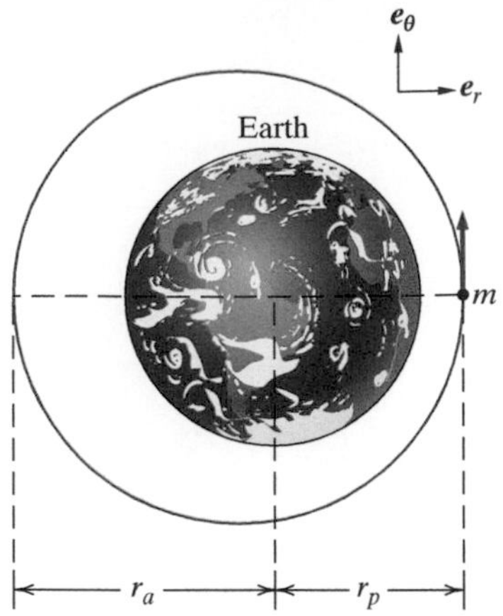

Figure 3.64 Elliptical earth orbit

A body of mass m has been carried into space on a rocket and placed into orbit as shown in **Figure 3.64**. At the instant illustrated, m's height above the earth is 160 km, and it has zero radial velocity and a speed in the $\boldsymbol{e}_\theta$ direction equal to 8500 m/s. The hope is that the orbit will be elliptical. Determine whether this is the case; if so, calculate the maximum height the mass attains above the earth and its associated speed at that height.

Goal Determine the orbit type and find r_a.

Given Position and velocity of an orbiting mass.

Assume Body is unpowered during trajectory.

Draw r_p and r_a correspond to the orbit's perigee (the least distance to the earth) and apogee (the greatest distance from the earth). **Figure 3.64** is drawn so that the body is initially at perigee, but that's only a guess. We don't know yet whether the orbit is actually elliptical and, if it is, whether the initial radius is the maximum or the minimum one.

Formulate Equations We'll start by finding A from (3.70). Because θ is defined to be zero at $r = r_p$, we have

$$A = \frac{1}{r_p} - \frac{Gm_E}{h^2} \tag{3.91}$$

Next, we'll use (3.72), $e = \frac{Ah^2}{Gm_E}$, to find e. Finally, if e indicates an elliptical orbit, we'll use (3.70) with $\theta = \pi$ to determine r_a and then, based on the fact that h is constant, will determine the body's speed at that point.

Solve The radius of the earth is 6.37×10^6 m, and so with the body 160 km above the earth's surface, $r_p = 6.37 \times 10^6 + 1.60 \times 10^5 = 6.53 \times 10^6$ m. We know that $h = r^2\dot{\theta}$, which, if we use $v_\theta = r\dot{\theta}$, becomes $h = rv_\theta$. $h = rv_\theta = (6.53 \times 10^6\,\text{m})(8500\,\text{m/s}) = 5.55 \times 10^{10}\,\text{m}^2/\text{s}$. Using this result in (3.91) yields $A = 2.37 \times 10^{-8}\,\text{m}^{-1}$.

We can now find e from (3.72) to be $e = 0.183$. Because e is between 0 and 1, we know the orbit is elliptical. To find r_a, all we need to do is use (3.70) with $\theta = \pi$, which yields $r_a = 9.45 \times 10^6$ m. The corresponding height above the earth is $(9.45 \times 10^6 - 6.37 \times 10^6)\,\text{m} = 3.08 \times 10^3\,\text{km}$.

Because h is constant, we have

$$5.55 \times 10^{10}\,\text{m}^2/\text{s} = (9.45 \times 10^6\,\text{m})v_a \quad \Rightarrow \quad v_a = 5.87 \times 10^3\,\text{m/s}$$

Check A good way to verify our result is to look at the total energy of the orbiting body. We haven't talked about energy yet, so I'm going to wait until we do (Chapter 4) and then revisit this problem.

EXAMPLE 3.22 DETERMINING CLOSEST APPROACH DISTANCE (Theory on pages 170 and 172)

A rocket is launched from the earth, and at the time the main engines stop firing, the rocket has a velocity $\boldsymbol{v} = (8610\boldsymbol{e}_\theta + 2249\boldsymbol{e}_r)$ m/s and is 910 km from the earth's surface. At its closest approach, how far from the earth's surface is the rocket?

Goal Calculate r for $\theta = 0$.

Given Rocket's initial position and velocity.

Assume Rocket is unpowered during trajectory.

Draw **Figure 3.65** indicates the initial conditions of the rocket. Its velocity has both angular and radial components, and its position corresponds to some unknown angle θ.

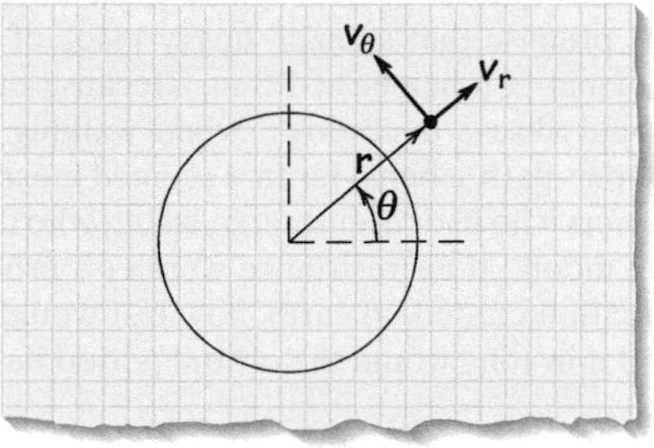

Figure 3.65 Orbital coordinates

Solve The rocket is 910 km from the earth's surface, and therefore its distance from the earth's center is $(6.37 \times 10^3 + 910)$ km $= 7.28 \times 10^6$ m. The initial angular speed ω_0 of the rocket about the earth is given by $\omega_0 = \frac{v_\theta}{r} = 1.18 \times 10^{-3}$ rad/s. h is found from $h = rv_\theta = 6.269 \times 10^{10}$ m^2/s.

Equation (3.77) gives us $\tan\theta = 1$, which means $\theta = 45°$. Thus we see that the entry conditions define a point 45° around from $\theta = 0$, where $\theta = 0$ is the point at which all our conic section constructions originate. Using (3.76) along with our known parameters yields $A = 5.07 \times 10^{-8}$ m^{-1} and so $C = A^{-1} = 1.97 \times 10^7$ m.

e is found from (3.72) to be $e = 0.50$. A value of 0.50 indicates that the orbit is elliptical.

We can use (3.78) to find r at $\theta = 0$:

$$r = \frac{Ce}{1+e} = (1.97 \times 10^7\ \text{m})\left(\frac{0.50}{1.50}\right) = 6.57 \times 10^6\ \text{m}$$

The rocket's distance above the surface of the earth at $\theta = 0$ is $(6.57 \times 10^6 - 6.37 \times 10^6)$ m $= 1.99 \times 10^5$ m $=$ 199 km. We know this is the minimum height because it's less than the initial height of 7280 km.

The two points on an elliptical orbit with zero radial velocity represent the minimum and maximum distances from the focus. All other points on the orbit have a radial distance from the focus that falls between these two extremes. Thus, if the distance at $\theta = 0$ had been greater than the initial distance, then that point would have to have been the point of maximum distance.

The corresponding velocity can be found from the conservation of angular momentum:

$$v_\theta = \frac{h}{r} = \frac{6.27 \times 10^{10}\ \text{m}^2/\text{s}}{6.57 \times 10^6\ \text{m}} = 9.54 \times 10^3\ \text{m/s}$$

EXERCISES 3.6

3.6.1. **[Level 1]** Approximate the moon as perfectly circular with a radius of 1700 km and a mass of 7.4×10^{22} kg. If you wanted a satellite to orbit at a height of 10 m above the moon's surface, how fast would it have to travel?

3.6.2. **[Level 1]** A science fiction story I once read had the premise that a small, heavy piece of rock could orbit a planet at a height of 1.5 m. Because of this, the natives would always duck down when walking across the small moon's path, a behavior that seemed inexplicable to the explorers who had recently arrived but who didn't know about the moon. If air resistance weren't a problem, and such a situation existed on earth, how fast would the moon travel and how long would the orbital period be?

3.6.3. **[Level 1]** Determine the orbital period of the earth assuming that it is in a circular orbit about the sun with a radius of 93 million miles.

3.6.4. **[Level 1]** Determine the orbital period of the moon around the earth. $\frac{m_{\text{moon}}}{m_{\text{earth}}} = 0.0123$ and $|r_{\text{moon/earth}}| = 3.84 \times 10^5$ km.

3.6.5. **[Level 1]** A satellite has zero radial velocity, is 161 km above the earth's surface, and is traveling at 19,300 km/h. You are asked to find the satellite's velocity when it reaches the halfway point of its orbit (on the opposite side of the earth). What you will find is that the value of A associated with these parameter values is negative. What does this imply about the orbit?

3.6.6. **[Level 1]** A spacecraft is in an elliptical orbit around the earth. When it reaches $\theta = \frac{\pi}{2}$ rad, lateral thrusters are fired toward the earth such that the craft's radial velocity goes to zero. What will the form of the resultant orbit be?

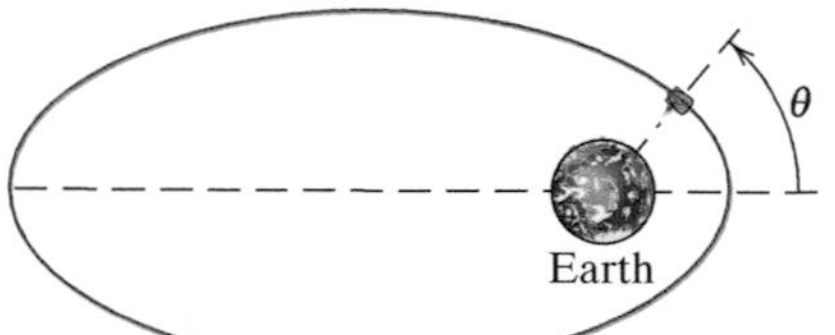

E3.6.6

3.6.7. **[Level 1]** Two satellites are orbiting the earth in orbits of 7000 km and 7000.1 km. Assume they both start at $\theta = 0$. What is the difference between the times at which they return to the $\theta = 0$ position?

3.6.8. **[Level 1]** Is the illustrated trajectory possible for a meteor traveling about the earth?

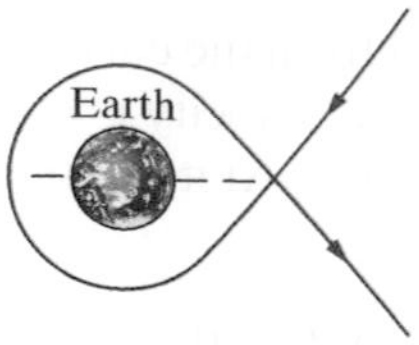

E3.6.8

3.6.9. **[Level 1]** How fast would a projectile need to be launched in order to impact the earth diametrically opposite the launch point if the launch angle was equal to 45°?

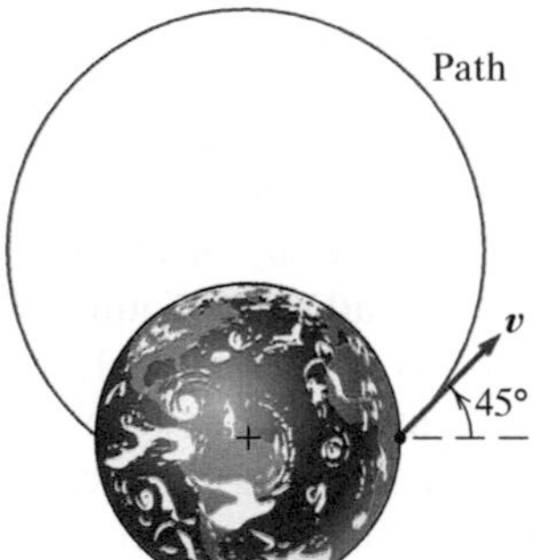

E3.6.9

3.6.10. **[Level 1]** Steve just received a rifle as a gift and his mother warned him not to "Not go firing that gun into the air—you might hurt someone." This got Steve curious. He wondered whether, if he fired the gun at a 45° angle, and it had enough force behind the bullet, and air resistance wasn't a problem, it would be possible to hurt himself by firing the bullet completely around the earth (as shown) and having it hit him from behind. Assuming that his assumptions are met (no air friction and very high muzzle velocity), could he actually injure himself in this way?

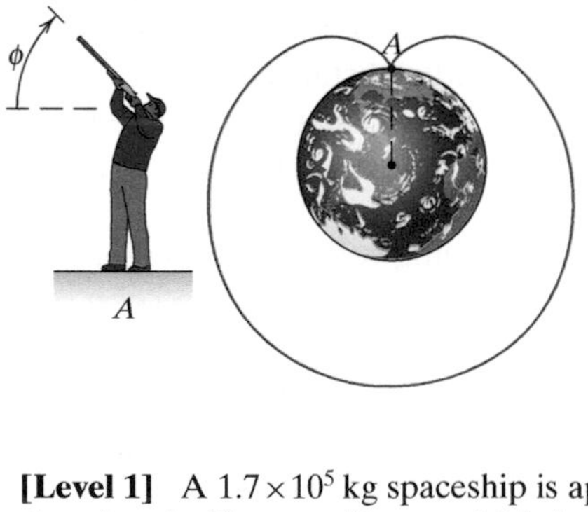

E3.6.10

3.6.11. **[Level 1]** A 1.7×10^5 kg spaceship is approaching the earth and at the illustrated instant (a) is $h = 250$ km above the earth's surface and is moving with velocity $\boldsymbol{v}_S = -(6{,}500\boldsymbol{i} + 5{,}200\boldsymbol{j})$ m/s. What linear impulse must be applied to the spaceship in order to put it into a circular orbit (b)?

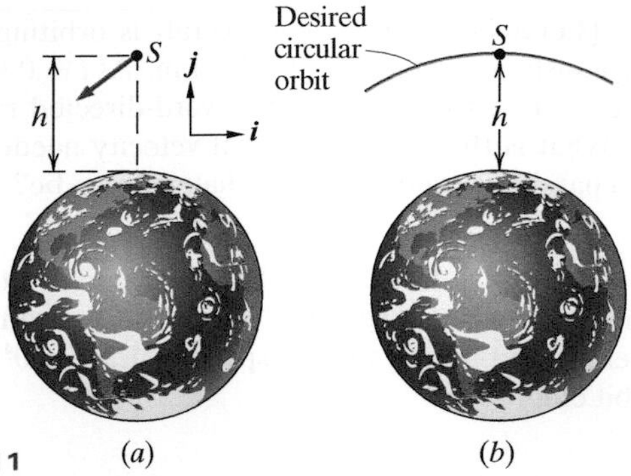

E3.6.11

3.6.12. **[Level 2]** A reconnaissance satellite is initially orbiting the moon with a radius of $r = 3.48 \times 10^3$ km. It then leaves lunar orbit with a speed v (no radial component) to return to the earth. When the satellite nears the earth, it gets pulled into an elliptical orbit at apogee ($r_a = 3.0 \times 10^4$ km) with the lunar orbit escape speed v (again, there is no radial component). The satellite is $r_p = 8.0 \times 10^3$ km from the earth's center at perigee. Suppose the satellite is programmed to transition to a circular orbit when it reaches perigee. What change in speed is needed to accomplish this change in trajectory?

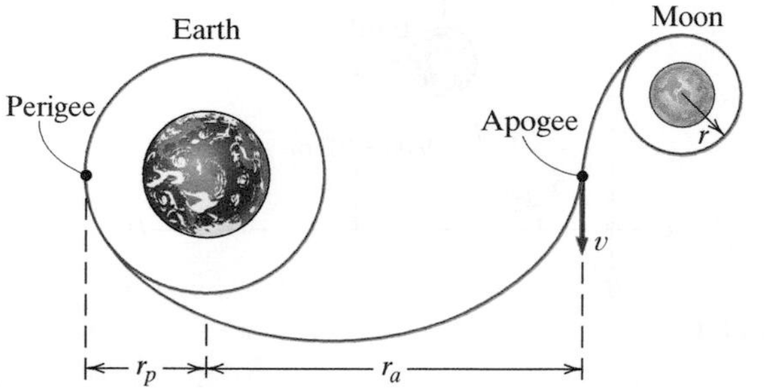

E3.6.12

3.6.13. **[Level 2]** A satellite is placed into an earth orbit and passes 161 km above the ground at its closest point while moving at 29,000 km/h. What is the furthest it gets from the surface, and how fast is it traveling at that point?

3.6.14. **[Level 2]** At $t = 0$, a spacecraft is 20,000 km from the earth's center and has no radial velocity. The captain fires the spacecraft's forward-facing engines to slow the craft sufficiently to put it into a circular orbit. The engines can produce a thrust of 4.50×10^5 N. The craft's speed at $t = 0$ is 5000 m/s and its mass is 5.00×10^4 kg. How long must the engines fire to bring the craft into a circular orbit?

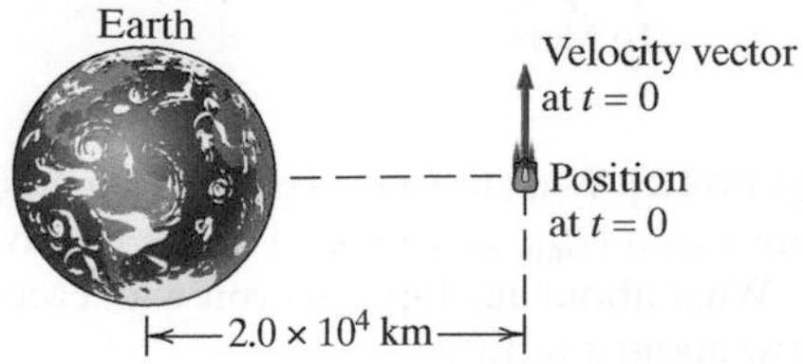

E3.6.14

3.6.15. **[Level 2]** If the speed v of a spaceship that's in a circular orbit around the earth (322 km above the ground) is increased by 10%, will its minimum speed along the orbit drop by 10% (to $0.9v$)?

3.6.16. **[Level 2]** Spacecrafts A and B are both orbiting the earth in different orbits. Unfortunately, both have the same perigee and at $t = 0$, spacecraft A slams into spacecraft B from behind. Just before impact, the 3×10^4 kg spacecraft A was moving at 9.00×10^3 m/s and the 4.50×10^4 kg spacecraft B was moving at 8.00×10^3 m/s. Both were 7000 km from the earth's center. Treat the wreckage as a single body and determine the maximum distance from the earth's center that it attains.

3.6.17. **[Level 2]** A 1.40×10^4 kg spacecraft is traveling around the earth in a circular orbit of radius 7000 km. When the spacecraft reaches point A the rearward-facing engines are fired, pushing the spacecraft into an elliptical orbit. When the craft reaches point B ($r_B = 8000$ km), the engines are fired again, putting the spacecraft into a new circular orbit. What is the magnitude of the impulse applied to the spacecraft in each of these maneuvers?

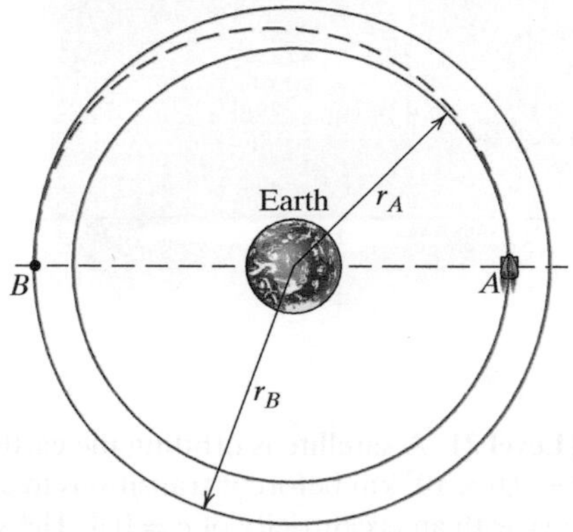

E3.6.17

3.6.18. **[Level 2]** By how much must a spacecraft's speed be increased to change a circular orbit of radius 8000 km into the illustrated elliptical one?

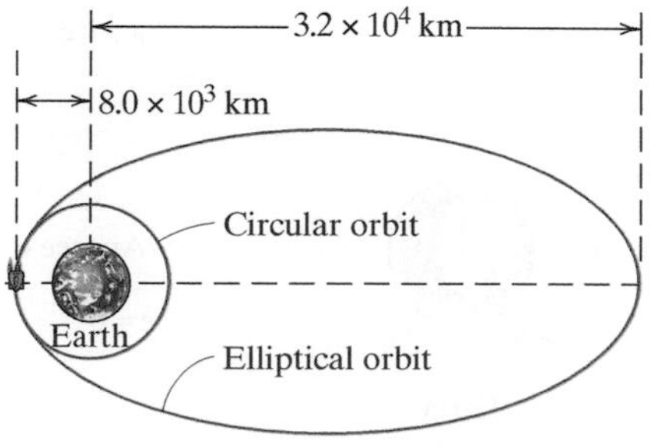

E3.6.18

3.6.19. **[Level 2]** A spacecraft and its cargo (a satellite) are circling the earth at an altitude of 600 km. Their combined mass is 1.20×10^5 kg. At $\theta = 0$, the captain ejects the 1.10×10^4 kg satellite out the back. The speed of the satellite relative to the spacecraft at the time of separation is 100 m/s. When the spacecraft gets to $\theta = \pi$, what will its altitude above the earth be?

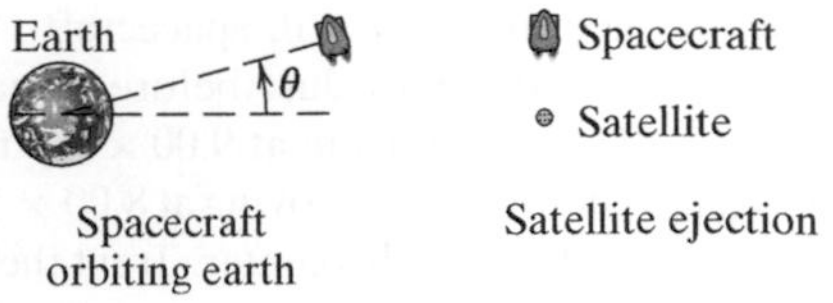

E3.6.19

3.6.20. **[Level 2]** A 6.00×10^4 kg spacecraft is traveling in a circular orbit of 6.70×10^6 m around the earth when a lateral rocket is fired. The rocket is fired for 60 s and produces a thrust of 7.50×10^3 N during that time. Ignore any position change during the firing interval and calculate the difference between its position if it had stayed in the circular orbit and r_{min}, its minimum distance from the earth's center in its new elliptical trajectory.

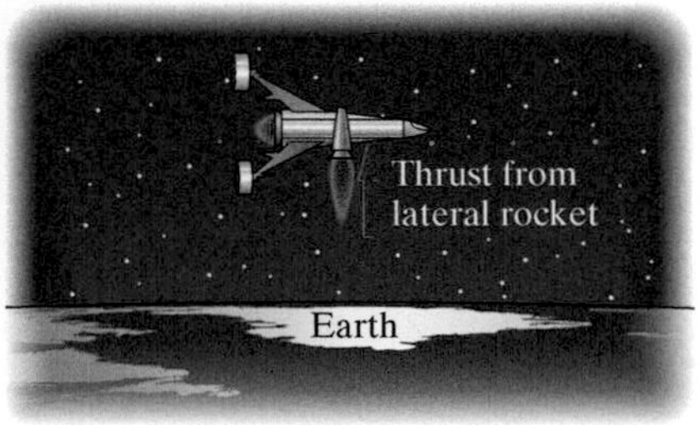

E3.6.20

3.6.21. **[Level 2]** A satellite is orbiting the earth with a radius of $R = 9.0 \times 10^3$ km before it transitions to an elliptical trajectory with an eccentricity of $e = 0.4$. The satellite then transitions to a circular orbit when it reaches apogee.

a. How much did the satellite's speed change when it transitioned from its original circular orbit to the elliptical one?

b. Find the change in speed needed to enter the larger circular orbit.

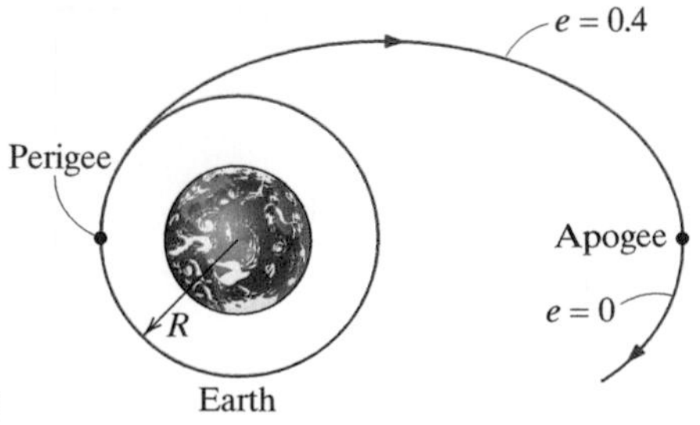

E3.6.21

3.6.22. **[Level 2]** Assume a spacecraft is orbiting the earth at a constant radius of 7.00×10^3 km. At $t = 0$ a lateral rocket is fired that imparts an inward-directed radial velocity. What is the minimum radial velocity needed to produce a parabolic trajectory, and what will r_{min} be?

3.6.23. **[Level 2]** A satellite is launched into orbit and at $t = 0$ it has zero radial speed, is 8.40×10^3 km from the earth's center, and has an angular speed of 1.20×10^4 m/s. Is the orbit elliptical?

3.6.24. **[Level 2]** A meteor approaches the earth from the direction of point A. (Consider A to be infinitely distant.) The trajectory of the meteor is such that it moves on a hyperbolic trajectory about the earth, its path asymptotic to $\overline{CD}$ as it moves away from the earth. What must the meteor's velocity be at point M?

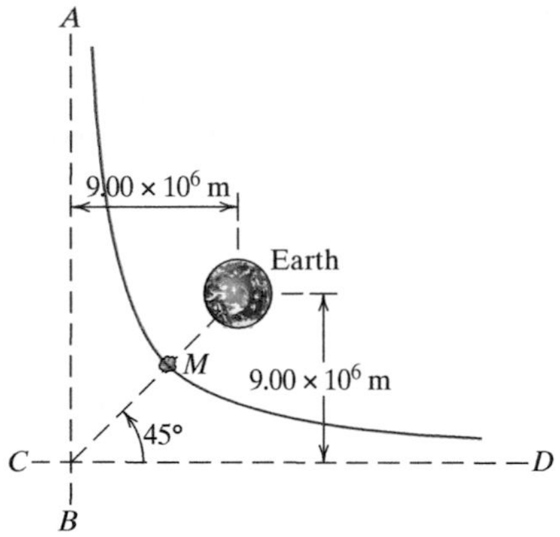

E3.6.24

3.6.25. **[Level 2]** A spacecraft, located at point A, fires its rocket so as to break its circular orbit, transition to a hyperbolic trajectory, and proceed to Mars. By how much must its velocity be increased to accomplish this? Its circular orbit has a radius of 7.50×10^6 m.

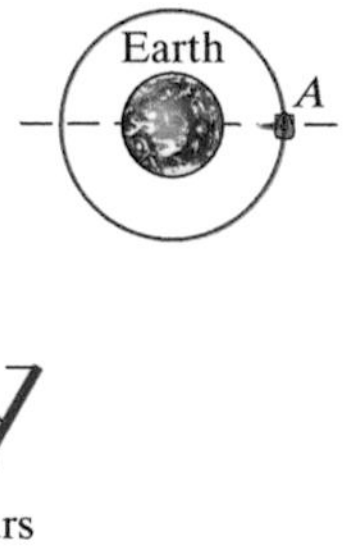

E3.6.25

3.6.26. **[Level 2]** What must the speed of a meteor be at its closest approach ($r_{min} = 6900$ km) if it is to have a circular orbit? What about an elliptical orbit with eccentricity $e = 0.5$? How about $e = 1.0$?

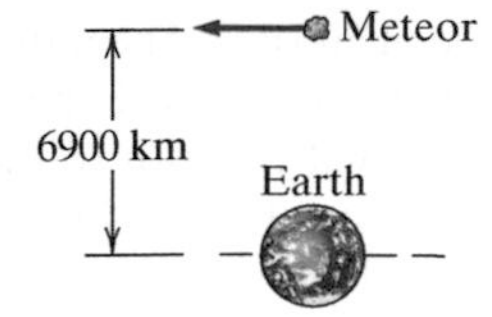

E3.6.26

3.6.27. **[Level 2]** A 7.20×10^4 kg spacecraft is in orbit around the earth with an eccentricity of $e = 0.20$. At the orbit's perigee, the spacecraft is slowed by the application of braking rockets, which produce a thrust of 8.70×10^4 N. The angular momentum of the spacecraft before the application of the braking thrusters is 4.08×10^{15} kg·m²/s. How long must the thrusters fire to put the spacecraft into a new orbit having an eccentricity equal to 0.18? Neglect the change of position of the spacecraft with respect to the earth during the thrusters' firing.

3.6.28. **[Level 2]** A meteor is located at the illustrated position and is traveling at a velocity of $3544\boldsymbol{i}$ m/s. Show that the meteor's orbit will cause it to just graze the earth's surface at its closest approach.

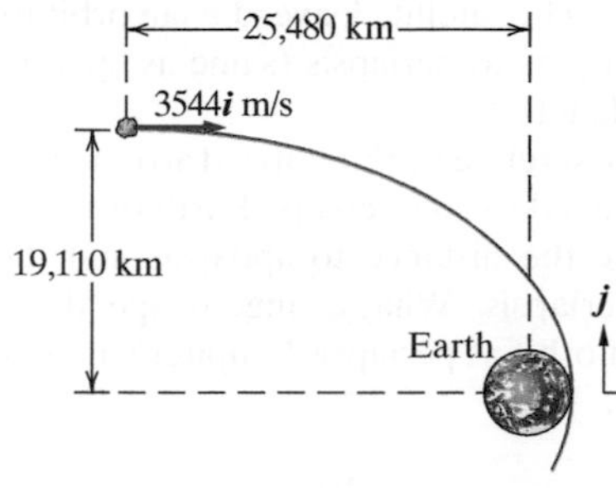

E3.6.28

3.6.29. **[Level 2] Computational** Use numerical integration to verify that the meteor of **Exercise 3.6.28** will just graze the earth's surface at its closest approach.

3.6.30. **[Level 2]** A meteor with mass m is traveling at 5160 m/s when it is at A. Will it hit the earth? $\theta = 18°$. Demonstrate your answer analytically.

E3.6.30

3.6.31. **[Level 2]** In *A Trip to the Moon*, Jules Verne imagined using a giant cannon to shoot people to the moon. Assume that such a device was built and fires a test projectile at 1500 m/s at an angle of 40°. How far around the earth will the projectile land, and how far above the earth's surface will it reach?

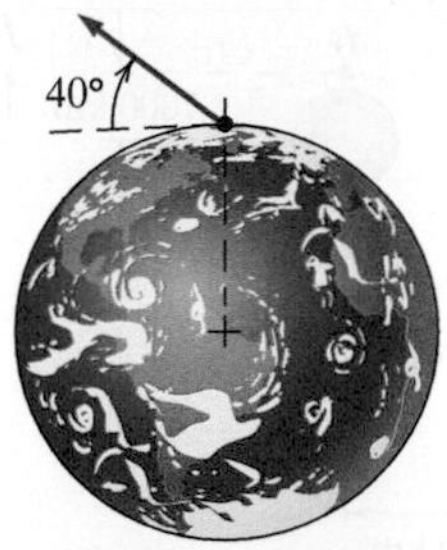

E3.6.31

3.6.32. **[Level 2]** An orbiting spacecraft is 8000 km from the earth's center, it has a speed of 4000 m/s, and its velocity vector is angled 45° from $\boldsymbol{e}_r$. What is the orbit's eccentricity?

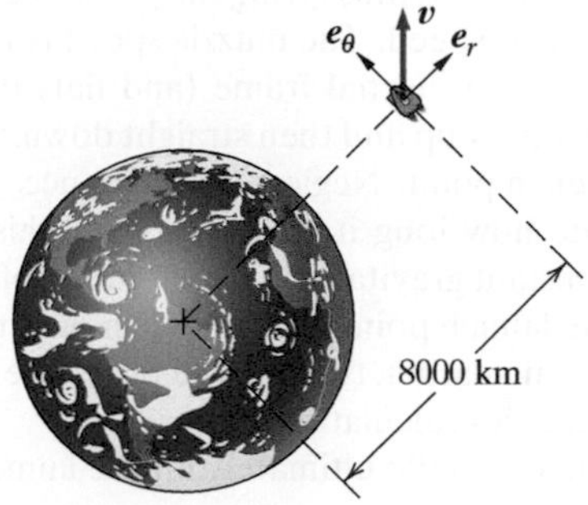

E3.6.32

3.6.33. **[Level 2]** A spacecraft is orbiting the earth at a radius of 6900 km. When at point A, it fires braking rockets that reduce its velocity by 1000 m/s. As a result, it impacts the earth's surface at B. Find the value of β at which it hits the earth.

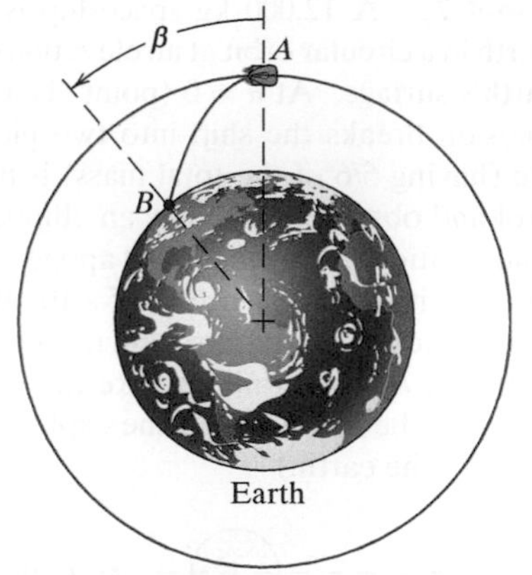

E3.6.33

3.6.34. **[Level 2]** A spacecraft is approaching the earth and is moving in a straight line. At A the engines are cut.

a. What must x equal so that the spacecraft's velocity is in the $\boldsymbol{i}$ direction at B?

b. What must the velocity be changed to at B in order for the subsequent orbit to be circular?

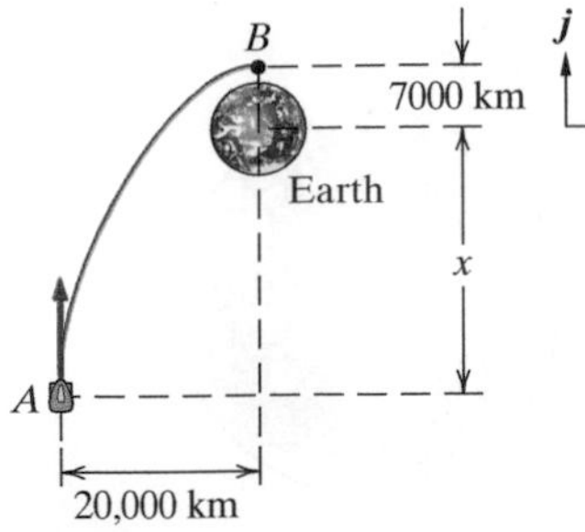

E3.6.34

3.6.35. [Level 2] Computational Let's examine the difference between treating a projectile problem as a true orbital mechanics problem and treating it as a rectilinear motion problem. Assume that a projectile is launched vertically from the equator, thus giving the projectile both a radial and an angular speed. The muzzle speed is 1600 km/h. If the earth were an inertial frame (and flat) the projectile would go straight up and then straight down, ultimately striking the launch point. Neglect air resistance.

a. Calculate how long it would take for this to occur (assuming a constant gravitational acceleration of 9.81 m/s^2, the value at the launch point) and then compare the results to an accurate calculation, treating the projectile as part of an orbital mechanics calculation.

b. Does the projectile ultimately hit the launch point?

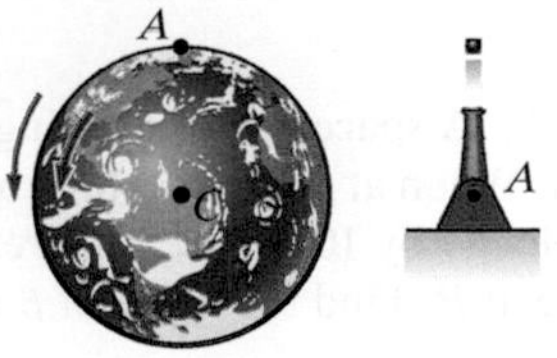

E3.6.35

3.6.36. [Level 2] A 12,000 kg spaceship is traveling around the earth in a circular orbit at an elevation of 300 km above the earth's surface. At $\theta = 0$ (point A) a powerful onboard explosion breaks the ship into two pieces. The forward piece (having 5/6 of the total mass) is tracked by ground control and observed to follow an elliptical trajectory, with A the position of its perigee. At apogee (the point B, for which $\theta = \pi$) it is at a distance 1.715×10^7 m from the earth's center. Its radial speed is zero and its angular speed is 3.609×10^3 m/s at B. Determine where the other piece of the spaceship will be 2 hours after the explosion. (*Note:* Ignore rotations of the earth.)

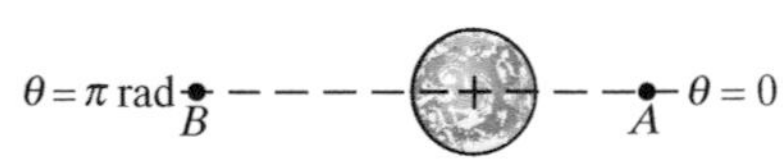

E3.6.36

3.6.37. [Level 2] Computational A spaceship is traveling around the earth in a circular orbit at an elevation of 300 km above the earth's surface. At $\theta = 0$ (point A) a powerful onboard explosion breaks the ship into two equal pieces. The inner piece has its velocity altered by the addition of a radial component directed directly toward the center of the earth with a magnitude of 700 m/s. The outer piece gains an outward velocity component, also with magnitude 700 m/s.

Both pieces will eventually return to the position they occupied at the time of the explosion but will do so at different times. Determine the time lapse between individual arrivals.

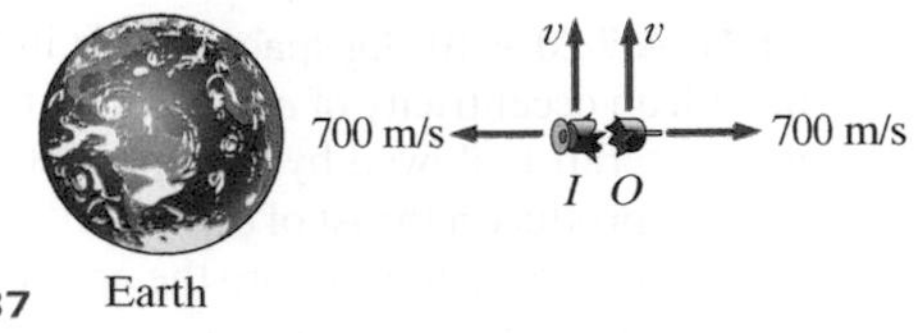

E3.6.37 Earth

3.6.38. [Level 2] A satellite headed to Jupiter on a reconnaissance mission is using a combination of boosters and gravity assist to gain speed for the return trip back to Earth. The satellite is initially launched into a circular orbit with radius $R = 7000$ km around Earth and then leaves Earth orbit for Jupiter. When the satellite nears Jupiter, it gets pulled into an elliptical orbit of $e = 0.5$ at apoapsis (which is "apogee" for planets other than Earth) with a speed of $v_a = 9500$ m/s. Assume that the satellite has no radial velocity at apoapsis. The satellite leaves Jovian orbit to return to Earth when it gets to periapsis (same as "perigee," but for non-Earth planets).

a. What is the satellite's orbital speed around Earth, and how fast does it need to go to escape Earth orbit?

b. Determine the distance to apoapsis and the satellite's speed at periapsis. What change in speed is needed to escape Jovian orbit at periapsis? Jupiter's mass is $m_J = 1.8986 \times 10^{27}$ kg.

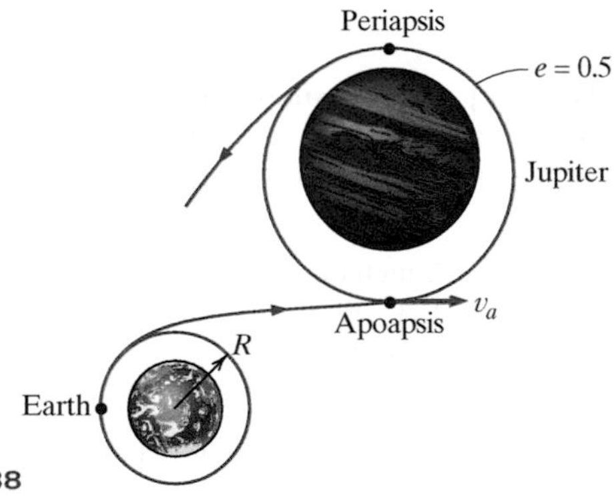

E3.6.38

3.6.39. [Level 2] A satellite approaches the moon with a speed $v = 1.0 \times 10^3$ m/s, and it gets pulled into an elliptical orbit at apoapsis (with zero radial speed), which is a distance $r_a = 5.22 \times 10^3$ km from the moon's center. Periapsis for the satellite's elliptical trajectory is $r_p = 3.48 \times 10^3$ km.

a. Find how much the satellite's speed needs to change to transition to a circular orbit at periapsis.

b. What change in speed is required to escape lunar orbit?

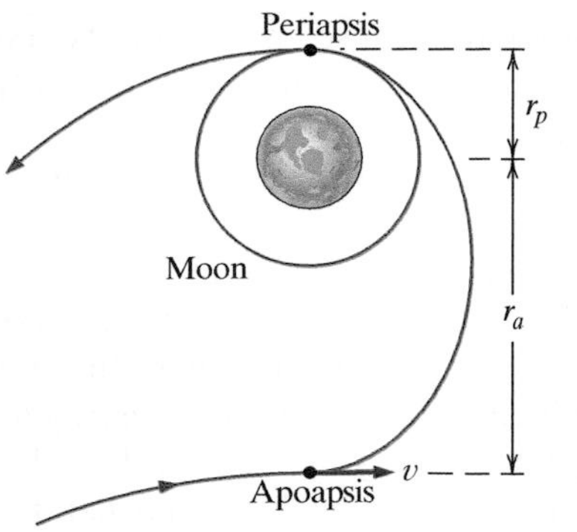

E3.6.39

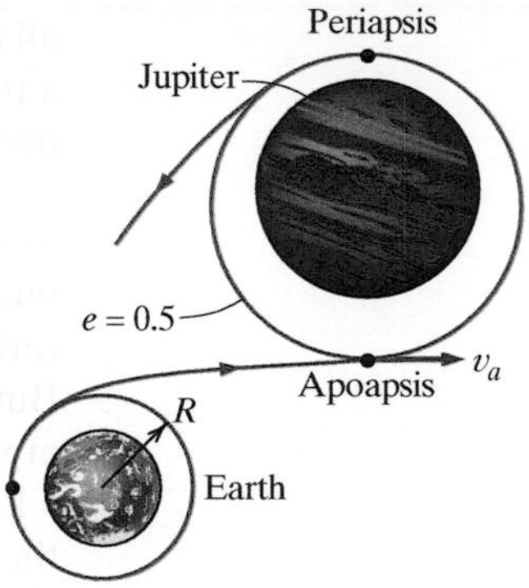

E3.6.40

3.6.40. [Level 2] A satellite is initially in circular orbit around Earth with a radius of $R = 7000$ km, and suppose it needs a speed of $\overline{v} = 11{,}000$ m/s to escape orbit and head to Jupiter. Once it gets to Jupiter, the satellite is pulled into an elliptical orbit with $e = 0.5$ at apoapsis at a speed of $v_a = 9500$ m/s (there is no radial velocity). The satellite is programmed to leave Jovian orbit at periapsis, and the necessary escape speed to do so is $\overline{v}_p = 33{,}000$ m/s. The satellite's boosters can generate an average thrust of $T = 5 \times 10^4$ N. How long must they fire to leave Earth orbit and Jovian orbit at periapsis? Jupiter's mass is $m_J = 1.8986 \times 10^{27}$ kg, and the satellite has a mass of $m_s = 700$ kg.

3.6.41. [Level 2] A satellite approaches Saturn in a hyperbolic trajectory. When it reaches periapsis at $r_p = 1.205 \times 10^8$ m, the satellite's speed is $v_p = 23{,}000$ m/s and enters an elliptical orbit with $e = 0.5$. Saturn's mass is $m_S = 5.6846 \times 10^{26}$ kg.

a. What change in speed is needed to enter an elliptical orbit at periapsis?

b. If the satellite then transitions to a circular orbit at apoapsis, determine the necessary speed change.

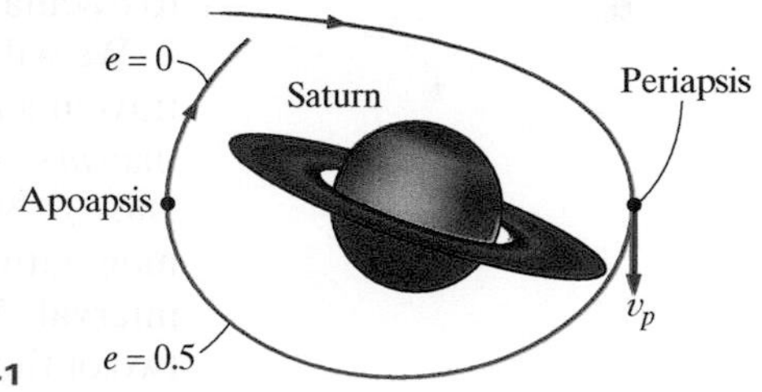

E3.6.41

3.7 IMPACT

This is a fun section because it deals with something that's depicted in a good 50% of Hollywood's movie releases—collisions. (We'll deal with explosions later.) Whether it's giant asteroids smashing into the earth or multiple cop cars crashing into each other, the key element is a loud and impressive collision.

There are two ways to approach collisions. There's the exact approach, in which we keep track of how the individual bodies deform as a result of the impact, employing sophisticated finite element codes to deal with the nonlinear, large deformation behavior of the bodies and complex numerical integration codes to make sure the interfacial contact conditions are appropriately handled. That's quite difficult. And then there's the easy approach, where we make a sweeping generalization, come up with an "impact coefficient," and try to convince ourselves that the analysis is still pretty good. So, which is it to be, the extraordinarily complex and realistic model or the pretty good and pretty easy model? That's right! As always, when faced with a tough choice, I'll opt for the easy way out. So it's the simple model for us. Actually, the simplified approach isn't really all that bad, and we're going to use it because, in all honesty, trying to do a truly realistic job of modeling an impact event is an extremely difficult undertaking, well beyond the scope of this text.

The "impact coefficient," called the **coefficient of restitution**, isn't really mysterious; rather, it is a single number that will give us a sense of how much energy is dissipated during a collision. Keep in mind that the restitution coefficient is not a precise energy measure—that will come later. But it will give us a good feel for what's happening with the system's energy. That's a key question with collisions. Is a lot of energy lost or just a little? Think about what happens when you drop a rubber ball. It hits the floor and then it bounces up, not quite back to your hand but pretty close. This means that not much energy was dissipated in the collision with the floor. The situation is quite different if you try to bounce a ball of clay. Once released, this ball splats against the floor and pretty much stays there. Clay doesn't tend to bounce. All the kinetic energy the clay had just before impact goes to zero as a result of the impact. And because clay isn't elastic, it doesn't store the energy in its deformed structure (as the rubber ball did) and therefore has nothing to "bounce back" with.

The two key pieces of information we will need are the linear momentum $\boldsymbol{L}$ of the two bodies involved in the collision and the coefficient of restitution between these bodies. With this information, we will be able to calculate how the velocities of the impacting bodies change.

We will start by examining **Figure 3.66**. Two particles are shown, both traveling in the $\boldsymbol{i}$ direction ($\boldsymbol{v}_1 = v_1\boldsymbol{i}, \boldsymbol{v}_2 = v_2\boldsymbol{i}$) but with m_1 traveling faster than m_2. Thus, eventually m_1 is going to find itself bumping into m_2 (**Figure 3.67*a***). Now remember what we just learned in Section 3.4—the linear momentum of a body changes if a force acts on the body over some time interval. When m_1 hits m_2, a force is going to be developed between the two of them. From Newton's third law we know that a pair of interacting forces will exist, equal in magnitude and opposed in direction, one acting on m_1 and the other on m_2. This is shown in **Figure 3.67*b***. These two forces exist as long as there is contact between the particles.

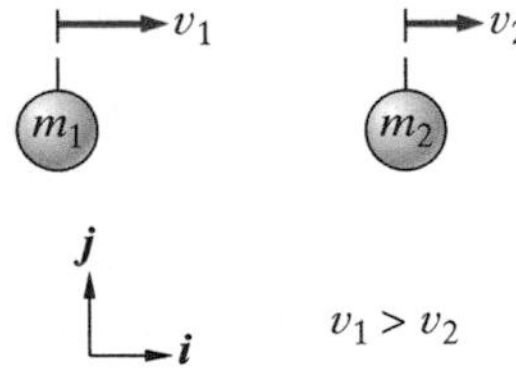

Figure 3.66 Two particles on a collision course

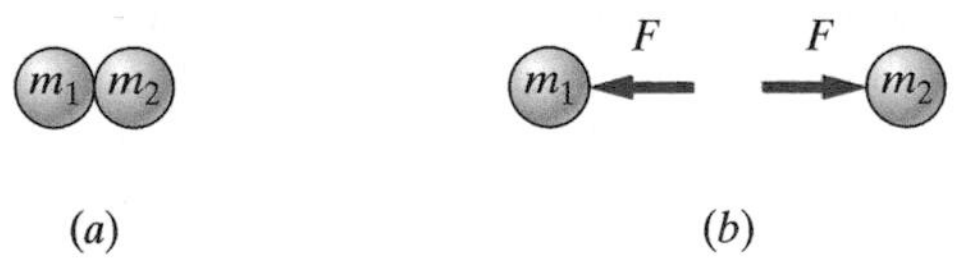

Figure 3.67 (*a*) Two particles that have just collided; (*b*) FBD for each particle

This is the point at which I have to wave my hands a little. Up to now I have said that we are dealing with infinitesimal point masses—masses having no dimensions at all. To get anywhere with our impact analysis, however, we are going to have to treat our colliding particles as having some finite dimensions. We are going to have to let them be deformable so that they can squash a little as they collide. **Figure 3.68** shows the complete situation.

Initially, the bodies haven't touched, and m_1 is approaching m_2 (**Figure 3.68*a***). A little later (**Figure 3.68*b***) they make first contact. This occurs at a

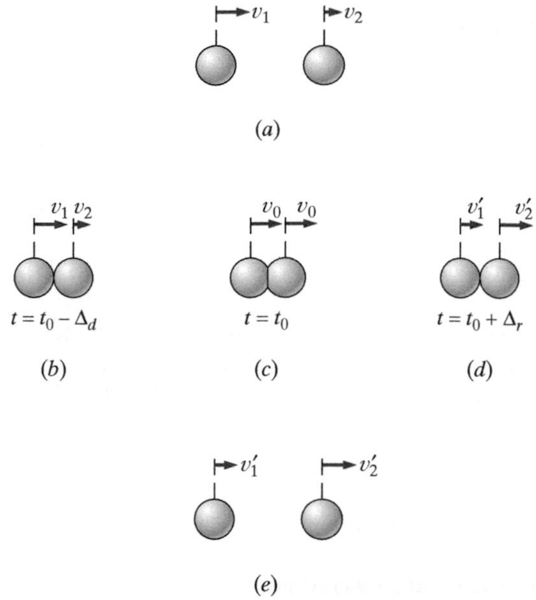

Figure 3.68 Complete collision scenario

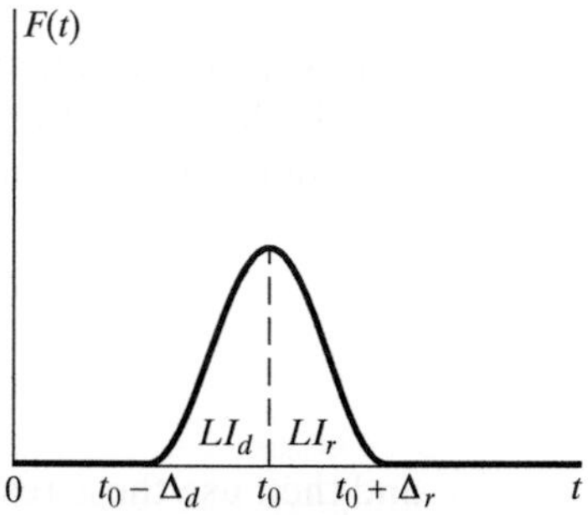

Figure 3.69 Force interaction during collision

time $t = t_0 - \Delta_d$ (d for deformation). At this instant, no deformation has taken place. Starting immediately, however, the bodies start to squash together because of their relative velocities, with the faster-moving m_1 trying to move through the slower-moving m_2. They continue to squash together until they reach a state of maximal squash at $t = t_0$ (**Figure 3.68*c***). At this point, their speeds are identical (v_0), the relative velocity difference between the two having gone to zero.

If you think about what happens to a basketball after it has been dropped to the ground, you will immediately realize what happens to our bodies after they reach maximal squash: they start to unsquash. The unsquashing phase continues until, at $t = t_0 + \Delta_r$, they again reach the state of just barely touching (**Figure 3.68*d***). (Note that Δ_r isn't necessarily the same as Δ_d.) Immediately after this point in time, the bodies either separate (if $v_2' > v_1'$) or continue moving together to the right with the same speed (if $v_2' = v_1'$) (**Figure 3.68*e***).

You will notice that I didn't allow the possibility of $v_2' < v_1'$. That's because it isn't possible. If it were true, it would mean that m_2 is moving more slowly than m_1. If that were the case, somehow m_1 would have to have moved through m_2 because m_1 started out behind m_2. That's clearly impossible, both bodies being solid objects, and so we needn't worry ourselves over the possibility.

The key to our analysis is what's happening from $t_0 - \Delta_d$ to t_0 and from t_0 to $t_0 + \Delta_r$. **Figure 3.69** shows a possible profile for force as a function of time during the collision. The interaction force $F(t)$ starts at zero, builds to a maximum at t_0, and then drops back to zero. Because this is a force-time plot, the integral (or area under the curve) is equal to the applied linear impulse. Let's use $\mathcal{LI}_d$ to denote the magnitude of the linear impulse from $t_0 - \Delta_d$ to t_0 and $\mathcal{LI}_r$ to denote the restoration impulse (which lasts from t_0 to $t_0 + \Delta_r$). Looking at the directions of the forces, we can form the individual linear momentum equations:

$$m_1v_1 - \mathcal{LI}_d = m_1v_0 \tag{3.92}$$

$$m_1v_0 - \mathcal{LI}_r = m_1v_1' \tag{3.93}$$

$$m_2v_2 + \mathcal{LI}_d = m_2v_0 \tag{3.94}$$

$$m_2v_0 + \mathcal{LI}_r = m_2v_2' \tag{3.95}$$

First I will show that the *system momentum* (where the system includes *both* bodies) is conserved in spite of the collision. We can find m_1v_0 and m_2v_0 in terms of $\mathcal{LI}$ values and the final momenta from (3.93) and (3.95):

$$m_1v_0 = \mathcal{LI}_r + m_1v_1' \tag{3.96}$$

$$m_2v_0 = -\mathcal{LI}_r + m_2v_2' \tag{3.97}$$

and then use these results in (3.92) and (3.94) to get

$$m_1v_1 - \mathcal{LI}_d = \mathcal{LI}_r + m_1v_1' \tag{3.98}$$
$$m_2v_2 + \mathcal{LI}_d = -\mathcal{LI}_r + m_2v_2' \tag{3.99}$$

These last two equations represent the change in momenta for both bodies as a function of the linear impulse due to the collision. Notice what happens when we add these equations together: the $\mathcal{LI}$ terms cancel and we are left with

$$m_1v_1 + m_2v_2 = m_1v_1' + m_2v_2' \tag{3.100}$$

which says that the *total* linear momentum of the system in the $\boldsymbol{i}$ direction (the sum of the two mv terms) is conserved.

Pretty neat! No matter how much energy is dissipated in the collision, the system momentum remains unchanged. This might (or might not) strike you as strange. After all, if you've got a huge collision and you know energy is being dissipated because of it, you might expect the velocities to change (they do) and might reasonably assume that the individual momenta will therefore change (they do) and thus conclude that the overall momentum of the system will change (it doesn't).

This fact is extremely important, and we will use it a great deal. So, just to repeat it one last time: *the total momentum of two colliding bodies does not change as a result of the collision*. This observation is a special case of a more general statement, namely, that the total linear momentum of a system of particles won't change unless some force completely independent of the system acts on the system. Collisions between the various particles inside a system, magnetic forces, gravity between the particles, and so forth—none of these *internal* forces can change the system's total linear momentum. We will look into this more deeply in Chapter 5, but for now let's keep looking at our collision problem.

We actually know *almost* all we need to know to figure out what happens to each body after a collision. The only missing piece is to figure out how *elastic* the collision is. And here's how we will do it. We have already seen that there are two phases to the collision: the compression phase (initial touch to maximal squash, involving $\mathcal{LI}_d$) and the expansion phase (maximal squash to barely touching, involving $\mathcal{LI}_r$).

What we will do is characterize the collision by the ratio $\frac{\mathcal{LI}_r}{\mathcal{LI}_d}$. If this ratio is 1, all the momentum change that went into deforming the bodies is given back during restoration. Physically, $\frac{\mathcal{LI}_r}{\mathcal{LI}_d} = 1$ means the impact is "bouncy," or *elastic*—purely elastic, in fact. We will see that there is zero energy loss in a collision of this sort. On the other hand, if $\frac{\mathcal{LI}_r}{\mathcal{LI}_d} = 0$, then there's no return of momentum. This corresponds to what we call a purely *plastic* deformation. It's what happens when you try to bounce clay. It just hits the ground with a thud—there is no restoration and therefore no bounce.

It is this ratio $\frac{\mathcal{LI}_r}{\mathcal{LI}_d}$ that is the coefficient of restitution mentioned previously. You will note that when I first mentioned the coefficient, I said it would give us a sense of how much energy is dissipated during a collision. This is true, but now you can see that it's giving us just a *sense* of how much, not a precise value, because it's associated with the system's change in momentum, not the system's energy. We'll use the symbol *e* for the coefficient of restitution, and the mathematical representation is

$$e = \frac{\mathcal{LI}_r}{\mathcal{LI}_d} \tag{3.101}$$

With this information, we can figure out everything we need to know in our collision problem. Notice that (3.92) and (3.93) can be solved to give us $\mathcal{LI}_d$ and $\mathcal{LI}_r$ for m_1, and (3.94) and (3.95) can be solved to give us $\mathcal{LI}_d$ and $\mathcal{LI}_r$ for m_2. If we do this and then in each case divide the expression for $\mathcal{LI}_r$ by the expression for $\mathcal{LI}_d$, we get

$$\frac{\mathcal{LI}_r}{\mathcal{LI}_d} = e = \frac{v_0 - v_1'}{v_1 - v_0} \tag{3.102}$$

$$\frac{\mathcal{LI}_r}{\mathcal{LI}_d} = e = \frac{v_2' - v_0}{v_0 - v_2} \tag{3.103}$$

Notice the totally fascinating fact that neither expression for *e* involves the masses of the colliding bodies.

We are still not where we want to be because both (3.102) and (3.103) involve v_0, the velocity of the two bodies in the middle of the collision. This isn't something we have any way of knowing, and so we have to get rid of it. Fortunately for us, we have two equations for *e* and so can say

$$\frac{v_0 - v_1'}{v_1 - v_0} = \frac{v_2' - v_0}{v_0 - v_2}$$

We can now cross-multiply and solve for v_0:

$$v_0 = \frac{v_2'v_1 - v_1'v_2}{v_1 - v_2 + v_2' - v_1'} \tag{3.104}$$

Arguably, this looks like a mess. And what I am going to suggest next might seem even worse. What we need to do now is substitute this expression for v_0 into either (3.102) or (3.103). It doesn't matter which; we'll get the same answer either way. You are probably thinking that the result of doing this will be horrendous—a huge mass of v_i's and v_i''s. And that's what you *will* get—at first. But if you start grouping terms, you will find that many of them cancel, and ultimately we are left with

$$e = \frac{v_2' - v_1'}{v_1 - v_2} \tag{3.105}$$

This is pretty amazing. When everything's said and done, e is simply the ratio of the relative velocity of the bodies *after* the collision to the relative velocity *before* the collision. The masses of the bodies don't matter—just the velocities. That's not to say the masses don't matter at all; they simply don't show up in this equation. They *do* show up in (3.100)—our equation for the total linear momentum of the system. And now let's write (3.100) and (3.105) together so that all is clear:

$$m_1 v_1 + m_2 v_2 = m_1 v_1' + m_2 v_2' \qquad e = \frac{v_2' - v_1'}{v_1 - v_2}$$

>>> **Check out Example 3.23 (page 189) and Example 3.24 (page 190) for applications of this material.**

EXAMPLE 3.23 DYNAMICS OF TWO POOL BALLS (Theory on pages 186 and 188)

We'll start by looking at the most common collision one sees in pool halls—two "particles" of identical mass m striking each other (**Figure 3.70**). For the present, we'll ignore the fact that pool balls are rolling spheres and simply treat them as particles. Initially, ball m_1 is moving at a velocity $v\boldsymbol{i}$ and ball m_2 is stationary. Assuming $e = 1$, what is the velocity of each ball after the collision?

Figure 3.70 Playing pool

Goal Find the final speed of each ball.

Given Initial velocities and coefficient of restitution.

Assume The coefficient of restitution is equal to 1.0.

Draw **Figure 3.71** shows the situation just before contact. Ball m_1 is headed in the $\boldsymbol{i}$ direction and is about to strike m_2.

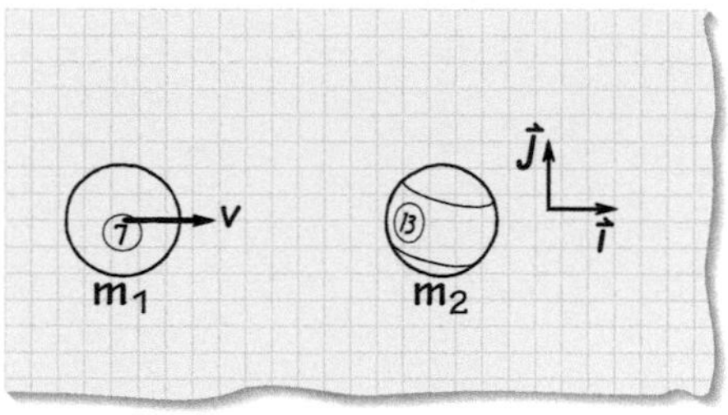

Figure 3.71 Configuration before collision

Formulate Equations (3.100) and (3.105), taken in the $\boldsymbol{i}$ direction:

Solve

$$mv + 0 = mv_1' + mv_2' \quad \Rightarrow \quad v_2' = v - v_1' \tag{3.106}$$

$$e = 1 = \frac{v_2' - v_1'}{v - 0} \quad \Rightarrow \quad v_1' = v_2' - v \tag{3.107}$$

Substituting (3.107) into (3.106), we get

$$v_2' = v - (v_2' - v) = 2v - v_2' \quad \Rightarrow \quad \boxed{v_2' = v}$$

which tells us that m_2's speed after the collision is equal to m_1's speed before the collision. Using $v_2' = v$ in (3.107) yields $\boxed{v_1' = 0}$. All the momentum that was initially in m_1 has been transferred to m_2.

Check The easiest check is to recall what happens when you strike one pool ball with another, assuming you've played pool at some time. Assuming that you weren't doing anything tricky with "English" on the cue ball, it pretty much came to a complete stop after colliding with a different ball, and that ball then started moving at about the same speed the cue ball had before the collision. That's exactly what we have just found here.

Another check is to realize that m_2 moves to the right after being hit by m_1 (which was also moving to the right). That's physically reasonable and so gives us confidence in the analysis. Don't think that this point is so obvious as to not be worth commenting on. I have often gotten test questions back from students that had exactly the opposite result—namely, that m_2 moves to the left after being hit by m_1. They got that result by dropping a sign somewhere in the analysis, which is an honest mistake. However, they should have noticed that such a result is physically implausible and then should have gone back to find their error, or at least commented on the test that the answer made no sense. So now there's no excuse for you to make the same error.

EXAMPLE 3.24 MORE POOL BALL DYNAMICS (Theory on pages 186 and 188)

Let's modify the previous example a bit by letting e equal 0.9. This means that the linear impulse "given back" after maximum deformation is less than the impulse originally "put in." The two pool balls still have the same mass m, the same initial conditions hold, and our task is again to find the speeds after the collision.

Goal Find the final speed of each ball.

Given Initial velocities and coefficient of restitution.

Assume No additional assumptions are needed.

Draw Same as Figure 3.71 in Example 3.23.

Formulate Equations We will again be using (3.100) and (3.105).

Solve The only change between this problem and the previous one is in the value of e, and so we have

$$mv = mv_1' + mv_2' \quad \Rightarrow \quad v_2' = v - v_1' \tag{3.108}$$

$$0.9 = \frac{v_2' - v_1'}{v} \quad \Rightarrow \quad v_1' = v_2' - 0.9v \tag{3.109}$$

Substituting (3.109) into (3.108), we get

$$v_2' = v - (v_2' - 0.9v) = 1.9v - v_2' \quad \Rightarrow \quad \boxed{v_2' = 0.95v}$$

and, using (3.109),

$$\boxed{v_1' = 0.05v}$$

Check Unlike the previous case, in this example both balls have positive final speed, and therefore both are moving to the right. As before, the fact that m_2 is going to the right makes sense from a physical standpoint, inasmuch as it was hit on its left side. It makes sense that it's moving more slowly than in the previous case ($0.95v$ instead of v) because less linear impulse is being applied.

We can go a little further by determining the velocity of both balls at maximum "squash" in both this example and the preceding one. If you calculate this velocity using (3.104), you will find that for both examples the balls were moving at $0.5v$ at maximal deformation. That means that the deformation linear impulse $\mathcal{LI}_d$ was the same for each case. Because $e = 1.0$ in the first case, we know that $\mathcal{LI}_r = \mathcal{LI}_d$ (because e is defined as their ratio), thus lowering m_1's speed to zero and raising m_2's to v. For $e < 1$, we can deduce that $\mathcal{LI}_r < \mathcal{LI}_d$, meaning that m_1's speed won't be able to drop all the way from $0.5v$ to zero and m_2's speed won't be able to reach v, as we just saw.

EXERCISES 3.7

3.7.1. **[Level 1]** A ball is released from rest at $H = 2\,\text{m}$ above the floor and directly over a small platform placed at a height $h = 1\,\text{m}$. The platform is removed once the ball makes contact with it, and the ball then bounces off the floor. The coefficients of restitution for the platform and floor are $e_1 = 0.6$ and $e_2 = 0.8$, respectively. How high does the ball bounce after hitting the floor?

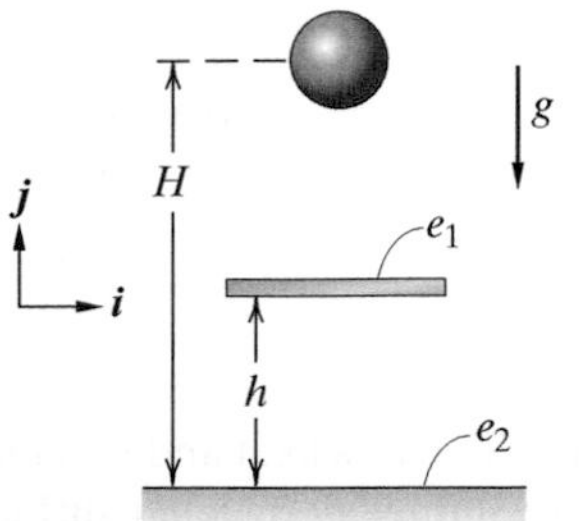

E3.7.1

3.7.2. **[Level 1]** Two masses slide along the ice, both traveling in the $\boldsymbol{i}$ direction, and impact each other. After the impact, the velocities are $\dot{\boldsymbol{x}}'_A = \dot{\boldsymbol{x}}'_B = 2.0\boldsymbol{i}$ m/s. If you're told $m_A = 10$ kg, $m_B = 20$ kg, and the initial velocity of A was $\dot{\boldsymbol{x}}_A = 5\boldsymbol{i}$ m/s, can you determine the coefficient of restitution and $\dot{\boldsymbol{x}}_B$?

3.7.3. **[Level 1]** A massless string connects the particle A (mass of 0.2 kg) to the ceiling. The particle is initially held in the position shown ($\theta = 30°$) and then released such that it is traveling to the right at 3 m/s when $\theta = 0°$. The 0.1 kg particle B is projected to the left so as to contact A when A is directly below O. The coefficient of restitution between A and B is 0.4. How fast must B be moving at the collision so that after the collision A remains stationary beneath O?

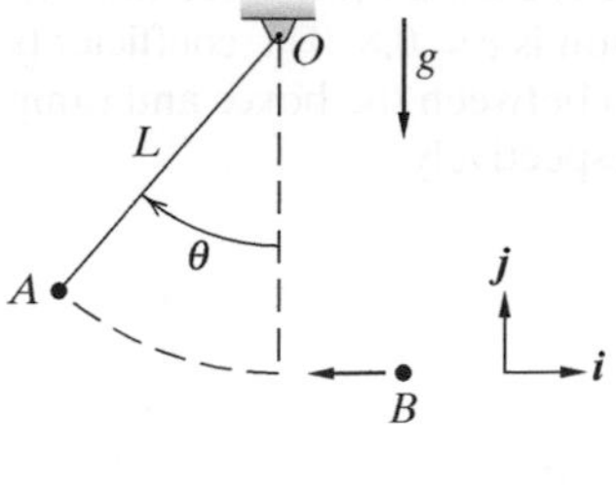

E3.7.3

3.7.4. **[Level 1]** A baseball traveling at 97 km/h hits the helmet of a player. If the baseball weighs 0.1 kg, the player's head weighs 11 kg, and the coefficient of restitution between the two is 0.2, what is the velocity of the player's head after the collision?

3.7.5. **[Level 1]** Two particles A and B strike each other as shown. $m_A = 3$ kg and $m_B = 2$ kg. Just before the collision $\boldsymbol{v}_A = 4\boldsymbol{i}$ m/s and $\boldsymbol{v}_B = -4\boldsymbol{i}$ m/s. $e = 0$. What are their velocities after the collision?

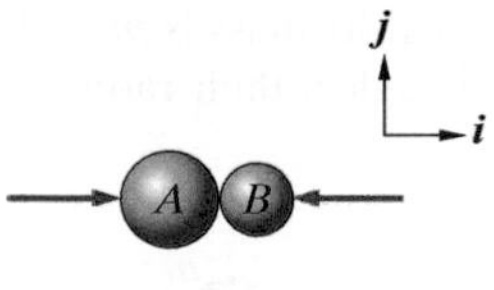

E3.7.5

3.7.6. **[Level 1]** Two particles A and B strike each other and then rebound. The coefficient of restitution between the two is 0.1. $m_A = 44$ kg and $m_B = 29$ kg. After the collision $\boldsymbol{v}'_A = 0.24\boldsymbol{i}$ m/s and $\boldsymbol{v}'_B = 0.3\boldsymbol{i}$ m/s. What are their velocities before the collision?

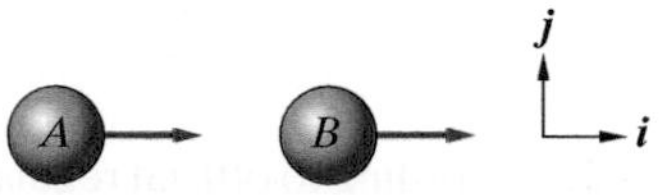

E3.7.6

3.7.7. **[Level 1]** Two particles A and B strike each other and then rebound. The coefficient of restitution between the two is 0.1. $m_A = 4\,\text{kg}$ and $m_B = 3\,\text{kg}$. Before the collision $\boldsymbol{v}_A = 2.0\boldsymbol{j}$ m/s and after $\boldsymbol{v}'_A = 1.0\boldsymbol{j}$ m/s. What are the pre- and post-collision velocities of particle B?

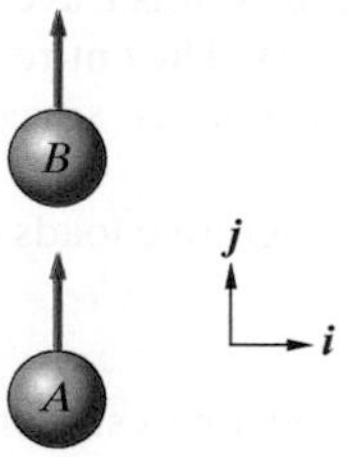

E3.7.7

3.7.8. **[Level 1]** A 6.8 kg bowling ball is projected down a lane at a constant $v_b = 3$ m/s to hit a single, stationary bowling pin that weighs 0.23 kg. If the coefficient of restitution is $e = 0.9$, what are the velocities of the bowling ball and pin immediately after impact?

E3.7.8

3.7.9. **[Level 2]** At some point in your childhood, you've probably either seen or done a neat trick that involves dropping a large and a small ball simultaneously, with the small ball just above the larger one. When the larger ball hits the ground and then collides with the smaller ball, it causes the small ball to bounce to a height several times greater than that from which the balls were originally released. Suppose the large ball ($M = 2$ kg) is dropped from $h = 1$ m above the ground, for which the coefficient of restitution is $e_1 = 0.9$.

The coefficient of restitution associated with the collision between the small and large balls is $e_2 = 0.7$. Find how high the small ball bounces if its mass is $m = 0.25$ kg. Treat the balls as particles, and neglect their radii.

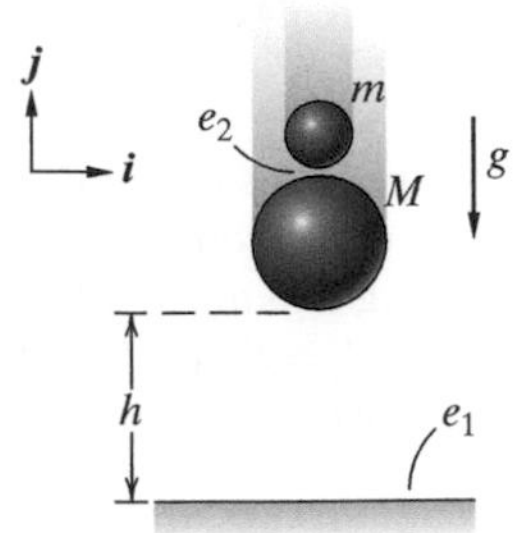

E3.7.9

3.7.10. **[Level 2]** According to official regulations, a tennis ball that's dropped from a height of 254 cm must rebound more than 135 cm and less than 147 cm. Calculate the range of e that this implies.

3.7.11. **[Level 2]** A 2268 kg SUV plows into the rear of a 998 kg sportscar. The coefficient of restitution is 0. Just before the collision the SUV was traveling at 32 km/h, and the sportscar was stationary. The entire collision takes 0.3 s.

a. What are the two vehicles' speeds immediately following the collision?

b. What average accelerative loads do the two vehicles experience?

3.7.12. **[Level 2]** A dynamics student is called on to investigate a two-car collision. It's known that car A struck car B. $m_A = 2000$ kg and $m_B = 1200$ kg. Both car A and car B then skidded 8.0 m (known from the skid marks on the street). It appears from the damage that car A struck car B at a relative speed of 12 m/s.

What was the speed of car A just before the collision? Assume a purely plastic collision ($e = 0$) and a coefficient of friction between the cars and the road of 0.7.

E3.7.12

3.7.13. **[Level 2]** Explain why, if the two leftmost balls are released from the position illustrated, the two rightmost balls swing up and attain a height equal to h_0 rather than some other result (such as the far-right ball swinging up to a height greater than h_0 or only balls 3–5 all swinging up to some height less than h_0). Assume $e = 1$. All five balls are identical.

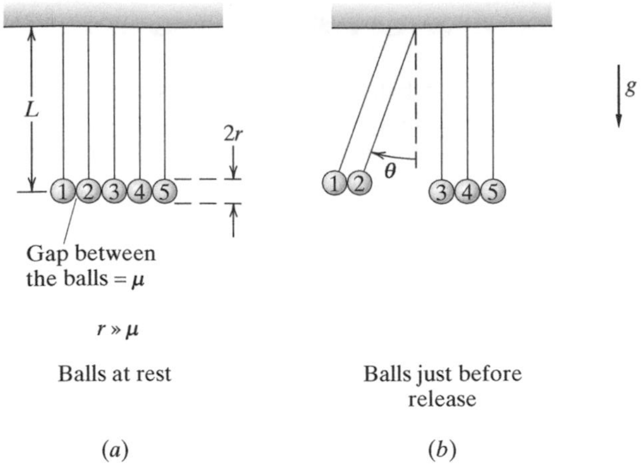

E3.7.13

3.7.14. **[Level 3]** Two blocks A and B of masses $m_A = 0.4$ kg, $m_B = 0.5$ kg are sliding along a flat surface. Just before the collision $\boldsymbol{v}_A = 5\boldsymbol{i}$ m/s and $\boldsymbol{v}_B = 2\boldsymbol{i}$ m/s. $e = 0.5$. How far will B travel after colliding with A?

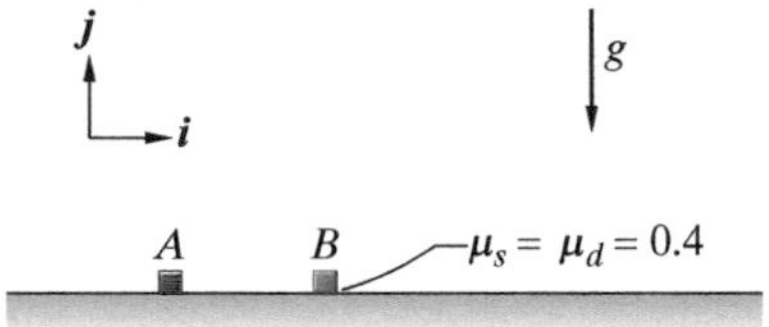

E3.7.14

3.7.15. **[Level 3]** Two boxes are on a rough ramp of length $L = 2$ m and angled at $\theta = 40°$ to the ground. Suppose box B ($m_B = 4$ kg) is first released from rest at the top of the ramp and allowed to slide 1 m down to C. Just before box B reaches this position, box A ($m_A = 2$ kg) is projected up the ramp from the base with an initial speed of $v_{A,0} = 5$ m/s so that it collides with box B at C. Determine the velocities of the boxes just after impact if the coefficient of restitution is $e = 0.8$. The coefficients of static and dynamic friction between the boxes and ramp are $\mu_s = 0.8$ and $\mu_d = 0.4$, respectively.

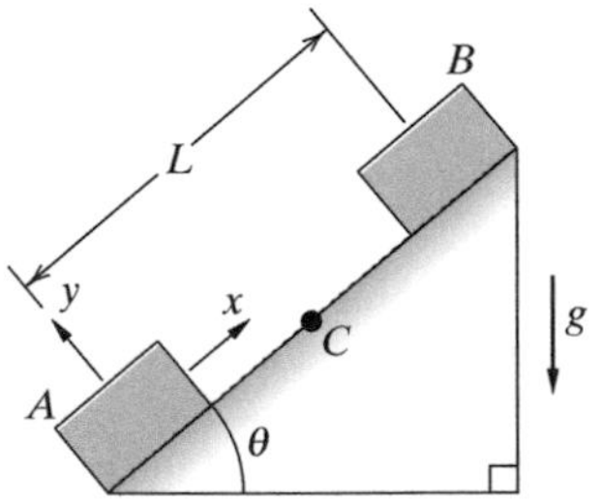

E3.7.15

3.7.16. **[Level 3]** A metal block (A) is slid across a floor and strikes a stationary block (B). $m_A = 0.1$ kg and $m_B = 0.15$ kg. The coefficient of dynamic friction between either block and the floor is equal to 0.6. How far from its initial position does block B end up if the initial velocity of block A is $\boldsymbol{v} = 10\boldsymbol{i}$ m/s and the distance between the blocks at the instant A is set into motion is 2 m?

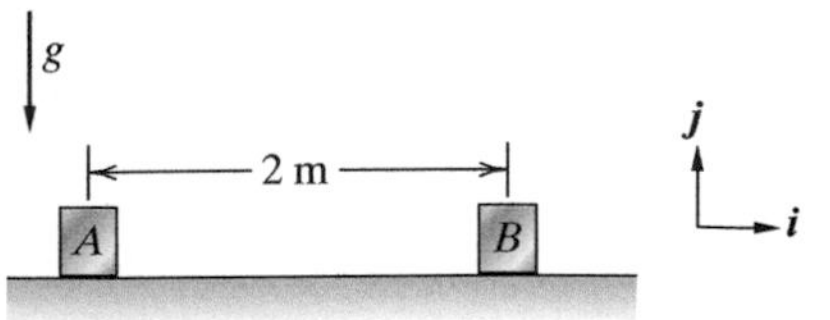

E3.7.16

3.7.17. **[Level 3]** How many bounces will it take until a ball that's dropped from 3 m rises up only 0.3 m or less (after striking the floor) if $e = 0.9$? Does it ever come completely to rest?

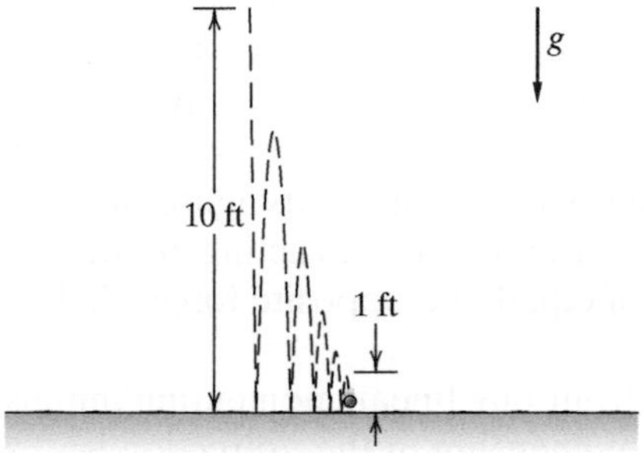

E3.7.17

3.7.18. **[Level 3]** Three particles are sliding along a flat, frictionless surface. $m_A = 0.4$ kg, $m_B = 0.5$ kg, and $m_C = 0.4$ kg. Before any collisions take place, $\boldsymbol{v}_A = 15\boldsymbol{i}$ m/s, $\boldsymbol{v}_B = 10\boldsymbol{i}$ m/s, and $\boldsymbol{v}_C = 5$ m/s. $e = 0.85$ between A and B and $e = 0.6$ between B and C. How fast will ball C be traveling after all collisions have ceased? At the illustrated, pre-impact position shown, $\boldsymbol{r}_{B_{/A}} = 2\boldsymbol{i}$ m and $\boldsymbol{r}_{C_{/B}} = 15\boldsymbol{i}$ m. Assume that A strikes B before B strikes C.

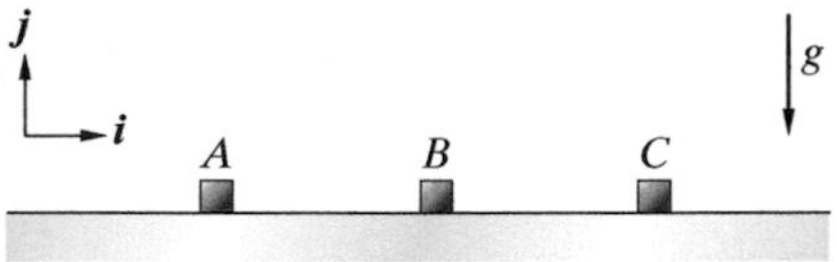

E3.7.18

3.8 OBLIQUE IMPACT

Fun as the preceding section was, it's still a little limiting.

For instance, if you wanted to write a computer program that lets you play pool on the screen, using what you have just learned, you could allow only a limited number of shots—no bank shots, no angled shots, nothing fancier than one ball smacking head-on into another ball. And if you were trying to analyze a vehicular collision, it would be the same thing: only head-ons allowed. Very boring. So obviously we have to spiff up the analysis so as to allow off-axis collisions, otherwise known as **oblique impacts**. **Figure 3.72** shows the kind of impact I'm talking about. It's clear that the two bodies are coming at each other at an angle, making the analysis more complicated than what we have been handling up to now.

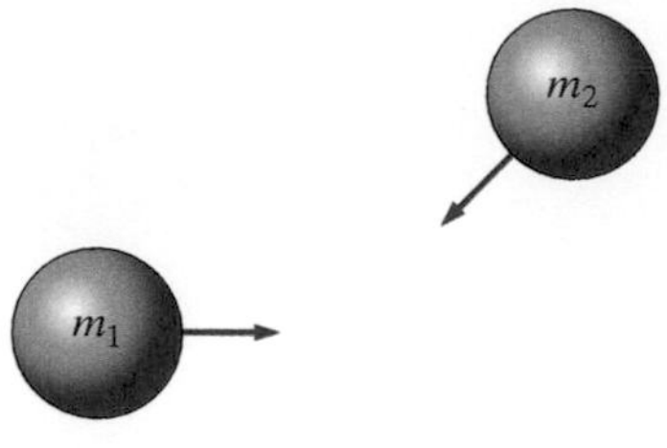

Figure 3.72 Two bodies approaching each other at an oblique angle

No worries. Although the problem looks much worse, the mathematics is going to be complicated only by the angles involved, nothing more. There's no new physics to worry about.

Figure 3.73 shows the two bodies just at the point of contact, along with their velocity vectors. Note that the vectors aren't aligned with each other in any particular way. Let's bring a bit more order to the picture by breaking the vectors into components, as shown in **Figure 3.74**. The t component direction is defined by the tangent to the two colliding bodies at their contact point, and the n component direction is defined to be normal to this. I have also drawn both bodies so that they're horizontally aligned.

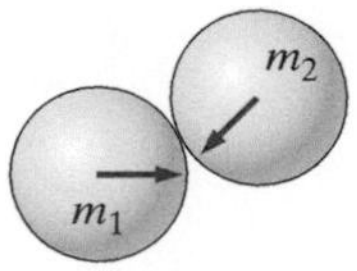

Figure 3.73 Two bodies at point of oblique contact

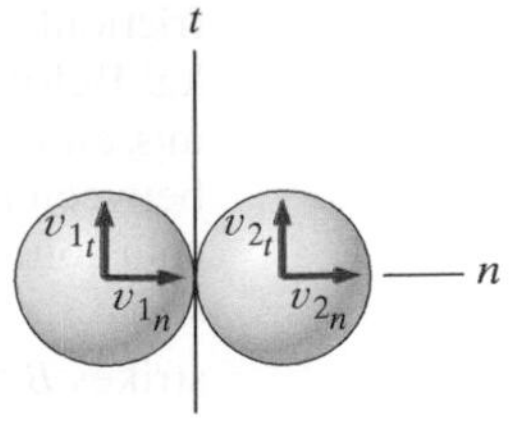

Figure 3.74 Velocity components along t and n axes

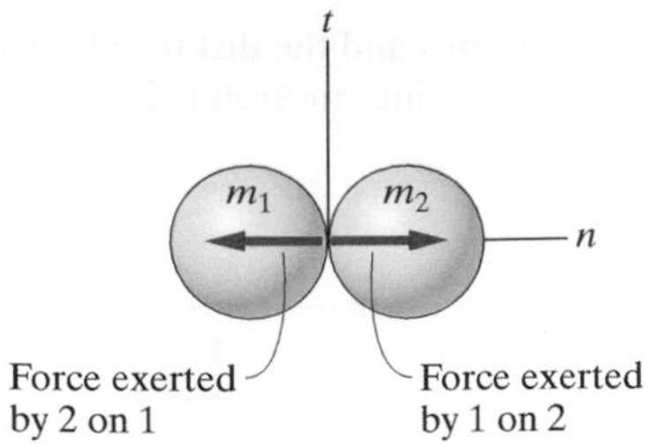

Figure 3.75 Force directions during impact

Because our interacting bodies are small, we will assume that no sliding friction is taking place. In other words, the bodies can't drag against each other as they collide. We make this obvious in the analysis by looking at the interaction forces and making sure there's zero force in the tangential direction. **Figure 3.75** illustrates what I mean. This figure shows the FBD for each body and indicates that the only forces acting on the two bodies during this collision are a set of equal and opposite forces that act in the n direction.

This is great! Remember from our linear momentum/impulse work that, being a vector quantity, momentum in the t direction isn't affected by momentum in the n direction—they're calculated independently. This means we can break our oblique-impact problem into two parts: the part in the n direction and the part in the t direction.

In the t direction, we have a particularly nice result. Because there's no interaction force in this direction, there's no change in momentum (no applied impulse means no momentum change). Therefore we know that the momentum of *each* body remains unchanged (in this direction, at least):

$$m_1 v_{1_t} = m_1 v'_{1_t}$$
$$m_2 v_{2_t} = m_2 v'_{2_t}$$

Eliminating the mass factors gives

$$v_{1_t} = v'_{1_t} \tag{3.110}$$

$$v_{2_t} = v'_{2_t} \tag{3.111}$$

How's that for uncomplicated? The result of the collision is that the velocity doesn't change at all along the tangential direction. What could be easier?

The n direction is just about as nice. What we have are two bodies that come toward each other, hit, and then separate. That's exactly the same situation as the on-axis impact problems we worked in the preceding section. So, using the results from that section tells us that the total system momentum is conserved in the n direction and that the coefficient of restitution is equal to the ratio of the relative velocities in the normal direction:

$$m_1 v_{1_n} + m_2 v_{2_n} = m_1 v'_{1_n} + m_2 v'_{2_n} \tag{3.112}$$

$$e = \frac{v'_{2_n} - v'_{1_n}}{v_{1_n} - v_{2_n}} \tag{3.113}$$

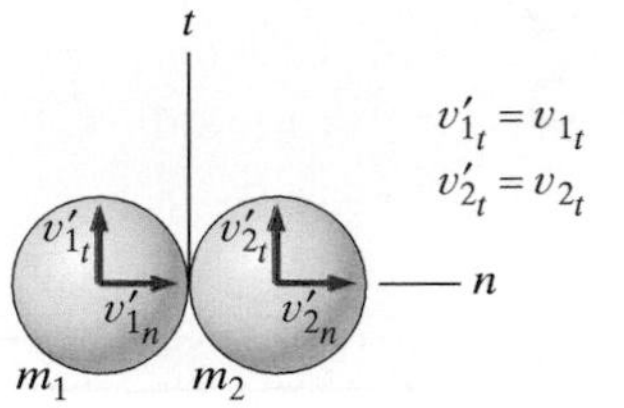

Figure 3.76 Post-impact velocities

Figure 3.76 shows the post-impact velocities used in these two formulas.

That's about all there is to it. In the normal direction, you treat the problem just like an on-axis impact, and in the tangential direction you leave the velocities of the bodies unchanged.

The only item left to take care of is how we set up the coordinates. In the most general case, the bodies collide at an angle and are not aligned horizontally, as illustrated in **Figure 3.73**. To show how we handle this, and also to get some experience in dealing with oblique collisions, let's solve a couple of examples. The first one won't require any realignment of our coordinates, but in the second one we will have to work a little to get the problem into a convenient form.

>>> **Check out Example 3.25 (page 196) and Example 3.26 (page 198) for applications of this material.**

EXAMPLE 3.25 OBLIQUE BILLIARD BALL COLLISION (Theory on page 195)

Figure 3.77 Two billiard balls on a billiard table

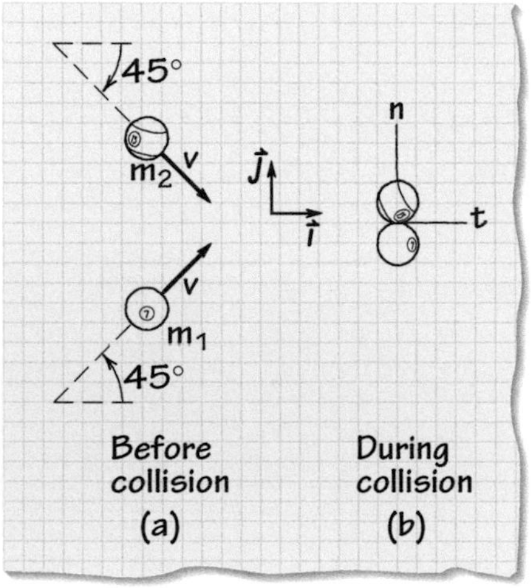

Figure 3.78 Approach and contact

Two billiard balls are coming toward each other as shown in **Figure 3.77**. **Figure 3.78*a*** shows a more schematic view of the two balls. The velocity vector for m_1 is oriented 45° up from $\boldsymbol{i}$, and m_2's velocity is oriented 45° down from $\boldsymbol{i}$. Both balls have the same mass m and the same speed v. As they collide (**Figure 3.78*b***), their centers both align with the $\boldsymbol{j}$ unit vector. Thus the tangential and normal directions are oriented along $\boldsymbol{i}$ and $\boldsymbol{j}$, respectively. Determine the velocities of the balls after they collide for $e = 0.5$ (which, although quite a bit too low for realistic billiard balls, will allow the velocities to change appreciably).

Goal Find the final velocity of each ball.

Given Initial velocities and coefficient of restitution.

Assume No additional assumptions are needed.

Draw **Figure 3.79** shows the configuration of each ball's velocity vector both before and after collision. Because the orientation of each ball is 45° from $\boldsymbol{i}$, we can write the incoming velocities as

$$\boldsymbol{v}_1 = \frac{v}{\sqrt{2}}\boldsymbol{i} + \frac{v}{\sqrt{2}}\boldsymbol{j}$$

$$\boldsymbol{v}_2 = \frac{v}{\sqrt{2}}\boldsymbol{i} - \frac{v}{\sqrt{2}}\boldsymbol{j}$$

as shown in **Figure 3.79*a***.

We can even label a couple of the outgoing velocity components. We know from (3.110) and (3.111) that the post-collision tangential velocity is the same as the pre-collision tangential velocity. Thus the component along the $\boldsymbol{i}$ direction is $\frac{v}{\sqrt{2}}$ for both m_1 and m_2, as shown in **Figure 3.79*b***.

Formulate Equations Having taken care of the tangential components, all we need to do now is determine the normal components. For this we need our equation for conservation of system momentum in the normal direction (3.112) and the impact equation, also in the normal direction (3.113):

$$m_1 v_{1_n} + m_2 v_{2_n} = m_1 v'_{1_n} + m_2 v'_{2_n}$$

$$e = \frac{v'_{2_n} - v'_{1_n}}{v_{1_n} - v_{2_n}}$$

Solve Substituting our known velocity components along with the given coefficient of restitution, and dropping all the mass factors because $m_1 = m_2 = m$, we get

$$m\frac{v}{\sqrt{2}} - m\frac{v}{\sqrt{2}} = mv'_{1_n} + mv'_{2_n} \quad \Rightarrow \quad v'_{1_n} = -v'_{2_n} \tag{3.114}$$

$$0.5 = \frac{v'_{2_n} - v'_{1_n}}{\frac{v}{\sqrt{2}} - \left(-\frac{v}{\sqrt{2}}\right)} \quad \Rightarrow \quad \frac{v}{\sqrt{2}} = v'_{2_n} - v'_{1_n} \tag{3.115}$$

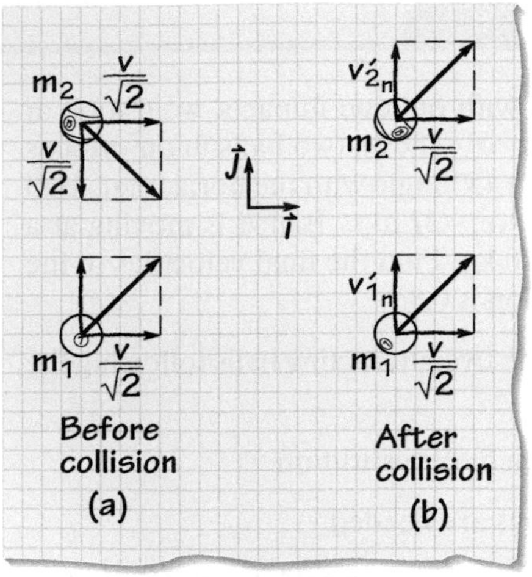

Figure 3.79 (*a*) Pre- and (*b*) post-collision velocities

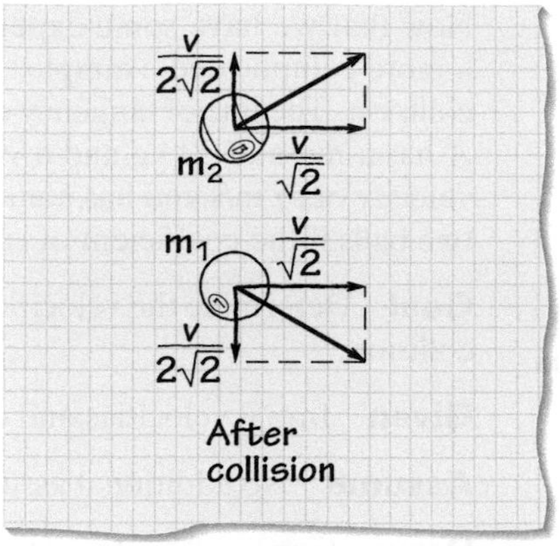

Figure 3.80 Post-collision velocities

$$(3.114) \rightarrow (3.115) \Rightarrow \qquad \frac{v}{\sqrt{2}} = v'_{2n} - (-v'_{2n}) \qquad \Rightarrow \qquad v'_{2n} = \frac{v}{2\sqrt{2}}$$

Finally, using this value for v'_{2n} in (3.114) gives us the value for our remaining velocity component:

$$v'_{1n} = -\frac{v}{2\sqrt{2}}$$

Check **Figure 3.80** shows the final velocity components. Because the problem was symmetrical at the start, we would expect it to be symmetrical at the end as well. And, happily, it is. The magnitudes of m_1 and m_2's velocity vectors are the same in each direction. The minus sign for v'_{1n} indicates that the balls bounce apart from each other after the collision, with m_2 moving up and m_1 moving down, as expected from intuition.

EXAMPLE 3.26 ANOTHER OBLIQUE COLLISION (Theory on page 195)

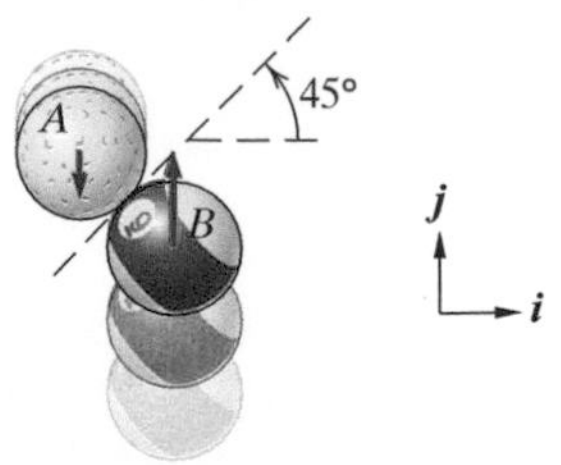

Figure 3.81 Two balls experiencing an off-axis collision

Now that we have some experience with oblique impacts, we'll look at a more complicated example. **Figure 3.81** shows two balls in an off-axis collision. The line of tangency makes a 45° angle with the horizontal. Ball A has a mass of 20 kg and a velocity of $-2\boldsymbol{j}$ m/s. Ball B is moving at a velocity of $5\boldsymbol{j}$ m/s and has a mass of 10 kg. Find the final velocities of the two balls if the coefficient of restitution is 0.8.

Goal Determine the velocities of the two interacting balls following the collision.

Given Initial velocities and coefficient of restitution.

Assume No additional assumptions are needed.

Draw **Figure 3.82*a*** shows the tangential and normal directions for this problem. Because this orientation is awkward to work from, I will rotate the entire system 45° clockwise to obtain **Figure 3.82*b***. (Everything will have to be rotated back before we are through. This rotation is just a way to make it simpler to deal with the problem.) Both the velocity vectors and their components along the t and n directions are shown.

Formulate Equations All we need to do is apply conservation of momentum for the system in the n direction, retain the individual velocities in the t direction, and apply our impact formula in the n direction. The t velocities can be read off **Figure 3.82*b***.

Our impact equation in the n direction is, from (3.113),

$$0.8 = \frac{v'_{A_n} - v'_{B_n}}{\left[\frac{5}{\sqrt{2}} - (-\sqrt{2})\right] \text{ m/s}} \tag{3.116}$$

and our expression for system momentum conservation, from (3.112), is

$$\left[10\left(\frac{5}{\sqrt{2}}\right) + 20\left(-\sqrt{2}\right)\right] \text{ m/s} = \left(10v'_{B_n} + 20v'_{A_n}\right) \text{ m/s} \tag{3.117}$$

Solve Equation (3.116) gives us

$$v'_{A_n} = v'_{B_n} + 3.96 \text{ m/s} \tag{3.118}$$

Substituting this into (3.117) yields

$$v'_{B_n} = -2.40 \text{ m/s}$$

and using (3.118) then gives

$$v'_{A_n} = 1.56 \text{ m/s}$$

All these velocity components are illustrated in **Figure 3.83*a***, along with the resultant velocities.

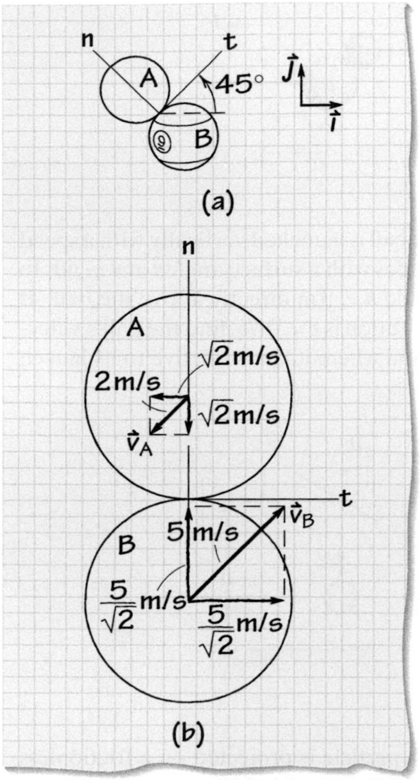

Figure 3.82 Coordinate sets—oblique impact

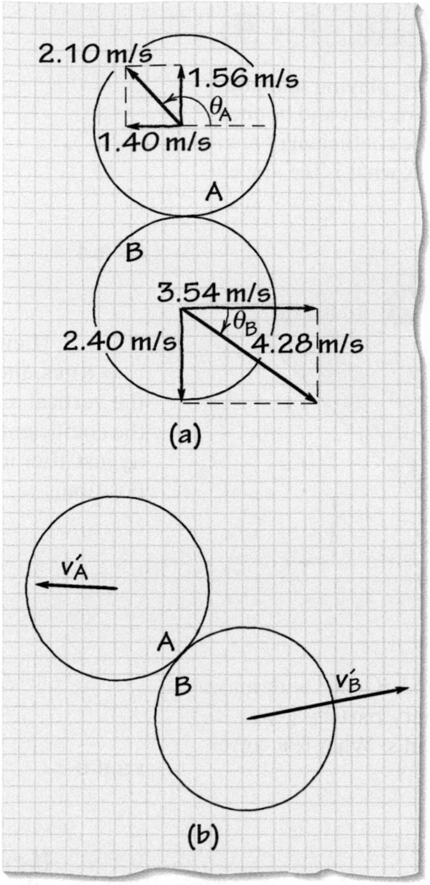

Figure 3.83 Resultant velocity components

Keep in mind, however, that we're not done because **Figure 3.83*a*** shows the rotated system, not the actual one. If we rotate the entire system counterclockwise by 45°, then θ_A, which was 132°, becomes 177°. Similarly, θ_B changes from −34.2° to 10.8°. The final results, shown in **Figure 3.83*b***, are

$$\boldsymbol{v}'_A = (-2.10\boldsymbol{i} + 0.10\boldsymbol{j})\text{ m/s}, \qquad |v'_A| = 2.10\text{ m/s}$$

$$\boldsymbol{v}'_B = (4.20\boldsymbol{i} + 0.80\boldsymbol{j})\text{ m/s}, \qquad |v'_B| = 4.28\text{ m/s}$$

Check One easy check is to use our calculated post-collision velocities and construct the system momentum. Doing so should exactly match the system momentum before collision. The initial momentum is equal to $(20\text{ kg})(-2\boldsymbol{i}\text{ m/s}) + (10\text{ kg})(5\boldsymbol{i}\text{ m/s}) = 10\boldsymbol{i}\text{ kg·m/s}$, and the final momentum is $(20\text{ kg})[(-2.10\boldsymbol{i} + 0.10\boldsymbol{j})\text{ m/s}] + (10\text{ kg})[(4.20\boldsymbol{i} + 0.80\boldsymbol{j})\text{ m/s}] = 10\boldsymbol{i}\text{ kg·m/s}$. Same result.

EXERCISES 3.8

3.8.1. **[Level 1]** Imagine that you're playing miniature golf. Your golf ball is aimed in the $\boldsymbol{i}$ direction and bounces off the inclined side wall as shown. If the coefficient of restitution between the ball and wall is 0.86, what is the angle θ at which the ball rebounds?

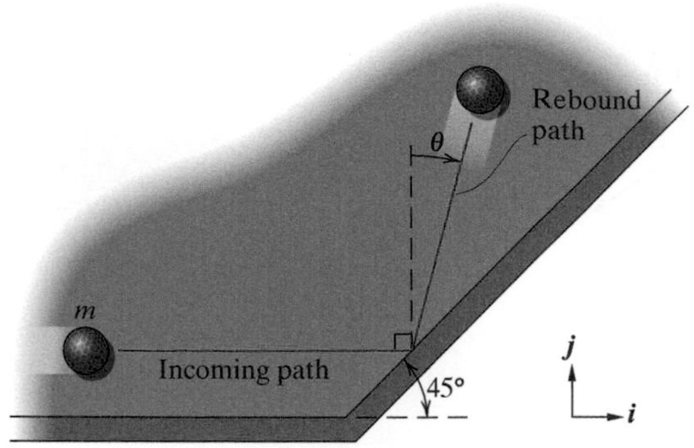

E3.8.1

3.8.2. **[Level 1]** Two particles A and B, each with mass m, travel diagonally toward a stationary particle C. $\boldsymbol{v}_A = (10\boldsymbol{i} - 10\boldsymbol{j})$ m/s and $\boldsymbol{v}_B = (10\boldsymbol{i} + 10\boldsymbol{j})$ m/s. What is $\boldsymbol{v}_C$ after the collision? $m_C = 2m$.

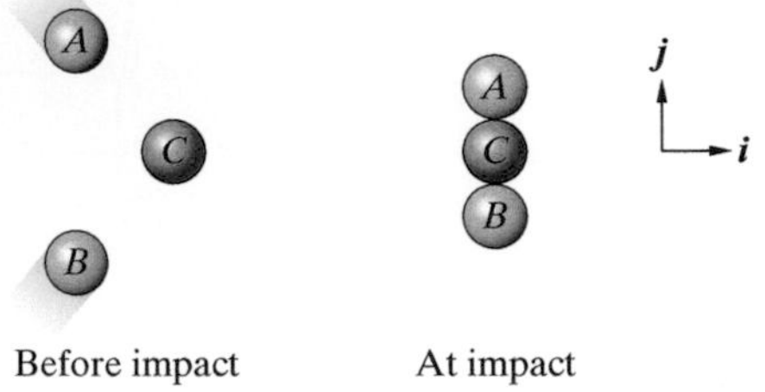

E3.8.2

3.8.3. **[Level 1]** Two particles A and B strike each other as shown. $m_A = 5$ kg and $m_B = 5$ kg. Just before the collision, $\boldsymbol{v}_A = -10\boldsymbol{j}$ m/s and $\boldsymbol{v}_B = (3\boldsymbol{i} + 3\boldsymbol{j})$ m/s. $e = 0.8$. What are their velocities after the collision?

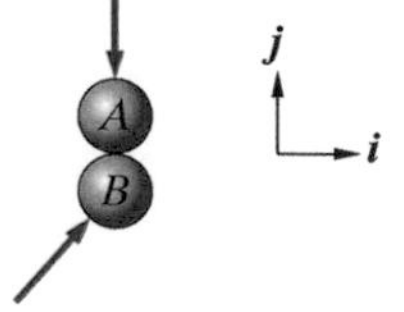

E3.8.3

3.8.4. **[Level 1]** Two particles A and B strike each other as shown. $m_A = 5$ kg and $m_B = 8$ kg. Just before the collision, $\boldsymbol{v}_A = 10\boldsymbol{i}$ m/s and $\boldsymbol{v}_B = -7\boldsymbol{i}$ m/s. $e = 0.4$. What are their velocities after the collision?

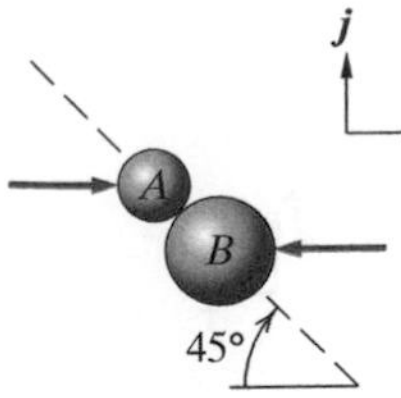

E3.8.4

3.8.5. **[Level 2]** A ball is launched vertically with a speed $v_0 = 8$ m/s, and it collides with a surface angled at $\theta = 60°$ to the vertical and located $h = 3$ m directly above the ball. If the coefficient of restitution is $e = 0.75$, find the ball's velocity just after impact and where it lands relative to its launch position.

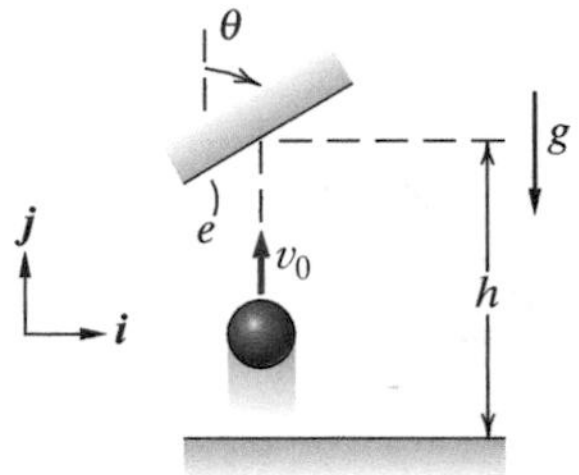

E3.8.5

3.8.6. **[Level 2]** A particle is projected on a frictionless surface in the $\boldsymbol{i}$ direction at a speed of 10 m/s. It bounces off a moving block that's moving at a constant speed of $\dot{x} = 5$ m/s. The coefficient of restitution between the particle and block is 0.8. Solve for θ. Neglect gravity.

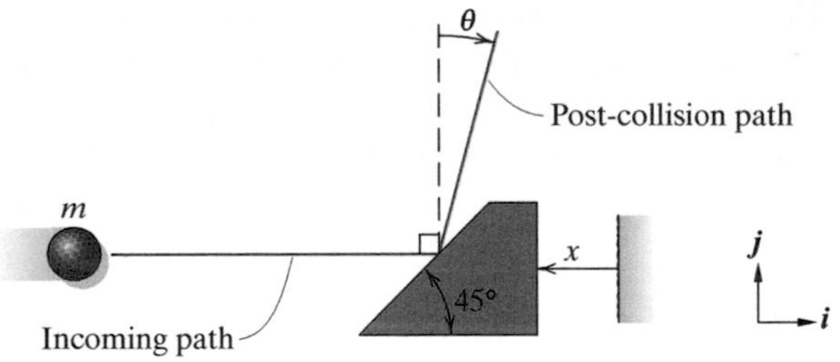

E3.8.6

3.8.7. **[Level 2]** Part of a pool table is shown. The first part of your task is to strike the cue ball so that it hits the other ball into the side pocket (it travels purely in the $\boldsymbol{j}$ direction). Your second task is to determine at what angle ϕ to strike the cue ball so that after the collision its velocity points in the direction $\cos(10°)\boldsymbol{i} + \sin(10°)\boldsymbol{j}$. $e = 0.93$.

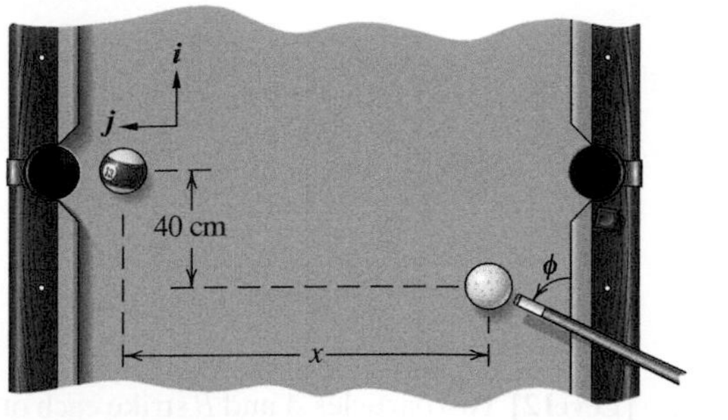

E3.8.7

3.8.8. **[Level 2]** A ball is dropped onto an inclined surface. The speed at impact is v_1. If you wish the ball to rebound so that its velocity immediately following impact is in the negative $\boldsymbol{i}$ direction, what must θ be? e for this exercise is 0.8.

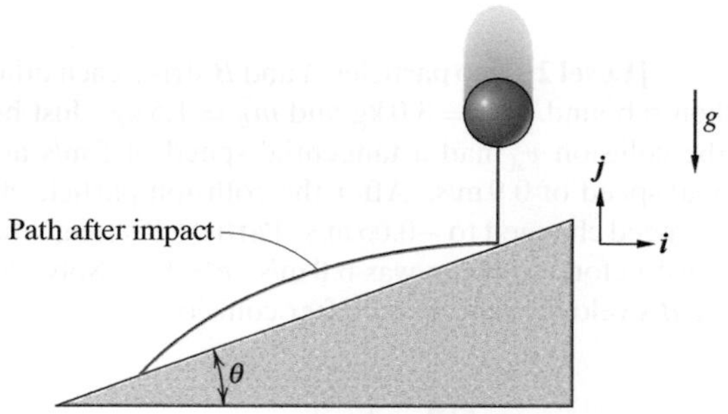

E3.8.8

3.8.9. **[Level 2]** Consider a billiard ball that strikes two other billiard balls as shown. The coefficient of restitution is 1.0. What will the final velocities and trajectories be for the balls? (Ignore rotations of the balls.) All the masses are equal to m and the impact speed is v_0.

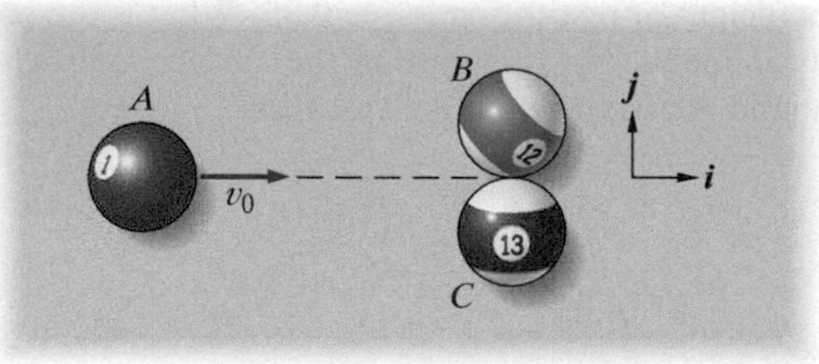

E3.8.9

3.8.10. **[Level 2]** A particular pool shot is illustrated. You want to strike ball A with the cue stick, sending it into ball B. Ball B needs to be propelled in the $\boldsymbol{j}$ direction and ball A has to move in the $\boldsymbol{i}$ direction after the collision. Both balls have a diameter of 0.08 m. What is the acceptable range of θ values for this problem? Assume that ball A can be positioned anywhere you would like along the line L and to the left of O. Specify where on B the contact has to occur. Is there a limitation on the coefficient of restitution?

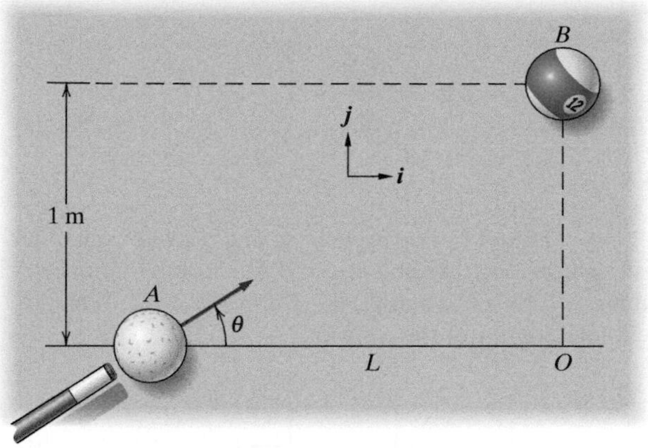

E3.8.10

3.8.11. **[Level 2]** Assume that you launch a small ball at an angle of 45° with an initial speed of 5 m/s. You want it to stop bouncing after traveling 2.8316 m from the starting position. What must the coefficient of restitution between the ball and the floor be for this to happen?

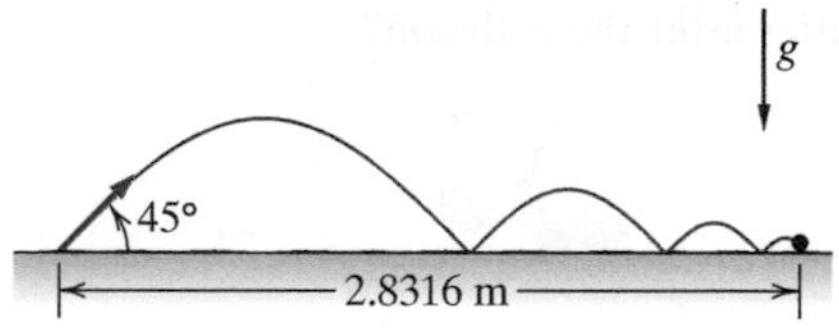

E3.8.11

3.8.12. **[Level 2]** In **Exercise 3.8.11** the task was to find e such that all bouncing of a ball ceased after the ball traveled horizontally for 2.8316 m. In this problem, look at the same situation—a bouncing ball having an initial velocity of $\left(\frac{5}{\sqrt{2}}\boldsymbol{i} + \frac{5}{\sqrt{2}}\boldsymbol{j}\right)$ m/s—and determine:

a. How far along the x axis does the bouncing stop if $e = 0.5$?

b. At what time $t = t_f$ does the bouncing stop?

c. What is the ball's behavior for $t > t_f$?

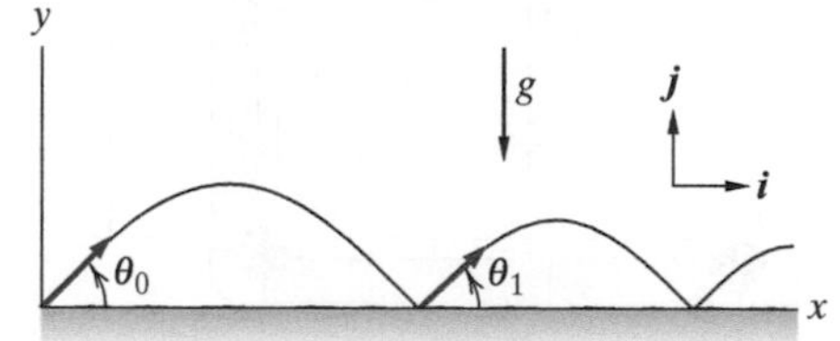

E3.8.12

3.8.13. **[Level 2]** Two particles A and B strike each other as shown. $m_A = 2$ kg and $m_B = 4$ kg. Just before the collision $\boldsymbol{v}_A = (2\boldsymbol{i} + 2\boldsymbol{j})$ m/s and $\boldsymbol{v}_B = (2\boldsymbol{i} - 2\boldsymbol{j})$ m/s. $e = 0.5$. What are their velocities after the collision?

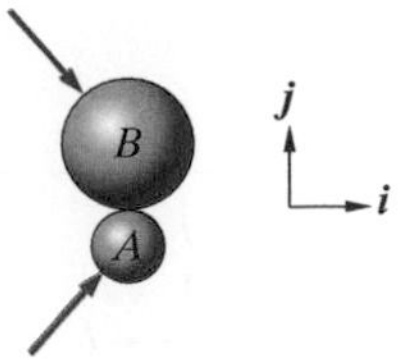

E3.8.13

3.8.14. **[Level 2]** Two particles A and B strike each other as shown. $m_A = 10$ kg and $m_B = 12$ kg. Just before the collision $\boldsymbol{v}_A = -20\boldsymbol{j}$ m/s and $\boldsymbol{v}_B = -13\boldsymbol{i}$ m/s. $e = 0.2$. What are their velocities after the collision?

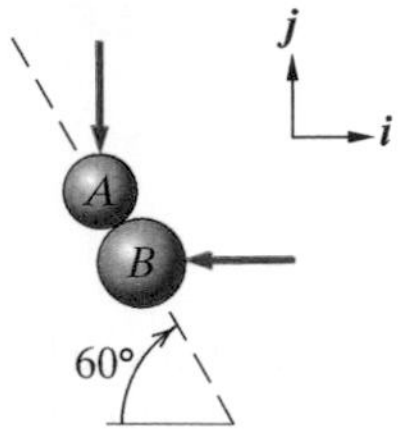

E3.8.14

3.8.15. **[Level 2]** Two particles A and B strike each other as shown. $m_A = 2$ kg and $m_B = 1.2$ kg. Just before the collision $\boldsymbol{v}_A = (5\boldsymbol{i} + 5\boldsymbol{j})$ m/s and $\boldsymbol{v}_B = 0$ m/s. $e = 1.0$. What are their velocities after the collision?

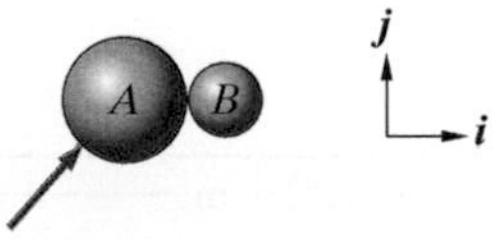

E3.8.15

3.8.16. **[Level 2]** A pool player wants to bank ball B off of point A into the side pocket O. The ball has a diameter $d = 5.72$ cm and a coefficient of restitution $e = 0.8$ when colliding with the side of the table. Find the location of point A on the table by calculating the length a. Treat the side of the table as immovable.

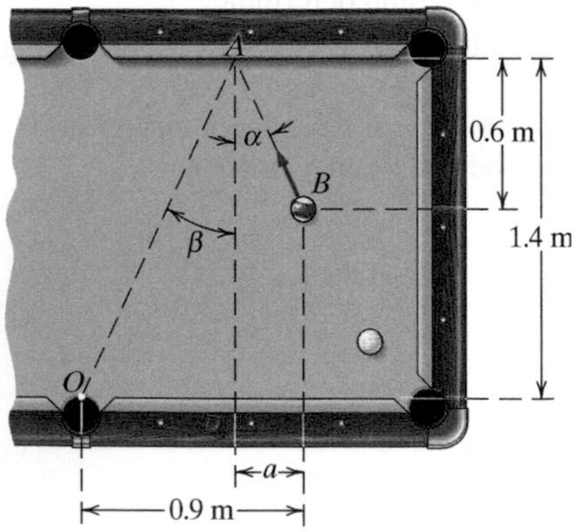

E3.8.16

3.8.17. **[Level 2]** Two particles A and B strike each other as shown. $m_A = 116$ kg and $m_B = 58$ kg. Just before the collision $\boldsymbol{v}_A = (-0.9\boldsymbol{i} + 0.9\boldsymbol{j})$ m/s and $\boldsymbol{v}_B = -(9\boldsymbol{i} + 9\boldsymbol{j})$ m/s. $e = 0.5$. What are their velocities after the collision?

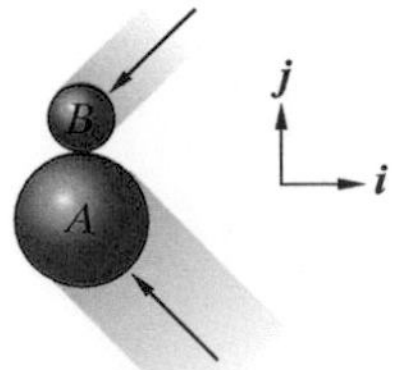

E3.8.17

3.8.18. **[Level 2]** Two particles A and B strike each other and then rebound. $m_A = 2$ kg and $m_B = 1$ kg. After the collision $\boldsymbol{v}'_A = (-3.0\boldsymbol{i} - 0.04\boldsymbol{j})$ m/s and $\boldsymbol{v}'_B = (-3.0\boldsymbol{i} + 0.68\boldsymbol{j})$ m/s. $e = 0.6$. What were their velocities before the collision?

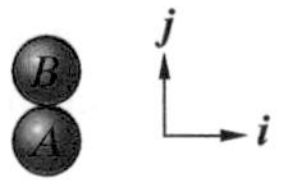

E3.8.18

3.8.19. **[Level 2]** Two particles A and B strike each other and then rebound. $m_A = 3.0$ kg and $m_B = 1.5$ kg. Just before the collision $\boldsymbol{v}_A$ had a tangential speed of 5 m/s and a normal speed of 0.9 m/s. After the collision particle A's normal speed changed to -0.06 m/s. Particle B's tangential speed just before collision was 6.0 m/s. $e = 0.6$. Solve for particle B's velocity before and after collision.

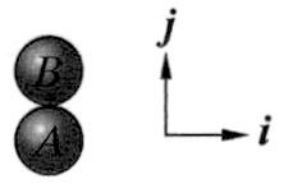

E3.8.19

3.8.20. **[Level 3]** As illustrated, a ball is projected with a speed $v_0 = 10$ m/s at an angle $\theta = 70°$ with respect to the horizontal, and it strikes a surface angled at $\beta = 60°$ from the horizontal right when it reaches the peak of its trajectory. The coefficient of restitution is $e = 0.85$. What is the ball's velocity immediately after hitting the surface, and where does it land relative to its launch location?

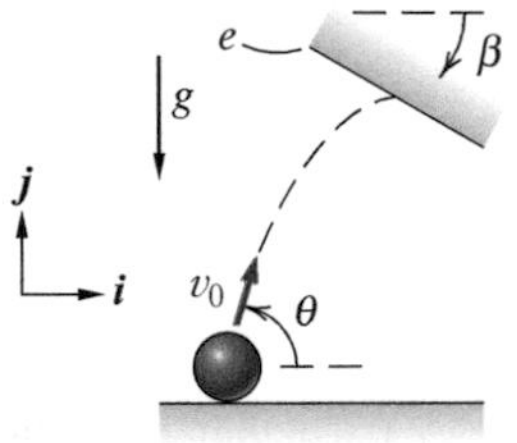

E3.8.20

3.8.21. **[Level 3]** Two round disks are free to move on a flat plane. The disk/surface interface is frictionless, and the coefficient of restitution between the disks is 0.8. What are the velocities of the two disks if before the collision disk A is stationary and disk B is traveling at $-5\boldsymbol{i}$ m/s? $m_A = 0.8$ kg and $m_B = 0.6$ kg.

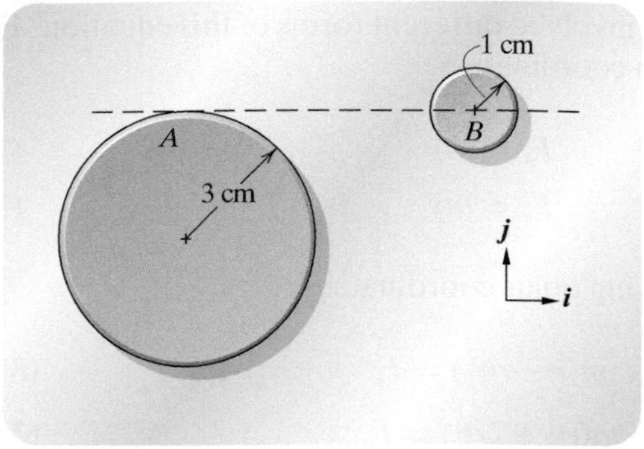

E3.8.21

3.8.22. **[Level 3]** Mark is playing a game of racquetball, and he hits the ball with a speed of $v_0 = 15$ m/s at an angle $\theta = 15°$ with respect to the court surface and from a height of $h = 0.3$ m. If the wall is $d = 4.6$ m from Mark and the coefficient of restitution is $e = 0.7$, how far from Mark does the ball hit the court surface after bouncing off the wall?

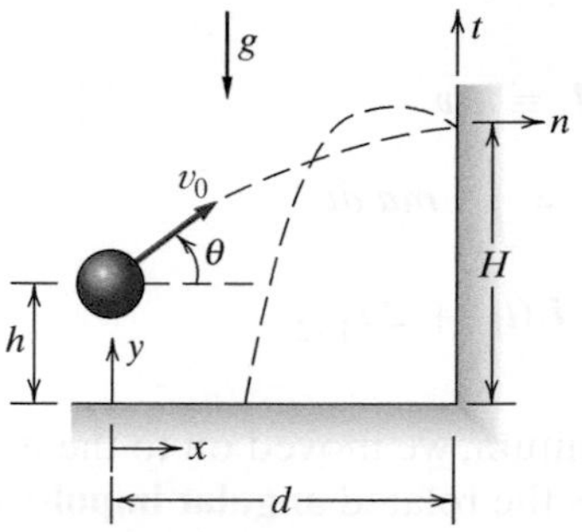

E3.8.22

3.8.23. **[Level 3]** Ball A ($m_A = 3$ kg) is released from rest at the top of a rough ramp angled at $\theta = 45°$ with respect to the ground. Just before reaching the end of the ramp, ball A hits ball B ($m_B = 2$ kg) such that the line of tangency makes an angle of 45° to the horizontal. Ball B is initially stationary on level ground and constrained to move horizontally in a frictionless chute of length $L = 3$ m. Take $e = 0$ in the chute so that ball B has only a horizontal component of velocity after impact with ball A. Ball A travels $s = 1$ m down the ramp, and $e = 0.8$. How long does it take for ball B to travel through the chute? The coefficients of static and dynamic friction between ball A and the ramp are $\mu_s = 0.8$ and $\mu_d = 0.25$, respectively.

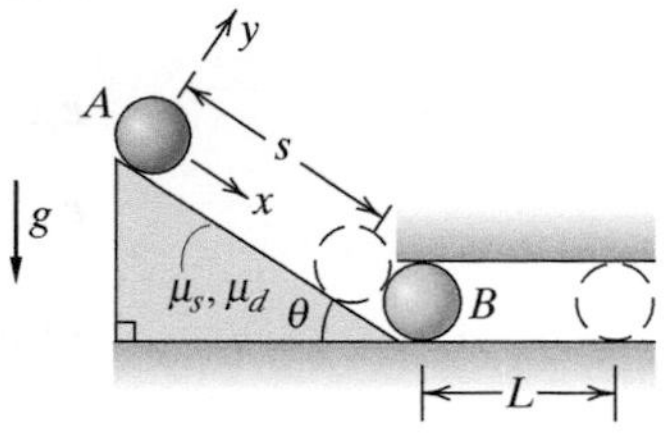

E3.8.23

3.8.24. **[Level 3]** In the Civil War there were numerous incidents in which the bullet fired from a Union soldier's rifle struck and fused with the bullet from a Confederate's rifle. Assume the impact scenario is as shown. The Union bullet (0.06 kg) has a speed of $v_U = 219$ m/s and its velocity is oriented at 2° to the horizontal. The Confederate bullet (0.07 kg) has a speed of $v_C = 213$ m/s and its velocity vector is oriented at 1° to the horizontal.

a. Where and when does the fused pair land?

b. How does this result differ from the case of oblique impact with zero force interaction in the vertical direction and $e = 0$?

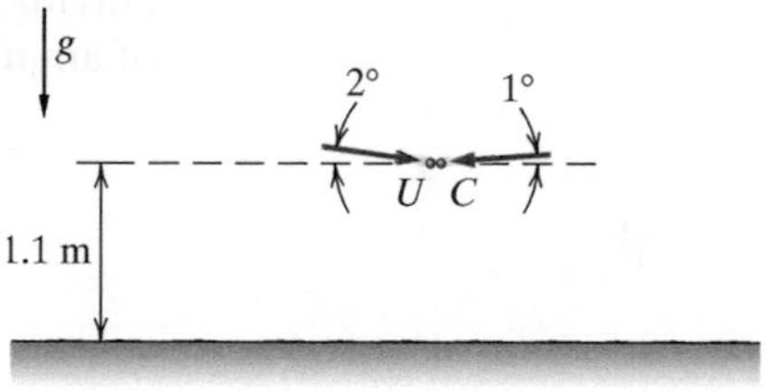

E3.8.24

3.9 JUST THE FACTS

This chapter dealt with **kinetics**: the study of forces and the dynamic response of systems to them. We applied $\boldsymbol{F} = m\boldsymbol{a}$ in a variety of coordinate systems and saw that the overall formulations were similar in character. By integrating in time, we created a momentum viewpoint with which we looked at both linear and angular motions. This allowed us to discover quite a bit about orbital mechanics. Finally, we delved into the topic of impacts—both normal and oblique.

The chapter opened with a statement of Newton's second law:

$$m\boldsymbol{a} = \sum_{i=1}^{n} \boldsymbol{F}_i \qquad (3.1)$$

and the rest of the chapter involved different forms of this equation. First up was a look at Cartesian coordinates:

$$\boldsymbol{i}: \quad F_x = m\ddot{x} \tag{3.5}$$

$$\boldsymbol{j}: \quad F_y = m\ddot{y} \tag{3.6}$$

and followed this by applying polar coordinates:

$$\boldsymbol{e}_r: \quad m(\ddot{r} - r\dot{\theta}^2) = F_r \tag{3.15}$$

$$\boldsymbol{e}_\theta: \quad m(r\ddot{\theta} + 2\dot{r}\dot{\theta}) = F_\theta \tag{3.16}$$

Our final set of coordinates were path coordinates:

$$\boldsymbol{e}_t: \quad F_t = m\dot{v} \tag{3.29}$$

$$\boldsymbol{e}_n: \quad F_n = m\frac{v^2}{r_C} \tag{3.30}$$

Next up was an investigation of **linear momentum** and **linear impulse**:

$$\boldsymbol{L} \equiv m\boldsymbol{v} \tag{3.36}$$

$$\mathcal{LI}_{1-2} = \int_{t_1}^{t_2} m\boldsymbol{a}\, dt \tag{3.38}$$

$$\boldsymbol{L}(t_2) = \boldsymbol{L}(t_1) + \mathcal{LI}_{1-2} \tag{3.39}$$

Having disposed of linear momentum, we moved on to the new concept of **angular momentum** (as well as the related **angular impulse**):

$$\boldsymbol{H}_O \equiv \boldsymbol{r} \times m\boldsymbol{v} \tag{3.45}$$

$$H_O = mvd \tag{3.47}$$

$$\dot{\boldsymbol{H}}_O = \boldsymbol{r} \times \boldsymbol{F} = \boldsymbol{M}_O \tag{3.49}$$

$$\mathcal{AI}_{O_{1-2}} \equiv \int_{t_1}^{t_2} \boldsymbol{M}_O\, dt \tag{3.51}$$

$$\boldsymbol{H}_O(t_2) = \boldsymbol{H}_O(t_1) + \mathcal{AI}_{O_{1-2}} \tag{3.52}$$

Having introduced momentum, we then proceeded to apply it to orbital problems, deriving the relevant equations for describing the motions of masses in orbit about larger bodies.

$$r^2\dot{\theta} = h \tag{3.58}$$

$$\frac{dA}{dt} = \frac{1}{2}(r^2\dot{\theta}) \tag{3.59}$$

$$p(\theta) = \frac{1}{r(\theta)} = \frac{Gm_S}{h^2} + A\cos\theta \tag{3.70}$$

$$e = \frac{Ah^2}{Gm_S} \tag{3.72}$$

$$x = \frac{c}{1+e} \tag{3.78}$$

$$A = \frac{1}{2}\left(\frac{1}{r_p} - \frac{1}{r_a}\right) \tag{3.79}$$

$$A = \frac{e}{r_p(1+e)} \tag{3.83}$$

$$\frac{1}{r_a} = A\left(\frac{1-e}{e}\right) \tag{3.85}$$

$$r_m = \sqrt{r_a r_p} \tag{3.86}$$

$$T_p = \frac{\pi(r_a + r_p)\sqrt{r_a r_p}}{h} \tag{3.89}$$

$$\frac{r_p}{r_a} = \frac{1-e}{1+e} \tag{3.90}$$

Next on the agenda was examining how two particles can interact during a collision. During this discussion we introduced e, the coefficient of restitution.

$$m_1 v_1 + m_2 v_2 = m_1 v_1' + m_2 v_2' \tag{3.100}$$

$$e = \frac{v_2' - v_1'}{v_1 - v_2} \tag{3.105}$$

Extending our analysis to **oblique impacts** gave us

$$v_{1_t} = v'_{1_t} \tag{3.110}$$

$$v_{2_t} = v'_{2_t} \tag{3.111}$$

$$m_1 v_{1_n} + m_2 v_{2_n} = m_1 v'_{1_n} + m_2 v'_{2_n} \tag{3.112}$$

$$e = \frac{v'_{2_n} - v'_{1_n}}{v_{1_n} - v_{2_n}} \tag{3.113}$$

SYSTEM ANALYSIS (SA) EXERCISES

SA3.1 Escape from Colditz

Colditz Castle (**Figure SA3.1.1*a***) in Germany was probably the most infamous of all German prisoner-of-war camps during World War II. The castle, with its dark granite walls, barred windows, and medieval towers, sat high on a cliff overlooking the small town of Colditz in Saxony, Germany. Over 300 escape attempts were made during the $5\frac{1}{2}$ year war. However, one escape plan stands apart, the Colditz Glider.

Flight Lieutenants Bill Goldfinch, Antony Rolt, and Jack Best conceived the idea of launching two prisoners in a glider from the castle roof across the Mulde River to freedom. The glider was built, along with a launching mechanism that included dropping a bathtub full of concrete 18 m to catapult the glider off the roof; see **Figure SA3.1.1*b***. However, the Allied prisoners of war were liberated in April 1945 before the glider could be launched.

a. A model of the simplified launching system is shown in **Figure SA3.1.1*b***. Assume that the glider with two prisoners weighs 230 kg and that the concrete-filled tub weighs 460 kg. In addition, assume that the glider rests on a freely rolling cart and that the pulley is frictionless and weightless. Using a force balance, calculate the initial acceleration of the glider when the concrete-filled tub is released.

b. Why is the glider's acceleration less than the acceleration of gravity (9.81 m/s^2)?

c. Is the glider's acceleration constant while the concrete-filled tub falls 18 m?

d. Assuming the glider has 18 m of runway, how long does it take the glider to reach the edge after the concrete-filled tub falls 18 m?

e. How fast is the glider going at the end of the runway in **d**?

Reuters/New Media Inc./Corbis Images

(*a*)

(*b*)

Figure SA3.1.1 (*a*) Colditz Castle; (*b*) escape glider

SA3.2 Kinetics of Variable Geometry Pulleys

Two different pulley arrangements are shown in **Figure SA3.2.1**. In case (*a*) the pulley ropes are vertical, whereas in (*b*) the rope segments are angled and the angle changes as block A rises. Assume that the pulleys are massless and neglect their dimensions. Assume further that a force of 60 N acts on the free end of the pulley rope (B), raising block A, and that block A has a mass of 10 kg. Plot the displacement of block A as a function of time for $0 < t < 0.7$ s. $a = 1$ m and b is initially equal to 5 m.

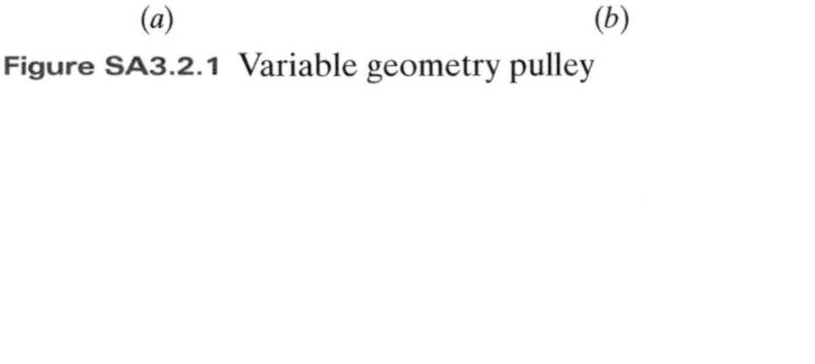

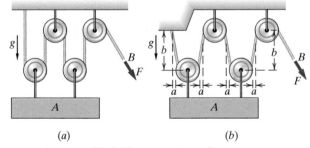

Figure SA3.2.1 Variable geometry pulley

SA3.3 The Somatogravic Illusion

The somatogravic illusion is one of the most common of all misperceptions experienced in the flight environment. If a pilot accelerates quickly enough, she or he will get a pitch-up sensation because, without other inputs, the vestibular system (part of the inner ear) is incapable of telling the difference between a pitch-up and a forward acceleration. If a pilot has poor visual references and does not check the plane's instruments, she or he may misinterpret the forward acceleration as a pitch-up. To counteract this perceived pitch-up, and maintain what she or he thought to be level flight, the pilot would push the stick forward, potentially flying the plane into the ground. This is what is known as an RBI, or "Really Bad Idea."

One of the first places that used this illusion for its entertainment value was DisneyWorld. The Disney engineers created a ride, called StarTours, in which the audience sat on a bench seat and watched a movie screen (**Figure SA3.3.1*a***). The weight of the person was perceived as a force N pointing straight up the spine. The seats were hydraulically controlled and could be made to pitch back, rise up, and so forth, just as a flight simulator would do. When the image on the screen showed a picture of stars zooming by, as if the audience were accelerating to warp speed, the seats rotated backwards, as shown in **Figure SA3.3.1*b***. This caused the normal force between person and seat, which had previously been pointing directly up the person's spine, to break into two parts (N_1 and N_2). The sum of these two forces was still equal to a vertical force, but the person sitting in the seat perceived the situation as a "sitting" force and a "seat pushing against me" force, just as would be experienced if the seat had been actually accelerating in the e_t direction, as shown in **Figure SA3.3.1*c***.

Because of the image on the screen, the brain interprets this "force at the back" sensation not as a pitching back but as an acceleration forward, making it seem as if the whole room is actually rocketing forward. And when the film shows an abrupt deceleration, the chair rotates forward, fooling the brain into thinking that the room is decelerating.

Your assignment is to create an in-flight profile that will simulate the somatogravic illusion for the F-16A Fighting Falcon. This plane weighs 9072 kg, has a maximum normal engine thrust of 66145 N, and can produce 106 kN of thrust if afterburners are used.

Assume that the flight begins at 103 m/s and model the passenger as a point mass.

a. How much thrust would be required to achieve a pitch-up sensation of 30°?

b. If the maximum airspeed of the aircraft is 309 m/s, how long can you maintain this sensation?

c. How much airspace (i.e., distance) would you need to accomplish this?

d. Assuming a starting speed of 103 m/s, create a plot of thrust versus perceived pitch sensation.

e. Think about and discuss what a person would sense as a result of a deceleration and what the implications might be.

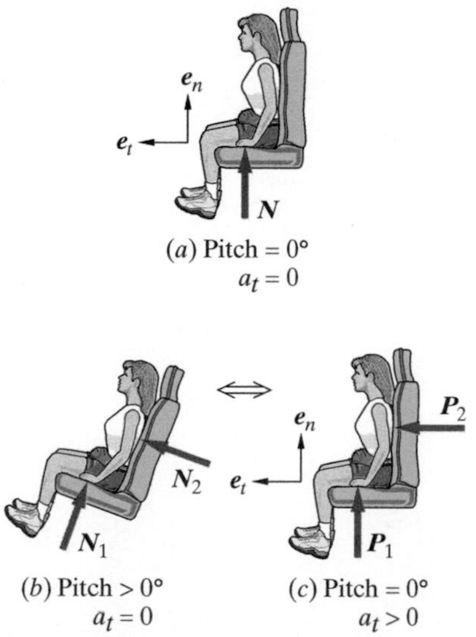

Figure SA3.3.1 Variation of normal force orientation with pitch

SA3.4 The Push-Pull Maneuver

Pulling positive g's means that the effective weight of a pilot exceeds his actual weight. One way for this to happen is for the plane to dive down and then pull up. The curved path causes a centripetal acceleration that acts to increase the normal force between pilot and seat beyond $1g$. One consequence of pulling positive g's is that the blood supply to the brain is compromised as the fluid flows toward the lower extremities. Although there are physiologic compensatory mechanisms to help combat these effects, they are not rapid enough to prevent g-induced loss of consciousness (GLOC).

With protective equipment and an effective anti-g straining maneuver, most pilots are able to withstand $9g$ turns fairly easily. Recently, however, some mishaps in the Canadian Air Force focused research on maneuvers that involved a push (less than $1g$ or even negative g acting on the pilot), followed by a pull (greater than $1g$ acting on the pilot). During a push maneuver, baroreceptors in the brain sense the extra pressure in the brain. This triggers a vasodilator response, which opens up the blood vessels in the neck to try to help drain some of the extra blood. If the pilot then pulls positive g's, blood rushes out of the head at an alarming rate. It has been found that GLOC after a push maneuver can occur at as little as $3g$'s. Physiologists measure the number of g's by looking at the normal force exerted on the pilot divided by the pilot's weight.

The test aircraft you will be analyzing is the F-16A Fighting Falcon, with a weight of 9072 kg. It has a maximum engine thrust of 66145 N, which increases to 106 kN when afterburners are used.

a. If the beginning airspeed is 103 m/s, what radius of curvature will be needed to achieve $0.5g$? $0g$? $-1g$? Note that the center of curvature is below the plane during this maneuver.

b. What aerodynamic force will the aircraft have to exert to achieve each of the conditions of **a**?

c. After the push maneuver, you will take 3 s to straighten out into level flight and accelerate before commencing your pull maneuver. You would like to perform a $4g$ climbing maneuver with a radius of 457 m. How much thrust will you need during the 3 s transition phase to achieve this (assuming constant acceleration)?

d. Graph the number of g's as a function of the radius of curvature and the speed of the plane (in knots). Let the radius vary from 457 to 1524 m, and let the speed vary from 51 to 309 m/s.

OBJECTIVES

Upon completion of this chapter you will be able to:

- Recognize and use the associated concepts of energy and work
- Deal with problems involving work, power, and efficiency
- Know when a problem is best approached from an energy standpoint as opposed to a force balance or momentum approach

CHAPTER 4

Energetics of Translating Bodies

JOHN PYLE/CSM/Landov

Having mastered $\boldsymbol{F} = m\boldsymbol{a}$ in a variety of forms, you probably feel pretty confident in approaching a broad range of dynamics problems. And you should, because it is not a trivial ability you have developed. There is, however, another approach to dynamic motion that, although derived from what we already know, looks at reality in quite a different way. For some problems, this alternative viewpoint is preferable to an application of Newton's laws. As the title of this chapter indicates, this approach deals with energy.

With Newton's laws, we can figure out an object's acceleration if we know the net force acting on the object and its mass. Then, to find the object's position at some later time, we have to integrate its equations of motion. By using an energy viewpoint, we will be able to figure what an object's speed and/or displacement is at some time in the future *without* needing to use numerical integration. This is clearly a nice capability, and, although it won't work on all problems, it works on enough of them to make it a worthwhile item for study.

(cont.)

(cont.)

Keep in mind that you will usually be using an energy approach when you're told something about a system *now* and will be asked to find out something about the system *later* (after it has moved to some other location). An example would be if I told you the position and speed of a mass attached to a vertical spring that's already stretched 3 cm. I might ask you to find the mass's speed after it had dropped another 2 cm under the influence of gravity. The tools we will develop will let you solve this problem in a straightforward way. If you *didn't* use an energy approach, you would have to find the system's equations of motion and integrate them with respect to time, watching all the while for the moment when the mass reached the desired position (5 cm). When the integration reached that point, you would stop and read the speed from the data output.

Even though integrating a system's equations of motion isn't difficult with the help of a computer, it's still a lot quicker simply to solve a single equation (which is what the energy approach lets us do). So, with this as motivation, let's see what energy can do for us.

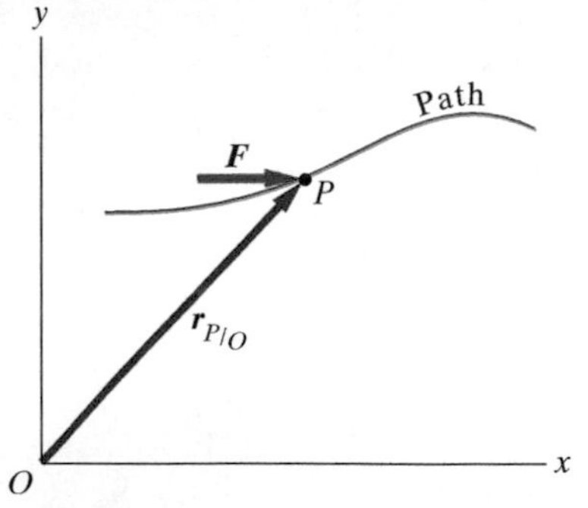

Figure 4.1 Force acting on a particle

4.1 KINETIC ENERGY

To start, consider a particle P in the x, y plane being acted on by a force $\boldsymbol{F}$, as shown in **Figure 4.1**. You probably remember from physics that the **work** expended in moving an object is equal to the force acting on the object times the displacement:

$$dW = \boldsymbol{F} \cdot d\boldsymbol{r} \tag{4.1}$$

In this case I have written the differential forms dW and $d\boldsymbol{r}$ because I want to look at small changes in work resulting from small changes in the position of particle P. We need to use the vector dot product because it's only the component of force along the direction of the particle's displacement that matters. The component normal to the displacement vector can't do any work because no motion takes place in that direction. **Figure 4.2** shows what I'm talking about. The particle moves a distance ds. The force $\boldsymbol{F}$ is made up of $\boldsymbol{F}_t$ tangent to the path (and therefore points along $\boldsymbol{r}$) and $\boldsymbol{F}_n$ normal to the path. You can see in the figure that $\boldsymbol{F}_n$ always points at right angles to the path and therefore can't do any work on m; there's no displacement associated with the direction in which it points. All of $\boldsymbol{F}_t$, however, points along the path; therefore the work done on m is given by

$$dW = F_t\, ds \tag{4.2}$$

Now let's push our particle along for a finite distance. Let s_1 and s_2 represent its positions at times t_1 and t_2, and let v_1 and v_2 represent its corresponding speeds. We can calculate the overall work expended over this distance by integrating (4.2):

$$W_{1-2} = \int_1^2 dW = \int_{s_1}^{s_2} F_t\, ds \tag{4.3}$$

This is just a repetition of what you encountered in physics class—namely, that W_{1-2}, the work done on a particle as the particle is moved from location 1 to location 2, is equal to the applied force times the displacement of the particle. Let's carry this analysis a bit further, shall we? The force in the direction of travel (F_t) has to equal $m\ddot{s}$ because Newton's second law still holds and $\ddot{s}$ is the acceleration along the path. Substituting $F_t = m\ddot{s}$

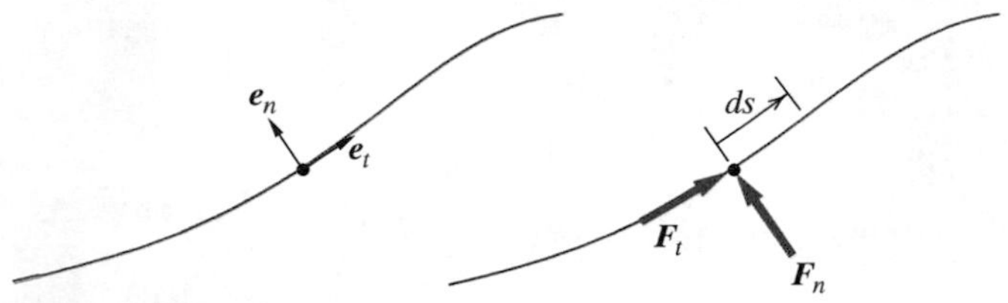

Figure 4.2 Forces acting over a differential motion

into (4.3) gives us

$$W_{1-2} = \int_{s_1}^{s_2} m\ddot{s}\, ds = m \int_{s_1}^{s_2} \ddot{s}\, ds \tag{4.4}$$

for a particle of constant mass m.

Remember the relationship among acceleration, speed, and displacement $a\, dx = v\, dv$ from Section 2.1? Using it in (4.4) gives us

$$W_{1-2} = m \int_{s_1}^{s_2} \ddot{s}\, ds = m \int_{v_1}^{v_2} \dot{s}\, d\dot{s} = m \left(\frac{v_2^2}{2} - \frac{v_1^2}{2} \right) \tag{4.5}$$

$$W_{1-2} = \frac{1}{2} m v_2^2 - \frac{1}{2} m v_1^2 \tag{4.6}$$

The term $\frac{1}{2}mv^2$ represents the **kinetic energy** of the particle. Let's use $\mathcal{KE}$ to denote the kinetic energy and rewrite (4.6) as

$$\mathcal{KE}\big|_2 = \mathcal{KE}\big|_1 + W_{1-2} \tag{4.7}$$

This is the key relationship we need for all our energy analyses. It shows us that the kinetic energy of a particle is altered when work is done on the particle. Because doing work involves exerting a force on the particle, we can also interpret this result as telling us that the kinetic energy of a particle is altered whenever a force is exerted on it. We already know from (3.35) that a particle's linear momentum changes when a force acts on the particle:

$$m\boldsymbol{v}_2 = m\boldsymbol{v}_1 + \int_{t_1}^{t_2} \boldsymbol{F}\, dt$$

Both of these relationships show what happens when we exert a force on a particle. The kinetic energy equation tells us that an application of work (tangential force component integrated over displacement) alters the particle's energy—a *scalar* quantity. The momentum equation tells us that an application of linear impulse (force vector integrated over time) alters the particle's linear momentum—a *vector* quantity. It is important to keep this distinction in mind: *energy* always involves *scalar* quantities, whereas *linear momentum* is always *vector*-based. Don't be fooled into trying to decompose an energy into different directions or into ignoring the independent vector directions that linear momentum possesses.

One last observation is in order. Because the kinetic energy relationship involves the square of the particle's velocity, there's no way to know in what direction the particle is moving. Only the magnitude of the velocity can be determined. As a concrete example, consider a particle moving along the x axis. Its kinetic energy is given by $\frac{1}{2}mv^2$, and without additional information it is impossible to know whether its velocity is $v\boldsymbol{i}$ or $-v\boldsymbol{i}$.

>>> **Check out Example 4.1 (page 212), Example 4.2 (page 213), and Example 4.3 (page 214) for applications of this material.**

EXAMPLE 4.1 SPEED OF AN ARROW (Theory on page 211)

Picture an archer releasing an arrow. His bowstring produces a force against a 5.67×10^{-2} kg arrow given by $f = ae^{-bx}$ where $a = 179$ N and $b = 10.5\,\text{m}^{-1}$. $x = 0$ when the arrow is pulled fully back and increases as the arrow moves forward. How fast is the arrow moving after traveling from $x = 0$ to $x = 0.49$ m?

Goal Find the speed of an arrow after it has moved 0.49 m.

Given Arrow's weight and force profile

Assume We'll assume that the arrow's speed is purely determined by the bowstring's force and isn't affected by air resistance or friction from the bow.

Draw The bowstring applies a force in the x direction, as shown in **Figure 4.3**.

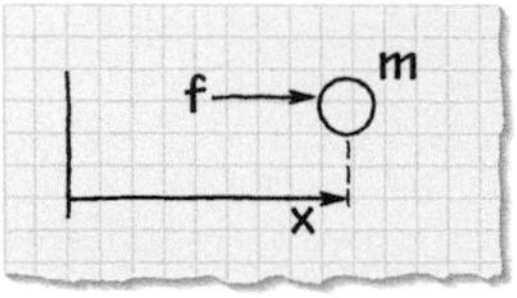

Figure 4.3 Force acting on arrow

Formulate Equations We'll need to employ our kinetic energy/work expression: $\mathcal{KE}|_2 = \mathcal{KE}|_1 + W_{1-2}$.

Solve The mass m of the arrow is 5.67×10^{-2} kg and our initial kinetic energy is 0. The work done by the bowstring is found from

$$W_{1-2} = \left[\int_0^{0.49} 179e^{-10.5x}dx\right] \text{N}\cdot\text{m} = \left[-\frac{179}{10.5}e^{-10.5x}\Big|_0^{0.49}\right] \text{N}\cdot\text{m} = 16.9\,\text{N}\cdot\text{m}$$

Our energy expression thus becomes

$$\frac{1}{2}m\dot{x}^2 = W_{1-2} \quad \Rightarrow \quad \frac{1}{2}(5.67 \times 10^{-2}\ \text{kg})\dot{x}^2 = 16.9\ \text{N}\cdot\text{m}$$

$$\dot{x} = 24\ \text{m/s}$$

Check An easy check is to use MATLAB to integrate $m\ddot{x} = 179e^{(-10.5x)}$ with a guess as to the appropriate `tspan`. Doing so, after some iteration, gives us the result that at $t = 0.0253$ s, the arrow has moved 0.49 m and has a speed of 24 m/s.

EXAMPLE 4.2 CHANGE IN SPEED DUE TO AN APPLIED FORCE (Theory on page 211)

Consider a car that enters a turn at 48 km/h (position A) and has a constant force of 4.89 kN acting in its direction of travel (**Figure 4.4**). We will neglect any drag forces for this example. Let's determine how fast it's going as it exits the turn 34 m later (at B). Assume that the car plus driver weighs 1460 kg.

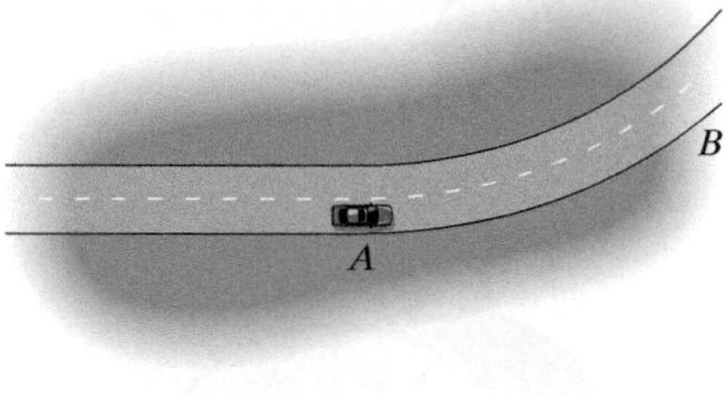

Figure 4.4 Car negotiating a turn

Goal Find the car's change in speed as it traverses a turn under the influence of a constant force.

Given Car's mass, applied force, and distance traveled.

Assume No additional assumptions are needed.

Draw The simplified system is shown in **Figure 4.5**. Our car is approximated as a lumped mass, the applied force always acts along the direction of travel, and s represents the distance traveled along the car's path.

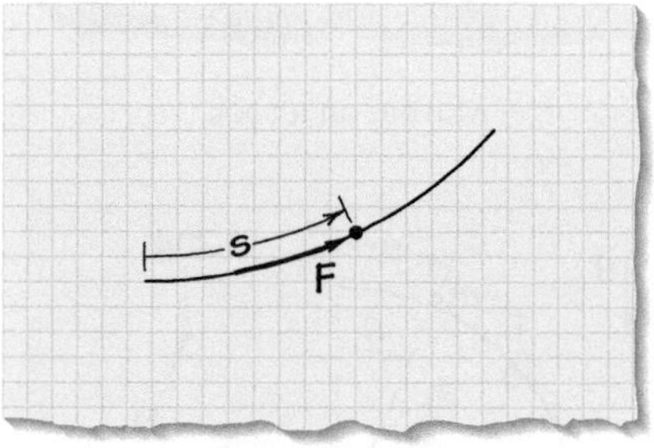

Figure 4.5 Force driving mass along path

Formulate Equations Equations (4.3) and (4.7) are all we need. A quick rearrangement yields

$$v_2 = \left[v_1^2 + \frac{2}{m} \int_{s_1}^{s_2} F\, ds \right]^{\frac{1}{2}} \tag{4.8}$$

Solve All that's needed is to express 48 km/h in m/s, determine the car's mass, and plug the appropriate values into (4.8):

$$v_2 = \left[(13.3\,\text{m/s})^2 + \frac{2}{1.46 \times 10^3\,\text{kg}} \int_0^{34\,\text{m}} 4.89 \times 10^3\,\frac{\text{kg}\cdot\text{m}}{\text{s}^2}\, ds \right]^{\frac{1}{2}}$$

$$v_2 = 20\,\text{m/s} = 72\,\text{km/h}$$

Check Is the result correct? In this case we can get an easy estimate. First, divide the force acting on the car by its mass to determine the acceleration:

$$a = \frac{F}{m} = \frac{4.89 \times 10^3\,\text{N}}{1.46 \times 10^3\,\text{kg}} = 3.35\,\text{m/s}^2$$

Starting at 13.3 m/s, and applying a constant acceleration of 3.35 m/s^2, the car will reach 20 m/s in t^* s:

$$20\,\text{m/s} = 13.3\,\text{m/s} + (3.35\,\text{m/s}^2)t^* \quad \Rightarrow \quad t^* = 2\,\text{s}$$

Now check how far the car would travel in 2 s at a constant acceleration of 3.35 m/s^2:

$$\Delta s = (13.3\,\text{m/s})(2\,\text{s}) + (3.35\,\text{m/s}^2)\frac{(2\,\text{s})^2}{2} = 33.3\,\text{m}$$

That's exactly how far it traveled through the corner, leading us to believe that we got it right.

EXAMPLE 4.3 CHANGE IN SPEED DUE TO SLIPPING (Theory on page 211)

Figure 4.6 Marble on a rock

Mary's been walking all day and stops to rest against a polished granite sculpture, the surface of which is in the shape of a semicircle of radius $r = 0.8$ m (**Figure 4.6**). As she rests, a 0.03 kg marble she's been holding in her hand slips free. The angle of release is $\theta = \frac{\pi}{24}$ rad. Consider it to have zero velocity at release and to slide without friction on the sculpture. (We're ignoring its dimensions and treating it as a particle.) How fast will it be going when it reaches $\theta = \frac{\pi}{12}$ rad?

Goal Find the marble's speed after it has moved from $\frac{\pi}{24}$ rad to $\frac{\pi}{12}$ rad.

Given Marble's mass and sculpture's surface shape.

Assume Assume that the marble slides without friction.

Draw The simplified system is shown in **Figure 4.7**, where s represents the distance the marble moves along the sculpture's surface.

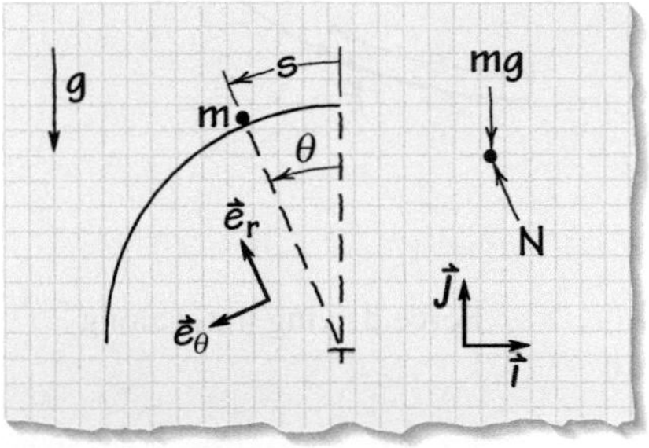

Figure 4.7 Mass moving on a semicircle

Formulate Equations The force acting on the marble is

$$\boldsymbol{F} = -mg\boldsymbol{j} + N\boldsymbol{e}_r = (-mg\cos\theta + N)\boldsymbol{e}_r + mg\sin\theta\,\boldsymbol{e}_\theta$$

Only the component of body force $mg\sin\theta$ that points along the $\boldsymbol{e}_\theta$ direction acts to accelerate the marble. We again need to invoke (4.3) and (4.7), but this time we'll need to think a bit more about how to handle the integral term. Our final speed can be found from

$$v_2 = \left[v_1^2 + \frac{2}{m}\int_{s_1}^{s_2} F\,ds\right]^{\frac{1}{2}} \tag{4.9}$$

Solve The difficulty is that the force in the $\boldsymbol{e}_\theta$ direction is known as a function of θ ($mg\sin\theta$), but the integral is expressed in terms of the displacement parameter s. What we can do is recall that the distance around a circle is equal to its radius multiplied by the subtended angle (measured in radians). Thus the total perimeter of a circle is equal to $2\pi r$. This lets us express the marble's position along the surface as $s = r\theta$, measuring from the top of the sculpture. This lets us re-express the differential ds as $ds = r\,d\theta$, and thus (4.9) can be rewritten as

$$v_2 = \left[0 + \frac{2}{m}\int_{\frac{\pi}{24}}^{\frac{\pi}{12}} mg\sin\theta\, r\,d\theta\right]^{\frac{1}{2}} = \boxed{0.633\text{ m/s}}$$

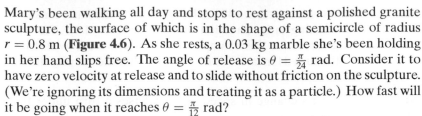

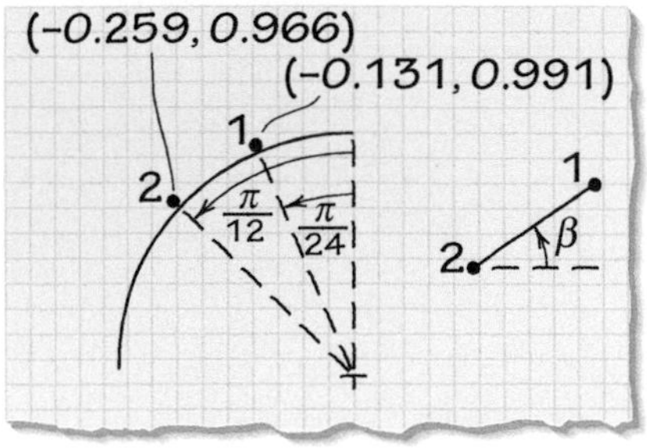

Figure 4.8 Marble on a rock

Check To check, we'll approximate the mass as moving from its initial position 1 to its final position 2 along a constant slope. Using x, y coordinates (units in meters), origin centered at the circle's center, we have $(x_1, y_1) = (-0.131, 0.991)$ and $(x_2, y_2) = (-0.259, 0.966)$ (see **Figure 4.8**). β, the inclination angle of the slope, is found from the arc tangent of the slope and is equal to 0.196 rad. The acceleration acting along the slope is given by $g\sin\beta$. The distance between the two points is 0.131 m. Using (2.8): $v\,dv = a\,ds \Rightarrow v_2^2 = 2a(s_2 - s_1) = 2(9.81\text{ m/s}^2)(\sin 0.196)(0.131\text{ m})$, which yields $v_2 = 0.71$ m/s. This is off by just 12% from our original result, indicating that we probably did it right the first time.

EXERCISES 4.1

4.1.1. **[Level 1]** A collar of mass $m = 2$ kg is constrained to move in the vertical plane along a rough rod of length $L = 2$ m angled at $\theta = 45°$ with respect to the horizontal. The coefficient of friction between the collar and rod is $\mu = 0.4$. The collar is acted on by a horizontal force $F = 15$ N, and it starts from rest at the rod's top A. Find the collar's speed v_B when it arrives at the rod's base B.

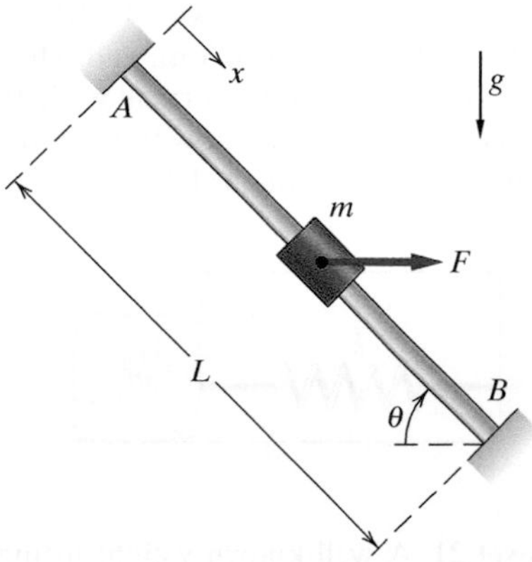

E4.1.1

4.1.2. **[Level 1]** A 1100 kg car, initially at rest, is raised on a hydraulic lift that accelerates upward at 0.9 m/s^2. What work does the lift perform on the car after it has been raised 2.0 m?

E4.1.2

4.1.3. **[Level 1]** The illustrated system models a children's toy. When the 0.2 kg block labeled A is pushed down, the linear spring is fully compressed and locked. After 10 s the spring unlocks, extends, and projects A upward. The force needed to compress the spring is given by $F = kx$, where x is the spring's compression (given in meters) and $k = 981$ N/m. The spring has an uncompressed length of 0.08 m. Assume that when fully compressed it has a length of zero. After the spring's release, what is the speed of block A at the point that the spring has extended itself back to 0.08 m in length?

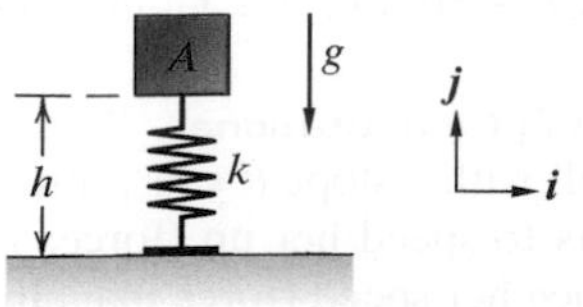

E4.1.3

4.1.4. **[Level 1]** A block with mass $m = 10$ kg is projected up a rough incline angled at $\theta = 40°$ to ground with a speed of $v = 5$ m/s. Just after release, a small booster strapped to the block ignites, providing a constant thrust of $T = 100$ N as the block travels up the incline. The coefficient of friction between the block and ramp is $\mu = 0.3$. If the booster cuts out at $d = 5$ m up the incline, what is the total distance that the block travels before coming to a stop?

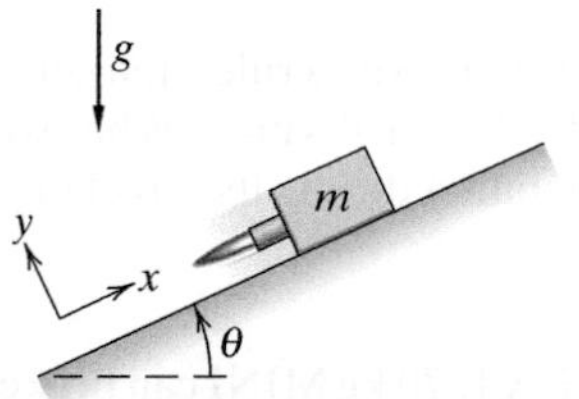

E4.1.4

4.1.5. **[Level 1]** A 23 kg block is slowly raised from the ground by pulling on an attached rope. How much work is done to raise the block to a height of 1.8 m above the ground? Assume that the block is stationary at the start and finish.

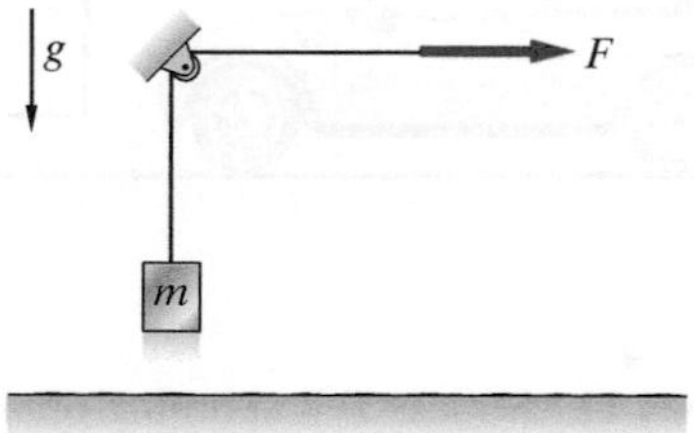

E4.1.5

4.1.6. **[Level 1]** A 5 kg rock is to be projected to a height of 10 m. The launcher produces a constant force against the rock that lasts until the rock leaves the launch tube. What must the force equal for the rock to just barely reach 10 m?

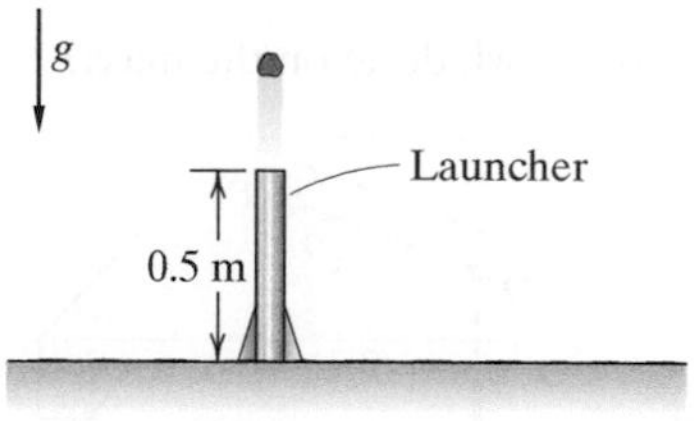

E4.1.6

4.1.7. **[Level 1]** A forklift has a bumper B attached to its rear to protect it against collisions. The force needed to compress the bumper varies with the compression of the bumper and is given by

$$F = ax^2$$

where x is the bumper's compression (measured in meters) and $a = 2.7 \times 10^7$ N/m^2. Assume that the 1000 kg forklift runs backwards into a wall at 1.5 m/s. How much will the bumper compress?

E4.1.7

4.1.8. **[Level 1]** As it exits a rifle's muzzle, a 0.028 kg bullet has a speed of 1.13×10^3 km/h. What was the average force acting on the bullet during its travel through the 1.2 m long barrel?

4.1.9. **[Level 1]** A 1270 kg MINI can brake from 96 km/h to 0 in 44 m. Assuming a coefficient of friction $\mu = 0.9$, determine the average braking force acting at the tire/road interface.

E4.1.9

4.1.10. **[Level 1]** A mass P is spinning on the end of a massless string on a horizontal, frictionless table. The mass is initially rotating at a rotational rate of ω_1 about O, and r is constant at r_1. Some time later (due to a string being pulled into a central hole in the table) r is again constant but at a new length r_2 ($r_2 < r_1$).

a. What is the rotational speed of the mass about O at this later time?

b. What was the work done on the mass?

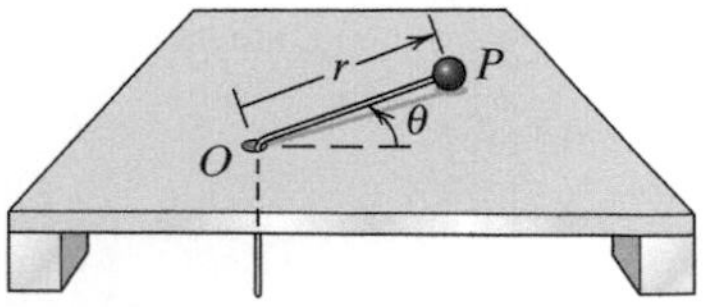

E4.1.10

4.1.11. **[Level 2]** A block of mass $m = 4$ kg is initially at rest on a rough incline angled at $\theta = 20°$ with respect to the horizontal. The block is compressing a spring of stiffness $k = 1100$ N/m by $\delta = 0.25$ m. Suppose the block loses contact with the spring platform when the spring returns to its uncompressed length. If the block travels an additional $\Delta x = 1$ m up the incline before coming to a stop after leaving the spring platform, find the coefficient of friction μ between it and the surface.

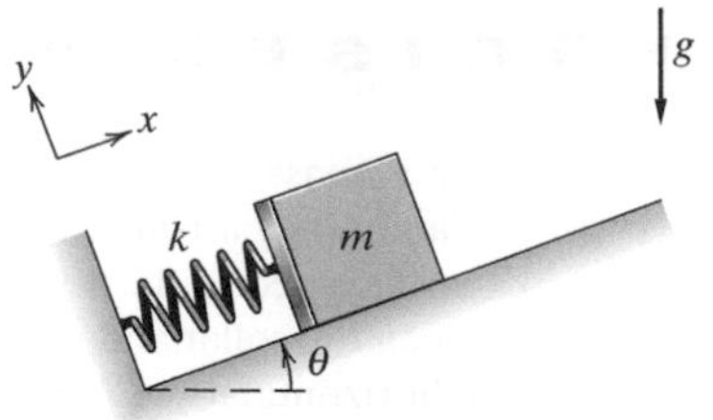

E4.1.11

4.1.12. **[Level 2]** A 0.5 kg mass is attached to a extensional spring with spring constant $k = 40$ N/m. The unstretched length of the spring is 0.3 m, and when initially released, the mass is 2 m from the vertical wall. If released from rest on a frictionless surface, how fast will the mass be moving when it is 0.6 m from the wall?

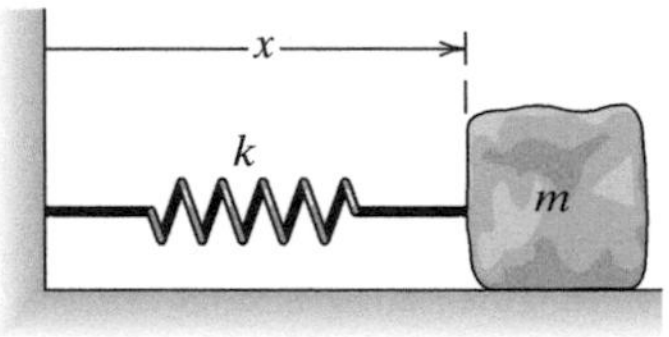

E4.1.12

4.1.13. **[Level 2]** A well-known weight-lifting move is the Clean and Jerk, where you bend down, grab hold of a barbell, and in a coordinated motion lift it over your head. Calculate the work done given the fact that the vertical acceleration of the 100 kg barbell is equal to $a_1 - a_2 y$, $a_1 = 4.0\ \text{m/s}^2$, $a_2 = \frac{4.0}{2.2}\ \text{s}^{-2}$, for $0 \le y \le 2.2$ m.

E4.1.13

4.1.14. **[Level 2] Computational** Use numerical integration (the MATLAB "quad" function will do nicely) to calculate the speed of a 50 kg mass that is initially moving to the right at 3 m/s and continues to the right for 2 more meters. The mass is acted on by a force directed to the left and given by

$$f = -a - b\frac{e^{-cx}}{d + x}$$

where $a = 100$ N, $b = 50$ N·m, $c = 1.1\ \text{m}^{-1}$, and $d = 2$ m.

4.1.15. **[Level 2] Computational** When a 50 kg cyclist coasts down a hill with a slope $\theta = 6°$, three forces act on her. Gravity acts to speed her up (force: $mg \sin\theta$), road drag acts to reduce her speed (force magnitude: 5 N), and aerodynamic drag also acts to reduce her speed (force magnitude: av^2, where $a = 0.04\ \mathrm{N{\cdot}s^2/m^2}$). Plot her speed and acceleration as a function of distance traveled, starting from rest and traveling a total of 50 m. How much work is done by the drag forces over this interval?

4.1.16. **[Level 2]** A 2 kg mass m is projected up a 30° slope with an initial speed of 10.2 m/s. The mass comes to rest at a point that's 5 m higher in elevation than its starting position. What must the coefficient of dynamic friction have been between the mass and the slope?

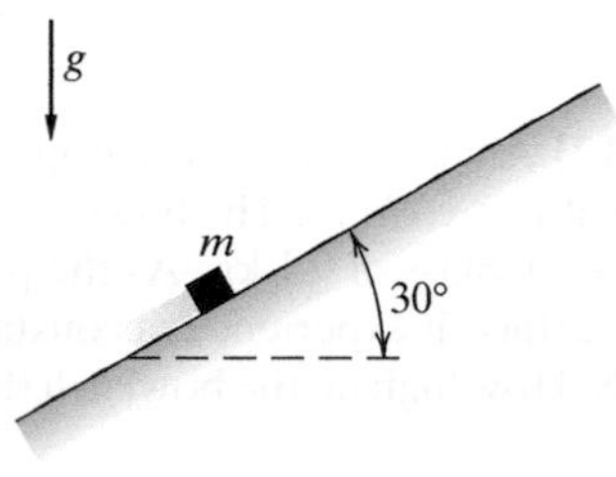

E4.1.16

4.1.17. **[Level 2]** How much slower will a mass be moving if it's released from rest and travels 10 m along a 45° slope with a coefficient of dynamic friction equal to 0.1 as compared to the case of a friction-free interface?

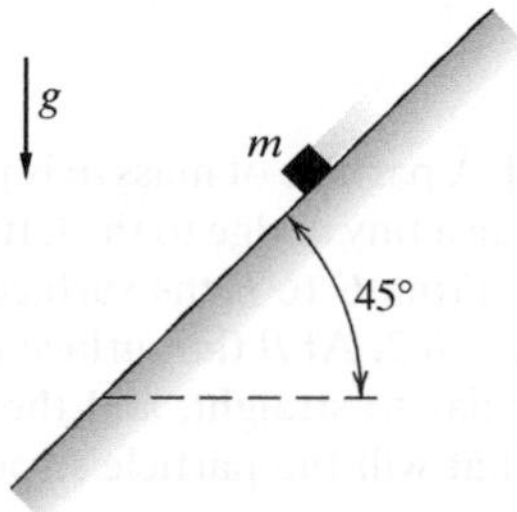

E4.1.17

4.1.18. **[Level 2]** A 0.3 kg pinecone strikes the surface of a snow-covered field after falling 30 m. The force of the snow resisting the pinecone's motion is given by

$$a + bx$$

where x indicates the distance the pinecone has penetrated the snow (given in meters), $a = 4$ N and $b = 14$ N/m. Calculate how deep the pinecone will bury itself in the snow.

4.1.19. **[Level 2]** A model catapult produces a force of 10 N against the 0.45 kg payload P, always oriented in the $\boldsymbol{e}_\theta$ direction. If the catapult is released from $\theta = 45°$ with zero velocity, what will the speed of the payload be at $\theta = 90°$?

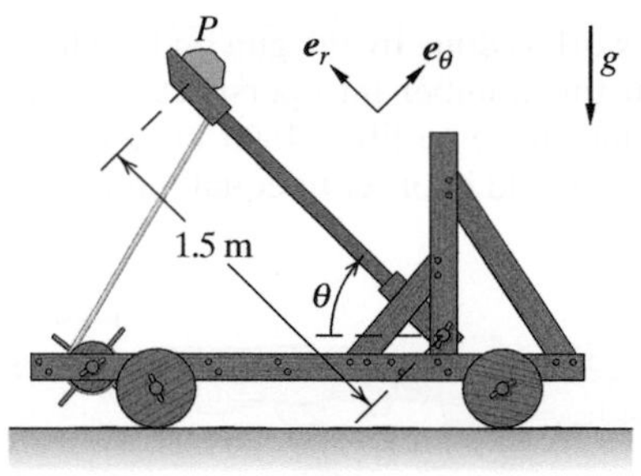

E4.1.19

4.1.20. **[Level 2]** A massless rope runs from block A (mass m_A), over a massless, frictionless pulley and attaches to block B (mass m_B). The coefficients of friction between block A and the table are $\mu_s = \mu_d = 0.6$. $m_A = 10$ kg and $m_B = 20$ kg. Assuming that the system is released from rest, what will block B's speed be after it has moved 0.5 m?

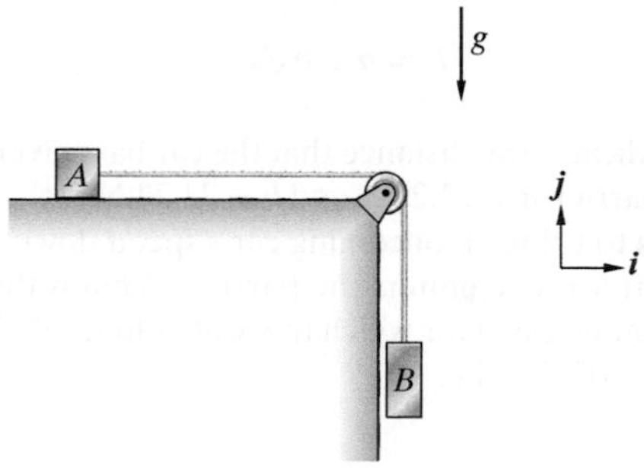

E4.1.20

4.1.21. **[Level 2]** This problem models an insertion process. Block A ($m_A = 2$ kg) is acted on by a horizontal force F ($F = 12$ N). As A moves to the right, it gets pressed more and more tightly in the channel. The tightening effect is modeled by a compression spring that is compressed linearly as the mass moves. The force exerted by the spring on A is given by $k(x_0 + \alpha x)$, $k = 100$ N/m, $x_0 = 0.05$ m, and $\alpha = 0.27$. The coefficients of friction between A and the ground are $\mu_s = \mu_d = 0.4$. Assuming that A starts moving from rest, at $x = 0$, what is its speed at the point that the force resisting its motion is equal in magnitude to F? Assume that the spring/wall interface is frictionless.

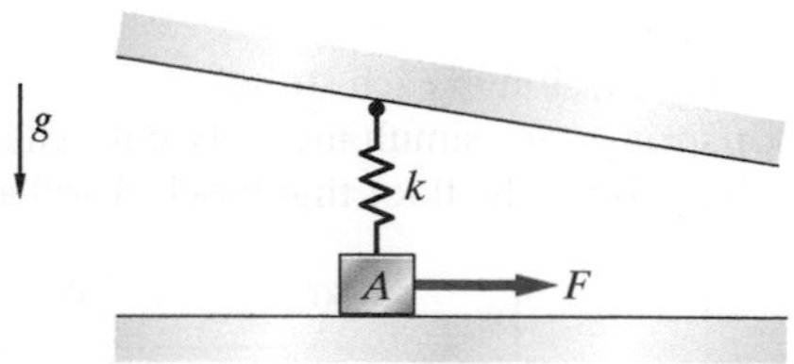

E4.1.21

4.1.22. **[Level 2]** In the movie *Eraser*, the bad guys had a gun that would shoot a 5 g aluminum slug at one quarter the speed of light. Let's assume that the slug is acted on by a constant force over the length of the 1 m long barrel. Further assume that the person shooting the gun is capable of keeping the gun stationary as the slug is fired.

a. What work is done by the gun on the slug?

b. To put this number into perspective, calculate how high you would have to lift a 1500 kg car in a constant 1 g gravitational field in order to equal the work calculated in **a**.

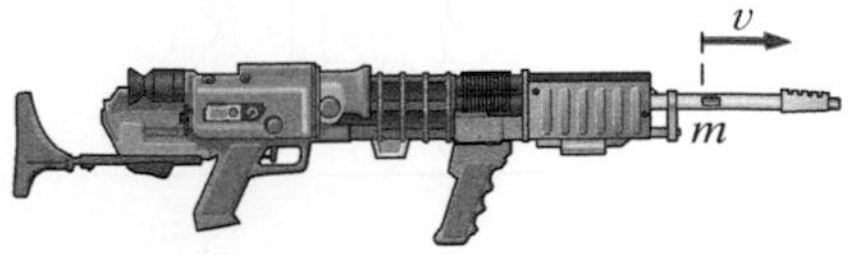

E4.1.22

4.1.23. [Level 2] Those rows of covered things that resemble garbage cans that you see along the highway are designed to stop speeding cars more safely than a concrete wall can. Assume that the force they exert against a car that's crashing into them can be approximated by

$$F = a + b\sqrt{x}$$

where x indicates the distance that the car has traveled into the crash barrier, $a = 22.2$ kN and $b = 24.2$ kN/m$^{0.5}$. The design goal is to bring an oncoming car's speed down to zero, 9 meters after it encounters the barrier. What is the maximum incoming speed for which this will hold true? Assume a car weight of 1724 kg.

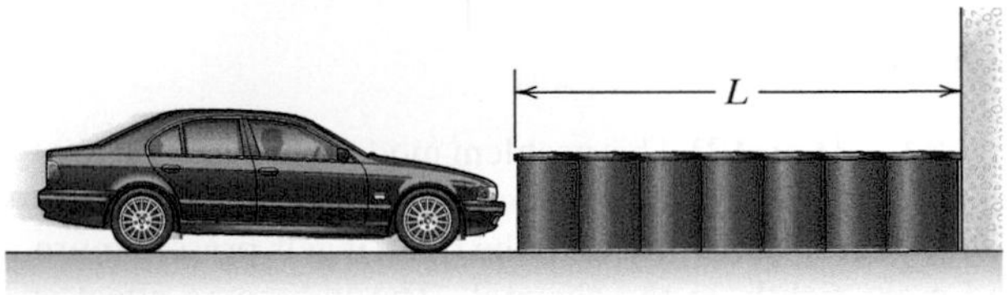

E4.1.23

4.1.24. [Level 2] Block A is supported by two springs. Each spring has a force characteristic of $F = kx$ where F is the force needed to compress the spring x m and $k = 500$ N/m. The springs, each having an unstretched length of $h = 0.06$ m, are compressed down to $h = 0.01$ m, and the mass is then restrained by two strings, with each string attached to the mass at one end and to the ground (at B and C) at the other end.

a. What is the tension in each string?

b. If both springs are simultaneously cut, what is the maximum height above the floor that block A will attain?

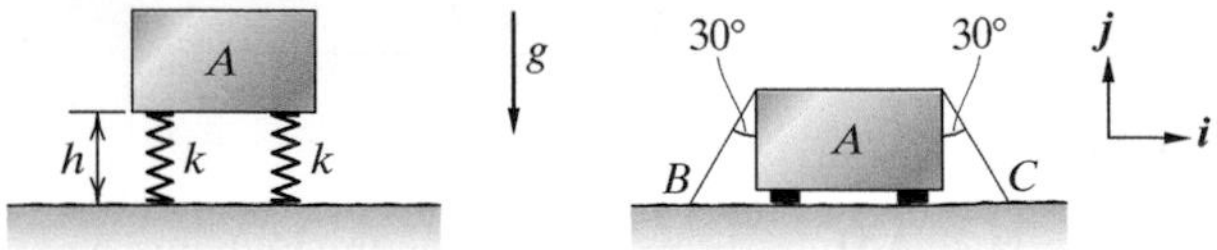

E4.1.24

4.1.25. [Level 2] The illustrated frictionless semicylinder has a radius of 1 m. A small mass $m = 0.5$ kg is slid upslope from $\theta = 20°$ at a speed of $v = 1.25$ m/s. What will be its velocity when it reaches $|\theta| = 30°$?

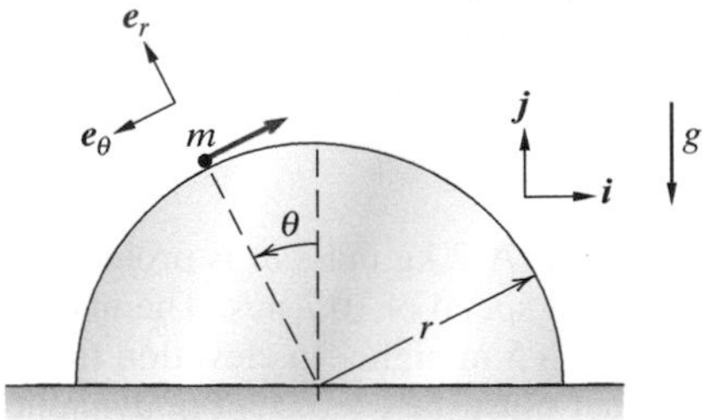

E4.1.25

4.1.26. [Level 2] A mass particle A is propelled from the bottom of a bowl at $\boldsymbol{v} = 1.2\boldsymbol{i}$ m/s. The bowl's radius is 1.2 m, and the particle has a mass of 0.1 kg. As the particle slides along the bowl's surface, it experiences a constant frictional resistance of 0.6 N. How high up the bowl will the mass rise?

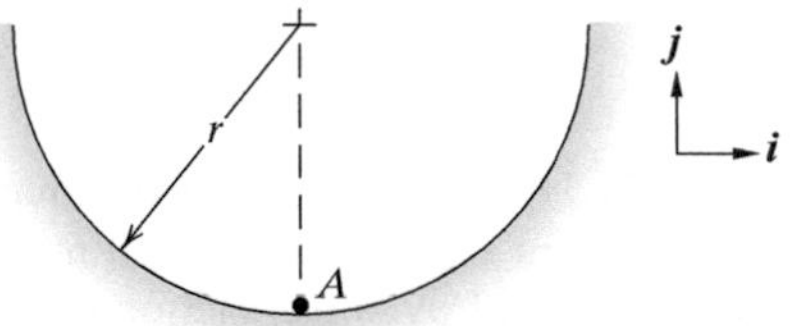

E4.1.26

4.1.27. [Level 2] A particle of mass m is placed at C, the top of a hill, and given a tiny nudge to the left so that it starts to slide downslope. From C to B the surface is frictionless, and from A to B $\mu_d = 0.2$. At B the surface profile changes smoothly from circular to straight, and the distance from A to B is 10 m. What will the particle's speed be when it reaches A?

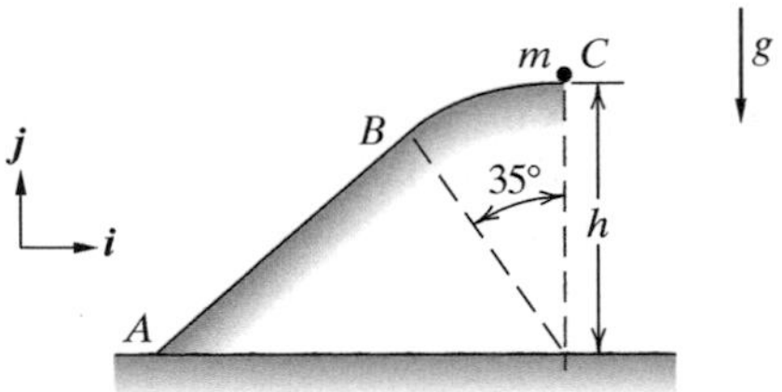

E4.1.27

4.1.28. [Level 2] A 60 kg skier begins her run from A. Assume a coefficient of friction from A to B of $\mu_d = 0.08$. The distance along the constant-slope path from A to B is 60 m. At B the track curves up, following a circular path with a radius of 100 m. For this portion of the track $\mu_d = 0.0$. What is the skier's speed at C?

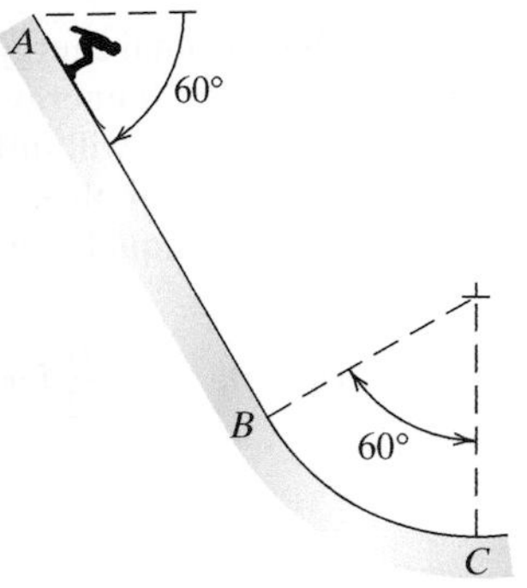

E4.1.28

4.1.29. **[Level 2]** The 3 kg mass m is released at $x = 0$ with velocity $\dot{x} = -5.0$ m/s. The coefficient of friction between the mass and the surface is $\mu = 0.4$. How fast is the mass traveling at $x = -5$ m?

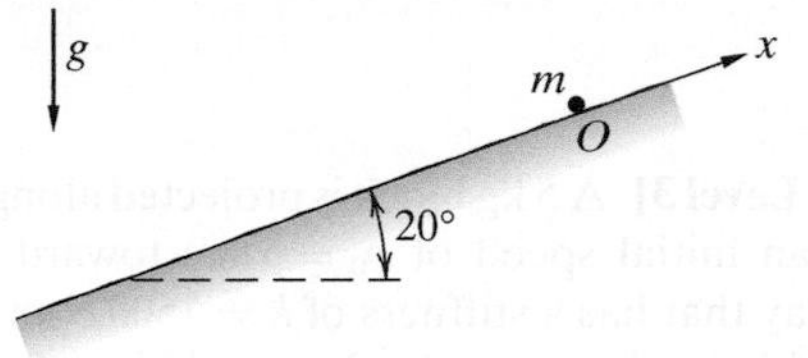

E4.1.29

4.1.30. **[Level 2]** A 0.1 kg mass P is spinning on the end of a massless string on a horizontal table. The other end of the string goes through a hole in the table at O and can be pulled further into the hole, as shown. A frictional coefficient μ exists between the mass and the table. At $t = 0$ the mass is located 1 m from the center hole and is set into rotation about O at an angular speed of $\dot{\theta} = 10$ rad/s. The string is withdrawn into the table at a constant rate $\dot{r} = v$. If r didn't change, the mass's rotational speed would decrease due to the frictional work done on it. But if we disregard friction for a moment, as the string is made shorter, the angular speed will increase due to conservation of angular momentum. With both effects acting it is unclear if the rotational speed will increase, decrease, or perhaps stay the same.

a. Numerically integrate the system's equation of motion with a coefficient of friction equal to 0.1. Experiment with different values of v (you can use -0.05 m/s as a good starting point) until you find the value of v for which the angular speed is equal to 10 rad/s from $t = 0$ to $t = 5$ s.

b. How much work was done on the mass during the motion?

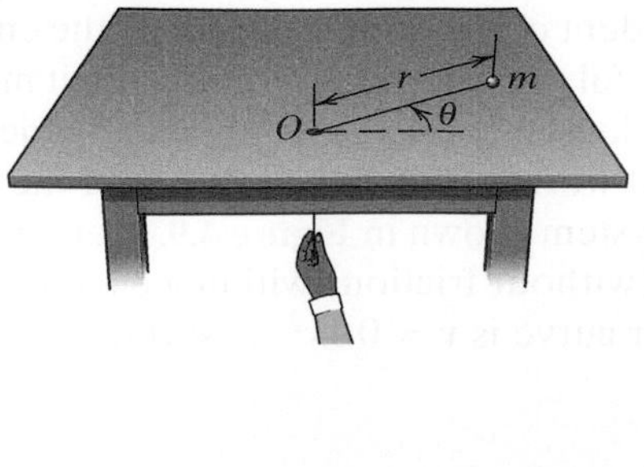

E4.1.30

4.1.31. **[Level 2]** A 10 kg crate is pushed along a rough floor with a force $F(x)$ applied at an angle $\theta = 20°$ to the ground. The applied force is described by $F(x) = F_0 + F_1 e^{ax}$, where $F_0 = 25$ N, $F_1 = 5$ N, $a = 0.1\ \text{m}^{-1}$, and x is the horizontal position of the crate. If the crate starts from rest, what is its speed after traveling $d = 10$ m? The coefficient of friction between the floor and crate is $\mu = 0.2$.

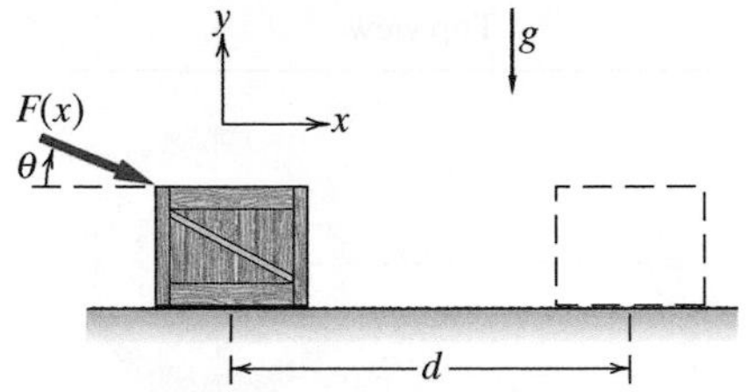

E4.1.31

4.1.32. **[Level 2]** A particle of mass $m = 2$ kg is acted upon by an external force and constrained to move in the horizontal plane along a frictionless path. The particle's position along the path relative to the origin O is given by $\boldsymbol{r}(t) = \overline{x}t^2\boldsymbol{i} + \overline{y}e^{-at}\boldsymbol{j}$, where $\overline{x} = 1\ \text{m/s}^2$, $\overline{y} = 4$ m, and $a = 0.5\ \text{s}^{-1}$. Find the work needed to move the particle from State 1 to State 2 if $t_1 = 0$ s and $t_2 = 2$ s.

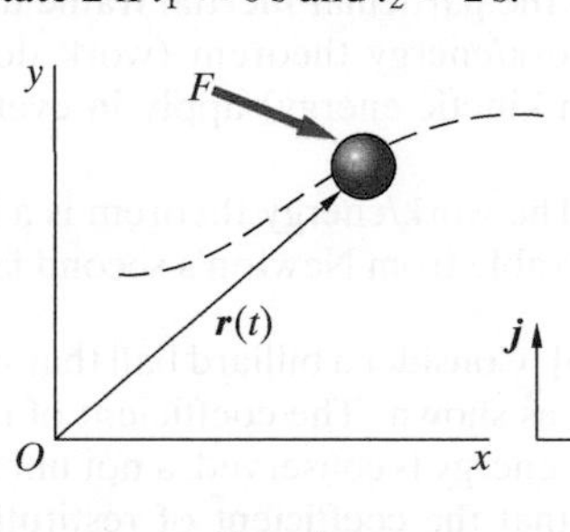

E4.1.32

4.1.33. **[Level 2]** A 45 kg cannonball is shot out of a cannon whose barrel is $d = 1.2$ m long and angled at $\theta = 45°$ with respect to the ground. The force applied to the cannonball while in the barrel is described by $F(s) = \overline{F}e^{as}$, where $\overline{F} = 44.5$ kN, $a = 2.30\ \text{m}^{-1}$, and s is the distance along the barrel. Find the maximum height H the cannonball achieves. Assume that the inside wall of the cannon barrel is frictionless and that the effects of gravity on the cannonball can be neglected when it is inside the barrel.

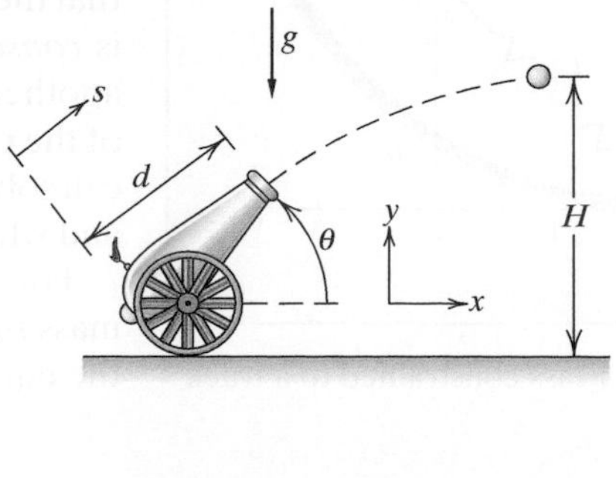

E4.1.33

4.1.34. **[Level 2]** Out of curiosity, a curling team decides to calculate the coefficient of friction μ between the curling stone and the playing surface (one of the players is an engineer). Suppose a team member launches the curling stone at $v_0 = 5.4$ m/s, and it travels in a straight-line path to the center of the target, a distance of $s_c = 28$ m, where it comes to a stop. If the curling stone has a mass of 18 kg, what is μ?

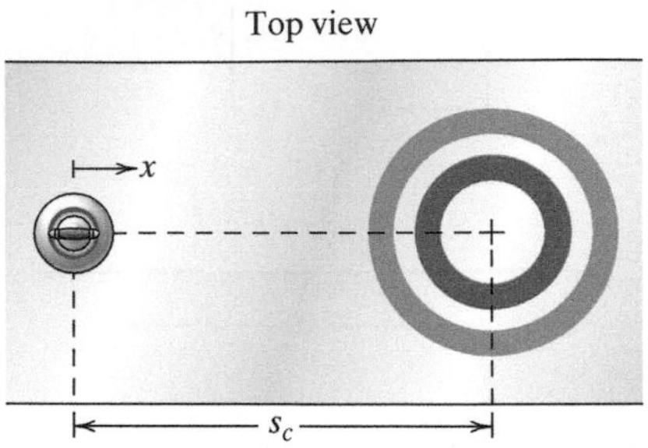

E4.1.34

4.1.35. **[Level 3]** A single particle is viewed from two different inertial reference frames.

a. Is the particle's kinetic energy dependent on the particular inertial reference frame used?

b. Is the work done by external forces acting on the particle dependent on the particular inertial frame used?

c. Does the work/energy theorem (work done by all forces = change in kinetic energy) apply in every inertial reference frame?

d. True/False: The work/energy theorem is a new principle that isn't derivable from Newton's second law.

4.1.36. **[Level 3]** Consider a billiard ball that strikes two other billiard balls as shown. The coefficient of restitution is 1.0. Assume that energy is conserved, a not unreasonable assumption given that the coefficient of restitution is 1.0. Solve for the final velocities of the balls by conserving momentum in the $\boldsymbol{i}$ direction, using symmetry to tell you that balls B and C go off at 30° angles (ball A's motion remaining in the $\boldsymbol{i}$ direction) and requiring that the final energy and the initial energy of the system be the same. If energy actually *isn't* conserved (and it isn't) your answers will differ from the correct results (shown below). Explain physically why the mismatch occurs. All the masses are equal to m and the impact speed is v_0.

Correct results for $e = 1$: $v'_A = 0$, $v'_B = \frac{v_0}{2}\boldsymbol{i} + \frac{\sqrt{3}v_0}{6}\boldsymbol{j}$, $v'_C = \frac{v_0}{2}\boldsymbol{i} - \frac{\sqrt{3}v_0}{6}\boldsymbol{j}$

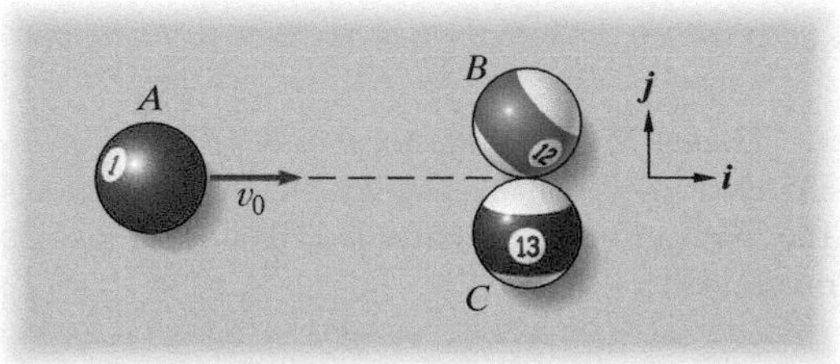

E4.1.36

4.1.37. **[Level 3]** A 5 kg block is projected along a rough floor with an initial speed of $v_0 = 6$ m/s toward a spring $d = 4$ m away that has a stiffness of $k = 1500$ N/m. The coefficient of friction between the floor and block is $\mu = 0.3$.

a. At what speed does the block make contact with the spring?

b. What is the maximum compression the spring experiences?

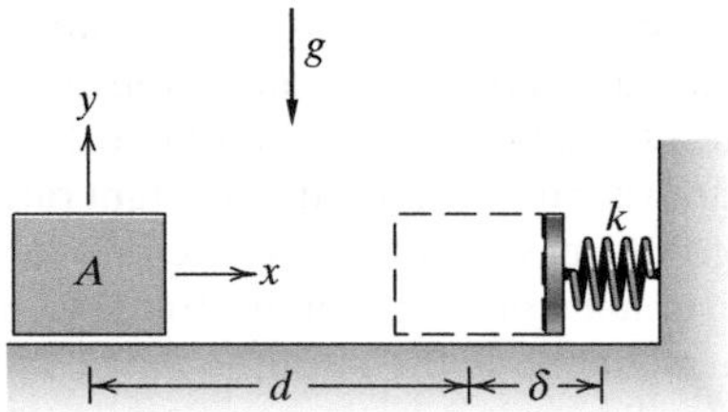

E4.1.37

4.2 POTENTIAL ENERGIES AND CONSERVATIVE FORCES

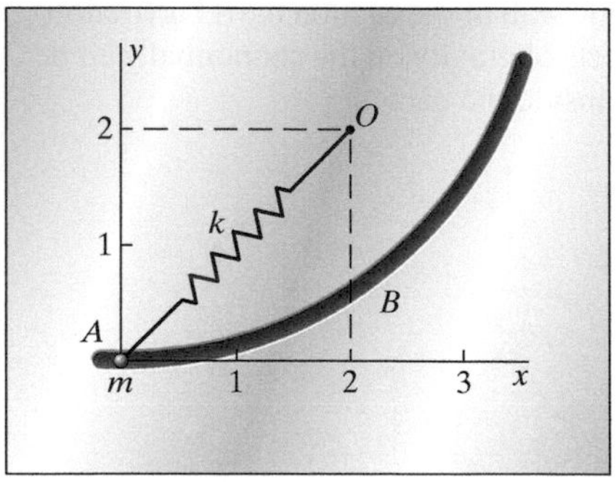

Figure 4.9 Mass constrained to a track

You might think it would pain someone who works at Berkeley (long considered the capital of academic liberalism) to talk about **conservative forces**, but in this case even the most diehard liberal would have to concede that they're a good thing. For our purposes, it's enough to say that if a force is *conservative*, the work it does in moving an object from one place to another is completely independent of the path taken. Only the endpoints of the path matter. This is a useful property of forces because it means we can solve a whole array of problems in which we know where objects start and where they end up without worrying about the details of the motion.

For instance, consider the system shown in **Figure 4.9**. Here we have a mass m constrained to move (without friction) within a curved slot, and the equation for this particular curve is $y = 0.1x^2$. A spring k is attached to the mass at one end and to the fixed point O at its other end. The unstretched length of the spring is 0.4 m. Given that the mass is released at point A and then zips along the slot to B at $(x, y) = (2, 0.4)$, find the speed of the mass at B.

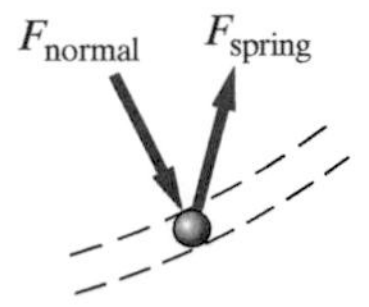

Figure 4.10 Forces acting on the mass particle

If we had to depend only on what we know so far, specifically (4.7), we would have to say that the kinetic energy at B is equal to the kinetic energy at A (zero in this case) plus the work done on the mass by the spring. To figure out how much work this is, we would have to do a free-body analysis (**Figure 4.10**), calculate the forces F_{normal} and F_{spring} acting on the mass for all positions along the slot, decompose them onto the path defined by the slot, multiply the tangential force component by the differential motion ds, and then integrate to find out how much work was done in moving the mass from A to B. This is not easy. Once we know about conservative forces, however, along with their associated potential energies, we can simply say that the kinetic energy plus potential energy at A is equal to the kinetic energy plus potential energy at B:

$$\underset{(\mathcal{KE}|_A)}{0} + \underset{(\mathcal{PE}|_A)}{\frac{1}{2}kx_1^2} = \underset{(\mathcal{KE}|_B)}{\frac{1}{2}mv^2} + \underset{(\mathcal{PE}|_B)}{\frac{1}{2}kx_2^2} \tag{4.10}$$

(Keep in mind that none of this material has yet been derived. I'm just showing you what the notion of potential energy will do for you, and once that's done, I will show you where it all comes from.)

Solving (4.10) gives us

$$v = \sqrt{\frac{2}{m}\left(\frac{1}{2}kx_1^2 - \frac{1}{2}kx_2^2\right)}$$

So you see that there's a good reason to add this ability to your arsenal of tools.

Simply put, a force that's conservative can do work (not surprising), but the total work done is independent of the path. As we just saw, the force exerted by a linear spring is one example of a conservative force. Another example is the force exerted by gravity. For instance, you might want to move a rock from point A to point B (**Figure 4.11**). Path 2 is a fancy path that loops around a bit and then ends up at B. Path 1 is a more mundane one in which you put off the heavy lifting until you're almost

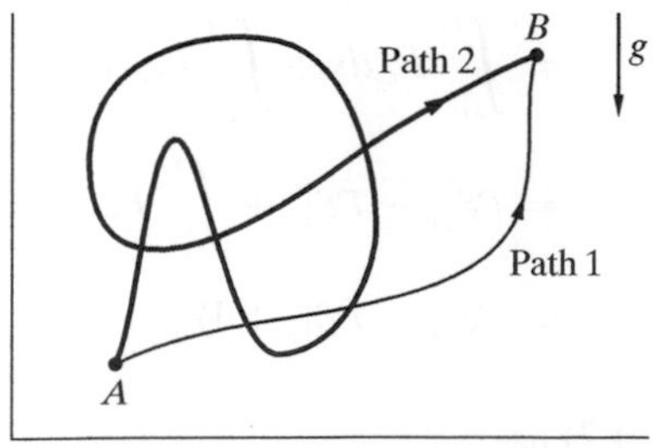

Figure 4.11 Two possible paths for moving a rock from A to B

under B and then lift up swiftly to B. *If* gravity is a conservative force, the work you do against this force will be the same for both paths. And since gravity *is* a conservative force, that's the result we in fact get. This is the case with all conservative forces—the work done is independent of the path.

The easy way to determine the work done by any conservative force is to use the special kind of energy that exists for conservative forces, called **potential energy**. The name reflects the fact that this energy is in the form of a **potential function**. You may or may not have come across potential functions in your physics course, but basically the situation is, given such a function, you can find the associated force by differentiating. Or, to go in the reverse direction, you can find a potential function by integrating the expression for the associated force. If only a single coordinate s is involved, the relationship is

$$F_c = -\frac{d\mathcal{PE}}{ds} \tag{4.11}$$

where $\mathcal{PE}$ represents the potential energy of the force F_c and the subscript on F_c indicates that the force produced by this potential energy is conservative. Keep in mind that both $d\mathcal{PE}$ and F_c are functions of s.

The integral relationship is also true:

$$\mathcal{PE}\big|_2 - \mathcal{PE}\big|_1 = -\int_{s_1}^{s_2} F_c\, ds \tag{4.12}$$

You will note that this formula looks like the expression for work done by a force (4.3). Using this similarity lets us update our work/energy formulation. Currently, we know from (4.7) that

$$\mathcal{KE}\big|_2 = \mathcal{KE}\big|_1 + W_{1-2}$$

where $W_{1-2} = \int_{s_1}^{s_2} F_t\, ds$ (4.3). What we will do now is re-express the general force F_t as being composed of a conservative portion F_c and a nonconservative portion F_{nc}:

$$\begin{aligned} W_{1-2} &= \int_{s_1}^{s_2} (F_c + F_{nc})\, ds \\ &= \int_{s_1}^{s_2} F_c\, ds + \int_{s_1}^{s_2} F_{nc}\, ds \\ \text{from (4.12),} \quad &= \mathcal{PE}\big|_1 - \mathcal{PE}\big|_2 + \int_{s_1}^{s_2} F_{nc}\, ds \\ &= \mathcal{PE}\big|_1 - \mathcal{PE}\big|_2 + W_{nc_{1-2}} \end{aligned}$$

which lets us rewrite (4.7) as

$$\mathcal{KE}\big|_2 = \mathcal{KE}\big|_1 + \mathcal{PE}\big|_1 - \mathcal{PE}\big|_2 + W_{nc_{1-2}}$$

or, more conveniently, as

$$\mathcal{KE}\big|_2 + \mathcal{PE}\big|_2 = \mathcal{KE}\big|_1 + \mathcal{PE}\big|_1 + W_{nc_{1-2}} \tag{4.13}$$

IMPORTANT NOTE! Most traditional dynamics texts represent potential energy with the symbol V. Because it's not much more work to use something that helps you remember what it is you're dealing with (namely, $\mathcal{PE}$ for potential energy), that's what I'll be using. Just keep this in mind if you're ever reading another dynamics text. Also, other texts often use T for kinetic energy, whereas I use $\mathcal{KE}$.

The potential energy due to gravity for a particle m is given by

$$\mathcal{PE}\big|_g = mgy \tag{4.14}$$

where y represents the positive vertical position of the particle, as shown in **Figure 4.12**. Differentiation of this expression with respect to y gives a form similar to (4.11):

$$F = -\frac{d}{dy}(mgy) = -mg \tag{4.15}$$

Here y took the place of the general variable s. As you can see, we got the right answer. The force produced by gravity on a particle m is indeed equal to mg and is oriented downward.

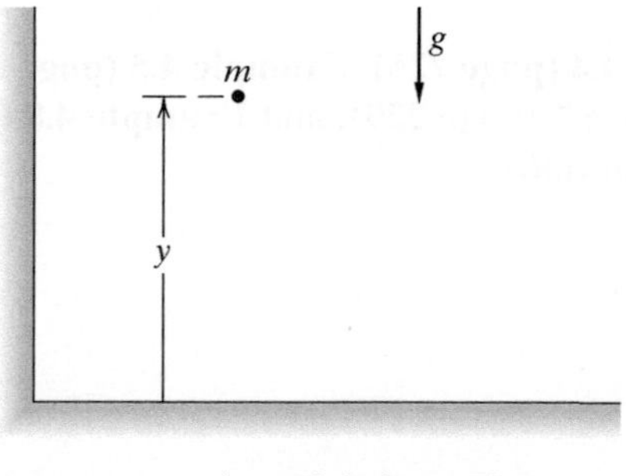

Figure 4.12 A particle located a distance y above the ground

Let's go in the opposite direction and find the potential energy of a spring, knowing that its force varies as a function of its deflection. **Figure 4.13*a*** shows a massless spring in the process of being stretched, and **Figure 4.13*b*** shows a force produced at the spring's free end. Assume an unstretched length L and a restoring force F_{sp} that's proportional to the spring's stretch:

$$F_{sp} = -kx \tag{4.16}$$

The spring is extended from a length L to a length $L + x$. Using (4.12) we have

$$\mathcal{PE}\big|_{sp} = -\int_0^x F_{sp}\,dx = -\int_0^x (-kx)\,dx = \frac{1}{2}kx^2$$

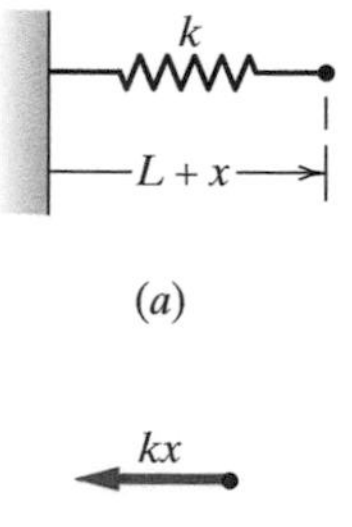

Figure 4.13 Forces acting on a spring

$$\mathcal{PE}\big|_{sp} = \frac{1}{2}kx^2 \qquad (4.17)$$

Thus the potential energy of a spring is a quadratic function of the displacement of the end (or ends in some cases) of the spring. This means that the potential energy grows very quickly as the spring is stretched.

As we have seen, both gravity and linear springs produce conservative forces, and we will be using their potential energies repeatedly when working energy problems that involve them. As you might guess, **nonconservative** forces are forces that aren't derivable from a potential function. When we encounter these forces, denoted by $\boldsymbol{F}_{nc}$, we will have to integrate over displacement in order to find the associated work.

If there are no nonconservative forces, then no nonconservative work is done and (4.13) becomes

$$\mathcal{KE}\big|_2 + \mathcal{PE}\big|_2 = \mathcal{KE}\big|_1 + \mathcal{PE}\big|_1 \qquad (4.18)$$

This relationship is often referred to as *conservation of mechanical energy*. It says that, in the absence of nonconservative forces, the total mechanical energy of the system (potential plus kinetic) is constant. A familiar example is given by a pendulum. In the absence of friction, a pendulum that has been displaced from equilibrium will oscillate back and forth without end. During the motion, both the kinetic energy and the potential energy will vary, but their sum remains constant.

>>> **Check out Example 4.4 (page 225), Example 4.5 (page 226), Example 4.6 (page 228), Example 4.7 (page 229), and Example 4.8 (page 230) for applications of this material.**

EXAMPLE 4.4 SPEED DUE TO A DROP (Theory on pages 222 and 224)

Use energy to determine what height building you would have to jump from in order to impact the ground at 56 km/h. This is a typical speed used in vehicle crash tests when evaluating the effectiveness of a car's restraints and crush characteristics.

Goal Find height h such that $v = 56$ km/h.

Given Speed at impact.

Assume Our only assumption will be to neglect wind drag.

Draw **Figure 4.14** shows our system.

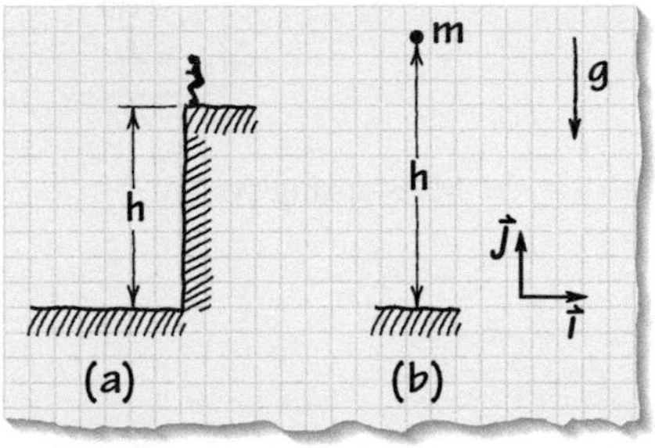

Figure 4.14 Crash test

Formulate Equations Because you're going to be in free-fall once you jump off the building, you need concern yourself only with your kinetic energy and your potential energy due to gravity. Using (4.18) we have

$$\mathcal{KE}\big|_2 + \mathcal{PE}_g\big|_2 = \mathcal{KE}\big|_1 + \mathcal{PE}_g\big|_1$$

where 1 indicates the initial system state and 2 the final system state.

Solve Your initial velocity is zero, and your initial height is h. If we define the potential energy at ground level to be zero, we have

State 1: $\quad \mathcal{KE}\big|_1 = 0, \quad \mathcal{PE}_g\big|_1 = mgh$

State 2: $\quad \mathcal{KE}\big|_2 = \frac{1}{2}mv^2, \quad \mathcal{PE}_g\big|_2 = 0$

Equating the energies at States 1 and 2 and then solving for h leave us with

$$\frac{1}{2}mv^2 = mgh \quad \Rightarrow \quad h = \frac{v^2}{2g}$$

After converting our desired value of 56 km/h to $v = 15.6$ m/s, we have

$$h = \frac{(15.6 \text{ m/s})^2}{2(9.81 \text{ m/s}^2)} = \boxed{12.4 \text{ m}}$$

Check There's not much to check. We could solve the problem a different way by realizing that the acceleration is constant and using our constant-acceleration formula to determine the distance traveled given the impact speed:

$$a\Delta y = \frac{\dot{y}_2^2}{2}$$

$$(9.81 \text{ m/s}^2)\Delta y = \frac{(15.6 \text{ m/s})^2}{2} \quad \Rightarrow \quad \Delta y = 12.4 \text{ m}$$

which clearly produces the same result.

EXAMPLE 4.5 DESIGNING A NUTCRACKER (Theory on pages 222 and 224)

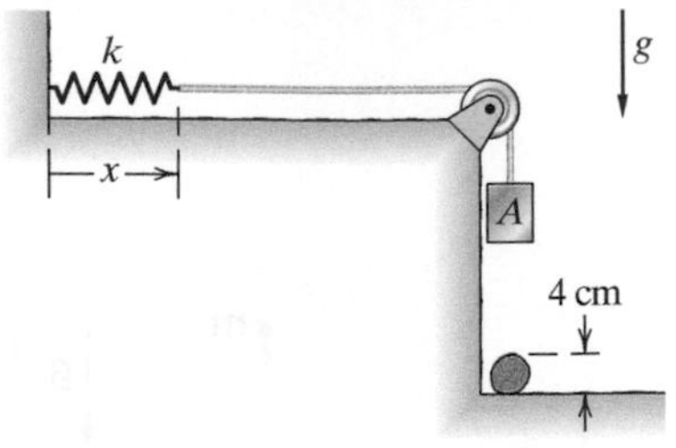

Figure 4.15 Mass-spring system

In this example we will look at both gravitational potential energy and spring potential energy. **Figure 4.15** shows a spring-restrained walnut cracker designed by a mechanical engineer with far too much time on her hands. It consists of block A ($m = 2.5$ kg) attached to a massless rope that is itself attached to a linear spring having a spring constant $k = 1000$ N/m. The position of the right end of the spring (the end that's attached to the rope) is given by x. The spring has an unstretched length of 0.1 m and at the instant pictured, the right end is at $x = 0.25$ m, indicating a stretch of 0.15 m. The bottom of block A is 0.16 m above the ground, and it is moving down at $\dot{y} = 4.22$ m/s. Determine the block's velocity when it is 4 cm from the floor (the height of a walnut).

Goal Find v when block A's bottom is 4 cm from the ground.

Given System parameters and initial configuration.

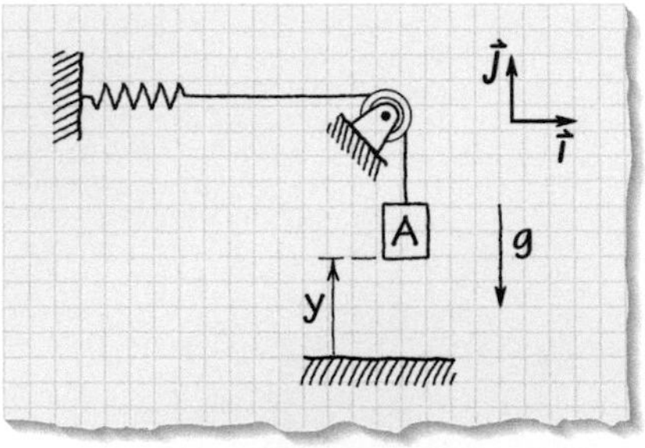

Figure 4.16 Labeling of spring-mass problem

Assume We will assume that there are no dissipative mechanisms present, such as air drag, and that the system's total energy is therefore conserved.

Draw **Figure 4.16** shows a simplified view of our system. The mass can move only up or down (in the direction $\boldsymbol{j}$), and y tracks its position.

Formulate Equations We will need to include both spring and gravitational potential energies this time. As always, the particular location from which we choose to reference the potential energy due to gravity is arbitrary, and for this problem we will let the ground define the zero gravitational potential energy state.

Our system energy will have three components: kinetic energy ($\frac{1}{2}mv^2$), gravitational potential energy (mgy), and spring potential energy ($\frac{1}{2}kx^2$). The spring is initally stretched $(0.25 - 0.1)$ m $= 0.15$ m and reaches a maximum stretch of $(0.25 + (0.16 - 0.04) - 0.1)$ m $= 0.27$ m. Our energy components at the start are given by

State 1:
$$\mathcal{KE}\big|_1 = \frac{1}{2}(2.5\,\text{kg})(-4.22\,\text{m/s})^2 = 5.28\,\text{kg}\cdot\text{m}^2/\text{s}^2$$
$$\mathcal{PE}_g\big|_1 = (2.5\,\text{kg})(9.81\,\text{m/s}^2)(0.16\,\text{m})$$
$$\mathcal{PE}_{sp}\big|_1 = \frac{1}{2}(1000\,\text{N/m})(0.15\,\text{m})^2$$

and at the final state are given by

State 2:
$$\mathcal{KE}\big|_2 = \frac{1}{2}(2.5\,\text{kg})v^2$$
$$\mathcal{PE}_g\big|_2 = (2.5\,\text{kg})(9.81\,\text{m/s}^2)(0.04\,\text{m})$$
$$\mathcal{PE}_{sp}\big|_2 = \frac{1}{2}(1000\,\text{N/m})(0.27\,\text{m})^2$$

Solve Equating the total energy at the two states gives us

$$\frac{1}{2}(2.5\,\text{kg})(-4.22\,\text{m/s})^2 + (2.5\,\text{kg})(9.81\,\text{m/s}^2)(0.16\,\text{m}) + \frac{1}{2}(1000\,\text{N/m})(0.15\,\text{m})^2$$
$$= \frac{1}{2}(2.5\,\text{kg})v^2 + (2.5\,\text{kg})(9.81\,\text{m/s}^2)(0.04\,\text{m}) + \frac{1}{2}(1000\,\text{N/m})(0.27\,\text{m})^2$$

When solved for v, this yields

$$v = \pm 0.053\,\text{m/s}$$

Realizing that the block was initially moving downward lets us correctly identify the negative root as the one we want:

$$v = -0.053\boldsymbol{j}\,\text{m/s}$$

The block started out moving at 4.22 m/s and at the specified height of 4 cm is almost stationary. The spring is strong enough to overcome the force of gravity and the block's initial momentum in order to bring it almost completely to a stop.

Check There's not a lot of intuitive checking that can be done here. We could obtain an independent verification of the results by numerically integrating the equation of motion from the given initial conditions. Doing so would produce the same result that we obtained through the energy approach. One thing that we can observe is that if we had considered a final height just slightly less than the one initially specified, we would have obtained a negative value for v^2, telling us that there is no real solution for v and that therefore the block can't go that low for the chosen initial conditions.

As a last observation, our analysis has shown that the current setup is a poor one for the intended purpose. Our engineer's design would not help much in the way of successfully cracking walnuts.

EXAMPLE 4.6 CHANGE IN SPEED USING POTENTIAL ENERGY (Theory on pages 222 and 224)

Figure 4.17 Marble on a rock

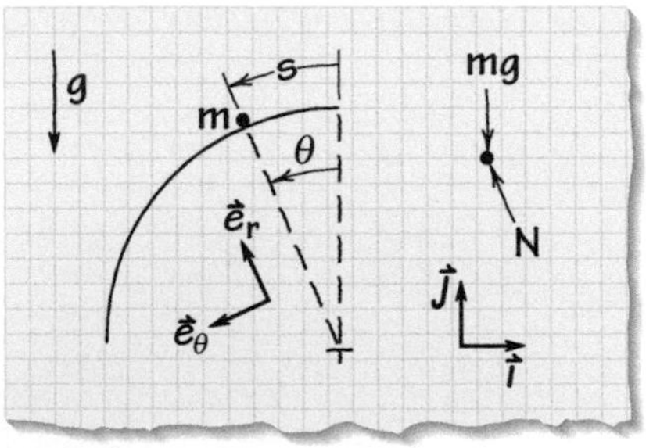

Figure 4.18 Mass moving on a semicircle

Let's revisit Example 4.3 and see how our knowledge of potential energy changes things. We'll again determine the speed of the 0.03 kg marble when it reaches $\theta = \frac{\pi}{12}$ rad (**Figure 4.17**).

Goal Find the marble's speed after it has moved from $\frac{\pi}{24}$ rad to $\frac{\pi}{12}$ rad.

Given $m = 0.03\,\text{kg}, r = 0.8\,\text{m}$

Assume Assume that the marble slides without friction and remains in contact with the surface.

Draw The simplified system is shown in **Figure 4.18**, where s represents the distance the marble moves along the sculpture's surface.

Formulate Equations The surface is frictionless, and thus no frictional (nonconservative) work is done. The only two energy elements we need to worry about are the kinetic energy of the moving mass and its potential energy change as it drops lower. The marble's velocity is given by

$$\boldsymbol{v}_m = r\dot{\theta}\boldsymbol{e}_\theta \quad \Rightarrow \quad \mathcal{KE} = \frac{1}{2}mr^2\dot{\theta}^2$$

If we define our zero potential energy position as the ground, then the potential energy at an angle θ is given by $mgh = mgr\cos\theta$.

$\mathcal{KE} + \mathcal{PE}$ at $\theta = \pi/24$: $\quad 0 + mgr\cos(\pi/24)$

$\mathcal{KE} + \mathcal{PE}$ at $\theta = \pi/12$: $\quad \frac{1}{2}mr^2\dot{\theta}^2 + mgr\cos(\pi/12)$

Equating the initial and final energies gives us

$$mgr\cos(\pi/24) = \frac{1}{2}mr^2\dot{\theta}^2 + mgr\cos(\pi/12)$$

$$\dot{\theta} = \sqrt{\frac{2g}{r}(\cos(\pi/24) - \cos(\pi/12))}$$

Using the given parameter values yields $v = 6.33\,\text{m/s}$.

Check We've gotten the same result as in Example 4.3, indicating all is well. Using potential energy clearly simplified our analysis.

EXAMPLE 4.7 FALLING ENCLOSURE (Theory on pages 222 and 224)

Protective packaging is an important area of design engineering. As one example, consider **Figure 4.19**. An enclosure, initially suspended above the floor, is released. Assume that at release the mass ($m = 0.02$ kg) is centered between the two identical positioning springs. The enclosure impacts the floor at 8 m/s and remains stationary against the ground. What is the minimum spring constant k for which the mass will not impact the enclosure floor?

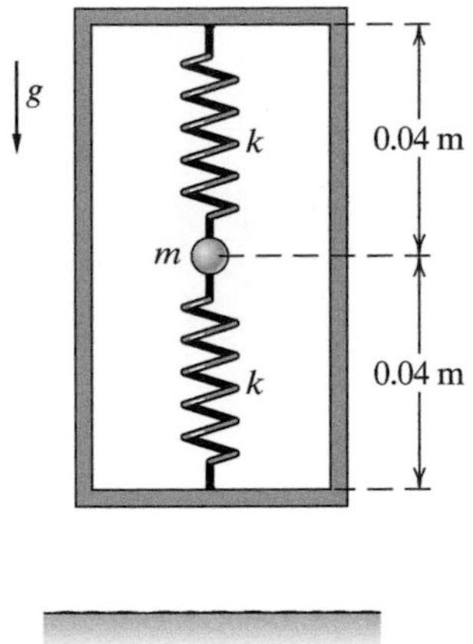

Figure 4.19 Dropped enclosure

Goal Determine the minimum spring constant k so that a mass does not impact the bottom of a dropped enclosure.

Given Size of enclosure, number of springs and arrangement, mass, and clearance in enclosure.

Assume We'll assume that the mass hits the floor perfectly aligned between the two springs and that the ground defines the zero energy state.

Draw **Figure 4.20** shows the system at the instant it strikes the floor and then at a later time with the mass just touching the enclosure's floor.

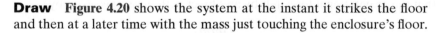

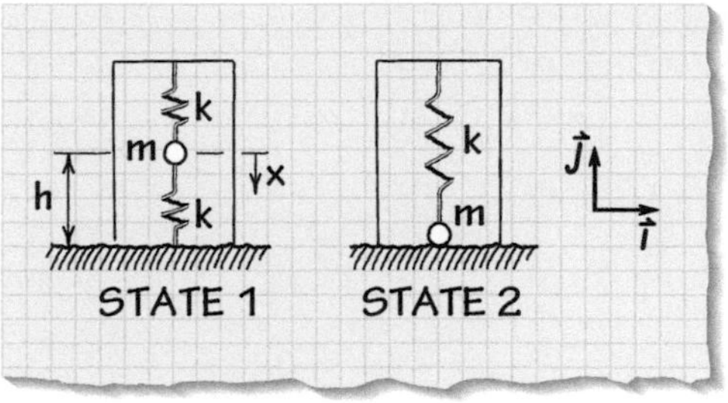

Figure 4.20 Initial contact (State 1) and post-contact (State 2)

Formulate Equations Conservation of energy from State 1 to State 2:

$$\mathcal{PE}|_1 + \mathcal{KE}|_1 = \mathcal{PE}|_2 + \mathcal{KE}|_2$$

Solve We're given that the speed of the enclosure and mass just before impact is 8 m/s. Thus the mass will still have this speed even though the enclosure is brought to an abrupt stop. At State 1 the mass has a finite kinetic energy and also a finite potential energy due to its height above the ground. At State 2 we assume zero speed, zero potential energy due to gravity, and a potential energy due to the extension/compression of the positioning springs.

$$mgh + \frac{1}{2}mv^2 = \frac{1}{2}(2k)h^2$$

$$(0.02\,\text{kg})(9.81\,\text{m/s}^2)(0.04\,\text{m}) + \frac{1}{2}(0.02\,\text{kg})(8\,\text{m/s})^2 = \frac{1}{2}(2k)(0.04\,\text{m})^2$$

$$k = 404.9\,\text{N/m}$$

Check As a numerical check we'll use MATLAB to integrate $m\ddot{y} = -2ky - mg$, with y tracking the mass m, positive upward, and defined to be zero at State 1. Integrating the equations of motion gives the results that at $t = 0.00783$ s, $y = -0.04$ m and $\dot{y} = 0$.

EXAMPLE 4.8 REEXAMINATION OF AN ORBITAL PROBLEM (Theory on pages 222 and 224)

In Chapter 3 we looked at the orbit of a mass around the earth (Example 3.21) and said we would use energy to verify the solution. So let's do it now.

Goal Verify the results of Example 3.21 by showing that the total energy is the same at the calculated apogee as at the given initial state (perigee).

Assume Because we're looking at an orbit in space, we can neglect any nonconservative forces arising from air drag.

Draw See **Figure 3.64**.

Formulate Equations We need to account for both kinetic and potential energy. The kinetic energy is simply $\frac{1}{2}m\|\boldsymbol{v}\|^2$. We can derive the potential energy by realizing that the relation between a conservative force and its potential energy is

$$\boldsymbol{F} = -\frac{d\mathcal{PE}}{dr}\boldsymbol{e}_r \tag{4.19}$$

where I've assumed a purely radially based potential. We already know that the gravitational force felt by an object with mass m is given by

$$\boldsymbol{F}_g = -\frac{Gmm_e}{r^2}\boldsymbol{e}_r \tag{4.20}$$

where r is the distance from the earth's center to the mass and m_e is the earth's mass. Integrating (4.19) and using the specific force relation given in (4.20) yield

$$\mathcal{PE} = -\frac{Gmm_e}{r} \tag{4.21}$$

Solve The kinetic energy is given by $\mathcal{KE} = \frac{1}{2}m\|\boldsymbol{v}\|^2$. The potential energy, as just derived, is given by $\mathcal{PE} = -\frac{Gmm_e}{r}$ and the total energy is simply $\mathcal{KE} + \mathcal{PE}$. At perigee (the given initial conditions for the orbit) we have a total energy of

$$\frac{1}{2}m(8500\,\text{m/s})^2 - \frac{\left(6.67\times10^{-11}\,\text{N}\cdot\text{m}^2/\text{kg}^2\right)(5.98\times10^{24}\,\text{kg})m}{6.53\times10^{6}\,\text{m}} = -(2.5\times10^{7}\,\text{m}^2/\text{s}^2)m$$

At apogee we have

$$\frac{1}{2}m(5.87\times10^{3}\,\text{m/s})^2 - \frac{\left(6.67\times10^{-11}\,\text{N}\cdot\text{m}^2/\text{kg}^2\right)(5.98\times10^{24}\,\text{kg})m}{9.45\times10^{6}\,\text{m}} = -(2.5\times10^{7}\,\text{m}^2/\text{s}^2)m$$

The end result is that we have the exact same total energy at both locations along the orbit. Although this doesn't guarantee our work is correct, it's a strong indication that it is. No nonconservative forces were acting, and thus we expect to have the same total energy at all points of the orbit.

Check There is no need to check a check!

EXERCISES 4.2

4.2.1. **[Level 1]** A mass particle P is projected up into a frictionless hollow U-shaped tube with a speed of 4.0 m/s. $h = 0.4$ m, $r = 0.2$ m. What is its speed at C? Verify that your result makes sense.

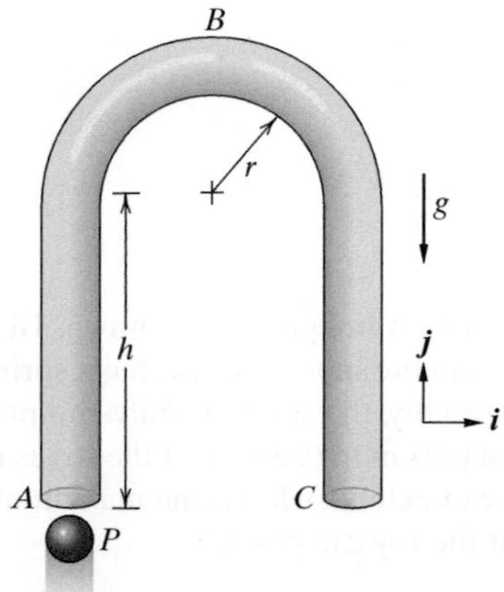

E4.2.1

4.2.2. **[Level 1]** Back in the dark ages researchers tested car bumpers with a swinging bar, similar to the one illustrated. The trapezoidal arrangement of the arms and weight ensured that the swinging bar contacted the bumper with a horizontal velocity (assuming it had been set up properly to impact with the arms vertical). Because of the trapezoidal geometry, the system can be viewed as a simple pendulum (lumped mass on the end of a pivoting arm). Each arm is 1.8 m long and the mass of the bar is 150 kg. If the arms are rotated up $\theta = 31°$ from vertical and then released, how fast will the bar be traveling at impact?

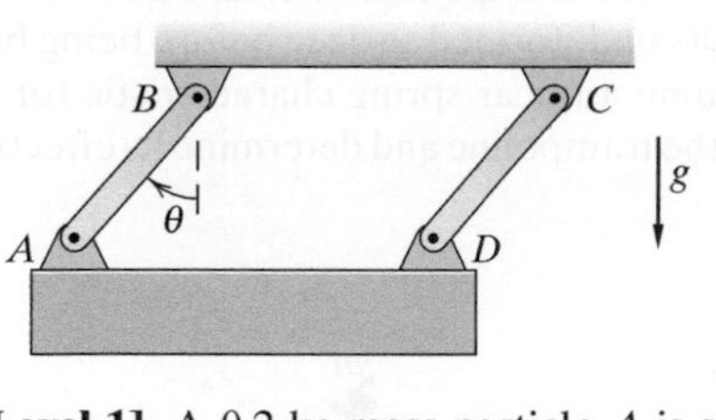

E4.2.2

4.2.3. **[Level 1]** A 0.2 kg mass particle A is suspended over the illustrated track. Once dropped from a height $h = 1.5$ m, it smoothly contacts the curved track ($r = 0.5$ m), slides along the ground, and finally impacts an elastic cushion B. The spring constant of the cushion is 40 N/m. What is the cushion's maximum compression?

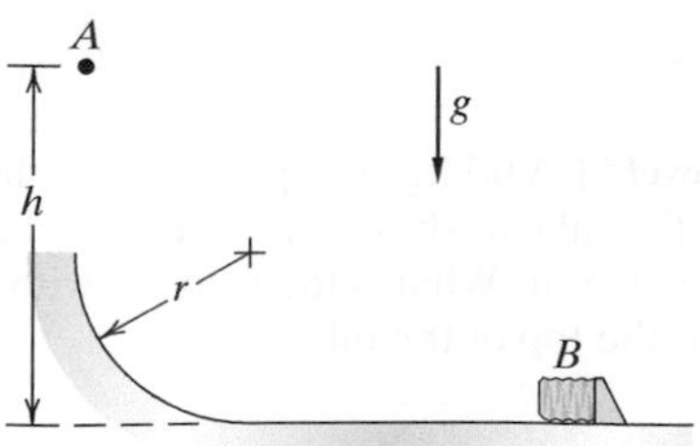

E4.2.3

4.2.4. **[Level 1]** A block of mass m is projected up a frictionless slope with speed $v = 8$ m/s and contacts a linear spring. What should the spring stiffness k be if the spring compression is to be limited to 10 cm? $m = 2$ kg, $L = 5$ m, and $\theta = 30°$.

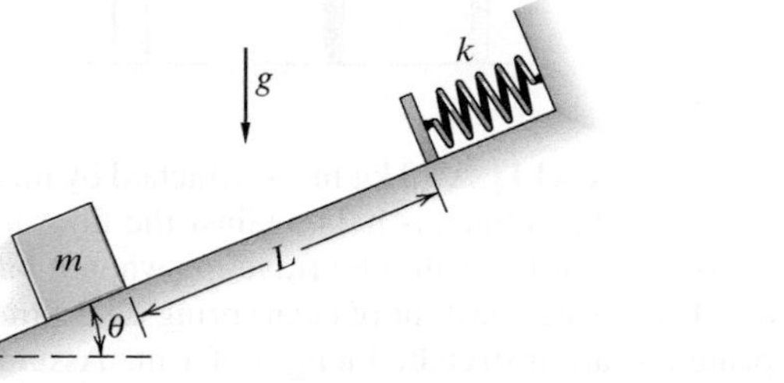

E4.2.4

4.2.5. **[Level 1]** A collar of mass $m = 5$ kg is constrained to move in the vertical plane along a smooth rod bent into a 90° arc of radius $r = 1$ m. A spring of stiffness $k = 500$ N/m and free length $L_0 = 0.25$ m is attached to the collar as depicted. If the collar is released from rest at the rod's top A ($\theta = 90°$), find its speed v halfway along the rod (i.e., at $\theta = 45°$).

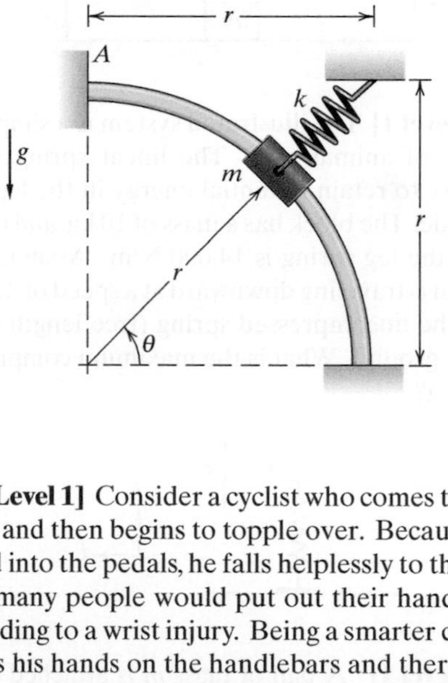

E4.2.5

4.2.6. **[Level 1]** Consider a cyclist who comes to a stop at a stop sign and then begins to topple over. Because his feet are clipped into the pedals, he falls helplessly to the side. At this point many people would put out their hand to break the fall, leading to a wrist injury. Being a smarter cyclist, our rider keeps his hands on the handlebars and therefore topples all the way over and finally strikes the ground with his upper torso. We will consider the simplest approximation to this problem and model the cyclist as a lumped mass m atop a massless rod of length h. The mass will contact the ground with a speed v after the rod has pivoted by $\frac{\pi}{2}$ rad. Find v for a height of 1 m.

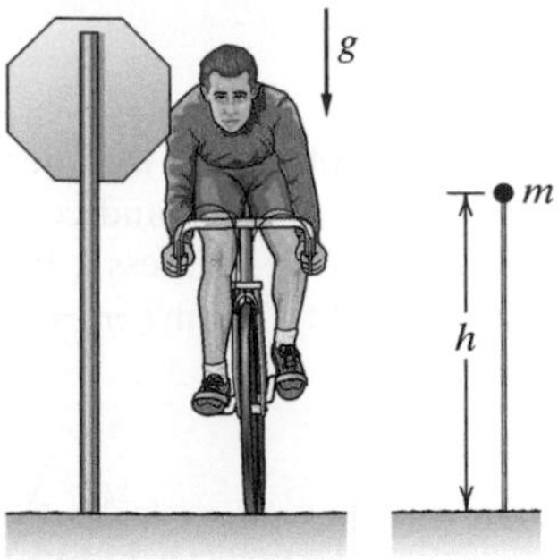

E4.2.6

4.2.7. **[Level 1]** A 10 kg mass, attached by means of two springs to the ceiling, is held against the floor and is then released. How fast will it be traveling when it hits the ceiling? The spring constant of each spring is 80 N/m, and each spring has an unstretched length of 1 m. Assume that the springs become loose and floppy once they're at their rest length.

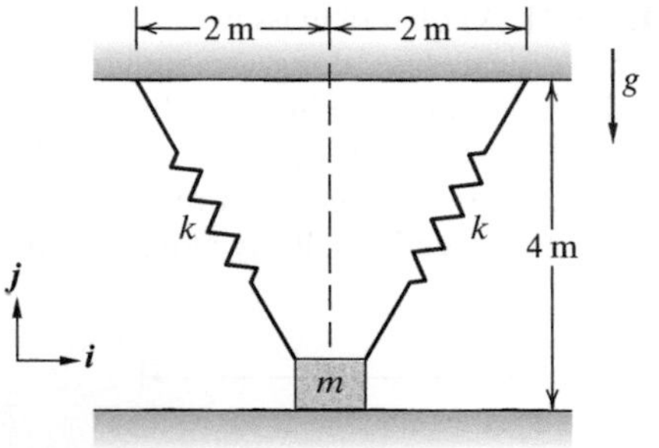

E4.2.7

4.2.8. **[Level 1]** The illustrated system is a simple model for analyses of animal gait. The linear spring models a body's ability to retain potential energy in the leg muscles during a stride. The block has a mass of 10 kg, and the spring constant of the leg spring is 14,000 N/m. Assume that the entire system is traveling downward at a speed of 2 m/s when the end of the uncompressed spring (free length of 0.5 m) contacts the ground. What is the maximum compression of the leg spring?

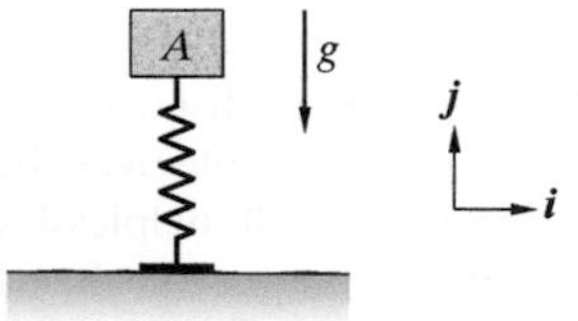

E4.2.8

4.2.9. **[Level 1]** A ball of mass m is attached to a rigid rod of length L and negligible mass that is constrained to rotate about the pivot O in the vertical plane. The ball is released from rest at $\theta = 90°$, where θ is the angle that the rod makes with the vertical, as illustrated. The ball strikes a vertical surface when it reaches the bottom of its swing. If the coefficient of restitution is $e = 0.7$, at what angle does the ball come to a stop after impact?

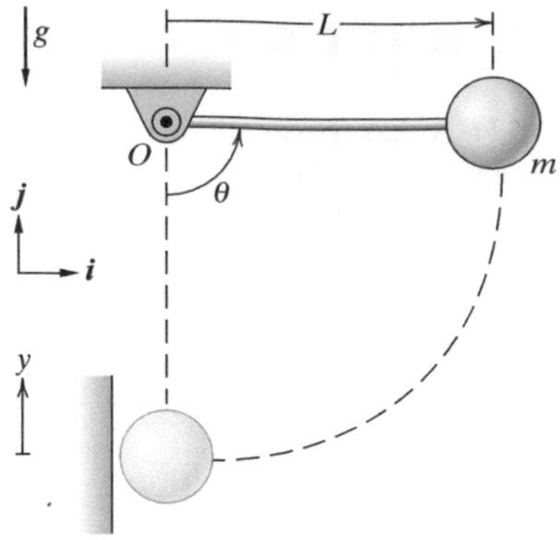

E4.2.9

4.2.10. **[Level 1]** A hopping toy is shown. Block C has a mass of 0.08 kg, and the support spring has a spring constant of 3.0 N/mm. Initially, the spring is fully compressed from its rest length of 0.04 m to 0.008 m. If the toy is released so that the spring can extend, what is the maximum height off the ground that the toy can reach?

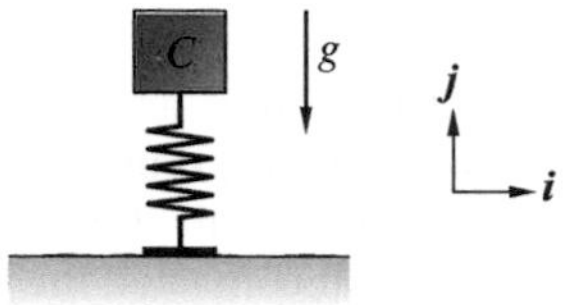

E4.2.10

4.2.11. **[Level 1]** A circus trapeze artist swings off the trapeze, arcs up, and finally lands on a trampoline. Neglect the lateral motion and concentrate only on vertical positions and speeds. At release the 68 kg man had an upward speed of 3.6 m/s and was 15 m above the trampoline's surface. After landing on the trampoline, he sank down 0.40 m below the level of its undeformed surface before being brought to a halt. Assume a linear spring characteristic for the compression of the trampoline and determine its effective spring constant.

E4.2.11

4.2.12. **[Level 1]** A 0.1 kg mass particle m is shot upward at 11 m/s into the tube as shown. The surface of the tube is frictionless. $r = 0.5$ m. What is the velocity with which the mass arrives at the top of the tube?

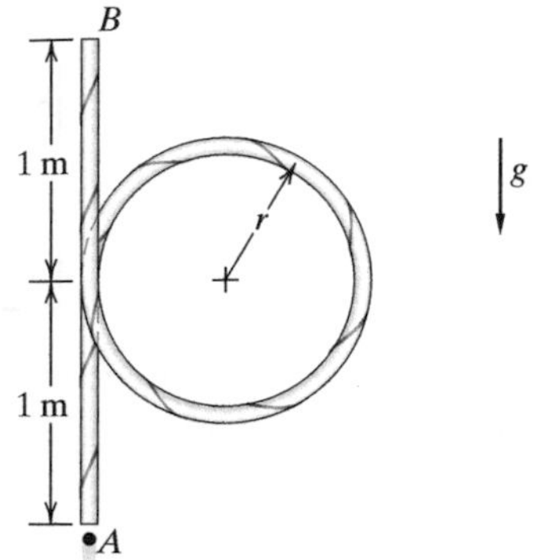

E4.2.12

4.2.13. [Level 1] A particle ($m = 0.5$ kg) is held in place between two linear springs ($k = 60$ N/m and unstretched length $L = 0.01$ m). The mass is then released from rest. What is v_m when the mass has dropped 0.04 m?

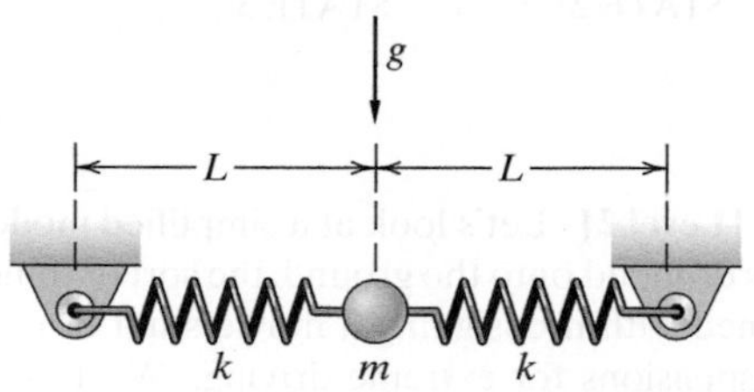

E4.2.13

4.2.14. [Level 1] A cyclist is riding along at 40 km/h and begins to coast just before encountering a hill with a 6% grade. How far up the hill can she coast before coming to a stop? Ignore air resistance and rolling friction.

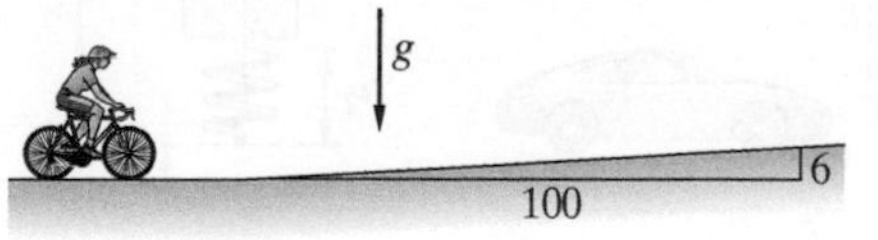

E4.2.14

4.2.15. [Level 1] A bowling ball with mass m is held by a bowler whose arm is modeled as a massless rod with length L. The ball is at one end of the rod, and the other is fixed at point O, the shoulder joint. The bowling ball starts from rest with the rod in the horizontal position. The bowler then starts swinging the ball in the vertical plane with a constant applied moment M at the shoulder joint. Calculate the ball's speed when it reaches the lowest point of its path.

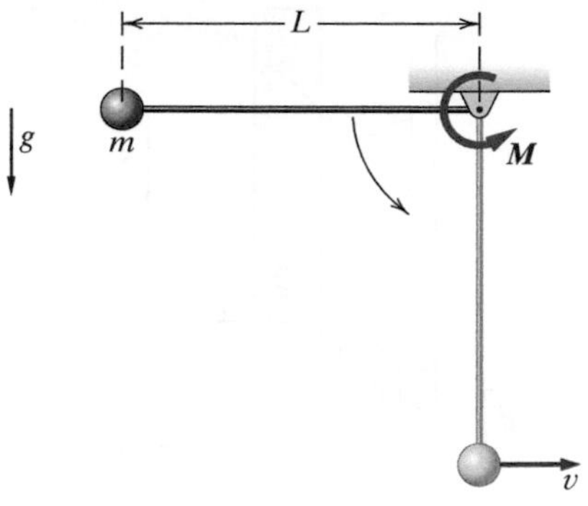

E4.2.15

4.2.16. [Level 1] If you've ever been to the docks, you know that the dock sides are often covered with elastic blocks that act to keep the docked boats from banging directly into the wooden dock structure. Real restraining blocks have significant damping, but we will neglect that in this exercise. Assume that a docked boat has a mass of 9000 kg and is moving toward the dock at a speed of 0.3 m/s. What linear spring constant k is needed if the elastic block is to deform a maximum of 5 cm? If that same spring constant was used in each of the four suspension springs of a 1400 kg car, how much would the car settle due to a static gravity load?

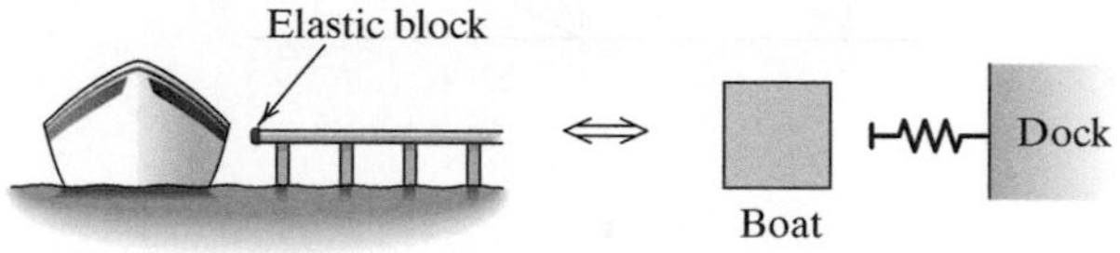

E4.2.16

4.2.17. [Level 1] The front end of a car is designed to crush during a collision, protecting the occupants from the worst effects of the crash. We will model the car as a lumped mass and the crush zone as an ideal spring. Calculate the maximum deformation of the vehicle structure (compression of the spring) if the car impacts an immovable wall at 64 km/h. The car weighs 1360 kg and $k = 1.17$ MN/m.

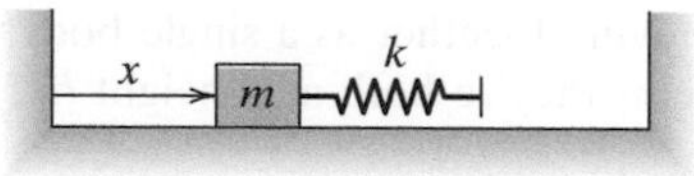

E4.2.17

4.2.18. [Level 1] The system shown is a simplified pile driver. The parallel links ensure that the main mass doesn't rotate and is traveling vertically at the instant it strikes the ground. Each guiding link has a length of 2 m, $m = 100$ kg, and at release $\theta = 30°$. What is the speed of the main mass just before impact?

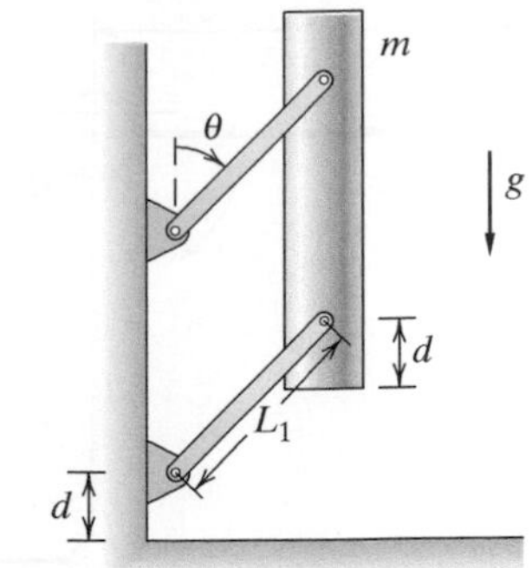

E4.2.18

4.2.19. **[Level 1]** A particle is placed just to the right of the top of a smooth semicircular hill. Because it's infinitesimally to the right and the surface is frictionless, it will start to slide down the right slope. Determine its speed when it's at an arbitrary value of θ (assume the particle is always in contact with the hill).

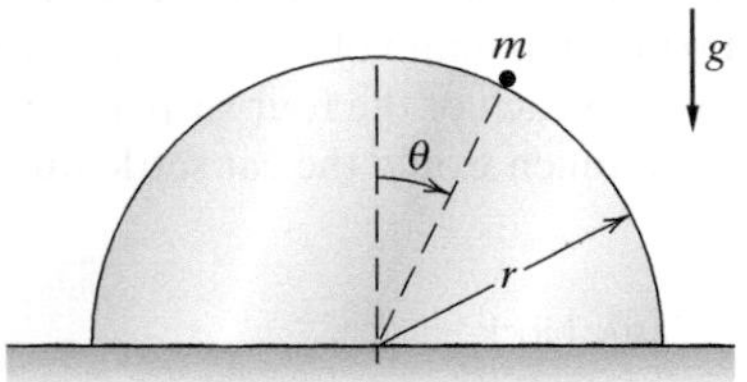

E4.2.19

4.2.20. **[Level 2]** Block A (with mass m) is released from rest in the configuration shown in State 1. The spring has an unstretched length of zero, and the equilibrium position for the spring/mass system is a distance $h/2$ below the spring's upper attachment point ($\frac{h}{2} = \frac{mg}{k}$). After release block A moves upward. B ($m_B = \frac{m}{2}$) is falling downward. State 2^- shows both blocks at $h/2$ above the ground, an instant before collision, with B moving down with speed v_B and A moving up with speed v_A.

State 2^+ shows the two blocks an instant after collision, both masses moving together as a single body with speed v. At this instant they're both at a height $h/2$ above the ground.

The two blocks remain joined after the collision. What we want is for the velocity of the $A + B$ combination to go to zero just as it reaches the ground (State 3).

Determine at what speed v_B must block B be moving when it impacts block A for this sequence to occur.

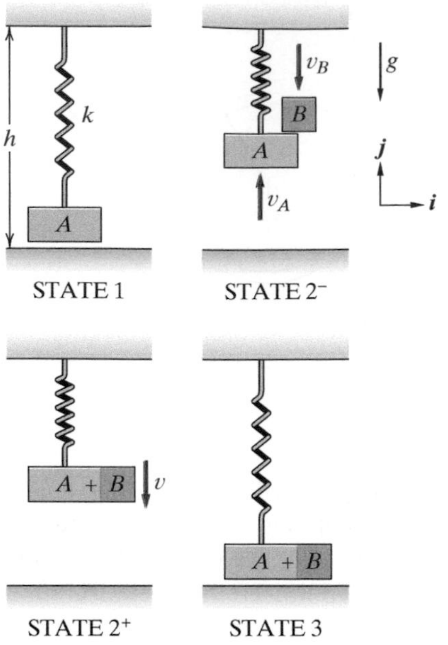

E4.2.20

4.2.21. **[Level 2]** Let's look at a simplified modeling of a car that's dropped onto the ground, the sort of thing you'd be concerned with if designing a movie stunt or designing vehicle suspensions for extreme driving. We'll model the suspension as a single linear spring and the rest of the vehicle as a single lumped mass. The car is dropped from a height of $h_1 = 0.9$ m. What is the maximum mass m for which the suspension does not bottom out? Assume a maximal allowable suspension travel of 10 cm and a spring constant of 2.4×10^6 N/ m.

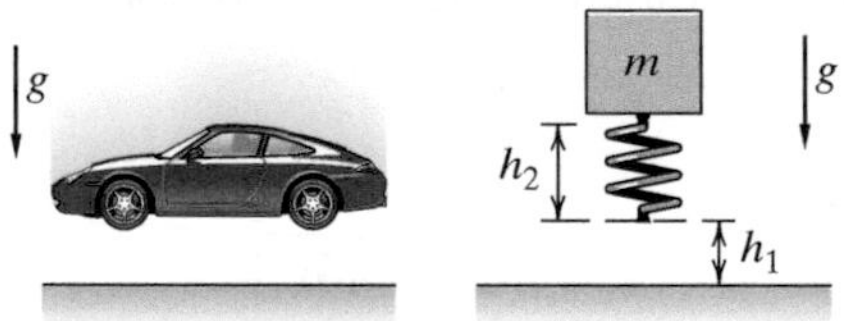

E4.2.21

4.2.22. **[Level 2]** Ball A ($m_A = 2$ kg) is released from rest and allowed to travel down a frictionless path given by a quarter circle with radius $R = 1$ m. At the end of the path, ball A encounters a spring with a stiffness of $k = 500$ N/m that is attached to block B, which is initially stationary and has a mass of $m_B = 15$ kg. Block B rests on top of a rubber mat nailed onto the ground with a coefficient of static friction equal to $\mu_s = 0.8$. Does block B move as a result of ball A colliding with the spring?

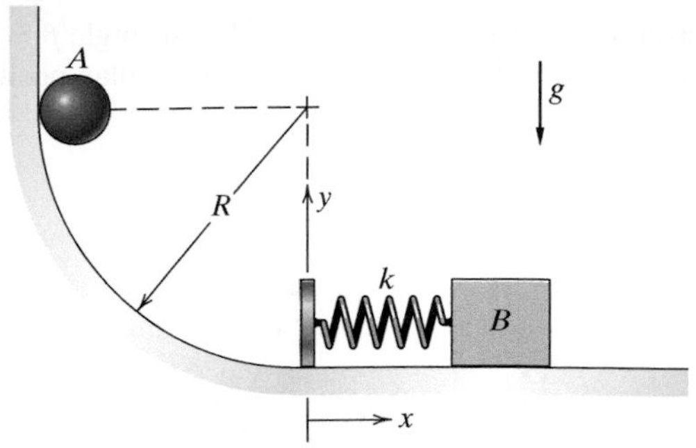

E4.2.22

4.2.23. **[Level 2]** The Charpy v-notch test provides a means of measuring the toughness of a material by examining how much energy it absorbs before fracture. The test apparatus involves a large, heavy, pivoted hammer that is released from rest at a prescribed angle and impacts a piece of test material with a small v-shaped notch cut into it at the bottom of its swing. After the hammer breaks the test piece, it is allowed to swing up and come to a stop, at which point the hammer's final orientation is noted. The amount of energy absorbed by the test piece (and thus its toughness) is related to the difference in the hammer's release height and the height at which it comes to rest. Suppose the hammerhead weighs 13.6 kg and the release angle is $\theta_0 = 60°$ as measured from the vertical. By means of a high-speed camera, it is found that the impact between the hammerhead and test piece occurs over $\Delta t = 0.01$ s. The hammer's post-impact swing makes an angle of 45° to the vertical when the hammer comes to a stop. Calculate the average force transmitted to the test piece during impact. The hammer's arm is $L = 0.9$ m long, and its mass is negligible compared to that of the hammerhead.

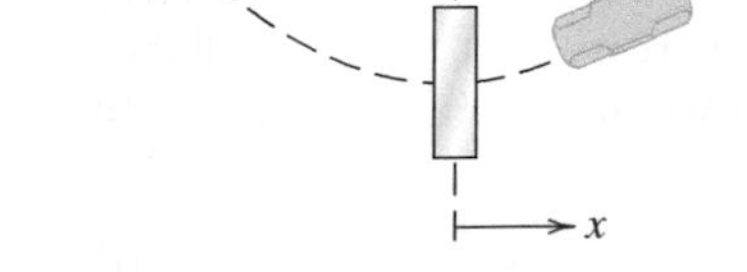

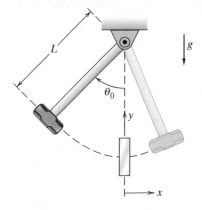

E4.2.23

4.2.24. **[Level 2]** Jake wants to determine the coefficient of dynamic friction μ_d for a $L = 2$ m long ramp angled at $\theta = 50°$ with respect to the horizontal. To do so, he releases a wooden block from the top of the ramp and lets it slide to the base, at which point he measures the block's speed to be $v = 4$ m/s. What is μ_d?

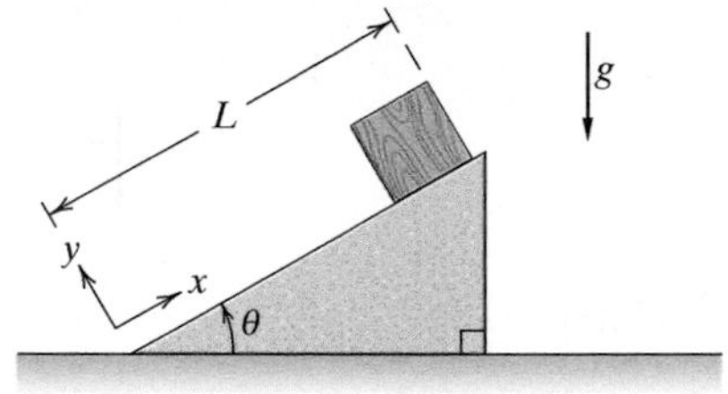

E4.2.24

4.2.25. **[Level 2]** Mass m_1 slides on a rough surface whose coefficient of dynamic friction is $\mu_d = 0.3$. Mass m_2 is suspended under a pulley and connected to m_1 by an inextensible rope that goes over the pulley. The system is initially at rest, and a spring with a stiffness of $k = 1000$ N/m is located $h = 3$ m directly below m_2. For $m_1 = 20$ kg and $m_2 = 12$ kg, m_1 slips and causes m_2 to fall toward the spring. Upon contact, m_2 compresses the spring by $\delta = 0.153$ m before the tension in the rope vanishes. Calculate the blocks' speed v at the instant the rope goes slack.

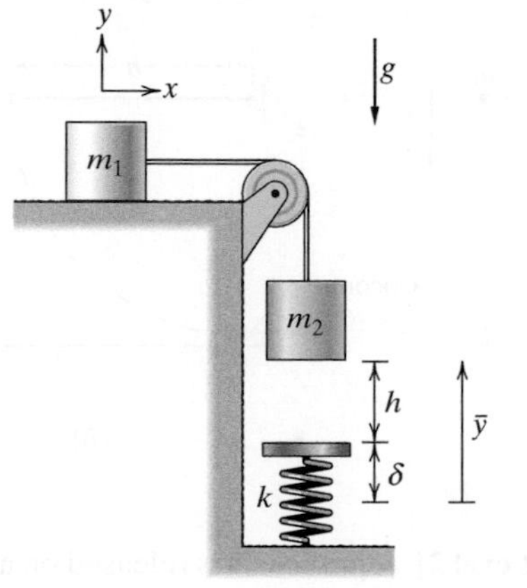

E4.2.25

4.2.26. **[Level 2]** A particle P is released from the position shown, slides down the frictionless track, and strikes the spring at $x = 2$ m. $m = 5$ kg and $k = 400$ N/m. Solve for the velocity corresponding to a spring compression of 0.8 m.

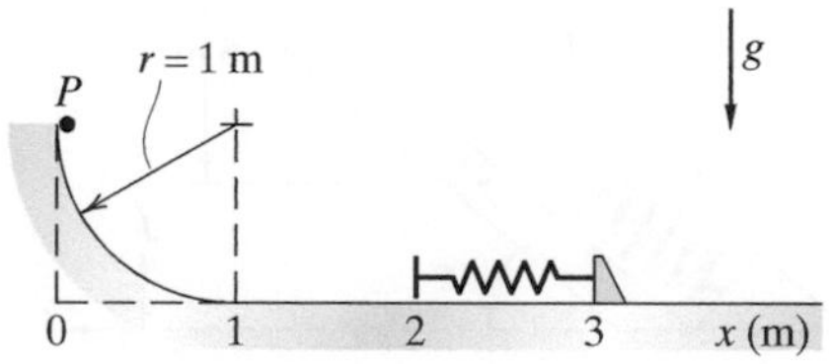

E4.2.26

4.2.27. **[Level 2]** A mass particle P is projected to the left at 7.95 m/s along the illustrated frictionless track. At what angle θ will it lose contact with the track? $r = 1.5$ m and $m = 4$ kg.

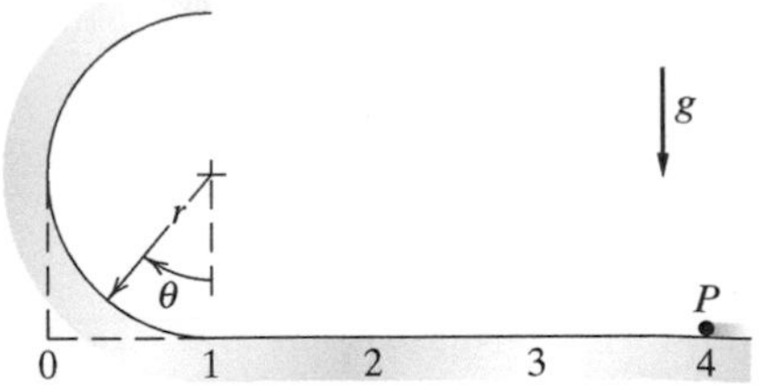

E4.2.27

4.2.28. **[Level 2]** You have been called in as a consultant for the North Dakota Coconut Company to evaluate two coconut-opening concepts, a problem that you may have already examined using the concepts of angular momentum. This time around you will use energy. The first concept, (*a*), consists of a mass m that's dropped from a height h onto a coconut. The second, (*b*), utilizes a mass m that's attached to the end of a pivoted, massless rod of length h. The rod is raised to a horizontal position and then released. In this case the mass strikes the coconut at the bottom of its swing, smashing it against a vertical wall. Is one device superior to the other in its coconut-cracking ability?

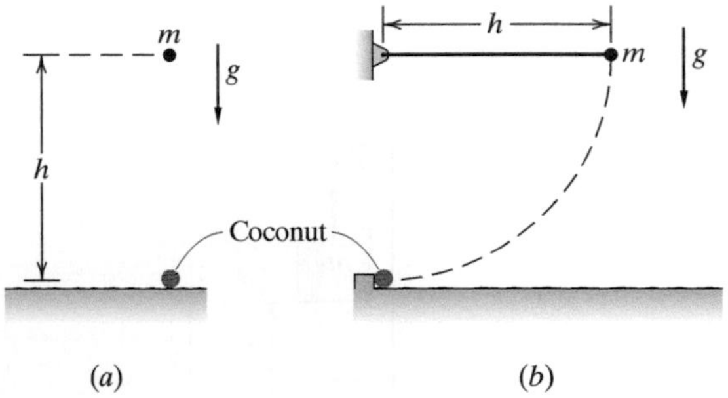

E4.2.28

4.2.29. **[Level 2]** A particle A is released on a 40° slope. The coefficient of friction along the slope is $\mu_s = \mu_d = 0.2$. The particle eventually contacts a massless pad attached to a linear spring having a spring constant of $k = 150$ N/mm. It bounces upslope and then returns. How fast is it traveling when it contacts the pad a second time? $m_A = 0.5$ kg.

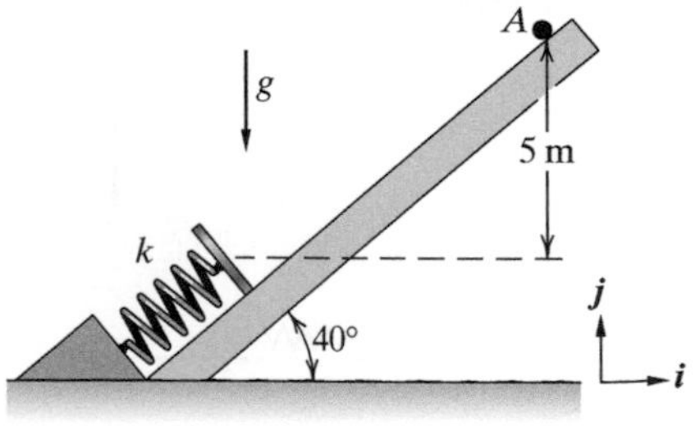

E4.2.29

4.2.30. **[Level 2]** As illustrated, a collar of mass $m = 1$ kg is constrained to move in the vertical plane along a smooth rod bent into a semicircle of radius $r = 0.5$ m. The collar starts from rest at the rod's top A ($\theta = 90°$) and is acted on by a constant force of $F = 20$ N applied at an angle $\beta = 45°$ with respect to the horizontal. At what angular position does the collar come to a stop?

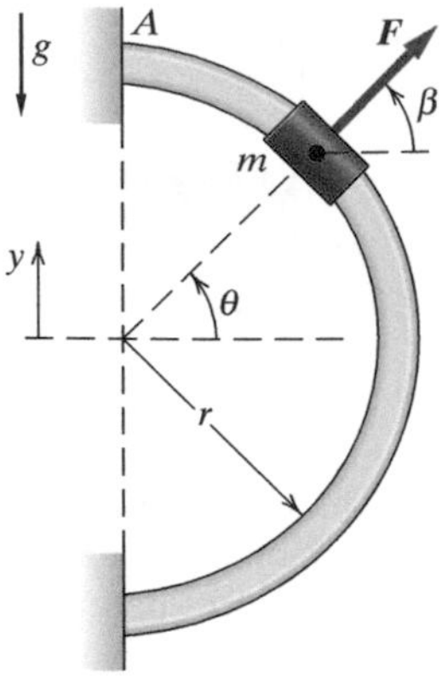

E4.2.30

4.2.31. **[Level 2]** The illustrated mechanism consists of a massless, rigid link $\overline{AB}$ of length 0.5 m that's connected by a frictionless hinge to the 0.4 m massless link $\overline{BC}$. The hinge at B has a mass of 10 kg. At C the link $\overline{BC}$ connects to a small, massless wheel that can roll along the horizontal surface. The axle of the wheel has a mass of 15 kg. Attached to the axle is a stretched linear spring with spring constant $k = 80$ N/m and an unstretched length of 0.2 m. At the instant illustrated $L = 1.2$ m. How fast will the hinge B be moving when $\theta = 30°$?

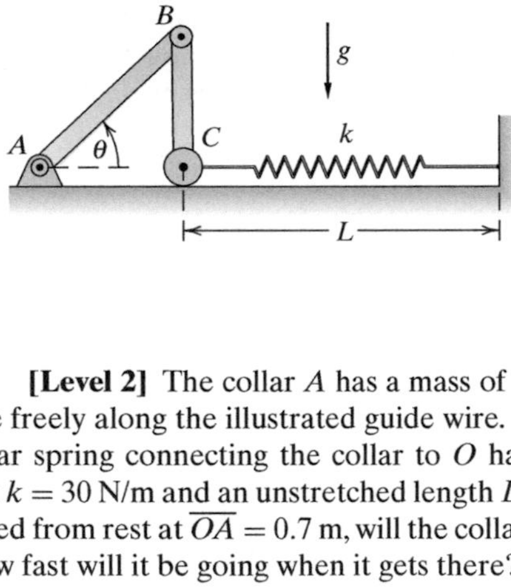

E4.2.31

4.2.32. **[Level 2]** The collar A has a mass of 0.5 kg and can slide freely along the illustrated guide wire. The massless linear spring connecting the collar to O has a spring constant $k = 30$ N/m and an unstretched length $L_0 = 0.1$ m. If released from rest at $\overline{OA} = 0.7$ m, will the collar reach B? If so, how fast will it be going when it gets there?

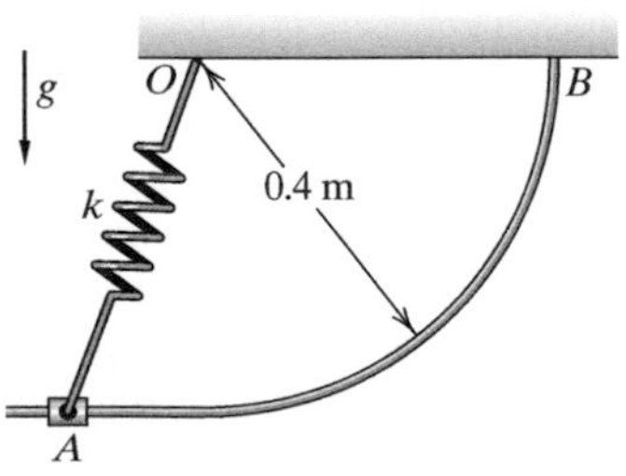

E4.2.32

4.2.33. **[Level 2]** A rigid, massless disk is freely pivoted about its center. A massless string is wound around the disk's periphery, hangs down the left side, and is attached to block B ($m_B = 0.5$ kg). A massless rod, fixed to the disk, extends out to a 1.1 kg mass at A. The system is released from rest with $\theta = 30°$. What is block B's velocity when $|\theta| = 45°$? $r = 0.08$ m and $L = 0.6$ m.

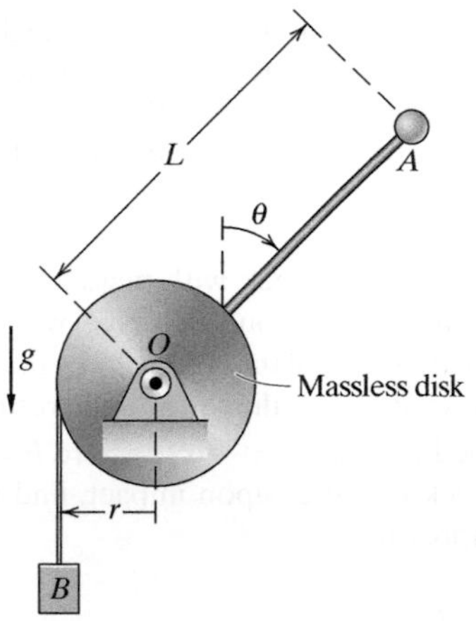

E4.2.33

4.2.34. **[Level 2]** Two identical, inverted pendulums are joined together by a spring ($k = 42$ N/m, unstretched length of 1.23 m). Mass A and mass B are each 2.0 kg, and $\overline{OA} = \overline{OB} = 1.2$ m. The two rods are pulled down so that $\theta = 60°$. If both are released from rest, what is $\boldsymbol{v}_B$ when $\theta = 45°$?

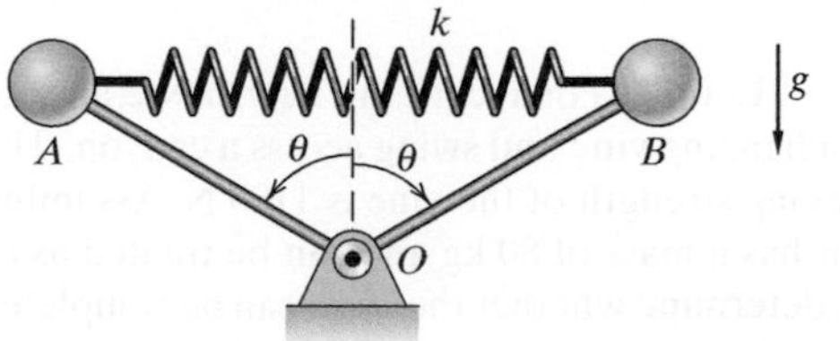

E4.2.34

4.2.35. **[Level 2]** A block with mass $m = 3$ kg is dropped from rest at a height $h = 2$ m above a spring. The spring is nonlinear, and the restoring spring force is given by $F_{sp}(x) = -ax^{\frac{1}{3}}$, where $a = 500\,\text{N/m}^{\frac{1}{3}}$.

a. Find the maximum spring compression δ_{max}.

b. Suppose the nonlinear spring is replaced by a linear one of stiffness $k = 1000$ N/m. From what height should the block be dropped such that the linear spring also has a maximum compression of δ_{max}?

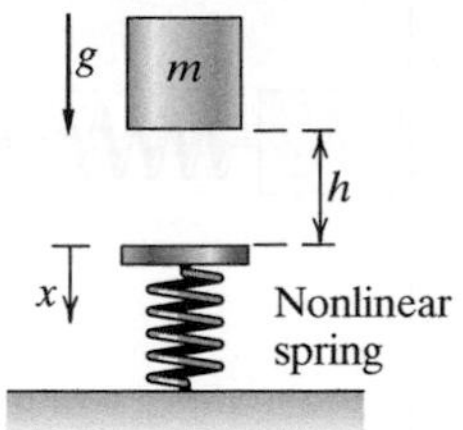

E4.2.35

4.2.36. **[Level 2]** Block A has a mass of 10 kg and is suspended in equilibrium by a linear spring ($k = 200$ N/m). Assume an unstretched length of zero for the spring. At $t = 0$ half the mass shears off of block A and falls to the floor. At what speed will the part of the block still attached to the spring contact the ceiling?

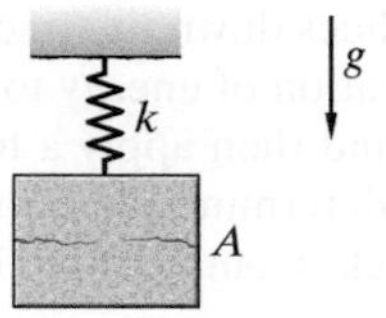

E4.2.36

4.2.37. **[Level 2]** A fishing pole is suspended as shown. The rod can be approximated as a massless, flexible beam with a tip mass A ($m_A = 0.04$ kg). A 10 kg fish (approximated by the lumped mass B) is attached to the end of the rod, causing the total end deflection to be $y = 0.2$ m. If the fish suddenly falls off the line, with what speed will A be traveling when $y = 0$?

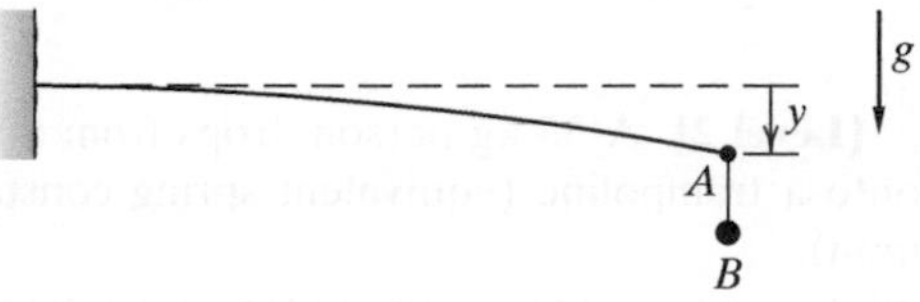

E4.2.37

4.2.38. **[Level 2]** A spring-loaded block of mass $m = 5$ kg is forced up against a rough wall at a height $h = 2$ m above another spring, where both springs have a stiffness of $k = 800$ N/m. The other end of the block's restraining spring is constrained to move vertically with the block within a smooth guide. The block is released from rest, slides down the wall, and eventually compresses the ground-fixed spring by $\Delta y = 0.3$ m before coming to a stop. If the compression of the block's restraining spring is a constant $\delta = 0.1$ m, what is the coefficient of friction μ between the block and the wall?

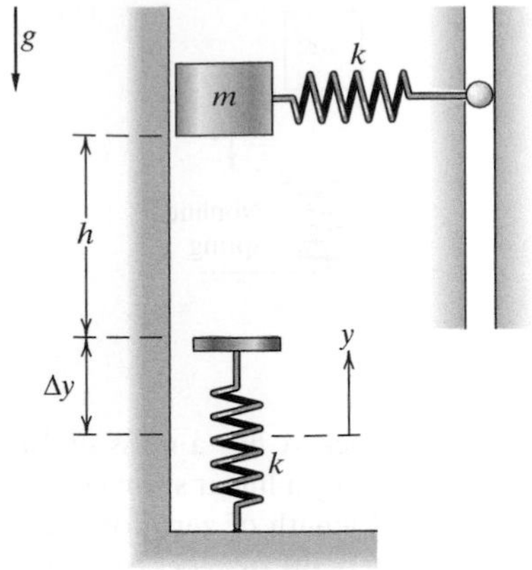

E4.2.38

4.2.39. [Level 2] Let's see what happens to a skateboard rider as she starts down the track. Treat the 32 kg rider as a mass particle that slides down the track in a friction-free manner. Use conservation of energy to determine her velocity when $\theta = 45°$ and then apply a force balance using polar coordinates to determine the normal force between the skater and the track. Comment on the result.

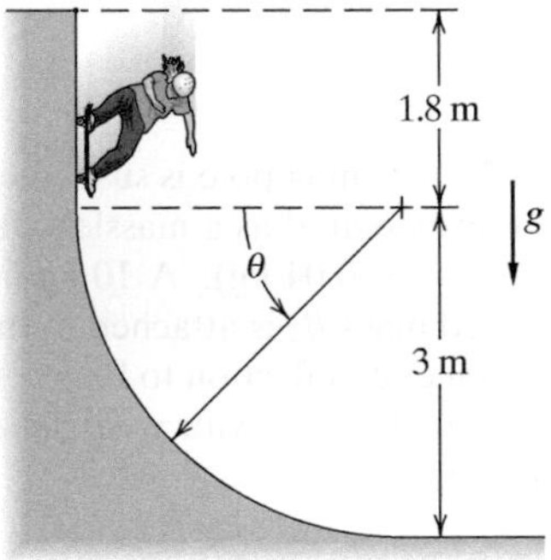

E4.2.39

4.2.40. [Level 2] A 55 kg person drops from a height of 6 m onto a trampoline (equivalent spring constant of $k = 7$ N/mm).

a. Will the trampoline bottom out (meaning the person contacts the ground)?

b. If it does bottom out, how high should the supports be made to avoid bottoming?

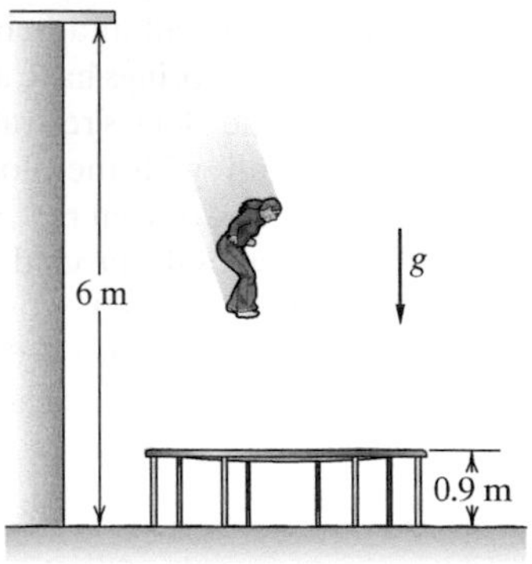

E4.2.40

4.2.41. [Level 2] Computational Let's now reconsider **Exercise 3.4.45**. You ran the code, but did you get the right result? Let's find out by using energy. Take your numerical time simulation from that problem and calculate the initial energy, the final energy, and the energy dissipated through friction. Do the results agree? Does the initial energy minus the frictional loss equal the final energy? Inquiring minds want to know. And, just because this is so cool, see if you can find a coefficient of kinetic friction for which the person will stop sliding when he reaches $\theta = 30°$ (measured from the vertical). After all, that's what Bond wants, right? He doesn't really *want* to fall off of the building, does he?

4.2.42. [Level 2] A block with mass $m = 6$ kg is projected over a smooth horizontal surface with a speed of $v = 15$ m/s toward a block of mass $M = 15$ kg that is attached to a spring. The spring is nonlinear, and the restoring spring force is governed by $F_{sp}(x) = -bx^{\frac{1}{3}}$, where $b = 700\ \text{N/m}^{\frac{1}{3}}$. If the blocks stick together upon impact, find the spring's maximum compression δ_{max}.

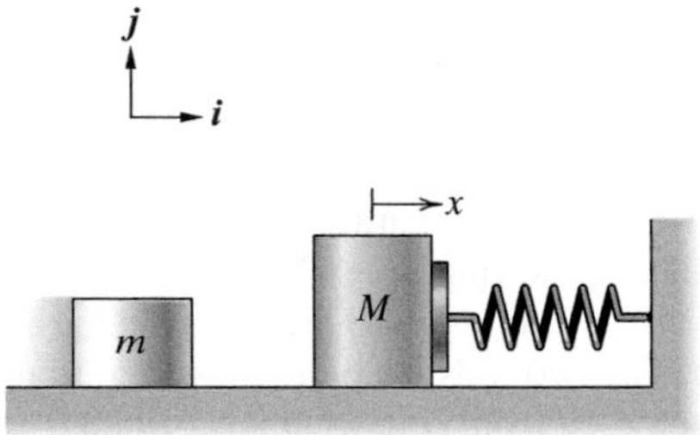

E4.2.42

4.2.43. [Level 2] For a scene in a new movie, an actor has to grab a hanging vine and swing across a canyon. The tensile breaking strength of the vine is 1100 N. Assuming that the actor has a mass of 80 kg and can be treated as a mass particle, determine whether the stunt can be completed successfully. Assume she begins the swing by stepping off the ledge with essentially zero velocity. $L = 10$ m.

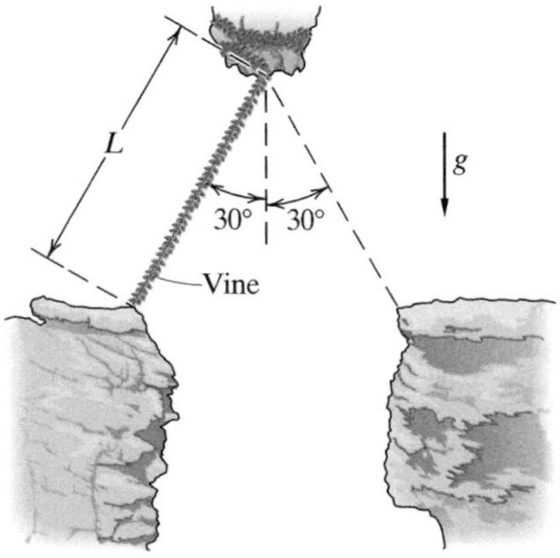

E4.2.43

4.2.44. **[Level 2]** A young boy thinks it would be fun to hang a rope from a tree branch and swing from A to B. He's fully capable of supporting his own weight, and tests with his younger sister proved that he could hold onto a stationary rope even when she was hanging from his shoulders. That was the limit, however; any more weight and he would be unable to maintain his grip. Approximate the system as a point mass hanging from a massless rope of length L and determine how large θ can be for him to successfully reach point B. His mass is 40 kg, his sister's is 25 kg, and $L = 4$ m. (He swings without his sister.)

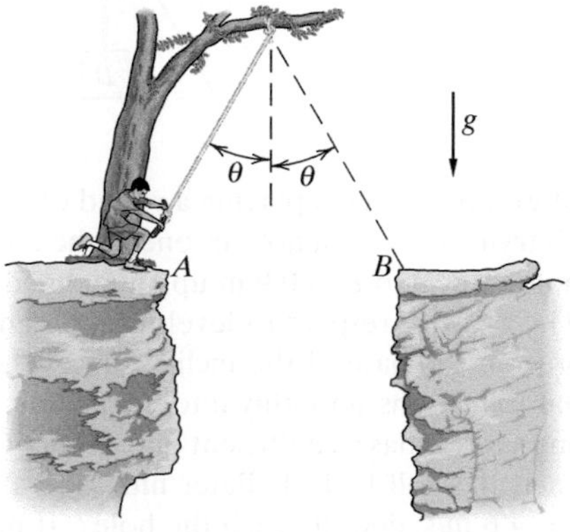

E4.2.44

4.2.45. **[Level 2]** A bungie jumper needs to calculate how much bungie cord to attach to herself so that it will bring her to rest 3 m above the ground. The spring constant of the bungie cord is 22 N/m, and she has a mass of 55 kg. Neglect the bungie cord's mass.

a. How long a bungie cord is required?

b. If she uses the length calculated in **a**, but the spring stiffness is 10% less than it was advertised to be, how fast will she hit the ground?

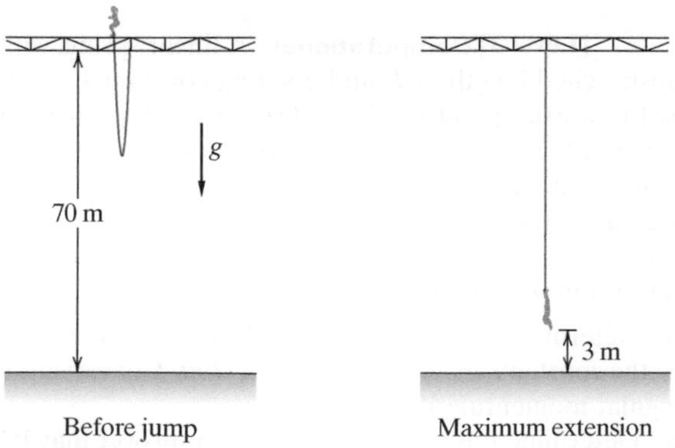

E4.2.45

4.2.46. **[Level 2]** A particle (mass m) is released from rest. A linear compression spring ($F = -kx$, where x is the spring's compression) is located a distance L from the mass and is initially unstretched. The slope on which the mass is released has an angle β, and the coefficient of friction is μ. $m = 1.1$ kg, $L = 2$ m, $\beta = 30°$, $k = 3500$ N/m, and $\mu = 0.1$. What is the spring's maximum deflection?

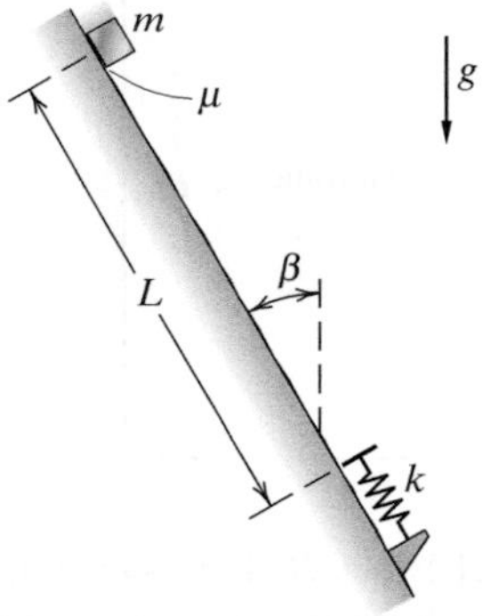

E4.2.46

4.2.47. **[Level 2]** A person on a pogo stick can be modeled as a mass m on top of a spring. The person and pogo stick have an initial velocity of $v_0 = 2.5\boldsymbol{j}$ m/s and an initial height of $h_0 = 1.5$ m. Assume that $k = 3000$ N/m, $m = 50$ kg, and the spring has a rest length (neither compressed nor extended) of $L = 1.2$ m. Calculate the maximum height reached and the maximum compression of the pogo stick.

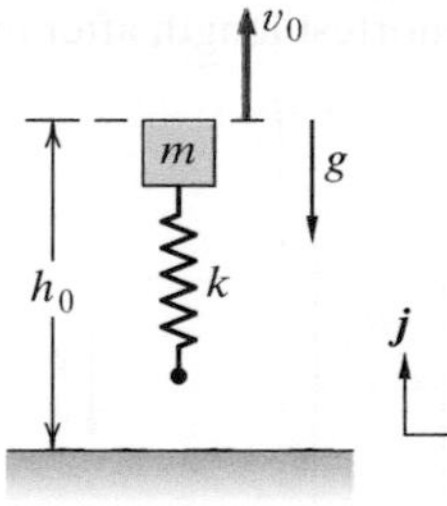

E4.2.47

4.2.48. **[Level 2]** In a recent action movie, the heroine is pushed over a railing, but she grabs the railing as she goes over, managing to hold on for a moment, her body dangling beneath her. Before her grip fails, she grabs the end of a conveniently placed elastic bungie cord. An instant later she falls and her co-star manages to secure the free end of the bungie cord to the railing. Assuming that she can hold on to her end of the bungie cord, will she contact the ground? Assume that the bungie cord going from the railing to the heroine has an unstretched length of 20 m and a spring constant of 63 N/m. The railing is 52.2 m above the ground, she has a mass of 57 kg, and the length from her hands to her feet is 2.2 m.

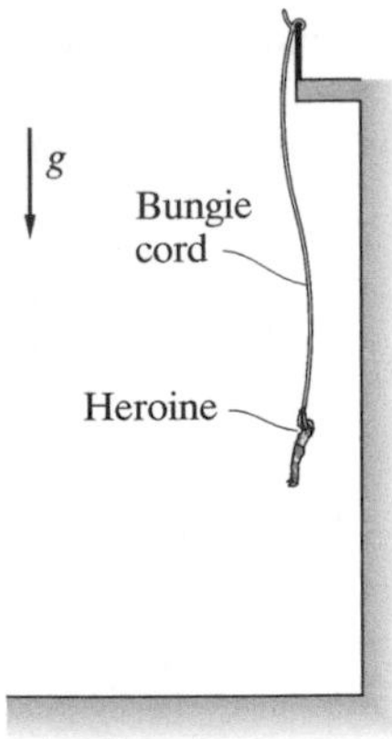

E4.2.48

4.2.49. **[Level 2]** Block B of mass m_B is falling down the illustrated chute and has a speed $-v_0\boldsymbol{j}$. At the instant shown it is h above the ground. At impact, block B sticks to block A, which is attached to a linear spring with uncompressed length L. A constant frictional force F_f acts between block B and the walls of the chute (total force from both walls $2F_f$). Answer the following in symbolic form using the given parameters.

a. Assume a (short) impact duration time Δ. Using the fact that the impact time is short, write down an expression for the magnitude of the force between blocks A and B during the impact.

b. Write down an expression that would allow you to solve for the spring's shortest length after impact.

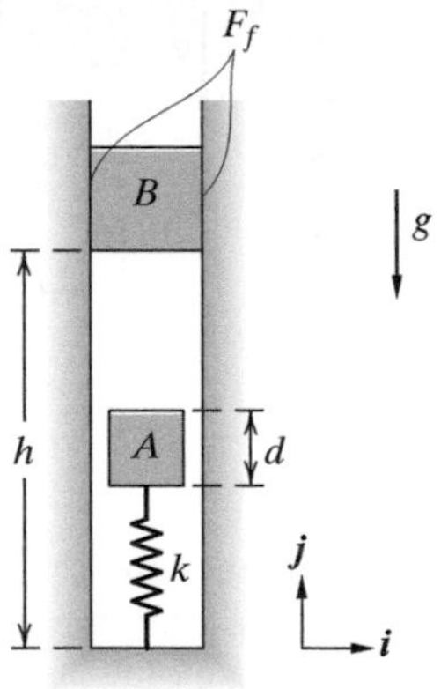

E4.2.49

4.2.50. **[Level 2]** Block A rests on a horizontal, frictionless surface and is attached by a massless rope to a vertical wall at P and by another massless rope to block B. Both A and B have identical masses. Block A is also connected to the fixed point O by means of an extensional spring with rest length L_0. The force applied by the spring to block A is given by

$$F_s = \frac{5mg}{L_0}(L - L_0)$$

where L is the spring's stretched length. At $t = 0$ the left rope snaps, allowing block A to slide to the right. Find block A's speed at the instant that it loses contact with the horizontal surface due to the upward force of the spring. $L_0 = 0.5$ m.

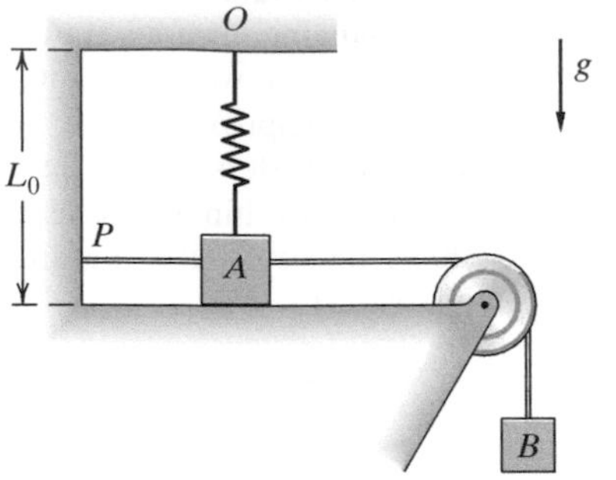

E4.2.50

4.2.51. **[Level 3]** Peter is playing a round of miniature golf, and he's trying to score a hole-in-one at the number 14 hole. The hole is located $L = 0.9$ m up an inclined surface angled at $\theta = 30°$ with respect to level ground, and Peter is $s = 0.9$ m from the base of the incline on level ground. Level ground transitions smoothly into the incline, and the entire playing surface has a coefficient of friction of $\mu = 0.4$ between it and the golf ball. If Peter hits the ball with a speed of $v_0 = 4.6$ m/s, does it reach the hole? If not, what speed does he need to hit the ball with so that it just makes it to the hole?

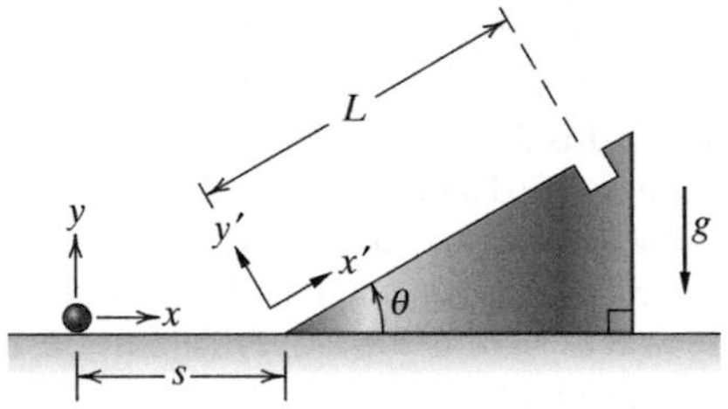

E4.2.51

4.2.52. **[Level 3] Computational** A linear spring, with an unstretched length of L and a spring constant k, is attached to a fixed point O ($F = -k(r - L)$). A particle of mass m is attached to the other end of the spring. The particle slides on a frictionless, horizontal plane. The particle is given an initial velocity, and at one point in its trajectory (point A), it passes by O with a speed v and a distance D. At this instant $\boldsymbol{r}_{m/O} = D\boldsymbol{i}$ and $\boldsymbol{v}_m = v\boldsymbol{j}$.

a. Determine $\dot{\theta}$ as a function of r. Your answer may involve the constants m, L, k, D, and v. (*Hint:* Conservation of angular momentum.)

b. Determine $\dot{r}$ as a function of r. Your answer may involve the constants m, L, k, D, and v. (*Hint:* Conservation of energy.)

c. Assume $L = 0$. What is r_{max} for the trajectory? Let $m = 1.1$ kg, $k = 50$ N/m, $D = 0.5$ m, and $v = 10$ m/s.

d. Verify your answer to **c** by numerically integrating the equation of motion.

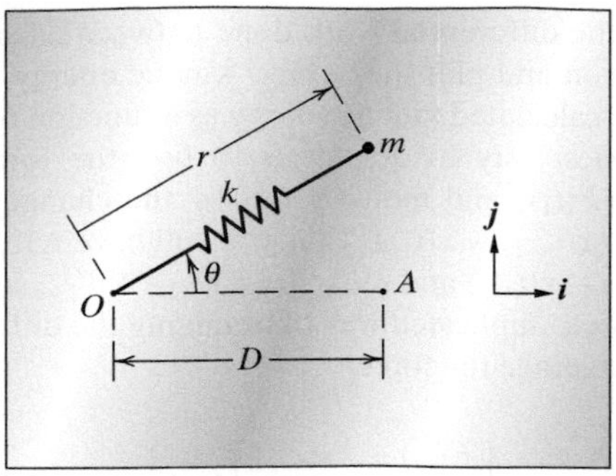

E4.2.52

4.2.53. [Level 3] Computational Pictured is a mass particle m within a circular hoop. If the mass particle has an initial speed of $v_m = 3\sqrt{gr} = 9.905$ m/s at $\theta = 0$, it will easily spin around the inside of the hoop, staying in contact the entire time, if no friction is present.

a. What is the minimum value of kinetic friction μ for which the mass will *not* stay in contact with the hoop for all values of θ and will instead lose contact at $\theta = \pi$ rad?

b. Show both analytically and through numerical simulation that if $\dot{\theta} = \sqrt{\frac{5g}{r}}$ at $\theta = 0$, then the mass will just barely make it around the hoop without losing contact if the mass/hoop interface is frictionless. Let $m = 1$ kg and $r = 0.9$ m.

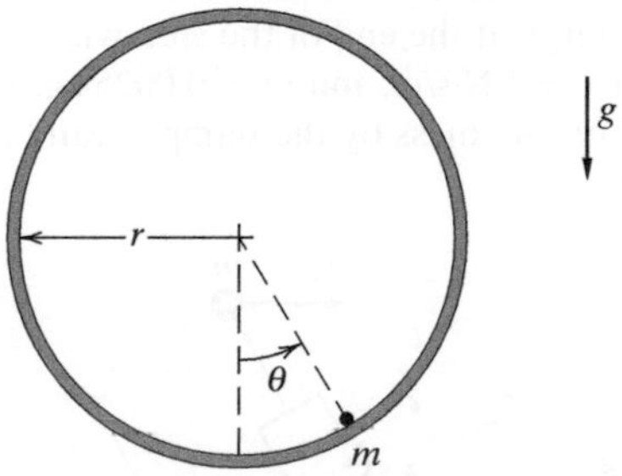

E4.2.53

4.2.54. [Level 3] Block A is pressed against a linear spring with spring constant k such that the spring is compressed 10 cm. Once released, how far along the slope will the mass travel? $k = 100$ N/m, $m_A = 0.05$ kg, $\mu_s = 0.6$, and $\mu_d = 0.3$.

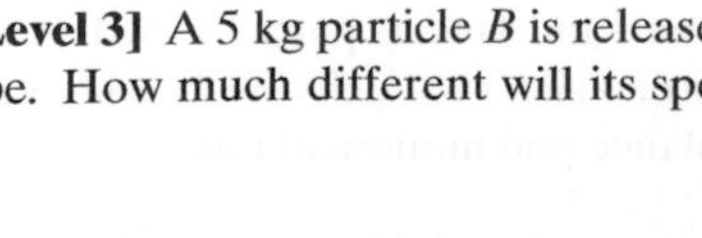

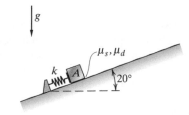

E4.2.54

4.2.55. [Level 3] A 5 kg particle B is released from rest on a 15° slope. How much different will its speed be after sliding 2 m if there's a coefficient of friction equal to 0.2 as opposed to a frictionless interface condition?

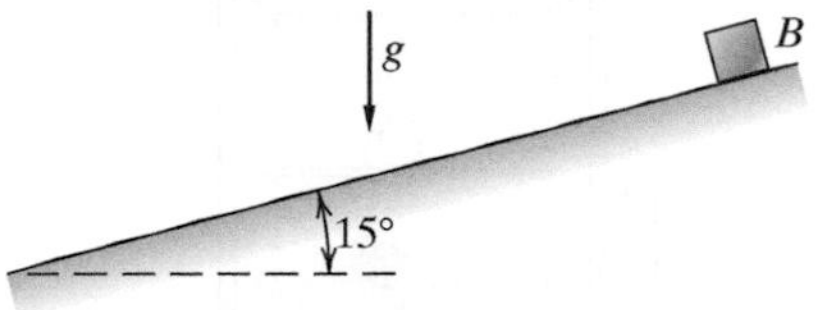

E4.2.55

4.2.56. [Level 3] Block A can slide on an inclined frictionless surface. The block is connected through a pulley to the mass B. B is rigidly attached to the end of a massless bar (length L), which is pivoted to the ground. A torsional spring ($M = -k_\theta \theta$, $\mathcal{PE} = \frac{1}{2} k_\theta \theta^2$) is connected between the bar and ground. $m_A = 5$ kg, $m_B = 10$ kg, $L = 0.8$ m, and $k_\theta = 10$ N·m/rad. If the system is released from rest and $\theta = \frac{\pi}{2}$ rad, what is the velocity of block A when B strikes the ground?

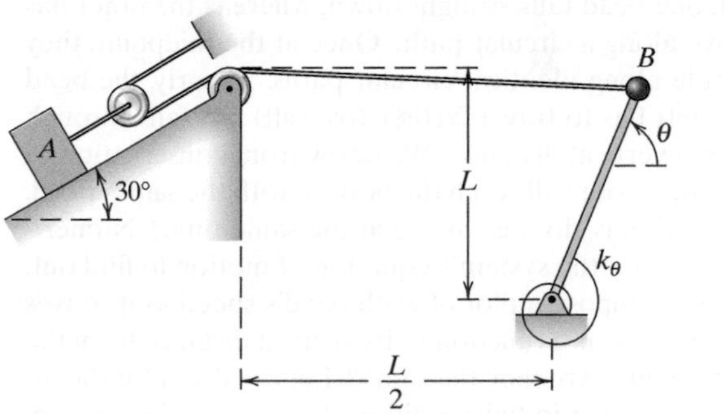

E4.2.56

4.2.57. [Level 3] Let's revisit the elevator problem in **Exercise 3.1.33**. Determine the maximum height with respect to the floor of the elevator that you can reach if you jump during the start of the deceleration phase (**E3.1.33**). Assume, as in the original exercise, that you weigh 36 kg, and further assume that you're able to leap 0.3 m in a stationary elevator. Proceed in the following manner. First determine how fast you need to be moving in order to reach 0.3 m as a maximum leap height. Next, find the (assumed constant) force that acts on your center of mass as you go from a crouch to an upright position. Assume that your center of mass moves 15 cm during the leg extension phase. This step involves work/energy from full crouch to liftoff. Assume that you produce the same (constant) force when leaping in a decelerating elevator. Using $\boldsymbol{F} = m\boldsymbol{a}$, calculate the velocity you will have at liftoff.

Finally, using this velocity, the elevator's known velocity (5.5 m/s), your acceleration in free-fall (-9.81 m/s^2), and the elevator's deceleration during the deceleration phase, determine the maximum height of your jump with respect to the elevator's floor.

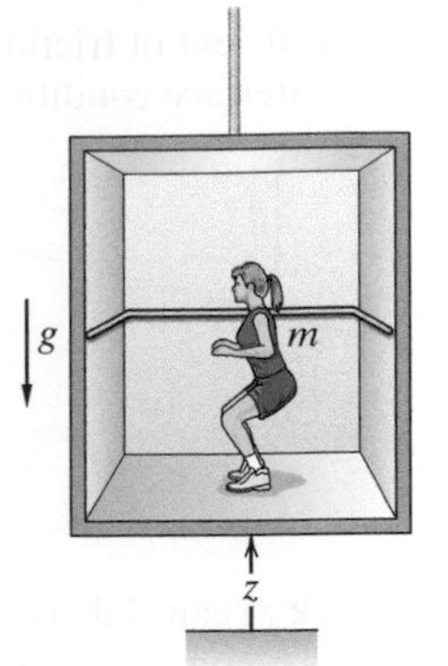

E4.2.57

4.2.58. **[Level 3] Computational** Shown are two curved wire tracks. On the left is a wire that's in the form of a semicircle. The wire on the right consists of a straight section which then flows into a quarter circle. A small bead of mass m is positioned at the top of each (points A and B). Assume that no friction exists between the beads and the wires. The difference in the two paths is that for the top half of the travel, one bead falls straight down, whereas the other has to move along a circular path. Once at the midpoint, they both ride along identical circular paths. Clearly, the bead on the left has to travel farther (overall) but falls through the same vertical distance. We know from conservation of energy that both will reach the bottom with the same speed. The question is, do they arrive at the same time? Numerically integrate the system's equation of motion to find out. Then superimpose a plot of each bead's speed as it moves down the wire as a function of its vertical distance from the starting point. Are they the same? Let $r = 0.4$ m for the integrations, and as initial conditions let each bead start from the top with a speed of 0.01 m/s.

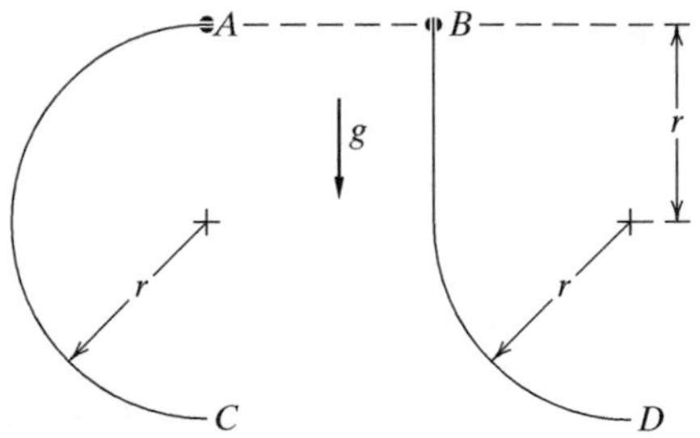

E4.2.58

4.2.59. **[Level 3] Computational** You've seen that doing work on a mass should increase its kinetic energy. In this exercise you will numerically verify this fact. Using MATLAB (or another routine of your choice), numerically integrate the response of a 0.5 kg mass acted on for 2 s by a force given by $F = \sin(\pi t)$ N. The governing equation is

$$(0.5 \text{ kg})\ddot{x} = \sin(\pi t) \text{ N}$$

Write an M-file that accepts as inputs the values of time, position, and speed calculated by the numerical integration. Calculate the differential work done between each calculated position and plot the "exact" kinetic energy ($\frac{1}{2}m\dot{x}^2$) against the calculated kinetic energy as a function of time.

The easiest way to do this is to find the force at a time step ($F(t_n)$) and multiply this by the change in displacement ($x(t_{n+1}) - x(t_n)$). Thus you will have $\Delta W_{n-n+1} = F(t_n(x(t_{n+1}) - x(t_n)))$ and $KE_{n+1} = KE_n + \Delta W_{n-n+1}$.

The more complicated way of obtaining the differential work is to average the force:

$$\Delta W_{n-n+1} = \left[\frac{F(t_n) + F(t_{n+1})}{2}\right](x(t_{n+1}) - x(t_n))$$

Implement both approaches and comment on the results.

4.2.60. **[Level 3] Computational** According to the latest research, animal legs can be modeled as inverted pendulums. A simple model of a cockroach leg is shown here. During the stance phase, the muscles in the leg are modeled as a massless spring and damper in parallel, with spring constant k and damping coefficient c, respectively. The animal body is modeled as a point mass m connected to the top of the leg, and the acceleration due to gravity g points downward as shown. The other end of the pendulum is fixed at point O. The initial velocity is $v_0 = -0.2\boldsymbol{i}$ m/s, the initial angle at touchdown is $\theta_0 = \frac{\pi}{3}$ rad, and the spring is initially neither stretched nor compressed. The spring's rest length is $L = 0.005$ m. Calculate the initial kinetic energy and the final kinetic energy at the end of the step when $\theta_1 = \frac{2\pi}{3}$rad. $k = 4000$ N/m, $c = 1$ N·s/m, and $m = 0.0025$ kg. (*Note:* The force exerted on the mass by the damper can be modeled as $\boldsymbol{F}_d = -c\dot{r}\boldsymbol{e}_r$.)

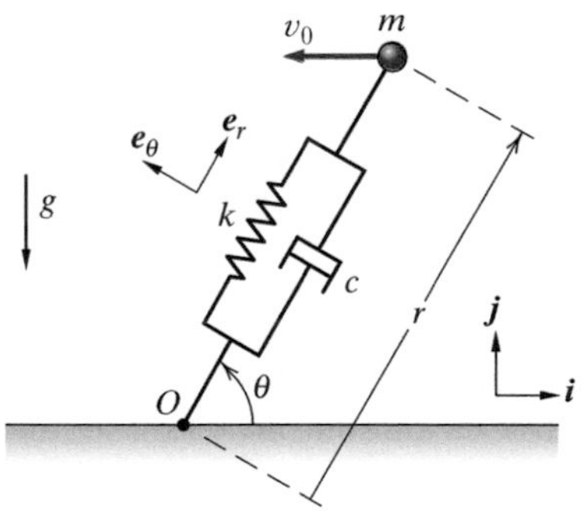

E4.2.60

4.2.61. **[Level 3]** Jane is sitting on a slowly sinking boat on the surface of a lake of quicksand. Tarzan hopes to save her by swinging on a vine, modeled as a spring with a rest length of 14 m and a spring constant k. Tarzan starts from a height h, and the vine is attached at a distance d directly above Jane. Tarzan does not want to get stuck in the quicksand and so his elevation should never go below its surface. The vine is initially stretched 1 m and Tarzan starts from rest. Treat both Tarzan and Jane as particles with zero length. If an energy approach isn't enough to solve the problem, then use a force balance and numerical integration.

a. Calculate the required spring constant k of the vine for Tarzan to successfully reach Jane without touching the quicksand. Assume that $m_T = 81$ kg, $d = 20$ m, $h = 12$ m, $L = 15$ m.

b. Plot Tarzan's swing path to Jane.

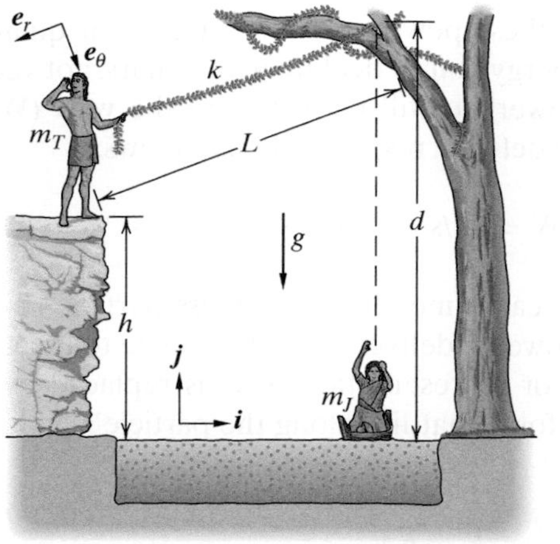

E4.2.61

4.2.62. **[Level 3]** A particle of mass m_1 is projected down with an initial speed v_0, as shown. The portion of the track from A to B is frictionless. m_1 is attached by a linear spring k to O. The unstretched length of the spring is L, and $L < r$. When m_1 reaches the bottom of the curve (assume that it does reach that position), it collides with a mass particle m_2 and the two particles stick together and both slide toward point C.

a. Assume that the duration of the impact was Δt seconds. What is the magnitude of the average force acting between the two particles during impact?

b. Assuming a constant friction force equal to F_C acting between the joined particles and the surface as they travel from B toward C, write an expression that can be solved for the longest distance traveled by the two-mass pair.

c. Assuming that a Coulomb friction model ($F = \mu_d N$) holds between B and C, write down an expression that can be solved for the longest distance traveled by the two-mass pair.

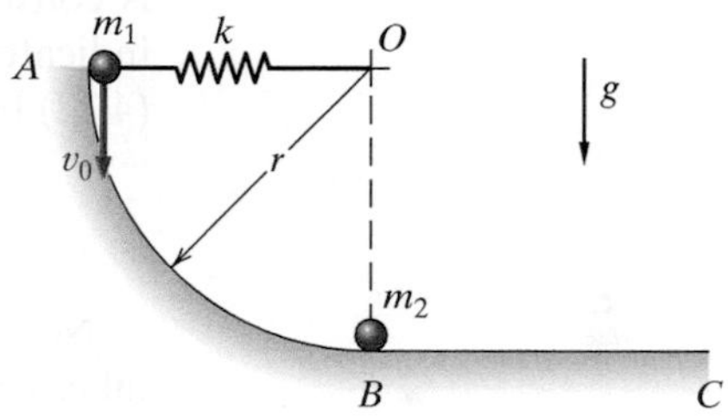

E4.2.62

4.3 POWER AND EFFICIENCY

If you've driven a car for a while, you know that going from 64 km/h to 80 km/h takes longer than going from 48 to 64, even though the change in speed (16 km/h) is the same in both cases. Although several factors affect automotive performance, you already have the necessary tools to examine one aspect of performance—the car's **power**.

What precisely is power? Simply stated, power is the *rate* at which work is being done. It takes a fixed amount of work to lift a 1 kg mass to a height of 10 m, regardless of how long it takes to do that work. The power needed, on the other hand, can be quite different from one situation to another. Raising the mass in 1 s, for instance, requires ten times the power needed to raise it in 10 s.

The power P developed by a force $\boldsymbol{F}$ is given by

$$P = \frac{dW}{dt} = \boldsymbol{F} \cdot \frac{d\boldsymbol{r}}{dt} = \boldsymbol{F} \cdot \boldsymbol{v} \tag{4.22}$$

Remember that the mathematical dot $\cdot$ is a vector dot product, indicating that it's the component of force oriented along the mass's path that matters, just as we saw when examining work.

Of course, you don't have to restrict yourself to translational forces. Otherwise how would we deal with the most common power-generating device around us—the car engine? Moving a torque through an angular displacement requires energy, and producing the torque at a given rate requires power:

$$P = \frac{dW}{dt} = T\frac{d\theta}{dt} = T\omega \tag{4.23}$$

Keep in mind that both of these power measures are scalar quantities: once you're dealing with energy you're dealing with scalars, not vectors.

The two measures of power we will be using are the **watt (W)** and **horsepower (hp)**, which are defined, respectively, as follows:

$$1\,\text{W} = 1\,\text{J/s} = 1\,\text{N}\cdot\text{m/s}$$

We will assume that our car is modeled by a mass particle (no rigid bodies yet!) and that the power P delivered by the engine to the ground is constant. Here the vector representation of $\boldsymbol{F}$ is replaced by F_t to indicate the component of force that lies along the particle's path. Thus (4.22) becomes

$$P = F_t v$$

Next I will re-express F_t in a way that will let me calculate how long it takes to change the speed of an object and how far the object moves while that change takes place, under the assumption that the applied power is constant. Replacing F_t by ma_t gives us

$$P = ma_t v \tag{4.24}$$

Recalling that a is $\frac{dv}{dt}$ lets me write

$$P = m\frac{dv}{dt}v$$

and multiplying both sides by $\frac{dt}{P}$ yields

$$dt = \frac{m}{P}v\,dv$$

Two exact differentials! Integrating them gives

$$\int_{t_1}^{t_2} dt = \frac{m}{P}\int_{v_1}^{v_2} v\,dv$$

$$t_2 - t_1 = \frac{m}{P}\left(\frac{v_2^2}{2} - \frac{v_1^2}{2}\right) \tag{4.25}$$

This expression lets us immediately determine the length of time that must elapse in order for the speed to change from v_1 to v_2 under the conditions of constant power P.

>>> **Check out Example 4.9 (page 247) and Example 4.10 (page 248) for applications of this material.**

Next, let's take another look at (4.24). Once again it's time for our old friend $a_t\,dx = v\,dv$. Rearranging it gives us

$$a_t = \frac{v\,dv}{dx}$$

which we can substitute into (4.24) to get

$$P = m\frac{v\,dv}{dx}v = mv^2\frac{dv}{dx}$$

Multiplying both sides by $\frac{dx}{P}$ yields

$$dx = \frac{m}{P}v^2\,dv \tag{4.26}$$

and integrating gives

$$\int_{x_1}^{x_2} dx = \frac{m}{P}\int_{v_1}^{v_2} v^2\,dv$$

which leads to

$$x_2 - x_1 = \frac{m}{P}\left(\frac{v_2^3}{3} - \frac{v_1^3}{3}\right) \tag{4.27}$$

And now we have it. With (4.25) and (4.27) we have the means to evaluate both the time required and the distance traveled by a car under the influence of a constant power.

Power is all well and good, but if you don't use it efficiently, it won't do much for you. And what exactly is efficiency? Glad you asked. Efficiency is simply a way of evaluating how completely power flows through whatever system it is that concerns you. Say a certain power flows into a device and the device changes that power into power of a different sort. For instance, the input might be mechanical power delivered by your legs to the cranks of a bicycle. As a result, the bicycle leaps forward. If you put in 149 W, you're going to see a speed change that corresponds not exactly to 149 W but to something more like 146 W. Almost the same but a bit less. (The loss is so small because bicycles are extremely efficient devices—the losses in the drivetrain are quite miniscule.)

We characterize the drivetrain loss through the parameter e_m, called the mechanical efficiency. e_m is defined as

$$e_m \equiv \frac{\text{mechanical power out}}{\text{mechanical power in}} \tag{4.28}$$

For our bicycle example the mechanical efficiency is

$$e_m = \frac{146}{149} = 0.98$$

Thus we would say that our bicycle is 98% efficient. As you might guess, there are other types of efficiency measures as well, such as electrical efficiency, e_e, and thermal efficiency, e_{th}.

The advantage of characterizing the efficiency of various energetic processes is that they can all be rolled into an overall efficiency measure simply by multiplying them. So, for example, let's say we have a two-component electromechanical machine that has a mechanical efficiency of 80% and an electrical efficiency of 92%. The overall efficiency of the total system is simply

$$e = (e_m)(e_e) = (0.80)(0.92) = 0.736$$

Thus our device has an overall efficiency of 73.6%.

Efficiency is an extremely important concept when you're dealing with energy in a practical sense. The overall mechanical efficiency of an automobile drivetrain, for instance, is around 80%. The combustion efficiency of the engine is only 28%. Thus, these two factors alone give us a combined efficiency of about 22%. If you could magically raise both of these efficiencies to 100% (which is quite impossible, by the way), then a car that previously got 8.50 km/lt would see its gas mileage improve to 37.8 km/lt. This is why automotive manufacturers put so much time and effort into increasing the efficiency of both their engines and their drivetrains.

Before leaving the topic of energy, I want to emphasize once again that, when solving problems in dynamics, whether or not you use an energy approach or a force approach depends on what you're being asked to find. If you're being asked to find the acceleration given a set of forces and/or constraints, then you want to use a force approach. This is because the force approach means you're finding the system's equations of motion ($\boldsymbol{F} = m\boldsymbol{a}$), which can then be used to determine the acceleration of the system.

If, on the other hand, you're told something about the system in one position (when a mass is 4 m above the table, a spring is stretched 0.5 m, and so on) and are then asked to calculate some dynamical quantity at a *later* time (not "at this instant" but an honest to goodness, finite interval of time later), then you can use an energy approach.

>>> **Check out Example 4.11 (page 249) for an application of this material.**

EXAMPLE 4.9 TIME NEEDED TO INCREASE SPEED (Theory on page 244)

Cars are commonly advertised as delivering a certain amount of horsepower, and sometimes in America more is better. So we're going to look into how a car performs if it's always putting out maximum horsepower. Assume that the maximum power available is 100 kW (100,000 W) and that this power level is always available, regardless of the car's speed. Calculate how long it takes the car to accelerate from 10 m/s to 20 m/s and from 20 m/s to 30 m/s. The car and driver have a combined mass of 1400 kg.

Goal Find the time Δ_1 needed to go from 10 to 20 m/s and the time Δ_2 needed to go from 20 to 30 m/s.

Given Car's mass and power output.

Assume Assume that there's no tire slippage.

Formulate Equations The equation we need is (4.25):

$$t_2 - t_1 = \frac{m}{P}\left(\frac{v_2^2}{2} - \frac{v_1^2}{2}\right)$$

Solve Using (4.25), we will find

$$\text{10–20 m/s:} \qquad \Delta t = \frac{1400\,\text{kg}}{2(100{,}000\,\text{N}\cdot\text{m/s})}\left[(20\,\text{m/s})^2 - (10\,\text{m/s})^2\right] = 2.1\,\text{s}$$

$$\text{20–30 m/s:} \qquad \Delta t = \frac{1400\,\text{kg}}{2(100{,}000\,\text{N}\cdot\text{m/s})}\left[(30\,\text{m/s})^2 - (20\,\text{m/s})^2\right] = 3.5\,\text{s}$$

We see that going from 10 to 20 m/s takes only 2.1 s, but going from 20 to 30 m/s takes 3.5 s, a substantially longer time.

Check Does this result mesh with your experience? If you floor the gas pedal in a car, does it take longer to increase your speed by some set amount if you're already traveling quickly than if you're going more slowly? The answer is definitely yes. Just pick up any *Road and Track* magazine and look at the car test sections where they show graphs of speed versus time. You will see that the slope is steepest at the beginning and decreases as the speed rises, as shown in **Figure 4.21**.

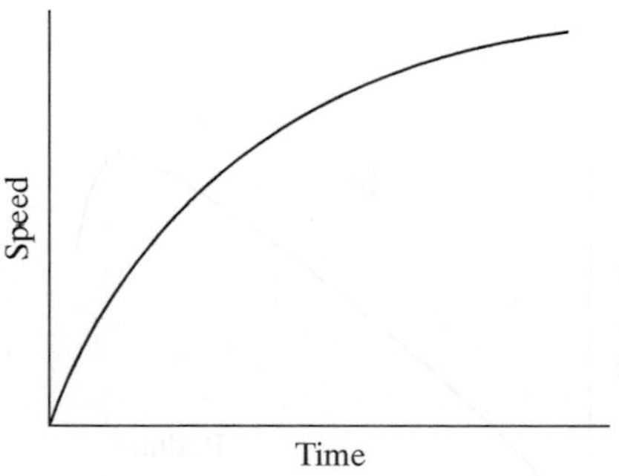

Figure 4.21 Speed versus time for an accelerating car

EXAMPLE 4.10 0 TO 60 TIME AT CONSTANT POWER (Theory on page 244)

In this example we will relate what we have just learned about speed and power to another real-world situation. In 1998 I purchased a Volkswagen Passat that had a 142 kW engine. The total weight of the car, some fuel, and a driver was 1542 kg, and the car could go from 0 to 96 km/h in 7 s. Determine how closely an assumption of constant maximum power matches these results.

Goal Determine the time needed to go from 0 to 96 km/h using a constant-power assumption.

Given Car's mass and power.

Assume We will assume roll without slip—that is, no power losses due to tire slippage.

Formulate Equations The relevant equation is again (4.25). To use it, we need the mass of the vehicle:

$$m = 1542 \text{ kg}$$

In addition, 96 km/h corresponds to 26.7 m/s, and because 1 kW equals $10^3 \frac{\text{N·m}}{s}$, we have a maximum power of 142 kW.

Solve

$$\Delta t = \frac{1542 \text{ kg}}{142 \times 10^3 \text{ N·m/s}} \frac{(26.7 \text{ m/s})^2}{2} = 3.9 \text{ s}$$

Our conclusion is that, if driven at maximum power over the entire test, the car would have a 0 to 60 time of 3.9 s. This is quite a bit quicker than the 7 s obtained in real-world testing.

Check Is this a sensible result? Yes. A car's horsepower curve generally reaches a maximum when the engine is spinning near its limit (i.e, near redline), as shown in **Figure 4.22**.

Because a 0 to 60 test starts from a dead stop, we would expect horsepower to be low at the start and to build as the car gets moving. Thus we're nowhere near a condition of constant maximum power during the run. In addition, to get from 0 to 60 requires at least one shift of gears (and often two), resulting in a drop in kilowatt as the shift takes place (because the shift drops the engine's speed as determined by the gear ratios). The time needed to complete the shifts also adds to the overall time.

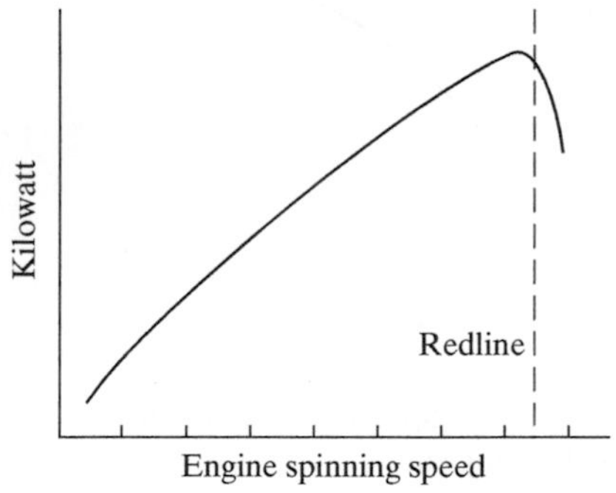

Figure 4.22 Kilowatt versus rpm

EXAMPLE 4.11 DETERMINING A CYCLIST'S ENERGY EFFICIENCY (Theory on page 245)

A bicyclist is being given a stress test as part of an overall physical. The doctor has determined that he's burning energy at a rate of 1.49 kJ/s. The bicycle has a transmission efficiency of 98%. Sensors at the crank reveal that the cyclist is pedaling at 10 rad/s and generating an average torque of 37.3 J. What is his body's efficiency e_b?

Goal Calculate a person's efficiency in producing power.

Given Power being used, pedal speed, and torque generation.

Assume We will simply assume that the only losses taking place are the ones mentioned.

Formulate Equations From (4.23) we know

$$P = T\omega$$

We're given the torque and rotational rate so that we can find the power.

Solve

$$P = (37.3\text{ J})(10\text{ rad/s}) = 373\text{ J/s}$$

The overall efficiency e is the product of the body's efficiency and the transmission efficiency:

$$e = e_b e_t = \frac{\text{power out}}{\text{power in}} = \frac{373\text{ J}}{1.49 \times 10^3\text{ J}} = 0.25$$

$$e_b = \frac{0.25}{e_t} = \frac{0.25}{0.98} = 0.255$$

The cyclist has an efficiency of 25.5%.

Check This isn't so much a check on your work as an interesting fact—namely, that most people have an efficiency of around 25%. A colleague of mine at Berkeley measured the metabolic efficiency of a variety of people, across age and fitness levels, and found that efficiency was consistently about 25%. Higher-performing athletes didn't have a higher efficiency than more average individuals but simply were capable of putting out more power. This is analogous to a 4-cylinder engine and a V8 having somewhat similar efficiencies but vastly different power outputs due to the size difference.

EXERCISES 4.3

4.3.1. [Level 1] A motor pulls a small block of mass $m = 6\,\text{kg}$ from rest along a smooth horizontal surface. The block's speed is $v = 5\,\text{m/s}$ after traveling a distance $\Delta x = 1\,\text{m}$. If the motor is 70% efficient, what is its power input?

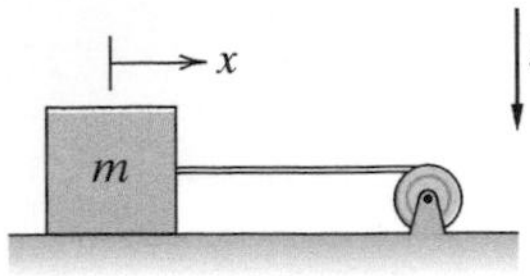

E4.3.1

4.3.2. [Level 1] I routinely trot up the stairs to get to my office on the sixth floor from my third floor classroom and encourage my colleagues to do so as well. There are 13 steps (17.8 cm per step) per flight and two flights per floor. Assume it takes my colleague 12 s to go up a single floor, and he weighs 76 kg. What is his power output in watts?

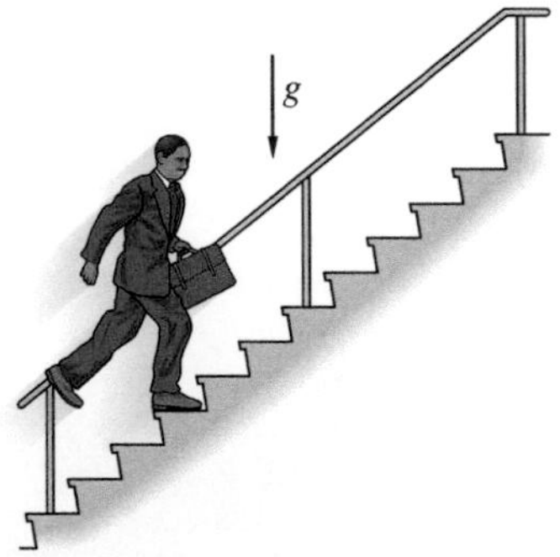

E4.3.2

4.3.3. [Level 1] Let's find out the impact of a sudden headwind on a cyclist. Assume the cyclist is initially traveling at a constant speed $v_0 = 10\,\text{m/s}$ on a flat road in still conditions. Wind drag is given by $F_W = av^2$, where v is the speed at which the air strikes the cyclist (not simply the cyclist's speed). Suddenly a headwind starts up, at $v_W = 10\,\text{m/s}$, directed directly toward the cyclist. What will the final steady-state speed of the cyclist be under the assumption that the power output stays the same for both cases?

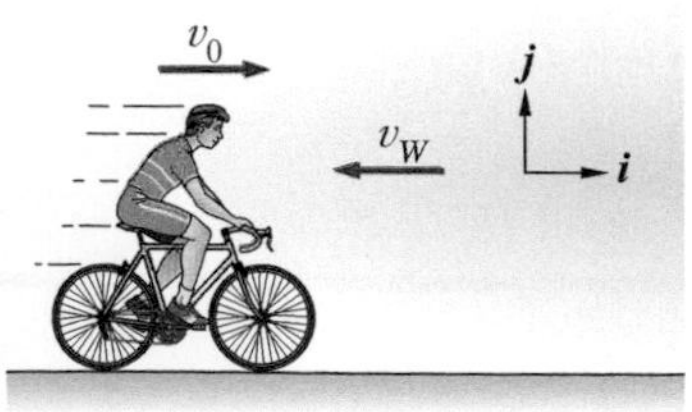

E4.3.3

4.3.4. [Level 1] A crate of mass $m = 50\,\text{kg}$ is pulled up a rough ramp at a constant speed v by a motor with a power input of $P_m = 2\,\text{kW}$. The ramp is angled at $\theta = 40°$ with respect to the horizontal, and the coefficient of friction between it and the crate is $\mu = 0.5$. Find the crate's speed v up the ramp if the motor is 60% efficient.

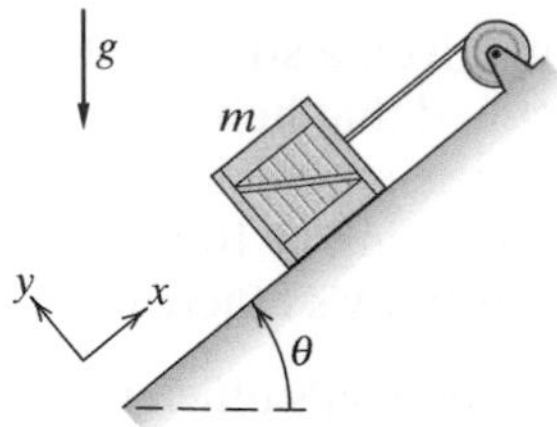

E4.3.4

4.3.5. [Level 1] A 1360 kg car can accelerate from 0 to 96 km/h in $\Delta t = 4\,\text{s}$. If the drivetrain efficiency is 85%, determine the power output from the engine that is needed to accomplish this acceleration.

E4.3.5

4.3.6. [Level 1] An electric motor lifts a 200 kg block up a distance of 4 m in 3 s, and it's observed that the electrical power drawn by the motor is equal to 3 kW. What is the motor's efficiency?

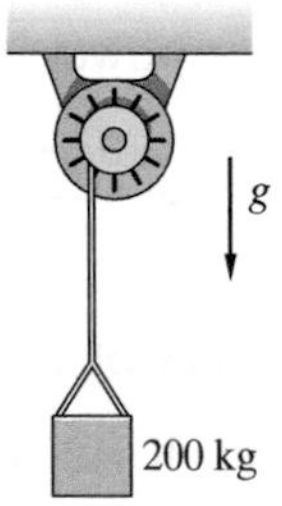

E4.3.6

4.3.7. [Level 1] A motor pulls the 125 kg block A up a frictionless surface. The block starts from rest and is accelerated uniformly to a speed of 2 m/s over 3 s. Plot the power acting on the block as a function of time for $0 \le t \le 2$ s.

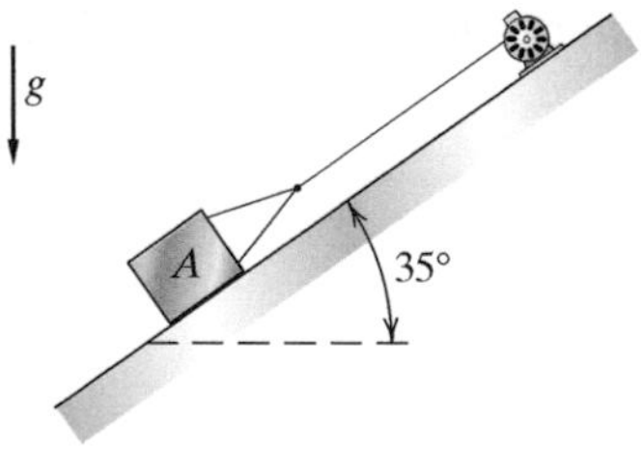

E4.3.7

4.3.8. [Level 1] How fast can a 1450 kg car with a 149 kW engine accelerate from 0 to 48 km/h? Neglect drivetrain losses and air and road drag.

4.3.9. **[Level 1]** A 1600 kg car, traveling at 100 km/hr, slows to rest at a constant deceleration in 4 s. Plot the rate at which energy is being removed from the car as a function of time.

4.3.10. **[Level 1]** A block with mass $m = 15\,\text{kg}$ is pulled along a rough horizontal surface at a constant speed $v = 4\,\text{m/s}$ by a motor with a power input and efficiency of $P_1 = 0.5\,\text{kW}$ and 80%, respectively.

a. What is the coefficient of friction μ between the block and the surface?

b. Suppose the motor is replaced by one with a power rating of $P_2 = 0.7\,\text{kW}$ and 65% efficiency. What is the fastest constant speed at which this motor can move the block?

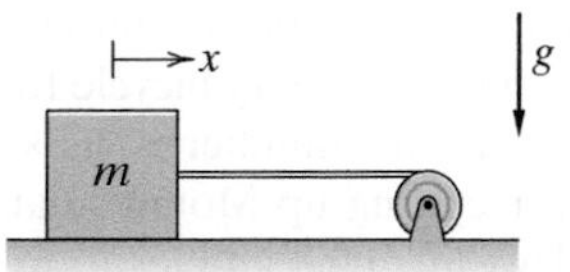

E4.3.10

4.3.11. **[Level 1]** An electric motor can raise the illustrated 50 kg mass a distance of 3 m in 5 s at a constant speed. Given that the motor's overall efficiency is 60%, calculate the power input to the motor.

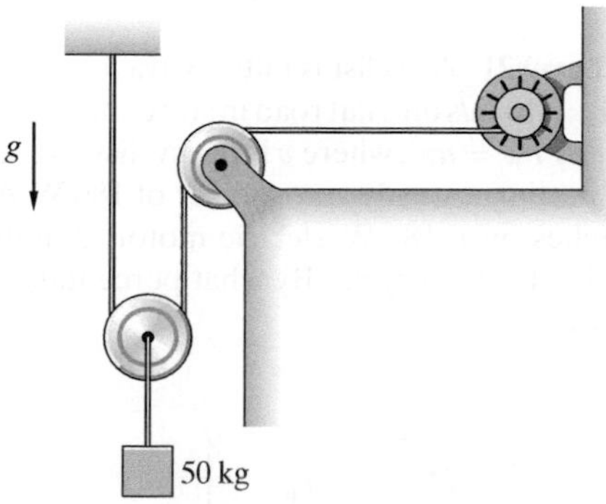

E4.3.11

4.3.12. **[Level 1]** What is the cost to pump 1000 liters of water to the top of a 30 m high water tank? Assume a pumping efficiency of 50% and an electrical cost of \$0.15 per kW·h.

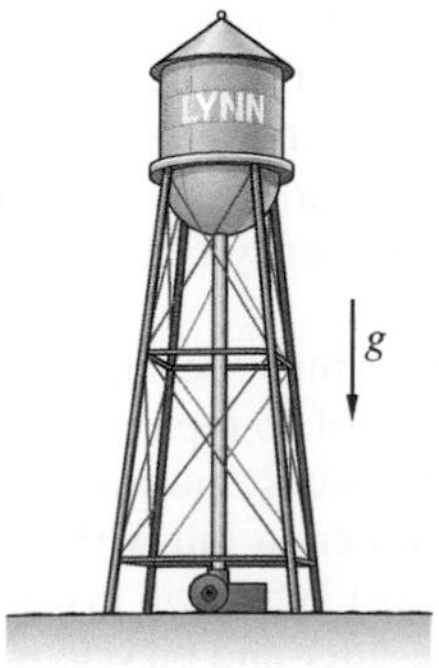

E4.3.12

4.3.13. **[Level 1]** A 1724 kg car is coming down a 10% grade at a constant speed of 48 km/h. How much power is being dissipated by the brakes to let this happen?

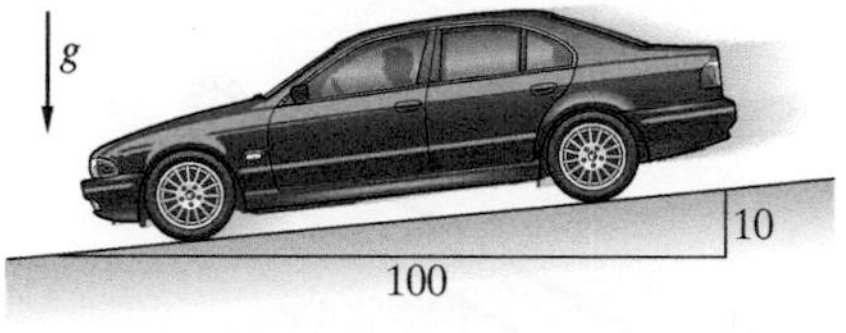

E4.3.13

4.3.14. **[Level 1]** How far will a cup of yogurt get you? Assume that the yogurt contains 628 J, your body has an overall efficiency of 25%, and you have a mass of 75 kg. How far up a mountain will that cup of yogurt get you?

4.3.15. **[Level 1]** The fastest that anyone has gotten to the top of a local mountain (elevation change of 1070 m) is 44 minutes. Assume a weight of 72.6 kg for the bike and rider. What was the average power output used for this ride?

4.3.16. **[Level 1]** Aerodynamic drag force increases with the square of a body's speed, $F_d = av^2$, where a is a scaling constant that depends on the body's cross-sectional area and the fluid being moved through, and v is the body's speed. Assume that this is the only force acting against a moving body, and determine by what percentage the body's power output must increase for the body's speed to double.

4.3.17. **[Level 1]** What power is necessary to move a 1200 kg car up a 5% grade at a constant 20 m/s? Neglect air and road drag.

E4.3.17

4.3.18. **[Level 1]** The equation of motion of a car moving against road and wind drag is given by $m\ddot{x} = -c_r mg - c_a\dot{x}^2 + F$ where m is the car's mass, c_r is the coefficient of friction with the road, $c_a\dot{x}^2$ is the aerodynamic drag force, and F is the driving force produced by the engine and transferred to the tires. The car weighs 1588 kg, $c_r = 0.02$, $c_a = 0.527\,\text{N·s}^2/\text{m}^2$, and $F = 1.56\,\text{kN}$. What is the maximum attainable speed of the car, and what kilowatt is it developing when traveling at that speed?

E4.3.18

4.3.19. **[Level 1]** A motor pulls the illustrated 1000 kg cart C up an inclined slope at a constant speed of 2 m/s. Doing so requires a power input of 18 kW. What is the motor's efficiency equal to?

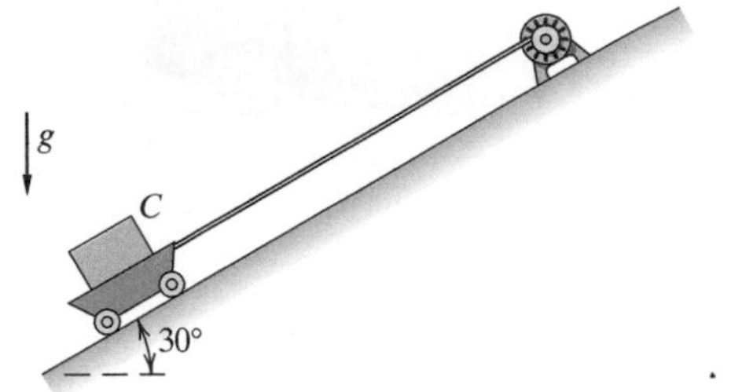

E4.3.19

4.3.20. **[Level 1]** A hopper that brings gravel to the roof of a house moves at 0.46 m/s. Assume that a load of 45.4 kg of gravel is moved from the ground to the roof and that the hopper itself weighs 13.6 kg. What is the power needed to accomplish this? What is the total work done?

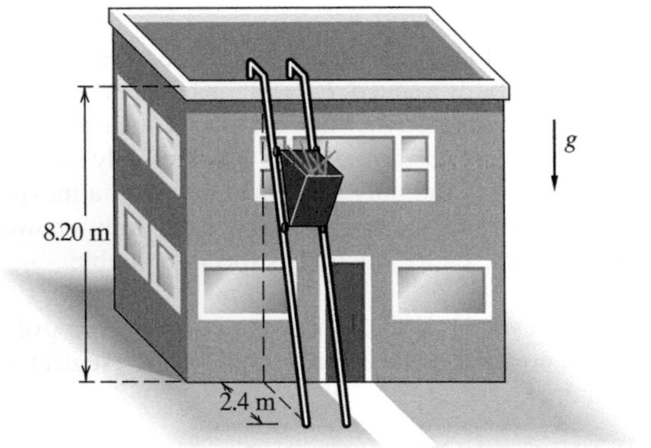

E4.3.20

4.3.21. **[Level 2]** Riding uphill on a bicycle can be tiring, but the ride down often compensates. However, does it compensate *enough*? Consider two situations, a ride up and down a mountain versus a flat ride of the same distance. The distance between points A and C is 20 km. The slope from A to B is 0.06 and from B to C is -0.06. Calculate the actual distance traveled and call this distance L. Assume that the only forces acting on the rider D are wind drag (with force F_W) and the force due to gravity. The wind drag is given by $F_W = av^2$, where a is a constant and v is the cyclist's speed. Assume further that the cyclist can output a constant power of 246 W and that the total mass of the cyclist plus the bicycle is 78 kg.

a. Assuming that the cyclist can maintain a maximum speed of 14 m/s when riding on level ground, determine a.

b. Calculate how long the cyclist needs to cover the distance L on level ground.

c. Using a FBD, determine the traction force the cyclist needs to apply on both the uphill and downhill portions of the mountain and, by assuming a constant power output of 246 W, determine the speed on both portions.

d. Calculate the total time needed to go from A to B to C and compare it to the time needed to cover the same distance but on level ground. Is the difference significant?

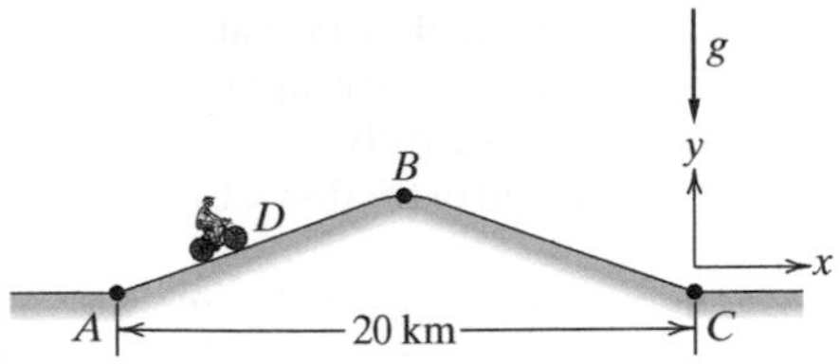

E4.3.21

4.3.22. **[Level 2]** In 2006 the *New York Times* reported that Floyd Landis was capable of sustaining a 1250 W power output while pedaling a stationary bicycle for 5 s. To understand how remarkable this number is, let's extrapolate it. The record time for cycling up Mount Diablo is a bit under 45 minutes. This involved an elevation gain of 990 m in 17.4 km of riding. Approximate the course as a constant grade. Assume a wind drag equal to av^2, where v is the cyclist's speed in m/s and $a = 0.23\,\text{N}\cdot\text{s}^2/\text{m}^2$. How fast would Mr. Landis have climbed the mountain if he could have sustained that 1250 W power output for the entire ride? Assume a total mass for rider plus bicycle of 78 kg.

4.3.23. **[Level 2]** A cyclist is initially traveling at a constant speed $v_0 = 10$ m/s on a flat road in still conditions. Wind drag is given by $F_W = av^2$, where v is the cyclist's speed. The cyclist is outputting a steady-state power of 186 W. Assume that he switches on a 186 W electric motor, doubling the power supplied to the bicycle. By what percentage will his speed increase?

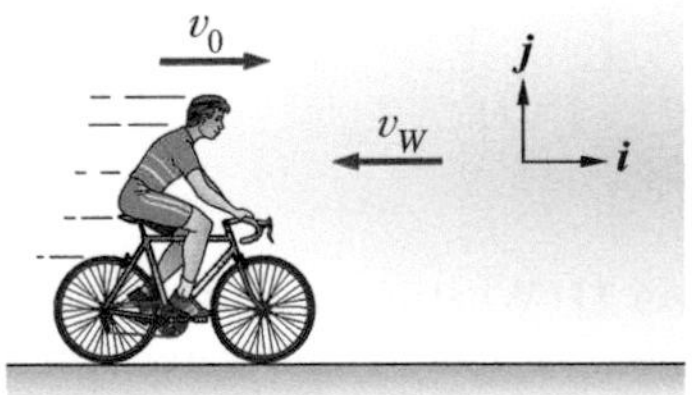

E4.3.23

4.3.24. **[Level 2]** The force acting against a bicyclist traveling on a flat road in still air is given by $F = a_1 + a_2v^2$ where a_1 is the force due to road friction, a_2v^2 is the force due to aerodynamic drag, and v is the velocity, expressed in units of m/s. F is given in newtons. $a_1 = 3.3$ N and $a_2 = 0.24\,\text{N}\cdot\text{s}^2/\text{m}^2$. Assume a drivetrain efficiency of 96%. (You will have to convert units.)

a. What horsepower must be supplied to the pedals to maintain a steady speed of 25 mph?

b. Assume that the cyclist, putting out the horsepower calculated in **a**, encounters a hill with a 5% grade. (Tangent of the slope is 0.05.) What will his speed drop to? The combined mass of the rider and bicycle is 85 kg.

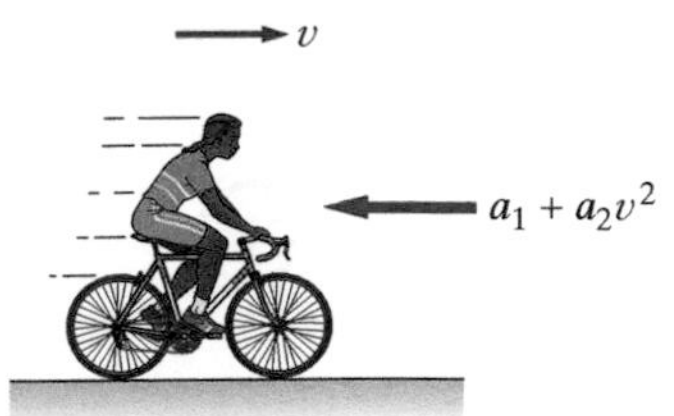

E4.3.24

4.3.25. **[Level 2]** When I ride my road bike up Mount Diablo, a local mountain, I average 18 km/h. The road has an average grade of 5.5%. My bike and I have the combined mass of 76 kg. The force from road drag is equal to 2.5 N, and the aerodynamic drag (in newtons) is equal to $(0.2\ \text{N·s}^2/\text{m}^2)v^2$ where v is my speed in m/s. What is my average power output? Neglect any drivetrain losses.

4.3.26. **[Level 2]** A variation of water skiing is barefoot skiing, where your feet are used to keep you on top of the water instead of skis. As you would expect, using your feet instead of buoyant skis to keep you afloat requires that you be moving at pretty high speeds, usually between 56 and 96 km. Suppose you're being pulled at a constant speed v via an inextensible cable by a boat whose drivetrain has an efficiency of 80%. The air is still, and the air drag acting on you is given by $F_a = av^2$, where $a = 0.25\,\text{N·s}^2/\text{m}^2$. The resistance between your feet and the water (which keeps you afloat) can be modeled as $F_w = bv^2$, where $b = 2\,\text{N·s}^2/\text{m}^2$. Your feet make an angle $\theta = 30°$ with respect to the water's surface, and F_w acts normal to your feet. If your mass is $m = 70$ kg, what is the speed v needed to keep you afloat and how much power does the boat's engine need to supply?

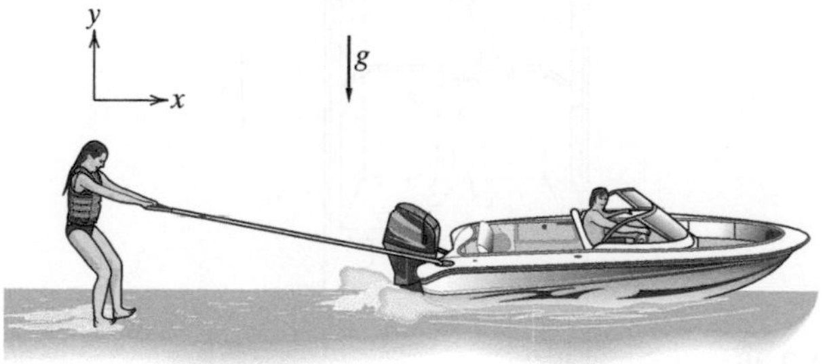

E4.3.26

4.3.27. **[Level 2]** Kelly is initially bicycling on level ground at a constant speed $v = 32$ km/h before encountering a hill with a 6% grade. She would like to maintain her speed v as she rides up the hill. Kelly and her bicycle have a combined mass of $m = 80$ kg, and the total drag acting on her is given by $F_d = a + bv^2$, where $a = 3$ N is the friction between her tires and the ground, and $b = 0.25\,\text{N·s}^2/\text{m}^2$ is the aerodynamic drag coefficient.

a. Find Kelly's power output while bicycling on level ground.

b. If her maximum power output is $P_{\text{max}} = 0.37$ kW, can Kelly maintain her initial speed up the hill? If not, what is the fastest she can go?

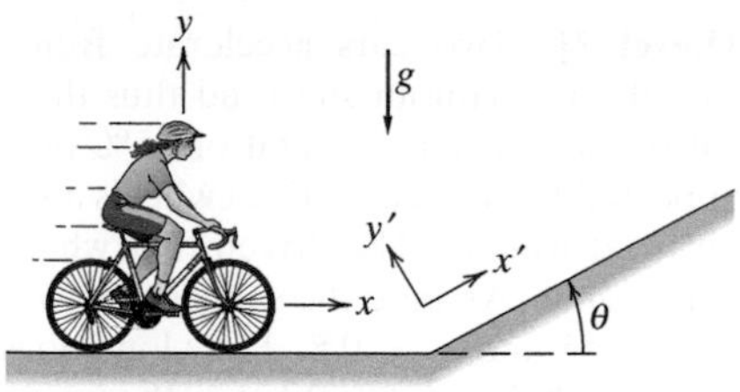

E4.3.27

4.3.28. **[Level 2]** Cyclists often pump up their tires to minimize rolling resistance (and thus go faster). Let's see how road drag affects my speed up Mount Diablo. The air drag is $(0.21\ \text{N·s}^2/\text{m}^2)v^2$, where v is given in m/s and the drag force is given in newtons. The road drag is equal to 2.5 N. The road grade is 6%. The mass of the bicycle and myself totals 76 kg. If I put out 0.23 kW, what will be my speed? How much will it change if the road drag is eliminated? Neglect drivetrain frictional power losses.

4.3.29. **[Level 2]** Given a gasoline usage of 10.2 lt per hour while traveling at 96 km/h, determine the coefficient of drag, c_d, for a passenger car. The drag force acting on the car is given by

$$F_d = 0.5\rho A c_d v^2$$

where ρ is the density of air (use 1.2 kg/m³), A is the car's cross-sectional area (use $A = 2\ \text{m}^2$), and v is the car's speed in m/s. Assume a drivetrain efficiency of 80% and an engine efficiency of 28%. The energy density of gasoline is 1.25×10^8 J/gal.

4.3.30. **[Level 2]** In a braking test, a 1588 kg car goes from 96 km/h to zero in 36 m with constant deceleration. Plot the rate at which energy was being extracted from the car from start to finish.

4.3.31. **[Level 2]** A 10 kg wheel, with radius $r = 0.1$ m, is driven by a motor at O and turns at a constant angular velocity $\boldsymbol{\omega} = -15\boldsymbol{k}$ rad/s. A mass ($m = 0.08$ kg) is in contact with the wheel and in perfect balance. The downward force of gravity is exactly counterbalanced by the frictional force generated as it slips relative to the wheel. θ remains constant. Assuming a coefficient of dynamic friction $\mu_d = 0.2$, calculate the power that needs to be supplied to the wheel.

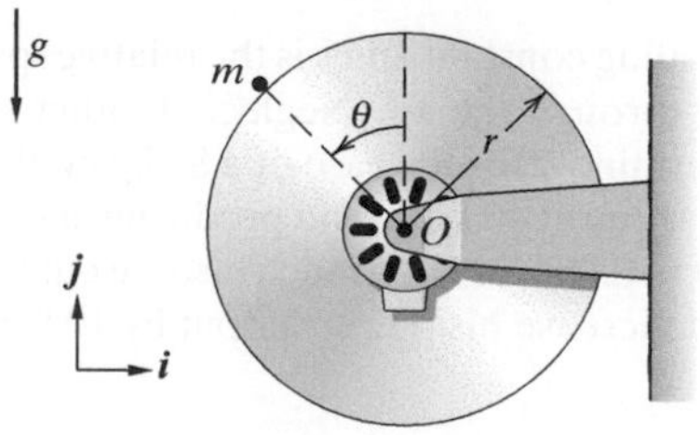

E4.3.31

4.3.32. **[Level 2]** Two cars accelerate from rest to 24 km/h. Car A rolls without slip (and thus the relevant coefficient of friction is μ_s). A total of 65% of the car's weight is supported by the rear (drive) wheels and 35% by the undriven front wheels. The driven rear wheels of car B slip the entire time. Assume the same 65/35 weight distribution. $\mu_d = 0.55$ and $\mu_s = 0.8$. How long does it take for each car to reach 24 km/h? What is the average kilowatt generated by the cars? What do the results tell you about the effectiveness of slip versus no-slip conditions in accelerating the car? Both cars weigh 1360 kg.

4.3.33. **[Level 2]** An all-wheel-drive vehicle is accelerating in a straight line. All tires are transmitting their maximum force without slipping, and therefore we would use the static coefficient of friction ($\mu_s = 0.9$) to determine the maximum traction force. The vehicle has a mass of 1500 kg. What is the horsepower developed by the vehicle after it travels for 3 s? What is the average horsepower over the distance traveled? Assume no drivetrain losses or external drag.

4.3.34. **[Level 2]** I can ride up a 5% grade at a constant speed of 16 km/h. If my power output remains constant and the grade increases to 6%, what will my speed drop to? Neglect air and road drag.

E4.3.34

4.3.35. **[Level 2]** I can pedal up a nearby mountain in 75 minutes. The total elevation change is 1070 m. Assuming a total weight of 80 kg, determine my average power output. Next, determine how much faster I could arrive at the top if I added a 9 kg, 246 W motor to my bicycle.

4.3.36. **[Level 2]** Wind drag can be approximated as being quadratic in speed:

$$F_d = av^2$$

where a is a scaling constant and v is the relative speed of the body moving through the air. Neglect all other sources of drag and determine a for the case of a 80 kg cyclist (including bicycle) cycling at 40 km/h and producing a steady-state power output of 0.26 kW. How much faster would the cyclist go if he could increase his power output by 10%?

E4.3.36

4.3.37. **[Level 2]** A 1450 kg car is coming down an 8% grade. By what percentage will the instantaneous power dissipated by the brakes change if the car goes from a steady speed of 48 km/h to a deceleration of 0.8 g?

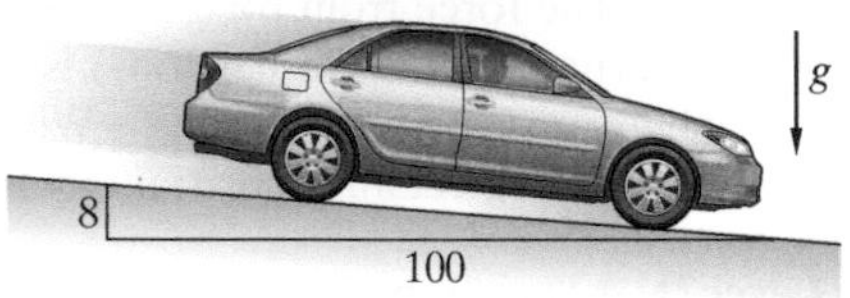

E4.3.37

4.3.38. **[Level 2]** How cheap is electricity? Let's work an example. I will assume a cost of \$0.12 per kW·h. Assume that you've constructed an elevator with an overall efficiency of 84%. What is the cost of using this elevator to raise itself (270 kg) and a single passenger (70 kg) up four stories (19 m)? What is the average power needed if it takes 20 s to complete the trip?

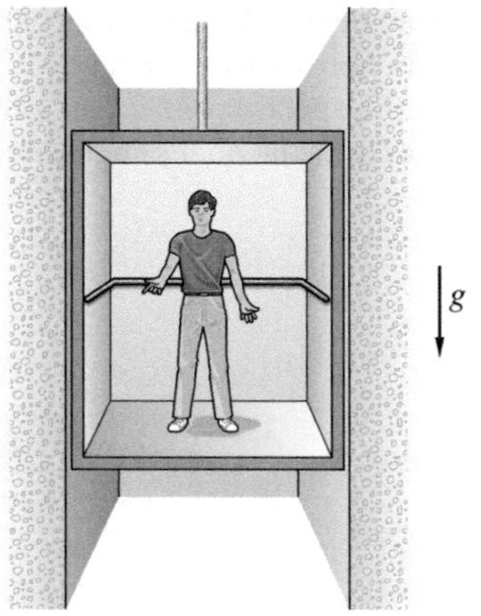

E4.3.38

4.3.39. **[Level 3] Computational** Let's assume you're riding your bicycle at 24 km/h. Your maximum power output is 0.30 kW. The road friction is 2.8 N, and the aerodynamic drag is $(0.22\ \text{N·s}^2/\text{m}^2)v^2$ (drag in N, and v in m/s). A person zooms by you, and you want to catch up.

a. What is the maximum acceleration you can achieve? The total mass of you and your bicycle is 82 kg. Neglect frictional losses in the bicycle's drivetrain.

b. Assume you exert maximum power as you try to catch the person. How long will it take and how far will you travel if you accelerate up to 32 km/h? Use numerical integration and plot both x and $\dot{x}$ versus t.

4.4 JUST THE FACTS

This chapter dealt with **energy**. In contrast to a force balance analysis, which is vector based, energy measures are scalar. We saw that through an energy analysis, we could determine how the speed of a mass would change when it reached a new location without having to integrate its equation of motion. We saw how the fundamental energy relation linked kinetic energy to **work**, and then we went beyond this to introduce the idea of a potential energy. Finally, by looking at the time rate of change of energy, we introduced the notion of power.

The chapter started with an examination of **kinetic energy**. We began by introducing differential work

$$dW = F_t\,ds \tag{4.2}$$

and then integrated to get a total work expression

$$W_{1-2} = \int_1^2 dW = \int_{s_1}^{s_2} F_t\,ds \tag{4.3}$$

The relationship between work and kinetic energy followed from there:

$$W_{1-2} = \frac{1}{2}mv_2^2 - \frac{1}{2}mv_1^2 \tag{4.6}$$

$$\mathcal{KE}\big|_2 = \mathcal{KE}\big|_1 + W_{1-2} \tag{4.7}$$

The notion of **potential energy** and **conservative forces** came next. First a differential form of the potential energy:

$$F_c = -\frac{d\mathcal{PE}}{ds} \tag{4.11}$$

followed by a finite form:

$$\mathcal{PE}\big|_2 - \mathcal{PE}\big|_1 = -\int_{s_1}^{s_2} F_c\,ds \tag{4.12}$$

Putting **nonconservative work**, kinetic energy, and potential energy together gave us:

$$\mathcal{KE}\big|_2 + \mathcal{PE}\big|_2 = \mathcal{KE}\big|_1 + \mathcal{PE}\big|_1 + W_{nc_{1-2}} \tag{4.13}$$

Specific types of potential energy included gravitational potential energy:

$$\mathcal{PE}\big|_g = mgy \tag{4.14}$$

and spring potential energy:

$$\mathcal{PE}\big|_{sp} = \frac{1}{2}kx^2 \tag{4.17}$$

Power from translational force was given by

$$P = \boldsymbol{F} \cdot \boldsymbol{v} \tag{4.22}$$

whereas power from a rotational force was seen to be

$$P = \frac{dW}{dt} = T\omega \tag{4.23}$$

If power was constant, then we found a time–speed relationship:

$$t_2 - t_1 = \frac{m}{P}\left(\frac{v_2^2}{2} - \frac{v_1^2}{2}\right) \tag{4.25}$$

as well as a displacement relationship:

$$x_2 - x_1 = \frac{m}{P}\left(\frac{v_2^3}{3} - \frac{v_1^3}{3}\right) \tag{4.27}$$

The chapter ended with a discussion of **efficiency**:

$$e_m \equiv \frac{\text{mechanical power out}}{\text{mechanical power in}} \tag{4.28}$$

SYSTEM ANALYSIS (SA) EXERCISES

SA4.1 Bungie Jump Energetics

As part of a circus act, two people, each with mass 65 kg, jump off a ledge with a long bungie cord attached to them (**Figure SA4.1.1**). They're harnessed together so that they fall as a unit. Eventually they stop bobbing up and down and hang motionless. At this point, one of the jumpers releases herself from the harness, dropping down into the safety net as the other person starts to rebound upward, pulled by the bungie cord. The hope is that the trapeze artist can swing by at just the right instant so as to grab her. Your task is to design the stunt.

a. How high does that person get at the peak of the rebound? Assume an unstretched length of 8 m for the bungie cord and a linear spring constant of 100 N/m.

b. How much higher will she go if the spring constant is increased by 10% (to 110 N/m)?

c. Investigate how a mass difference (two people with different masses) affects the results.

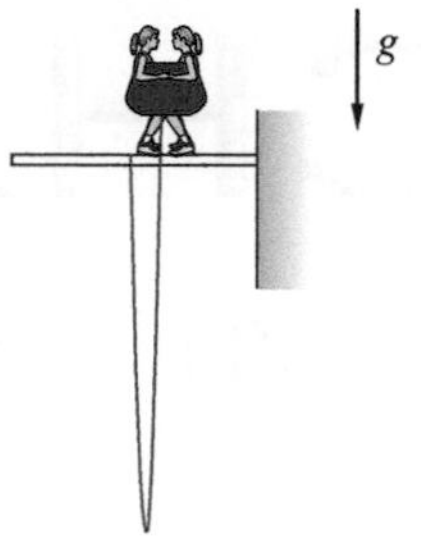

Figure SA4.1.1 Bungie jumping/aerial act

SA4.2 Escape from Colditz—Take Two

Read **SA3.1** as background information.

a. A model of the simplified launching system is shown in **Figure SA4.2.1*a***. Assume the glider with two prisoners weighs 227 kg and the concrete-filled tub weighs 454 kg. In addition, assume the glider rests on a free-rolling cart. Using the concepts of work and energy, calculate the glider's velocity after the concrete-filled tub falls 18 m.

b. Graph the relationship between the weight of the concrete-filled tub and the final velocity of the glider. Assume the concrete-filled tub can weigh between 454 and 227 kg. Discuss your results.

c. A double pulley will alter the system performance. Two examples of double pulleys are shown in (*b*) and (*c*). Double Pulley 1 (*b*) has the glider cable attached to the outer radius and the tub attached to the inner radius. Double Pulley 2 (*c*) has the glider cable attached to the inner radius and the tub cable to the outer radius. Which double pulley do you recommend? Support your answer with calculations.

d. Assume the outer radius is 30 cm and the inner radius is 15 cm for both double pulleys. Recalculate the glider's velocity in **a** for both pulleys.

e. Recommend another physical change to the system to increase the velocity of the glider. Support your recommendation with calculations.

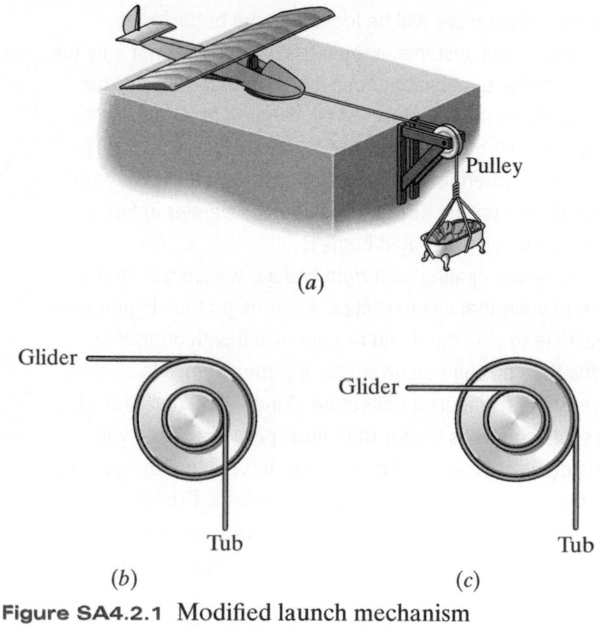

Figure SA4.2.1 Modified launch mechanism

OBJECTIVES

Upon completion of this chapter you will be able to:

- Deal with the kinematics and kinetics of multiple particles
- Apply momentum concepts to multiple-particle problems
- Use an energy approach when dealing with a system of multiple particles

CHAPTER 5

MULTIBODY SYSTEMS

In this chapter we will be looking at the behavior of multiparticle systems, a topic that's important not only for its own sake, but also because it will springboard us into rigid-body dynamics. How? Well, if you think about it, a rigid body is a collection of particles (a very large collection) held together by intermolecular forces. The equations we come up with here will be directly usable (and simpler in form) when we move on to rigid bodies.

Even when dealing with rigid bodies, we will sometimes want to view them as particles. A car might look bigger than a particle to you, but if you're concerned with analyzing traffic flow through an urban environment, you'll likely want to view the traffic as a collection of individual particles (the cars). The earth is bigger than most particles, but if you consider what it looks like from the viewpoint of the sun, it's a rather small ball of rock, a particle in fact. Orbital mechanics can actually be thought of as an analysis of multiple particles—the sun, the earth, an Apollo spacecraft, and so forth.

Thus you can view this chapter as either an extension of the single-particle analysis we have been doing up until now or an extension of rigid-body mechanics in which we wish to study several rigid bodies at once and from a viewpoint for which the dimensions of the bodies are negligible.

I will follow the same approach used in the previous chapters—namely, start off with a force balance ($\boldsymbol{F} = m\boldsymbol{a}$), integrate to get some linear momentum results, then do a moment balance followed by some angular momentum, and finish off with some energy observations.

Benson Tongue

5.1 FORCE BALANCE AND LINEAR MOMENTUM

Figure 5.1 shows what I like to think of as the "dynamical potato." The potato-like outline indicates that we're looking at a collection of bodies, all contained within the outline and free to move about as they will. There's nothing rigid about this outline; it simply indicates what bodies are part of the system being examined. There can be as many bodies as you like in there—it doesn't matter. I've drawn in three—m_1, m_2, and m_i, along with their position vectors (referenced to the origin O). As before, we'll view each body as being well approximated by a single particle.

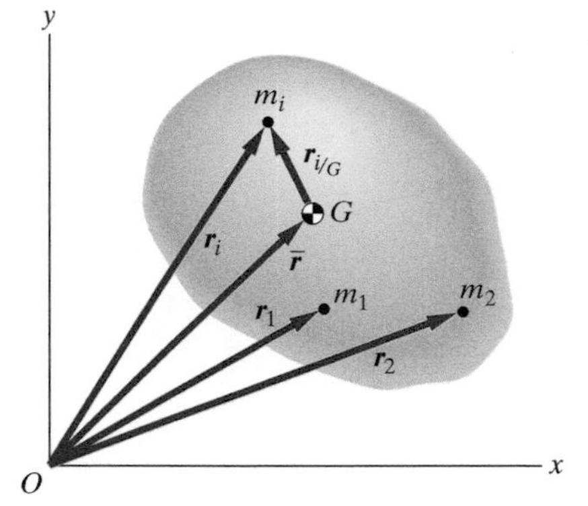

Figure 5.1 General multiparticle system

I have also drawn in the system's center of mass and labeled it G. Whenever you see anything labeled G throughout the rest of the book, remember that it will always refer to a center of mass. In addition to using G exclusively in reference to the center of mass, I will also use an overbar to indicate a property associated with a system's mass center. Thus in **Figure 5.1** I've labeled the position vector of the mass center $\bar{\boldsymbol{r}}$. I could label it $\boldsymbol{r}_G$, but our equations will be a little less messy with $\bar{\boldsymbol{r}}$, and so that's what I'll use. Also I'll use a little BMW logo to indicate the center of mass. Great piece of product placement, isn't it?

How can we start to analyze our system? First, we can be explicit about how we find the center of mass. As you may recall, the mass center has a clear physical meaning. If you could somehow hang the collection of particles like a big mobile (imagine massless, rigid rods connecting them all), then this mobile would be perfectly balanced if you hung it from the mass center. There would be no gravitational torques that would act to spin it. We find the location of the mass center by using the expression

$$m\bar{\boldsymbol{r}} = \sum_{i=1}^{n} m_i \boldsymbol{r}_i, \quad \text{where } m \equiv \sum_{i=1}^{n} m_i \tag{5.1}$$

where n is the total number of particles in the potato. Rearranging (5.1) gives us

$$\bar{\boldsymbol{r}} = \frac{\sum_{i=1}^{n} m_i \boldsymbol{r}_i}{m} \tag{5.2}$$

>>> **Check out Example 5.1 (page 263) and Example 5.3 (page 265) for applications of this material.**

This expression is going to be useful in our work, but we will also find it helpful to look at a related expression. In **Figure 5.1** I have drawn in the vector $\boldsymbol{r}_{i/G}$, which is in the position vector of an arbitrary particle m_i with respect to the mass center G. What we will do now is see how this position vector figures into the calculation for locating a mass center. You can see from **Figure 5.1** that the position vector $\boldsymbol{r}_i$ for the mass particle m_i can just as well be expressed as $\bar{\boldsymbol{r}} + \boldsymbol{r}_{i/G}$. If we substitute this into (5.1), we will get

$$
\begin{aligned}
m\bar{\boldsymbol{r}} &= \sum_{i=1}^{n} m_i \left(\bar{\boldsymbol{r}} + \boldsymbol{r}_{i_{/G}}\right) \\
&= \bar{\boldsymbol{r}} \sum_{i=1}^{n} m_i + \sum_{i=1}^{n} m_i \boldsymbol{r}_{i_{/G}} \\
&= m\bar{\boldsymbol{r}} + \sum_{i=1}^{n} m_i \boldsymbol{r}_{i_{/G}}
\end{aligned}
$$

which leaves us with the conclusion that

$$
\sum_{i=1}^{n} m_i \boldsymbol{r}_{i_{/G}} = 0 \tag{5.3}
$$

This result makes loads of sense, assuming that you bought into (5.2), which tells us where to go, if we start at the origin O, in order to reach the mass center. If we *start* from the mass center in the first place, then we don't want to go anywhere. And that's what (5.3) tells us. The summation of all the masses times their displacements from the center of mass is what ultimately tells us how far to go to get to the mass center. And if we're already there, we will obviously want to go zero distance—hence the summation equals zero.

Now that we know about the center of mass, what's next? Well, we know there could be lots of forces acting on each particle in our system, and we also know how to write down the individual equations of motion. So let's do that. There will be two basic kinds of forces for us to be concerned with: forces that are internal to the system and forces that are external. **Internal forces** are the forces that the particles exert on one another. The gravitational attraction of one particle to another is an example of an internal force, and intra-atomic forces are another example. There might be springs connecting the particles, in which case the spring forces would be internal forces because they act between the particles of the system. The particles could also be bumping into one another, as happens when planetesimals are accreting interstellar dust; these contact forces are another type of internal force.

Not surprisingly, **external forces** are forces that originate outside the system. If you push on a system of particles with your hand, then that's an external force. If our system is defined to be the solar system, then the gravitational attraction of the rest of the Milky Way on the solar system is an external force.

To keep these two types of forces separate in our minds, we will designate the internal forces as $\tilde{\boldsymbol{F}}_{ji}$ and the external forces as $\boldsymbol{F}_{ki}$. The ji subscript indicates that $\tilde{\boldsymbol{F}}_{ji}$ is the internal force exerted by particle m_j on particle m_i. In the case of $\boldsymbol{F}_{ki}$ I am showing that several external forces may be acting on each particle; thus $\boldsymbol{F}_{12}$ is the first external force acting on particle m_2, $\boldsymbol{F}_{22}$ is the second force acting on m_2, and so on.

We can write the equation of motion for the particle m_i as

$$\sum_{j=1}^{n} \tilde{\boldsymbol{F}}_{ji} + \sum_{k=1}^{p_i} \boldsymbol{F}_{ki} = m_i \ddot{\boldsymbol{r}}_i \tag{5.4}$$

where p_i indicates the number of external forces acting on the ith particle. Try not to be put off by the variables; all I'm saying is that each particle in the system can potentially exert an internal force on the ith particle, as can a collection of external forces. Add up all these forces, and the sum is equal to the particle's mass times its acceleration.

The summation notation is convenient, but as written it introduces the possibility that a particle exerts a force on itself. We eliminate this possibility by saying that

$$\tilde{\boldsymbol{F}}_{ij} \equiv 0 \qquad i = j$$

So, all well and good but not very enlightening. But we haven't used our center of mass concept yet, have we? Let's now do so. We will calculate the sums of (5.4) for each particle in the system and add up all the results. This will give us a summation over all the particles, and we will therefore have a double summation in front of the forces:

$$\sum_{i=1}^{n}\sum_{j=1}^{n} \tilde{\boldsymbol{F}}_{ji} + \sum_{i=1}^{n}\sum_{k=1}^{p_i} \boldsymbol{F}_{ki} = \sum_{i=1}^{n} m_i \ddot{\boldsymbol{r}}_i \tag{5.5}$$

Because the double differentiation in the $\ddot{\boldsymbol{r}}_i$ term doesn't depend on i, we can pull the differentiation out of the summation, which gives us

$$\sum_{i=1}^{n}\sum_{j=1}^{n} \tilde{\boldsymbol{F}}_{ji} + \sum_{i=1}^{n}\sum_{k=1}^{p_i} \boldsymbol{F}_{ki} = \frac{d^2}{dt^2}\left(\sum_{i=1}^{n} m_i \boldsymbol{r}_i\right) \tag{5.6}$$

Ah ha! What's that quantity in parentheses? It's $m\bar{\boldsymbol{r}}$ from (5.1), that's what it is. And take a look at the double summations. The one on the left is a summation of all internal forces in the system, and you know that Newton's third law states that for every force there's an equal and *opposite* reaction force. So when we add up all the internal forces, we will get the force exerted by m_i on m_j and the equal and opposite force exerted by m_j on m_i, which cancel. In fact, the entire summation simply pops out of existence, done in by the pairings of equal and opposite forces.

This leaves us with the second double summation, which is simply the summation of all external forces acting on the system. In order to keep the subscripts in some kind of control, allow me to take all the external forces that might be acting on particle m_i, add them vectorially, and replace them with a resultant vector which I will call $\boldsymbol{F}_i$, so that $\sum_{k=1}^{p_i} \boldsymbol{F}_{ki}$ becomes simply $\boldsymbol{F}_i$. Taking all this into account, we can now write (5.6) as

$$\sum_{i=1}^{n} \boldsymbol{F}_i = m\ddot{\bar{\boldsymbol{r}}} = m\bar{\boldsymbol{a}} \tag{5.7}$$

What did this get us? Quite a bit, actually. What (5.7) tells us is that in at least one sense we can treat the motion of a collection of particles as if it were a single particle. Here is the analogue that will make it as clear as possible:

$$\boldsymbol{F} = m\boldsymbol{a} \quad \Longleftrightarrow \quad \sum_{i=1}^{n} \boldsymbol{F}_i = m\bar{\boldsymbol{a}}$$

These two expressions look almost identical. The difference is that in the left equation we are including all the forces acting on a single particle, but in the right equation we are including all the forces acting on *all* the particles in the system. In the left equation, m is the mass of one particle and $\boldsymbol{a}$ is its acceleration; in the right equation, m is the mass of all the system particles and $\bar{\boldsymbol{a}}$ is the acceleration of the system's center of mass.

Now let's examine linear momentum. Each particle in our system has a linear momentum

$$\boldsymbol{L}_i = m_i \dot{\boldsymbol{r}}_i$$

As before, we will sum all these linear momenta and see what that gets us.

$$\begin{aligned}
\boldsymbol{L} &= \sum_{i=1}^{n} \boldsymbol{L}_i = \sum_{i=1}^{n} m_i \dot{\boldsymbol{r}}_i = \sum_{i=1}^{n} m_i \left(\dot{\bar{\boldsymbol{r}}} + \dot{\boldsymbol{r}}_{i_{/G}}\right) \\
&= \dot{\bar{\boldsymbol{r}}} \sum_{i=1}^{n} m_i + \frac{d}{dt} \overbrace{\sum_{i=1}^{n} m_i \boldsymbol{r}_{i_{/G}}}^{0 \text{ from } (5.3)} \\
&= m\dot{\bar{\boldsymbol{r}}}
\end{aligned}$$

$$\boxed{\boldsymbol{L} = m\bar{\boldsymbol{v}}} \tag{5.8}$$

Here's the first place where knowing that $\sum_{i=1}^{n} m_i \boldsymbol{r}_{i_{/G}} = 0$ has proved helpful. Because of this fact, we have gotten our second basic result: The linear momentum of a system of particles is equal to the total mass times the velocity of the mass center. Again, this is almost the same relationship we had when dealing with a single particle—momentum equals mass times velocity.

We can differentiate (5.8) with respect to time to get

$$\boxed{\dot{\boldsymbol{L}} = m\dot{\bar{\boldsymbol{v}}} = m\bar{\boldsymbol{a}}} \tag{5.9}$$

This is the $\boldsymbol{F} = m\boldsymbol{a}$ relationship we derived earlier in this section, just as you would expect.

If there are no external forces acting on our system, then we have

$$\boxed{\dot{\boldsymbol{L}} = 0} \tag{5.10}$$

which is a statement of **conservation of linear momentum** for the system. This very important equation tells us that as long as no external forces act on our system, the system's linear momentum is constant.

Integrating (5.10) gives us $\boldsymbol{L} = \text{constant}$.

>>> **Check out Example 5.2 (page 264), Example 5.4 (page 266), and Example 5.5 (page 267) for applications of this material.**

EXAMPLE 5.1 FINDING A MASS CENTER (Theory on page 259)

Let's start by finding the mass center of the three-particle system shown in **Figure 5.2**. $m_A = 4$ kg, $m_B = 6$ kg, $m_C = 2$ kg, $\boldsymbol{r}_{A/O} = 1\boldsymbol{i}$ m, $\boldsymbol{r}_{B/O} = (2\boldsymbol{i} + 2\boldsymbol{j})$ m, and $\boldsymbol{r}_{C/O} = 4\boldsymbol{i}$ m.

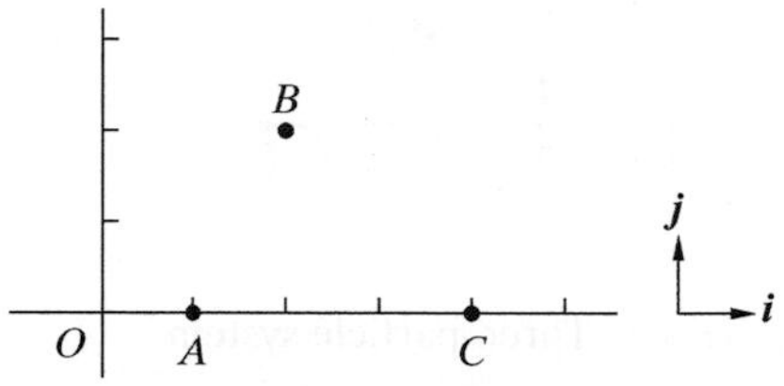

Figure 5.2 Three-particle system

Goal Determine the position of the mass center of a three-mass system with respect to the origin O.

Assume No additional assumptions are needed.

Draw **Figure 5.3** shows $\boldsymbol{r}_{G/O}$, the vector we're after (also known as $\bar{\boldsymbol{r}}$).

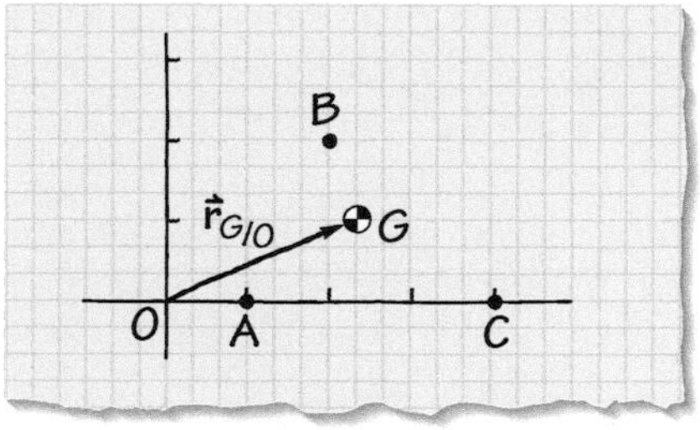

Figure 5.3 Mass center of three particles

Formulate Equations We will need to apply (5.2).

Solve Using (5.2) gives us

$$\bar{\boldsymbol{r}} = \frac{\sum_{i=1}^{n} m_i \boldsymbol{r}_i}{m} = \frac{(4\,\text{kg})(1\boldsymbol{i}\,\text{m}) + (6\,\text{kg})[(2\boldsymbol{i} + 2\boldsymbol{j})\,\text{m}] + (2\,\text{kg})(4\boldsymbol{i}\,\text{m})}{12\,\text{kg}}$$

$$\boxed{\bar{\boldsymbol{r}} = (2\boldsymbol{i} + 1\boldsymbol{j})\,\text{m}}$$

Check A nice check is to use (5.3). If we get zero, then we can be sure we found $\bar{\boldsymbol{r}}$ correctly. Our relative position vectors are $\boldsymbol{r}_{A/G} = (-1\boldsymbol{i} - 1\boldsymbol{j})$ m, $\boldsymbol{r}_{B/G} = 1\boldsymbol{j}$ m, and $\boldsymbol{r}_{C/G} = (2\boldsymbol{i} - 1\boldsymbol{j})$ m. Using (5.3) gives us

$$\sum_{i=1}^{n} m_i \boldsymbol{r}_{i/G} = (4\,\text{kg})[(-1\boldsymbol{i} - 1\boldsymbol{j})\,\text{m}] + (6\,\text{kg})(1\boldsymbol{j}\,\text{m}) + (2\,\text{kg})[(2\boldsymbol{i} - 1\boldsymbol{j})\,\text{m}] = 0$$

as expected, validating our original result.

EXAMPLE 5.2 **FINDING A SYSTEM'S LINEAR MOMENTUM** (Theory on page 262)

Now let's go a little further and consider the same system of particles we just saw in Example 5.1. As before, $m_A = 4$ kg, $m_B = 6$ kg, $m_C = 2$ kg, $\boldsymbol{r}_{A/O} = 1\boldsymbol{i}$ m, $\boldsymbol{r}_{B/O} = (2\boldsymbol{i} + 2\boldsymbol{j})$ m, and $\boldsymbol{r}_{C/O} = 4\boldsymbol{i}$ m. In addition, we are also given the particles' velocities: $\boldsymbol{v}_A = 4\boldsymbol{j}$ m/s, $\boldsymbol{v}_B = (3\boldsymbol{i} + 3\boldsymbol{j})$ m/s, and $\boldsymbol{v}_C = -2\boldsymbol{i}$ m/s. Find the system's linear momentum.

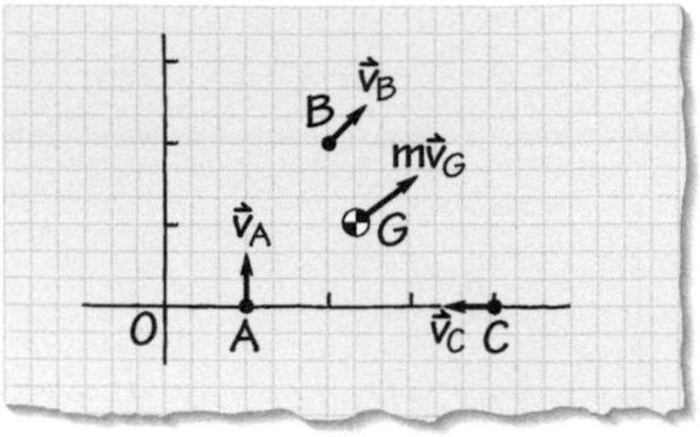

Figure 5.4 Three-particle system

Goal Determine the linear momentum of a three-particle system.

Assume No additional assumptions are needed.

Draw **Figure 5.4** shows the velocity vectors of the three points, as well as the system's linear momentum vector (direction and magnitude still undetermined).

Formulate Equations For this problem all we need is (5.8).

Solve To apply (5.8) we need the velocity of the system's mass center, and an easy way to get that is simply to differentiate (5.2):

$$\overline{\boldsymbol{v}} = \frac{\sum_{i=1}^{n} m_i \boldsymbol{v}_i}{m} = \frac{(4\,\text{kg})(4\boldsymbol{j}\,\text{m/s}) + (6\,\text{kg})[(3\boldsymbol{i} + 3\boldsymbol{j})\,\text{m/s}] + (2\,\text{kg})(-2\boldsymbol{i}\,\text{m/s})}{12\,\text{kg}}$$

Using this in (5.8) gives us

$$\boldsymbol{L} = m\overline{\boldsymbol{v}} = (14\boldsymbol{i} + 34\boldsymbol{j})\,\text{kg}\cdot\text{m/s}$$

Check An equivalent way to express $\boldsymbol{L}$ is simply as the sum of the individual linear momenta. Thus you can calculate

$$\boldsymbol{L} = m_A \boldsymbol{v}_A + m_B \boldsymbol{v}_B + m_C \boldsymbol{v}_C$$

and if you do so, you will get the same result as we did a few lines up.

EXAMPLE 5.3 MOTION OF A TWO-PARTICLE SYSTEM (Theory on page 259)

An unpowered, 100 kg missile A is projected upward at an inclination of 45° and a velocity of $\boldsymbol{v} = (98.1\boldsymbol{i} + 98.1\boldsymbol{j})$ m/s. At the top of its trajectory it explodes (**Figure 5.5**) such that a 75 kg piece receives a linear impulse in the $-\boldsymbol{i}$ direction and the remaining 25 kg piece gets an equal and opposite linear impulse (in the $\boldsymbol{i}$ direction). The 75 kg piece is later found at $x = x_A = 1862$ m. At what position x_B should you expect to find the other piece?

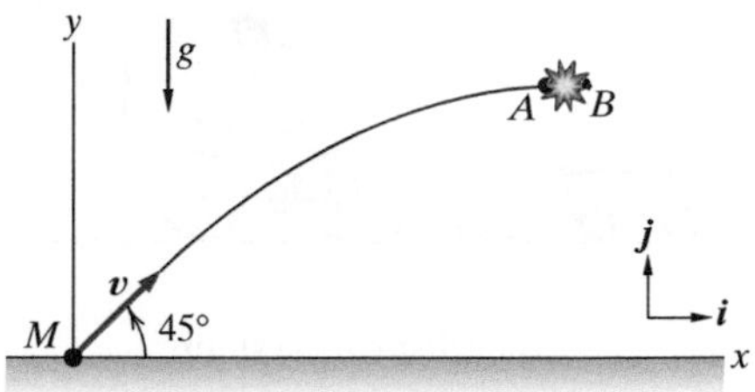

Figure 5.5 Unpowered missile launch

Goal Determine where a piece of a missile will come to earth, knowing the masses of the two pieces and the landing position of one piece (x_A).

Assume We will neglect any effects due to variable air resistance; in fact, we will neglect air resistance entirely.

Draw **Figure 5.6** shows the trajectory of the mass center, as well as the two pieces after breakup.

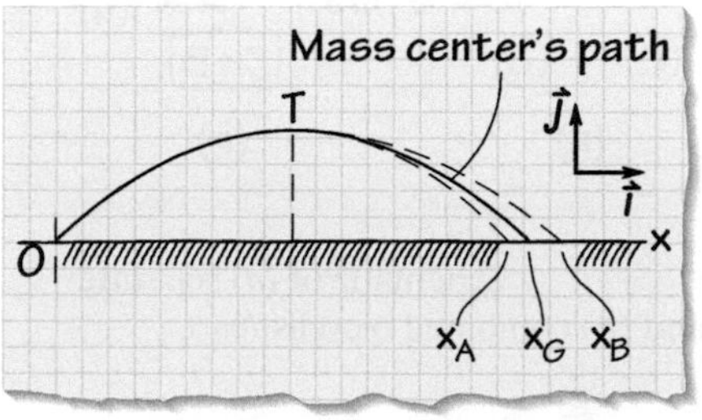

Figure 5.6 Trajectories of mass center and missile pieces

Formulate Equations The key here is to find the position of the mass center on contact (5.7), as well as the relationship between the mass center and the individual parts of the missile (5.2).

Solve The missile has no external forces acting on it other than that due to gravity, and thus the mass center's trajectory will be parabolic. The position of the mass center G is given by

$$\boldsymbol{r}_G = \left[(98.1\boldsymbol{i}\ \text{m/s})t + \left((98.1\ \text{m/s})t - \frac{(9.81\ \text{m/s}^2)t^2}{2}\right)\boldsymbol{j}\right] \text{m}$$

Differentiating the missile's height with respect to time and setting the result to zero tell us that the peak of the trajectory is reached in 10 s, and at this time the missile is at $x = 981$ m, $y = 490.5$ m. Waiting 20 s brings the mass center along the next half of the parabolic arc and back to the ground:

$$\boldsymbol{r}_G = 1962\boldsymbol{i}\ \text{m} \quad \text{at} \quad t = 20\ \text{s} \quad \Rightarrow \quad x_G = 1962\ \text{m}$$

Now that we have x_G we can put (5.2) to use.

$$x_G = \frac{(75\ \text{kg})(1862\ \text{m}) + (25\ \text{kg})(x_B)}{100\ \text{kg}} \quad \Rightarrow \quad x_B = 2262\ \text{m}$$

Check Let's look at the positions of the two masses with respect to the mass center and see if they satisfy (5.3).

$$m_A \boldsymbol{r}_{A/G} + m_B \boldsymbol{r}_{B/G} = (75\ \text{kg})(-100\ \text{m}) + (25\ \text{kg})(300\ \text{m}) = 0$$

Excellent—we get zero, just as we should. Thus we can conclude that the 25 kg piece of the missile will be found 2262 m beyond the launch point.

EXAMPLE 5.4 FINDING SPEED OF A BICYCLIST/CART (Theory on page 262)

Figure 5.7 Bicyclist/cart system

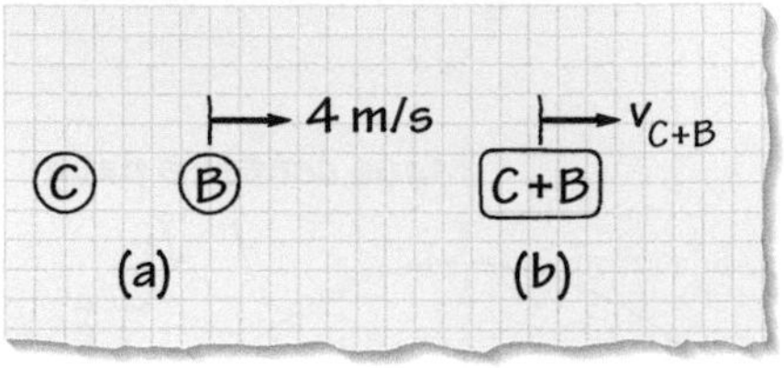

Figure 5.8 Schematic of (*a*) separate and (*b*) combined bicyclist/cart

An 80 kg bicyclist (mass of cyclist plus bicycle) moves forward at 4 m/s and, after he travels a couple of meters, a rope attached between the bicycle and a 20 kg wheeled cart goes taut, causing the bicycle and cart to move as a single unit (**Figure 5.7**). What is the speed of the bicyclist/cart combination just after the rope goes taut?

Goal Determine the speed of this combined system (bicyclist and cart) after the two begin to move as a unit.

Assume No additional assumptions are needed.

Draw **Figure 5.8*a*** shows a schematic of the system before the rope goes taut, in which bicyclist B is moving at 4 m/s and cart C is stationary. In **Figure 5.8*b*** the entire system $B + C$ is moving as a unit with a speed v.

Formulate Equations All forces are internal (due to the rope), and thus (5.10) is applicable.

Solve The initial momentum $\boldsymbol{L}_1$ is given by

$$\begin{aligned} \boldsymbol{L}_1 &= m_B \boldsymbol{v}_B + m_C \boldsymbol{v}_C \\ &= (80\,\text{kg})(4\boldsymbol{i}\,\text{m/s}) + (20\,\text{kg})(0\boldsymbol{i}\,\text{m/s}) \\ &= 320\boldsymbol{i}\,\text{kg}\cdot\text{m/s} \end{aligned}$$

After the rope goes taut, the system moves as a unit with mass $m_B + m_C$ and velocity $v\boldsymbol{i}$:

$$\boldsymbol{L}_2 = (m_B + m_C)\boldsymbol{v} = (100\,\text{kg})v\boldsymbol{i}$$

We can now apply conservation of linear momentum.

$$(5.10) \Rightarrow \qquad \boldsymbol{L}_1 = \boldsymbol{L}_2$$

$$320\boldsymbol{i}\,\text{kg}\cdot\text{m/s} = (100\,\text{kg})v\boldsymbol{i}$$

$$\boxed{v = 3.2\,\text{m/s}}$$

Check For this case we'll just do a "Does it seem reasonable?" check. The bicyclist outweighed the cart by four to one. Thus the extra 25% of mass should have reduced his speed, but not by a huge amount. In fact, we saw that the initial speed of 4 m/s was reduced to 3.2 m/s, which is very much in line with expectations.

EXAMPLE 5.5 MOMENTUM OF A THREE-MASS SYSTEM (Theory on page 262)

An astronaut (B) was working outside a space station and accidentally sent a tool (C) off to the right. She quickly tethered herself to her robotic repair pod (A) and, with a brief burst from her backpack reaction jets, sent herself in pursuit of the tool (**Figure 5.9**). At the instant shown the repair pod is stationary, $v_B = 2\,\text{m/s}$, and $v_C = 0.5\,\text{m/s}$. Eventually, she catches up with the tool, after which the rope between the pod and herself goes taut. Assume that the rope is inextensible and massless. After the rope becomes taut, what is the velocity of the repair pod? $m_A = 200\,\text{kg}$, $m_B = 80\,\text{kg}$, and $m_C = 5\,\text{kg}$.

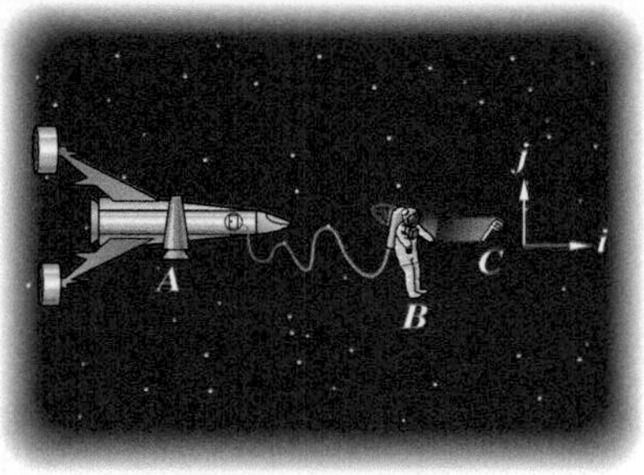

Figure 5.9 Astronaut chasing a tool

Goal Find the velocity of the repair pod after the astronaut has recovered her lost tool.

Given The masses and initial velocities of the pod, astronaut, and tool.

Assume Because the entire event unfolds in space, we're safe in neglecting any air friction. In addition, we will completely ignore gravity. We're neglecting any elasticity in the tether rope as well. In an actual system we would expect the rope to have some give, but in our model it's inextensible and therefore couples the pod to the astronaut such that they move together as a single unit.

Draw **Figure 5.10** shows a schematic of the three masses before the astronaut has caught up with the tool and an equivalent mass equal to the sum of the individual masses that represents the system after tool recovery has taken place.

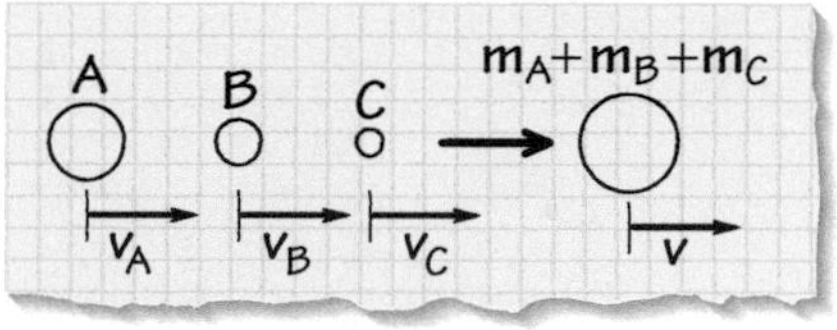

Figure 5.10 Schematic of system before and after tool recovery

Formulate Equations The key for this problem is to apply conservation of linear momentum. The only forces to consider are the impact forces between the astronaut and the tool as she grabs it and the tension in her tether rope when it goes taut. Both of these forces are internal to the system, and therefore the total linear momentum is conserved.

Solve

(5.10) ⇒ $$m_A v_A + m_A v_B + m_A v_C = (m_A + m_B + m_C)v$$

$$0 + (80\,\text{kg})(2\,\text{m/s}) + (5\,\text{kg})(0.5\,\text{m/s}) = (285\,\text{kg})v$$

$$v = 0.570\,\text{m/s}$$

Check Does this answer seem sensible? The main source of linear momentum is the astronaut. She's moving faster than the tool and is much more massive; therefore her contribution to the linear momentum is far greater. So a simplified analysis would ignore the tool entirely. In this case we would have

$$0 + (80\,\text{kg})(2\,\text{m/s}) = (280\,\text{kg})v$$

and thus $v = 0.571\,\text{m/s}$, a speed that's almost equal to the original result. Thus our approximate analysis is in line with the more complete one, giving us some confidence in the work.

EXERCISES 5.1

5.1.1. **[Level 1]** Find the illustrated system's linear momentum vector: $m_A = 5\,\text{kg}$, $m_B = 8\,\text{kg}$, $m_C = 2\,\text{kg}$, $m_D = 7\,\text{kg}$, $\boldsymbol{v}_A = (2\boldsymbol{i} + 3\boldsymbol{j})\,\text{m/s}$, $\boldsymbol{v}_B = (3\boldsymbol{i} - 4\boldsymbol{j})\,\text{m/s}$, $\boldsymbol{v}_C = 5\boldsymbol{i}\,\text{m/s}$, and $\boldsymbol{v}_D = -4\boldsymbol{j}\,\text{m/s}$.

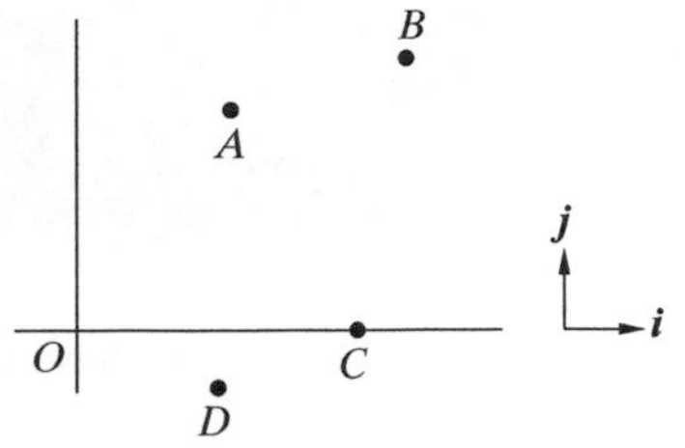

E5.1.1

5.1.2. **[Level 1]** Find the illustrated system's center of mass and linear momentum vector: $m_A = 15$ kg, $m_B = 45$ kg, $m_C = 60$ kg, $\boldsymbol{r}_A = 0.9\boldsymbol{k}$ m, $\boldsymbol{r}_B = (0.6\boldsymbol{i} + 0.9\boldsymbol{j} - 0.3\boldsymbol{k})$ m, $\boldsymbol{r}_C = (-0.3\boldsymbol{i} - 0.6\boldsymbol{j} + 0.6\boldsymbol{k})$ m, $\boldsymbol{v}_A = (-0.3\boldsymbol{i} + 0.3\boldsymbol{j})$ m/s, $\boldsymbol{v}_B = -0.9\boldsymbol{k}$ m/s, and $\boldsymbol{v}_C = -0.6\boldsymbol{j}$ m/s.

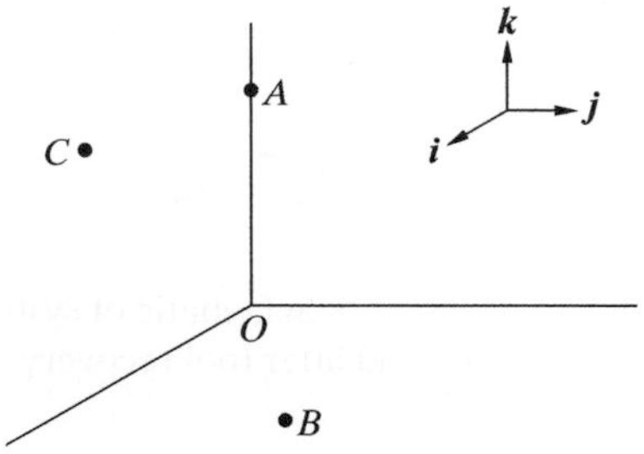

E5.1.2

5.1.3. **[Level 1]** Find the illustrated system's center of mass. $m_A = 2$ kg, $m_B = 3$ kg, $\boldsymbol{r}_{A/O} = 4\boldsymbol{j}$ m, and $\boldsymbol{r}_{B/O} = 4\boldsymbol{i}$ m.

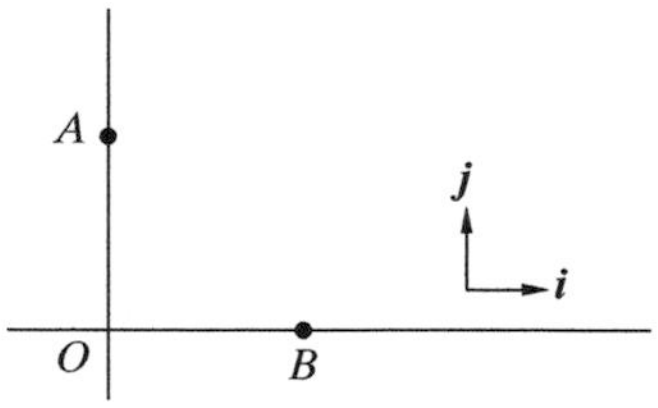

E5.1.3

5.1.4. **[Level 1]** Find the illustrated system's center of mass. $m_A = 73$ kg, $m_B = 146$ kg, $m_C = 292$ kg, $m_D = 146$ kg, $\boldsymbol{r}_{A/O} = 0.6\boldsymbol{j}$ m, $\boldsymbol{r}_{B/O} = (0.6\boldsymbol{i} + 0.6\boldsymbol{j})$ m, $\boldsymbol{r}_{C/O} = 0.6\boldsymbol{i}$ m, and $\boldsymbol{r}_{D/O} = 0.9\boldsymbol{i}$ m.

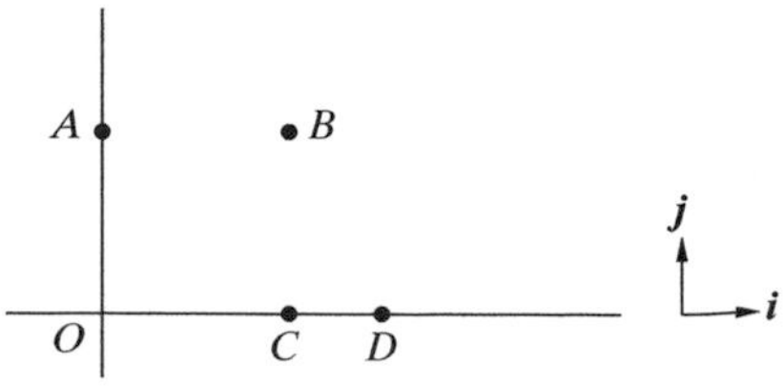

E5.1.4

5.1.5. **[Level 1]** Two carts (A and B) are initially traveling with a constant velocity $\boldsymbol{v}_0 = 6\boldsymbol{i}\,\text{m/s}$, separated by a distance $d = 3\,\text{m}$ but connected by an inextensible rope of length $L = 5\,\text{m}$. Suppose the brakes on cart B are suddenly applied, and the cart experiences an absolute deceleration of $\boldsymbol{a}_B = -4\boldsymbol{i}\,\text{m/s}^2$. If $m_A = 10\,\text{kg}$ and $m_B = 15\,\text{kg}$, find the speed of the two carts immediately after the rope between them becomes taut.

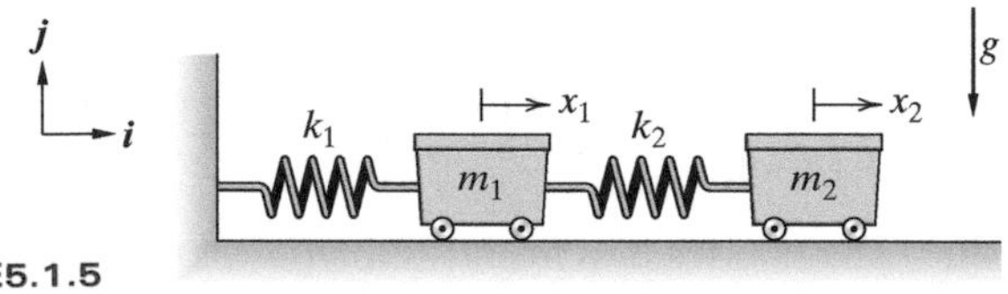

E5.1.5

5.1.6. **[Level 1]** Find the illustrated system's center of mass. $m_A = 2$ kg, $m_B = 2$ kg, $m_C = 2$ kg, $m_D = 2$ kg, $\boldsymbol{r}_{A/O} = 0.5\boldsymbol{i}$ m, $\boldsymbol{r}_{B/O} = 1.2\boldsymbol{j}$ m, $\boldsymbol{r}_{C/O} = -0.5\boldsymbol{k}$ m, and $\boldsymbol{r}_{D/O} = (-1.2\boldsymbol{j} + 1.5\boldsymbol{k})$ m.

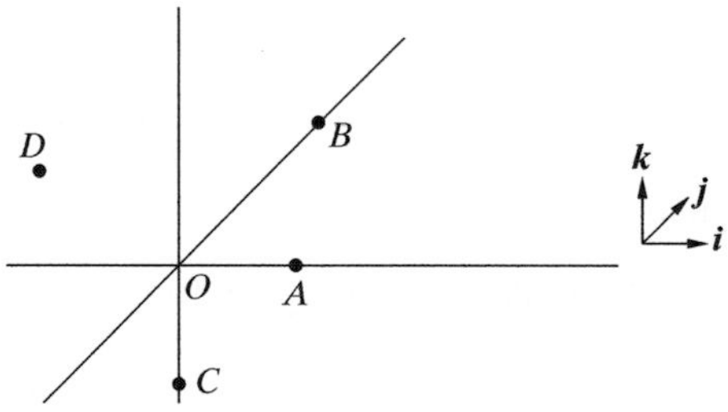

E5.1.6

5.1.7. **[Level 1]** Find the illustrated system's center of mass. $m_A = 5$ kg, $m_B = 2$ kg, $m_C = 6$ kg, $\boldsymbol{r}_{A/O} = -0.2\boldsymbol{j}$ m, $\boldsymbol{r}_{B/O} = (-0.2\boldsymbol{j} + 0.25\boldsymbol{k})$ m, and $\boldsymbol{r}_{C/O} = (0.3\boldsymbol{i} - 0.2\boldsymbol{k})$ m.

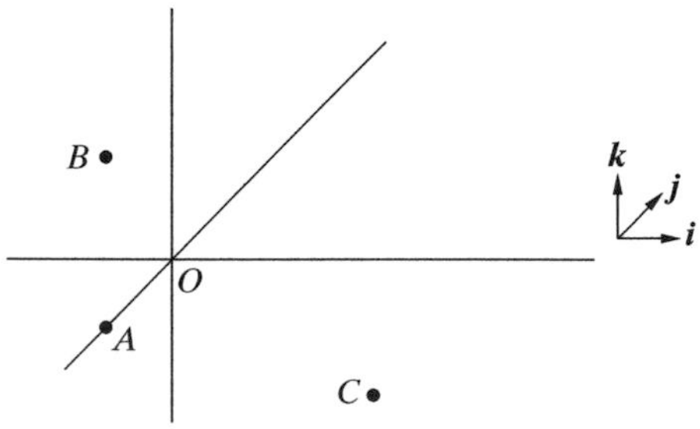

E5.1.7

5.1.8. **[Level 1]** Find the illustrated system's linear momentum vector. $m_A = 30$ kg, $m_B = 45$ kg, $m_C = 60$ kg, $m_D = 60$ kg, $\boldsymbol{r}_{A/O} = 0.6\boldsymbol{i}$ m, $\boldsymbol{r}_{B/O} = 0.7\boldsymbol{j}$ m, $\boldsymbol{r}_{C/O} = -0.75\boldsymbol{i}$ m, $\boldsymbol{r}_{D/O} = -0.7\boldsymbol{j}$ m, $\boldsymbol{v}_A = 1.5\boldsymbol{j}$ m/s, $\boldsymbol{v}_B = 1.8\boldsymbol{i}$ m/s, $\boldsymbol{v}_C = -1.5\boldsymbol{i}$ m/s, and $\boldsymbol{v}_D = -1.8\boldsymbol{j}$ m/s.

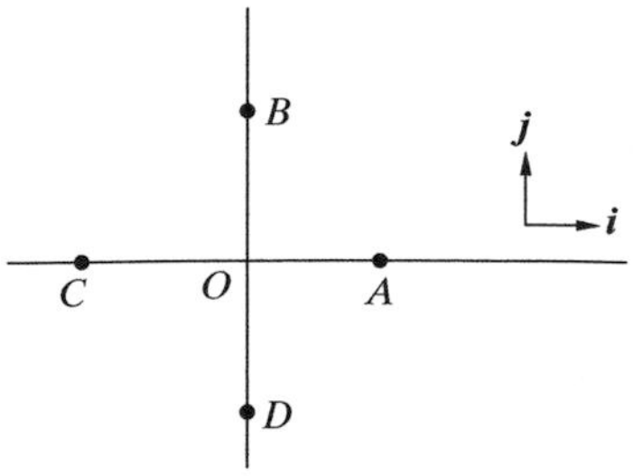

E5.1.8

5.1.9. **[Level 1]** Three balls ($A, B,$ and C) of equal mass m are projected toward each other in the horizontal plane such that they all impact simultaneously with the given velocities: $\boldsymbol{v}_A = 10\boldsymbol{i}$ m/s, $\boldsymbol{v}_B = -5\boldsymbol{i}$ m/s, and $\boldsymbol{v}_C = (-11\boldsymbol{i} - 9\boldsymbol{j})$ m/s. If the three balls stick together upon collision, what is the system's velocity just after impact?

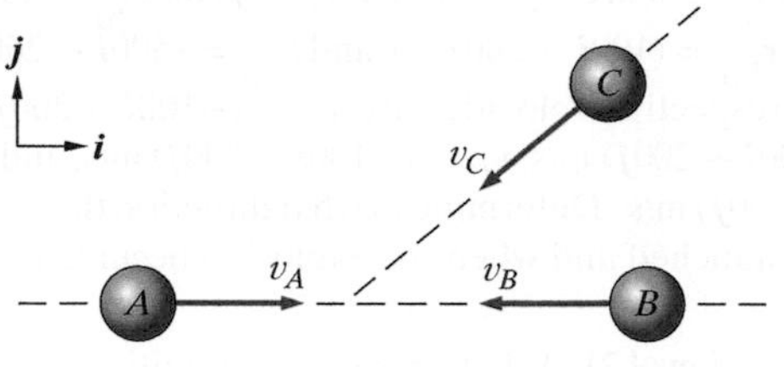

E5.1.9

5.1.10. **[Level 1]** Find the illustrated system's linear momentum vector. $m_A = 2$ kg, $m_B = 3$ kg, $m_C = 10$ kg, $\boldsymbol{r}_{A/O} = 0.2\boldsymbol{j}$ m, $\boldsymbol{r}_{B/O} = 0.4\boldsymbol{j}$ m, $\boldsymbol{r}_{C/O} = -0.6\boldsymbol{j}$ m, $\boldsymbol{v}_A = 10\boldsymbol{j}$ m/s, $\boldsymbol{v}_B = 20\boldsymbol{j}$ m/s, and $\boldsymbol{v}_C = -10\boldsymbol{j}$ m/s.

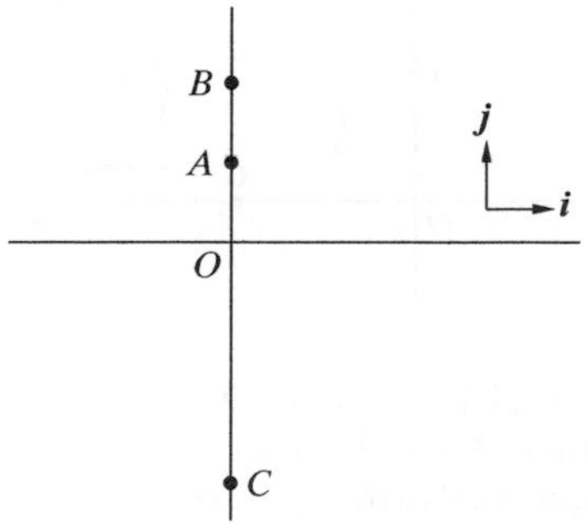

E5.1.10

5.1.11. **[Level 1]** When designing a dance platform, the builder needs to consider several factors, one of which is the load created by people moving along the floor and thereby inducing a horizontal load upon it. As an example, consider the situation shown in the figure. Two dancers are accelerating to the left at 1.8 m/s^2, while two are standing still upon the floor. What is the total horizontal force being applied to the floor? $\boldsymbol{v}_A = \boldsymbol{a}_A = \boldsymbol{v}_B = \boldsymbol{a}_B = 0$. $\boldsymbol{v}_C = \boldsymbol{v}_D = 1.5\boldsymbol{i}$ m/s and $\boldsymbol{a}_C = \boldsymbol{a}_D = -1.8\boldsymbol{i}$ m/s^2. Dancers A and B weigh 54 kg, dancer C weighs 50 kg, and dancer D weighs 68 kg.

E5.1.11

5.1.12. **[Level 1]** Four cars traveling at different velocities become involved in a collision while traveling in Minnesota during an ice storm. What is the velocity of the four cars' center of mass after they have collided (and are moving as a single, large mass)?

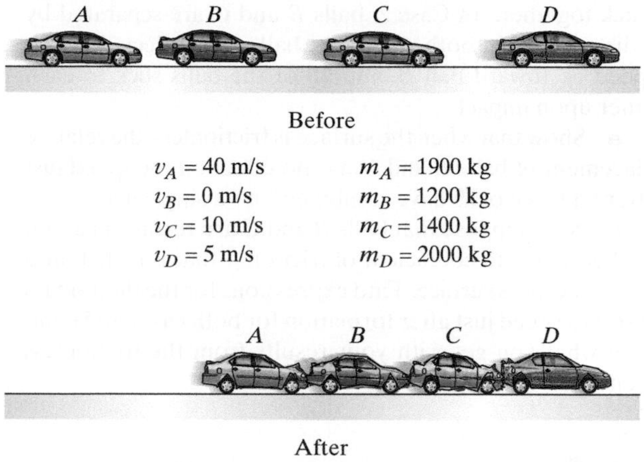
A
B
C
D
Before
$v_A = 40$ m/s
$v_B = 0$ m/s
$v_C = 10$ m/s
$v_D = 5$ m/s
$m_A = 1900$ kg
$m_B = 1200$ kg
$m_C = 1400$ kg
$m_D = 2000$ kg
A
B
C
D
After

E5.1.12

5.1.13. **[Level 1]** The illustrated mechanism is an extending set of drawers. Assume that they can slide without friction and are on a frictionless surface. At $t = 0$ the drawer A is moving to the right at 3 m/s, and the other two pieces are stationary. Drawer A retracts fully into B, and then both move to the right as a single piece. A/B fully retracts into C and hits C's rightmost wall. What is the speed of the completed retracted set of drawers immediately after this final collision? $m_A = 1$ kg, $m_B = 2$ kg, and $m_C = 4$ kg.

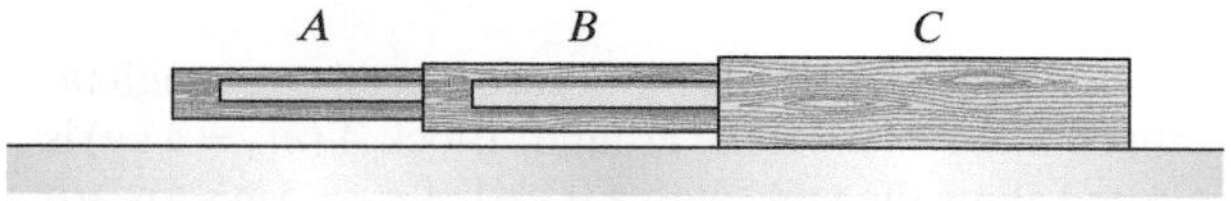

E5.1.13

5.1.14. **[Level 1]** A woman of mass m is standing on a rigid platform. Show that, if the platform has zero mass and the person is holding herself in equilibrium (platform off the ground), the normal force between the platform and her feet is equal to $\frac{2}{3}mg$.

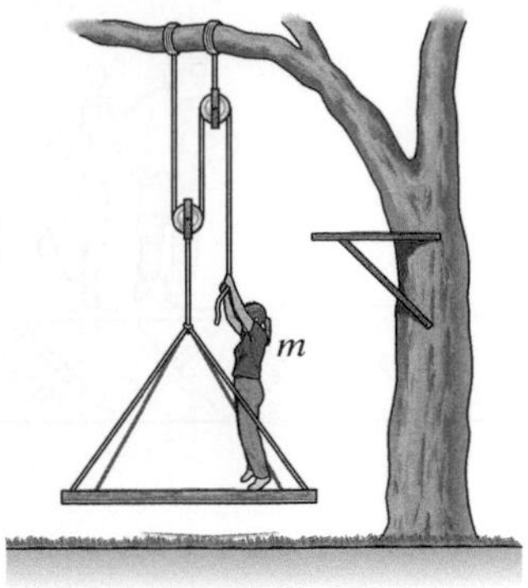

E5.1.14

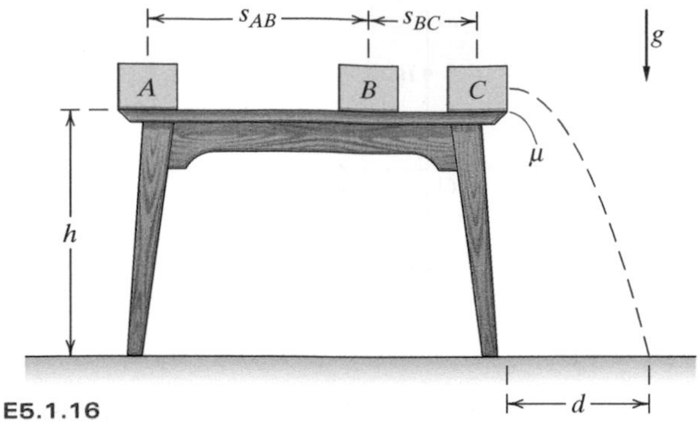

E5.1.16

5.1.15. **[Level 2]** Consider the following two cases involving three balls (A, B, and C) on a horizontal surface. In Case 1, balls B and C are right next to each other and stuck together. In Case 2, balls B and C are separated by a distance d. In both situations, ball A is projected with a speed v_A toward ball B, and all of the balls stick to each other upon impact.

a. Show that when the surface is frictionless, the relative placement of balls B and C has no effect on the speed just after all three balls have combined into a single mass.

b. Now suppose that balls B and C are resting on a thin rubber mat with coefficient of friction μ that is nailed onto the frictionless surface. Find expressions for the three-mass system's speed just after formation for both cases, and compare what you get with your results from the frictionless scenario.

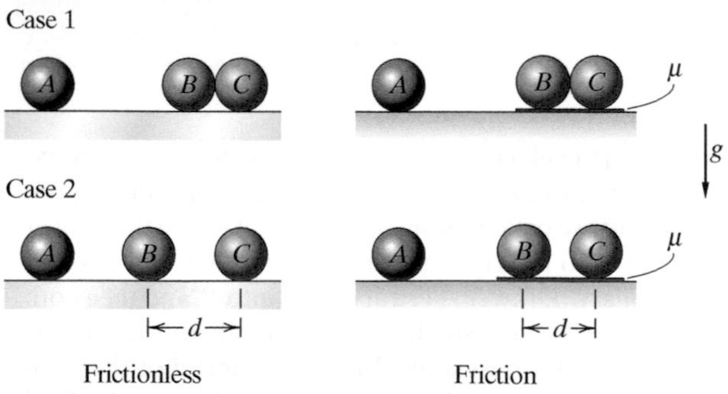

E5.1.15

5.1.16. **[Level 2]** Three blocks are placed on a rough table that is $h = 1.25$ m off the ground. Block A ($m_A = 4$ kg) is projected along the table with a speed of $v_A = 5$ m/s toward block B ($m_B = 1$ kg), which is initially at rest and $s_{AB} = 1$ m away from A. When blocks A and B collide, they stick together and travel $s_{BC} = 0.5$ m before impacting with and sticking to the stationary block C ($m_C = 2$ kg) located at the edge of the table. If the coefficient of friction between the blocks and table is $\mu = 0.2$, how far from the table's ledge does the three-block system land?

5.1.17. **[Level 2]** A sled is launched across a frozen lake from a position O with an initial speed v. At time t_1 the body explodes into four pieces (A, B, C, and D), each with equal mass. The positions of the four pieces at $t = 3$ s are $\boldsymbol{r}_{A/O} = (100\boldsymbol{i} + 200\boldsymbol{j})$ m, $\boldsymbol{r}_{B/O} = (500\boldsymbol{i} + 200\boldsymbol{j})$ m, $\boldsymbol{r}_{C/O} = (100\boldsymbol{i} - 200\boldsymbol{j})$ m, and $\boldsymbol{r}_{D/O} = (500\boldsymbol{i} - 200\boldsymbol{j})$ m, and the respective velocities are $\boldsymbol{v}_A = (-100\boldsymbol{i} + 200\boldsymbol{j})$ m/s, $\boldsymbol{v}_B = (300\boldsymbol{i} + 200\boldsymbol{j})$ m/s, $\boldsymbol{v}_C = (-100\boldsymbol{i} - 200\boldsymbol{j})$ m/s, and $\boldsymbol{v}_D = (300\boldsymbol{i} - 200\boldsymbol{j})$ m/s. Determine in what direction the sled was initially launched and when the explosion occurred.

5.1.18. **[Level 2]** A 45 kg rigid body is initially at rest at O on a frozen lake. Is it possible that once it has exploded into the three pieces, A, B, C, with masses $m_A = 20$ kg, $m_B = 15$ kg, and $m_C = 10$ kg, it can occupy the illustrated configuration $\boldsymbol{r}_{A/O} = 50\boldsymbol{j}$ m, $\boldsymbol{r}_{B/O} = (30\boldsymbol{i} + 20\boldsymbol{j})$ m, and $\boldsymbol{r}_{C/O} = 70\boldsymbol{i}$ m, at some later time?

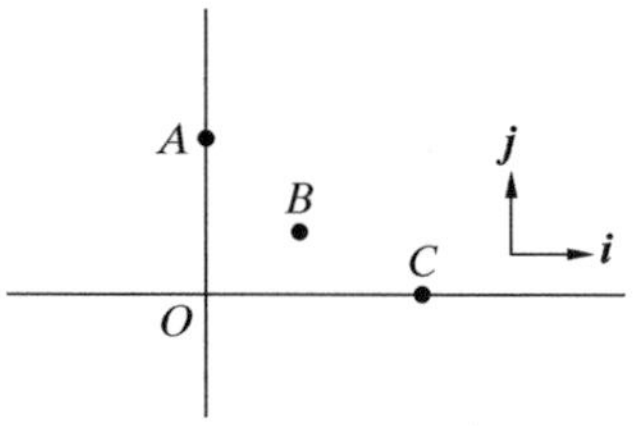

E5.1.18

5.1.19. **[Level 2]** Two blocks rest on a horizontal surface, one atop the other. What is their acceleration if acted on by a force $T = 30$ N, applied to the upper right corner of block A? Assume that they can move only horizontally. $\mu_{1_s} = 0.4$, $\mu_{1_d} = 0.35$, $\mu_{2_s} = 0.2$, $\mu_{2_d} = 0.15$, $m_A = 10$ kg, $m_B = 1$ kg.

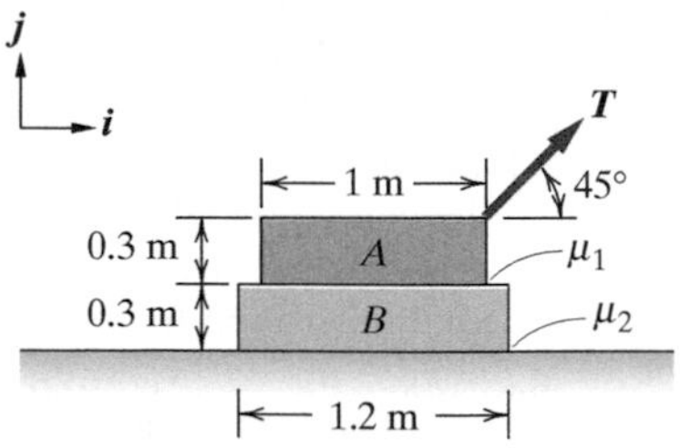

E5.1.19

5.1.20. **[Level 2]** A boat, initially stationary in the water, fires a shell in the $\boldsymbol{i}$ direction. The muzzle speed of the shell was 1000 m/s. The shell's speed at impact with its target was 990 m/s. The shell was in the gun for 0.1 s, and the force exerted on the shell decreased linearly from the initial force to zero over the firing interval. The shell's mass was 10 kg, and the boat's mass was 990 kg. Find the maximum force that acted on the shell and calculate the final velocity of the boat.

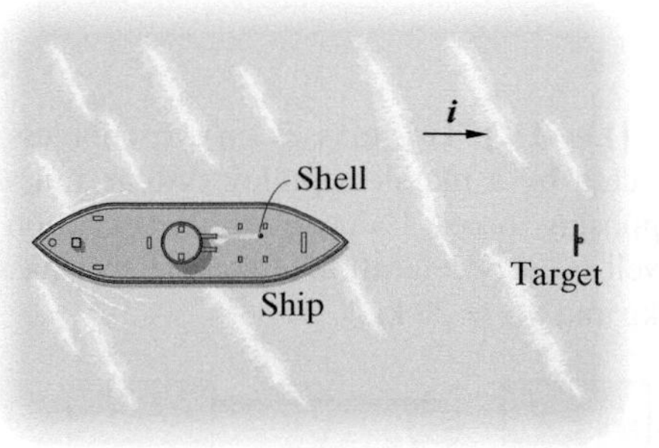

E5.1.20

5.1.21. **[Level 2]** The two masses m_1 and m_2, connected by a massless string, are moving with velocity $-10\boldsymbol{j}$ m/s on a frictionless, horizontal plane. Initially stationary, m_3 is contacted by the string, causing m_1 and m_2 to move around it as shown. When m_1 and m_2 eventually collide, what are the velocities of m_1, m_2, and m_3? $m_1 = m_2 = 2$ kg and $m_3 = 4$ kg.

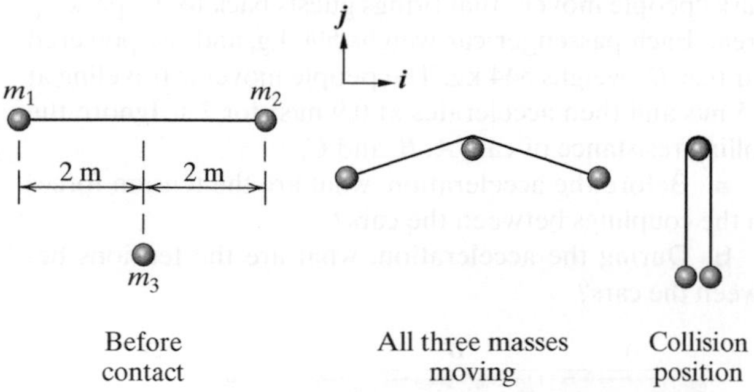

E5.1.21

5.1.22. **[Level 2]** A shuttle astronaut (modeled as a particle of mass m_2) is at the bottom of a ladder (m_1), and both are moving down ($-\boldsymbol{j}$ direction) at 1 m/s. When she reaches a position 2 m below the lower edge of the shuttle's hatch, she starts to climb the ladder. How fast (constant speed) must she climb (velocity taken relative to the ladder) so that she can reach the lower edge of the main hatch before the ladder moves completely beyond it? $m_1 = 75$ kg and $m_2 = 75$ kg.

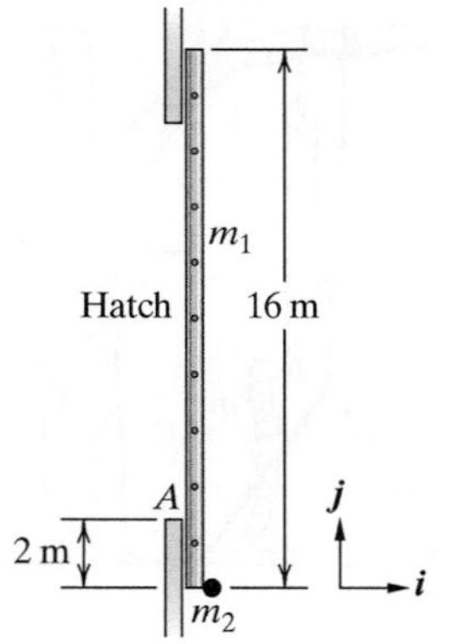

E5.1.22

5.1.23. **[Level 2]** The mass m_2 represents a child standing on a platform m_1 that itself is lying on a frozen lake. The coefficient of friction between m_1 and the ice is zero. The child uniformly accelerates toward the far end (A) (measured relative to the platform). In 2 s she reaches A. $m_1 = 40$ kg, $m_2 = 60$ kg, and $L = 6$ m.

a. What is the platform's absolute velocity when the child reaches A?

b. What was the force between the child and the platform during the motion?

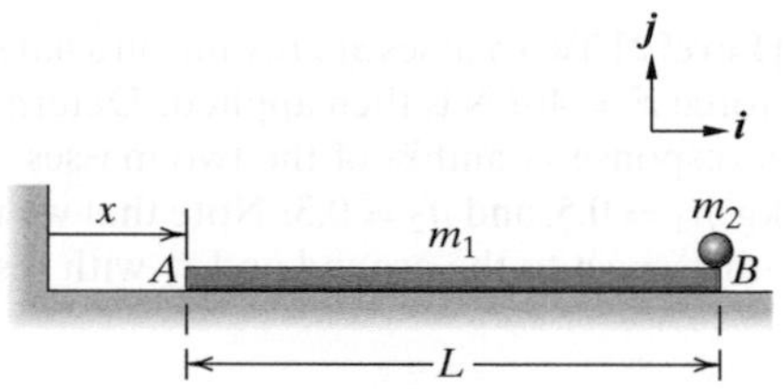

E5.1.23

5.1.24. **[Level 2]** Two particles, m_1 and m_2, slide on a frictionless, horizontal plane. The particles are connected by an ideal spring (zero mass and $F = -kx$). Explain whether the following is true or false: During a motion, the total work done by the spring (the sum of the work done by the spring on m_1 and the work done by the spring on m_2) is equal to *zero* because the spring forces are internal to the m_1, m_2 system and hence cancel in pairs.

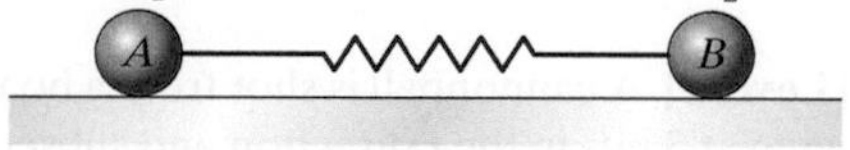

E5.1.24

5.1.25. **[Level 2]** A woman of mass m is standing on a rigid platform of mass m_P. She is pulling down on the illustrated pulley rope in an attempt to raise the platform and herself. Assume she exerts a force of S on the pulley rope. What conditions must be met for the woman to remain in contact with the platform while the platform is held above the ground in an equilibrium condition (zero velocity)?

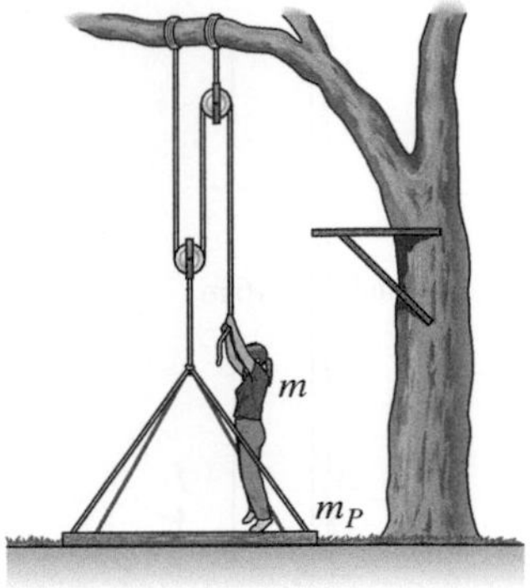

E5.1.25

5.1.26. **[Level 2]** Reconsider **Exercise 5.1.25**. If the rope is pulled in at a rate of 1.2 m/s^2, what is the maximal platform mass for which the person stays in contact with the platform? $m = 60$ kg. Note that the rate at which the rope is pulled in depends on the absolute acceleration of both the rope and the platform.

5.1.27. **[Level 2]** Two masses are resting on a flat surface. A constant force $\boldsymbol{F} = 40\boldsymbol{i}$ N is then applied. Determine the acceleration response $\ddot{x}_1$ and $\ddot{x}_2$ of the two masses. $m_1 = 4$ kg, $m_2 = 5$ kg, $\mu_1 = 0.5$, and $\mu_2 = 0.3$. Note that we're measuring x_2 with respect to the ground and x_1 with respect to mass m_2.

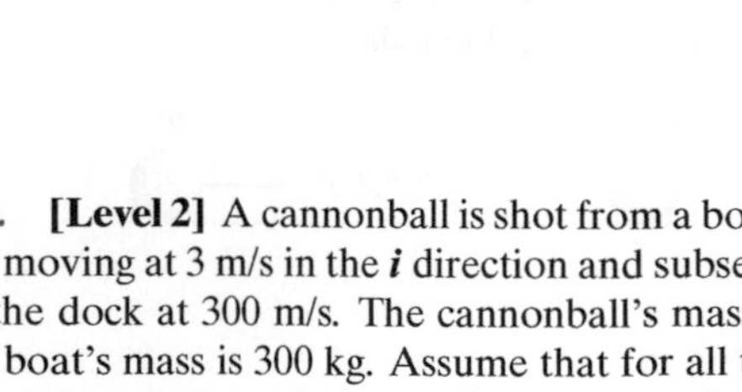

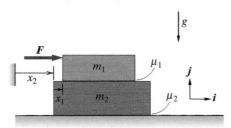

E5.1.27

5.1.28. **[Level 2]** A cannonball is shot from a boat that's initially moving at 3 m/s in the $\boldsymbol{i}$ direction and subsequently strikes the dock at 300 m/s. The cannonball's mass is 5 kg and the boat's mass is 300 kg. Assume that for all the time that the cannonball was being projected through the cannon's bore it was acted on by a constant force (due to the expanding gases) and that when it exited the cannon the force immediately went to zero. The cannon's bore is 1.5 m long. Find Δt, the time that the cannonball takes to travel through the cannon, as well as the final velocity of the boat.

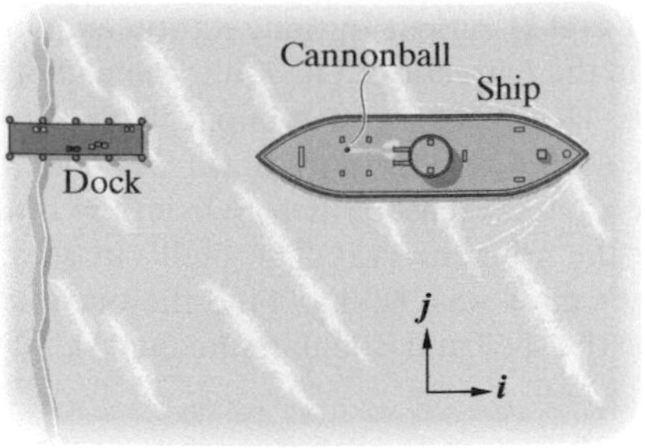

E5.1.28

5.1.29. **[Level 2]** Two masses on a frictionless surface are connected by a massless pulley system and acted on by two opposing forces. Calculate the acceleration of each mass, as well as the acceleration of the system's mass center. $m_1 = 10$ kg and $m_2 = 20$ kg.

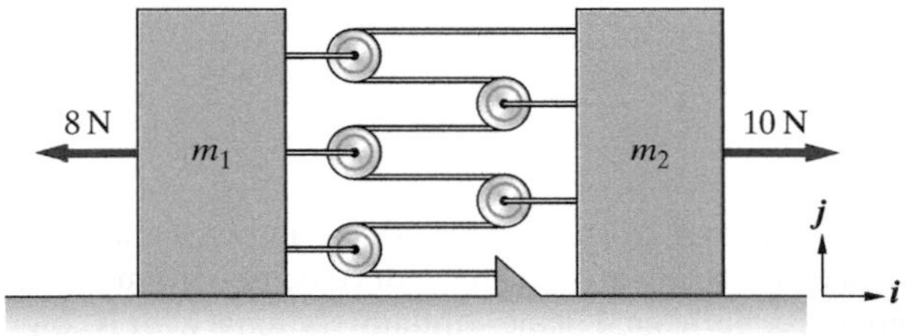

E5.1.29

5.1.30. **[Level 2]** The illustrated system is an amusement park "people mover" that brings guests back to the parking area. Each passenger car weighs 590 kg, and the powered car (car D) weighs 544 kg. The people mover is traveling at 1.5 m/s and then accelerates at 0.9 m/s^2 for 2 s. Ignore the rolling resistance of cars A, B, and C.

a. Before the acceleration, what are the tension forces in the couplings between the cars?

b. During the acceleration, what are the tensions between the cars?

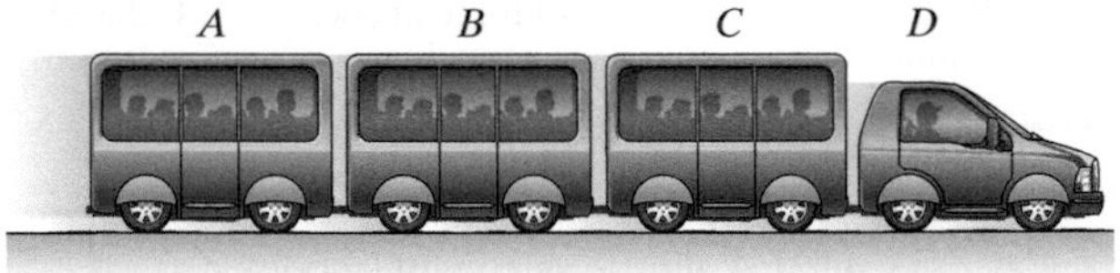

E5.1.30

5.1.31. **[Level 3]** The illustrated body m_2 is free to roll along a horizontal guide. Mass m_1 is attached to the end of a rigid, massless rod, which is itself attached to m_2 at O through a frictionless pivot. When the system is released at $t = 0$, the following conditions hold: $\dot{x}(0) = 0$, $\theta = 45°$, and $\dot{\theta} = 0$. How far will m_2 have moved when $\theta = 0°$ and how fast will it be moving? $m_1 = 2$ kg, $m_2 = 4$ kg, and $L = 3$ m. Use $v_{m_1}\boldsymbol{i} = v_{m_2}\boldsymbol{i} + L\dot{\theta}\boldsymbol{i}$ when $\theta = 0°$.

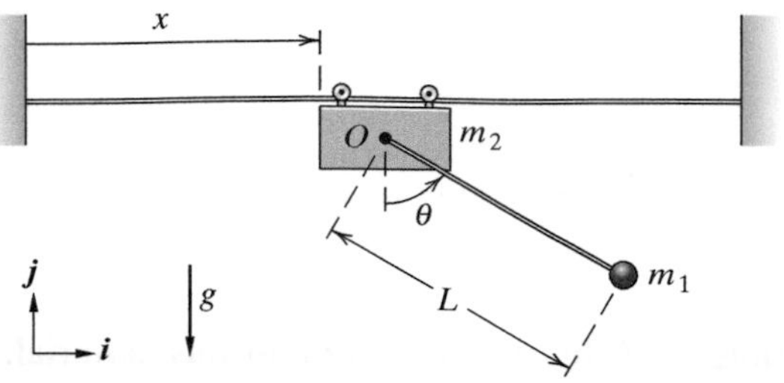

E5.1.31

5.1.32. **[Level 3]** At $t = 0$ s a satellite S is ejected at $\boldsymbol{v} = 20\boldsymbol{i}$ m/s from O, an ejection port on a spaceship. Three seconds after being ejected, it explodes into two pieces. The smaller piece (A) has a mass of 100 kg, and the larger piece (B) has a mass of 500 kg. At $t = 5$ s, piece B is located at $\boldsymbol{r}_{B/O} = (100\boldsymbol{i} + 10\boldsymbol{j})$ m. Where is piece A, and how fast is it traveling?

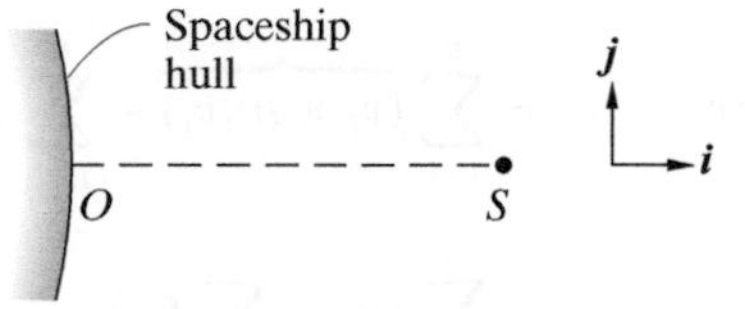

E5.1.32

5.1.33. **[Level 3]** You have been brought into a court case as an expert witness. A satellite is known to have exploded into three pieces, A, B, and C. $m_A = 1000$ kg, $m_B = 500$ kg, and $m_C = 500$ kg. Telemetry data showed that at a particular time the positions of the satellite with respect to a known reference position O were $\boldsymbol{r}_{A/O} = (400\boldsymbol{i} - 1000\boldsymbol{j})$ m, $\boldsymbol{r}_{B/O} = 1000\boldsymbol{j}$ m, and $\boldsymbol{r}_{C/O} = (800\boldsymbol{i} + 1000\boldsymbol{j})$ m. The respective velocities were $\boldsymbol{v}_A = -500\boldsymbol{j}$ m/s, $\boldsymbol{v}_B = (-400\boldsymbol{i} + 500\boldsymbol{j})$ m/s, and $\boldsymbol{v}_C = (400\boldsymbol{i} + 500\boldsymbol{j})$ m/s. The defense claims that a single explosion caused the described configuration, whereas the prosecution contends that at least two separate explosions were responsible. Who is correct?

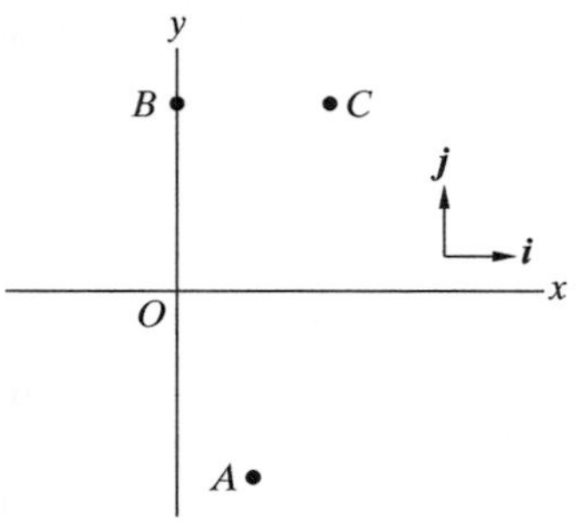

E5.1.33

5.1.34. **[Level 3]** Two blocks are arranged so that block A lies atop block B. $m_A = 20$ kg and $m_B = 100$ kg. The coefficient of friction between blocks A and B is 0.5, and the coefficient of friction between block B and the ground is 0.05.

a. What are the blocks' accelerations for $F = 40$ N?
b. What are the blocks' accelerations for $F = 80$ N?

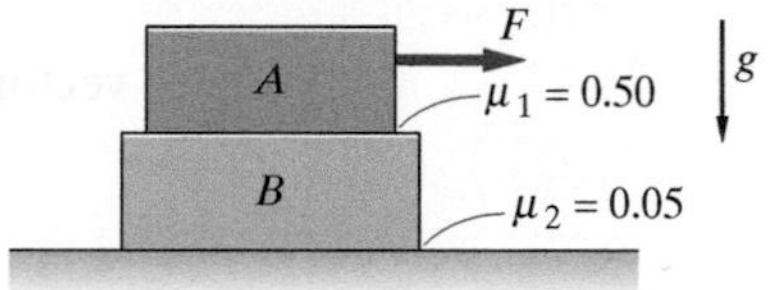

E5.1.34

5.1.35. **[Level 3]** Two blocks are shown that may or may not slide on each other, depending on friction. Assume that both move together as a unit ($a_A = a_B$) due to F, and then show from an examination of the system's FBD that this assumption is an invalid one. $F = 30$ N and $m_A = m_B = 10$ kg. (Assume $a_A \neq 0$.)

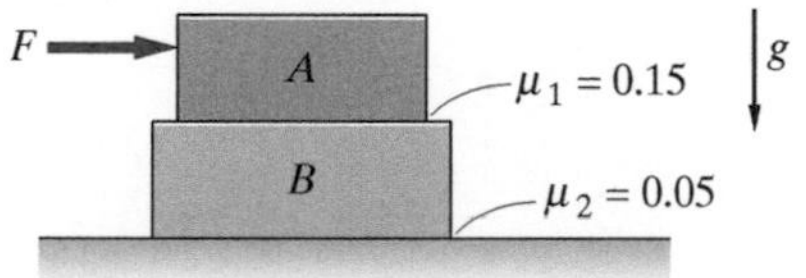

E5.1.35

5.2 ANGULAR MOMENTUM

We have now seen that both $\boldsymbol{F} = m\boldsymbol{a}$ and $\boldsymbol{L} = m\boldsymbol{v}$ for a system of particles are identical in form to the respective expressions for single particles, but can this simplicity hold with angular momentum? In this case we will have particles flying all over the place, each having a different angular momentum with respect to some fixed point. And what if our reference point isn't fixed? Can we still come up with useful angular momentum relationships? Well, read on and see.

Let's start by determining, for each particle in **Figure 5.1**, the angular momentum about O: $\boldsymbol{H}_{i/O} = \boldsymbol{r}_i \times m_i\boldsymbol{v}_i$ (3.45). We will then sum the individual contributions to get $\boldsymbol{H}_O$, the angular momentum about O for the system:

$$\boldsymbol{H}_O = \sum_{i=1}^{n} (\boldsymbol{r}_i \times m_i \boldsymbol{v}_i) \tag{5.11}$$

⋙ **Check out Example 5.6 (page 277) for an application of this material.**

Differentiating with respect to time gives us

$$\dot{\boldsymbol{H}}_O = \frac{d}{dt} \sum_{i=1}^{n} (\boldsymbol{r}_i \times m_i \boldsymbol{v}_i)$$

product rule:

$$= \sum_{i=1}^{n} (\dot{\boldsymbol{r}}_i \times m_i \boldsymbol{v}_i) + \sum_{i=1}^{n} (\boldsymbol{r}_i \times m_i \dot{\boldsymbol{v}}_i)$$

vector crossed with itself is zero:

$$= \sum_{i=1}^{n} \overbrace{(\boldsymbol{v}_i \times m_i \boldsymbol{v}_i)}^{0} + \sum_{i=1}^{n} (\boldsymbol{r}_i \times m_i \boldsymbol{a}_i)$$

$$= \sum_{i=1}^{n} \left(\boldsymbol{r}_i \times \sum \boldsymbol{F}_i\right) \tag{5.12}$$

Note that there's no problem in using the product rule with vector cross products; it works in just the same way as when we are dealing with the products of scalar functions. The term in parentheses in (5.12) is simply the total applied moment about O. Thus we've arrived at our moment balance:

$$\dot{\boldsymbol{H}}_O = \sum_{i=1}^{n} \boldsymbol{M}_{O_i} \tag{5.13}$$

Looks a lot like (3.49), doesn't it? So, once again, we're getting relatively simple relationships, even though we're dealing with a lot of particles. In this case, however, the simplicity is a bit of a deception. To evaluate (5.13) we would need to look at each particle individually, figure out its velocity to get its angular momentum contribution, and then differentiate. So it's not as nice as it looks. Luckily, once we hit rigid bodies, almost all the complexity will evaporate because the particles making up the rigid body can't move anywhere they please. They're constrained by the fact that they're all part of the same rigid body, unlike the more independent-minded particles in our system of particles.

The next way we're going to look at our system of particles wouldn't make any sense if we had only a single particle because we're going to use G as our reference point. If we had only one particle, then this discussion would be a major waste of time because the particle has zero angular momentum about itself ($\boldsymbol{r} \times m\boldsymbol{v}$ is zero because $\boldsymbol{r}$ is zero).

When we have a large number of particles, however, it works just fine to use $\boldsymbol{r}_{i_{/G}}$ crossed with individual particle linear momenta. So let's do just that:

$$\boldsymbol{H}_G = \sum_{i=1}^{n}\left(\boldsymbol{r}_{i_{/G}} \times m_i \boldsymbol{v}_i\right)$$

re-expressing $\boldsymbol{v}_i$:

$$= \sum_{i=1}^{n}\left[\boldsymbol{r}_{i_{/G}} \times m_i\left(\overline{\boldsymbol{v}} + \boldsymbol{v}_{i_{/G}}\right)\right]$$

$$= \sum_{i=1}^{n}\left(\boldsymbol{r}_{i_{/G}} \times m_i \overline{\boldsymbol{v}}\right) + \sum_{i=1}^{n}\left(\boldsymbol{r}_{i_{/G}} \times m_i \boldsymbol{v}_{i_{/G}}\right)$$

velocity pulled out of summation:

$$= -\overline{\boldsymbol{v}} \times \overbrace{\sum_{i=1}^{n} m_i \boldsymbol{r}_{i_{/G}}}^{0 \text{ from } (5.3)} + \sum_{i=1}^{n}\left(\boldsymbol{r}_{i_{/G}} \times m_i \boldsymbol{v}_{i_{/G}}\right)$$

$$\boldsymbol{H}_G = \sum_{i=1}^{n}\left(\boldsymbol{r}_{i_{/G}} \times m_i \boldsymbol{v}_{i_{/G}}\right) \tag{5.14}$$

Now take a second to look at the term on the right in this cross product. It looks a lot like what we started the derivation with, except when we began we were dealing with the actual linear momentum of each particle ($m_i\boldsymbol{v}_i$), whereas what we ended up with was the *relative* momentum of each particle. The momentum term $m_i\boldsymbol{v}_{i_{/G}}$ isn't the "real" momentum because it involves the linear velocity *relative to G*, and not the actual linear velocity that's defined with respect to a fixed Newtonian reference frame.

This is amazing! No, I'm serious, it really is. We started with the angular momentum about G (the "real" angular momentum) and ended with the relative momentum. Thus it would appear that, as far as the angular momentum is concerned, it doesn't matter whether we use the actual velocities of the particles or just the relative velocities. As long as we're referencing to the center of mass, we will get the same answer. Thus we can write

$$\boldsymbol{H}_G = \boldsymbol{H}_G\big|_{rel} \tag{5.15}$$

where *rel* indicates that the linear velocity in the momentum term is a relative velocity, not an absolute one. This will be of great help in the future because it is easier to obtain the relative velocity of a rigid body than the absolute velocity. Thus (5.15) tells us we can take the easy route and end up with the same final result.

⋙ **Check out Example 5.7 (page 278) for an application of this material.**

We saw with (5.13) what happens when we differentiate $\boldsymbol{H}_O$, and so it's time to see what we get when we do the same with $\boldsymbol{H}_G$:

$$\dot{\boldsymbol{H}}_G = \frac{d}{dt}\sum_{i=1}^{n}\left(\boldsymbol{r}_{i_{/G}} \times m_i \boldsymbol{v}_i\right)$$

product rule:

$$= \sum_{i=1}^{n}\left(\boldsymbol{v}_{i_{/G}} \times m_i \boldsymbol{v}_i\right) + \sum_{i=1}^{n}\left(\boldsymbol{r}_{i_{/G}} \times m_i \boldsymbol{a}_i\right)$$

re-expressing $\boldsymbol{v}_i$ and $m_i\boldsymbol{a}_i$:

$$= \sum_{i=1}^{n}\left[\boldsymbol{v}_{i_{/G}} \times m_i\left(\overline{\boldsymbol{v}} + \boldsymbol{v}_{i_{/G}}\right)\right] + \sum_{i=1}^{n}\left(\boldsymbol{r}_{i_{/G}} \times \sum \boldsymbol{F}_i\right)$$

vector crossed with itself is zero:

$$= -\overline{\boldsymbol{v}} \times \sum_{i=1}^{n} m_i \boldsymbol{v}_{i_{/G}} + \overbrace{\sum_{i=1}^{n}\left(\boldsymbol{v}_{i_{/G}} \times m_i \boldsymbol{v}_{i_{/G}}\right)}^{0}$$

definition of mass center:

$$= -\overline{\boldsymbol{v}} \times \overbrace{\sum_{i=1}^{n} m_i \boldsymbol{v}_{i_{/G}}}^{0 \text{ from } (5.3)} + \sum_{i=1}^{n}\left(\boldsymbol{r}_{i_{/G}} \times \sum \boldsymbol{F}_i\right)$$

$$= \sum_{i=1}^{n}\left(\boldsymbol{r}_{i_{/G}} \times \sum \boldsymbol{F}_i\right)$$

$$\dot{\boldsymbol{H}}_G = \sum_{i=1}^{n} \boldsymbol{M}_{G_i} \tag{5.16}$$

That took a while but look what it got us. The time rate of change of the angular momentum about G is equal to the applied moments about G. You have no idea how useful this relationship is going to be when we get to rigid bodies. It'll be one of your most frequently used equations. I guarantee it.

As in the single-particle case, we can integrate our moment balance about O (5.13) to get

$$\int_{t_1}^{t_2} \dot{\boldsymbol{H}}_O\, dt = \int_{t_1}^{t_2} \sum_{i=1}^{n} \boldsymbol{M}_{O_i}\, dt \tag{5.17}$$

or, defining the angular impulse $\mathcal{AI}_{1-2}$ (Section 3.5) as

$$\mathcal{AI}_{O_{1-2}} \equiv \int_{t_1}^{t_2} \sum_{i=1}^{n} \boldsymbol{M}_{O_i}\, dt \tag{5.18}$$

we have

$$\boldsymbol{H}_{O_2} = \boldsymbol{H}_{O_1} + \mathcal{AI}_{O_{1-2}} \tag{5.19}$$

In the same way, we can find our momentum relationships with respect to G from (5.18):

$$\int_{t_1}^{t_2} \dot{\boldsymbol{H}}_G \, dt = \int_{t_1}^{t_2} \sum_{i=1}^{n} \boldsymbol{M}_{G_i} \, dt \tag{5.20}$$

to be

$$\boldsymbol{H}_{G_2} = \boldsymbol{H}_{G_1} + \mathcal{AI}_{G_{1-2}} \tag{5.21}$$

where

$$\mathcal{AI}_{G_{1-2}} \equiv \int_{t_1}^{t_2} \sum_{i=1}^{n} \boldsymbol{M}_{G_i} \, dt \tag{5.22}$$

EXAMPLE 5.6 ANGULAR MOMENTUM OF THREE PARTICLES (Theory on page 274)

In **Figure 5.11**, three particles, A, B, and C, located at $\boldsymbol{r}_{A/O} = 0.6\boldsymbol{i}$ m, $\boldsymbol{r}_{B/O} = 0.9\boldsymbol{j}$ m, and $\boldsymbol{r}_{C/O} = -0.6\boldsymbol{j}$ m, have masses $m_A = 30$ kg, $m_B = 45$ kg, and $m_C = 30$ kg, and velocities $\boldsymbol{v}_A = 0.6\boldsymbol{j}$ m/s, $\boldsymbol{v}_B = (0.6\boldsymbol{i} + 0.42\boldsymbol{j})$ m/s, and $\boldsymbol{v}_C = (-0.06\boldsymbol{i} + 0.9\boldsymbol{j})$ m/s. Find the system's total angular momentum about O.

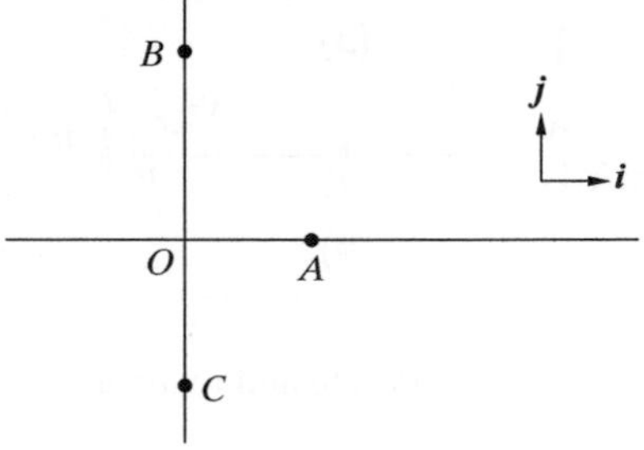

Figure 5.11 Three particles

Goal Find the angular momentum about O of three particles.

Assume No additional assumptions are needed.

Draw **Figure 5.12** shows the system's position and velocity vectors.

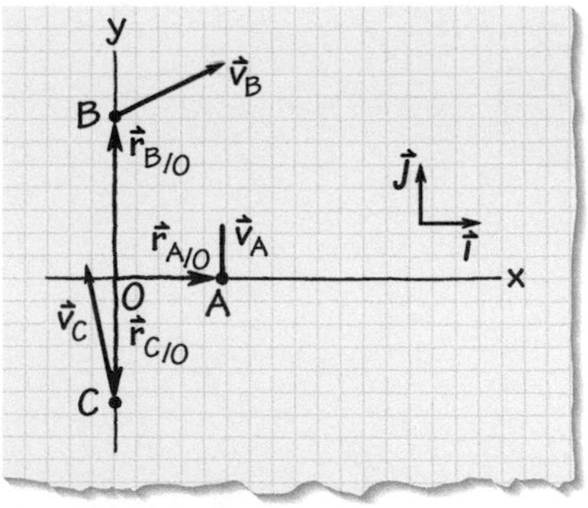

Figure 5.12 Three particles showing position and velocity vectors

Formulate Equations All we need to do is apply (5.11) with the given parameter values.

Solve

$$\begin{aligned}\boldsymbol{H}_O = {} & (0.6\boldsymbol{i}\text{ m}) \times (30\text{ kg})(0.6\boldsymbol{j}\text{ m/s}) \\ & + (0.9\boldsymbol{j}\text{ m}) \times (45\text{ kg})[(0.6\boldsymbol{i} + 0.42\boldsymbol{j})\text{ m/s}] \\ & + (-0.6\boldsymbol{j}\text{ m}) \times (30\text{ kg})[(-0.06\boldsymbol{i} + 0.9\boldsymbol{j})\text{ m/s}]\end{aligned}$$

$$\boldsymbol{H}_O = -14\text{ kg}\cdot\text{m}^2\text{/s}$$

Check There's not much to check here beyond the correctness of the calculations.

EXAMPLE 5.7 ANGULAR MOMENTUM ABOUT A SYSTEM'S MASS CENTER (Theory on page 275)

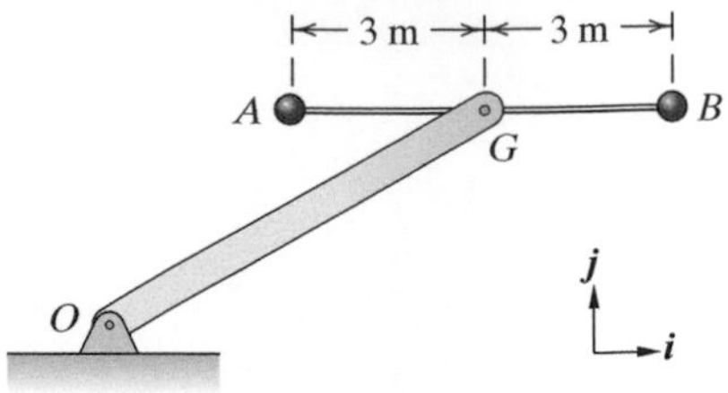

Figure 5.13 Rotating masses

Let's verify (5.15) by working out a physical example and seeing how the expression for angular momentum of a system of particles about its mass center can use either absolute or relative velocities. **Figure 5.13** shows the system we will be considering—two particles of mass m located at A and B. The rigid arm connecting the two is rotating about G, and the larger arm $\overline{OG}$ is rotating about O. $\boldsymbol{v}_A = (-10\boldsymbol{i} + 8\boldsymbol{j})$ m/s, $\boldsymbol{v}_B = (-10\boldsymbol{i} + 12\boldsymbol{j})$ m/s, $\boldsymbol{v}_G = (-10\boldsymbol{i} + 10\boldsymbol{j})$ m/s, $\boldsymbol{r}_{A/G} = -3\boldsymbol{i}$ m, $\boldsymbol{r}_{B/G} = 3\boldsymbol{i}$ m, and $|r_{G/O}| = 10$ m.

Goal Find the angular momentum of the two masses about their mass center using absolute and relative velocities.

Assume No additional assumptions are needed.

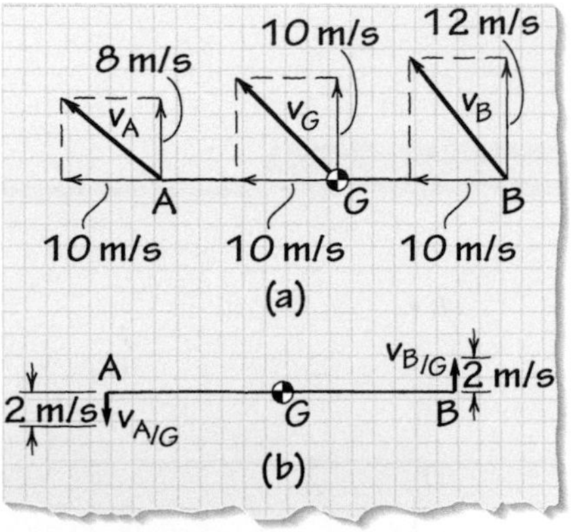

Figure 5.14 Absolute and relative velocities

Draw **Figure 5.14*a*** shows the bar $\overline{AB}$ and the absolute velocities of points A, G, and B, while **Figure 5.14*b*** illustrates the velocities *relative* to G. The magnitudes of the velocity components along the $\boldsymbol{i}, \boldsymbol{j}$ directions are shown as well.

Formulate Equations A and B are equal in mass; thus the mass center is located at G, halfway between them. Using absolute velocities, we have

$$\boldsymbol{H}_G = \boldsymbol{r}_{A/G} \times m\boldsymbol{v}_A + \boldsymbol{r}_{B/G} \times m\boldsymbol{v}_B \tag{5.23}$$

and for relative velocities we have

$$\boldsymbol{H}_G\big|_{rel} = \boldsymbol{r}_{A/G} \times m\boldsymbol{v}_{A/G} + \boldsymbol{r}_{B/G} \times m\boldsymbol{v}_{B/G} \tag{5.24}$$

Solve Substituting the given values into (5.23) gives

$$\begin{aligned}\boldsymbol{H}_G &= (-3\boldsymbol{i}\text{ m}) \times m[(-10\boldsymbol{i} + 8\boldsymbol{j})\text{ m/s}] + (3\boldsymbol{i}\text{ m}) \times m(-10\boldsymbol{i} + 12\boldsymbol{j})\text{ m/s}\\ &= (-24\text{m}\boldsymbol{k} + 36m\boldsymbol{k})\text{ m/s} = \boxed{12m\boldsymbol{k}\text{ m/s}}\end{aligned}$$

To find the relative angular momentum, we need the relative velocities.

$$\begin{aligned}\boldsymbol{v}_A = \boldsymbol{v}_G + \boldsymbol{v}_{A/G} \quad &\Rightarrow \quad \boldsymbol{v}_{A/G} = \boldsymbol{v}_A - \boldsymbol{v}_G\\ \boldsymbol{v}_B = \boldsymbol{v}_G + \boldsymbol{v}_{B/G} \quad &\Rightarrow \quad \boldsymbol{v}_{B/G} = \boldsymbol{v}_B - \boldsymbol{v}_G\end{aligned}$$

Using the expressions for relative velocity and the given data in (5.24) gives

$$\begin{aligned}\boldsymbol{H}_G\big|_{rel} &= \boldsymbol{r}_{A/G} \times m\boldsymbol{v}_{A/G} + \boldsymbol{r}_{B/G} \times m\boldsymbol{v}_{B/G}\\ &= \boldsymbol{r}_{A/G} \times m(\boldsymbol{v}_A - \boldsymbol{v}_G) + \boldsymbol{r}_{B/G} \times m(\boldsymbol{v}_B - \boldsymbol{v}_G)\\ &= (-3\boldsymbol{i}\text{ m}) \times m(-2\boldsymbol{j}\text{ m/s}) + (3\boldsymbol{i}\text{ m}) \times m(-2\boldsymbol{j}\text{ m/s})\\ &= (6m\boldsymbol{k} + 6m\boldsymbol{k})\text{ m/s} = \boxed{12m\boldsymbol{k}\text{ m/s}}\end{aligned}$$

Check The point of this example was to check (5.15), which we've done.

EXERCISES 5.2

5.2.1. **[Level 1]** Calculate the angular momentum of the system of particles about O. $m_A = 2$ kg, $m_B = 4$ kg, $m_C = 10$ kg, $\boldsymbol{r}_{A/O} = (2\boldsymbol{j} + 2\boldsymbol{k})$ m, $\boldsymbol{r}_{B/O} = (3\boldsymbol{i} + 2\boldsymbol{j})$ m, $\boldsymbol{r}_{C/O} = (2\boldsymbol{i} + 2\boldsymbol{k})$ m, $\boldsymbol{v}_A = -10\boldsymbol{k}$ m/s, $\boldsymbol{v}_B = (3\boldsymbol{i} + 3\boldsymbol{j})$ m/s, and $\boldsymbol{v}_C = (-4\boldsymbol{i} - 4\boldsymbol{k})$ m/s.

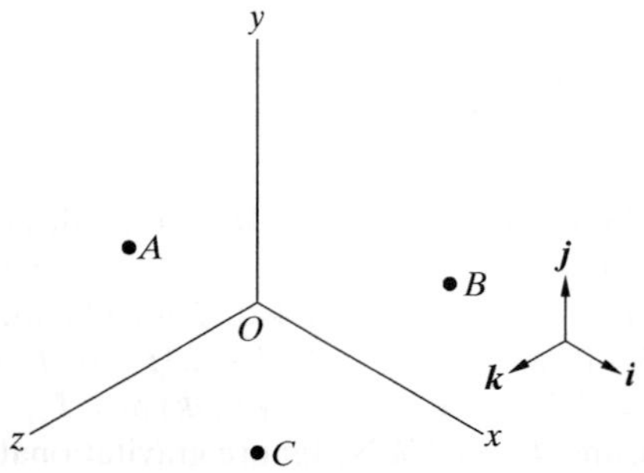

E5.2.1

5.2.2. **[Level 1]** Calculate the angular momentum of the system of particles about point P. $\boldsymbol{r}_{P/O} = (-10\boldsymbol{i} + 20\boldsymbol{j} + 30\boldsymbol{k})$ cm. $m_A = 0.2$ kg, $m_B = 0.4$ kg, $m_C = 0.1$ kg, $\boldsymbol{r}_{A/O} = \boldsymbol{r}_{B/O} = \boldsymbol{r}_{C/O} = (50\boldsymbol{i} + 40\boldsymbol{k})$ cm, $\boldsymbol{v}_A = (10\boldsymbol{i} + 5\boldsymbol{j})$ cm/s, $\boldsymbol{v}_B = (2\boldsymbol{i} + 2\boldsymbol{k})$ cm/s, and $\boldsymbol{v}_C = (-5\boldsymbol{j} - 5\boldsymbol{k})$ cm/s.

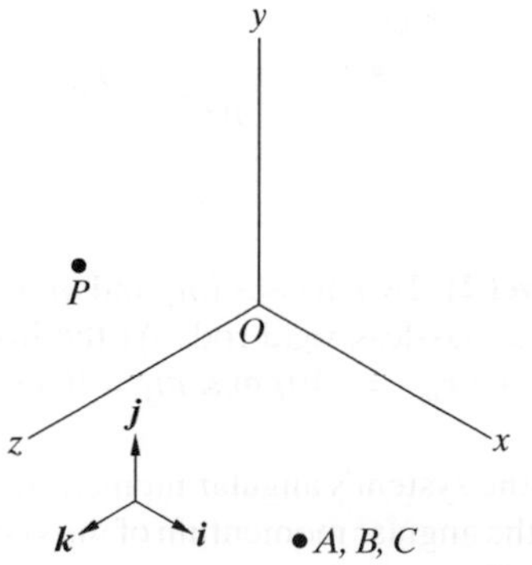

E5.2.2

5.2.3. **[Level 1]** What is $\boldsymbol{H}_O$ for the three masses A, B, and C? $m_A = 0.2$ kg, $m_B = 0.4$ kg, $m_C = 0.3$ kg, $\boldsymbol{r}_{A/O} = 5\boldsymbol{i}$ cm, $\boldsymbol{r}_{B/O} = 10\boldsymbol{i}$ cm, $\boldsymbol{r}_{C/O} = 15\boldsymbol{i}$ cm, $\boldsymbol{v}_A = -3\boldsymbol{j}$ cm/s, $\boldsymbol{v}_B = 3\boldsymbol{k}$ cm/s, and $\boldsymbol{v}_C = -3\boldsymbol{j}$ cm/s.

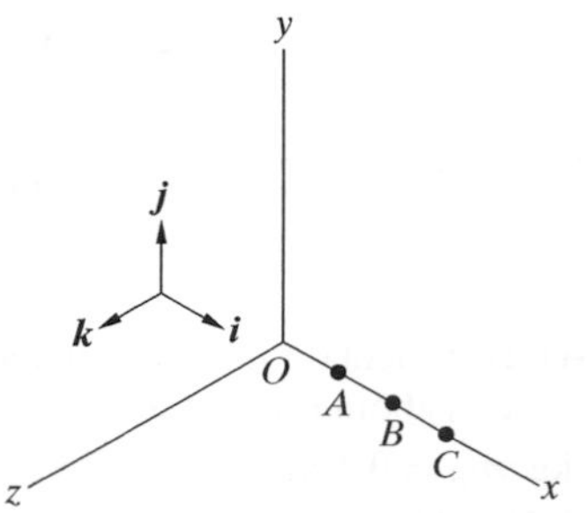

E5.2.3

5.2.4. **[Level 1]** What is $\boldsymbol{H}_P$ for the three masses A, B, and C? $m_A = m_B = m_C = 0.2$ kg, and all three masses are clustered at the origin O of the x, y, z axes. $\boldsymbol{r}_{P/O} = 10\boldsymbol{k}$ cm, $\boldsymbol{v}_A = 5\boldsymbol{i}$ cm/s, $\boldsymbol{v}_B = 5\boldsymbol{j}$ cm/s, and $\boldsymbol{v}_C = 5\boldsymbol{k}$ cm/s.

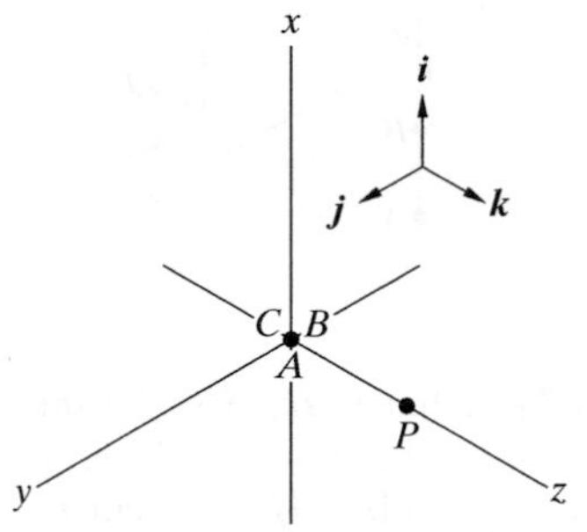

E5.2.4

5.2.5. **[Level 1]** What is $\boldsymbol{H}_O$ for the two masses A and B? m_A and m_B have a mass of 0.5 kg, $\boldsymbol{r}_{A/O} = 6\boldsymbol{j}$ cm, $\boldsymbol{r}_{B/O} = 6\boldsymbol{k}$ cm, $\boldsymbol{v}_A = 5\boldsymbol{j}$ cm/s, and $\boldsymbol{v}_B = 5\boldsymbol{j}$ cm/s.

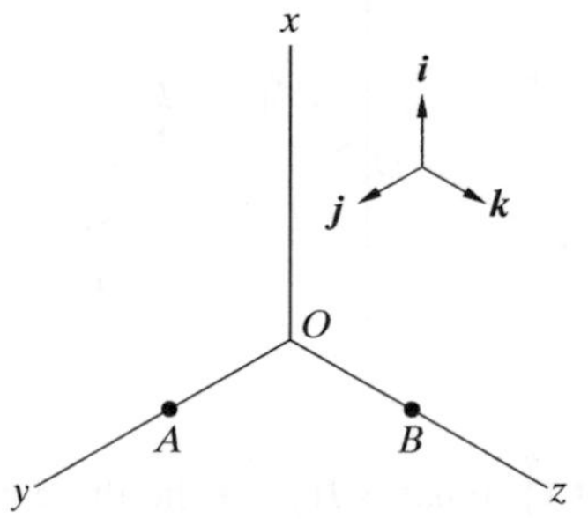

E5.2.5

5.2.6. **[Level 1]** What is $\boldsymbol{H}_O$ for the four masses A, B, C, and D? $m_A = m_B = 0.8$ kg and $m_C = m_D = 1.6$ kg. $\boldsymbol{r}_{A/O} = -6\boldsymbol{i}$ cm, $\boldsymbol{r}_{B/O} = -4\boldsymbol{i}$ cm, $\boldsymbol{r}_{C/O} = -4\boldsymbol{j}$ cm, $\boldsymbol{r}_{D/O} = 4\boldsymbol{i}$ cm, $\boldsymbol{v}_A = 5\boldsymbol{i}$ cm/s, $\boldsymbol{v}_B = 15\boldsymbol{i}$ cm/s, $\boldsymbol{v}_C = 25\boldsymbol{j}$ cm/s, and $\boldsymbol{v}_D = 5\boldsymbol{i}$ cm/s.

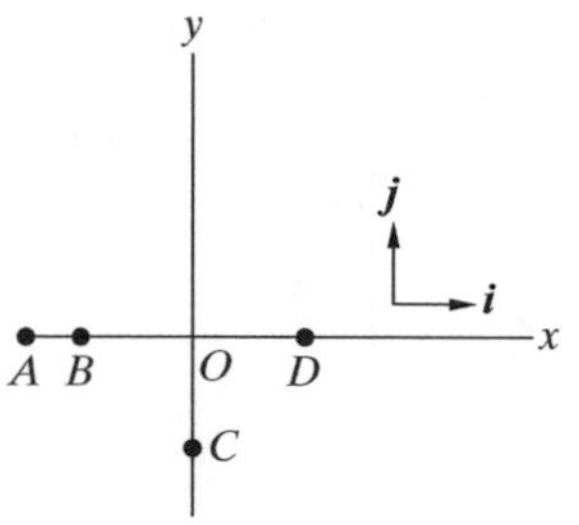

E5.2.6

5.2.7. **[Level 1]** What is $\boldsymbol{H}_P$ for the two masses A and B? $m_A = 0.2$ kg and $m_B = 0.3$ kg. $\boldsymbol{r}_{A/O} = -8\boldsymbol{k}$ cm, $\boldsymbol{r}_{B/O} = -4\boldsymbol{k}$ cm, $\boldsymbol{v}_A = (4\boldsymbol{i} + 6\boldsymbol{j} + 4\boldsymbol{k})$ cm/s, $\boldsymbol{v}_B = (5\boldsymbol{i} + 5\boldsymbol{j} + 5\boldsymbol{k})$ cm/s, and $\boldsymbol{r}_{P/O} = (4\boldsymbol{i} + 4\boldsymbol{j} + 4\boldsymbol{k})$ cm.

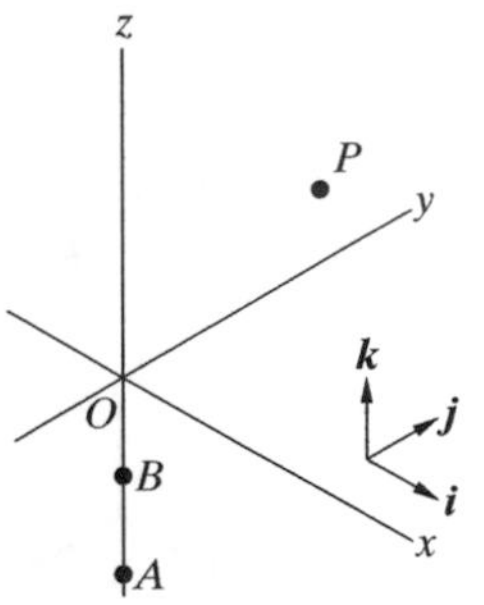

E5.2.7

5.2.8. **[Level 1]** What is $\boldsymbol{H}_O$ for the three masses A, B, and C? $m_A = 0.05$ kg, $m_B = 0.04$ kg, and $m_C = 0.01$ kg. $\boldsymbol{r}_{A/O} = 4\boldsymbol{i}$ cm, $\boldsymbol{r}_{B/O} = (4\boldsymbol{i} + 4\boldsymbol{j} + 6\boldsymbol{k})$ cm, $\boldsymbol{r}_{C/O} = (-1.5\boldsymbol{i} - 3.2\boldsymbol{j} - 2\boldsymbol{k})$ cm, $\boldsymbol{v}_A = 5\boldsymbol{k}$ cm/s, $\boldsymbol{v}_B = (9\boldsymbol{i} + 4\boldsymbol{j} + 3\boldsymbol{k})$ cm/s, and $\boldsymbol{v}_C = -6\boldsymbol{k}$ cm/s.

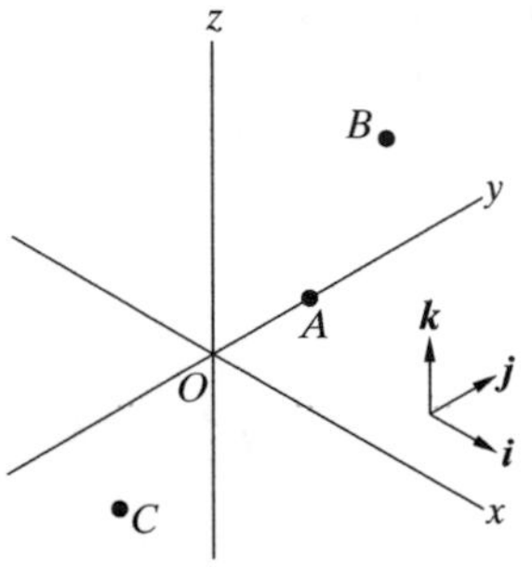

E5.2.8

5.2.9. **[Level 1]** What is $\boldsymbol{H}_P$ for the three masses A, B, and C? $m_A = 0.2$ kg, $m_B = 0.3$ kg, and $m_C = 0.5$ kg. $\boldsymbol{r}_{A/O} = (-3\boldsymbol{i} + 2\boldsymbol{j})$ cm, $\boldsymbol{r}_{B/O} = (3\boldsymbol{i} + 4\boldsymbol{j})$ cm, $\boldsymbol{r}_{C/O} = (2\boldsymbol{i} - 3\boldsymbol{j})$ cm, $\boldsymbol{v}_A = (2\boldsymbol{i} + 2\boldsymbol{j})$ cm/s, $\boldsymbol{v}_B = (5\boldsymbol{i} + 5\boldsymbol{j})$ cm/s, $\boldsymbol{v}_C = 5\boldsymbol{j}$ cm/s, and $\boldsymbol{r}_{P/O} = (-4\boldsymbol{i} - 4\boldsymbol{j})$ cm.

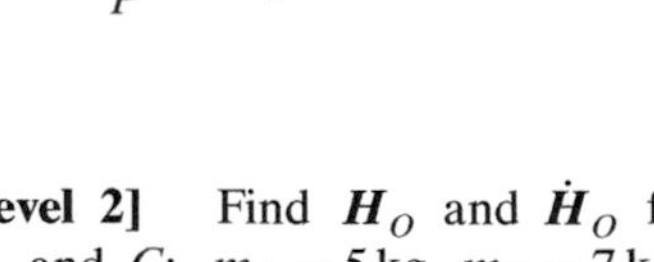

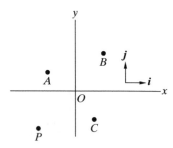

E5.2.9

5.2.10. **[Level 2]** Find $\boldsymbol{H}_O$ and $\dot{\boldsymbol{H}}_O$ for the three masses A, B, and C: $m_A = 5$ kg, $m_B = 7$ kg, $m_C = 2$ kg, $\boldsymbol{r}_{A/O} = (-2\boldsymbol{i} + 3\boldsymbol{j} + \boldsymbol{k})$ m, $\boldsymbol{r}_{B/O} = (3\boldsymbol{i} + 7\boldsymbol{j})$ m, $\boldsymbol{r}_{C/O} = (2\boldsymbol{i} + \boldsymbol{j} - 4\boldsymbol{k})$ m, $\boldsymbol{v}_A = 10\boldsymbol{k}$ m/s, $\boldsymbol{v}_B = (-3\boldsymbol{i} - 5\boldsymbol{j})$ m/s, $\boldsymbol{v}_C = 4\boldsymbol{k}$ m/s, $\boldsymbol{F}_A = 8\boldsymbol{k}$ N, $\boldsymbol{F}_B = -9\boldsymbol{i}$ N, and $\boldsymbol{F}_C = 7\boldsymbol{k}$ N. Ignore gravitational effects.

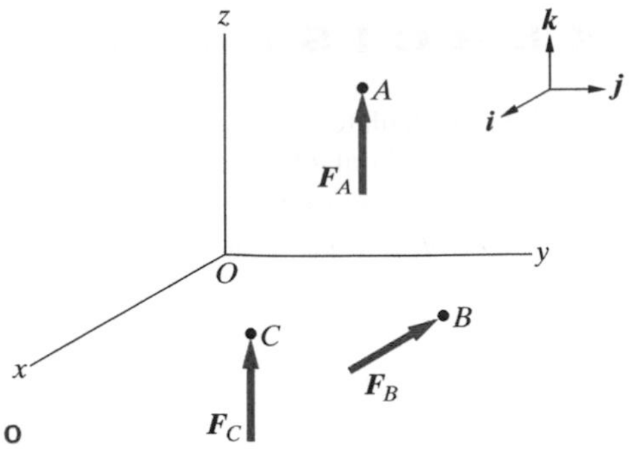

E5.2.10

5.2.11. **[Level 2]** Find $\boldsymbol{H}_P$ and $\dot{\boldsymbol{H}}_P$ for the three masses A, B, and C: $m_A = 30$ kg, $m_B = 75$ kg, $m_C = 45$ kg, $\boldsymbol{r}_{P/O} = (10.6\boldsymbol{i} + 0.9\boldsymbol{j})$ m, $\boldsymbol{r}_{A/O} = (-0.9\boldsymbol{i} + 0.3\boldsymbol{j} + 1.8\boldsymbol{k})$ m, $\boldsymbol{r}_{B/O} = (1.2\boldsymbol{i} + 2.1\boldsymbol{j} - 0.3\boldsymbol{k})$ m, $\boldsymbol{r}_{C/O} = (0.6\boldsymbol{i} - 0.3\boldsymbol{j} + 0.9\boldsymbol{k})$ m, $\boldsymbol{v}_A = 2.1\boldsymbol{j}$ m/s, $\boldsymbol{v}_B = 2.4\boldsymbol{i}$ m/s, $\boldsymbol{v}_C = (1.5\boldsymbol{i} + 0.9\boldsymbol{k})$ m/s, $\boldsymbol{F}_A = 44\boldsymbol{j}$ N, $\boldsymbol{F}_B = 67\boldsymbol{i}$ N, and $\boldsymbol{F}_C = 27\boldsymbol{k}$ N. Ignore gravitational effects.

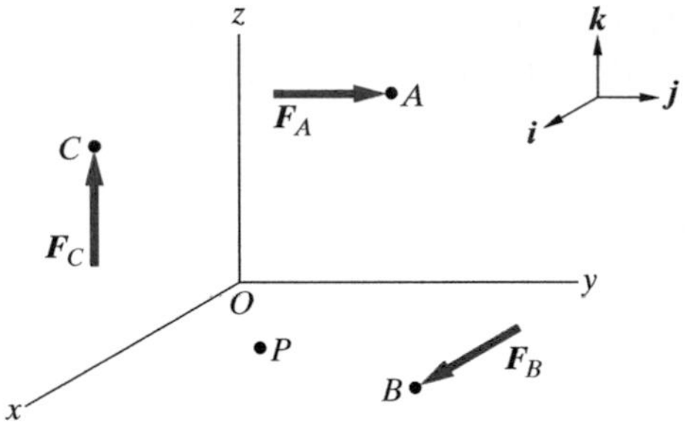

E5.2.11

5.2.12. **[Level 2]** Two masses (m_1 and m_2) are attached to the ends of a massless, rigid rod. At the instant shown, $\boldsymbol{v}_{m_1} = -5\boldsymbol{j}$ m/s and $\boldsymbol{v}_{m_2} = -10\boldsymbol{j}$ m/s. $m_1 = 10$ kg, $m_2 = 40$ kg, and $L = 0.5$ m.

a. What is the system's angular momentum about O?

b. What is the angular momentum of the system's center of mass about O?

c. Why do the results of **a** and **b** differ?

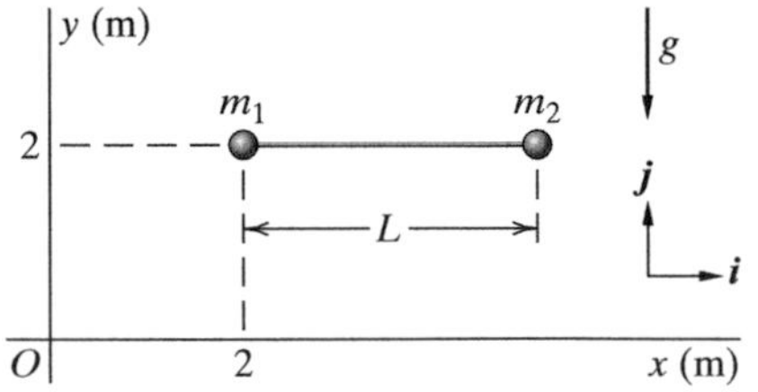

E5.2.12

5.2.13. **[Level 2]** Calculate the angular momentum of the system of particles about point P. $\boldsymbol{r}_{P/O} = (2\boldsymbol{i} + 2\boldsymbol{j} + 2\boldsymbol{k})$ cm. $m_A = 0.5$ kg, $m_B = 0.4$ kg, $m_C = 1.0$ kg, $\boldsymbol{r}_{A/O} = (3\boldsymbol{i} + 3\boldsymbol{j})$ cm, $\boldsymbol{r}_{B/O} = (1.5\boldsymbol{i} + 3\boldsymbol{j})$ cm, $\boldsymbol{r}_{C/O} = 3\boldsymbol{i}$ cm, $\boldsymbol{v}_A = -10\boldsymbol{k}$ cm/s, $\boldsymbol{v}_B = 3\boldsymbol{k}$ cm/s, and $\boldsymbol{v}_C = (-2\boldsymbol{i} - 2\boldsymbol{k})$ cm/s.

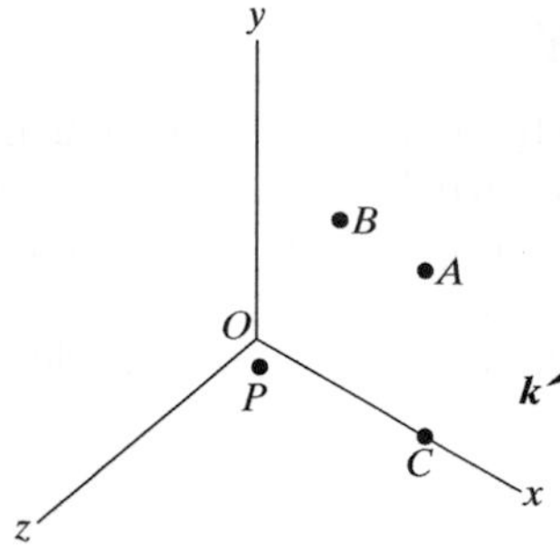

E5.2.13

5.2.14. **[Level 2]** The illustrated system is free to pivot about the fixed point O in the horizontal plane, where $m = 1\,\text{kg}$, $M = 3\,\text{kg}$, and $r = 0.5\,\text{m}$. Suppose the system has a counterclockwise angular speed $\omega = 2\,\text{rad/s}$ when a constant torque $T = 20\,\text{N·m}$ is applied at O in the indicated direction. If the surface over which the system rotates has a coefficient of friction of $\mu = 0.3$, how fast is the system spinning about O after $\Delta t = 4\,\text{s}$?

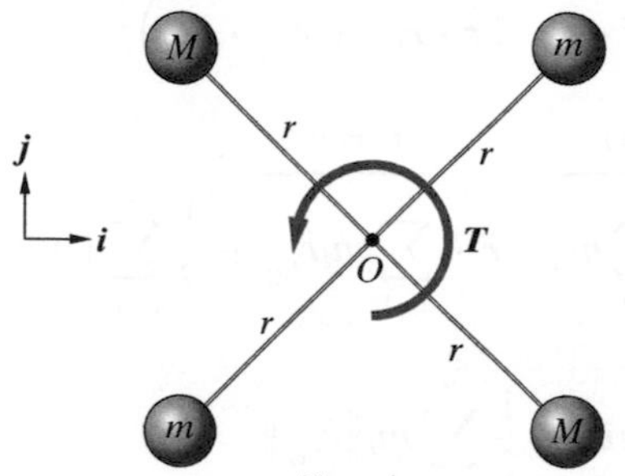

E5.2.14 Top view

5.2.15. **[Level 2]** In the depths of gravity-free space, an asteroid of mass m_0 is observed at $t = 0$ to pass through the coordinates $(0, 0, 0)$ with velocity $10\boldsymbol{j}$ m/s. Later, an explosive charge is detonated within the asteroid, breaking it into three distinct pieces—A, B, C—with masses $m_A = 0.5m_0$, $m_B = 0.3m_0$, and $m_C = 0.2m_0$. At $t = t_1$, $\boldsymbol{v}_A = (-20\boldsymbol{i} + 10\boldsymbol{j})$ m/s, $\boldsymbol{v}_B = (33.\bar{3}\boldsymbol{i} + 10\boldsymbol{j})$ m/s, $\boldsymbol{v}_C = (v_{C_1}\boldsymbol{i} + v_{C_2}\boldsymbol{j} + v_{C_3}\boldsymbol{k})$ m/s, $\boldsymbol{r}_{A/O} = (-20\boldsymbol{i} + y_A\boldsymbol{j})$ m, $\boldsymbol{r}_{B/O} = (33.\bar{3}\boldsymbol{i} + 20\boldsymbol{j})$ m, and $\boldsymbol{r}_{C/O} = 20\boldsymbol{j}$ m.

a. Use mass center and linear momentum concepts to determine $\boldsymbol{v}_C$ and y_A.

b. Verify that the angular momentum about O is zero, even after the asteroid has split up.

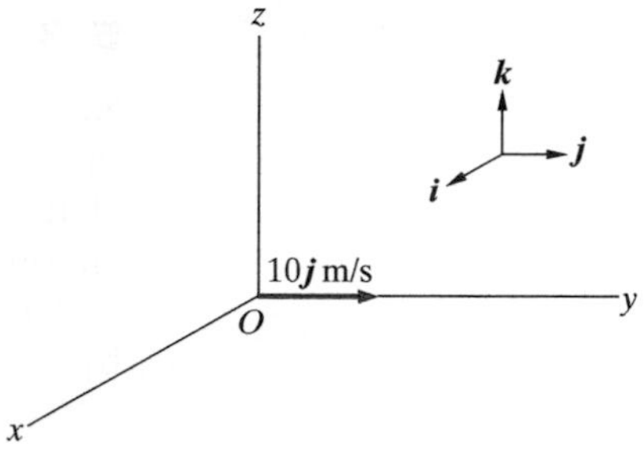

E5.2.15

5.2.16. **[Level 2]** The depicted system is constrained to rotate about the pivot O in the horizontal plane, where $m = 2\,\text{kg}$, $M = 5\,\text{kg}$, and $r = 0.6\,\text{m}$. Suppose the system is given an initial angular speed of $\omega = 180\,\text{rpm}$ in the counterclockwise direction, and the appended masses are made of different materials such that the coefficients of friction for m and M are $\mu_m = 0.2$ and $\mu_M = 0.25$, respectively. Find how long it takes for the system to come to a stop.

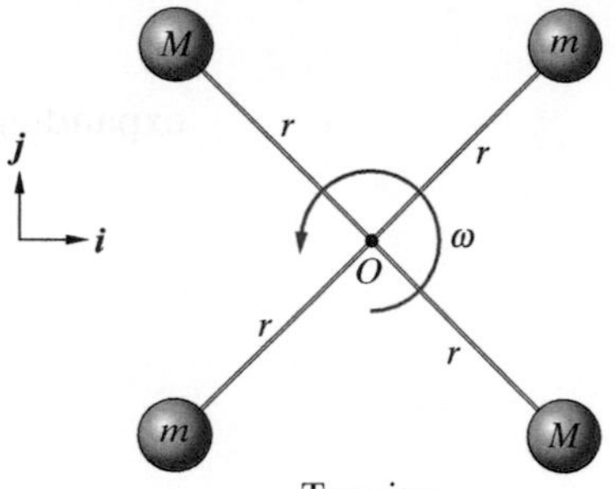

E5.2.16 Top view

5.2.17. **[Level 3]** Find $\boldsymbol{H}_G$ for the masses A, B, and C. Then find $\boldsymbol{H}_O$. Finally, show that

$$\boldsymbol{H}_O = \boldsymbol{H}_G + \boldsymbol{r}_{G/O} \times m\boldsymbol{v}_G$$

$m_A = 0.2$ kg, $m_B = 0.4$ kg, and $m_C = 0.4$ kg. $\boldsymbol{r}_{A/O} = (7\boldsymbol{j} + 7\boldsymbol{k})$ cm, $\boldsymbol{r}_{B/O} = (7\boldsymbol{i} + 4\boldsymbol{j})$ cm, $\boldsymbol{r}_{C/O} = (6\boldsymbol{i} + 5\boldsymbol{k})$ cm, $\boldsymbol{v}_A = -3\boldsymbol{i}$ cm/s, $\boldsymbol{v}_B = 1\boldsymbol{i}$ cm/s, and $\boldsymbol{v}_C = 5\boldsymbol{j}$ cm/s.

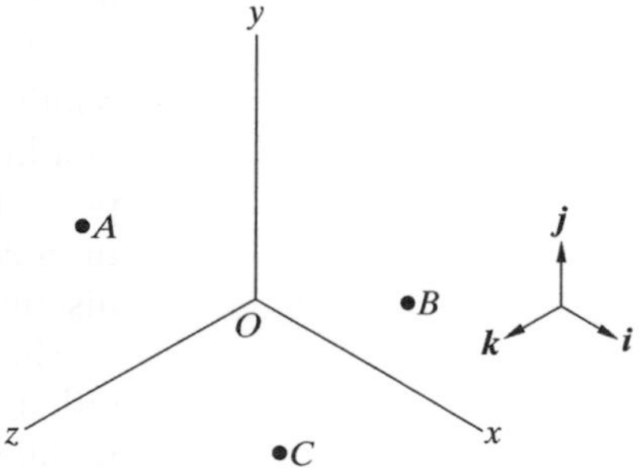

E5.2.17

5.3 WORK AND ENERGY

Not much about this section is going to be shocking—pretty much all we'll say is that when you have a system of particles, you have to account for all their energies. It's all pretty uninteresting with one exception—kinetic energy. Here we will see some nice results.

The kinetic energy of the ith particle in a system of particles is given by

$$\mathcal{KE}_i = \frac{1}{2} m_i \|\dot{\boldsymbol{r}}_i\|^2$$

The total kinetic energy is found by summing over all the particles:

$$\mathcal{KE} = \frac{1}{2}\sum_{i=1}^{n} m_i \|\dot{\boldsymbol{r}}_i\|^2 = \frac{1}{2}\sum_{i=1}^{n} (m_i \dot{\boldsymbol{r}}_i \cdot \dot{\boldsymbol{r}}_i)$$

re-expressing $\dot{\boldsymbol{r}}_i$:

$$= \frac{1}{2}\sum_{i=1}^{n} m_i \left[\left(\dot{\bar{\boldsymbol{r}}} + \dot{\boldsymbol{r}}_{i_{/G}}\right) \cdot \left(\dot{\bar{\boldsymbol{r}}} + \dot{\boldsymbol{r}}_{i_{/G}}\right)\right]$$

expanding:

$$= \frac{1}{2}\sum_{i=1}^{n} m_i \left(\dot{\bar{\boldsymbol{r}}} \cdot \dot{\bar{\boldsymbol{r}}} + 2\dot{\bar{\boldsymbol{r}}} \cdot \dot{\boldsymbol{r}}_{i_{/G}} + \dot{\boldsymbol{r}}_{i_{/G}} \cdot \dot{\boldsymbol{r}}_{i_{/G}}\right)$$

$$= \frac{1}{2}(\dot{\bar{\boldsymbol{r}}} \cdot \dot{\bar{\boldsymbol{r}}}) \overbrace{\sum_{i=1}^{n} m_i}^{m} + \dot{\bar{\boldsymbol{r}}} \cdot \overbrace{\sum_{i=1}^{n} m_i \dot{\boldsymbol{r}}_{i_{/G}}}^{0 \text{ from } (5.3)} + \frac{1}{2}\sum_{i=1}^{n} \left(m_i \dot{\boldsymbol{r}}_{i_{/G}} \cdot \dot{\boldsymbol{r}}_{i_{/G}}\right)$$

$$\mathcal{KE} = \frac{1}{2} m \bar{v}^2 + \frac{1}{2}\sum_{i=1}^{n} m_i \left\|\dot{\boldsymbol{r}}_{i_{/G}}\right\|^2 \tag{5.25}$$

>>> **Check out Example 5.8 (page 284), Example 5.9 (page 285), and Example 5.10 (page 286) for applications of this material.**

You can see that there are two distinctly different parts to the total kinetic energy of the system. The first term is the analogue of single-particle linear motion and represents the kinetic energy of the system's mass center. The second term contains a summation over all the remaining particles and, as you can see, involves $\dot{\boldsymbol{r}}_{i_{/G}}$, the time rate of change of their *relative* positions. To see what this second term describes, imagine you're floating at the center of mass of a group of particles. Think about their possible motions *relative to you*. Each one could be moving toward you or moving away from you, or circling around you at some constant radial distance. Anything else? Waiting ...

The answer is that there's no other possibility. Each particle can move either toward or away from the center of mass or circle around it. When we finally get to rigid bodies (we're almost there, I promise), the possibility of the particle moving toward or away from the mass center will disappear. After all, how can you have a rigid body when pieces of it are moving away from the center of mass? That would imply that the object is breaking up or, at the very least, stretching. Similarly, parts of the body can't move toward the mass center unless the body is compressing. No, the only thing that can happen in a rigid body is that the pieces that compose it can rotate around the center of mass. That's one of the simplifications that will benefit us when we are dealing with rigid bodies: the kinetic energy will have only two terms—one due to translation of the body's center of mass and one due to rotation about it.

Now that we have finished with kinetic energy, examining the potential energy of our system of particles and the work done on or by the system won't take long. The work done on each particle going from State 1 to State 2 is equal to the change in kinetic energy of the particle:

$$W_{1-2_i} = \Delta\mathcal{KE}_i$$

So if we sum over all the particles, we will get the total work done on the system:

$$W_{1-2} = \Delta\mathcal{KE} \tag{5.26}$$

If no work is being done on the system, then

$$\Delta\mathcal{KE} = 0 \tag{5.27}$$

that is, energy is conserved $\mathcal{KE}_1 = \mathcal{KE}_2$.

Just as we did in the single-particle case, we can simplify our lives by using potential as well as kinetic energy. If no external work is being done on the system, then

$$\Delta\mathcal{KE} + \Delta\mathcal{PE}\big|_{sp} + \Delta\mathcal{PE}\big|_{g} = 0 \tag{5.28}$$

where *sp* refers to spring forces and *g* refers to gravitational forces. For the multiparticle case we simply have to account for all the potential energy changes in our system and sum them. If we're dealing with a conservative system (no energy dissipation and no external nonconservative work being done), then the sum of all the potential and kinetic energies is constant in time.

If external work is done (nonconservative work such as that due to friction), then we have to account for it:

$$\Delta\mathcal{KE} + \Delta\mathcal{PE}\big|_{sp} + \Delta\mathcal{PE}\big|_{g} = W_{nc_{1-2}} \tag{5.29}$$

In this equation $W_{nc_{1-2}}$ accounts for all the external nonconservative work done as the particles move from their initial to their final states.

⋙ **Check out Example 5.10 (page 286) for an application of this material.**

EXAMPLE 5.8 KINETIC ENERGY OF A MODIFIED BATON (Theory on page 282)

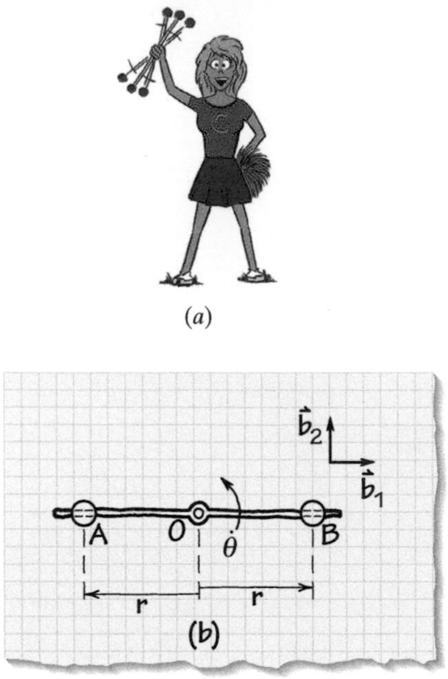

Figure 5.15 (*a*) Majorette with baton; (*b*) simplified picture of baton

A school marching band wouldn't be complete without majorettes, and majorettes wouldn't be complete without batons (**Figure 5.15*a***). The only modification I've made is that, rather than being fixed at the end of the connecting rod, the masses in this baton are free to move. The rod they ride on is massless and rigid. We'll analyze the system under the assumption that the center of the rod stays fixed in space (the majorette is simply twirling the baton and is not tossing it up) and we'll also assume that the masses are identical and stay the same distance (r) away from the point of rotation. In this way the system's center of mass and the center of rotation are the same, stationary point. Let's calculate the kinetic energy components of the system—the component that represents translation of the mass center and the component that represents the motion of the masses relative to the mass center.

Goal Examine the translational and rotational kinetic energy components of the twirling baton.

Draw Examine the simplified model of the baton shown in **Figure 5.15*b***. The $\boldsymbol{b}_1, \boldsymbol{b}_2$ unit vectors are fixed to the rotating baton, with $\boldsymbol{b}_1$ always pointing in the directions of motion away from or toward the rod center and $\boldsymbol{b}_2$ always pointing in the direction of motion around the rod center.

Formulate Equations I will tell you the velocities of the baton masses here since we haven't gone far enough yet to actually derive them. Once you hit Section 6.4 you will see how I did it. The two velocities are given by

$$\begin{aligned} \boldsymbol{v}_A &= -\dot{r}\boldsymbol{b}_1 - r\dot{\theta}\boldsymbol{b}_2 \\ \boldsymbol{v}_B &= \dot{r}\boldsymbol{b}_1 + r\dot{\theta}\boldsymbol{b}_2 \end{aligned}$$

where $\dot{\theta}$ is the rotational speed of the baton.

The total kinetic energy of the baton is given by

$$\mathcal{KE} = \frac{1}{2}mv_A^2 + \frac{1}{2}mv_B^2$$

Solve Using our velocity expressions in the kinetic energy expression yields

$$\mathcal{KE} = \frac{1}{2}m(\dot{r}^2 + r^2\dot{\theta}^2) + \frac{1}{2}m(\dot{r}^2 + r^2\dot{\theta}^2) = \frac{1}{2}(2m)\dot{r}^2 + \frac{1}{2}(2mr^2)\dot{\theta}^2$$

The term $\frac{1}{2}(2mr^2)\dot{\theta}^2$ is the kinetic energy associated with the rotational motion of either weight *around* the baton's center of mass. The term $2mr^2$ is what we will be calling the mass moment of inertia (when we get to that point in Chapter 7). The term $\frac{1}{2}(2m)\dot{r}^2$ represents the kinetic energy associated with translational motion *away* from the mass center. Note that the two terms are completely decoupled—one involves angular rotation and the other involves the sliding of the weights along the baton's length.

EXAMPLE 5.9 KINETIC ENERGY OF A TRANSLATING MODIFIED BATON (Theory on page 282)

Example 5.8 looked at the twirling aspect of baton twirling, but that's not the most dynamic part. To use a baton correctly, you have to toss it up into the air while it's turning and (hopefully) catch it on the way down. This motion is quite different from the simple twirling motion we looked at in that the baton will now be translating as well as rotating, as shown in **Figure 5.16*a***. The baton is again a modified one in which the end weights can slide along the massless connecting rod. We will also again assume that the weights, if they move, stay equidistant from the center of the rod. Thus the physical center of the baton will also be the mass center. The center of the rod will be thrown directly upward at a speed v_0, and we will calculate the kinetic energy components that represent the motion of the mass center and the motions of the end weights with respect to the mass center.

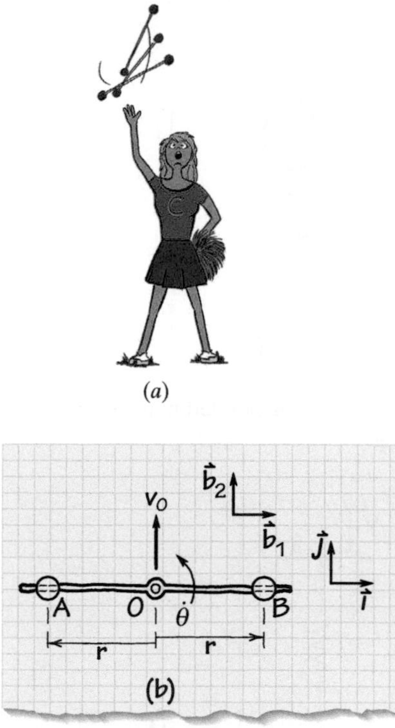

Figure 5.16 (*a*) Majorette tossing baton; (*b*) simplified baton model

Goal Find the translational and rotational kinetic energy components of a twirling baton tossed into the air.

Draw **Figure 5.16*b*** shows our simplified baton model. The $\boldsymbol{b}_1, \boldsymbol{b}_2$ unit vectors are again fixed to the rotating baton and point in the directions of motion away from and toward the center of mass and around it; the $\boldsymbol{i}, \boldsymbol{j}$ unit vectors define fixed coordinate directions. For the instant under consideration, the $\boldsymbol{i}, \boldsymbol{j}$ and $\boldsymbol{b}_1, \boldsymbol{b}_2$ unit vectors align with one another.

Formulate Equations The velocity of m_A and m_B are given by

$$\begin{aligned} \boldsymbol{v}_A &= -\dot{r}\boldsymbol{b}_1 + (v_0 - r\dot{\theta})\boldsymbol{b}_2 \\ \boldsymbol{v}_B &= \dot{r}\boldsymbol{b}_1 + (v_0 + r\dot{\theta})\boldsymbol{b}_2 \end{aligned}$$

and the total kinetic energy of the baton is $\frac{1}{2}mv_A^2 + \frac{1}{2}mv_B^2$.

Solve Substituting our velocity expressions in the kinetic energy expression produces

$$\begin{aligned} \mathcal{KE} &= \frac{1}{2}m(v_0^2 - 2v_0r\dot{\theta} + r^2\dot{\theta}^2 + \dot{r}^2) + \frac{1}{2}m(v_0^2 + 2v_0r\dot{\theta} + r^2\dot{\theta}^2 + \dot{r}^2) \\ &= \frac{1}{2}(2\,m)v_0^2 + \frac{1}{2}(2mr^2)\dot{\theta}^2 + \frac{1}{2}(2m)\dot{r}^2 \end{aligned}$$

More is going on here than in the simple twirling example. In addition to the terms we have already seen ($\frac{1}{2}(2mr^2)\dot{\theta}^2$ for motion around the center of mass and $\frac{1}{2}(2m)\dot{r}^2$ for motion away from it), we now have the term $\frac{1}{2}(2\,m)\,v_0^2$, which is the same kinetic energy we would have gotten if the entire baton had been compressed into a ball and thrown up into the air at speed v_0. As predicted, the system's center of mass "acts like" a single particle.

EXAMPLE 5.10 SPRING-MASS SYSTEM (Theory on page 282 and 283)

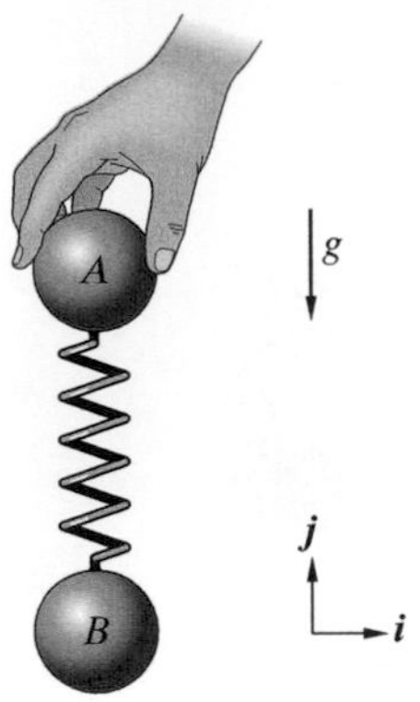

Figure 5.17 Suspended masses and spring

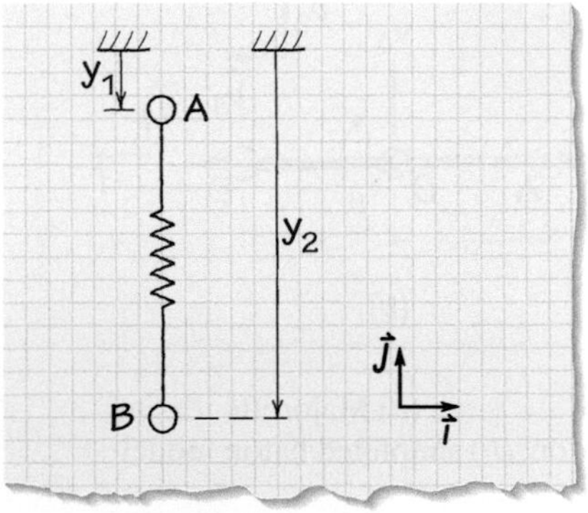

Figure 5.18 Coordinates for spring/mass system

Figure 5.17 shows two balls, each with mass 1.1 kg. Ball A is held firmly and ball B is suspended in static equilibrium 3 m below ball A. The linear spring connecting the balls is modeled as having an unstretched length of zero. At $t = 0$ ball A is released and at t^* the two balls hit each other, ball A moving at $\boldsymbol{v}_A = -16.5\boldsymbol{j}$ m/s and ball B moving at $\boldsymbol{v}_B = 13.5\boldsymbol{j}$ m/s. Determine t^* and the spring constant k.

Goal Calculate t^*, the time of collision of the balls, and determine k.

Assume Air drag will be neglected.

Draw **Figure 5.18** shows the coordinates we will be using. y_1 gives the vertical position of m_A and y_2 tracks m_B. Both are positive downward.

Formulate Equations We'll use linear momentum and energy:

$$m\overline{\boldsymbol{v}} = m_A\boldsymbol{v}_A + m_B\boldsymbol{v}_B \tag{5.30}$$

$$\mathcal{KE}_1 + \mathcal{PE}_1 = \mathcal{KE}_2 + \mathcal{PE}_2 \tag{5.31}$$

Solve Once the assemblage is dropped, the center of mass will accelerate at 9.81 m/s^2 because gravity provides the only external force acting on the system. Our equation of motion for the mass center is given by $m\ddot{\overline{y}} = mg$, which has the solution $\overline{\boldsymbol{v}} = -gt + \overline{\boldsymbol{v}}_0$.

Using the fact that the mass center is initially stationary and evaluating at t^* give us $\overline{\boldsymbol{v}} = -(9.81\text{ m/s}^2)t^*\boldsymbol{j}$. Using the given velocity data, we can calculate the velocity of the mass center. Both balls have the same mass, and thus the mass center's velocity will be the average of the two: $\overline{\boldsymbol{v}} = (\boldsymbol{v}_A + \boldsymbol{v}_B)/2 = -1.54\boldsymbol{j}$ m/s. Equating the two values for $\overline{\boldsymbol{v}}$ gives $t^* = 0.153$ s.

Now we can deal with finding k. We can calculate the distance that the center of mass dropped by integrating the equation of motion twice:

$$\Delta\overline{y} = \frac{g}{2}t^{*2} = \frac{9.81\text{ m/s}^2}{2}(0.153\text{ s})^2 = 0.115\text{ m}$$

There's no spring energy at t^* (collision time), and thus the only terms we need to consider are the final kinetic energy, the initial spring potential energy, and the change in potential energy due to gravity. Applying (5.31) gives us

$$0 + \frac{1}{2}k\,(y_2(0) - y_1(0))^2 + mg\Delta\overline{y} = \frac{1}{2}m_A v_A^2 + \frac{1}{2}m_B v_B^2 + 0 + 0$$

$$\frac{1}{2}k(3\text{ m})^2 + (2.2\text{ kg})(9.81\text{ m/s}^2)(0.115\text{ m}) = \frac{1}{2}(1.1\text{ kg})(16.5\text{ m/s})^2 + \frac{1}{2}(1.1\text{ kg})(13.5\text{ m/s})^2$$

$$k = 55\text{ N/m}$$

EXERCISES 5.3

5.3.1. [Level 1] Consider two elastically constrained carts that are free to oscillate in the vertical plane over a smooth horizontal surface. As illustrated, the cart of mass m_1 is attached to a wall via a spring with stiffness k_1, and the coordinate x_1 measures the cart's distance away from its equilibrium position. Cart m_2 is connected to m_1 by a spring of stiffness k_2, where x_2 describes the cart's oscillation about equilibrium. Derive expressions for the system's total kinetic and potential energies.

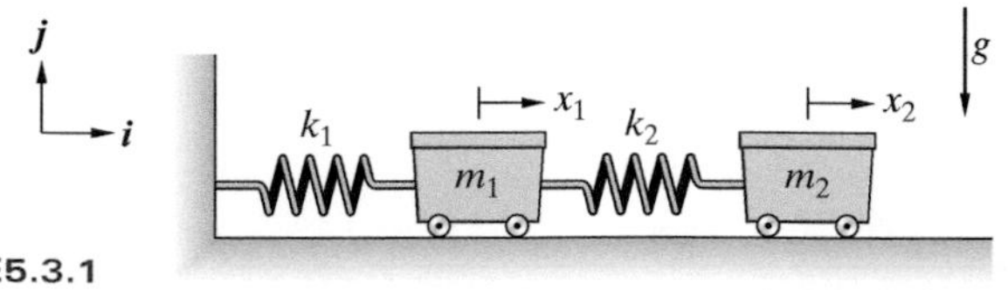

E5.3.1

5.3.2. [Level 2] The system shown consists of two rigid bodies. Both bodies are constructed of square "bricks," the bricks being equal in mass. Each brick is 0.1 m on a side. Both bodies are joined together by a spring that has an unstretched length of 0.2 m. The interface between body A and body B is frictionless, as is the interface between body B and the ground. Each brick has a mass of 0.1 kg, the total mass is equal to the number of visible bricks (8 bricks and 16 bricks) and $k = 120$ N/m. The two bodies are released from rest.

a. Where will the rightmost edge of body A be when it strikes the stop S? The origin of our horizontal axis is indicated as O in the figure.

b. How fast will body B be traveling when the collision occurs?

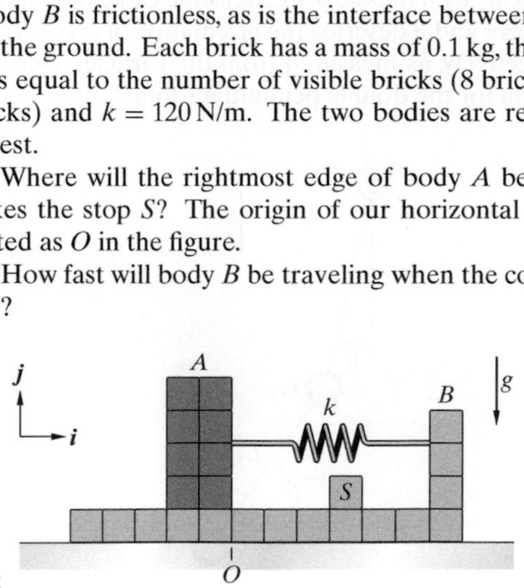

E5.3.2

5.3.3. [Level 2] A system consisting of two elastically constrained masses m_1 and m_2 is free to slide on a frictionless rod that rotates about the pivot point O. Mass m_1 is attached to O along the rod by a spring of stiffness k_1 with a free length of l_1, and m_1 and m_2 are connected by a spring whose stiffness and free length are k_2 and l_2, respectively. The angle θ describes the rod's orientation relative to the horizontal, and the positions of m_1 and m_2 along the rod are given by L_1 and L_2, respectively. Find expressions for the system's total kinetic and potential energies.

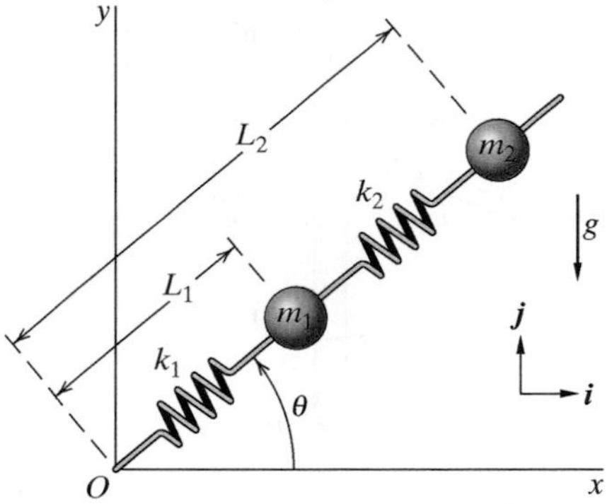

E5.3.3

5.3.4. [Level 2] A 4 kg triangular body m_2 rests on a frictionless, horizontal surface. At $t = 0$, a 1 kg mass m_1 is released from rest as shown. It's located just slightly to the left of center so that it can slide down the left side. Assume zero friction between m_1 and m_2.

a. Where will the mass contact the ground?

b. How fast will the mass be moving when it hits the ground?

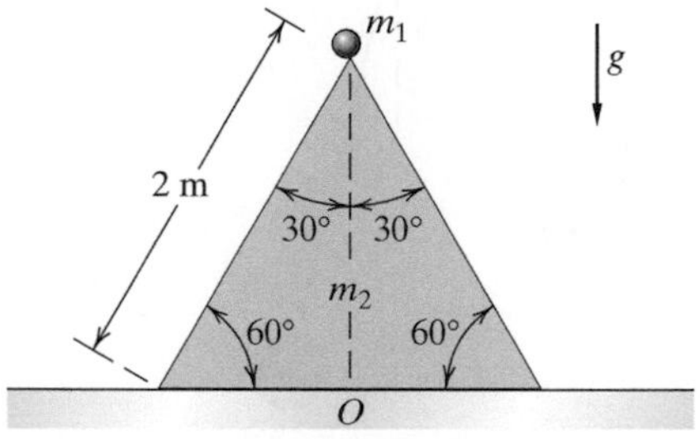

E5.3.4

5.3.5. [Level 2] The device shown in the figure has four massless arms attached to a central (massless) rod that's free to rotate in its base B. The four masses can slide along the arms up to a distance of 0.4 m from the center. Initially, they are a distance of 0.2 m from the center. $m = 0.5$ kg and the initial rotation rate is equal to 10 rad/s. At $t = 0$ the masses are released with zero velocity relative to the arm they're attached to. The tangential speed of the masses is equal to the rotation rate of the arm ($\dot{\theta}$) multiplied by the distance of the mass from the center of rotation (r).

a. Determine the rotation rate of the overall system after the masses have reached the stop at the end of the arms.

b. What is the kinetic energy change?

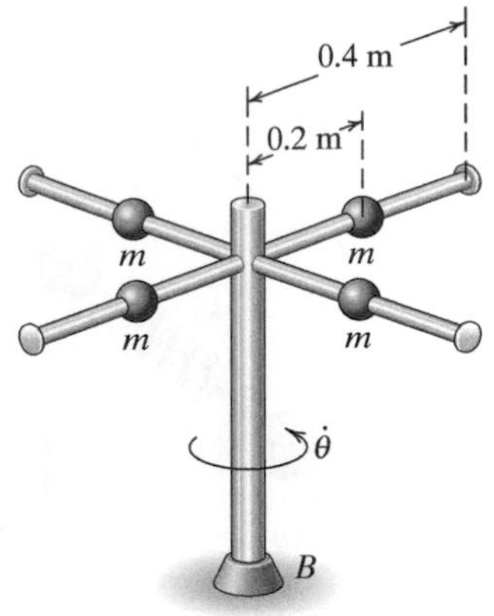

E5.3.5

5.3.6. [Level 2] As depicted, a simple pendulum of length L and mass M is free to pivot about the point O on a collar of mass m that is constrained to move along a smooth horizontal rail. Let x denote the collar's horizontal position along the rail, and take θ to be the angle that the pendulum makes with the vertical. Write expressions for the system's total kinetic and potential energies.

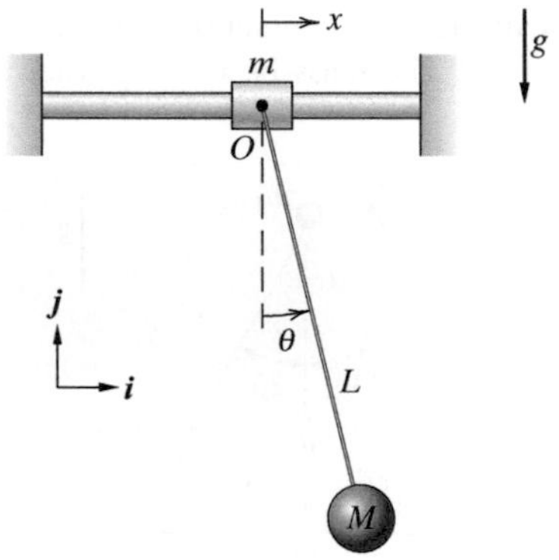

E5.3.6

5.3.7. [Level 2] An able-bodied seaman is aloft on the mainyard of a Napoleonic-era warship. The task at hand is to raise a bucket of paint up from the deck to where he's sitting. This involves lifting the 5.0 kg bucket up 12 m. The rope he uses weighs 0.15 kg/meter. Calculate the work he does in accomplishing the task, including both the bucket and the continuous span of rope.

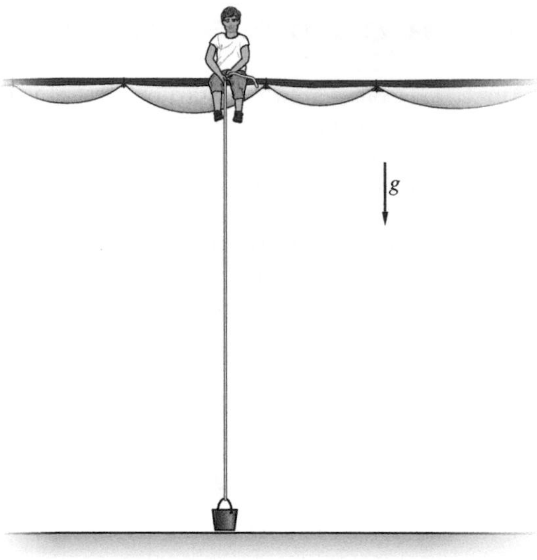

E5.3.7

5.3.8. [Level 2] Consider the illustrated double pendulum. Mass m_1 is connected to the pivot point O via a massless, extensible cable of length L_1, and mass m_2 is attached to m_1 by a massless, inextensible cable of length L_2. The angles θ and β describe the orientation of cables L_1 and L_2, respectively, as measured from the vertical. Derive an expression for the double pendulum's total kinetic energy.

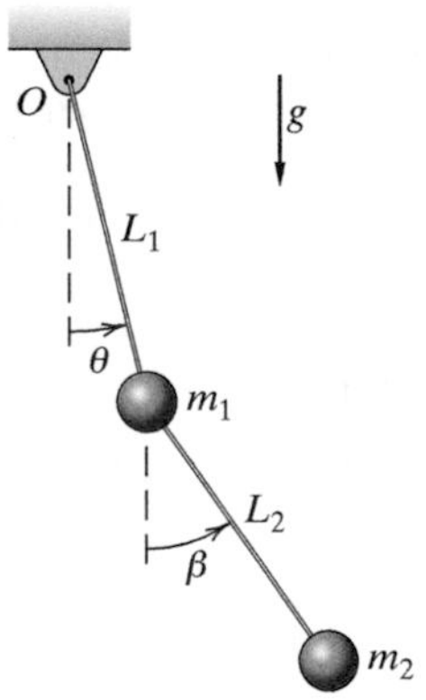

E5.3.8

5.4 STATIONARY ENCLOSURES WITH MASS INFLOW AND OUTFLOW

We do not always need to analyze a single mass or even a finite collection of masses; instead, we might be faced with a continuous stream of mass. Perhaps we need to determine the force necessary to hold a garden hose as water shoots through it. Or we might be asked to determine the acceleration of a rocket due to the exhaust gases being ejected from it.

The primary conceptual problem we will face is determining how to construct meaningful free-body diagrams. For solid bodies we have the advantage that the whole body moves as one rigid piece. When we're considering fluids, this is no longer the case. We can have a pipe that gets narrower as the fluid moves along it, which means that the fluid velocity has to increase. (Volumetric flow is equal to the area times the velocity, and if the volumetric flow is constant and the area decreases, then the velocity increases.) Do we pick a chunk of fluid and follow it along, or do we look at a particular location in the pipe and keep track of the fluid that enters and exits? Each approach is viable, and in a fluids class you will discuss both. Our approach in this section will be to identify the solid object that's reacting to the flow and define that as the object of interest. Our concern will then center on the fluid entering, exiting, and/or accumulating within the body.

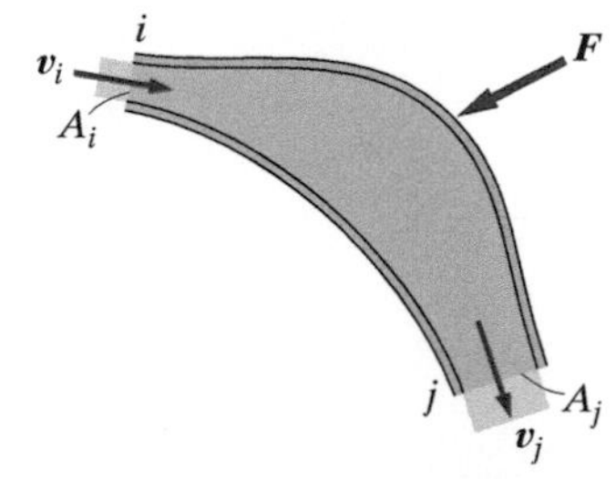

Figure 5.19 Enclosure with mass entering and leaving

Figure 5.19 shows an enclosure filled with fluid. More fluid is entering at position i, and at the same time fluid is leaving at j. Interaction forces exist between the fluid within the enclosure walls and the enclosure itself, but they won't affect our overall force balance because they're all internal to the fluid/enclosure system. I have drawn in a vector that represents the sum of all the external forces (the ones that hold the enclosure in place) and have labeled it $\boldsymbol{F}$.

The approach we'll use to find the overall force $\boldsymbol{F}$ acting on the enclosure will parallel the way in which we originally defined velocity and acceleration. We'll look at the system at some time t, then examine it at a later time $t + \Delta t$, identify the difference, divide by Δt, and take the limit as Δt goes to zero.

The cross-sectional area is A_i at i and A_j at j. The fluid is moving at right angles to these areas as it passes through. Assume that the fluid entering at i has density ρ_i and speed v_i. The fluid leaving at j has density ρ_j and speed v_j. The rate at which fluid is entering (or leaving) is given by the volumetric flow rate (area times speed) times the density and is denoted by $\frac{dm}{dt}$. If we assume that the same amount leaves at j as enters at i, then

$$\rho_i A_i v_i = \rho_j A_j v_j = \frac{dm}{dt}$$

We now know how to characterize how much fluid enters or leaves our container. Let's say that the mass of fluid entering over Δt seconds is given by Δm. At t, the enclosure is full of fluid. At $t + \Delta t$ an amount Δm has left at j (with velocity $\boldsymbol{v}_j$) and an amount Δm has entered with velocity $\boldsymbol{v}_i$. What's the momentum change of the system?

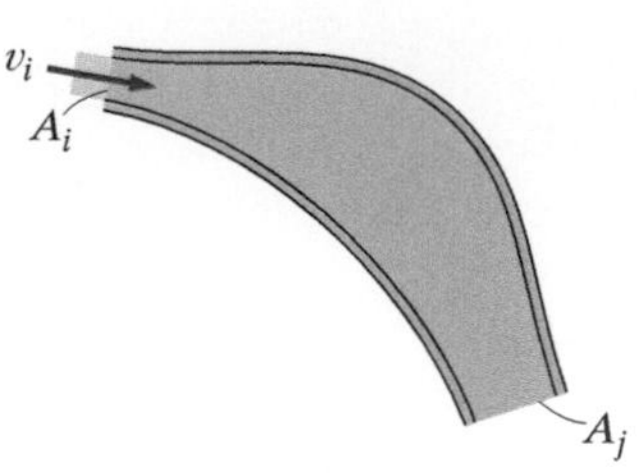

Figure 5.20 Configuration at t

When we started this section, I said that the enclosure and the fluid within it would constitute our system. But now I'm going to stretch that definition a little. I'm going to define our system as the enclosure, the fluid within it, and the tiny chunk shown in **Figure 5.20** that's about to enter at i. Keep in mind that I'm going to take the limit as this chunk goes to zero, so what I'm ultimately looking at is simply the enclosure and the fluid within it. The reason I'm bothering with the little chunk of fluid to the left is because it actually affects (in a major way) the overall system momentum. This chunk of fluid is moving at velocity $\boldsymbol{v}_i$, and so its momentum is $\Delta m \boldsymbol{v}_i$.

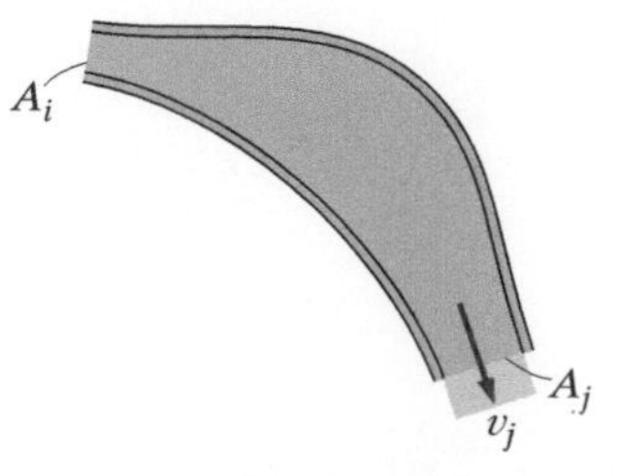

Figure 5.21 Configuration at $t + \Delta t$

Δt seconds later the situation is the one shown in **Figure 5.21**. The chunk of fluid at i has entered the enclosure, and a new chunk has left at

j. The overall mass in the system doesn't change, and so the mass that just left at j is equal to Δm, and its linear momentum is $\Delta m\boldsymbol{v}_j$.

Put it all together and we have

$$\Delta \boldsymbol{L} = \Delta m(\boldsymbol{v}_j - \boldsymbol{v}_i)$$

Dividing by Δt and taking the limit as Δt goes to zero give us the time rate of change of linear momentum:

$$\dot{\boldsymbol{L}} = \lim_{\Delta t \to 0} \frac{\Delta m(\boldsymbol{v}_j - \boldsymbol{v}_i)}{\Delta t} = \dot{m}(\boldsymbol{v}_j - \boldsymbol{v}_i)$$

We already know that the time rate of change of linear momentum is equal to the sum of the external forces being applied to the system (the $\boldsymbol{F}$ in **Figure 5.19**), and thus we have

$$\boldsymbol{F} = \dot{m}(\boldsymbol{v}_j - \boldsymbol{v}_i) \tag{5.32}$$

>>> **Check out Example 5.11 (page 291) and Example 5.12 (page 292) for applications of this material.**

If the velocity at j is the same in magnitude and direction as at i, then from (5.32) we can see that there's no momentum change. But if the speeds and/or the directions differ, then there is a momentum change and therefore a resulting force.

We can find the time rate of change of angular momentum as well. **Figure 5.22** shows our original system, along with an arbitrarily placed fixed point O. We've already learned that the sum of the moments about a fixed point is equal to the time rate of change of a system's angular momentum about that point. In our case we have

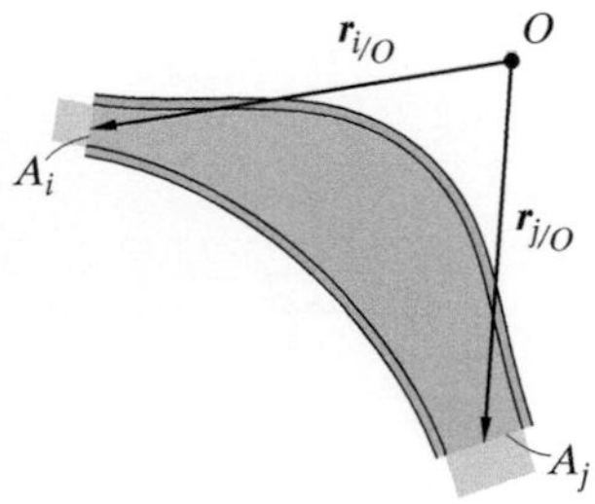

Figure 5.22 Moment about O

$$\Delta \boldsymbol{H}_O = \boldsymbol{r}_{j/O} \times \Delta m\boldsymbol{v}_j - \boldsymbol{r}_{i/O} \times \Delta m\boldsymbol{v}_i$$

Dividing by Δt and taking the limit as Δt goes to zero give us

$$\dot{\boldsymbol{H}}_O = \lim_{\Delta t \to 0} \boldsymbol{r}_{j/O} \times \frac{\Delta m}{\Delta t}\boldsymbol{v}_j - \boldsymbol{r}_{i/O} \times \frac{\Delta m}{\Delta t}\boldsymbol{v}_i$$

$$\dot{\boldsymbol{H}}_O = \dot{m}\left(\boldsymbol{r}_{j/O} \times \boldsymbol{v}_j - \boldsymbol{r}_{i/O} \times \boldsymbol{v}_i\right) \tag{5.33}$$

And there you have it: two formulas that enable you to calculate the forces and moments applied to something with fluid moving through it.

EXAMPLE 5.11 WATER JET IMPINGING ON STATIONARY VANE (Theory on page 290)

A jet of water impinges on a stationary vane and is evenly bisected into two streams oriented at 45° to the incoming stream, as shown in **Figure 5.23**. The volume flow rate is $Q = 0.13\ \text{m}^3/\text{s}$ and the speed of the incoming stream is $v = 35\ \text{m/s}$. Determine the force required to hold the vane stationary.

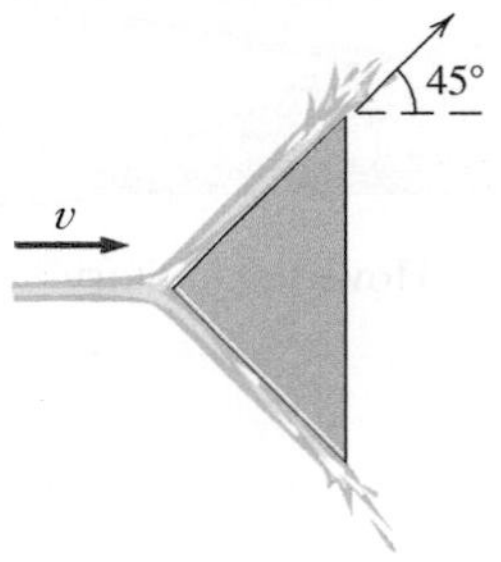

Figure 5.23 Water jet impinging on stationary block

Goal Find the force necessary to hold the deflection vane stationary as it redirects a jet of water.

Given $Q = 0.13\ \text{m}^3/\text{s}$, $v = 35\ \text{m/s}$, and water stream deflected at 45°.

Assume We'll assume that the water jets exit the vane at the same speed with which they entered and that the density of the water jet is $\rho = 1000\ \text{kg/m}^3$.

Draw A box has been drawn in **Figure 5.24** that represents the "body" having an equal inflow and outflow of mass.

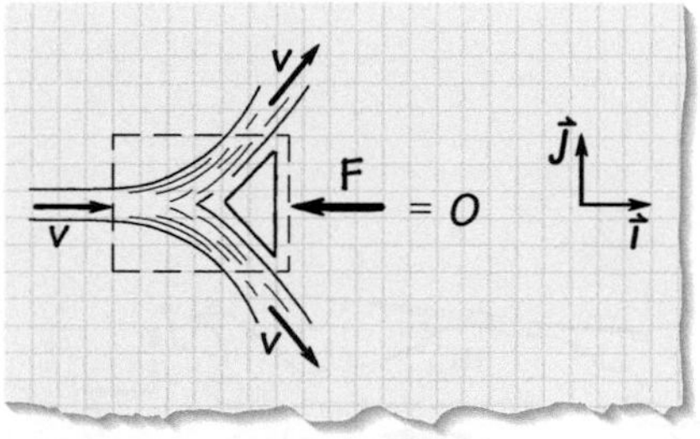

Figure 5.24 FBD=IRD for our system

Formulate Equations We'll apply (5.32). From symmetry we see that we need only apply this equation in the $\boldsymbol{i}$ direction—the force components of the two resultant streams in the $\boldsymbol{j}$ direction cancel each other out. Mathematically, we can dot our vector equation with $\boldsymbol{i}$ to pull out the horizontal components, giving us

$$-F = \dot{m}(\boldsymbol{v}_j - \boldsymbol{v}_i) \cdot \boldsymbol{i}$$

$$-F = \dot{m}(\boldsymbol{v}_j \cdot \boldsymbol{i} - v) \tag{5.34}$$

where $v = 35$ m/s is the given speed of the fluid entering the box.

Solve The mass flow rate of the jet is the volume flow rate multiplied by its density:

$$\dot{m} = \rho Q = (1000\ \text{kg/m}^3)(0.13\ \text{m}^3/\text{s}) = 130\ \text{kg/s}$$

Now apply equation (5.34).

$$-F = \dot{m}\,(v \cos 45° - v)$$

$$= (130\ \text{kg/s})[(35\ \text{m/s}) \cos 45° - 35\ \text{m/s}]$$

$$F = 1.33 \times 10^3\ \text{N}$$

EXAMPLE 5.12 FORCE DUE TO A STREAM OF MASS PARTICLES (Theory on page 290)

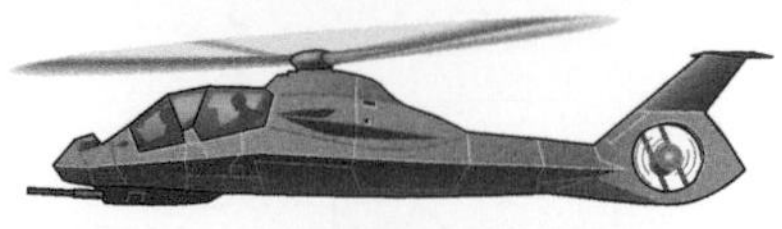

Figure 5.25 Hovering military helicopter

Let's look at a system in which the fluid stream isn't really a continuous stream of liquid but rather a stream of mass particles. We'll consider the average flow rate and be able to apply our fluid equations without any problem. The system we'll consider is a hovering military helicopter. The 6000 kg helicopter (**Figure 5.25**) is equipped with a single Vulcan machine gun, mounted on the underside of the helicopter and protruding just beyond the nose. The maximum rate of fire is 4000 shots per minute, firing 0.11 kg bullets at a muzzle velocity of 1200 m/s. If the gun were fired continuously at its maximum rate, what compensating thrust would be required to keep the helicopter hovering in a stationary position?

Goal Determine the effects that a firing machine gun will have on the motion of a hovering helicopter on which it is mounted.

Given Mass of helicopter, mass of bullets, rates of fire, and muzzle velocity.

Assume We'll assume that the mass of the exiting bullets does not significantly affect the total mass of the helicopter; that is, the total change in mass due to the lost bullets is negligible in comparison with the helicopter's overall mass.

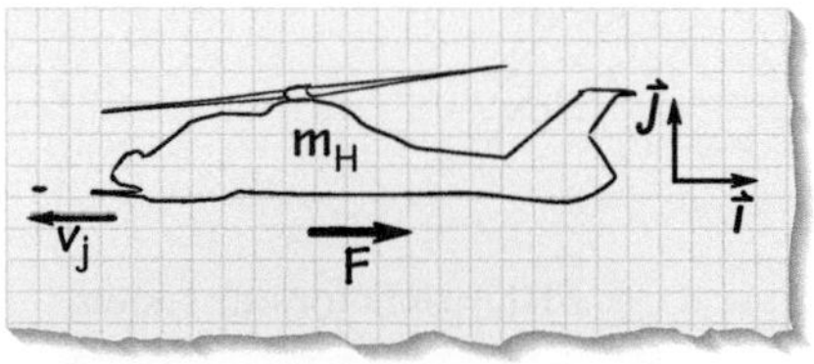

Figure 5.26 Schematic of helicopter and exiting bullets

Draw As **Figure 5.26** shows, our system has an outflow of mass but no inflow.

Formulate Equations All we need is an application of (5.32), with the inflow of mass set to zero:

$$\boldsymbol{F} = \dot{m}\boldsymbol{v}_j \tag{5.35}$$

Solve First, we'll compute $\dot{m}$ for the maximum rate of fire:

$$\dot{m} = \frac{4000 \text{ shots/min}}{60 \text{ sec/min}}(0.11 \text{ kg/shot}) = 7.33 \text{ kg/s}$$

We're given that each bullet's exit speed is 1200 m/s. Using this information along with the $\dot{m}$ in (5.35) gives us:

$$\boldsymbol{F} = \dot{m}\boldsymbol{v}_j = (7.33\,\text{kg/s})(-1200\boldsymbol{i}\text{ m/s}) = \boxed{-8800\boldsymbol{i}\text{ N}}$$

Check We can check that this force makes sense by looking at the effect of a single bullet and then calculating the total force at the given rate of firing. Every minute 4000 bullets are fired, which means that the time between bullets is $\frac{60\,\text{s}}{4000} = 0.015$ s. To get a 0.11 kg bullet moving from 0 m/s to 1200 m/s means we have a momentum change of $(0.11\,\text{kg})(1200\,\text{m/s}) = 132\,\text{kg·m/s}$. The applied linear impulse is therefore $\frac{132\,\text{kg·m/s}}{0.015\,\text{s}} = 8800$ N. Clearly, the force is acting to the left in order to have the bullets travel left. Thus we see that the discrete analysis on the individual bullets gives us the same result as the continuous analysis of the bullet stream.

EXERCISES 5.4

5.4.1. **[Level 1]** A stream of water is directed from an opening in a fire hydrant onto a stationary washing machine (waiting for delivery). The water is flowing at 8 l/s. Assume that the incoming stream of water has a cross-sectional area of 4 cm^2. What force acts on the washing machine?

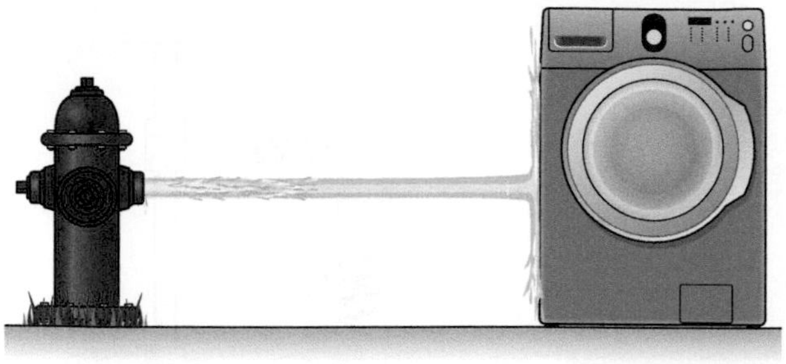

E5.4.1

5.4.2. **[Level 1]** A jet of water impinges on a vane and is deflected as shown. The volume flow rate is $Q = 0.13\ m^3/s$ and the nozzle velocity is $v = 35$ m/s. Determine the force required to hold the vane stationary.

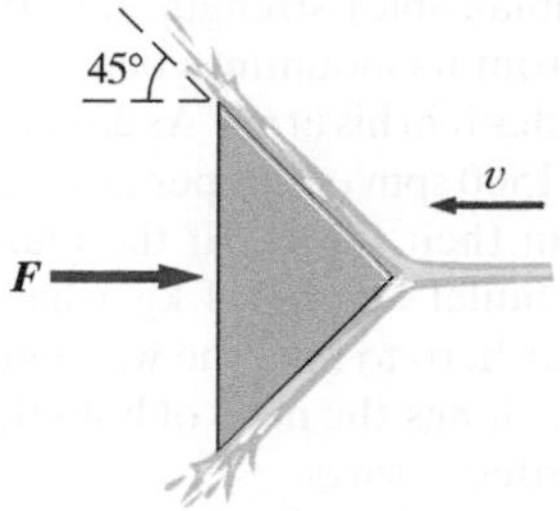

E5.4.2

5.4.3. **[Level 1]** A jet of water impinges on a vane and is deflected as shown. The volume flow rate is $Q = 0.13\ m^3/s$ and the nozzle velocity is $v = 35$ m/s. Determine the force required to hold the vane stationary.

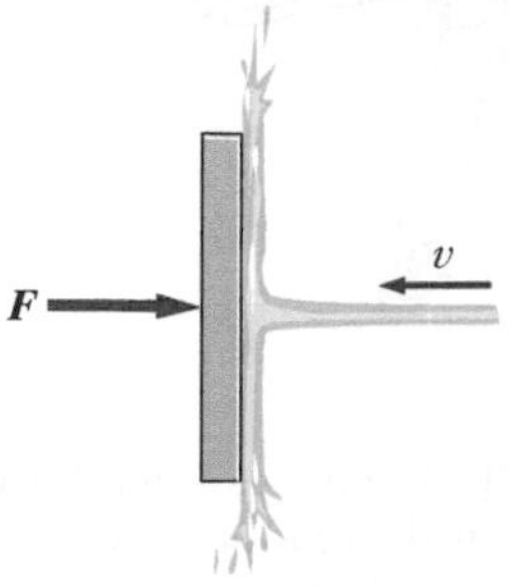

E5.4.3

5.4.4. **[Level 1]** A jet of water impinges on a vane and is deflected as shown. The volume flow rate is $Q = 0.13\ m^3/s$ and the nozzle velocity is $v = 35$ m/s. Determine the force required to hold the vane stationary.

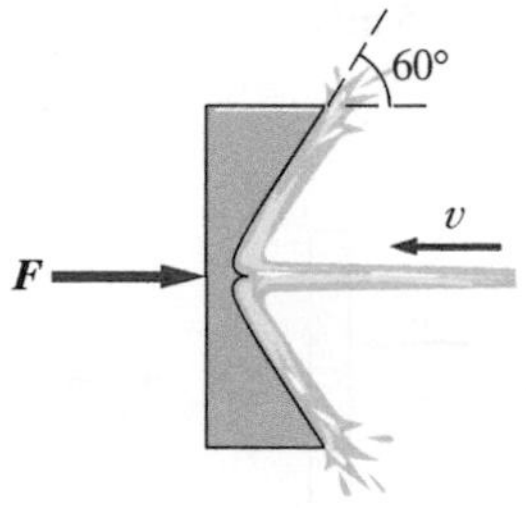

E5.4.4

5.4.5. **[Level 1]** A jet of water impinges on a vane and is deflected as shown. The volume flow rate is $Q = 0.13\ m^3/s$ and the nozzle velocity is $v = 35$ m/s. Determine the force required to hold the vane stationary.

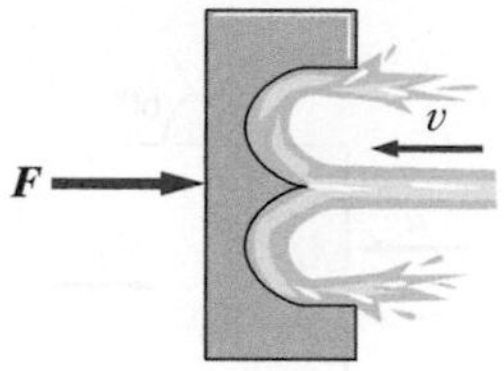

E5.4.5

5.4.6. **[Level 1]** What force must be exerted on the illustrated vane if it is to move directly away from the jet at a constant speed of $u = 10$ m/s?

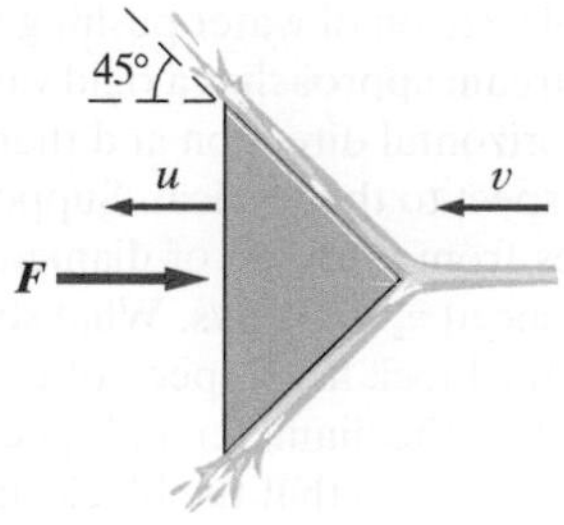

E5.4.6

5.4.7. **[Level 1]** What force must be exerted on the illustrated vane if it is to move directly away from the jet at a constant speed of $u = 10$ m/s?

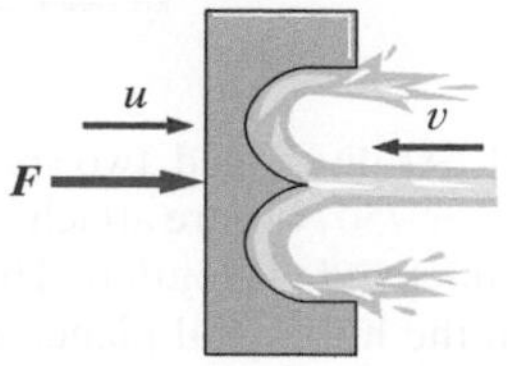

E5.4.7

5.4.8. **[Level 1]** What force must be exerted on the illustrated vane if it is to move directly toward the jet at a constant speed of $u = 15$ m/s?

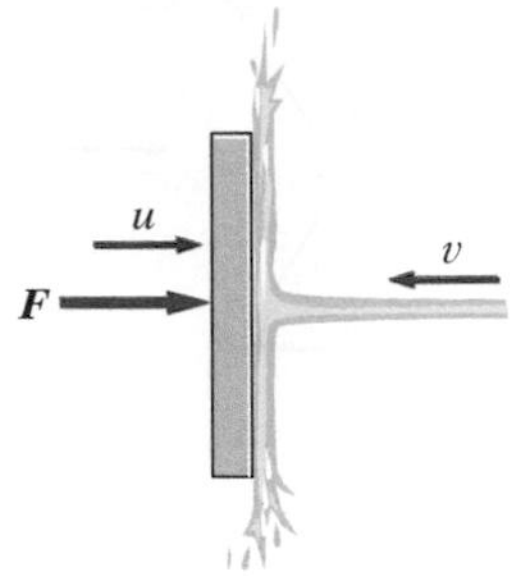

E5.4.8

5.4.9. **[Level 1]** What force must be exerted on the illustrated vane if it is to move directly toward the jet at a constant speed of $u = 15$ m/s?

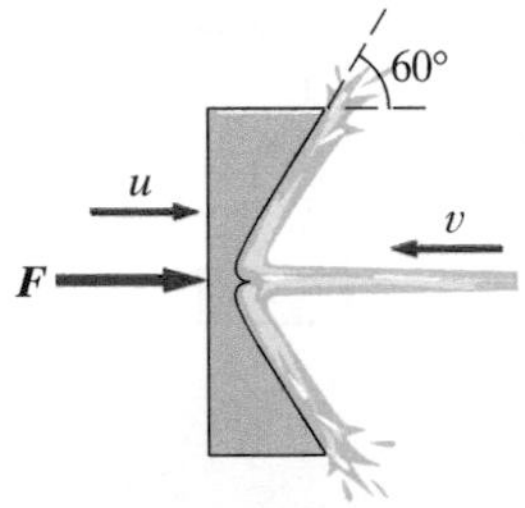

E5.4.9

5.4.10. **[Level 2]** A block with mass $m = 20$ kg is constrained to slide within a smooth slot in the horizontal plane under the action of a steam of water pushing on it. As illustrated, the water stream approaches a rigid vane attached to the block in the horizontal direction and then shoots off at an angle θ with respect to the vertical. Suppose the stream of water originates from a nozzle of diameter $d = 0.05$ m, and it exits with a speed $v_w = 30$ m/s. What should the vane angle θ be so that the block has a speed of $v' = 20$ m/s after $\Delta t = 4$ s? Assume that the diameter and speed of the water stream remain constant and that the block starts from rest.

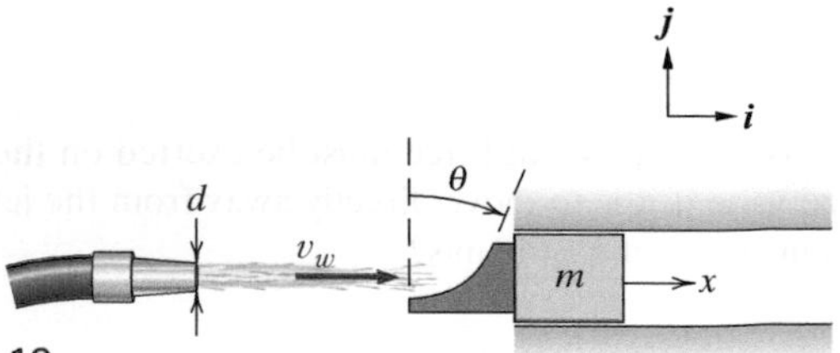

E5.4.10

5.4.11. **[Level 2]** As illustrated, two springs of stiffness $k_1 = 1000$ N/m and $k_2 = 750$ N/m are attached to a light, rigid enclosure with one inlet and two outlets. The system is constrained to move in the horizontal plane. The enclosure's inlet has a radius of $r_1 = 1$ cm, and the radii of the outlets are $r_2 = 0.6$ cm and $r_3 = 0.4$ cm. A steady stream of water is then injected into the enclosure with an inlet speed of $v_1 = 15$ m/s, and water exits through one outlet at a speed $v_2 = 25$ m/s and an angle $\theta = 45°$ with respect to the vertical. What are the deflections of the springs when the system has reached static equilibrium?

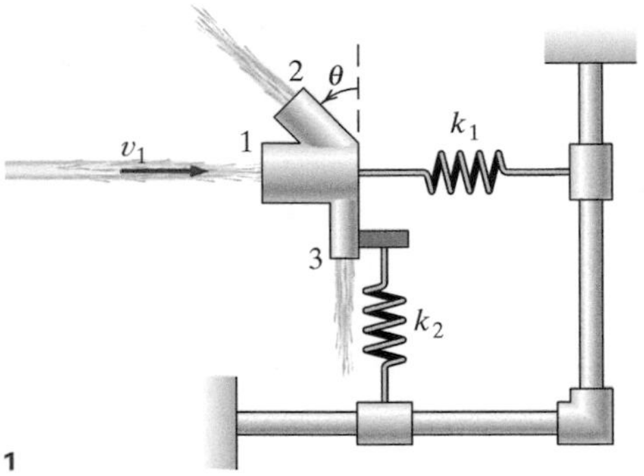

E5.4.11

5.4.12. **[Level 2]** After a disastrous descent into enemy territory, our hero has found himself completely weaponless, yet only slightly disheveled. In a feat of inconceivable (perhaps even implausible) strength, he pries a perfectly intact chain gun from its mountings beneath his chopper's wreckage and cradles it in his arms. As enemy troops swarm in, he unleashes a 1500 spm (shots per minute) barrage that stops them dead in their tracks. If the muzzle velocity is 975 m/s and each bullet weighs 0.1 kg, what average force F is required of our hero to hold the weapon in place? Neglect, among other things, the mass of both the ammunition belt and the discarded casings.

E5.4.12

5.4.13. **[Level 2]** A 160 kg swamp boat is powered by twin fans mounted at the rear of the craft. Each fan has a 0.9 m wide slipstream and, at maximum throttle, accelerates the entering air from zero velocity to a velocity $u = 17$ m/s relative to the boat. What is the acceleration of the boat when $v = 0$? The hydrodynamic drag force is approximated by $f = -cv$, where v is expressed in m/s. If the boat's top speed is 40 km/h, what is c? The specific weight of air is 1.22 kg/m^3.

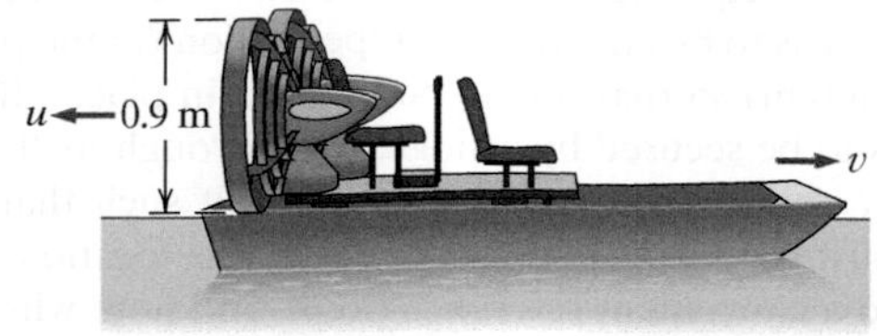

E5.4.13

5.4.14. **[Level 2]** A grain combine equipped with a 6 m wide cutting platform travels at 4.8 km/h as it harvests crops. The harvested grain ($\rho = 7.21 \times 10^{-1}$ kg/mm^3) is separated, cleaned, and then carried up a 4.2 m, 45° elevator to the loading auger. The elevator belt rotates at 38 rpm. A force plate mounted at 20° to the horizontal lies just beyond the top of the elevator and is used to measure the force exerted by the grain as it is deflected toward the auger. Over a 3.0 s interval, the average recorded force is 110 N, perpendicular to the plate. If the collision between the grain and force plate may be modeled as perfectly elastic, what is the farmer's yield (in mm^3/m^2)? Assume that the velocity of the ejected grain is the same as that of the elevator belt and that the distance between the elevator and force plate is small enough that gravitational forces may be neglected.

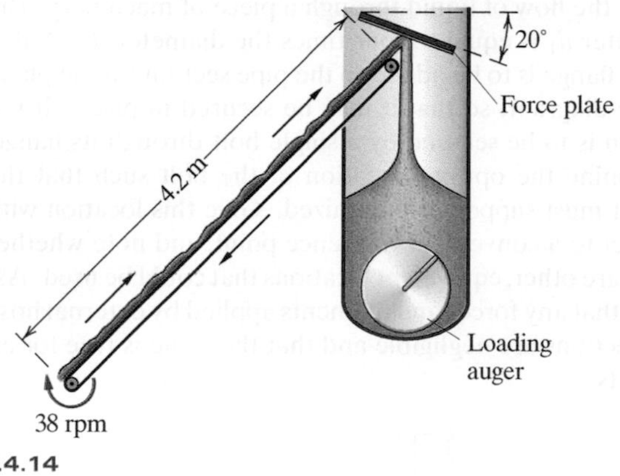

E5.4.14

5.4.15. **[Level 2]** Not long ago there was a popular yard toy that used a novel means of showering a play area with water. When the faucet was turned on, a stream of water shot from the base (a clown's head) and lifted a 0.18 kg hollow conical shell (the clown's hat) into the air. The diameter of the base nozzle is $d = 0.90$ cm. If the volume flow rate is $Q = 0.52$ l/s, and if the water is assumed to deflect radially at an angle $\theta = 25°$ and speed $v_2 = 0.4v_1$, what is the steady height h between the tip of the hat and the base? (Any changes in the diameter of the rising stream are negligible.)

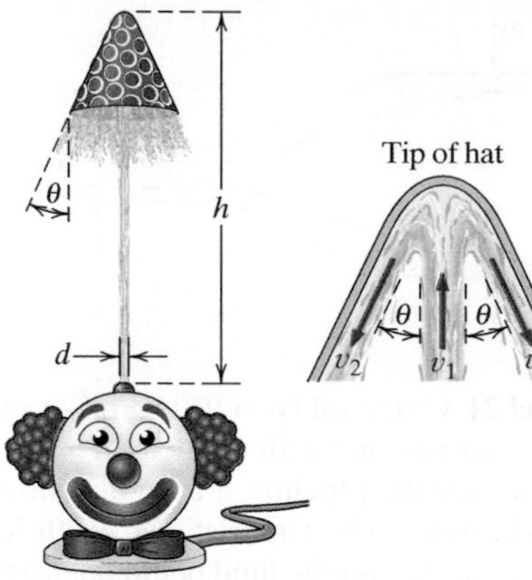

E5.4.15

5.4.16. **[Level 2]** A 65 kg snowboarder (including gear) is gliding down an 8° slope when she brings herself to a stop by turning both feet perpendicular to her motion and "plowing" loose, unpacked snow ($\rho = 140$ kg/m^3) with the long edge of her snowboard. The effective contact length of the board is 120 cm, and it carves a 7.5 cm layer of snow from the surface as she skids to a halt. The carving action of the board's edge generates an additional 130 N of resistive force per unit length (meters) of the contact edge. If her initial velocity is $v = 54$ km/h, how far does she travel before coming to rest? Assume that the velocity component of the plowed snow, in the direction parallel to the slope, is equal to that of the snowboard at the instant the snow is carved.

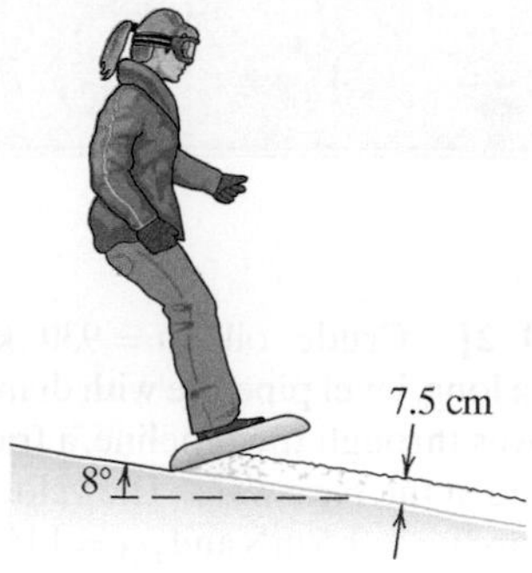

E5.4.16

5.4.17. **[Level 2]** Water is shown flowing through a gradual constriction in a pipe. The upstream diameter of the pipe is $d_1 = 4$ cm, and the downstream diameter is $d_2 = 2$ cm. The pressure drop across the contraction is $\Delta P = 50$ kPa, and the flow rate is $Q = 3$ l/s. A pressure gauge measures a pressure downstream from the contraction of $P_2 = 100$ kPa. What force does the water exert on the area of pipe contraction?

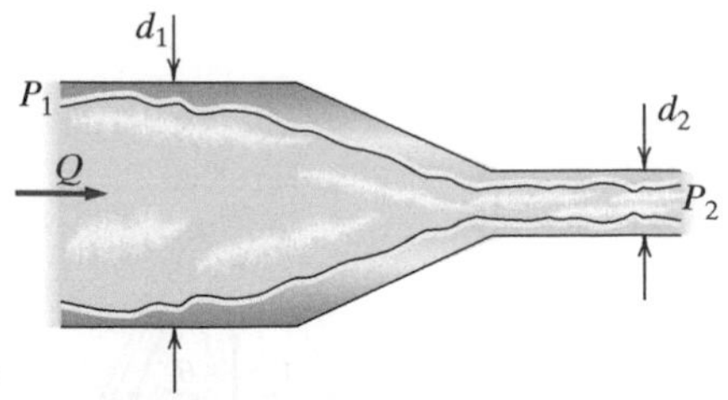

E5.4.17

5.4.18. **[Level 2]** Crude oil ($\rho = 930$ kg/m^3) is pumped through a long, level pipeline with diameter $d = 1.0$ m. As the oil moves through the pipeline, a frictional interaction is developed at the pipe walls. The frictional coefficient, expressing the force exerted on the fluid per unit length of the pipeline, is $\mu = 45$ N/m. Neglect gravity and assume that all variables have constant, uniform values at any given cross section.

a. Draw a free-body diagram of the fluid volume between points A and B and label all forces acting on this volume.

b. If the flow velocity is $v_A = 2.8$ m/s at cross section A, determine the velocity v_B and mass flow rate $\dot{m}_B$ at B.

c. What is the net force acting on this fluid volume in the direction parallel to the length of pipe (the horizontal direction)? If necessary, reexamine your diagram and answers in **a** and **b**.

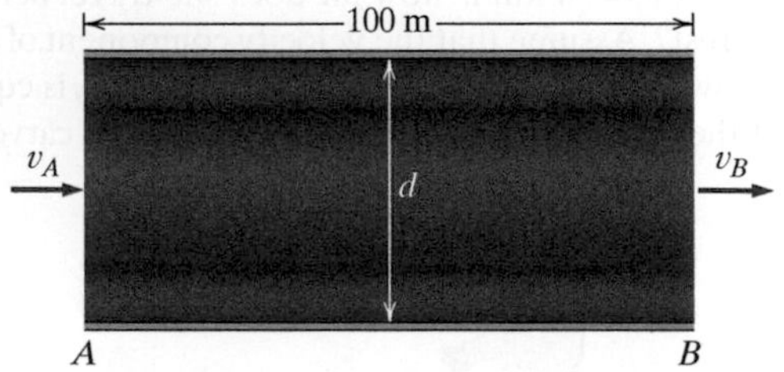

E5.4.18

5.4.19. **[Level 2]** Crude oil ($\rho = 930$ kg/m^3) flows steadily through a long, level pipeline with diameter $d = 1.0$ m. As the oil moves through the pipeline, a frictional interaction is developed at the pipe walls. The velocity and pressure at section A are $v_A = 3.5$ m/s and $p_A = 144.5$ kPa. If the pressure 100 m down the pipe, at section B, is $p_B = 138.0$ kPa, determine the frictional force exerted on the oil per unit length (meters) of the pipeline. Neglect gravity and any variation in flow velocity over a cross section.

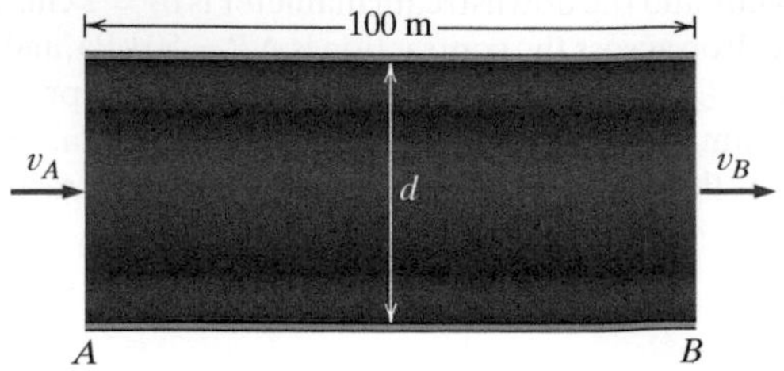

E5.4.19

5.4.20. **[Level 2]** This pipe section is used to alter and divert the flow of liquid through a piece of machinery. The diameter d_1 is equal to four times the diameter d_2. A flat, metal flange is to be added to the pipe section (in the plane of the diagram) so that it may be secured in place. If the section is to be secured by a single bolt through its flange, determine the optimal location of the bolt such that the load it must support is minimized. Give this location with respect to a convenient reference point, and note whether there are other equivalent locations that could be used. Assume that any forces and moments applied by external hose connections are negligible and that the same is true for all weights.

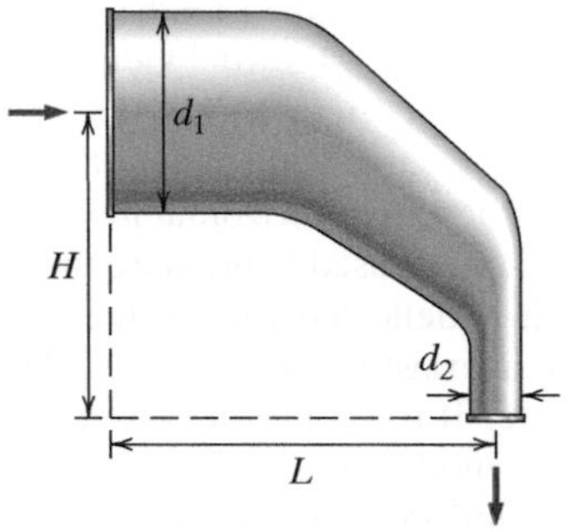

E5.4.20

5.4.21. **[Level 2]** This pipe section is used to alter and divert the flow of liquid through a piece of machinery. The diameter d_1 is equal to four times the diameter d_2. A flat, metal flange is to be added to the pipe section (in the plane of the diagram) so that it may be secured in place. If the section is to be secured by a single bolt through its flange, determine the optimal location of the bolt such that the load it must support is minimized. Give this location with respect to a convenient reference point, and note whether there are other, equivalent locations that could be used. Assume that any forces and moments applied by external hose connections are negligible and that the same is true for all weights.

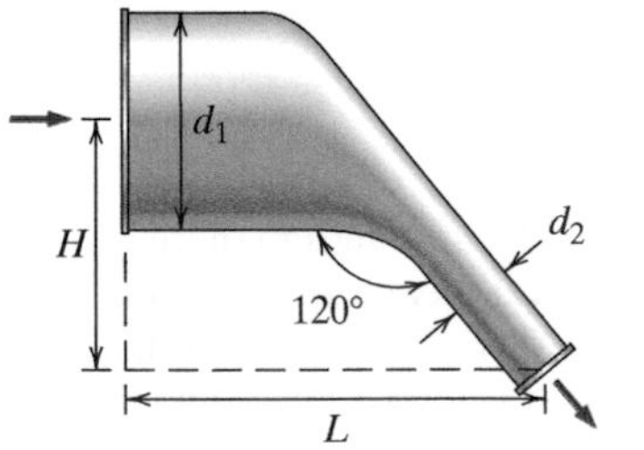

E5.4.21

5.4.22. **[Level 2]** A heavy metal plate of mass $M = 20$ kg and length $L = 40$ cm is supported from above by means of a frictionless, horizontal hinge. A horizontal jet of water is directed at the plate and strikes it at a vertical distance d from the center of the hinge. If $d = 8.0$ cm, and the jet velocity and mass flow rate are $v = 32$ m/s and $\dot{m} = 12$ kg/s, calculate the angle θ that the plate forms with the vertical, once a steady position has been established. If $\dot{m}$ is increased to 24 kg/s, what can you conclude about the new position of the plate?

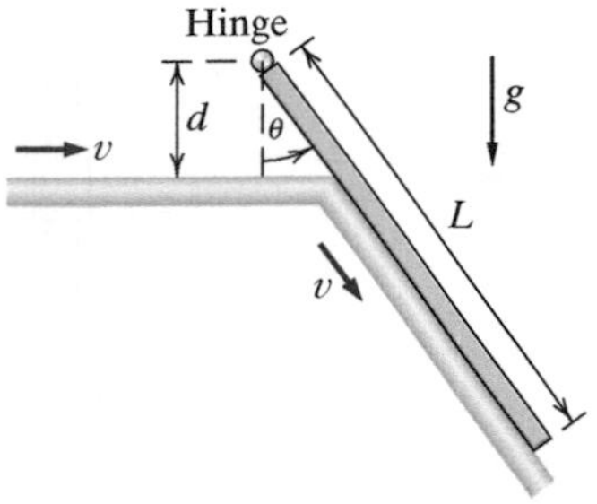

E5.4.22

5.4.23. **[Level 2]** A pressurized water spray is used to clean a hard, flat surface. The spray is emitted with a nozzle velocity of 60 m/s through an area $A = 6.7$ mm^2 in the pattern shown in the figure (a 5 mm thick circular wedge, spanning 75°). If the maximum safe pressure to avoid surface damage is 23 kPa, what is the minimum distance d that the nozzle should be held from the surface? Assume that the water is evenly distributed over the dispersal area and that the deflected velocity is negligible.

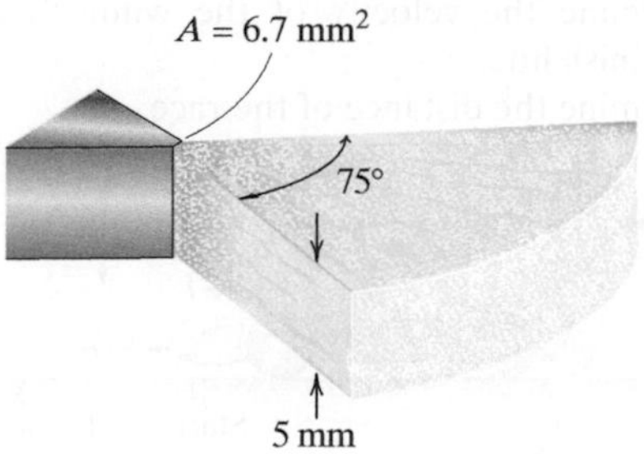

E5.4.23

5.4.24. **[Level 2]** A new turbojet engine is being tested in the laboratory. The air intake is 74.8 kg/s through a cross-sectional area $A = 6720$ cm^2, with a (gage) pressure $p = -5.2$ kPa over the intake area. The exhaust velocity of the fuel–air mixture is $v_e = 1180$ m/s at atmospheric pressure. The measured forward thrust of the engine is 86.46 kN. What is the fuel-to-air (mass) ratio of the exhaust? Assume the density of air drawn into the engine is 1.18 kg/m^3.

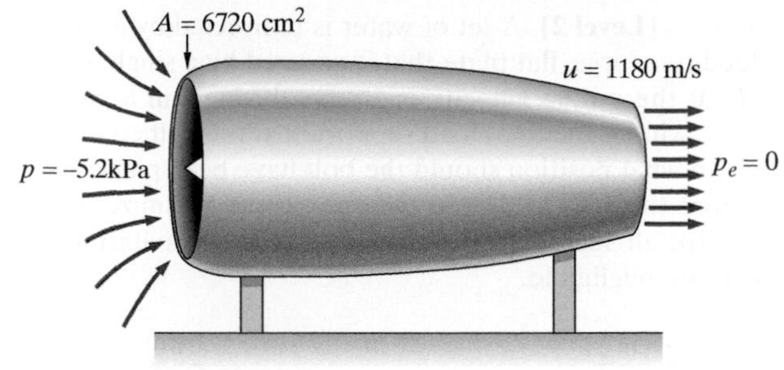

E5.4.24

5.4.25. **[Level 2]** Water flows into a circular pipe of radius $R = 10.0$ cm. The inlet velocity v_1 is uniform across the entrance cross section. The velocity profile at a downstream cross section is a function of radius $v_2(r) = c(1 - r^2/R^2)$, where $c = 5.0$ m/s. Neglect frictional resistance.

a. Determine the uniform inlet velocity v_1 and the mass flow rate $\dot{m}_2$. (*Hint:* Let $dA = 2\pi r\,dr$.)

b. The net force acting on the fluid volume between points 1 and 2 is $F = p_1A_1 - p_2A_2$, where p and A are the corresponding pressures and areas. Determine the pressure drop $\Delta p = p_1 - p_2$. (*Hint:* Modify a familiar equation so that it is expressed in terms of A; then follow a procedure similar to that used in **a.**)

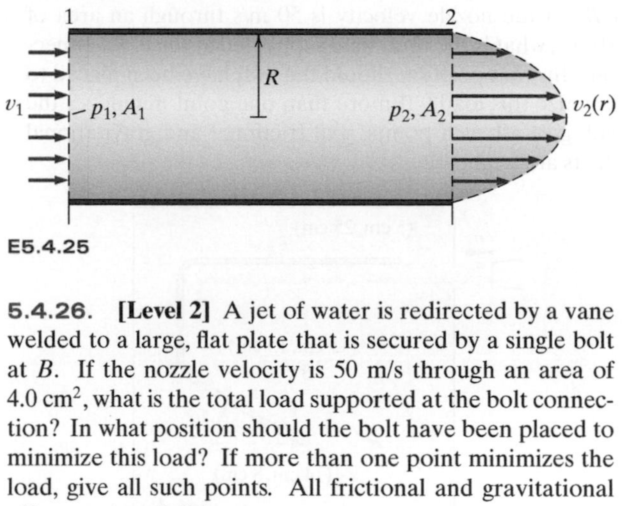

E5.4.25

5.4.26. **[Level 2]** A jet of water is redirected by a vane welded to a large, flat plate that is secured by a single bolt at B. If the nozzle velocity is 50 m/s through an area of 4.0 cm^2, what is the total load supported at the bolt connection? In what position should the bolt have been placed to minimize this load? If more than one point minimizes the load, give all such points. All frictional and gravitational effects are negligible.

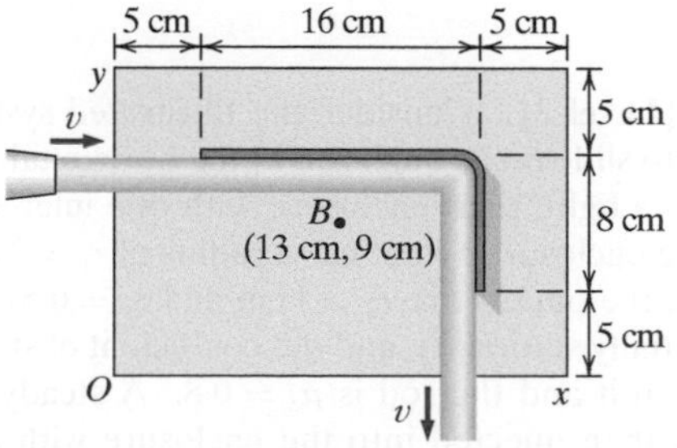

E5.4.26

5.4.27. **[Level 2]** A jet of water is redirected by a vane welded to a large, flat plate that is secured by a single bolt at B. If the nozzle velocity is 50 m/s through an area of 4.0 cm^2, what is the total load supported at the bolt connection? In what position should the bolt have been placed to minimize this load? If more than one point minimizes the load, give all such points. All frictional and gravitational effects are negligible.

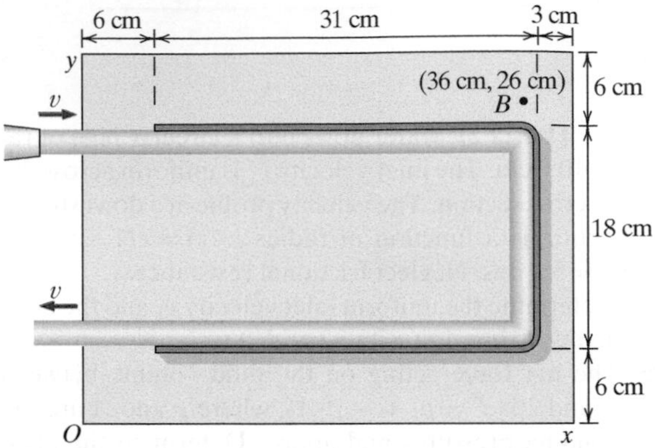

E5.4.27

5.4.28. **[Level 2]** A jet of water is redirected by a vane welded to a large, flat plate that is secured by a single bolt at B. If the nozzle velocity is 50 m/s through an area of 4.0 cm^2, what is the total load supported at the bolt connection? In what position should the bolt have been placed to minimize this load? If more than one point minimizes the load, give all such points. All frictional and gravitational effects are negligible.

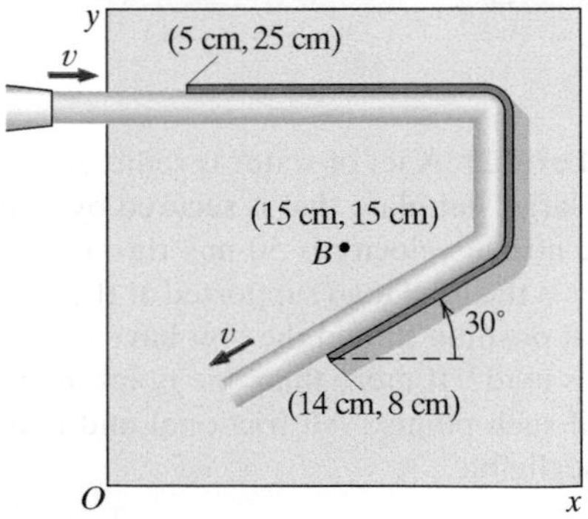

E5.4.28

5.4.29. **[Level 3]** Consider the illustrated system. A collar free to slide on a rough rod in the horizontal plane is attached to a light, rigid enclosure with one inlet and two outlets. The enclosure's inlet has a radius of $r_1 = 2$ cm, and the radii of the outlets are $r_2 = 1$ cm and $r_3 = 0.5$ cm. The collar is initially stationary, and the coefficient of static friction between it and the rod is $\mu_s = 0.8$. A steady stream of water is then injected into the enclosure with an inlet speed of $v_1 = 10$ m/s. It is observed that water exits through one outlet at a speed $v_2 = 30$ m/s. The outlets are angled at $\theta = 60°$ to the horizontal. Does the stream of water cause the collar to slip? If not, what would μ_s need to be for it to do so?

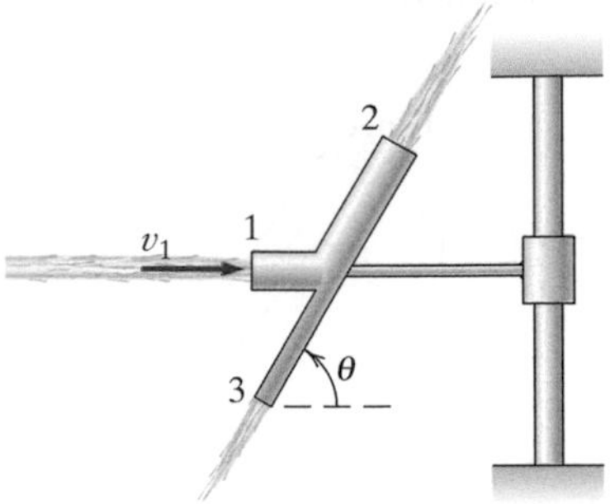

E5.4.29

5.4.30. **[Level 3]** Guests at a carnival game race small model cars using pressurized water guns to propel them forward. When properly aimed, the jets impact the cars horizontally and are deflected downward by a $\theta = 20°$ vane on the rear of the car. Each car has a mass of 3.2 kg and is at rest when the race begins. The flow rate of the water jet is $Q = 0.06$ l/s, and the nozzle exit velocity is 38 m/s. All frictional and gravitational effects are negligible. The winner maintained perfect aim throughout the race, and the winning time was 3.9 s.

a. Determine the initial acceleration of the car.

b. Determine the acceleration when $v_{car} = 1.0$ m/s.

c. Determine the velocity of the winner's car as it crossed the finish line.

d. Determine the distance of the race.

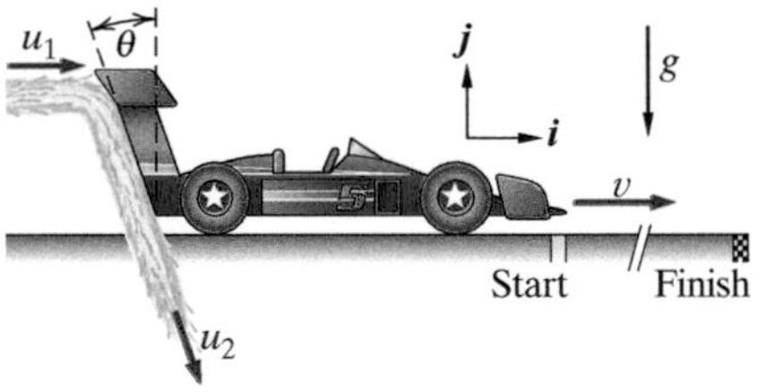

E5.4.30

5.4.31. **[Level 3]** Water is discharged through a lawn sprinkler at a volume flow rate $Q = 34$ L/min. The nozzle area of each sprinkler head is $A = 25.8\ mm^2$, and each nozzle is tilted up at a 40° angle to the horizontal. Each sprinkler arm has length $l = 152$ mm. If $\omega = 10$ rad/s during normal operation, what is the torsional friction T_f in the sprinkler's shaft, and what is the relative velocity of the emitted water with respect to the ground? If tinkering with the sprinkler reduces the friction by 80%, what are the new angular velocity ω and the new velocity of the emitted spray (and what has happened to the coverage area)? Neglect the length of the nozzle regions and assume that the water is ejected with no radial velocity component.

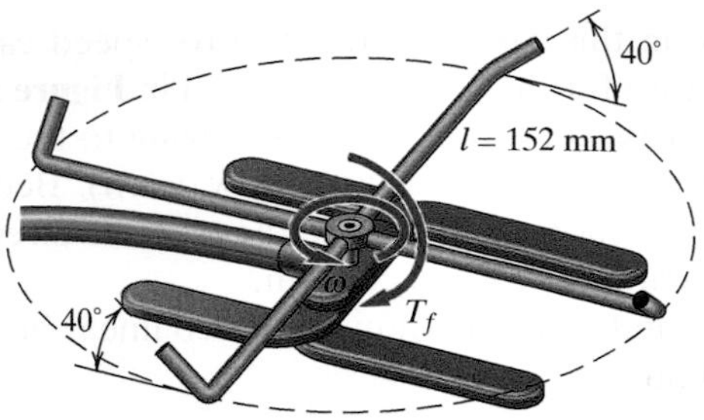

E5.4.31

5.5 NONCONSTANT MASS SYSTEMS

In the previous section we identified the forces and moments that act on a system that has a constant mass (the enclosure and the fluid inside of it) but has a continuous inflow and outflow of liquid. Now we'll look at a slightly different problem: one in which the mass of our system actually changes over time. What could this apply to? Think of the classic arcade game *Ms. Pacman*. Ms. Pacman runs along the corridors, chomping up dots as she goes. I would have to assume that she's gaining mass (although I notice that she never gets bigger). What, you want a more realistic example? Okay, how about a rocket? It moves because it's expelling exhaust gases out from its bottom and the reaction pushes the rocket up. And then there's the ram scoop as used on *Star Trek*. You open it up and suck in interstellar gas as the *Enterprise* moves through it. Very useful device.

Now that you're convinced of the real-world and cinematic advantages of systems that change in mass, let's begin analyzing them. First, we'll examine a rocket, shown in **Figure 5.27**, and will apply a momentum approach in determining the relevant equation of motion. The analysis is relatively straightforward. We'll start with the rocket's momentum at a particular time t, add what happens from t to $t + \Delta t$ due to any applied impulse, and end with the final momentum at $t + \Delta t$. By taking the difference of the final and initial momentum, dividing by Δt, and taking the limit as Δt goes to zero, we will capture the correct equation of motion. This approach will be referred to as a **lumped-mass momentum analysis**.

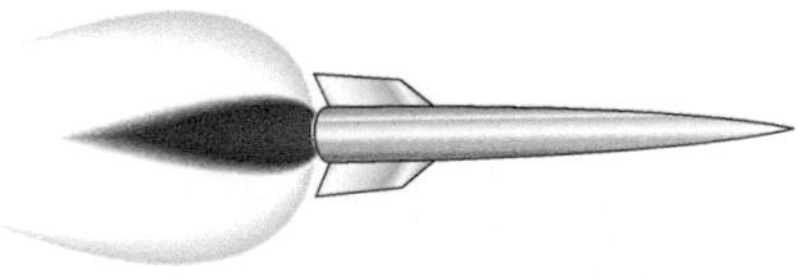

Figure 5.27 Rocket expelling mass

The leftmost element of **Figure 5.28** depicts our enclosure (the rocket body, m_E) and the enclosed fluid (the liquid fuel within the rocket, m_F), both traveling at a speed v. These represent an initial momentum of $(m_E + m_F)v$.

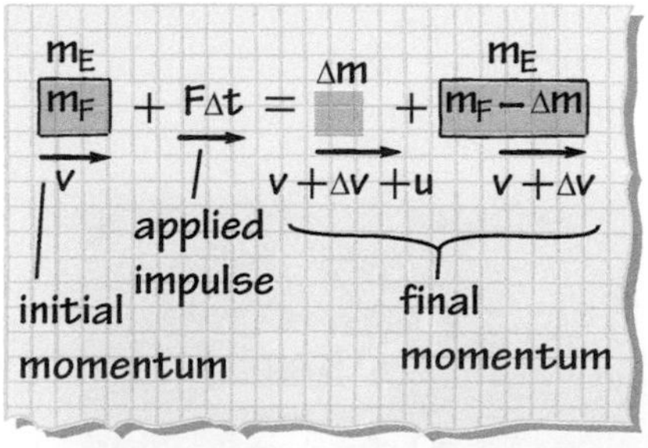

Figure 5.28 Lumped-mass momentum analysis

Moving to the right in **Figure 5.28**, we see the applied linear impulse due to the external forces, $F\Delta t$. To the right of the equals sign we have two bodies. The first is the mass that's been expelled from the rocket, labeled Δm. To the right of it is the rocket, which now comprises the enclosure and a bit less fluid $(m_F - \Delta m)$ due to some having just been ejected. The main body of the rocket, having experienced a change in speed due to the applied forces and the expelled fluid, is now traveling at a new speed $v + \Delta v$. Note that the change in speed may be positive or negative, depending on whether the external forces were acting in the direction of travel or against it. Finally, the ejected mass is moving at a speed equal to the rocket's *plus* the relative speed at which it was ejected from the rocket. Be aware in this case that the relative speed can be negative (this would be the case for the rocket illustrated in **Figure 5.27**) or positive (as would be the case for a rocket that's moving to the right and firing a braking thruster to the right in order to slow down). Because we're only analyzing horizontal motion, we'll take the positive direction to the right and not bother writing in $\boldsymbol{i}$ for each term.

Expressing **Figure 5.28** as "initial momentum + applied linear impulse = final momentum," we obtain

$$(m_E + m_F)v + F\Delta t = \Delta m(v + \Delta v + u) + (m_E + m_F - \Delta m)(v + \Delta v)$$

Canceling the $(m_E + m_F)v$ term on each side leaves us with

$$F\Delta t = \Delta m(v + u) + \Delta m \Delta v + (m_E + m_F)\Delta v - \Delta m \Delta v - \Delta m v$$

The two Δmv terms and the two $\Delta m\Delta v$ terms cancel, giving us

$$F\Delta t = \Delta m u + (m_E + m_F)\Delta v$$

and dividing by Δt then gives us

$$F = \frac{\Delta m}{\Delta t}u + (m_E + m_F)\frac{\Delta v}{\Delta t}$$

If we now take the limit as Δt goes to zero, the $\frac{\Delta v}{\Delta t}$ terms go to $\dot{v}$, $\frac{\Delta m}{\Delta t}$ goes to $\dot{m}$, and we can rename $m_E + m_F$ as m, the total mass of the moving body. This leaves us with

$$F = m\dot{v} + \dot{m}u \tag{5.36}$$

Does this make sense? Let's think of some particular cases. Assume that the rocket isn't expelling any fluid but is being acted on by an external force F. Then we would expect our old friend $F = ma$, and that's what we get: $F = m\dot{v}$. Next, assume there are no external forces. We can rearrange (5.36) to give us

$$m\dot{v} = -\dot{m}u$$

Let's assume that the fluid is being ejected from the back of the rocket, as in **Figure 5.27**. This means that u is negative. If it weren't, the fluid couldn't get out of the rocket body. So what we have is an equation that tells us that $m\dot{v}$ for the rocket is equal to a positive number, implying that the rocket is accelerating to the right. That's good; it's what we expect to see happen.

Figure 5.29 Spaceship firing braking thrusters

Now what if u were positive? That would mean the fluid is being ejected from the front of the rocket, as in the case shown in **Figure 5.29**, in which a spaceship traveling to the right is firing braking thrusters (also to the right) in order to slow down. We still have

$$m\dot{v} = -\dot{m}u$$

but now u is positive, not negative. Thus our $m\dot{v}$ is equal to a negative number, implying deceleration.

So, as we can see, all the physical scenarios match what our equations give us, and thus we are now free to examine the opposite case. Rather than ejecting material from the rocket, we will allow material to be taken in, thereby increasing the rocket's mass. **Figure 5.30** shows our rocket, again traveling to the right, along with a chunk of mass Δm. These then meet up and join, which they can do in two ways. The first is if the rocket is hit from behind by a stream of mass that's traveling faster than the rocket. The other possibility is that the rocket catches up to and engulfs a stream of mass that's moving more slowly than the rocket. The only difference in our equations will be the sign of u, the relative speed between the chunk of mass and the rocket.

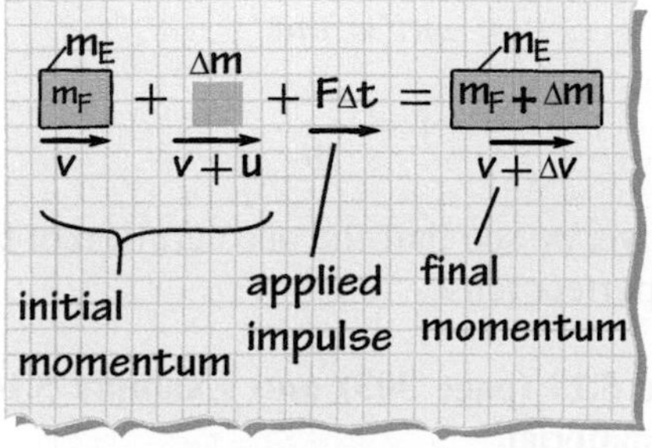

Figure 5.30 Body taking in mass

In words, **Figure 5.30** is saying "initial linear momentum of both rocket and mass chunk + applied linear impulse = final linear momentum of the rocket/mass chunk." In equation form this is

$$(m_E + m_F)v + \Delta m(v + u) + F\Delta t = (m_E + m_F + \Delta m)(v + \Delta v)$$

Canceling terms and dividing by Δt give us

$$\frac{\Delta m}{\Delta t}u + F = (m_E + m_F)\frac{\Delta v}{\Delta t} + \Delta m\frac{\Delta v}{\Delta t}$$

In the limit of $\Delta t \to 0$, the last term goes to zero (clearly, Δm goes to zero as Δt goes to zero), and, again labeling $m_E + m_F$ as m, we're left with

$$\dot{m}u + F = m\dot{v}$$

Putting this into the same order as (5.36) produces

$$F = m\dot{v} - \dot{m}u \tag{5.37}$$

As you can see, the only difference between (5.36) and (5.37) is the sign on the $\dot{m}u$ term.

Let's do a mental "makes sense" test on (5.37) just as we did for (5.36). We will assume that the enclosure is traveling to the right at a constant speed and engulfing mass; thus u is negative. The constant speed means that $\dot{v}$ is zero, and so we're left with

$$F = -\dot{m}u$$

Realizing that u being negative means that F is positive, we need to exert an external force to the right on the enclosure to keep it moving to the right. This certainly makes sense, especially for those who live in snowy climes. Every winter snow has to be shoveled from the sidewalk, and as those who do it can testify, moving the snow shovel through the snow requires a strong force in the direction in which the shovel is being pushed.

Let's keep the snow shovel analogy and this time assume that the shovel has been shoved forward into the snow and then released. For this case the force F is zero and our equation of motion gives us

$$m\dot{v} = \dot{m}u$$

$$\dot{v} = \frac{\dot{m}u}{m}$$

Since u is still negative, we see that the acceleration of the shovel is negative, which makes perfect sense.

>>> **Check out Example 5.13 (page 303) and Example 5.14 (page 304) for applications of this material.**

EXAMPLE 5.13 MOTION OF A TOY ROCKET (Theory on page 300)

Figure 5.31 shows a classic toy—the bottle rocket. The rocket is filled with water and compressed air. Once released, the water shoots out the bottom and the rocket takes off. Calculate the rocket's vertical acceleration, accounting for gravity and ignoring aerodynamic drag, and then plot the rocket's vertical position as a function of time from launch to when it runs out of water. The rocket body has a mass of 0.06 kg and an initial load of 0.15 l of water. The exit nozzle has a cross-sectional area of 9 mm^2, and water flows out at a steady rate of 0.2 l/s.

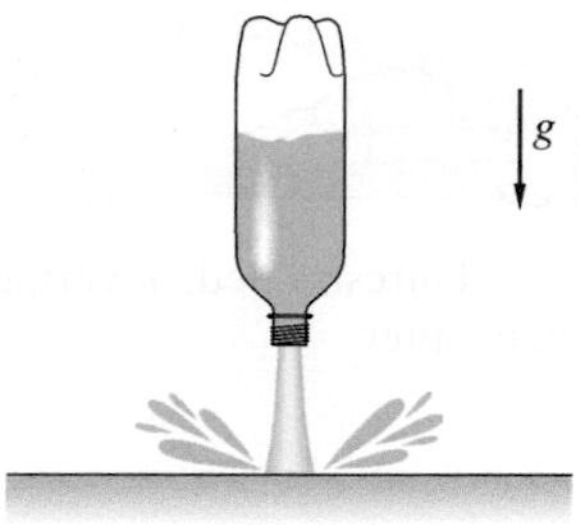

Figure 5.31 Bottle rocket

Goal Plot the rocket's vertical position as a function of time.

Assume We will ignore air drag.

Draw **Figure 5.32** shows our system. All we have to resist the upward reaction from the expelled water is the force of gravity.

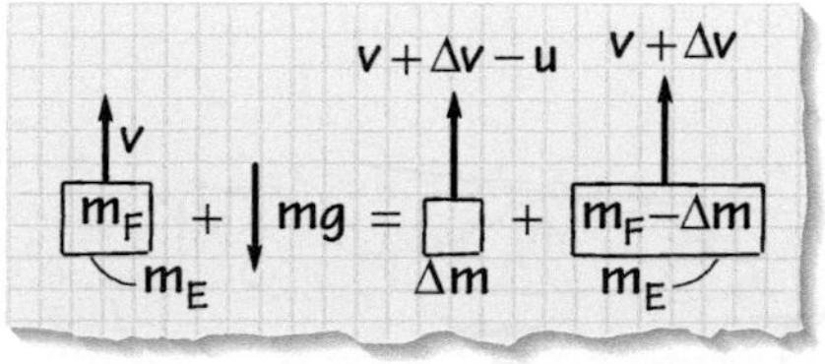

Figure 5.32 Lumped-mass momentum diagram

Formulate Equations Note in **Figure 5.32** that u has a negative sign in front of it. This is intentional so that you can see how easy it is to set up your own lumped-mass momentum analysis. The reason for the negative sign is that we are going to exploit the system's physical setup. We know that the water is ejected backward from the rocket and so we'll use that from the start. A positive u indicates that the water is directed downward. The fluid has a density ρ and is flowing through a cross-sectional area A at a speed u. Thus we have $\dot{m} = \rho u A$. Recalling that 1 kg of water takes up 1 liter of volume, lets us re-express the volumetric flow rate of 0.2 l/s as a mass flow rate of 0.2 kg/s. Taken together, this all means that

$$u = \frac{\dot{m}}{\rho A} = \frac{0.2\,\text{kg/s}}{(1000\,\text{kg/m}^3)(9\times 10^{-6}\,\text{m}^2)} = 22.2\,\text{m/s}$$

The mass of the rocket and water is given by

$$m = (0.15 + 0.06)\,\text{kg} - (0.2\,\text{l/s})(1\,\text{kg/l})t = [0.21\,\text{kg} - (0.2\,\text{kg/s})t]\,\text{kg}$$

Solve **Figure 5.32** in equation form is

$$(m_F + m_E)v - mg\Delta t = \Delta m(v + \Delta v - u) + (m_F - \Delta m + m_E)(v + \Delta v)$$

Canceling terms, dividing by Δt, and taking the limit as Δt goes to zero give us

$$-mg = -\dot{m}u + m\dot{v} \quad \Rightarrow \quad \dot{v} = \frac{\dot{m}u}{m} - g$$

with m, $\dot{m}$, and u given above.

The constant flow rate is 0.2 l/s and the rocket holds 0.15 l of water. Hence the total time of powered flight is $t_p = \frac{0.15\,\text{l}}{0.2\,\text{l/s}} = 0.75$ s. Integrating $\dot{v}$ on MATLAB for 0.75 s with initial conditions of $y(0) = 0, \dot{y}(0) = 0$ gives us the height versus time plot shown in **Figure 5.33**.

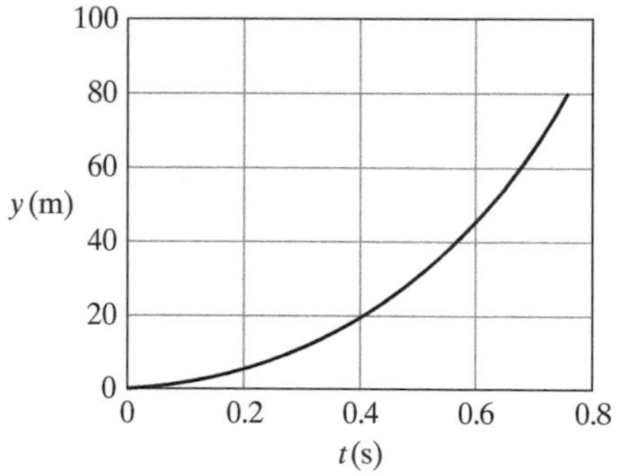

Figure 5.33 Height versus time for powered flight

EXAMPLE 5.14 HELICOPTER RESPONSE TO A STREAM OF BULLETS (Theory on page 300)

Figure 5.34 Unrestrained, hovering military helicopter

In Example 5.12 we found the forward thrust needed to keep the helicopter stationary. Now let's drop that restriction and determine the helicopter's (**Figure 5.34**) acceleration response to the bullet stream when there's no counterthrust applied.

Goal Determine the uncontrolled helicopter's acceleration.

Given Mass of helicopter, mass of bullets, rates of fire, and muzzle velocity.

Assume We'll again neglect the mass of exiting bullets in comparison with the total mass of the helicopter. The speed of the helicopter will be viewed as negligible compared to the muzzle speed of the bullets. Thus we'll take the muzzle velocity to be the same as the bullet's absolute velocity with respect to a ground frame.

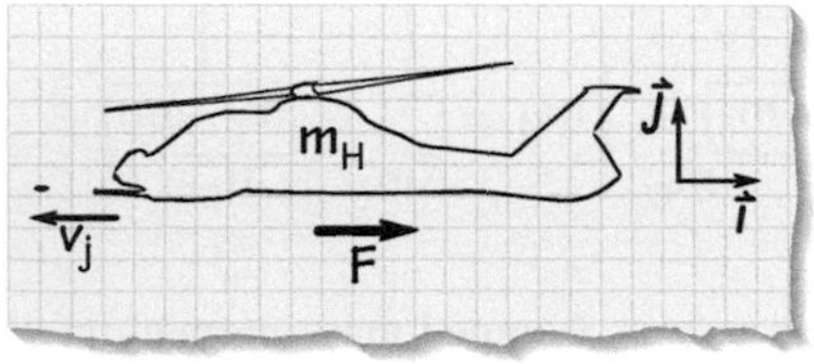

Figure 5.35 Schematic of hovering helicopter

Draw **Figure 5.35** shows our system.

Formulate Equations All we need is an application of (5.36):

$$F = m\dot{v} + \dot{m}u \tag{5.38}$$

Solve From Example 5.12 we know that the maximum rate of fire is $\dot{m} = 7.33$ kg/s. With no compensation applied, we set $F = 0$ in (5.38):

$$0 = m\dot{v} + \dot{m}u \quad \Rightarrow \quad \dot{v} = -\frac{\dot{m}u}{m} = \frac{8800\text{ N}}{6000\text{ kg}} = \boxed{1.5\text{ m/s}^2}$$

Check Did we get the signs right? Apparently so, because the bullets are being fired to the left and we would therefore expect the helicopter to be pushed to the right, as it was. Is the acceleration reasonable? At the highest firing rate we have an acceleration of $\frac{14.7\,\text{m/s}}{10\,\text{s}} = 1.47\,\text{m/s}^2$. This is about $0.15g$, or about a seventh of a g. Though certainly not negligible, it seems reasonable given the large amount of linear impulse being applied. Let's do a lumped analysis just to be sure. Rather than considering 4000 shots per minute at 0.11 kg per bullet, we will lump it into one huge bullet of $4000(0.11\,\text{kg}) = 440\,\text{kg}$. To launch this 440 kg bullet at 1200 m/s, we would need to apply a linear impulse of $(440\,\text{kg})(1200\,\text{m/s}) = 5.28 \times 10^5\,\text{kg·m/s}$. Each of these super-bullets is launched every minute, giving us an average applied force of $\frac{5.28\times10^5\,\text{kg·m/s}}{60\,\text{s}} = 8800\,\text{N}$. We can conclude that what matters for this example is the rate at which the total linear impulse is applied to the bullets, and whether we launch the bullets all at once or in many little shots seems to make no difference. Is this *really* true? Well, consider that we neglected the loss of mass due to the bullets being fired in our analyses. If we account for the time-varying mass of the helicopter over time, then we can expect to see some differences between the cases of many small bullets versus a few large ones.

EXERCISES 5.5

5.5.1. [Level 1] The figure shows a new design for an automated snow plow. The plow moves forward at a constant speed v and snow enters the plow at a mass rate of $\dot{m}$. Half the snow remains in the plow and the other half is ejected at a 45° angle from the plow. Upon ejection the snow is moving at a speed $2v$ relative the plow (clearly there's some machinery in the plow to do this). Determine what the force is that acts on the plow (supplied by the track in back of the plow) for it to move at a constant speed v. Consider only motion/force in the $\boldsymbol{i}$ direction.

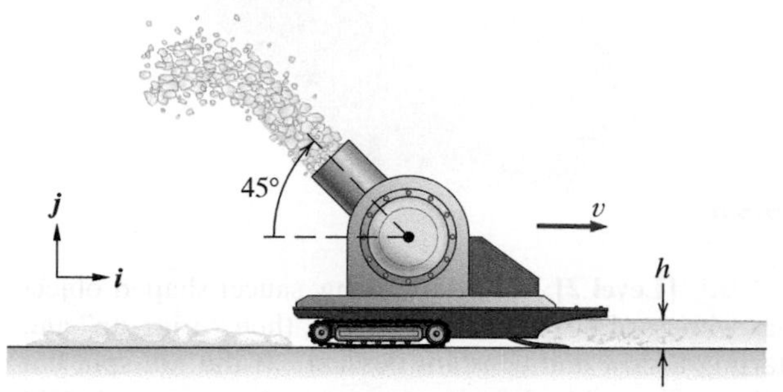

E5.5.1

5.5.2. [Level 2] At time $t = 0$ a sleigh is sliding to the right at a speed $v = 20$ m/s. A snow-gun, mounted on a motorized snow-wagon, is firing $m_b = 1$ kg snowballs at the sleigh with a muzzle speed (speed at which the snow exits the gun, taken with respect to the gun) of $v_m = 100$ m/s. It fires one snowball every 0.1 s. The snow-wagon is moving to the right at a constant speed of $v_w = 40$ m/s. Let the mass of the sleigh and whatever snow is within it at the start of our analysis be denoted m_s and assume a friction-free interface between sleigh and snow.

a. Determine the average rate at which mass is increasing in the sleigh.

b. Solve for the initial acceleration of the sleigh in the limit of infinitesimally small snowballs having the same mass flow as the average rate for part (b). Express in terms of $m_s, v, \dot{m}$, etc. Assume that the "snow-stream" already exists from snow-cannon to sleigh from the very start of the problem; i.e., there's no start-up transient as the first bit of snow travels from the snow-cannon to the sleigh.

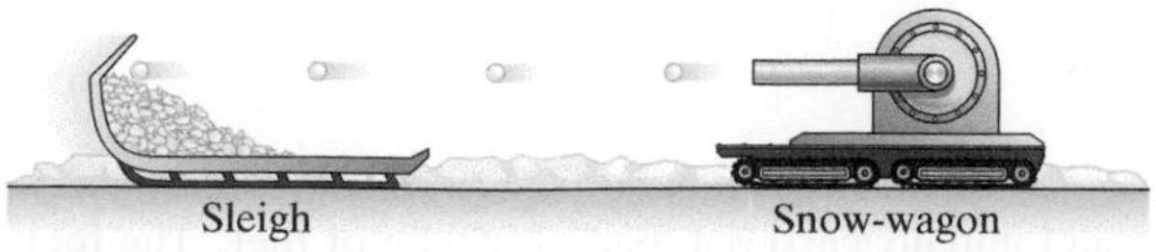

E5.5.2

5.5.3. [Level 2] Computational At time $t = 0$ a sleigh is sliding to the right at a speed $v = 20$ m/s. Assume that the mass of the sleigh and any snow within it is 120 kg at this instant. A snowthrower is projecting a continuous stream of snow to the left with a speed of $v_{st} = \left[100 - 20(1 - \cos[(1.0s^{-1})t]\right]$ m/s (i.e., the speed of the snow varies with time). The snow (with a density of 1000 kg/m^3) is projected through a nozzle with a cross-sectional area of 1 cm^2. Assume a friction-free interface between sleigh and snow.

Find the acceleration of the sleigh in terms of the system parameters and numerically integrate and plot the sleigh's position and speed for $0 \le t \le 3$ s.

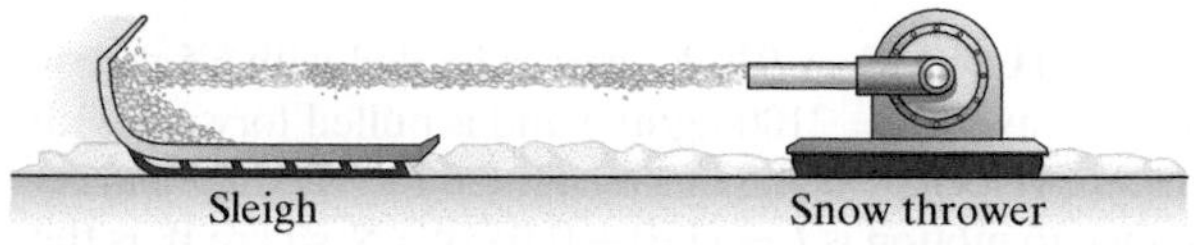

E5.5.3

5.5.4. [Level 2] The figure shows a cart being propelled to the right by a right-moving stream of fluid. Assume that the originating mass flow rate of the fluid stream is $Q = 20$ kg/s, it has a speed $u = 12$ m/s, and the cart's speed is given by v. At the given instant $v = 2$ m/s. All the fluid that enters the cart also leaves (zero mass accumulation) and the mass of the cart and on-board fluid is given by $m_C = 10$ kg. $\theta = 36.87°$, (3/4/5 triangle). Determine the total vertical force acting on the ground and the acceleration of the cart. Do this first in terms of the system variables (i.e., a general solution) and then evaluate for the given values.

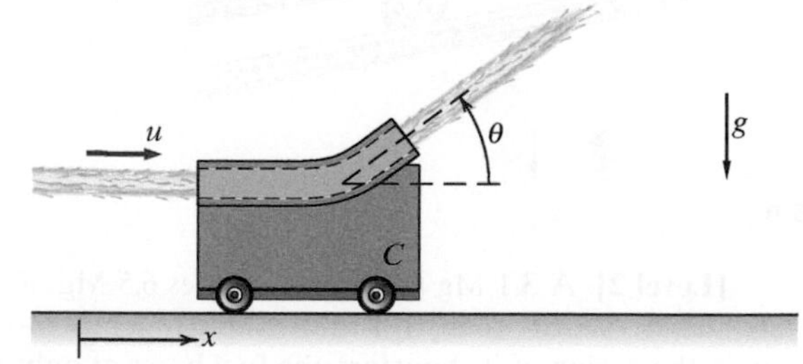

E5.5.4

5.5.5. [Level 2] At time $t = 0$ a sleigh is sliding to the right at a speed $v = 20$ m/s. It's encountering a stream of 1 kg snowballs being directed toward the sleigh at a speed of 100 m/s. Let the mass of the sleigh and whatever snow is within it at the start of our analysis be denoted m_s. The spacing between the snowballs in the stream is 10 m. Assume a friction-free interface between sleigh and snow.

a. Draw a lumped-mass momentum diagram for the sleigh/snowballs.

b. Determine the average rate at which mass is increasing in the sleigh.

c. Construct a lumped-mass momentum balance equation and solve for the acceleration of the sleigh in the limit of infinitesimally small snowballs having the same mass flow as the average rate for **b**. Express in terms of $m_s, v, \dot{m}$, etc.

Assume that the "snow-stream" already exists from snow-cannon to sleigh from the very start of the exercise; i.e., there's no start-up transient as the first bit of snow travels from the snow-cannon to the sleigh.

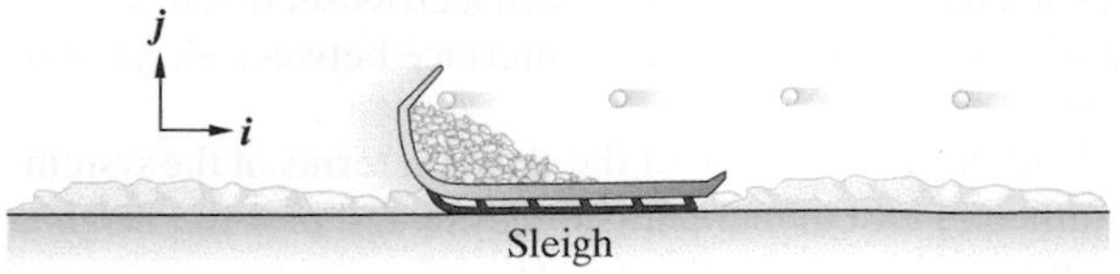

E5.5.5

5.5.6. [Level 2] A 930 kg cart is loaded with 6.5 metric tons of gravel ($\rho = 2100$ kg/m^3) and is pulled forward with constant velocity $v = 4.8$ m/s by a force F. The frictional resistance to motion is $f = (110 + 0.035W)$ N, where W is the total weight of the cart in newtons. The gravel is released through a vertical chute in the bottom of the cart at 0.62 m^3/s. Determine the force F after the chute has been open for 2.2 s.

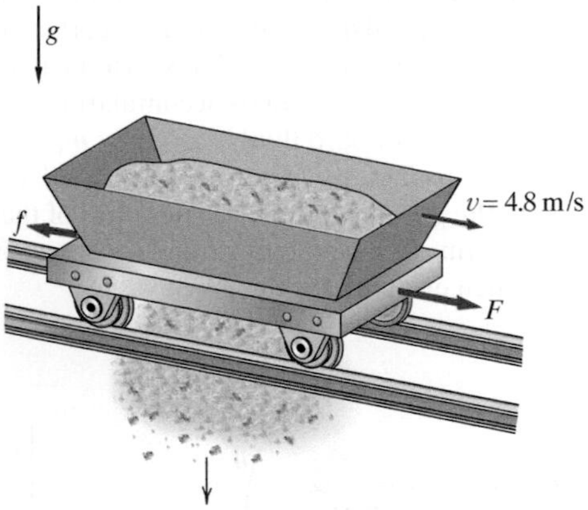

E5.5.6

5.5.7. [Level 2] A 3.1 Mg test rocket carries 6.5 Mg of fuel and stands vertical, ready for liftoff. Due to a miscalculation in the design of its boosters, the fuel burns at only 66 kg/s, and the exhaust velocity is a mere 450 m/s relative to the rocket. Will the rocket ever leave the ground? If so, at what time t will liftoff occur, and what maximum speed will be attained by the rocket? Assume gravitational acceleration is constant, $g = 9.81$ m/s^2.

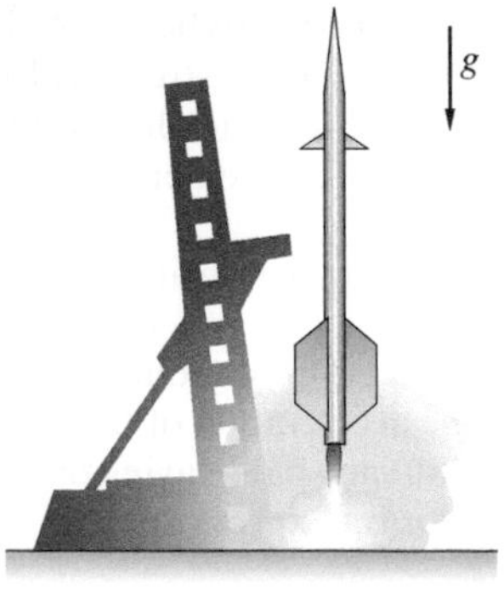

E5.5.7

5.5.8. [Level 2] A 1200 kg rocket is in deep space when its second-stage boosters are ignited. The boosters contain 300 kg of fuel, which is burned at a constant rate with an exhaust velocity of 1600 m/s relative to the rocket. What change in velocity will the rocket have experienced once all the fuel is expended?

E5.5.8

5.5.9. [Level 2] An odd-looking, saucer-shaped object was witnessed hovering above earth, then "firing up" and starting a slow and constant descent. It did not spin but rather appeared to draw its support from an array of eight "blasters" (rocket boosters) scattered about its perimeter. With an estimated weight of "a couple big rigs" (1 big rig $= 1.36 \times 10^4$ kg) and approximate fuel ejection speeds "like 10 big tornados" (1 big tornado = 193 km/h), what rate of fuel consumption, $\dot{m}(t)$, would have been required at each booster to maintain the saucer's constant rate of descent? From eyewitness descriptions of size, it is guessed that the booster fuel may have been 20% of the saucer's total initial mass. It has also been reported that the descent lasted at least five minutes. Is this possible under the above assumptions?

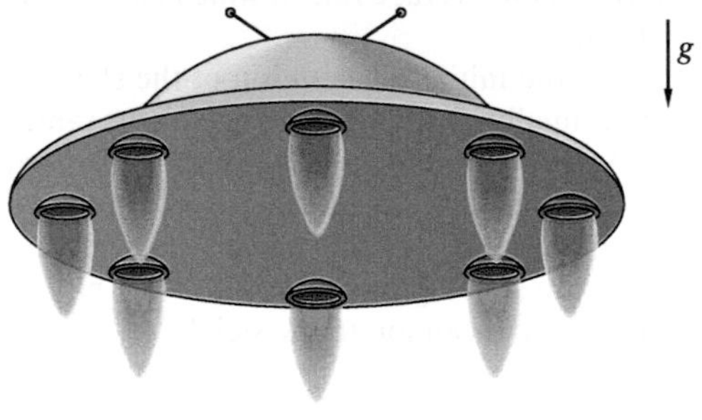

E5.5.9

5.5.10. [Level 2] A rocket sled has an empty mass of 2.5 Mg and initially carries a load of 850 kg of fuel. The fuel is consumed at a rate of 22 kg/s, with an exhaust velocity of 540 m/s relative to the sled. Write an expression for the sled's acceleration as a function of time, and determine its burnout velocity. Neglect all frictional effects.

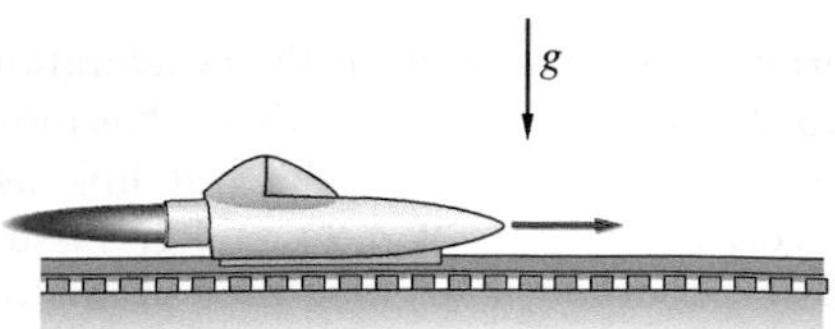

E5.5.10

5.5.11. **[Level 2]** A neatly coiled, inextensible string with linear density ρ lies on a smooth, horizontal surface. A force F is applied to the end of the string such that it unravels with constant velocity v. Determine the force F, the work W done by the force F, and the kinetic energy $\mathcal{KE}$ of the unraveled length of string. Can you account for any difference found between W and $\mathcal{KE}$?

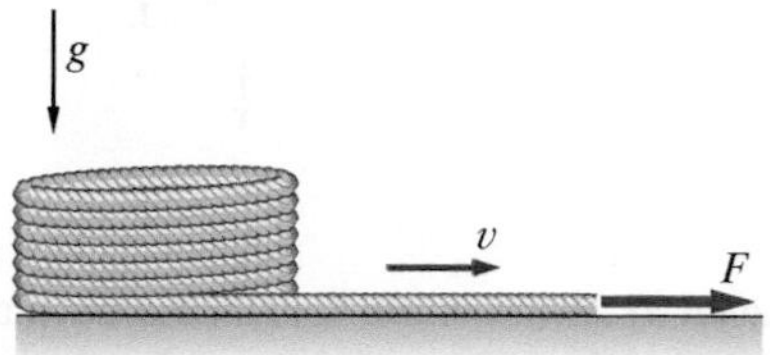

E5.5.11

5.5.12. **[Level 2]** Consider a small carriage that has been equipped with a winch and cable and glides freely on an inclined rail. The cable has been hooked into a solid wall at the end of the incline, and the winch motor draws cable either in or out to adjust the position of the carriage along the rail. The total length of the cable is L and the carriage is initially a distance d down the incline. Let ρ be the linear density of the cable, and M the total mass of the carriage and winch (excluding the cable). Determine the tension T in the cable that the winch should generate at the carriage if a constant acceleration a is desired in the direction up the incline. The carriage is initially at rest at time $t = 0$.

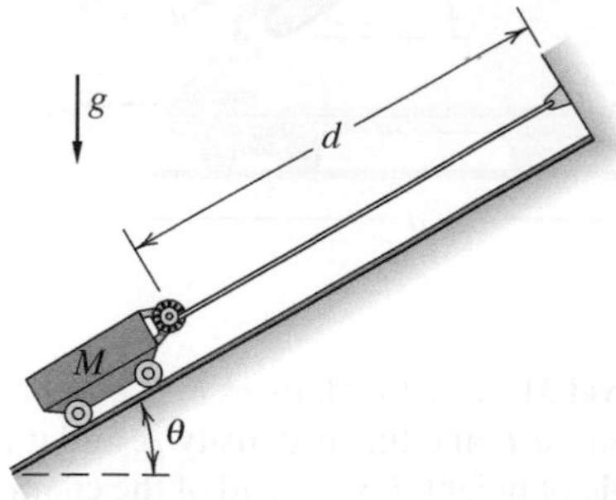

E5.5.12

5.5.13. **[Level 2]** The Bombardier CL-415 amphibious aircraft is used as an aerial firefighter to battle large-scale wildfires. The plane can hold 6.0 kL of water, and it takes just $\Delta t = 12$ s to collect this much while skimming the ocean's surface for $d = 410$ m at $v_p = 130$ km/h. The plane weighs $W_p = 1.28 \times 10^4$ kg without any water in its holding tank, and it is powered by two Pratt & Whitney turboprop engines with a maximum power output of 1.5 MW each. If the plane accelerates while collecting water to a final speed of $\overline{v}_p = 160$ km/h, what average power output is needed from each engine? Neglect any frictional effects, and assume that the water is still. Take the relative speed u between the water and plane to be the average of the plane's initial and final speeds.

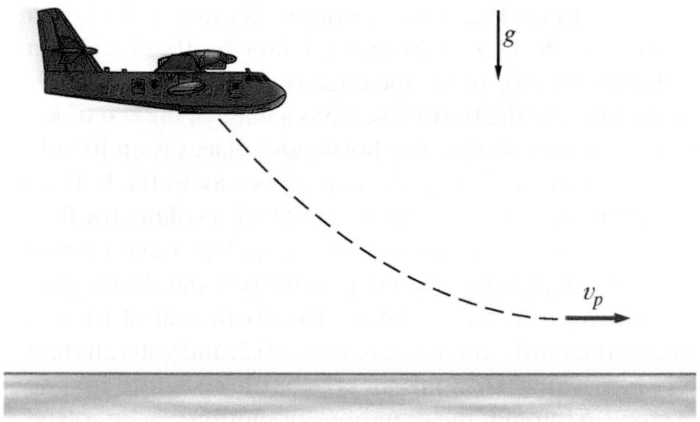

E5.5.13

5.5.14. **[Level 2]** A 3.5 kg rocket body, containing 1.5 kg of fuel, is attached to a light, rigid rod of length $R = 0.5$ m. The rod pivots freely in the horizontal plane (so you can neglect gravity) about the fixed point O. Assume that the rocket burns fuel at a rate $\dot{m} = 0.25$ kg/s, with an exhaust velocity $u = 20$ m/s relative to the rocket, and determine the angular velocity ω and angular acceleration α of the rocket, and the tension T in the rod at $t = 6$ s.

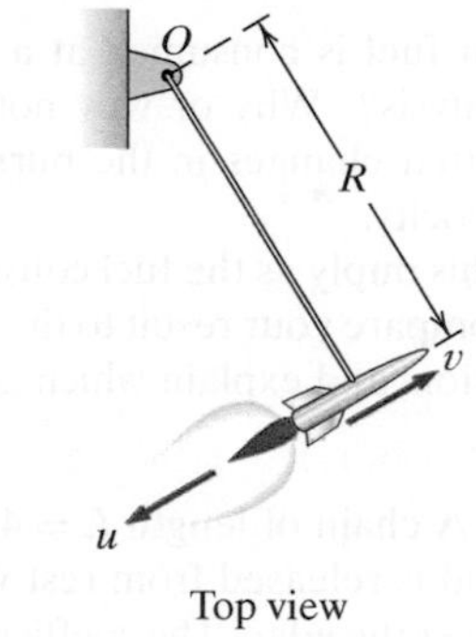

E5.5.14

5.5.15. **[Level 3]** A cart of mass $m_c = 0.1$ kg is filled with $m_w = 0.2$ kg of water and placed at the bottom of a frictionless ramp angled at $\theta = 20°$ with respect to the ground. Pressurized air is then injected into the cart, causing the water to gush out the back at a rate $\dot{m}_w = 0.5$ kg/s through an orifice of radius $r = 0.7$ cm. How fast is the cart moving when all of the water has been expelled, and how much farther from this point does the cart travel before coming to a stop? Solve the equations of motion analytically.

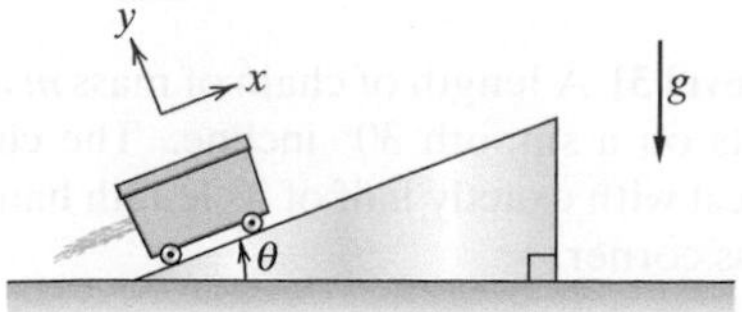

E5.5.15

5.5.16. **[Level 3]** Revisit all parts of **Exercise 5.5.7** and assume that the design miscalculations led to a greater exhaust velocity, at 710 m/s relative to the rocket, but a lesser burn rate, at 55 kg/s.

5.5.17. **[Level 3] Computational** Ryan is really bored one day and decides to propel a 1 liter bottle of soda on the skating surface of his local ice rink. The soda bottle is initially full, and the bottle itself has a mass of $m_b = 0.07$ kg. Ryan vigorously shakes the bottle and places it on its side on the ice surface. When the cap gives way to the bottle's internal pressure, the soda gushes out with a volumetric flow rate of $Q_s = 0.75$ l/s. The bottle's spout has a diameter of $d = 2.5$ cm. Just as the soda begins to spew out, Ryan gives the bottle a push at $v_0 = 3$ m/s. The coefficient of friction between the bottle and ice rink is $\mu = 0.2$, and you can treat the soda as water. How far does the bottle go before coming to a stop? Although the equations of motion can be solved analytically, use numerical integration.

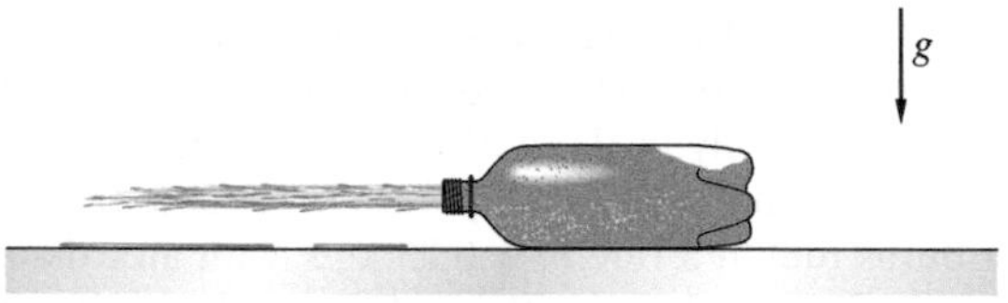

E5.5.17

5.5.18. **[Level 3]** Return to **Exercise 5.5.8** and answer the following:

a. Is the fact that fuel is consumed at a *constant* rate important in the analysis? Why or why not? (Consider $\dot{m} = f(t)$.) Assume that changes in the burn rate do not affect the exhaust velocity.

b. What should this imply as the fuel consumption time approaches zero? Compare your result to the analysis of an instantaneous explosion, and explain which analysis seems more realistic to you.

5.5.19. **[Level 3]** A chain of length $L = 4.0$ m lies on a horizontal surface and is released from rest with an initial length $L/8$ hanging over the edge. The coefficient of friction between the chain and the surface is $\mu = 0.12$. Determine the velocity of the chain as a function of the distance x that it has moved, and calculate this velocity at the moment when the last link just reaches the edge. Assume that a small, massless, toothed pulley exists at the corner, which takes the horizontally moving chain and directs it downward.

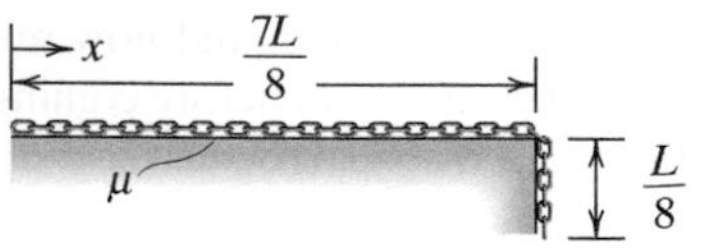

E5.5.19

5.5.20. **[Level 3]** A length of chain of mass m and linear density ρ rests on a smooth 30° incline. The chain is released from rest with exactly half of its length hanging over the frictionless corner.

a. Determine an expression for the acceleration of the chain, as a function of the distance x that it has moved.

b. Determine the velocity of the last link as it just reaches the corner, in terms of ρ, m, and g. Assume that a small, massless, toothed pulley is located at the top of the incline and redirects the chain that's moving up the slope straight down.

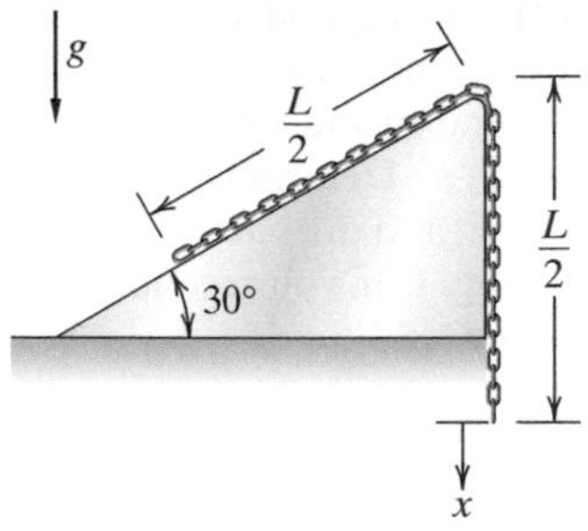

E5.5.20

5.5.21. **[Level 3]** A 35° chute loads coal onto a 1500 kg flatcar of length $d = 5$ m. The coal leaves the chute at a rate of 100 kg/s with a velocity $u = 4.0$ m/s. The initial velocity of the flatcar is $v_0 = 1.0$ m/s as it heads under the chute. Determine the velocity of the flatcar 3 s after loading begins and once it is fully loaded and past the chute. How much coal will the flatcar be carrying as it leaves the loading area? Neglect the rolling friction in the wheels and assume that any amount of coal that falls from the front or back of the car is negligible.

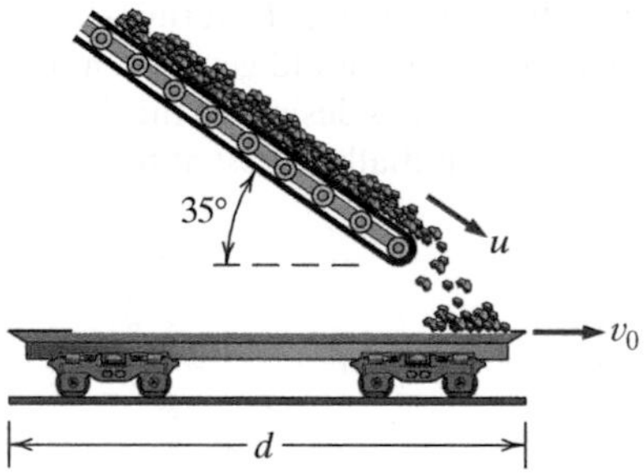

E5.5.21

5.5.22. **[Level 3]** A cart with mass M has been attached to a chain of length L and linear density ρ, and it rests on a frictionless table of height d. One end of the chain has been pinned to the underside of the cart, at a point directly below its center. The other end hangs through a smooth hole in the table surface, just barely in contact with the floor. The cart glides freely over the hole. If the system is released from rest, determine the velocity of the cart when it is at the position directly above the hole, and the speed at which it strikes a bumper a distance $(L - d)/2$ beyond this position. If the collision with the bumper is perfectly elastic, what will be the position of the cart the next time it comes to rest? $M = 20$ kg, $L = 5$ m, $d = 2$ m, and $\rho = 2$ kg/m.

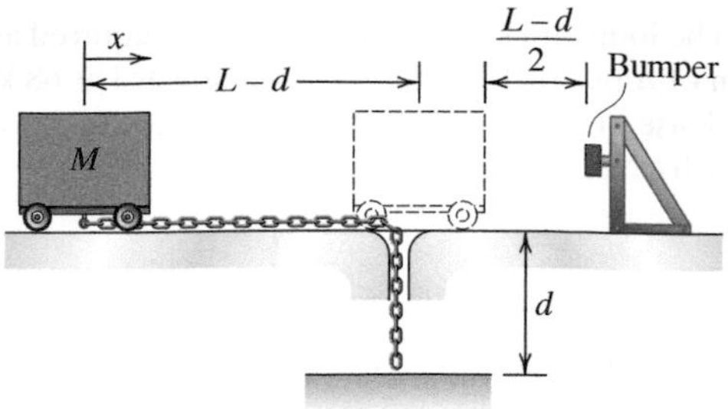

E5.5.22

5.5.23. **[Level 3] Computational** A mass M is connected to the end of a loose pile of chain and given an initial velocity v_0 across a smooth, horizontal surface. At the moment the mass is released, a length of chain L has been pulled from the pile and moves with the same velocity as the mass. The linear density of the chain is ρ.

a. Determine the velocity of the mass as a function of the distance x that it has traveled after release.

b. Computational Add a frictional coefficient $\mu = 0.15$ between the chain and surface, and allow the mass M to roll without friction. Let $L = 1.0$ m, $M = 10$ kg, $v_0 = 2$ m/s, and $\rho = 5$ kg/m. How far does the mass travel, and at what time t after the release does it come to rest? On a single graph, plot the velocity v and position x of the mass versus time. On a second graph, plot v versus x. Comment on your results and label any significant points in the motion.

c. As a final task, briefly discuss how the release distance L affects the stopping distance of the mass, assuming all other parameters remain fixed. If necessary, create some plots illustrating your conclusions. Compare this situation qualitatively to the scenario where the chain is initially at rest and then suddenly latches onto the moving mass.

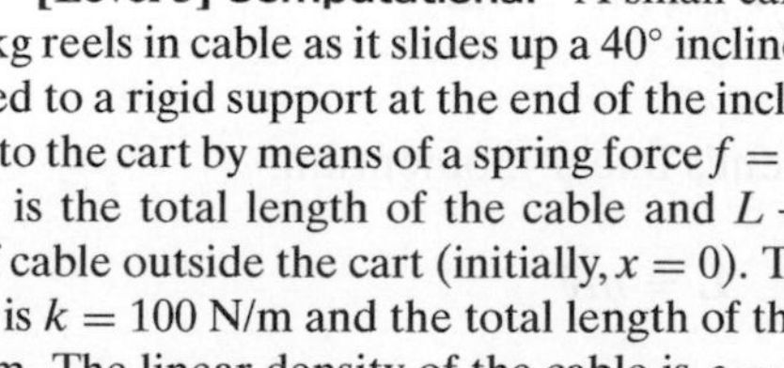

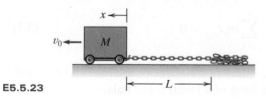

E5.5.23

5.5.24. **[Level 3] Computational** A small cart of mass $M = 20$ kg reels in cable as it slides up a 40° incline. The cable is fixed to a rigid support at the end of the incline and is drawn into the cart by means of a spring force $f = k(L - x)$, where L is the total length of the cable and $L - x$ is the length of cable outside the cart (initially, $x = 0$). The spring constant is $k = 100$ N/m and the total length of the cable is $L = 3.5$ m. The linear density of the cable is $\rho = 2.8$ kg/m. If the coefficients of static and dynamic friction between the cart and incline are equal, $\mu = 0.08$, determine how far up the incline the cart will have reached when it first comes to rest. Will the cart remain in this position or will it begin to slide back down?

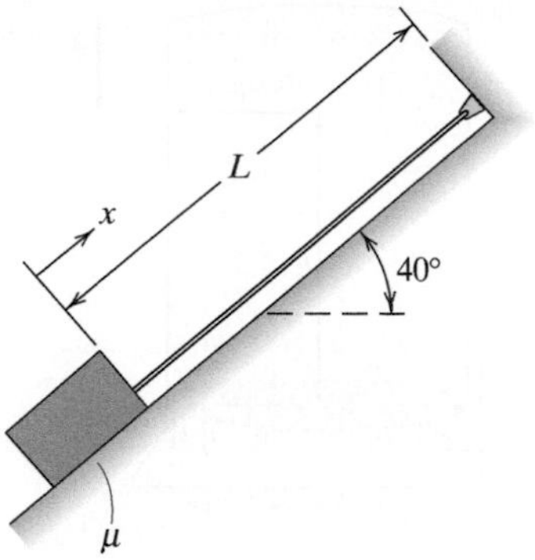

E5.5.24

5.5.25. **[Level 3]** Consider a 2 kg (total mass) model rocket placed on a smooth, horizontal track. The drag force acting on the rocket is $f = 0.0051v^2$ N, where v has units of m/s. If the rocket's 1.5 kg of propellant is evenly burned over a 12 s time span, with an ejection speed of 45 m/s relative to the rocket, what is the rocket's maximum attained speed?

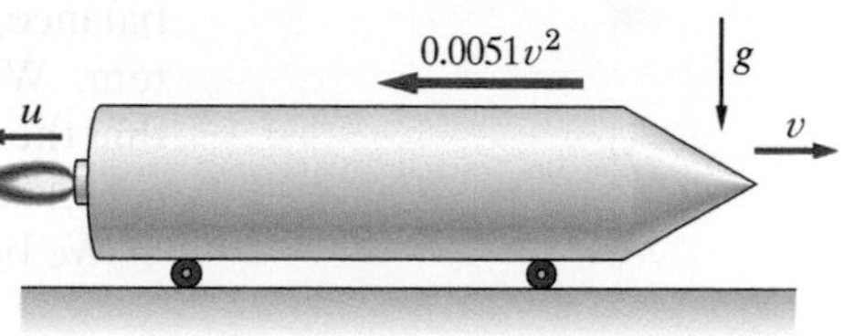

E5.5.25

5.5.26. **[Level 3] Computational** In a class competition, students were asked to build "water-powered" carts, harnessing the potential energy stored in a tank of water to propel their carts forward. One such design is shown. The mass of the empty cart is 9.4 kg, and the initial height of the water is 41 cm. The nozzle diameter is $d = 8.0$ cm, and the radius of the water tank is $r = 0.19$ m. In all calculations, neglect the small amount of water in the nozzle inlet area (below $h = 0$). The nozzle velocity of the water can be approximated as a function of the height of the water column remaining in the tank, $u(h) = 0.95 \times \sqrt{2g(h + d/2)}$, and the frictional resistance of the cart to horizontal motion is $f = 0.009W$, where W is the total weight of the cart.

a. Determine the initial acceleration of the cart.

b. Determine the acceleration when $t = 3.6$ s. (*Hint:* How are u and $\dot{h}$ related?)

c. Computational Determine the velocity of the cart at the instant the water level reaches $h = 0$. (Although you may be able to find the analytical solution if you look hard enough in tabled solutions, solve using numerical techniques.)

d. Determine the total distance the cart will travel before coming to rest.

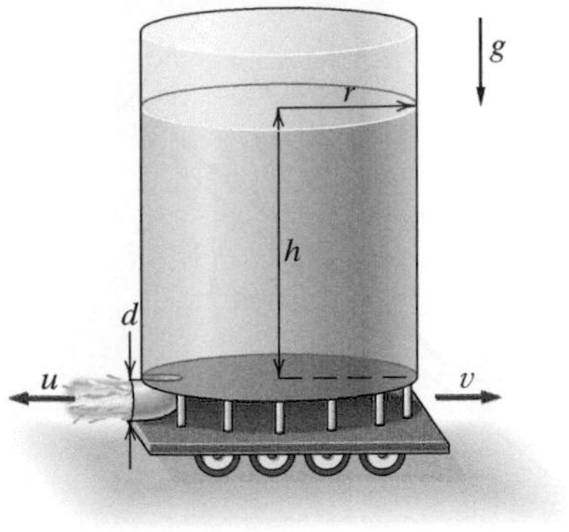

E5.5.26

5.5.27. **[Level 3] Computational** A 0.20 kg rock is tied to the end of a neatly coiled, 23 m length of rope weighing 0.54 kg. The loop and knot holding the rock required an extra 35 cm of rope. If the rock is hurled upward at 68 km/h, with a release height of 1.9 m, how high above the ground will it reach? Neglect air resistance.

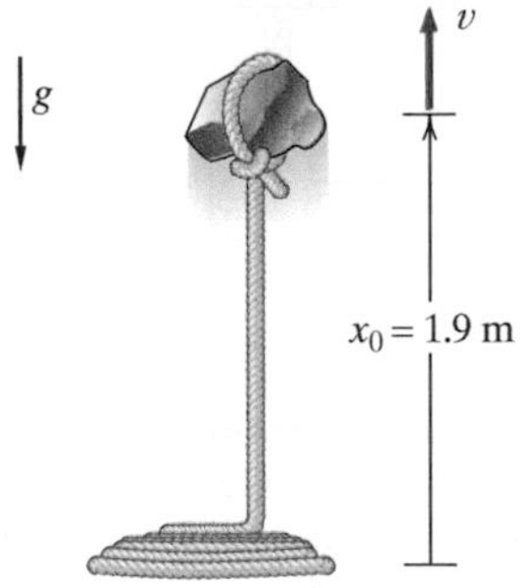

E5.5.27

5.6 JUST THE FACTS

This chapter dealt with systems of particles. We saw how the concepts we've already learned with regard to a single particle (force and moment balance, momentum, and work/energy) carry over to a multiparticle system. When focusing our attention on the system's mass center, we saw that the associated equations often took a particularly simple form, highly reminiscent of the single-particle case.

We began by defining the position of the system's center of mass:

$$\bar{\boldsymbol{r}} = \frac{\sum_{i=1}^{n} m_i \boldsymbol{r}_i}{m} \tag{5.2}$$

and then derived the useful result that

$$\sum_{i=1}^{n} m_i \boldsymbol{r}_{i_{/G}} = 0 \tag{5.3}$$

We next began to look at the effect of forces, both **internal** and **external**, and came up with the system form of a linear force balance:

$$\sum_{i=1}^{n} \boldsymbol{F}_i = m\ddot{\bar{\boldsymbol{r}}} = m\bar{\boldsymbol{a}} \tag{5.7}$$

and an expression for the system's linear momentum:

$$\boldsymbol{L} = m\bar{\boldsymbol{v}} \tag{5.8}$$

Putting the two together gave us the result that the time rate of change of the system's linear momentum is equal to the system's mass times the acceleration of its center of mass:

$$\dot{\boldsymbol{L}} = m\dot{\bar{\boldsymbol{v}}} = m\bar{\boldsymbol{a}} \tag{5.9}$$

Conservation of linear momentum holds when there are no external forces acting on the system:

$$\dot{\boldsymbol{L}} = 0 \tag{5.10}$$

The next task was to consider angular momentum relationships, and we started by defining the angular momentum about a point O for a system of particles:

$$\boldsymbol{H}_O = \sum_{i=1}^{n} (\boldsymbol{r}_i \times m_i \boldsymbol{v}_i) \tag{5.11}$$

Differentiating this gave us our moment balance about a fixed point:

$$\dot{\boldsymbol{H}}_O = \sum_{i=1}^{n} \boldsymbol{M}_{O_i} \tag{5.13}$$

Next we saw what the moment about the system's mass center looked like

$$\boldsymbol{H}_G = \sum_{i=1}^{n} \left(\boldsymbol{r}_{i/G} \times m_i \boldsymbol{v}_{i/G} \right) \tag{5.14}$$

and differentiating this led to

$$\dot{\boldsymbol{H}}_G = \sum_{i=1}^{n} \boldsymbol{M}_{G_i} \tag{5.16}$$

Integrating both our moment balance expressions gave us the expected result that the respective angular momentums at some later time are equal to the angular momentum at an earlier time plus the applied impulse over the system of particles:

$$\boldsymbol{H}_{O_2} = \boldsymbol{H}_{O_1} + \mathcal{AI}_{O_{1-2}} \tag{5.19}$$

$$\boldsymbol{H}_{G_2} = \boldsymbol{H}_{G_1} + \mathcal{AI}_{G_{1-2}} \tag{5.21}$$

Having dispensed with the force/moment-based analysis, we then proceeded to look at work/energy.

The kinetic energy for a system of particles had an interesting form in which one part accounted for the kinetic energy of the mass center and the other parts dealt with motions of individual particles with respect to the mass center:

$$\mathcal{KE} = \frac{1}{2} m \bar{v}^2 + \frac{1}{2} \sum_{i=1}^{n} m_i \left\| \dot{\boldsymbol{r}}_{i/G} \right\|^2 \tag{5.25}$$

In an exact parallel to our single-particle results, we saw that the work done on a system of particles will alter its kinetic energy:

$$W_{1-2} = \Delta \mathcal{KE} \tag{5.26}$$

and if no work is done, then there's no change in the system's kinetic energy:

$$\Delta\mathcal{KE} = 0 \tag{5.27}$$

Introducing the concept of potential energy made life easier for cases in which a potential energy function could be identified. For no external work we had

$$\Delta\mathcal{KE} + \Delta\mathcal{PE}\big|_{sp} + \Delta\mathcal{PE}\big|_{g} = 0 \tag{5.28}$$

which became slightly more complex when an external work term was present:

$$\Delta\mathcal{KE} + \Delta\mathcal{PE}\big|_{sp} + \Delta\mathcal{PE}\big|_{g} = W_{nc_{1-2}} \tag{5.29}$$

Allowing a stream of material to enter and then leave a body required an extension to our familiar $\boldsymbol{F} = m\boldsymbol{a}$ formula:

$$\boldsymbol{F} = \dot{m}(\boldsymbol{v}_j - \boldsymbol{v}_i) \tag{5.32}$$

We obtained the related moment balance formula as well:

$$\dot{\boldsymbol{H}}_O = \dot{m}\left(\boldsymbol{r}_{j/O} \times \boldsymbol{v}_j - \boldsymbol{r}_{i/O} \times \boldsymbol{v}_i\right) \tag{5.33}$$

Our final formulas pertained to a system for which the mass was changing in time. The first considered a system that loses mass:

$$F = m\dot{v} + \dot{m}u \tag{5.36}$$

If the mass was increasing, we had

$$F = m\dot{v} - \dot{m}u \tag{5.37}$$

SYSTEM ANALYSIS (SA) EXERCISES

SA5.1 Multi-Station Disorientation Demonstrator

Seasickness, carsickness, and airsickness—these are the names for the nausea that can affect people traveling by ship, car, or plane, and their root cause is the acceleration that occurs when traveling in these vehicles. Eventually, the sensations pass when the travelers become acclimated to the new motions. Thus it is that passengers on board an oceangoing ship are often seasick for the first day or so and then "get their sea legs."

The U.S. Navy utilizes the Multi-Station Disorientation Demonstrator (MSDD) to help acclimate large numbers of aircrew in the most efficient means possible. The MSDD (**Figure SA5.1.1**) consists of 12 individual 150 kg stations on a rotating platform (*a*). Assume that each station is 5 m away from the center of rotation and the maximum rotation rate is limited to 22 rpm. Neglect the platform's mass.

a. Determine the total kinetic energy of the system at maximum rotation.

b. Determine the total linear momentum of the system.

c. Determine the angular momentum about the center of rotation.

d. If you were going to spin the device up to speed by applying a tangential force to each of the stations, how much force would you have to apply to reach the maximum rpm within four revolutions?

e. Finally, the Navy is considering modifying the MSDD so that half of the stations are at a radius of 3 m (*b*) as a result of the observation that the tighter radius can often result in more severe motion sickness. How would this change the total kinetic energy and total angular momentum about the center of the device?

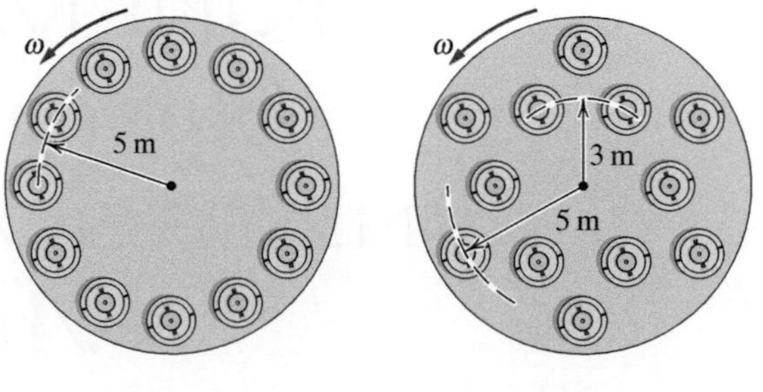

Figure SA5.1.1 Multi-Station Disorientation Demonstrator

SA5.2 Sand Loader

A large bin is to be loaded with 1.5 m^3 of sand ($\rho = 1540$ kg/m^3) as it sits on a cargo scale (**Figure SA5.2.1**). The bin is 2 m tall with a 1 m diameter. It is initially empty with a scale reading of zero. The sand drops at a constant rate of 300 l/s from a chute 3 m above the bottom of the bin and has an initial downward velocity $v_0 = 0.50$ m/s as it leaves the chute.

a. Determine the rate $\dot{m}(s)$ at which the sand collects in the bin. (*Hint:* Be careful to note the dependence on the height s of the sand.)

b. Determine the downward force exerted on the scale: (i) at the instant just after sand has begun collecting in the bottom of the bin ($s = 0$), (ii) when $s = 1.0$ m, and (iii) at the instant just before it is filled to the desired amount.

c. Plot this downward force versus s and determine the maximum downward force exerted on the scale.

d. One method for approximating $\dot{m}$ would be to multiply the density of the sand by its volume flow rate through the chute. What error is neglected in this computation? Use this approximation to compute the downward force exerted on the truck bed at the same three instants and the resulting errors.

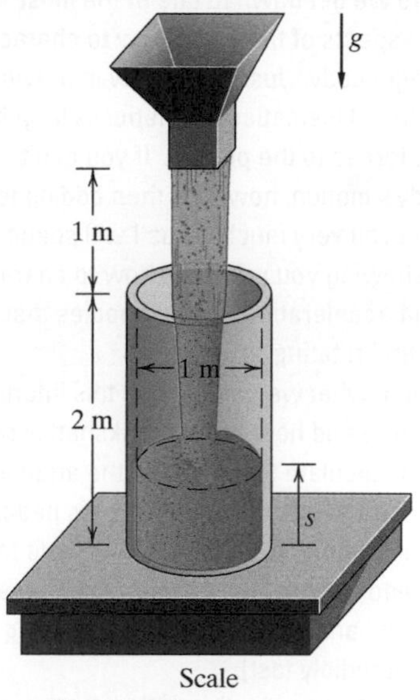

Figure SA5.2.1 Sand loader

OBJECTIVES

Upon completion of this chapter you will be able to:

- Determine the velocity/acceleration of one part of a solid body once you know the velocity/acceleration of a different point and the body's rotational motion
- Deal with rotating reference frames as well as fixed frames of reference

CHAPTER 6

Kinematics of Rigid Bodies Undergoing Planar Motion

Here's where we get down to one of the most important aspects of the book: how to characterize the motion of a rigid body. Just as we saw in previous chapters, going from kinematics to kinetics is largely a matter of adding forces to the picture. If you can't describe the body's motion, however, then adding forces isn't going to help you very much. Thus I will spend a fair amount of time showing you precisely how to characterize the velocities and accelerations of rigid bodies that are both translating and rotating.

There's no end to what we can do with this information. For example, if we're told how fast a crankshaft is rotating, we will be able to calculate the speed of the attached pistons. Or, given the rpm at which a cyclist is pedaling, as well as the particular gear he's in, we will be able to determine his speed. I used this technique just recently to determine how fast Lance Armstrong was traveling in Le Tour de France (incredibly fast).

We will spend some time dealing with rotating reference frames as well. You will learn how they affect our approach to finding a system's kinematic equations. This will let us determine the velocity/acceleration of a point that's moving on a body that is itself translating and rotating. Correctly accounting for all the acceleration terms is very important and requires a bit of insight, insight that you will gain after finishing the chapter.

Benson Tongue

6.1 RELATIVE VELOCITIES ON A RIGID BODY

Figure 6.1 shows several aerial snapshots of a sports car model that illustrates what happens when a driver comes around a corner too quickly and spins the car. This complex motion is the most general case that will concern us here. As you can see, the car's center of mass follows a curved path, indicated by the black line, the car body's orientation is changing continuously, and the rate at which the car's angular orientation is changing increases dramatically toward the end of the trajectory. (Skilled drivers will often allow their cars to do some sliding, and stunt drivers do a lot of it for the movies. But it's not something you should try on your own unless you like large body shop repair bills.)

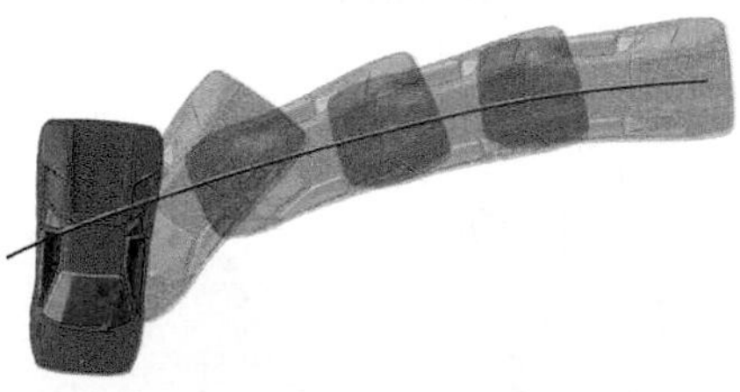

Figure 6.1 General planar motion

Figure 6.2 shows motion that is much simpler—a body rotating about a single, fixed point. In this case it's the passenger door that pivots upward about a fixed point of rotation. (Yes, spend enough money and your door can do this too.) Although the door's orientation is changing, it isn't both translating and rotating as it moves through space in the manner that the car in **Figure 6.1** is. Every point on the door moves in a purely circular arc about the point of rotation.

Figure 6.2 Rotation about a fixed point

Both of these examples require an examination of the rotational motion of the body in order to account for the orientation change. This doesn't have to be the case, however. **Figure 6.3** shows three positions of an automobile floor jack (with a box on top to help you visualize the top's motion). Note that the top of the jack remains horizontal at all times. It moves almost directly upward at first and then moves both back and up. But at all times the box maintains the same horizontal orientation. A dynamic analysis of the box would only require us to account for the acceleration of its mass center, and we could safely ignore any angular accelerations, which are all zero by virtue of the fact that the box isn't rotating.

We will examine all these cases in due time. To begin, however, let's look at **Figure 6.4**. A general body is shown at two instants, $t = 0$ and $t = \Delta t$, and it's clear from the positions of points A and B that the body has both translated and rotated. Thus we're looking at the most general motion of a rigid body that's moving on a plane. This general motion can be divided into two conceptual parts: translation and rotation.

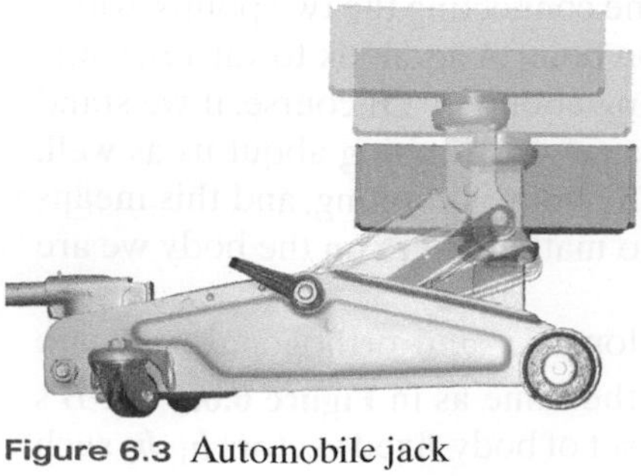

Figure 6.3 Automobile jack

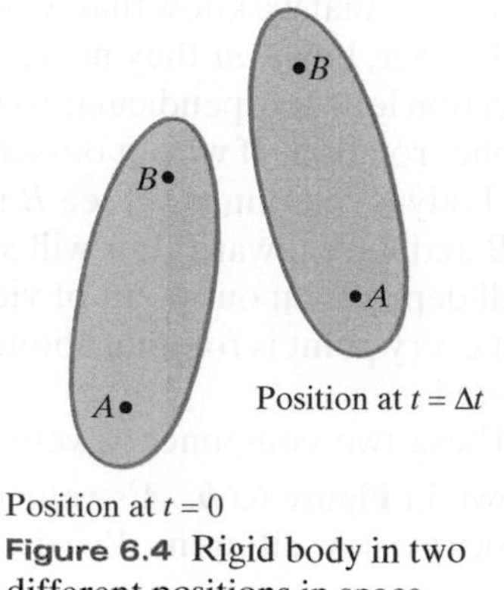

Figure 6.4 Rigid body in two different positions in space

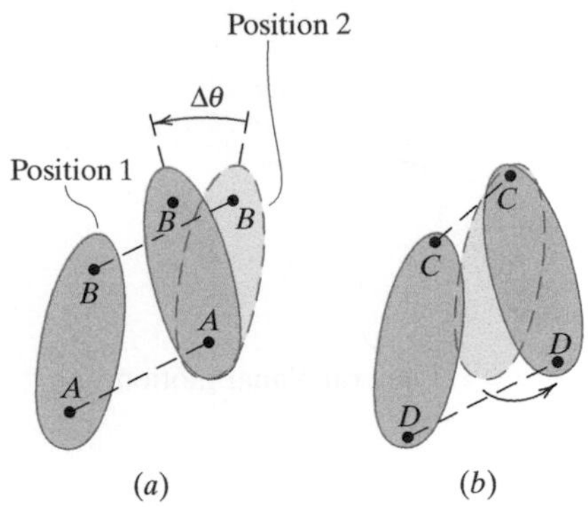

Figure 6.5 General motion in a plane

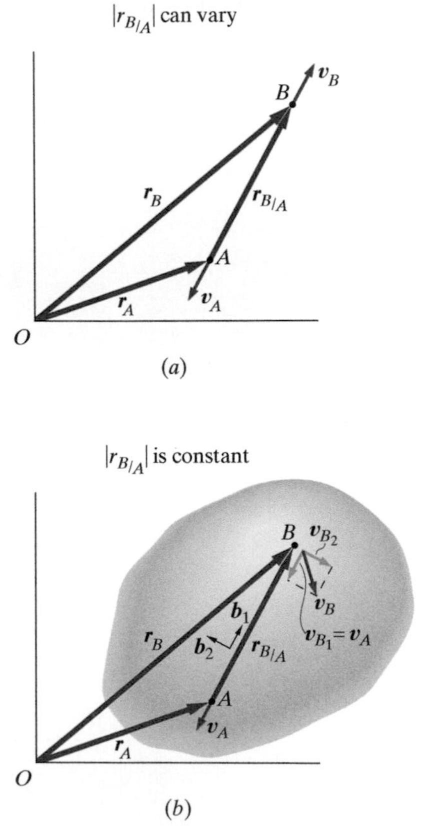

Figure 6.6 Constraints of rigid-body motion

We can handle the translation aspect rather easily by imagining that every point on the body moves the same distance at the same inclination, as was shown in **Figure 6.3**. To make this clear, I've added an additional outline (**Figure 6.5*a***). To get to the final position, the body first translates (Position 1 to Position 2) and then, once there, rotates about A by an amount $\Delta\theta$ to reach its final position.

This procedure is always valid. All general displacements of a rigid body can be decomposed into a translation and a rotation. Note that there was nothing special about choosing the point A. Any point will suffice, and **Figure 6.5*b*** shows another way to get the same result with the rotation about point C this time. As you can imagine, there are literally an infinite number of other possibilities.

We'll next do something similar to our approach in Chapter 2, where we answered the question of how to define velocities. You will recall that we took a particle's position at time $t + \Delta t$, subtracted from that quantity the particle's position at t, and then divided the difference by Δt. The limit as Δt went to zero gave us the velocity. In our current case we'll be looking at two effects—translation and rotation—in an attempt to relate the velocity of point B to the velocity of point A on the same body.

Figure 6.6 shows how the rigid-body case differs from the kinematics familiar to us. **Figure 6.6*a*** is a figure that could have come straight out of Chapter 2. Two points A and B are illustrated, along with their absolute position vectors, their relative position vector, and their velocities. The key point to realize is that there is *no* constraint on their relative velocity. In fact, the two velocity vectors $\boldsymbol{v}_B$ and $\boldsymbol{v}_A$ I've drawn emphasize this fact. You will note that they are both aligned with $\boldsymbol{r}_{B/A}$ and point in opposite directions. This means that A and B are moving away from each other, which is a perfectly fine situation for two independent particles. But what's fine for two independent particles is most definitely *not* fine for a rigid body. The velocity vectors shown in **Figure 6.6*a*** would be impossible if both A and B were fixed points on the same rigid body. Their moving apart would mean that the body was stretching, something rigid bodies cannot, by definition, do. Similarly, they cannot move toward each other because their doing so would mean the body was compressing.

This is the critical difference between our kinematics in Chapter 2 and the kinematics we're developing here. For points on rigid bodies, there can't be *any* motion toward or away from each other. This fact holds for every pair of points on the rigid body.

So, now that we know that two points on a rigid body can't move toward each other, how *can* they move? If you think about it, there's only one direction left—perpendicular to the line connecting the two points, which implies rotation. If we put ourselves on point A and look toward B, while the body is rotating, we'll see B rotating about us. Of course, if we stand at B and look toward A, it will seem as if A is rotating about us as well. It all depends on our point of view. The body is rotating, and this means that every point is rotating about us no matter where on the body we are located.

These two components, velocity along $\boldsymbol{r}_{B/A}$ and orthogonal to it, are shown in **Figure 6.6*b***. A's velocity is the same as in **Figure 6.6*a***, but B's velocity is now different. I've drawn a set of body-fixed vectors $\boldsymbol{b}_1, \boldsymbol{b}_2$ such that $\boldsymbol{b}_1$ points from A to B and $\boldsymbol{b}_2$ points in the direction in which $\boldsymbol{b}_1$ would move if the body were to rotate in the counterclockwise direction. Thus $\boldsymbol{r}_{B/A} = r_{B/A}\boldsymbol{b}_1$. The velocity vector $\boldsymbol{v}_B$ is broken down into two orthogonal components:

$$\boldsymbol{v}_B = \boldsymbol{v}_{B_1} + \boldsymbol{v}_{B_2}$$

You will notice that $\boldsymbol{v}_{B_1}$ has the same length as $\boldsymbol{v}_A$. That's good: it means that B is moving toward A at the same speed at which A is moving away from B. As a result, there is no change in distance between them, which satisfies our main criterion for rigid bodies. The other component, $\boldsymbol{v}_{B_2}$, accounts for the rotation of B about A.

In a moment we will see that this motion of B with respect to A is expressible as

$$\boldsymbol{v}_{B/A} = \boldsymbol{\omega} \times \boldsymbol{r}_{B/A}$$

In this expression $\boldsymbol{\omega}$ represents the angular velocity of the body, and $\boldsymbol{r}_{B/A}$ is, as usual, the position of B with respect to A. Thus we will see that the relative velocity $\boldsymbol{v}_{B/A}$ depends on how fast the body is spinning and how far B is from A. Showing this won't be too hard. Here's how.

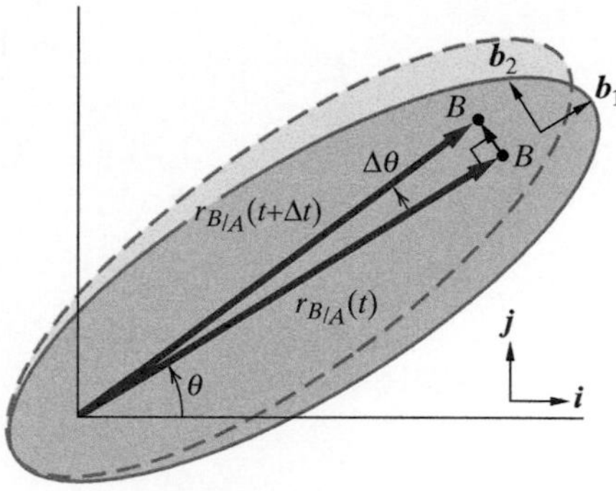

Figure 6.7 Rigid body, slightly rotated

Because we're concerned only with B's relative velocity, we can sketch out B's initial position and its position a short time Δt later (*with respect to A*), as shown in **Figure 6.7**. The difference between these two positions, divided by the elapsed time, is what gives us the velocity if we take the limit as the elapsed time goes toward zero.

Simply imagine that you're sitting at A and looking out toward B. All B can do is rotate about A, and the position vector $\boldsymbol{r}_{B/A}$ has changed orientation by the angle $\Delta\theta$ because of this rotation. We already looked at problems like this one in Section 2.3, when we determined the value of $\dot{\boldsymbol{e}}_r$. It was the same situation—a vector rotated slightly as a result of an angular velocity. Although I could follow the same approach as in Section 2.3, I'll keep things interesting by using a different method this time.

You can see from **Figure 6.7** that

$$\boldsymbol{r}_{B/A}(t) = r_{B/A}\boldsymbol{b}_1 = r_{B/A}[\cos\theta\boldsymbol{i} + \sin\theta\boldsymbol{j}]$$

After $\boldsymbol{r}_{B/A}$ has moved into its displaced position (at time $t + \Delta t$), we have

$$\boldsymbol{r}_{B/A}(t+\Delta t) = r_{B/A}[\cos(\theta+\Delta\theta)\boldsymbol{i} + \sin(\theta+\Delta\theta)\boldsymbol{j}]$$

We can expand these trigonometric terms using the identities

$$\cos(a+b) = \cos a\cos b - \sin a\sin b$$

and

$$\sin(a+b) = \sin a\cos b + \cos a\sin b$$

Using these relations gives us

$$\boldsymbol{r}_{B/A}(t+\Delta t) = r_{B/A}[\cos\theta\cos(\Delta\theta) - \sin\theta\sin(\Delta\theta)]\boldsymbol{i} + r_{B/A}[\sin\theta\cos(\Delta\theta) + \cos\theta\sin(\Delta\theta)]\boldsymbol{j}$$

Because $\Delta\theta$ is going to be small, we can linearize this expression, which yields

$$\boldsymbol{r}_{B/A}(t+\Delta t) = r_{B/A}[\cos\theta - \sin\theta(\Delta\theta)]\boldsymbol{i} + r_{B/A}[\sin\theta + \cos\theta(\Delta\theta)]\boldsymbol{j}$$

Now we can get the velocity of B relative to A by subtracting $\boldsymbol{r}_{B/A}(t)$ from $\boldsymbol{r}_{B/A}(t+\Delta t)$, dividing by Δt, and taking the limit as Δt goes to zero:

$$\boldsymbol{v}_{B/A} = \lim_{\Delta t\to 0}\frac{\boldsymbol{r}_{B/A}(t+\Delta t) - \boldsymbol{r}_{B/A}(t)}{\Delta t}$$

$$= \lim_{\Delta t\to 0}\frac{r_{B/A}[\cos\theta - \sin\theta(\Delta\theta)]\boldsymbol{i} + r_{B/A}[\sin\theta + \cos\theta(\Delta\theta)]\boldsymbol{j} - r_{B/A}[\cos\theta\boldsymbol{i} + \sin\theta\boldsymbol{j}]}{\Delta t}$$

Factoring out $\Delta\theta$:

$$= \lim_{\Delta t\to 0}\left(\frac{\Delta\theta}{\Delta t}\right) r_{B/A}[\cos\theta\boldsymbol{j} - \sin\theta\boldsymbol{i}]$$

$$= \dot{\theta} r_{B/A}[\cos\theta\boldsymbol{j} - \sin\theta\boldsymbol{i}] \tag{6.1}$$

Ah, we're so close I can taste it. Let's think about angular speed $\dot{\theta} = \omega$ for a second. I'm sure you recall from Physics class that bodies rotating on a plane have an angular velocity vector that is normal to the plane. Thus, if our plane were defined by $\boldsymbol{i}$ and $\boldsymbol{j}$, our angular velocity would be given by $\dot{\theta}\boldsymbol{k}$. Well, what happens if we cross this angular velocity with our relative position vector? We get

$$\begin{aligned}\boldsymbol{\omega} \times \boldsymbol{r}_{B/A} &= \dot{\theta}\boldsymbol{k} \times r_{B/A}[\cos\theta\boldsymbol{i} + \sin\theta\boldsymbol{j}] \\ &= \dot{\theta} r_{B/A}[\cos\theta\boldsymbol{j} - \sin\theta\boldsymbol{i}]\end{aligned} \tag{6.2}$$

And where have we seen this before? Just look up a few lines, at (6.1). It's the velocity of B with respect to A. As predicted, we have found the relationship

$$\boldsymbol{v}_{B/A} = \boldsymbol{\omega} \times \boldsymbol{r}_{B/A} \tag{6.3}$$

where $\boldsymbol{\omega} = \dot{\theta}\boldsymbol{k}$.

Because we now know $\boldsymbol{v}_{B/A}$, we can go back to our relative velocity formulation $\boldsymbol{v}_B = \boldsymbol{v}_A + \boldsymbol{v}_{B/A}$ (from 2.65) to obtain

$$\boldsymbol{v}_B = \boldsymbol{v}_A + \boldsymbol{\omega} \times \boldsymbol{r}_{B/A} \tag{6.4}$$

>>> **Check out Example 6.1 (page 320), Example 6.2 (page 321), Example 6.3 (page 322), Example 6.4 (page 323), and Example 6.5 (page 324) for applications of this material.**

This formula will singlehandedly allow us to solve a whole array of problems. For instance, consider the robotic arm illustrated in **Figure 6.8**. We have three links, each of which has a torque actuator at its base. Perhaps we would like to know the velocity of the arm's tip. If we know how fast each link is rotating $(\dot{\theta}_1, \dot{\theta}_2, \dot{\theta}_3)$, then we can easily compute the tip's velocity by writing

$$\boldsymbol{v}_A = \boldsymbol{v}_O + \dot{\theta}_1 \boldsymbol{k} \times \boldsymbol{r}_{A/O}$$

$$\boldsymbol{v}_B = \boldsymbol{v}_A + \dot{\theta}_2 \boldsymbol{k} \times \boldsymbol{r}_{B/A}$$

$$\boldsymbol{v}_C = \boldsymbol{v}_B + \dot{\theta}_3 \boldsymbol{k} \times \boldsymbol{r}_{C/B}$$

or, combining these equations into one,

$$\boldsymbol{v}_C = \boldsymbol{v}_O + \dot{\theta}_1 \boldsymbol{k} \times \boldsymbol{r}_{A/O} + \dot{\theta}_2 \boldsymbol{k} \times \boldsymbol{r}_{B/A} + \dot{\theta}_3 \boldsymbol{k} \times \boldsymbol{r}_{C/B}$$

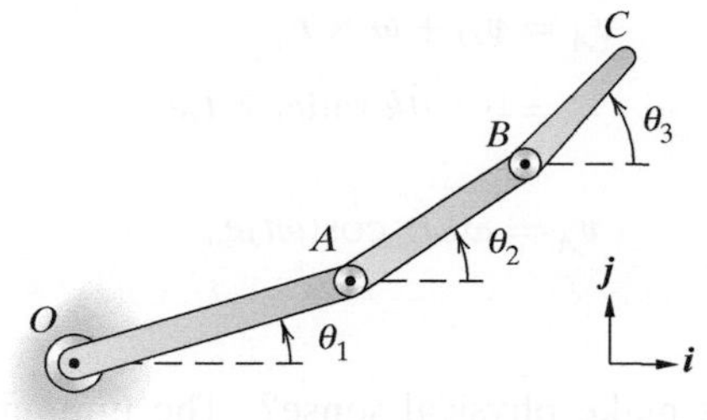

Figure 6.8 Three-link, robotic arm

EXAMPLE 6.1 VELOCITY OF A PENDULUM (Theory on page 318)

Figure 6.9 Grandfather clock

The grandfather clock shown in **Figure 6.9** has a pendulum with length L and the pendulum's angular deflection from vertical is given by $\theta = \theta_0 \sin(\omega t)$. Determine the velocity of the pendulum's tip as a function of time.

Goal Find the velocity of a pendulum's tip for any value of time.

Given Length of pendulum and its angular deflection as a function of time.

Assume No additional assumptions are needed.

Draw A simplified pendulum is shown in **Figure 6.10**, pivoted at O and with its tip labeled as A. Cylindrical coordinates are the logical choice for this example.

Formulate Equations All we need is (6.4), along with an expression for $\boldsymbol{\omega}$.

Solve Differentiating θ with respect to time gives us $\dot{\theta} = \omega\theta_0 \cos(\omega t)$. Using (6.4) gives us

$$\begin{aligned} \boldsymbol{v}_A &= \boldsymbol{v}_O + \boldsymbol{\omega} \times \boldsymbol{r}_{A/O} \\ &= 0 + \dot{\theta}\boldsymbol{k} \text{ rad/s} \times L\boldsymbol{e}_r \end{aligned}$$

$$\boldsymbol{v}_A = \omega L\theta_0 \cos(\omega t)\boldsymbol{e}_\theta$$

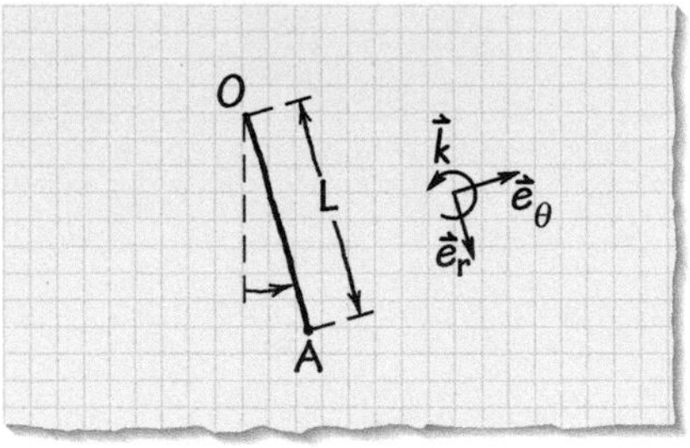

Figure 6.10 Simplified pendulum

Check Does this make physical sense? The units are correct—m/s. And the way the variables enter the solution makes sense as well. If the pendulum gets longer but keeps ticking at the same rate, we'd expect the tip's speed to increase. This effect is accounted for by the L in the solution. Likewise, increasing the rate at which it ticks should make all parts of the pendulum move faster, something that's accounted for by the angular speed ω appearing explicitly in the solution. Finally, the θ_0 term indicates that if the pendulum has to sweep out a larger angle, the speed of the tip will increase. This explanation makes sense: if the tip has to move farther in the same period of time, then we'd expect its speed to increase.

EXAMPLE 6.2 VELOCITY OF A CONSTRAINED LINK (Theory on page 318)

Figure 6.11 shows a single-link mechanism. The ends of the link are free to move in straight guides, A along the $\boldsymbol{i}$ direction and B along the $\boldsymbol{j}$ direction. The length of the link $\overline{AB}$ is 0.5 m. If A is being pushed to the right at 4 m/s, determine $\boldsymbol{v}_B$ when θ is 40°.

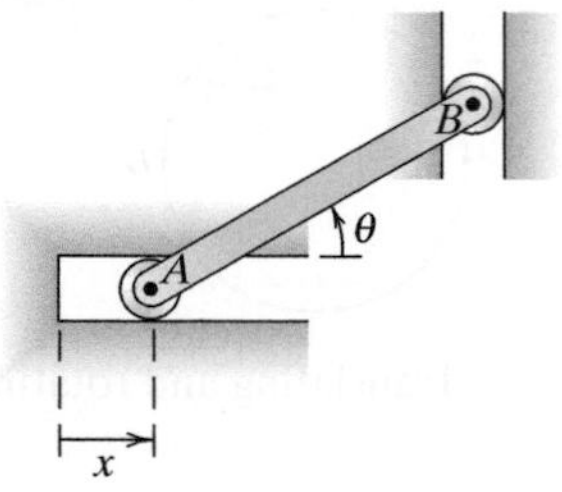

Figure 6.11 Constrained link

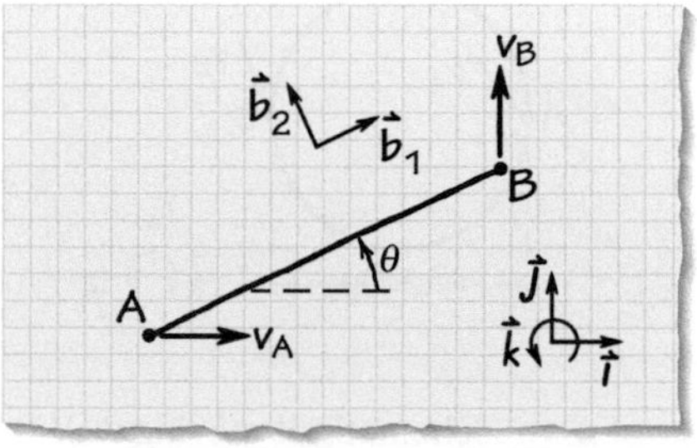

Figure 6.12 Simplified diagram of constrained link

Goal Find $\boldsymbol{v}_B$ when $\theta = 40°$.

Given Velocity of A and length of link.

Draw **Figure 6.12** shows a simplified representation of our system. Our coordinate transformation array is

	$\boldsymbol{i}$	$\boldsymbol{j}$
$\boldsymbol{b}_1$	$\cos\theta$	$\sin\theta$
$\boldsymbol{b}_2$	$-\sin\theta$	$\cos\theta$

Assume Our system is definitely constrained because A and B aren't free to move wherever they please. This means that

$$\boldsymbol{v}_A = v_A\boldsymbol{i} \quad \text{and} \quad \boldsymbol{v}_B = v_B\boldsymbol{j}$$

Formulate Equations Our governing equation is given by (6.4):

$$\boldsymbol{v}_B = \boldsymbol{v}_A + \boldsymbol{\omega} \times \boldsymbol{r}_{B/A}$$

It would be lovely if we knew $\boldsymbol{\omega}$ because then it would be a simple matter to plug our known values in and get the answer. Unfortunately, we don't know $\boldsymbol{\omega}$. Are we lost? Of course not. If we don't know it explicitly, then there must be constraints that will allow us to solve for it. And these constraints have already been mentioned in our **Assume** step.

Solve Using our constraints ($\boldsymbol{v}_A = v_A\boldsymbol{i}$, $\boldsymbol{v}_B = v_B\boldsymbol{j}$) in (6.2) yields

$$\begin{aligned} v_B\boldsymbol{j} &= v_A\boldsymbol{i} + \dot{\theta}\boldsymbol{k} \times 0.5\boldsymbol{b}_1 = 4\boldsymbol{i}\ \text{m/s} + \dot{\theta}\boldsymbol{k} \times 0.5(\cos\theta\boldsymbol{i} + \sin\theta\boldsymbol{j})\ \text{m} \\ &= 4\boldsymbol{i}\ \text{m/s} + 0.5\dot{\theta}(\cos\theta\boldsymbol{j} - \sin\theta\boldsymbol{i})\ \text{m} \end{aligned}$$

$\boldsymbol{i}$: $$0 = 4\ \text{m} - 0.5\dot{\theta}\sin\theta\ \text{m} \tag{6.5}$$

$\boldsymbol{j}$: $$v_B = 0.5\dot{\theta}\cos\theta\ \text{m} \tag{6.6}$$

$$(6.5), (6.6) \Rightarrow \quad \dot{\theta} = 12.4\ \text{rad/s} \quad \text{and} \quad v_B = 4.77\ \text{m/s} \quad \Rightarrow \quad \boxed{\boldsymbol{v}_B = 4.77\boldsymbol{j}\ \text{m/s}}$$

Check We can go backwards (from B to A), using the results we have just derived, and see whether A is moving in a purely horizontal manner.

$$\begin{aligned} \boldsymbol{v}_A &= \boldsymbol{v}_B + \dot{\theta}\boldsymbol{k} \times (-0.5\boldsymbol{b}_1\ \text{m}) \\ &= (0\boldsymbol{j} + 4\boldsymbol{i})\ \text{m/s} \end{aligned}$$

Thus we find $\boldsymbol{v}_A = 4\boldsymbol{i}$ m/s, a perfect match with our initial data.

EXAMPLE 6.3 ANGULAR SPEED OF A SPINNING DISK (Theory on page 318)

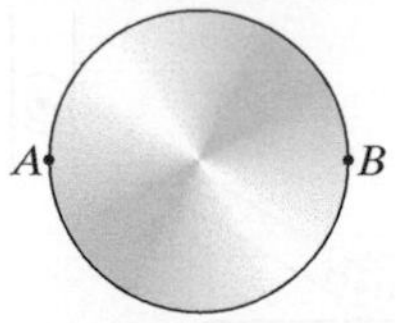

Figure 6.13 Translating and rotating disk

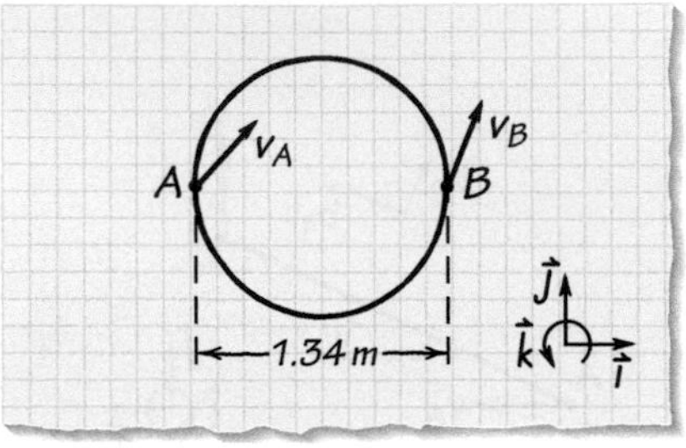

Figure 6.14 Schematic of moving disk

Figure 6.13 shows a circular disk. The velocities of points A and B ($\boldsymbol{r}_{B/A} = 1.34\boldsymbol{i}$ m) are known: $\boldsymbol{v}_A = (0.6\boldsymbol{i} + 0.6\boldsymbol{j})$ m/s, $\boldsymbol{v}_B = (0.6\boldsymbol{i} + 1.2\boldsymbol{j})$ m/s. Determine the angular speed of the disk.

Goal Calculate the angular speed of a solid disk.

Given Velocities of two points on the disk and the distance between them.

Assume No additional assumptions are needed.

Draw **Figure 6.14** shows the velocities of A and B and the relevant unit vectors.

Formulate Equations All we'll need is a judicious application of (6.4).

Solve Reordering (6.4) gives us

$$\boldsymbol{\omega} \times \boldsymbol{r}_{B/A} = \boldsymbol{v}_B - \boldsymbol{v}_A$$

$$\omega\boldsymbol{k} \times 1.34\boldsymbol{i}\text{ m} = (0.6\boldsymbol{i} + 1.2\boldsymbol{j})\text{ m/s} - (0.6\boldsymbol{i} + 0.6\boldsymbol{j})\text{ m/s}$$

$$1.34\boldsymbol{j}\text{ m} = 0.6\boldsymbol{j}\text{ m/s}$$

$$\boxed{\omega = 0.45\text{ rad/s}}$$

Check Let's check our work by finding $\boldsymbol{v}_B$ knowing $\boldsymbol{v}_A$ and ω:

$$\boldsymbol{v}_B = \boldsymbol{v}_A + \boldsymbol{\omega} \times \boldsymbol{r}_{B/A}$$
$$= (0.6\boldsymbol{i} + 0.6\boldsymbol{j})\text{ m/s} + 0.45\boldsymbol{k}\text{ rad/s} \times 1.34\boldsymbol{i}\text{ m} = (0.6\boldsymbol{i} + 1.2\boldsymbol{j})\text{ m/s}$$

exactly as expected.

EXAMPLE 6.4 VELOCITY OF LINK-CONSTRAINED BODY (Theory on page 318)

Figure 6.15 shows a body that's constrained by two attached links. The link $\overline{AB}$ is being driven at $\omega_{\overline{AB}} = 10\,\text{rad/s}$. $\boldsymbol{r}_{B/A} = 0.6\boldsymbol{i}\,\text{m}$, $\boldsymbol{r}_{C/B} = 0.3\boldsymbol{i}\,\text{m}$, $\boldsymbol{r}_{D/C} = (0.4\boldsymbol{i} + 0.4\boldsymbol{j})\,\text{m}$ and $\boldsymbol{r}_{E/C} = 0.5\boldsymbol{i}\,\text{m}$. Determine the velocity of E.

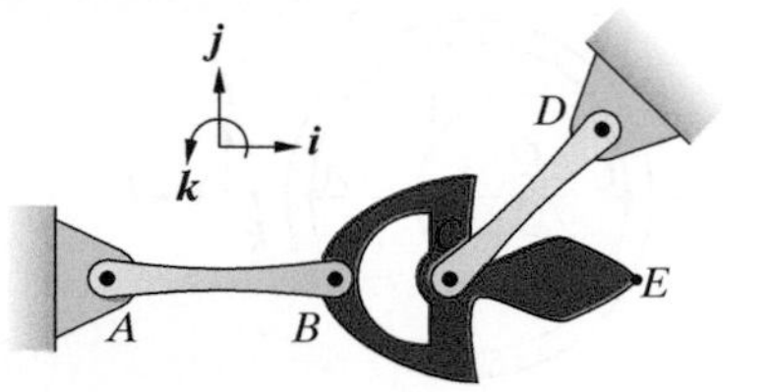

Figure 6.15 Body constrained by two rigid links

Goal Determine $\boldsymbol{v}_E$.

Given Dimensions and angular velocity of one member.

Assume It's important to note that all that matters is where the relevant points are, not the actual look of the structure. Hence when we sketch it, there's no need to include any of the body's interesting details.

Draw **Figure 6.16** shows a schematic of the two gears.

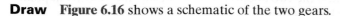

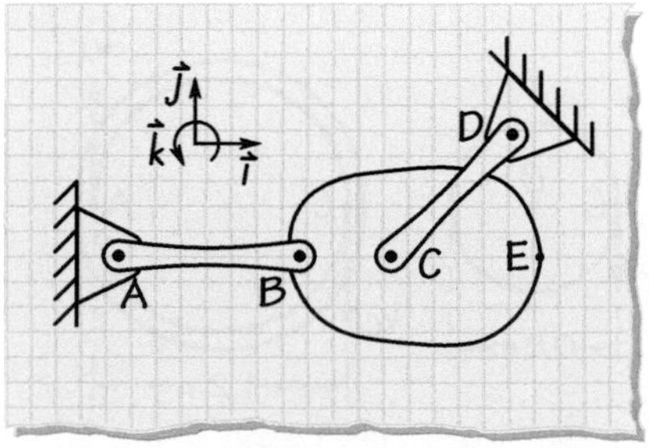

Figure 6.16 Mechanism schematic

Formulate Equations To solve this problem we'll need to apply (6.4).

Solve

(6.4) between B and $A \Rightarrow$ $$\boldsymbol{v}_B = \boldsymbol{v}_A + \boldsymbol{\omega}_{AB} \times \boldsymbol{r}_{B/A} \qquad (6.7)$$

(6.7) $\Rightarrow$ $$\boldsymbol{v}_B = 0 + 10\boldsymbol{k}\,\text{rad/s} \times 0.6\boldsymbol{i}\,\text{m} \quad \Rightarrow \quad \boldsymbol{v}_B = 6\boldsymbol{j}\,\text{m/s} \qquad (6.8)$$

(6.4) between C and $B \Rightarrow$ $$\boldsymbol{v}_C = \boldsymbol{v}_B + \boldsymbol{\omega}_{BC} \times \boldsymbol{r}_{C/B} \qquad (6.9)$$

(6.9) $\Rightarrow$ $$\boldsymbol{v}_C = 6\boldsymbol{j}\,\text{m/s} + \omega_{\overline{BC}}\boldsymbol{k} \times 0.3\boldsymbol{i}\,\text{m} \quad \Rightarrow \quad \boldsymbol{v}_C = (6\,\text{m/s} + (0.3\,\text{m})\omega_{\overline{BC}})\boldsymbol{j} \qquad (6.10)$$

(6.4) between C and $D \Rightarrow$ $$\boldsymbol{v}_C = \boldsymbol{v}_D + \boldsymbol{\omega}_{CD} \times \boldsymbol{r}_{C/D} \qquad (6.11)$$

(6.11) $\Rightarrow$ $$\boldsymbol{v}_C = 0 + \omega_{\overline{CD}}\boldsymbol{k} \times (-0.4\boldsymbol{i}\,\text{m} - 0.4\boldsymbol{j}\,\text{m}) \quad \Rightarrow \quad \boldsymbol{v}_C = (0.4\,\text{m})\omega_{\overline{CD}}\boldsymbol{i} - (0.4\,\text{m})\omega_{\overline{CD}}\boldsymbol{j} \qquad (6.12)$$

(6.10), (6.12) $\Rightarrow$ $$\omega_{\overline{CD}} = 0, \qquad \omega_{\overline{BC}} = -20\,\text{rad/s}$$

(6.4) between E and $B \Rightarrow$ $$\boldsymbol{v}_E = \boldsymbol{v}_B + \boldsymbol{\omega}_{\overline{BC}} \times \boldsymbol{r}_{E/B} \qquad (6.13)$$

(6.13) $\Rightarrow$ $$\boldsymbol{v}_E = 6\boldsymbol{j}\,\text{m/s} - 20\boldsymbol{k}\,\text{rad/s} \times 0.8\boldsymbol{i}\,\text{m} = -10\boldsymbol{j}\,\text{m/s}$$

Check We'll learn a nice way to check this problem in Example 6.7.

EXAMPLE 6.5 RELATIVE ANGULAR VELOCITY (Theory on page 318)

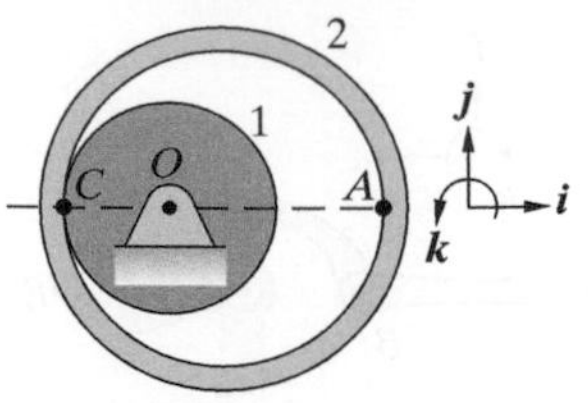

Figure 6.17 Ring gear rolling on inner gear

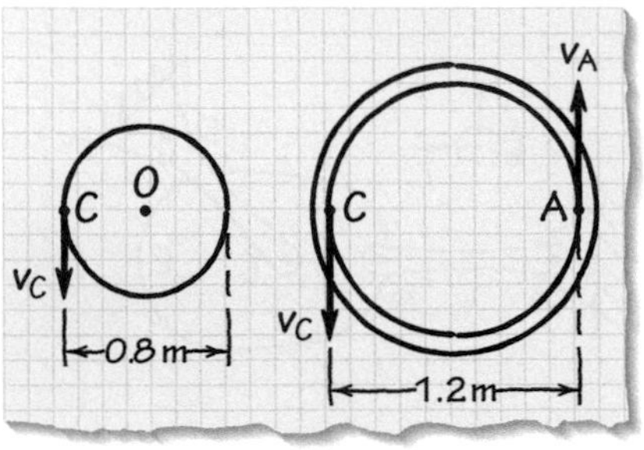

Figure 6.18 Schematic of the two gears

Body 1 is a gear that's rotating about O at a constant angular velocity of $5\boldsymbol{k}$ rad/s. The inner surface of the ring gear (Body 2) rolls without slip along the outer surface of Body 1 (**Figure 6.17**). The radius of Body 1 is 0.4 m, and the inner diameter of Body 2 is 1.2 m. Given that $\boldsymbol{v}_A = 30\boldsymbol{j}$ m/s, what is the angular velocity of Body 2 *with respect to* Body 1?

Goal Find relative angular velocity of Body 2 with respect to Body 1.

Given The angular velocity of Body 1, the gear dimensions, and the velocity of a point on Body 2.

Assume Because the two bodies roll without slip, the velocity at point C is the *same* for both bodies.

Draw **Figure 6.18** shows a schematic of the two gears.

Formulate Equations We'll use the no-slip assumption along with (6.3).

Solve Let ω_1 represent the rotational velocity of Body 1 and ω_2 be the rotational velocity of Body 2.

$$(6.3) \Rightarrow \quad \boldsymbol{v}_C = \boldsymbol{v}_O + \boldsymbol{\omega}_1 \times \boldsymbol{r}_{C/O} = 0 + 5\boldsymbol{k}\text{ rad/s} \times (-0.4\boldsymbol{i}\text{ m}) = -2\boldsymbol{j}\text{ m/s} \tag{6.14}$$

$$\boldsymbol{v}_A = 30\boldsymbol{j}\text{ m/s} = \boldsymbol{v}_C + \boldsymbol{\omega}_2 \times \boldsymbol{r}_{A/C} \underbrace{=}_{(6.14)} -2\boldsymbol{j}\text{ m/s} + \omega_2\boldsymbol{k} \times 1.2\boldsymbol{i}\text{ m}$$

$$30\boldsymbol{j}\text{ m/s} = -2\boldsymbol{j}\text{ m/s} + 1.2\omega_2\boldsymbol{j}\text{ m} \quad \Rightarrow \quad \omega_2 = 26.7\text{ rad/s}$$

Using $\boldsymbol{\omega}_2 = \boldsymbol{\omega}_1 + \boldsymbol{\omega}_{2/1}$ gives us

$$\boldsymbol{\omega}_{2/1} = \boldsymbol{\omega}_2 - \boldsymbol{\omega}_1 = 26.7\boldsymbol{k}\text{ rad/s} - 5\boldsymbol{k}\text{ rad/s} = \boxed{21.7\boldsymbol{k}\text{ rad/s}}$$

Check We can verify our analysis from a physical standpoint. Point C is moving down at 2 m/s, as previously determined, and point A is moving up at 30 m/s. The motion of a rigid body can be viewed as a superposition of a translation and a rotation. If we take the translational speed to be 2 m/s (down), then by adding 2 m/s (up) to both points we will make C's speed zero and A's speed 32 m/s. Now we can view A as rotating about C, and we know that its speed will be its distance from C times Body 2's angular speed:

$$32\text{ m/s} = (1.2\text{ m})\omega_2 \quad \Rightarrow \quad \omega_2 = 26.7\text{ rad/s}$$

Physically, if we're standing on Body 1 and therefore rotating with it, everything around us will seem to be rotating in the clockwise direction at the same rate that we're rotating counterclockwise. Body 1 is rotating at 5 rad/s, and we have just seen that Body 2 is rotating at 26.7 rad/s; therefore Body 2 will seem to be rotating at only

$$26.7\text{ rad/s} - 5\text{ rad/s} = 21.7\text{ rad/s}$$

EXERCISES 6.1

6.1.1. [Level 1] This exercise looks at how the real world differs from our idealized models. Tom, a car enthusiast, knows how to read the nomenclature on a tire and has purchased a 225/45/432 Z rated set of tires for his car. The 432 indicates the diameter of the wheel (in millimeters), the 225 indicates the width of the tire (in millimeters), and the 45 indicates the aspect ratio of the tire: The radial distance from the outside of the wheel to the outside of the rubber tire (the tire's "height") is equal to its width (225 mm) times 0.45. Thus in this case the distance from the center of the wheel to the outside of the tire would be $(\frac{432}{2}\text{ mm}) + (225)(0.45)\text{ mm} = 317\text{ mm}$. He mounts the tires and then gets in and starts driving. He notices from the tachometer that the engine is turning at 3000 rpm. He knows that speedometers are notoriously inaccurate and so he uses his GPS to determine that his car is moving at 27 km/h. He knows that first gear involves a gear ratio of 4.21 and the final drive ratio is 3.07. Thus his tires are rotating 12.9 times slower than the engine. Some mental math shows him that the car's actual speed doesn't match the speed he finds from applying $v = \omega r$ (28 km/h), the value he expected. He knows the car isn't slipping and he's also confident that his tachometer and GPS signals are highly accurate, as is his value for the car's gear ratios. Why does he observe this discrepancy between his predicted speed and the observed speed?

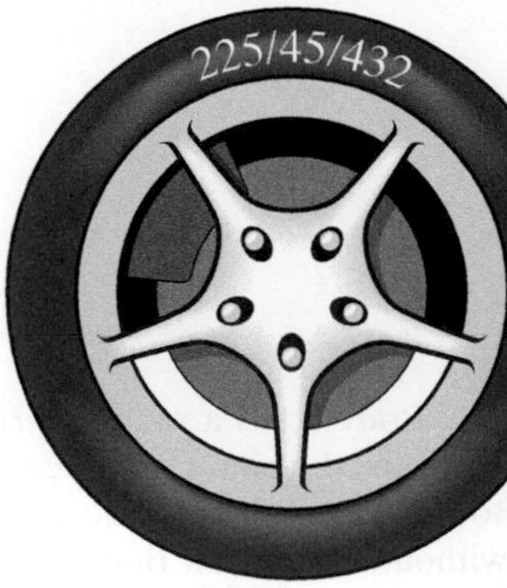

E6.1.1

6.1.2. [Level 1] A rally-car wheel has been instrumented so as to produce velocity data as the car moves through the race course. While traveling to the left $(-\boldsymbol{i})$ the telemetry data indicates that the point A has a velocity of $(102\boldsymbol{i} - 408\boldsymbol{j})$ cm/s. $\boldsymbol{r}_{A/C} = (-20\boldsymbol{i} + 15\boldsymbol{j})$ cm. What is the slip speed at C and the speed of the wheel's center G? The wheel/tire has a radius of 30 cm.

E6.1.2

6.1.3. [Level 1] A traditional set of cycling rollers has two identical, parallel cylinders in the rear of the device that the rear tire of the bicycle rests on. Assume that the rear tire is rotating at $\boldsymbol{\omega} = 24\boldsymbol{k}$ rad/s. What are the angular velocities of the two cylinders?

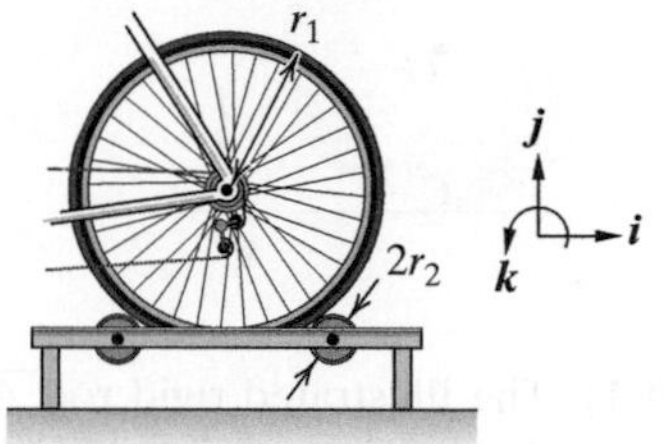

E6.1.3

6.1.4. [Level 1] My microwave oven has a rotating plate that helps ensure uniform cooking. The plate takes 10 s to complete one revolution and is driven by three rotating wheels. Assuming no slip between the wheels and plate, determine the rotational speed of the rollers.

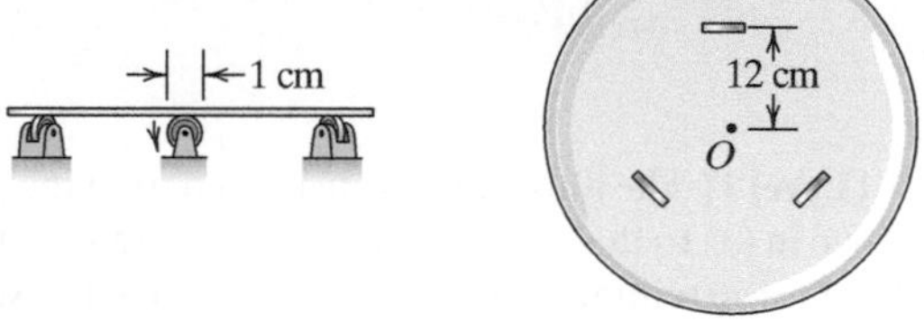

E6.1.4

6.1.5. [Level 1] A uniform cylinder of radius $R = 0.3$ m rolls without slip on a horizontal rail with a constant angular velocity of $\boldsymbol{\omega} = 4\boldsymbol{k}$ rad/s. The end A of a link with length $L = 0.5$ m is pinned near the edge of the cylinder, and its free end B is allowed to rotate about A with a constant angular velocity of $\boldsymbol{\omega}_{\overline{AB}} = -2\boldsymbol{k}$ rad/s. At the illustrated instant, A is located directly above the cylinder's center O, and the link is oriented at $\theta = 30°$ with respect to the vertical. What is the speed of the link's free end B?

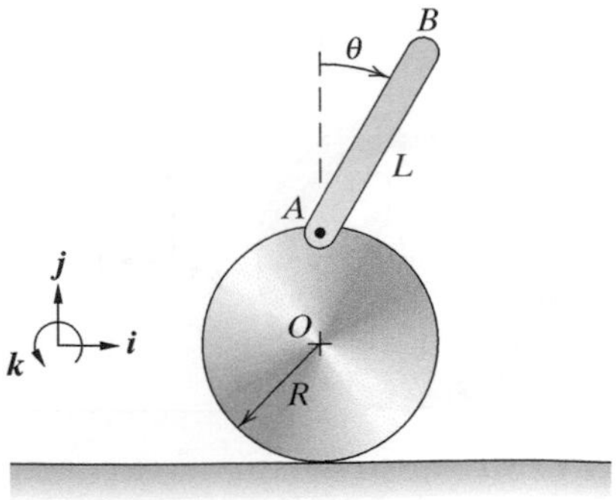

E6.1.5

6.1.6. **[Level 1]** The illustrated wheel has a radius of 0.2 m, an angular velocity of $10\boldsymbol{k}$ rad/s, and $\boldsymbol{v}_G = -1.5\boldsymbol{i}$ m/s. What can you say about the wheel/ground interface condition at C?

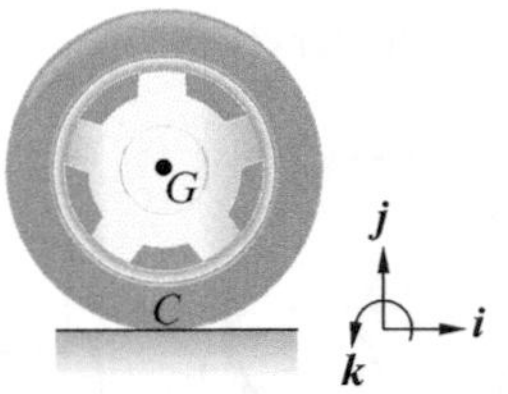

E6.1.6

6.1.7. **[Level 1]** The illustrated rigid rod $\overline{AB}$ is sliding within a circular guide of radius $r = 2$ m. At the illustrated instant $\boldsymbol{v}_A = 3\boldsymbol{i}$ m/s. Find $\boldsymbol{v}_B$.

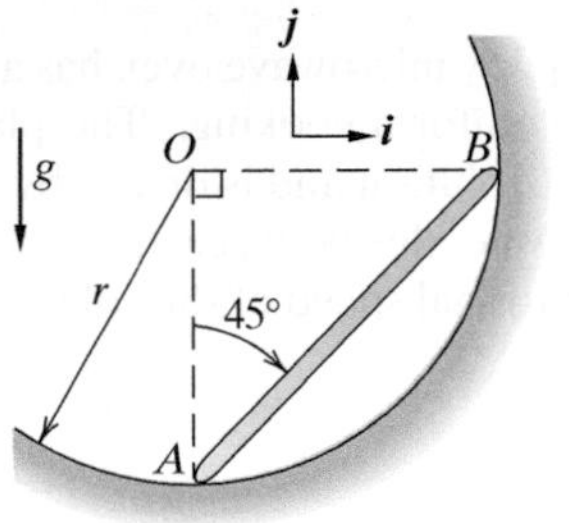

E6.1.7

6.1.8. **[Level 1]** The illustrated crate moved from the position shown in (*a*) to that in (*b*) over the span of 15 seconds.

a. What is the minimum average angular velocity experienced by the crate?

b. Why was it necessary for me to specify the minimum average angular velocity, rather than simply the average angular velocity?

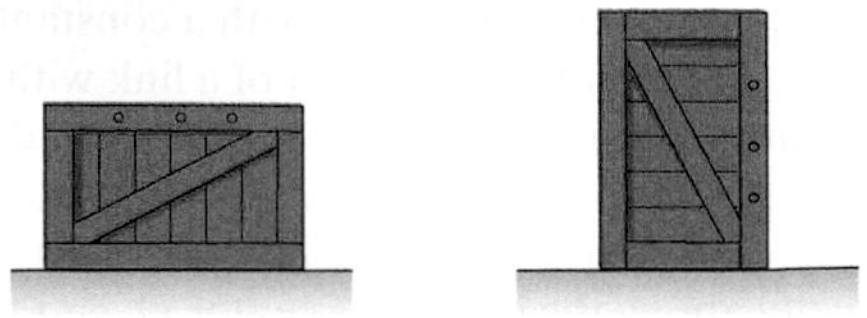

E6.1.8 (*a*) (*b*)

6.1.9. **[Level 1]** An old-fashioned type of bicycle, called an ordinary or a penny farthing, is shown. This bicycle had no chain—the cranks were connected rigidly to the front wheel. Thus, in order to get a reasonable speed out of it, the front wheel had to be large. Determine how large D must be for the bicycle to travel at 16 km/h with a pedaling cadence of 60 rpm. Assume that the bicycle's speed is equal to the rotation rate of the wheel in rad/s times the wheel's radius.

E6.1.9

6.1.10. **[Level 1]** The illustrated ring gear (Body 2) with an inner diameter of 0.48 m meshes with the outside of a fixed circular gear (Body 1) of radius 0.20 m. If the angular velocity of Body 2 is $10\boldsymbol{k}$ rad/s, what is $\boldsymbol{v}_A$?

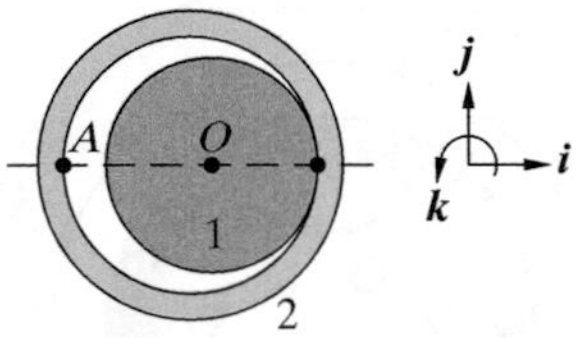

E6.1.10

6.1.11. **[Level 2]** Body 1 has a radius of $r_1 = 0.25$ m and rotates about the fixed point O at a constant rate $\omega_1 = 15\boldsymbol{k}$ rad/s. Body 2 is attached to the rigid link $\overline{AO}$ at A, and it rolls without slip against Body 1. The radius of Body 2 is $r_2 = 0.1$ m. If link $\overline{AO}$ has an angular velocity of $\boldsymbol{\omega}_{\overline{AO}} = -4\boldsymbol{k}$ rad/s at the illustrated instant, what are the angular velocity of Body 2 and $\boldsymbol{v}_B$?

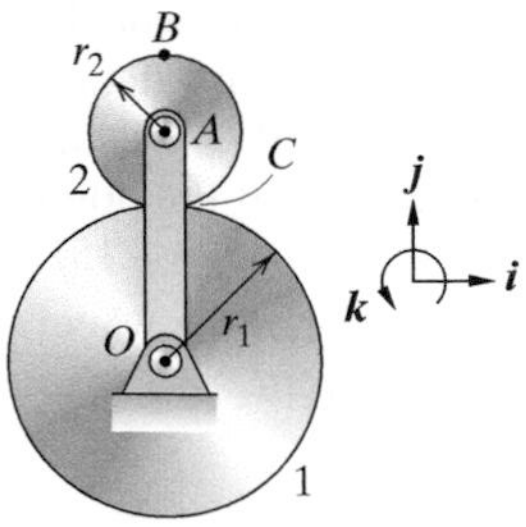

E6.1.11

6.1.12. [Level 2] An overhead fan accelerates from rest such that $\theta = (1 - e^{at})$ rad, where t is in seconds and $a = -0.8\,\text{s}^{-1}$. At $t = 3$ s, what is $\|\boldsymbol{v}_A\|$ equal to? ($r_{A/O} = 1$ m.)

E6.1.12

6.1.13. [Level 2] Calculate ω_{BCD} and $\boldsymbol{v}_C$ for the given device: $\boldsymbol{\omega}_{\overline{AB}} = 5\boldsymbol{k}$ rad/s, $\boldsymbol{r}_{B/A} = (-0.2\boldsymbol{i} + 0.5\boldsymbol{j})$ m, $\boldsymbol{r}_{C/B} = (-0.1\boldsymbol{i} + 0.2\boldsymbol{j})$ m, $\boldsymbol{r}_{D/B} = 0.5\boldsymbol{i}$ m, and $\boldsymbol{r}_{D/E} = (0.1\boldsymbol{i} + 0.1\boldsymbol{j})$ m.

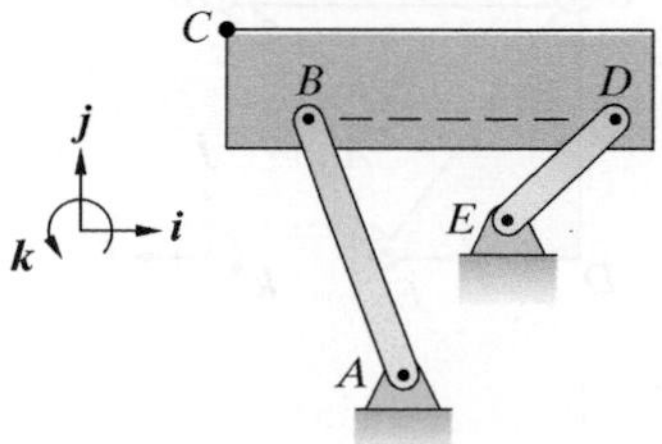

E6.1.13

6.1.14. [Level 2] The rack on the left (Rack A) is moved up at a speed v_A. The rack causes the two gears to rotate, which in turn causes the right rack (Rack B) to translate. Find the direction and speed with which Rack B moves.

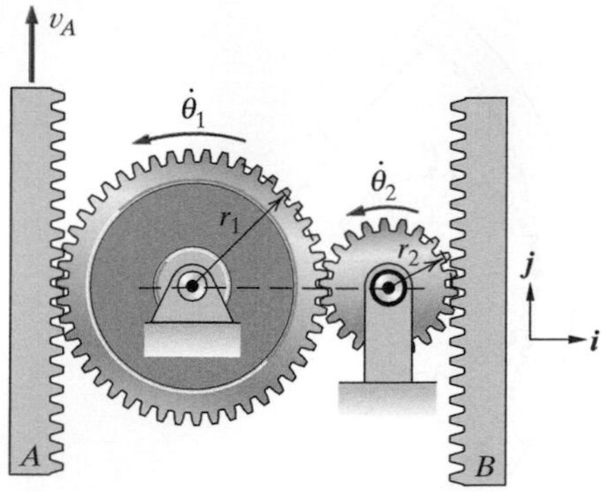

E6.1.14

6.1.15. [Level 2] A fixed-gear bicycle transmission is shown. When the rider pedals, he rotates the crank and hence the chain ring (labeled A). A chain connects the chain ring to a toothed gear (B) rigidly attached to the rear tire. $r_2 = 10.8$ cm, $r_1 = 3.8$ cm, $r_3 = 34$ cm, and $r_4 = 18$ cm. At what rpm must the rider pedal if the rear tire is to rotate at 26 rad/s?

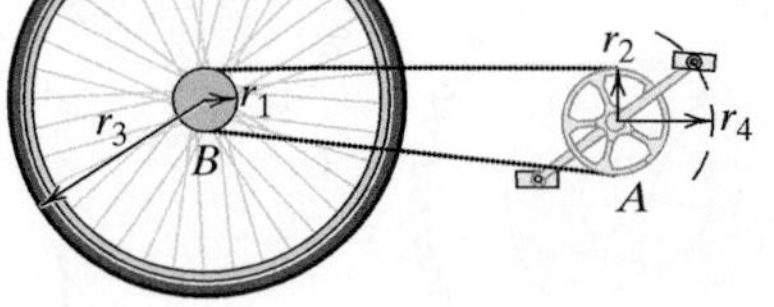

E6.1.15

6.1.16. [Level 2] The triangular plate ADC is connected to the triangular plate EFG by the link $\overline{BE}$. The angular velocity of plate EFG is $2.0\boldsymbol{b}_3$ rad/s. At the illustrated instant $\boldsymbol{r}_{B/A} = (2.0\boldsymbol{b}_1 + 2.0\boldsymbol{b}_2)$ m, $\boldsymbol{r}_{E/G} = (1.0\boldsymbol{b}_1 + 1.0\boldsymbol{b}_2)$ m, $\boldsymbol{r}_{G/A} = 1.0\boldsymbol{b}_1$ m. What is the angular velocity of the link $\overline{BE}$?

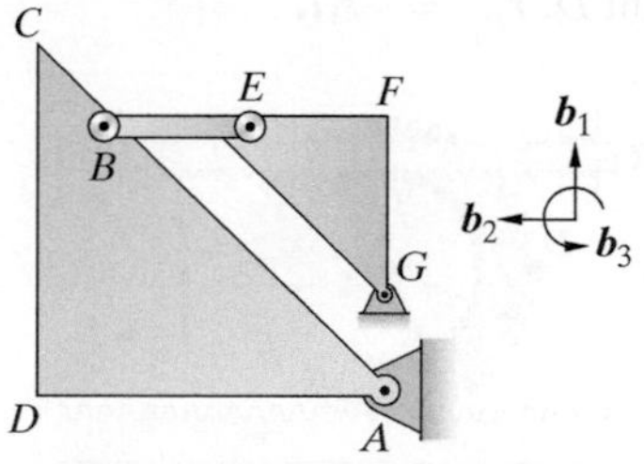

E6.1.16

6.1.17. [Level 2] Let's analyze the gear range of a typical bicycle. Assume an effective tire radius of 34 cm from the hub center to the ground. The two front chain rings have 39 and 52 teeth, respectively. The rear gear cluster has a range of cogs from 12 teeth up to 28 teeth. Assuming that you pedal at 75 rpm, calculate the maximum and minimum rotational speed of the rear wheel.

6.1.18. [Level 2] Gear 1 rotates about the fixed point O and meshes with Rack A. At the center of Gear 1 is a pulley of radius $r_1 = 0.05$ m, around which an inextensible cable C is wound. The other end of the cable is wound around another pulley of radius $r_2 = 0.03$ m, which spins with Gear 2 about the fixed point P. Gear 2 meshes with Rack B, and both racks are constrained to move in the horizontal direction. Gears 1 and 2 have a radius of $R_1 = 0.08$ m and $R_2 = 0.1$ m, respectively. There is no slip between the cable C and the pulleys. Calculate the velocity of Rack B when Rack A moves at $\boldsymbol{v}_A = -5\boldsymbol{i}$ m/s.

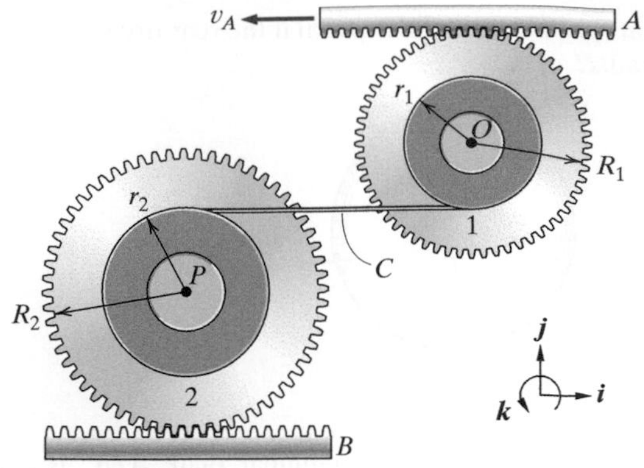

E6.1.18

6.1.19. **[Level 2]** The illustrated double gear rolls on the stationary lower rack, and the velocity of its center is given by $\boldsymbol{v}_G = 1.2\boldsymbol{i}$ m/s. $r_1 = 0.15$ m and $r_2 = 0.1$ m. Determine the angular velocity of the gear and the velocities of the upper rack R and point D. $\boldsymbol{r}_{D/G} = -r_1\boldsymbol{i}$.

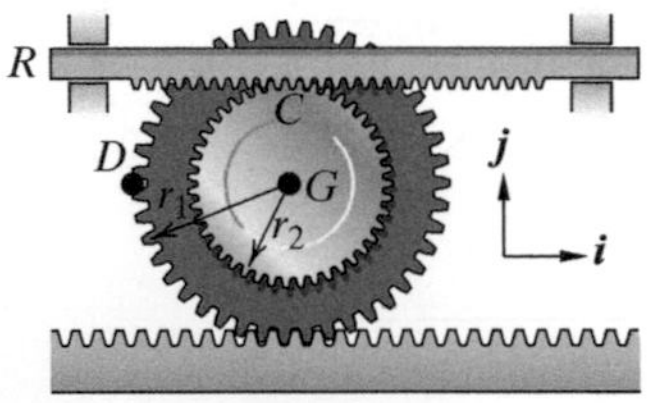

E6.1.19

6.1.20. **[Level 2]** A reel of massless string is shown, having an inner and outer radius of r_1 and r_2, respectively. The free end of the string is pulled to the right for each case at a constant rate $\dot{x} = v_0$.

a. Which scenario will induce a higher velocity of the reel's center of mass, (a) or (b)? Assume that the reel moves to the right for each case and that the point of the reel in contact with the ground has zero velocity.

b. Does the reel unroll or roll up the string?

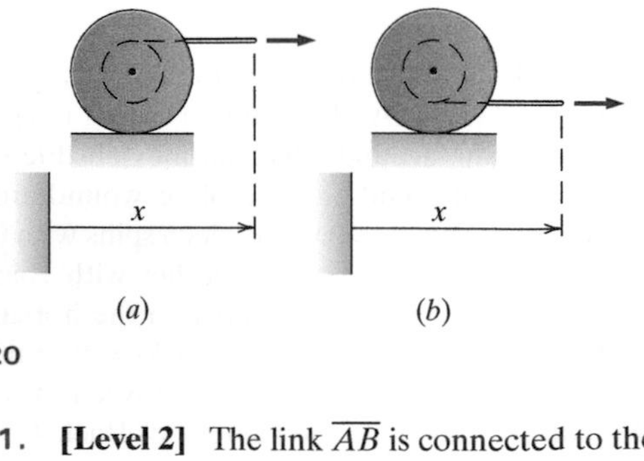

E6.1.20

6.1.21. **[Level 2]** The link $\overline{AB}$ is connected to the rectangular plate $BCDE$, which is itself connected to ground via the link $\overline{EF}$. The angular velocity of the link EF is $1.0\boldsymbol{k}$ rad/s. At the illustrated instant $\boldsymbol{r}_{B/A} = (-0.12\boldsymbol{i} - 0.12\boldsymbol{j})$ m, $\boldsymbol{r}_{E/F} = (0.06\boldsymbol{i} + 0.06\boldsymbol{j})$ m, $\boldsymbol{r}_{E/B} = -0.06\boldsymbol{j}$ m. What is the angular velocity of the rectangular plate $BCDE$?

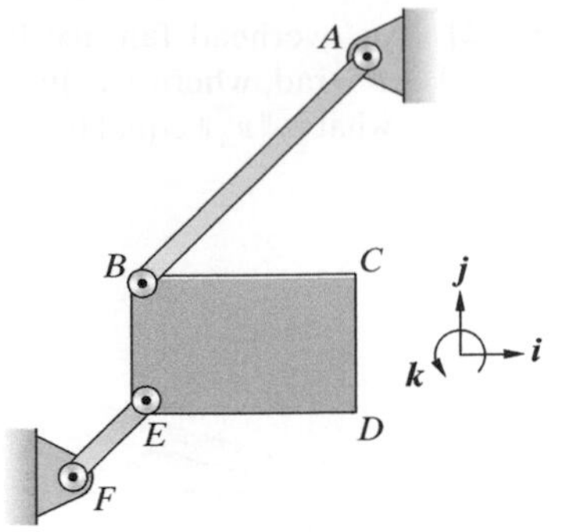

E6.1.21

6.1.22. **[Level 2]** The triangular plate CDE is controlled by the two links $\overline{AB}$ and $\overline{BC}$. At the illustrated instant $\boldsymbol{r}_{A/B} = (-0.06\boldsymbol{i} - 0.03\boldsymbol{j})$ m, $\boldsymbol{r}_{E/C} = (0.06\boldsymbol{i} - 0.06\boldsymbol{j})$ m, and $\boldsymbol{r}_{C/B} = -0.12\boldsymbol{i}$ m. The link $\overline{AB}$ is rotating with an angular velocity of 12 rad/s. Determine the velocity of the point C and the angular velocity of the plate CDE.

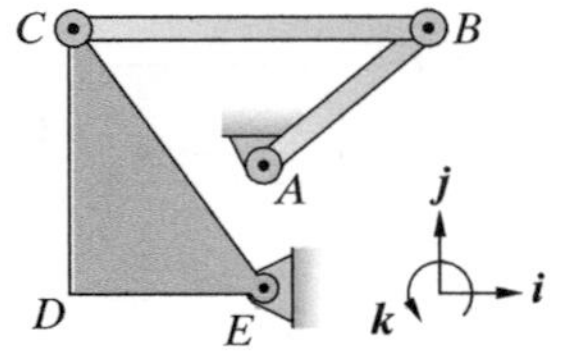

E6.1.22

6.1.23. **[Level 2]** A gear of radius r_1 (Body 1) rolls without slip along the inner radius of Body 2. The rigid link $\overline{OA}$ is restrained by a fixed pivot at O. $\boldsymbol{\omega}_{\overline{OA}} = 4\boldsymbol{k}$ rad/s. $r_1 = 0.1$ m and $r_2 = 0.3$ m. Find the angular velocity of Body 1 and $\boldsymbol{v}_B$ for the illustrated instant.

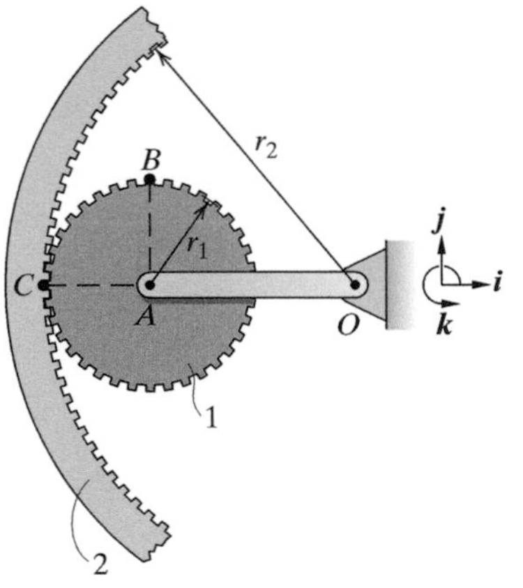

E6.1.23

6.1.24. **[Level 2]** Point O is fixed in space. Gear 1 has an angular velocity of $20\boldsymbol{k}$ rad/s, and Gear 2 has an angular velocity of $-10\boldsymbol{k}$ rad/s. $r_1 = 0.3$ m and $r_2 = 0.2$ m. What is the angular velocity of the connecting arm $\overline{OA}$, and what is the speed of point A?

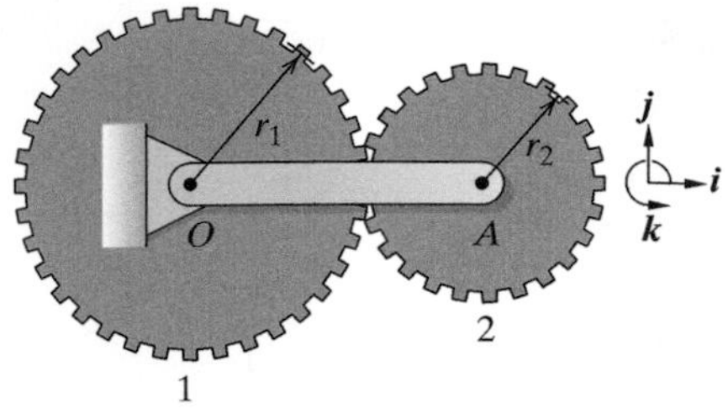

E6.1.24

6.1.25. [Level 2] Part of a pulley is shown. An inextensible rope, attached to the ceiling at O, wraps halfway around Body 1. The free end of the rope can move up or down, its position given by x. The pulley's radius, r_1, is equal to 0.2 m. At a particular instant, $\dot{x} = 2$ m/s. Find the velocity of points A, B, C, and D. There exists no slip between the rope and Body 1.

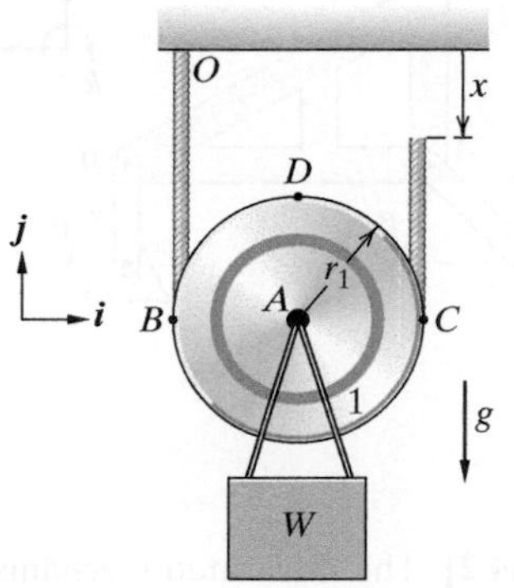

E6.1.25

6.1.26. [Level 2] Shown is a common toy in which the solid axle of a plastic wheel rolls along metal guide rails. To the right is a simplified picture of the system. r_1, the radius of the axle, is 0.127 cm. r_2 is 3.8 cm. $|v_A| = 30$ cm/s. What is $\boldsymbol{v}_B$? Assume counterclockwise rotation.

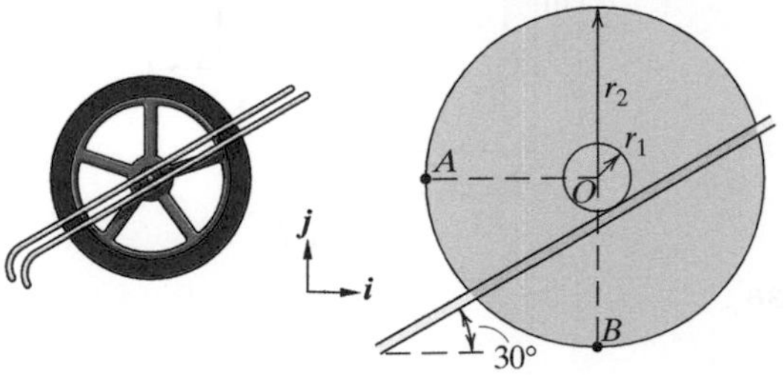

E6.1.26

6.1.27. [Level 2] At the illustrated instant, Body 1 is rolling without slip against Body 2. $r_1 = 0.1$ m, $r_2 = 0.08$ m, $\overline{OA} = 0.2$ m, and $\overline{BC} = 0.16$ m. $\boldsymbol{\omega}_{\overline{OA}} = 10\boldsymbol{k}$ rad/s, ω_1 (angular velocity of Body 1) $= -20\boldsymbol{k}$ rad/s, and $\boldsymbol{\omega}_{\overline{BC}} = 30\boldsymbol{k}$ rad/s. Find the angular velocity of Body 2.

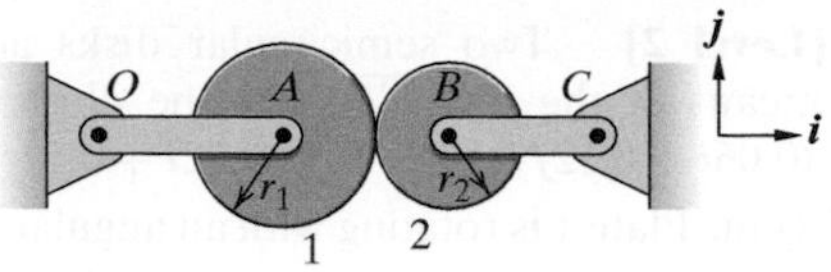

E6.1.27

6.1.28. [Level 2] Body 1 is attached to the rigid link $\overline{AB}$ at A, and Body 2 is attached at B. The bodies roll without slip against each other. The link $\overline{AB}$ rotates at a constant rate $\boldsymbol{\omega}_{\overline{AB}} = -4\boldsymbol{k}$ rad/s. Body 1 is also in contact with a circular inner ring at D. The radius of this inner ring is 0.25 m. $r_1 = 0.15$ m, $r_2 = 0.1$ m, and $h = 0.1$ m. What are the angular velocities of Bodies 1 and 2?

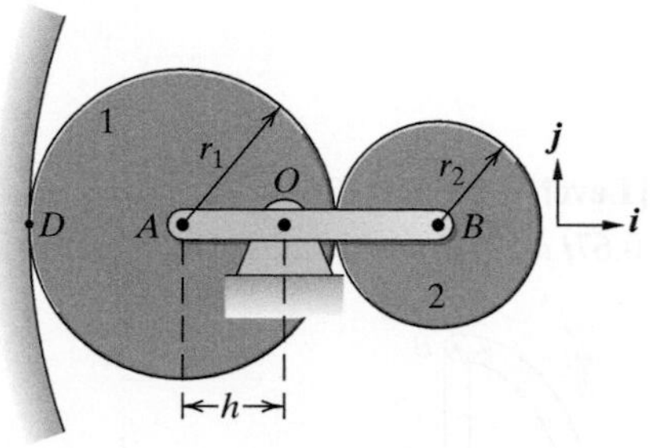

E6.1.28

6.1.29. [Level 2] $\boldsymbol{r}_{B/A} = -0.6\boldsymbol{j}$ m, $\boldsymbol{r}_{D/B} = 0.7\boldsymbol{i}$ m, and $\boldsymbol{r}_{E/D} = (0.5\boldsymbol{i} + 0.5\boldsymbol{j})$ m. $\boldsymbol{\omega}_{\overline{AB}} = 4\boldsymbol{k}$ rad/s. Determine $\boldsymbol{\omega}_{\overline{BCD}}$ and $\boldsymbol{v}_D$.

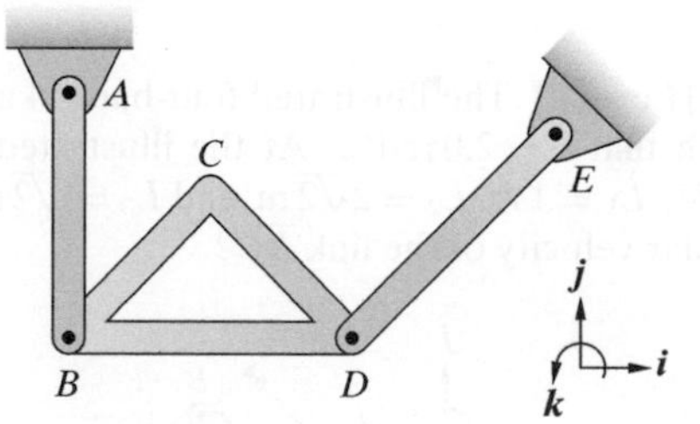

E6.1.29

6.1.30. [Level 2] $\boldsymbol{r}_{C/B} = -0.4\boldsymbol{i}$ m, $\boldsymbol{r}_{A/B} = (-0.2\boldsymbol{i} - 0.1\boldsymbol{j})$ m, and $\boldsymbol{r}_{D/C} = (0.2\boldsymbol{i} - 0.2\boldsymbol{j})$ m. $\boldsymbol{\omega}_{\overline{AB}} = -2\boldsymbol{k}$ rad/s. Determine $\boldsymbol{v}_C$ and $\boldsymbol{\omega}_{\overline{CD}}$.

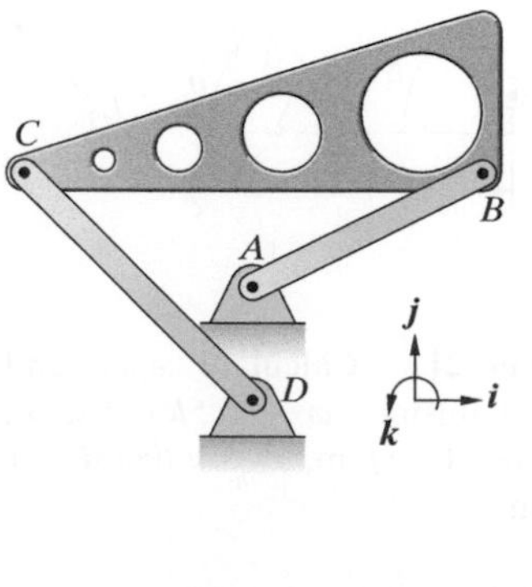

E6.1.30

6.1.31. **[Level 2]** Two semicircular disks are positioned by means of the link $\overline{CD}$. At the illustrated instant $\boldsymbol{r}_{B_{/A}} = (0.06\boldsymbol{i} + 0.12\boldsymbol{j})$ m, $\boldsymbol{r}_{D_{/C}} = (-0.12\boldsymbol{i} + 0.12\boldsymbol{j})$ m, and $\boldsymbol{r}_{C_{/B}} = -0.24\boldsymbol{j}$ m. Plate 1 is rotating with an angular velocity of $4\boldsymbol{k}$ rad/s. Determine $\boldsymbol{v}_C$ and the angular velocity of link $\overline{CD}$.

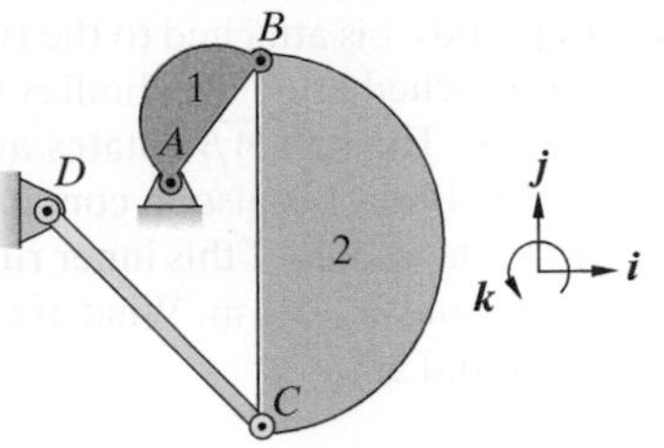

E6.1.31

6.1.32. **[Level 2]** $\boldsymbol{r}_{B_{/A}} = (0.36\boldsymbol{i} + 72\boldsymbol{j})$ m, $\boldsymbol{r}_{C_{/B}} = -0.55\boldsymbol{j}$ m, and $\boldsymbol{r}_{D_{/C}} = 0.67\boldsymbol{i}$ m. $\boldsymbol{\omega}_{\overline{BC}} = 5\boldsymbol{k}$ rad/s. Determine $\boldsymbol{v}_B$ and $\boldsymbol{v}_C$.

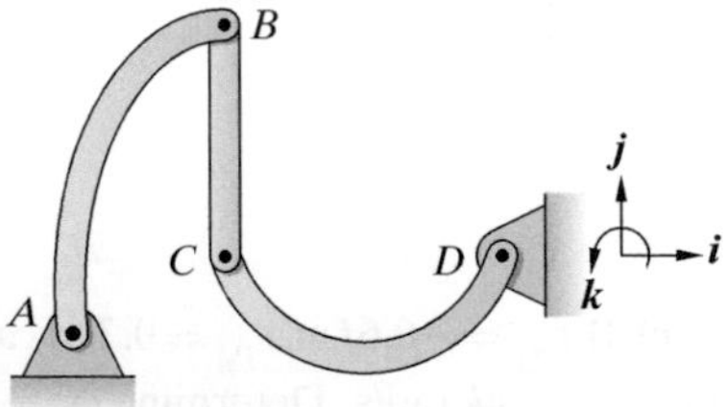

E6.1.32

6.1.33. **[Level 2]** The illustrated four-bar linkage is actuated such that $\dot{\theta} = 2.0$ rad/s. At the illustrated instant, $\theta = \beta = 45°$. $L_1 = 1$ m, $L_2 = 2\sqrt{2}$ m, and $L_3 = \sqrt{2}$ m. What is the angular velocity of the link $\overline{BC}$?

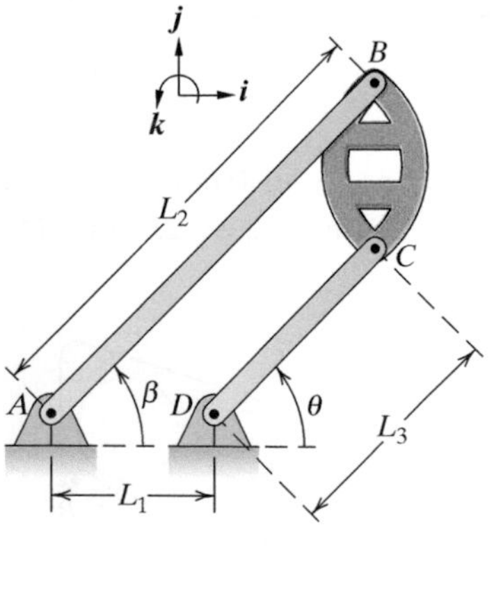

E6.1.33

6.1.34. **[Level 2]** Calculate ω_{BCE} and $\boldsymbol{v}_E$ for the illustrated mechanism: $\boldsymbol{\omega}_{\overline{AB}} = 5\boldsymbol{k}$ rad/s, $\boldsymbol{r}_{B_{/A}} = (0.21\boldsymbol{i} - 0.40\boldsymbol{j})$ m, $\boldsymbol{r}_{C_{/D}} = -0.55\boldsymbol{j}$ m, $\boldsymbol{r}_{C_{/B}} = (0.45\boldsymbol{i} - 0.15\boldsymbol{j})$ m, and $\boldsymbol{r}_{E_{/B}} = -0.15\boldsymbol{j}$ m.

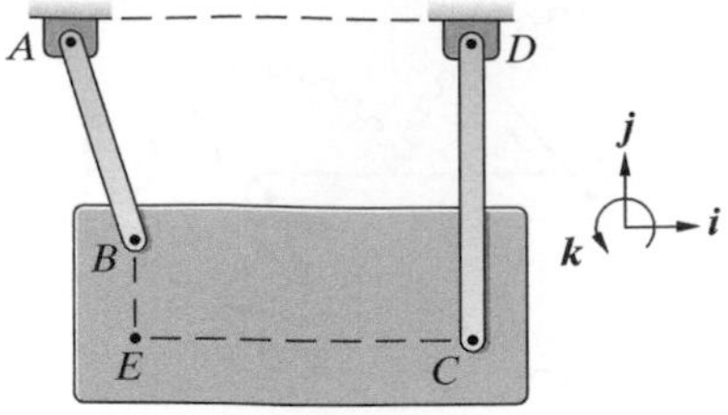

E6.1.34

6.1.35. **[Level 2]** The illustrated triangular body changes orientation as the supporting links rotate. At the moment shown, $\theta = 90°$, $\dot{\theta} = 5$ rad/s, and $\beta = 45°$. Determine $\boldsymbol{v}_A$ and the angular velocity of the triangular body. $r_{B_{/E}} = 2\boldsymbol{j}$ cm and $r_{B_{/A}} = 5\boldsymbol{i}$ cm.

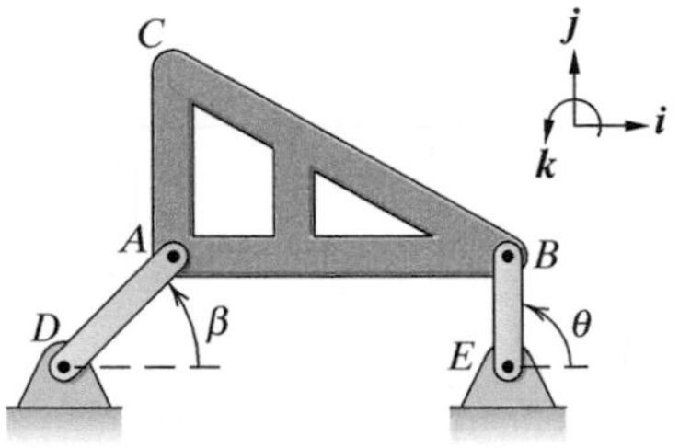

E6.1.35

6.1.36. **[Level 2]** The illustrated mechanism, hinged at B and D, is controlled by the motion of the two links $\overline{AB}$ and $\overline{DE}$. Determine the velocity of C. $\boldsymbol{v}_B = 30\boldsymbol{j}$ m/s at the instant shown.

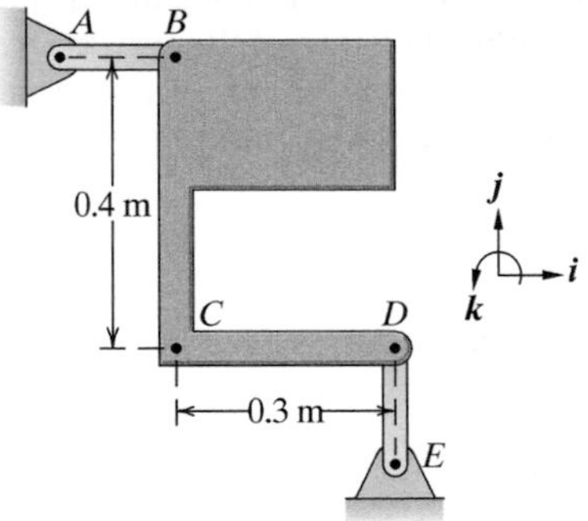

E6.1.36

6.1.37. **[Level 2]** Two square plates (1 and 2) are connected by a rigid link $\overline{BC}$. At the illustrated instant $\boldsymbol{r}_{A_{/B}} = (0.09\boldsymbol{b}_1 + 0.18\boldsymbol{b}_2)$ m, $\boldsymbol{r}_{D_{/C}} = (-0.18\boldsymbol{b}_1 + 0.18\boldsymbol{b}_2)$ m, and $\boldsymbol{r}_{C_{/B}} = -0.36\boldsymbol{b}_2$ m. Plate 1 is rotating with an angular velocity of $-4\boldsymbol{b}_3$ rad/s. Determine the angular velocity of Plate 2.

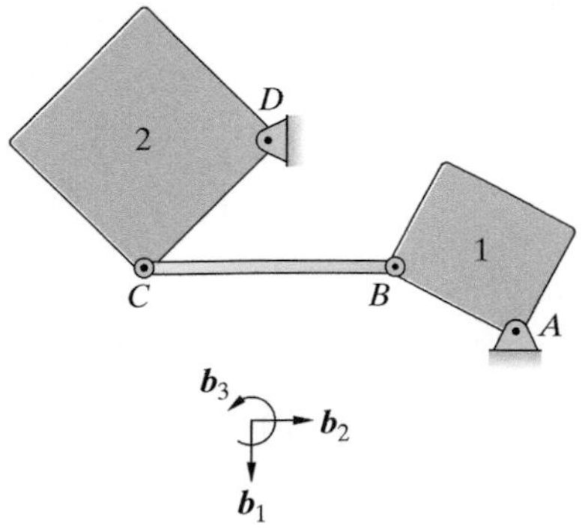

E6.1.37

6.1.38. **[Level 2]** The body R is supported by two links: $\overline{CA}$ and $\overline{BD}$. Is it possible that the links are rotating with angular velocities $\boldsymbol{\omega}_{\overline{CA}} = 10\boldsymbol{k}$ rad/s and $\boldsymbol{\omega}_{\overline{BD}} = 5\boldsymbol{k}$ rad/s?

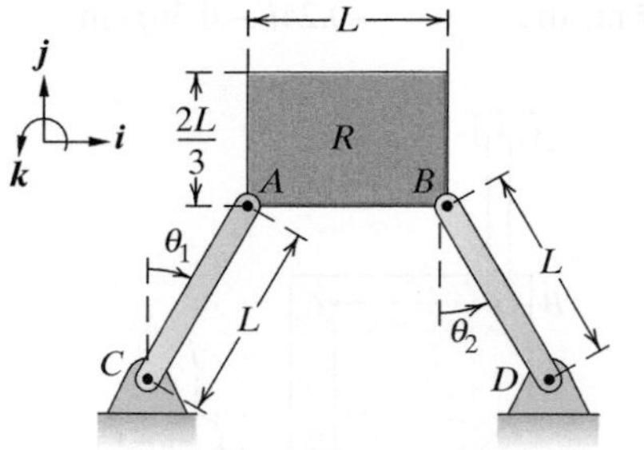

E6.1.38

6.1.39. **[Level 2]** A and B are points on a triangular rigid body. $\boldsymbol{v}_A = (0.3\boldsymbol{i} + 0.3\boldsymbol{j})$ m/s and $\boldsymbol{v}_B = (-5.8\boldsymbol{i} + 4.9\boldsymbol{j})$ m/s.

a. What must θ be equal to?

b. What is the distance from A to B?

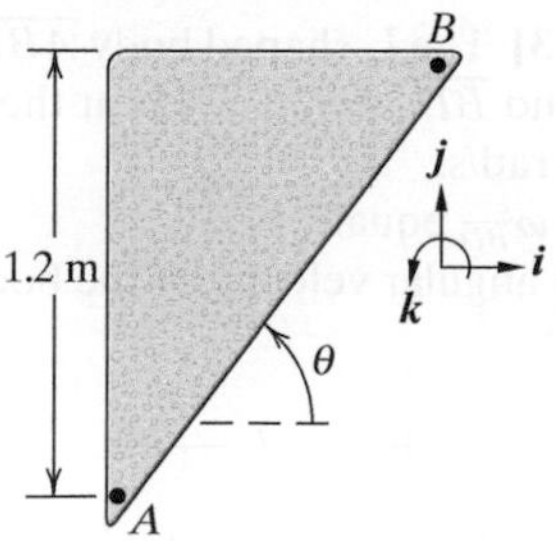

E6.1.39

6.1.40. **[Level 2]** The velocities of three points A, B, and C on a rigid body are given. Determine the rotational velocity $\boldsymbol{\omega}$ of the body as well as x_2, y_1, and y_2.

$$\boldsymbol{v}_A = (3\boldsymbol{i} + 2\boldsymbol{j})\text{ m/s}$$
$$\boldsymbol{v}_B = 4\boldsymbol{i}\text{ m/s}$$
$$\boldsymbol{v}_C = -4\boldsymbol{i}\text{ m/s}$$
$$\boldsymbol{r}_{A/B} = 1\boldsymbol{i}\text{ m} + y_1\boldsymbol{j}$$
$$\boldsymbol{r}_{C/B} = x_2\boldsymbol{i} + y_2\boldsymbol{j}$$

6.1.41. **[Level 2]** A multilink robotic arm is shown.

a. Determine the velocity of C in terms of the $\boldsymbol{c}_1, \boldsymbol{c}_2$ frame as well as $\dot{\theta}$.

b. What is the velocity of C relative to an observer at O, who rotates with the link $\overline{AB}$? Express your answer in terms of $\boldsymbol{b}_1, \boldsymbol{b}_2$. $\theta = 0, \phi = 30°, \boldsymbol{v}_B = 6\boldsymbol{j}$ m/s, and $\dot{\phi} = 10$ rad/s.

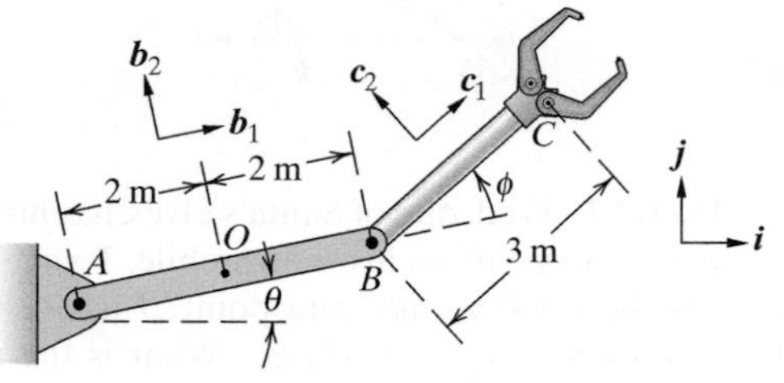

E6.1.41

6.1.42. **[Level 2]** Find $\boldsymbol{\omega}_{\overline{BC}}$ and $\boldsymbol{v}_B$ for the depicted apparatus: $\boldsymbol{\omega}_{\overline{CD}} = -6\boldsymbol{k}$ rad/s, $\boldsymbol{r}_{B/A} = 0.2\boldsymbol{j}$ m, $\boldsymbol{r}_{C/D} = (0.6\boldsymbol{i} + 0.7\boldsymbol{j})$ m, and $\boldsymbol{r}_{C/B} = (0.6\boldsymbol{i} - 0.1\boldsymbol{j})$ m.

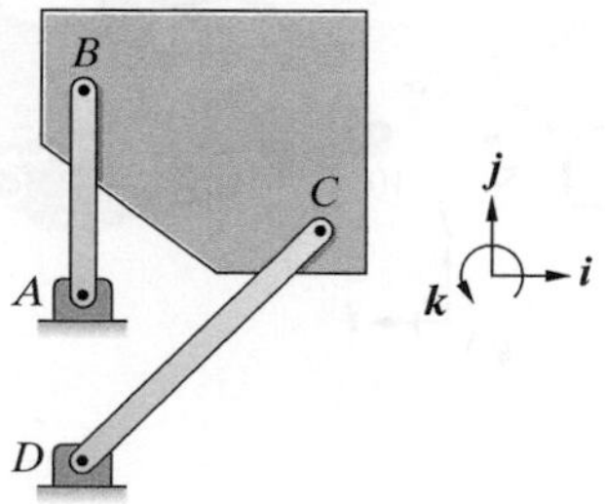

E6.1.42

6.1.43. **[Level 2]** The illustrated two-link system is actuated so that $\ddot{\theta}_1 = 6$ rad/s² and $\ddot{\theta}_2 = -10$ rad/s², starting from the configuration $\theta_1 = \theta_2 = \dot{\theta}_1 = \dot{\theta}_2 = 0$. When the link $\overline{BC}$ becomes vertically oriented (C below B), what is the velocity of C in terms of $\boldsymbol{i}$ and $\boldsymbol{j}$? $|\overline{AB}| = 4$ m, $|\overline{BC}| = 2$ m.

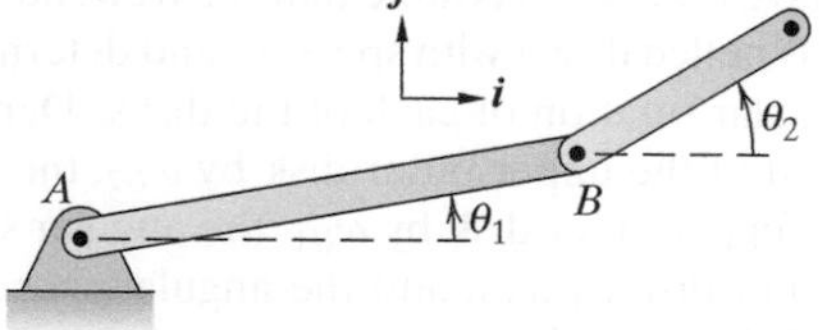

E6.1.43

6.1.44. **[Level 2]** Shown is a piston P in a cylinder, along with a connecting rod and crank arm. A is a fixed point of rotation. As the crankshaft rotates, it causes the connecting rod $\overline{BC}$ to oscillate, causing P to move up and down. Calculate the velocity of P when $\boldsymbol{r}_{B/A} = 0.06\boldsymbol{j}$ m. $\boldsymbol{\omega}_{AB} = 3000\boldsymbol{k}$ rpm and $r_{B/C} = 0.10$ m.

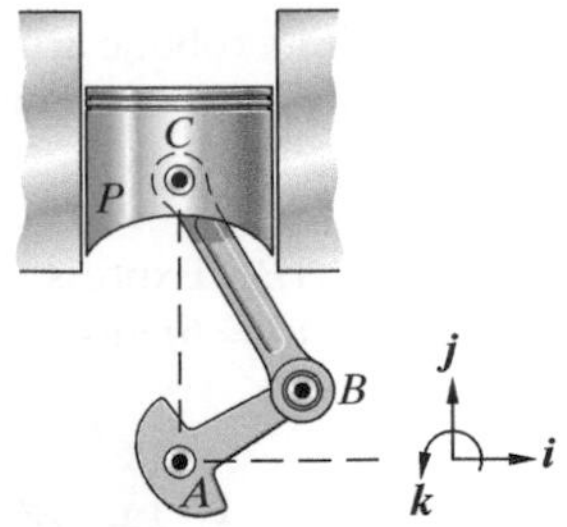

E6.1.44

6.1.45. **[Level 2]** Fred, one of Santa's elves, has just run off the edge of a snow cliff on his snowmobile. Point A has velocity $(-0.45\boldsymbol{b}_1 + 0.45\boldsymbol{b}_2)$ m/s, and point B has velocity $(-0.45\boldsymbol{b}_1 - 0.6\boldsymbol{b}_2)$ m/s. $\boldsymbol{r}_{B/A} = 0.7\boldsymbol{b}_1$ m. What is the rotational velocity of the snowmobile?

E6.1.45

6.1.46. **[Level 2]** Consider **Exercise 6.1.45**. Is it possible for the snowmobile to have the velocities $\boldsymbol{v}_A = (-0.37\boldsymbol{b}_1 + 0.53\boldsymbol{b}_2)$ m/s and $\boldsymbol{v}_B = (-0.4\boldsymbol{b}_1 - 0.53\boldsymbol{b}_2)$ m/s?

6.1.47. **[Level 3]** The illustrated pulley consists of four rotating pulley disks. In both the upper and the lower assembly, the outer disks both have radius r_2 and the inner disks both have radius r_1. Assume that the free end B of the pulley rope is pulled down with speed v_1 and determine the speed of angular rotation of each of the disks. Denote the angular speed of the upper outer disk by ω_{U2}, the angular speed of the upper inner disk by ω_{U1}, the angular speed of the lower outer disk by ω_{L2}, and the angular speed of the lower inner disk by ω_{L1}. Verify your answer by showing that the other end of the rope has zero speed (as it should, being attached to the center of the upper assembly). Assume for simplicity that all ropes are oriented vertically.

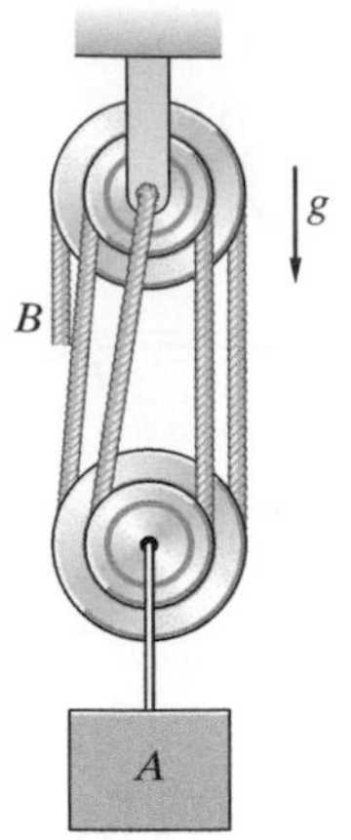

E6.1.47

6.1.48. **[Level 3]** Find $\boldsymbol{\omega}_{\overline{BCD}}$ and $\boldsymbol{v}_C$ for the given system: $\boldsymbol{\omega}_{\overline{DE}} = -7\boldsymbol{k}$ rad/s, $\boldsymbol{r}_{B/A} = -0.18\boldsymbol{j}$ m, $\boldsymbol{r}_{D/E} = (-0.3\boldsymbol{i} + 0.3\boldsymbol{j})$ m, $\boldsymbol{r}_{C/B} = 0.24\boldsymbol{i}$ m, and $\boldsymbol{r}_{D/B} = (-0.24\boldsymbol{i} - 0.36\boldsymbol{j})$ m.

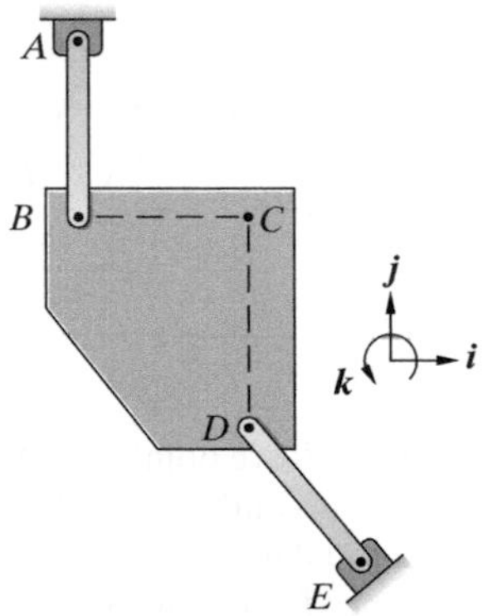

E6.1.48

6.1.49. **[Level 3]** The L-shaped body $\overline{ABE}$ is supported by the links $\overline{CA}$ and $\overline{BD}$. $\theta = \beta = 45°$ at the illustrated instant. $\boldsymbol{\omega}_{\overline{AC}} = 10\boldsymbol{k}$ rad/s.

a. What must $\boldsymbol{\omega}_{\overline{BD}}$ equal?

b. What is the angular velocity of the body $\overline{ABE}$?

c. What is $\boldsymbol{v}_E$?

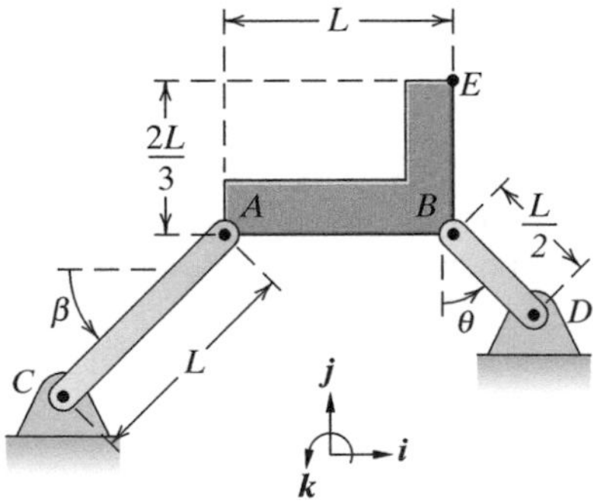

E6.1.49

6.1.50. **[Level 3]** Body 1 is an automotive piston, which moves vertically within its cylinder. The crank $\overline{OA}$ rotates at a constant angular velocity $\omega_{\overline{OA}} = 100\boldsymbol{k}$ rad/s about O, $|\overline{OA}| = 7.6$ cm, and $|\overline{AB}| = 13$ cm. Plot the speed of Body 1 versus time for one complete cycle of motion. You will see that the velocity profile looks sinusoidal. Try to match the response to a true sinusoidal response. Assume $v_B = A\cos(\theta - \phi)$ and determine A and ϕ from your numerical data. Plot both $v_B(t)$ and the approximation $A\cos(\theta - \phi)$. Is the match precise? Why or why not?

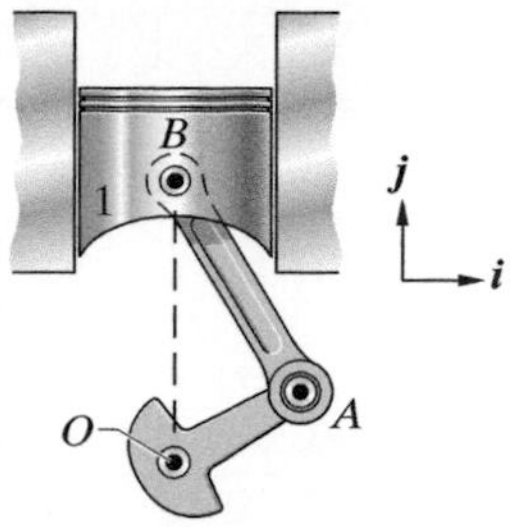

E6.1.50

6.2 INSTANTANEOUS CENTER OF ROTATION (ICR)

Now that we have seen how to find the velocity of an arbitrary point on a rigid body, it's time to learn an interesting fact that's often helpful. For any rigid body moving in a plane, there *always* exists a point that has zero velocity and is therefore a center of rotation, for at least an instant of time. This point is called the **instantaneous center of rotation (ICR)** or, alternatively, the **instantaneous center of zero velocity**. The simplest case is one in which the body is pivoting about some fixed point. In that case the instantaneous center of rotation is fixed in space and is just the pivot point. Even for bodies in general motion, however, a point of rotation exists at every instant of time. The notion of an instantaneous center of rotation is a specific case of a more general finding—that the motion of any rigid body can be viewed as being made up of a rigid translation and a rotation about a single axis.

How can we find the instantaneous center of rotation? It's actually pretty easy to do. Look at **Figure 6.19**, which shows a body that's rotating about a fixed pivot point at O. Three representative velocity vectors are shown, along with the position vectors from the pivot point to three points on the body. Note how each velocity is at a right angle to its associated position vector. This makes sense because the velocity of point A, for instance, is given by

$$\boldsymbol{v}_A = \boldsymbol{v}_O + \boldsymbol{\omega} \times \boldsymbol{r}_{A/O}$$

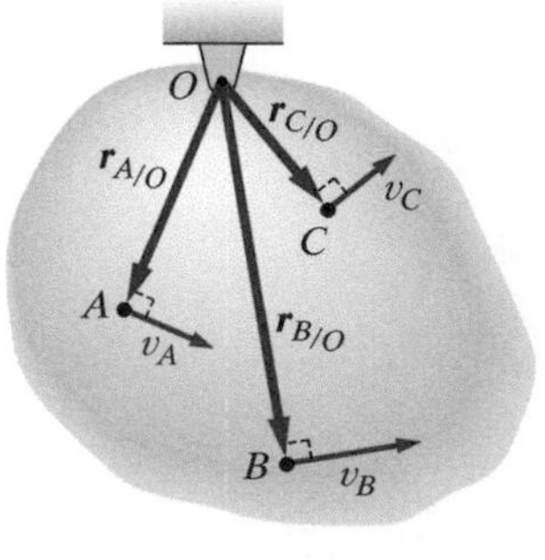

Figure 6.19 Body rotating about a fixed point

Because $\boldsymbol{v}_O$ is zero, all we have is the cross product of angular velocity with the direction vector $\boldsymbol{r}_{A/O}$.

What we get out of this observation is the fact that the velocity of any point on a rigid body is always going to be directed at right angles to the position vector going from the instantaneous center of rotation to that point. Thus, to find the instantaneous center of rotation, all we need to do is find two points, sketch their velocity vectors, draw lines at right angles to these vectors, and find where they intersect. Because both lines have to go through the instantaneous center of rotation and because this is a unique point, the intersection of the two lines defines that point.

Once you've found the ICR of a body B, you can find the velocity of any arbitrary point A on the body from

$$\boldsymbol{v}_A = \boldsymbol{\omega}_B \times \boldsymbol{r}_{A/ICR} \tag{6.15}$$

I drew **Figure 6.19** with a physical hinge simply for illustrative purposes. All rotating rigid bodies, whether or not they're hinged, have an instantaneous center of rotation.

>>> **Check out Example 6.6 (page 335), Example 6.7 (page 336), Example 6.8 (page 337), Example 6.9 (page 338), and Example 6.10 (page 340) for applications of this material.**

EXAMPLE 6.6 ANGULAR SPEED DETERMINATION VIA ICR (Theory on page 334)

Figure 6.20 shows a two-dimensional body that's connected to the ground by two links $\overline{AB}$ and $\overline{CD}$, which are both free to rotate in the plane of the figure so that the angles θ and β can change. $|\overline{AB}| = |\overline{CD}| = 1.1$ m. For $\boldsymbol{v}_B = -10\boldsymbol{j}$ m/s, $\beta = 90°$, and $\theta = 45°$, find $\dot{\phi}$, the angular speed of the body.

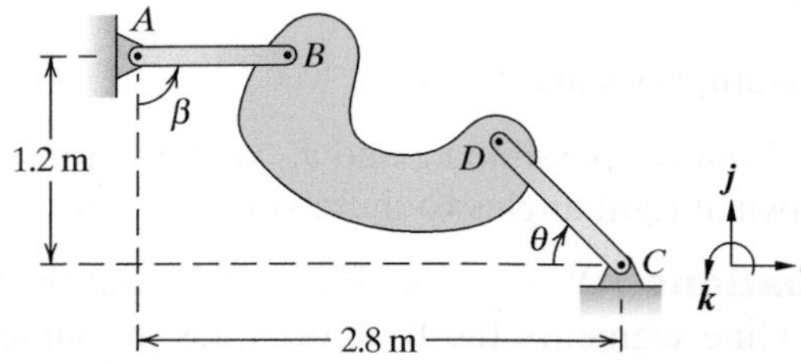

Figure 6.20 Rigid body constrained by two links

Goal Find the instantaneous value of $\dot{\phi}$ for the illustrated constrained body.

Given $|\overline{AB}| = |\overline{CD}| = 1.1$ m, $\boldsymbol{v}_B = -10\boldsymbol{j}$ m/s, $\beta = 90°$, and $\theta = 45°$

Assume No assumptions are needed.

Draw **Figure 6.21** expands upon **Figure 6.20** by adding $\boldsymbol{v}_B$ and $\boldsymbol{v}_D$ to the sketch, along with line segments at right angles to these velocity vectors. The intersection of these two line segments, labeled *ICR*, is the instantaneous center of rotation. From geometry, we can find L_1, the distance from the end B of the link AB to the *ICR*, to be 0.5 m.

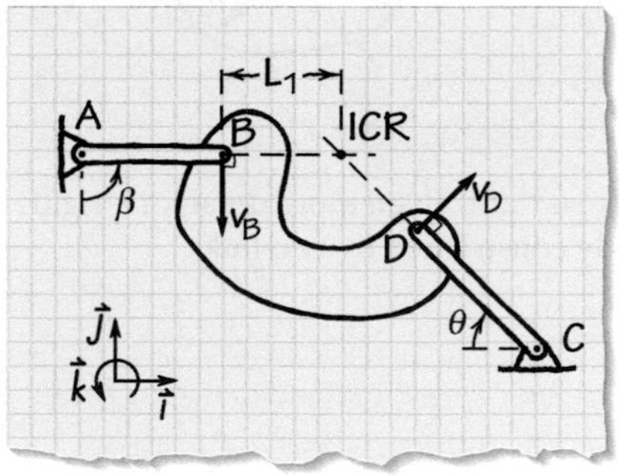

Figure 6.21 More detail of a rigid body constrained by two links

Formulate Equations Our relevant equation is (6.15):

$$\boldsymbol{v}_B = \dot{\phi}\boldsymbol{k} \times \boldsymbol{r}_{B/ICR}$$

Solve

$$\boldsymbol{v}_B = \dot{\phi}\boldsymbol{k} \times \boldsymbol{r}_{B/ICR}$$

$$-10\boldsymbol{j}\,\text{m/s} = \dot{\phi}\boldsymbol{k} \times (-L_1\boldsymbol{i})$$

$$-10\boldsymbol{j}\,\text{m/s} = -(0.5\,\text{m})\dot{\phi}\boldsymbol{j}$$

This last equation is easily solved:

$$\dot{\phi} = 20\ \text{rad/s}$$

Check For this example we will just check on the correctness of the rotation direction. From **Figure 6.21** we see that $\boldsymbol{v}_B$ is oriented downward and thus indicates a counterclockwise rotation about the *ICR*. This means we should expect a positive value of $\dot{\phi}$, exactly as we just obtained.

EXAMPLE 6.7 VELOCITY ON A CONSTRAINED BODY VIA ICR (Theory on page 334)

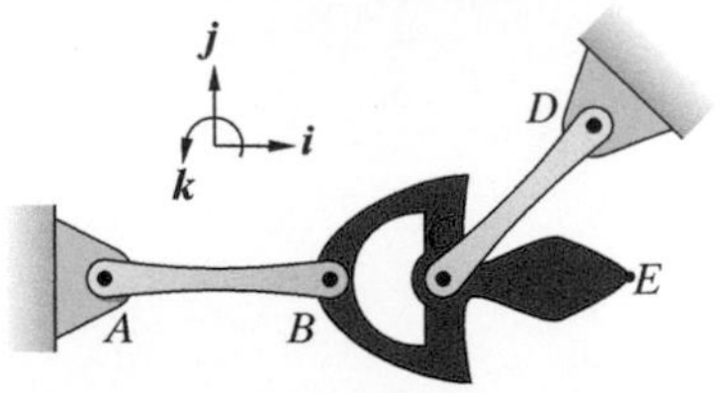

Figure 6.22 More detail of a rigid body constrained by two links

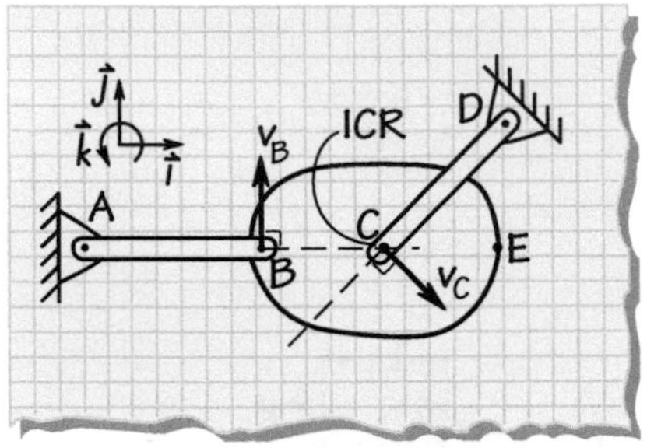

Figure 6.23 Even more detail of a rigid body constrained by two links

Figure 6.22 shows the system we encountered in Example 6.4. We'll verify our previous solution using the ICR concept.

Goal Determine E's velocity.

Given $\omega_{\overline{AB}} = 10\,\text{rad/s}$, $\boldsymbol{r}_{B/A} = 0.6\boldsymbol{i}\,\text{m}$, $\boldsymbol{r}_{C/B} = 0.3\boldsymbol{i}\,\text{m}$, $\boldsymbol{r}_{D/C} = (0.4\boldsymbol{i} + 0.4\boldsymbol{j})\,\text{m}$, and $\boldsymbol{r}_{E/C} = 0.5\boldsymbol{i}\,\text{m}$.

Assume No assumptions are needed.

Draw **Figure 6.23** shows possible $\boldsymbol{v}_B$ and $\boldsymbol{v}_C$ vectors, along with dashed line segments drawn at right angles to these velocity vectors.

Formulate Equations We see from **Figure 6.23** that the intersection of the two dashed line segments, the ICR, occupies the same position as C. Thus we can view the body as rotating about C. Note that this also means that the length of the $\boldsymbol{v}_C$ vector, drawn arbitrarily long in the figure, actually has a length of zero.

Solve Because our body can be viewed as rotating about C, we have

$$\omega_{\overline{BC}} = \frac{|v_B|}{|r_{B/C}|} = \frac{|v_E|}{|r_{E/C}|} \quad \Rightarrow \quad |v_E| = |v_B|\frac{|r_{E/C}|}{|r_{B/C}|}$$

$$|v_E| = (6\,\text{m/s})\left(\frac{0.5\,\text{m}}{0.3\,\text{m}}\right) = 10\,\text{m/s}$$

From geometry, we see that if $\boldsymbol{v}_B$ is moving in the $\boldsymbol{j}$ direction, then $\boldsymbol{v}_E$ must be moving in the $-\boldsymbol{j}$ direction.

$$\boldsymbol{v}_E = -10\boldsymbol{j}\,\text{m/s}$$

Check Our result matches the result calculated in Example 6.4.

EXAMPLE 6.8 VELOCITY OF THE CONTACT POINT DURING ROLL WITHOUT SLIP (Theory on page 318)

Figure 6.24 shows the front wheel and tire of a car. We'll approximate the wheel/tire as a circular disk of radius r. The car is traveling to the left, and the disk's center has a velocity of $-v\boldsymbol{i}$. Determine the velocity of the point of the disk that's in contact with the floor.

Figure 6.24 Rolling wheel/tire

Goal Determine the velocity of the contact point of a rolling disk.

Given Speed of the disk's center and the disk's dimensions.

Assume We'll assume that the tire (disk) is perfectly circular.

Draw I've sketched just three positions of the disk in **Figure 6.25**, along with labels for the disk's center G and the contact point C.

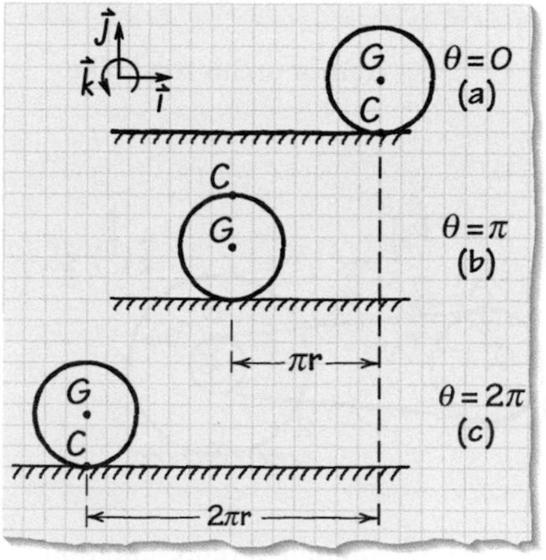

Figure 6.25 Snapshots of rolling wheel

Formulate Equations The relevant equation is again (6.4). We're given the fact that the disk's center is moving to the left with speed v: $v_G = -v\boldsymbol{i}$. An easy way to relate the rotational speed to the disk's translational speed is to assume that v is constant and let the disk roll for one full revolution, as shown in **Figure 6.25**. You know from geometry that the perimeter of the disk is $2\pi r$, and therefore G, the disk's center, has moved $2\pi r$ to the left. The time it took to do this is equal to the distance traveled divided by the speed: $\Delta t = 2\pi r/v$.

Finding the rotational speed ω is straightforward—just divide the angle rotated (2π) by the time it took (Δt):

$$\omega = \frac{2\pi}{\Delta t} = \frac{v}{r}$$

Recalling that the disk is moving to the left, we have $\boldsymbol{v}_G = -\omega r\boldsymbol{i}$.

Solve In **Figure 6.25*a*** point C is in contact with the ground, and you can see that $\boldsymbol{r}_{C/G} = -r\boldsymbol{j}$. Applying (6.4) gives us

$$\boldsymbol{v}_C = \boldsymbol{v}_G + \boldsymbol{\omega}\times\boldsymbol{r}_{C/G} = -\omega r\boldsymbol{i} + \omega\boldsymbol{k}\times(-r\boldsymbol{j}) = -\omega r\boldsymbol{i} + \omega r\boldsymbol{i} \quad\Rightarrow\quad \boxed{\boldsymbol{v}_C = 0}$$

Thus we see that the *velocity of the point of contact between a disk rolling without slip and the ground is zero—the contact point is the wheel's instantaneous center of rotation.*

Check We can go backwards to check, from ground to G. If the contact point C has zero velocity, then we can view the disk as rotating about that point. $\boldsymbol{v}_G = 0 + \omega\boldsymbol{k}\times\boldsymbol{r}_{G/C} = -\omega r\boldsymbol{i}$. This matches our original data, namely that the center of the disk is moving to the left at a speed $v = \omega r$.

EXAMPLE 6.9 PEDALING CADENCE AND BICYCLE SPEED (Theory on pages 318 and 337)

Figure 6.26 Author on his road bike

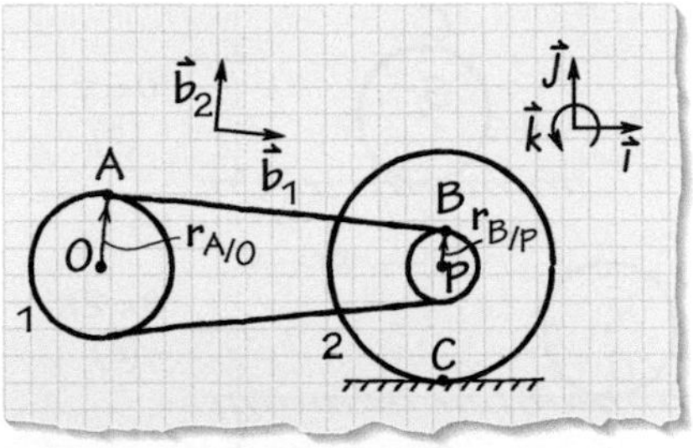

Figure 6.27 Simplified representation of moving bicycle

Figure 6.26 shows a shot of me cruising along on my Air Friday, a high-performance folding bicycle. Assume that I'm pedaling at 90 rpm and my chain runs between the 12 cm radius front chain ring and the 3 cm radius rear cog. The wheels have a radius of 25 cm. Determine how fast I'm traveling.

Goal Determine the translational velocity of a cyclist.

Given Pedaling cadence and bicycle dimensions.

Assume Assume that the cyclist is traveling in a no-slip condition.

Draw In **Figure 6.27** the drivetrain is drawn as a single speed (dropping the multiple rear cogs and derailleur). Body 1 (the front chain ring) is coupled to Body 2 (the rear tire/wheel/cogs, all of which move as a single unit) by the chain. The vectors $\boldsymbol{r}_{A/O}$ and $\boldsymbol{r}_{B/P}$ go from the centers of their respective bodies to the chain and are at right angles to it.

Formulate Equations This example will require (6.4) as well as the roll-without-slip condition. We'll move from the ground to the front chain ring and determine how the front chain ring's rotational speed affects the bike's translational velocity.

Solve We've already seen that the point of contact (during roll without slip) has zero velocity, and knowing the angular speed of the wheel lets us move from C to P using (6.4):

$$\begin{aligned} \boldsymbol{v}_P &= \boldsymbol{v}_C + \boldsymbol{\omega}_2 \times \boldsymbol{r}_{P/C} = 0 + \omega_2 \boldsymbol{k} \times r_{P/C}\boldsymbol{j} \\ &= -\omega_2 r_{P/C}\boldsymbol{i} \end{aligned} \tag{6.16}$$

Now we know the bike's translational velocity ($-\omega_2 r_{P/C}\boldsymbol{i}$), but it's not in a fully usable form because we don't yet know how to relate ω_2 to our given data. To go further, jump from P to B, the point at which the chain leaves the rear cog:

$$\begin{aligned} \boldsymbol{v}_B &= \boldsymbol{v}_P + \boldsymbol{\omega}_2 \times \boldsymbol{r}_{B/P} = -\omega_2 r_{P/C}\boldsymbol{i} + \omega_2 \boldsymbol{k} \times r_{B/P}\boldsymbol{b}_2 \\ &= -\omega_2 r_{P/C}\boldsymbol{i} - \omega_2 r_{B/P}\boldsymbol{b}_1 \end{aligned}$$

Now we jump from B to A, the point at which the chain gets taken up by the front chain ring. The whole point of a chain is that, even though it's flexible when bent, it acts like a rigid link with regard to longitudinal motion when under tension. Thus $\boldsymbol{v}_A = \boldsymbol{v}_B$. This means that we now know the velocity of A, a point on the periphery of the front chain ring, and we can use (6.4) to get O's velocity:

$$\begin{aligned} \boldsymbol{v}_O &= \boldsymbol{v}_A + \boldsymbol{\omega}_1 \times \boldsymbol{r}_{O/A} = \boldsymbol{v}_A + \omega_1 \boldsymbol{k} \times (-r_{O/A}\boldsymbol{b}_2) \\ &= \boldsymbol{v}_A + \omega_1 r_{O/A}\boldsymbol{b}_1 = -\omega_2 r_{P/C}\boldsymbol{i} - \omega_2 r_{B/P}\boldsymbol{b}_1 + \omega_1 r_{O/A}\boldsymbol{b}_1 \end{aligned} \tag{6.17}$$

According to (6.17), $\boldsymbol{v}_O$ has a velocity with components in the $\boldsymbol{i}$ and $\boldsymbol{b}_1$ directions. But we know from physical considerations that O simply moves horizontally, along with the rest of the bicycle frame. Thus the $\boldsymbol{b}_1$ component must be zero, and so we require

$$-\omega_2 r_{B_{/P}} + \omega_1 r_{O_{/A}} = 0$$

$$\omega_2 = \frac{\omega_1 r_{O_{/A}}}{r_{B_{/P}}} \tag{6.18}$$

This is the key result we need. Using both (6.16) and (6.18) gives us

$$\boldsymbol{v}_P = -\frac{\omega_1 r_{O_{/A}} r_{P_{/C}}}{r_{B_{/P}}}\boldsymbol{i} \tag{6.19}$$

From here it's just plug and chug. We know that the front chainwheel is being turned at 90 rpm, or 3π rad/s, and that the different radii have all been specified. Using the given data yields

$$\boldsymbol{v}_P = -\frac{(3\pi \text{ rad/s})(0.12 \text{ m})(0.25 \text{ m})}{0.03 \text{ m}}\boldsymbol{i} = -9.42\boldsymbol{i} \text{ m/s}$$

Check Let's just do a physical reasonableness check for this one. The data given are all realistic and come from my actual bicycle: 90 rpm is a racing cadence, the given chain ring is my "big" chain ring, and the cog radius corresponds to one of my smaller rear cogs. Thus I would expect the speed to be fairly high, around 32–40 km/h. In line with my expectations, 9.42 m/s corresponds to 34 km/h. So I can reasonably conclude that the answer is likely to be accurate.

EXAMPLE 6.10 ROTATION RATE OF AN UNWINDING REEL VIA ICR (Theory on page 334)

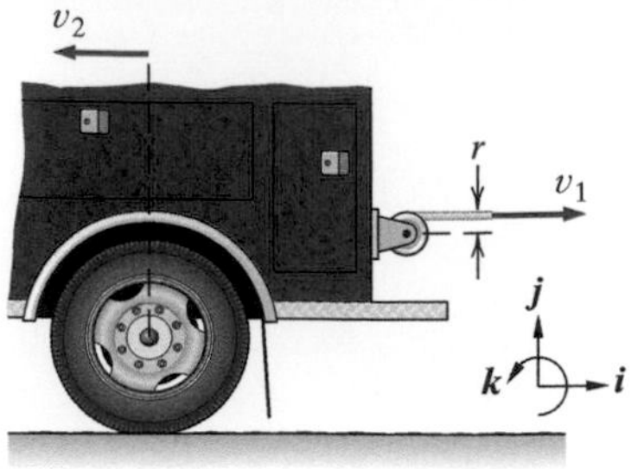

Figure 6.28 Rope unreeling from moving truck

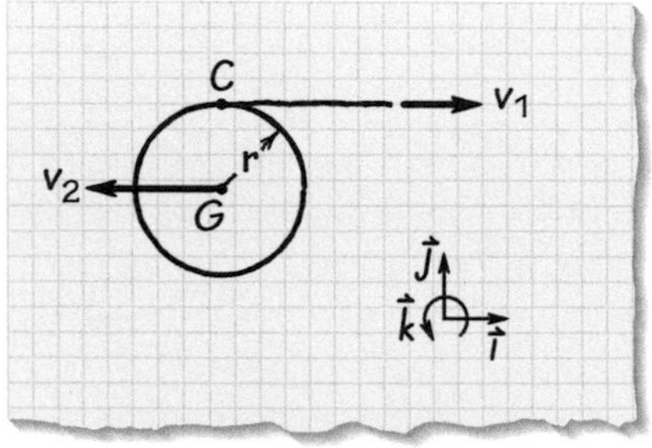

Figure 6.29 Schematic of the reel

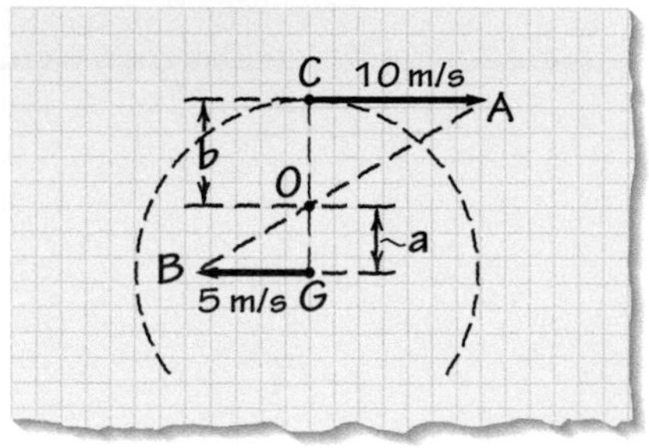

Figure 6.30 Position of instantaneous center of rotation

A tow rope attached to the rear of an emergency vehicle is being unreeled as the truck moves to the left (**Figure 6.28**). The truck's velocity is $-v_2\boldsymbol{i}$, and the velocity of the rope's free end with respect to the ground is $v_1\boldsymbol{i}$. Find the angular velocity of the reel and the position of its instantaneous center of rotation ($r = 0.3$ m, $v_1 = 10$ m/s, and $v_2 = 5$ m/s).

Goal Find a reel's instantaneous center of rotation and angular velocity.

Given Velocity of two points on the moving reel.

Assume No additional assumptions are needed.

Draw **Figure 6.29** shows the reel and the velocities imposed by the truck and the rope, with the reel center defined as the point G.

Formulate Equations The only equations we will need in this case are

$$\boldsymbol{v}_1 = \boldsymbol{\omega}_R \times \boldsymbol{r}_{C/O} \quad \text{and} \quad \boldsymbol{v}_2 = \boldsymbol{\omega}_R \times \boldsymbol{r}_{G/O}$$

where O represents the instantaneous center of rotation and $\boldsymbol{\omega}_R$ is the reel's angular velocity.

Solve **Figure 6.30** shows the instantaneous center of rotation (labeled O) along with the points C and G and their associated velocity vectors, with their tips labeled A and B. The two triangles $\overline{OCA}$ and $\overline{OGB}$ are similar, and therefore the ratio $\frac{|CA|}{|GB|}$ is the same as $\frac{|OC|}{|OG|}$. Thus we have

$$\frac{b}{a} = \frac{10}{5} = 2 \quad \Rightarrow \quad b = 2a \tag{6.20}$$

We know that the total radius of the reel is 0.3 m and so

$$r = a + b \underbrace{=}_{(6.20)} 3a = 0.3\,\text{m} \quad \Rightarrow \quad a = 0.1\,\text{m} \quad \Rightarrow \quad \boxed{\boldsymbol{r}_{O/G} = 0.1\boldsymbol{j}\,\text{m}}$$

We note that the reel's rotation is clockwise, and from the lower triangle, we have

$$\boldsymbol{v}_2 = \boldsymbol{\omega}_R \times \boldsymbol{r}_{G/O} \quad \Rightarrow \quad -5\boldsymbol{i}\,\text{m/s} = \omega_R\boldsymbol{k} \times (-0.1\boldsymbol{j}\,\text{m}) \quad \Rightarrow \quad \boxed{\boldsymbol{\omega}_R = -50\boldsymbol{k}\,\text{rad/s}}$$

Check We can verify the accuracy of these answers by using our rigid-body velocity relationships. First, we can relate $\boldsymbol{v}_1$ to $\boldsymbol{v}_2$:

$$\boldsymbol{v}_1 = \boldsymbol{v}_2 + \boldsymbol{\omega}_R \times \boldsymbol{r}_{C/G} \quad \Rightarrow \quad 10\boldsymbol{i}\,\text{m/s} = -5\boldsymbol{i}\,\text{m/s} + \omega_R\boldsymbol{k} \times 0.3\boldsymbol{j}\,\text{m} \quad \Rightarrow \quad \boldsymbol{\omega}_R = -50\boldsymbol{k}\,\text{rad/s}$$

$\boldsymbol{v}_2 + \boldsymbol{\omega}_R \times r_{O/G}\boldsymbol{j} = \boldsymbol{v}_O = 0 \Rightarrow a = 0.1\,\text{m}$. Thus we have verified that $\boldsymbol{r}_{O/G} = 0.1\boldsymbol{j}\,\text{m}$.

EXERCISES 6.2

6.2.1. [Level 1] Find $\omega_{\overline{BC}}$ and v_B. $\boldsymbol{\omega}_{\overline{CD}} = -6\boldsymbol{k}$ rad/s, $\boldsymbol{r}_{B_{/A}} = 0.2\boldsymbol{j}$ m, $\boldsymbol{r}_{C_{/D}} = (0.6\boldsymbol{i} + 0.7\boldsymbol{j})$ m, and $\boldsymbol{r}_{C_{/B}} = (0.6\boldsymbol{i} - 0.1\boldsymbol{j})$ m.

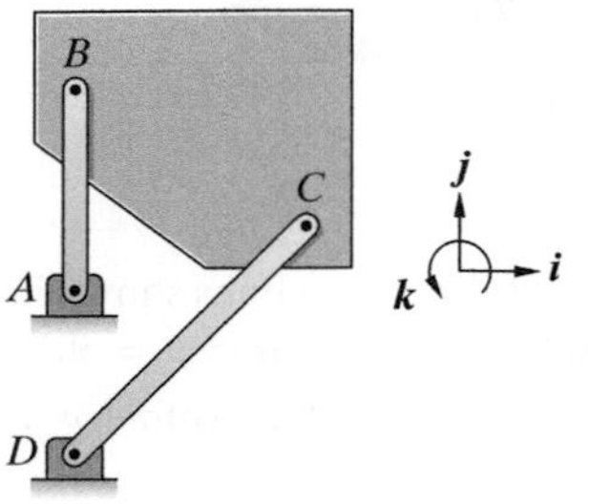

E6.2.1

6.2.2. [Level 1] The link $\overline{AB}$ connects the rotating link $\overline{OA}$ with the vertically translating link $\overline{BC}$ through rotational joints. $\theta = 30°$, $\dot{\theta} = 4$ rad/s, $\ddot{\theta} = 0$, $L_1 = 0.8$ m, and $L_2 = 0.462$ m. What is the angular velocity of the link $\overline{AB}$?

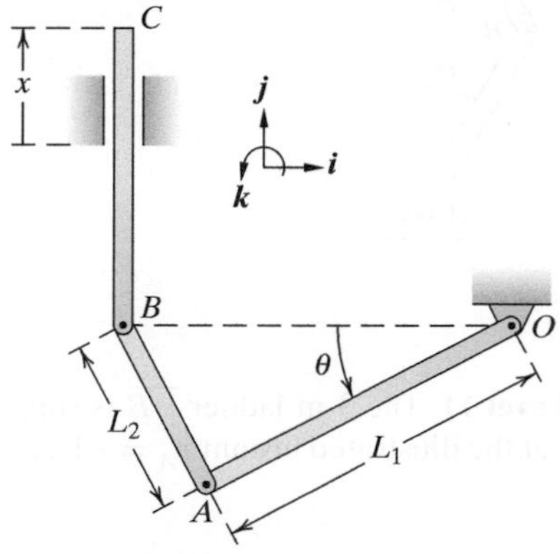

E6.2.2

6.2.3. [Level 1] Find v_O using the instantaneous center of rotation. $v_A = 11\boldsymbol{i}$ m/s and $v_B = -3.6\boldsymbol{j}$ m/s.

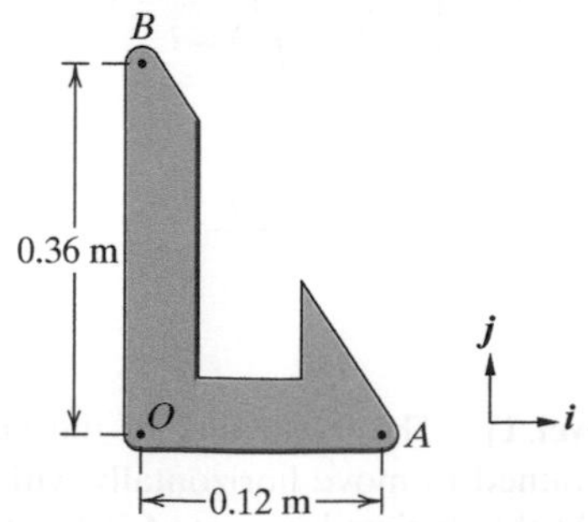

E6.2.3

6.2.4. [Level 1] At the given instant, link $\overline{AB}$ is angled at $\theta = 35°$ with respect to the horizontal, where its left end A is attached to the center of a uniform cylinder with radius $r = 0.25$ m that is rolling without slip at a constant angular speed of $\omega = 10$ rad/s counterclockwise. The other end B is connected to a collar that is constrained to move along a smooth vertical rail. If link $\overline{AB}$ has a length of $L = 1$ m, what are $\omega_{\overline{AB}}$ and v_B?

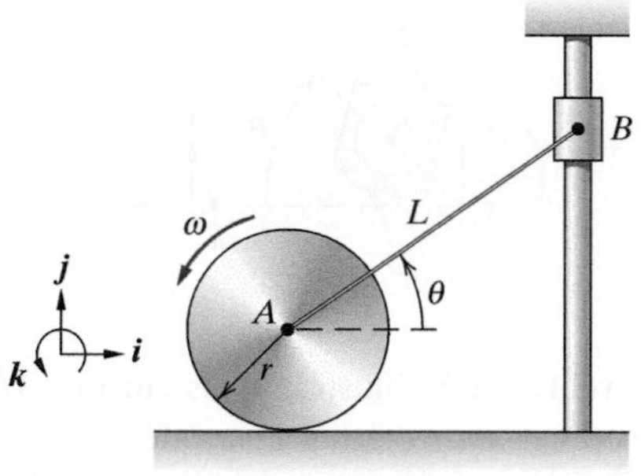

E6.2.4

6.2.5. [Level 1] Solve **Exercise 6.1.36** using the instantaneous center of rotation.

6.2.6. [Level 1] An actuator drives the end A of the rod $\overline{AC}$ purely in the $\boldsymbol{i}$ direction. The motion of the rod is further constrained by the fact that it passes through a small hole in a flat plate at B. Note that B indicates the location of the hole in the constraining plate, not a fixed point on the rod. Determine the rod's rotation rate when $\theta = 30°$, $|\boldsymbol{r}_{AB}| = 0.2$ m, and $\boldsymbol{v}_A = 4\boldsymbol{i}$ m/s.

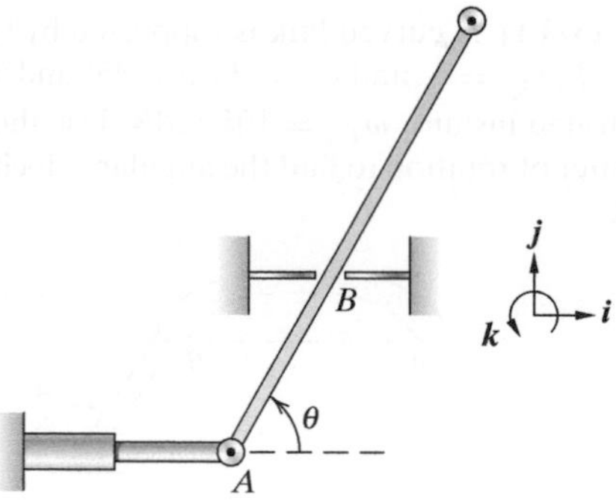

E6.2.6

6.2.7. [Level 1] A cross section of a piston P in a cylinder is shown, along with a connecting rod and crank arm. As the crankshaft rotates, it causes the connecting rod $\overline{BC}$ to oscillate, inducing a reciprocating, vertical motion on P. The crank arm is rotating at $\boldsymbol{\omega}_{\overline{AB}} = 3000\boldsymbol{k}$ rpm. Calculate the velocity of P when $\boldsymbol{r}_{B_{/A}} = 0.06\boldsymbol{j}$ m and $r_{C_{/B}} = 0.10\boldsymbol{j}$ m.

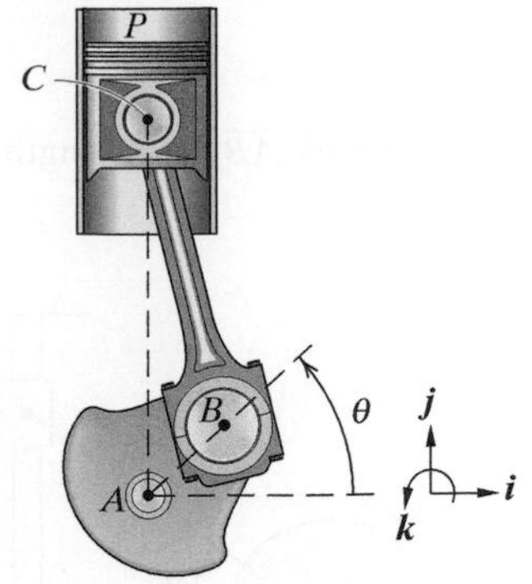

E6.2.7

6.2.8. **[Level 1]** At the illustrated instant, link $\overline{AB}$ ($r_{B/A} = 0.18$ m) is vertical and rotating at a rate $\boldsymbol{\omega}_{\overline{AB}} = 9\boldsymbol{k}$ rad/s, and link $\overline{CD}$ ($r_{C/D} = 0.21$ m) is horizontal. If link $\overline{BC}$ ($r_{C/B} = 0.27$ m) is angled at $\theta = 60°$ with respect to the vertical, what are $\omega_{\overline{BC}}$ and $\omega_{\overline{CD}}$?

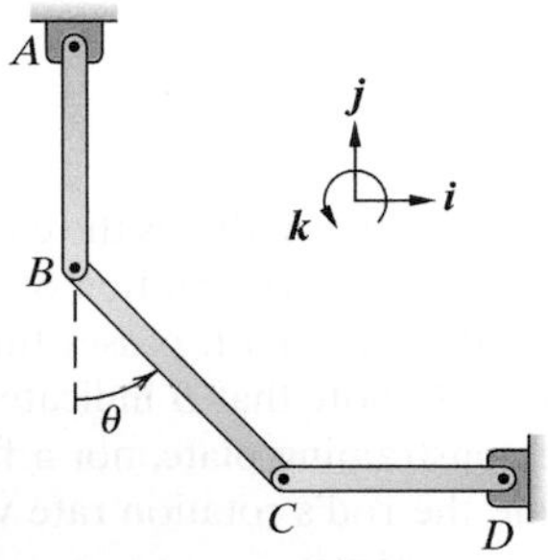

E6.2.8

6.2.9. **[Level 1]** A curved link is supported by two rigid links. $r_{C/A} = L$, $r_{B/D} = \frac{L}{2}$, and $r_{B/A} = L$. $\beta = 45°$ and $\theta = 135°$ at the illustrated instant. $\boldsymbol{\omega}_{\overline{CA}} = 10\boldsymbol{k}$ rad/s. Use the instantaneous center of rotation to find the angular velocity of the curved link.

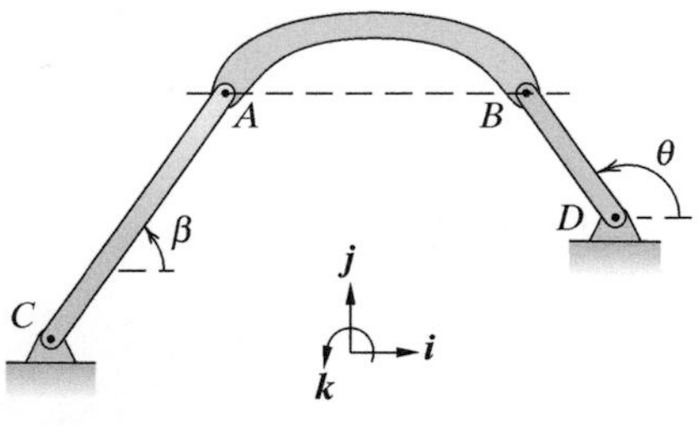

E6.2.9

6.2.10. **[Level 1]** A wheel is instrumented as part of a car manufacturer's vehicle-testing program so as to produce velocity information on a new mini-car that's undergoing winter testing. A series of brake tests are being run under icy conditions, and thus slippage may occur. When the tests are run in a straight line in the $-\boldsymbol{i}$ direction, the telemetry data indicate that the point A has a velocity of $(-102\boldsymbol{i} - 76\boldsymbol{j})$ cm/s. $\boldsymbol{r}_{A/G} = (-20\boldsymbol{i} - 15\boldsymbol{j})$ cm. What is the velocity of the point B?

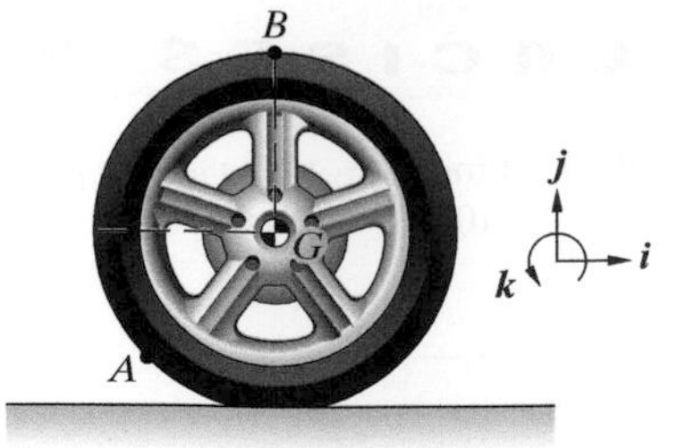

E6.2.10

6.2.11. **[Level 1]** The three links shown in the figure are oriented such that $\theta = 90°$ and $\alpha = \beta = 30°$. $r_{D/C} = 4L$ and $r_{C/B} = r_{B/A} = 3L$. $\dot{\theta} = 2$ rad/s. Determine $\boldsymbol{\omega}_{\overline{BC}}$.

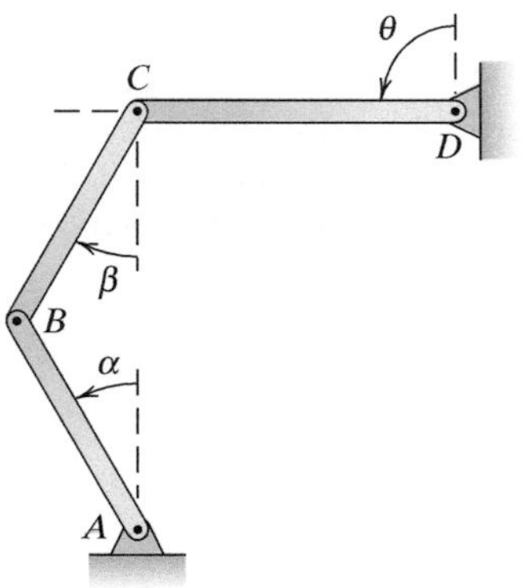

E6.2.11

6.2.12. **[Level 1]** The 3 m ladder $\overline{AB}$ is slipping down the wall, and at the illustrated instant $\boldsymbol{v}_A = -1.2\boldsymbol{j}$ m/s. What is $\boldsymbol{\omega}_{\overline{AB}}$?

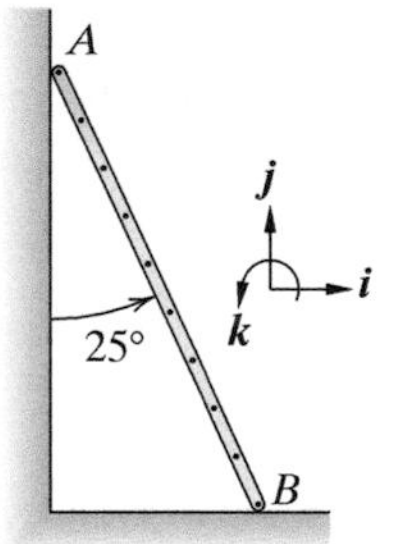

E6.2.12

6.2.13. **[Level 1]** The lower link of the given mechanism is constrained to move horizontally with a velocity $\boldsymbol{v}_A = 3\boldsymbol{i}$ m/s. At the depicted instant, A is located directly beneath the fixed pivot point C, and links $\overline{AB}$ and $\overline{BC}$ are orthogonal and of equal length, $r_{B/A} = r_{B/C} = 0.36$ m. Find $\omega_{\overline{AB}}$ and v_B.

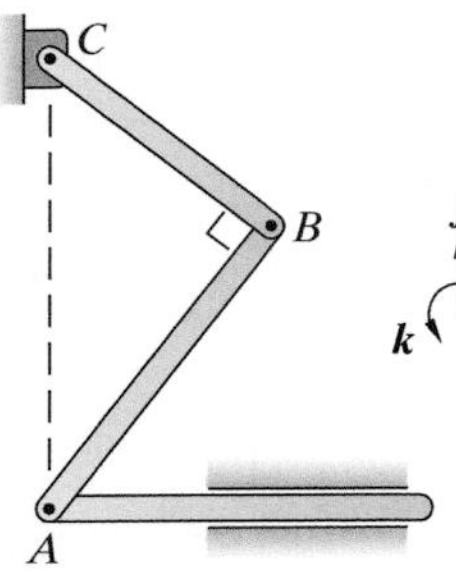

E6.2.13

6.2.14. **[Level 1]** Two cylinders (A and B) of identical height are resting on their side on a smooth horizontal surface. As illustrated, their centers are connected by a rigid rod of length L, and the cylinders are constrained to roll without slip against two vertical surfaces that meet at a right angle. At a certain instant, the rod makes an angle of $\theta = 40°$ as depicted, and cylinder B rotates at $\omega_B = 3$ rad/s clockwise. If the radii of the cylinders are $r_A = 0.15$ m and $r_B = 0.1$ m, find the rotational speed ω_A for cylinder A.

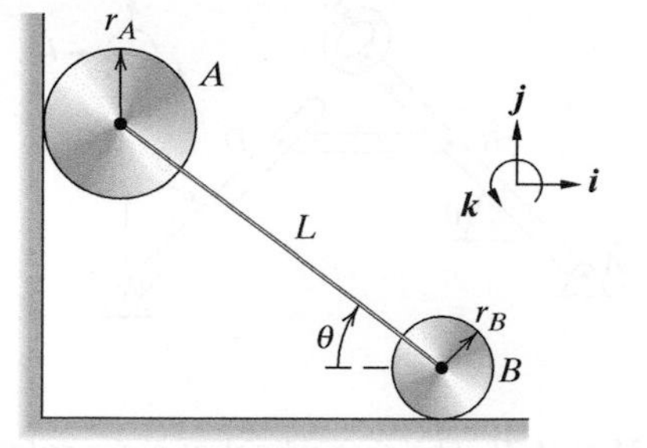

E6.2.14 Top view

6.2.15. **[Level 1]** The rotating propeller is moving to the left with speed v. $\dot{\theta} = \omega$ and $\ddot{\theta} = 0$. Determine where the instantaneous center of rotation is along the propeller when $\theta = 90°$.

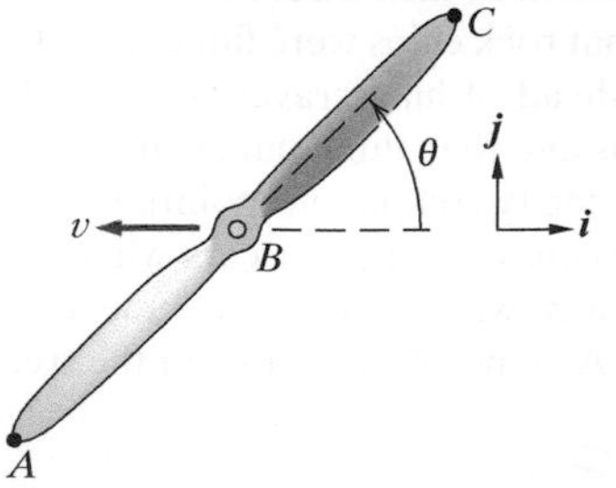

E6.2.15

6.2.16. **[Level 1]** Two straight links control the position of a half-disk. $|r_{C/A}| = |r_{B/A}| = h$, $|r_{B/D}| = \frac{h}{2}$. $\theta = \beta = 45°$ at the illustrated instant. $\omega_{\overline{CA}} = 20\boldsymbol{k}$ rad/s. Use the instantaneous center of rotation to find the angular velocity of the half-disk. $\overline{AB}$ is horizontal.

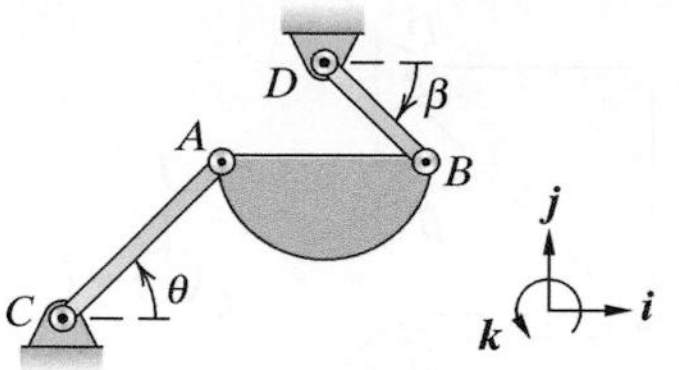

E6.2.16

6.2.17. **[Level 1]** Link $\overline{AB}$ of the illustrated system is $r_{B/A} = 0.75$ m long and angled at $\theta = 30°$ with respect to the horizontal. Link $\overline{CD}$ has a length of $r_{C/D} = 0.5$ m and is oriented at $\beta = 60°$ to the horizontal. Link $\overline{BC}$ is horizontal and $r_{C/B} = 1$ m long, and $\boldsymbol{\omega}_{\overline{AB}} = -4\boldsymbol{k}$ rad/s at the given instant. Calculate $\omega_{\overline{BC}}$ and $\omega_{\overline{CD}}$.

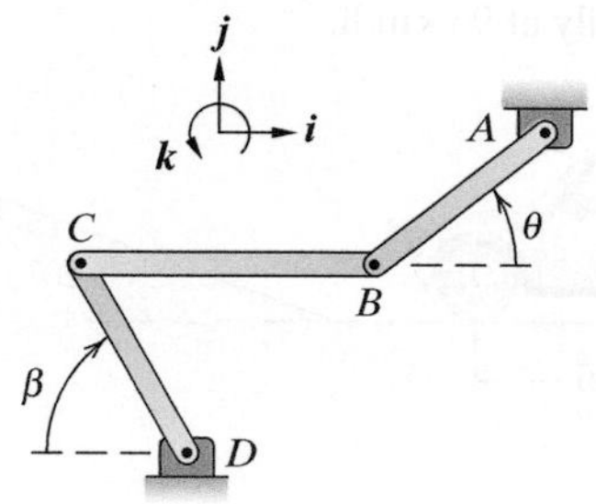

E6.2.17

6.2.18. **[Level 1]** The illustrated symmetric trapezoidal linkage is pulled upward at C at 0.015 m/s. All links are 0.36 m long. Determine $\boldsymbol{\omega}_{\overline{BC}}$ when $\theta = 34°$.

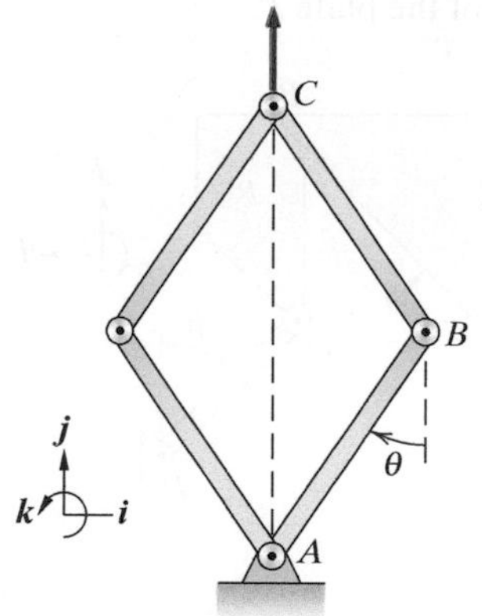

E6.2.18

6.2.19. **[Level 1]** A triangular plate is connected to ground through two links. $|r_{C/D}| = 1.2$ m, $|r_{C/B}| = |r_{B/A}| = 0.6$ m. $\beta = 45°$ and $\gamma = 135°$ at the illustrated instant. $\omega_{\overline{CD}} = 2\boldsymbol{k}$ rad/s. $\overline{BC}$ is horizontal. Use the instantaneous center of rotation to find the angular velocity of the connecting link $\overline{BC}$.

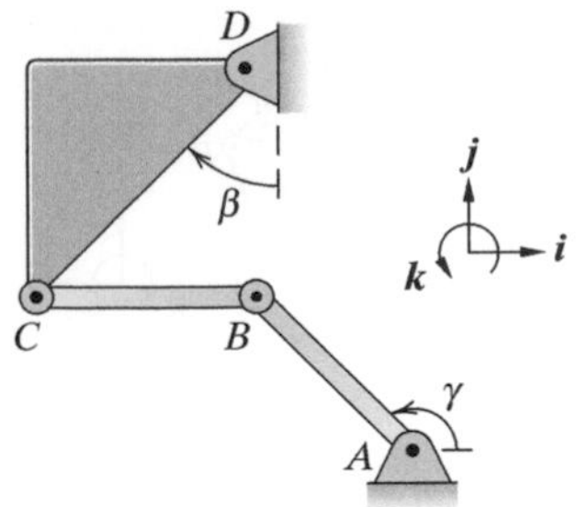

E6.2.19

6.2.20. **[Level 1]** A car is traveling toward a 20° slope at 96 km/h. The wheels have a radius of 30 cm, and the car's wheelbase is 254 cm. What is the angular rotation rate of the car immediately after the front wheel contacts the slope? Assume that A, the center of the rear wheel, continues to move horizontally at 96 km/h.

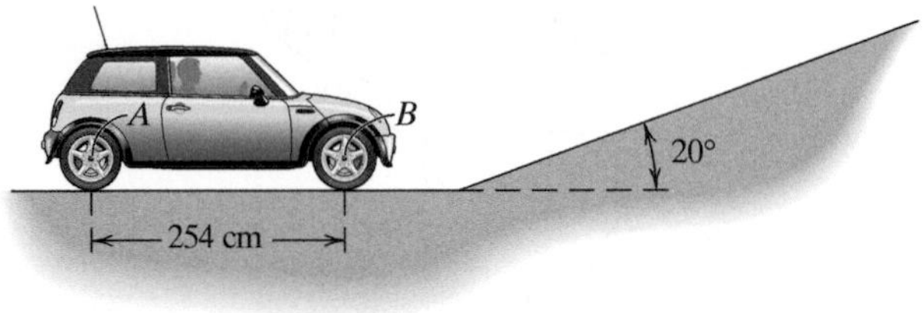

E6.2.20

6.2.21. **[Level 1]** Two plates are positioned by the link $\overline{CD}$. At the illustrated instant, $\theta = \beta = 45°$ and B is directly above C. $|r_{C/D}| = |r_{C/B}| = 0.4\,\text{m}$, $|r_{B/A}| = 0.6\,\text{m}$. $\omega_{\overline{CD}} = -12\boldsymbol{k}$ rad/s. Use the instantaneous center of rotation to find the angular velocity of the plate P.

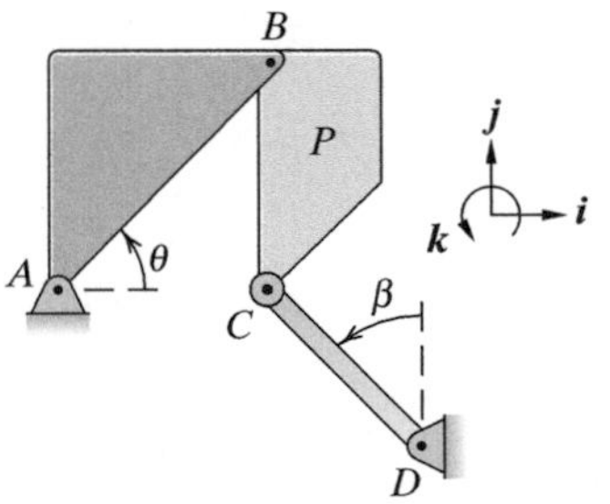

E6.2.21

6.2.22. **[Level 1]** Consider the system of **SA6.2** at the end of this chapter. Is there ever a configuration of the system for which $\omega_{\overline{AB}}$ is zero?

6.2.23. **[Level 1]** At the depicted instant, link $\overline{AB}$ is vertical, link $\overline{BC}$ is horizontal, and link $\overline{CD}$ is angled at $\theta = 50°$ with respect to the horizontal. Links $\overline{AB}$ and $\overline{BC}$ are $r_{B/A} = 0.3\,\text{m}$ and $r_{C/B} = 0.6\,\text{m}$ long, respectively. Find $\omega_{\overline{BC}}$ and v_C if link $\overline{AB}$ is rotating at $\boldsymbol{\omega}_{\overline{AB}} = 9\boldsymbol{k}$ rad/s.

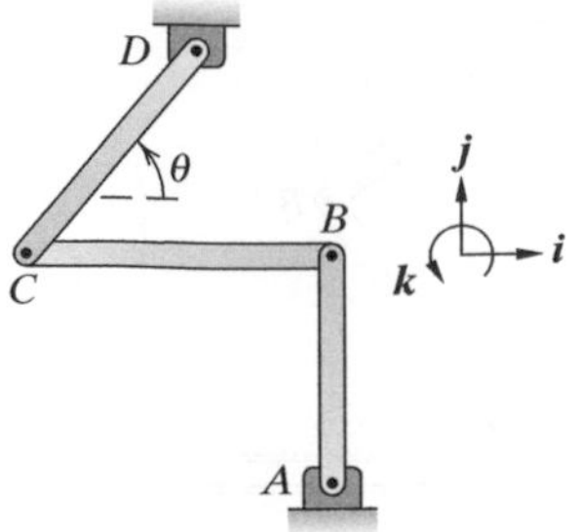

E6.2.23

6.2.24. **[Level 2]** Solve **Exercise 6.1.19** using the instantaneous center of rotation.

6.2.25. **[Level 2]** A gripper arm $\overline{AC}$ is slaved to the control arm $\overline{DE}$ by means of the connecting arm $\overline{BD}$. $\overline{BD}$ is horizontal. $|r_{B/A}| = 0.3\,\text{m}$, $|r_{D/B}| = 0.2\,\text{m}$, $|r_{D/E}| = 0.5\,\text{m}$. $\beta = \theta = 45°$ at the illustrated instant. $\boldsymbol{\omega}_{\overline{DE}} = 10\boldsymbol{k}$ rad/s. Use the instantaneous center of rotation to find the angular velocity of the gripper arm $\overline{AC}$.

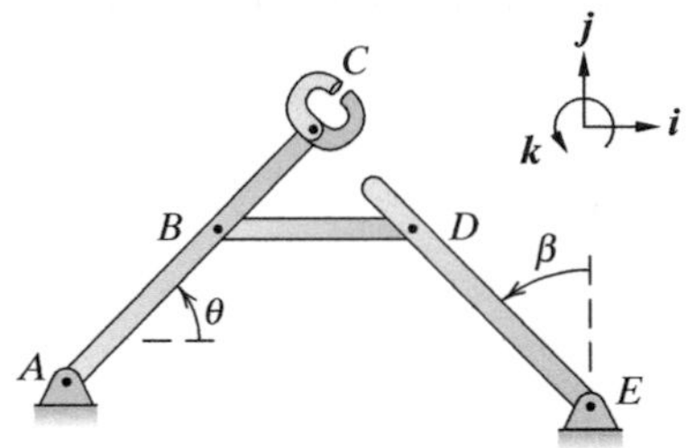

E6.2.25

6.2.26. **[Level 2]** The rear wheel of a car rotates counterclockwise at a constant angular velocity of 88 rad/s and rolls without slip.

a. Calculate the velocity vectors for the points G, A, and B.

b. Show that there are no points on the tire for which the $\boldsymbol{i}$ velocity component is positive.

c. You showed in **b** that a tire rolling to the left has no velocity component in the $\boldsymbol{i}$ direction (to the right). My father thought that rock chips were flung up at his windshield from the car ahead of him, gravel being wedged between the tires' treads and then flung out as the tire spun around. If there's no velocity component pointing toward him, and yet stones do seem to strike our cars when we are driving down the highway, what do you conclude about the origin of the stones? Are they flung out from the treads or not?

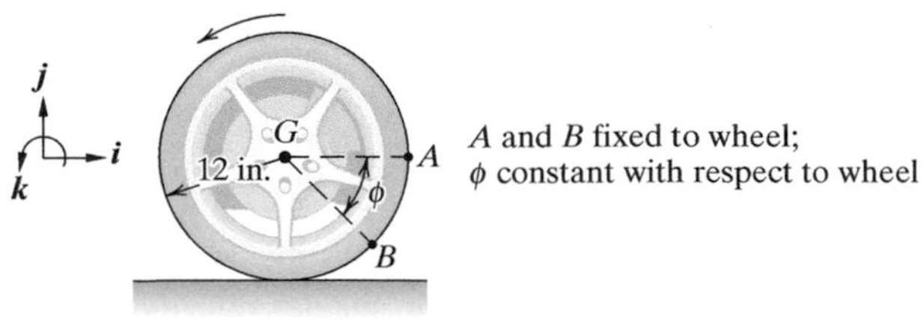

E6.2.26

6.2.27. **[Level 2]** Solve **Exercise 6.1.38** using the instantaneous center of rotation.

6.2.28. **[Level 2]** Solve **Exercise 6.1.39** using the instantaneous center of rotation.

6.2.29. **[Level 2]** Consider the system of **Exercise 6.2.7.** What is $\omega_{\overline{BC}}$ when $\theta = 45°$?

6.2.30. **[Level 2]** At the illustrated instant $\theta = \frac{\pi}{2}$ rad, $\gamma = \frac{\pi}{3}$ rad, and A, C, and D are horizontally aligned. All links have length $h = 1.0$ m. $\dot{\gamma} = 10$ rad/s. Determine $\boldsymbol{\omega}_{\overline{BC}}$.

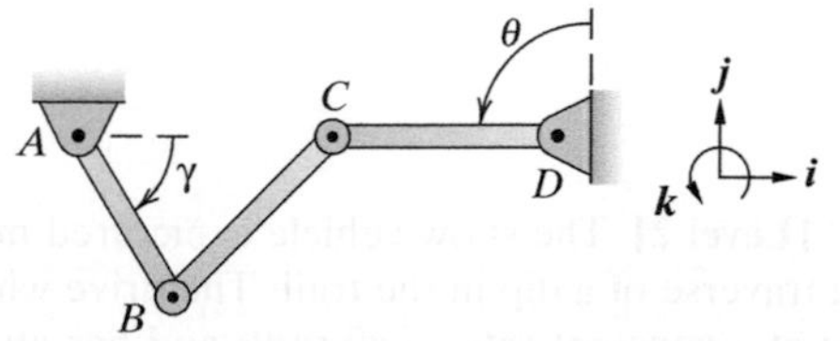

E6.2.30

6.2.31. **[Level 2]** The illustrated rigid body is composed of two cylindrical pieces, and the smaller-diameter cylinder rolls on a moving support. The figure shows a side view (a) and a perspective view (b) for clarity. The inner cylinder has a radius of r_i and the outer cylinder a radius of r_o. The inner cylinder rolls without slip on the surface that's moving with velocity $v\boldsymbol{i}$ and is also rotating with angular velocity $\omega\boldsymbol{k}$. What are the velocities of points A and B?

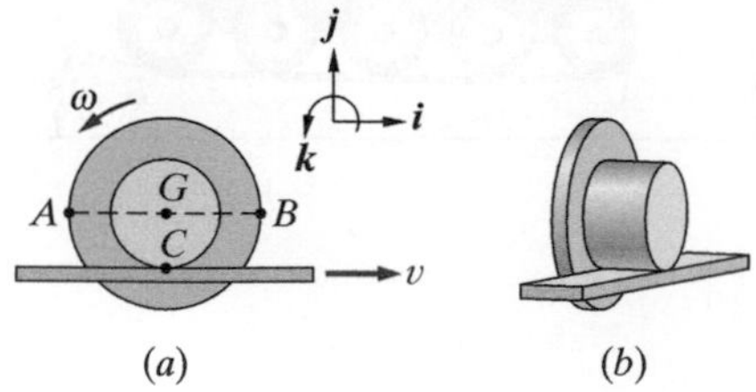

E6.2.31

6.2.32. **[Level 2]** The illustrated system shows a ring gear (Body 3) that contains a rigid frame (OAB) that rotates about O and that has two gears attached. Gear 1 contacts the ring gear at C, and Gear 2 contacts it at D. The body OAB contains a motor at B that drives Gear 2, and another motor at O drives the frame. Gear 1 rotates freely. At the illustrated instant $\overline{OB}$ makes a 45° angle with the horizontal ($\boldsymbol{r}_{B/O} = r_{B/O}(\frac{1}{\sqrt{2}}\boldsymbol{i} + \frac{1}{\sqrt{2}}\boldsymbol{j})$). The inner radius of the ring gear is 0.5 m, $r_1 = 0.1$ m, and $r_2 = 0.16$ m. The angular velocity of Body OAB is $10\boldsymbol{k}$ rad/s, and Gear 2 is being driven with an angular velocity of $4\boldsymbol{k}$ rad/s *with respect to Body OAB*.

a. Determine the angular velocity of Gear 1.

b. Determine how far the instantaneous center of rotation of Gear 1 is from C.

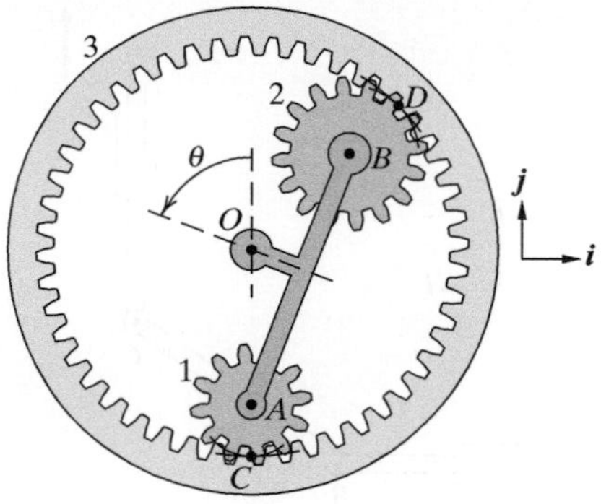

E6.2.32

6.2.33. **[Level 2]** The three links are oriented such that $\theta = 45°$ and $\gamma = \beta = 90°$. $r_{D/C} = 2\sqrt{2}L$ and $r_{C/B} = r_{B/A} = 2L$. $\dot{\gamma} = 10$ rad/s. Determine $\boldsymbol{\omega}_{\overline{CD}}$ and $\boldsymbol{\omega}_{\overline{BC}}$.

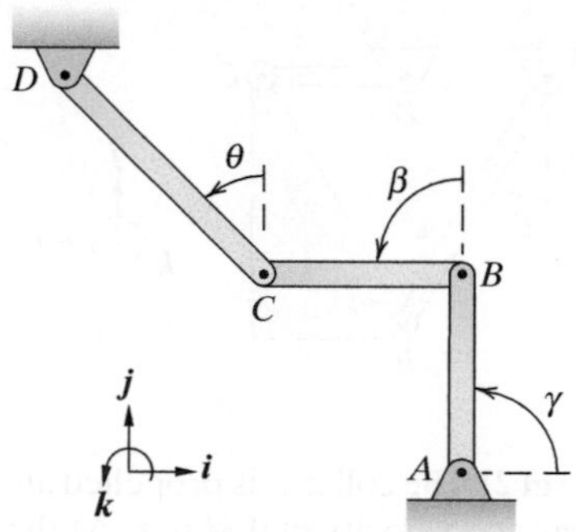

E6.2.33

6.2.34. **[Level 2]** The lower link is moved to the right with a speed v. $r_{B/A} = 4L$ and $r_{C/B} = 2L$. Determine $\boldsymbol{\omega}_{\overline{AB}}$ and $\boldsymbol{\omega}_{\overline{BC}}$. Link $\overline{BC}$ is horizontal at this instant.

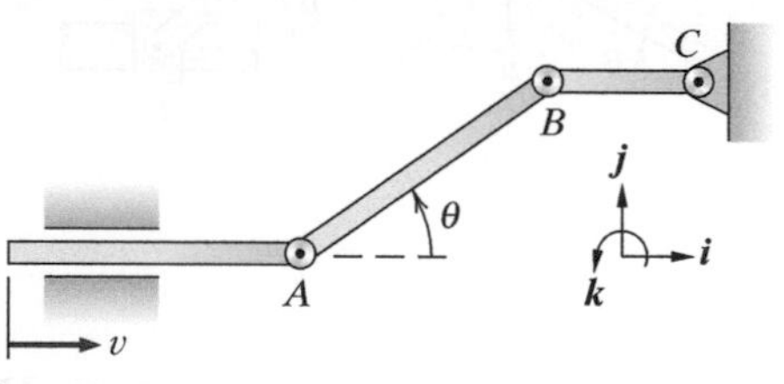

E6.2.34

6.2.35. **[Level 2]** The link $\overline{AB}$ moves to the right with speed $v_1 = 2$ m/s. $r_{C/B} = 1$ m and $\theta = 30°$. Show that it's not possible for link $\overline{CD}$ to be moving up at 3 m/s.

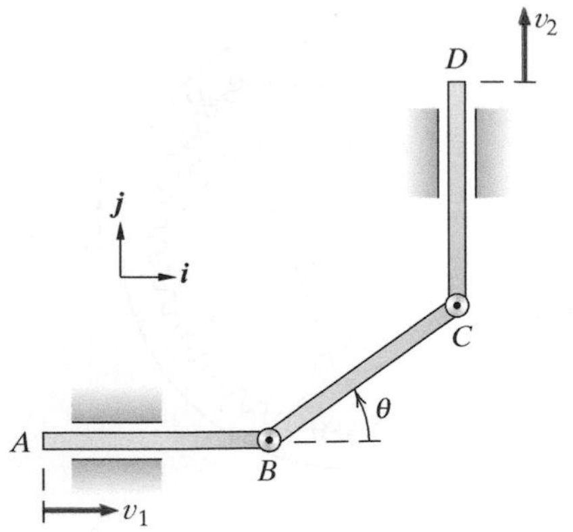

E6.2.35

6.2.36. **[Level 2]** The driven link $\overline{AB}$ is rotating at $20\boldsymbol{k}$ rad/s. $\boldsymbol{r}_{B/A} = (0.15\boldsymbol{i} - 0.26\boldsymbol{j})$ m, $\boldsymbol{r}_{C/B} = (0.15\boldsymbol{i} + 0.26\boldsymbol{j})$ m, and $\boldsymbol{r}_{D/C} = -0.15\boldsymbol{i}$ m. Determine the rotational velocity of the triangular plate 1.

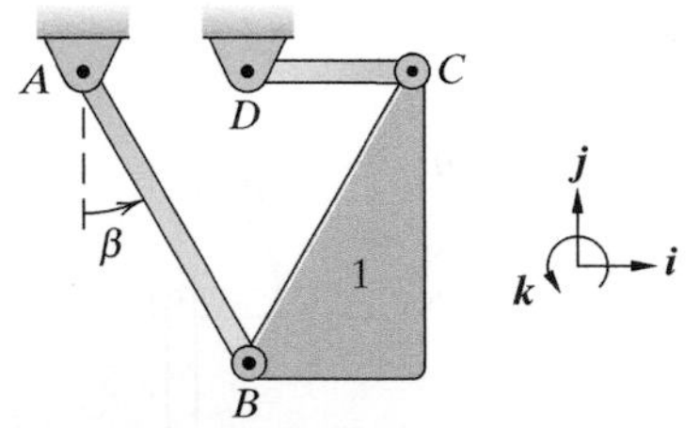

E6.2.36

6.2.37. **[Level 2]** The collar C is propelled along the horizontal guidebar at a velocity of $0.3\boldsymbol{i}$ m/s. At the illustrated instant $\theta = 30°$ and $\beta = 40°$. $r_{B/A} = 0.8$ m and $r_{C/B} = 0.5$ m. (Note that a is nonzero.) Determine $\boldsymbol{\omega}_{\overline{BC}}$.

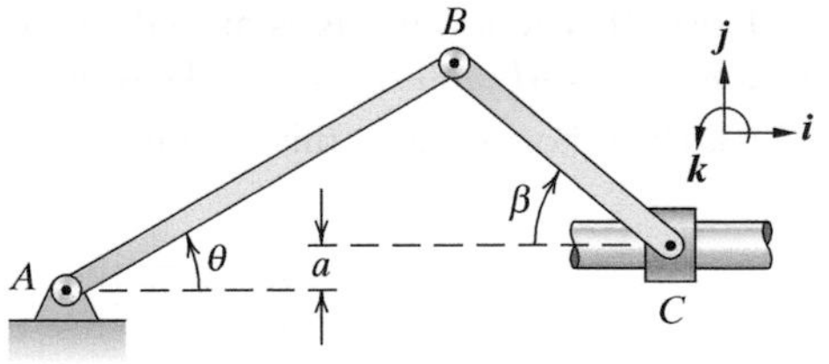

E6.2.37

6.2.38. **[Level 2]** The bar $\overline{AB}$ is moved to the left at 0.3 m/s. The bar $\overline{CE}$ can move only vertically, constraining point C on link $\overline{BD}$ to move vertically as well. $\theta = 30°$ at the illustrated instant, $r_{C/B} = 0.2$ m, and $r_{D/B} = 0.6$ m. Determine the speed of the plunger P.

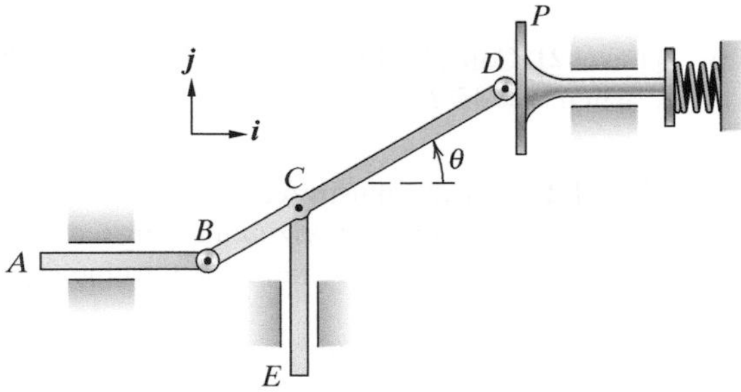

E6.2.38

6.2.39. **[Level 2]** The snow vehicle is pictured midway through a traverse of a dip in the trail. The drive wheel W is rotating at a constant rate of 40 rad/s and has an effective radius (including track) of 0.1 m; the vehicle is moving without slip. $\boldsymbol{r}_{B/A} = 4\boldsymbol{i}$ m. Determine the vehicle's angular velocity and its instantaneous center of rotation.

E6.2.39

6.3 ROTATING REFERENCE FRAMES AND RIGID-BODY ACCELERATIONS

Now that you've seen how rotating bodies work, it is time to introduce rotating reference frames. These frames are introduced because when a body rotates, any reference frames attached to it do so as well. It turns out that we can come up with a formula that gives us the derivative of any vector associated with a rotating body that we would like to examine. Position, velocity, acceleration, whatever—if you have a vector that's associated with a rotating body, then this formula will quickly and easily give you the vector's absolute time derivative. For clarity, I'll be using a position vector in the following discussion because that's the easiest one to visualize.

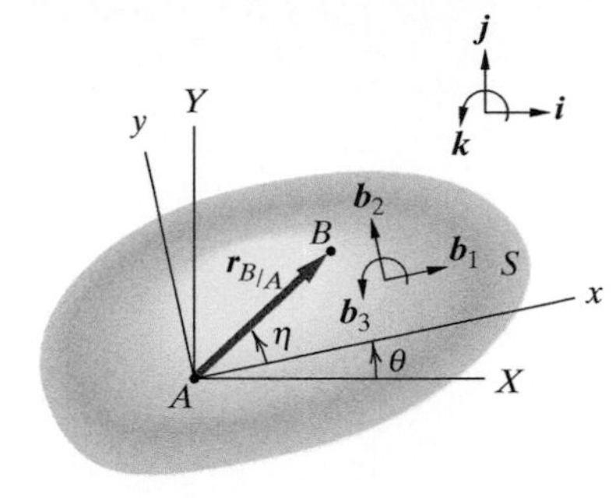

Figure 6.31 Reference frames for rotational analysis

Figure 6.31 shows the basic setup. Our inertial reference frame—call it the N frame—is represented by the X–Y axes and the $\boldsymbol{i}, \boldsymbol{j}$ unit vectors. As usual, $\boldsymbol{i}$ and $\boldsymbol{j}$ don't rotate. In addition to this frame, we also have a frame represented by the x–y axes. This frame rotates with respect to X–Y, and the angle of rotation is θ. The $\boldsymbol{b}_1, \boldsymbol{b}_2$ unit vectors rotate at the same rate as the x–y axes. The pivot point for the rotation is A. *Brief recap*: X–Y are fixed in space, x–y rotate, $\boldsymbol{i}, \boldsymbol{j}$ are fixed in space, and $\boldsymbol{b}_1, \boldsymbol{b}_2$ rotate along with x–y.

I've also drawn a body S in **Figure 6.31**. Let's say S is rotating and its rate of rotation is the same as that of the x–y axes. This means that if you were rotating with x–y and looking down at the body S, you wouldn't notice it moving at all. For those of you who have seen the movie *2001: A Space Odyssey*, recall the scene where the Pan Am shuttle from earth is docking with the space station. (For those of you who haven't seen it—go rent it, it's a classic!) Remember how from an external shot you could see the space station turning and the shuttle rotating to match it? And then the point of view switched to the one from inside the shuttle, looking toward the space station. From the spaceship's viewpoint, the space station looked completely stationary (but now the stars seemed to be rotating).

That's the same sort of thing I'm talking about here. If B is a point that's fixed to the body S, then B's position won't change with respect to x–y as the body turns. Thus $\boldsymbol{v}_{B/A}$ is zero *with respect to the x–y axes*. The vector $\boldsymbol{r}_{B/A}$ is certainly moving with respect to the inertial frame X–Y because the tip of the vector is rotating about A, and thus there is a velocity associated with that rotation. It's all wrapped up in the point of view. To a passenger in a speeding car the driver doesn't seem to be moving, but to an outside observer the driver appears to be zipping along.

What we need to find is the time derivative of $\boldsymbol{r}_{B/A}$ with respect to the ground frame, and so that's the frame we will use for our $\boldsymbol{F} = m\boldsymbol{a}$ analyses. For convenience I'll refer to this as the N frame. Let's now express $\boldsymbol{r}_{B/A}$ as

$$\boldsymbol{r}_{B/A} = r_1\boldsymbol{b}_1 + r_2\boldsymbol{b}_2$$

Differentiating with respect to the N frame gives

$$\left.\frac{d}{dt}\right|_N \boldsymbol{r}_{B/A} = \left.\frac{d}{dt}\right|_N (r_1\boldsymbol{b}_1 + r_2\boldsymbol{b}_2) = \dot{r}_1\boldsymbol{b}_1 + \dot{r}_2\boldsymbol{b}_2 + r_1\dot{\boldsymbol{b}}_1 + r_2\dot{\boldsymbol{b}}_2 \tag{6.21}$$

We figured out what $\dot{\boldsymbol{b}}_1$ and $\dot{\boldsymbol{b}}_2$ are equal to back in Section 2.3. There we were using $\boldsymbol{e}_r$ and $\boldsymbol{e}_\theta$ (because we were talking about polar coordinates), but since they were just a set of unit vectors rotating counterclockwise and were governed by θ (just like the current situation), we can simply swap $\boldsymbol{b}_1$ for $\boldsymbol{e}_r$ and $\boldsymbol{b}_2$ for $\boldsymbol{e}_\theta$. Our results back then (2.48) were

$$\dot{\boldsymbol{e}}_r = \dot{\theta}\boldsymbol{e}_\theta \quad \text{and} \quad \dot{\boldsymbol{e}}_\theta = -\dot{\theta}\boldsymbol{e}_r$$

which means that for the present case, we have

$$\dot{\boldsymbol{b}}_1 = \dot{\theta}\boldsymbol{b}_2 \quad \text{and} \quad \dot{\boldsymbol{b}}_2 = -\dot{\theta}\boldsymbol{b}_1$$

Using this in (6.21) gives us

$$\left.\frac{d}{dt}\right|_N \boldsymbol{r}_{B/A} = \dot{r}_1\boldsymbol{b}_1 + \dot{r}_2\boldsymbol{b}_2 + \dot{\theta}(r_1\boldsymbol{b}_2 - r_2\boldsymbol{b}_1) \tag{6.22}$$

We already know that $\boldsymbol{\omega} = \dot{\theta}\boldsymbol{b}_3$ for this problem, and so it would be nice to hook up the $\dot{\theta}$ in (6.22) with a $\boldsymbol{k}$, thus letting us form $\boldsymbol{\omega}$. It's no problem doing that, as long as we deal correctly with the cross products on the righthand side of the equation. Rewriting (6.22) as

$$\left.\frac{d}{dt}\right|_N \boldsymbol{r}_{B/A} = \dot{r}_1\boldsymbol{b}_1 + \dot{r}_2\boldsymbol{b}_2 + \dot{\theta}\boldsymbol{b}_3 \times (r_1\boldsymbol{b}_1 + r_2\boldsymbol{b}_2) \tag{6.23}$$

accomplishes the task nicely.

It's no secret that the last term is just $\boldsymbol{r}_{B/A}$, and $\dot{\theta}\boldsymbol{b}_3 = \boldsymbol{\omega}$, so

$$\left.\frac{d}{dt}\right|_N \boldsymbol{r}_{B/A} = \dot{r}_1\boldsymbol{b}_1 + \dot{r}_2\boldsymbol{b}_2 + \boldsymbol{\omega} \times \boldsymbol{r}_{B/A}$$

How about the term $\dot{r}_1\boldsymbol{b}_1 + \dot{r}_2\boldsymbol{b}_2$? Is that anything in particular? You bet it is! It's the time derivative of $\boldsymbol{r}_{B/A}$ *with respect to the body S*. There are two possibilities here. If B is fixed to S, then if you are standing on S and looking at B, you don't see B move (because it's fixed to S), which means $\dot{r}_1$ and $\dot{r}_2$ must be zero. The more general case is where B *isn't* fixed to S, in which case $\dot{r}_1$ and $\dot{r}_2$ aren't necessarily zero.

Our final equation is therefore

$$\left.\frac{d}{dt}\right|_N \boldsymbol{r}_{B/A} = \left.\frac{d}{dt}\right|_S \boldsymbol{r}_{B/A} + \boldsymbol{\omega} \times \boldsymbol{r}_{B/A}$$

or, being even more general and applying this expression to some arbitrary vector $\boldsymbol{p}$,

$$\left.\frac{d}{dt}\right|_N \boldsymbol{p} = \left.\frac{d}{dt}\right|_S \boldsymbol{p} + \boldsymbol{\omega} \times \boldsymbol{p} \tag{6.24}$$

where S represents our rotating reference frame. In words, the derivative of $\boldsymbol{p}$ with respect to an inertial frame (the N frame) is equal to its derivative with respect to a rotating frame (the S frame) *plus* that rotating frame's angular velocity crossed with the vector $\boldsymbol{p}$ itself.

This allows us to verify (6.4). We begin with our basic relative motion expression $\boldsymbol{r}_B = \boldsymbol{r}_A + \boldsymbol{r}_{B/A}$ and take the time derivative:

$$\frac{d}{dt}\boldsymbol{r}_B = \frac{d}{dt}\boldsymbol{r}_A + \frac{d}{dt}\boldsymbol{r}_{B/A}$$

We can use (6.24) to evaluate $\frac{d}{dt}\boldsymbol{r}_{B/A}$:

$$\frac{d}{dt}\boldsymbol{r}_{B/A} = \overbrace{\left.\frac{d}{dt}\right|_S \boldsymbol{r}_{B/A}}^{0} + \boldsymbol{\omega}\times\boldsymbol{r}_{B/A} = \boldsymbol{\omega}\times\boldsymbol{r}_{B/A}$$

As mentioned in the derivation, if the vector under consideration doesn't move with respect to the rotating frame, then its derivative with respect to that frame is zero. Hence our first term in the above expression drops out. We're left with $\boldsymbol{v}_B = \boldsymbol{v}_A + \boldsymbol{\omega}\times\boldsymbol{r}_{B/A}$, which is (6.4), giving us the same result we derived through a different approach in Section 6.1.

I know that right about now you're saying, "If you think this formula is so great—prove it. Find the acceleration of B using it." All right, I will.

All we have to do is apply (6.24) to our velocity vector (6.4):

$$\left.\frac{d}{dt}\right|_N \boldsymbol{v}_B = \left.\frac{d}{dt}\right|_N \boldsymbol{v}_A + \left.\frac{d}{dt}\right|_S \boldsymbol{v}_{B/A} + \boldsymbol{\omega}\times\boldsymbol{v}_{B/A}$$

$$= \boldsymbol{a}_A + \left.\frac{d}{dt}\right|_S \left(\boldsymbol{\omega}\times\boldsymbol{r}_{B/A}\right) + \boldsymbol{\omega}\times\left(\boldsymbol{\omega}\times\boldsymbol{r}_{B/A}\right)$$

$$= \boldsymbol{a}_A + \overbrace{\left(\left.\frac{d}{dt}\right|_S \boldsymbol{\omega}\right)}^{\boldsymbol{\alpha}}\times\boldsymbol{r}_{B/A} + \boldsymbol{\omega}\times\overbrace{\left.\frac{d}{dt}\right|_S \boldsymbol{r}_{B/A}}^{0} + \boldsymbol{\omega}\times\left(\boldsymbol{\omega}\times\boldsymbol{r}_{B/A}\right)$$

$$= \boldsymbol{a}_A + \boldsymbol{\alpha}\times\boldsymbol{r}_{B/A} + \boldsymbol{\omega}\times\left(\boldsymbol{\omega}\times\boldsymbol{r}_{B/A}\right)$$

And there you have it. The acceleration of point B on a moving body is given by

$$\boldsymbol{a}_B = \boldsymbol{a}_A + \boldsymbol{\alpha}\times\boldsymbol{r}_{B/A} + \boldsymbol{\omega}\times\left(\boldsymbol{\omega}\times\boldsymbol{r}_{B/A}\right) \tag{6.25}$$

In words, what we have here is the following. To find the acceleration of point B on a moving body, we first need to get onto the body. That's what $\boldsymbol{a}_A$ does for us—it's the translational accleration of a particular point A on the body. Then we need to get the acceleration of B with respect to A, which is given by $\boldsymbol{\alpha}\times\boldsymbol{r}_{B/A}$ and $\boldsymbol{\omega}\times(\boldsymbol{\omega}\times\boldsymbol{r}_{B/A})$, where $\boldsymbol{\alpha}$ is the angular acceleration and $\boldsymbol{\omega}$ is the angular velocity of the body. The second term in (6.25) corresponds to $r\ddot{\theta}\boldsymbol{e}_\theta$ (if we were using a polar representation), whereas the third term corresponds to $-r\dot{\theta}^2\boldsymbol{e}_r$.

>>> **Check out Example 6.11 (page 350), Example 6.12 (page 351), Example 6.13 (page 352), Example 6.14 (page 354), and Example 6.15 (page 355) for applications of this material.**

EXAMPLE 6.11 ACCELERATION OF A PEDAL SPINDLE (Theory on page 349)

Figure 6.32 Cyclist on rollers

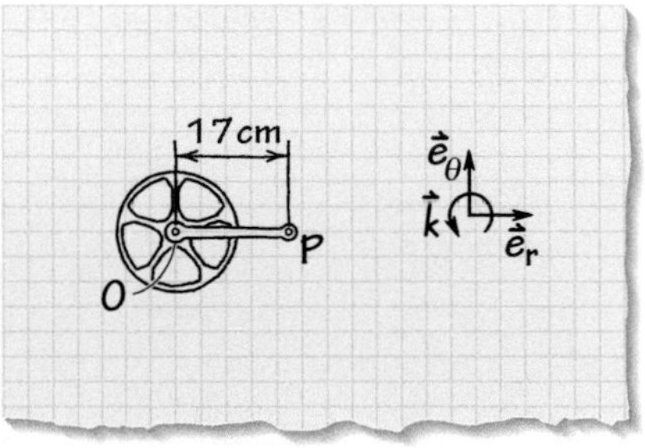

Figure 6.33 Crank and chain ring

To keep in shape during the winter, cyclists often put their bikes on rollers, which allows them to pedal as hard as they wish without going anywhere (**Figure 6.32**). Assume that the cyclist is pedaling at 4 rad/s and then begins to accelerate so that the angular acceleration of the main crank assembly is equal to 1.2 rad/s^2. After accelerating at this rate for 2 s, what is the acceleration of the pedal spindle P? The distance from O, the center of the bottom bracket, to the pedal spindle is 17 cm, and the bottom bracket center O is fixed in space.

Goal Find the acceleration of a bike's pedal spindle.

Given System dimensions and acceleration, along with initial rotation rate.

Assume No additional assumptions are needed.

Draw **Figure 6.33** shows the key elements of our system—the center of the bottom bracket O, the center of the pedal spindle P, and a set of cylindrical unit vectors.

Formulate Equations We need to break the problem down into two parts. First we will need to determine the rotation rate after 2 s by integrating with respect to time. Then we will need to apply (6.25).

Solve Integrating for 2 s gives us an angular speed of

$$\dot{\theta}(0) + (\ddot{\theta})(2\,\text{s}) = 4\,\text{rad/s} + (1.2\,\text{rad/s}^2)(2\,\text{s}) = 6.4\,\text{rad/s}$$

Applying (6.25) gives us

$$\begin{aligned}
\boldsymbol{a}_P &= \boldsymbol{a}_O + \boldsymbol{\alpha} \times \boldsymbol{r}_{P/O} + \boldsymbol{\omega} \times \left(\boldsymbol{\omega} \times \boldsymbol{r}_{P/O}\right) \\
&= 0 + 1.2\boldsymbol{k}\,\text{rad/s}^2 \times 0.17\boldsymbol{e}_r\,\text{m} + 6.4\boldsymbol{k}\,\text{rad/s} \times [(6.4\boldsymbol{k}\,\text{rad/s}) \times (0.17\boldsymbol{e}_r\,\text{m})] \\
&= (0.204\boldsymbol{e}_\theta - 6.96\boldsymbol{e}_r)\,\text{m/s}^2
\end{aligned} \tag{6.26}$$

$$\boldsymbol{a}_P = (0.204\boldsymbol{e}_\theta - 6.96\boldsymbol{e}_r)\,\text{m/s}^2$$

EXAMPLE 6.12 ACCELERATION DURING ROLL WITHOUT SLIP (Theory on page 349)

In Example 6.8 we saw that the point of contact between a rolling disk and the ground has zero velocity. The obvious question that then arises is, "Does the contact point of a rolling disk have zero acceleration?" Well, let's find out. **Figure 6.34** shows a typical rolling wheel/tire of the sort we'll be considering. As in the past, we'll idealize the wheel/tire as a disk of radius r, one that's rolling to the left with speed v and linear acceleration a.

Benson Tongue

Figure 6.34 Rolling wheel/tire

Goal Determine the acceleration of the contact point of a rolling disk.

Given Speed and acceleration of the disk's center.

Assume As in Example 6.24, we'll assume roll without slip.

Draw **Figure 6.35** shows our system.

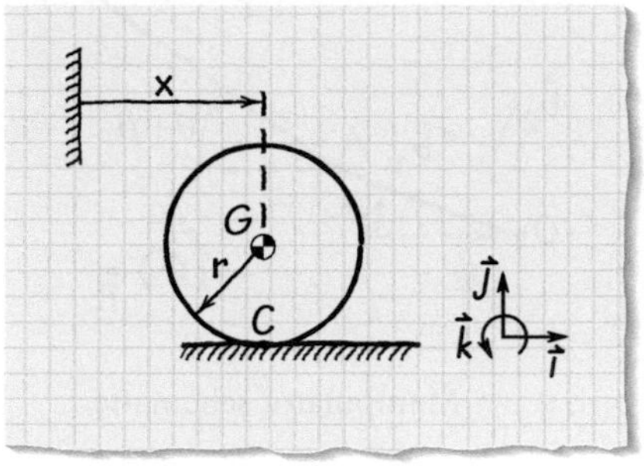

Figure 6.35 Rolling disk schematic

Formulate Equations We'll use (6.25) and the disk's velocity ($\boldsymbol{v}_G = -\omega r\boldsymbol{i}$) found in Example 6.8.

Solve $\boldsymbol{a}_G = \frac{d}{dt}(-\omega r\boldsymbol{i}) = -\alpha r\boldsymbol{i}$. Knowing $\boldsymbol{a}_G$, we can use (6.25) to find $\boldsymbol{a}_C$:

$$\begin{aligned}\boldsymbol{a}_C &= \boldsymbol{a}_G + \boldsymbol{\alpha}\times\boldsymbol{r}_{C/G} + \boldsymbol{\omega}\times\left(\boldsymbol{\omega}\times\boldsymbol{r}_{C/G}\right)\\ &= -\alpha r\boldsymbol{i} + \alpha\boldsymbol{k}\times(-r\boldsymbol{j}) + \omega\boldsymbol{k}\times[\omega\boldsymbol{k}\times(-r\boldsymbol{j})]\\ &= -\alpha r\boldsymbol{i} + \alpha r\boldsymbol{i} + \omega^2 r\boldsymbol{j} \quad\Rightarrow\quad \boxed{\boldsymbol{a}_C = \omega^2 r\boldsymbol{j}}\end{aligned}$$

This says the contact point has no acceleration in the $\boldsymbol{i}$ direction but has a large acceleration component vertically ($\boldsymbol{j}$ direction).

Check We can derive the same result by looking at the vertical position of C as the disk rolls. The height is given by $y = r(1 - \cos\theta)$ (where $\theta = \omega t$ indicates the rotation of the disk.) Differentiating twice gives us $\ddot{y} = \omega^2 r\cos\theta$. At $\theta = 0$ this gives us an acceleration of $\omega^2 r$, matching our earlier result.

EXAMPLE 6.13 TIP ACCELERATION OF A TWO-LINK MANIPULATOR (Theory on page 349)

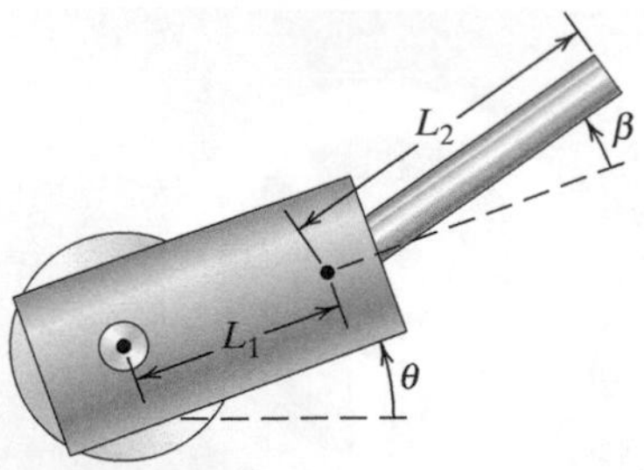

Figure 6.36 Two-link manipulator

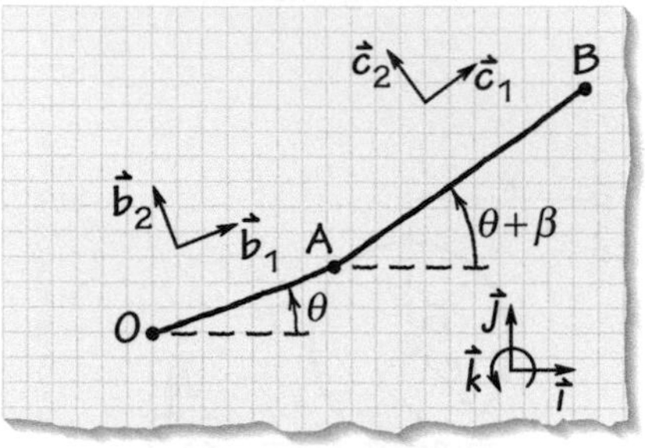

Figure 6.37 Manipulator schematic

Figure 6.36 shows a two-link manipulator. L_1 and L_2 are held constant, and θ, $\dot{\theta}$, $\ddot{\theta}$, β, $\dot{\beta}$, and $\ddot{\beta}$ are assumed known. The angle θ gives the orientation of the base arm with respect to ground, and β tells us what the deflection of the secondary arm is with respect to the base arm. The task is to find the acceleration of the outmost tip of the manipulator.

Goal Find the acceleration at the tip of a two-link robotic arm.

Given Arm configuration and motion variables.

Assume L_1 and L_2 are constant; that is, the robot arms aren't extending or retracting.

Draw A simplified drawing of our system is illustrated in **Figure 6.37**. The $\boldsymbol{b}_1, \boldsymbol{b}_2$ unit vectors move with the base arm, and $\boldsymbol{c}_1, \boldsymbol{c}_2$ move with the secondary arm.

Formulate Equations The key observation has already been shown in **Figure 6.37**. Although β, the angle between the base arm and secondary arm, is given, that's not the angle we care about. All our rigid-body kinematic analyses have been done with angles that are measured with respect to ground—absolute angles and not relative. Thus, to apply our kinematic formulas we need to express everything with respect to ground, and so the inclination of the secondary arm is given by $\theta + \beta$ and the tranformation arrays are given by

	$\boldsymbol{i}$	$\boldsymbol{j}$
$\boldsymbol{b}_1$	$\cos(\theta)$	$\sin(\theta)$
$\boldsymbol{b}_2$	$-\sin(\theta)$	$\cos(\theta)$

	$\boldsymbol{i}$	$\boldsymbol{j}$
$\boldsymbol{c}_1$	$\cos(\theta+\beta)$	$\sin(\theta+\beta)$
$\boldsymbol{c}_2$	$-\sin(\theta+\beta)$	$\cos(\theta+\beta)$

Differentiating its angular displacement with respect to time gives us the secondary arm's rotational speed and acceleration as $\dot{\theta} + \dot{\beta}$ and $\ddot{\theta} + \ddot{\beta}$, respectively. Beyond this, all we'll need is (6.25).

Solve First we'll go from O to A, using (6.4) and (6.25) to calculate $\boldsymbol{v}_A$ and $\boldsymbol{a}_B$, respectively.

$$\begin{aligned} \boldsymbol{v}_A &= \boldsymbol{v}_O + \boldsymbol{\omega}_{OA} \times \boldsymbol{r}_{A/O} \\ &= 0 + \dot{\theta}\boldsymbol{b}_3 \times L_1\boldsymbol{b}_1 \\ &= L_1\dot{\theta}\boldsymbol{b}_2 \end{aligned}$$

$$\begin{aligned} \boldsymbol{a}_A &= \boldsymbol{a}_O + \boldsymbol{\alpha}_{OA} \times \boldsymbol{r}_{A/O} + \boldsymbol{\omega}_{OA} \times \left(\boldsymbol{\omega}_{OA} \times \boldsymbol{r}_{A/O}\right) \\ &= 0 + \ddot{\theta}\boldsymbol{b}_3 \times L_1\boldsymbol{b}_1 + \dot{\theta}\boldsymbol{b}_3 \times (\dot{\theta}\boldsymbol{b}_3 \times L_1\boldsymbol{b}_1) \\ &= L_1\ddot{\theta}\boldsymbol{b}_2 - L_1\dot{\theta}^2\boldsymbol{b}_1 \end{aligned}$$

Now we simply jump along the final link, going from A to B:

$$\begin{aligned}\boldsymbol{v}_B &= \boldsymbol{v}_A + \boldsymbol{\omega}_{AB} \times \boldsymbol{r}_{B/A}\\ &= L_1\dot{\theta}\boldsymbol{b}_2 + (\dot{\theta} + \dot{\beta})\boldsymbol{c}_3 \times L_2\boldsymbol{c}_1\end{aligned}$$

$$\boldsymbol{v}_B = L_1\dot{\theta}\boldsymbol{b}_2 + L_2(\dot{\theta} + \dot{\beta})\boldsymbol{c}_2 \tag{6.27}$$

$$\begin{aligned}\boldsymbol{a}_B &= \boldsymbol{a}_A + \boldsymbol{\alpha}_{AB} \times \boldsymbol{r}_{B/A} + \boldsymbol{\omega}_{AB} \times \left(\boldsymbol{\omega}_{AB} \times \boldsymbol{r}_{B/A}\right)\\ &= L_1\ddot{\theta}\boldsymbol{b}_2 - L_1\dot{\theta}^2\boldsymbol{b}_1 + (\ddot{\theta} + \ddot{\beta})\boldsymbol{c}_3 \times L_2\boldsymbol{c}_1 + (\dot{\theta} + \dot{\beta})\boldsymbol{c}_3 \times [(\dot{\theta} + \dot{\beta})\boldsymbol{c}_3 \times L_2\boldsymbol{c}_1]\end{aligned}$$

$$\boldsymbol{a}_B = L_1\ddot{\theta}\boldsymbol{b}_2 - L_1\dot{\theta}^2\boldsymbol{b}_1 + L_2(\ddot{\theta} + \ddot{\beta})\boldsymbol{c}_2 - L_2(\dot{\theta} + \dot{\beta})^2\boldsymbol{c}_1 \tag{6.28}$$

Note that by moving beyond a single link, we have opened the door to a good deal of complication. I wish it weren't the case, but this is a problem that won't go away. As soon as you have interconnections of multiple bodies, you will find that the number and complexity of the governing equations increase.

Check The easiest check to make in this example is to set the angles to something that produces a simple system, one that we can easily solve, and verify that the results match. You can do this in many ways; here we will just look at one. Let $\theta = \beta = 0$ rad, $\dot{\beta} = 0$ rad/s, and $\ddot{\beta} = 0$ rad/s^2. This should lock the secondary arm to the base arm and reduce the system to a single link with length $L_1 + L_2$ and angular speed/acceleration of $\dot{\theta}/\ddot{\theta}$. In this case we would expect an acceleration of

$$\boldsymbol{a}_B = (L_1 + L_2)\ddot{\theta}\boldsymbol{j} - (L_1 + L_2)\dot{\theta}^2\boldsymbol{i}$$

Using (6.28) and letting $\beta = 0$ in the coordinate transformation arrays give us

$$\begin{aligned}\boldsymbol{a}_B &= L_1\ddot{\theta}\boldsymbol{j} - L_1\dot{\theta}^2\boldsymbol{i} + L_2\ddot{\theta}\boldsymbol{j} - L_2\dot{\theta}^2\boldsymbol{i}\\ &= (L_1 + L_2)\ddot{\theta}\boldsymbol{j} - (L_1 + L_2)\dot{\theta}^2\boldsymbol{i}\end{aligned}$$

exactly as expected.

EXAMPLE 6.14 ACCELERATION OF A POINT ON A COG OF A MOVING BICYCLE (Theory on page 349)

Benson Tongue

Figure 6.38 Cyclist traveling in a straight line

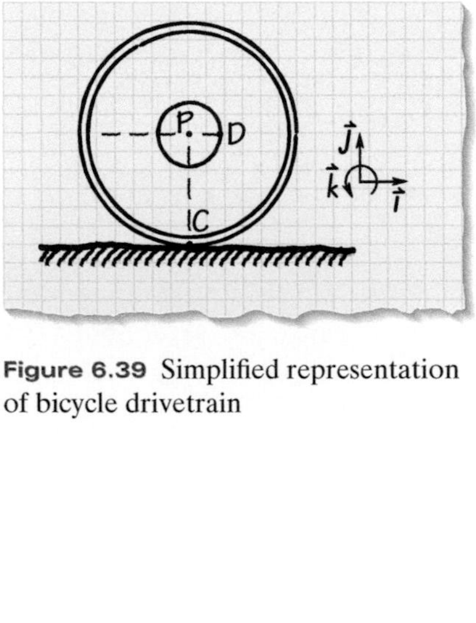

Figure 6.39 Simplified representation of bicycle drivetrain

As in Example 6.9, **Figure 6.38** shows a picture of a cyclist. We will assume that the same parameter values hold; that is, a 3 cm radius rear cog and a wheel radius of 25 cm. The cyclist is traveling at 12 m/s and decelerating at 2 m/s^2. Determine the linear acceleration of point D on the periphery of the rear cog.

Goal Find the acceleration of a point on the periphery of a bicycle's rear cog.

Given Pedaling cadence and bicycle dimensions.

Assume Assume that the cyclist is traveling in a no-slip condition.

Draw **Figure 6.39** shows a simplified view of our system.

Formulate Equations As we're looking for accelerations, we will need to apply (6.25). We will make use of the instantaneous center of rotation to find ω and apply the results of Example 6.8 to find α, the wheel's angular acceleration.

Solve We're given that $\boldsymbol{v}_P = -12\boldsymbol{i}$ m/s and $\boldsymbol{a}_P = 2\boldsymbol{i}$ m/s^2. Because C is an instantaneous center of rotation, we have

$$\boldsymbol{v}_P = \omega\boldsymbol{k} \times \boldsymbol{r}_{P/C} \quad \Rightarrow \quad -12\boldsymbol{i}\text{ m/s} = \omega\boldsymbol{k} \times r_{P/C}\boldsymbol{j} = -\omega r_{P/C}\boldsymbol{i}$$

$$\omega = \frac{12\text{ m/s}}{r_{P/C}} = \frac{12\text{ m/s}}{0.25\text{ m/s}} = 48\text{ rad/s}$$

Now let's find α, the tire's angular acceleration. We saw in Example 6.8 that there's no acceleration in the $\boldsymbol{i}$ direction at C. An application of (6.25) from P to C will produce

$$\boldsymbol{a}_C = \boldsymbol{a}_P + \boldsymbol{\alpha} \times \boldsymbol{r}_{C/P} + \boldsymbol{\omega} \times \left(\boldsymbol{\omega} \times \boldsymbol{r}_{C/P}\right)$$

$$(0\boldsymbol{i} + a_C\boldsymbol{j})\text{ m/s}^2 = 2\boldsymbol{i}\text{ m/s}^2 + \alpha\boldsymbol{k} \times \left(-r_{C/P}\boldsymbol{j}\right) + \omega\boldsymbol{k} \times \left[\omega\boldsymbol{k} \times \left(-r_{C/P}\boldsymbol{j}\right)\right]$$

$$\boldsymbol{i}: \qquad 0\text{ m/s}^2 = 2\text{ m/s}^2 + \alpha r_{C/P} \quad \Rightarrow \quad \alpha = \frac{-2\text{ m/s}^2}{0.25\text{ m}} = -8\text{ m/s}^2$$

We now have all we need to apply (6.25) from P to D:

$$\boldsymbol{a}_D = \boldsymbol{a}_P + \boldsymbol{\alpha} \times \boldsymbol{r}_{D/P} + \boldsymbol{\omega} \times \left(\boldsymbol{\omega} \times \boldsymbol{r}_{D/P}\right) = 2\boldsymbol{i}\text{ m/s}^2 + \alpha\boldsymbol{k} \times r_{D/P}\boldsymbol{i} + \omega\boldsymbol{k} \times \left(\omega\boldsymbol{k} \times r_{D/P}\boldsymbol{i}\right)$$

$$= 2\boldsymbol{i}\text{ m/s}^2 + (-8\boldsymbol{k}\text{ rad/s}^2) \times 0.03\boldsymbol{i}\text{ m} + 48\boldsymbol{k}\text{ rad/s} \times (48\boldsymbol{k}\text{ rad/s} \times 0.03\boldsymbol{i}\text{ m})$$

$$= (2.0 - 69.12)\boldsymbol{i}\text{ m/s}^2 - 0.24\boldsymbol{j}\text{ m/s}^2 \quad \Rightarrow \quad \boxed{\boldsymbol{a}_D = (-67.12\boldsymbol{i} - 0.24\boldsymbol{j})\text{ m/s}^2}$$

Check The bike is moving to the left and decelerating; hence the positive sign for the 2 m/s^2 deceleration is fine. Because the acceleration is to the right, we need a negative angular acceleration, which we have (−8 rad/s^2). This is clockwise, and thus the angular acceleration component of $\boldsymbol{a}_D$, that due to $\boldsymbol{\alpha} \times \boldsymbol{r}_{D/P}$, should be negative, as it turns out to be.

EXAMPLE 6.15 PATH OF POINT ON ROLLING DISK (Theory on page 349)

We've now looked at some velocity and acceleration results for a disk rolling without slip. Let's look a bit deeper. Determine the path that the initial point of contact follows as the disk (**Figure 6.40**) rolls, and discuss any interesting characteristics.

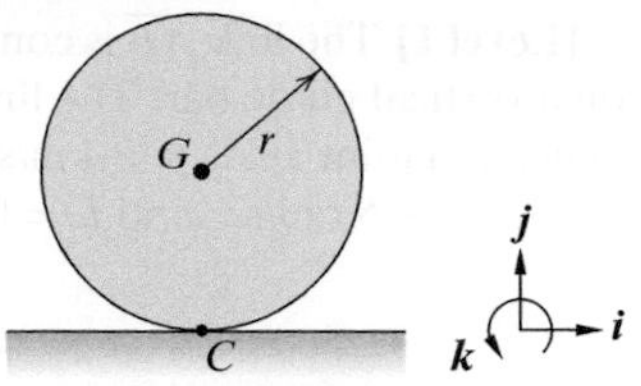

Figure 6.40 Rolling disk

Goal Plot the path of an initial contact point and discuss results.

Given Disk rolls without slip.

Assume Assume that the disk is perfectly circular.

Draw **Figure 6.41** shows the variables we'll be using. x and y show the position of a point on the periphery of the disk, originally the point of contact with the ground.

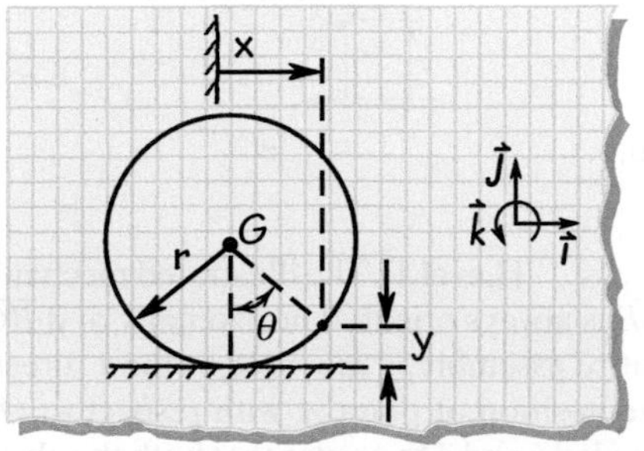

Figure 6.41 Parameters for rolling disk

As the disk rolls this point moves in space as illustrated in **Figure 6.42**. The outline of the rolling disk is shown five times. The disk is rolling to the left and each disk snapshot is rotated 90 degrees counterclockwise with respect to the one to its immediate right. To aid in visualizing this, I've put a red line across the rightmost disk, with a green dot at the ground contact point. As you can see, this line/dot rotates in 90 degree increments as you move left.

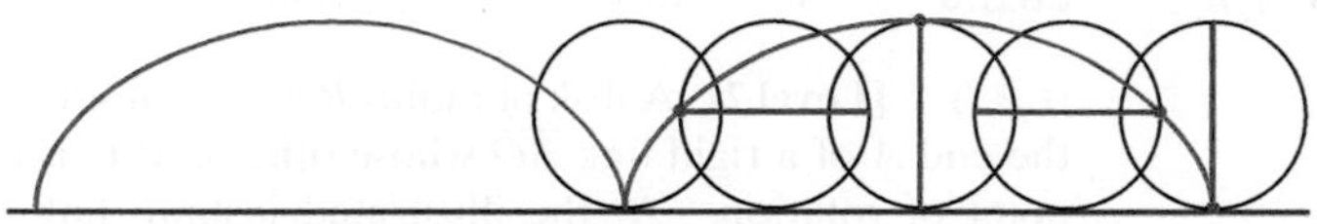

Figure 6.42 Rolling disk schematic

The green curve is a locus of all the points in space visited by the initial contact point. As expected, this line goes through the green dots since the position of the green dot (for *all*) positions of the disk is what defines the curve.

Formulate Equations and Solve The line was plotted by formulating the position of the contact point:

$$x = -r\theta + r\sin\theta, \qquad y = r(1 - \cos\theta)$$

We can derive the speed and acceleration components by differentiating with respect to time:

$$\dot{x} = -r\dot{\theta} + r\dot{\theta}\cos\theta, \qquad \dot{y} = r\dot{\theta}\sin\theta$$

$$\ddot{x} = -r\ddot{\theta} + r\ddot{\theta}\cos\theta - r\dot{\theta}^2\sin\theta, \qquad \ddot{y} = r\ddot{\theta}\sin\theta + r\dot{\theta}^2\cos\theta$$

Note the similarity in these terms to what we've seen with particle motion in polar coordinates, acceleration terms such as $r\dot{\theta}^2, r\ddot{\theta}$, and so on. What we have here are simply these familiar terms projected into their horizontal and vertical components (hence the $\sin\theta$ and $\cos\theta$ terms).

Check We can check these results by comparing to a prior example (Example 6.12). The solution for the acceleration of the contact point between the tire and road was $\omega^2 r\boldsymbol{j}$. Evaluating our equations for $\theta = 0$ gives us $\ddot{x} = -r\ddot{\theta} + r\ddot{\theta} = 0$ and $\ddot{y} = r\dot{\theta}^2$, a precise match since $\boldsymbol{j}$ indicates motion in the y direction and $\omega = \dot{\theta}$.

EXERCISES 6.3

6.3.1. **[Level 1]** The link $\overline{AB}$ is connected to a collar that moves on a vertical guide bar. The link is joined to the collar through a pin joint at A. $\dot{y} = 4$ m/s, $\ddot{y} = 10$ m/s^2, $\theta = 30°$, $\dot{\theta} = 14$ rad/s, $\ddot{\theta} = -5$ rad/s ^{2}, and $L = 0.75$ m. Determine $\boldsymbol{v}_B$ and $\boldsymbol{a}_B$.

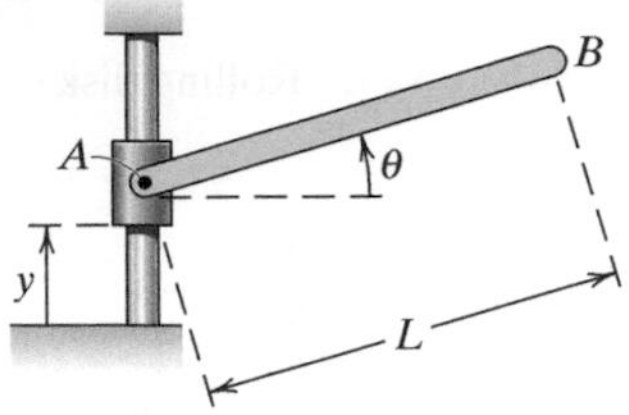

E6.3.1

6.3.2. **[Level 1]** The illustrated mechanism has two links. $\overline{AB}$ is always oriented horizontally, and $\overline{BC}$ can rotate. The entire assembly is movable in the $\boldsymbol{i}$ direction, x indicating the position of point A with respect to O. Determine the needed $\dot{x}$ and $\ddot{x}$ to ensure that both the absolute velocity and acceleration of C are purely vertical (in the $\boldsymbol{j}$ direction) for given θ, $\dot{\theta}$, and $\ddot{\theta}$. Express your answer in terms of r, θ, $\dot{\theta}$, and $\ddot{\theta}$.

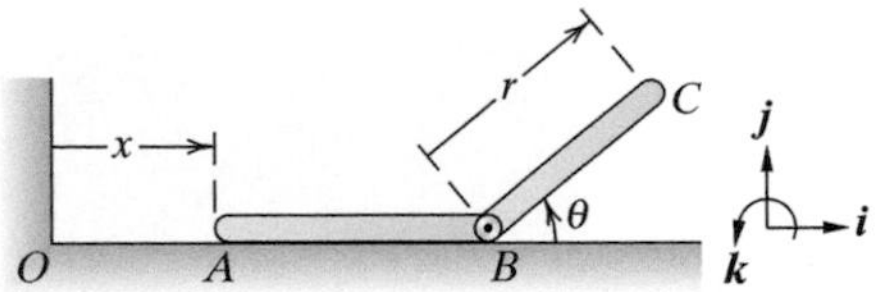

E6.3.2

6.3.3. **[Level 2]** A car is being tested while on a lift. Currently, the lift is stationary, and the tire is experiencing an angular velocity of $50\boldsymbol{k}$ rad/s and an angular acceleration of $-10\boldsymbol{k}$ rad/s^2. Determine the velocity and acceleration of points B and C.

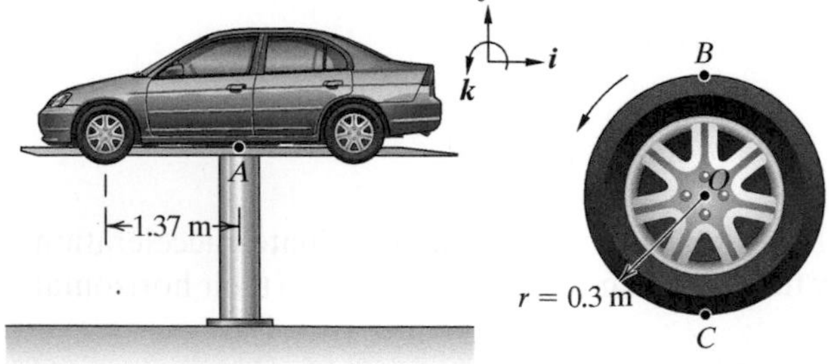

E6.3.3

6.3.4. **[Level 2]** Point A has an acceleration of $\boldsymbol{a}_A = (-1131\boldsymbol{i} - 1697\boldsymbol{j})$ m/s^2 when $\theta = 45°$. Determine the disk's angular velocity and angular acceleration. $r = 0.2$ m.

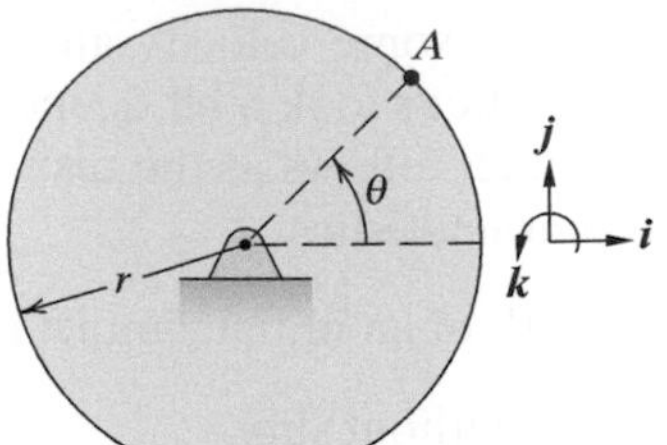

E6.3.4

6.3.5. **[Level 2]** Two opposing points on the ends of a propeller are shown as A and B.

a. What is their acceleration if the propeller has an angular speed of 300 rpm and an angular acceleration of -1000 rad/s^2?

b. What is the sum of their accelerations $(\boldsymbol{a}_A + \boldsymbol{a}_B)$? Does this make physical sense to you?

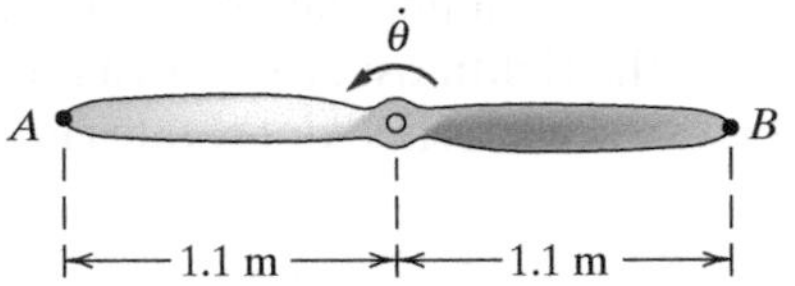

E6.3.5

6.3.6. **[Level 2]** A disk of radius $R = 0.12$ m spins about the end A of a rigid link $\overline{AO}$ whose other end O is a fixed point of rotation. At the illustrated instant, link $\overline{AO}$ is spinning at $\boldsymbol{\omega}_{\overline{AO}} = 7\boldsymbol{k}$ rad/s with an angular acceleration of $\boldsymbol{\alpha}_{\overline{AO}} = 2\boldsymbol{k}$ rad/s^2. Also, $\boldsymbol{r}_{A/O} = 0.40\boldsymbol{j}$ m, and point B on the disk is directly above A. The disk spins at a constant angular velocity of $\boldsymbol{\omega}_{\overline{AB}} = -3\boldsymbol{k}$ rad/s. What are $\boldsymbol{v}_B$ and $\boldsymbol{a}_B$?

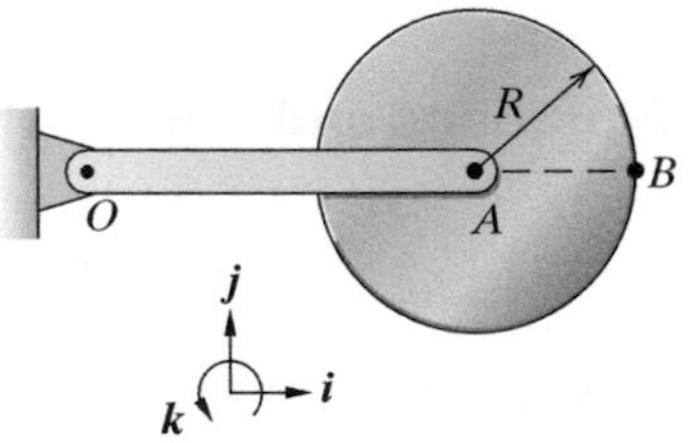

E6.3.6

6.3.7. **[Level 2]** The front chain ring (G_1), rear cog (G_2), and chain that form a bicycle's drivetrain are shown. $r_1 = 10$ cm and $r_2 = 5$ cm. If the cyclist is pedaling at a constant rate, find the ratio of the magnitude of $\boldsymbol{a}_B$ to $\boldsymbol{a}_A$ (i.e., $\frac{\|\boldsymbol{a}_B\|}{\|\boldsymbol{a}_A\|}$).

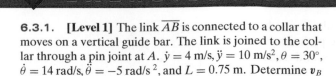

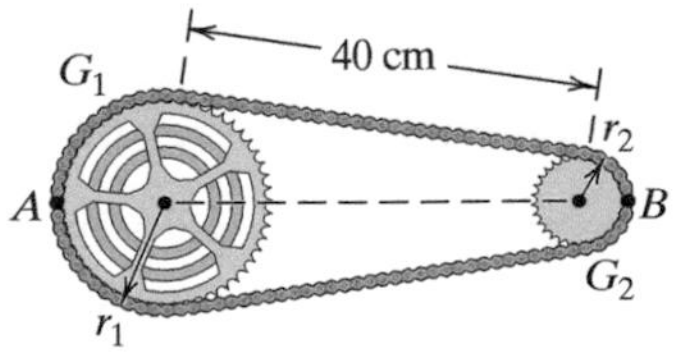

E6.3.7

6.3.8. **[Level 2]** Two meshing gears contact each other at a point. C_1 is attached to Gear 1, and C_2 is attached to Gear 2. Both of these points currently occupy the same position—the point of contact between the two gears. What are the velocity and acceleration of C_1 and C_2? $\dot{\theta}_1$ and $\ddot{\theta}_1$ are both nonzero. (Note that the $\dot{\theta}_2$ and $\dot{\theta}_1$ indicated in the figure are simply showing the positive rotation direction—physically one of the gears must be rotating in a clockwise manner if the other is rotating counterclockwise.)

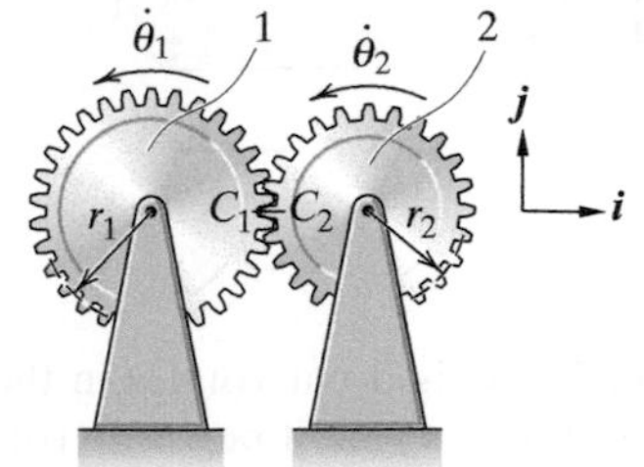

E6.3.8

6.3.9. **[Level 2]** The illustrated wheel is rolling without slip such that $\boldsymbol{v}_A = (-0.3\boldsymbol{i} + 0.3\boldsymbol{j})$ m/s and $\boldsymbol{a}_A = (-0.85\boldsymbol{i} + 0.4\boldsymbol{j})$ m/s^2. $\boldsymbol{r}_{G/C} = 0.2\boldsymbol{j}$ m and $\boldsymbol{r}_{A/G} = 0.2\boldsymbol{i}$ m. Determine $\boldsymbol{v}_B$ and $\boldsymbol{a}_B$.

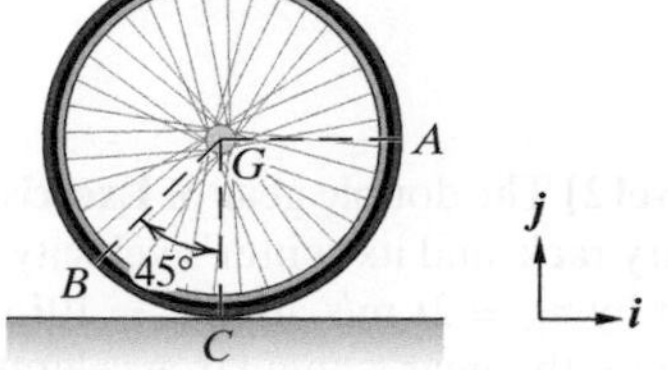

E6.3.9

6.3.10. **[Level 2]** An accessory drive belt for an automotive engine is shown. $r_1 = 7$ cm and $r_2 = 4$ cm. The belt is moving at 120 cm/s in the counterclockwise direction and is slowing at 30 cm/s^2. Determine the ratio $\frac{\|\boldsymbol{a}_A\|}{\|\boldsymbol{a}_B\|}$.

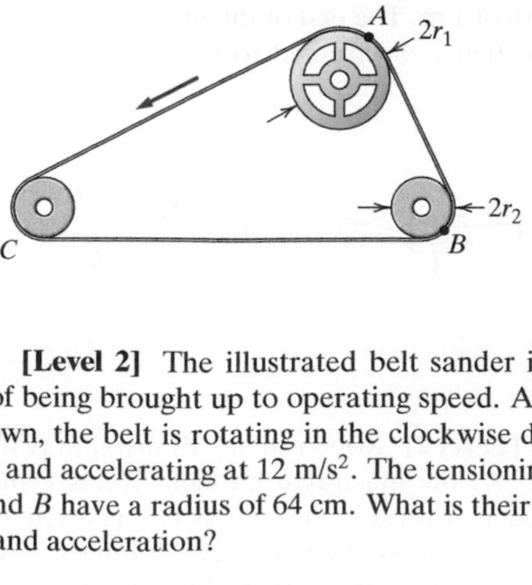

E6.3.10

6.3.11. **[Level 2]** The illustrated belt sander is in the process of being brought up to operating speed. At the instant shown, the belt is rotating in the clockwise direction at 12 m/s and accelerating at 12 m/s^2. The tensioning cylinders A and B have a radius of 64 cm. What is their angular velocity and acceleration?

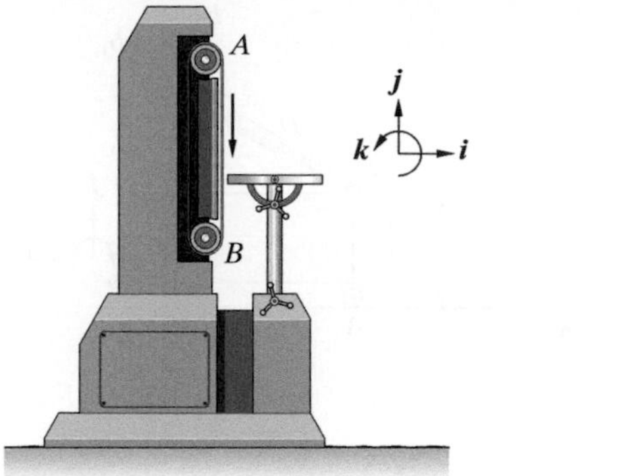

E6.3.11

6.3.12. **[Level 2]** A 4 m long bar ($\overline{AB}$) is being raised by the illustrated rope. The diameter of the reel C that's taking up the rope is $d = 20$ cm. The reel is turning with an angular velocity of $-10\boldsymbol{k}$ rad/s and an angular acceleration of $-0.5\boldsymbol{k}$ rad/s^2. Determine the acceleration of B for the illustrated instant.

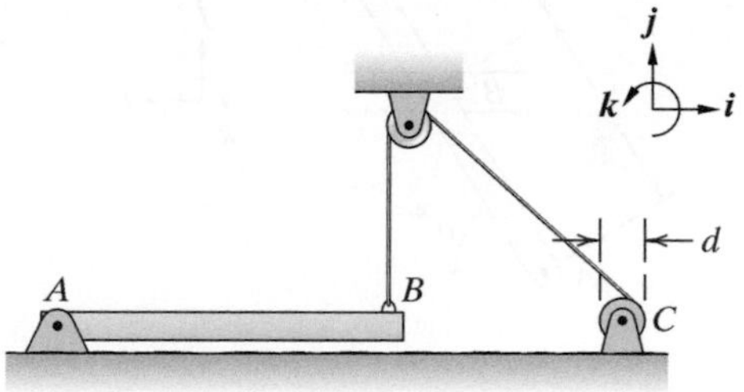

E6.3.12

6.3.13. **[Level 2]** A cylinder with three attached arms is rolling without slip down a sloped surface. The arms overhang the edge of the sloped surface, and thus the whole system rolls freely. The arms are equally spaced around the cylinder and are all of length $L = 0.5$ m. At the illustrated instant, C is directly above the cylinder's center O. $\theta = 15°$ and $d = 0.2$ m. The center O has a velocity $0.8\boldsymbol{b}_1$ m/s and acceleration $1.2\boldsymbol{b}_1$ m/s^2. What is the acceleration of C?

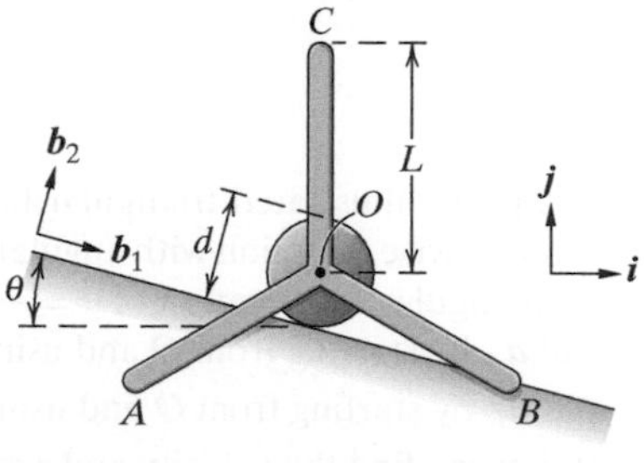

E6.3.13

6.3.14. **[Level 2]** A retractable arm $\overline{CD}$ is attached to the illustrated wheeled cart at C. The arm pulls back into its housing, causing C's velocity and acceleration to be $1.2\boldsymbol{b}_1$ m/s and $0.8\boldsymbol{b}_1$ m/s^2, respectively. $h = 0.5$ m, $r_{B/A} = 0.8$ m, and $\boldsymbol{b}_1 = \cos 40°\boldsymbol{i} + \sin 40°\boldsymbol{j}$. Determine the angular velocity and acceleration of the cart.

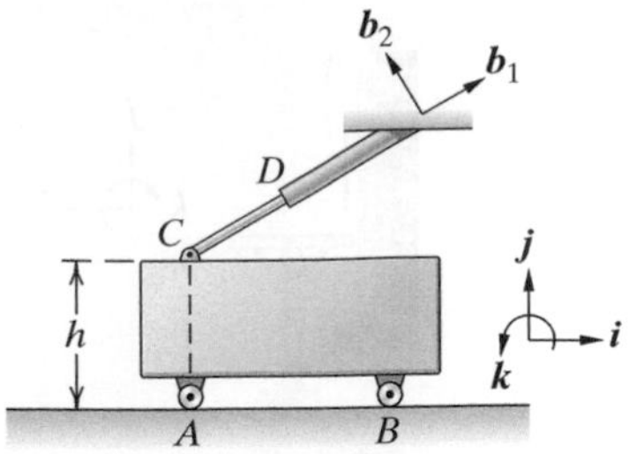

E6.3.14

6.3.15. [Level 2] Block C is driven within the vertical channel such that $\boldsymbol{v}_C = 0.4\boldsymbol{j}$ m/s and $\boldsymbol{a}_C = -0.2\boldsymbol{j}$ m/s^2. $L = 1.8$ m and $\theta = 60°$. Find $\boldsymbol{a}_A$.

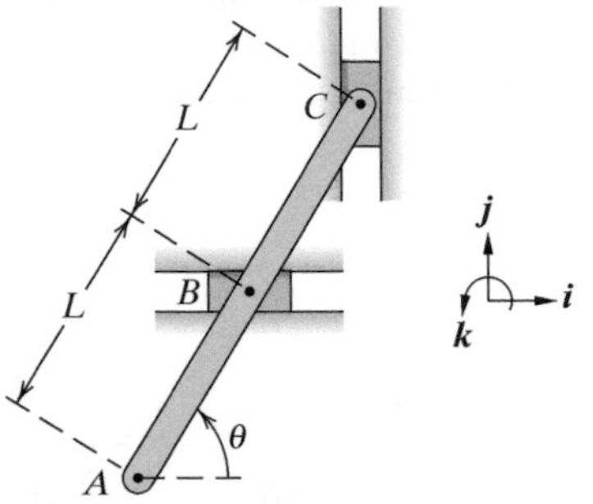

E6.3.15

6.3.16. [Level 2] The illustrated mechanism is free to rotate about C. $\boldsymbol{v}_A = (-0.3\boldsymbol{i} + 0.4\boldsymbol{j})$ m/s and $\boldsymbol{a}_A = (-4.0\boldsymbol{i} - 3.0\boldsymbol{j})$ m/s^2. Determine $\boldsymbol{v}_B$ and $\boldsymbol{a}_B$.

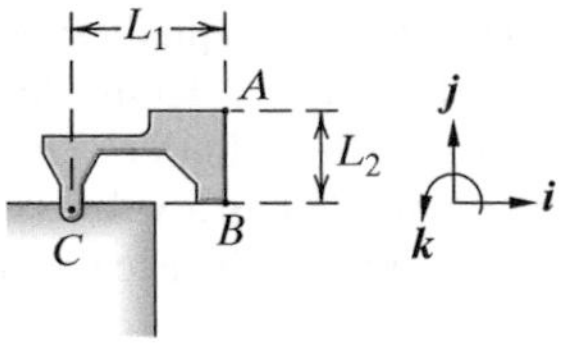

E6.3.16

6.3.17. [Level 2] The illustrated triangular body is rotating in a counterclockwise direction with angular speed of $\dot{\theta} = 10$ rad/s and an angular acceleration of $\ddot{\theta} = -3$ rad/s^2.

a. Find $\boldsymbol{v}_A$ and $\boldsymbol{a}_A$ by starting from O and using $\boldsymbol{r}_{A/O}$.

b. Find $\boldsymbol{v}_B$ and $\boldsymbol{a}_B$ by starting from O and using $\boldsymbol{r}_{B/O}$.

c. Show that you can find the velocity and acceleration at an arbitrary point in a body from some other point if you know $\dot{\theta}$ and $\ddot{\theta}$ by starting from A (use the results of **a**) and using $\boldsymbol{r}_{B/A}$, $\dot{\theta}$, and $\ddot{\theta}$ to find $\boldsymbol{v}_B$ and $\boldsymbol{a}_B$. Verify your answer using the results of **b**.

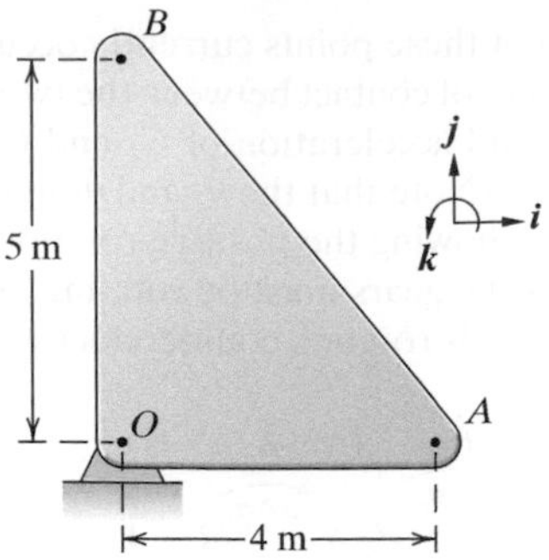

E6.3.17

6.3.18. [Level 2] A rigid bar rotates in the horizontal plane about O. Initially at rest, it begins to rotate counterclockwise with a constant angular acceleration of 15 rad/s^2. How long does it take for the magnitude of A's normal acceleration to equal the magnitude of its tangential acceleration?

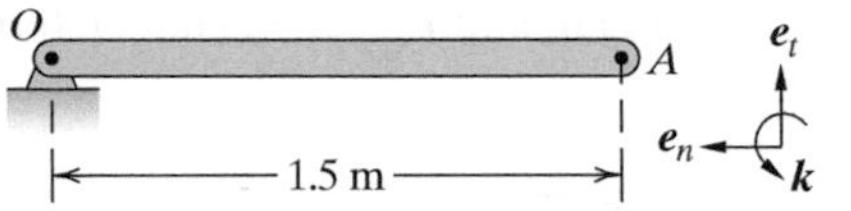

E6.3.18

6.3.19. [Level 2] The double gear of **Exercise 6.1.19** rolls on the stationary rack, and its center's velocity and acceleration are given by $\boldsymbol{v}_G = 2\boldsymbol{i}$ m/s and $\boldsymbol{a}_G = 10\boldsymbol{i}$ m/s^2, respectively. Determine the gear's angular acceleration and the acceleration of the upper rack.

6.3.20. [Level 2] For the dimensions and angles given, determine $\boldsymbol{a}_A$ if $\overline{OB}$ is rotating counterclockwise at a constant angular speed ω. $\overline{AB}$ has a length of $\sqrt{2}$ m, and $\overline{OB}$ has a length of 1 m. The end of the link $\overline{AB}$ labeled A rides on a collar that's constrained to move in the $\boldsymbol{i}$ direction. ($\theta = 45°$)

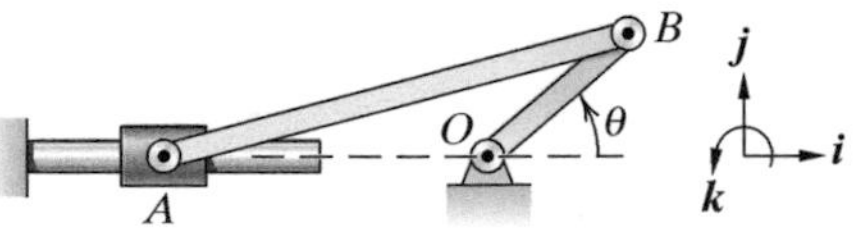

E6.3.20

6.3.21. [Level 2] Rope is being drawn into two motorized reels. At the illustrated moment Reels 1 and 2 are taking rope in at rates v_1 m/s and v_2 m/s, respectively. The rate is constant for Reel 1 but is accelerating at a_2 m/s^2 for Reel 2. What is the acceleration of point C located on the top of the pulley?

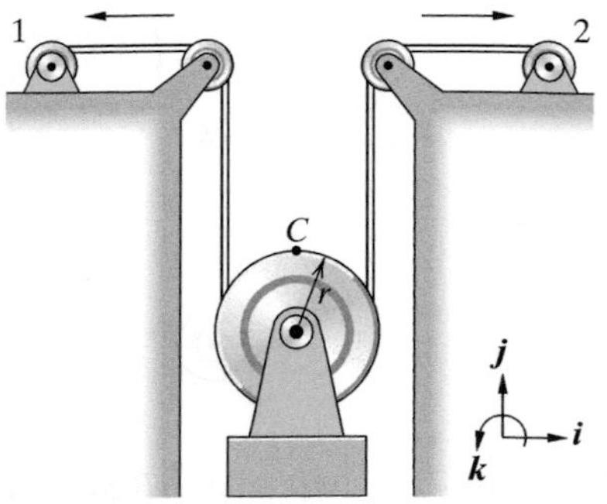

E6.3.21

6.3.22. [Level 3] The illustrated T-bar is positioned by the two links $\overline{AD}$ and $\overline{BE}$. At the illustrated instant, $\boldsymbol{r}_{A/D} = 2L\boldsymbol{b}_2$, $\boldsymbol{r}_{B/E} = L\boldsymbol{b}_2$, $\boldsymbol{r}_{A/B} = \sqrt{3}L\boldsymbol{b}_1 + L\boldsymbol{b}_2$, and $\boldsymbol{r}_{C/A} = -\sqrt{3}L\boldsymbol{b}_1 + L\boldsymbol{b}_2$, $\dot{\theta}$ is nonzero and $\ddot{\theta} = 0$. Determine $\boldsymbol{a}_C$ in terms of the given parameters.

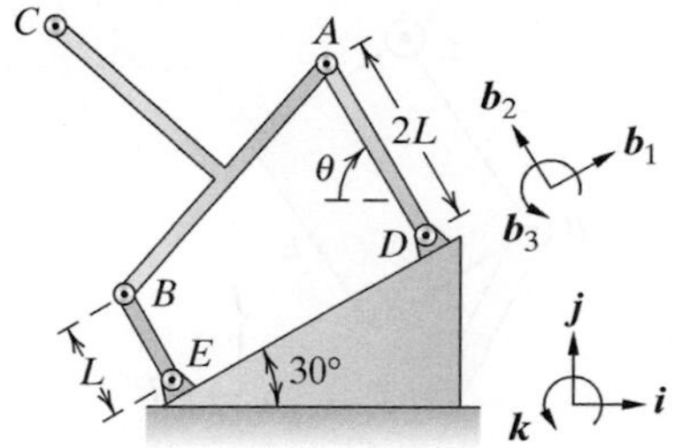

E6.3.22

6.3.23. [Level 3] A T-shaped bracket rotates about the fixed point O at a constant angular speed. When the bracket is oriented as depicted, point A has a velocity of $\boldsymbol{v}_A = (0.18\boldsymbol{i} - 0.46\boldsymbol{j})$ m/s. The dimensions of the bracket are as follows: $L_1 = 0.3$ m, $L_2 = 0.12$ m, and $L_3 = 0.06$ m. Find $\boldsymbol{v}_B$, $\boldsymbol{a}_A$, and $\boldsymbol{a}_B$.

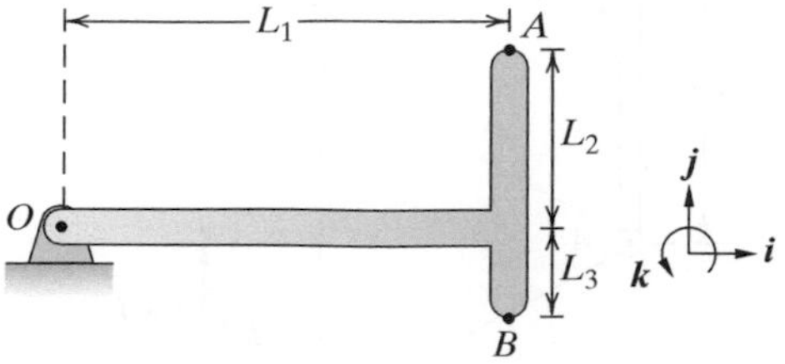

E6.3.23

6.3.24. [Level 3] The illustrated penny farthing is traveling at 16 km/h to the left and accelerating at 1.2 m/s^2. The front wheel has a diameter of 127 cm, and the rear has a diameter of 46 cm. Find $\frac{\|\boldsymbol{a}_Q\|}{\|\boldsymbol{a}_R\|}$.

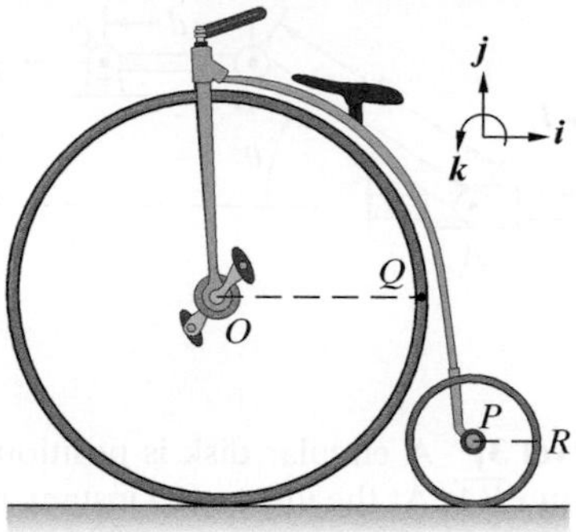

E6.3.24

6.3.25. [Level 3] A trapezoidal plate is positioned by the two links $\overline{BD}$ and $\overline{AC}$. At the illustrated instant, $\boldsymbol{r}_{D/B} = h\boldsymbol{b}_2$, $\boldsymbol{r}_{A/C} = 2h\boldsymbol{b}_2$, $\boldsymbol{r}_{B/A} = 2h\boldsymbol{i}$, $\boldsymbol{r}_{E/C} = 4h\boldsymbol{b}_2$, and $\ddot{\gamma} = 0$. $\boldsymbol{b}_1 = \frac{\sqrt{3}}{2}\boldsymbol{i} - \frac{1}{2}\boldsymbol{j}$. Determine $\boldsymbol{a}_E$ in terms of h and $\dot{\gamma}$.

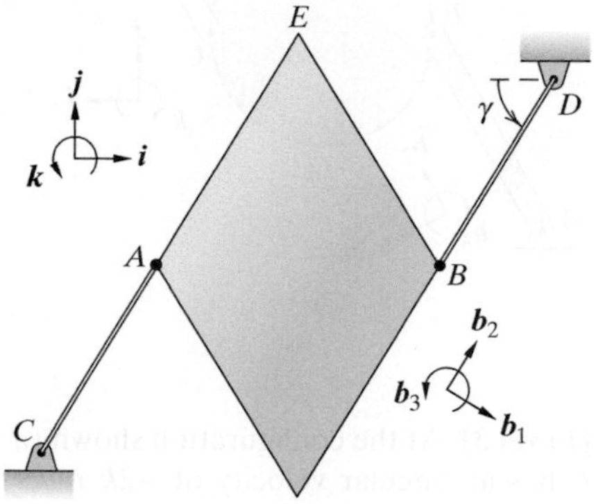

E6.3.25

6.3.26. [Level 3] Link $\overline{AB}$ of the illustrated system is $r_{B/A} = 0.75$ m long and angled at $\theta = 30°$ with respect to the horizontal. Link $\overline{CD}$ has a length of $r_{C/D} = 0.5$ m and is oriented at $\beta = 60°$ to the horizontal. Link $\overline{BC}$ is horizontal and $r_{C/B} = 1$ m long, and $\boldsymbol{\omega}_{\overline{AB}} = -4\boldsymbol{k}$ rad/s at the given instant. If link $\overline{CD}$ is spinning at a constant angular speed, what are $\boldsymbol{\omega}_{\overline{BC}}$, $\boldsymbol{\alpha}_{\overline{BC}}$, and $\boldsymbol{\alpha}_{\overline{AB}}$?

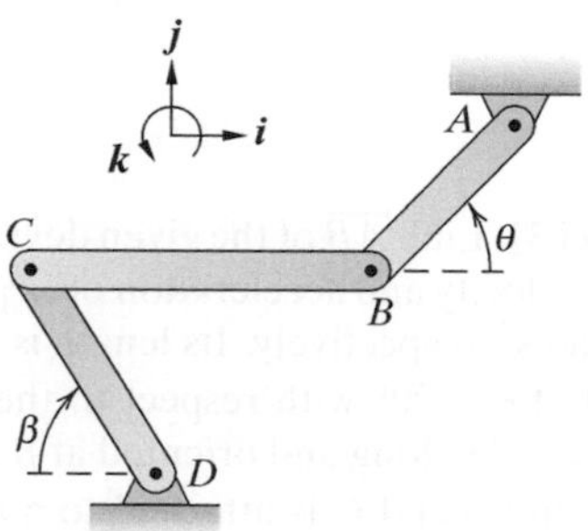

E6.3.26

6.3.27. [Level 3] The link $\overline{BC}$ is rotated at a constant rate of $12\boldsymbol{k}$ rad/s. Derive an expression for $\ddot{\theta}$ and an expression for the acceleration of the collar at A.

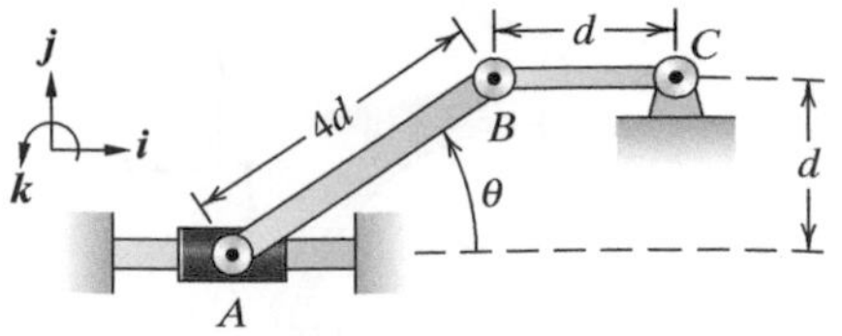

E6.3.27

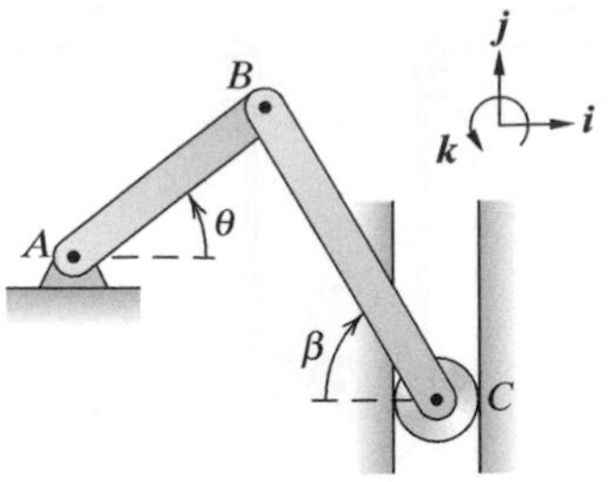

6.3.28. **[Level 3]** A circular disk is positioned by the two links $\overline{CD}$ and $\overline{BA}$. At the illustrated instant, $\mathbf{r}_{D/C} = h\mathbf{b}_2$, $\mathbf{r}_{B/A} = 2h\mathbf{b}_2$, $\mathbf{r}_{C/B} = 2h\mathbf{i}$, $\mathbf{r}_{E/C} = -h(\mathbf{i}+\mathbf{j})$, and $\ddot{\beta} = 0$. $\mathbf{b}_1 = \frac{\sqrt{3}}{2}\mathbf{i} - \frac{1}{2}\mathbf{j}$. Determine $\mathbf{a}_E$ in terms of h and $\dot{\beta}$.

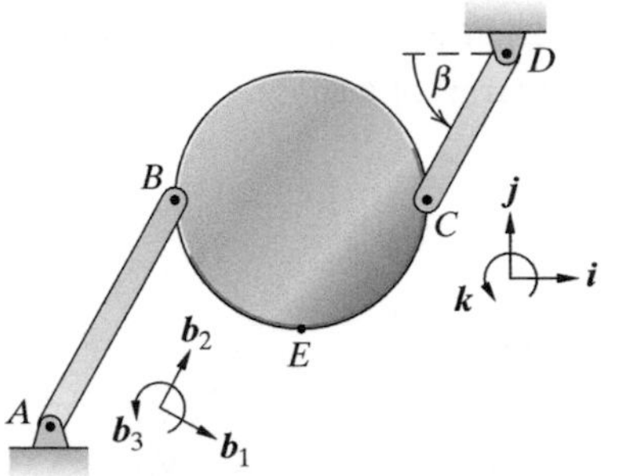

E6.3.28

6.3.29. **[Level 3]** At the configuration shown ($\eta = 30°$), the link $\overline{AB}$ has an angular velocity of $-2\mathbf{k}$ rad/s and an angular acceleration of zero. Find $\mathbf{a}_C$ and the angular acceleration of $\overline{BC}$. C slides freely on the horizontal surface.

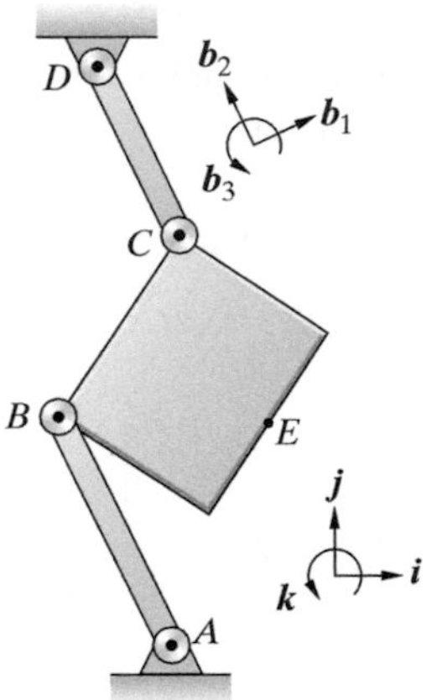

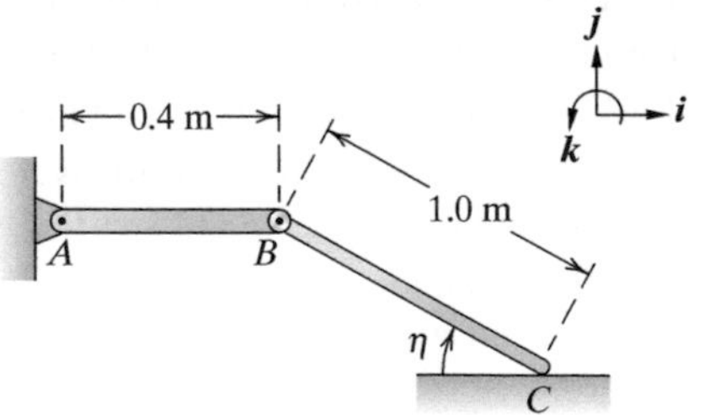

E6.3.29

6.3.30. **[Level 3]** Link $\overline{AB}$ of the given device is rotating with an angular velocity and acceleration of $\boldsymbol{\omega}_{\overline{AB}} = 6\mathbf{k}$ rad/s and $\boldsymbol{\alpha}_{\overline{AB}} = 3\mathbf{k}$ rad/s², respectively. Its length is $r_{B/A} = 0.9$ m, and it is angled at $\theta = 20°$ with respect to the horizontal. Link $\overline{BC}$ is $r_{B/C} = 1.3$ m long and oriented at $\beta = 70°$ to the horizontal. The link's end C is attached to a roller that is constrained to move in a vertical slot. Find $\boldsymbol{\omega}_{\overline{BC}}$ and $\boldsymbol{\alpha}_{\overline{BC}}$.

E6.3.30

6.3.31. **[Level 3]** The two links $\overline{AB}$ and $\overline{CD}$ control the position of a rectangular plate. At the illustrated instant, $\mathbf{r}_{D/C} = \mathbf{r}_{B/A} = 2d\mathbf{b}_2$, $\mathbf{r}_{C/B} = \sqrt{3}d\mathbf{b}_1 + d\mathbf{b}_2$, $\mathbf{r}_{E/C} = \frac{d}{2}[(1-\sqrt{3})\mathbf{b}_1 - (1+\sqrt{3})\mathbf{b}_2]$, $\dot{\theta} = 5$ rad/s, $\ddot{\theta} = 0$, and $d = 0.15$ m. $\mathbf{b}_1 = \frac{\sqrt{3}}{2}\mathbf{i} + \frac{1}{2}\mathbf{j}$. Determine $\mathbf{a}_E$.

E6.3.31

6.3.32. **[Level 3]** The illustrated equilateral triangle is supported by two links. $d = 0.5$ m. At the illustrated position, $\dot{\theta} = 4$ rad/s and $\ddot{\theta} = 0$. Find $\mathbf{a}_C$.

E6.3.32

6.3.33. **[Level 3]** At the illustrated instant, link $\overline{AB}$ ($r_{B/A} = 0.18$ m) is vertical and rotating at a constant rate $\boldsymbol{\omega}_{\overline{AB}} = 9\mathbf{k}$ rad/s, and link $\overline{CD}$ ($r_{C/D} = 0.21$ m) is horizontal. If link $\overline{BC}$ ($r_{C/B} = 0.27$ m) is angled at $\theta = 60°$ with respect to the vertical, what are $\boldsymbol{\alpha}_{\overline{BC}}$ and $\boldsymbol{\alpha}_{\overline{CD}}$?

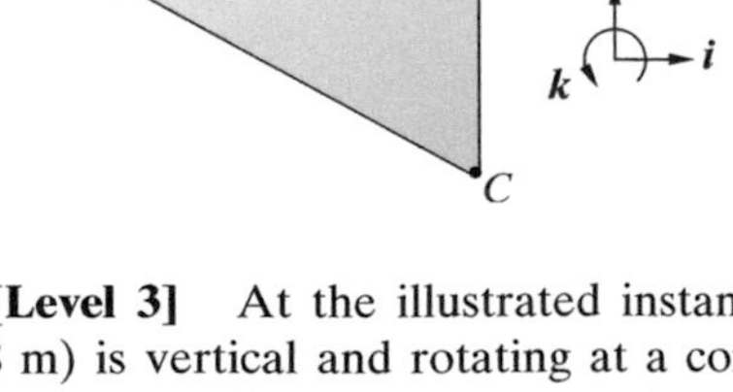

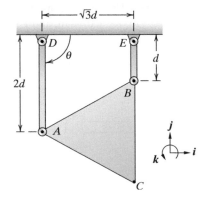

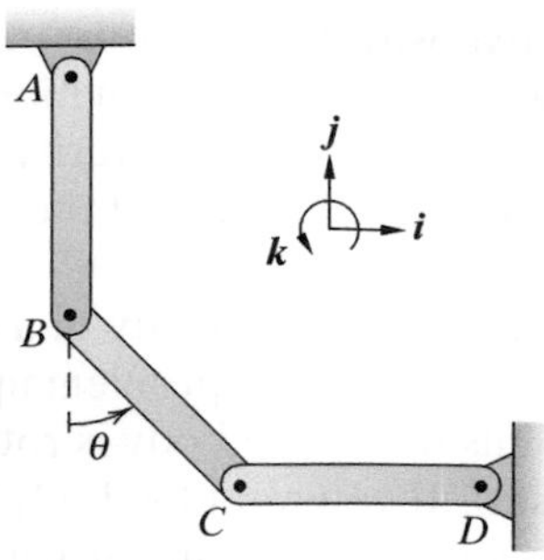

E6.3.33

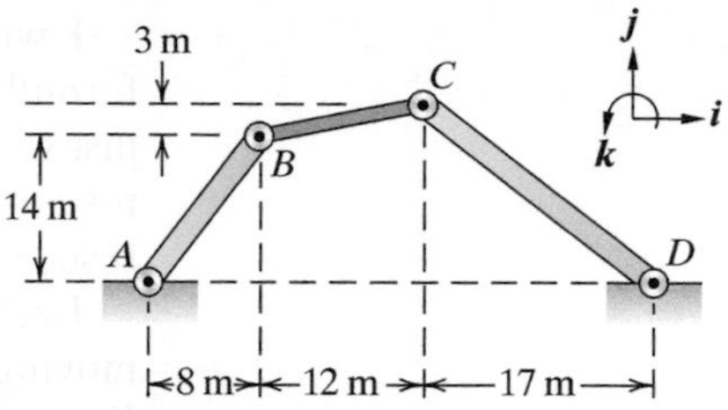

E6.3.34

6.3.34. [Level 3] The illustrated linkage is controlled by motions of crank $\overline{AB}$. For the position shown, with $\boldsymbol{\omega}_{\overline{AB}} = 20\boldsymbol{k}$ rad/s and $\boldsymbol{\alpha}_{\overline{AB}} = 0$, determine the angular velocities and accelerations of the connecting rods $\overline{BC}$ and $\overline{CD}$.

6.3.35. [Level 3] Computational Consider the system of **SA6.3**. The same parameters hold: $\boldsymbol{r}_{C/O} = 4.5\boldsymbol{b}_1$ m and $\boldsymbol{r}_{A/C} = 1.5\boldsymbol{c}_1$ m. The primary arm is rotating at a constant rate of $0.8\boldsymbol{k}$ rad/s, and the secondary arm is rotating with respect to the primary arm at a constant rate of $2\boldsymbol{k}$ rad/s. Plot the components of acceleration in the $\boldsymbol{c}_1$ and $\boldsymbol{c}_2$ directions of the tip A. Start from $\theta = \beta = 0$ and plot for one complete rotation of the primary arm.

6.4 RELATIVE MOTION ON A RIGID BODY

This topic, though lots of fun, is one of the hardest to grasp, at least at first. We looked at all the possible two-dimensional motions of a rigid body in Section 6.3 and by now have a good sense of how to find the velocity/acceleration at one place in the body if we're given the velocity/acceleration of another point as well as the body's rotational behavior. In this section we're going to go further and consider the possibility of some point moving *on* the body, like an ant running along the surface of a rolling orange or a passenger running up the aisle in an airplane that's taking off. In both of these cases, we have a rigid body in motion (the orange and the airplane), and we also have some point that's moving with respect to this body (the ant and the passenger). Finding the velocity or acceleration of this moving point won't be as easy as simply tacking on a relative velocity or acceleration term (as we did in Section 2.5). Actually, for the velocity case it is this simple, but the acceleration case will cause us some problems. Here's what you would *want* to have happen, if life were being kind to us:

$$\boldsymbol{v}_B = \boldsymbol{v}_A + \boldsymbol{\omega} \times \boldsymbol{r}_{B/A} + \boldsymbol{v}_{rel} \qquad \text{(actually correct!)}$$

and

$$\boldsymbol{a}_B = \boldsymbol{a}_A + \boldsymbol{\omega} \times \left(\boldsymbol{\omega} \times \boldsymbol{r}_{B/A}\right) + \boldsymbol{\alpha} \times \boldsymbol{r}_{B/A} + \boldsymbol{a}_{rel} \qquad \text{(sadly, incorrect)}$$

where $\boldsymbol{v}_{rel}$ and $\boldsymbol{a}_{rel}$ are the velocity and acceleration, respectively, of the moving point with respect to the body it's moving on. All I did was take our rigid-body results [(6.4) and (6.25)] and tack on a correction term to account for the relative motion ($\boldsymbol{v}_{B/A}$ and $\boldsymbol{a}_{B/A}$). As we will see right now, the velocity expression is correct as written, but the acceleration formula needs one additional term. So don't memorize the acceleration formula just yet—it's not quite correct.

I will ultimately use the derivative-with-respect-to-a-rotating-frame formula of (6.24), but first I want to go through the situation physically, just so you can get a good sense of what's going on. I'll verify the velocity results through the rotating-frame formula and then get the acceleration results directly.

Let's start by finding out how to express the velocity of a point that's moving *on* a rotating body. **Figure 6.43** breaks the problem up into steps. I've set $\boldsymbol{v}_A$ to zero for simplicity. Thus the entire body is rotating about this point (like a playground carousel). If you want the body to translate as well as rotate, then you simply need to add whatever velocity vector you want A to have, just as we saw in the previous section.

At any instant, the moving point B is touching some particular point on the rotating body, *and* B is also moving relative to the body. B's velocity due to the rotation is shown in **Figure 6.43*a***, which ignores B's velocity relative to the body. **Figure 6.43*b*** shows the velocity vector for B that is due to the motion of B relative to the rotating body. This velocity has nothing to do with the fact that the body is rotating. It's the term you would get if someone on the carousel watched someone else (also on the carousel) get off one horse and walk toward another.

We can come up with the overall velocity of B by simply adding the two terms together, as shown in **Figure 6.43*c***. Again, if it's not clear, just think about it physically. If you were standing in the aisle of a plane that was taxiing down the runway, your velocity would be the same as the point of the plane directly under your feet. But if you then started walking, you would be generating a relative velocity with respect to the airplane that would have to be added to the airplane's velocity. If you're in contact with a moving body, you will have the same velocity as the contact point, *and* if you yourself are moving with respect to the body, then you also have this additional velocity term.

The correct way to account for motion relative to a moving body is therefore given by

$$\boldsymbol{v}_B = \boldsymbol{v}_A + \boldsymbol{\omega} \times \boldsymbol{r}_{B/A} + \boldsymbol{v}_{rel} \tag{6.29}$$

just as we had hoped it would be.

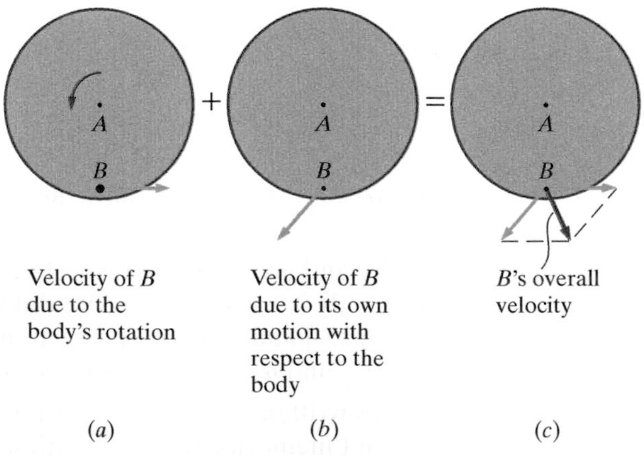

Figure 6.43 Velocity components of a point moving on a body

>>> **Check out Example 6.16 (page 365) and Example 6.17 (page 366) for applications of this material.**

Now that we've convinced ourselves that this is the right answer, let's get it in one step from our rotating-reference-frame formulation,

$$\left.\frac{d}{dt}\right|_N \boldsymbol{p} = \left.\frac{d}{dt}\right|_S \boldsymbol{p} + \boldsymbol{\omega}\times\boldsymbol{p}$$

for any vector $\boldsymbol{p}$ on the rotating body. In our case, our vector will be the position vector $\boldsymbol{r}_{B/A}$:

$$\left.\frac{d}{dt}\right|_N \boldsymbol{r}_{B/A} = \left.\frac{d}{dt}\right|_S \boldsymbol{r}_{B/A} + \boldsymbol{\omega}\times\boldsymbol{r}_{B/A}$$

The time derivative with respect to the rotating frame is just $\boldsymbol{v}_{rel}$, the velocity of point B on the body measured with respect to the rotating reference frame S. Thus we have

$$\boldsymbol{v}_{B/A} = \boldsymbol{v}_{rel} + \boldsymbol{\omega}\times\boldsymbol{r}_{B/A}$$

and our complete expression for the velocity of B is

$$\boldsymbol{v}_B = \boldsymbol{v}_A + \boldsymbol{v}_{B/A} = \boldsymbol{v}_A + \boldsymbol{\omega}\times\boldsymbol{r}_{B/A} + \boldsymbol{v}_{rel} \tag{6.30}$$

exactly the same result as (6.29).

Why is the situation more complex for acceleration? Recall what the general expression for acceleration is (2.50) for a particle in a polar frame:

$$\boldsymbol{a} = (\ddot{r} - r\dot{\theta}^2)\boldsymbol{e}_r + (r\ddot{\theta} + 2\dot{r}\dot{\theta})\boldsymbol{e}_\theta \tag{6.31}$$

Next, recall what the expression is (6.25) to get the acceleration of a point B on a rigid body that's pivoting about a point A (for which $\boldsymbol{a}_A = 0$):

$$\boldsymbol{a}_B = \boldsymbol{\omega}\times\left(\boldsymbol{\omega}\times\boldsymbol{r}_{B/A}\right) + \boldsymbol{\alpha}\times\boldsymbol{r}_{B/A} \tag{6.32}$$

If you compare (6.31) and (6.32), you can make some matches. The $\boldsymbol{\omega}\times(\boldsymbol{\omega}\times\boldsymbol{r}_{B/A})$ term corresponds to $-r\omega^2$, and the $\boldsymbol{\alpha}\times\boldsymbol{r}_{B/A}$ term matches up with $r\ddot{\theta}$. But what about the $\ddot{r}$ term in (6.32)? It doesn't show up on a rigid body (because it presumes a time-varying r), but if we had a point moving on a rigid body, then $\ddot{r}$ would be the $\boldsymbol{a}_{rel}$ term we speculated about a few lines up.

This leaves us to identify $2\dot{r}\dot{\theta}$ in (6.31). This is the nonintuitive term, just as it was nonintuitive when we ran into it the first time. It's an acceleration term (the coriolis acceleration) that depends on both rotational motion and translational motion. Can we conclude that for our problem, this term $2\dot{r}\dot{\theta}$ involves $\boldsymbol{\omega}$ and $\boldsymbol{v}_{rel}$? Absolutely. If the rigid body isn't rotating, then this term won't be present, and if our point of interest isn't moving relative to the body, the $2\dot{r}\dot{\theta}$ term is absent as well. But when both terms show up, then so does this acceleration component, and it depends on the body's angular velocity $\boldsymbol{\omega}$ and the moving points velocity with respect to the body $\boldsymbol{v}_{rel}$.

With that as a motivation as to where we're going, let's see how we can derive this result by using (6.29) in (6.24). To get the acceleration, we need to differentiate our velocity term:

$$\boldsymbol{a}_B = \left.\frac{d}{dt}\right|_N \left(\boldsymbol{v}_A + \boldsymbol{\omega} \times \boldsymbol{r}_{B/A} + \boldsymbol{v}_{rel}\right)$$

$$= \left.\frac{d}{dt}\right|_N \boldsymbol{v}_A + \left.\frac{d}{dt}\right|_N \left(\boldsymbol{\omega} \times \boldsymbol{r}_{B/A} + \boldsymbol{v}_{rel}\right)$$

$$(6.24) \Rightarrow \quad = \boldsymbol{a}_A + \left.\frac{d}{dt}\right|_S \left(\boldsymbol{\omega} \times \boldsymbol{r}_{B/A} + \boldsymbol{v}_{rel}\right) + \boldsymbol{\omega} \times \left(\boldsymbol{\omega} \times \boldsymbol{r}_{B/A} + \boldsymbol{v}_{rel}\right)$$

$$= \boldsymbol{a}_A + \overbrace{\boldsymbol{\alpha} \times \boldsymbol{r}_{B/A} + \boldsymbol{\omega} \times \boldsymbol{v}_{rel}}^{\left.\frac{d}{dt}\right|_S \left(\boldsymbol{\omega} \times \boldsymbol{r}_{B/A}\right)} + \overbrace{\boldsymbol{a}_{rel}}^{\left.\frac{d}{dt}\right|_S \boldsymbol{v}_{rel}} + \boldsymbol{\omega} \times \left(\boldsymbol{\omega} \times \boldsymbol{r}_{B/A}\right) + \boldsymbol{\omega} \times \boldsymbol{v}_{rel}$$

$$= \boldsymbol{a}_A + \boldsymbol{\alpha} \times \boldsymbol{r}_{B/A} + \boldsymbol{\omega} \times \left(\boldsymbol{\omega} \times \boldsymbol{r}_{B/A}\right) + 2\boldsymbol{\omega} \times \boldsymbol{v}_{rel} + \boldsymbol{a}_{rel} \tag{6.33}$$

Let's go over this slowly so that the origin of each term is clear. Remember that when we want to differentiate a rotating vector, we can break the problem up into two parts: time rates of change that are taking place with respect to the rotating frame and an $\boldsymbol{\omega} \times \ldots$ term. I've put braces over two of the terms in (6.33) to show where they came from. The first braced term derives from applying the chain rule of differentiation to the $\boldsymbol{\omega} \times \boldsymbol{r}_{B/A}$ term *with respect to the rotating frame*. The angular speed ω differentiates to become an angular acceleration α, and the position vector $\boldsymbol{r}_{B/A}$ of B with respect to A becomes the relative velocity $\boldsymbol{v}_{rel}$. The second braced term is pretty reasonable—it's just the time rate of change of the relative velocity with respect to the rotating frame, that is, the relative acceleration.

Rearranging a bit yields

$$\boldsymbol{a}_B = \boldsymbol{a}_A + \boldsymbol{\alpha} \times \boldsymbol{r}_{B/A} + \boldsymbol{\omega} \times \left(\boldsymbol{\omega} \times \boldsymbol{r}_{B/A}\right) + \boldsymbol{a}_{rel} + 2\boldsymbol{\omega} \times \boldsymbol{v}_{rel} \tag{6.34}$$

And there you have it: the correct way to deal with accelerating objects on a rotating body. It's helpful to remember the analogues to a polar representation of a single particle:

Rigid Body		Polar
$\boldsymbol{\alpha} \times \boldsymbol{r}_{B/A}$	$\rightleftharpoons$	$r\ddot{\theta}$
$\boldsymbol{\omega} \times \left(\boldsymbol{\omega} \times \boldsymbol{r}_{B/A}\right)$	$\rightleftharpoons$	$-r\dot{\theta}^2$
$\boldsymbol{a}_{rel}$	$\rightleftharpoons$	$\ddot{r}$
$2\boldsymbol{\omega} \times \boldsymbol{v}_{rel}$	$\rightleftharpoons$	$2\dot{r}\dot{\theta}$

>>> **Check out Example 6.18 (page 367), Example 6.19 (page 368), and Example 6.20 (page 370) for applications of this material.**

EXAMPLE 6.16 ABSOLUTE VELOCITY OF A SPECIMEN IN A CENTRIFUGE (Theory on page 362)

Figure 6.44 shows a specimen A in a centrifuge. The tube it's in is spinning counterclockwise about O at 10 rps, and at the instant illustrated it's 10 cm from O and moving outward at 16 mm/s. What is the absolute velocity of the specimen?

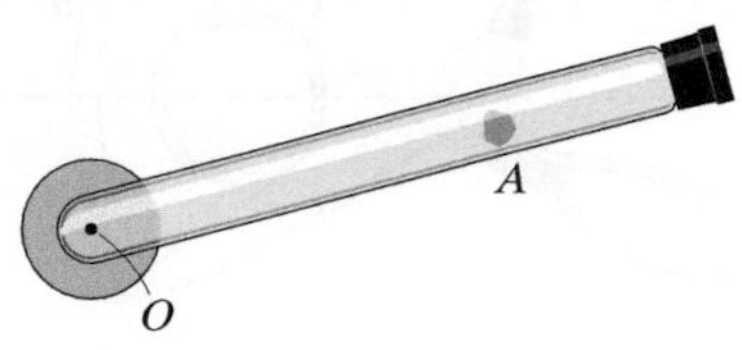

Figure 6.44 Centrifuge specimen

Goal Calculate the absolute velocity of A.

Given Rotational speed of enclosing body, relative position, and speed of enclosed specimen.

Assume The specimen is constrained to remain within the tube.

Draw In **Figure 6.45** we see the basic system along with appropriate unit vectors and motion variables.

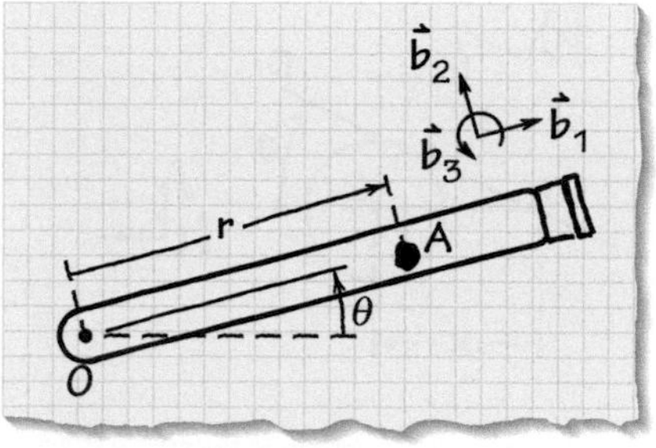

Figure 6.45 Schematic of centrifuge

Formulate Equations The specimen's velocity is found by using (6.29):

$$\boldsymbol{v}_A = \boldsymbol{v}_O + \boldsymbol{\omega} \times r\boldsymbol{b}_1 + \boldsymbol{v}_{rel} \quad (6.35)$$

Solve We first have to find the centrifuge's rotational speed. It spins at 10 rps, and knowing it rotates through 2π rad for each revolution gives us a rotational speed of

$$\omega = (10\,\text{rps})(2\pi\,\text{rad/rev}) = 20\pi\,\text{rad/s}$$

Using the given data and ω in (6.35) gives us

$$\boldsymbol{v}_A = 0 + 20\pi\,\boldsymbol{b}_3\,\text{rad/s} \times (0.10\,\text{m})\boldsymbol{b}_1 + 0.016\boldsymbol{b}_1\,\text{m/s}$$

$$\boldsymbol{v}_A = (2\pi\,\boldsymbol{b}_2 + 0.016\boldsymbol{b}_1)\,\text{m/s}$$

EXAMPLE 6.17 VELOCITY CONSTRAINTS—CLOSING SCISSORS (Theory on page 362)

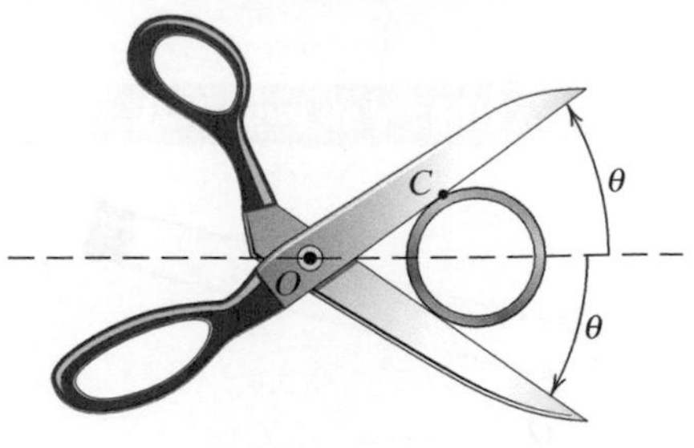

Figure 6.46 Cutting a pipe with scissors

The scissors shown are closing (**Figure 6.46**), trying to cut through a cylindrical pipe of radius r. The pipe is too hard to cut and so, as the scissors' jaws close, the pipe translates to the right. As the pipe moves, the points contacting the scissors scrape along the jaws. Determine how fast the contact point moves along the upper jaw. The pivot point O of the scissors remains fixed in space.

Goal Determine the relative speed of a contact point.

Given Physical configuration of the system.

Assume We've already been told that the pipe remains an impenetrable body and that O remains fixed in space. No additional assumptions are needed.

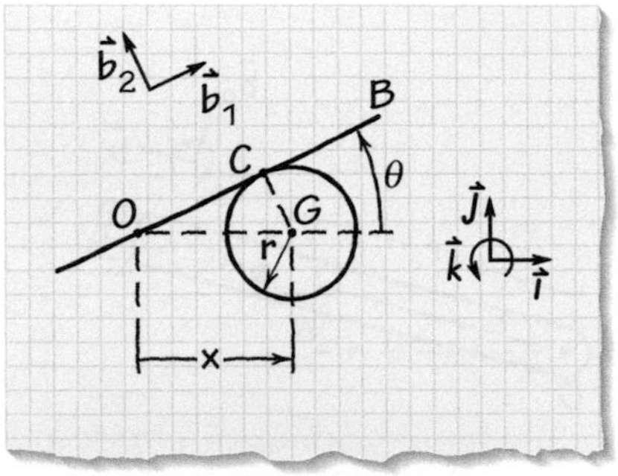

Figure 6.47 Scissors schematic

Draw **Figure 6.47** shows a model of our system. The upper jaw pivots about O and extends to B. The upper jaw is tangent to the pipe at the point of contact C, and from geometry we know that the triangle OCG is a right triangle.

Formulate Equations We'll first view the point C as a point that's moving with respect to a rotating body (the upper jaw). Then we'll consider C to be a point that moves with respect to the translating cylindrical pipe. Finally, we'll equate the velocities found from each of these two approaches and determine the relative speed of C along the upper jaw in terms of x, θ, r, and $\dot{\theta}$.

Viewing C as moving along the upper jaw gives us

$$\begin{aligned} \boldsymbol{v}_C &= \boldsymbol{v}_O + \boldsymbol{w}_{\overline{OB}} \times \boldsymbol{r}_{C/O} + \boldsymbol{v}_{C/OB} = 0 + \dot{\theta}\boldsymbol{b}_3 \times (x\cos\theta)\boldsymbol{b}_1 + v_{C/OB}\boldsymbol{b}_1 \\ &= (x\dot{\theta}\cos\theta)\boldsymbol{b}_2 + v_{C/OB}\boldsymbol{b}_1 \end{aligned} \quad (6.36)$$

Now we express C with respect to O as $\boldsymbol{r}_{C/O} = \boldsymbol{r}_{G/O} + \boldsymbol{r}_{C/G} = x\boldsymbol{i} + r\boldsymbol{b}_2$ and then differentiate with respect to time to obtain $\boldsymbol{v}_C$.

$$\boldsymbol{v}_C = \dot{x}\boldsymbol{i} + \dot{r}\boldsymbol{b}_2 + r\dot{\boldsymbol{b}}_2 = \dot{x}\boldsymbol{i} - r\dot{\theta}\boldsymbol{b}_1 \quad (6.37)$$

Solve

(6.36), (6.37) ⇒ $v_{C/\overline{OB}}\boldsymbol{b}_1 + x\dot{\theta}\cos\theta\boldsymbol{b}_2 = \boldsymbol{b}_1(\dot{x}\cos\theta - r\dot{\theta}) + \boldsymbol{b}_2(-\dot{x}\sin\theta)$

$\boldsymbol{b}_1$: $$v_{C/\overline{OB}} = \dot{x}\cos\theta - r\dot{\theta} \quad (6.38)$$

$\boldsymbol{b}_2$: $$x\dot{\theta}\cos\theta = -\dot{x}\sin\theta \quad \Rightarrow \quad \dot{x} = -x\dot{\theta}\cos\theta/\sin\theta \quad (6.39)$$

(6.49)→(6.38) ⇒ $$v_{C/\overline{OB}} = -\dot{\theta}\left[x\frac{(\cos\theta)^2}{\sin\theta} + r\right]$$

Check The units check out correctly. Physically, if the jaws are closing, $\dot{\theta} < 0$ and the pipe moves rightward. The quantity in the square brackets is positive and is multiplied by $-\dot{\theta}$. Hence the overall expression is positive, matching our physical intuition.

EXAMPLE 6.18 VELOCITY AND ACCELERATION IN A TUBE (Theory on page 362 and 364)

Having mentioned the space station in *2001: A Space Odyssey*, let's use it again to illustrate motion on a moving body. **Figure 6.48** shows our system. A personnel pod m rides in one of the four transportation tubes that lie within the arms that join the central docking area with the outer ring. At the instant illustrated, the pod is traveling outward at 30 m/s and decelerating at 1.1 m/s^2. It's also 80 m from the center of the space station. The station itself is rotating with a constant angular speed of 0.3 rad/s. Calculate the velocity and acceleration of the pod.

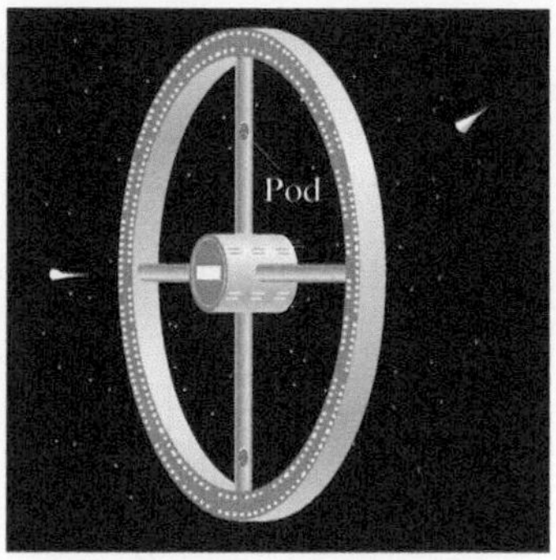

Figure 6.48 Space station

Goal Calculate the velocity and acceleration of m.

Given The space station is rotating about its center O at 0.3 rad/s. The pod is 80 m from O, traveling outward at 30 m/s and decelerating at 1.1 m/s^2.

Draw **Figure 6.49** shows that m is constrained to move in the $\boldsymbol{b}_1$ direction, and the $\boldsymbol{b}_1, \boldsymbol{b}_2$ unit vectors rotate with the body.

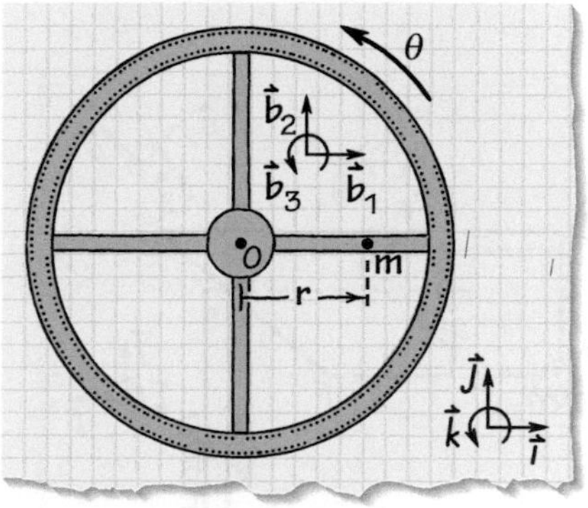

Figure 6.49 Space station schematic

Formulate Equations Our velocity expression is given by (6.29):

$$\boldsymbol{v}_m = \boldsymbol{v}_O + \boldsymbol{\omega} \times r\boldsymbol{b}_1 + \boldsymbol{v}_{rel}$$

and our acceleration by (6.34):

$$\boldsymbol{a}_m = \boldsymbol{a}_O + \boldsymbol{\alpha} \times r\boldsymbol{b}_1 + \boldsymbol{\omega} \times (\boldsymbol{\omega} \times r\boldsymbol{b}_1) + \boldsymbol{a}_{rel} + 2\boldsymbol{\omega} \times \boldsymbol{v}_{rel}$$

Assume The motion of the pod m is constrained to run along the transportation tube. Because r indicates the position of the pod relative to the station's center O, we have $v_{rel} = \dot{r}$ and $a_{rel} = \ddot{r}$. $\boldsymbol{v}_O = \boldsymbol{a}_O = 0$.

Solve Using the given kinematic constraints in our equations gives us

$$\boldsymbol{v}_m = \dot{\theta}\boldsymbol{b}_3 \times r\boldsymbol{b}_1 + \dot{r}\boldsymbol{b}_1 = r\dot{\theta}\boldsymbol{b}_2 + \dot{r}\boldsymbol{b}_1$$

$$\begin{aligned} \boldsymbol{a}_m &= \dot{\theta}\boldsymbol{b}_3 \times (\dot{\theta}\boldsymbol{b}_3 \times r\boldsymbol{b}_1) + \ddot{r}\boldsymbol{b}_1 + 2\dot{\theta}\boldsymbol{b}_3 \times \dot{r}\boldsymbol{b}_1 \\ &= (\ddot{r} - r\dot{\theta}^2)\boldsymbol{b}_1 + 2\dot{r}\dot{\theta}\boldsymbol{b}_2 \end{aligned}$$

Using the given parameter values yields

$$\boldsymbol{v}_m = (30\boldsymbol{b}_1 + 24\boldsymbol{b}_2)\text{ m/s} \quad \Rightarrow \quad \boldsymbol{a}_m = (-8.3\boldsymbol{b}_1 + 18\boldsymbol{b}_2)\text{ m/s}^2$$

Check Note the high value of lateral acceleration (acceleration at right angles to the path) induced by the pod's radial motion. 18 m/s^2 is almost 2 g's—quite a substantial acceleration. Clearly, the designers of the personnel transport system aren't as free to use high velocities as we are on earth. The time needed to get from the central docking area to the outer ring is limited not just by the length of the transport tube but also by the maximal acceleration the people can tolerate. A smaller $\dot{r}$ means lower accelerations but also a longer trip.

EXAMPLE 6.19 ANGULAR ACCELERATION OF A CONSTRAINED BODY (Theory on page 362 and 364)

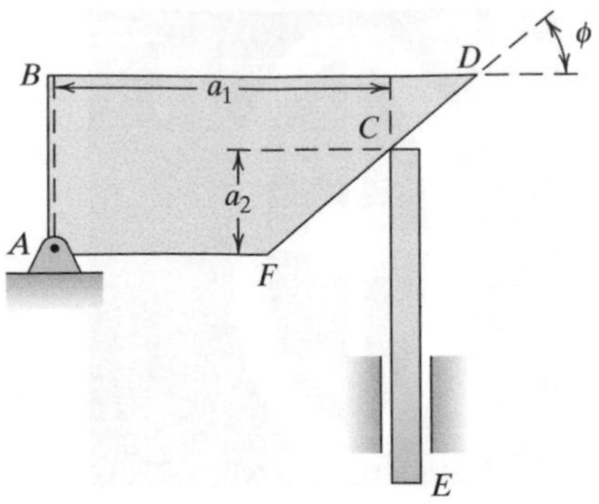

Figure 6.50 Pivoting link actuated by moving rod

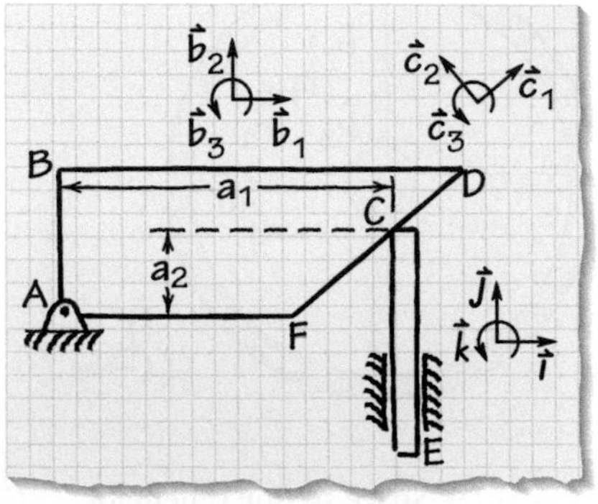

Figure 6.51 System labeling

A body that's hinged at A is positioned by the vertical guide rod $\overline{CE}$ (**Figure 6.50**). At the instant shown, the guide rod $\overline{CE}$ is moving upward at 0.9 m/s and accelerating at 1.2 m/s^2. $a_1 = 0.4$ m, $a_2 = 0.15$ m, and $\phi = 40°$. What is the angular acceleration $\boldsymbol{\alpha}$ of the body at this instant?

Goal Find the rigid body's angular acceleration at the instant shown.

Draw **Figure 6.51** shows our system along with three sets of unit vectors.

Formulate Equations The coordinate transformation

	$\boldsymbol{b}_1$	$\boldsymbol{b}_2$
$\boldsymbol{c}_1$	$\cos\phi$	$\sin\phi$
$\boldsymbol{c}_2$	$-\sin\phi$	$\cos\phi$

lets us relate $\boldsymbol{b}_1, \boldsymbol{b}_2$ to $\boldsymbol{c}_1, \boldsymbol{c}_2$. $\boldsymbol{b}_1, \boldsymbol{b}_2$ are fixed to the rotating body and are initially aligned with the ground-fixed $\underline{\boldsymbol{i}}, \underline{\boldsymbol{j}}$; $\boldsymbol{c}_1, \boldsymbol{c}_2$ are also body-fixed but are oriented such that $\boldsymbol{c}_1$ is parallel to $\overline{FD}$.

Let $\dot{\theta}\boldsymbol{b}_3, \ddot{\theta}\boldsymbol{b}_3$ be the angular velocity and angular acceleration, respectively, of the pivoting body. v_{rel}, a_{rel} are the relative speed and acceleration at which the vertically moving rod slides along the angled face $\overline{FD}$.

$$(6.29) \Rightarrow \quad \boldsymbol{v}_C = \boldsymbol{v}_A + \dot{\theta}\boldsymbol{b}_3 \times (a_1\boldsymbol{b}_1 + a_2\boldsymbol{b}_2) + v_{rel}\boldsymbol{c}_1$$

$$= 0 + \dot{\theta}\boldsymbol{b}_3 \times (a_1\boldsymbol{b}_1 + a_2\boldsymbol{b}_2) + v_{rel}(\cos\phi\boldsymbol{b}_1 + \sin\phi\boldsymbol{b}_2)$$

$$= (-a_2\dot{\theta} + v_{rel}\cos\phi)\boldsymbol{b}_1 + (a_1\dot{\theta} + v_{rel}\sin\phi)\boldsymbol{b}_2 = 0.9\boldsymbol{b}_2 \text{ m/s}$$

$\boldsymbol{b}_1$:
$$v_{rel} = \frac{a_2\dot{\theta}}{\cos\phi} \tag{6.40}$$

$\boldsymbol{b}_2$:
$$a_1\dot{\theta} + v_{rel}\sin\phi = 0.9 \text{ m/s} \tag{6.41}$$

$$(6.34) \Rightarrow \quad \boldsymbol{a}_C = \boldsymbol{a}_A + \dot{\theta}\boldsymbol{b}_3 \times [\dot{\theta}\boldsymbol{b}_3 \times (a_1\boldsymbol{b}_1 + a_2\boldsymbol{b}_2)] + \ddot{\theta}\boldsymbol{b}_3 \times (a_1\boldsymbol{b}_1 + a_2\boldsymbol{b}_2)$$

$$+ 2\dot{\theta}\boldsymbol{b}_3 \times v_{rel}(\cos\phi\boldsymbol{b}_1 + \sin\phi\boldsymbol{b}_2) + a_{rel}(\cos\phi\boldsymbol{b}_1 + \sin\phi\boldsymbol{b}_2)$$

$$= \boldsymbol{b}_1(-a_1\dot{\theta}^2 - a_2\ddot{\theta} - 2\sin\phi\dot{\theta}v_{rel} + a_{rel}\cos\phi)$$

$$+ \boldsymbol{b}_2(-a_2\dot{\theta}^2 + a_1\ddot{\theta} + 2\cos\phi\dot{\theta}v_{rel} + a_{rel}\sin\phi) = 1.2\boldsymbol{b}_2 \text{ m/s}^2$$

$\boldsymbol{b}_1$:
$$-a_1\dot{\theta}^2 - a_2\ddot{\theta} - 2\sin\phi\dot{\theta}v_{rel} + a_{rel}\cos\phi = 0 \tag{6.42}$$

$\boldsymbol{b}_2$:
$$-a_2\dot{\theta}^2 + a_1\ddot{\theta} + 2\cos\phi\dot{\theta}v_{rel} + a_{rel}\sin\phi = 1.2 \text{ m/s}^2 \tag{6.43}$$

Solve

(6.40)→(6.41) ⇒ $$(a_1 + a_2 \tan\phi)\dot{\theta} = 0.9\,\text{m/s}$$

$$\dot{\theta} = \frac{0.9\,\text{m/s}}{0.4\,\text{m} + (0.15\,\text{m})\tan 40^\circ}$$

$$\dot{\theta} = 1.71\,\text{rad/s} \qquad (6.44)$$

(6.44)→(6.40) ⇒ $$v_{rel} = 0.335\,\text{m/s}$$

(6.42), (6.43) ⇒ $$\begin{bmatrix} -a_2 & \cos\phi \\ a_1 & \sin\phi \end{bmatrix} \begin{bmatrix} \ddot{\theta} \\ a_{rel} \end{bmatrix} = \begin{bmatrix} a_1\dot{\theta}^2 + 2\sin\phi\dot{\theta}v_{rel} \\ a_2\dot{\theta}^2 - 2\cos\phi\dot{\theta}v_{rel} + 1.2\,\text{m/s}^2 \end{bmatrix}$$

$$\begin{bmatrix} -0.15\,\text{m} & \cos 40^\circ \\ 0.4\,\text{m} & \sin 40^\circ \end{bmatrix} \begin{bmatrix} \ddot{\theta} \\ a_{rel} \end{bmatrix} = \begin{bmatrix} 1.91\,\text{m/s}^2 \\ 0.761\,\text{m/s}^2 \end{bmatrix}$$

Solving these two equations in the two unknowns $\ddot{\theta}, a_{rel}$ yields

$$\ddot{\theta} = -1.60\,\text{rad/s}^2$$

$$a_{rel} = 2.18\,\text{m/s}^2$$

$$\boldsymbol{\alpha} = -1.60\boldsymbol{b}_3\,\text{rad/s}^2$$

Check Let's do a "Does it look at least reasonable?" check. The guide rod is moving up, and therefore we would expect the body to rotate in a counterclockwise manner and the relative velocity to be in the positive $\boldsymbol{c}_1$ direction. Both of these expectations are borne out by the results. The acceleration results aren't as straightforward to see on physical grounds because we'll have angular acceleration components due to the rod's velocity *and* its acceleration. So we'll take comfort in the fact that the velocity terms make immediate sense and hope that the math worked out for the acceleration as well.

EXAMPLE 6.20 ANGULAR ACCELERATION (Theory on page 362 and 364)

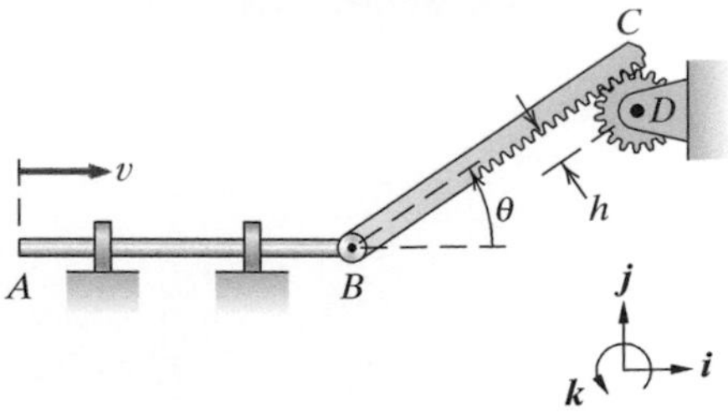

Figure 6.52 Gear rack coupled to circular gear

The link $\overline{AB}$ in **Figure 6.52** moves horizontally and at B connects via a pin joint to the gear rack $\overline{BC}$. The teeth of the rack engage with a gear G pivoted at D with radius h. At the illustrated moment $\boldsymbol{r}_{C/B} = (3\boldsymbol{i} + 2\boldsymbol{j})$ m, $\boldsymbol{v}_A = 2\boldsymbol{i}$ m/s, and $\boldsymbol{a}_A = 3\boldsymbol{i}$ m/s^2. $h \ll \overline{BD}$. What are $\boldsymbol{\omega}_G$ and $\boldsymbol{\alpha}_G$?

Goal Find $\boldsymbol{\omega}_D$ and $\boldsymbol{\alpha}_D$ of the gear.

Given $\boldsymbol{r}_{C/B} = (3\boldsymbol{i} + 2\boldsymbol{j})$ m, $\boldsymbol{v}_A = 2\boldsymbol{i}$ m/s, $\boldsymbol{a}_A = 3\boldsymbol{i}$ m/s.

Assume Assume h is negligible compared to $\overline{BD}$.

Draw **Figure 6.53** shows a schematic of our system.

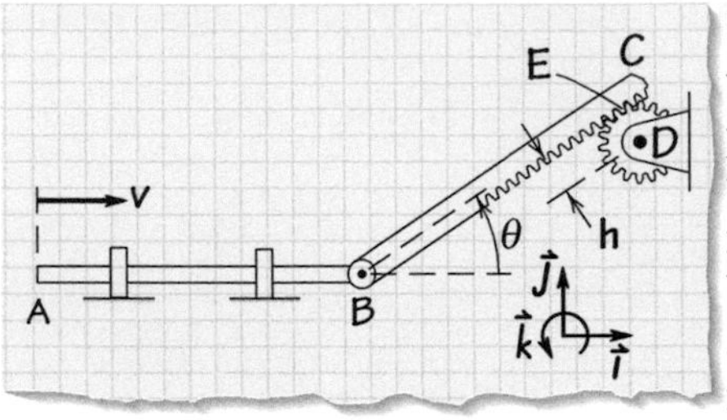

Figure 6.53 System schematic

Formulate Equations We'll ignore the dimensions of the teeth and view the gear and rack as a smooth link sliding along a fixed point. This point is shown as E in **Figure 6.52**. This means that we're neglecting the dimension h in comparison with the rest of the mechanism. Hence for our velocity/acceleration calculations we'll approximate E and D as being coincident. Once we've determined v_{rel} and $\boldsymbol{a}_{rel}$ (the speed and acceleration with which the arm moves past E), we can then consider h in order to calculate the gear's rotational velocity and acceleration.

$$\boldsymbol{v}_E = 0 = \boldsymbol{v}_B + \dot{\theta}\boldsymbol{k} \times \boldsymbol{r}_{E/B} + v_{rel}\boldsymbol{b}_1 \tag{6.45}$$

$$\boldsymbol{a}_E = 0 = \boldsymbol{a}_B + \left(\ddot{\theta}\boldsymbol{k} \times \boldsymbol{r}_{E/B}\right) + \dot{\theta}\boldsymbol{k} \times \left(\dot{\theta}\boldsymbol{k} \times \boldsymbol{r}_{E/B}\right) + a_{rel}\boldsymbol{b}_1 + 2\dot{\theta}\boldsymbol{k} \times v_{rel}\boldsymbol{b}_1 \tag{6.46}$$

where $\theta = \tan^{-1}(2/3) = 33.69°$, $\boldsymbol{v}_B = \boldsymbol{v}_A = (v_A \cos\theta \boldsymbol{b}_1 - v_A \sin\theta \boldsymbol{b}_2)$, $\boldsymbol{a}_B = \boldsymbol{a}_A = (a_A \cos\theta \boldsymbol{b}_1 - a_A \sin\theta \boldsymbol{b}_2)$, and $\boldsymbol{r}_{E/B} = [(3\text{ m})\cos\theta + (2\text{ m})\sin\theta]\boldsymbol{b}_I = 3.61\boldsymbol{b}_1$ m.

The angular velocity and angular acceleration of the gear are given by

$$-v_{rel}\boldsymbol{b}_1 = \omega_G\boldsymbol{k} \times h\boldsymbol{b}_2 = -\omega_G h\boldsymbol{b}_1 \quad \Rightarrow \quad \omega_G = \frac{v_{rel}}{h} \tag{6.47}$$

$$-a_{rel}\boldsymbol{b}_1 = \alpha_G\boldsymbol{k} \times h\boldsymbol{b}_2 = -\alpha_G h\boldsymbol{b}_1 \quad \Rightarrow \quad \alpha_G = \frac{a_{rel}}{h} \tag{6.48}$$

Solve

(6.45) $\Rightarrow$
$$0 = (v_A \cos\theta \boldsymbol{b}_1 - v_A \sin\theta \boldsymbol{b}_2) + (3.61\text{ m})\dot{\theta}\,\boldsymbol{b}_2 + v_{rel}\boldsymbol{b}_1 \tag{6.49}$$

$\boldsymbol{b}_1$:
$$0 = v_A \cos\theta + v_{rel} \quad \Rightarrow \quad v_{rel} = -v_A \cos\theta = -1.66\text{ m/s} \tag{6.50}$$

$\boldsymbol{b}_2$:
$$0 = -v_A \sin\theta + (3.61\text{ m})\dot{\theta} \quad \Rightarrow \quad \dot{\theta} = \frac{v_A \sin\theta}{3.61\text{ m}} = 0.31\text{ rad/s} \tag{6.51}$$

(6.50) $\rightarrow$ (6.47) $\Rightarrow$
$$\omega_G = \frac{-1.66\text{ m/s}}{h} \quad \Rightarrow \quad \boxed{\boldsymbol{\omega}_G = -\frac{1.66\text{ m/s}}{h}\boldsymbol{k}} \tag{6.52}$$

(6.46) $\Rightarrow$
$$0 = (a_A \cos\theta \boldsymbol{b}_1 - a_A \sin\theta \boldsymbol{b}_2) + \ddot{\theta}(3.61\text{ m/s})\boldsymbol{b}_2 - \dot{\theta}^2(3.61\text{ m/s})\boldsymbol{b}_1 + a_{rel}\boldsymbol{b}_1 + 2\dot{\theta}(-1.66\text{ m/s})\boldsymbol{b}_2$$

$\boldsymbol{b}_1$:
$$0 = a_A \cos\theta - \dot{\theta}^2(3.61\text{ m/s}) + a_{rel} \tag{6.53}$$

(6.52) → (6.53) ⇒
$$a_{rel} = \dot{\theta}^2(3.61\,\text{m/s}) - a_A \cos\theta = -2.15\,\text{m/s}^2 \quad (6.54)$$

(6.48), (6.54) ⇒
$$\alpha_G = \frac{-2.15\,\text{m/s}^2}{h}$$

$$\boldsymbol{\alpha}_G = \frac{-2.15\,\text{m/s}^2}{h}\boldsymbol{k}$$

EXERCISES 6.4

6.4.1. [Level 1] As illustrated, a peg P is constrained to slide within both a smooth horizontal chute and the slotted link $\overline{OQ}$, which is free to rotate about its pivot O with an angular speed and acceleration of ω and α, respectively. At a particular instant, the peg is a distance $r_{P_{/O}} = 0.9\,\text{m}$ from the pivot O, and the link is angled at $\theta = 65°$ with respect to the horizontal. If the peg's velocity in the chute is a constant $\boldsymbol{v}_P = 5\boldsymbol{i}$ m/s, what are $\boldsymbol{v}_{rel}$ and $\boldsymbol{a}_{rel}$? Express your answers in the $\{\boldsymbol{b}_1, \boldsymbol{b}_2, \boldsymbol{b}_3\}$ basis.

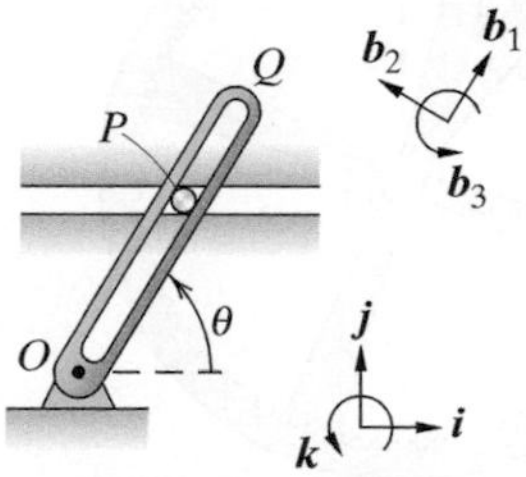

E6.4.1

6.4.2. [Level 1] The illustrated simplified tire is moving to the left at a constant velocity: $\boldsymbol{v}_G = -13\boldsymbol{i}$ m/s. The tire's rotation velocity is $7\boldsymbol{k}$ rad/s, and its radius is 0.3 m. What is the relative velocity of the ground with respect to the tire at point C, the tire/ground interface?

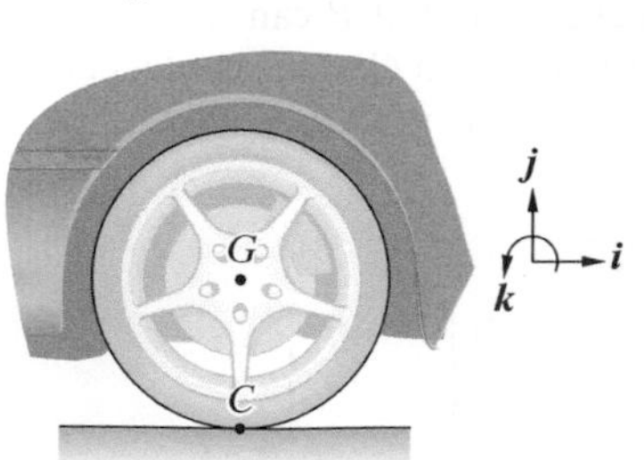

E6.4.2

6.4.3. [Level 1] A rigid bar $\overline{OA}$, hinged at O, is being drawn upright by a rope attached to the bar at A and being reeled into the wall at B. Determine $\frac{d}{dt}\overline{BA}$ in terms of the given parameters.

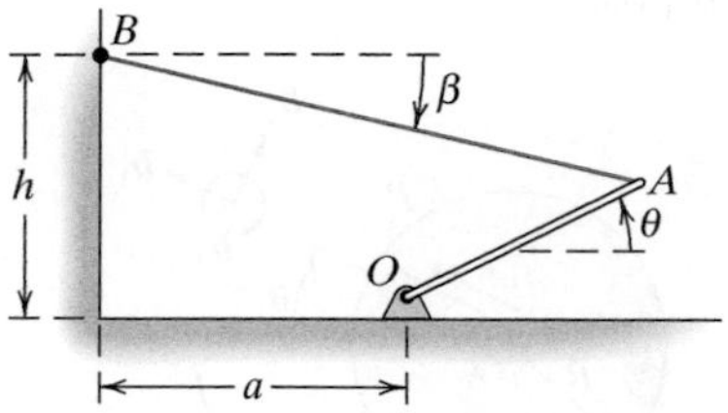

E6.4.3

6.4.4. [Level 1] The rotating arm $\overline{OA}$ is connected to the fixed guide by means of a pin P that rides within both the slot in the arm and the channel in the fixed guide. The channel in the fixed guide is circular, with radius b. Show that if $\dot{\theta}$ is a constant, the acceleration of the pin is constant in magnitude with respect to ground ($\boldsymbol{i}, \boldsymbol{j}$ frame).

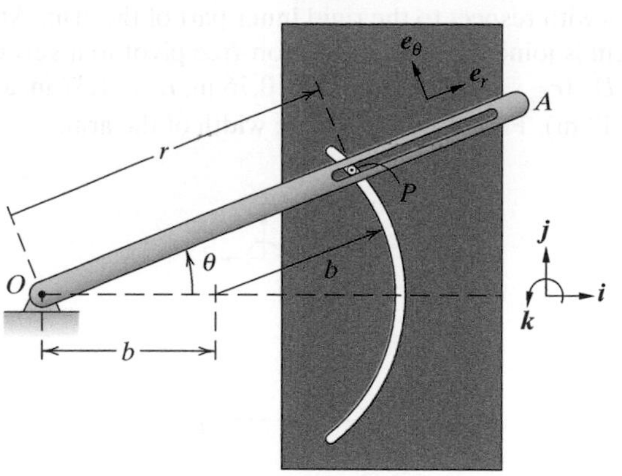

E6.4.4

6.4.5. [Level 1] Part of an experimental theme park ride is shown. The car (shown in a front view) consists of a top piece that contains the passengers and is mounted on two hydraulic supports, B and C. These supports ride on a rigid bed that can pivot at D. Determine the acceleration felt at A, if the bed is given a constant angular velocity of $3\boldsymbol{k}$ rad/s and the hydraulic supports B and C both extend with speed 0.9 m/s and acceleration 2.4 m/s^2.

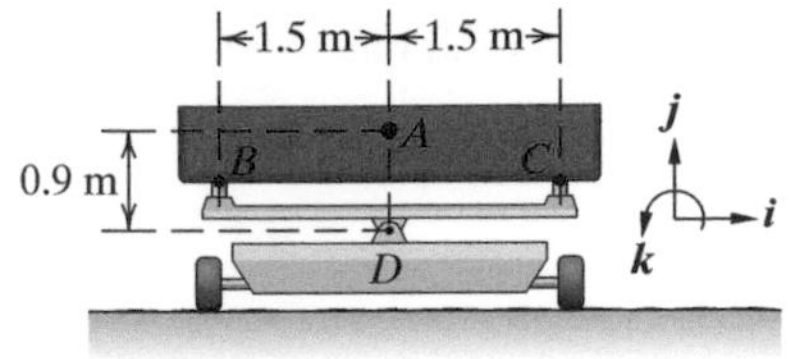

E6.4.5

6.4.6. [Level 2] A cylinder of radius $R = 0.8$ m is rolling without slip with an angular velocity and acceleration of $\boldsymbol{\omega} = 7\boldsymbol{k}$ rad/s and $\boldsymbol{\alpha} = 2\boldsymbol{k}$ rad/s^2, respectively. As illustrated, a peg P is constrained to slide within a radial slot in the cylinder. At this instant, the peg is a distance $r = 0.4$ m from the cylinder's center O, and the slot is angled at $\theta = 30°$ with respect to the horizontal. The peg is moving outward with a speed $v_{rel} = 1.2$ m/s and acceleration $a_{rel} = 0.6$ m/s^2 with respect to the cylinder. Find $\boldsymbol{v}_P$ and $\boldsymbol{a}_P$, and express them in the $\{\boldsymbol{b}_1, \boldsymbol{b}_2, \boldsymbol{b}_3\}$ basis.

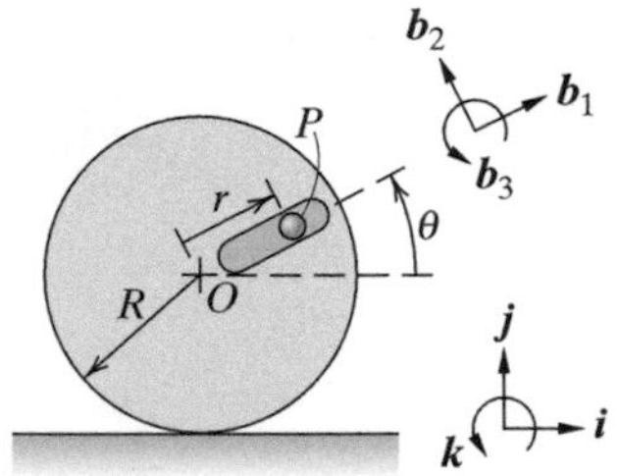

E6.4.6

6.4.7. [Level 2] The arm $\overline{AC}$ is similar to a chainsaw in that the outer surface comprises a moving belt of gear teeth. They mesh at B with a rigidly fixed gear (it doesn't rotate) with radius r. The moving belt is rotating in a counterclockwise fashion, and an individual gear tooth has a speed of 1.2 m/s with respect to the rigid inner part of the arm. At C the arm is joined through a friction-free pivot to a vertical bar $\overline{CD}$. ($\boldsymbol{r}_{B/C} = L_1\boldsymbol{i} - L_2\boldsymbol{j}$. $L_1 = 0.76$ m, $L_2 = 1.37$ m, and $r = 0.12$ m). Find $\boldsymbol{v}_C$. Neglect the width of the arm.

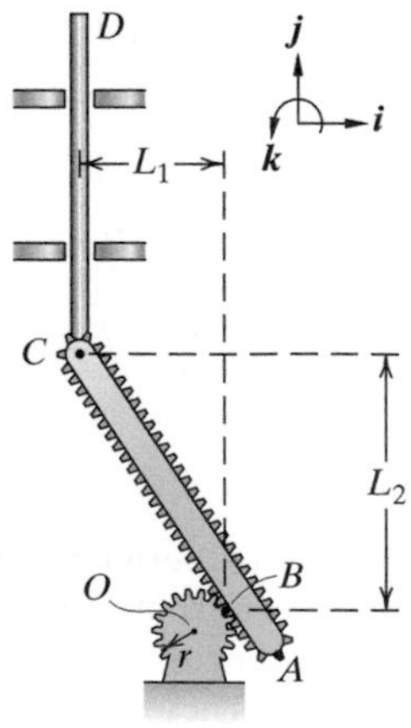

E6.4.7

6.4.8. [Level 2] The illustrated system is a ring gear in a framework with an arm attached to its center. At the end of the arm is a gear (G) that meshes with the inside surface of the ring gear. The outer surface of the ring gear is smooth, and the ring gear rolls without slip on the ground. **E6.4.8a** shows the system with arm and wheel aligned, and **E6.4.8b** shows the system at a displaced orientation. The absolute rotation of the ring is given by θ and the orientation of the arm $\overline{OA}$ with respect to the ring is given by β. $\boldsymbol{v}_O = -5\boldsymbol{i}$ m/s and $L \equiv |\overline{OA}| = 0.8$ m. The radius of the smaller gear is $r = 0.2$ m and the thickness of the ring gear's rim is $h = 0.2$ m. The gear G rotates at $\boldsymbol{\omega}'_G = 10\boldsymbol{k}$ rad/s with respect to the arm $\overline{OA}$. Determine the angular velocity of the arm $\overline{OA}$.

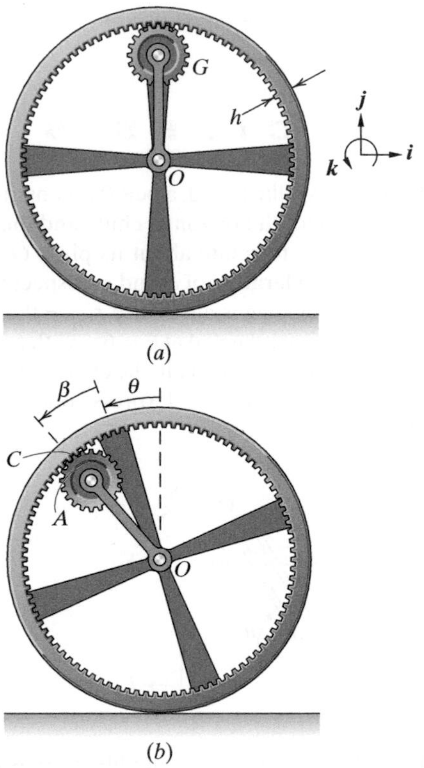

E6.4.8

6.4.9. [Level 2] The illustrated mechanism consists of a hinged rigid rod $\overline{OC}$ that can control the position of the rod $\overline{AB}$. The wheel centered at B can rotate without friction, and L is the distance from O to the contact point D. Relate the angular speed of the rod $\overline{OC}(\dot{\theta})$ to the speed of $A(\dot{x})$.

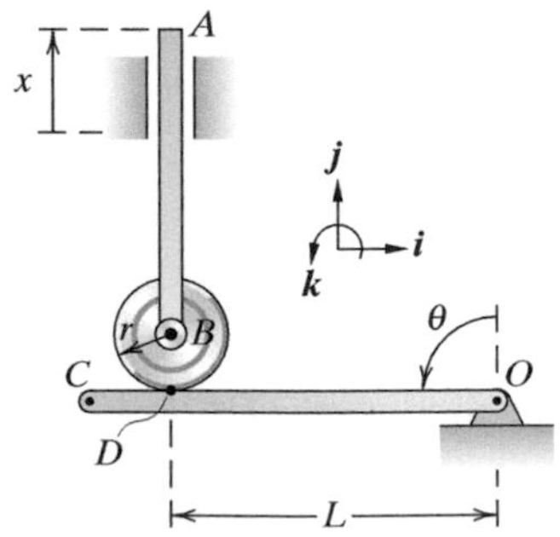

E6.4.9

6.4.10. **[Level 2]** A circular body is rolling without slip, and at the same time a mass m is moving along the illustrated guide, its position relative to the slot given by x. At the instant illustrated, $\boldsymbol{b}_1$ and $\boldsymbol{b}_2$ (body-fixed) line up with $\boldsymbol{i}$ and $\boldsymbol{j}$ (ground-fixed).

a. Find $\boldsymbol{v}_m$ as a function of $r, x, \dot{\theta}$, and $\dot{x}$.

b. Find $\ddot{\theta}$ as a function of $\dot{\theta}, x, r$, and $\dot{x}$ so that the lateral acceleration of the mass m ($\boldsymbol{i}$ direction) is zero for the illustrated configuration.

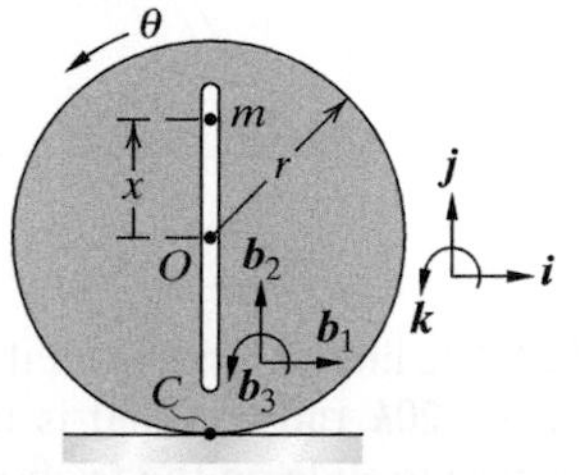

E6.4.10

6.4.11. **[Level 2]** A slot S is cut into the illustrated rotating disk, and a pin P is free to move within the slot. The pin also rides along a horizontal track T, separate from the disk. Thus, as the disk rotates about O, the pin will move with respect to the disk as well as with respect to the ground. Express the displacement of the pin in $\boldsymbol{b}_1$, $\boldsymbol{b}_2$ coordinates, transform to $\boldsymbol{i}, \boldsymbol{j}$ coordinates, and then write the constraint equation that ensures that P remains within the horizontal track. Finally, differentiate this constraint equation to show that

$$\dot{u} = -L_1(1 + \tan^2\theta)\dot{\theta}$$

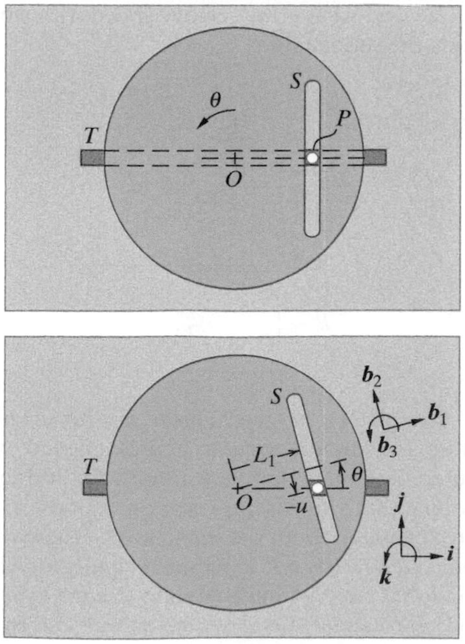

E6.4.11

6.4.12. **[Level 2]** A rigid bar $\overline{AB}$, hinged at A, is being drawn upright by a rope attached to the bar at B and being pulled at the other end by someone standing at O, directly below a pulley at C. $\overline{AB} = \overline{OC} = \overline{OA} = 2$ m. Rope is being reeled in at 0.5 m/s. Find $\dot{\theta}$ when $\theta = 30°$.

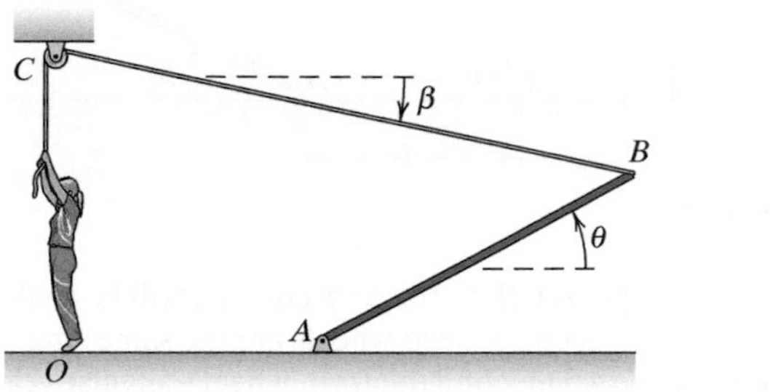

E6.4.12

6.4.13. **[Level 2]** Consider the same system as that of **Exercise 6.4.11.** Find the same result in a different manner. Find $\boldsymbol{v}_P$, taking into account the constraint that $\boldsymbol{v}_P$ has no vertical component. From $\boldsymbol{v}_P$ deduce that

$$\dot{u} = -L_1(1 + \tan^2\theta)\dot{\theta}$$

6.4.14. **[Level 2]** Consider the illustrated rigid, rectangular body. $\boldsymbol{b}_1, \boldsymbol{b}_2$ are body-fixed and $\boldsymbol{i}, \boldsymbol{j}$ are ground-fixed.

$$\boldsymbol{v}_A = (0.9\boldsymbol{b}_1 + 2.1\boldsymbol{b}_2) \text{ m/s}$$
$$\boldsymbol{v}_B = (1.5\boldsymbol{b}_1 + 1.5\boldsymbol{b}_2) \text{ m/s}$$

Determine the dimension h and the angular velocity of the rigid body with respect to the ground. Determine the velocity of point C relative to ground.

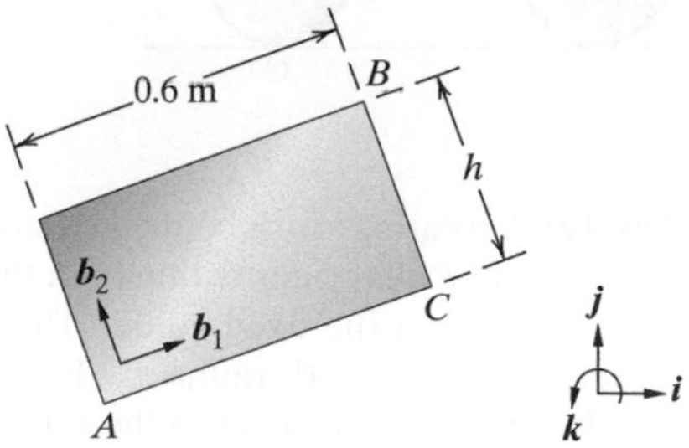

E6.4.14

6.4.15. **[Level 2]** A two-link mechanism is shown. The link $\overline{AB}$ is pivoted at A and connects at the hinge B to the second link. This link has an inner track that encloses a pin at C. Thus as β increases, the link $\overline{BD}$ is pushed so that it slides along the pin C. Determine $\dot{\gamma}$ as a function of $\beta, \gamma, \dot{\beta}$, $\overline{AB}$ and $\overline{CB}$.

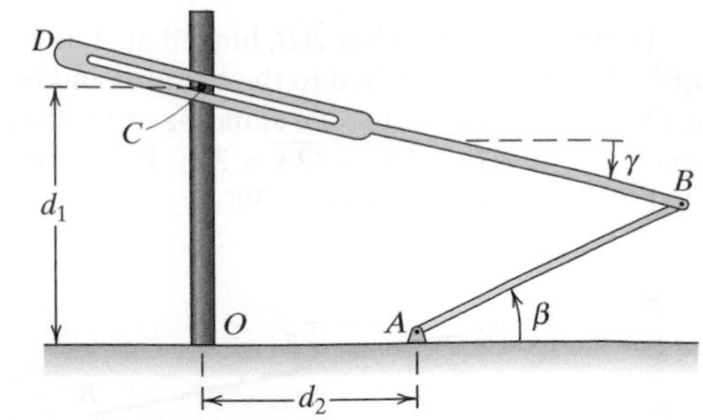

E6.4.15

6.4.16. **[Level 2]** In this exercise you will be analyzing a tricycle for which all three wheels are the same size: $r = 0.5$ m. For $v_G = 4.5$ m/s, find the front wheel's angular velocity if it's rolling without slipping, as well as the angular velocity of the tricycle body. Does it matter if the front wheel is being steered ($\dot{\phi} \neq 0$)? The rear wheels can rotate independently of each other.

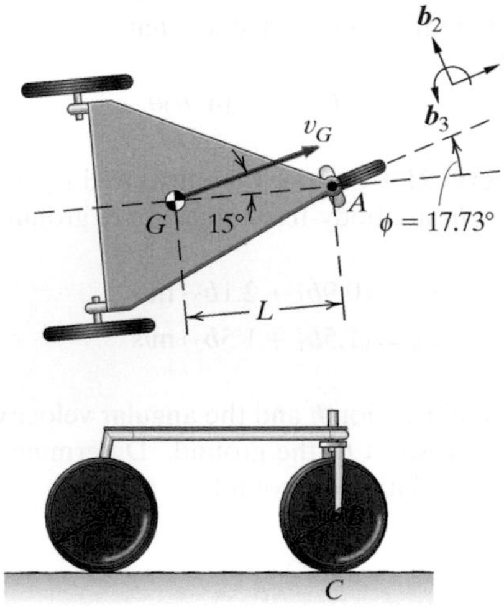

E6.4.16

6.4.17. **[Level 2]** A rotating arm is connected to the fixed guide by means of a pin P that rides within both the slot in the arm and the channel in the fixed guide. The channel in the fixed guide is circular, with radius b. Find a_{rel}, the acceleration of the pin P with respect to the rotating arm. Let $\ddot{\theta} = 0$.

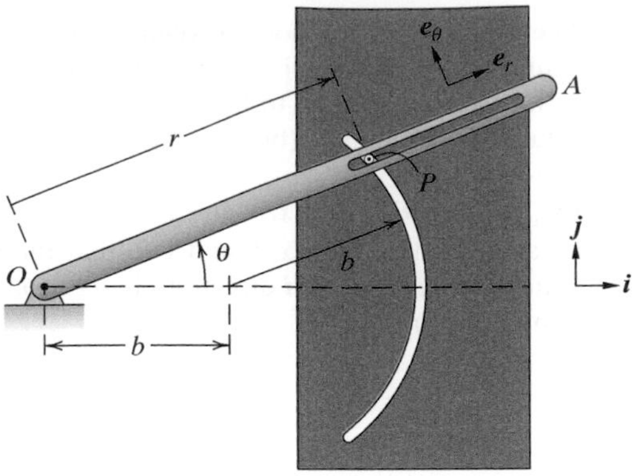

E6.4.17

6.4.18. **[Level 2]** The link $\overline{OA}$, horizontal at this instant, is rotating at $\omega_{\overline{OA}} = -20\boldsymbol{k}$ rad/s. What is the velocity of Body 1? Body 1 is free to slide on the frictionless horizontal supporting surface and $|\overline{OA}| = 0.4$ m.

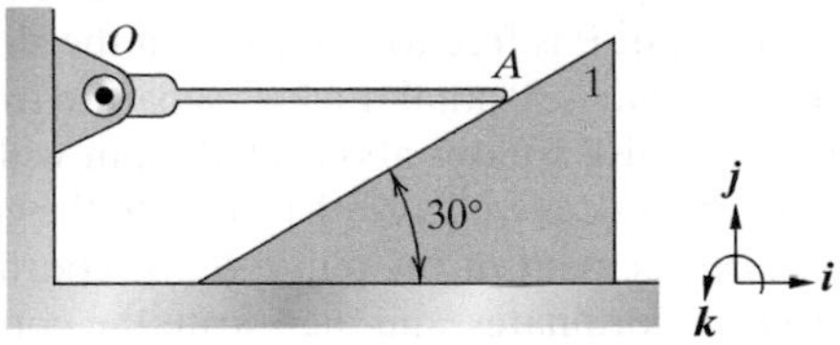

E6.4.18

6.4.19. **[Level 2]** The link $\overline{ED}$ oscillates about the horizontal position, inducing an oscillation in body ABC. A compression spring ensures that the two bodies are always in contact with each other. If the rotational speed of $\overline{ED}$ is ω_1 at the illustrated configuration, what is the rotational speed of ABC and what is the relative speed with which D moves along the surface $\overline{BC}$?

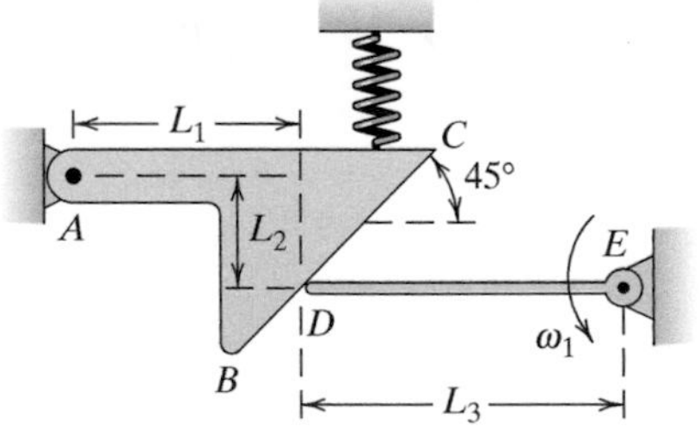

E6.4.19

6.4.20. **[Level 2]** A mechanism is shown in two configurations—both rectangular bodies aligned with a ground reference frame and both tilted (by different angles). Assume that you wish to track the motion of Body A as Body B rotates about the hinge at O. The two bodies are connected through P, a pin that's rigidly attached to Body B and that rides within a straight slot cut into Body A. As Body B rotates, Body A rotates as well, but by a different amount. Let counterclockwise rotations of Body A

and Body B be given by η and ϕ, respectively. The pin P slides in the slot, and s_{rel} tracks how far it moves from its $\phi = \eta = 0$ position. Use the $\boldsymbol{b}_1, \boldsymbol{b}_2$ reference frame to find $\boldsymbol{r}_{P_{/Q}}$. Next, use the $\boldsymbol{c}_1, \boldsymbol{c}_2$ reference frame to formulate $\boldsymbol{r}_{P_{/O}}$. Then use the $\boldsymbol{i}, \boldsymbol{j}$ frame to find $\boldsymbol{r}_{O_{/Q}}$. These last two vectors can be added to give $\boldsymbol{r}_{P_{/Q}}$. This gives you two different forms for $\boldsymbol{r}_{P_{/Q}}$. Set them equal and re-express all the unit vectors in terms of $\boldsymbol{i}, \boldsymbol{j}$ to find the two equations in two unknowns that would have to be solved in order to find s_{rel} and η in terms of an input ϕ. The point here is to illustrate how complex it can be to track the motions of interconnected bodies.

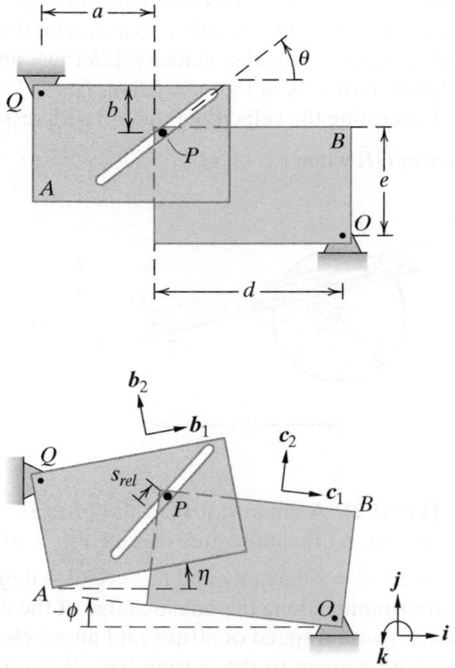

E6.4.20

6.4.21. [Level 2] Find the speed at which the collar at B moves along $\overline{AD}$ if $\theta = \frac{\pi}{4}$ rad, $\beta = \frac{\pi}{6}$ rad, and $\dot{\beta} = 0.6$ rad/s. At this instant, $|\boldsymbol{r}_{B_{/A}}| = 2$ m and $|\boldsymbol{r}_{C_{/B}}| = 2\sqrt{2}$ m.

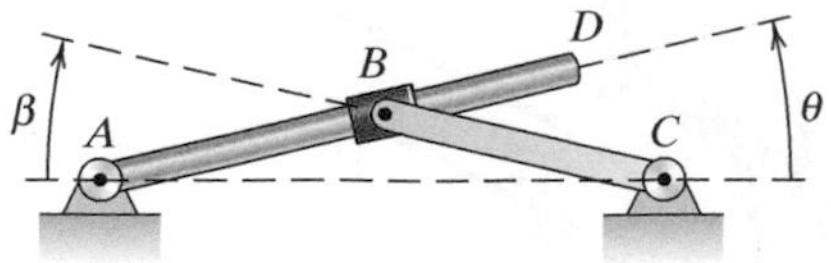

E6.4.21

6.4.22. [Level 2] The illustrated links $\overline{AB}$ and $\overline{CD}$ are connected through the pin E. The pin is rigidly attached to $\overline{AB}$ and slides within a slot cut into $\overline{CD}$. θ varies according to

$$\theta(t) = \frac{\pi}{6}\sin(5t)$$

Find $\beta(\theta)$ and $\dot{\beta}(\theta, \dot{\theta}, \beta)$.

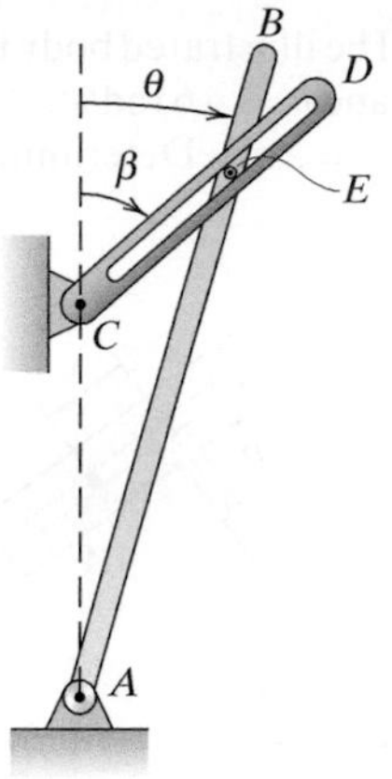

E6.4.22

6.4.23. [Level 2] A lawn sprinkler is shown with four identical, spinning arms. At the illustrated configuration, the shape of arm 1 is given by $y = ax^2$, where $a = 3.6\,\text{m}^{-1}$ and $0 \le x \le 0.17$ m. Assume that water is flowing through the arms at 15 m/s and that the speed of flow is increasing at 3 m/s^2. The whole assembly is rotating with angular velocity $-25\boldsymbol{k}$ rad/s and angular acceleration $-4\boldsymbol{k}$ rad/s^2. What is the total acceleration of a particle of water for the current configuration if the particle is in arm 1 at a position corresponding to $x = 0.15$ m?

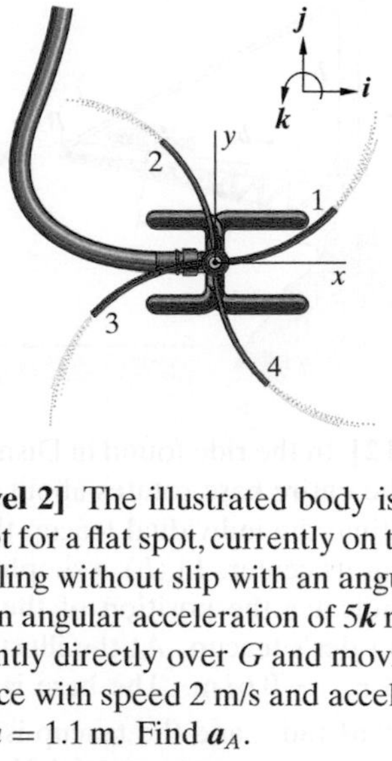

E6.4.23

6.4.24. [Level 2] The illustrated body is circular (with radius r), except for a flat spot, currently on top of the body. The body is rolling without slip with an angular velocity of $10\boldsymbol{k}$ rad/s and an angular acceleration of $5\boldsymbol{k}$ rad/s^2. The particle A is currently directly over G and moving to the right along the surface with speed 2 m/s and acceleration 3 m/s^2. $r = 1.5$ m and $a = 1.1$ m. Find $\boldsymbol{a}_A$.

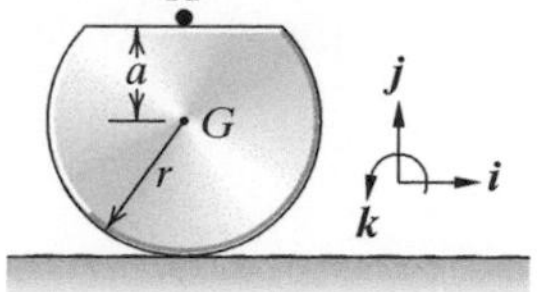

E6.4.24

6.4.25. **[Level 2]** The illustrated body is rotating about O such that $\dot{\theta} = 4\,\text{rad/s}$ and $\ddot{\theta} = -6\,\text{rad/s}^2$. $x = 1\,\text{m}, \dot{x} = 10\,\text{m/s}$, and $\ddot{x} = -5\,\text{m/s}^2$. $r_{P_{/O}} = 2\,\text{m}$. Determine the velocity and acceleration of A.

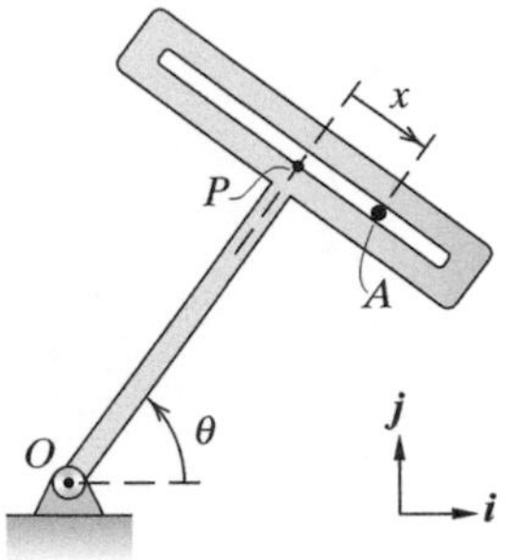

E6.4.25

6.4.26. **[Level 2] Computational** A secret agent in pursuit of international smugglers has driven his car onto a drawbridge that is in the process of opening. Her task is to jump the gap and get safely onto the other side of the river. Our task is to determine the acceleration of her car. The bridge is rotating at a constant rate of 0.1 rad/s. The car enters the bridge at $t = 0$ s with a speed of 64 km/h and accelerates at a constant rate of 7.6 m/s^2 up to the end of the 27 m long bridge. Plot the absolute acceleration of the car as a function of its position along the bridge section.

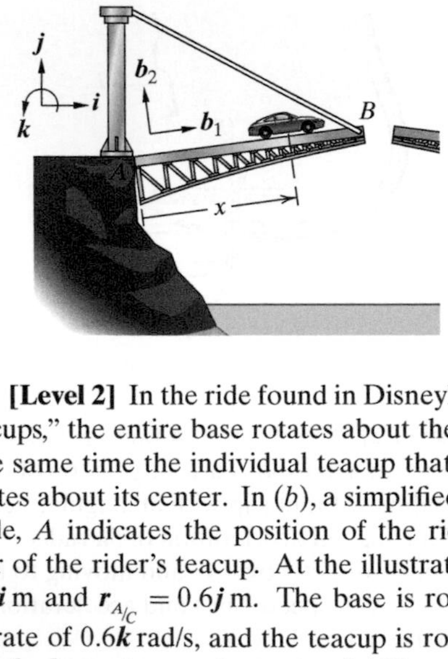

E6.4.26

6.4.27. **[Level 2]** In the ride found in Disneyland called "The Teacups," the entire base rotates about the center O, and at the same time the individual teacup that the riders sit in rotates about its center. In (b), a simplified overview of the ride, A indicates the position of the rider and C the center of the rider's teacup. At the illustrated instant, $\boldsymbol{r}_{C_{/O}} = 1.8\boldsymbol{i}\,\text{m}$ and $\boldsymbol{r}_{A_{/C}} = 0.6\boldsymbol{j}\,\text{m}$. The base is rotating at a constant rate of $0.6\boldsymbol{k}$ rad/s, and the teacup is rotating *with respect to the base* at a constant rate of $4.2\boldsymbol{k}$ rad/s. Determine $\boldsymbol{a}_A$.

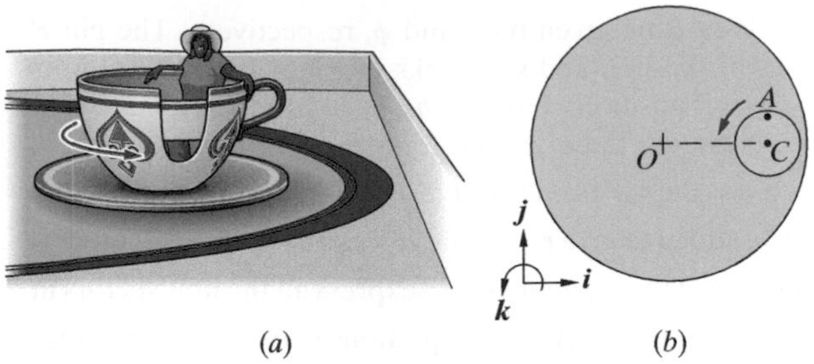

E6.4.27

6.4.28. **[Level 2]** The arm $\overline{OR}$ is freely hinged at O and constrained by the pin P that rides within the illustrated slot. $|\overline{PQ}| = r$. The pin P is rigidly attached to a circular disk that spins about Q with an angular velocity $12\boldsymbol{k}$ rad/s and angular acceleration of $-2\boldsymbol{k}$ rad/s^2. $r = 0.1\,\text{m}$, $L = 0.3\,\text{m}$, and $\boldsymbol{r}_{O_{/Q}} = L\boldsymbol{i}$. Determine the velocity of the pin with respect to the slot in arm $\overline{OR}$ when $\boldsymbol{r}_{P_{/Q}} = r\boldsymbol{j}$.

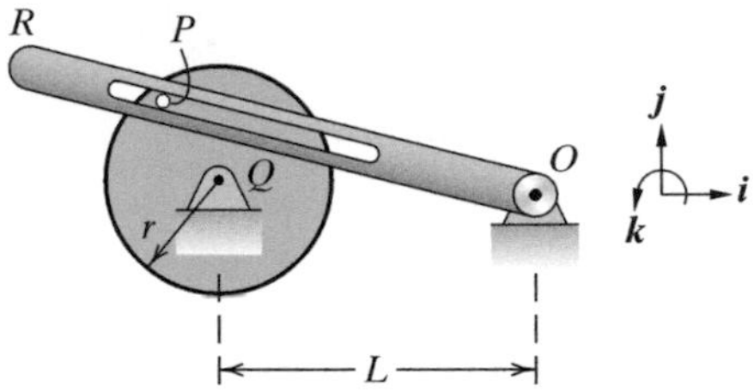

E6.4.28

6.4.29. **[Level 2]** A chain saw is in the process of cutting through a log. At the illustrated instant, $\boldsymbol{r}_{B_{/A}} = (0.36\boldsymbol{i} + 0.15\boldsymbol{j})\,\text{m}, \theta = 27°, \dot{\theta} = -0.3\,\text{rad/s}$, and $\boldsymbol{v}_A = (0.6\boldsymbol{i} + 0.6\boldsymbol{j})\,\text{m/s}$. The teeth are running along the outside edge of the cutting bar (clockwise) with a speed of 30 m/s and an acceleration of -3.6 m/s with respect to the cutting bar. What are the velocity and acceleration of tooth B? $\boldsymbol{a}_A = 0$.

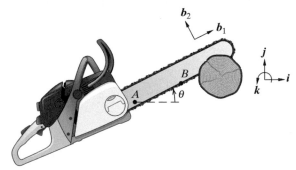

E6.4.29

6.4.30. **[Level 2]** While attempting to cut through a wooden rod, a tooth of the illustrated chainsaw hit a nail in the wooden rod (A), causing the tooth to come to an abrupt stop relative to the rod. Unfortunately, the tooth continued to move relative to the cutting bar, flinging the saw blade back because of reaction forces. At the instant under consideration, point O has velocity $\boldsymbol{v}_O = -3\boldsymbol{i}$ m/s, $\boldsymbol{r}_{B/O} = 0.24\boldsymbol{b}_1$ m, $\theta = 20°$, $\dot{\theta} = 18\boldsymbol{k}$ rad/s, and $\ddot{\theta} = 0$. Tooth B is moving with a speed relative to the cutting bar of 5.5 m/s (in the $-\boldsymbol{b}_1$ direction) and an acceleration (also relative to the cutting bar) of $-1.5\boldsymbol{b}_1$ m/s^2. What is the absolute acceleration of tooth B? $\boldsymbol{a}_O = 0$.

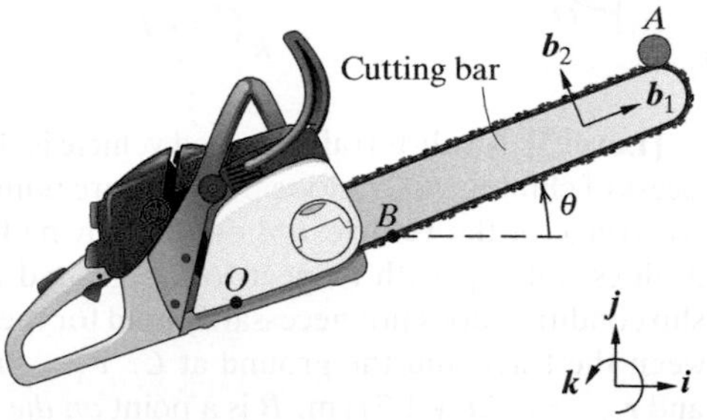

E6.4.30

6.4.31. **[Level 2]** One of Santa's elves (Fred) decided to test out some of the more exotic toys, such as the illustrated Snow-Speeder. As we join Fred, he has driven off the edge of a precipice and has started on his parabolic trajectory down. The Snow-Speeder is propelled by twin tracks that rotate clockwise, driving the vehicle over the snow. Just before going off the edge, Fred was moving at 5.5 m/s, and the tracks are still moving at the same constant speed as when they were in contact with the snow. The entire system (Fred plus Snow-Speeder) is rotating about G with an angular velocity of $-1.1\boldsymbol{k}$ rad/s and an angular acceleration of $0.5\boldsymbol{k}$ rad/s^2. $\boldsymbol{r}_{A/G} = -0.46\boldsymbol{b}_2$ m, and the Snow-Speeder is inclined at 10°, as shown. What is the absolute acceleration of point A, a part of the moving track?

E6.4.31

6.4.32. **[Level 3]** A peg P slides within a slotted link $\overline{OQ}$ that rotates about its fixed end O. The peg is also constrained to move within a chute C angled at $\beta = 45°$ with respect to the horizontal. At the illustrated instant, the slotted link $\overline{OQ}$ is at $\theta = 15°$ to the horizontal, and the peg is $r_{P/O} = 0.6$ m from the fixed point of rotation O. The peg is traveling up through the chute with a speed and acceleration of $v_P = 10$ m/s and $a_P = 3$ m/s^2, respectively, at this instant.

a. What is the velocity of the peg with respect to the slotted link $\overline{OQ}$, and how fast is the link rotating? Express the relative velocity in the $\{\boldsymbol{b}_1, \boldsymbol{b}_2, \vec{b}_3\}$ basis.

b. Find the peg's acceleration relative to link $\overline{OQ}$ (in terms of the $\{\boldsymbol{b}_1, \boldsymbol{b}_2, \vec{b}_3\}$ basis) and the link's angular acceleration.

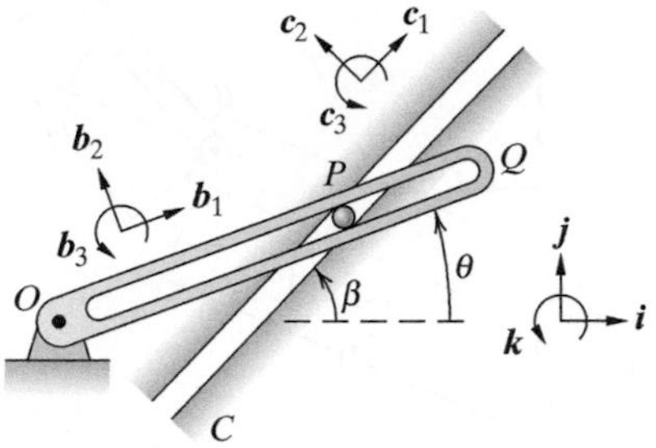

E6.4.32

6.4.33. **[Level 3]** The link $\overline{BC}$ is driven upward at a constant speed $\dot{x}$. What are $\dot{\theta}$ and $\ddot{\theta}$ as functions of L, θ, and $\dot{x}$?

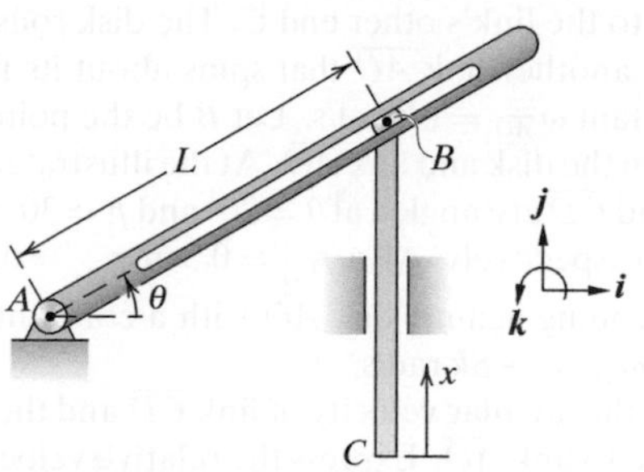

E6.4.33

6.4.34. **[Level 3]** The center A of a disk is constrained to slide within a slotted link $\overline{OP}$ with a constant speed $v_{rel,A} = 2.4$ m/s. The disk has a small radial slot in which a marble B is free to slide, and the link $\overline{OP}$ rotates about its fixed end O. At the given instant, link $\overline{OP}$ is rotating with an angular velocity and acceleration of $\boldsymbol{\omega}_{\overline{OP}} = -7\boldsymbol{k}$ rad/s and $\boldsymbol{\alpha}_{\overline{OP}} = 4\boldsymbol{k}$ rad/s^2, respectively, and the disk is spinning at a constant $\boldsymbol{\omega}_d = 20\boldsymbol{k}$ rad/s as its center A moves away from O. The marble is sliding within the disk's slot with a speed $v_{rel,B} = 1.2$ m/s and acceleration $a_{rel,B} = 0.6$ m/s^2 away from A, and the disk's slot makes an angle $\beta = 20°$ with respect to link $\overline{OP}$. If $r_{A/O} = 0.24$ m and $r_{B/A} = 0.09$ m, what are the absolute velocity and acceleration of the marble in terms of the $\{\boldsymbol{b}_1, \boldsymbol{b}_2, \boldsymbol{b}_3\}$ basis?

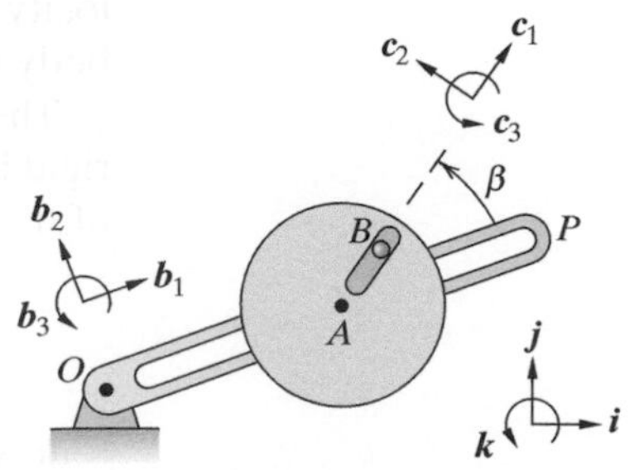

E6.4.34

6.4.35. [Level 3] A "mixer," used to completely mix chemical solutions, is depicted. $\ddot{\theta} = \ddot{\beta} = 0$. Consider the main arm AB to be a rigid body and C the point moving with respect to the body, and determine $\boldsymbol{v}_C$ and $\boldsymbol{a}_C$ in terms of $\theta, \dot{\theta}, \beta, \dot{\beta}, r_1, r_2, \dot{r}_2$, and $\ddot{r}_2$.

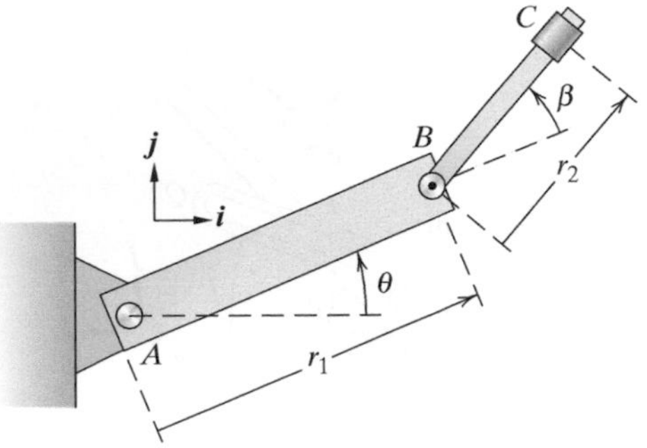

E6.4.35

6.4.36. [Level 3] As depicted, link $\overline{CD}$ rotates about its fixed end D, and the center of a disk with radius $R = 0.1$ m is attached to the link's other end C. The disk rolls without slip against another link $\overline{AO}$ that spins about its fixed end O at a constant $\boldsymbol{\omega}_{\overline{AO}} = 6\boldsymbol{k}$ rad/s. Let B be the point of contact between the disk and link $\overline{AO}$. At the illustrated instant, links $\overline{AO}$ and $\overline{CD}$ are angled at $\theta = 20°$ and $\beta = 30°$ from the horizontal, respectively. Also, $r_{B_{/O}} = 0.3$ m, $r_{C_{/D}} = 0.9$ m, and the disk is rolling against link $\overline{AO}$ with a constant angular velocity of $\boldsymbol{\omega}_{\overline{BC}} = -8\boldsymbol{k}$ rad/s.

a. Find the angular velocity of link $\overline{CD}$ and the velocity of B relative to link $\overline{AO}$. Express the relative velocity in the $\{\boldsymbol{b}_1, \boldsymbol{b}_2, \boldsymbol{b}_3\}$ basis.

b. What are the angular acceleration of link $\overline{CD}$ and the acceleration of B with respect to link $\overline{AO}$ (in terms of the $\{\boldsymbol{b}_1, \boldsymbol{b}_2, \boldsymbol{b}_3\}$ basis)?

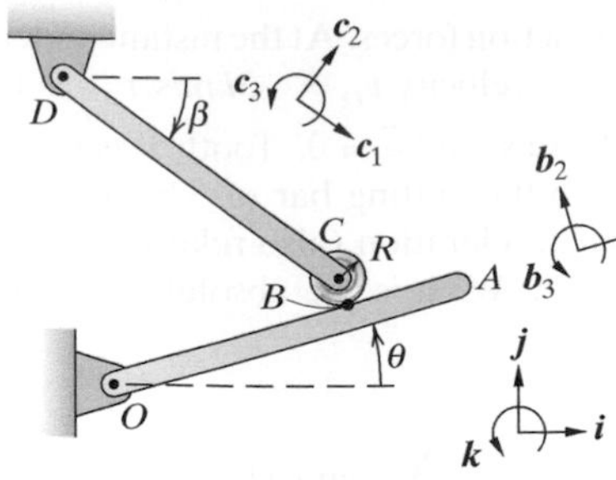

E6.4.36

6.4.37. [Level 3] An all-terrain, tracked vehicle is shown in the process of climbing a 45° slope. A and C are points on the undercarriage of the vehicle, *not* on the moving track. The track does not slip with respect to the ground at A. This no-slip condition does not necessarily hold for the contact between the track and the ground at C. $\boldsymbol{v}_A = 4\boldsymbol{i}$ m/s, $\boldsymbol{a}_A = 0$, and $\boldsymbol{r}_{C_{/A}} = (3.2\boldsymbol{i} + 1.2\boldsymbol{j})$ m. B is a point *on the track* (it moves relative to the vehicle body) and is located at $\boldsymbol{r}_{B_{/A}} = (1.6\boldsymbol{i} + 0.6\boldsymbol{j})$ m. Neglect the width of the track itself and find $\boldsymbol{a}_B$.

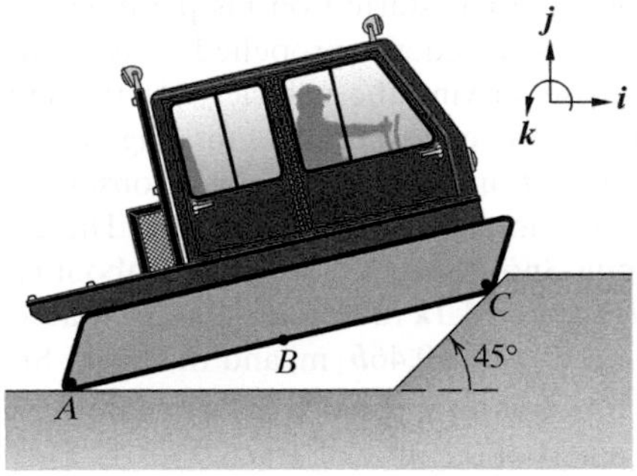

E6.4.37

6.5 JUST THE FACTS

This chapter dealt with the kinematics of extended bodies. We looked at both velocities and accelerations between different points of the same rigid body and derived the formulas that enabled us to compute the absolute velocity and acceleration of a point on a rigid body relative to an inertial reference frame. We looked at rotating reference frames (often useful for kinematic analyses) and then saw how to characterize the velocity and acceleration of points that are moving with respect to a rigid body rather than being fixed to it.

The first item on our agenda was to find the velocity of one point on a rigid body, given the velocity of a different point and the angular velocity of the body:

$$\boldsymbol{v}_{B_{/A}} = \boldsymbol{\omega} \times \boldsymbol{r}_{B_{/A}} \tag{6.3}$$

followed by an expression for the absolute velocity of point B on the rigid body:

$$\boldsymbol{v}_B = \boldsymbol{v}_A + \boldsymbol{\omega} \times \boldsymbol{r}_{B/A} \tag{6.4}$$

The highly useful formula with which to express velocities and accelerations with rotating reference frames then made a much appreciated appearance:

$$\left.\frac{d}{dt}\right|_N \boldsymbol{p} = \left.\frac{d}{dt}\right|_S \boldsymbol{p} + \boldsymbol{\omega} \times \boldsymbol{p} \tag{6.24}$$

and allowed us to get an expression for the acceleration of point B:

$$\boldsymbol{a}_B = \boldsymbol{a}_A + \boldsymbol{\alpha} \times \boldsymbol{r}_{B/A} + \boldsymbol{\omega} \times \left(\boldsymbol{\omega} \times \boldsymbol{r}_{B/A}\right) \tag{6.25}$$

Moving beyond fixed points, we found out how to express the velocity of a point that's moving with respect to a rigid body:

$$\boldsymbol{v}_B = \boldsymbol{v}_A + \boldsymbol{\omega} \times \boldsymbol{r}_{B/A} + \boldsymbol{v}_{rel} \tag{6.29}$$

as well as its acceleration:

$$\boldsymbol{a}_B = \boldsymbol{a}_A + \boldsymbol{\alpha} \times \boldsymbol{r}_{B/A} + \boldsymbol{\omega} \times \left(\boldsymbol{\omega} \times \boldsymbol{r}_{B/A}\right) + \boldsymbol{a}_{rel} + 2\boldsymbol{\omega} \times \boldsymbol{v}_{rel} \tag{6.34}$$

SYSTEM ANALYSIS (SA) EXERCISES

SA6.1 Evaluating Head Rotation Effects

The Vertical Drop Tower (VDT) at Wright-Patterson Air Force Base is used to examine accelerations that could occur during an ejection from an aircraft. The person (or test dummy) sits on a carriage that is dropped from a predetermined height. At the end of the fall, a plunger on the bottom of the carriage contacts a fluid-filled chamber. This deceleration mimics the positive acceleration encountered during the catapult phase of an ejection. The magnitude and duration of the deceleration pulse are controlled by the shape of the plunger, thus allowing control over the magnitude and duration of the deceleration profile.

There has been a recent trend to add more equipment to the pilot's helmet (**Figure SA6.1.1**), including night vision goggles, target acquisition systems (e.g., the Joint Helmet Mounted Cueing System), and other display symbology/information systems. This additional weight may shift the center of gravity of the head system forward, which can result in a forward pitching motion during an ejection pulse. In addition, improper positioning of the headrest can exacerbate the problem.

Recent tests have been performed at Wright-Patterson Air Force Base to determine the effects of moving the headrest forward and adding more weight to the front of the helmet. It's important to examine the range of velocity and acceleration that the mount experiences so that we can be confident that the structure has been designed robustly enough that it will not inadvertently release. When the initial acceleration impulse occurs, the head rotates forward with rotational acceleration peaks of up to 300 rad/s^2.

One set of tests was performed with a plunger that caused the sled's acceleration to vary as

$$\ddot{z} = -\left(10\,\text{s}^{-1}\right)\dot{z}$$

Model the angular acceleration of the head with the expression

$$\ddot{\theta} = (100\,\text{rad/s}^2)[1 - 400(t - 0.05\,\text{s})^2]$$

Assume that the head rotates about the occipital condyles, which are located at the top of the spine in the neck. There is a mount for night vision goggles located 10 cm forward and 20 cm above the condyle (which we will assume moves with the same rate as the chair). Find the linear acceleration of the mount as a function of time during the first 100 ms of the test. Initially $v = -3$ m/s.

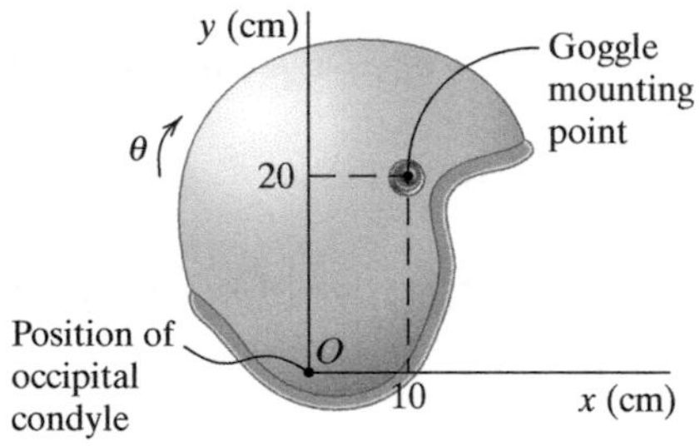

Figure SA6.1.1 Head/helmet model

SA6.2 Design of a Shooting Gallery Game

Carnival shooting galleries have a variety of targets that move back and forth or up and down. The device illustrated in **Figure SA6.2.1** is a mechanism that moves a wooden cutout of a rabbit back and forth as part of a carnival shooting gallery. $\dot{\beta} = 8$ rad/s, $\boldsymbol{r}_{B/A} = (2.4\boldsymbol{i} + 1.2\boldsymbol{j})$ m, $\boldsymbol{r}_{B/C} = (0.6\boldsymbol{i} + 0.6\boldsymbol{j})$ m, and $\boldsymbol{r}_{C/A} = (1.8\boldsymbol{i} + 0.6\boldsymbol{j})$ m at the illustrated instant. Determine $\boldsymbol{\omega}_{\overline{AB}}$.

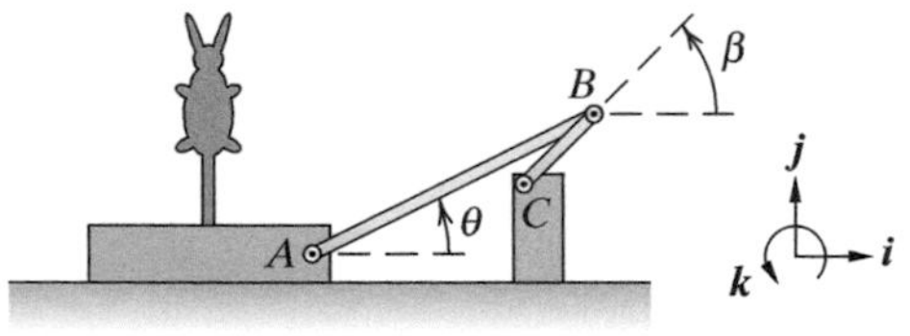

Figure SA6.2.1 Shooting gallery mechanism

SA6.3 Design of a "Snap-the-Whip" Ride

A ride commonly found in amusement parks is called Snap the Whip. Three primary arms extend from a central base and rotate about the base (**Figure SA6.3.1*a***). Attached to the end of each arm are three secondary arms, each terminating in a passenger pod. The secondary arms rotate about the end of the base arms. **Figure SA6.3.1*b*** shows a simplified view of a single primary and secondary arm. $\boldsymbol{r}_{C/O} = 4.5\boldsymbol{b}_1$ m and $\boldsymbol{r}_{A/C} = 1.5\boldsymbol{c}_1$ m. The primary arm is rotating at a constant rate of $0.8\boldsymbol{k}$ rad/s, and the secondary arm is rotating with respect to the primary arm at a constant rate of $2\boldsymbol{k}$ rad/s. $\theta = 45°$ and $\beta = 15°$.

a. Find $\boldsymbol{a}_A$ with respect to the $\boldsymbol{c}_1, \boldsymbol{c}_2$ unit vectors.
b. Investigate how altering the lengths of the arms while

keeping the same rotational speeds affects the acceleration felt at the rider's pod.

c. The derivative of acceleration with respect to time is called *jerk*, and it is a good way of quantifying how comfortable a ride feels to a passenger. To give you a sense of what jerk measures, think about alternately hitting the accelerator and brake in a car. The driver and passenger will be flung back and forward, and it won't take long for a passenger to complain about it. This quick change of acceleration causes a high level of jerk.

Calculate the jerk levels for this ride for the given nominal parameters and plot the jerk versus time for both the individual components along and orthogonal to the secondary arm (the ones aligned with the passenger) as well as the overall magnitude of the jerk. Comment on how these levels compare to the analogy suggested of a car being alternately braked and accelerated.

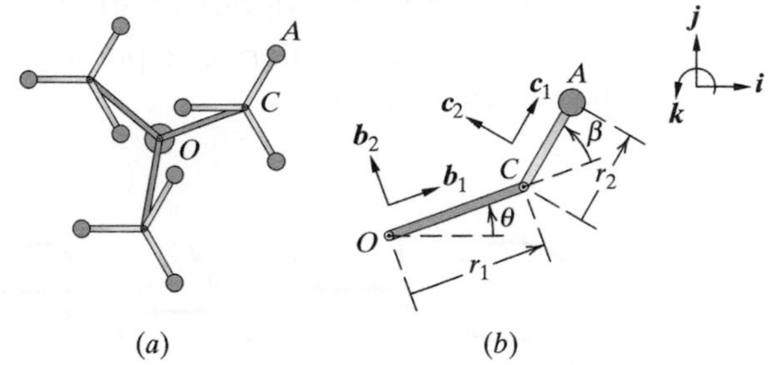

Figure SA6.3.1 Snap-the-Whip design schematic

SA6.4 Kinematics of Catapults

Ancient siege engines provided military commanders with the ability to engage an enemy from a distance; essentially they were the artillery of the ancients. Unfortunately, what is known of these medieval siege engines is limited to crude artist's renditions and manuscript references. Hence, the hypothetical analysis of siege engines is still intriguing and challenging. Let us analyze two simple siege engines: the catapult in **Figure SA6.4.1*a*** and the trebuchet in **Figure SA6.4.1*b***. Both siege engines fire missiles using the energy gained from a dropping weight and the advantage of a lever arm. The trebuchet adds a sling at the end of the throwing arm.

Assume the following for the catapult and the trebuchet at the point of missile release (45° from the horizontal):

$$L_{arm} = 4.5\,\text{m}$$
$$L_{sling} = 3\,\text{m}$$
$$\omega_{arm} = 2\,\text{rad/s}$$
$$\omega_{sling} = 3\,\text{rad/s}$$

a. Calculate the release velocity of the catapult and trebuchet missiles.

b. Compare your results from the catapult and the trebuchet. Do the answers make sense? Why?

c. Calculate the missile range for the catapult and trebuchet. Assume the support pin for the throwing arm is 3.6 m above the ground, and ignore air resistance on the missile.

d. Graph the relationship between angle of release and missile range for any missile firing and landing at the same elevation. Ignore air resistance on the missile.

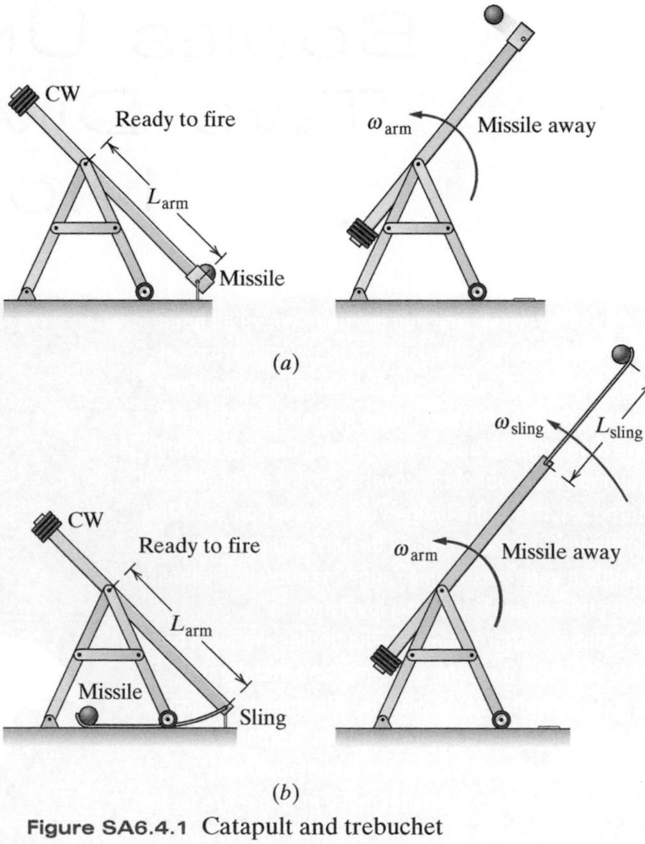

Figure SA6.4.1 Catapult and trebuchet

OBJECTIVES

Upon completion of this chapter you will be able to:

- Analyze the time-dependent behavior of a particle: where it is and how it is moving
- Describe a particle's motion in a variety of coordinate systems
- Switch from one description to another
- Deal with relative motion (one particle that's moving with respect to another) as well as constrained motion

CHAPTER 7

KINETICS OF RIGID BODIES UNDERGOING TWO-DIMENSIONAL MOTION

At last we've reached the point we've all been waiting for—the kinetics of two-dimensional rigid bodies. Of course, three-dimensional motion is even more fun (you can't have gyroscopic motion in two dimensions), but it's also a topic that's probably beyond what your class covers. So, for all intents and purposes, this is the main event of your course. After finishing this chapter, you'll be fully equipped to handle a very reasonable range of problems. What sort of problems? Anything involving two-dimensional motion of rigid bodies. You will be able to model a car, for instance, and include the engine, drivetrain, and wheels. (But planar motion only. If you want to model a thrown football's wobble, then you'll need a three-dimensional analysis.) Robotic arms that move in-plane are on the menu, as are automatic door openers, rudders on airplanes, and bicycle gearchains.

(cont.)

Benson Tongue

(cont.)

All we'll need to solve the two-dimensional problems of this chapter are three scalar equations (along with some constraints). The first two equations come from a force balance, $\sum \boldsymbol{F} = m\bar{\boldsymbol{a}}$. By breaking this vector relationship into components along two orthogonal directions ($\boldsymbol{e}_r$-$\boldsymbol{e}_\theta$, $\boldsymbol{i}$-$\boldsymbol{j}$, $\boldsymbol{e}_t$-$\boldsymbol{e}_n$, and so forth), we will get two scalar equations. The third equation is our moment balance, and it tells us how the rotational acceleration of the body is related to the applied forces and moments. Here are the forms of this equation that we'll be using:

Form 1: $$\sum \boldsymbol{M}_G = I_G\ddot{\theta}\boldsymbol{b}_3$$

Form 2: $$\sum \boldsymbol{M}_O = I_O\ddot{\theta}\boldsymbol{b}_3$$

Form 3: $$\sum \boldsymbol{M}_A = I_G\ddot{\theta}\boldsymbol{b}_3 + \boldsymbol{r}_{G_{/A}} \times m\boldsymbol{a}_G$$

Form 4: $$\sum \boldsymbol{M}_A = I_A\ddot{\theta}\boldsymbol{b}_3 + \boldsymbol{r}_{G_{/A}} \times m\boldsymbol{a}_A$$

In Form 1 we're summing moments about the body's mass center G, whereas in Form 2 we're summing moments about a fixed point of rotation O. In Forms 3 and 4 we're summing moments about a general point A (not necessarily fixed and not necessarily the mass center). Any of these forms will work, and the one we choose will depend on which seems most convenient for the problem at hand. So let's reiterate: three equations and three unknowns (a_x, a_y, and $\ddot{\theta}$, for instance). Drill this into your consciousness! That's all you're going to need. If you're lucky, you can sometimes get away with using just one or two of the equations but at most, you will need only three.

Although we could jump right into fully general two-dimensional motion, it probably makes more sense to ease our way into it. So, with this in mind, we'll start off by looking at systems that don't involve any rotation of the rigid body; that is, all we have to deal with is curvilinear translation. The simplest case, motion in a straight line, is shown in **Figure 7.1*a***. Every point in the rectangular body follows the same straight trajectory. One simplification this constraint implies is that there's not going to be any kinetic energy due to rotation—just kinetic energy arising from translation. When we try to apply our balance laws, this lack of rotation means that the only "active" laws are those involving translation of the mass center. The moment balance is a static equation in this case. **Figure 6.3** from the previous chapter demonstrated this motion for the first part of the jack's travel—the supported box essentially moved straight up. The next step up in complexity is for the body to follow a curved path but still not rotate; this is shown schematically in **Figure 7.1*b*** and is demonstrated by the box's complete path in **Figure 6.3**.

Moving beyond curvilinear motion, we will consider bodies that rotate about a fixed point in the plane, like that illustrated in **Figure 7.1*c***. Because the center of mass is moving, we will have a translational kinetic energy term, and because the body is rotating as well, we will also have a rotational kinetic energy term. The same goes for our balance laws: we will have fully operational balances of applied forces as well as moments. Often, however, we won't need to use a force balance; the rotational relationship will be sufficient. **Figure 6.2** from the previous chapter showed this motion as well.

The final case we'll examine will be the fully general case—motion that involves both translation and rotation. You will recall that at the start of Chapter 6 we saw how a car can spin out (**Figure 6.1**), something that's schematically represented in **Figure 7.1*d***. There won't be any additional terms in the kinetic energy or balance equations for this case; the only difference between this case and the previous one is that for pure rotation

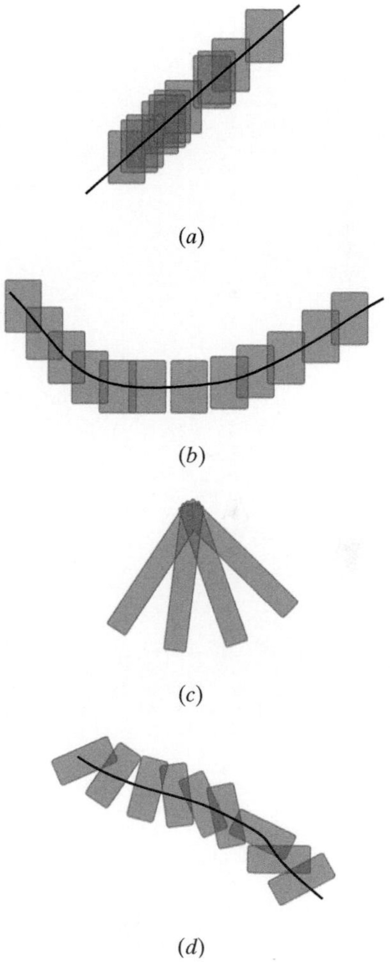

(*a*) (*b*) (*c*) (*d*)

Figure 7.1 Types of two-dimensional motion

there's a definite relationship between the body's rotation rate and the velocity of its mass center, whereas in the general case these quantities are pretty much independent of each other.

In all cases we'll follow the procedure you're already accustomed to—draw a free-body diagram and inertial response diagram, determine any constraints, apply the balance laws (forces and moments), and then solve the equations.

7.1 CURVILINEAR TRANSLATION

Don't you just love the title? Curvilinear translation. Sounds like something that happens to the *Enterprise* when it encounters a deep-space anomaly. Sadly, it's not really that exotic. Curvilinear translation simply means that a rigid body can move along a curving path (the curvilinear part) but doesn't rotate (hence it's just translating), like the motion shown in **Figure 7.1*b***. Because the body never rotates, the angular velocity and angular acceleration are always equal to zero, a fact that simplifies our life when we're trying to compute the system's dynamical response. For the case of planar motion this gives us two force balances:

$$\sum F_x = m\ddot{x} \tag{7.1}$$

$$\sum F_y = m\ddot{y} \tag{7.2}$$

where the accelerations $\ddot{x}$ and $\ddot{y}$ are associated with the center of mass of the body we're considering.

As already mentioned, the body experiences no rotation. This means that our moment balance equation is the one you're already familiar with from statics—the sum of the moments about the body's mass center is zero:

$$\sum M_G = 0 \tag{7.3}$$

>>> **Check out Example 7.1 (page 385), Example 7.2 (page 386), Example 7.3 (page 388), and Example 7.4 (page 390) for applications of this material.**

In the next sections we will be deriving the moment balance equations quite carefully, and you can easily verify that when the rotation is zero, we simply end up with (7.3). Thus we won't derive this relationship now but will take it as a given and see what sort of results we can get with it.

EXAMPLE 7.1 DETERMINING THE ACCELERATION OF A TRANSLATING BODY (Theory on page 384)

Consider the Batsled, as shown in **Figure 7.2**. At $t = 0$ its rocket engine is fired, producing a thrust $T = 4.3 \times 10^4$ N. The mass of the Batsled is $m = 1600$ kg, and the coefficient of friction between the snow and the sled is $\mu = 0.2$. Find the sled's acceleration.

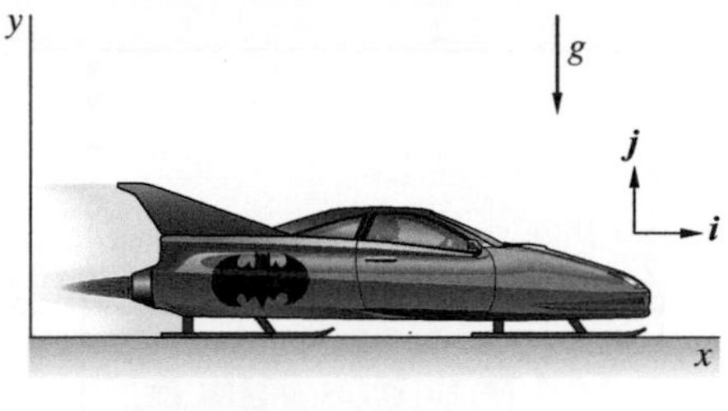

Figure 7.2 The Batsled

Goal Determine the Batsled's acceleration.

Draw **Figure 7.3** shows both the free-body diagram and the inertial response diagrams for our system, with $\boldsymbol{i}$ pointing in the Batsled's direction of travel.

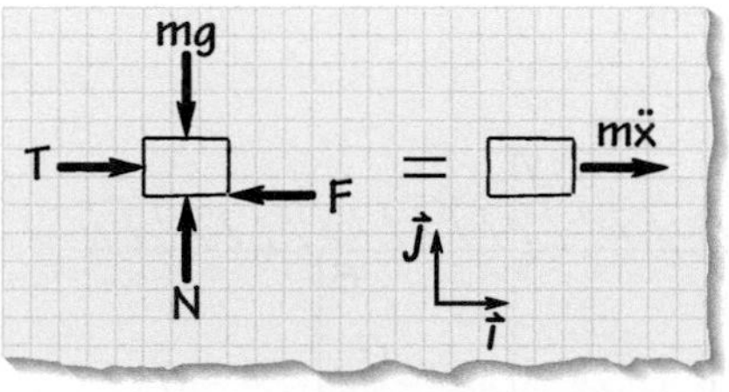

Figure 7.3 FBD=IRD for the Batsled

Formulate Equations Because the Batsled can move only horizontally ($y = \dot{y} = \ddot{y} = 0$), the only forces we need consider in order to determine the normal force N are those in the $\boldsymbol{j}$ direction. Applying (7.1) and (7.2) yields

$$\boldsymbol{i}: \quad m\ddot{x} = T - F$$

$$\boldsymbol{j}: \quad 0 = N - mg$$

Assume So far we have two equations and three unknowns—$\ddot{x}$, N, and F. To reduce the number of unknowns, we have to make an assumption—either the thrust is enough to cause slipping to occur, or it's insufficient and the Batsled ends up sitting in the same spot, its rocket engine firing pitifully as the bad guys run away laughing. Because the second scenario is too horrible to consider, let's start by assuming that the thrust is enough to overcome the friction developed between the sled and the snow. We will have to check this assumption, of course.

Solve Under the assumption of slipping, we know that F will attain the maximum value possible, which is μN. Rewriting our equations with the given parameter values yields

$$\boldsymbol{i}: \quad (1600\,\text{kg})\ddot{x} = 43{,}000\,\text{N} - 0.2N \tag{7.4}$$

$$\boldsymbol{j}: \quad 0 = N - (1600\,\text{kg})(9.81\,\text{m/s}^2) \tag{7.5}$$

Equation (7.5) tells us that $N = 1.57 \times 10^4$ N. Using this in (7.4) gives

$$\ddot{x} = \frac{4.3 \times 10^4\,\text{N} - 0.2(1.57 \times 10^4\,\text{N})}{1600\,\text{kg}} = 24.9\,\text{m/s}^2 = 2.54g$$

Check We can tell that our slip assumption is correct from the fact that the acceleration is positive to the right; that is, the amount of thrust was enough to counteract the maximal friction force μN and yield a positive acceleration. We can also provide a qualitative check on the work by noting that the acceleration is about two and a half g's—well in line with the acceleration expected of a superhero's vehicle.

EXAMPLE 7.2 TENSION IN SUPPORT CHAINS (Theory on page 384)

Figure 7.4 Welcome sign on massless chains

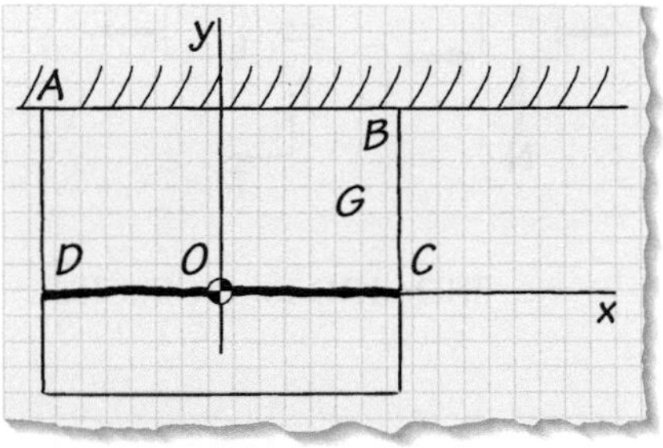

Figure 7.5 Position of the system in its rest postion

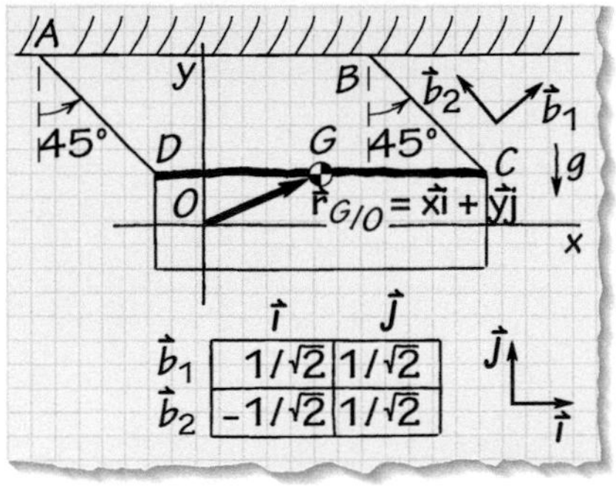

Figure 7.6 Position of the system in its displaced position

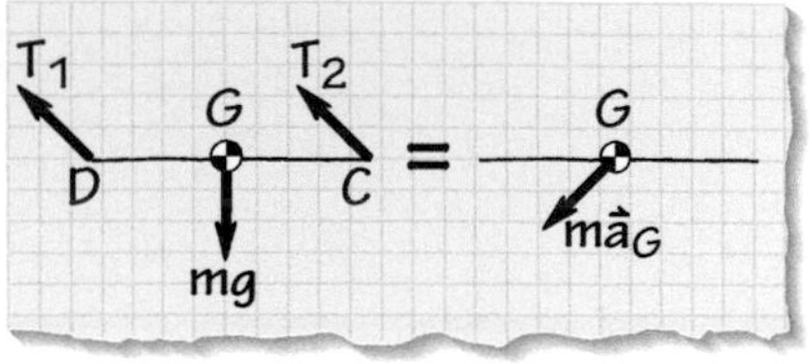

Figure 7.7 FBD=IRD

Let's examine a welcome sign (**Figure 7.4**) that's been pulled to the right and then released. At the instant of release, the sign is so far from its equilibrium position that the supporting (massless) chains are at 45° to the horizontal. The velocity at this instant is zero. The sign is $2L$ long and $\frac{L}{2}$ high and has a mass m. Essentially all of the sign's mass is concentrated in a horizontal bar along the top of the sign. We will therefore assume that the mass center is located directly in the middle of the top edge. Each flexible supporting chain has length L. Find the tension in the chains at the instant of release.

Goal Find the chain tension at the instant of release.

Draw **Figure 7.5** shows the system in diagrammatic form. The sign is shown in its rest position in **Figure 7.5** and in its displaced position in **Figure 7.6**.

Two sets of unit vectors are illustrated: $\boldsymbol{i}, \boldsymbol{j}$ and $\boldsymbol{b}_1, \boldsymbol{b}_2$. The $\boldsymbol{i}, \boldsymbol{j}$ set is fixed to ground, and the $\boldsymbol{b}_1, \boldsymbol{b}_2$ set aligns with the supporting chains. O, the origin of our coordinate axes, corresponds to G's position when the sign is hanging straight down. (The coordinate axes in **Figure 7.5** are *not* attached to the sign: they're ground fixed.) When disturbed from its vertical position, the location of the sign's mass center with respect to O is given by $\boldsymbol{r}_{G/O} = x\boldsymbol{i} + y\boldsymbol{j}$.

Assume The chains can only supply a tension force, and thus the sign will experience a force at each end that's directed along the chains, as well as a force due to gravity. The sign is released from rest and therefore has zero velocity at the point of release. **Figure 7.7** shows the associated free-body diagram and inertial response diagram.

Formulate Equations A force balance in the $\boldsymbol{i}$ and $\boldsymbol{j}$ directions gives us

$$\boldsymbol{i}: \qquad m\ddot{x} = -T_1\frac{1}{\sqrt{2}} - T_2\frac{1}{\sqrt{2}} \tag{7.6}$$

$$\boldsymbol{j}: \qquad m\ddot{y} = T_1\frac{1}{\sqrt{2}} + T_2\frac{1}{\sqrt{2}} - mg \tag{7.7}$$

and a moment balance about the mass center (7.3) yields

$$\boldsymbol{k}: \qquad 0 = \boldsymbol{r}_{D/G} \times T_1\boldsymbol{b}_2 + \boldsymbol{r}_{C/G} \times T_2\boldsymbol{b}_2$$

$$0 = -L\boldsymbol{i} \times T_1\boldsymbol{b}_2 + L\boldsymbol{i} \times T_2\boldsymbol{b}_2$$

$$0 = -L\boldsymbol{i} \times T_1\left(-\frac{1}{\sqrt{2}}\boldsymbol{i} + \frac{1}{\sqrt{2}}\boldsymbol{j}\right) + L\boldsymbol{i} \times T_2\left(-\frac{1}{\sqrt{2}}\boldsymbol{i} + \frac{1}{\sqrt{2}}\boldsymbol{j}\right)$$

$$0 = -\frac{T_1L}{\sqrt{2}} + \frac{T_2L}{\sqrt{2}} \tag{7.8}$$

This last equation, (7.8), tells us that T_1 is equal to T_2. So let's denote both tensions with T. Where does that leave us?

An examination of (7.6) and (7.7) tells us that we're going to have to come up with another equation to solve this problem because we have three unknowns ($\ddot{x}, \ddot{y}$, and T) and only these two equations. The key in this case is geometry. You can see from **Figure 7.5** that the endpoints of the chains define a trapezoid ($ABCD$). The bottom edge of the sign always remains horizontal, and all points within the sign have the same-shaped trajectory. **Figure 7.8** shows what the velocity vectors of different points on the sign will look like as soon as the sign begins to move. They're all aligned and all pointing at a downward 45° angle.

The speeds in the x and y directions are the same at this instant: $\dot{x} = \dot{y}$. The acceleration of the mass center, $\boldsymbol{a}_G$, is oriented down by 45°, as shown in **Figure 7.7**.

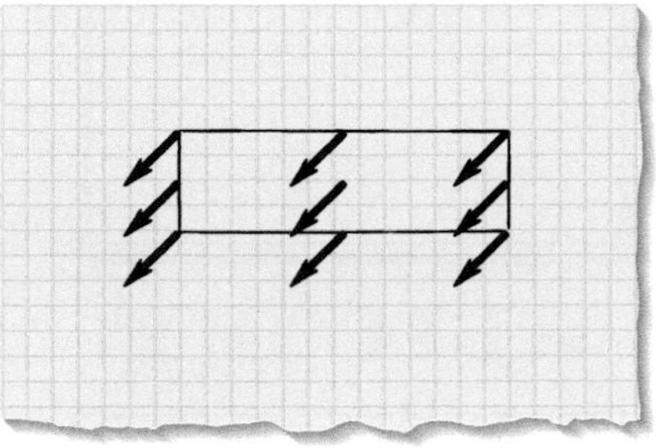

Figure 7.8 Velocity vectors at release

Solve Using $\ddot{x} = \ddot{y}$ in (7.6) and (7.7) gives us

$$i: \quad m\ddot{x} = -T\frac{1}{\sqrt{2}} - T\frac{1}{\sqrt{2}} = -\sqrt{2}T \tag{7.9}$$

$$j: \quad m\ddot{x} = T\frac{1}{\sqrt{2}} + T\frac{1}{\sqrt{2}} - mg = \sqrt{2}T - mg \tag{7.10}$$

which is easily solved. Equation (7.9) yields

$$\ddot{x} = -\frac{\sqrt{2}T}{m} \tag{7.11}$$

$$(7.11) \rightarrow (7.10) \Rightarrow \qquad -\sqrt{2}T = \sqrt{2}T - mg \quad \Rightarrow \quad \boxed{T = \frac{mg}{2\sqrt{2}}}$$

Check This is a great example of how dynamics differs from statics. If the welcome sign had been hanging straight down (**Figure 7.5**) and we had solved for the tension in the chains, we would have gotten $\frac{mg}{2}$ in each—that is, each chain supporting half the gravitational load. Now, because the system is dynamic, we get tensions that, combined, are only 71% of the static load. Where's the rest of the gravitational load gone? Into accelerating the sign! Equation (7.11) shows that

$$\ddot{x} = -\frac{\sqrt{2}T}{m} = -\frac{g}{2}$$

which is the same as $\ddot{y}$ at release. The sign immediately begins to accelerate down and to the left. This is why it's dangerous to base estimates of forces on a static load analysis when the system is dynamic. The world is full of instances when this has been done. The analysis of the Tacoma Narrows bridge was the best *static* analysis the builders knew how to do. According to the static load projections, the bridge should have been more than strong enough to do its job. Unfortunately, a relatively small dynamic force (due to the wind) was enough ultimately to destroy the bridge. Our own government is currently basing its vehicle safety evaluations with respect to rollovers on a static analysis that simply looks at the vehicle's wheelbase and the position of its mass center. All the important dynamic effects that occur during driving are ignored. Something to think about as you're driving down the road.

EXAMPLE 7.3 GENERAL MOTION OF A SWINGING SIGN (Theory on page 384)

Figure 7.9 Welcome sign on massless chains

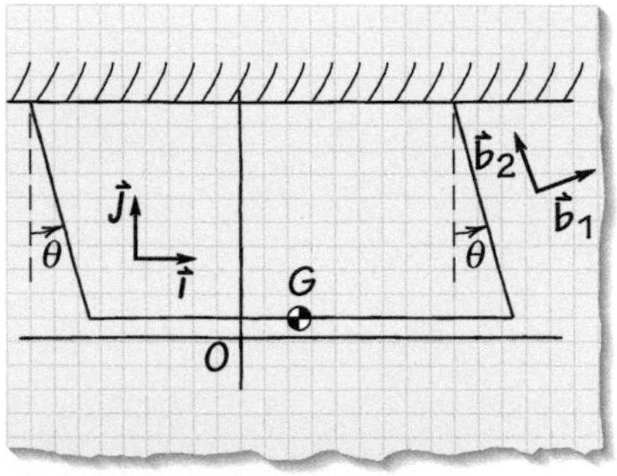

Figure 7.10 Labeling for sign

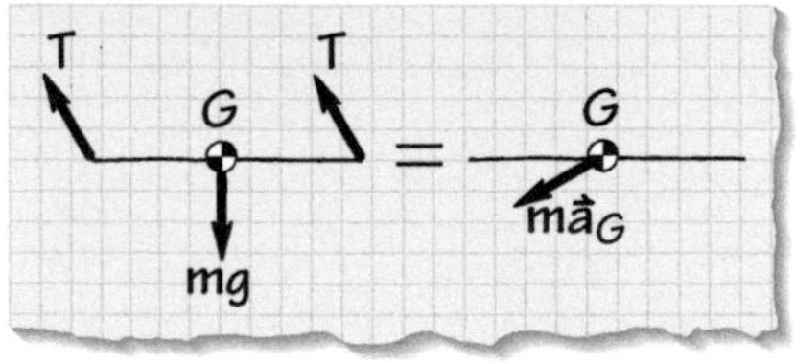

Figure 7.11 FBD=IRD for sign

Now that we've looked at the tension in the chains supporting a welcome sign (Example 7.2), let's up the ante and try to determine the general equations of motion that will enable us to calculate the position of the sign (**Figure 7.9**) at any time after release from an arbitrary angle θ. Once we have done so, we will plot out the horizontal position of the sign's mass center as a function of time for a release from $\theta = 30°$.

As before, the sign is $2L$ long and $\frac{L}{2}$ high and has a mass m. $L = 0.5\,\text{m}$ and $m = 1.2\,\text{kg}$.

Goal Find the equations of motion for a freely swinging sign, and plot the position of the mass center along the horizontal axis as a function of time.

Draw **Figure 7.10** shows our system in diagrammatic form with two sets of unit vectors: $\boldsymbol{i}, \boldsymbol{j}$ and $\boldsymbol{b}_1, \boldsymbol{b}_2$. The $\boldsymbol{i}, \boldsymbol{j}$ unit vectors are fixed to ground, and $\boldsymbol{b}_1, \boldsymbol{b}_2$ are oriented with the supporting chains. The chains can supply only a tension force, and thus the sign will experience a force at each end that's directed along the chains, as well as a force due to gravity, all shown in **Figure 7.11**.

Our transformation array is given by

	$\boldsymbol{i}$	$\boldsymbol{j}$
$\boldsymbol{b}_1$	$\cos\theta$	$\sin\theta$
$\boldsymbol{b}_2$	$-\sin\theta$	$\cos\theta$

Assume As in Example 7.2, the endpoints of the chains define a parallelogram ($ABCD$), the bottom edge of the sign always remains horizontal, and all points on the sign have the same-shaped trajectory (not the same trajectory—just the same shape). Two typical trajectories are shown in **Figure 7.12*a***, one starting from point A (and going to A') and the other starting from an arbitrary point E (going to E'). Their paths both trace circular arcs, with radius L, constrained to be so by the two equal-length, supporting chains.

Formulate Equations The key point for our analysis is that points on the welcome sign that travel along their circular arcs will develop acceleration components in two orthogonal directions. It's most convenient to use the $\boldsymbol{b}_1$ and $\boldsymbol{b}_2$ unit vectors, and a quick kinematic analysis tells us that the acceleration of the mass center G is given by

$$\boldsymbol{a}_G = L\ddot{\theta}\boldsymbol{b}_1 + L\dot{\theta}^2\boldsymbol{b}_2$$

Figure 7.12*b* shows the two components, along with their projections in the $\boldsymbol{i}, \boldsymbol{j}$ directions. As you can see, unlike the case of Example 7.2, this time we can't determine a simple relationship between $\ddot{x}$ and $\ddot{y}$. Both $\ddot{x}$ and $\ddot{y}$ depend on $\theta, \dot{\theta}$, and $\ddot{\theta}$:

$$\ddot{x} = L\ddot{\theta}\cos\theta - L\dot{\theta}^2\sin\theta$$
$$\ddot{y} = L\ddot{\theta}\sin\theta + L\dot{\theta}^2\cos\theta$$

A force balance in the $\boldsymbol{b}_1$ and $\boldsymbol{b}_2$ directions gives us

$\boldsymbol{b}_1$: $$mL\ddot{\theta} = -mg\sin\theta \tag{7.12}$$

$\boldsymbol{b}_2$: $$mL\dot{\theta}^2 = 2T - mg\cos\theta \tag{7.13}$$

Solve We need only consider (7.12) to find the system's time response. The system's equation of motion is

$$L\ddot{\theta} + g\sin\theta = 0$$

and we can use this to find $\ddot{\theta}$:

$$\ddot{\theta} = -\frac{g}{L}\sin\theta = -\frac{9.81\ \text{m/s}^2}{0.5\ \text{m}}\sin\theta$$

To find the time response we will need to integrate this expression. We first put it into state variable form (`y1=`θ`,y2=`$\dot{\theta}$) and then we execute the MATLAB command

```
[t,y]=ode45('welcome',tspan,y0)
```

with `y0=[0.524 0]` and `tspan=[0 2]`.

This will give us both θ and $\dot{\theta}$ for $0 \le t \le 2$ s.

We're asked for the horizontal motion of the mass center. Assuming that we have $x = 0$ when $\theta = 0$, we need to apply the transformation

$$x = L\sin\theta$$

which in MATLAB is done via

```
x=L*sin(y(:,1));
```

Plotting and gridding with the commands `plot[t,x]` and `grid` produce **Figure 7.13**, the requested plot of the sign's mass center horizontal position as a function of time. As usual, the relevant `M-file`, named `welcome.m`, is found on this book's webpage.

Check We will just do a physical reasonableness check here. You will notice that the mass center starts out positive, 0.25 m to the right of the origin. It then begins moving to the left and goes just as far left as it started right (−0.25 m). Then it returns to its starting position and begins the cycle again. This is just how you would expect a pendulum to behave, ticking back and forth. And that's precisely how the sign is behaving. The whole sign behaves as a large pendulum, going right, then left, and then back. If you look at the governing equation, you will see the reason—it's exactly the same equation as you would find if all the mass had been concentrated at a point that was supported by a single chain. The fact that the body has dimensions didn't really affect the overall behavior.

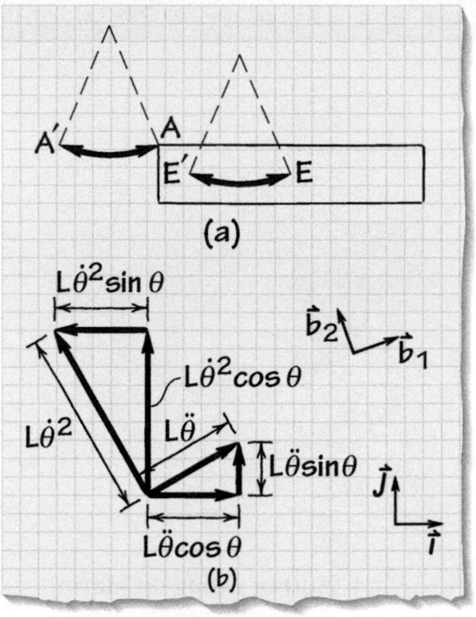

Figure 7.12 Two circular paths and mass center's acceleration

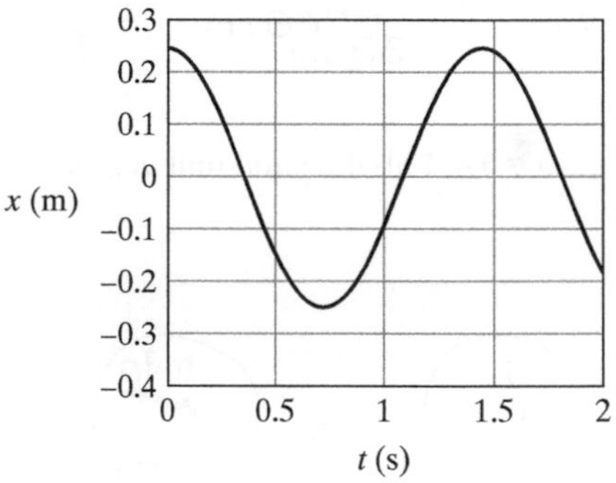

Figure 7.13 Horizontal motion of mass center

EXAMPLE 7.4 NORMAL FORCES ON A STEEP HILL (Theory on page 384)

Benson Tongue

Figure 7.14 Author riding up Marin Avenue

There's an incredibly steep street near where I work (Marin Avenue), which has a 22% grade (rise over run). (Note to those who don't cycle: the toughest climbs in the Tour de France are 11% grades; that gives you some idea!) Assume that I'm cycling up at a constant 6 km/h, I weigh 67 kg, my bike is 10 kg, and my assorted gear weighs 4.5 kg. **Figure 7.14** shows the bicycle, slope, and dimensions. The illustrated mass center is the mass center of the overall system (bike/rider).

(a) What are the normal forces at the tire/road interface?

(b) What do your results imply about how I usually ride up Marin (in- or out-of-saddle)?

Goal Determine the normal forces between the bicycle's tires and the road, and deduce something about the likely riding position.

Given Marin Avenue has a 22% grade, the speed is constant at 4 mph, and the total weight is 81.5 kg.

Assume We will assume that the tires don't slip as the bicycle moves uphill and that the center of mass stays a constant distance above the road surface. We're also implicitly neglecting the fact that the wheels are actually rotating. Later on in the chapter we will look at the wheels and body as separable units, but for now we will treat the system as a single body.

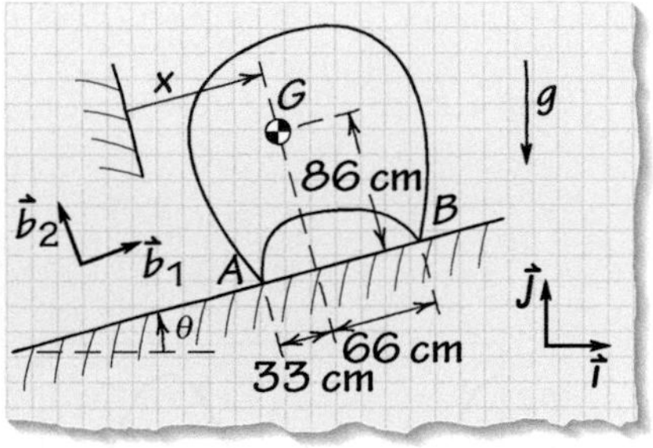

Figure 7.15 Labeling and unit vectors

Draw **Figure 7.15** shows two sets of unit vectors and the relevant dimensions of the problem, and **Figure 7.16** presents a general free-body diagram and inertial response diagram. The inertial response $m\boldsymbol{a}_G$ is drawn in the $\boldsymbol{b}_1$ direction, in line with our assumption that the mass center stays a fixed distance from the road surface.

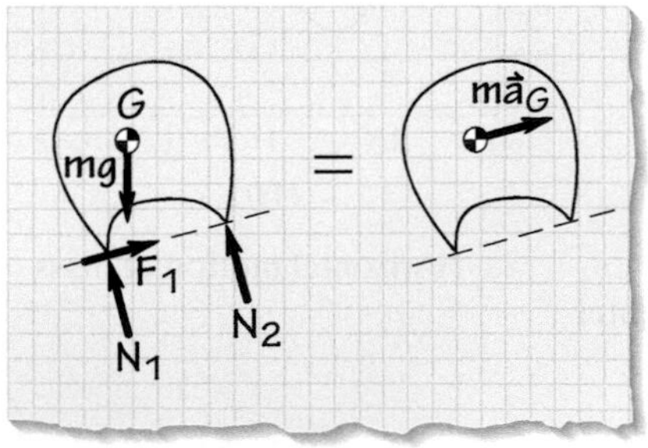

Figure 7.16 FBD=IRD

Formulate Equations The coordinate transformation array

	$\boldsymbol{i}$	$\boldsymbol{j}$
$\boldsymbol{b}_1$	$\cos\theta$	$\sin\theta$
$\boldsymbol{b}_2$	$-\sin\theta$	$\cos\theta$

lets us express all the relevant quantities in terms of a single unit-vector set, in this problem the $\boldsymbol{b}_1$, $\boldsymbol{b}_2$ set. We will make use of all three of our balance equations: (7.1), (7.2), and (7.3).

Notice that the force vector $\boldsymbol{F}_1$ goes through both contact points—A and B. For this problem the physical fact that the drive force is coming from the rear wheel makes no difference; it could just as well be an all-wheel drive bicycle as far as our analysis is concerned.

Solve **(a)** We can find the angle of the slope by taking the inverse tangent of the grade:

$$\theta = \tan^{-1} 0.22 = 12.4^\circ$$

The velocity is constant, and so the linear acceleration is zero. Knowing that the center of mass remains at a fixed height implies no angular acceleration as well. Thus our force balance is

$$F_1\boldsymbol{b}_1 - mg\boldsymbol{j} + (N_1 + N_2)\boldsymbol{b}_2 = 0$$
$$F_1\boldsymbol{b}_1 - mg(\sin\theta\boldsymbol{b}_1 + \cos\theta\boldsymbol{b}_2) + (N_1 + N_2)\boldsymbol{b}_2 = 0$$
$$\boldsymbol{b}_1(F_1 - mg\sin\theta) + \boldsymbol{b}_2(-mg\cos\theta + N_1 + N_2) = 0$$

$\boldsymbol{b}_1$: $$F_1 = mg\sin\theta \tag{7.14}$$

$\boldsymbol{b}_2$: $$N_1 + N_2 = mg\cos\theta \tag{7.15}$$

A moment balance gives us

$$\sum \boldsymbol{M}_G = 0$$

$\boldsymbol{b}_3$: $$66N_2 - 33N_1 + 86F_1 = 0 \tag{7.16}$$

(7.14) → (7.16) ⇒ $$66N_2 - 33N_1 + 86mg\sin\theta = 0$$

$$N_2 = -\frac{86mg\sin\theta}{66} + \frac{N_1}{2} \tag{7.17}$$

(7.17) → (7.15) ⇒ $$1.5N_1 - \frac{86mg\sin\theta}{66} = mg\cos\theta$$

$$N_1 = \frac{mg\left[\cos\theta + \frac{86}{66}\sin\theta\right]}{1.5} = \frac{(81.5\,\text{kg})(1.257)}{1.5} = 68\,\text{kg} \tag{7.18}$$

(7.18) → (7.15) ⇒ $$N_2 = 11.6\,\text{kg} \tag{7.19}$$

(b) The implication of these results is that I don't usually ride up Marin in a seated position. The grade is so steep that my front tire is pressed very lightly onto the road, and this severe lack of normal force interaction makes steering very problematic, especially given the uneven pedal forces that happen when climbing slowly. Thus, when I'm on this climb, I quickly get out of the saddle to shift more weight over the front axle.

Check This isn't a check but rather an interesting observation about everyday forces. The force due to gravity that's making it so difficult to climb the hill is $mg\sin 12.4° = 17.5\,\text{kg}$. That might not sound impossibly large; it's about the weight of a medium dog. But it's just about the upper limit for someone on a bike to pull. Just an interesting fact to remember.

EXERCISES 7.1

7.1.1. **[Level 1]** The illustrated rear-wheel-driven sports car could theoretically pull a wheelie (lift its front tires off the ground) if it accelerated strongly enough. What acceleration would be necessary for this to occur? Is it reasonable to expect that this could be achieved in real life? $L_1 = L_2 = 127$ cm and $h = 60$ cm.

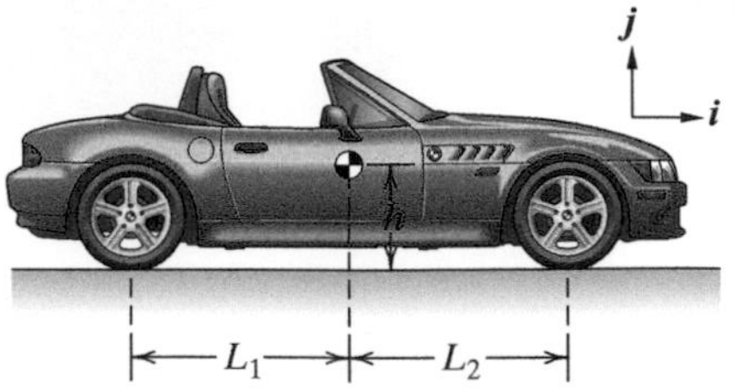

E7.1.1

7.1.2. **[Level 1]** In this exercise we'll consider the effect of height on a vehicle's braking capabilities. We'll consider two cases: a vehicle with a mass center at $h = 0.8$ m and a "flat" vehicle (zero height). $l = 2.6$ m, $w = 1.6$ m, and $m = 1500$ kg. The car's mass center is centered between the four tires. The maximum braking force is equal to the normal force multiplied by the coefficient of friction and $\mu = 0.9$. Assume that this force is supplied at all times.

Compare the time required to come to rest, and the distance traveled, for the case of finite height and that of zero height if maximum braking is applied from a speed of 100 km/hr and the maximum braking force is developed at the tire contact points at all times.

7.1.3. **[Level 1]** The thin plate $\overline{ABCD}$ has mass 10 kg and is held in the position shown by a wire $\overline{BH}$ and the two massless links $\overline{AF}$ and $\overline{DE}$. What is the acceleration of the plate and the forces in each link immediately after the wire is cut? The dimensions given are in millimeters. $\phi = 30°$.

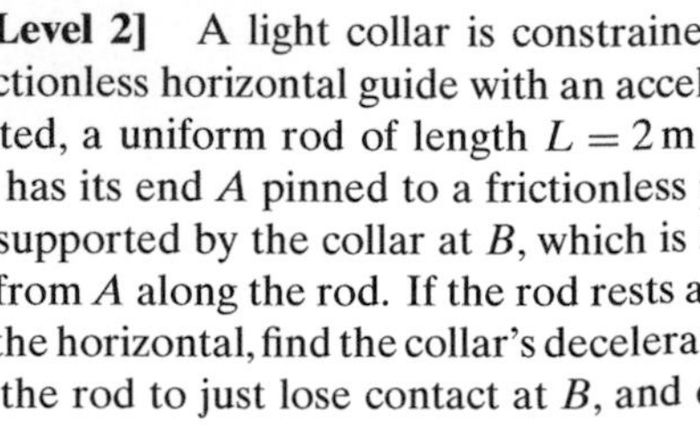

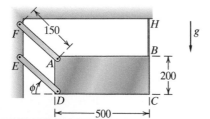

E7.1.3

7.1.4. **[Level 2]** A light collar is constrained to slide along a frictionless horizontal guide with an acceleration $\ddot{x}$. As illustrated, a uniform rod of length $L = 2$ m and mass $m = 10$ kg has its end A pinned to a frictionless pivot, and the rod is supported by the collar at B, which is a distance $d = 0.7$ m from A along the rod. If the rod rests at an angle $\theta = 40°$ to the horizontal, find the collar's deceleration $\ddot{x}$ that will cause the rod to just lose contact at B, and determine the corresponding resultant reaction force F_A at A.

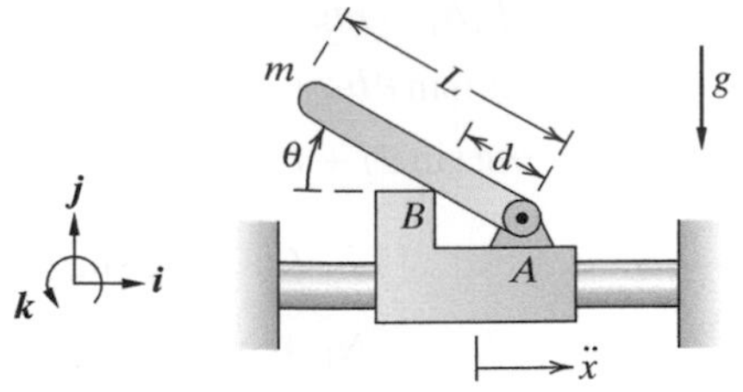

E7.1.4

7.1.5. **[Level 2]** The weight of the illustrated car is 1860 kg. The coefficient of friction between the tires and the ground is 0.85.

a. Assuming an engine that can deliver as much power/torque as demanded, calculate the maximum attainable acceleration $\ddot{x}$ and the normal forces at each tire.

b. Which tires (front or back) feel the most normal force?

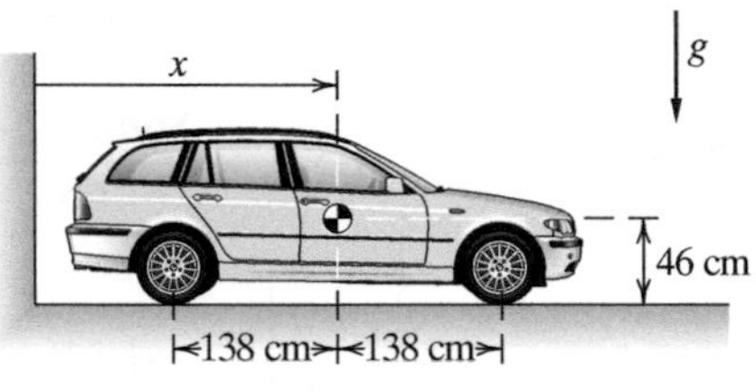

E7.1.5

7.1.6. **[Level 2]** In this exercise we'll see how much effect a partially iced road has on the braking performance of a modern car. We'll approximate the car as a rectangular block with zero height and with $l = 2.6$ m, $w = 1.6$ m, and $m = 1500$ kg. In Case 1 the car operates on a dry road, with a coefficient of friction equal to 0.9 for all four tires. In Case 2 the left tires (C and D) are in contact with dry road but the right tires (A and B) are in contact with a sheet of ice having a coefficient of friction equal to 0.1.

Because the car is perfectly balanced and has zero height, the vertical force at each tire contact patch is equal to one-quarter of the car's weight for all times during the deceleration. To simplify the analysis, we'll neglect any lateral forces at the contact patch and assume that all the forces point backwards, decelerating the car.

Compare the time required to come to rest, and the distance traveled, for the two cases if maximum braking is applied from a speed of 100 km/hr and the maximum braking force is developed at the tire contact points at all times.

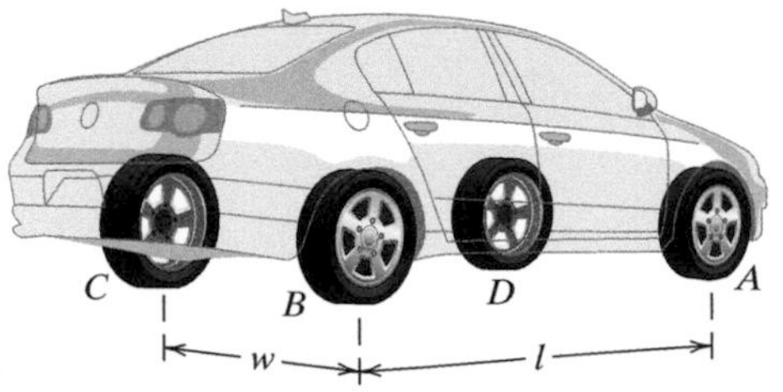

E7.1.6

7.1.7. **[Level 2]** When a car accelerates, the normal force between the tires and the ground changes. Assume that the car has a 50/50 weight distribution (equal weight on the front and rear tires) when at rest. By what percentage will the normal forces change if the vehicle accelerates forward at $0.25g$? $m = 1300$ kg, $h = 0.7$ m, and $L = 1.35$ m.

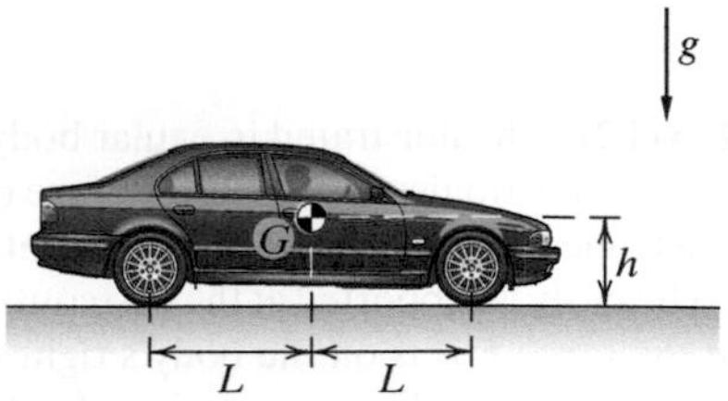

E7.1.7

7.1.8. **[Level 2]** In this exercise we'll extend our analysis of **Exercise 7.1.6**. As in **Exercise 7.1.6**, we have two cases: a dry road and a road for which the left tires are in contact with the dry road ($\mu = 0.9$) and the right are on ice ($\mu = 0.0$). We'll approximate the car as a rectangular block with zero height and with length $l = 2.6$ m, width $w = 1.6$ m, and $m = 1500$ kg. Because the car is perfectly balanced and has zero height, the vertical force at each tire contact patch is equal to one-quarter of the car's weight for all times during the deceleration. You must account for the lateral forces that are needed to keep the car from rotating. Assume that the magnitudes of both the lateral and the longitudinal forces are the same for the front and rear tires (the same magnitude front to back but different side to side). Apply a moment balance about the non-rotating vehicle's mass center and set it to zero in order to determine the level of lateral force needed to maintain a non-rotating state.

Compare the time required to come to rest, and the distance traveled, for the two cases if maximum braking is applied from a speed of 100 km/hr and the maximum braking force is developed at the tire contact points at all times.

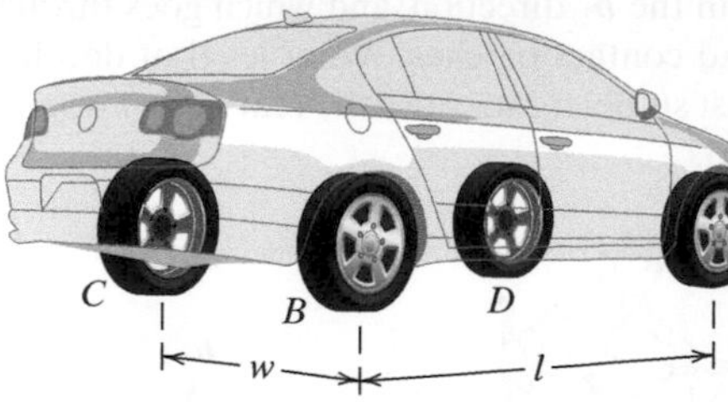

E7.1.8

7.1.9. **[Level 2]** In this exercise you will see, from a different point of view, why the weight transfers toward the back when a car accelerates forward. If the car accelerates forward at $0.2g$, then a force S must act, equal to $0.2mg$. Imagine that the car, rather than riding on the ground, is traveling through space. In this case there are no positional constraints. We can ignore gravity since the car is in a zero g environment. Now imagine that a force S acts on the bottom of the tires, as shown, so that the center of mass feels an acceleration equal to $0.2g\boldsymbol{i}$. In the absence of gravity and the ground, calculate the response of the car, treating it as a single rigid body. What forces N_1 and N_2 must be exerted on the car to maintain zero rotational acceleration and zero vertical acceleration? How do these forces affect a car that's in contact with the ground?

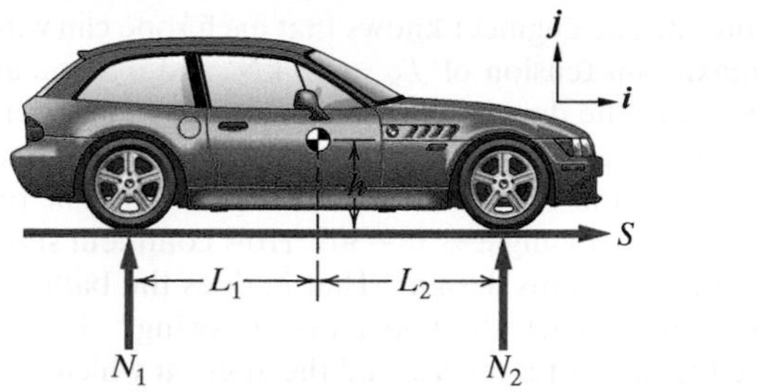

E7.1.9

7.1.10. **[Level 2]** A cabinet of width $2d$ and mass m is placed on a shake table that oscillates horizontally according to $x_0(t) = \bar{x}_0 \sin \omega t$. The shake table's surface is rough, and the coefficient of static friction between it and the cabinet is μ_s. The cabinet is symmetrical and its mass center G is located a distance h above the shake table surface.

a. At least how long should d be so that the cabinet never tips?

b. Suppose d is chosen such that the cabinet just barely tips. What is the minimum frictional coefficient μ_s needed so that the cabinet will not slip while tipping?

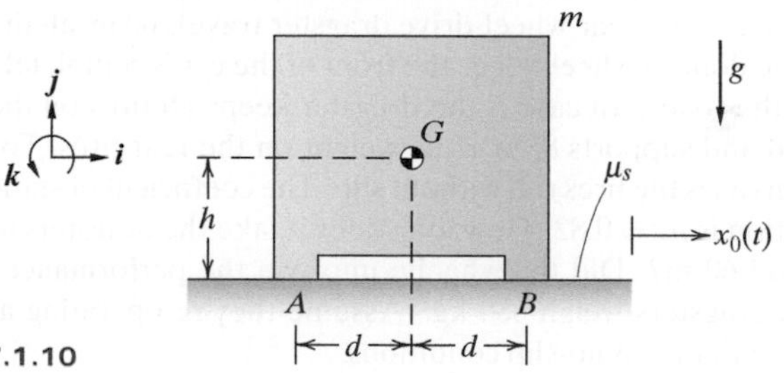

E7.1.10

7.1.11. **[Level 2]** What is the maximum inclination θ for which the car can ascend (assuming an arbitrarily large coefficient of friction) without flipping over due to gravity? Assume a constant speed of travel. If the car developed the same force when traveling horizontally at 96 km/h, how much power would it be producing? The car weighs 1360 kg.

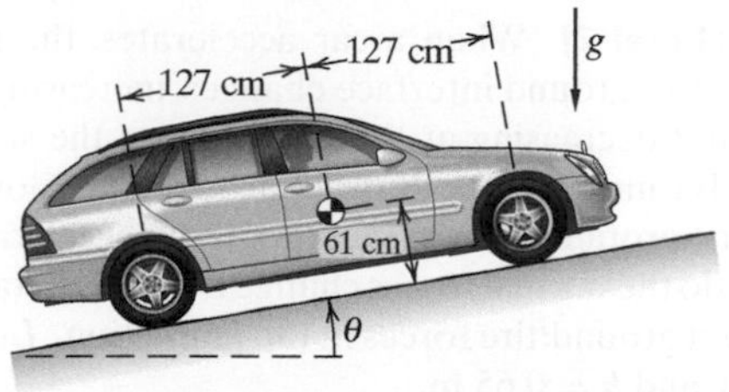

E7.1.11

7.1.12. [Level 2] Computational A medieval engineer builds a battering ram that consists of a tree trunk of weight $W = 6.8 \times 10^3$ kg that is supported at its ends by two light ropes of length $L = 1.2$ m, where θ is the angle that each rope makes with respect to the vertical. The tree trunk has a width and length of $2h$ and $2d$, respectively, where $h = 0.3$ m and $d = 3$ m, and the mass of the tree trunk is uniformly distributed. The engineer knows that each rope can withstand a maximum tension of $T_\infty = 42$ kN, and a static analysis (these are the days before dynamics was well understood) tells him that the ropes are strong enough to support the battering ram. Suppose the battering ram is to be released from rest at an angle $\theta_0 = -30°$. How confident should the engineer be in his design? That is, does the battering ram successfully reach the bottom of its swing? If it doesn't, give the time after release and the angle at which structural failure occurs. Neglect the log's diameter.

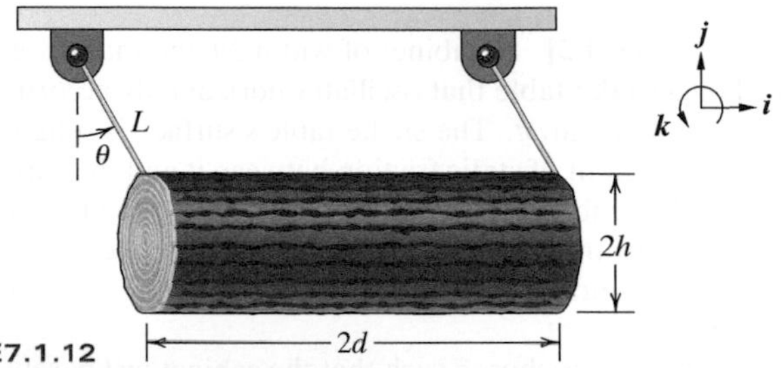

E7.1.12

7.1.13. [Level 2] Can wheelies improve performance? In case A the rear-wheel-drive dragster travels 60 m, all the while doing a wheelie (i.e., the front of the car is completely off the road). In case B the dragster keeps all tires on the road and supports 85% of its weight on the rear tires. For both cases the tires roll without slip. The coefficient of static friction is $\mu_s = 0.82$. How long does it take the dragsters to travel 60 m? Did the wheelie improve the performance? The dragsters weigh 680 kg. Assume they're operating at the limit of slip/no-slip conditions.

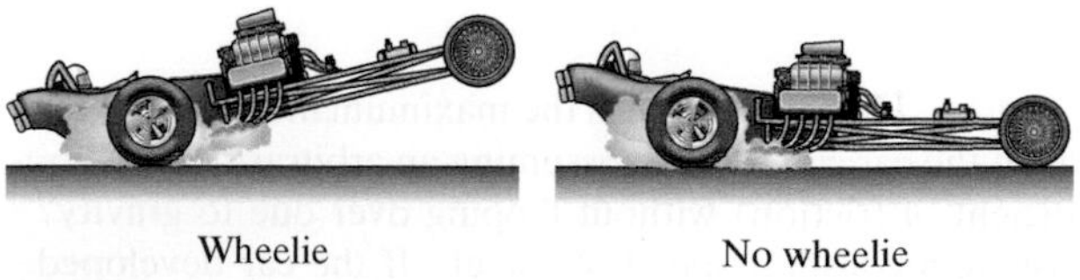

E7.1.13

7.1.14. [Level 2] When a car accelerates, the normal force at the tire/ground interface changes, increasing at the rear tires and decreasing at the front. Does the same occur for the Batmobile? The Batmobile is shown, along with its jet-engine propulsion system. Assume an acceleration of $0.9g$. How do the normal forces change from their static values? Neglect ground/tire forces in the $\boldsymbol{i}$ direction. $L_1 = 1$ m, $L_2 = 1.8$ m, and $h = 0.65$ m.

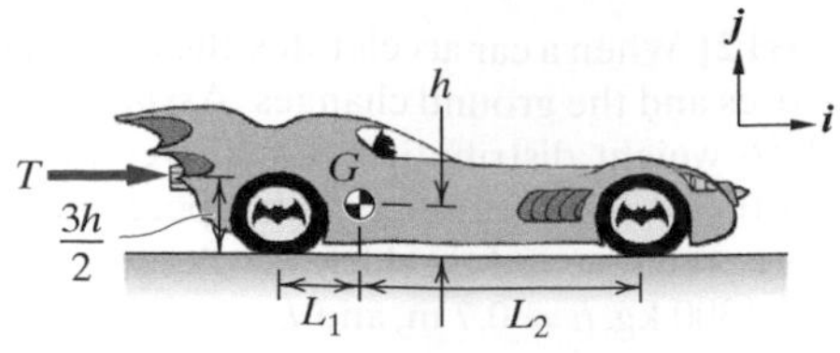

E7.1.14

7.1.15. [Level 2] The illustrated irregular body of mass m is pushed along a frictionless horizontal surface by a force F applied at an angle θ with respect to the horizontal, where $0° \le \theta < 90°$. The body is supported at the two contact points A and B, and the force F acts on the body's right edge at a height h from the surface. Find expressions for the acceleration $\ddot{x}$ that just causes the body to lose contact with the surface at A and B. What major difference between these two expressions do you notice?

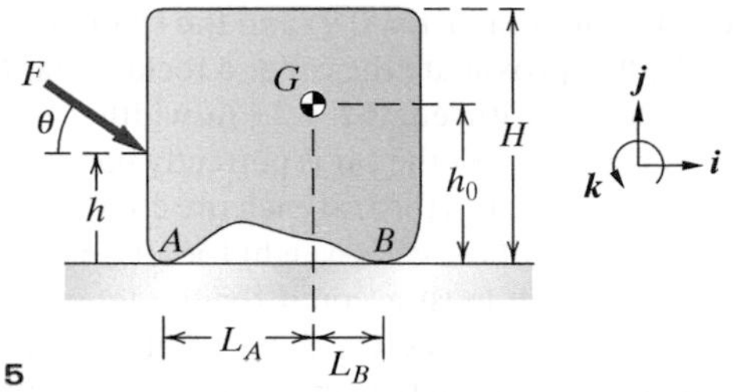

E7.1.15

7.1.16. [Level 2] When I was a kid, I was once riding down a hill and then abruptly applied my front brake (dumb but true). After finishing a 360° rotation through the air and subsequent meeting with the pavement, I pondered the meaning of deceleration. Let's analyze this problem a bit. Assume a rider is moving down a hill at a constant velocity and at $t = 0$ uses the brakes to decelerate. Assume that the result of squeezing the brake levers is to produce a force that points in the $\boldsymbol{b}_1$ direction, and which goes through the two tire/road contact patches. What level of deceleration would be just sufficient to cause the rear tire's normal force to go to zero?

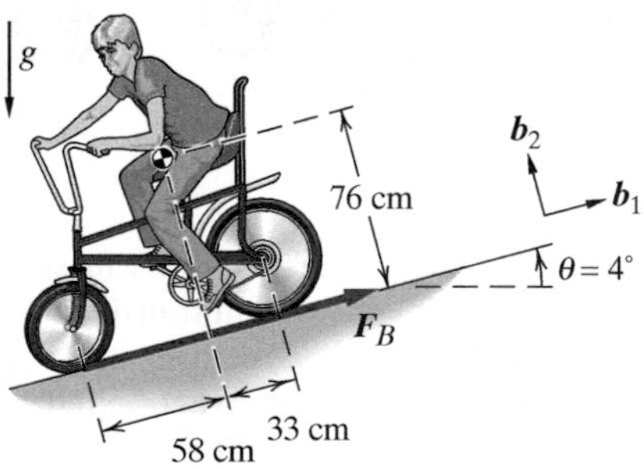

E7.1.16

7.1.17. [Level 2] The illustrated mechanism is a sensor in which the contact will open if $\ddot{x}$ exceeds a predetermined limit. The spring is prestressed so that it pulls with a force of 0.25 N at B, forcing the upper end against the mechanical stop P. The linear density ρ of the L-shaped bar is 0.1 kg/m and $m_1 = 1.1 \times 10^{-2}$ kg. $L_1 = 0.005$ m, $L_2 = 0.01$ m, and $L_3 = 0.01$ m. Neglect the width of the L_1, L_2, L_3 elements. What is the critical value of acceleration $\ddot{x}$ for which the contact will open?

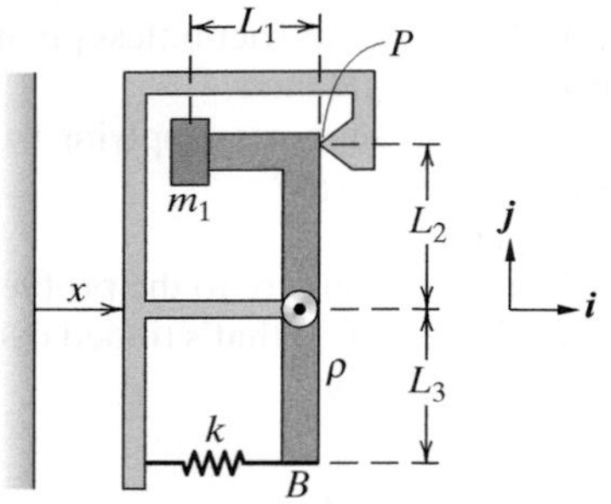

E7.1.17

7.1.18. **[Level 2]** Illustrated is a simplified model of a cyclist in a turn. The cyclist is approximated as a point mass, and the bicycle as a massless, rigid connection to ground. The turn is circular with radius 12 m. During the turn, the cyclist's lean angle is constant at 20°. What is the cyclist's speed?

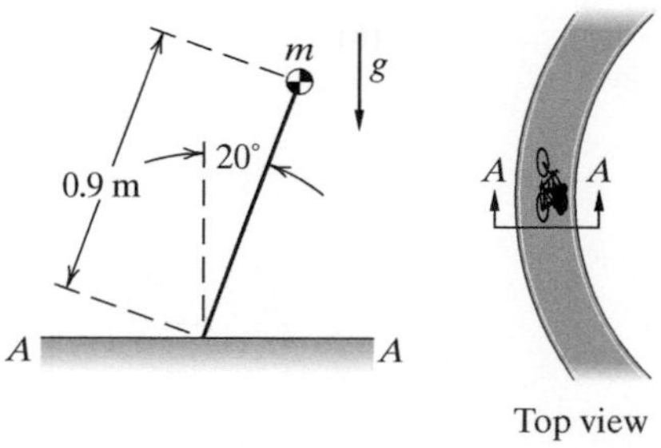

E7.1.18

7.1.19. **[Level 2]** Occasionally, superheroes have to hold on to the rear of a car to keep the bad guy from getting away. Assume that Captain Insanity is gripping the rear of the illustrated vehicle so that it can't move forward or backward. The rear tire (it's a rear-wheel-drive car) is spinning at 88 rad/s, and the coefficient of friction between the tire and the road is 0.8. We will model the tires as perfectly circular with a diameter of 60 cm. The mass of the car is 1500 kg, and it has a 50/50 weight distribution (center of mass directly between the front and rear axles). It has a wheelbase of 270 cm, and the mass center is positioned 90 cm from the ground. Assume that Captain Insanity will grip the car within the range of 50 to 75 cm from the ground. Where should he grip the car in order to exert the least force? What grip position would require the most force? What's happening physically to cause the difference?

E7.1.19

7.1.20. **[Level 2]** A uniform, square body is suspended from the ceiling by two massless cords. At the instant shown, $\dot{\theta} = 0$ and $\theta = \frac{\pi}{3}$ rad. $L = 1.1$ m and $m = 20$ kg. $|\overline{AB}| = |\overline{CD}| = 2$ m. Find the tension in each of the cords.

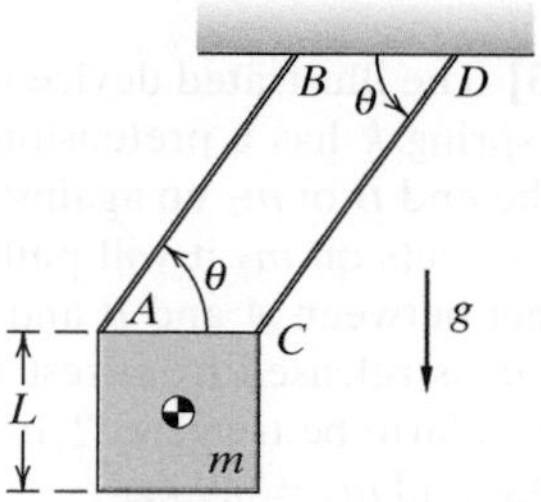

E7.1.20

7.1.21. **[Level 3]** **Computational** In this exercise we'll consider the case of a car that's undergoing braking in which the left tires are in contact with a dry road surface ($\mu = 0.9$) and the right are on ice ($\mu = 0.0$). We'll approximate the car as a rectangular block with zero height and with length $l = 2.6$ m, width $w = 1.6$ m, and $m = 1500$ kg. Because the car is perfectly balanced and has zero height, the vertical force at each tire contact patch is equal to one-quarter of the car's weight for all times during the deceleration. You'll have to determine the particular in-plane forces acting at each tire. At both left tires there will be a force component directed backwards, slowing the car, and a lateral force as well, acting to keep the car moving straight without rotating. Assume that each tire develops an overall in-plane force equal to the coefficient of friction at that tire times the vertical force acting between the tire and the road.

There exist a number of ways that the lateral force resisting car rotation can come about. One way is for the front left tire to provide all the lateral force necessary, leaving the left rear to simply brake. Or the reverse could be true, the rear supplying all the needed lateral force. Or, more generally, both could provide some lateral force.

Plot the total braking force developed as a function of the lateral force at the left front tire. Let the limits of this lateral force run between zero at the low end and the maximum needed to keep the car rotating (on the assumption that the rear tire is providing no lateral force at all) at the high end. Can you deduce anything about what's happening with the system by looking at the particular force values that correspond to the maximum overall braking force?

7.1.22. **[Level 3]** Two pinned links are attached to a horizontally translating block. $\overline{AB} = \overline{BC} = 1.2$ m. Each link has a mass of 5 kg. There exists a stable configuration for which both links simply translate to the right, each at a constant inclination angle, when $\ddot{x}$ is a constant. What is this configuration when $\ddot{x} = \frac{g}{2}$?

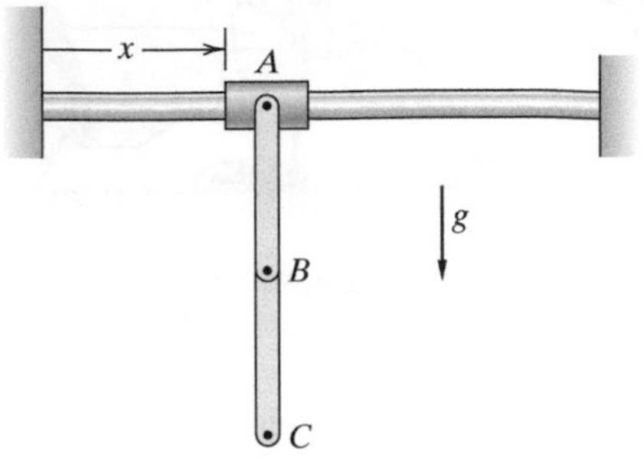

E7.1.22

7.1.23. **[Level 3]** The illustrated device is a mechanical force alarm. The spring k has a pretension of 10 N. This pretension holds the end B of m_2 up against the stop A. If a great enough force acts on m_2 it will pull away from A, breaking the contact between A and B and setting off the alarm. If a mass m_1 is released from rest in the position shown here, will the alarm be triggered? $\mu_1 = 0.3$, $\mu_2 = 0$, $\theta = 28°$, $m_1 = 1.5$ kg, and $m_2 = 0.9$ kg.

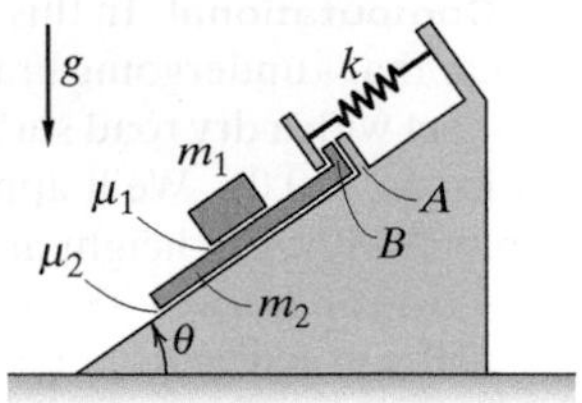

E7.1.23

7.1.24. **[Level 3]** The illustration shows a hunting stand that a bow hunter has built from which to target deer. A couple of civic-minded squirrels removed the nails holding the stand to the tree, and since hunters are poor carpenters in general, the insufficiently reinforced stand immediately began falling to the side. The illustration shows the stand when $\theta = 60°$ and $\dot{\theta} = -0.936$ rad/s. $m = 15$ kg. Model the connections at $A, B, C,$ and D as frictionless pivots and treat the supports as massless, rigid links.

a. What forces are the supports supplying to the stand's floor at the instant shown in (b)?

b. What is $\ddot{\theta}$ equal to?

c. How do the results compare to the problem of an inverted pendulum (shown in (c)) that's tipped over from the upright (unstable) equilibrium?

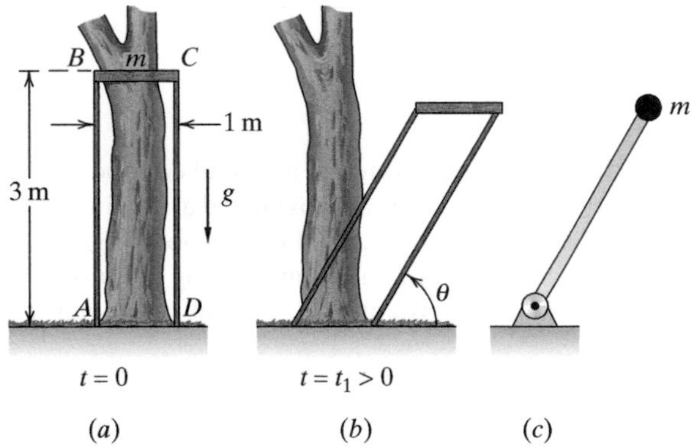

E7.1.24

7.2 ROTATION ABOUT A FIXED POINT

In this section we will derive the balance-of-angular-momentum expressions presented at the start of the chapter. When a body is undergoing pure rotation, we benefit from the fact that once we know the rotation rate, we can find the linear velocity of any point on the body. The same holds true for acceleration. Before we go any further, however, we must come up with another equation, one that allows us to tackle rotation. The relationship $\boldsymbol{F} = m\boldsymbol{a}$ is great, but we need more if we're going to deal with rotating bodies. Luckily, most of the work's already been done. In Section 5.2 we derived two moment balance equations—one with respect to a fixed origin O and the other with respect to a system's center of mass G:

$$\dot{\boldsymbol{H}}_O = \sum \boldsymbol{M}_O \quad \text{and} \quad \dot{\boldsymbol{H}}_G = \sum \boldsymbol{M}_G$$

where $\boldsymbol{H}$ represents the system's angular momentum and $\boldsymbol{M}$ represents the applied moments. In Section 5.2 these expressions weren't all that easy to deal with because of the generality of the angular momentum term. That is no longer the case.

As you might guess, we need to know where G is if we're going to use a moment balance that refers to it. So let's determine where the mass center is for a rigid body. In Section 5.1 we used the formula (5.2)

$$\boldsymbol{r}_{G/O} = \frac{\sum_{i=1}^{n} m_i \boldsymbol{r}_i}{m}$$

to determine the mass center of a collection of particles. The difference for us now is that we're integrating over a continuous body rather than adding up a finite number of masses. The summation becomes an integration, and we have

$$\boldsymbol{r}_{G/O} = \frac{\int_{\text{Body}} \boldsymbol{r}_{dm/O}\, dm}{m} \tag{7.20}$$

where dm is a differential mass element, as shown in **Figure 7.17**.

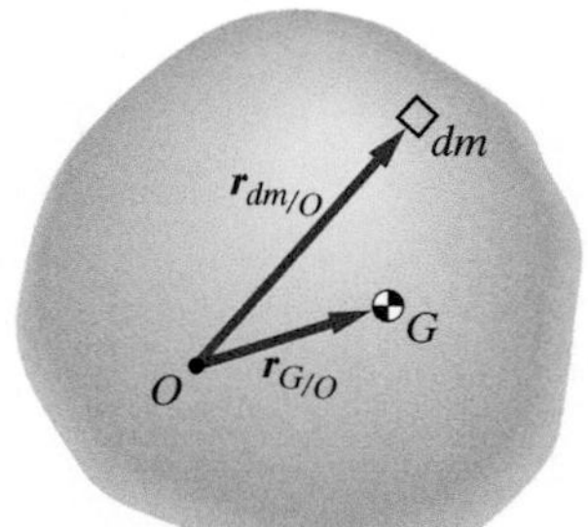

Figure 7.17 Mass center of a rigid body

>>> **Check out Example 7.10 (page 410) for an application of this material.**

Now that that's out of the way, let's do some moment balancing. We'll start with $\dot{\boldsymbol{H}}_O$ because this is going to be the easiest of all our moment balance equations. The formula we're about to derive will apply when we have a rigid body that's pivoting about a fixed point O, such as the one shown in **Figure 7.18**.

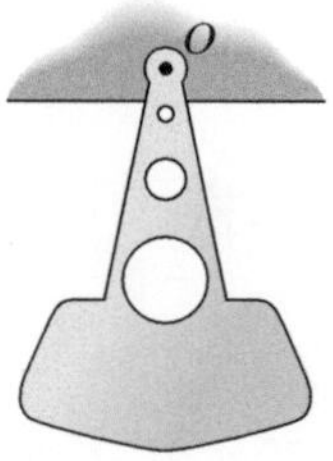

Figure 7.18 Body pivoted about a fixed point O

If you take a second to glance back at Section 5.2, you'll see that we derived our moment balance laws for many individual particles (equations (5.13) and (5.16)). We can take two approaches to get from multiple particles to continuous, rigid bodies. The first is to keep a multiparticle outlook but insist that all the particles are joined together in a rigid mass. Then, in order to have a continuous solid, we have to let the number of particles increase (to infinity) and their size decrease (toward zero). The individual masses become differential masses, and the summations become integrals. That's a fine way to go about it, but instead we will take the second approach and look at the problem as a continuum from the start. This means that the angular momentum about O is given by

$$\boldsymbol{H}_O = \int_{\text{Body}} \left(\boldsymbol{r}_{dm/O} \times \boldsymbol{v}_{dm}\right) dm \tag{7.21}$$

The $/O$ is to remind you that the position vector is referenced to the fixed point of rotation O. The dm is the bit of differential mass that $\boldsymbol{r}_{dm/O}$ is pointing toward, and $\boldsymbol{v}_{dm}$ is its absolute velocity. Finally, the body is there to remind you that the integral is a spatial one taken over the entire body being examined. **Figure 7.19*a*** shows a pivoting body of the sort being examined. I've included a set of unit vectors $\boldsymbol{b}_1, \boldsymbol{b}_2, \boldsymbol{b}_3$ that are aligned with $\boldsymbol{r}_{dm/O}$ in order to clarify the following discussion.

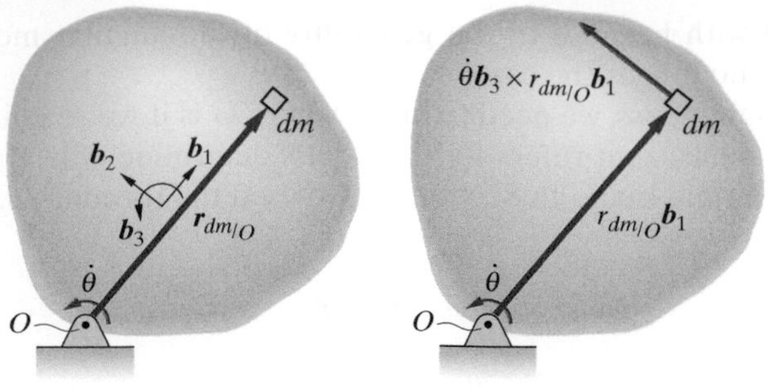

Figure 7.19 Body pivoting about O

We start with our well-known vector relation

$$\boldsymbol{v}_{dm} = \boldsymbol{v}_O + \boldsymbol{v}_{dm/_O}$$

$\boldsymbol{v}_O$ is zero (O is a fixed point of rotation), and because we're dealing with a rigid body, $\boldsymbol{v}_{dm}$ is given by

$$\boldsymbol{v}_{dm} = \boldsymbol{\omega} \times \boldsymbol{r}_{dm/_O} \tag{7.22}$$

Using (7.22) in (7.21) gives us

$$\begin{aligned} \boldsymbol{H}_O &= \int_{\text{Body}} \boldsymbol{r}_{dm/_O} \times \left(\boldsymbol{\omega} \times \boldsymbol{r}_{dm/_O}\right)\, dm \\ &= \int_{\text{Body}} r_{dm/_O}\boldsymbol{b}_1 \times \left(\dot{\theta}\boldsymbol{b}_3 \times r_{dm/_O}\boldsymbol{b}_1\right)\, dm \\ &= \int_{\text{Body}} r_{dm/_O}\boldsymbol{b}_1 \times \left(\dot{\theta} r_{dm/_O}\boldsymbol{b}_2\right)\, dm \end{aligned}$$

Figure 7.19*b* gives a graphical idea of how these terms are oriented on the body. You can see that because we have $\boldsymbol{b}_1$ crossed with $\boldsymbol{b}_2$, we will end up with a resultant that's in the $\boldsymbol{b}_3$ direction:

$$\boldsymbol{H}_O = \int_{\text{Body}} \dot{\theta} \left|r_{dm/_O}\right|^2 dm\, \boldsymbol{b}_3$$

This is a good result! $\dot{\theta}$ doesn't have anything to do with the mass distribution over the body and can thus be taken out of the integral, giving us

$$\boldsymbol{H}_O = \dot{\theta} \int_{\text{Body}} \left|r_{dm/_O}\right|^2 dm\, \boldsymbol{b}_3 \tag{7.23}$$

or, in scalar form,

$$H_O = \dot{\theta} \int_{\text{Body}} \left|r_{dm/_O}\right|^2 dm \tag{7.24}$$

Is there anything particularly fancy about the integral now? It's just a spatial integral, over the body, of the squared distance of the differential mass from O multiplied by the associated differential mass. What we have just derived is called the **mass moment of inertia** about O. You can take the mass moment of inertia about any point in a body, in which case you would have the mass moment of inertia about A or B or whatever. The symbol for mass moment of inertia is I, and we will subscript it to show what point it's taken about:

$$I_O \equiv \int_{\text{Body}} r^2_{dm/O}\, dm \qquad (7.25)$$

This lets us write our angular momentum equation (7.24) as

$$H_O = I_O\dot{\theta} \qquad (7.26)$$

A related term that is often used when dealing with mass moments of inertia is the **radius of gyration**, most often represented by k, where

$$mk^2 = I$$

>>> **Check out Example 7.5 (page 401) and Example 7.6 (page 402) for applications of this material.**

What someone would do in this case is give you the radius of gyration of an object, along with its mass, and then leave it to you to calculate the mass moment of inertia. Physically, what k represents is the distance from O that you would have to put a lumped mass (equal to the mass of the body) in order to produce the body's mass moment of inertia. Most commonly the radius of gyration is referenced to the body's mass center, but it could be referenced to any point as long as the tabulation makes clear what point is being used. This value is typically found experimentally.

Let's talk a little more about what I really represents. The mass moment of inertia is the rotational equivalent of translational mass. You know that it's harder to throw a boulder than a baseball, and you also know this is a consequence of the vastly greater mass of the boulder as compared to the baseball. In the same way, it's much easier to spin a wheel like the one shown in **Figure 7.20*a*** about O than it is to spin the one shown in **Figure 7.20*b***, even though they have the same mass. They differ in their mass moments of inertia, and you can see why from (7.25). Each piece of differential mass contributes linearly to the mass part of the moment of inertia, but the effect of its distance from the point of rotation goes as the square of that distance. So objects with their mass concentrated near the center of rotation are easier to spin than objects with the mass far from that center. If you're into cars, you've probably heard that you always want to minimize spinning mass and that, when choosing wheels, you should get wheels that have as little mass out at the rim as possible. Now you know why. It's to minimize the rotational inertia—to make the system "lighter" with regard to spinning.

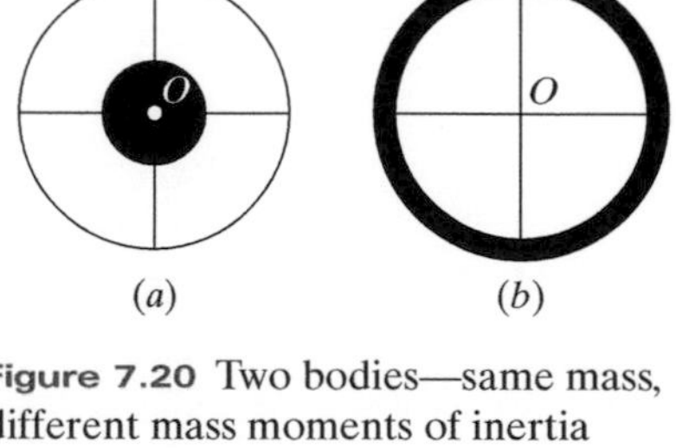

Figure 7.20 Two bodies—same mass, different mass moments of inertia

It's now time to get our final equation—the one that actually governs this spin. You'll note that the mass moment of inertia is independent of time and depends only on the object's mass distribution. Thus differentiating (7.26) with respect to time gives us

$$\dot{H}_O = I_O\ddot{\theta} \tag{7.27}$$

As mentioned at the start of the section, we have already derived the fact that $\dot{\boldsymbol{H}}_O = \sum \boldsymbol{M}_O$, and thus we have

$$I_O\ddot{\theta}\boldsymbol{b}_3 = \sum \boldsymbol{M}_O \tag{7.28}$$

or, in scalar form,

$$I_O\ddot{\theta} = \sum M_O \tag{7.29}$$

where M_O is the magnitude of the applied moments about O. To give you a realistic example, assume that the body we're examining is the crankshaft of a car. O indicates the crankshaft's center of rotation, and $\sum M_O$ is the moment generated by the engine and exerted on the crankshaft. If we know I_O, then we can use this equation to calculate the angular acceleration of the crankshaft. Of course, in a real car other bits need to be accounted for as well, such as the drivetrain and the car body. Never fear, we'll be able to handle them all in a short while.

This formula is great, and there would be no need to go any further if everything in the world moved around a fixed point. Since that's not the case, however, we'll have to keep working. But if you're analyzing a body that *is* moving around a fixed point, this is the formula you'll most likely want to use.

⋙ **Check out Example 7.8 (page 406) and Example 7.9 (page 408) for applications of this material.**

EXAMPLE 7.5 MASS MOMENT OF INERTIA OF A RECTANGULAR PLATE (Theory on page 399)

We'll gain some experience with calculating mass moments of inertia by examining a relatively simple case—that of a rectangular plate (**Figure 7.21**). The plate has an **areal density** ρ (mass per unit area) and sides of length a and b. We will solve for the mass moment of inertia about its mass center, located at the geometric center of the plate.

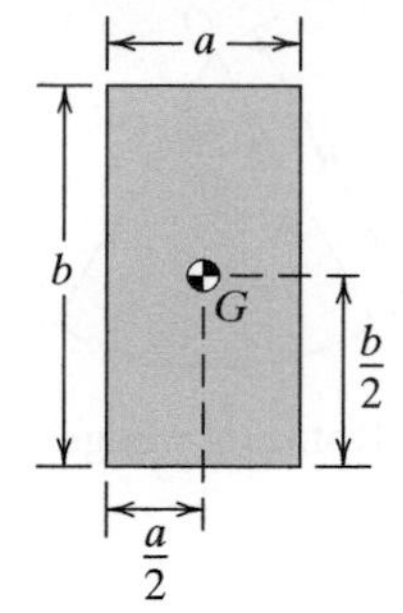

Figure 7.21 Rectangular body

Goal Find the mass moment of inertia of a rectangular body.

Given Body's dimensions and density.

Assume No additional assumptions are needed.

Draw **Figure 7.22** shows what the differential element that we will use in the integration looks like.

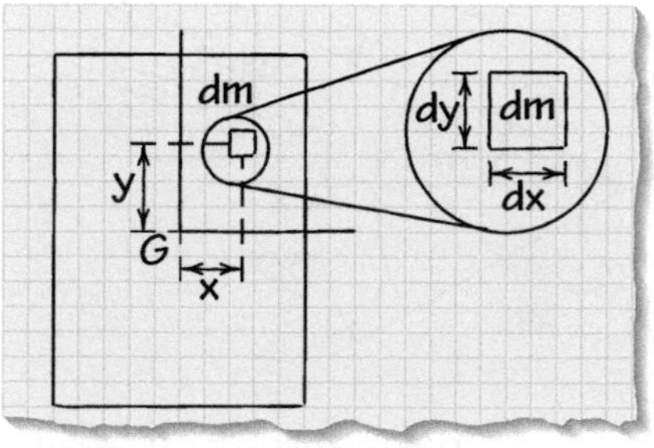

Figure 7.22 Differential mass element on plate

Formulate Equations The only equation needed is (7.25), referenced to the plate's center.

Solve Using the given integration limits in (7.25) gives us

$$I_G = \int_{\text{Body}} r^2_{dm/G}\, dm = \int_{-\frac{b}{2}}^{\frac{b}{2}} \int_{-\frac{a}{2}}^{\frac{a}{2}} (x^2 + y^2)\rho\, dx\, dy$$

$$= \rho \int_{-\frac{b}{2}}^{\frac{b}{2}} \left(\frac{x^3}{3} + xy^2\right)\Bigg|_{-\frac{a}{2}}^{\frac{a}{2}} dy = \rho \int_{-\frac{b}{2}}^{\frac{b}{2}} \left(\frac{a^3}{12} + ay^2\right) dy$$

$$= \rho \left(\frac{a^3 y}{12} + \frac{ay^3}{3}\right)\Bigg|_{-\frac{b}{2}}^{\frac{b}{2}} = \rho \left(\frac{a^3 b}{12} + \frac{ab^3}{12}\right) = \frac{m(a^2 + b^2)}{12}$$

Check One way to check results is to look at simpler limiting cases and check that we can obtain the same results from our more complex equation. For our case what we'll do is squash the plate along the y axis so that it's compressed into a bar. **Figure 7.23** shows our new bar once everything has been squashed onto the x axis. Finding the mass moment of inertia of such a body is very quick. Let ρ_1 be the linear density of the bar ($m = \rho_1 a$). Our mass moment of inertia is given by

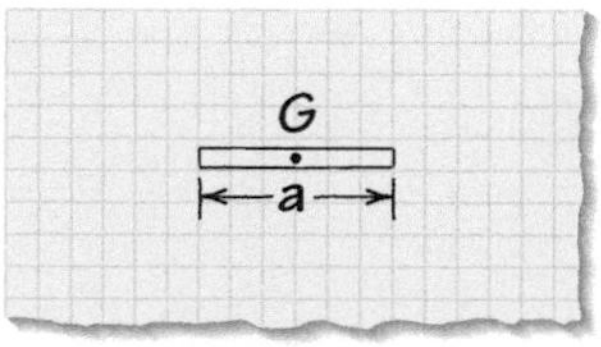

Figure 7.23 Uniform bar limit

$$I_G = \rho_1 \int_{-\frac{a}{2}}^{\frac{a}{2}} x^2\, dx = \rho_1 \frac{x^3}{3}\Bigg|_{-\frac{a}{2}}^{\frac{a}{2}} = \rho_1 \frac{a^3}{12}$$

Thus we have $I_G = \frac{\rho_1 a^3}{12}$ or, in terms of m, $I_G = \frac{ma^2}{12}$.

Does this agree with our plate results? For a plate we found that $I_G = \frac{m(a^2+b^2)}{12}$. If we let $b = 0$ (squashing down the plate), this gives us $I_G = \frac{ma^2}{12}$, an exact match with our bar result. Thus we can have some confidence that our integration was correct.

EXAMPLE 7.6 MASS MOMENT OF INERTIA OF A CIRCULAR SECTOR (Theory on page 399)

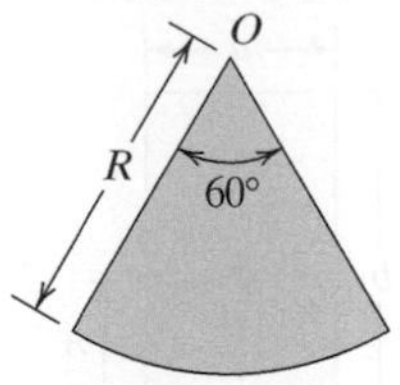

Figure 7.24 Rotating wedge

The rotating element that drives automatic watches (the ones that self-wind due to the motions of the wearer) is a wedge-shaped piece of steel, similar in shape to the one shown in **Figure 7.24**. The wedge has an areal density ρ (mass per unit area), a length R, and an enclosed angle of 60°. Solve for the mass moment of inertia about the point O.

Goal Find the mass moment of inertia of a wedge-shaped body about a point that isn't its mass center.

Given Body's dimensions and density.

Assume No additional assumptions are needed.

Draw **Figure 7.25** shows what the differential element that we will use in the integration looks like, as well as indicating the coordinates we'll use.

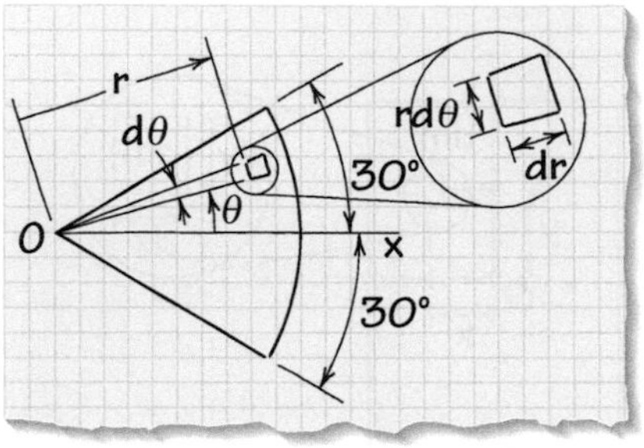

Figure 7.25 Differential mass element on wedge

Formulate Equations The only equation needed is (7.25), referenced to the point O.

Solve Unlike Example 7.5, in this case we will use a polar coordinate system to perform the needed integrations.

$$I_O = \int_{\text{Body}} r^2_{dm/O}\, dm = \int_{-\frac{\pi}{6}}^{\frac{\pi}{6}} \int_0^R r^2 \rho\, (dr)(r\, d\theta) = \rho \int_{-\frac{\pi}{6}}^{\frac{\pi}{6}} \int_0^R r^3\, dr\, d\theta$$

$$= \rho \int_{-\frac{\pi}{6}}^{\frac{\pi}{6}} \left. \frac{r^4}{4} \right|_0^R d\theta = \rho \int_{-\frac{\pi}{6}}^{\frac{\pi}{6}} \frac{R^4}{4}\, d\theta$$

$$= \left. \frac{\rho R^4 \theta}{4} \right|_{-\frac{\pi}{6}}^{\frac{\pi}{6}} = \frac{\rho \pi R^4}{12} = \boxed{\frac{\rho \pi R^4}{12} = \frac{mR^2}{2}}$$

Check We will do just a rough check this time. If all the mass were concentrated a distance R away from O, then I_O would be mR^2. That's the maximum possible mass moment of inertia, and it's twice as large as our answer. So at least we know we didn't do something in the calculations that caused a result that was bigger than the theoretical maximum. If all the mass were concentrated at $R/2$, then we would have an I_O equal to $mR^2/4$. Our result is double this, which also seems reasonable, because we can see from inspection that the mass center of the wedge is farther out from O than $R/2$. And that's all we're going to do. We have verified that the result isn't obviously incorrect, and we will rely on our mathematical skills to convince us that it is, in fact, correct.

We've now seen that we can determine the angular acceleration of rigid bodies that are rotating about a fixed point by using $I_O\ddot{\theta} = \sum M_O$ and have been told we can do the same for bodies in general motion with $I_G\ddot{\theta} = \sum M_G$. One clear requirement of either approach is that we have the body's mass moment of inertia about the appropriate point. Does this mean that we have to calculate mass moments of inertia for each problem we encounter? It would be nice if someone had already compiled a table that lists common mass moments of inertia (*Hint:* Someone already has. See Appendix B.) But how could a table list the mass moments of inertia about any point we might care about? The center of mass is unique, so that's an easy one to compile results for. But the body might be pivoting about any of an infinite number of locations O, and to deal with those situations we would like to have I_O. It turns out that there's a convenient formula that helps us overcome this problem, and that's the next topic of discussion.

We'll start with the assumption that we already know the mass moment of inertia about G. **Figure 7.26** shows what we're dealing with. From the definition of mass moment of inertia, we know that

$$I_O = \int_{\text{Body}} r^2_{dm/O}\, dm$$

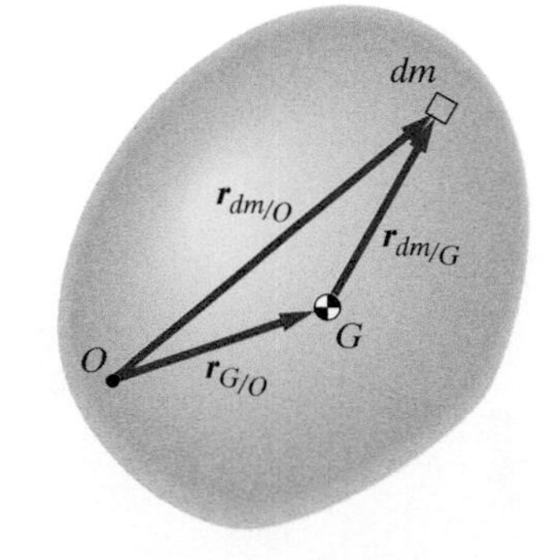

Figure 7.26 Finding moment of inertia about two different points

We can also see that the vector from O to our differential mass element dm can be expressed as

$$\boldsymbol{r}_{dm/O} = \boldsymbol{r}_{G/O} + \boldsymbol{r}_{dm/G}$$

So let's substitute this relationship into our expression for I_O and see what it gets us:

$$I_O = \int_{\text{Body}} \left(\boldsymbol{r}_{G/O} + \boldsymbol{r}_{dm/G}\right) \cdot \left(\boldsymbol{r}_{G/O} + \boldsymbol{r}_{dm/G}\right) dm$$

(Just a quick reminder in case you've forgotten: $\boldsymbol{r} \cdot \boldsymbol{r}$ is equal to r^2, the square of $\boldsymbol{r}$'s magnitude.)

Expanding this last expression yields

$$\begin{aligned}
I_O &= \int_{\text{Body}} \left(r^2_{G/O} + 2\boldsymbol{r}_{G/O} \cdot \boldsymbol{r}_{dm/G} + r^2_{dm/G}\right) dm \\
&= \int_{\text{Body}} r^2_{G/O}\, dm + 2\,\boldsymbol{r}_{G/O} \cdot \int_{\text{Body}} \boldsymbol{r}_{dm/G}\, dm + \overbrace{\int_{\text{Body}} r^2_{dm/G}\, dm}^{I_G} \\
&= r^2_{G/O} \overbrace{\int_{\text{Body}} dm}^{m} + 2\boldsymbol{r}_{G/O} \cdot \int_{\text{Body}} \boldsymbol{r}_{dm/G}\, dm + I_G \\
&= r^2_{G/O} m + 2\boldsymbol{r}_{G/O} \cdot \int_{\text{Body}} \boldsymbol{r}_{dm/G}\, dm + I_G
\end{aligned} \tag{7.30}$$

The only annoying term left is the middle one. Remember how we had a convenient relationship back in the multiple-particle chapter—namely, that the distance to the center of mass was zero if you were already there? It looked like

$$\sum_{i=1}^{n} m_i \boldsymbol{r}_{i_{/G}} = 0$$

What would this summation become if we took the limits of lots of little particles? The summation would morph into an integral over the body, the m_i would become a differential dm, and the discrete position vector $\boldsymbol{r}_{i_{/G}}$ would change into $\boldsymbol{r}_{dm_{/G}}$, which means we would end up with

$$\int_{\text{Body}} \boldsymbol{r}_{dm_{/G}} \, dm$$

which is exactly what we've got in the middle term of (7.30). So that middle term is equal to zero, and we're left with

$$I_O = I_G + m r_{G_{/O}}^2 \tag{7.31}$$

This is the very useful relationship I mentioned earlier. It goes by the name of the **parallel axis theorem**, and it comes in handy much of the time. Whenever you need to find the mass moment of inertia for some body and already know what it is with respect to the body's mass center, all you need do is apply this formula to find what it is with respect to any other point. Just find out how far the new rotation point is from G, square it, multiply by the total mass, and then add this to I_G, and the problem is solved.

WARNING! DON'T try to use this in the opposite manner. If you already have I about some other point, say B, and you wish to find I about some point A, do NOT attempt to use

$$I_A = I_B + m r_{A_{/B}}^2$$

It won't work. You have to start from G and then go to the new point. Always start from G.

⋙ **Check out Example 7.7 (page 405) for an application of this material.**

EXAMPLE 7.7 MASS MOMENT OF INERTIA OF A COMPLEX DISK (Theory on page 399 and 404)

In this example we'll look at a body that's a bit more complex: a circular disk with a hole that's eccentrically placed (**Figure 7.27**). The disk has an **areal density** ρ (mass per unit area) and $r_1 = \frac{r_2}{4}$. Express the mass moment of inertia (in-plane rotation), about O, in terms of ρ and r_2.

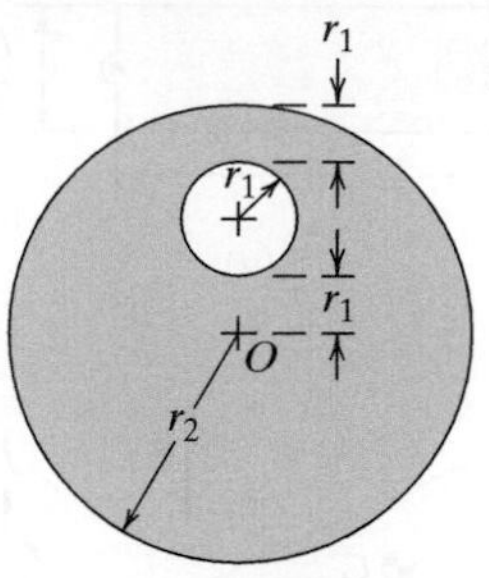

Figure 7.27 Complex disk

Goal Find the mass moment of inertia of a disk.

Given Body's dimensions and density.

Assume No additional assumptions are needed.

Draw **Figure 7.28** shows how we'll be viewing the system: Disk 2 (our system) plus "the hole" (disk 3) equals a uniform disk (disk 1).

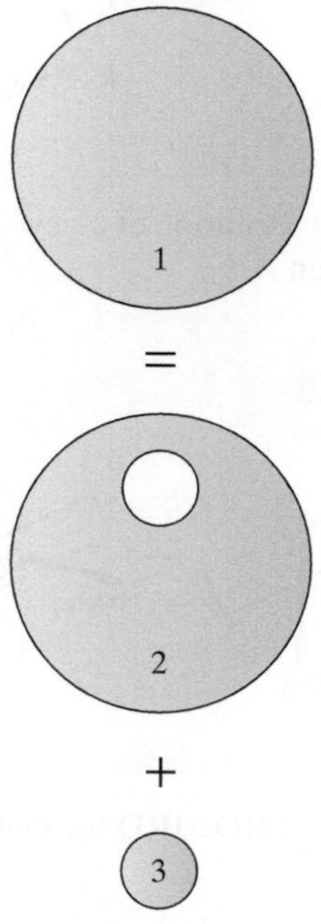

Figure 7.28 Disk deconstruction

Formulate Equations We'll use (7.25) and (7.31):

$$I_O = \int r^2_{dm_{/O}}\, dm, \qquad I_O = I_G + m r^2_{O_{/G}}$$

Solve A circular disk of radius R has a mass moment of inertia of

$$\bar{I} = \int_0^R \tau^2 \rho(2\pi\tau)d\tau = 2\rho\pi \int_0^R \tau^3 d\tau = \frac{\rho\pi R^4}{2}$$

where τ is a dummy variable of integration for radial position.

From **Figure 7.28** we see that disk 2 is equal to disk 1 minus its "hole" (disk 3). So we can calculate the relevant mass moment of inertia for disk 3:

$$I_{hole_{/O}} = \frac{\rho\pi r_1^4}{2} + m_{hole}(2r_1)^2$$

and subtract it from that of disk 1:

$$I_O = \frac{\rho\pi {r_2}^4}{2} - \left(\frac{\rho\pi r_1^4}{2} + \rho\pi {r_1}^2(2r_1)^2\right)$$

$$r_1 = \frac{r_2}{4} \Rightarrow \qquad I_O = \frac{256\rho\pi {r_2}^4}{512} - \frac{9\rho\pi {r_2}^4}{512} = \boxed{\frac{247\rho\pi {r_2}^4}{512}}$$

EXAMPLE 7.8 ANALYSIS OF A ROTATING BODY (Theory on page 400)

(a)

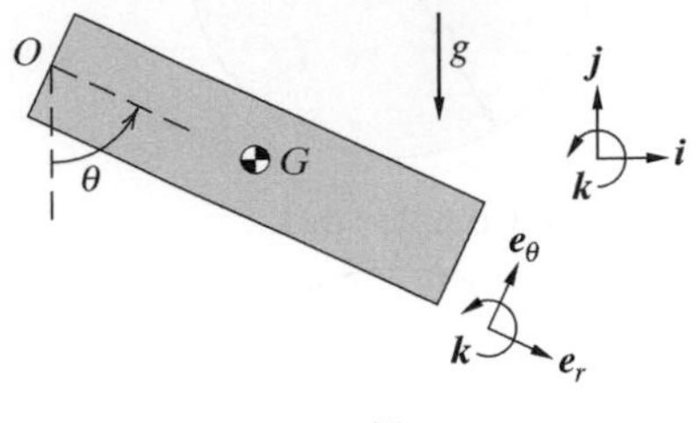

(b)

Figure 7.29 Rotation of a new welcome sign

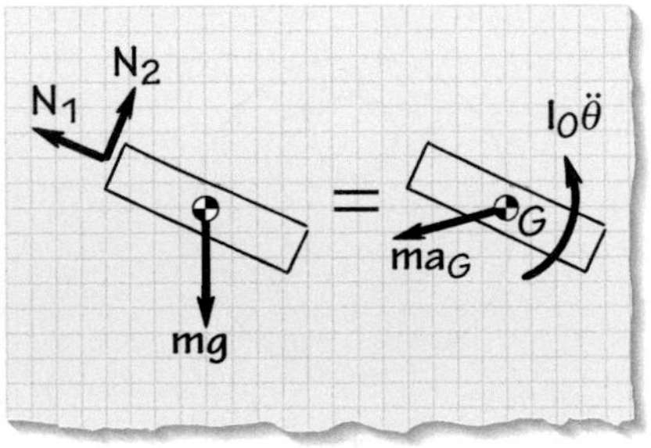

Figure 7.30 FBD=IRD for swinging sign

In this example we'll be looking at a different kind of welcome sign (**Figure 7.29*a***). Unlike the sign in Example 7.4, our current sign (mass m) is attached to a wall by two nails. As a result of ridiculously poor workmanship, the right nail has fallen out, and, as pictured in **Figure 7.29*b***, the sign is in the process of swinging around the left nail, (labeled O). Assume that the sign pivots freely about the nail, and determine the equation of motion and the rotational speed the sign will have when $\theta = 30°$. Treat the sign as a rectangular solid. Originally the sign was horizontal ($\theta = 90°$). Express your answers in terms of m, g, and L.

Goal Find the sign's equation of motion and $\dot{\theta}$ for $\theta = 30°$.

Draw **Figure 7.29*b*** shows two sets of unit vectors—one ground-based $(\boldsymbol{i}, \boldsymbol{j}, \boldsymbol{k})$ and the other attached to the rotating sign $(\boldsymbol{e}_r, \boldsymbol{e}_\theta, \boldsymbol{k})$. The FBD=IRD of **Figure 7.30** shows that the sign has only three forces acting on it—two reaction forces at the pivot and a gravity force. Because the sign freely pivots about O, no reaction moment is present.

Formulate Equations This example is about as polar as one could wish, and so we will use the $\boldsymbol{e}_r$, $\boldsymbol{e}_\theta$ coordinates for our final equations. From (7.28) we have

$$\boldsymbol{k}: \qquad I_O\ddot{\theta}\boldsymbol{k} = L\boldsymbol{e}_r \times (-mg)\boldsymbol{j} \tag{7.32}$$

Solve You'll note that by summing moments around O we completely avoided having to determine the reaction forces N_1 and N_2. This is the obvious advantage of purely rotational motion: you need only one equation of motion in order to determine the object's trajectory.

We already know I_G from Example 7.5; $I_G = \frac{m(a^2+b^2)}{12}$, where a and b are the lengths of the rectangle's sides. Thus for the particular values of our example, we have

$$I_G = \frac{m\left[(2L)^2 + (L/2)^2\right]}{12} = \frac{17mL^2}{48}$$

Using (7.31) then gives us

$$I_O = I_G + mr^2_{G/O} = \frac{17mL^2}{48} + mL^2 = \frac{65mL^2}{48} \tag{7.33}$$

In order to solve our equation of motion, all we have to do is whip up a coordinate transformation matrix between the $\boldsymbol{i}, \boldsymbol{j}$ and the $\boldsymbol{e}_r, \boldsymbol{e}_\theta$ systems:

	$\boldsymbol{i}$	$\boldsymbol{j}$
$\boldsymbol{e}_r$	$\sin\theta$	$-\cos\theta$
$\boldsymbol{e}_\theta$	$\cos\theta$	$\sin\theta$

These transformation relations enable us to rewrite (7.32) as

$$I_O\ddot{\theta}\boldsymbol{k} = -mgL\boldsymbol{e}_r \times (-\cos\theta\boldsymbol{e}_r + \sin\theta\boldsymbol{e}_\theta)$$

$$= -mgL\sin\theta\boldsymbol{k}$$

Our equation of motion is therefore

$$I_O\ddot{\theta} = -mgL\sin\theta$$

Now we need to find the $\dot{\theta}$ that corresponds to a particular angle of θ. Dividing our equation of motion by I_O gives us an explicit expression for $\ddot{\theta}$:

$$\ddot{\theta} = \frac{-mgL}{I_O}\sin\theta \tag{7.34}$$

We're in luck because this expression depends on θ. We saw in Section 2.1 that this is easily integrated:

$$\ddot{\theta}\,d\theta = \dot{\theta}\,d\dot{\theta} \quad \Rightarrow \quad \int_{\theta_1}^{\theta_2}\ddot{\theta}\,d\theta = \frac{1}{2}(\dot{\theta}_2^2 - \dot{\theta}_1^2)$$

For our example, State 1 refers to $\theta = \frac{\pi}{2}$ rad and State 2 corresponds to $\theta = \frac{\pi}{6}$ rad. The integral expression becomes

$$\int_{\frac{\pi}{2}}^{\frac{\pi}{6}}\ddot{\theta}\,d\theta \underbrace{=}_{(7.34)} -\frac{mgL}{I_O}\int_{\frac{\pi}{2}}^{\frac{\pi}{6}}\sin\theta\,d\theta = \frac{mgL}{I_O}\cos\theta\Big|_{\frac{\pi}{2}}^{\frac{\pi}{6}} = \frac{\sqrt{3}mgL}{2I_O}$$

We know that the sign starts with zero rotational velocity ($\dot{\theta}_1 = 0$), and so we have

$$\frac{1}{2}\dot{\theta}_2^2 = \frac{\sqrt{3}mgL}{2I_O}$$

$$\dot{\theta}_2 = \sqrt{\frac{\sqrt{3}mgL}{I_O}} \underbrace{=}_{(7.33)} \sqrt{\frac{48\sqrt{3}\,g}{65L}}$$

Check A dimensional check on our final answer for $\dot{\theta}_2$ yields

$$\sqrt{\frac{(\cancel{\text{kg}})\left(\frac{\cancel{\text{m}}}{\text{s}^2}\right)(\cancel{\text{m}})}{(\cancel{\text{kg}})(\cancel{\text{m}})^2}} = \frac{1}{\text{s}}$$

which is the correct result for a rotational velocity.

EXAMPLE 7.9 FORCES ACTING AT PIVOT OF FIREWORKS DISPLAY (Theory on page 400)

Figure 7.31 Fireworks pinwheel

Figure 7.31 shows a type of firework called a pinwheel. A rocket is attached to the end of a freely pivoting board, and when ignited, the rocket spins the board, showing sparks in a circular display. Unfortunately for the fans, this particular rocket got wet and is fated for failure as a display. Once ignited from the position shown, the thrust produced by the rocket is given by $T = (50\ \text{N/s}^2)t^2e^{-3t}$ where T is in newtons and t is in seconds. Find the equation of motion for the device, integrate with respect to time, and determine the forces acting at the pivot at $t = 1$ s. Describe what the system is doing at that time. The supporting board has a mass of 0.8 kg and a length of 1.2 m. $I_G = 0.24\ \text{kg·m}^2$ for the board/rocket system, and the center of mass is located 0.9 m from the pivot. The rocket has a mass of 0.8 kg, assumed constant.

Goal Find the forces acting at the pivot of a fireworks display after its rocket has been firing for 1 s, and describe the physical behavior of the system.

Given Mass moment of inertia, masses, dimensions, and rocket's thrust.

Assume The rocket maintains a constant mass.

Draw **Figure 7.32** shows the variables we will use to describe the system. Note that the device is oriented vertically, and therefore we will have to account for the effect of gravity on it as it rotates. Free-body and inertial-response diagrams are shown in **Figure 7.33**.

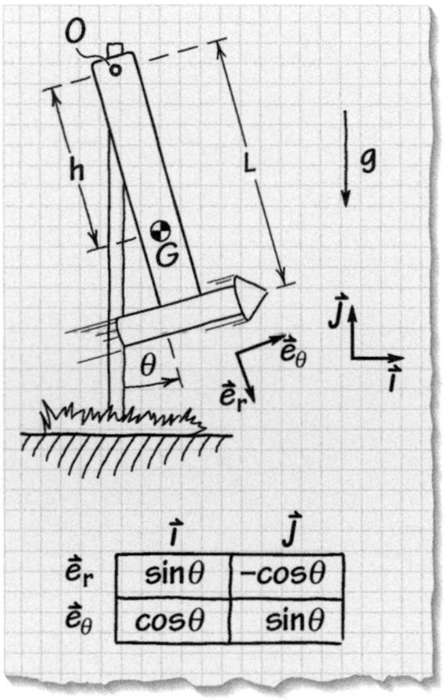

Figure 7.32 Pinwheel labeling

Formulate Equations We will use (7.28) for our moment balance:

$$I_O\ddot{\theta}\boldsymbol{k} = \boldsymbol{r}_{A/O} \times T\boldsymbol{e}_\theta + \boldsymbol{r}_{G/O} \times (-mg\boldsymbol{j})$$

$\boldsymbol{k}$:
$$I_O\ddot{\theta} = TL - mgh\sin\theta \tag{7.35}$$

and apply $\boldsymbol{F} = m\boldsymbol{a}_G$ along the $\boldsymbol{e}_r$ and $\boldsymbol{e}_\theta$ directions:

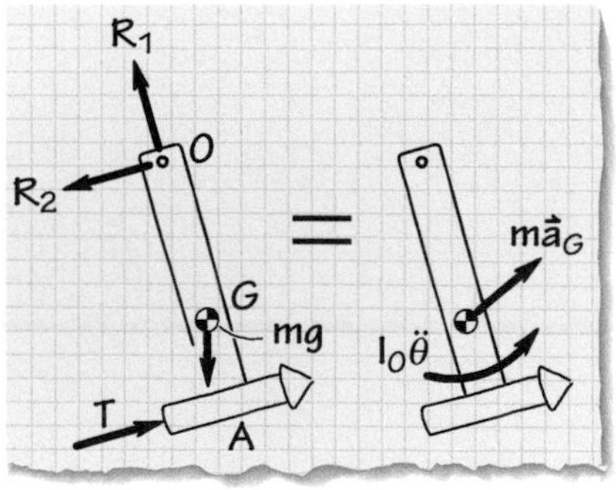

Figure 7.33 Pinwheel FBD=IRD

$\boldsymbol{e}_r$:
$$-mr_{G/O}\dot{\theta}^2 = -R_1 + mg\cos\theta \tag{7.36}$$

$\boldsymbol{e}_\theta$:
$$mr_{G/O}\ddot{\theta} = -R_2 - mg\sin\theta + T \tag{7.37}$$

In order to evaluate (7.35), we will need to find I_O. Luckily, that's not too difficult now that we've got (7.31) to help. Applying it gives us

$$I_O = I_G + mr_{G/O}^2 = 0.24\ \text{kg·m}^2 + (1.6\ \text{kg})(0.9\ \text{m})^2 = 1.536\ \text{kg·m}^2$$

Solve Our equations are in a particularly convenient form, thanks to our having expressed the force balances with respect to $\boldsymbol{e}_r$ and $\boldsymbol{e}_\theta$ rather than $\boldsymbol{i}, \boldsymbol{j}$. Equations (7.36) and (7.37) can be rearranged to give us the desired pivot forces as functions of our system parameters, $\theta, \dot{\theta}$, and $\ddot{\theta}$:

(7.36) ⇒
$$R_1 = mr_{G/O}\dot{\theta}^2 + mg\cos\theta \tag{7.38}$$

(7.37) ⇒
$$R_2 = -mr_{G/O}\ddot{\theta} - mg\sin\theta + T \tag{7.39}$$

and (7.35) gives us our equation of motion:

$$(7.35) \Rightarrow \qquad I_O\ddot{\theta} + mgh\sin\theta = TL \qquad (7.40)$$

We can't evaluate the forces at $t = 1$ s without knowing the values of θ, $\dot{\theta}$, and $\ddot{\theta}$ at that time, so that means we have to integrate our equation of motion. Let $y1=\theta$, $y2=\dot{\theta}$ and integrate the equations

```
ẏ1=y2
ẏ2=(50*t∧2*exp(-3*t)*L-m*g*h*sin(y1))/I0
```

using `tspan=[0 1]`, `y0=[0 0]`, `L=1.2`, `h=0.9`, `g=9.81`, `m=1.6`, and `I0=1.536`. The appropriate M-file is found on this book's website as `pinwheel.m`.

The integration gives $t = 1$ s values of $\theta = 0.394$ rad and $\dot{\theta} = 0.526$ rad/s. Evaluating (7.40) at $t = 1$ s produces

$$\ddot{\theta} = \frac{(2.49\,\text{N})(1.2\,\text{m}) - (1.6\,\text{kg})(9.81\,\text{m/s}^2)(0.9\,\text{m})(\sin 0.394)}{1.536\,\text{kg·m}^2} = -1.59\,\text{rad/s}^2$$

Using these values in (7.38) and (7.39) gives us the desired forces at the pivot:

$$R_1 = 14.9\,\text{N}$$

$$R_2 = -1.25\,\text{N}$$

Plotting the time behavior of θ and $\dot{\theta}$ (**Figure 7.34**) shows us that the fizzling rocket doesn't even have enough energy to get the board spinning around. The angle θ increased to a maximum value of about 0.5 rad at $t = 1.2$ s and then began to decrease. A most unimpressive fireworks display.

Check Let's use MATLAB to help us further convince ourselves that our equations are correct. The rocket's thrust builds from zero to a maximum value and then goes back to zero, as shown in **Figure 7.35**.

This plot shows that after 4 s, the rocket is essentially dead. So what should we expect? The rocket kicked the board up to about 0.5 rad, the board starts rotating back, and the rocket quickly sputters out. That leaves us with an unpowered pendulum, hinged at O. All it can do is rock back and forth at some steady amplitude. To see if this is right, all we need to do is integrate a bit further in time. Changing `tspan` to `[0 8]` gives us longer-term responses, which are plotted in **Figure 7.36**. As you can see, the system behaves as expected. Thus we can have some confidence in the mathematics.

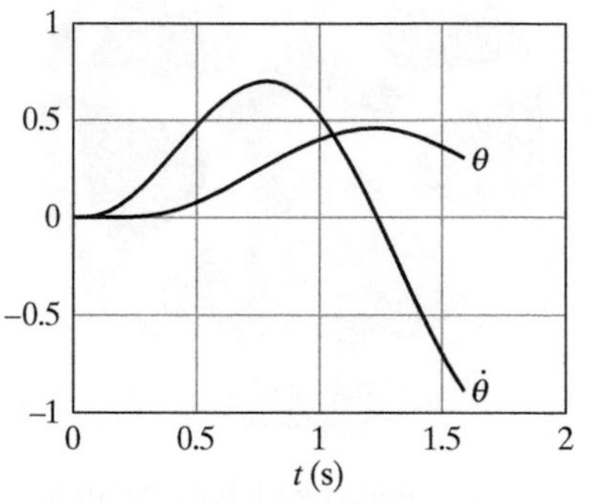

Figure 7.34 θ and $\dot{\theta}$ versus t

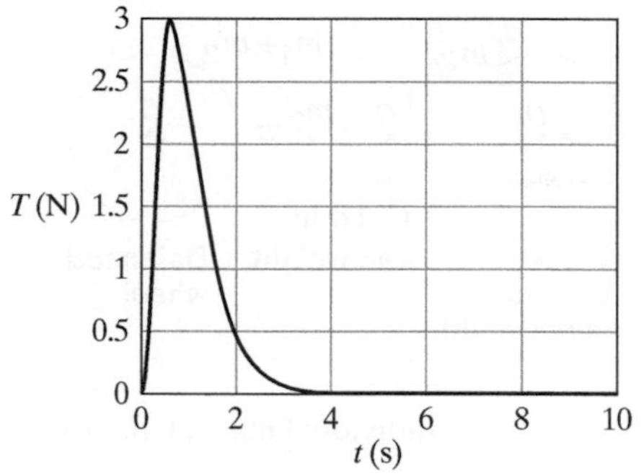

Figure 7.35 Rocket's thrust characteristics

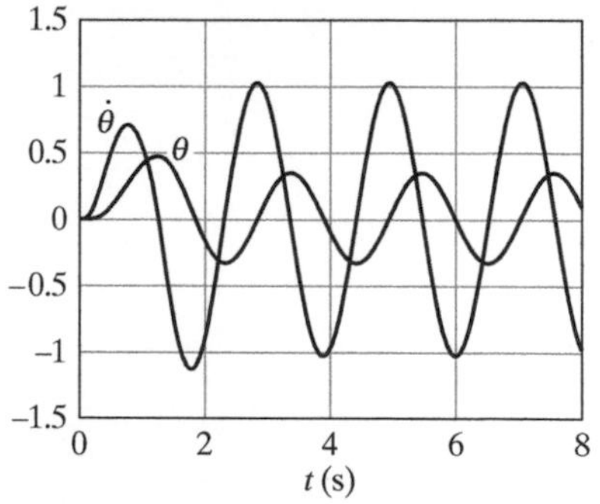

Figure 7.36 Rocket's long-term behavior

EXAMPLE 7.10 DETERMINING A WHEEL'S IMBALANCE ECCENTRICITY (Theory on page 397)

Benson Tongue

Figure 7.37 Wheel with lead balance weight

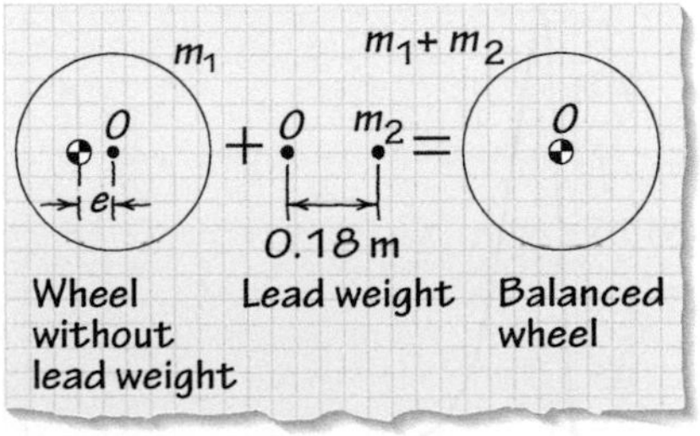

Figure 7.38 Individual mass elements

Oscillatory forces due to rotating imbalances are ubiquitous sources of vibrations. A commonly seen example is the imbalanced car wheel, leading to a noticeable vibration at certain speeds. **Imbalance eccentricity** is the term used to characterize the degree of imbalance and is defined as the distance between the body's center of rotation and its center of mass. It's the center of mass rotating about the center of rotation that leads to the vibrations. **Figure 7.37** shows a 43 cm wheel that's had a 0.034 kg lead weight added to the inner rim 0.18 m from the wheel's center in order to position the center of mass exactly at the wheel's geometric center. Before balancing, the wheel weighed 8 kg (which, by the way, is quite a light weight for an alloy wheel this size). What was the imbalance eccentricity before balancing?

Goal Find the imbalance eccentricity of a wheel knowing the balancing weight, its placement, and the wheel's weight before balancing.

Draw **Figure 7.38** shows how the individual pieces make up the total balanced wheel. At the left is the imbalanced wheel (with mass m_1), and its mass center is shown as being a distance e to the left of its geometric center (labeled O). e is the imbalance eccentricity. We add to this the lead weight (mass m_2) positioned 0.18 m from O. This gives us the balanced wheel shown above, for which the total mass is $m_1 + m_2$ and for which its mass center coincides with its geometric center.

Formulate Equations We will need to use (7.20), the definition of the mass center.

Solve Luckily for us, we don't have to actually compute any integrals. We know that after balancing, the mass center is at the center of rotation. Thus $r_{G_{/O}} = 0$. Equation (7.20) therefore becomes

$$0 = \frac{\int_{\text{Body}} \boldsymbol{r}_{dm_{/O}}\, dm + (0.18\,\text{m})(0.034\,\text{kg})\boldsymbol{i}}{m_1 + m_2} = \frac{-(8\,\text{kg})e\boldsymbol{i} + (0.18\,\text{m})(0.034\,\text{kg})\boldsymbol{i}}{8.034\,\text{kg}}$$

$$8e = (0.18\,\text{m})(0.034) \quad \Rightarrow \quad e = 7.65 \times 10^{-4}\,\text{m} = 0.0765\,\text{cm}$$

Check The force generated by the imbalance will be due to the centrifugal acceleration, equal to $e\dot{\theta}^2$. Let's consider a car traveling at 97 km/h, with a 0.3 m radius tire. This implies a rotation rate of 88 rad/s. For our problem the imbalance force is therefore $\frac{8\,\text{kg}}{9.82\,\text{m/s}^2}(7.65 \times 10^{-4}\,\text{m})(88\,\text{s}^{-1})^2 = 4.83\,\text{kg}$. The result is a 4.83 kg force that's pushing up and down at the car body in a continuously alternating fashion. This is certainly enough to be noticeable.

Note that in this analysis we didn't include the tire. In reality, both tires and wheels can contribute to imbalances, and for that reason a wheel is never balanced alone; the wheel and tire are balanced together.

EXERCISES 7.2

MASS MOMENTS OF INERTIA EXERCISES

7.2.1. **[Level 1]** As shown, a slender rod of mass m and length $3L$ is free to rotate in the horizontal plane about the pivot O, which is located a distance L from the rod's right end B. A lumped body with mass $2m$ is attached at B and at the rod's left end A. What is the system's in-plane rotational inertia I_O?

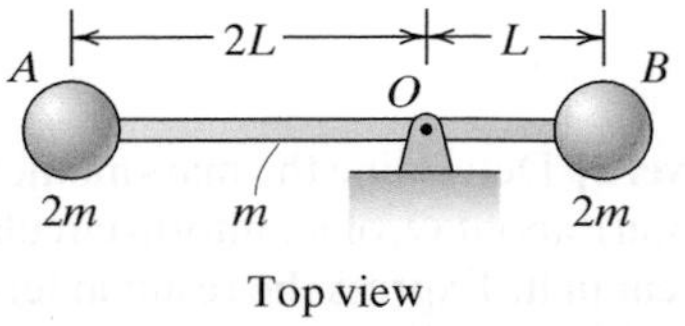

E7.2.1

7.2.2. **[Level 1]** What is the mass moment of inertia of the illustrated body about the x axis? Assume an areal density of ρ. Treat h as negligible (much smaller than a, b, or c). Express your answer in terms of ρ, not the total mass.

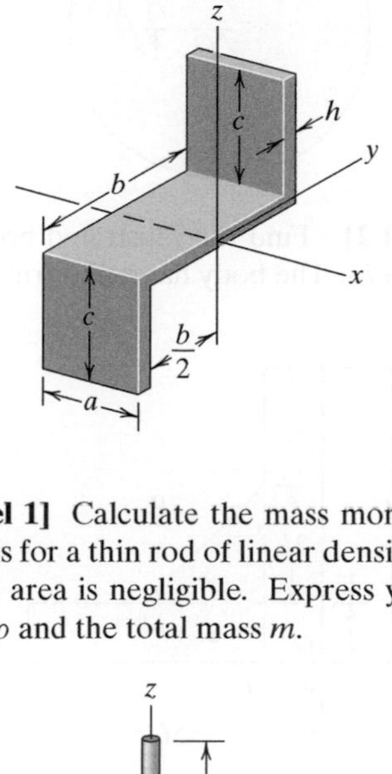

E7.2.2

7.2.3. **[Level 1]** Calculate the mass moment of inertia about the x axis for a thin rod of linear density ρ kg/m. The cross-sectional area is negligible. Express your answer in terms of both ρ and the total mass m.

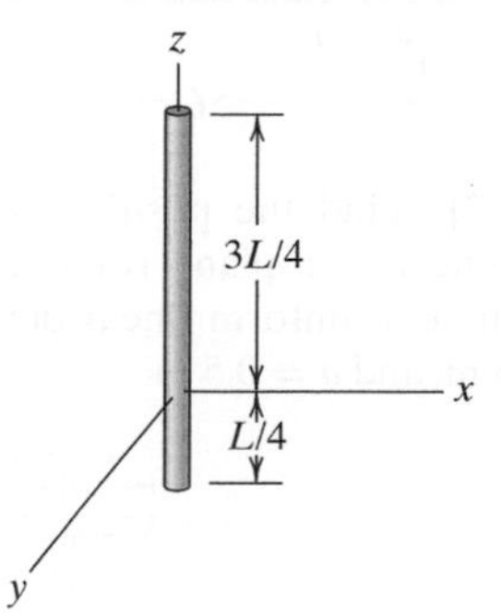

E7.2.3

7.2.4. **[Level 1]** The illustrated body is free to rotate about the pivot O in the vertical plane. If the body has a uniform linear density ρ, what is its in-plane moment of inertia I_O?

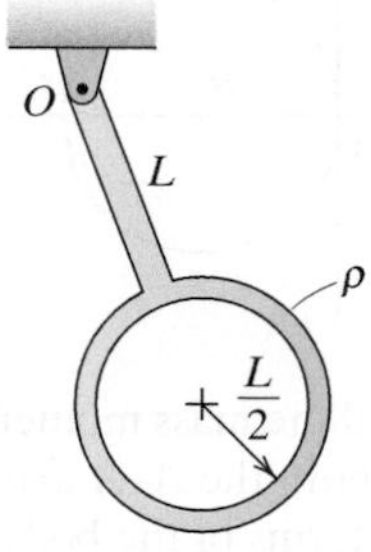

E7.2.4

7.2.5. **[Level 1]** A thin plate (nonuniform) has a total mass of 20 kg. Dimensions are given in meters. The mass center is marked, as are points A and B. I_A is the mass moment of inertia (in-plane rotation) at A and $I_A = 140\ \text{kg}\cdot\text{m}^2$. What is I_B?

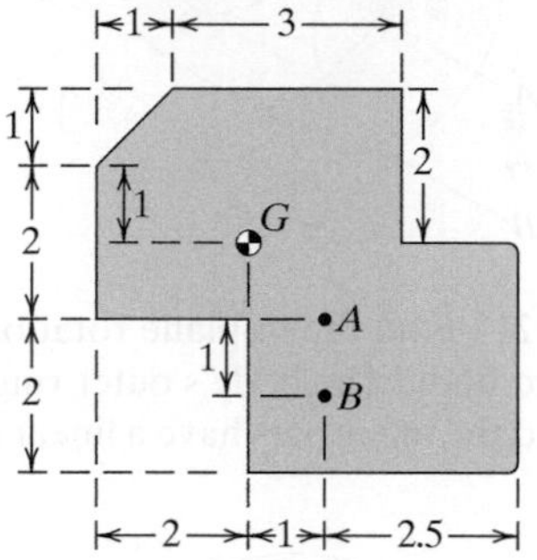

E7.2.5

7.2.6. **[Level 1]** As depicted, a thin ring of uniform linear density ρ is resting in the horizontal plane, where G denotes the body's mass center. Derive an expression for the ring's in-plane rotational inertia I_G in terms of its mass m and radius r.

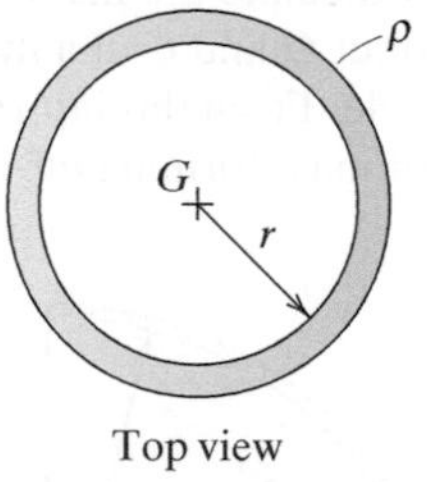

E7.2.6

7.2.7. **[Level 2]** Determine the moment of inertia, about A, of a perforated circular disk. The areal density of the disk is ρ, the outer radius is R, and the radius of hole is $\frac{R}{2}$. Express your result in terms of ρ and R.

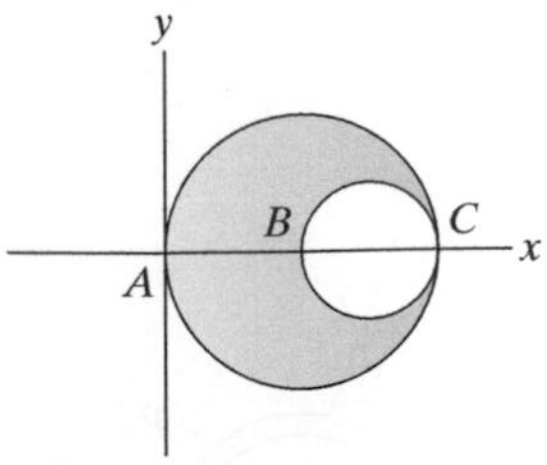

E7.2.7

7.2.8. **[Level 2]** Find the mass moments of inertia of the illustrated tube about both the A–A axis and the B–B axis. Express your result in terms of the body's density ρ and in terms of its mass m.

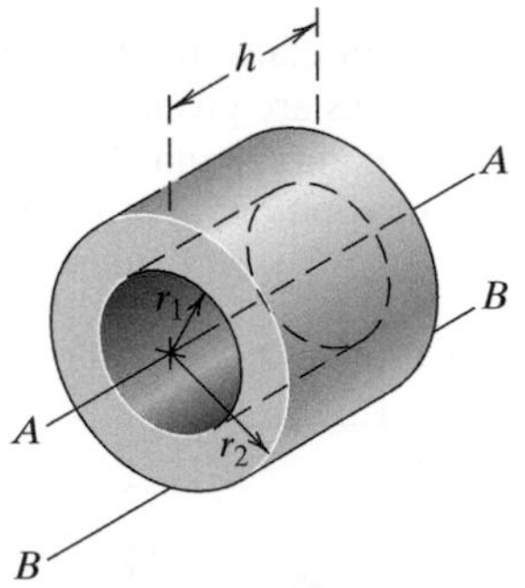

E7.2.8

7.2.9. **[Level 2]** Find the in-plane rotational inertia I_O for the illustrated body. The body's outer ring has an areal density of ρ_A, and the inner bars have a linear density of ρ_L.

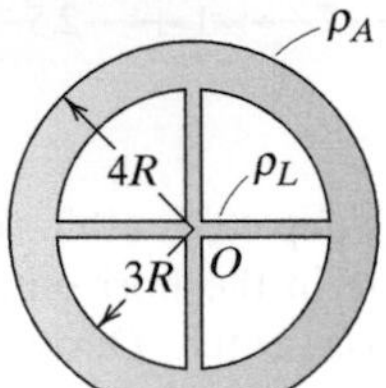

E7.2.9

7.2.10. **[Level 2]** Calculate the mass moment of inertia (in-plane rotation) about O and C of a thin rod that's been bent into a quarter-circle. From this information (and without using the center of mass formula $m\bar{r} = \int r\,dm$), deduce what $\bar{r}$ must be.

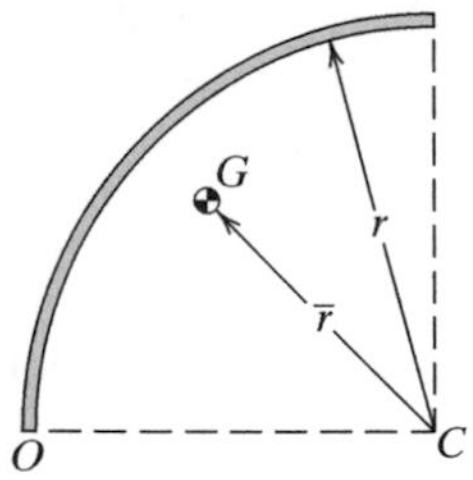

E7.2.10

7.2.11. **[Level 2]** The illustrated solid of revolution is formed by rotating the curve $z = ay^{\frac{1}{3}}$ about the z axis and then filling in the resulting shell. The body has density ρ and mass m, and it extends for $0 \leq z \leq h$, where the body's radius is r at $z = h$. Derive the body's mass moment of inertia I_{zz} about the z axis, and express it in terms of m and r.

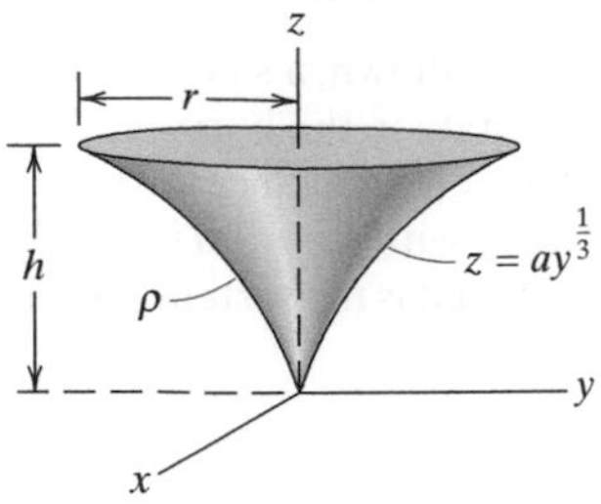

E7.2.11

7.2.12. **[Level 2]** Determine the mass moment of inertia (in-plane rotation), about O, of a uniform circular disk with a circular hole cut in it. Express the result in terms of ρ, the areal density, and r_2. $r_1 = \frac{r_2}{4}$.

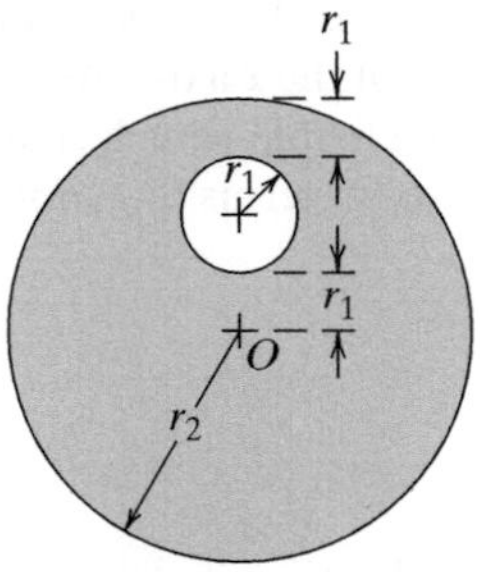

E7.2.12

7.2.13. **[Level 2]** Find the illustrated body's in-plane rotational inertia I_O. The body has a uniform areal density of ρ.

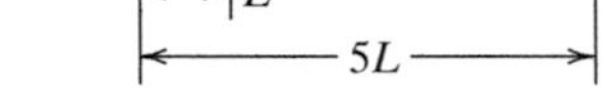

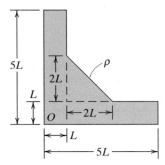

E7.2.13

7.2.14. **[Level 2]** Find the position of the center of mass, and determine I_O (in-plane rotation) for the illustrated body. Assume a uniform linear density of 5 kg/m. $c = 1.8$ m, $b = 0.3$ m, and $a = 0.5$ m.

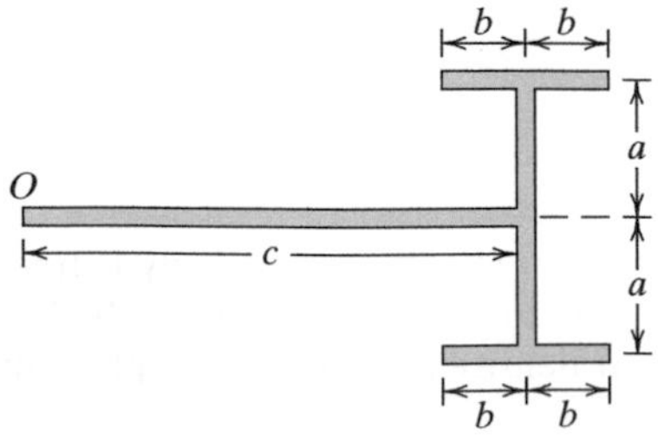

E7.2.14

7.2.15. **[Level 2]** The depicted solid of revolution is generated by rotating the curve $z = r_0 e^{-ay}$ about the y-axis and then filling in the resulting shell. The body has density ρ and mass m, and it extends for $0 \le y \le h$. The body has an initial radius r_0 at $y = 0$, and its radius at $y = h$ is r. Derive the body's mass moment of inertia I_{yy} about the y-axis, and express it in terms of m, r_0, and r.

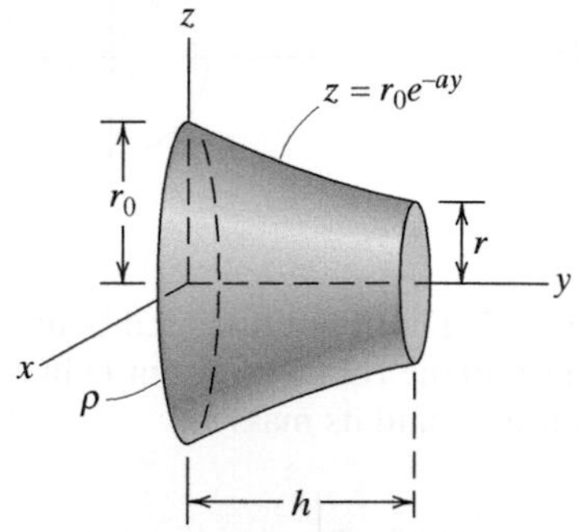

E7.2.15

7.2.16. **[Level 2]** The illustrated body is made up of two pieces, one a 45°/45°/90° triangle and the other a semicircle. $\rho_1 = 10\,\text{kg/m}^2$ and $\rho_2 = 4\,\text{kg/m}^2$. $L = 0.5\,\text{m}$. Find I_O for in-plane rotation.

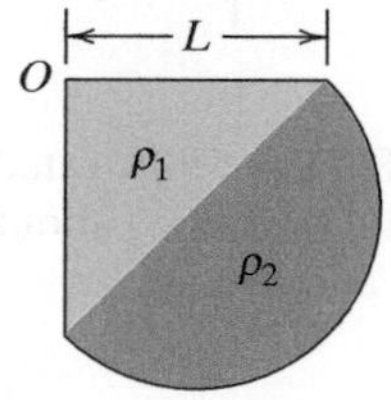

E7.2.16

7.2.17. **[Level 2]** Two plates, each made of a different material, have been bonded together. The areal density of plate A is $1\,\text{kg/m}^2$. The center of mass has been experimentally determined to be located at

$$\bar{\boldsymbol{r}}_{G_{/O}} = (4.739\boldsymbol{i} + 3.544\boldsymbol{j})\text{ cm}$$

What is the density of plate B?

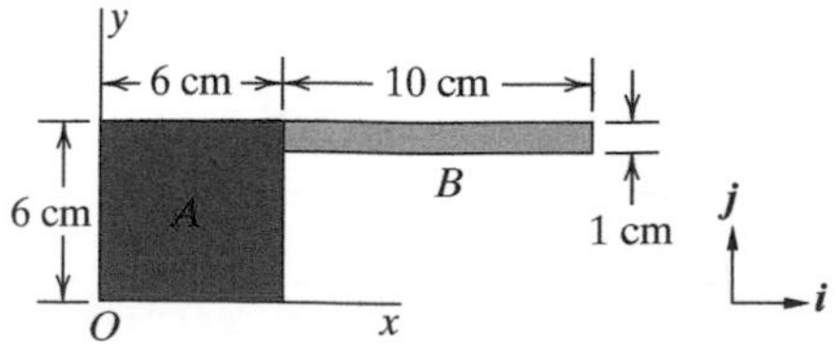

E7.2.17

7.2.18. **[Level 2]** Find the mass moments of inertia (in-plane rotation) of the illustrated body (three thin rigid bars welded together) about the points A and G. Express your result in terms of the body's linear density ρ and in terms of its mass m. The thickness of each bar is negligible compared to its length.

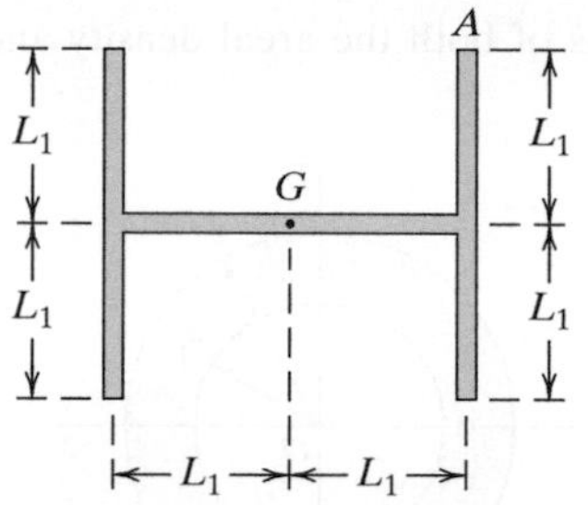

E7.2.18

7.2.19. **[Level 2]** Determine the in-plane rotational inertia I_G for the illustrated system. $r = 0.2\,\text{m}$. The two wheels (pinned to the bar at A and B) can rotate about their pivots with no friction. Each wheel has a mass of 0.8 kg, and the bar has a mass of 1.8 kg.

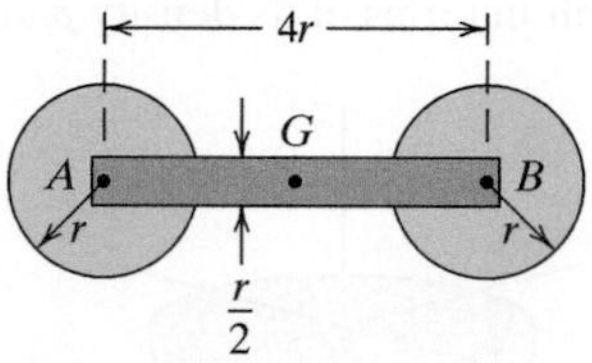

E7.2.19

7.2.20. **[Level 2]** Determine the moment of inertia, about A, of a half-disk. The distance from B to C is R and the mass of the disk is m. Express your answer in terms of m and R.

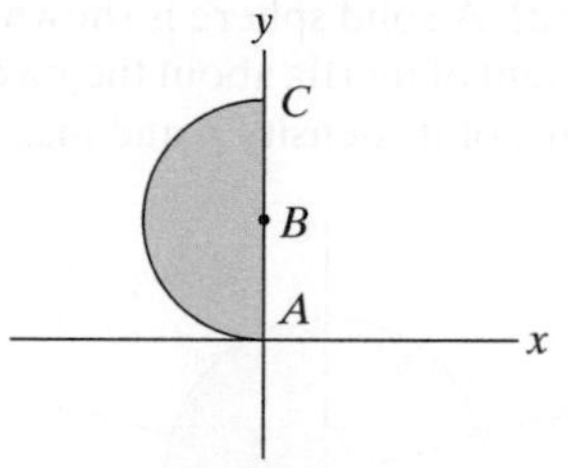

E7.2.20

7.2.21. **[Level 2]** The curved surface of the illustrated solid of revolution is found by rotating the curve $z = ax^2$ about the z axis and then filling the resulting shell. Determine the mass moment of inertia about the z axis if the solid extends for $0 \le z \le z_0$. Express your answer in terms of the solid's density ρ, as well as in terms of its mass m.

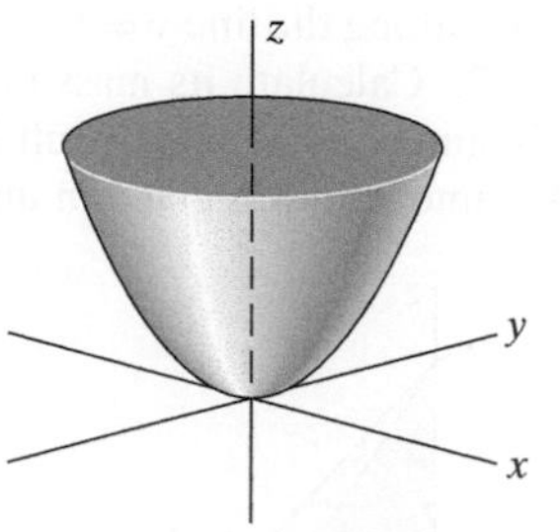

E7.2.21

7.2.22. **[Level 2]** A flat washer is illustrated with areal density ρ. Determine its mass moment of inertia about the out-of-plane z axis and also about the y axis. Express your answer in terms of both the areal density and the body's mass.

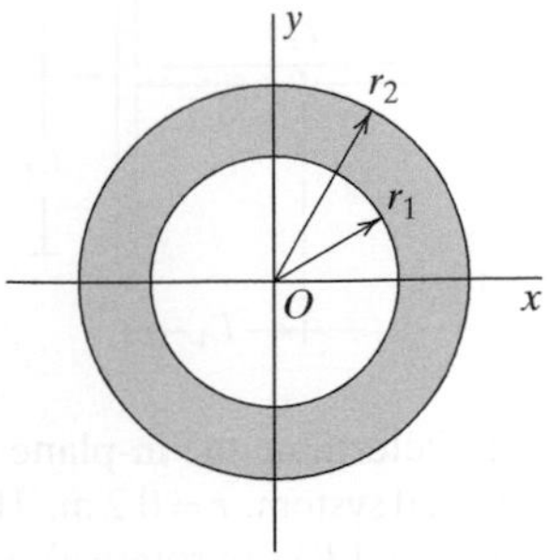

E7.2.22

7.2.23. **[Level 2]** A solid half-sphere is shown, with radius r. Find its mass moment of inertia about the z axis, and express your result in terms of its density ρ and its mass m.

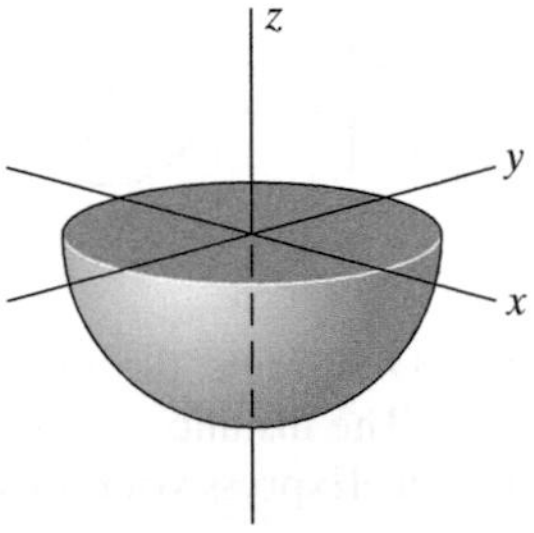

E7.2.23

7.2.24. **[Level 2]** A solid sphere is shown, with radius r. Find its mass moment of inertia about the x axis, and express your result in terms of its density ρ and mass m.

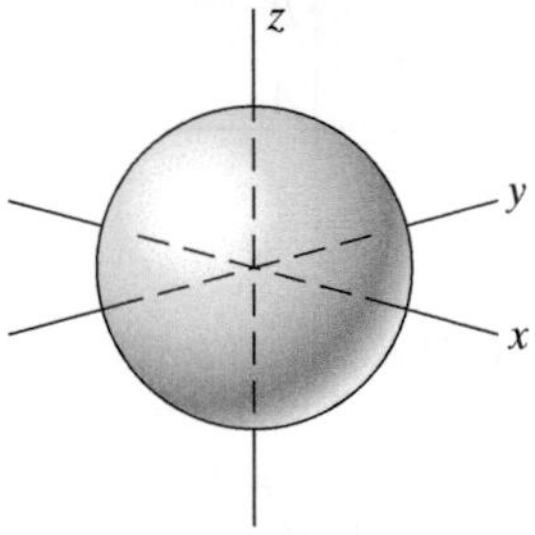

E7.2.24

7.2.25. **[Level 2]** A triangular pyramid is shown. One of the base edges lies along the x axis, one lies along the y axis, and the third lies along the line $y = 1 - x$. The pyramid has a height of $z = 2$. Calculate its mass moment of inertia about the z axis, and express your result in terms of the body's mass m. Assume length is given in units of meter.

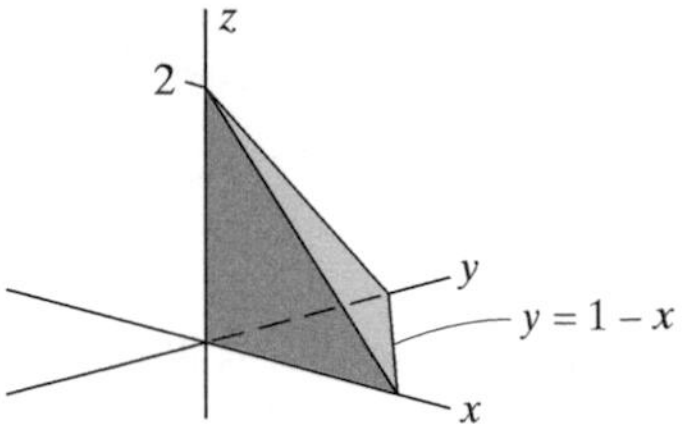

E7.2.25

7.2.26. **[Level 2]** Determine the moment of inertia about the z axis of the illustrated composite solid cone/half-sphere. Both have density ρ. Express you answer in terms of ρ.

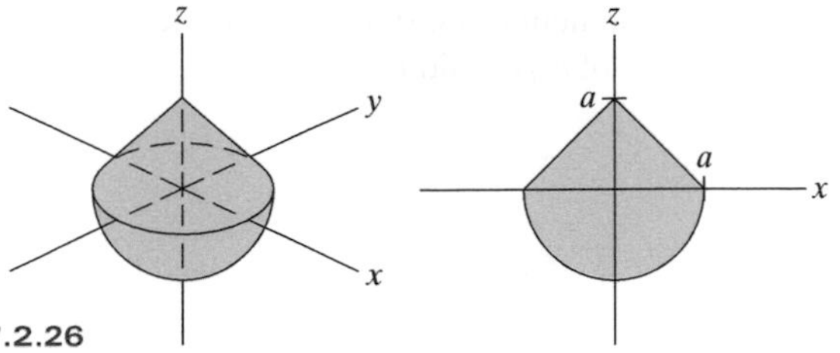

E7.2.26

7.2.27. **[Level 2]** Find the illustrated body's mass moment of inertia (in-plane rotation) about O in terms of the body's areal density ρ and its mass m.

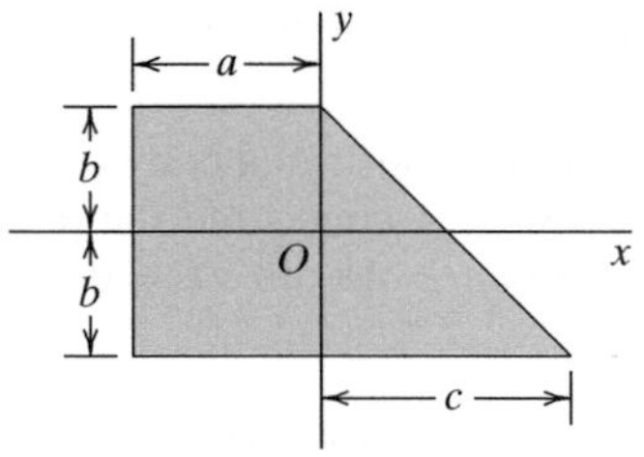

E7.2.27

7.2.28. **[Level 2]** Find the illustrated body's mass moment of inertia (in-plane-rotation) about O in terms of the body's areal density ρ and its mass m.

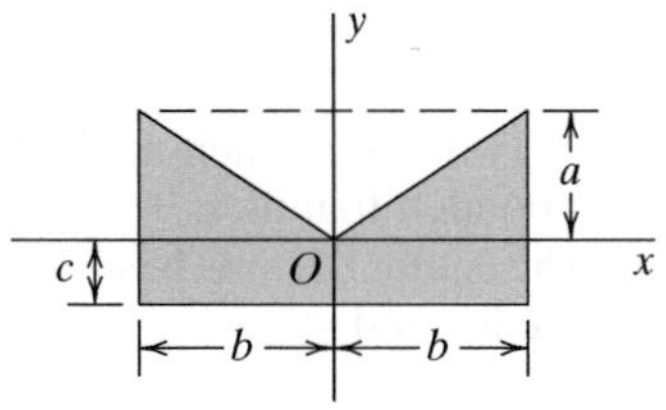

E7.2.28

7.2.29. **[Level 2]** Find the illustrated body's mass moment of inertia (in-plane rotation) about O in terms of the body's areal density ρ and its mass m.

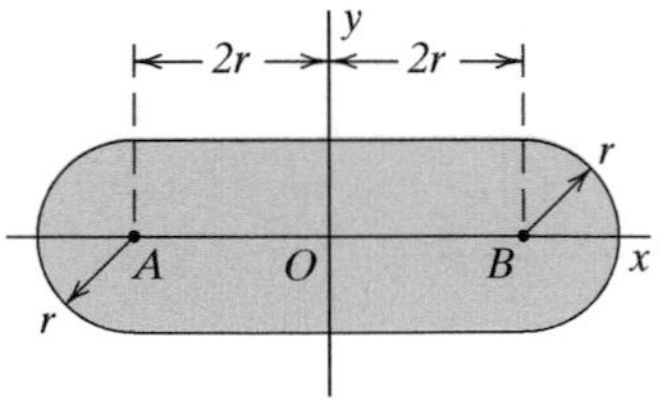

E7.2.29

7.2.30. **[Level 2]** The body illustrated is a circular disk with a hemispherical cavity in the middle of its top surface. Find the body's mass moment of inertia about the x axis in terms of the body's density ρ and its mass m.

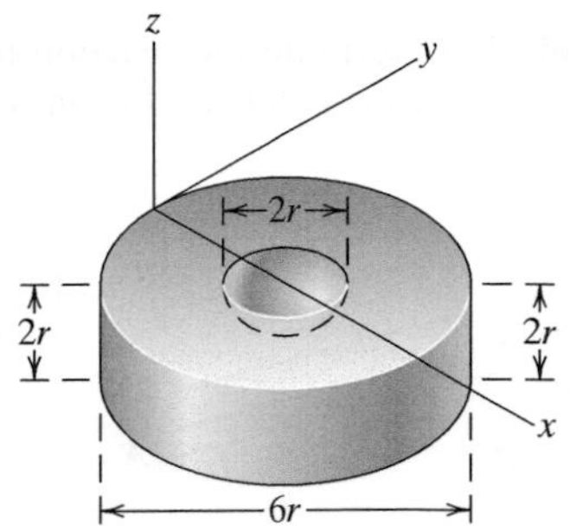

E7.2.30

7.2.31. [Level 3] Find the location of the illustrated body's mass center and its in-plane moment of inertia I_O. Assume a uniform linear density of $\rho = 2\,\text{kg/m}$. Let $a = 0.2\,\text{m}$ and $m = 2\,\text{kg}$ for each lumped mass.

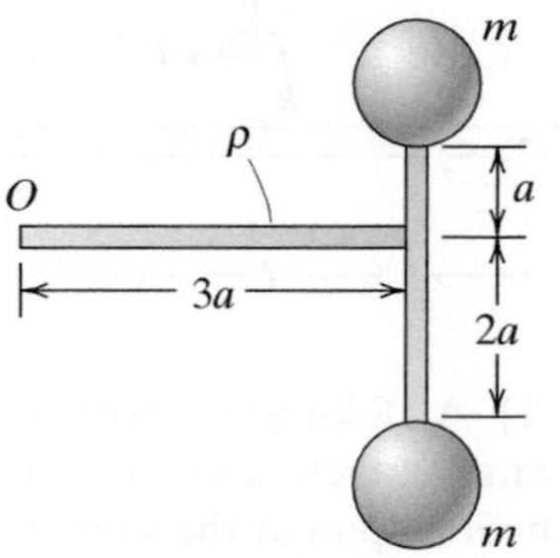

E7.2.31

7.2.32. [Level 3] A half-spherical shell is shown, with radius r. Find its mass moment of inertia about the x' axis and express your result in terms of its areal density ρ and mass m. The x' axis goes through the point $x = y = 0$ and $z = -r$.

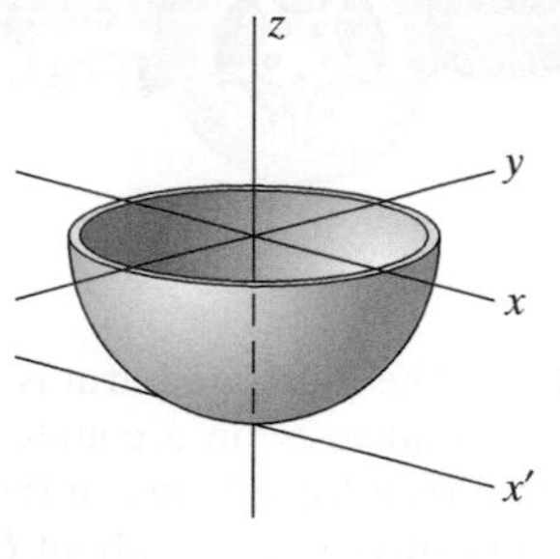

E7.2.32

7.2.33. [Level 3] A thin shell is formed by rotating the curve $z = ax^2$ about the z axis. Determine the mass moment of inertia of this shell about the z axis if the shell extends for $0 \le z \le z_0$. Express your answer in terms of the shell's areal density ρ, as well as in terms of its mass m.

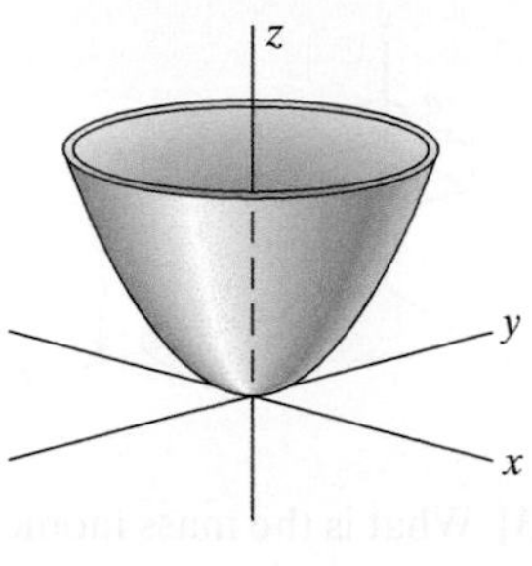

E7.2.33

7.2.34. [Level 3] Find the position of the center of mass and determine I_A (in-plane rotation) for the illustrated body. The outer edge of the equilateral triangle has a linear mass density of 4 kg/m, and the interior portion has an areal density of 25 kg/m^2. $L = 0.8$ m.

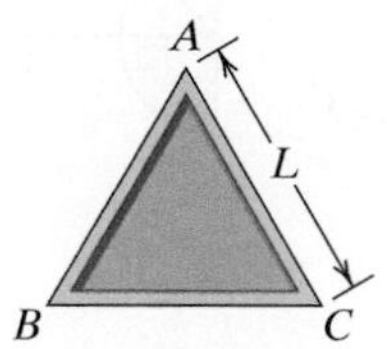

E7.2.34

7.2.35. [Level 3] Find the in-plane rotational inertia I_A. Assume a constant areal density of $\rho = 12\,\text{kg/m}^2$. $L = 1.6$ m.

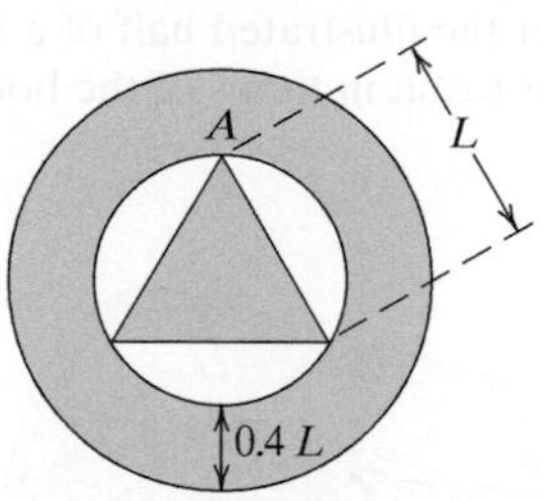

E7.2.35

7.2.36. [Level 3] The body illustrated is a simplified model of a bicycle's chain ring. Determine its mass moment of inertia (in-plane rotation) about O. $\rho = 8.6 \times 10^{-4}\,\text{kg/cm}^2$, $a = 2.5$ cm.

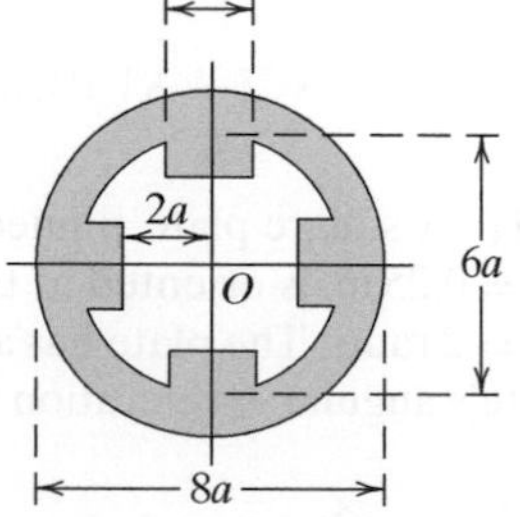

E7.2.36

7.2.37. [Level 3] The body illustrated is a rectangular piece of sheet steel that's been bent into a channel shape. Determine the body's mass moment of inertia about both the y and z axes. Express your answer in terms of the body's areal density ρ and its mass m.

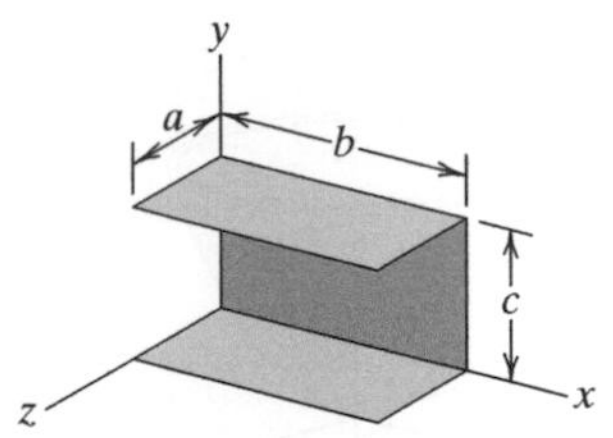

E7.2.37

7.2.38. **[Level 3]** What is the mass moment of inertia of the illustrated half-sphere about the y axis and about the y' axis? Express your answer in terms of both the density ρ and the total mass m.

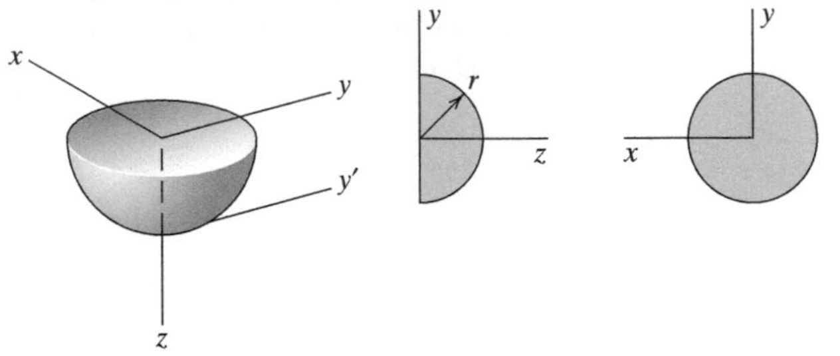

E7.2.38

7.2.39. **[Level 3]** Derive the mass moment of inertia about the x axis for the illustrated half of a thin cylindrical shell. Express your result in terms of the body's mass m.

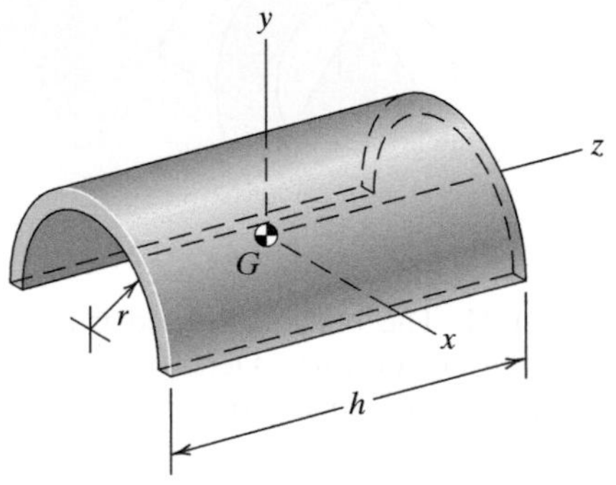

E7.2.39

FORCE BALANCE EXERCISES

7.2.40. **[Level 1]** A square plate, pinned at A and with edge dimension $b = 0.25\,\text{m}$, is oriented at this instant such that $\theta = 45°$ and $\dot{\theta} = 2\,\text{rad/s}$. The plate has a mass of 2.5 kg. Determine the plate's angular acceleration at this instant.

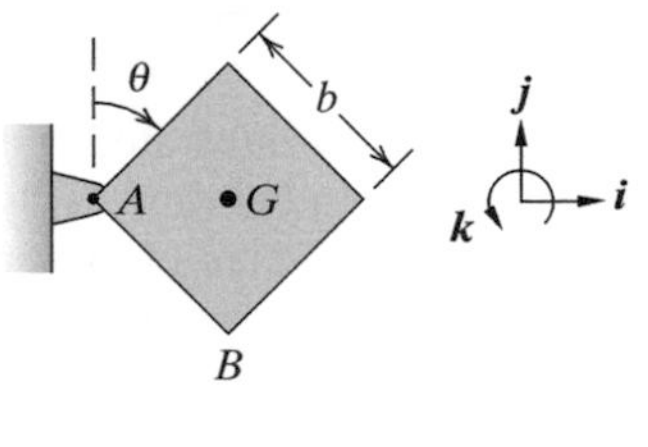

E7.2.40

7.2.41. **[Level 1]** Determine the reaction forces at the hinge O if the bar (mass m) is released from rest at $\theta = 55$ degrees.

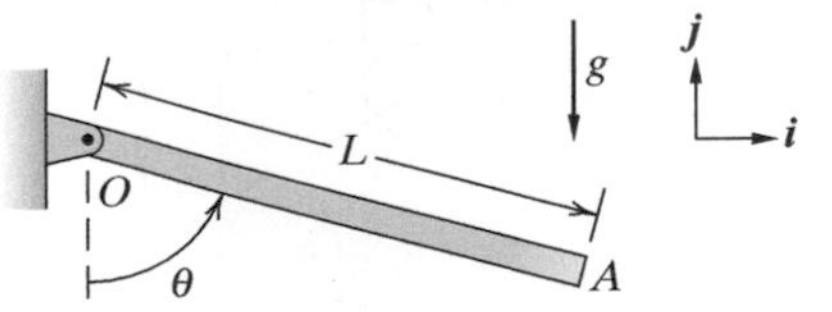

E7.2.41

7.2.42. **[Level 1]** The rigid rod of length L is free to rotate in a horizontal plane. A moment $M\boldsymbol{k}$ is applied to the vertical shaft that connects to the rod at O. What are the reaction forces acting at O in terms of $\dot{\theta}$, the rod's angular speed, M, m, the mass of the rod, and L?

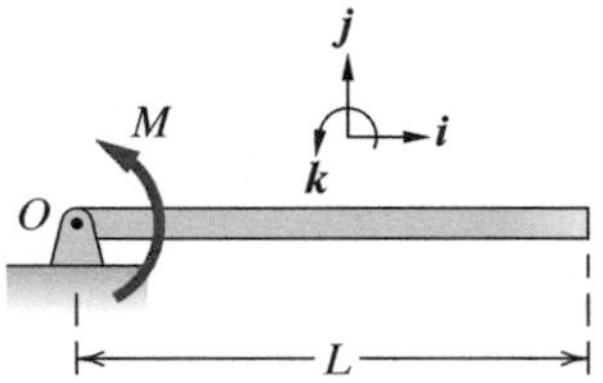

E7.2.42

7.2.43. **[Level 1]** A 18 kg wheel with an imbalance eccentricity of 0.13 cm is attached to a car being tested on a simulator. The wheel is spun at the same rate as if the car were traveling at 97 km/h. Determine the imbalance force acting on the vehicle body. The distance from the axle center to ground during normal operating conditions is 33 cm.

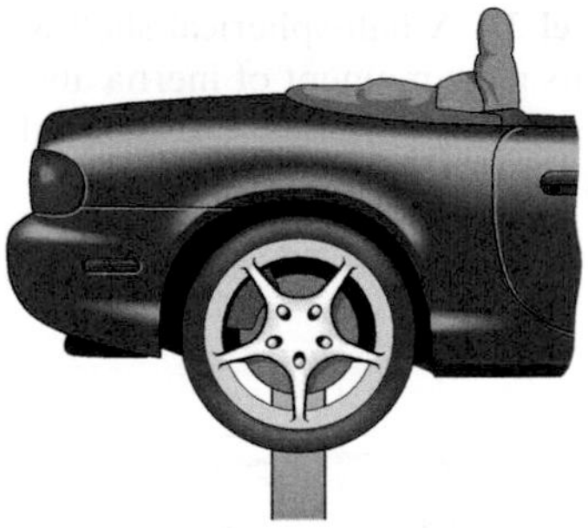

E7.2.43

7.2.44. **[Level 2]** The illustrated arm is freely pivoted about O, and its far end rests on a counterclockwise rotating disk. The frictional force between the disk and the arm's tip induces a positive moment about O. A torsional spring at O acts to counteract this moment. The moment applied by the spring to the arm is given by $M_{sp} = -k_\theta \theta$, where $k_\theta = 1.92\,\text{N}\cdot\text{m/rad}$ and θ is measured in radians. The mass moment of inertia of the arm about O is given by $I_O = 0.048\,\text{kg}\cdot\text{m}^2$, and the coefficient of friction between the end of the arm and the disk is 0.8. The normal force between the end of the arm and the disk is 3 N. The distance from O to the end of the arm is 0.2 m.

a. Determine the steady-state angle between the arm's position with an unstretched spring and the position it moves to under the influence of the frictional force.

b. If the torsional spring suddenly breaks (and the arm is at its steady-state position), what angular acceleration does the arm initially experience?

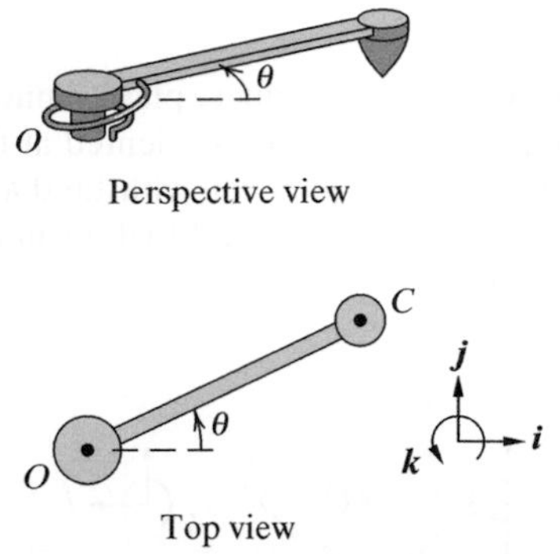

E7.2.44

7.2.45. **[Level 2]** When a pendulum is allowed to freely swing through small angles of its rotational coordinate θ with no damping, its equation of motion is of the form $\ddot{\theta} + \omega_n^2\theta = 0$, where ω_n is the pendulum's natural frequency. Consider the illustrated compound pendulum constrained to rotate about the fixed, frictionless pivot O in the vertical plane. Its rod has length L and mass m, and the lumped body on its free end has a mass of $M = 2m$.

a. Find the compound pendulum's natural frequency ω_n.

b. Suppose the system is approximated as a simple pendulum, for which the natural frequency is given by $\omega_n^* = \sqrt{\frac{g}{L}}$. What is the percent error e in ω_n associated with this approximation?

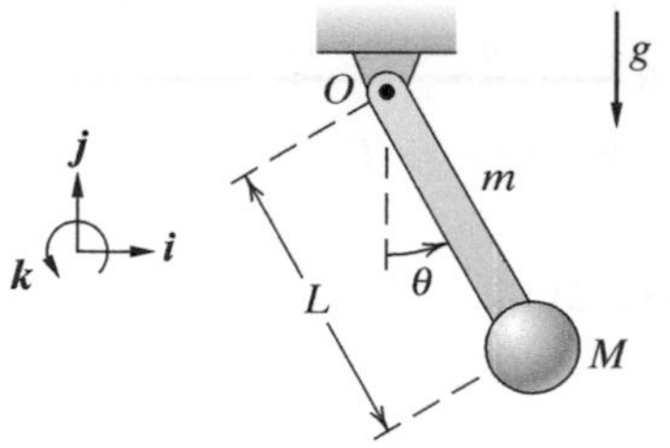

E7.2.45

7.2.46. **[Level 2]** **Computational** Two bars are hinged at their center and have mass moments of inertia about this center point of $\bar{I}_1 = 0.1\,\text{kg}\cdot\text{m}^2$ and $\bar{I}_2 = 0.2\,\text{kg}\cdot\text{m}^2$. They are also connected to each other via a linear torsional spring with rotational spring constant $k_\theta = 10.5\,\text{N}\cdot\text{m/rad}$. The bars are lying on a flat, friction-free surface, and the spring is unstretched in the position illustrated. Denote rotations from the equilibrium positions by θ_1 and θ_2, as shown. Bar 1 is held fixed and bar 2 is rotated 90° so as to align the two bars. Upon release the two bars will oscillate. Numerically integrate the two equations of motion to find θ_1 and θ_2 as functions of time. Plot θ_2–θ_1 vs. time for 1.0 s. What function does the resulting plot resemble?

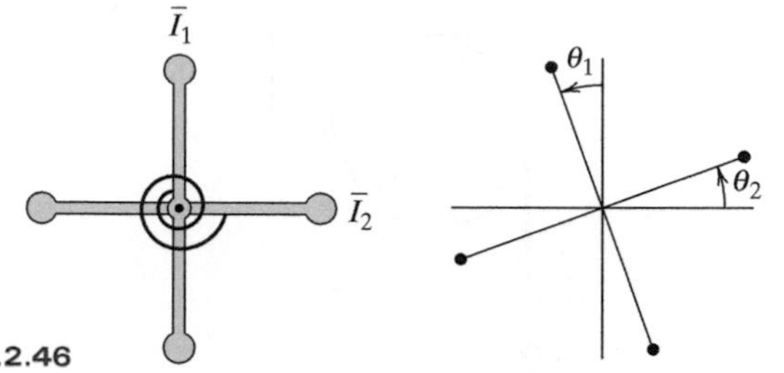

E7.2.46

7.2.47. **[Level 2]** Two situations are illustrated: a moment $M\boldsymbol{k}$ that's being directly applied to gear 2 (Case 1) and a simple transmission (Case 2) in which the moment M is applied to a small gear (gear 1), which then acts on gear 2. $\bar{I}_1 = 0$. Gear 2 has mass m and mass moment of inertia $\bar{I}_2$.

a. Calculate the acceleration of gear 2 for Case 1.

b. Calculate the acceleration of gear 2 for Case 2.

c. Discuss the results of **a** and **b**.

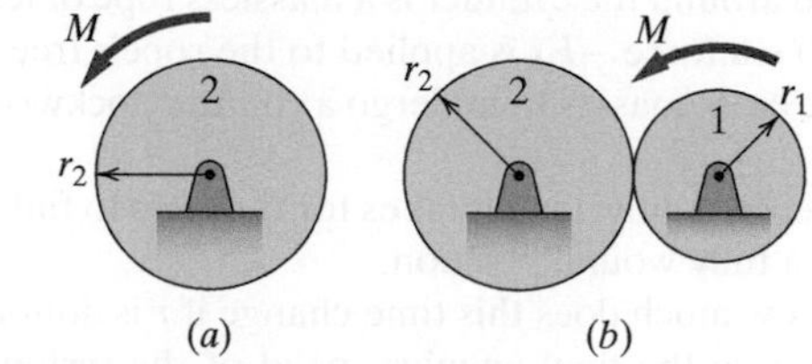

E7.2.47

7.2.48. **[Level 2]** As depicted, a plate of mass $m = 2\,\text{kg}$ in the shape of an isosceles triangle is supported at its base in the vertical plane by a frictionless pin joint at the left end A and by a string at the right end B. The plate's base is $a = 0.5\,\text{m}$ long, and its height is $b = 0.8\,\text{m}$. If the string attached at B is cut, find the plate's initial angular acceleration $\ddot{\theta}$ about A and the resultant reaction force F_A at the pivot A.

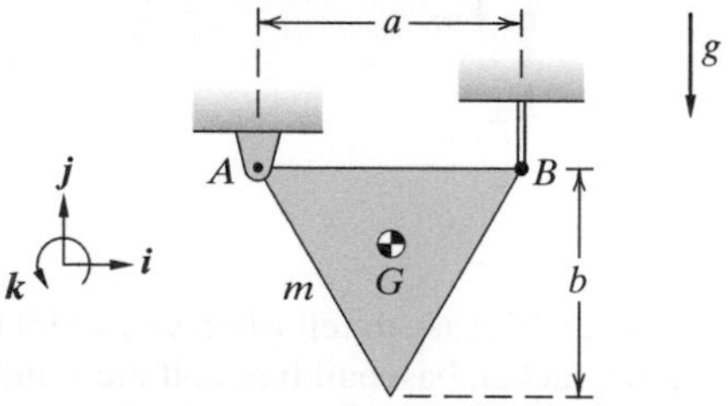

E7.2.48

7.2.49. **[Level 2]** In a new action movie, California Sam (our hero) grabs hold of the free end of a hanging cable. 50 m of cable is wrapped around a heavy spool S, and 2 m of cable was hanging free of the spool at the time Sam grabbed on. The spool has a mass of 80 kg and a radius of 1.1 m. Model the spool as a uniform cylindrical body. The cable has a linear density of 0.5 kg/m. The axle/spool interface has enough friction that 11 N·m of torque is needed to start the spool turning and remains at this level once the spool is rotating. Sam's mass is 75 kg. What is Sam's vertical acceleration after he grabs hold of the cable?

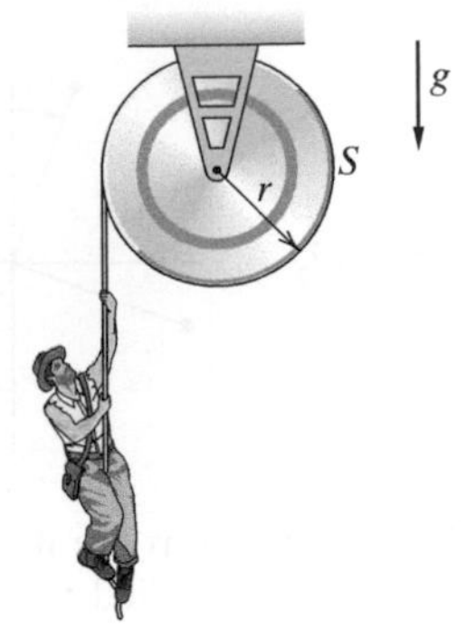

E7.2.49

7.2.50. **[Level 2]** Four identical masses are rigidly attached to a massless cross that is free to rotate about its center O. Also attached to the cross is a cylinder of radius r. Wrapped around the cylinder is a massless rope of length L. At $t = 0$ s, a force $-F\boldsymbol{j}$ is applied to the rope's free end, causing the cross/masses to undergo a counterclockwise rotation.

a. Determine how long it takes for the rope to fully unwind from a fully wound position.

b. By how much does this time change if r is doubled?

c. Compare the final angular speed of the system for both cases. Did changing r make much of a change in the final angular speed?

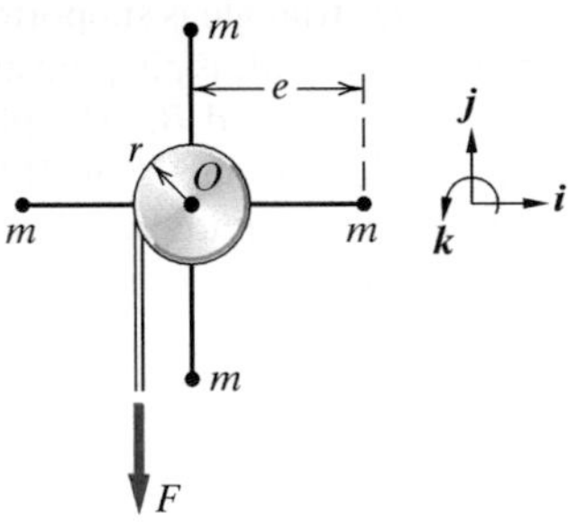

E7.2.50

7.2.51. **[Level 2]** It's easy to tell when you've hit the ball well with a tennis racket, baseball bat, golf club, and so on. The ball just seems to fly off the racket (bat, club) and you hardly feel the impact. A bad hit, on the other hand, sends a strong shock into your hand. When you do it right, it's said that you've hit the "sweet spot." Let's solve for this spot. Model the racket (bat, club) as a uniform bar, freely pivoted at O. (This is where you would be holding the bar.) A force $-F\boldsymbol{j}$ is applied a distance r from O. Examine the reaction forces at O and determine what r should be to minimize them. Ignore gravity and assume that the bar is rotating at ω rad/s when struck.

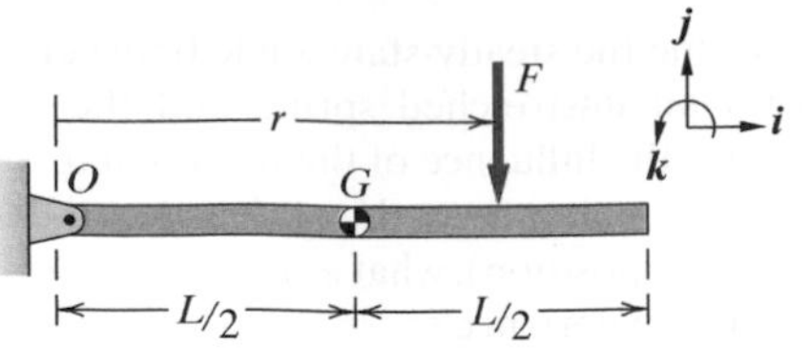

E7.2.51

7.2.52. **[Level 2]** A 2.5 kg square plate, pinned at A and with edge dimension $b = 0.25$ m, is oriented at this instant such that $\theta = 45°$. The pin joint can withstand a maximum force of 10.0 N. What is the greatest that $\dot{\theta}$ can be without causing a joint failure?

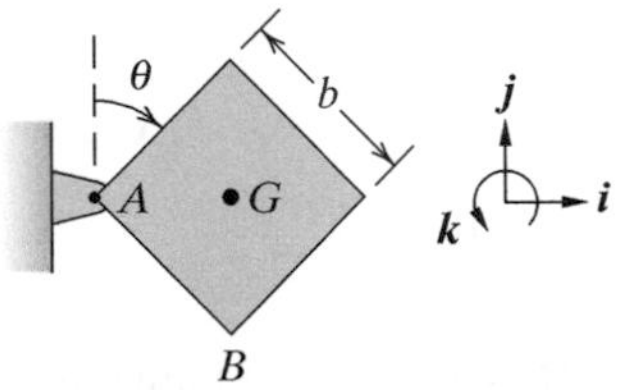

E7.2.52

7.2.53. **[Level 2]** Two mass particles can be positioned along the massless rod $\overline{BD}$. The rod is suspended by massless strings running from B to C and from D to E. At $t = 0$ the string $\overline{BC}$ is cut. Determine which mass configuration produces the smallest tension in the string $\overline{DE}$ immediately after the cut: both masses fixed in the center of the rod or one fixed at each end.

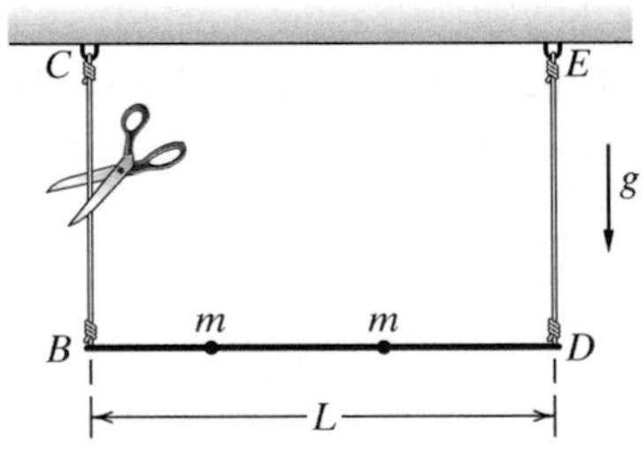

E7.2.53

7.2.54. **[Level 2]** Show that using $I_A = I_B + mr^2_{A/B}$, where neither A nor B corresponds to the mass center, doesn't work.

7.2.55. **[Level 2]** The bearing at O is not frictionless but has a frictional moment (static and kinetic) of 2 N·m. This means that if there is no motion, the bearing can resist a moment up to 2 N·m in either direction. If there is rotation, then regardless of the speed of rotation, the bearing exerts a moment of 2 N·m that resists the rotation. Determine the magnitude and direction of angular acceleration of the pulley (if it does, in fact, move) after it is released from rest. The geometric center of the pulley is the mass center, $\bar{I} = 0.4$ kg·m^2, and the mass of the pulley is 10 kg. $r_1 = 0.2$ m and $r_2 = 0.3$ m.

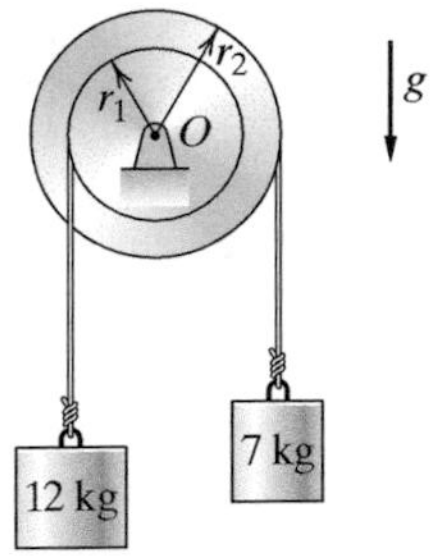

E7.2.55

7.2.56. **[Level 2]** A simplified disk brake assembly is shown. A force $F = 133$ N is applied to the lower end of the massless brake lever at a distance d from the lever's pivot O. The brake pad is located $L = 0.06$ m from O at the upper end of the lever, and the coefficient of friction between the brake pad and disk is $\mu = 0.7$. The disk has a weight and radius of $W = 54$ kg and $R = 0.076$ m, respectively, and it is initially spinning at a constant $\omega_0 = 1000$ rpm. What should d be if the disk is to come to a stop in $\Delta t = 1$ s?

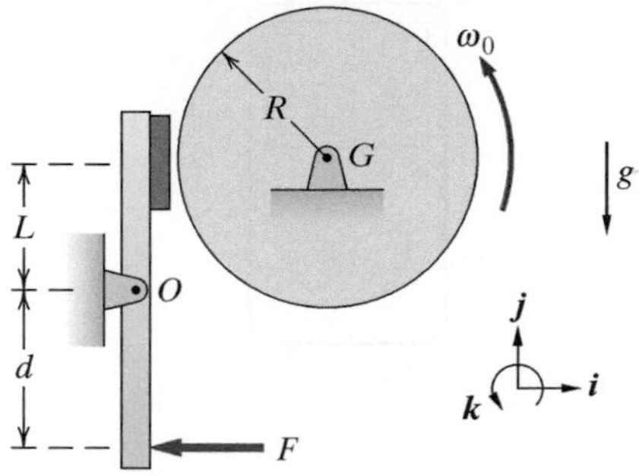

E7.2.56

7.2.57. **[Level 2]** Consider a plate (0.12 kg) with dimensions 0.06 m by 0.08 m, suspended from two massless pins at A and B, and pin B is suddenly removed.

a. Determine the angular acceleration of the plate.
b. Determine the reaction force components at A.

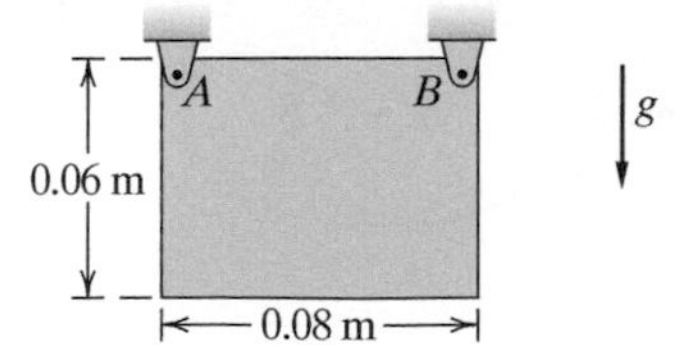

E7.2.57

7.2.58. **[Level 2]** A block of mass $M = 15$ kg is attached to a rope that winds around a large reel of radius $R = 0.6$ m and mass $m = 5$ kg that is constrained to rotate about its mass center G, which is located high above the ground. Dry friction in the reel's bearings generates a frictional torque $M_f = 50$ N·m that impedes the reel's rotation. Suppose the block is released from rest at a height $h = 4$ m above the ground. How long Δt does it take for the block to reach the ground, and with what speed v does it hit? Assume that the rope does not slip over the reel, and the block's mass is large enough so that the reel will spin after release.

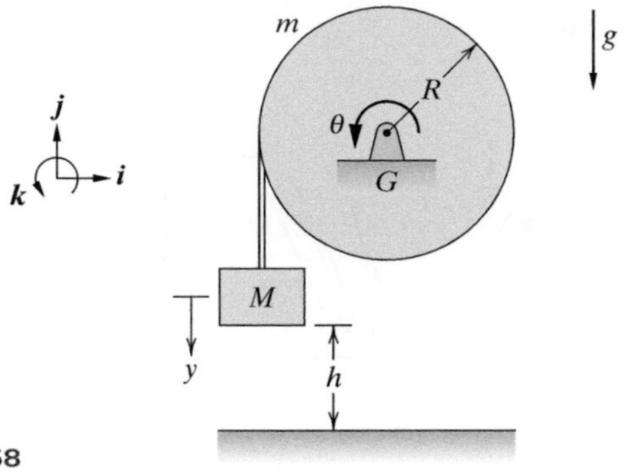

E7.2.58

7.2.59. **[Level 2]** A horizontal log is being raised by pulling the free end of the massless rope at D at 0.6 m/s. Assume that the log pivots freely at A, has a length of 3.0 m, and has a weight of 181 kg. Point C is positioned 3.0 m above point A. What is the tension in the rope?

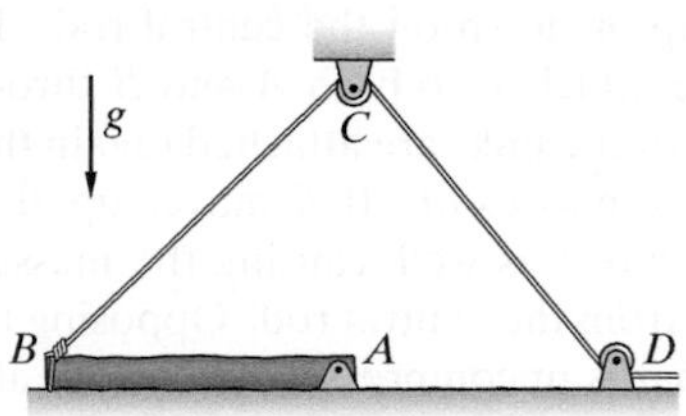

E7.2.59

7.2.60. **[Level 2]** Let's revisit **Exercise 3.1.1**. In that exercise we found the deceleration limit for a no-slip condition, treating the box as a sliding particle. Now we will treat the box as a solid body. What is the maximum magnitude of deceleration that can be withstood before the box begins to rotate? Take point A as the point around which the box will rotate (if the deceleration is severe enough). To model double-sided tape, assume that it supplies a uniformly distributed, vertical force of 96.5 Pa that resists upward motion. The depth of the box is 127 cm, $d = 71$ cm, and $h = 97$ cm. The mass weighs 41 kg. Assume that the car's roof is flat and fully supports the box.

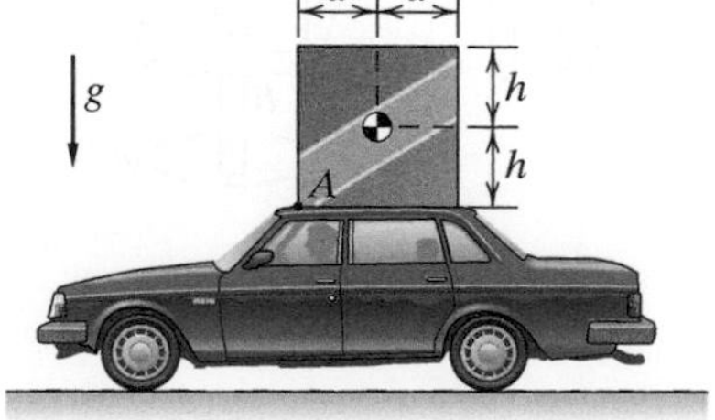

E7.2.60

7.2.61. **[Level 2]** An inverted compound pendulum of length L rotates about its fixed pivot O, where the rod's rotation θ is measured from the horizontal. The mass of the pendulum's rod is m, and a lumped body of mass $M = 2m$

is attached to the rod's free end. Let x and y describe the horizontal and vertical displacements, respectively, of the pendulum's mass center. Find expressions for $\ddot{\theta}, \ddot{x}$, and $\ddot{y}$ in terms of θ and $\dot{\theta}$.

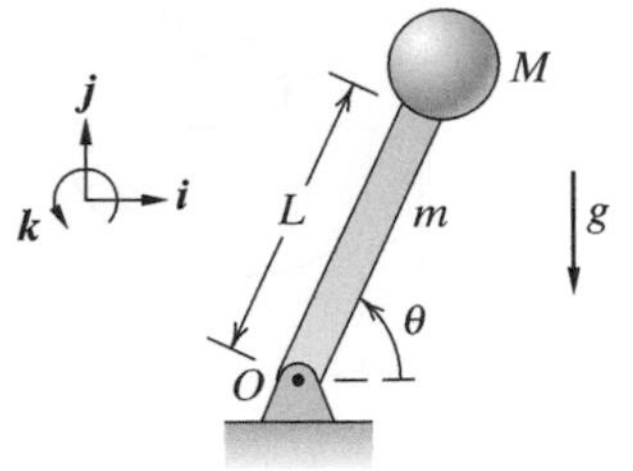

E7.2.61

7.2.62. [Level 2] A fly ball governor is shown. This is a mechanical device that you've often seen in movies and TV programs that show technology of the late 1800s. The center rod rotates, and the cap piece (A), being rigidly attached to it, rotates at the same rate. B is a movable sleeve that can ride up or down on the central rod. Rigid links of length L are attached to both A and B through pivots. The other ends of the links are attached (again through pivots) to two large masses m. If B moves up, the links are constrained to move as well, causing the masses to move outward, away from the central rod. Opposing this motion is a spring k that is uncompressed when B is at its lowest position. Determine the speed ω at which the system can rotate in equilibrium ($\dot{\theta} = \ddot{\theta} = 0$ with the arms inclined at a constant angle $\overline{\theta}$). Neglect the mass of the arms. Break the system up into six rigid bodies (four links and two masses). Use symmetry to reduce the problem to two links and one mass.

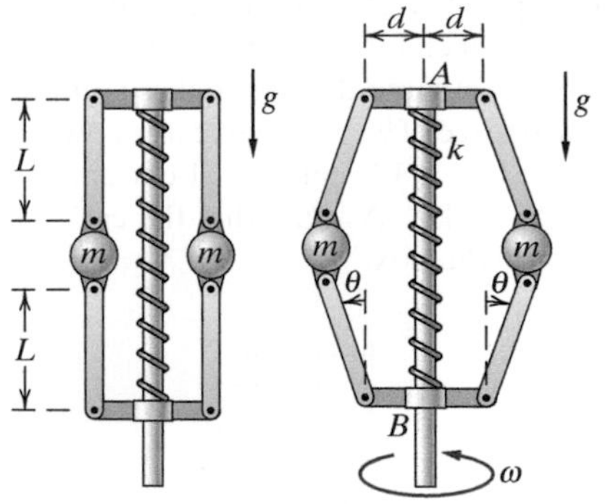

E7.2.62

7.2.63. [Level 2] A uniform cylinder of radius R is spun about its central axis at an angular speed of ω_0 and then placed into a corner. The coefficient of friction between the corner walls and the cylinder is equal to μ. How many turns will the cylinder undergo before it stops turning?

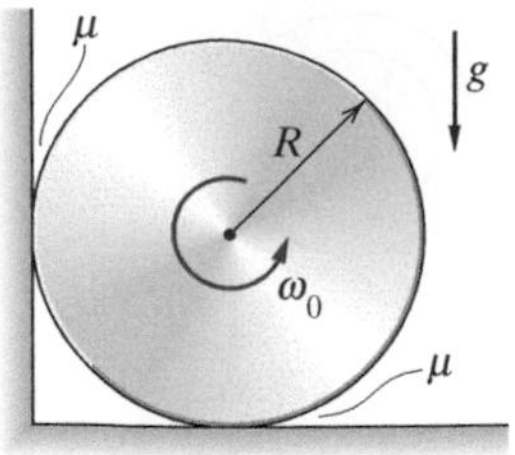

E7.2.63

7.2.64. [Level 2] My iPod's hard drive takes $\Delta t = 3.5$ s to spin up from rest to its operating speed of $\omega = 4200$ rpm. The hard disk has a mass and radius of $m = 48$ g and $r = 22.9$ mm, respectively.

a. Suppose the hard drive's motor provides a constant torque T to spin up the hard disk. What must T be to achieve the given specifications?

b. Now assume that the motor's torque varies according to $T(t) = T_0 t^2$. What is the value of the coefficient T_0?

E7.2.64

7.2.65. [Level 2] A uniform disk of radius R is spun at an angular speed ω_0 and then carefully placed, flat side down, on a horizontal surface. How long will the disk be rotating on the surface if the coefficient of dynamic friction is equal to μ? The pressure exerted by the disk on the surface should be regarded as uniform.

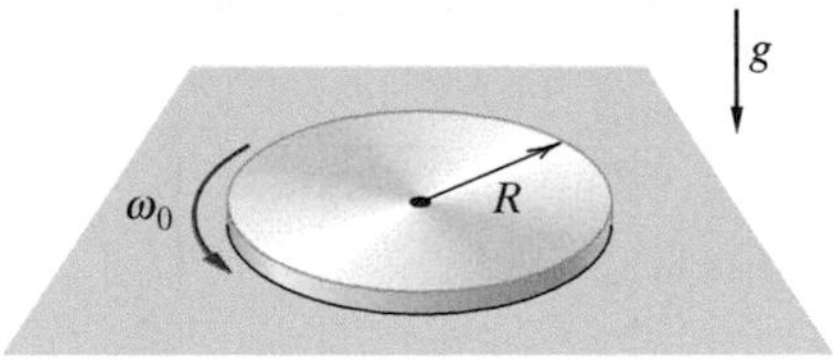

E7.2.65

7.2.66. [Level 3] Computational The illustrated body is a solid right triangle with a semicircular rod welded to it at A and C. At the midspan of the rod is a lumped mass m. The triangular body has a mass of 4 kg, the rod has a linear density of 2 kg/m, $L = 1.2$ m, and $m = 4$ kg. Defining the current position as $\theta = 0$, rotate the body 90 degrees counterclockwise and then release it from that position. Plot θ versus t for 5 s.

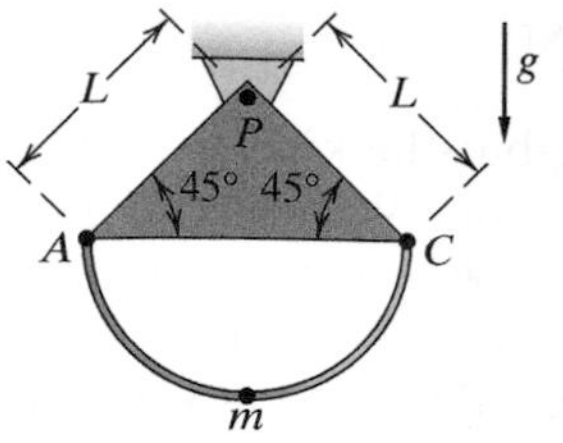

E7.2.66

7.2.67. **[Level 3]** Find the equation of motion for the illustrated system. Let the rotation of the bar be given by θ, and consider only small angles of rotation ($\sin\theta \approx \theta$). The spring k_θ is a torsional spring: $M = -k_\theta \theta$, and ρ is the linear density of the bar. Neglect gravity and nonlinear terms ($\dot{\theta}^2, \ldots$).

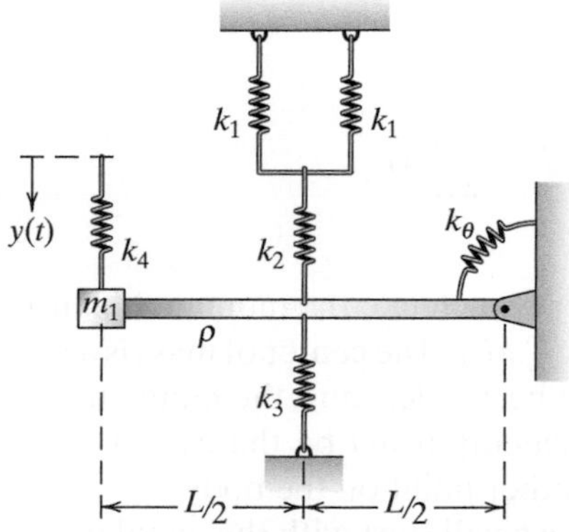

E7.2.67

7.2.68. **[Level 3]** A moment $M\boldsymbol{k}$ acts on the quarter-circle-shaped bar shown. As a result of this moment, a reaction force equal to $(0.944\boldsymbol{i} + 3.46\boldsymbol{j})$ N acts on the frictionless pivot O. Determine the acceleration of the point A and the value of M. $r = 2.0$ m and $\rho = 0.5$ kg/m. Neglect gravity.

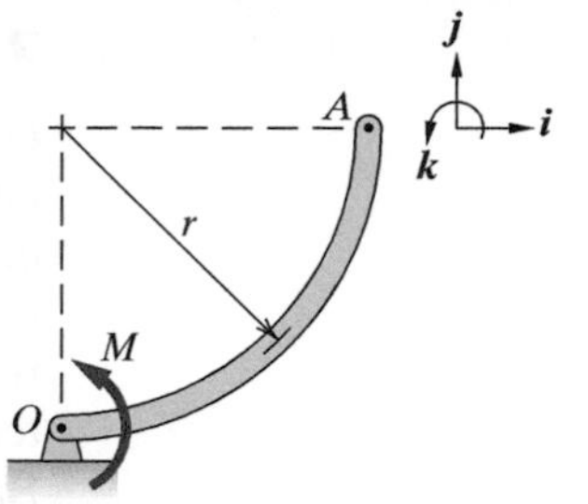

E7.2.68

7.2.69. **[Level 3]** **Computational** Consider a reel (modeled as a uniform cylinder with mass moment of inertia about the axle of 2 kg·m^2 and radius 0.81 m) with 56 m of flexible rope wrapped around it. The linear density of the rope is 0.1 kg/m. At the start of the motion, all the rope is wrapped up on the reel, $\theta(0) = 0$, and $\dot{\theta}(0) = 1$ rad/s, θ representing the reel's counterclockwise rotation angle. Immediately after the start, rope begins to unwind from the reel.

a. Find the exact equation for $\ddot{\theta}(t)$.

b. Find the equation for $\ddot{\theta}(t)$ under the assumption that the center of mass of the reel, plus the rope still wrapped around it, is located at the reel's geometric center (coincident with the axle).

c. Does the rope leave the reel at the same time for both cases?

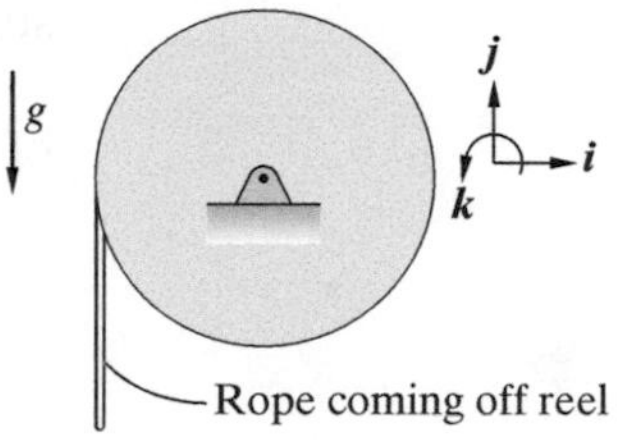

E7.2.69

7.3 GENERAL MOTION

General motion means everything-but-the-kitchen-sink motion: translation plus rotation, everything moving. We will definitely need all the equations we've got at our disposal: $\boldsymbol{F} = m\boldsymbol{a}$ applied at the mass center to deal with translation plus a moment balance equation to handle rotation. What you will see as we go along is that some choices are better than others with respect to which moment balance equation we use. Basically, you will be trying to decide if you should sum moments about some contact point, about the center of mass, or wherever. Although you can ultimately get everything you need with any of the moment balance equations we will be deriving, choosing wisely can save you time and effort.

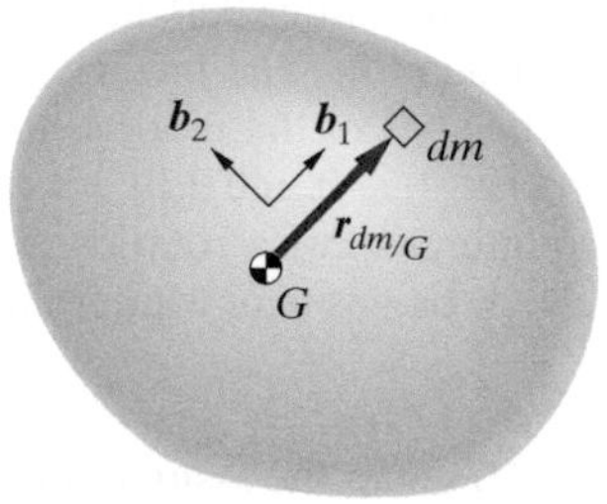

Figure 7.39 Body in general motion

We will start by deriving a moment balance about G, the body's mass center (**Figure 7.39**). This particular moment balance formula will be extremely simple in form:

$$I_G\ddot{\theta} = \sum M_G$$

You will note that this looks very much like the moment balance derived in the previous section—$I_O\ddot{\theta} = \sum M_O$. The center of mass is a very special point, and even though it might be accelerating, the moment balance still has this simple form. This decidedly *won't* be the case if you consider summing moments about any other point on the body.

As in the preceding section, we will start with the angular momentum of the body about G:

$$\boldsymbol{H}_G = \int_{\text{Body}} \left(\boldsymbol{r}_{dm/G} \times \boldsymbol{v}_{dm}\right) dm \tag{7.41}$$

Remember (5.15)? That was the formula that told us that the angular momentum $\boldsymbol{H}_G$ about the center of mass is the same whether you use absolute velocities or relative velocities; that is,

$$\boldsymbol{H}_G = \boldsymbol{H}_G\Big|_{rel}$$

Here's where this fact will come in handy. It's not necessary that we use it, but it will save us a few lines of work. Rewrite (7.41) as

$$\boldsymbol{H}_G = \int_{\text{Body}} \left(\boldsymbol{r}_{dm/G} \times \boldsymbol{v}_{dm/G}\right) dm$$

using the relative velocity $\boldsymbol{v}_{dm/G}$ instead of the absolute velocity. Differentiating then gets us

$$\dot{\boldsymbol{H}}_G = \int\limits_{\text{Body}} \left[\overbrace{\left(\boldsymbol{v}_{dm/G} \times \boldsymbol{v}_{dm/G}\right)}^{0} + \left(\boldsymbol{r}_{dm/G} \times \boldsymbol{a}_{dm/G}\right) \right] dm$$
$$= \int\limits_{\text{Body}} \left(\boldsymbol{r}_{dm/G} \times \boldsymbol{a}_{dm/G}\right) dm \tag{7.42}$$

This is okay, but we can do much better. Because we're dealing with a rigid body, we know something about $\boldsymbol{a}_{dm/G}$. In fact, we know more than something—we know a lot. For any two points A and B on a rigid body undergoing planar motion, we have

$$\boldsymbol{a}_B = \boldsymbol{a}_A + \boldsymbol{a}_{B/A}$$

and, furthermore, from Section 6.3 we know the form of $\boldsymbol{a}_{B/A}$, given in (6.25):

$$\boldsymbol{a}_{B/A} = \boldsymbol{\alpha} \times \boldsymbol{r}_{B/A} + \boldsymbol{\omega} \times \left(\boldsymbol{\omega} \times \boldsymbol{r}_{B/A}\right)$$

which in our case becomes

$$\boldsymbol{a}_{dm/G} = \boldsymbol{\alpha} \times \boldsymbol{r}_{dm/G} + \boldsymbol{\omega} \times \left(\boldsymbol{\omega} \times \boldsymbol{r}_{dm/G}\right)$$

(replacing A with G and B with dm).

In **Figure 7.39** I've aligned my unit vectors so that $\boldsymbol{b}_1$ points from G to dm, which means that $\boldsymbol{r}_{dm/G} = r_{dm/G}\boldsymbol{b}_1$. I don't have to do this—any set of unit vectors would suffice—but locating them in any other orientation would just mean a little more work taking cross products and give us the same end result.

In terms of these quantities, and with the substitutions $\boldsymbol{\omega} = \dot{\theta}\boldsymbol{b}_3$ and $\boldsymbol{\alpha} = \ddot{\theta}\boldsymbol{b}_3$, (7.42) becomes

$$\dot{\boldsymbol{H}}_G = \int\limits_{\text{Body}} \left\{ \overbrace{r_{dm/G}\boldsymbol{b}_1}^{\boldsymbol{r}_{dm/G}} \times \left[\overbrace{\ddot{\theta}\boldsymbol{b}_3}^{\boldsymbol{\alpha}} \times \overbrace{r_{dm/G}\boldsymbol{b}_1}^{\boldsymbol{r}_{dm/G}} + \overbrace{\dot{\theta}\boldsymbol{b}_3}^{\boldsymbol{\omega}} \times \left(\overbrace{\dot{\theta}\boldsymbol{b}_3}^{\boldsymbol{\omega}} \times \overbrace{r_{dm/G}\boldsymbol{b}_1}^{\boldsymbol{r}_{dm/G}} \right) \right] \right\} dm$$

Interestingly, because $\boldsymbol{b}_1 \times \boldsymbol{b}_1 = 0$, the second term is equal to zero. This implies that the rotational velocity of the rigid body doesn't enter into the moment balance equation. Carrying out the cross products for the first term yields

$$\dot{\boldsymbol{H}}_G = \int\limits_{\text{Body}} \ddot{\theta} r_{dm/G}^2 \, dm \, \boldsymbol{b}_3 \tag{7.43}$$

Dispensing with the unit vector (the momentum quantities are always in the $\boldsymbol{b}_3$ direction for planar motion), let's write the equation as

$$\dot{H}_G = \int_{\text{Body}} \ddot{\theta} r_{dm/G}^2 \, dm$$

If we pull the angular acceleration out of the integral, we're left with

$$\dot{H}_G = \ddot{\theta} \int_{\text{Body}} r_{dm/G}^2 \, dm$$

This should look familiar, for the integral term is just about the same as the one we derived in the previous section, except in that case the mass element was referenced to a fixed point of rotation O and now it's referenced to the mass center G. What we've just derived is the mass moment of inertia about G:

$$I_G \equiv \int_{\text{Body}} r_{dm/G}^2 \, dm \tag{7.44}$$

As previously mentioned, we will occasionally use $\overline{I}$ instead of I_G if it makes the equations cleaner. Recalling that $\dot{\boldsymbol{H}}_G = \sum \boldsymbol{M}_G$ and using (7.44), we obtain

$$I_G \ddot{\theta} \boldsymbol{b}_3 = \sum \boldsymbol{M}_G \tag{7.45}$$

where $\sum \boldsymbol{M}_G$ is the sum of all the moments applied about G. Dropping the out-of-plane vectors from the equation gives us the scalar form:

$$I_G \ddot{\theta} = \sum M_G \tag{7.46}$$

This is the formula you are most likely to reach for when you're doing a general planar kinetics problem that involves a body that is both rotating and translating. It's simple in form, it's easy to use, and it's always going to give you the right answer. There are, however, other formulas that can prove superior in certain problems. We will discuss these formulas later in the chapter.

>>> **Check out Example 7.11 (page 425), Example 7.12 (page 428), Example 7.14 (page 432), Example 7.15 (page 434), Example 7.16 (page 437) Example 7.17 (page 438), Example 7.18 (page 440), and Example 7.19 (page 442) for applications of this material.**

EXAMPLE 7.11 ACCELERATION RESPONSE OF AN UNRESTRAINED BODY (Theory on page 424)

Figure 7.40 shows a space capsule that's moving to the left as it approaches the earth. Two thrusters are shown, A and B. The astronauts inside the capsule attempt to reduce the capsule's speed slightly by firing them both simultaneously, but only the single thruster B fires, producing a thrust T. What is the resultant angular acceleration of the capsule? Assume that the capsule's dimensions and I_G are known.

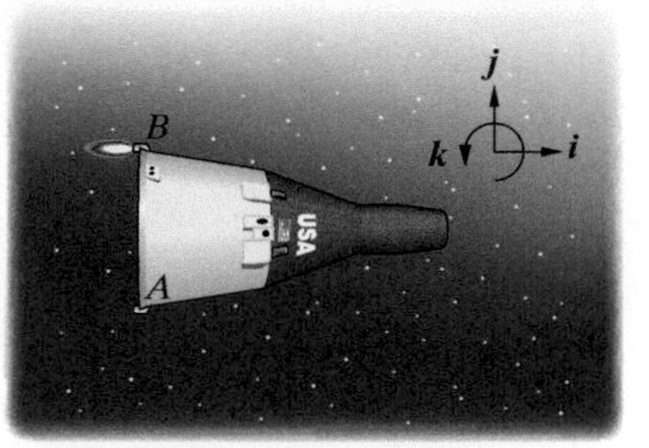

Figure 7.40 Space capsule

Goal Find the angular acceleration of a space capsule upon the application of an off-axis force.

Given Capsule dimensions, mass moment of inertia, and thrust.

Assume Because the capsule is in space, we can safely assume that there aren't any significant forces that resist the craft's motion. In other words, it's completely unrestrained.

Draw **Figure 7.41** shows the only force acting on the capsule, as well as the distances needed to determine the applied moment about G.

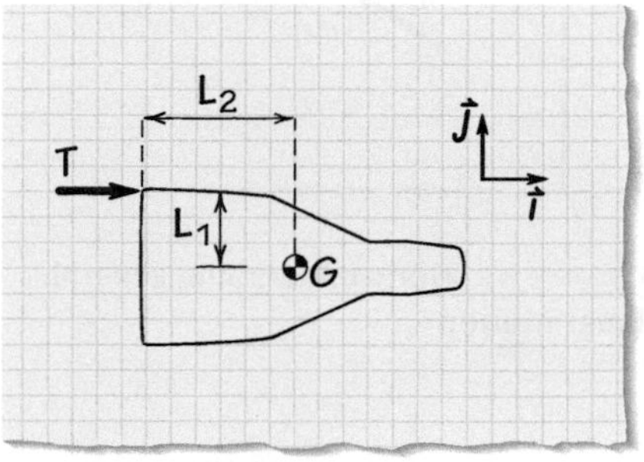

Figure 7.41 Dimensions/FBD for capsule

Formulate Equations All that's needed is to invoke (7.45):

$$I_G\ddot{\theta}\boldsymbol{k} = (-L_2\boldsymbol{i} + L_1\boldsymbol{j}) \times T\boldsymbol{i} \quad (7.47)$$

Solve Carrying through the cross products in (7.47) gives us

$$I_G\ddot{\theta}\boldsymbol{k} = -L_1T\boldsymbol{k}$$

$$I_G\ddot{\theta} + L_1T = 0$$

$$\ddot{\theta} = -\frac{L_1T}{I_G}$$

Check Let's do a logic check. First, the angular acceleration is negative. That certainly makes physical sense; the thrust is producing a clockwise moment on the capsule. It also tells us that the angular acceleration scales linearly with both L_1 and T. This is sensible as well. We would expect that a doubling of thrust would double the angular acceleration, and if you double the moment arm L_1 you would likewise expect a doubling of the applied moment and therefore of the angular acceleration. Finally, it tells us that increasing I_G will decrease the angular acceleration, a finding that's totally in line with our expectations. Increasing the mass moment of inertia of a body will make it harder to turn, and so we would expect to see a reduced angular acceleration.

We have now seen the moment balance equations for a rotating body with reference to a fixed point of rotation O and with reference to the body's center of mass G. Next we will derive a moment balance equation with respect to some arbitrary point A (not necessarily O or G) on the body and get two forms of the equation because sometimes one form will be most convenient and sometimes the other will be. The point A has no restrictions on it. It can be a fixed point of rotation, it can be moving at a uniform velocity, or it can be accelerating—anything at all.

Figure 7.42 shows a representative body along with some relevant labeling. You will recall from statics that the effect of a set of forces acting on a rigid body is equivalent to that produced by applying the sum of those forces to a point on the body and adding the appropriate couple. In line with this observation, the sum of forces is shown going through G, and $\sum M_G$ represents the appropriate moment sum about G.

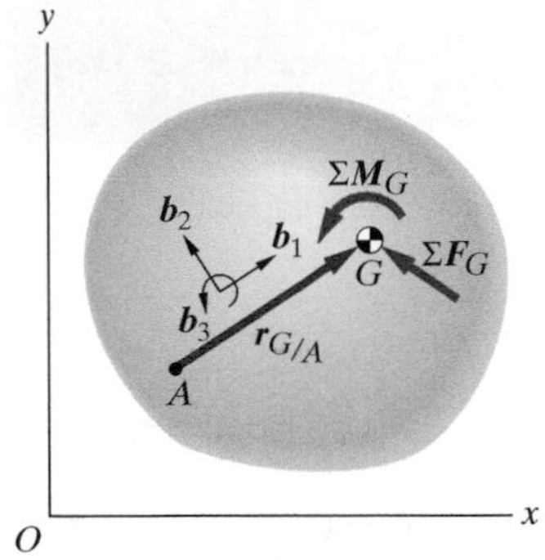

Figure 7.42 Body undergoing general planar motion

We will start by taking a moment sum about A and see where that leads us:

$$\sum \boldsymbol{M}_A = \sum \boldsymbol{M}_G + \boldsymbol{r}_{G/A} \times \sum_{i=1}^{n} \boldsymbol{F}_G$$

Using $\sum \boldsymbol{F}_G = m\boldsymbol{a}_G$ and $\sum M_G = I_G\ddot{\theta}$ gives us

$$\sum \boldsymbol{M}_A = I_G\ddot{\theta}\boldsymbol{b}_3 + \boldsymbol{r}_{G/A} \times m\boldsymbol{a}_G \tag{7.48}$$

where $\boldsymbol{b}_3$ is a unit vector that's orthogonal to the plane of rotation, as usual.

This is the first form of the moment balance equation about an arbitrary point on the body. It's a little odd, however, because it involves not the mass moment of inertia about A (the point we're taking moments about) but rather the inertia moment about G, the center of mass. It will take a bit of manipulation, but we can get it into a form that involves I_A as well. Before that, however, here's a nice way to remember this formula. **Figure 7.43** shows a body, along with $\boldsymbol{r}_{G/A}$, $m\boldsymbol{a}_G$, and the distance d, the minimum distance from A to an extension of $m\boldsymbol{a}_G$. This lets us write the scalar form of (7.48) as

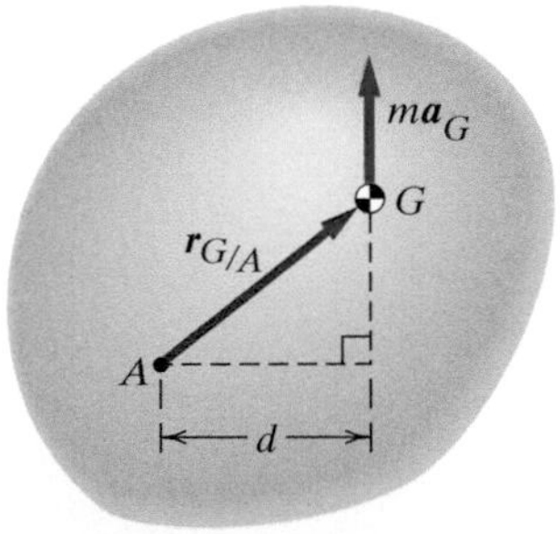

Figure 7.43 "mad" representation

$$\sum M_A = I_G\ddot{\theta} + mad$$

where a is the magnitude of $\boldsymbol{a}_G$. Here's how you remember it. It makes you *mad* that you have to deal with this complicated formula. The first two letters of mad are m and a, and our moments are given by M_A. The last term (*mad*) is clearly a mad term. As for the $I_G\ddot{\theta}$, you will just have to remember that part.

Okay, back to our derivations. Because both G and A are on the body, they can be related through

$$\boldsymbol{a}_G = \boldsymbol{a}_A + \boldsymbol{\alpha} \times \boldsymbol{r}_{G/A} + \boldsymbol{\omega} \times \left(\boldsymbol{\omega} \times \boldsymbol{r}_{G/A}\right) \tag{7.49}$$

which gives us

$$\sum \boldsymbol{M}_A = I_G\ddot{\theta}\boldsymbol{b}_3 + \boldsymbol{r}_{G/A} \times m\left[\boldsymbol{a}_A + \boldsymbol{\alpha} \times \boldsymbol{r}_{G/A} + \boldsymbol{\omega} \times \left(\boldsymbol{\omega} \times \boldsymbol{r}_{G/A}\right)\right]$$

The final term, $\boldsymbol{r}_{G/A} \times [\boldsymbol{\omega} \times (\boldsymbol{\omega} \times \boldsymbol{r}_{G/A})]$, is equal to zero for the simple reason that $\boldsymbol{a} \times [\boldsymbol{b} \times (\boldsymbol{b} \times \boldsymbol{a})]$ for two orthogonal vectors $\boldsymbol{a}$ and $\boldsymbol{b}$ is *always* zero, which leaves us with

$$\sum \boldsymbol{M}_A = I_G \ddot{\theta} \boldsymbol{b}_3 + \boldsymbol{r}_{G/A} \times m\boldsymbol{a}_A + m\boldsymbol{r}_{G/A} \times \left(\boldsymbol{\alpha} \times \boldsymbol{r}_{G/A}\right) \tag{7.50}$$

The last term is straightforward. I've included the unit vectors $\boldsymbol{b}_1$ and $\boldsymbol{b}_2$ on **Figure 7.42** and aligned $\boldsymbol{b}_1$ so that $\boldsymbol{r}_{G/A} = r_{G/A}\boldsymbol{b}_1$. The last term can therefore be written as

$$\begin{aligned} m\boldsymbol{r}_{G/A} \times \left(\boldsymbol{\alpha} \times \boldsymbol{r}_{G/A}\right) &= m r_{G/A} \boldsymbol{b}_1 \times \left(\ddot{\theta}\boldsymbol{b}_3 \times r_{G/A}\boldsymbol{b}_1\right) \\ &= m r_{G/A}^2 \ddot{\theta} \boldsymbol{b}_3 \end{aligned}$$

which means that (7.50) can now be expressed as

$$\sum \boldsymbol{M}_A = \left(I_G + m r_{G/A}^2\right) \ddot{\theta} \boldsymbol{b}_3 + \boldsymbol{r}_{G/A} \times m\boldsymbol{a}_A$$

or, using $I_A = I_G + m r_{G/A}^2$ from the parallel axis theorem,

$$\sum \boldsymbol{M}_A = I_A \ddot{\theta} \boldsymbol{b}_3 + \boldsymbol{r}_{G/A} \times m\boldsymbol{a}_A \tag{7.51}$$

>>> **Check out Example 7.13 (page 430) for an application of this material.**

Both (7.48) and (7.51) are more general than $\sum M_O = I_O\ddot{\theta}$ or $\sum M_G = I_G\ddot{\theta}$ and we can easily derive the simpler expressions as special cases.

For instance, take (7.51) and let A be a fixed point. In that case $\boldsymbol{a}_A = 0$, and the equation becomes

$$\sum M_A = I_A \ddot{\theta}$$

which is the same form as (7.28), the expression we derived for a fixed point O. What if we let A be the center of mass? Then $\boldsymbol{r}_{G/A} = 0$, and we get

$$\sum M_A = I_A \ddot{\theta}$$

which again is the same form as (7.45), the expression we derived for the center of mass.

There's one other time that (7.51) simplifies, and this simplified form comes in handy when you're looking at rolling wheels. It's when $\boldsymbol{r}_{G/A}$ and $\boldsymbol{a}_A$ are collinear. In this case their cross product is zero, and we're again left with

$$\sum M_A = I_A \ddot{\theta}$$

This isn't going to happen all the time, but when it does, this is a nice fact to remember.

EXAMPLE 7.12 RESPONSE OF A FALLING ROD (Theory on page 424)

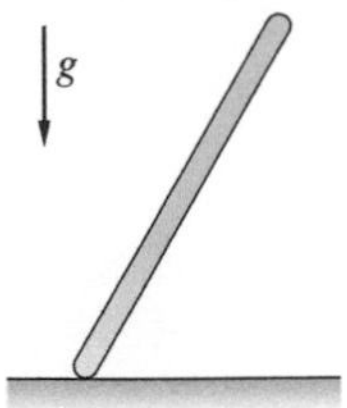

Figure 7.44 Imbalanced rod

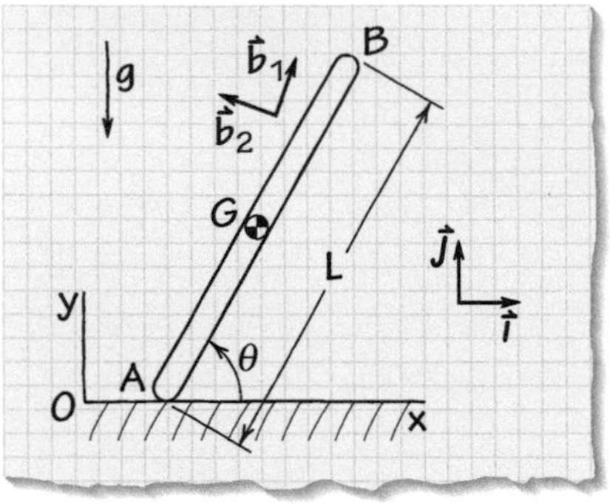

Figure 7.45 Labeling for imbalanced rod

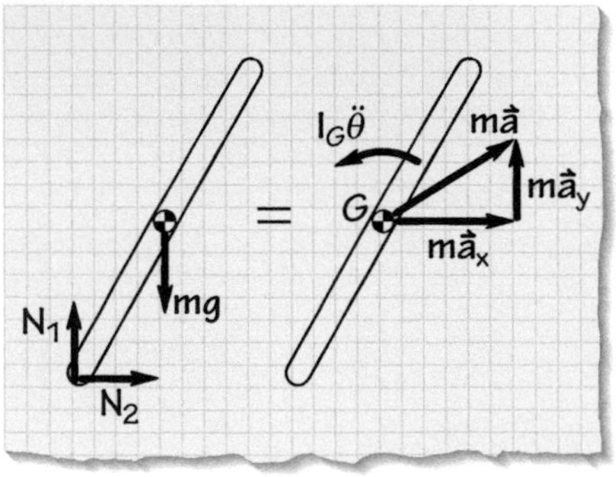

Figure 7.46 FBD=IRD for imbalanced rod

A uniform, rigid rod is positioned on a frictionless surface as shown in **Figure 7.44** and then released from that position at $t = 0$. Determine the rod's equation of motion.

Goal Find the rod's equation of motion.

Draw **Figure 7.45** shows the labeling for our system (the rod). θ measures the inclination of the rod with respect to horizontal (thus $\theta = 0$ indicates the rod at rest on its side). A fully general FBD and IRD are shown in **Figure 7.46**. Two reaction forces are shown at A, the bottom end of the rod, and a gravitational force is shown acting through the center of mass. No moment can be applied by the ground to the rod at A, and therefore none is shown in the free-body diagram. The inertial response, $m\bar{\boldsymbol{a}}$, is shown as an as-yet unconstrained vector.

Formulate Equations To start, let's take the safest and "easiest" set—that is, the set in which everything is expressed with respect to the mass center. We'll examine what an alternative choice can do for us in Example 7.13. Our three force/moment balance equations are

$\boldsymbol{i}$: $$N_2 = m\bar{a}_x$$

$\boldsymbol{j}$: $$N_1 - mg = m\bar{a}_y$$

$\boldsymbol{k}$: $$I_G\ddot{\theta} = N_2\frac{L}{2}\sin\theta - N_1\frac{L}{2}\cos\theta$$

Assume Because our three equations have five unknowns—$\ddot{\theta}$, $\bar{a}_x$, $\bar{a}_y$, N_1, and N_2—something's clearly got to give in order for us to get anywhere. Getting rid of one unknown is easy: N_2 is zero because the problem statement told us that there's no friction. Getting rid of another unknown is a bit tougher, and to do it we're going to have to apply some physical understanding to the problem. Look at end A. What we can use is the fact that it's constrained by the ground to move only horizontally. Thus its acceleration can be expressed as $\boldsymbol{a}_A = a_A\boldsymbol{i}$.

Of course, we don't know what a_A is at this point, and so it may seem that I've introduced another unknown for no purpose, but actually I haven't. We already have as unknowns the total acceleration of the rod's center of mass $\bar{\boldsymbol{a}}_x$, $\bar{\boldsymbol{a}}_y$ as well as the angular acceleration $\ddot{\theta}$. Given this information, we can easily solve for the acceleration of any point on the rod by using

$$\boldsymbol{a}_A = \boldsymbol{a}_G + \ddot{\theta}\boldsymbol{k} \times \boldsymbol{r}_{A/G} + \dot{\theta}\boldsymbol{k} \times \left(\dot{\theta}\boldsymbol{k} \times \boldsymbol{r}_{A/G}\right)$$

In general, this acceleration would have components in both the $\boldsymbol{i}$ and $\boldsymbol{j}$ directions. Thus, by explicitly stating that the $\boldsymbol{j}$ component is zero, we've constrained the problem.

Solve Let's keep a_A as an unknown and see what we can do. If we reverse the analysis just presented, we can jump from A to G and express the acceleration of the center of mass in terms of a_A and $\ddot{\theta}$:

$$\boldsymbol{a}_G = a_A\boldsymbol{i} + \ddot{\theta}\boldsymbol{k} \times \boldsymbol{r}_{G/A} + \dot{\theta}\boldsymbol{k} \times \left(\dot{\theta}\boldsymbol{k} \times \boldsymbol{r}_{G/A}\right) = a_A\boldsymbol{i} + \ddot{\theta}\boldsymbol{k} \times \frac{L}{2}\boldsymbol{b}_1 + \dot{\theta}\boldsymbol{k} \times \left(\dot{\theta}\boldsymbol{k} \times \frac{L}{2}\boldsymbol{b}_1\right)$$

A little coordinate transformation matrix would be of use right about now:

	$\boldsymbol{i}$	$\boldsymbol{j}$
$\boldsymbol{b}_1$	$\cos\theta$	$\sin\theta$
$\boldsymbol{b}_2$	$-\sin\theta$	$\cos\theta$

Using this in our equation for $\boldsymbol{a}_G$ gives us

$$\boldsymbol{a}_G = \frac{1}{2}\left(2a_A - L\ddot{\theta}\sin\theta - L\dot{\theta}^2\cos\theta\right)\boldsymbol{i} + \left(L\ddot{\theta}\cos\theta - L\dot{\theta}^2\sin\theta\right)\boldsymbol{j} \tag{7.52}$$

Interesting. The entire acceleration vector of the center of mass is determined by the linear acceleration a_A of A and the values of $\theta, \dot{\theta}$, and $\ddot{\theta}$. This means that our sum total of unknowns is now N_1, a_A, and $\ddot{\theta}$—three unknowns and three equations. Using these results in our equations of motion yields

$\boldsymbol{i}$:
$$0 = \frac{m}{2}\left(2a_A - L\ddot{\theta}\sin\theta - L\dot{\theta}^2\cos\theta\right) \tag{7.53}$$

$\boldsymbol{j}$:
$$N_1 - mg = \frac{m}{2}\left(L\ddot{\theta}\cos\theta - L\dot{\theta}^2\sin\theta\right) \tag{7.54}$$

$\boldsymbol{k}$:
$$I_G\ddot{\theta} = -\frac{N_1 L\cos\theta}{2} \tag{7.55}$$

(7.53) ⇒
$$a_A = \frac{1}{2}\left(L\ddot{\theta}\sin\theta + L\dot{\theta}^2\cos\theta\right) \tag{7.56}$$

(7.57) → (7.52) ⇒
$$\boldsymbol{a}_G = \frac{1}{2}\left(L\ddot{\theta}\cos\theta - L\dot{\theta}^2\sin\theta\right)\boldsymbol{j} \tag{7.57}$$

Thus, as a result of the sliding constraint at A, the acceleration of the body's center of mass is completely known once we know $\theta, \dot{\theta}$, and $\ddot{\theta}$.

Using (7.54) is next on the agenda:

(7.54) ⇒
$$N_1 = mg + m\left(\frac{L\ddot{\theta}\cos\theta}{2} - \frac{L\dot{\theta}^2\sin\theta}{2}\right) \tag{7.58}$$

(7.58) → (7.55) ⇒
$$I_G\ddot{\theta} = -\frac{Lm\cos\theta}{2}\left(g + \frac{L\ddot{\theta}\cos\theta - L\dot{\theta}^2\sin\theta}{2}\right)$$

which gives us the rod's equation of motion:

$$\left(I_G + \frac{mL^2}{4}\cos^2\theta\right)\ddot{\theta} = -\frac{mgL\cos\theta}{2} + \frac{mL^2\dot{\theta}^2\sin\theta\cos\theta}{4} \tag{7.59}$$

Check Our check will be to move on to Example 7.13.

EXAMPLE 7.13 MORE RESPONSE OF A FALLING ROD (Theory on page 427)

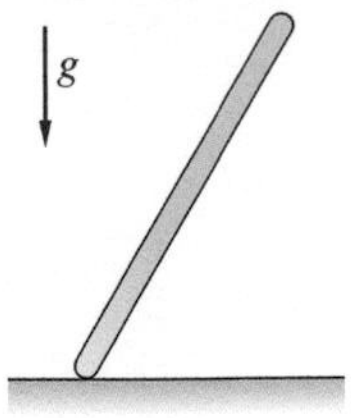

Figure 7.47 Imbalanced rod

In Example 7.13 we found the equations of motion for a falling rod (**Figure 7.47**). Let's extend the analysis. First, we'll find the rod's equation of motion in a new way. Then we'll form an expression for the rod's angular acceleration, and solve for and plot the rod's response (θ versus t) from an initial condition of $\theta(0) = 80°, \dot{\theta}(0) = 0$ to $\theta = 45°$. Finally, we'll determine the corresponding value of $\dot{\theta}$ as well.

Goal Find the rod's equation of motion, find its angular acceleration, plot a time response from $\theta = 80°$ to $\theta = 45°$, and determine $\dot{\theta}$ when $\theta = 45°$.

Assume We'll assume that the end A of the rod remains in contact with the floor.

Draw **Figure 7.48** shows the labeling for our system (the rod). A fully general free-body diagram and inertial response diagram are shown in **Figure 7.49**.

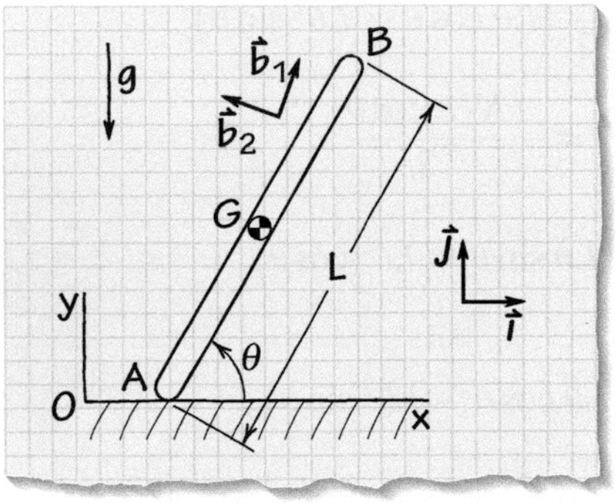

Figure 7.48 Labeling for imbalanced rod

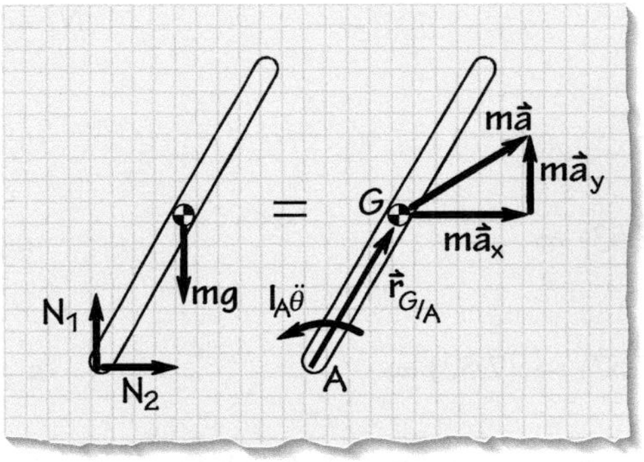

Figure 7.49 FBD=IRD for imbalanced rod

Formulate Equations To get our equation of motion more directly, we will have to use one of our angular momentum expressions. The one to choose is (7.51):

$$\sum \boldsymbol{M}_A = I_A\ddot{\theta}\boldsymbol{k} + \left(\boldsymbol{r}_{G_{/A}} \times m\boldsymbol{a}_A\right)$$

Applying this formula to our problem gives us

$$-\frac{mgL\cos\theta}{2}\boldsymbol{k} = \left(I_G + \frac{mL^2}{4}\right)\ddot{\theta}\boldsymbol{k} + \left[\left(\frac{L}{2}\right)\boldsymbol{b}_1 \times ma_A\boldsymbol{i}\right]$$

$$= \left(I_G + \frac{mL^2}{4}\right)\ddot{\theta}\boldsymbol{k} + \left[ma_A\left(\frac{L}{2}\right)(\cos\theta\boldsymbol{i} + \sin\theta\boldsymbol{j}) \times \boldsymbol{i}\right]$$

Using our the expression for a_A (7.56) gives us

$$-\frac{mgL\cos\theta}{2}\boldsymbol{k} = \left(I_G + \frac{mL^2}{4}\right)\ddot{\theta}\boldsymbol{k} - \frac{mL^2\sin\theta}{4}(\ddot{\theta}\sin\theta + \dot{\theta}^2\cos\theta)\boldsymbol{k}$$

$$= I_G\ddot{\theta}\boldsymbol{k} + (1-\sin^2\theta)\frac{mL^2}{4}\ddot{\theta}\boldsymbol{k} - \frac{mL^2\dot{\theta}^2\sin\theta\cos\theta}{4}\boldsymbol{k}$$

$$\left(I_G + \frac{mL^2}{4}\cos^2\theta\right)\ddot{\theta} = -\frac{mgL\cos\theta}{2} + \frac{mL^2\dot{\theta}^2\sin\theta\cos\theta}{4}$$

exactly the same equation of motion as in Example 7.13.

Solve It's a simple matter to solve for $\ddot{\theta}$:

$$\ddot{\theta} = \frac{mL\cos\theta(L\dot{\theta}^2\sin\theta - 2g)}{4\left(I_G + \frac{mL^2}{4}\cos^2\theta\right)}$$

At this point, all we need to do is use the given parameter values and apply MATLAB's integration code ode45. Doing so will produce **Figure 7.50**. The time interval used was $0 \le t \le 0.6667$ s, and the initial conditions for the integration were $\theta = 1.3963$ rad, $\dot{\theta}(0) = 0$. The final values from the integration were $\theta(0.6667) = 0.7856$ rad and $\dot{\theta}(0.6667) = -1.808$ rad/s.

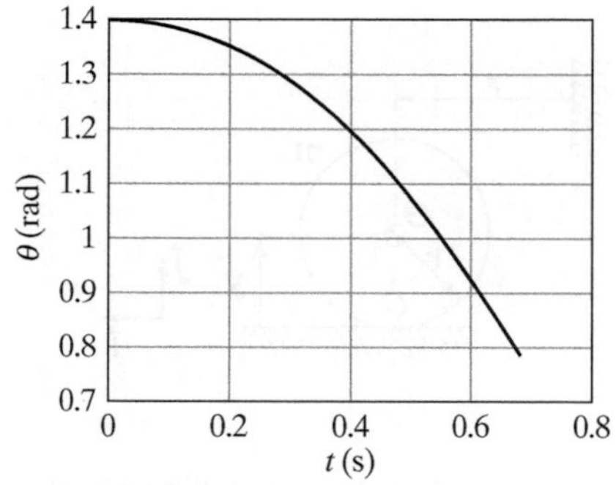

Figure 7.50 θ versus t for falling rod

Check We obtained the same equation of motion through two different equations, a good indication that our work is correct. By choosing the best moment balance equation, we got our answer without having to use a force balance.

EXAMPLE 7.14 ACCELERATION RESPONSE OF A DRIVEN WHEEL (Theory on page 424)

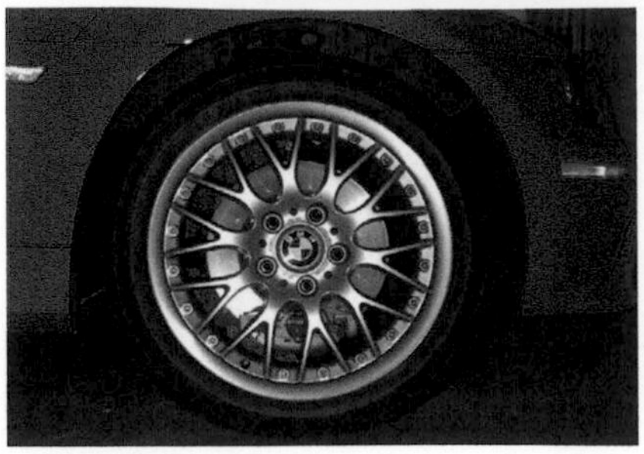
Benson Tongue

Figure 7.51 Wheel/tire

Figure 7.51 shows a wheel/tire (which we'll approximate as a uniform cylinder) that can roll on a horizontal surface. For now we'll ignore the fact that there's a car involved as well and simply consider the wheel as being driven by a moment about its central axis. Find the acceleration of the wheel's center of mass, the tractive force (the force between the ground and wheel that "makes the wheel go"), and the wheel's angular acceleration. The relevant parameters are a clockwise moment $M = 85$ N·m, $m = 23$ kg, $\mu = 0.7$, and $r = 0.3$ m.

Goal Find the linear acceleration of the center of mass, the tractive force between wheel and ground, and the wheel's angular acceleration.

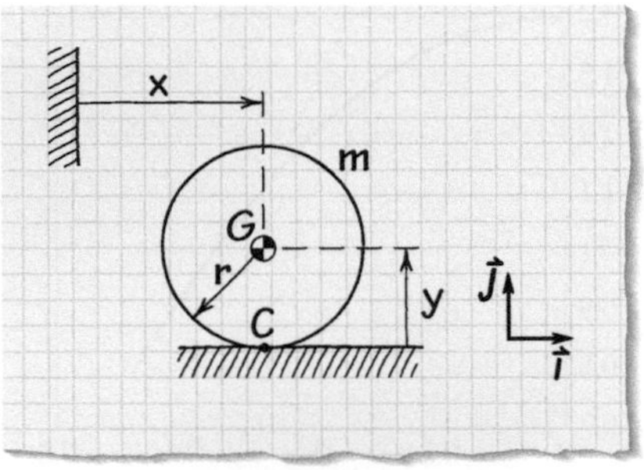

Figure 7.52 Labeling for rolling wheel

Draw **Figure 7.52** shows the basics of our system. As usual, G represents the center of mass, and C is introduced to represent the contact point between the wheel and the ground. As **Figure 7.53** shows, three forces and one moment are applied to the wheel: the ground interaction forces N_1 and N_2, the gravity force mg, and the applied moment M, applied by some external agency.

Formulate Equations We have the equation set

$\boldsymbol{i}$: $$N_2 = m\bar{a}_x$$

$\boldsymbol{j}$: $$N_1 - mg = m\bar{a}_y$$

$\boldsymbol{k}$: $$I_G\ddot{\theta} = rN_2 - M$$

where x indicates the horizontal position of the mass center ($\boldsymbol{i}$ direction), y indicates the vertical position ($\boldsymbol{j}$ direction), and I_G is the mass moment of inertia about the wheel's mass center.

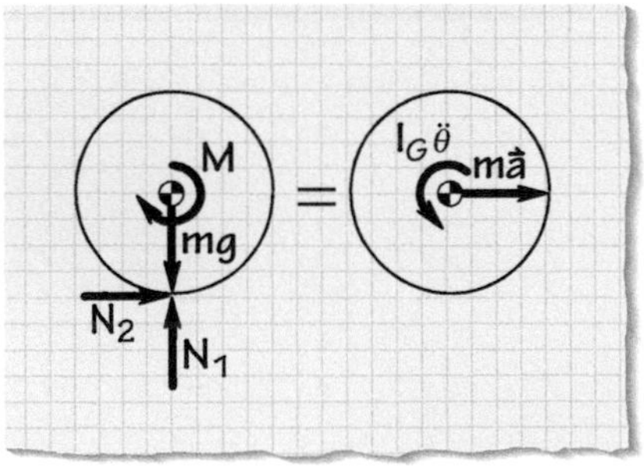

Figure 7.53 FBD=IRD for rolling wheel

Assume Once again, there are two more unknowns than equations. Because we don't know $\bar{a}_x$, $\bar{a}_y$, $\ddot{\theta}$, N_1, or N_2, it's clear that we need some constraints. The first is pretty obvious—the wheel is going to remain in contact with the ground. This means $\bar{a}_y$ is zero, effectively eliminating it as an unknown. What else can we look at? Well, consider the movies. Sometimes cars just start up and drive (roll without slip) and sometimes there's a lot of tire spin with lots of burning rubber (roll with slip). What's going to happen for our example? There's really no way to know at this point. Either of these conditions gives us the additional constraints we need, and to proceed further we have to choose one or the other, solve the problem, and then check whether our assumption was justified.

If we've got pure rolling, then the acceleration of the center of mass can be expressed in terms of $\ddot{\theta}$. There's no need to keep translational acceleration and angular acceleration as separate unknowns because knowing one immediately implies the other. Because the wheel can move only horizontally, the only component of linear acceleration that the center of mass possesses is the component in the $\boldsymbol{i}$ direction, and we can find it from (6.25):

$$\bar{\boldsymbol{a}} = \boldsymbol{a}_C + \boldsymbol{\alpha} \times \boldsymbol{r}_{G/C} + \boldsymbol{\omega} \times \left(\boldsymbol{\omega} \times \boldsymbol{r}_{G/C}\right) = \ddot{\theta}\boldsymbol{k} \times r\boldsymbol{j}$$

which gives us a lateral acceleration of $\bar{a}_x = -r\ddot{\theta}$. Thus we've lost the independent unknown $\bar{a}_x$ and are free to solve the problem. Note that $\dot{\theta}$ doesn't enter into our acceleration equation because the mass center is moving horizontally. There's no centripetal acceleration component and thus no need for the $\boldsymbol{\omega} \times (\boldsymbol{\omega} \times \boldsymbol{r}_{G/C})$ term of (6.25).

That's one scenario. The other scenario is the case in which we have slip. For this case there's no direct correlation between $\bar{a}_x$ and $\ddot{\theta}$ as there is in the pure rolling case. What we *do* know, however, is that in order for slip to take place, the contact force requirements for pure rolling motion must exceed the frictional force limit μN_1. When this happens, the tractive force N_2 takes on this maximal value, which means N_2 is known. Therefore we again have reduced the unknowns to three and can solve the problem.

Solve Now we choose one of our two possibilities (pure roll or roll with slip), solve the problem based on this choice, and then check to make sure the choice is validated by our results. Because there's no particular way to determine what's going to happen without some experience with the system, I will guess that roll with slip occurs.

$\boldsymbol{i}$: $$\mu N_1 = m\bar{a}_x$$

$\boldsymbol{j}$: $$N_1 - mg = 0$$

$\boldsymbol{k}$: $$I_G\ddot{\theta} = \mu r N_1 - M$$

where I've used $\bar{a}_y = 0$ and $N_2 = \mu N_1$. Because $N_1 = mg$, we can re-express our $\boldsymbol{i}$ and $\boldsymbol{k}$ equations as

$\boldsymbol{i}$: $$\mu g = \bar{a}_x$$

$\boldsymbol{k}$: $$I_G\ddot{\theta} = \mu rmg - M$$

$$\bar{a}_x = \mu g = 0.7(9.81\,\text{m/s}^2) = 6.87\,\text{m/s}^2 \qquad \ddot{\theta} = \frac{\mu rmg - M}{I_G} = -36.3\,\text{rad/s}^2$$

where I used $I_G = \frac{mr^2}{2} = 0.5(23\,\text{kg})(0.3\,\text{m})^2 = 1.035\,\text{kg}\cdot\text{m}^2$, the mass moment of inertia for a uniform disk. Because $N_2 = \mu N_1$ we also have $N_2 = \mu mg = 0.7(23\,\text{kg})(9.81\,\text{m/s}^2) = 158\,\text{N}$.

Check How can we most easily check the validity of our analysis? The two main assumptions made were that the center of mass stays a distance r from the ground and that the wheel slips. Checking if the wheel stays put is simply a matter of checking N_1's sign. If it's positive, then the wheel must be in contact with the ground. We have already seen that N_1 is simply mg, which is clearly positive. Thus that's an okay assumption. How can we determine if the slip condition is correct? The most complete way is to assume the opposite and show that that can't be true. So let's assume that slip *doesn't* occur. That means we have pure rolling motion and the equations

$$N_2 = m(-r\ddot{\theta}), \qquad N_1 - mg = 0, \qquad I_G\ddot{\theta} = rN_2 - M$$

The first and third equation give us $\ddot{\theta} = -\frac{M}{I_G + mr^2} = -27.4\,\text{rad/s}^2$. Because $N_2 = -mr\ddot{\theta}$ we have

$$N_2 = -(23\,\text{kg})(0.3\,\text{m})(-27.4\,\text{rad/s}^2) = 189\,\text{N}$$

Is this possible? We have already seen that the maximum force that can exist between wheel and ground is $\mu mg = 158$ N, and 189 N exceeds this limit by a comfortable margin. Thus we *can't* have pure rolling because there isn't enough friction to support it. So our assumption of slip was a good one.

EXAMPLE 7.15 ACCELERATION RESPONSE OF A DRIVEN WHEEL—TAKE TWO (Theory on page 424)

In Example 7.14 a wheel slipped as it accelerated forward, just as a powerful car slips when gunned in first gear. One way to increase a car's torque is by installing a supercharger, the hope being to increase the car's acceleration. Let's examine what effect this would have on the wheel of Example 7.14. We will assume that the available torque from the engine has doubled. How does that affect the forward acceleration? Use the same parameters as in Example 7.14, except increase the applied moment M to 170 N·m.

Goal Find the forward acceleration of the wheel and compare the result with the result of Example 7.14.

Assume Because the wheel slipped with 85 N·m being applied, it's rather certain that it will still slip if we increase the applied torque. So we will assume slip conditions.

Draw Same as in Example 7.14.

Formulate Equations Same as in Example 7.12.

Solve The slip condition equations of motion are

$$\boldsymbol{i}: \quad \mu N_1 = m\bar{a}_x$$

$$\boldsymbol{j}: \quad N_1 - mg = 0$$

$$\boldsymbol{k}: \quad I_G\ddot{\theta} = \mu r N_1 - M$$

The second and third equations let us solve for $\ddot{\theta}$:

$$\ddot{\theta} = \frac{\mu rmg - M}{I_G} = \frac{0.7(0.3\,\text{m})(23\,\text{kg})(9.81\,\text{m/s}^2) - 170\,\text{N}\cdot\text{m}}{1.035\,\text{kg}\cdot\text{m}^2} = -118.5\,\text{rad/s}^2$$

This is much bigger in magnitude than the $-36.3\ \text{rad/s}^2$ we obtained in the previous example. So it's clear that doubling the torque does something. However, $\ddot{\theta}$ isn't directly related to the forward acceleration, because we're not in a pure rolling condition. We have to solve for the forward acceleration using our $\boldsymbol{i}$ expression:

$$\mu N_1 = m\bar{a}_x$$

$$\bar{a}_x = \frac{\mu N_1}{m} = \mu g = 6.87\,\text{m/s}^2$$

We see that doubling the torque didn't increase acceleration at all! The reason is that the maximal force that can be developed between the ground and the wheel, at least according to our model, is the coefficient of friction times the normal force. This quantity isn't going to change when the applied moment changes—the moment doesn't even enter into the equation. Increasing the car's torque will spin its tires more quickly, but it won't move the car any quicker. An important point to remember if you ever plan to participate in a streetlight drag race.

- Note that this result isn't meant to imply that there's no benefit to doing burnouts. They create a lot of tire smoke and noise, and they lay down a covering of rubber on the road surface, enhancing your ability to get up to speed quickly the next time around.

Check We've already done all the checking we need in Example 7.14. There's no need to repeat it.

After thinking about the dynamics of rolling wheels, many students start to get confused as they think more carefully about the ground interaction force. For instance, consider **Figure 7.54**, which shows the free-body diagram for a wheel being acted on by an applied moment M. What students often find confusing is the fact that the ground seems to be accelerating the wheel via the ground/tire traction force $N_2\boldsymbol{i}$. The equation of motion seems clear on this point:

$$m\ddot{x} = N_2$$

where m is the mass of the wheel. But how can this force be the one pushing the wheel? If it were, wouldn't the wheel rotate counterclockwise instead of clockwise as it did in Examples 7.14 and 7.15? And how can the ground "push" in the first place? Isn't the ground just sitting there? How can it exert a force that moves with the wheel?

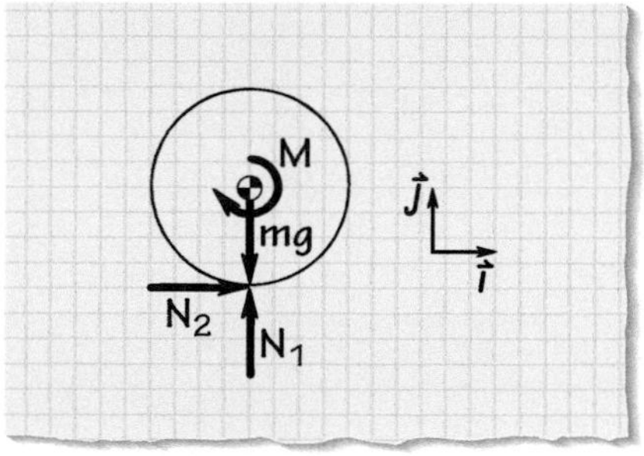

Figure 7.54 Rolling wheel FBD

These are all good questions. To answer them, I suggest that, instead of a wheel with an applied moment, we look at a more physically accessible system: someone trying to paddle a canoe across dry land. This usually happens only in cartoons, and so Captain Insanity, previously seen in **Exercise 7.1.19**, has been drafted to illustrate what's going on. **Figure 7.55** shows three snapshots of Captain Insanity, with time increasing from (*a*) to (*c*). In **Figure 7.55*a*** he has touched his mystic sword to the ground and has started to pull. In (*b*) he has reached the halfway point of the stroke, and in (*c*) he is just finishing the stroke.

This situation provides a great analogue to a torque-driven wheel. Look at **Figure 7.56**, which shows the forces acting on the sword for the configuration of **Figure 7.55*b***. Captain Insanity is exerting two forces on the sword: $F_1\boldsymbol{i}$ and $-F_2\boldsymbol{i}$. He is using his left hand to push forward against the top of the sword ($F_1\boldsymbol{i}$), and with his right hand he's pulling the sword back ($-F_2\boldsymbol{i}$). The other end of the sword is stuck into the ground and is being pulled back against the dirt. I've even drawn in a little mound of dirt to illustrate how the sword tip is pushing against it. What's naturally going to pop up where you're pushing against something? An equal and

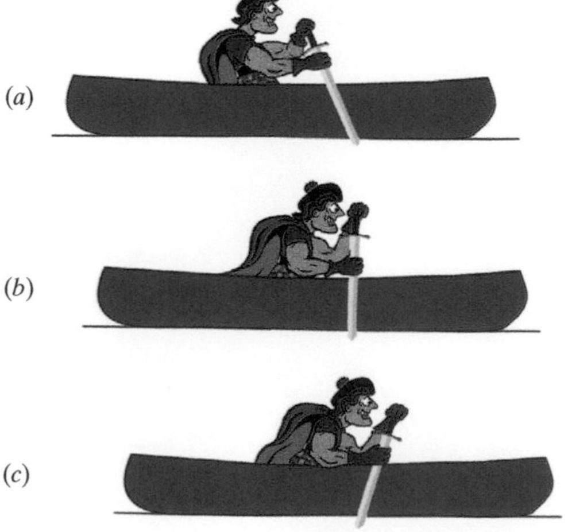

Figure 7.55 A rowing analogue of rolling

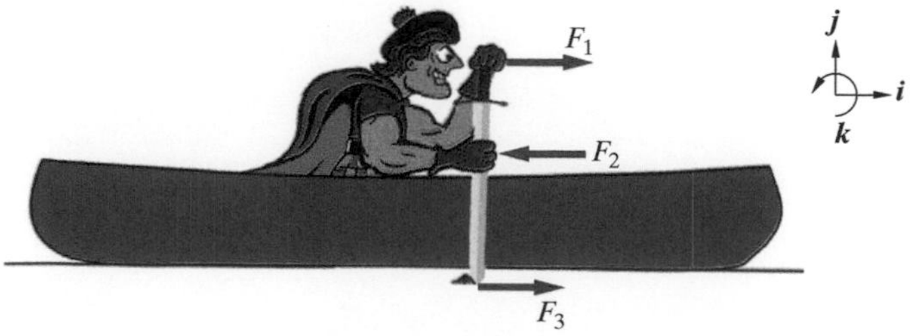

Figure 7.56 FBD of rower

opposite force, of course. And that's what's shown in **Figure 7.56**, the reaction of the ground against the sword: $F_3\boldsymbol{i}$.

If we compare **Figure 7.56** with **Figure 7.54**, we can make some immediate connections. The force $N_2\boldsymbol{i}$ on the wheel serves the same purpose as $F_3\boldsymbol{i}$ does for the sword/canoe. It's the reaction of the ground to the vehicle moving on it. In **Figure 7.54** we have an applied moment $-M\boldsymbol{k}$. Do we have something similar in **Figure 7.56**? Sure. The two forces $F_1\boldsymbol{i}$ and $-F_2\boldsymbol{i}$ produce a moment. If we could move Captain Insanity's hands closer and closer together while keeping the forces they exert equal in magnitude, they would produce a pure clockwise couple.

This canoe example holds only in an approximate sense, of course, because it isn't rotating. However, what you see in the canoe approximates what's happening with the rolling wheel as it rotates through some small angle. The applied moment acts to twist the wheel around the ground contact point, and the ground acts as a stable surface for the wheel to push against.

This point of view also shows what happens during wheel slip. Imagine that the ground isn't packed firmly enough to resist the sword tip. If this were the case, the sword would push across the dirt, losing efficiency as it slid backwards instead of remaining fixed in place. The same thing happens to a wheel. If the coefficient of friction isn't sufficiently large to supply the reaction force needed by the wheel, then the wheel slips—an analogue to the sword slipping across the ground.

EXAMPLE 7.16 FALLING SPOOL (Theory on page 424)

The solid spool shown in **Figure 7.57** is made up of an inner cylinder with radius R and two outer cylinders, each with radius r. A block A is attached by a massless string to the inner cylinder. The two outer cylinders are attached by massless strings to the ceiling. The mass moment of inertia of the spool about its longitudinal central axis is I_S, the mass of the spool is m_S, and the block's mass is I_B. Determine the angular acceleration of the spool immediately after the system is released from rest.

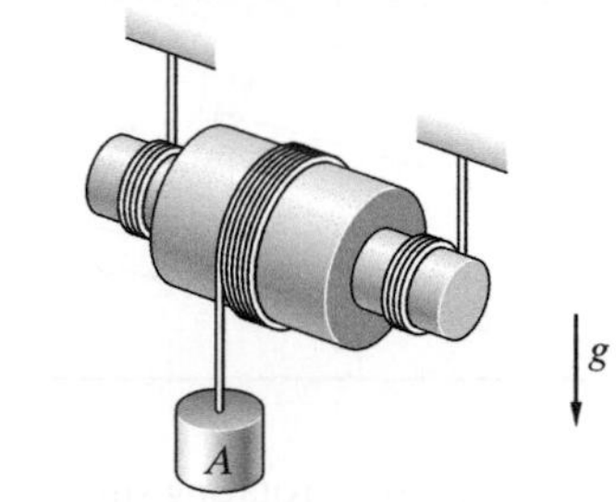

Figure 7.57 Falling spool

Goal Find the acceleration of mass A with respect to the ground.

Given Spool's mass and moment of inertia, block's mass, system dimensions.

Assume All the forces in this example are vertical and therefore the bodies can only rotate and translate downward.

Draw **Figure 7.58** shows a labeled sketch and **Figure 7.59** shows a FBD=IRD sketch.

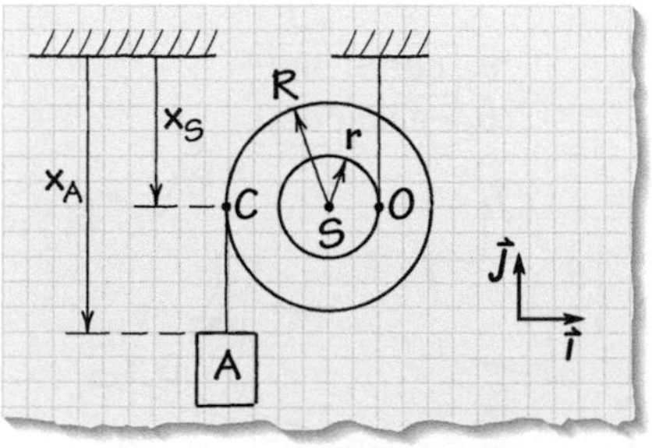

Figure 7.58 Labeling for falling spool

Formulate Equations Because the string is inextensible and all forces are oriented downward, the point labeled O can be viewed as a stationary pivot point:

$$\ddot{x}_S = \ddot{\theta} r \tag{7.61}$$

where θ, the rotation angle of the spool, is taken as positive in the clockwise direction. The acceleration of the block is the same as the vertical acceleration of point C on the spool:

(7.61) ⇒
$$\ddot{x}_A = \ddot{x}_C = \ddot{x}_S + R\ddot{\theta} = \ddot{\theta}(r + R) \tag{7.62}$$

Force balance, block:
$$m_A g - T_2 = m_A \ddot{x}_A \tag{7.63}$$

Force balance, spool:
$$m_S g + T_2 - 2T_1 = m_S \ddot{x}_S \tag{7.64}$$

Moment balance, spool:
$$R T_2 + 2r T_1 = I_S \ddot{\theta} \tag{7.65}$$

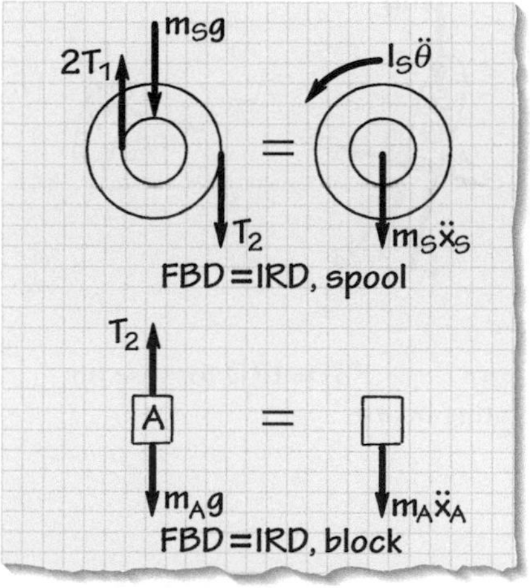

Figure 7.59 FBD=IRD for falling spool

Solve

(7.61)→(7.65) ⇒
$$\ddot{\theta} = \frac{[m_S r + m_A(r + R)]g}{r^2 m_S + (r + R)^2 m_A + I_S}$$

Check All the units are consistent, a good sign. Letting $m_S = I_S = 0$ gives us $\ddot{\theta} = \frac{g}{r+R}$, which leads to $\ddot{x}_A = g$, matching physical expectations.

EXAMPLE 7.17 TIPPING OF A MING VASE (Theory on page 424)

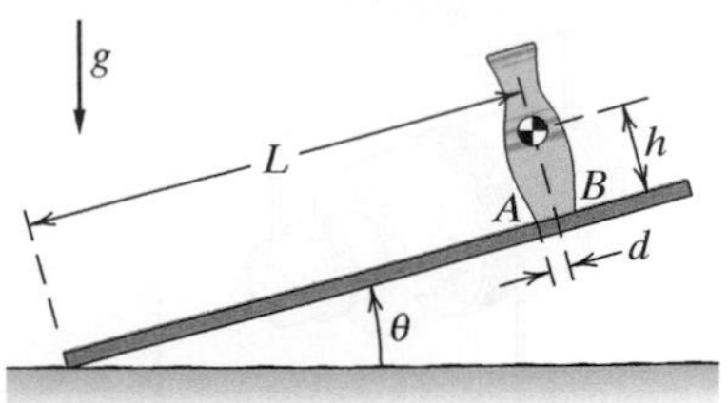

Figure 7.60 Vase on a rotating surface

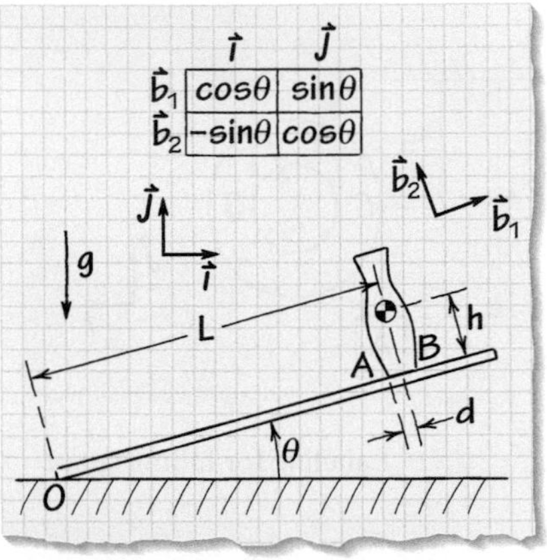

Figure 7.61 Vase dimensions and parameters

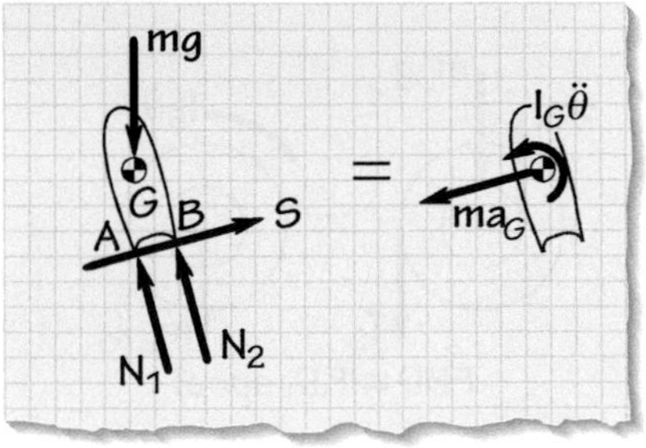

Figure 7.62 FBD=IRD for vase on rotating board

Rather than the squat bowl of Example 3.6, we now consider a tall Ming vase (**Figure 7.60**). The vase is initially stationary with respect to the supporting board, its center of mass a distance L from the hinge O (along the slope), and the board is rotating at a constant $\dot{\theta} = \omega_0$. We will include a static coefficient of friction μ_s, μ_s chosen large enough so that the vase doesn't *ever* slip. It can, however, tip over. Because the board is rotating, the vase feels a centripetal acceleration even before it starts tipping. Model the vase as contacting the board at two points, A and B, find the angle at which tip begins. Is this angle greater or less than the angle associated with a static imbalance? $\omega_0 = 0.4$ rad/s, $L = 2.0$ m, $d = 0.04$ m, and $h = 0.09$ m.

Goal Formulate the equation of motion for rotation of the vase and determine when tipping occurs. Compare results to the static case.

Given Slip will not occur.

Draw **Figure 7.61** shows the coordinates to be used. All that matters to the analysis are the contact points with the board and the position of the mass center relative to them. The FBD=IRD is shown in **Figure 7.62**.

Formulate Equations For tipping to occur, one of the normal forces in **Figure 7.62** has to drop to zero. In the static case, the force at B becomes zero when the board's inclination is too great, and we will continue to focus on this with our rotating board.

Note that the frictional force holding the vase in place is represented by a single force S. There's no need to know the specifics of how much force acts at A and how much at B, because the lines of action are the same.

The most convenient approach is to sum moments about A. In this way, the normal force N_1 and the frictional force S drop out of the moment equation because they both act through A. Notice that we will have to account for the fact that A is accelerating—it's not a fixed point. Thus we will use (7.48).

The acceleration of the vase's center of mass, assuming no tipping yet, is

$$\boldsymbol{a}_G = \boldsymbol{\omega}_0 \times \left(\boldsymbol{\omega}_0 \times \boldsymbol{r}_{G_{/O}}\right) = -\omega_0^2(L\boldsymbol{b}_1 + h\boldsymbol{b}_2)$$

Summing moments about A gives

$$\sum \boldsymbol{M}_A = I_G\ddot{\theta}\boldsymbol{b}_3 + \left(\boldsymbol{r}_{G_{/A}} \times m\boldsymbol{a}_G\right)$$

$$\overbrace{(d\boldsymbol{b}_1 + h\boldsymbol{b}_2)\times(-mg\boldsymbol{j})}^{\text{moment due to gravity}} + \overbrace{2d\boldsymbol{b}_1 \times N_2\boldsymbol{b}_2}^{\text{moment due to } N_2} = I_G\ddot{\theta}\boldsymbol{b}_3 + \overbrace{(d\boldsymbol{b}_1 + h\boldsymbol{b}_2)}^{\boldsymbol{r}_{G_{/A}}} \times \overbrace{(-m\omega_0^2)(L\boldsymbol{b}_1 + h\boldsymbol{b}_2)}^{m\boldsymbol{a}_G}$$

$$mg(h\sin\theta - d\cos\theta) + 2dN_2 = I_G\ddot{\theta} + hm\omega_0^2(L - d) \qquad (7.60)$$

Solve Before tipping, $\ddot{\theta} = 0$, and at the limit just before tipping occurs we have $N_2 = 0$. Thus (7.60) becomes

$$g(h\sin\theta - d\cos\theta) = h\omega_0^2(L-d) \quad \Rightarrow \quad \sin\theta - \left(\frac{d}{h}\right)\cos\theta = \frac{\omega_0^2}{g}(L-d) \tag{7.61}$$

Solving (7.61) will take some thought. What we're looking for is the value of θ that satisfies the equation for the given parameter values. Often this will mean using some numerical root finder. In this particular case, however, we can get a closed-form solution by recalling that $P\sin(\theta - \phi) = P\sin\theta\cos\phi - P\cos\theta\sin\phi$. Set this form equal to $\sin\theta - \frac{d}{h}\cos\theta$ to match up with (7.61): $P\sin\theta\cos\phi - P\cos\theta\sin\phi = \sin\theta - (d/h)\cos\theta$. Equating the coefficients of the $\sin\theta$ and $\cos\theta$ terms on both sides of the equation gives us

$$P\cos\phi = 1 \quad \text{and} \quad P\sin\phi = \frac{d}{h}$$

Squaring both expressions and adding give us $P = \sqrt{1 + \left(\frac{d}{h}\right)^2}$ and dividing the two expressions leads to

$$\tan\phi = \frac{d}{h} \quad \Rightarrow \quad \phi = \tan^{-1}\left(\frac{d}{h}\right)$$

Thus we can write (7.61) as $\sqrt{1 + \left(\frac{d}{h}\right)^2}\,\sin(\theta - \phi) = \frac{\omega_0^2}{g}(L-d)$, or, more simply

$$\sin(\theta - \phi) = \frac{\omega_0^2(L-d)}{g\sqrt{1 + \left(\frac{d}{h}\right)^2}}$$

Solving now becomes straightforward. We can evaluate the righthand side, take the inverse sine, and then add ϕ to find θ:

$$\sin(\theta - \phi) = \frac{(0.4\,\text{rad/s})^2(2\,\text{m} - 0.04\,\text{m})}{(9.81\,\text{m/s}^2)\sqrt{1 + \left(\frac{0.04}{0.09}\right)^2}} = 0.0292$$

$$\theta - \tan^{-1}\left(\frac{0.04}{0.09}\right) = \sin^{-1}(0.0292)$$

$$\theta = 0.418\,\text{rad} + 0.029\,\text{rad} = 0.447\,\text{rad}$$

Thus the vase tips when $\theta = 0.447\,\text{rad} = 25.6°$. In the static case, the vase tips when the point A is directly under the mass center. This occurs when $\theta = \tan^{-1}\left(\frac{0.04}{0.09}\right) = 0.418\,\text{rad} = 24.0°$.

Check The rotation of the board has served to delay tipping, allowing the board to incline about 7% farther than the static case. This increase is a direct result of the centripetal acceleration imposed on the vase by the board's rotation.

EXAMPLE 7.18 EQUATIONS OF MOTION FOR A SIMPLE CAR MODEL (Theory on page 424)

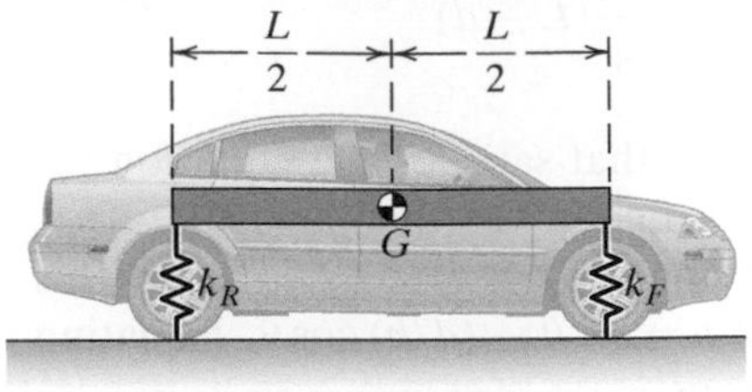

Figure 7.63 Half-car model

A common dynamic system you have all encountered in your day-to-day life is a car. We have spent some time already looking at the car's wheels, so now it's time to think about the body/suspension. The suspension is a complicated system of spring and damper elements that supports the body as the car is driven up, down, and around. Small imperfections in the road transmit vibrations through the suspension to the body, and traveling along the road causes the body to sway, tilt, and so on. If we *really* wanted to complicate matters, we wouldn't even view the body as being rigid but would instead model it as a flexible structure that can flex in a variety of ways [4].

Shown in **Figure 7.63** is a simple approximate model overlaid on a car profile so you can see what the simplified pieces correspond to. This is called a **half-car model** because we're flattening the entire vehicle (four wheels) into a "flat" car model (two wheels). All the mass of the car and its contents have been replaced by a rigid bar of mass m, and the suspension system and tires have been replaced by simple linear springs, one in back and one in front. For this analysis we will consider the car to be stationary and will neglect any fore/aft motion. All we're concerned with is vertical motion and rotation. Assume that the springs are already compressed due to gravity and that, in their steady-state condition, the bar is horizontal. For this to be so, we will need a compression of $\frac{mg}{2k_R}$ at the rear and $\frac{mg}{2k_F}$ at the front, where k_R and k_F are the effective rear and front linear spring constants, respectively. (You can verify this with a static analysis if you wish.) What we want to do is find the equations of motion that govern the car's dynamic response.

Goal Determine the car's equations of motion.

Given Simplified model consisting of a rigid bar and two linear springs.

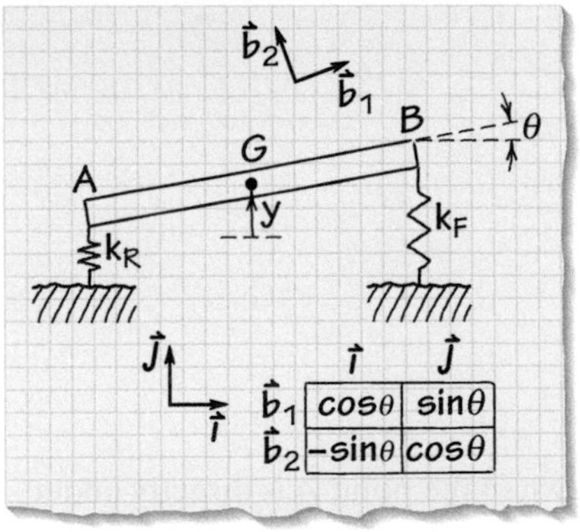

Figure 7.64 Labeling for car model

Assume To simplify the analysis, assume that the vertical motions and rotation angles are small. This means that the springs will stay in their linear range, and we can safely assume that the forces acting on the ends of the bar are always vertical. We'll use y to represent vertical motion of the mass center away from its equilibrium position and θ to represent counterclockwise rotations of the bar with respect to a horizontal orientation.

Draw **Figure 7.64** shows the bar in a displaced position, along with the needed labeling, and **Figure 7.65** shows the relevant free-body and inertial response diagrams.

Formulate Equations To find the force exerted by each spring, we need to find the spring displacement and then multiply by the appropriate spring constant. Referencing from G, the positions of A and B are initially $\boldsymbol{r}_R = -L/2\boldsymbol{i}$, $\boldsymbol{r}_F = L/2\boldsymbol{i}$. After the bar has been displaced, we have

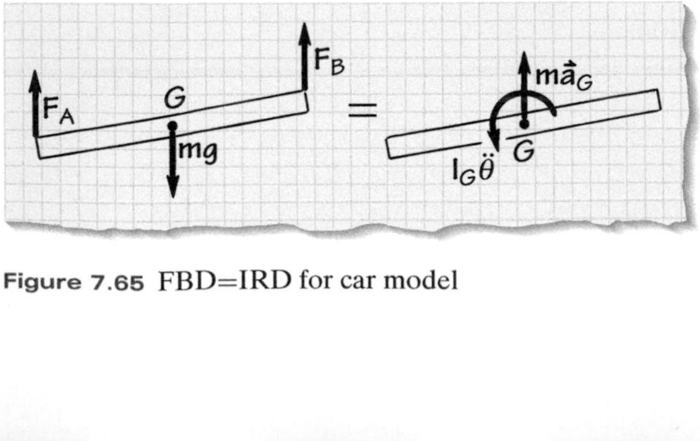

Figure 7.65 FBD=IRD for car model

$$\boldsymbol{r}_R = y\boldsymbol{j} - (L/2)\boldsymbol{b}_1 = (y - (L/2)\sin\theta)\,\boldsymbol{j} - (L/2)\cos\theta\,\boldsymbol{i}$$
$$\boldsymbol{r}_F = y\boldsymbol{j} + (L/2)\boldsymbol{b}_1 = (y + (L/2)\sin\theta)\,\boldsymbol{j} + (L/2)\cos\theta\,\boldsymbol{i}$$

Subtracting the initial position from the displaced position gives us our end deflections:

$$\Delta\boldsymbol{r}_R = (y - (L/2)\sin\theta)\,\boldsymbol{j} - (L/2)(\cos\theta - 1)\boldsymbol{i}$$
$$\Delta\boldsymbol{r}_F = (y + (L/2)\sin\theta)\,\boldsymbol{j} + (L/2)(\cos\theta - 1)\boldsymbol{i}$$

Because we're looking only at small values of y and θ, these expressions simplify (via the approximations $\sin\theta \approx \theta$ and $\cos\theta \approx 1$ for small θ) to

$$\Delta\boldsymbol{r}_R = y\boldsymbol{j} - (L/2)\boldsymbol{b}_1 = (y - (L/2)\theta)\,\boldsymbol{j}$$
$$\Delta\boldsymbol{r}_F = y\boldsymbol{j} + (L/2)\boldsymbol{b}_1 = (y + (L/2)\theta)\,\boldsymbol{j}$$

The forces F_A and F_B can now be calculated:

$$F_A = k_R\left(\frac{mg}{2k_R} - y + (L/2)\theta\right)$$

$$F_B = k_F\left(\frac{mg}{2k_F} - y - (L/2)\theta\right)$$

Examine **Figure 7.65** to see that all the terms in our force equations make physical sense. Let's look at F_R to see how. There's a positive component $\frac{mg}{k_R}$, the upward force produced by the spring's static loading, which counteracts the downward gravitational force. Because the bar moves up as y increases, the spring is stretched, producing a restoring force $-k_R y$. Finally, the bar can rotate. If the rotation is positive (counterclockwise), the left end of the bar dips down, increasing the spring's compression and therefore increasing the force $(k_R\frac{L}{2}\theta)$. A similar analysis holds for the right end of the bar, the only real difference being that a positive rotation of the bar decreases the force $k_F\frac{L}{2}\theta$ rather than increasing it.

Our final equations are therefore

$\sum \boldsymbol{F}$:
$$\left[k_R\left(\frac{mg}{2k_R} - y + (L/2)\theta\right) + k_F\left(\frac{mg}{2k_F} - y - (L/2)\theta\right) - mg\right]\boldsymbol{j} = m\ddot{y}\boldsymbol{j}$$

$$m\ddot{y} + (k_R + k_F)y + (k_F - k_R)(L/2)\theta = 0$$

$\sum \boldsymbol{M}_G$:
$$\boldsymbol{r}_{A/G} \times \boldsymbol{F}_A + \boldsymbol{r}_{B/G} \times \boldsymbol{F}_B = \bar{I}\ddot{\theta}$$

$$-\frac{Lk_R}{2}\left(\frac{mg}{2k_R} - y + \frac{L\theta}{2}\right) + \frac{Lk_F}{2}\left(\frac{mg}{2k_F} - y - \frac{L\theta}{2}\right) = \bar{I}\ddot{\theta}$$

$$\bar{I}\ddot{\theta} + \left[\frac{(k_R + k_F)L^2}{4}\right]\theta + \left[\frac{(k_F - k_R)L}{2}\right]y = 0$$

Check What's interesting about these equations is that gravity doesn't show up. You might have expected it to appear, inasmuch as it's acting along the entire bar. Recall that we based our analysis on motions that deviate from the static equilibrium position. Gravity definitely affects the static equilibrium, and the car will sag until the force exerted by the springs is enough to counteract the steady pull of gravity. After that, however, gravity has nothing to contribute. All of the motions that occur about the equilibrium position are affected only by the physical springs and masses.

EXAMPLE 7.19 ANALYSIS OF A SIMPLE TRANSMISSION (Theory on page 424)

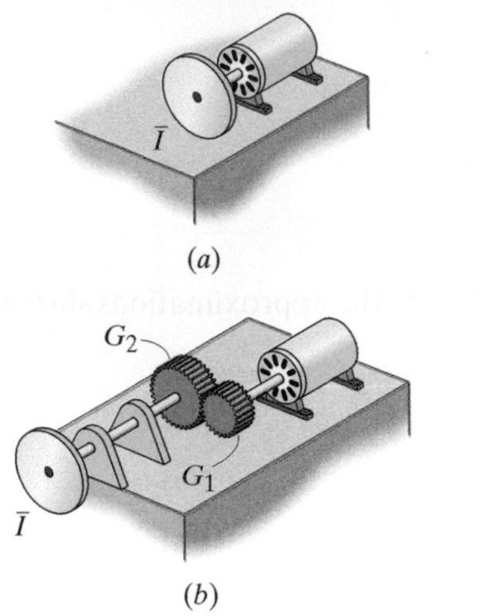

Figure 7.66 Motor-driven rotating disk

Figure 7.66*a* shows a motor that produces a constant moment M and that drives a solid disk with inertia about its rotational axis of $\overline{I}$. After examining this system, you decide that a more elegant arrangement would have the disk connect to the motor via two meshing gears (**Figure 7.66*b***). Calculate how this would change the disk's rotational acceleration. Assume that only the rotating disk has any mass.

G_1 and G_2 have effective radii r_1 and r_2, respectively, with $r_2 > r_1$.

Goal Determine how the acceleration of the solid disk varies between a geared drive and a direct drive connection.

Given Moment supplied by motor, mass moment of inertia of disk, and gear sizes.

Draw **Figure 7.67** shows the forces and moment acting on the disk, and **Figure 7.68** shows the same for the disk-gear system.

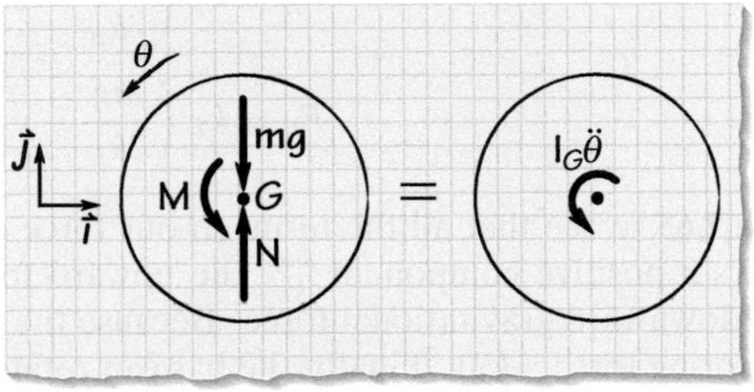

Figure 7.67 Direct-drive connection, FBD=IRD

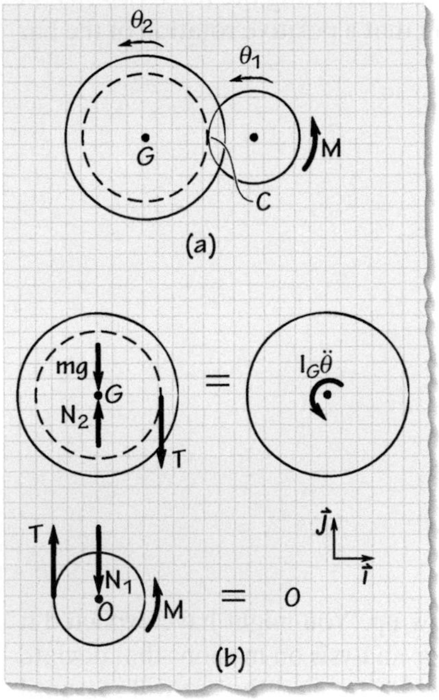

Figure 7.68 Geared connection, FBD=IRD

Assume You can see from **Figure 7.67** that only the applied moment M will affect the disk's rotation. The disk's weight is counteracted by a reaction force supplied by the shaft, and because both go through the disk's center, neither will affect the angular acceleration. The geared case (**Figure 7.68*a***), however, is more complex. For this case we have two separate bodies that are coupled together at their point of contact C. If we break the bodies apart, we have the two free-body diagrams shown in **Figure 7.68*b***. Each body has forces that go through their respective points of rotation and therefore don't affect their angular acceleration. In addition, the interaction force T provides a moment both to the motor and to the disk. We will assume that any side forces in between the two systems (***i*** direction) are negligible.

Formulate Equations We'll set up the simple direct-drive system first. The rotation of the disk is given by θ, and a moment sum about G gives us

$$\overline{I}\ddot{\theta} = M \tag{7.62}$$

The two-gear case requires two equations—a moment balance for each body. We don't have to bother with a vertical force balance because all that would do is let us solve for N_1 and N_2, something that's not necessary to solve the problem. The two moment balances about the bodies' respective centers are

$$M - r_1 T = 0 \tag{7.63}$$

$$-r_2 T = \bar{I}\ddot{\theta}_2 \tag{7.64}$$

The reason for the "= 0" in (7.63) is that the mass moment of inertia is zero (due to the mass being zero).

Solve Equation (7.62) can be solved immediately:

$$\ddot{\theta} = \frac{M}{\bar{I}}$$

Combining (7.63) and (7.64) gives us

$$\ddot{\theta}_2 = -\frac{r_2 M}{r_1 \bar{I}}$$

Check This result is worth some discussion. First, notice that the sign of the angular accelerations is opposite for the two cases. This makes sense if you look at the system. A counterclockwise moment for the first case will certainly turn the disk in a counterclockwise direction. For the two-gear case, however, if the smaller gear turns counterclockwise, the larger gear to its left has to turn clockwise in order to ensure that the point of contact has the same velocity for each gear. This is important because the gear teeth are meshing at that point, and so they *must* go in the same direction.

The bigger deal is the factor $\frac{r_2}{r_1}$ that shows up for the two-gear case. Let's say that G_2 has twice the radius of G_1 (small gear turning a big one). This would result in a *doubling* of the angular acceleration. As far as acceleration is concerned, it's as if we had doubled the effective moment being applied.

All this has solid real-world implications. What we have just done is uncover what goes on in a car's transmission. The engine is connected to the wheels via a series of gears. If the gear ratio from wheel to engine is large (large gear at the wheels, small gear at the engine), that translates into a large angular acceleration and hence a large linear acceleration of the car (the wheels are in contact with the ground). So now, when you hear people talk about final drive ratios and switching the ratios of their transmission, you'll know what they're talking about.

By the way, lest you think that all you need do is swap in some huge gears in order to give your car some monster amount of torque, you have to also realize that those large gears mean the engine is going to have to spin faster. The rotational speed of the wheel is defined by the car's translational speed and the wheel's radius, so the speed at which the wheel spins can't be changed to accommodate the new gearing. Thus any changes in the gear ratios have to be reflected in the engine's speed. Car engines can spin only so fast before they hit their redline; thus realistically you can change the amount of torque your car puts to the ground by only a limited amount.

EXERCISES 7.3

7.3.1. [Level 1] A solid reel (inner cylinder attached to two outer cylinders) is positioned on a slope. The reel has mass m and mass moment of inertia I_G about its center. An inextensible string, wrapped around the inner cylinder, connects to a massless spring (spring constant k). Assuming roll without slip, construct an appropriate FBD=IRD diagram, and write down the governing equation of motion (good for $t \geq 0$) for the reel. Assume that at $t = 0$ the reel is released from rest and that the spring is initially unstretched.

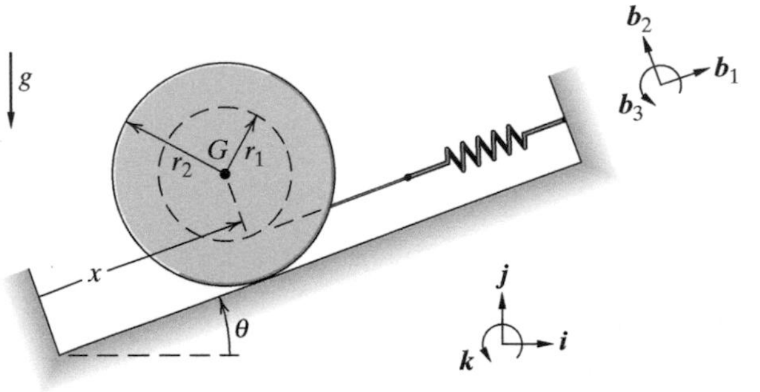

E7.3.1

7.3.2. [Level 1] A simple model of a rocket having two engines capable of applying thrust (N_1 and N_2) is shown. Each engine is located 0.5 m from the rocket's centerline. The rocket has mass m and mass moment of inertia $\bar{I}$. Treat the rocket as a uniform bar of length L. What are the rocket's equations of motion?

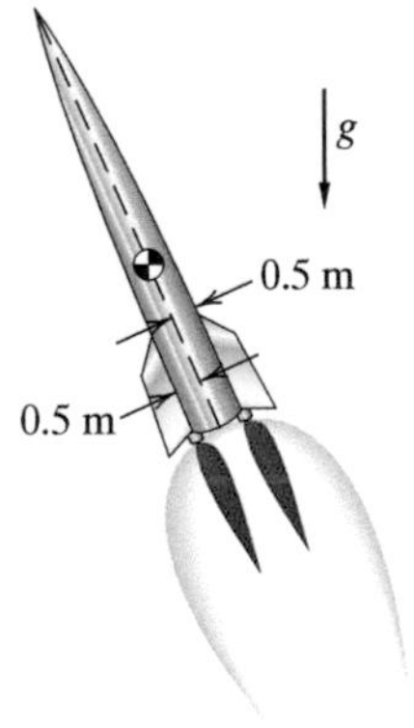

E7.3.2

7.3.3. [Level 1] A uniform sphere of mass m and radius r is released from rest at the top of a rough surface inclined at an angle β with respect to the horizontal. If the coefficient of friction between the sphere and surface is μ, find the maximum angle for which the sphere will roll without slip.

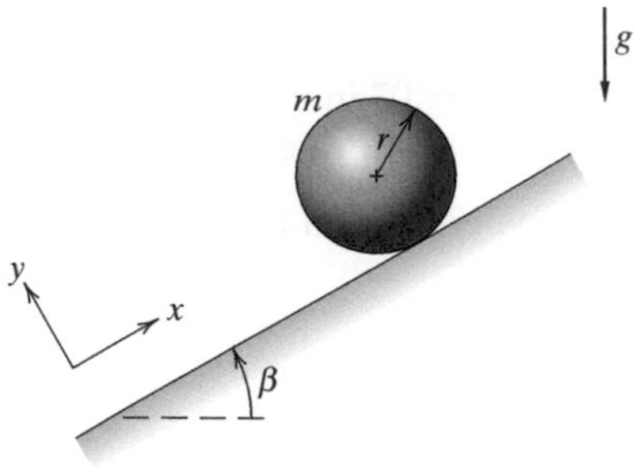

E7.3.3

7.3.4. [Level 1] A uniform cylindrical body is connected to the ceiling by means of a linear spring with spring constant k and through a massless rope that's wrapped around the body and that attaches to the ceiling at A. Assume that the spring is stretched 0.06 m from its rest length in the position shown. The cylinder has a weight of 7.0 kg, it has a radius of 0.15 m, and $k = 15$ kg/m. Upon release, what is the angular acceleration of the cylinder?

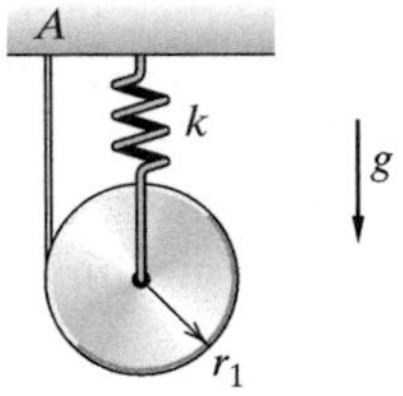

E7.3.4

7.3.5. [Level 1] The illustrated system is a wheel with an attached rod. A motor in the rod applies a moment $M_0\boldsymbol{k}$ to the wheel, which rolls without slip on the ground. $M_0 = 9.50\,\text{N}\cdot\text{m}$. The wheel is accelerating at a constant rate to the left and $\beta = 20°, \dot{\beta} = \ddot{\beta} = 0$. $m_{rod} = 10\,\text{kg}$, $R = 0.5\,\text{m}$, and $|\overline{OA}| = 0.5\,\text{m}$. Determine $\ddot{x}\boldsymbol{i}$ (the acceleration of the wheel's center of mass).

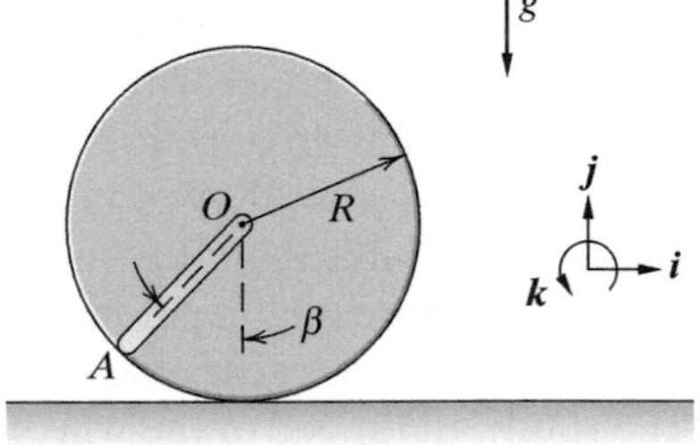

E7.3.5

7.3.6. [Level 1] In this exercise we will calculate the reaction force at a batter's hand. Model the bat as shown below, a rigid, uniform bar swung around the point O. The batter grips the bat at A, and the bat is struck by the baseball at B. Assume that the baseball imparts a force F to the bat, acting at B. Find the reaction force at A due to this impact force. Assume a constant angular speed ω of the bat prior to impact.

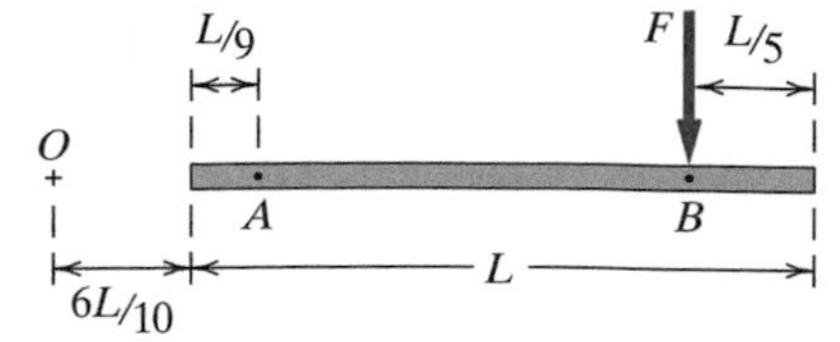

E7.3.6

7.3.7. **[Level 1]** Determine the equations of motion for the illustrated system using the illustrated coordinates y and θ. Consider only small values of y and θ and assume that, when in equilibrium, the bar is horizontal. Neglect gravity and nonlinear terms ($\dot{\theta}^2, \ldots$).

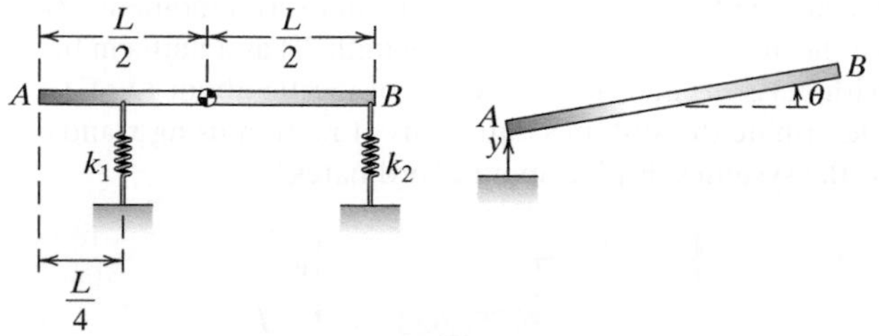

E7.3.7

7.3.8. **[Level 1]** In this exercise we will analyze the motion of a yo-yo, consisting of a central axle of radius r_1 and a main body (two disks of radius r_2). A massless cord is wrapped around the central axle, and the free end is pulled upward with a force F. The axle has a mass $m_1 = 0.015$ kg, and each disk has a mass of $m_2 = 0.1$ kg. $r_1 = 0.003$ m, $r_2 = 0.02$ m, and $F = 3$ N. What are the angular and vertical accelerations of the yo-yo?

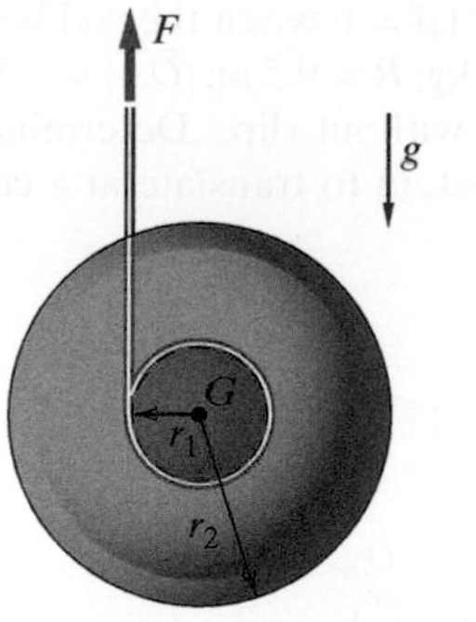

E7.3.8

7.3.9. **[Level 2] Computational** A circular disk (weight of 1.4 kg and radius of gyration about its mass center equal to 0.054 m) is attached to the ceiling via a linear spring (spring constant of 4.5 kg/m and an unstretched length of 0.15 m) and a massless string that's wrapped around the cylinder and connected at A to the ceiling. $r_1 = 0.076$ m. If the cylinder is released with the spring unstretched, how fast will its center be traveling after falling for 0.339 s?

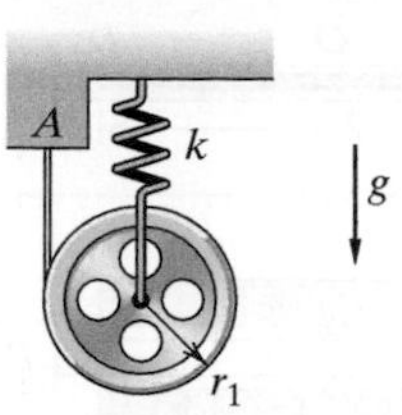

E7.3.9

7.3.10. **[Level 2]** The illustrated system is a wheel with an attached rod. A motor in the rod applies a moment $M_0\boldsymbol{k}$ to the 20 kg wheel. $M_0 = 9.50$ N$\cdot$ m. The inclination of the rod from vertical (β) is constant. $m_{rod} = 10$ kg, $R = 0.5$ m, and $|\overline{OA}| = 0.5$ m. Treat the wheel as a uniform cylinder. What is the minimum coefficient of friction that will allow roll without slip and a constant acceleration of the system to the left?

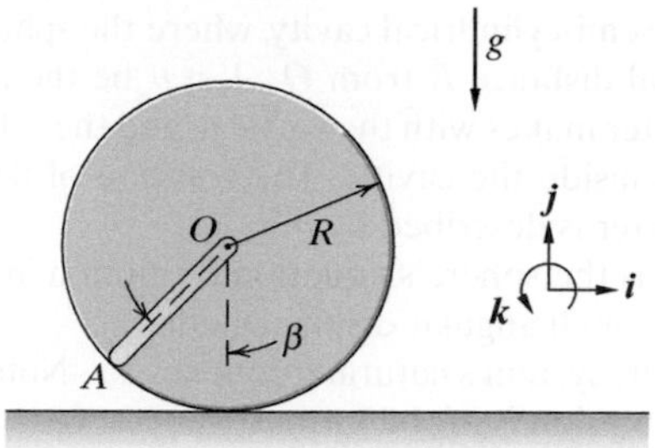

E7.3.10

7.3.11. **[Level 2]** The illustrated figure shows a simplified model of a satellite. The satellite consists of two parts, joined together at O by a pivot. Two identical outer links $\overline{CD}$ are rigidly connected to each other. A motor (part of main body) can provide a torque to rotate the outer links with respect to the main satellite body $\overline{AB}$. At $t = 0$ the motor starts to exert a constant torque of 24 N·m. The masses at the ends of the satellite are both 4 kg, the rigid bar connecting them is 8 kg, and each outer link is 12 kg. Treat the links $\overline{CD}$ and the 8 kg inner bar (without the tip masses) as uniform bars.

a. How long will it be before the outer links first rotate to make an angle of 90° with respect to the satellite body?

b. How long will it be before the outer links are again aligned east to west (with D to the right of C, as it initially is)?

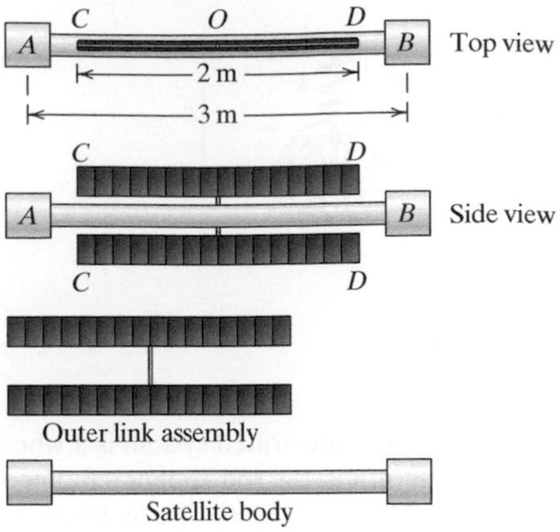

E7.3.11

7.3.12. **[Level 2]** As illustrated, a uniform sphere with mass m and radius r is constrained to roll along the rough surface of a semi-cylindrical cavity, where the sphere's center is a radial distance R from O. Let θ be the angle the sphere's center makes with the vertical, and the sphere rolls without slip inside the cavity. The rotation of the sphere about its center is described by β.

a. What is the sphere's equation of motion in terms of θ? Consider small angular displacements.

b. Find the system's natural frequency ω_n. Note that the equation of motion is of the form $\ddot{\theta} + \omega_n^2\theta = 0$.

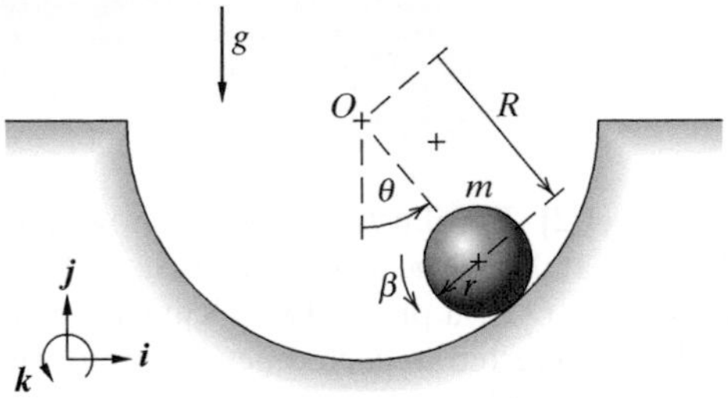

E7.3.12

7.3.13. **[Level 2]** A simplified model of a Lunar Lander is shown. Unfortunately, the Lander's legs were damaged during landing, and are now free to pivot at A and B. Luckily, the Lander landed on a rock R and thus is in static equilibrium in the position shown.

Using the given parameters, determine the needed thrust T for the legs to lose contact with the ground immediately upon application of the thrust. The mass of the Lander body is m_1, and the mass of each of the two legs is m_2. Consider the leg/moon interface to be frictionless.

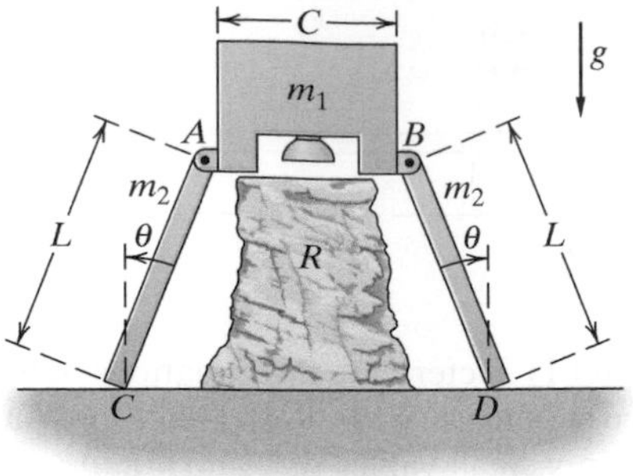

E7.3.13

7.3.14. **[Level 2]** The illustrated cart has mass m_c and is acted on by a force F. The cart can exert a moment $M\boldsymbol{k}$ on the arm at O. Treat the arm (length $2l$) as a uniform bar with mass m_a and mass moment of inertia about O of I_O. Determine the system's equations of motion using x and θ as the system's displacement coordinates.

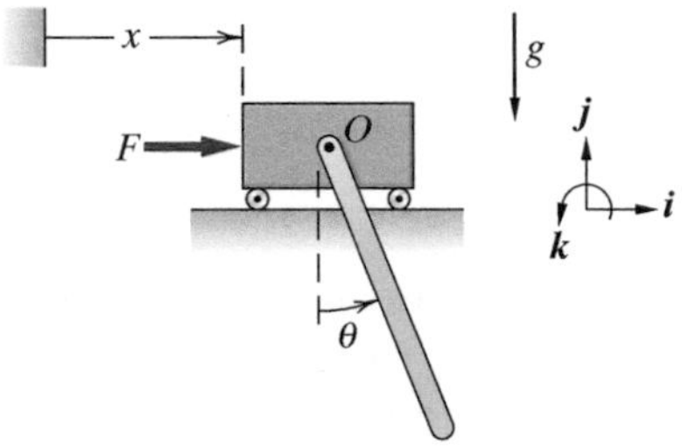

E7.3.14

7.3.15. **[Level 2]** A uniform cylinder rolls on a flat surface due to the moment applied by an attached rod (a motor in the rod applies a moment $M_0\boldsymbol{k}$ to the 20 kg cylinder). $M_0 = 9.50\,\text{N}\cdot\text{m}$. The inclination of the rod from vertical is given by β ($\beta = 0$ when the rod is oriented straight down). $m_{rod} = 10\,\text{kg}$, $R = 0.5\,\text{m}$, $|\overline{OA}| = 0.5\,\text{m}$. Assume that the cylinder rolls without slip. Determine the constant β needed for the system to translate at a constant acceleration.

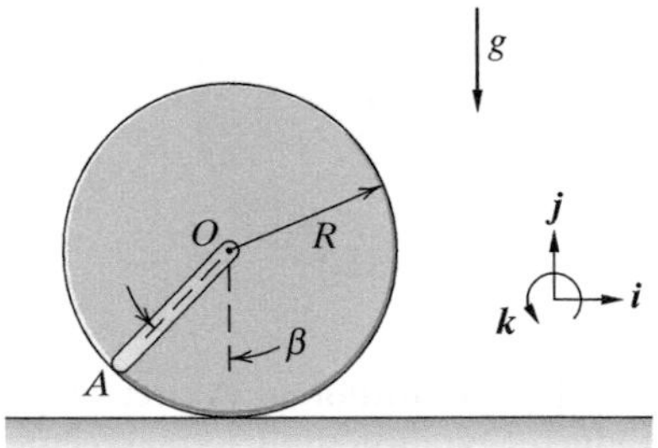

E7.3.15

7.3.16. **[Level 2]** When running at high speeds, cockroaches are sometimes observed to run only on their two hind legs. The running cockroach is modeled as a thin, uniform rod of mass m and length L, supported by a massless, rigid leg of length r. During each step of the run, a cockroach maintains a constant angle ϕ with the horizontal. What is the required angular acceleration $\ddot{\theta}$ of the leg OA in order for the angle ϕ to be constant? Express in terms of $r, \dot{\theta}, \theta,$ ϕ, and g.

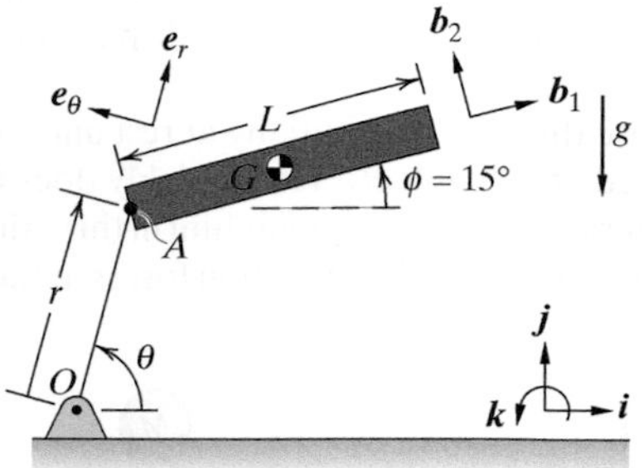

E7.3.16

7.3.17. **[Level 2]** A rigid rod of mass m is supported by two massless strings. At $t = 0$ the right string is cut. What are the resultant translational and angular acceleration of the rod? Show FBDs for pre- and post-cut conditions.

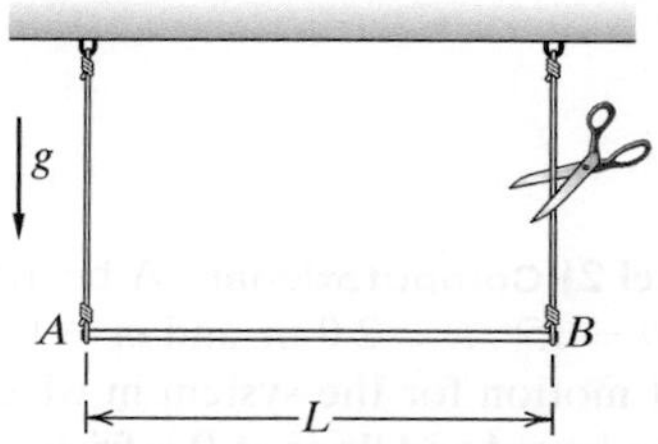

E7.3.17

7.3.18. **[Level 2]** A heavy crate ($m = 60$ kg) is being lifted, and by accident, when the left end has been lifted up (with the right end still on the ground), the workman lost his grip. Assume that when the workman lost his grip, the bottom of the crate was oriented at an angle of 30° to the ground and the crate was initially stationary. What is the angular acceleration of the crate immediately after the workman's grip was lost? The coefficient of friction between crate and ground is $\mu = 0.4, a = 0.7$ m, and $b = 2$ m.

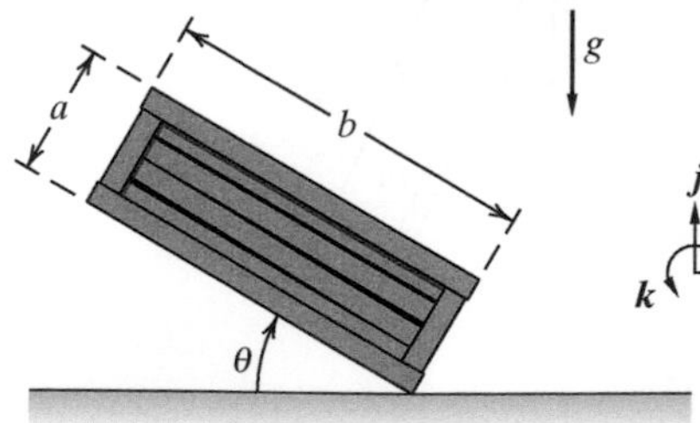

E7.3.18

7.3.19. **[Level 2]** A rigid rod of mass m is supported in equilibrium by two identical, massless, linear springs. At $t = 0$ the right spring breaks. Find the resulting translational and angular acceleration of the rod. Show FBDs for the system immediately before and after the break.

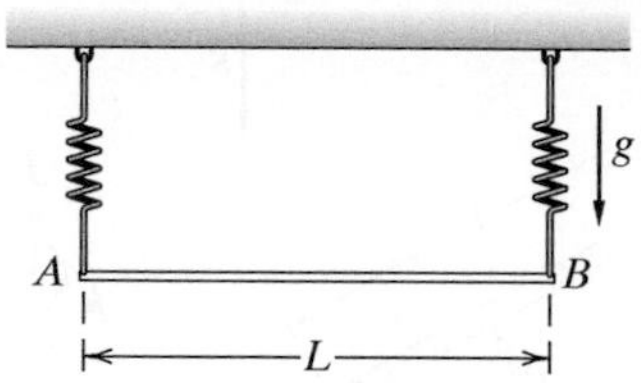

E7.3.19

7.3.20. **[Level 2]** Let's extend Example 7.18. We will assume an unequal weight distribution so that the car's mass center is closer to the front than the rear, as would be the case for a front-wheel-drive car. Find the system equations of motion in terms of y, vertical translation of the mass center, and θ, counterclockwise rotation of the bar. Assume that the bar is horizontal when at rest, and assume small values of y and θ. Does gravity drop out of the equations as it did in Example 7.18? Neglect nonlinear terms ($\dot{\theta}^2, \ldots$).

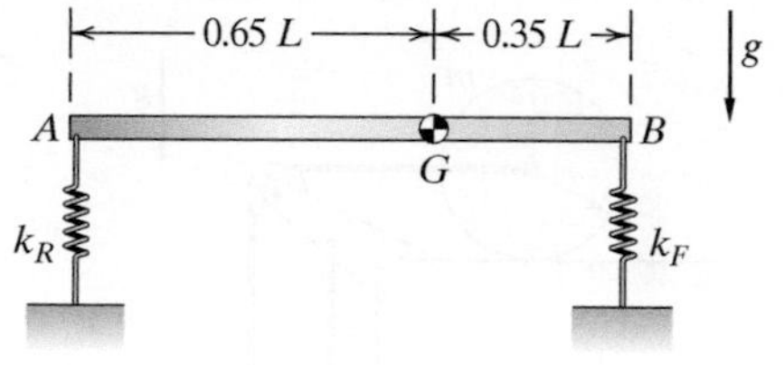

E7.3.20

7.3.21. **[Level 2]** You're in charge of analyzing a car and have been given the illustrated model. You're required to use the illustrated coordinates. Assume that the bar is horizontal when the system is in equilibrium and that y and θ motions are small. What are the system's two equations of motion? The body is modeled as a uniform bar of mass m with two lumped masses m_a and m_b. $m_A = \frac{m}{2}$ and $m_B = \frac{m}{4}$. Neglect gravity and nonlinear terms ($\dot{\theta}^2, \ldots$).

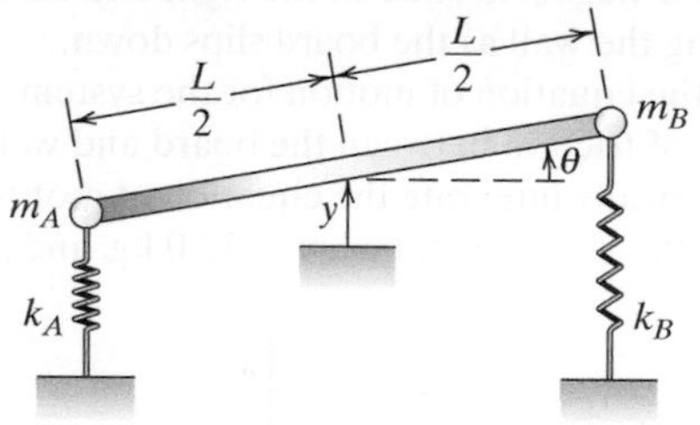

E7.3.21

7.3.22. **[Level 2]** A steel plate ($m = 100$ kg) is being lifted, and by accident, when the left end has been lifted up (with the right end still on the ground), the left end is released. Assume that at the moment of release, the plate had a 30° orientation with respect to the ground. What is the angular acceleration of the plate immediately after release? The coefficient of friction between the plate and ground is $\mu = 0.5, l = 2.1$ m.

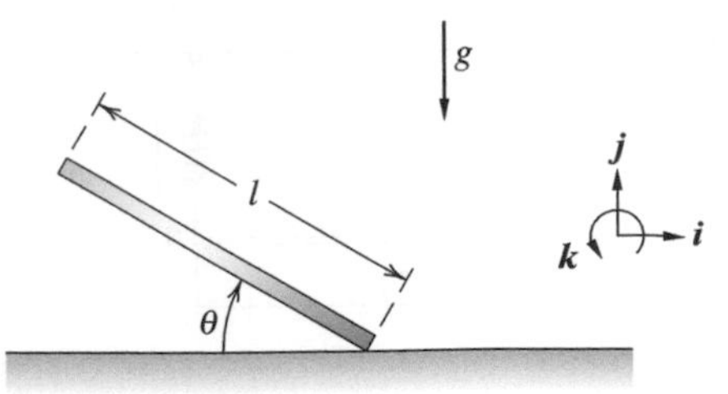

E7.3.22

7.3.23. **[Level 2]** As illustrated, a uniform cylinder is pulled along a rough horizontal surface by means of a falling block supported by an inextensible cable wound around a system of massless pulleys. The cylinder's mass and radius are m and r, respectively, and the block has a mass of $M = 4m$. The coefficient of friction between the cylinder and surface is μ. What is the minimum coefficient of friction for which the cylinder rolls without slip?

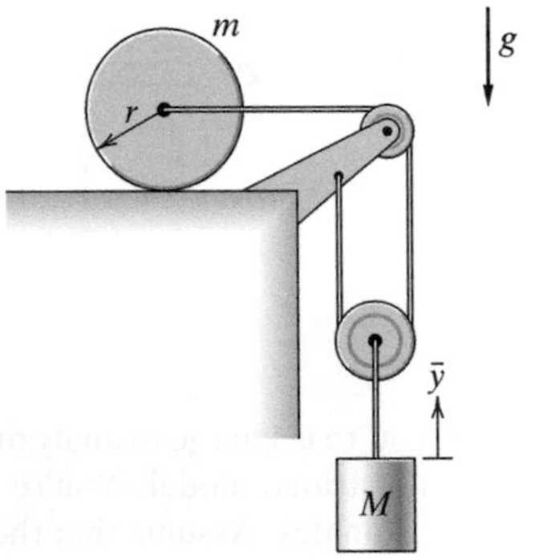

E7.3.23

7.3.24. **[Level 2] Computational** A board is leaning against a wall. θ is initially equal to 15°. At $t = 0$, some oil is spilled on the ground, causing the coefficient of friction between the board and the ground to drop to zero. Assume that the end B begins to slide to the right and that the end A drags along the wall as the board slips down.

a. Find the equation of motion for the system. Assume a coefficient of friction between the board and wall of μ.

b. Numerically integrate the equation of motion to find $\dot{\theta}$ when $\theta = 45°$. Let $L = 2.0$ m, $m = 10.0$ kg, and $\mu = 0.3$.

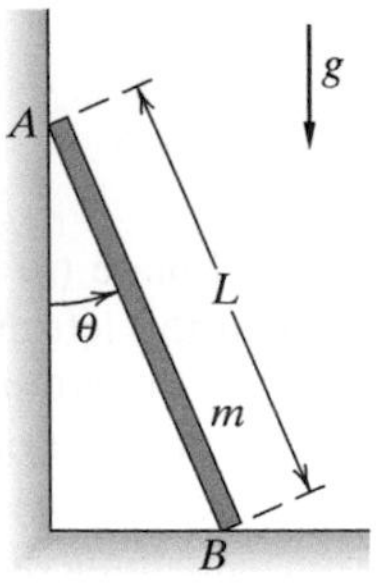

E7.3.24

7.3.25. **[Level 2]** An inverted compound pendulum of length L rotates about its pivot O, which is attached to a massless collar that is free to slide on a frictionless horizontal rail. The pendulum's rod has a mass of m, and a lumped body of mass $M = 3m$ is connected to the rod's free end. The rod's orientation with respect to the horizontal is given by θ, and x_O describes the horizontal displacement of the collar.

a. Suppose the system is initially at rest and the pendulum is oriented at an angle θ_0. How quickly does the collar need to accelerate to keep the pendulum in this orientation?

b. Find the corresponding reaction forces at the pivot O.

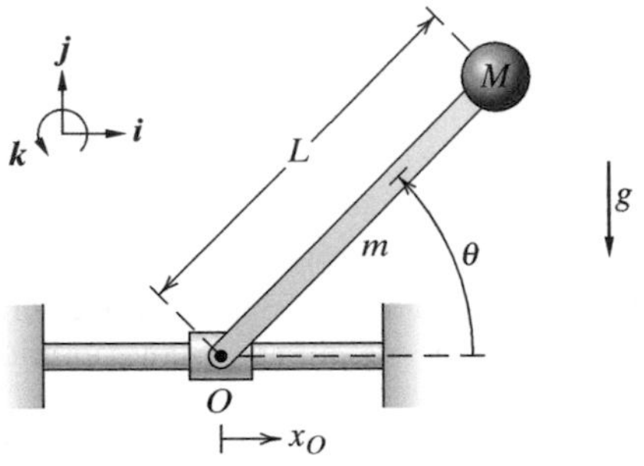

E7.3.25

7.3.26. **[Level 2] Computational** A board is released from rest with $\theta = 15°$, $L = 2.0$ m, and $m = 10$ kg. Derive the equation of motion for the system in which the interface between the board and floor at B is friction-free, while the interface at A has a coefficient of friction μ. Determine how long it takes for the board to reach an angle of 45° and plot this time versus μ for μ going from 0 to 1.0 in 0.1 increments. Also, record the value of $\dot{\theta}$ corresponding to $\theta = 45°$ for the same μ values. Plot $\dot{\theta}|_{\theta=45°}$ versus μ, and $\dot{\theta}|_{\theta=45°}$ versus $t|_{\theta=45°}$. Do the results make sense?

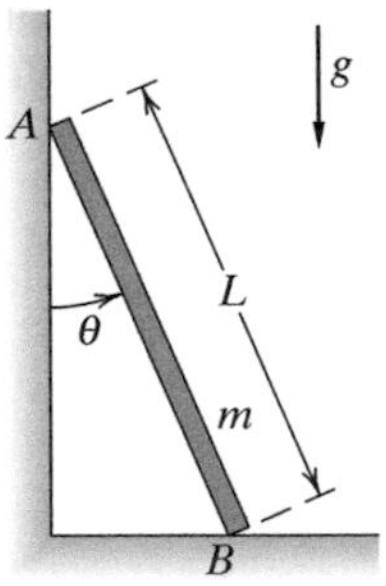

E7.3.26

7.3.27. **[Level 2]** Find the equations of motion for the illustrated system. Assume that the bar is horizontal when the system is in equilibrium, and consider small angles of rotation. Neglect gravitational effects.

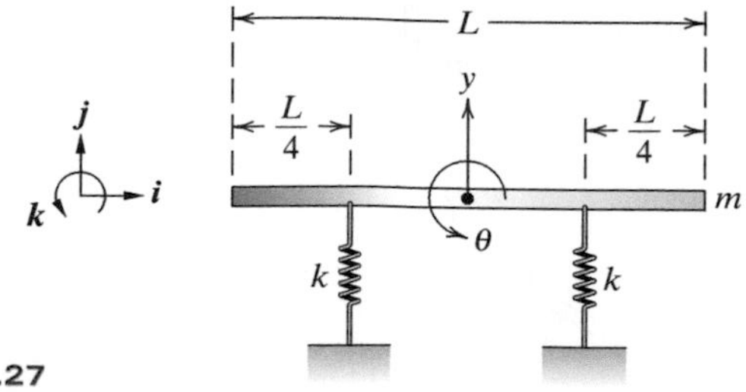

E7.3.27

7.3.28. **[Level 2]** Find the equations of motion for the illustrated system. The bar has mass m_2 and has an attached mass m_1. Let the angle of the bar be given by θ and consider only small angles of rotation ($\sin\theta \approx \theta$). Use y and θ as your coordinates. Neglect gravity and nonlinear terms ($\dot{\theta}^2, \ldots$).

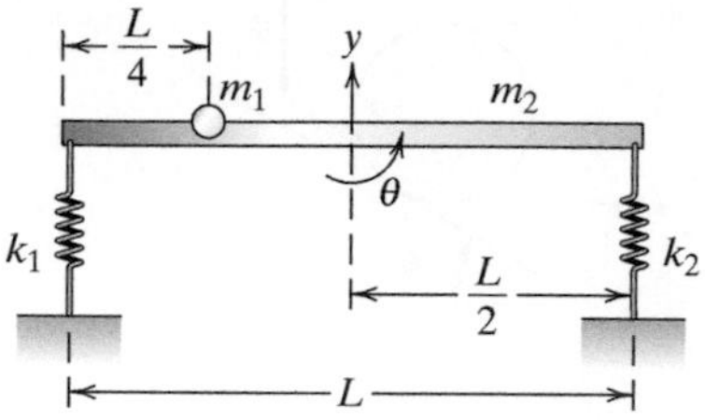

E7.3.28

7.3.29. **[Level 2]** Find the equations of motion for the illustrated system. Consider small angles of rotation ($\sin\theta \approx \theta$). The bar itself has a mass of ρL and in addition has three masses attached. Neglect gravity and nonlinear terms ($\dot{\theta}^2, \ldots$).

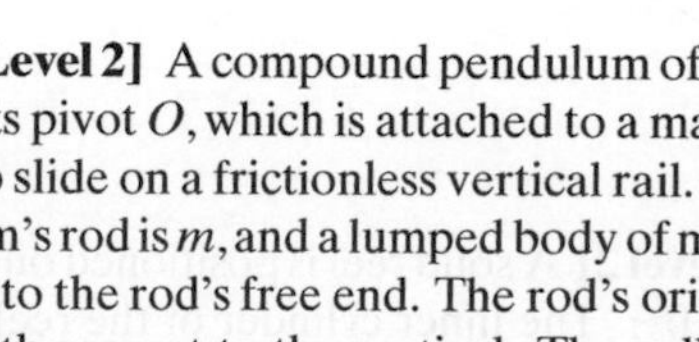

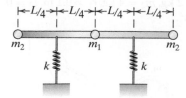

E7.3.29

7.3.30. **[Level 2]** A compound pendulum of length L rotates about its pivot O, which is attached to a massless collar that is free to slide on a frictionless vertical rail. The mass of the pendulum's rod is m, and a lumped body of mass $M = 4m$ is connected to the rod's free end. The rod's orientation θ is measured with respect to the vertical. The collar is moved up and down the rail according to $y_O(t) = \overline{y}_O \sin(\omega t)$, where $\overline{y}_O$ and ω are the amplitude and frequency of oscillation, respectively. What is the pendulum's equation of motion?

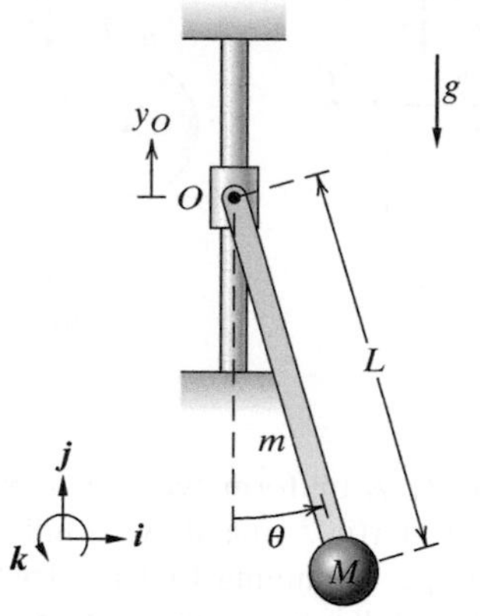

E7.3.30

7.3.31. **[Level 2]** The illustrated body is a 18 kg equilateral triangle made out of welded steel bars. $L = 0.6$ m. What are the body's linear and angular acceleration immediately following an application of the two forces shown in the figure? $F = 45$ N.

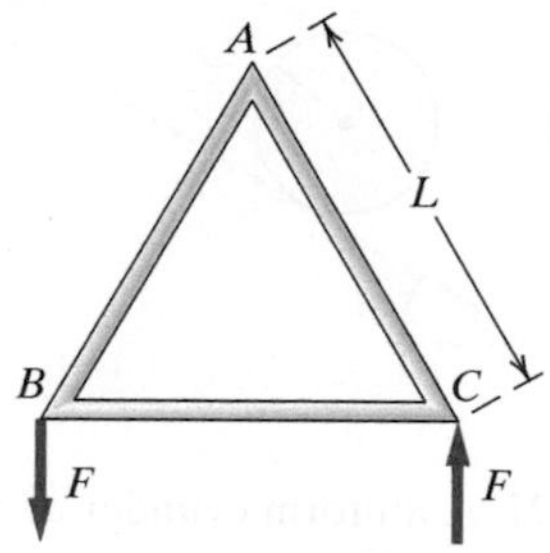

E7.3.31

7.3.32. **[Level 2]** In both of the illustrated cases, a reel of massless string, initially at rest, is pulled to the right. Assume a frictionless reel/floor interface. Will the point of the reel contacting the floor accelerate to the right or left? Is only one answer possible for each case? The inner radius is equal to r_1 and the outer radius is equal to r_2.

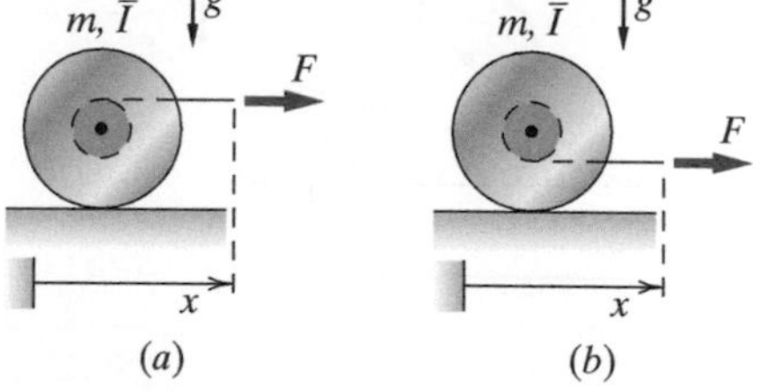

E7.3.32

7.3.33. **[Level 2]** In (a) a reel of massless string is being pulled to the right, and the string unrolls from the top of the inner cylinder. In (b) it is again being pulled to the right, but in this case the string unrolls from the bottom of the spool. Solve for the spool's acceleration for each case. Assume pure rolling. The inner radius is equal to r_1 and the outer radius is equal to r_2.

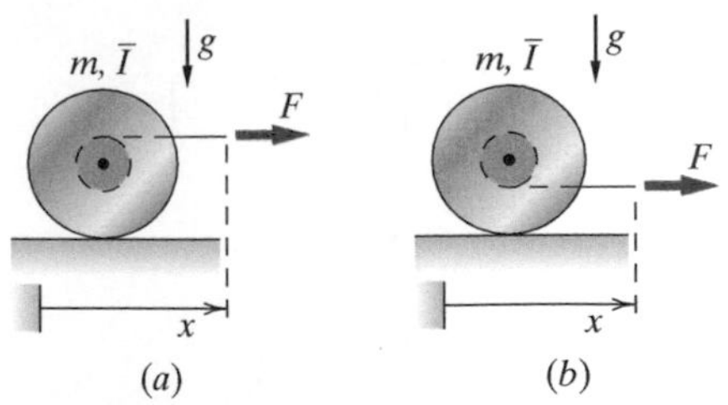

E7.3.33

7.3.34. **[Level 2]** A uniform cylinder of mass m, radius r, and moment of inertia $\bar{I}$ rolls down a hill having an angle of inclination β. Assuming that the friction between the cylinder and the hill is large enough to prevent slipping, compute:

a. The friction force that acts on the cylinder during a free downhill roll.

b. The acceleration of the cylinder's center of mass.

c. The minimum coefficient of static friction to prevent slipping.

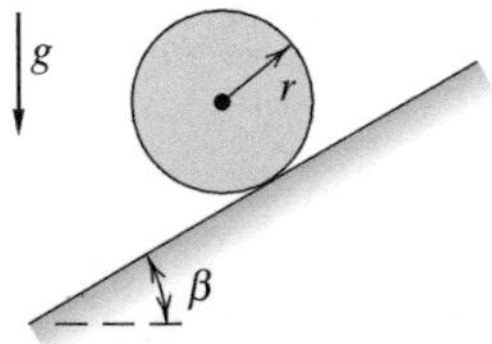

E7.3.34

7.3.35. **[Level 2]** A uniform cylinder of mass m, radius r, and moment of inertia $\frac{mr^2}{2}$ rolls down an inclined surface. The surface is simultaneously accelerated upward at a constant rate, $\ddot{z} = a_z$.

a. Find the acceleration of the cylinder's center of mass under no-slip conditions.

b. What is the minimum coefficient of static friction for which no-slip conditions occur?

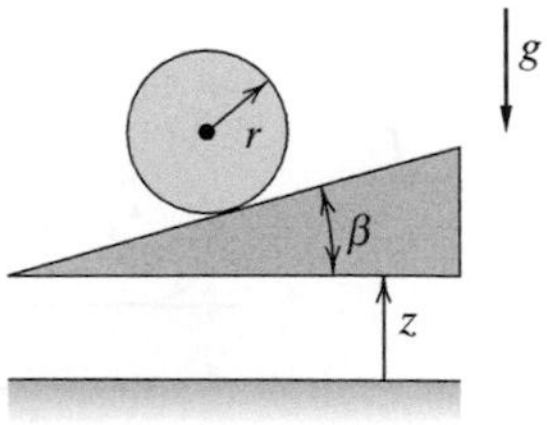

E7.3.35

7.3.36. **[Level 2] Computational** A rotating, three-arm body ($\bar{I} = 0.5\ \text{kg·m}^2$) has three identical masses ($m = 0.5$ kg) that can slide along the arms. The entire system initially is rotating at $\dot{\theta} = 10$ rad/s, $r = 0.6$ m, and $\dot{r} = 0$. At $t = 0$, the three masses are simultaneously released and slide without friction to the ends of the rods. Ignore the dimensions of each mass.

What are $\dot{\theta}$ and $\dot{r}$ just as the masses reach the ends of the rods ($r = 1.2$ m)?

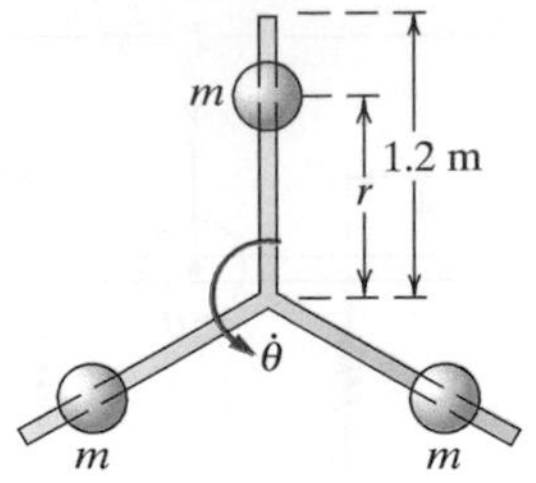

E7.3.36

7.3.37. **[Level 2]** The figure shows a solid cylinder of radius r placed on an inclined slope. The coefficient of friction is 0.4. Assume the cylinder is released from rest and determine the critical value of θ above which the cylinder slips rather than rolling without slip.

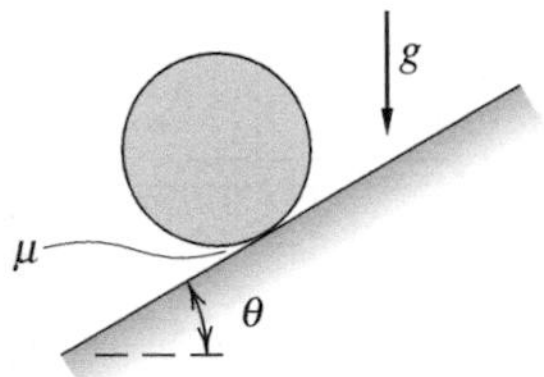

E7.3.37

7.3.38. **[Level 2]** The illustrated spool is released from rest on a slope. A massless string is wrapped around the inner radius, extends up the slope over a massless pulley, and then extends downward to support a mass m_2. How long will it take for the spool to rotate one full turn? Does the spool move upslope or downslope? Assume roll without slip. $m = 2$ kg, $m_2 = 4$ kg, $r_1 = 0.1$ m, $r_2 = 0.15$ m, and $\bar{I} = 0.02\ \text{kg·m}^2$.

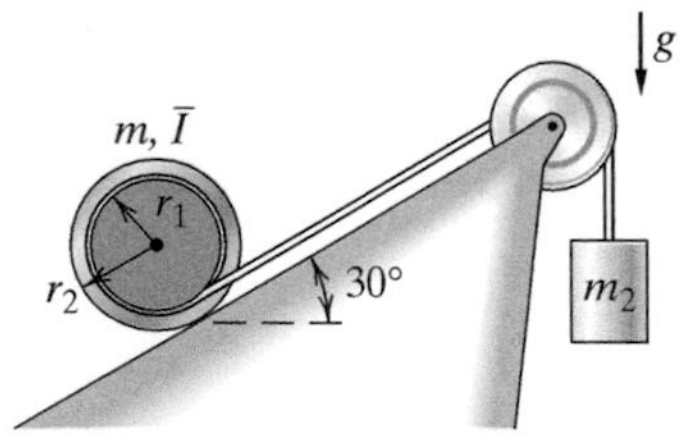

E7.3.38

7.3.39. **[Level 2]** A solid reel is positioned on an inclined surface ($\theta = 10°$). The inner cylinder of the reel has radius $r_1 = 0.5$ m, and the outer cylinders have radius $r_2 = 0.7$ m. The reel's mass is 20 kg, and its mass moment of inertia about its center is 6.8 kg·m^2. A massless string runs from the inner cylinder (wrapped several times around the cylinder) to a fixed attachment point A. What is the acceleration of the reel's center if it is released from rest? Assume roll without slip.

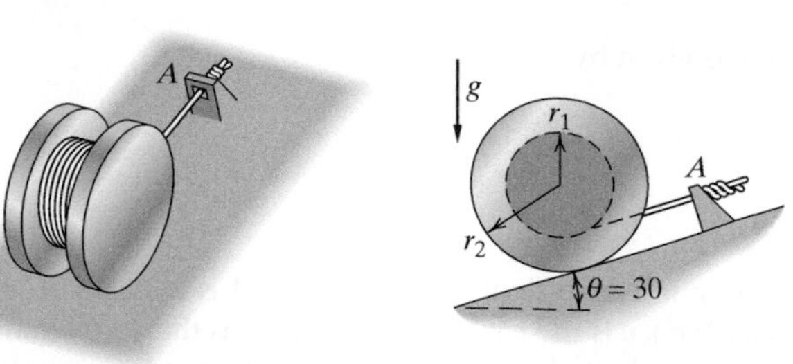

Reel on a slope   Side view

E7.3.39

7.3.40. [Level 2] A uniform, slender bar, of mass m and length L, pivots about a point O. A rigid body B slides, without friction, along the bar. The mass center of the rigid body is located a distance d away from the bar. The mass moment of inertia of the rigid body about its mass center is $\bar{I}_B$, and its mass is m_B. A free-body diagram is shown for the rigid body B. Note that all the forces of interaction between B and the bar have been combined into a normal force N and a moment C. You will have to draw the free-body diagram of the bar. Four equations of motion are shown. Each is complete but includes several possibilities within itself. You must decide which term is correct.

$$\begin{bmatrix}1\\0\\-1\end{bmatrix} C + m_B g\begin{bmatrix}\cos\theta\\ \sin\theta\end{bmatrix} + \begin{bmatrix}1\\0\\-1\end{bmatrix} N = \begin{bmatrix}m_B\\ m_L\end{bmatrix}\left(2\dot{r}\dot{\theta} + r\begin{bmatrix}\dot{\theta}^2\\ \ddot{\theta}\end{bmatrix} + \begin{bmatrix}d\dot{\theta}^2\\ d\ddot{\theta}\\ -d\dot{\theta}^2\\ -d\ddot{\theta}\end{bmatrix}\right) \quad (1)$$

$$\begin{bmatrix}C\\ Nr\\ Nd\cos\theta\end{bmatrix} = \begin{bmatrix}\frac{1}{3}m_L L^2\\ \bar{I}_B\\ \bar{I}_B + m_B d^2\\ \bar{I}_B - m_B d^2\\ \frac{1}{12}m_L L^2\end{bmatrix}\ddot{\theta} + \begin{bmatrix}1\\0\\-1\end{bmatrix} m\ddot{r}r \quad (2)$$

$$\begin{bmatrix}m_L g\frac{L}{2}\cos\theta\\ m_L g\frac{L}{2}\sin\theta\\ m_L gL\sin\theta\\ m_L gL\cos\theta\end{bmatrix} + \begin{bmatrix}1\\0\\-1\end{bmatrix} C + \begin{bmatrix}Nr\\ Nd\cos\theta\end{bmatrix} = \begin{bmatrix}\frac{1}{3}m_L L^2\\ \bar{I}_B\\ \bar{I}_B + m_B d^2\\ \bar{I}_B - m_B d^2\\ \frac{1}{12}m_L L^2\end{bmatrix}\ddot{\theta} \quad (3)$$

$$\begin{bmatrix}1\\0\\-1\end{bmatrix} N + m_B g\begin{bmatrix}\cos\theta\\ \sin\theta\end{bmatrix} = m_B\left(\ddot{r} + \begin{bmatrix}-r\dot{\theta}^2\\ r\ddot{\theta}\\ 2\dot{r}\dot{\theta}\end{bmatrix} + \begin{bmatrix}d\dot{\theta}^2\\ d\ddot{\theta}\\ -d\dot{\theta}^2\\ -d\ddot{\theta}\end{bmatrix}\right) \quad (4)$$

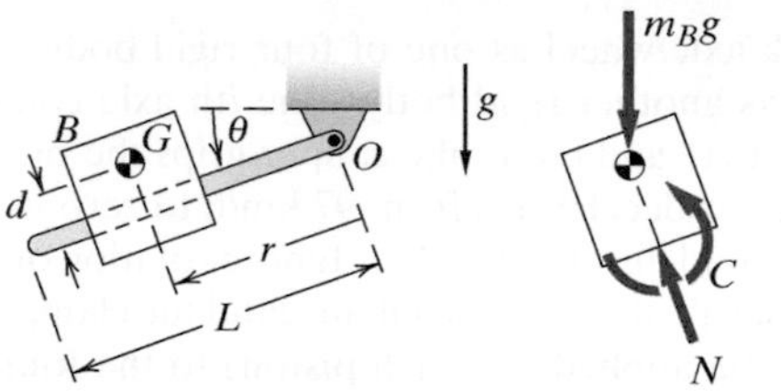

E7.3.40

7.3.41. [Level 2] A uniform cylinder (radius r and mass m_1) has, attached to its axle, a rigid rod of length L and mass m_2. A motor at O in the rod produces a constant torque M that acts on the cylinder, which rolls without slip. The system is released from rest with the bar at $\theta = 0°$ and $\dot{\theta} = 0$. What must M be for the system to translate to the right with the bar remaining fixed at 0 degrees? What's the angular acceleration of the cylinder?

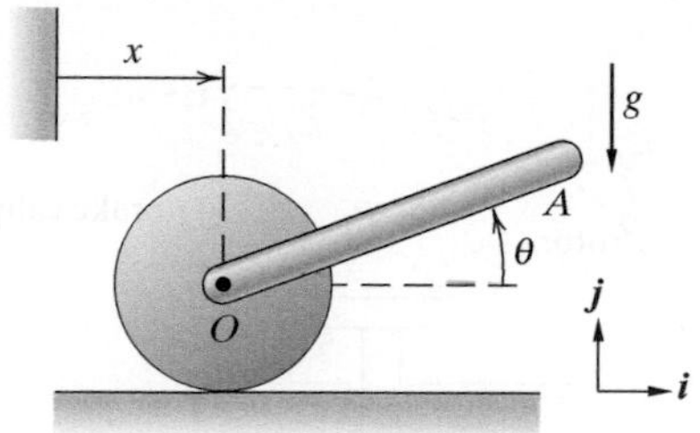

E7.3.41

7.3.42. [Level 2] A hoop of mass m and radius r travels down a rough incline angled at β with respect to the horizontal. If the rotational inertia about the hoop's mass center G is $I_G = mr^2$, what is the minimum coefficient of friction μ for which the hoop rolls without slip down the incline?

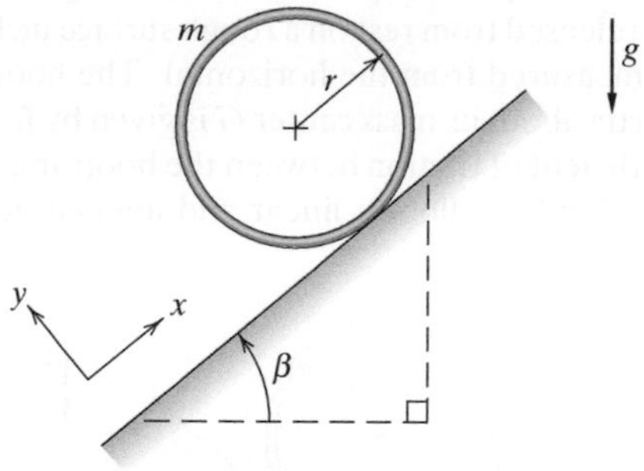

E7.3.42

7.3.43. [Level 2] In **Exercise 3.1.8** we saw that to brake a 1315 kg car from 97 km/h to zero in 2 s required a force of 17,628 N. Now let's look more closely at the problem. Assume that the car has four disk brakes, each with a pair of pistons that clamp the brake disk. Assume that the brake pads are located 13 cm from the axle axis. A simplified model is shown in the accompanying figure. Consider the brake disk/axle/wheel as one of four rigid bodies and the car body as another rigid body. The ith axle contacts the car body at A_i, and the brake caliper grips the brake rotor at B_i. The car decelerates from 97 km/h to zero in 2 s. The axle-to-ground distance is 0.3 m. Ignore rotation of the car's body; it doesn't affect the problem. Find the clamping force needed to be applied (by each piston) to the four brakes. Assume that all brake/tire combinations experience identical conditions (same normal force magnitude, rotation rate, etc.). Let the mass of the brake/tires be zero.

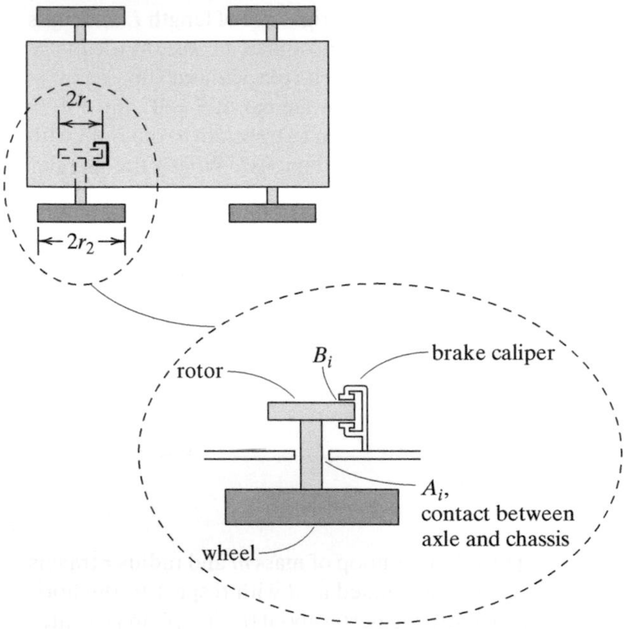

E7.3.43

7.3.44. [Level 2] A hoop of mass $m = 50$ kg and radius $r = 0.7$ m is released from rest on a rough surface inclined at $\beta = 40°$ as measured from the horizontal. The hoop's moment of inertia about its mass center G is given by $I_G = mr^2$, and the coefficient of friction between the hoop and surface is $\mu = 0.35$. Find the hoop's linear and angular accelerations.

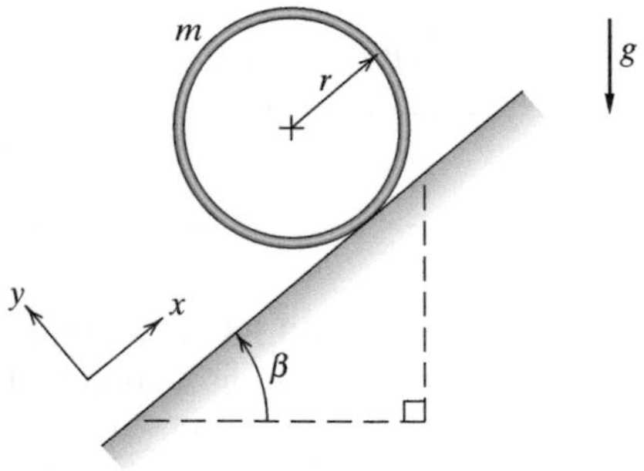

E7.3.44

7.3.45. [Level 2] The goal of this exercise is to predict the top speed of a new sports sedan. The aerodynamic drag force is given by

$$F_{\text{aero}} = \frac{1}{2}\rho c_a A v^2$$

where ρ is the air density (1.2 kg/m^3), A is the car's frontal area ($2m^2$), v is the car's velocity, and c_a is the aerodynamic drag coefficient. The drag force from the road is given by

$$F_{\text{road}} = c_r mg$$

where m is the car's mass and c_r is the rolling resistance coefficient. The vehicle we will be looking at has a mass of 1450 kg, and its engine develops a constant torque of 300 N·m. The distance from the axles to the ground is 0.3 m. The gear ratio from the engine to the wheels is 2.74. Using an aerodynamic resistance coefficient of 0.02, calculate the car's maximum attainable speed. Assume a 15% loss in force from the engine due to drivetrain losses.

7.3.46. [Level 3] Suppose a hoop and uniform cylinder, both with mass m and radius r, are simultaneously released from rest at the top of a rough surface angled at β with respect to the horizontal. If the coefficient of friction between the two bodies and the surface is μ, prove that the hoop can never reach the bottom of the incline before the cylinder. The moment of inertia for a hoop about its mass center G is given by $I_G = mr^2$.

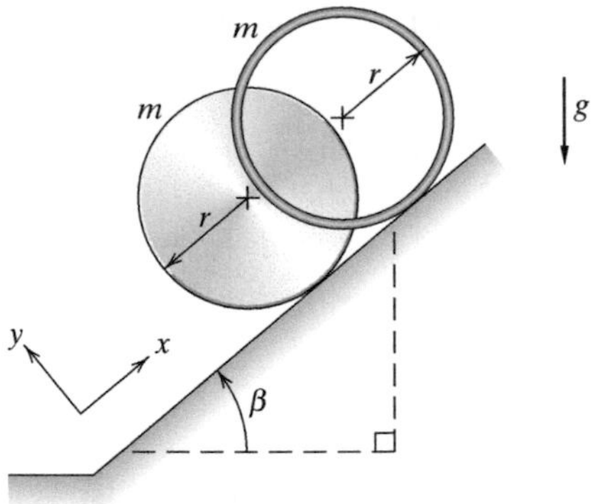

E7.3.46

7.3.47. [Level 3] A planetary gear system is shown with the gear teeth omitted for simplicity. The spider ($OABC$) has mass m_s, and each arm is of length L. Its mass moment of inertia about O is $\overline{I}_s$. Each gear rotates without slip on the fixed outer ring D. If a counterclockwise moment M is applied to the spider, what is the angular acceleration of the spider? Each gear has a radius r, mass m_g, and mass moment of inertia about its center of $\overline{I}_g$.

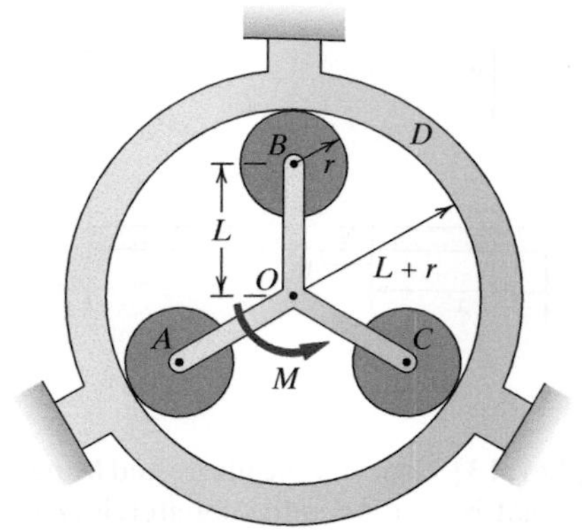

E7.3.47

7.3.48. **[Level 3]** A force $\boldsymbol{F} = 10\boldsymbol{j}$ N is applied to the illustrated stationary body. What is the body's acceleration (linear and angular) immediately following the application of the force? $a = 0.8$ m, $b = 0.5$ m, and $c = 2.2$ m. The body has a mass of 40 kg.

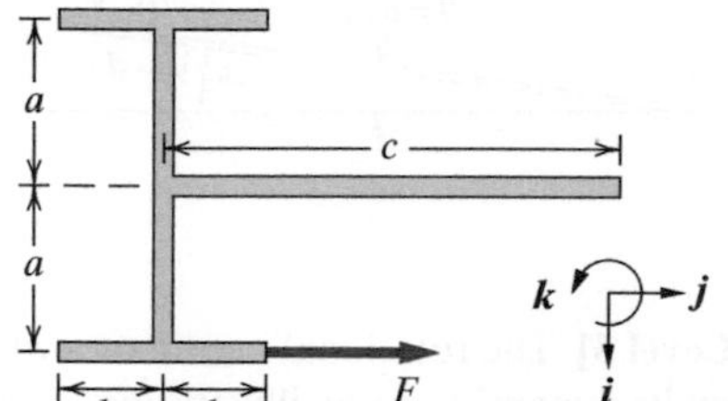

E7.3.48

7.3.49. **[Level 3]** A rod of mass m and length L is initially held in the depicted configuration, where $\theta = 60°$ with respect to the horizontal. The rod is supported by a rough horizontal surface at its end A, and a lumped body of mass $2m$ is attached to its free end B. The coefficient of friction between the rod and surface is $\mu = 0.4$. Suppose the rod is released from rest and allowed to fall. If $m = 4$ kg and $L = 1.3$ m, what is the rod's initial angular acceleration $\ddot{\theta}$?

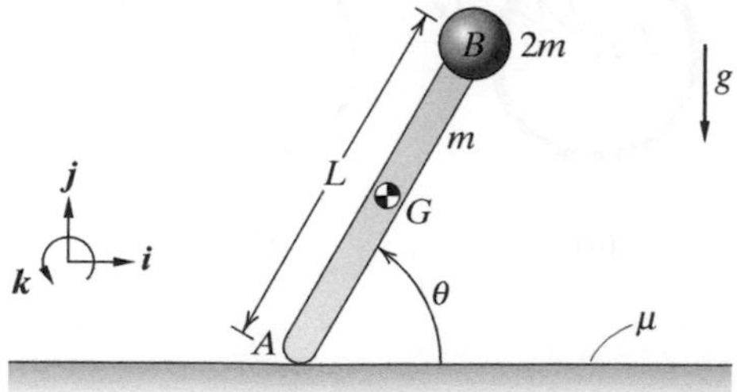

E7.3.49

7.3.50. **[Level 3] Computational** At $t = 0$, a uniform bar is released. Assume that the springs have an unstretched length of 0.5 m and a spring constant $k = 10$ N/m. The bar has a mass of 2 kg and a length of 2 m.

a. What are $\overline{\boldsymbol{a}}$ and $\boldsymbol{\alpha}$ just after release?

b. Determine $\overline{\boldsymbol{a}}$ and $\boldsymbol{\omega}$ in configuration (b) via numerical integration.

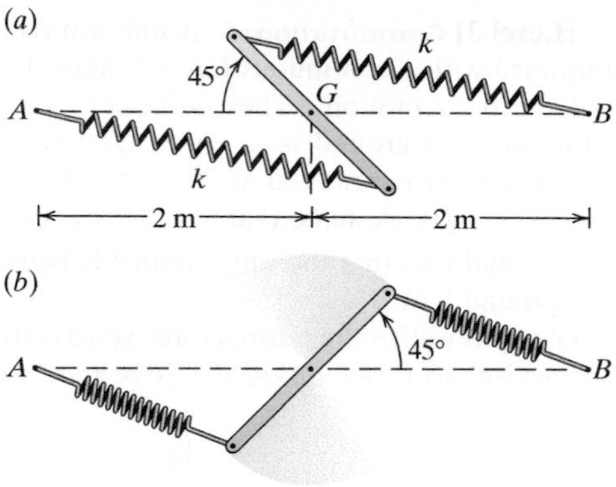

E7.3.50

7.3.51. **[Level 3] Computational** A drawbridge is controlled by a hanging weight m_2. Calculate the time needed to raise the drawbridge from a closed position ($\theta = 0°$) to $\theta = 60°$. $m_1 = 200$ kg, $m_2 = 150$ kg, and $L = 4$ m. Assume that the drawbridge pivots freely at O and that the rope connecting m_2 to the bridge is massless. Plot θ (the inclination angle of the bridge) versus time.

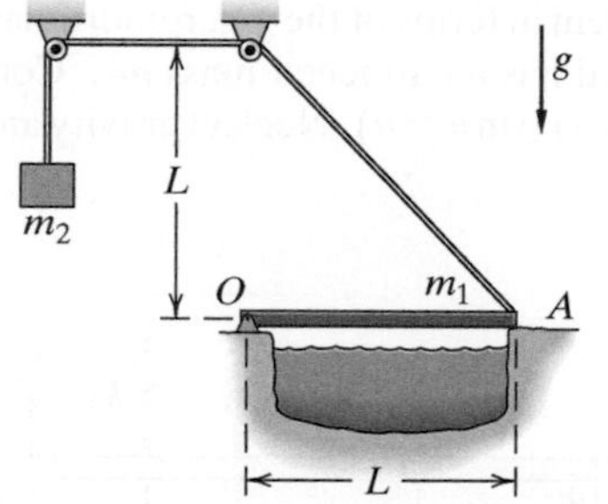

E7.3.51

7.3.52. **[Level 3] Computational** Assume that you have taken the job of theme park designer and have to construct a trapdoor that opens at a predetermined rate. The massless rope that permits the door to open has to run over the guide A at a constant speed of 1.1 m/s. What is the force T that has to be exerted on the end of the rope from a fully closed position to $\theta = 80°$? The trapdoor is 5 m long and has a mass of 600 kg. Plot T versus time and T versus θ. What is the time at which $\theta = 80°$?

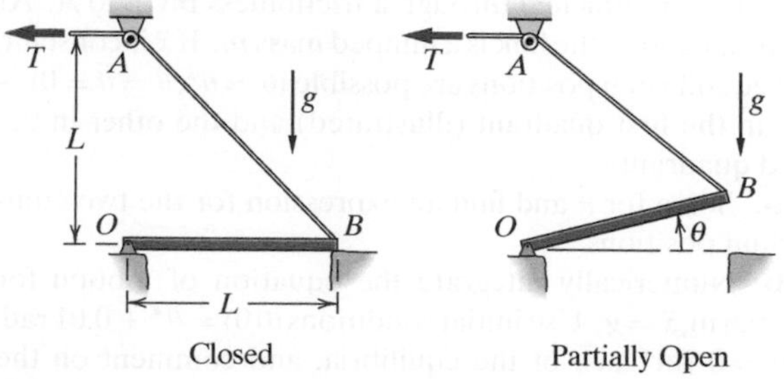

E7.3.52

7.3.53. [Level 3] Computational A uniform rigid bar is pulled upward with a continuous force T, angled at 45° throughout the bar's motion. The coefficient of friction between the bar and ground is $\mu = 0.5$. Determine the time at which the bar is oriented at 30° to the horizontal $[\boldsymbol{r}_{A/B} = L(-\frac{\sqrt{3}}{2}\boldsymbol{i} + \frac{1}{2}\boldsymbol{j})]$. Assume that at $t = 0$, the end A is raised infinitesimally so that the only contact between the bar and the ground is at B.

Plot θ versus t as well as the normal force versus t (to verify that contact isn't lost). $m = 2$ kg, $T = 20$ N, and $L = 1$ m.

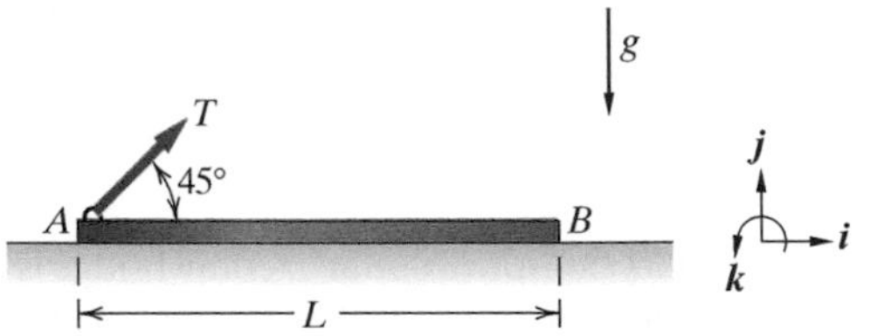

E7.3.53

7.3.54. [Level 3] Find the equations of motion for the illustrated system in terms of the given coordinates. The bar has mass m and has an attached mass m_1. Consider small angles of rotation ($\sin\theta \approx \theta$). Neglect gravity and nonlinear terms ($\dot{\theta}^2, \ldots$).

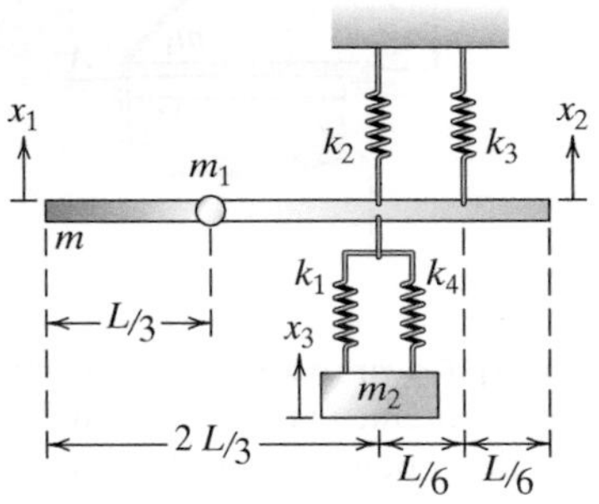

E7.3.54

7.3.55. [Level 3] Computational The system shown is a collar A that runs along a horizontal bar. A massless link (length L) is attached through a frictionless pivot to A. At the other end of the link is a lumped mass m. If $\ddot{x}$ is constant, two equilibrium positions are possible ($\theta = \theta^*, \dot{\theta} = \ddot{\theta} = 0$)—one in the first quadrant (illustrated) and the other in the third quadrant.

a. Solve for $\ddot{\theta}$ and find an expression for the two equilibrium positions.

b. Numerically integrate the equation of motion for $L = 0.6$ m, $\ddot{x} = g$. Use initial conditions $\theta(0) = \theta^* + 0.01$ rad, $\dot{\theta}(0) = 0$ for each of the equilibria, and comment on the results.

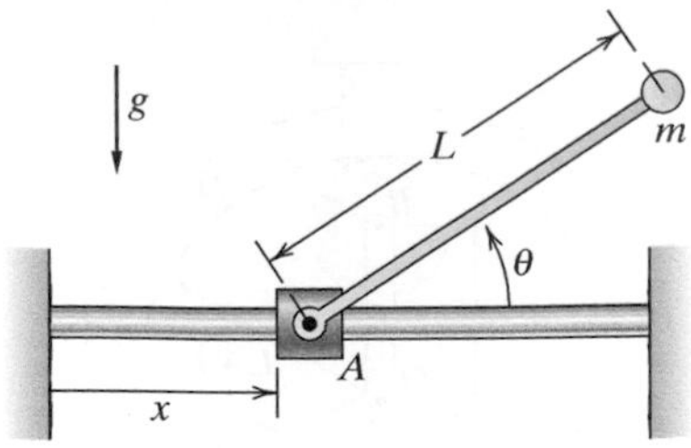

E7.3.55

7.3.56. [Level 3] Consider the illustrated Ming vase. It's on a surface that is rotating in the counterclockwise direction at a constant rate ω_0. Determine whether the vase will tip or slip first. $\mu = 0.7$, $\omega_0 = 0.8$ rad/s, $L = 2$ m, $d = 0.04$ m, and $h = 0.09$ m.

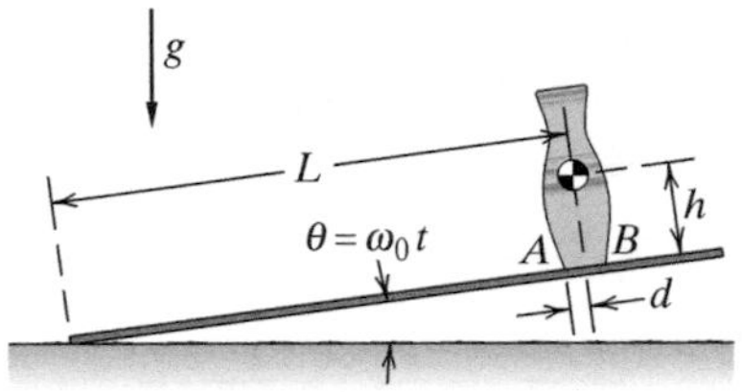

E7.3.56

7.3.57. [Level 3] The rotational inertia of a wheel has a large effect on its dynamics. As an illustration, consider two simplified models of a bicycle wheel. In (a) all the wheel's mass has been distributed to the outer edge (massless hub and spokes), whereas in (b) the mass is concentrated at the hub (massless rim and spokes).

a. Calculate the acceleration of the mass center for each case under the assumption of pure rolling.

b. Determine the conditions needed to ensure that slip doesn't occur.

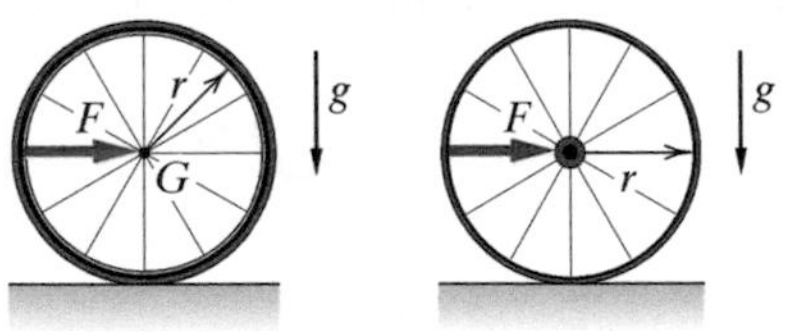

E7.3.57 (*a*) (*b*)

7.3.58. [Level 3] The uniform cylinder of mass $m = 100$ kg and radius $r = 0.5$ m is released from rest on a sloped surface. The coefficient of friction between the cylinder and the sloped surface is μ.

a. Find the cylinder's angular and linear acceleration if $\mu = 0.1$.

b. Find the cylinder's angular and linear acceleration if $\mu = 0.3$.

c. Comment on the results of **a** and **b**.

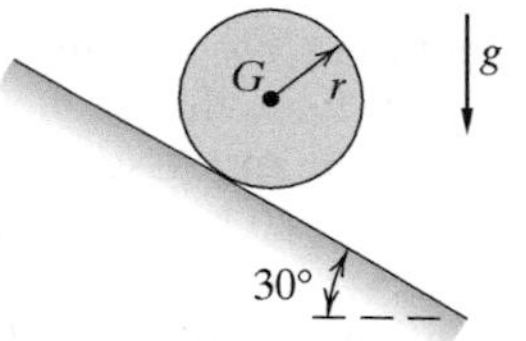

E7.3.58

7.3.59. [Level 3] A massless string is wound around the inner cylinder of a reel R (mass m_1), goes over a freely pivoted cylinder C (mass m_2), and connects to block A (mass m_3). Cylinder C moves with the string (no-slip). $\bar{I}_1 = 0.015\ \text{kg·m}^2$, $\bar{I}_2 = 0.018\ \text{kg·m}^2$, $m_1 = 4$ kg, $m_2 = 20$ kg, $m_3 = 5$ kg, $r_1 = 7$ cm, $r_2 = 10$ cm, and $r_3 = 3$ cm. Assume roll without slip on the left slope and a frictionless surface on the right. Determine the angular acceleration of cylinder C if the entire system is released from rest.

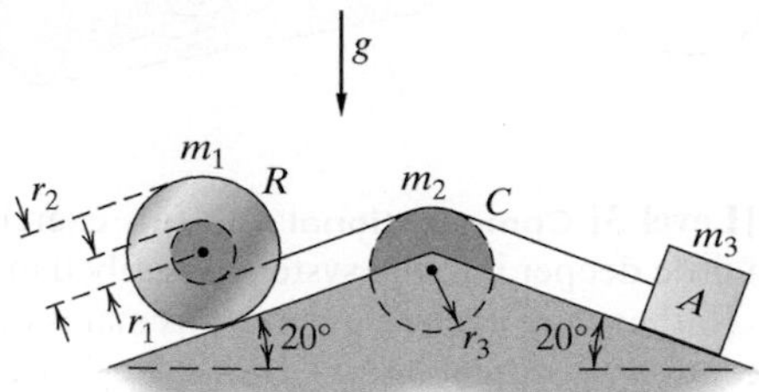

E7.3.59

7.3.60. [Level 3] Links $\overline{AB}$ and $\overline{BC}$ are joined by a friction-free pivot and are kept from slipping by friction between the ground and their ends A and C. Each bar has a mass of 2 kg and a length of 0.9 m.

a. For $F = 0$, find the minimum coefficient of friction μ that supports a no-slip condition.

b. Let μ equal the value found in **a**. Assume slip isn't occurring and find the maximum F that supports this case.

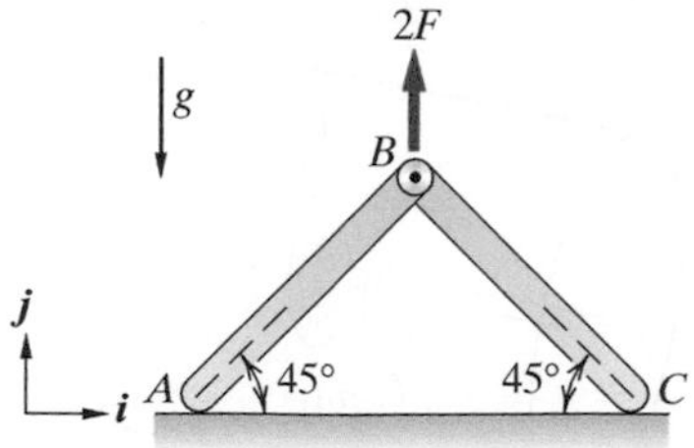

E7.3.60

7.3.61. [Level 3] A simplified model of a person riding a unicycle uphill is shown. The cyclist's body is modeled as a rigid rod of length L and mass m_1. The unicycle wheel is modeled as a uniform hoop with mass m_2 and radius r. Neglect the mass of the hub and spokes, as well as the mass of the pedals, cranks, and all the other pesky details. The rod maintains a constant inclination β. Consider M, the moment applied by the cyclist to the wheel, to be specified; assume it is sufficient to cause the system to move upslope (as opposed to down), and furthermore assume that a no-slip condition holds.

a. Derive the equations of motion for the system.

b. Derive the equations that must be satisfied for the system to move upslope at a constant acceleration with the person (rod) oriented at a constant angle with respect to the ground.

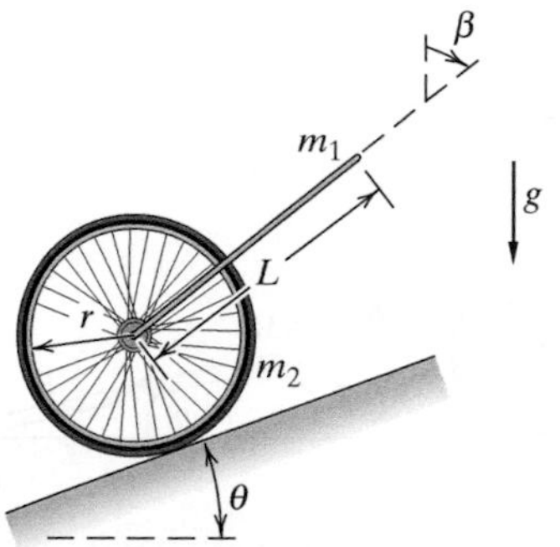

E7.3.61

7.3.62. [Level 3] The system shown consists of a wheel (m_w) with radius r, a "car body" (m_b) of length L, and a massless drag that keeps the body level. A moment M is applied from the link $\overline{BG}$ to the wheel. Assume that no friction exists between the drag and the road surface. Make an assumption regarding slip, solve for the vehicle's acceleration, and then verify your assumption. $r = 0.3$ m, $L = 1.4$ m, $m_b = 10$ kg, $m_w = 14$ kg, $\mu = 0.7$, and $M = 50$ N·m. Treat the wheel as a uniform circular disk.

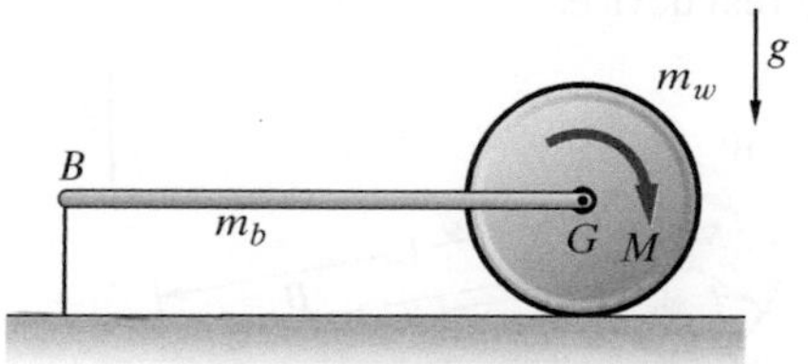

E7.3.62

7.3.63. [Level 3] Shown is a simple vehicle, consisting of a driven wheel W (treat it as a uniform cylinder) attached at O to a rigid link B. An unmodeled motor in B applies a moment $M(t)$ to the wheel W. Solve for the linear acceleration ($\ddot{x}$) experienced by the vehicle and the angular acceleration of the wheel. $m_W = 10$ kg, $m_B = 40$ kg, $L = 1$ m, $r = 0.5$ m, and $M = 2$ N·m. Neglect any friction between the right end of body B (point A) and the ground. Assume that W rolls without slip and that both W and B remain in contact with the ground.

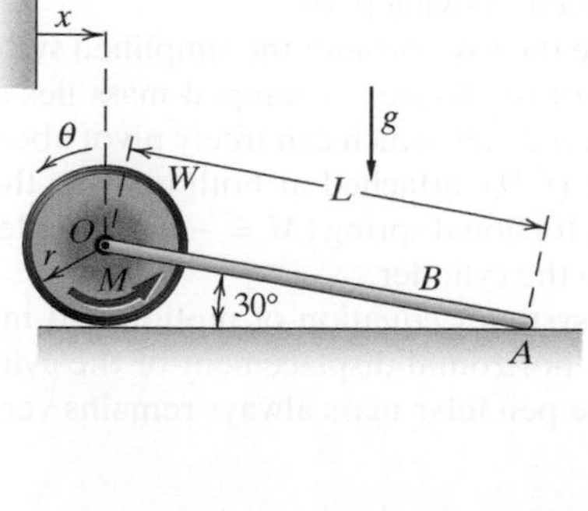

E7.3.63

7.3.64. **[Level 3]** Repeat **Exercise 7.3.63** but consider two values of applied moment M: 2 N·m and 200 N·m. Calculate $\ddot{x}$, $\ddot{\theta}$, R, S, Q, N, and T for both cases. Discuss the difference in their values for the two cases and explain why the differences make physical sense. The interaction forces are shown.

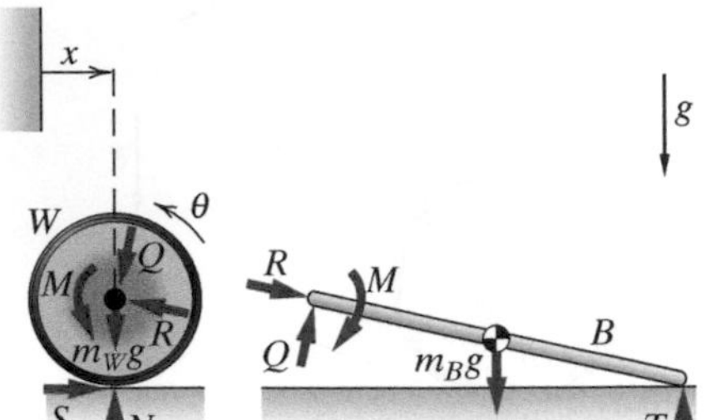

E7.3.64

7.3.65. **[Level 3]** Can a sufficient positive moment M be applied from the hinged bar B to the wheel W for the system shown to experience wheel liftoff (body W loses contact with the ground)? Assume roll without slip for body W and frictionless contact between the rigid link (body B) and the ground. $m_W = 10$ kg, $m_B = 40$ kg, $r = 0.5$ m, and $L = 1$ m. The rigid link B exerts the moment M on the wheel. Solve the problem first by assuming that a tractive force will exist between the wheel and the ground regardless of the normal force. Next, comment on the real-world chances of seeing liftoff for a real device.

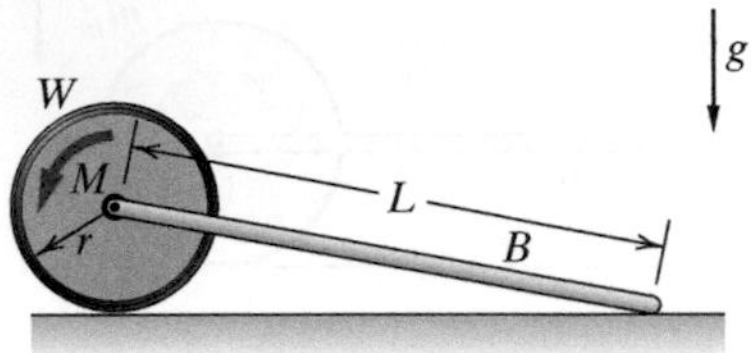

E7.3.65

7.3.66. **[Level 3] Computational** In this exercise we will analyze a toy I designed several years ago. It consists of three pieces—a cylindrical body, a metal strip, and a rubber band. When the toy is resting on the ground, the strip hangs straight down. If you then roll the cylinder away from yourself, the rubber bands wind up because the strip tends to remain hanging down as the cylinder rotates about it. Eventually, the moment from the twisting is enough to stop the cylinder from moving, and the toy then begins to roll back, gaining speed as it approaches. To the user this is quite mysterious because all she sees is a plain cylinder with no apparent moving parts.

To analyze the toy, consider the simplified system shown on the right of the figure. A lumped mass lies at the end of a massless rod $\overline{AB}$, which can freely pivot about another massless rod ($\overline{CD}$), attached at both ends to the cylinder (radius r). A torsional spring ($M = -k_\theta \theta$) couples the pendulum $\overline{AB}$ to the cylinder.

Find the system's equation of motion and integrate to find $x(t)$, the horizontal displacement of the cylinder. Assume that the pendular mass always remains vertical. This isn't precisely true, but the error in using such an assumption will be small. This reduces the problem to one having just a single degree of freedom. Assume that the cylinder rolls without slipping. $m_1 = 0.3$ kg, $m_2 = 5$ kg, $r = 0.1$ m, $k_\theta = 0.01$ N·m/rad, and $I_{\overline{CD}} = m_1 r^2$. Integrate from $t = 0$ to $t = 3$ s and use initial conditions of $\theta(0) = 0$ and $\dot{\theta}(0) = 3.0$ rad/s. Does the resulting motion make sense?

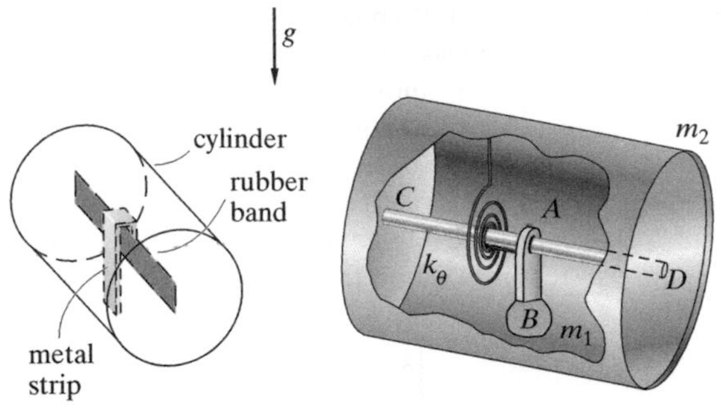

E7.3.66

7.3.67. **[Level 3] Computational** In this exercise we will delve a little deeper into the system described in **Exercise 7.3.66**. Rather than assuming that m_1 remains directly below the centrally positioned rod $\overline{CD}$ of the cylinder, let's perform a more exact calculation and allow the pendulum $\overline{AB}$ to rotate.

a. Derive the two governing equations of motion (for $\ddot{\theta}$ and $\ddot{\beta}$). Assume roll without slip. Let θ indicate rotation of the outer cylinder, and let β show the rotation of the pendulum $\overline{AB}$ with respect to the cylinder. The moment generated by the torsional spring is given by $M = -k_\beta \beta$.

b. Numerically integrate the equations using $e = 0.08$ m, $r = 0.10$ m, $k = 9.07 \times 10^{-4}$ N·m/rad, $m_1 = 0.2$ kg, $m_2 = 0.1$ kg, $\theta(0) = 0, \dot{\theta}(0) = 10$ rad/s, $\beta(0) = 0, \dot{\beta}(0) = 0$, and $0 \le t \le 10$ s, and show that the device works as claimed.

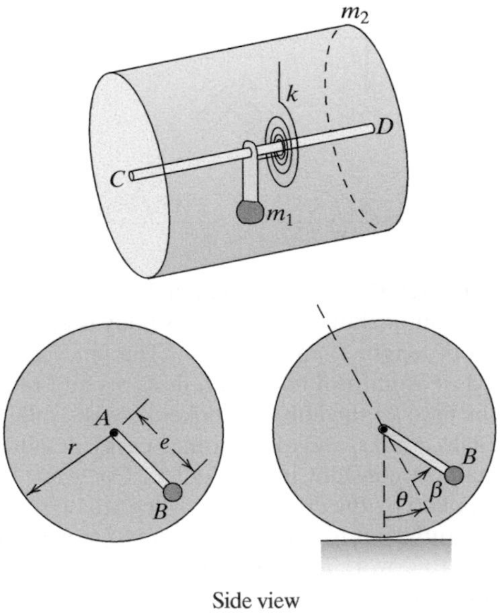

E7.3.67

7.3.68. **[Level 3]** The illustrated pulley system has four freely pivoting pulley disks with masses m_1, m_2, m_3, and m_4 and rotational inertias I_1, I_2, I_3, and I_4. Pulleys 1 and 2 share a common axle, as do pulleys 3 and 4. The pulley disks have radii $r_1 = r_3 = 0.07$ m and $r_2 = r_4 = 0.1$ m, masses $m_1 = m_3 = 0.5$ kg and $m_2 = m_4 = 0.75$ kg, and rotational inertias $I_1 = I_3 = 0.001$ kg·m^2 and $I_2 = I_4 = 0.0035$ kg·m^2. If the reel R pulls in rope at a rate of 0.7 m/s, the disks have rotation rates of $\omega_1 = -5$ rad/s, $\omega_2 = -7$ rad/s, $\omega_3 = -2.5$ rad/s, and $\omega_4 = -5.25$ rad/s, and block A ($m_A = 10$ kg) moves up at 0.175 m/s. Assume that reel R can exert a constant force of 50 N as it reels in the pulley rope. Assume that all of the pulley ropes are oriented vertically. (This keeps you from dealing with trigonometric nonlinearities.)

a. Determine the acceleration of block A and the tensions in the ropes.

b. As a check on your work, let the masses and inertias be close to zero and determine the tensions and acceleration of block A. Under zero pulley mass conditions, the acceleration of the block should be 10.19 m/s^2. Do you get this result?

c. Do the changes from zero pulley mass to finite pulley mass make sense?

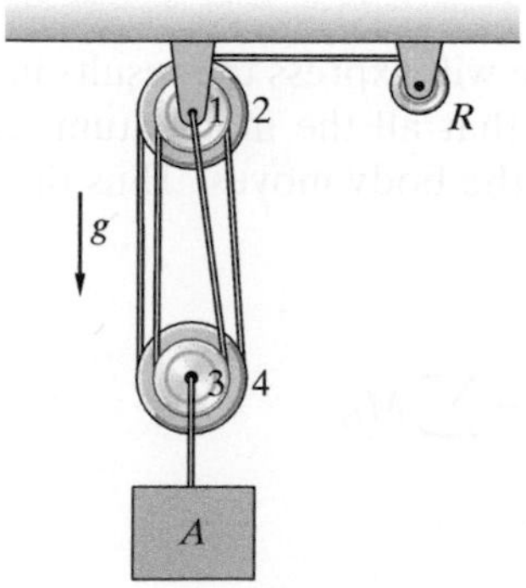

E7.3.68

7.3.69. **[Level 3]** Find the equation of motion for the illustrated car model. The axles are massless. Assume that all motion occurs in the x direction and that the car rolls without slip.

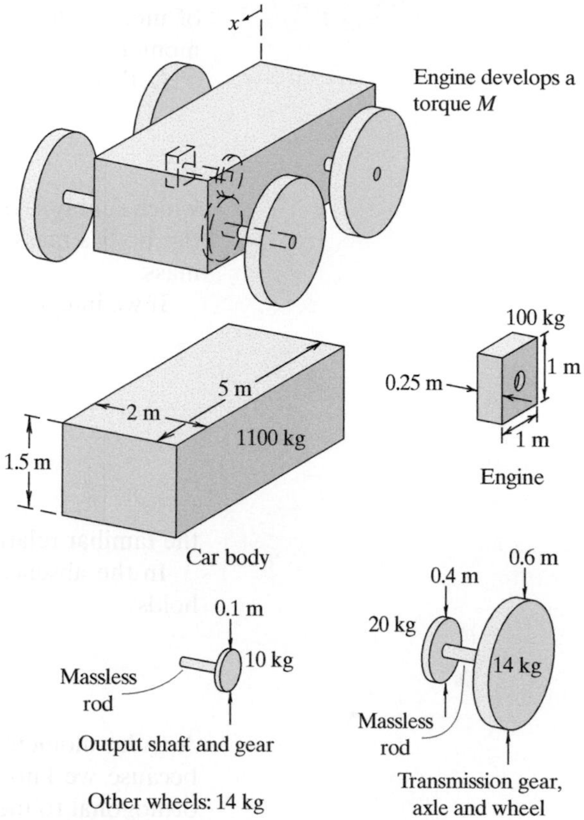

E7.3.69

7.3.70. **[Level 3]** Two identical uniform circular disks are joined by a rigid rod able to freely pivot at each end. Determine the angular acceleration of disk A. Assume roll without slip. $r = 1.0$ m, $m_A = m_B = 50$ kg, $m_C = 0$, and $M = 60$ N·m.

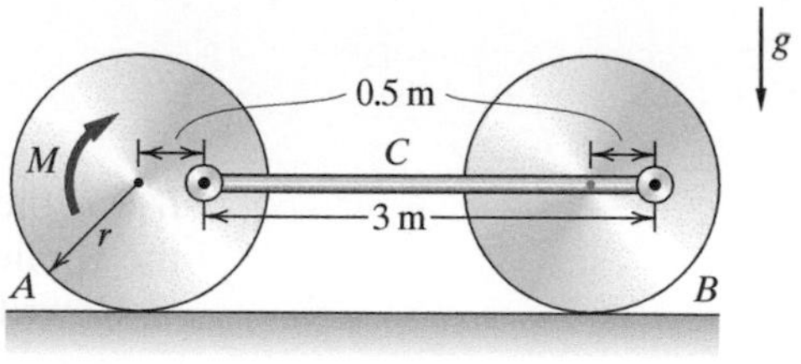

E7.3.70

7.4 LINEAR/ANGULAR MOMENTUM OF TWO-DIMENSIONAL RIGID BODIES

We find the linear and angular momentum relationships for rigid bodies in the same way as we did for single and multiple particles—we just integrate our force and moment balance equations with respect to time. The results will be just the same: the linear/angular momentum at some later time will equal the initial linear/angular momentum plus an applied linear/angular impulse. If there isn't any applied impulse, then we have conservation of momentum (which can apply to the linear momentum, the angular momentum, or both).

Let's start with the linear momentum. The relevant force balance is

$$\sum \boldsymbol{F} = m\bar{\boldsymbol{a}}$$

which simply says that the sum of the forces acting on the body will equal the body's mass multiplied by the acceleration of the body's center of mass.

If we integrate from t_1 to t_2 we will obtain

$$m\boldsymbol{v}(t_2) = m\boldsymbol{v}(t_1) + \int_{t_1}^{t_2} \sum \boldsymbol{F}\,dt$$

$$\boldsymbol{L}(t_2) = \boldsymbol{L}(t_1) + \mathcal{LI}_{1\text{-}2} \tag{7.70}$$

the familiar relation from our prior work.

In the absence of external forces, conservation of linear momentum holds:

$$\boldsymbol{L}(t_2) = \boldsymbol{L}(t_1) \tag{7.71}$$

Angular momentum is similar. We will express the results in scalar form because we know from the start that all the momentum terms will be orthogonal to the plane in which the body moves. Thus the only vector direction possible is $\boldsymbol{k}$.

Starting from

$$I_G\ddot{\theta} = \sum M_G$$

we can integrate in time to obtain

$$I_G\dot{\theta}(t_2) - I_G\dot{\theta}(t_1) = \int_{t_1}^{t_2} \sum M_G\,dt$$

$I_G\dot{\theta}$ is the angular momentum of the body about G, and $\int_{t_1}^{t_2} \sum M_G\,dt$ is the applied angular impulse.

Rewritten in slightly different form, this gives us

$$H_G(t_2) = H_G(t_1) + \mathcal{AI}_{G_{1\text{-}2}} \tag{7.72}$$

where I've written $I_G\dot{\theta}$ as H_G and used $\mathcal{AI}_{G_{1\text{-}2}}$ to represent the applied angular impulse from t_1 to t_2.

As in the linear case, if there's no applied angular impulse, then we have conservation of angular momentum about G:

$$H_G(t_2) = H_G(t_1) \tag{7.73}$$

Because the angular momentum balance about a fixed point O has exactly the same form as that about G ($\dot{H}_O = \sum M_O$ versus $\dot{H}_G = \sum M_G$), you can derive the formula for angular momentum about a fixed point O in exactly the same way, which yields

$$H_O(t_2) = H_O(t_1) + \mathcal{AI}_{O_{1-2}} \tag{7.74}$$

Finally, as in the previous case, a lack of applied angular impulses gives us an expression reflecting a conservation of angular momentum about O:

$$H_O(t_2) = H_O(t_1) \tag{7.75}$$

>>> **Check out Example 7.20 (page 460) and Example 7.21 (page 461) for applications of this material.**

EXAMPLE 7.20 ANGULAR IMPULSE APPLIED TO SPACE STATION (Theory on page 459)

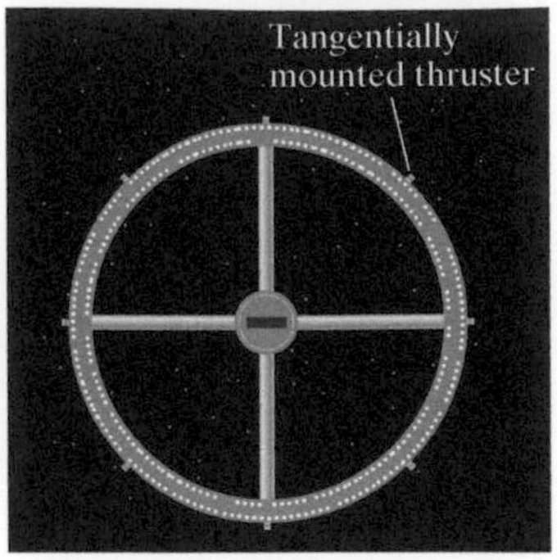

Figure 7.69 Rotating space station

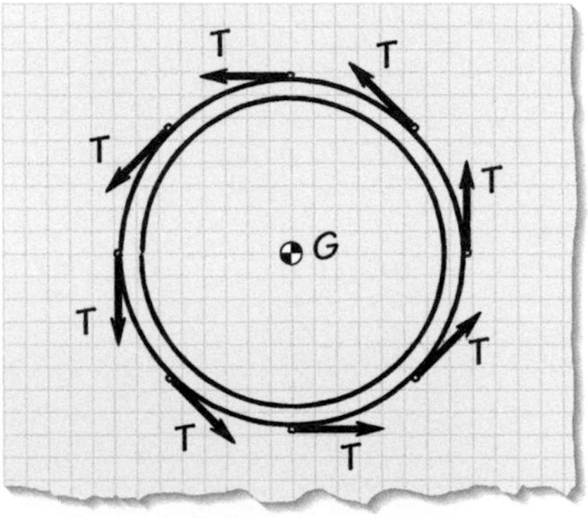

Figure 7.70 Eight thrusters acting on space station

Determine how long a group of eight thrusters must fire in order to bring the space station illustrated in **Figure 7.69** from rest to a rotational speed for which the simulated gravity (due to centripetal acceleration) is equal to 1g. The station will be modeled as a hoop, having a radius $r = 600$ m and mass $m = 8.0 \times 10^5$ kg. Each thruster produces a thrust of 300 N and fires tangent to the space station's surface.

Goal Determine the time needed for the thrusters to produce a perceived 1g acceleration at the station's rim.

Draw **Figure 7.70** shows the hoop model of our space station, as well as the eight thruster forces, each labeled T.

Formulate Equations All we need to consider is how the angular momentum changes over time:

$$H_G(t_2) = H_G(t_1) + \mathcal{AI}_{G_{1-2}} \tag{7.76}$$

The mass moment of inertia about the mass center G is given by

$$\bar{I} = mr^2 = (8.0 \times 10^5\,\text{kg})(600\,\text{m})^2 = 2.88 \times 10^{11}\,\text{kg·m}^2$$

The applied angular impulse is given by

$$\begin{aligned}\mathcal{AI}_{G_{1-2}} &= 8rT(t_2 - t_1) = 8(600\,\text{m})(300\,\text{N})(t_2 - t_1)\\ &= (1.44 \times 10^6\,\text{N·m})(t_2 - t_1)\end{aligned}$$

The final piece of information we need is the rotational speed that will create an artificial gravity environment equivalent to that on the earth's surface. Because the centripetal acceleration is equal to $r\dot{\theta}^2$, we have

$$(600\,\text{m})\dot{\theta}^2 = 9.81\,\text{m/s}^2 \quad \Rightarrow \quad \dot{\theta} = 0.128\,\text{rad/s}$$

Solve Because the initial momentum of the system is zero, we have, from (7.76),

$$\bar{I}\dot{\theta}(t_2) = \mathcal{AI}_{G_{1-2}}$$

$$(2.88 \times 10^{11}\,\text{kg·m}^2)(0.128\,\text{rad/s}) = (1.44 \times 10^6\,\text{N·m})(t_2 - t_1)$$

$$t_2 - t_1 = 2.56 \times 10^4\,\text{s}$$

Thus the time necessary to spin the station up to speed is a little over 7 hours.

EXAMPLE 7.21 IMPACT BETWEEN A PIVOTED ROD AND A MOVING PARTICLE (Theory on page 459)

We now have got enough information to move beyond simple particle collisions and look at collisions involving rigid bodies. As an example, consider the system shown in **Figure 7.71**. A mass particle A (mass m, speed v), traveling in the $\boldsymbol{b}_1$ direction, collides with the end of the illustrated bar (length L, mass m). The bar can pivot in the horizontal plane and is initially stationary. Immediately following the collision, the bar will have a rotational speed ω. Assume a coefficient of restitution equal to 1.0 and solve for the post-collision rotational speed.

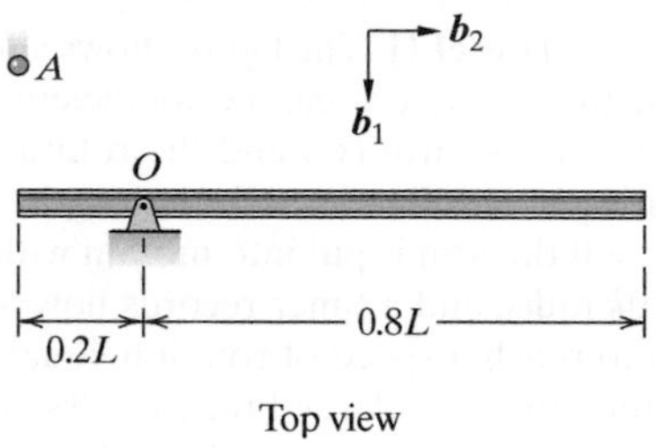

Figure 7.71 Hinged bar and impacting mass

Goal Find the rotational speed of a pinned bar after it has been impacted by a moving particle.

Draw **Figure 7.72** shows the reaction forces due to the collision.

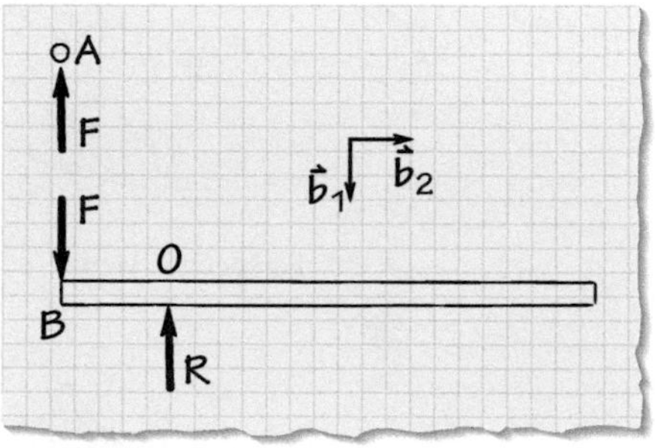

Figure 7.72 Forces acting on bar and mass

Formulate Equations During the collision, there exist forces between the bar and the mass (F) as well as a reaction force (R) that acts on the bar at the pivot location. The two equations we will need are the impact equation and conservation of angular momentum about the pivot. By applying conservation of angular momentum about O, we ensure that the reaction force R won't appear in our equations—a nice bonus. Use $v'\boldsymbol{b}_1$ to denote the particle's velocity after the collision and $v_B\boldsymbol{b}_1$ to indicate the velocity of the bar end B just after the impact. As in Section 3.7, we define our coefficient of restitution as the ratio of the separation speed to the approach speed.

$$e = \frac{v_B - v'}{v} = 1 \qquad (7.77)$$

Before collision, the system's angular momentum arises solely from the mass particle: $H_O = 0.2mLv$. After the collision, the angular momentum involves both the mass particle and the bar: $H_O = 0.2mLv' + I_O\omega$.

Equating the two momenta gives us

$$0.2mLv = 0.2mLv' + I_O\omega \qquad (7.78)$$

Assume Because the bar is a rigid body, v_B is not independent of ω:

$$v_B = 0.2\omega L \qquad (7.79)$$

Solve We need to invoke the parallel axis theorem to find I_O:

$$I_O = \bar{I} + m(0.3L)^2 = 0.17\bar{3}mL^2 \qquad (7.80)$$

$$(7.79) \rightarrow (7.77) \Rightarrow \quad v' = 0.2\omega L - v \qquad (7.81)$$

$$(7.80), (7.81) \rightarrow (7.78) \Rightarrow \quad 0.2v = 0.2(0.2\omega L - v) + 0.17\bar{3}L\omega \qquad (7.82)$$

$$\omega = \frac{1.875v}{L}\ \text{rad/s}$$

EXERCISES 7.4

7.4.1. **[Level 1]** The figure shows a testing device that is to be used to determine the coefficient of friction between the tip of the arm (C) and the rotating disk. The disk is rotating at $-20\boldsymbol{k}$ rad/s and the length of the arm is 0.3 m. At $t = 0$ the arm is put into motion with a rotational speed of $10\boldsymbol{k}$ rad/s, and a timer records how long it takes for the arm to reach a speed of rotation equal to that of the disk. Assume that the elapsed time is 2.5 s and calculate the coefficient of friction this implies. The normal force between the tip and the disk is 2 N and $I_O = 0.015\ \text{kg}\cdot\text{m}^2$.

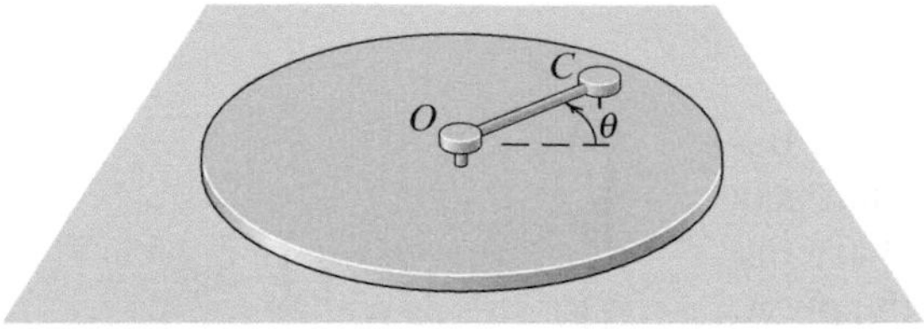

Perspective view

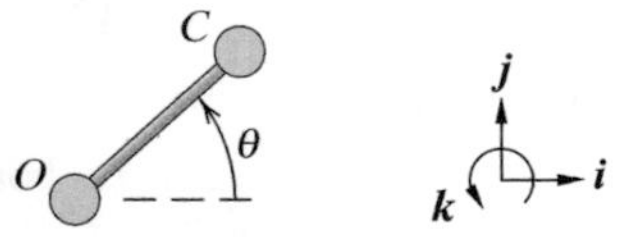

Top view

E7.4.1

7.4.2. **[Level 1]** An 80 kg man is initially stationary on a 20 kg, uniform, stationary platform. The man is initially centered on the platform, located just above the platform's center of mass and also located at an absolute position of $x = 0$ m. The platform lies on a frictionless surface. At $t = 0$ the man begins to walk to the right. What is the minimum length (L) of the platform such that the man can reach the absolute position $x = 2$ m without leaving the platform?

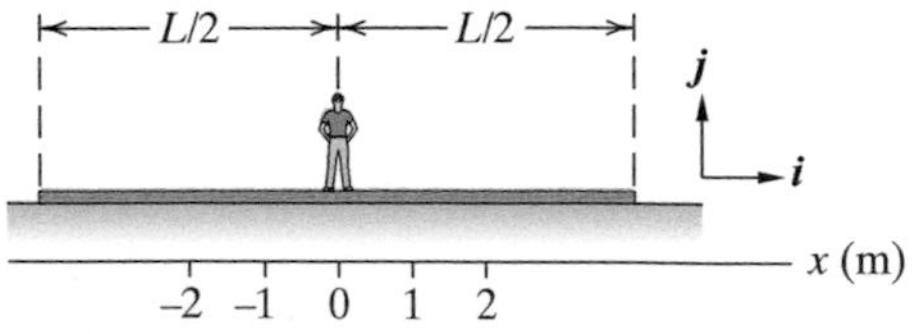

E7.4.2

7.4.3. **[Level 1]** A person rolls a $W = 4.5$ kg bowling ball with radius r from rest by repeatedly brushing the palms of his hands over the top of the ball. We'll assume that the resulting traction acting on the ball's top can be modeled as a continuous and constant horizontal force of $F = 4.45$ N. If the bowling ball rolls without slip, how fast v will it be moving after $\Delta t = 2$ s? Treat the bowling ball as a uniform sphere.

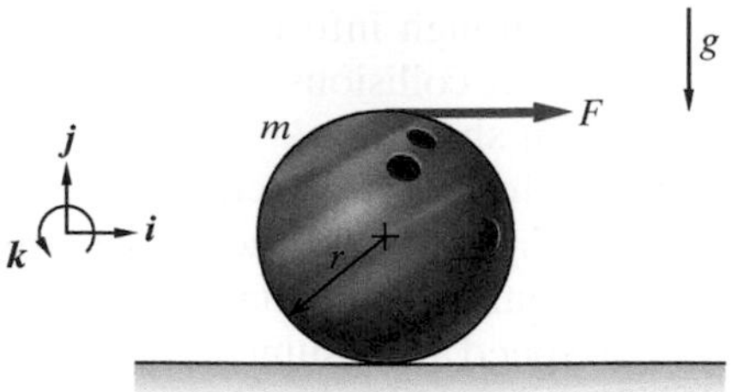

E7.4.3

7.4.4. **[Level 1]** A uniform cylinder with mass $m = 6$ kg and radius r is pulled from rest along level ground by a constant force F that acts on its center at an angle $\beta = 40°$ with respect to the horizontal. The cylinder rolls without slip, and its speed is $v = 5$ m/s after $\Delta t = 3$ s. How much force F is being applied?

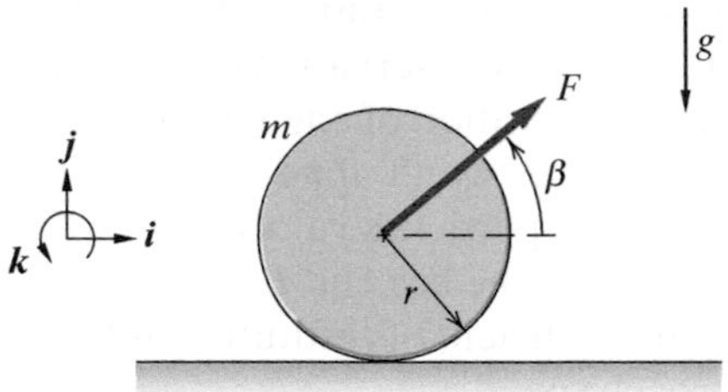

E7.4.4

7.4.5. **[Level 1]** A uniform disk A (mass $m_A = 10$ kg) rolls without slip on body B (mass $m_B = 5$ kg), and body B slides without friction along the ground. $r = 0.2$ m. Determine the rotational speed of disk A if $\boldsymbol{v}_O = 5\boldsymbol{i}$ m/s and the mass center of the total system is stationary.

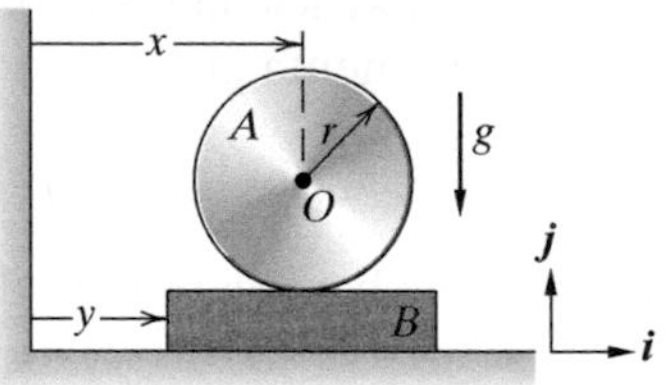

E7.4.5

7.4.6. **[Level 1]** At $\theta = 0$ the illustrated system is spinning about a vertical, frictionless axle at A (thus the motion is in a horizontal plane). At $\theta = 0$, $\dot{\theta} = 5$ rad/s, and $t = 0$. The rigid, uniform disk is not rotating with respect to the bar $\overline{GA}$ because a massless clamp holds the two bodies tightly together. Besides the clamp, the bodies are joined through a friction-free pivot at G.

When $\theta = 90$ degrees ($t = t_1$), the clamp unlatches and a negative moment is applied to $\overline{AG}$ at A, which is meant to slow its rotation to zero. Once the bar becomes stationary, the moment goes to zero ($t = t_2$). $r = 0.3$ m, $L = 1.5$ m, $m_d = 10$ kg, $m_b = 20$ kg, and $\boldsymbol{M} = -100\boldsymbol{k}$ N·m.

a. What is the absolute rotational velocity of the disk at the instant the moment goes to zero?

b. What is the behavior of the system after t_2?

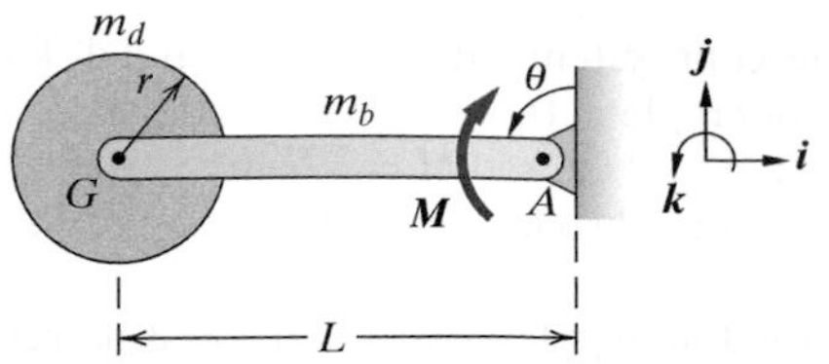

E7.4.6

7.4.7. [Level 1] Solve **Exercise 7.4.3** using an angular impulse-momentum analysis.

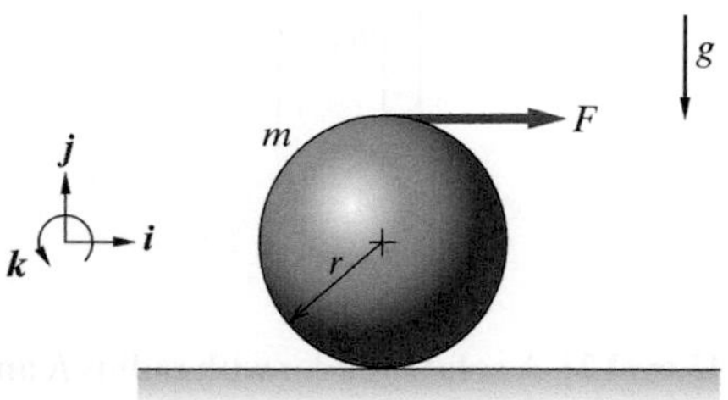

E7.4.7

7.4.8. [Level 2] A uniform cylinder of mass m and radius r is pulled from rest along a rough horizontal surface by a constant force F acting on its center. If the cylinder rolls without slip, find how long it takes to reach a speed v using a linear impulse-momentum analysis.

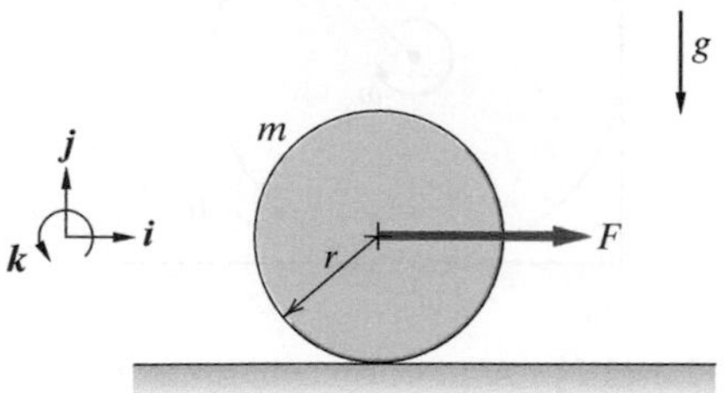

E7.4.8

7.4.9. [Level 2] Computational Two bars are lying on a flat, friction-free surface, are hinged at their center, and have mass moments of inertia about this center point of $\bar{I}_1 = 0.1\,\text{kg}\cdot\text{m}^2$ and $\bar{I}_2 = 0.2\,\text{kg}\cdot\text{m}^2$. They are also connected to each other via a linear torsional spring with rotational spring constant $k_\theta = 10.5\,\text{N}\cdot\text{m/rad}$. θ_1 and θ_2 indicate rotations away from the bars' equilibrium positions, positions at which the spring is unstretched. Bar 1 is held fixed and bar 2 is rotated 90° so as to align the two bars. Upon release the two bars will oscillate. Numerically integrate the two equations of motion to find θ_1 and verify that the total angular momentum of the system is constant. Plot the angular momentum and the individual angular speeds of each bar versus time for $0 \le t \le 1.0$.

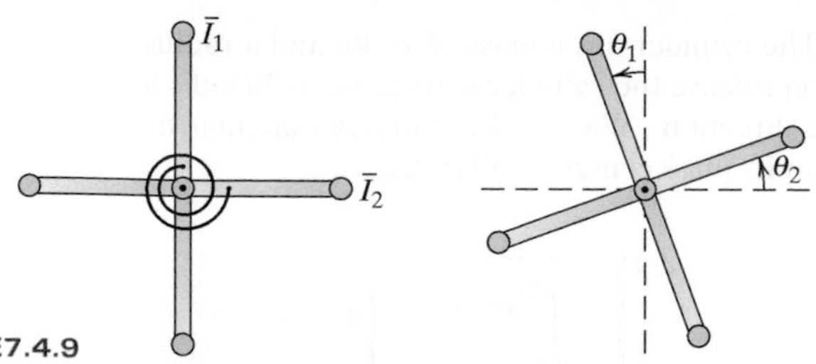

E7.4.9

7.4.10. [Level 2] The system shown in the figure consists of two main pieces, each floating outside of a space station. The main body can be approximated as a uniform bar of length $L = 4$ m and mass m_2. The wheeled cart (mass m_1) is designed to run along the main body by means of two counter-rotating wheels (radius r). We will be treating this as a lumped mass. Shown below is the initial configuration, with the cart at $x = 4$ m. At $t = 0$ the two guide wheels begin to rotate, the top wheel's rotation rate given by $\dot{\theta}$ and the bottom wheel's by $-\dot{\theta}$. The wheels' angular acceleration is constant ($\ddot{\theta} = \alpha_0$) for 2 s and then constant but with opposite sign ($\ddot{\theta} = -\alpha_0$) for the next 2 s. At $t = 4$ s, the cart is at A.

a. What is the velocity of B at $t = 4$ s?

b. What is the position of A with respect to the space station at $t = 4$ s?

c. What is the force exerted on the main body by the moving cart as a function of m_1 and m_2?

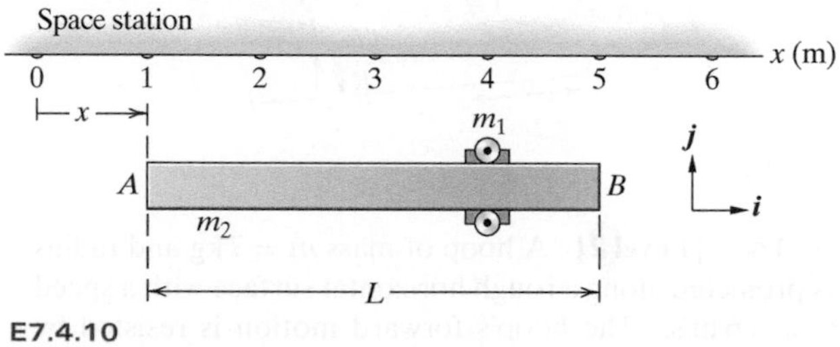

E7.4.10

7.4.11. [Level 2] A uniform disk A (mass $m_A = 16$ kg, radius $r = 0.5$ m) rolls without slip on block B ($m_B = 8$ kg), which lies on a frictionless surface. A force $\boldsymbol{F} = 10\boldsymbol{i}$ N is applied to the system that is initially at rest. After 5 s, the velocity of block B is $3.75\boldsymbol{i}$ m/s. What is the rotational velocity of disk A?

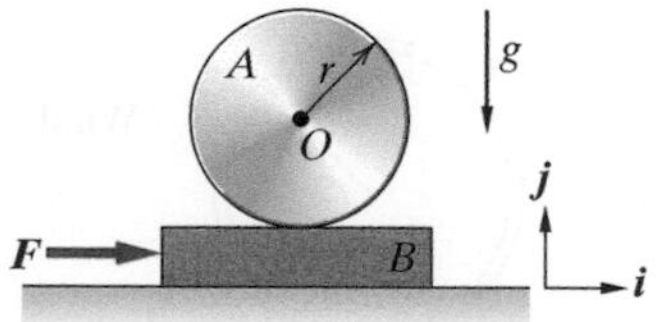

E7.4.11

7.4.12. [Level 2] A uniform cylindrical body is constrained in two ways. Its axle (about which it rotates freely) feels a constant 2 N force, directed upward, due to an actively controlled reel at B. In addition, a massless rope is wrapped around the cylinder and attaches to the ceiling at

A. The cylinder has a mass of 10 kg and a radius of 0.2 m. Upon release the cylinder is rotating at 20 rad/s in a clockwise direction. What is the cylinder's angular momentum about its mass center 3 s after release?

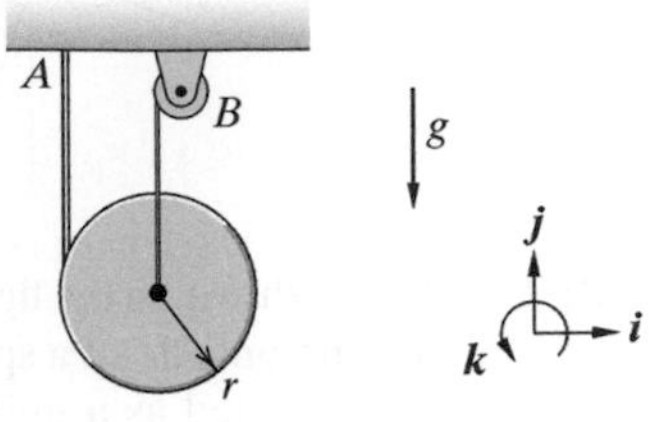

E7.4.12

7.4.13. **[Level 2]** The illustrated system consists of a motor with an attached gear that meshes with a linearly translating rack that moves on a frictionless surface. The motor produces a torque of 90 N·m, the gear has a mass moment of inertia about its center of 0.05 kg·m^2 and a radius of 0.1 m, and the rack has a mass of 10 kg. Calculate the rack's translational speed at $t = 0.3$ s if its initial velocity at $t = 0$ is $-5\boldsymbol{i}$ m/s.

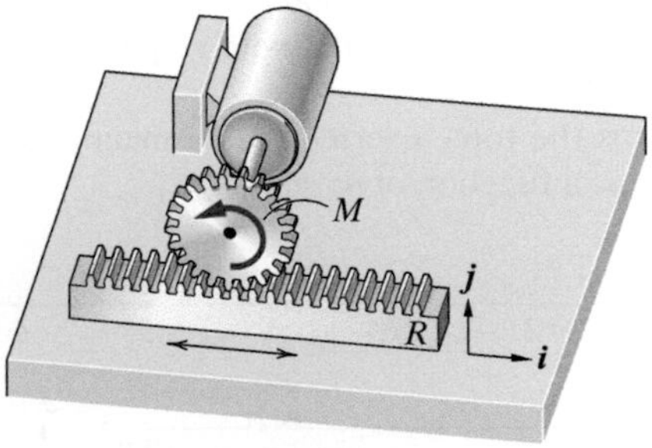

E7.4.13

7.4.14. **[Level 2]** A hoop of mass $m = 7$ kg and radius r is projected along a rough horizontal surface with a speed of $v_0 = 6$ m/s. The hoop's forward motion is resisted by a horizontal wind that can be modeled as a time-varying force $F_w(t)$ that acts on the hoop's center and is governed by $F_w(t) = \overline{F}_w\sqrt{t}$, where $\overline{F}_w = 15\,\text{N/s}^{\frac{1}{2}}$. If the hoop rolls without slip, calculate how long Δt it takes for the hoop to come to a stop. The rotational inertia about a hoop's mass center G is $I_G = mr^2$.

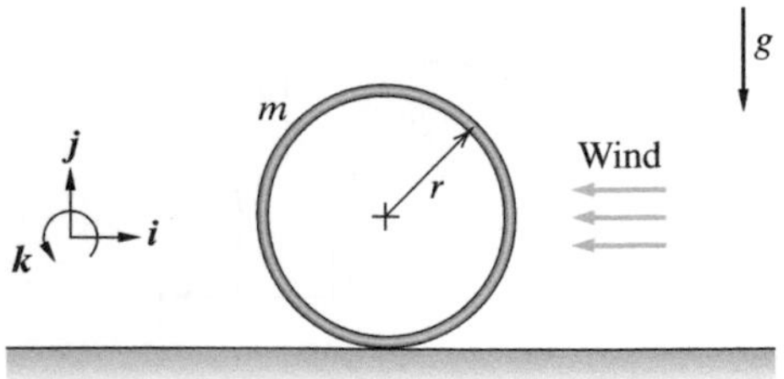

E7.4.14

7.4.15. **[Level 2]** A solid disk is mounted to a rigid bar $\overline{AG}$ through a frictionless pivot at G. The bar is pivoted to the ground at A. An external moment is applied to the initially quiescent system, causing the bar and disk to move in the *horizontal* plane (no gravity). $r = 0.1$ m, $L = 0.5$ m, $m_d = 1$ kg, $m_b = 1.2$ kg, and $M = 3$ N·m.

a. What is the absolute angular velocity of the disk at $t = 3$ s?

b. What is the angular velocity of the disk, relative to the bar $\overline{AG}$, at $t = 3$ s?

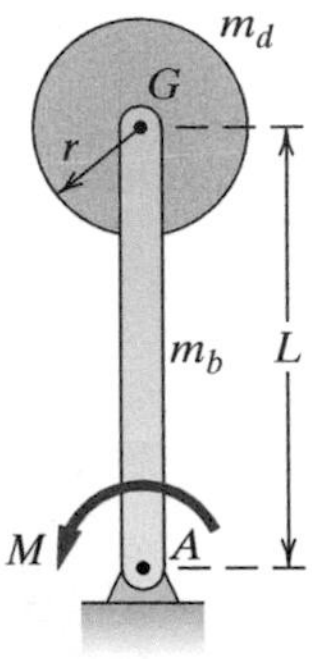

E7.4.15

7.4.16. **[Level 2]** A solid cylinder with radius R and mass m has rolled across the floor and into a wall. When it strikes, it has an angular speed ω_0. The coefficient of friction between the cylinder and both the floor and wall is μ. How long until the cylinder comes to rest? Assume zero rebound from the wall.

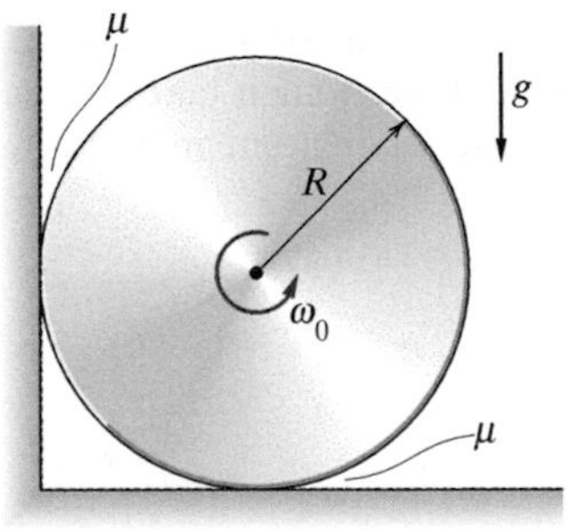

E7.4.16

7.4.17. **[Level 2]** A uniform cylindrical body (mass of 12 kg, radius of 0.25 m) is being pulled to the left with a constant force of 4 N. Assume roll without slip. What is the cylinder's angular momentum about its mass center 4 s after release?

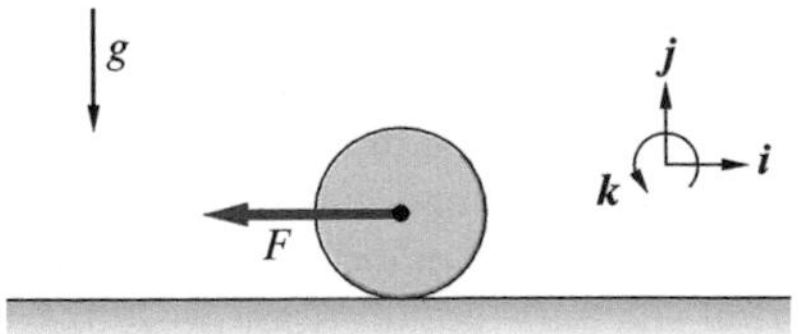

E7.4.17

7.4.18. **[Level 2]** A flat disk of radius R and mass m is placed on a horizontal table and then spun at an angular speed ω_0. How long will it take for it to come to a complete stop? The coefficient of dynamic friction is equal to μ. Assume a uniform pressure between the table and the disk.

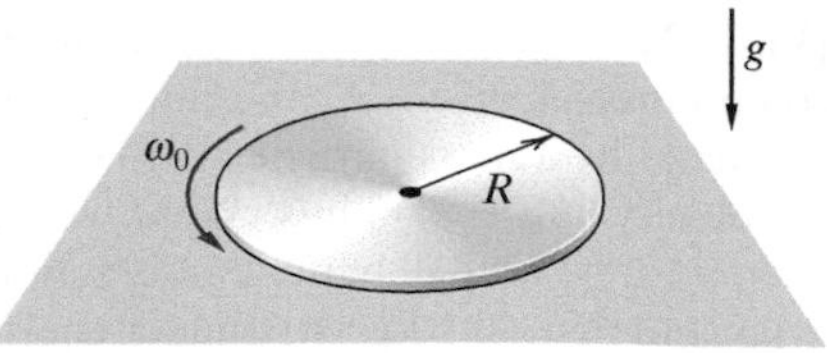

E7.4.18

7.4.19. **[Level 2]** To reduce some of the pressures of his job, Captain Kirk used to amuse himself by spinning in Uhura's chair whenever he was alone on the bridge. He found that if he spun himself around with his legs and arms extended, with a heavy Starfleet manual in each hand, and then drew his arms and legs close to his body, his rotation rate would increase markedly. Let's analyze this problem. The chair and his torso (plus his thighs) have a rotational inertia of 5.42 kg·m^2. In the illustration, A indicates the end of his leg, B is his knee joint, C is his hand, and D is his shoulder joint. B is located 41 cm from the chair's center of rotation, $\overline{AB} = 66$ cm, D is 23 cm from the chair's center of rotation, and $\overline{CD} = 66$ cm. Each of his arms ($\overline{CD}$) weighs 5 kg, each leg segment ($\overline{AB}$) weighs 7 kg, and each Starfleet manual weighs 2.2 kg. If his initial rotation rate is 5 rad/s (arms and legs extended), what is it with his arms and legs brought in? Treat his arms and legs as uniform thin rods. Treat the Starfleet manuals as point masses.

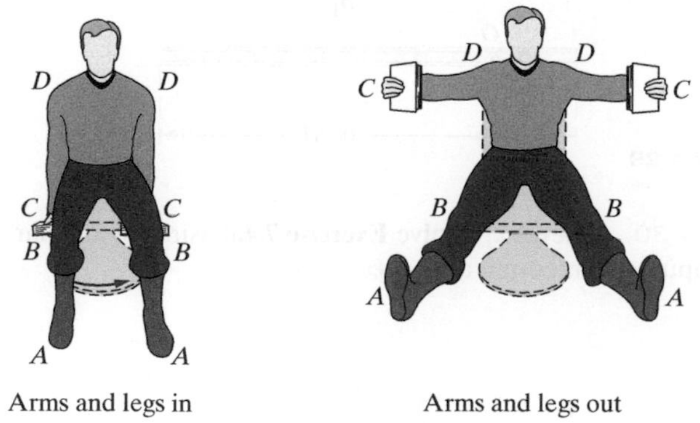

E7.4.19

7.4.20. **[Level 2]** Two uniform, rotating disks are brought together, immediately bonding to each other. Before they're joined, $\boldsymbol{\omega}_1 = -50\boldsymbol{k}$ rad/s and $\boldsymbol{\omega}_2 = 100\boldsymbol{k}$ rad/s. Their physical parameters are $r_1 = 5$ cm, $r_2 = 8$ cm, $m_1 = 1.1$ kg, and $m_2 = 3$ kg. What is the rotational velocity of each disk after they're joined?

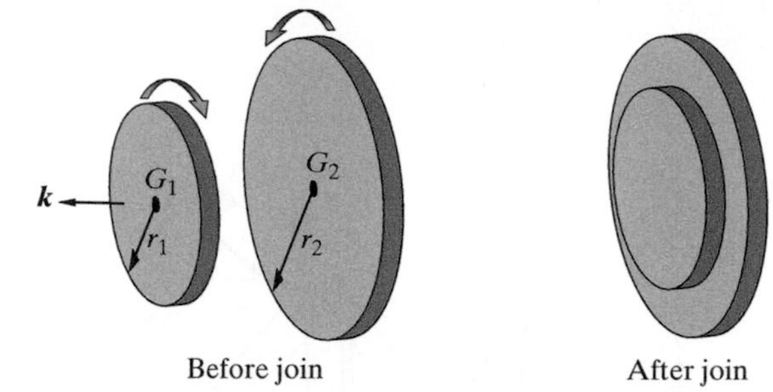

E7.4.20

7.4.21. **[Level 2]** By striking a pool ball below its midline, you can cause it to translate to the right while rotating counterclockwise (slipping). Assume a coefficient of friction equal to 0.65, a diameter of 7 cm, and a mass of 0.11 kg. Calculate how far the ball travels before it transitions to pure rolling (no slip). $v_G = 5$ m/s, $\dot{\theta} = 20$ rad/s.

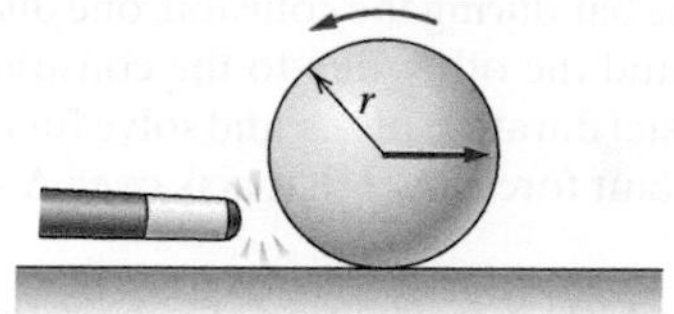

E7.4.21

7.4.22. **[Level 2]** A uniform cylindrical body (mass of 12 kg, radius of 0.25 m) is attached to the wall by a linear spring (initially unstretched) with spring constant $k = 72$ N/m. The cylinder is put into motion with an angular velocity of $-10\boldsymbol{k}$ rad/s. Assume roll without slip. What is the cylinder's angular momentum about its mass center $\frac{\pi}{8}$ s after release?

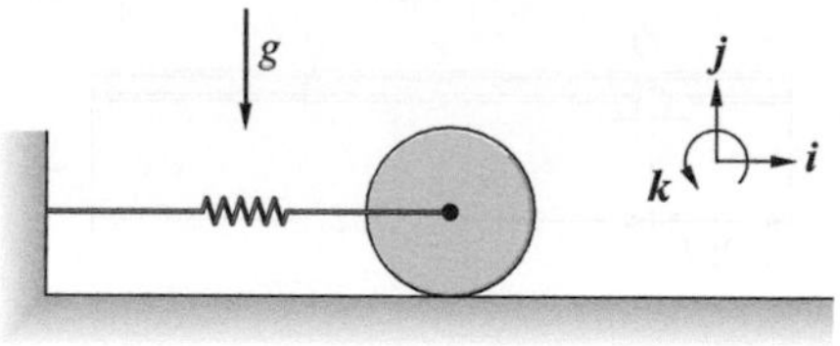

E7.4.22

7.4.23. **[Level 2]** Find the velocity of the mass particle of Example 7.21 after the collision.

7.4.24. **[Level 2]** A gymnast competing at the Olympics is performing a routine on the uneven bars. After completing a flip, she approaches the higher of the two bars with a speed v_0 at an angle β with respect to the ground. Assume that her body is aligned with the horizontal at approach. The gymnast has a mass of m, and her body has a length of L with her hands and legs stretched out. What is the gymnast's angular speed just after grabbing on to the bar?

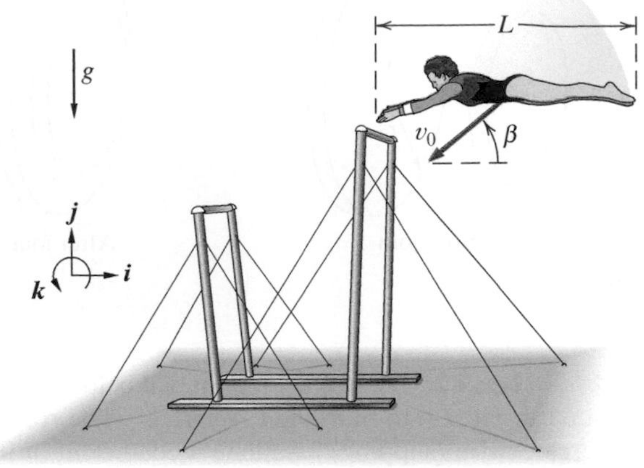

E7.4.24

7.4.25. **[Level 2]** As mentioned in Example 7.21, two forces act on the bar during the collision, one due to the colliding particle and the other due to the constraining pivot. Assume an impact duration of Δ s, and solve for these forces. Assume a constant force level that acts over Δ s.

7.4.26. **[Level 2]** A freely hinged bar (initially stationary) is impacted by a mass particle A moving with velocity $v\boldsymbol{b}_1$. Let the mass of the bar be m and the mass of the particle be m. Assume a coefficient of restitution equal to 0.5 and determine the velocity of the mass particle immediately following the collision.

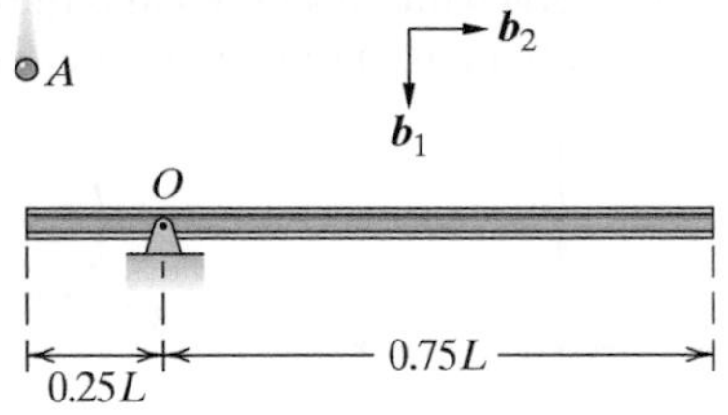

E7.4.26

7.4.27. **[Level 2]** A hoop of mass m and radius r rolls without slip with a constant speed v_0 on level ground. Find the hoop's angular speed immediately after it strikes an incline angled at β with respect to the horizontal. The rotational inertia for a hoop about its mass center G is given by $I_G = mr^2$, and assume zero rebound.

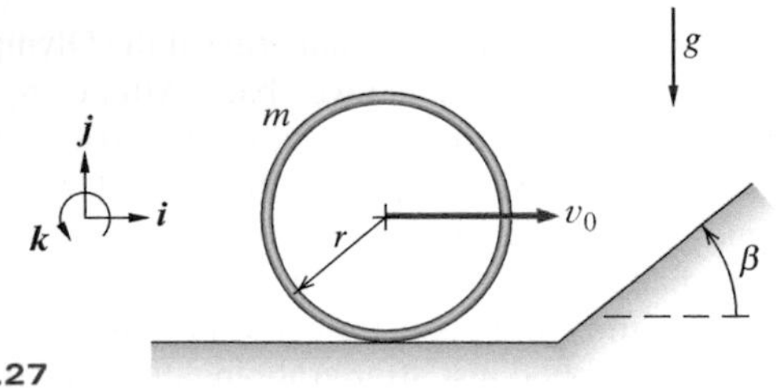

E7.4.27

7.4.28. **[Level 2]** A bar (mass m_1) is rotating around O at 25 rad/s (counterclockwise). At the illustrated instant, it is being impacted by a mass particle A (mass m_2) that's moving with velocity $v\boldsymbol{b}_1$, $v = 40$ m/s. What we have here is a simple model for a baseball bat/ball interaction. $L = 1$ m, $m_1 = 3.5$ kg, and $m_2 = 0.1$ kg. Assume a coefficient of restitution equal to 0.75, and determine the velocity of the mass particle immediately following the collision.

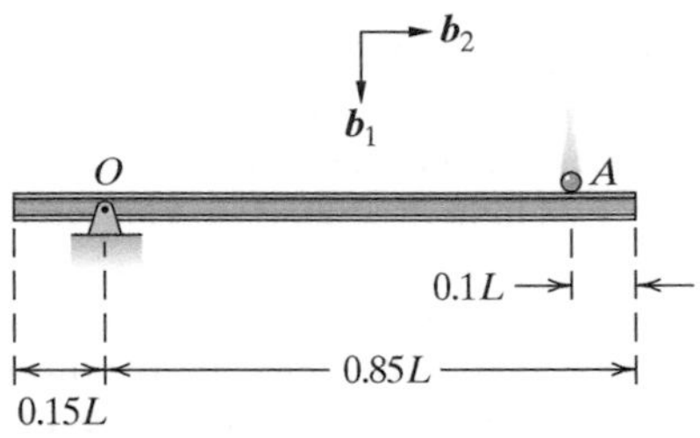

E7.4.28

7.4.29. **[Level 2]** An initially stationary bar (mass m_1) is struck by a mass particle (mass m_2) traveling with velocity $v\boldsymbol{b}_1$. Derive an expression for the rotational speed of the bar after impact as a function of the given parameters and e, the coefficient of restitution. Which causes a larger rotational speed, low or high values of e?

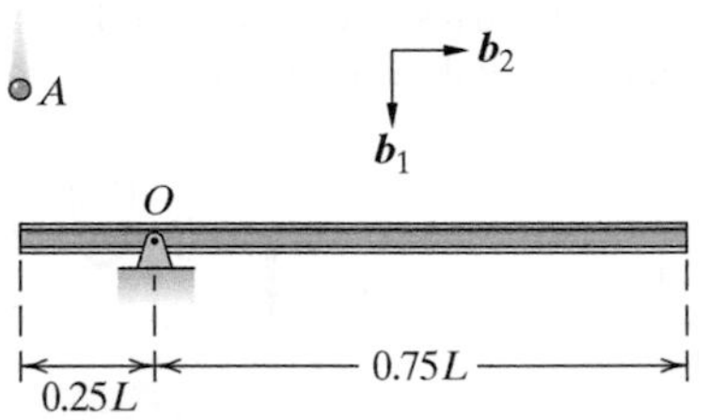

E7.4.29

7.4.30. **[Level 2]** Solve **Exercise 7.4.8** using an angular impulse-momentum analysis.

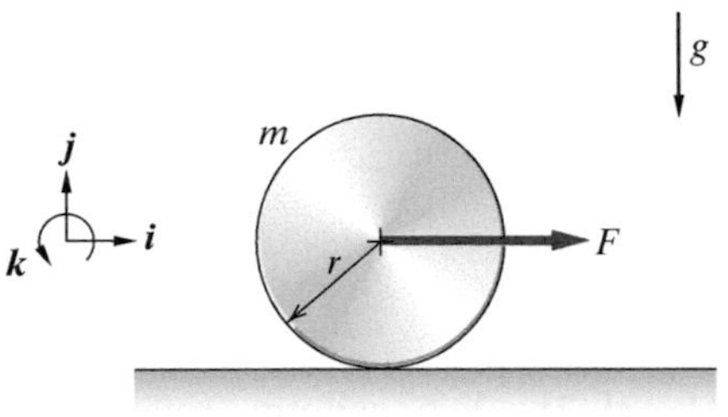

E7.4.30

7.4.31. **[Level 2]** Two uniform bars (A and B) are joined together at their left ends by a hinge joint. A massless motor in body A applies a moment M to body B. Initially, the system is at rest on a horizontal, frictionless surface, and is aligned as shown. At $t = 0$ the system is allowed to move as a result of the applied moment M. How large a circle, centered at $x = 0$, $y = 0$, must be drawn so that the entire mechanism stays within the circle for all time $t > 0$? $L = 0.3$ m, $m_A = 20$ kg, $m_B = 60$ kg, and $M = 10$ N·m.

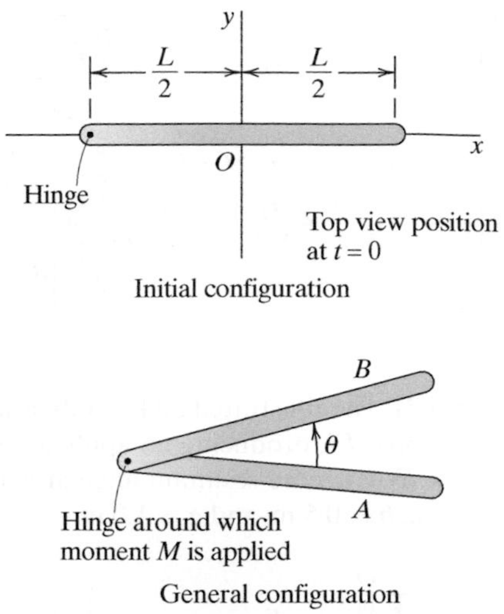

E7.4.31

7.4.32. **[Level 2]** The illustrated body of uniform linear density $\rho = 2\,\text{kg/m}$ is free to rotate in the horizontal plane about its pivoted end A. The body is initially at rest at an angle $\theta = 25^\circ$ with respect to the horizontal when a small ball of mass $m = 9\,\text{kg}$ strikes its free end B with a velocity $\boldsymbol{v} = 15\boldsymbol{j}\,\text{m/s}$. The ball sticks to the body at B upon impact, and $L = 0.5\,\text{m}$. Find the system's angular speed ω immediately after collision.

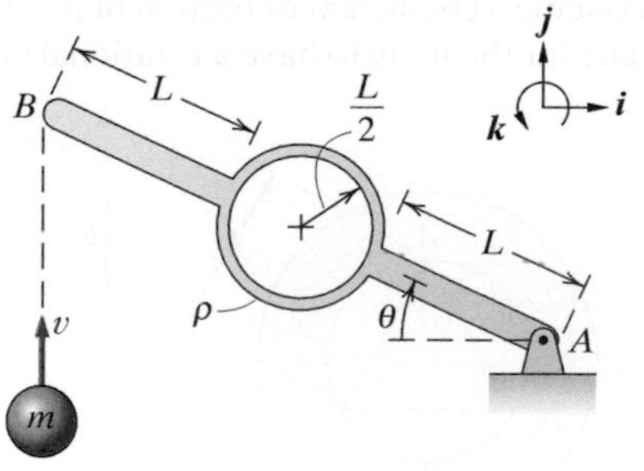

E7.4.32

7.4.33. **[Level 2]** A pulley of inner radius $r = 0.4\,\text{m}$ and outer radius $R = 0.7\,\text{m}$ is allowed to spin about its mass center G in the vertical plane, where the pulley's rotational inertia about G is given by $I_G = 7\,\text{kg·m}^2$. A block with mass $m = 6\,\text{kg}$ is pulled from rest along a rough horizontal surface by a rope that winds around the pulley's inner radius. The coefficient of friction between the block and surface is $\mu = 0.4$. Another rope winds around the pulley's outer radius, and it is pulled down with a constant force of $F = 25\,\text{N}$. Assume no slip between the ropes and pulley. What is the pulley's angular speed ω after $\Delta t = 5\,\text{s}$?

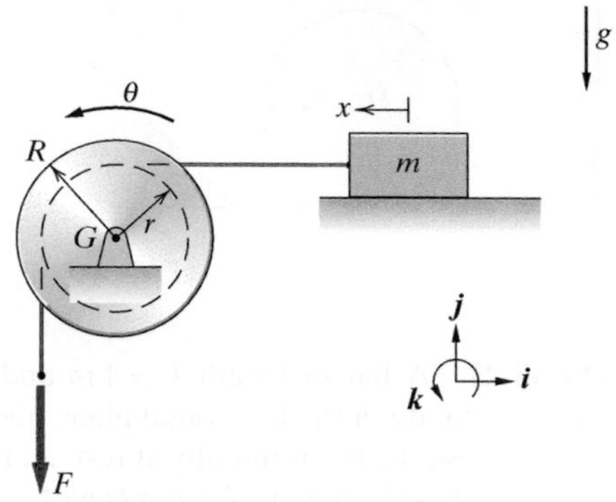

E7.4.33

7.4.34. **[Level 2]** The illustrated 9 kg uniform body (in the shape of an equilateral triangle) is acted on by an impulsive force $F\boldsymbol{i}$ that produces a linear impulse of 178 N·s in the $\boldsymbol{i}$ direction. What is the body's angular velocity after the impulse has been applied?

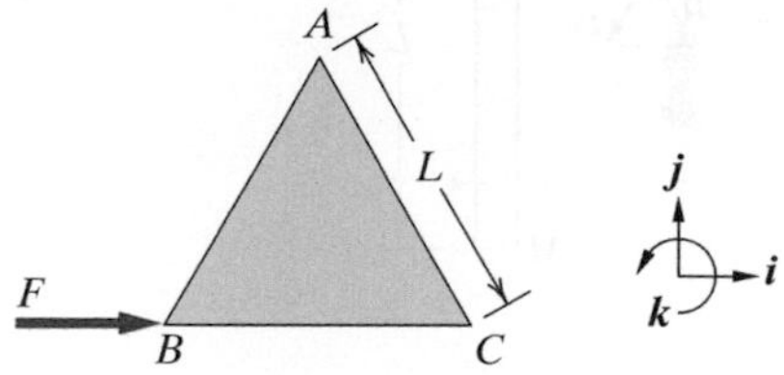

E7.4.34

7.4.35. **[Level 2]** A simplified disk brake assembly is shown. A force $F = 133\,\text{N}$ is applied to the lower end of the massless brake lever at a distance d from the lever's pivot O. The brake pad is located $L = 0.06\,\text{m}$ from O at the upper end of the lever, and the coefficient of friction between the brake pad and disk is $\mu = 0.7$. The disk has a weight and radius of $W = 54\,\text{kg}$ and $R = 0.076\,\text{m}$, respectively, and it is initially spinning at a constant $\omega_0 = 1000\,\text{rpm}$. What should d be if the disk is to come to a stop in $\Delta t = 1\,\text{s}$?

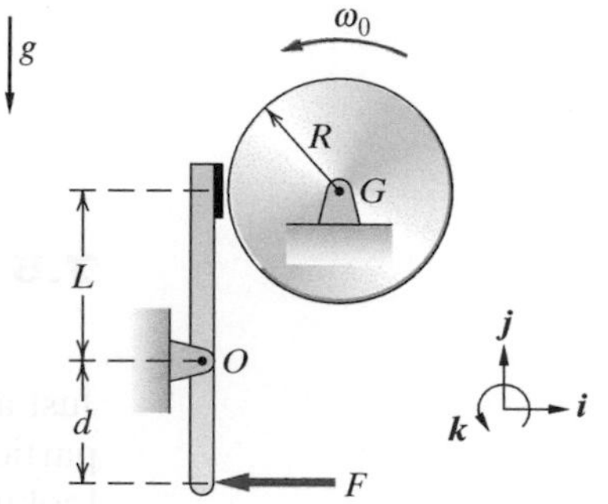

E7.4.35

7.4.36. **[Level 2]** A uniform disk A rolls without slip with velocity $v\boldsymbol{i}$. What is its angular velocity after striking a 45° sloped surface (assume zero rebound).

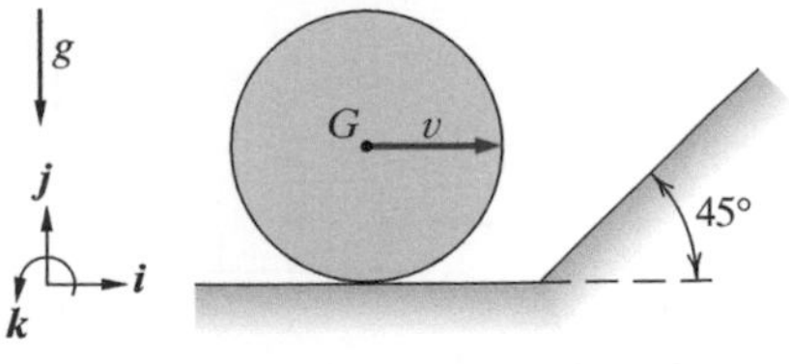

E7.4.36

7.4.37. **[Level 2]** A bar of length $L = 1$ m and mass $M = 8$ kg is free to rotate in the horizontal plane about its fixed pivot O. Suppose the bar is initially at rest, and a ball of mass $m = 3$ kg strikes the bar at $\frac{3}{4}L$ from O with a speed $v_0 = 10$ m/s at an angle $\beta = 60°$ from the horizontal, as illustrated. Find the bar's angular speed ω just after impact if the coefficient of restitution is $e = 0.7$.

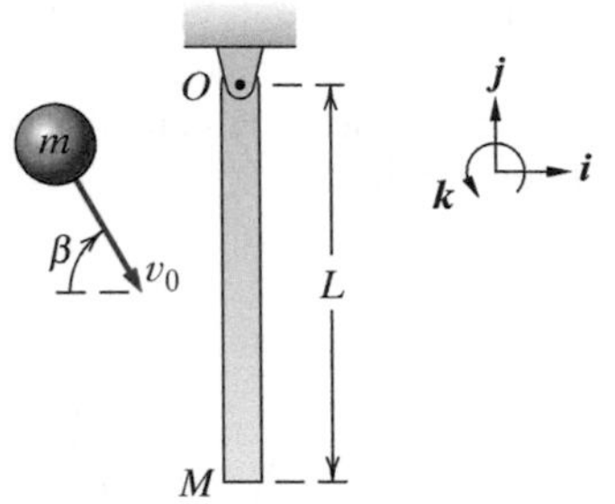

E7.4.37

7.4.38. **[Level 3]** The illustrated system consists of three intermeshing gears, all pivoting about their respective centers. Treat the gears as circular disks. The two racks, A and B, are constrained to move in the $\boldsymbol{j}$ direction by the two fixed pins within the illustrated slots. $m_A = 10$ kg, $m_B = 20$ kg, $m_C = 6$ kg, $m_D = 12$ kg, $m_E = 6$ kg, $r_C = 0.2$ m, $r_D = 0.4$ m, and $r_E = 0.2$ m. The system is initially at rest, and then a constant moment of 200 N·m is applied. How fast are the racks moving after 0.5 s?

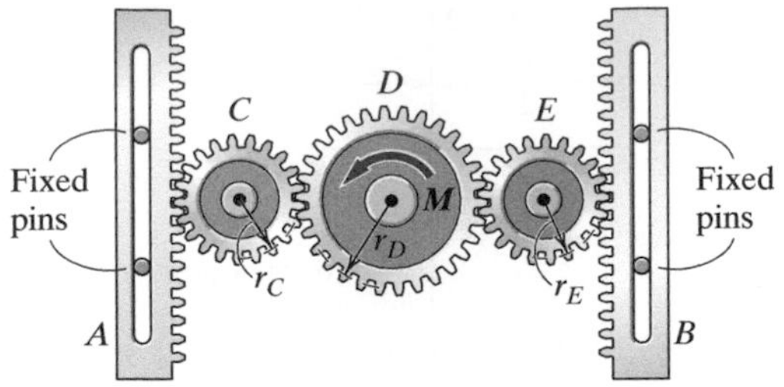

E7.4.38

7.4.39. **[Level 3]** The illustrated 50 kg body is acted on by an impulsive force $F\boldsymbol{i}$, producing an applied linear impulse of $100\boldsymbol{i}$ N·s. What is the resultant angular velocity of the body? $c = 2$ m, $b = 0.5$ m, and $a = 1.0$ m.

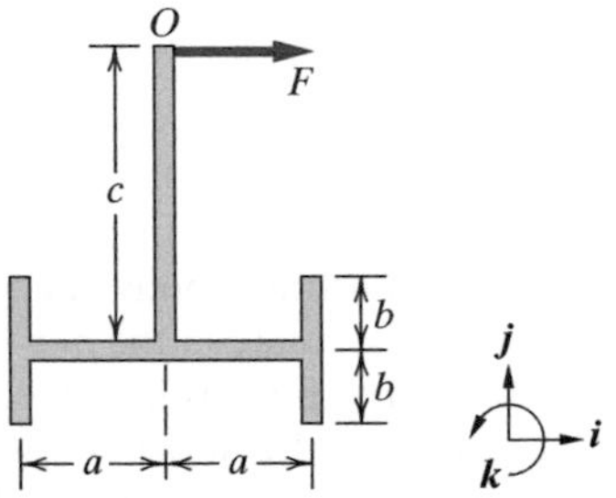

E7.4.39

7.4.40. **[Level 3]** The illustrated wheel is projected onto a flat surface with no rotational velocity and a translational velocity of $9\boldsymbol{i}$ m/s. The body has a total weight of 14 kg and $L = 0.3$ m. Assume a coefficient of friction of $\mu = 0.6$. How long will it take for the body to have a rotational velocity of $-10\boldsymbol{k}$ rad/s?

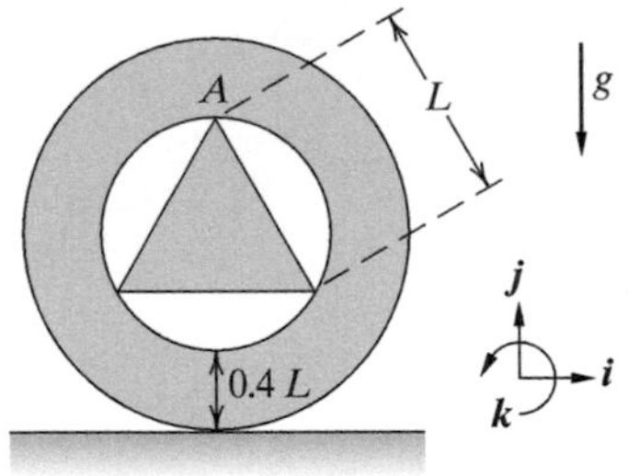

E7.4.40

7.5 WORK/ENERGY OF TWO-DIMENSIONAL RIGID BODIES

Just as a rigid body viewpoint simplified our lives with respect to multi-particle force and moment balances, it will make things easier when we're looking at energy. Recall what the kinetic energy of n particles (5.25) looked like:

$$\mathcal{KE} = \frac{1}{2}m\overline{v}^2 + \frac{1}{2}\sum_{i=1}^{n} m_i \left\| \dot{\boldsymbol{r}}_{i_{/G}} \right\|^2 = \frac{1}{2}m\overline{v}^2 + \frac{1}{2}\sum_{i=1}^{n} m_i \left\| \boldsymbol{v}_{i_{/G}} \right\|^2 \tag{7.83}$$

The first term is the kinetic energy that a single mass would have if it had the same mass as the system of particles and had the same speed as the system's center of mass. The second term has to do with the additional energy associated with motions of the individual masses *with respect to the mass center*. As long as the particles are free to move independently of each other, this term is going to remain somewhat complicated in form. Once we demand that the particles be part of a rigid body, however, everything simplifies.

To begin, imagine that our system of particles is locked together by rigid, massless rods. Although the entire mass can translate and rotate, the massless rods ensure that the relative orientation of the masses with respect to one another remains fixed. What we have done is transform our system of independent particles into a rigid structure, thus enabling us to use our rigid-body formulas to good effect. Thus we know that the velocity of the ith particle is given by

$$\begin{aligned} \boldsymbol{v}_i &= \boldsymbol{v}_G + \boldsymbol{v}_{i_{/G}} \\ &= \boldsymbol{v}_G + \boldsymbol{\omega} \times \boldsymbol{r}_{i_{/G}} \end{aligned}$$

Because we're now looking at a rigid body, the relative velocity of the ith particle with respect to G can be directed only *around* G, not toward or away from it. The kinetic energy can therefore be written as

$$\mathcal{KE} = \frac{1}{2} m \bar{v}^2 + \frac{1}{2} \sum_{i=1}^{n} m_i \left\| \boldsymbol{\omega} \times \boldsymbol{r}_{i_{/G}} \right\|^2 \tag{7.84}$$

To see how this further simplifies, write $\boldsymbol{\omega}$ as $\dot{\theta}\boldsymbol{k}$ and $\boldsymbol{r}_{i_{/G}}$ as $r_1\boldsymbol{i} + r_2\boldsymbol{j}$. The cross product $\boldsymbol{\omega} \times \boldsymbol{r}_{i_{/G}}$ is equal to $\dot{\theta}(r_1\boldsymbol{j} - r_2\boldsymbol{i})$. Squaring this (which remember just means taking the sum of the squares of the two components) gives us $(r_1^2 + r_2^2)\dot{\theta}^2$. This is just the magnitude squared of $\boldsymbol{r}_{i_{/G}}$ multiplied by $\dot{\theta}^2$—that is, $\dot{\theta}^2 r_{i_{/G}}^2$.

Equation (7.84) can now be rewritten as

$$\begin{aligned} \mathcal{KE} &= \frac{1}{2} m \bar{v}^2 + \frac{1}{2} \sum_{i=1}^{n} m_i \dot{\theta}^2 r_{i_{/G}}^2 \\ &= \frac{1}{2} m \bar{v}^2 + \frac{1}{2} \left(\sum_{i=1}^{n} m_i r_{i_{/G}}^2 \right) \dot{\theta}^2 \end{aligned} \tag{7.85}$$

As it's now written, we can see that two distinct parts make up the kinetic energy. The first is the kinetic energy associated with translation of the system's center of mass, and the second is the kinetic energy associated with the system's rotation. $\sum_{i=1}^{n} m_i r_{i_{/G}}^2$ is a constant term that depends on the system's mass distribution. Because the configuration of the masses is fixed, this quantity won't vary over time. What matters is the mass of a particular mass particle and its distance from the center of rotation (m_i and $r_{i_{/G}}^2$). All that we're doing is summing the kinetic energy contribution of each mass. If we had a lot of little masses (say an infinite number of infinitesimal masses), then the summation would mutate into an integration over the body:

$$\sum_{i=1}^{n} m_i r_{i/G}^2 \quad \rightarrow \quad \int_{\text{Body}} r^2\, dm$$

and the kinetic energy would become

$$\mathcal{KE} = \frac{1}{2} m \overline{v}^2 + \frac{1}{2} \overline{I} \dot{\theta}^2 \tag{7.86}$$

where $\overline{I}$ is defined as

$$\overline{I} \equiv \int_{\text{Body}} r^2\, dm$$

This is the same $\overline{I}$ that we met while examining the angular momentum of rigid bodies. Once again we see that it serves the purpose of a "rotational mass." This viewpoint is especially clear in (7.86), in which the terms making up the body's kinetic energy are the "translational mass" m, the "rotational mass" $\overline{I}$, the translational speed $\overline{v}$, and the rotational speed $\dot{\theta}$.

The various potential energies are the same for the rigid-body case as for the multiple-particle case—gravitational potential of the center of mass and any spring potentials. The work terms are the same as well—the external work done is simply the path integral of the applied forces. Thus, for the typical rigid-body energy problem, we would track a body from an initial state (State 1) to a later state (State 2) and say

$$\mathcal{PE}\big|_1 + \mathcal{KE}\big|_1 + W_{1-2} = \mathcal{PE}\big|_2 + \mathcal{KE}\big|_2 \tag{7.87}$$

As always, if no external work is done on the system, then energy is conserved:

$$\mathcal{PE}\big|_1 + \mathcal{KE}\big|_1 = \mathcal{PE}\big|_2 + \mathcal{KE}\big|_2 \tag{7.88}$$

>>> **Check out Example 7.22 (page 471), Example 7.23 (page 472), and Example 7.24 (page 474) for applications of this material.**

EXAMPLE 7.22 ANGULAR SPEED OF A HINGED TWO-DIMENSIONAL BODY (Theory on page 470)

Find the angular speed of the illustrated body at $\theta = 90°$ after it is released from rest at $\theta = 0°$. The body consists of a rigid bar ($m_1 = 10$ kg) that's welded to another bar ($m_2 = 4$ kg), as shown in **Figure 7.73**.

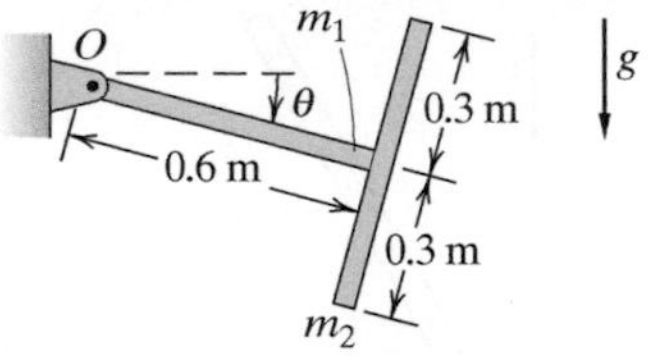

Figure 7.73 Hinged T-bar

Goal Determine the T-bar's angular velocity at $\theta = 90°$.

Given Body's dimensions and mass.

Draw **Figure 7.74** show the system's initial and final positions.

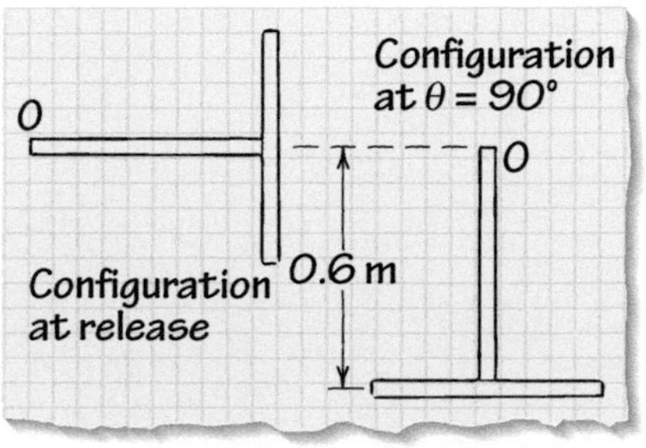

Figure 7.74 T-bar at $\theta = 0°$ and $90°$

Formulate Equations Because there are no external work terms, we have conservation of energy between State 1 ($t = t_1$, $\theta = 0°$) and State 2 ($t = t_2$, $\theta = 90°$). Thus we will apply (7.88): $\mathcal{KE}|_1 + \mathcal{PE}|_1 = \mathcal{KE}|_2 + \mathcal{PE}|_2$.

Assume The system is rotating about the pivot at O, which means we can express the entire kinetic energy in terms of I_O and $\dot{\theta}$: $\mathcal{KE} = \frac{1}{2}I_O\dot{\theta}^2$. Define the potential energy to be zero at the initial position. Note that we're not going to look at the potential energy of the overall mass center but, rather, keep track of each individual mass center associated with each bar (which is equivalent).

$$\overbrace{\frac{1}{2}I_O\dot{\theta}^2(t_1)}^{0} + \overbrace{m_1gh_1(t_1)}^{0} + \overbrace{m_2gh_2(t_1)}^{0} = \frac{1}{2}I_O\dot{\theta}^2(t_2) + m_1gh_1(t_2) + m_2gh_2(t_2) \tag{7.89}$$

where h_1, h_2 indicate the height of the mass center for the two bars.

We now must calculate I_O for the body. Break the system into two parts: the bar m_1 and the bar m_2. The mass moment of inertia of m_1 about O is given by $\frac{1}{3}m_1(0.6\,\text{m})^2$. For m_2, we can take its mass moment of inertia about its center of mass, $\frac{1}{12}m_2(0.6\,\text{m})^2$, and then use the parallel axis theorem to find its moment of inertia about O by adding $m_2(0.6\,\text{m})^2$. Taking this in total, and using the given values for the masses, give us

$$I_O = \frac{1}{3}m_1(0.6\,\text{m})^2 + \frac{1}{12}m_2(0.6\,\text{m})^2 + m_2(0.6\,\text{m})^2 = 2.76\,\text{kg}\cdot\text{m}^2$$

Notice that m_2 is more important than m_1 in determining I_O, even though it's 60% less massive ($1.56\,\text{kg}\cdot\text{m}^2$ as compared to $1.2\,\text{kg}\cdot\text{m}^2$). This illustrates the importance that distance from the center of rotation plays in mass moment of inertia calculations.

Solve Now that we have I_O we can solve (7.89): $[\dot{\theta}(t_2)]^2 = 38.4(\text{rad/s})^2$.

We know from physical considerations that the system will be swinging down in the clockwise direction. Thus $\dot{\theta}$ is negative, and we have the final result:

$$\dot{\theta}(t_2) = -6.20\,\text{rad/s}$$

EXAMPLE 7.23 RESPONSE OF A FALLING ROD VIA ENERGY (Theory on page 470)

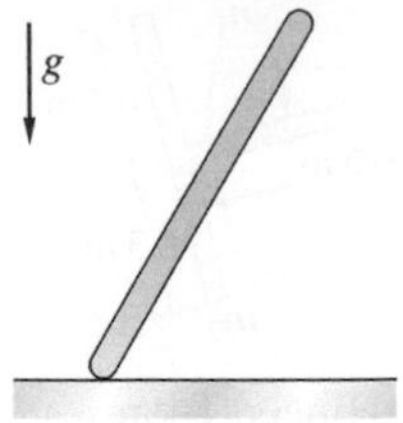

Figure 7.75 Rod on a frictionless surface

Now's a great time to look back at Example 7.12. You'll recall that we spent our time examining how a rigid rod (**Figure 7.75**) would fall if released on a frictionless surface. One of the tasks was to find $\dot{\theta}$ when it had fallen from $\theta = 80°$ to $\theta = 45°$. Now that we've learned about energy methods, we should be able to get this particular answer very easily. As in the prior example, $L = 4\,\text{m}$ and $m = 23\,\text{kg}$.

Goal Find the $\dot{\theta}$ associated with $\theta = 45°$ when the rod is released from rest at $\theta = 80°$.

Given Mass and length data as well as initial conditions.

Assume To solve this example we need a key assumption, or, more precisely, a key observation. The rod/ground interface is frictionless; therefore no horizontal force acts on the rod at any time. Thus we can conclude that the mass center of the rod doesn't move right or left. This fact is shown in **Figure 7.76*a***, in which you can see the mass center G moving vertically down.

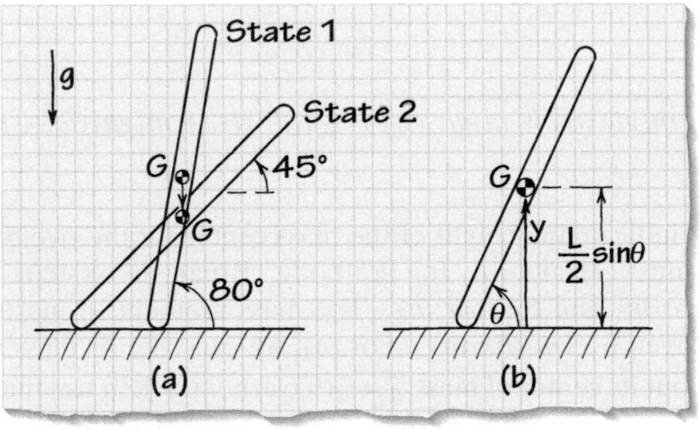

Figure 7.76 Rod falling on a frictionless surface

Draw **Figure 7.76*a*** shows the rod at the two relevant states: at release (State 1) and at 45° (State 2). **Figure 7.76*b*** shows the coordinate y with which we will track the height of the rod's mass center. As you can see, geometry tells us that $y = \frac{L}{2}\sin\theta$.

Formulate Equations We will need to apply (7.88): $\mathcal{PE}|_1 + \mathcal{KE}|_1 = \mathcal{PE}|_2 + \mathcal{KE}|_2$ where our potential energy comes from gravity and our kinetic energy involves both translational and rotational components. The general expression for the energy associated with a particular θ and $\dot{\theta}$ is $mgy + \frac{1}{2}m\overline{v}^2 + \frac{1}{2}\overline{I}\dot{\theta}^2$.

Solve At State 1 we have no motion of the body and thus only a potential energy term: $\frac{mgL\sin 80°}{2}$. (The zero energy potential state is referenced to $y = 0$ in this example.) At $\theta = 45°$ (State 2) we have a total energy of

$$\frac{mgL\sin 45°}{2} + \frac{1}{2}m\dot{y}^2 + \frac{1}{2}\overline{I}\dot{\theta}^2$$

Equating these two expressions gives us

$$\frac{mgL\sin 80^\circ}{2} = \frac{mgL\sin 45^\circ}{2} + \frac{1}{2}m\dot{y}^2 + \frac{1}{2}\overline{I}\dot{\theta}^2$$

It has already been mentioned that $y = \frac{L}{2}\sin\theta$, and by differentiating with respect to time and using this in our energy expression, we get

$$\frac{mgL\sin 80^\circ}{2} = \frac{mgL\sin 45^\circ}{2} + \frac{1}{2}m\left(\frac{L\dot{\theta}\cos\theta}{2}\right)^2 + \frac{1}{2}\overline{I}\dot{\theta}^2$$

Using the fact that $\overline{I} = \frac{mL^2}{12}$ lets us simplify this to

$$\frac{g}{2}(\sin 80^\circ - \sin 45^\circ) = L\left(\frac{\cos\theta^2}{8} + \frac{1}{24}\right)\dot{\theta}^2$$

$$1.362\,\text{m/s}^2 = (0.41\overline{6}\,\text{m})\dot{\theta}^2$$

$$\dot{\theta} = \pm 1.808\,\text{rad/s}$$

Physical considerations tell us that the negative solution is the physically reasonable one for a falling rod, and thus we have

$$\dot{\theta} = -1.808\,\text{rad/s}$$

Check We have an easy check for this problem—just look at what we got in Example 7.12 when we numerically integrated the rod's equation of motion. That answer was identical to what we just found here. Note that we didn't have to do any numerical integration in this example, which is an advantage over Example 7.12. But it should also be noticed that for this example we have no idea at what time the rod reaches $\theta = 45^\circ$; we know only that when it does, $\dot{\theta}$ is -1.808 rad/s. This is the normal tradeoff. Energy approaches don't tell you *when*—just *where* and *how fast*.

EXAMPLE 7.24 DESIGN OF A SPRING-CONTROLLED DRAWBRIDGE (Theory on page 470)

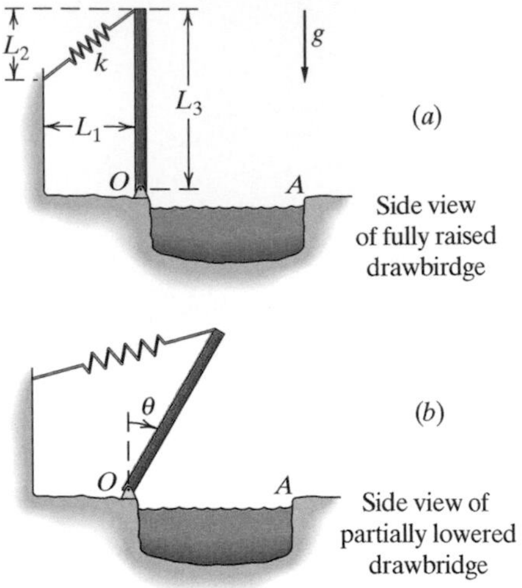

Figure 7.77 Side view of spring-controlled drawbridge

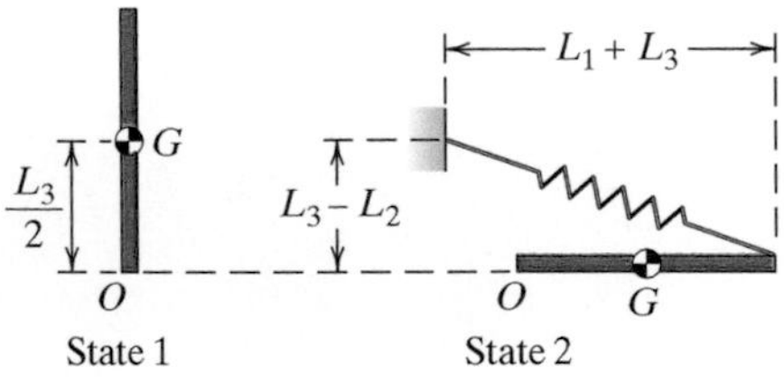

Figure 7.78 Start and end positions of platform

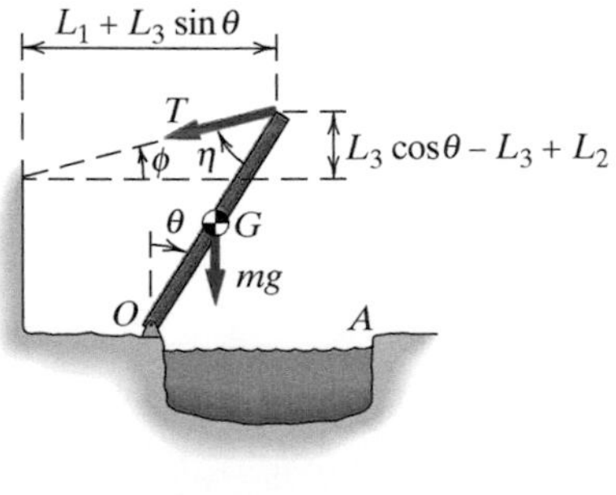

Figure 7.79 Forces causing moments about O

Being a progressive king, you've decided to update the drawbridge on your castle. Rather than controlling the bridge purely by ropes, you've attached two springs on either side of the movable platform, each with stiffness k, as shown in **Figure 7.77*a***. The springs are unstretched in the position shown. The idea is that once the top of the vertical platform is nudged to the right, it will rotate about O and fall into position to serve as a bridge between O and A (**Figure 7.77*b***). Because of the two springs, it won't crash into place but will touch down at a moderate rotational speed. Once horizontal, it's held in place by hidden clamps. Determine the value of k such that the rotational speed at impact is 1.0 rad/s. $L_1 = 2$ m, $L_2 = 2$ m, $L_3 = 5$ m, and m, the mass of the platform, is 400 kg.

Goal Determine the spring constant k that will ensure a rotational speed of 1.0 rad/s when a hinged platform falls from a vertical orientation to a horizontal one.

Given Dimensions and mass of the device.

Assume No additional assumptions are needed.

Draw **Figure 7.78** shows the platform in its raised (State 1) and lowered (State 2) positions.

Formulate Equations We will need to use (7.88) with both gravitation and spring potential energy terms: $\mathcal{PE}|_1 + \mathcal{KE}|_1 = \mathcal{PE}|_2 + \mathcal{KE}|_2$.

Solve We'll define the zero gravitational potential to be at the floor level. Thus we will have a finite gravitational potential at the start and none at the end. The springs are initially unstretched (zero spring potential energy) and are maximally stretched when the platform touches down. Finally, the platform has zero kinetic energy initially and a finite amount at touchdown. Putting this into equation form gives us

$$mg\frac{L_3}{2} = \frac{1}{2}(2k)\left[\sqrt{(L_3 - L_2)^2 + (L_3 + L_1)^2} - 2\sqrt{2}\,\text{m}\right]^2 + \frac{1}{2}\left(\frac{mL_3^2}{3}\right)\dot{\theta}^2$$

$$k = 355\ \text{N/m}$$

Check A good way to check on this result is to numerically integrate the system's equation of motion. If we sum moments about O, all we will need to worry about are the forces due to the springs (T) and due to gravity (mg). **Figure 7.79** shows these forces as well as the new angles ϕ and η.

From geometry we see that

$$\phi = \tan^{-1}\left(\frac{L_3\cos\theta - L_3 + L_2}{L_1 + L_3\sin\theta}\right), \quad \eta = \frac{\pi}{2} - \theta - \phi$$

Summing moments about O gives us the equation of motion

$$I_O\ddot{\theta} - \frac{mgL_3\sin\theta}{2} + L_3T\sin\eta = 0$$

Using MATLAB to integrate for θ and $\dot{\theta}$ using `y0=[0 0.01]` produced the θ versus t plot shown in **Figure 7.80**. As you can see, the drawbridge is moving at 1.0 rad/s when it reaches $\theta = \frac{\pi}{2}$ rad, just as predicted by the energy analysis.

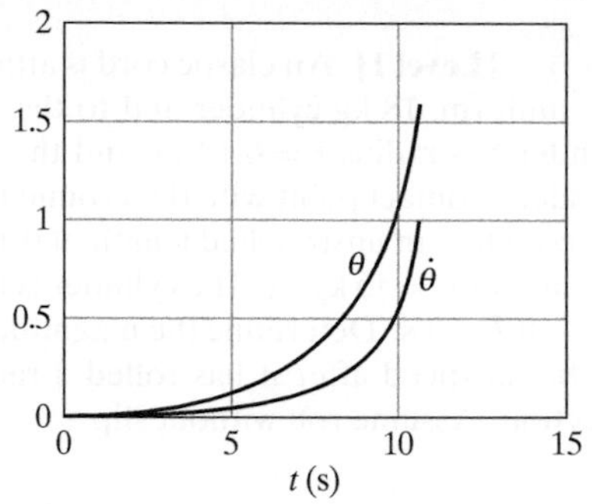

Figure 7.80 $\theta, \dot{\theta}$ versus t

EXERCISES 7.5

7.5.1. **[Level 1]** A uniform cylinder is spring restrained and free to roll without slip. State 1 corresponds to an unstretched spring and the cylinder rotating at $\boldsymbol{\omega} = -10\boldsymbol{k}$ rad/s. Determine the magnitude of the cylinder's rotational speed after it has rolled 2π rad in the clockwise direction. $m = 12$ kg, $r = 0.25$ m, and $k = 20$ N/m.

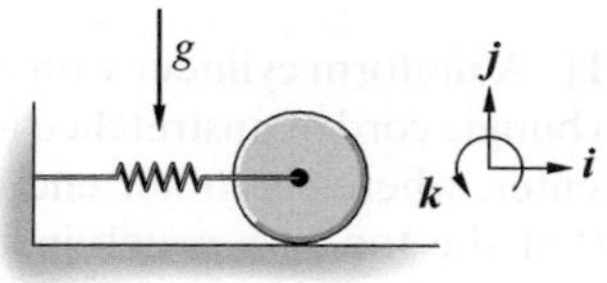

E7.5.1

7.5.2. **[Level 1]** A uniform sphere with mass m and radius r is rolled up a rough incline angled at $\theta = 30°$ to the horizontal such that the initial speed of its mass center is $v_0 = 5$ m/s. The sphere rolls without slip.

a. Find the distance d the sphere travels along the incline before coming to a stop.

b. What initial speed v_0^* does the sphere need to travel $d^* = 6$ m before stopping?

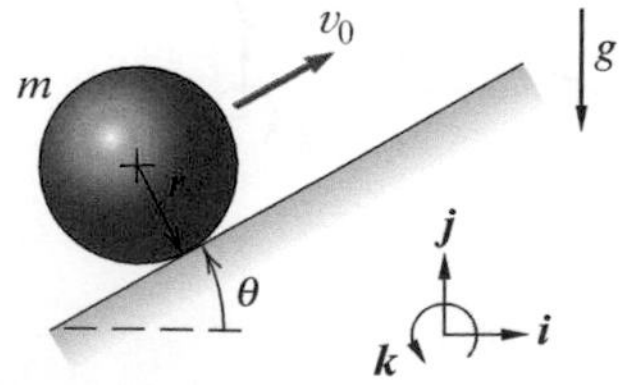

E7.5.2

7.5.3. **[Level 1]** In the sequel to *Search for the Lost What-ever*, California Sam (our hero) leaps from a ledge onto a chain that's wrapped around a large gear. The gear is rigidly attached to a large cylinder, and the other end of the chain is attached to a drawbridge (not shown). Assume that the load of the drawbridge is a constant force of 500 N and that Sam contacts the chain with zero velocity. The mass moment of inertia of the gear/cylinder about its axle is 240 kg·m², the radius r of the gear is 1.1 m, and Sam has a mass of 72 kg. After Sam has dropped 5 m, how fast is he moving? Neglect the chain's mass.

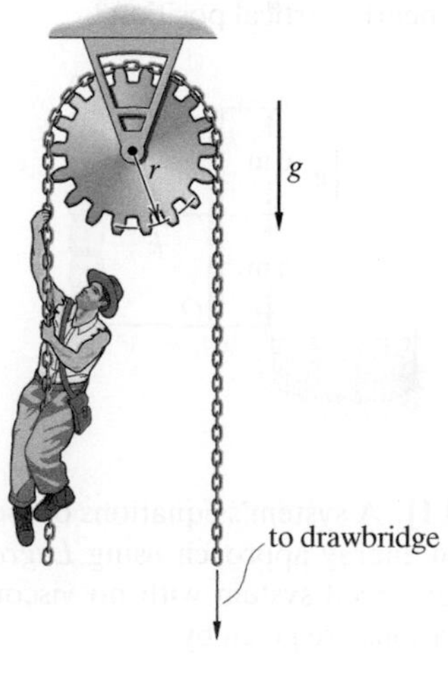

E7.5.3

7.5.4. **[Level 1]** A uniform cylinder is free to roll without slip. At all times a horizontal force (directed toward the left) acts on the cylinder's mass center with a magnitude of 4 N. State 1 corresponds to an initial state for which the cylinder is rotating at $\boldsymbol{\omega} = -12\boldsymbol{k}$ rad/s. Determine the magnitude of the cylinder's rotational speed after it has rolled 2π rad in the clockwise direction. $m = 10$ kg and $r = 0.20$ m.

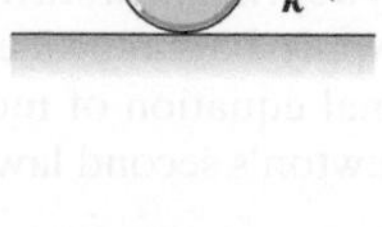

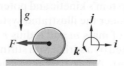

E7.5.4

7.5.5. **[Level 1]** An elastic cord is attached to the center of a uniform, 18 kg cylinder and to the ground at A. The cylinder has radius $r = 0.12$ m, and the distance from the cylinder's contact point with the ground to A is also 0.12 m. The cord has an unstretched length of 0.17 m and an elastic constant of $k = 15$ kg/m. The cylinder is initially rotating at $\boldsymbol{\omega} = -10\boldsymbol{k}$ rad/s. Determine the magnitude of the cylinder's rotational speed after it has rolled π rad in the clockwise direction. Assume roll without slip.

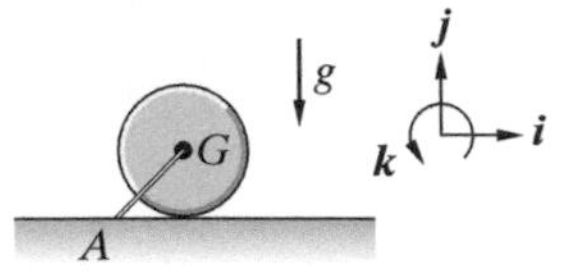

E7.5.5

7.5.6. **[Level 1]** A thoughtful king once decided to update the drawbridge of his castle by attaching the middle of the drawbridge to the castle wall via a single linear spring with an unstretched length of 6 m. The drawbridge is 12 m long, has a rotational inertia of about O of 21,600 kg·m^2, and has a mass of 450 kg. The king has decreed that the spring constant shall be 690 N/m. What's the velocity with which the end of the drawbridge will strike the ground if it's released from a nearly vertical position?

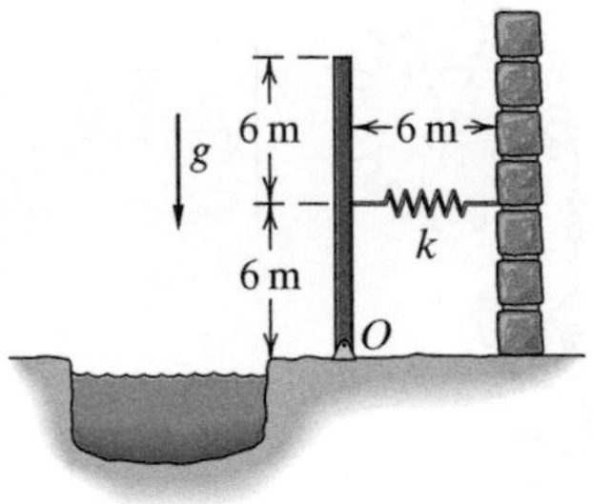

E7.5.6

7.5.7. **[Level 1]** A system's equations of motion can be derived from an energy approach using *Lagrange's equations*. For an unforced system with no viscous damping, Lagrange's equations are given by

$$\frac{d}{dt}\left(\frac{\partial L}{\partial \dot{x}_i}\right) - \frac{\partial L}{\partial x_i} = 0$$

where x_i are the coordinates describing the system's motion and L is the *Lagrangian*, which is defined as the difference between the system's kinetic and potential energies, or $L = \mathcal{KE} - \mathcal{PE}$. Consider the illustrated system consisting of a uniform cylinder of mass m and radius r whose center is elastically constrained by a spring with stiffness k. The cylinder rolls without slip. Apply Lagrange's equations to find the cylinder's translational equation of motion, and verify your answer by using Newton's second law.

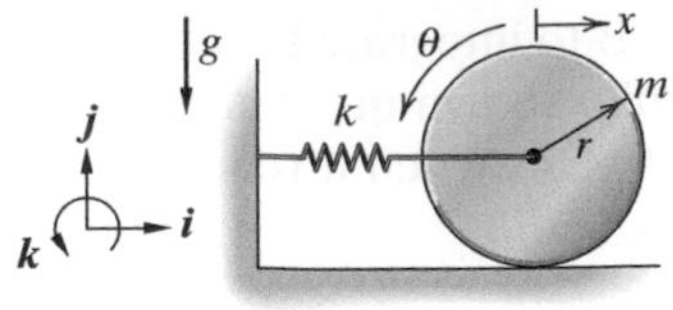

E7.5.7

7.5.8. **[Level 1]** Two uniform cylinders are connected by means of a linear spring. At the illustrated instant, the centers of A and B are separated by $4r$, where r is the radius of each cylinder. What will the speed of the centers of the cylinders be just before they strike each other? Assume they roll without slip and the spring has an unstretched length equal to r.

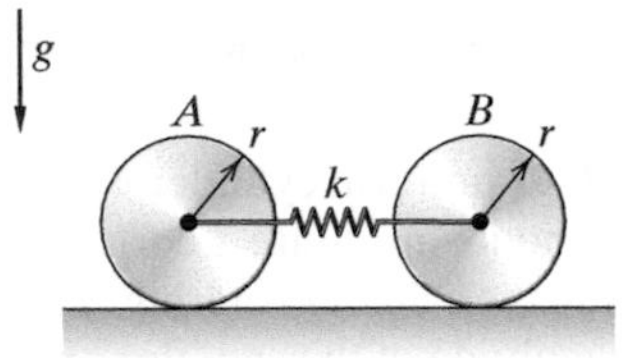

E7.5.8

7.5.9. **[Level 1]** A uniform cylinder with mass $m = 9$ kg and radius r has a bungie cord of unstretched length $L = 4$ m attached to its center, where the other end of the cord is fixed to a post P at the top of a rough incline angled at $\theta = 25°$ to the horizontal. Suppose the cylinder is rolled down the incline from P so that its mass center's initial speed is $v_0 = 3$ m/s. The cylinder rolls without slip, and it comes to a stop when the bungie cord is stretched $\delta = 1$ m. What is the bungie cord's stiffness k?

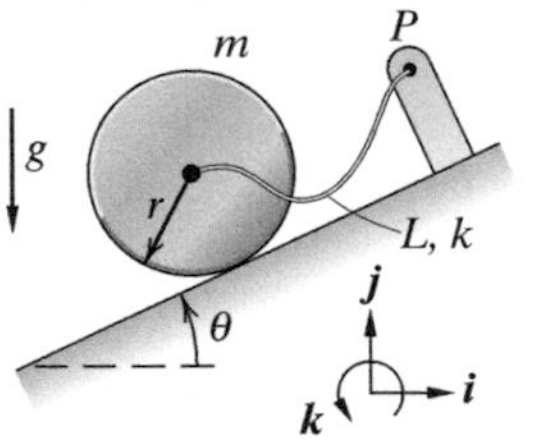

E7.5.9

7.5.10. **[Level 1]** A hoop with mass $m = 10$ kg and radius $r = 0.4$ m is rolled down a rough surface toward a spring of stiffness $k = 1500$ N/m. The surface is angled at $\theta = 45°$ with respect to the horizontal, and the hoop's mass center G is initially $d = 4$ m from the spring. The hoop rolls without slip down the incline, and its moment of inertia about G is $I_G = mr^2$. Find the initial speed v_0 of the hoop's mass center if the maximum compression of the spring is $\delta = 0.7$ m.

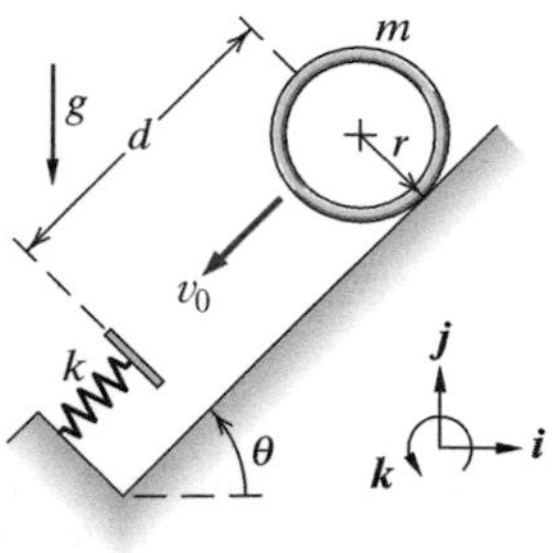

E7.5.10

7.5.11. **[Level 2]** As illustrated, a uniform bar with mass m and length $L = 1$ m has a small ball of mass $2m$ welded to its free end C, and the system is constrained to rotate about its end A in the vertical plane. A spring with stiffness $k = 1000$ N/m is attached to the rod at B, and it has a free length of $\frac{L}{2}$. The system is released from rest in a horizontal position, and the speed of its end C is $v = 3$ m/s when it reaches the vertical. What is the value of m?

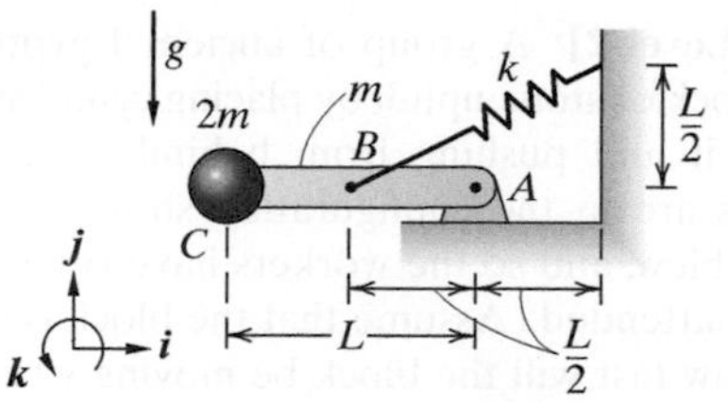

E7.5.11

7.5.12. **[Level 2]** A bottle is laid down on a slightly tilted table top and slowly begins to roll to the edge. Model the bottle as a uniform hollow cylinder of mass m, radius r, and length L. Assume the bottle is just barely rotating as it reaches the table edge. The coefficient of friction between the bottle and table is $\mu = 0.34$. Determine the angle θ at which the cylinder starts to slip.

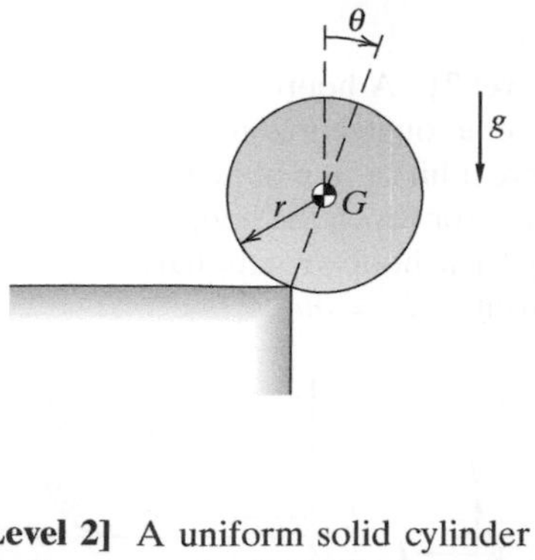

E7.5.12

7.5.13. **[Level 2]** A uniform solid cylinder of radius r rolls without slip over a horizontal plane that then joins with an inclined plane forming an angle α with the horizontal. Find the maximum velocity v_0 for which the cylinder will roll onto the inclined plane section without leaving the surface.

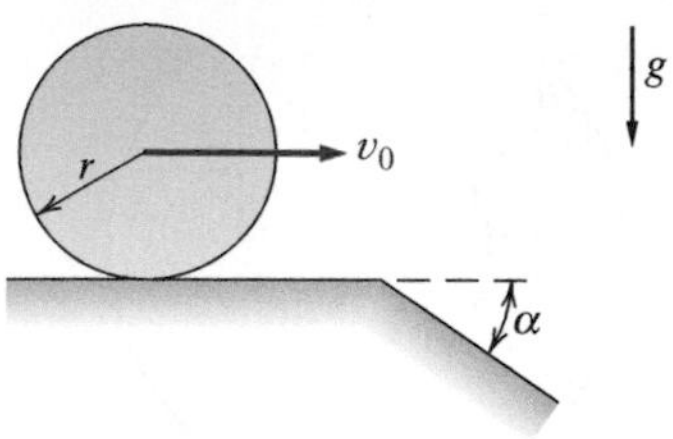

E7.5.13

7.5.14. **[Level 2]** A uniform sphere of mass $m = 7$ kg and radius r rolls without slip toward a large crate with mass $M = 60$ kg that rests on top of a rubber mat nailed into the ground. The coefficient of friction between the crate and mat is given by μ. A spring with stiffness k is attached to the crate at a height $h = 0.8$ m above the mat, and the crate's mass center G is located a distance $d = 0.5$ m from either of its supports A and B. The sphere makes contact with the spring with a speed of $v = 4$ m/s. Find the spring stiffness k so that the crate just barely tips without slipping, and solve for the minimum frictional coefficient μ needed to keep the crate from slipping when it tips.

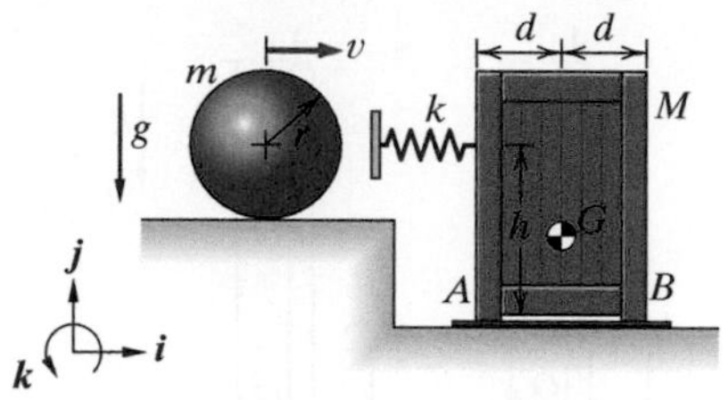

E7.5.14

7.5.15. **[Level 2]** A rod of length L and mass m_1, having a lumped mass m_2 at its tip, is freely hinged at O. The tip is pressed against a spring having spring constant $k = 2 \times 10^5$ N/m and is compressed 0.1 m. The rod is initially horizontal. The rod is then released and is propelled upward. $m_1 = 10$ kg, $m_2 = 5$ kg, and $L = 2.8$ m. What is the rod's angular velocity, and what are the reaction forces at the hinge when the rod is completely vertical?

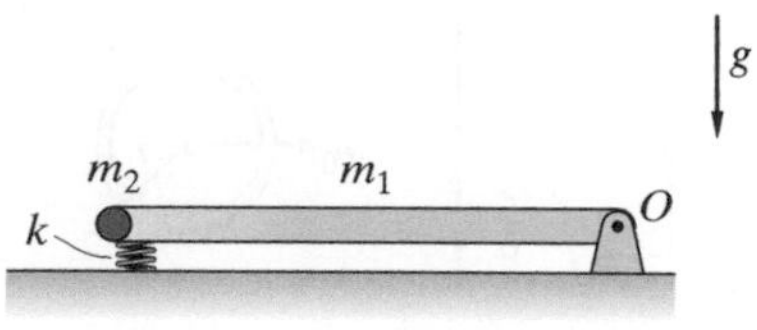

E7.5.15

7.5.16. **[Level 2]** A vertical rod $\overline{AB}$ of length 0.9 m is released from the illustrated vertical orientation within a frictionless, circular guide. What will $\overline{AB}$'s angular speed be when it achieves a horizontal orientation?

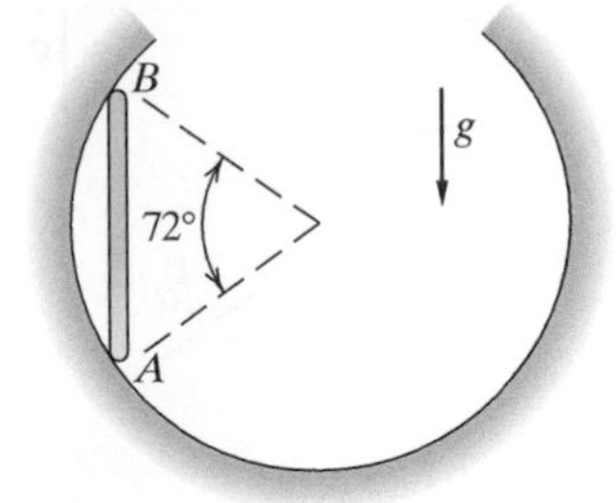

E7.5.16

7.5.17. **[Level 2]** The figures illustrate two scenarios that serve as *extremely* simplified models for human fall dynamics. In (a) we model the human as a rigid bar of length L and mass m that's pivoted at O. This lets us approximate the fall as a rotation about the feet. In (b) we model the falling body as a rigid bar (mass m and length L) that's free to slide without friction along the ground. This lets us model a person slipping on ice.

a. Calculate the velocity with which the end A of the rod, the person's "head," strikes the ground for each case.

b. What is the body's rotation rate just before impact?

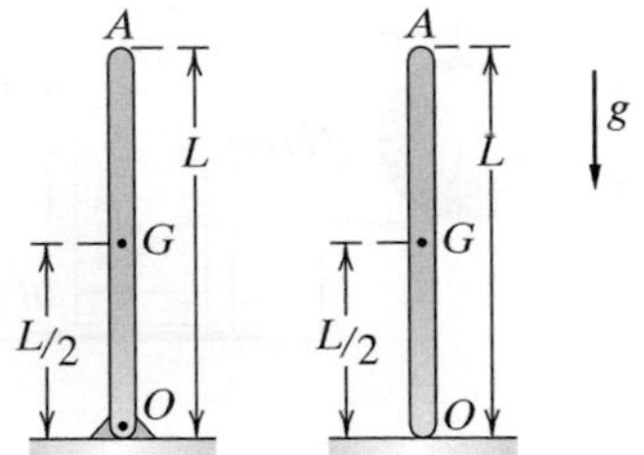

E7.5.17 (a) (b)

7.5.18. **[Level 2]** A hoop of mass m and radius r rolls without slip over a rough horizontal surface and has a speed of v_0 just before it hits a step of height $\frac{r}{6}$. What is the maximum value of v_0 for which the hoop will roll up the step without losing contact? The hoop's rotational inertia about its mass center G is given by $I_G = mr^2$.

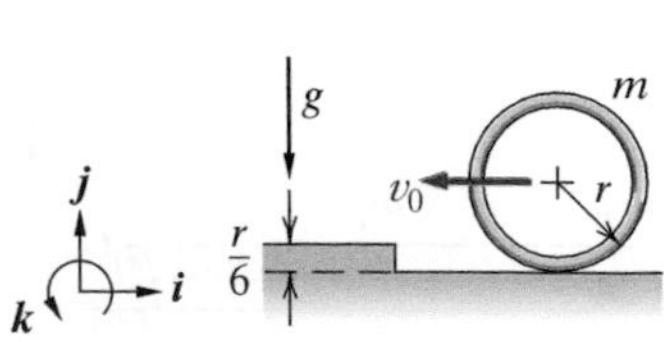

E7.5.18

7.5.19. **[Level 2]** Let the angular velocity of $\overline{OA}$ be $\omega\boldsymbol{k}$ and the angular acceleration be $\alpha\boldsymbol{k}$. What is the kinetic energy of the three-link system at the instant shown? The masses of the three links are $m_{\overline{OA}} = 2m$, $m_{\overline{AB}} = 3m$, and $m_{\overline{BC}} = m$.

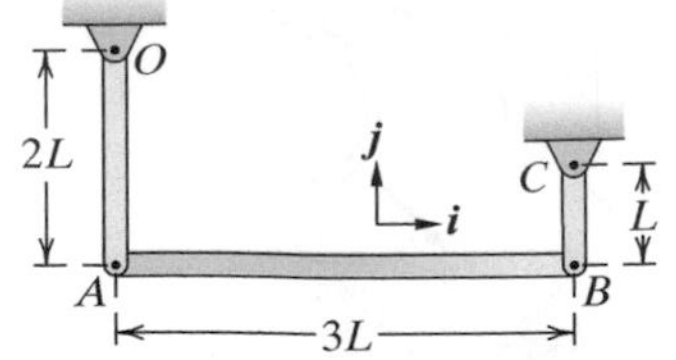

E7.5.19

7.5.20. **[Level 2]** Two uniform links, each of length L and mass m_L, are pinned together at B. The left link is pinned to ground at A, and at C the right link is pinned to a mass m_C, which can slide without friction in a horizontal guide. What is the kinetic energy of the system (two links and translational mass) in terms of β and $\dot{\beta}$?

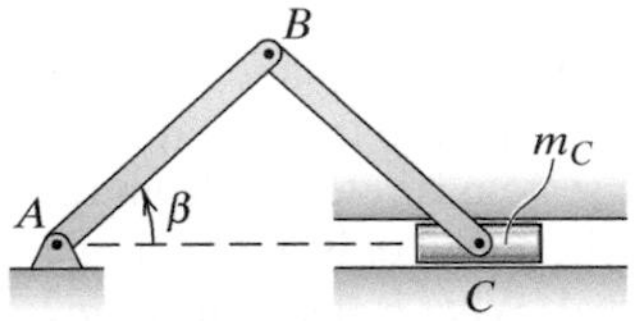

E7.5.20

7.5.21. **[Level 2]** A group of ancient Egyptians were moving a block of stone uphill by placing cylindrical rollers in front of it and pushing from behind. At $t = 0$ the block/rollers are in the configuration shown. The lunch whistle just blew, and so the workers have run off, leaving the block unattended. Assume that the block is stationary at $t = 0$. How fast will the block be moving when roller 1 has reached the midpoint of the block? $m_{\text{block}} = 700$ kg, $m_{\text{roller}} = 30$ kg, $d = 0.2$ m, $L = 2$ m, $h = 0.2$ m, and $\theta = 12°$.

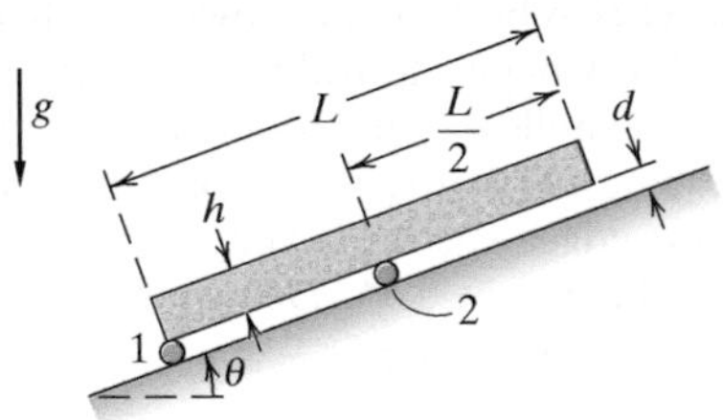

E7.5.21

7.5.22. **[Level 2]** A hoop of mass m and radius r rolls without slip over a rough horizontal surface and has a speed of v_0 just before it hits a step of height $\frac{r}{6}$. What is the minimum value of v_0 for which the hoop will just make it to the top of the step? The hoop's rotational inertia about its mass center G is given by $I_G = mr^2$.

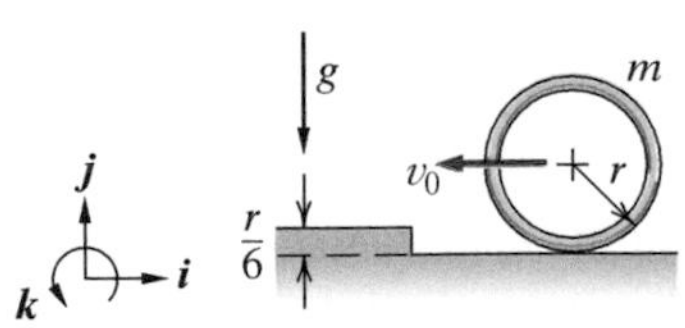

E7.5.22

7.5.23. **[Level 2]** While cleaning my road bike, I draped the chain over the front crank's outer chain ring. Unfortunately for me, I didn't center the chain, and 5 cm more of chain hung down on the left side than on the right. Consequently, the crank began to rotate. How fast was it rotating when the chain on the left had moved down 15 cm? The rotational inertia of the crank, pedal, and chain ring combination about the axle is 0.015 kg·m^2; the chain weighs 0.33 kg and has a length of 142 cm, and $r_1 = 10$ cm.

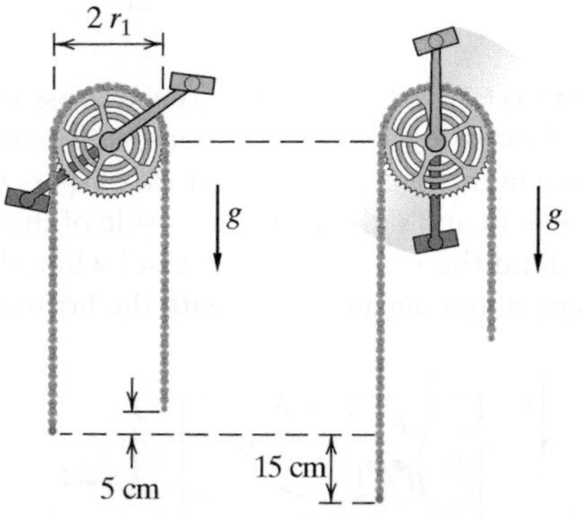

E7.5.23

7.5.24. **[Level 2]** The illustrated system consists of a rigid bar $\overline{OB}$, pivoted at O. A massless wire goes from B, around a frictionless pulley at A, over to C, and down to mass m_1. The dimensions of the pulleys are negligible. What is the minimum mass m_1 that will get the bar upright? $m_2 = 200$ kg and $L = 3$ m.

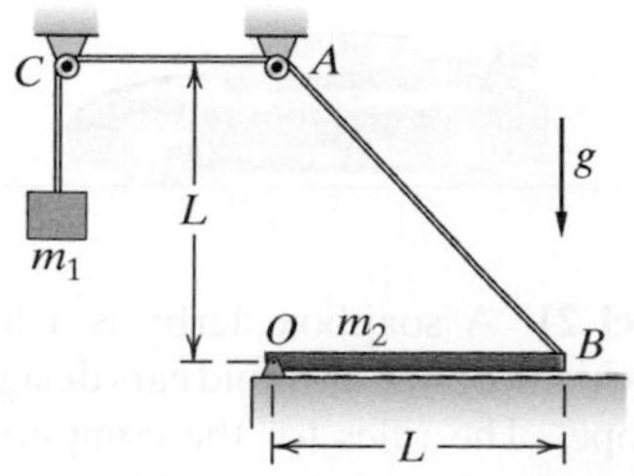

E7.5.24

7.5.25. **[Level 2]** A uniform cylinder with mass $m = 5$ kg and radius r is rolling without slip on a rough horizontal surface toward a spring of stiffness $k = 1000$ N/m attached to a heavy block resting on a rubber mat nailed into the surface. The block's mass is $M = 50$ kg, and the coefficient of friction between it and the rubber mat is $\mu = 0.8$. Suppose the cylinder impacts the spring with a speed of $v = 5$ m/s. Will the block move?

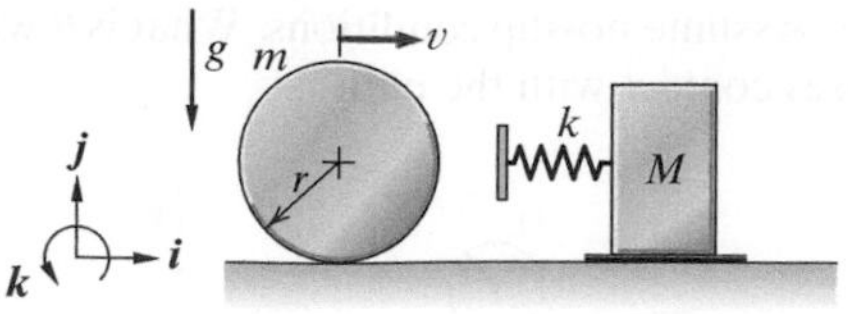

E7.5.25

7.5.26. **[Level 2]** The illustrated system consists of a rigid bar $\overline{OB}$, pivoted at O. A massless wire goes from B, around a frictionless pulley at A, over to C, and down to mass m_1. The dimensions of the pulleys are negligible. $m_1 = 70$ kg, $m_2 = 100$ kg, and $L = 2$ m. What is the acceleration of m_1 as the bar passes through the upright vertical position?

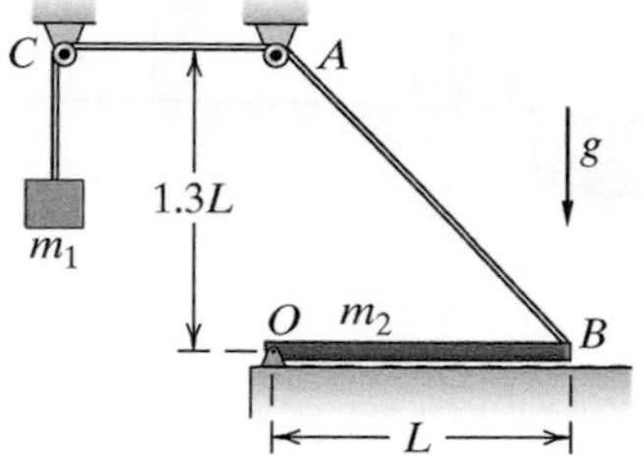

E7.5.26

7.5.27. **[Level 2]** Illustrated is a solid disk that's spring restrained as well as string restrained. A massless string wraps around it and attaches to the ceiling, and a linear spring runs from the disk's center to the ceiling. The spring stiffness is 4.5 kg/m, it has an unstretched length of 0.15 m, the radius of the disk is 0.076 m, and it weighs 1.36 kg. Assume that the disk is held 0.15 m below the ceiling (spring unstretched) and then released. How fast is its center moving once it has fallen 0.3 m?

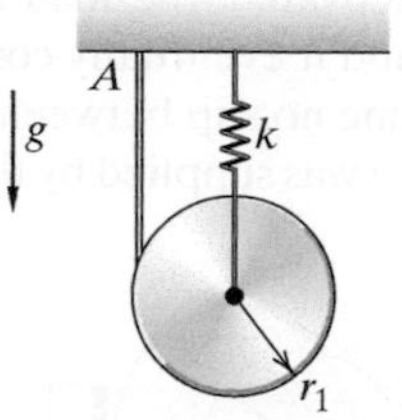

E7.5.27

7.5.28. **[Level 2]** As your first job out of college, you have been assigned the task of helping design a drawbridge for your eccentric employer. The drawbridge, freely hinged at O, has a length of 6 m, a mass of 450 kg, and a mass moment of inertia about O of $I_O = 5400$ kg·m^2. A massless rope is attached to the end of the drawbridge, passes through the castle wall, then passes over a massless pulley, and is attached to a spring. In the bridge's up position the spring is unextended. A colleague has already constructed most of it, and all you need to do is finish up the final details.

a. Determine the value of k necessary so that if the drawbridge is released from the vertical position (with a tiny nudge to get it going), it will hit the ground with a rotational speed of 0.5 rad/s.

b. Determine whether, when built with the spring constant you have just specified, the drawbridge will "work"; that is, when nudged from the vertical, will it continue down rather than being pulled back up to the vertical position due to the action of the stretched spring?

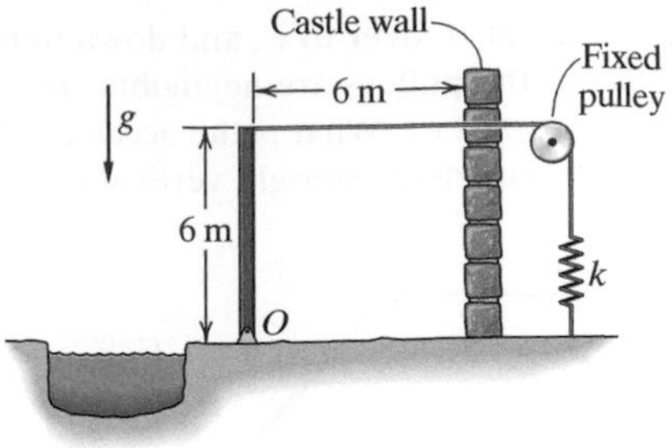

E7.5.28

7.5.29. [Level 2] The illustrated system consists of a pulley of radius $r = 0.06$ m that raises a load of weight $W_L = 27$ kg by means of a motor at the pulley's fixed pivot O. Attached to the pulley is a disk of radius $R = 0.12$ m that also rotates about O, and the pulley-disk unit has a radius of gyration and weight of $k = 0.076$ m and $W_d = 4.5$ kg, respectively. The disk is part of an emergency braking system, which also consists of a brake lever controlled by a linear actuator located a distance $d = 0.27$ kg below the lever's fixed pivot P. The brake pad is $L = 0.09$ m above P, and the coefficient of friction between it and the disk is $\mu = 0.7$. Suppose the pulley is raising the load and the motor suddenly stops working, causing the load to fall. When the emergency brake is activated, the load is traveling downward at $v = 4.5$ m/s, and it eventually comes to a halt after falling $h = 3$ m. Assume no slip between the pulley and its cable. How much force was supplied by the linear actuator?

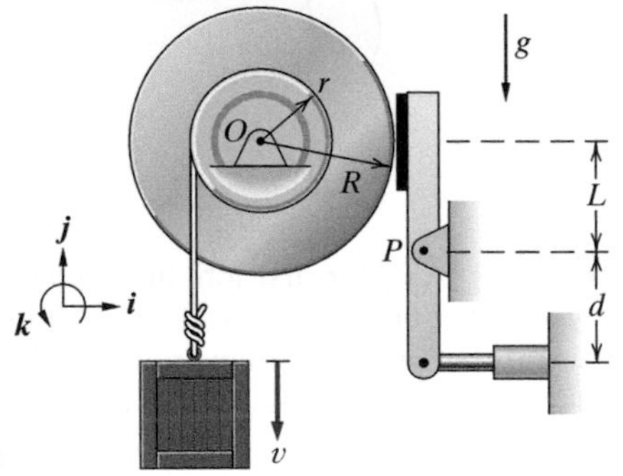

E7.5.29

7.5.30. [Level 2] The illustrated system is a spring-restrained cylinder that's connected to block A through a massless string. The string goes over another cylindrical body that rotates as the string moves (no slip between the body and the string). Both cylindrical bodies have a radius of 0.6 m, a mass of 5 kg, and a mass moment of inertia about their geometric centers of 0.9 kg·m^2. $m_A = 10$ kg and $k = 10$ N/m. The system is released from rest with the spring stretched 0.02 m from its rest length. What is the speed of block A after it has fallen 2 m?

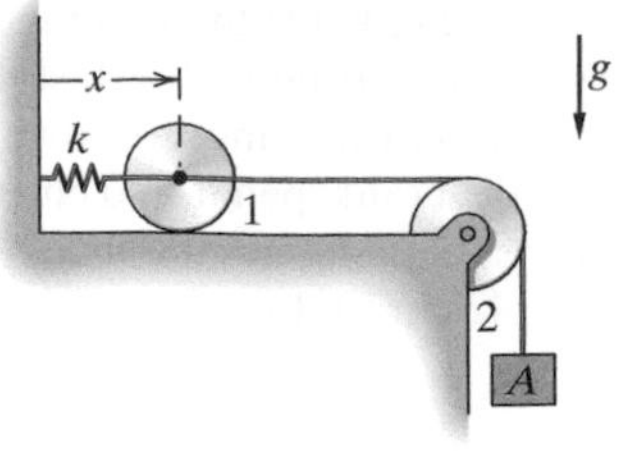

E7.5.30

7.5.31. [Level 2] A bowl with a circular cross section of radius R and mass m_2 lies on a frictionless horizontal surface as shown in the figure. A small block of mass m_1 starts at point A ($\theta = 0$) and slides down the inside of the frictionless vessel. Find the velocity of the vessel when the block reaches point B making an angle θ with the horizontal.

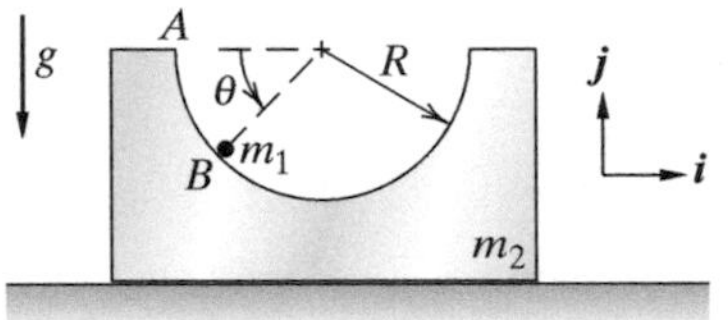

E7.5.31

7.5.32. [Level 2] A 1134 kg car is traveling at 97 km/h. Each of its four 23 kg wheels has a radius of gyration of 25 cm. The distance from the wheel's center of rotation to the road is 30 cm. Determine what percentage of the vehicle's kinetic energy is stored in the wheels. Assume roll without slip.

E7.5.32

7.5.33. [Level 2] A soapbox derby is a long-running competition in which youngsters build cars designed to coast down a long slope. The rules for the competition specify the radius of the wheels and the combined weight of the car and driver. Consider the option of using heavier wheels or lighter wheels with ballast added to the body. Make a recommendation with regard to which of the cars will be moving faster at the end of the race.

7.5.34. [Level 3] A cylinder is released from rest from point A. As it rolls past B it moves from a constant slope of $-8°$ and onto a circular path. The distance from A to B is 0.8 m, $r_1 = 0.15$ m, $r_2 = 1.5$ m, and the mass of the cylinder is $m = 4$ kg. Assume no-slip conditions. What is θ when the cylinder loses contact with the path?

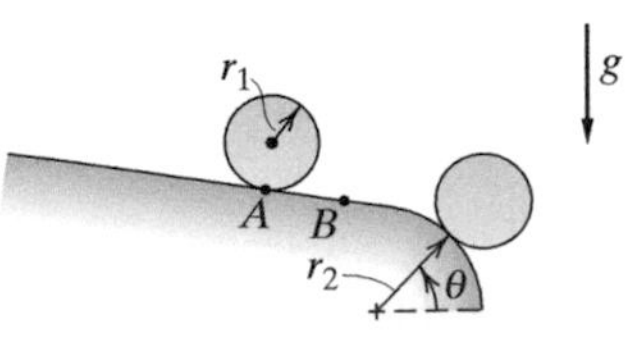

E7.5.34

7.5.35. **[Level 3]** Illustrated are two cases in which a uniform, square block is allowed to fall. It will either rotate about a frictionless hinge (*a*) or, alternatively, slide on a frictionless surface (*b*). The block has mass m. Initially each block is oriented as shown, positioned such that G is just slightly to the right of being directly above O and will thus fall in a clockwise direction.

a. With what velocity does the corner A hit the ground?

b. For both cases, what is the angular velocity of the block just before impact?

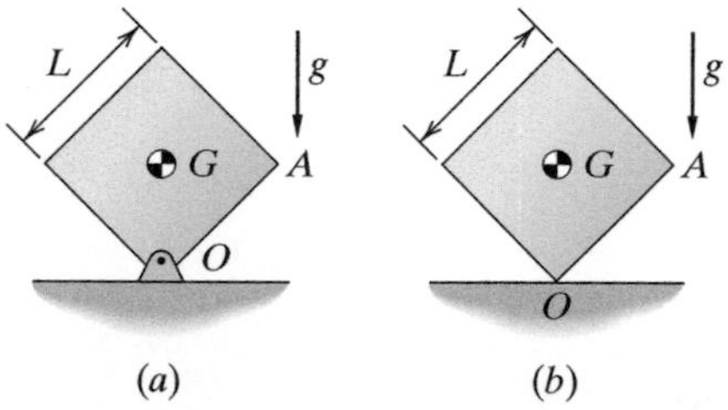

E7.5.35

7.5.36. **[Level 3]** **Computational** A board is released from rest in the configuration shown. θ is initially equal to 15°. Assume a friction-free interface at both A and B between the board and the rigid surfaces.

a. Use an energy formulation to determine the angular velocity of the board when $\theta = 45°$. $L = 2$ m and $m = 10$ kg.

b. Verify your answer by forming the system's equation of motion and integrating from $\theta = 15°$ to $\theta = 45°$.

c. How long does it take for θ to reach 45°?

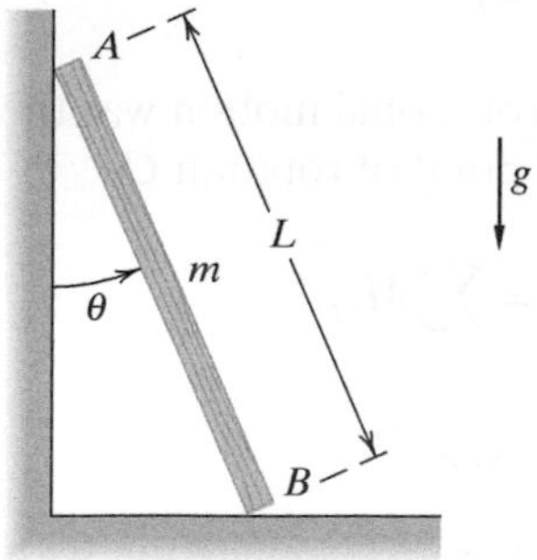

E7.5.36

7.5.37. **[Level 3]** A uniform cylinder is allowed to roll down the illustrated track.

a. What is the minimum value of h such that the cylinder never leaves the surface of the track? $r_1 = 0.9$ m and $r_2 = 0.04$ m. Assume the cylinder rolls without slip.

b. How does your answer change if you assume a frictionless interface between the cylinder and the track? Does the change make sense?

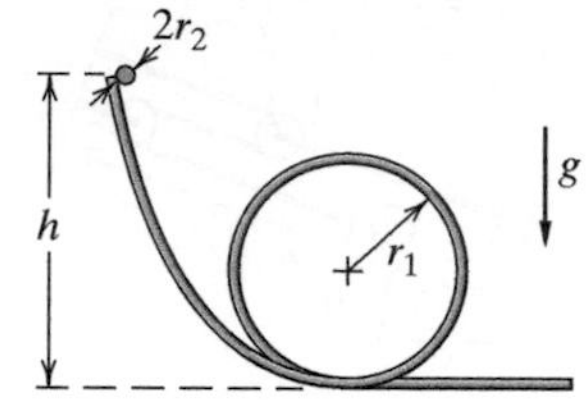

E7.5.37

7.5.38. **[Level 3]** **Computational** A circular disk (radius $r = 0.2$ m, $m_1 = 2$ kg) turns at a steady rate of $10\boldsymbol{k}$ rad/s. The pin P, rigidly attached to the disk at a point $e = 0.15$ m from G, rides in a grooved track. The pin causes a reciprocating motion in the illustrated link. Treat the track as a uniform bar of length $L = 0.5$ m and mass $m_2 = 0.5$ kg. At the track's far end is an attached block (mass $m_3 = 0.2$ kg, $a = 0.03$ m, $b = 0.15$ m). Plot the total energy of the system as a function of the disk's rotation angle θ, starting from a position $r_{P_{/G}} = 0.15\boldsymbol{i}$ m.

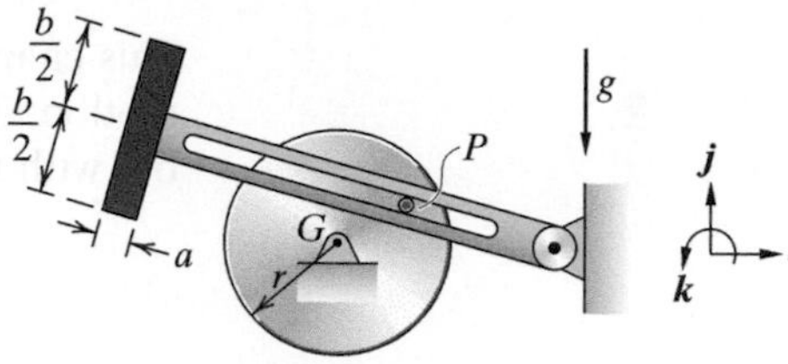

E7.5.38

7.5.39. **[Level 3]** **Computational** A uniform disk A (mass $m_A = 20$ kg, radius $r = 0.25$ m) can roll without slip, and a rigid link B (length 0.22 m, $m_B = 1$ kg) is attached via a frictionless pivot to the disk at its center O. At the illustrated instant the disk is motionless, and the link has a rotational velocity of $12\boldsymbol{k}$ rad/s. Plot the total energy of the system as well as the individual total energies of the link and disk as a function of the link's inclination angle (zero at the illustrated instant). End the plot when the link is again oriented vertically downward.

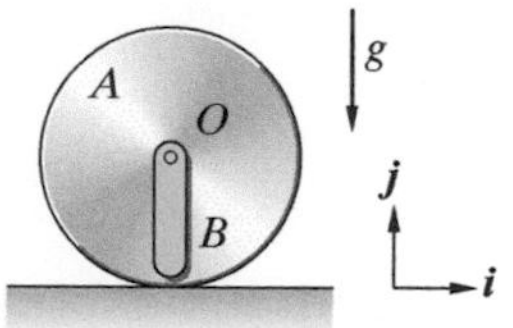

E7.5.39

7.5.40. **[Level 3]** Each individual pulley has a mass $m = 4$ kg and radius r of 5 cm. Blocks A and B have masses of 10 kg and 8 kg, respectively. Assume a frictionless interface and massless ropes. If the system is released from rest, what's the velocity of block B once block A has moved 1.5 m? By how much does your answer change if the pulleys are approximated as being massless?

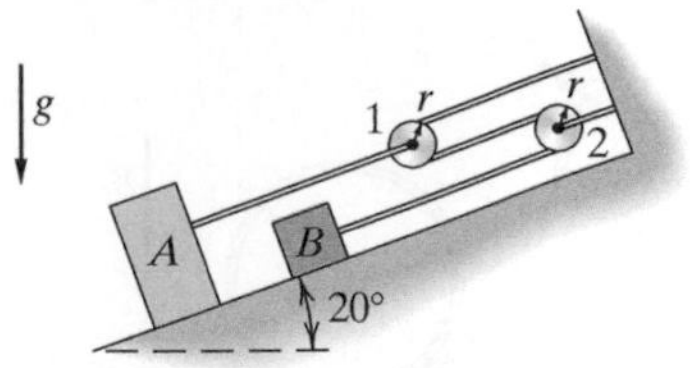

E7.5.40

7.5.41. **[Level 3]** The 2 m long, 20 kg slender bar is constrained to move in the illustrated guide. Point B is confined to move in the vertical guide, and A moves in the horizontal guide. At the instant shown, the spring is uncompressed, and B has a speed of 2 m/s directed downward. Although the horizontal guide is frictionless, the vertical guide resists motion of B within it with a force of 100 N. $k = 100$ N/m. What is the velocity with which B contacts the horizontal guide?

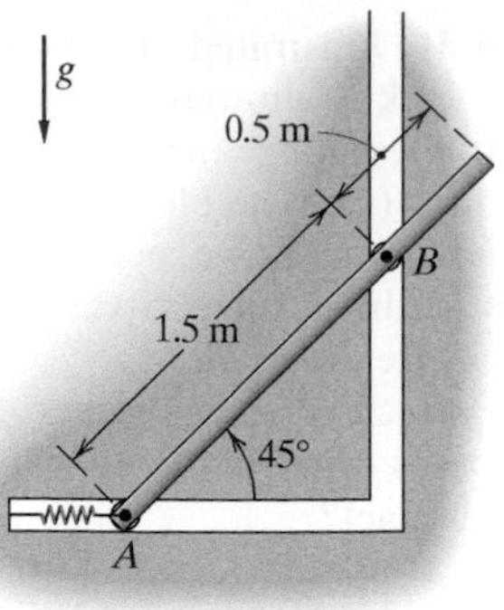

E7.5.41

7.6 JUST THE FACTS

This chapter dealt with the kinetics of rigid bodies in two-dimensional motion. The simplest systems involved motion along curved trajectories but with no rotations:

$$\sum F_x = m\ddot{x} \tag{7.1}$$

$$\sum F_y = m\ddot{y} \tag{7.2}$$

We also made use of the static moment balance,

$$\sum M_G = 0 \tag{7.3}$$

The first equation that considered rotational motion was the special case of a moment balance about a fixed point of rotation O:

$$I_O\ddot{\theta}\boldsymbol{b}_3 = \sum \boldsymbol{M}_O \tag{7.28}$$

$$I_O\ddot{\theta} = \sum M_O \tag{7.29}$$

Similar in form but much more general in application was the moment balance about the body's mass center. This formula was applicable to fully general planar motion.

$$I_G\,\ddot{\theta}\boldsymbol{b}_3 = \sum \boldsymbol{M}_G \tag{7.45}$$

$$I_G\,\ddot{\theta} = \sum M_G \tag{7.46}$$

An additional relationship that proved very useful in formulating moment balances was the **parallel axis theorem**:

$$I_O = I_G + mr^2_{G_{/O}} \tag{7.31}$$

Finally, we formulated moment balance equations about an arbitrary point A on the body:

$$\sum \boldsymbol{M}_A = I_G\ddot{\theta}\boldsymbol{b}_3 + \boldsymbol{r}_{G/A} \times m\boldsymbol{a}_G \tag{7.48}$$

$$\sum \boldsymbol{M}_A = I_A\ddot{\theta}\boldsymbol{b}_3 + \boldsymbol{r}_{G/A} \times m\boldsymbol{a}_A \tag{7.51}$$

Integrating our force balance between two distinct times gave us our linear momentum/linear impulse equation:

$$\boldsymbol{L}(t_2) = \boldsymbol{L}(t_1) + \mathcal{LI}_{1-2} \tag{7.70}$$

and a lack of external forces led to an expression for conservation of linear momentum.

$$\boldsymbol{L}(t_2) = \boldsymbol{L}(t_1) \tag{7.71}$$

Integrating our moment balance equation about the mass center led to

$$H_G(t_2) = H_G(t_1) + \mathcal{AI}_{G_{1-2}} \tag{7.72}$$

which without applied angular impulses gave us a conservation of angular momentum expression

$$H_G(t_2) = H_G(t_1) \tag{7.73}$$

Integrating about a fixed point O produced the angular momentum expression

$$H_O(t_2) = H_O(t_1) + \mathcal{AI}_{O_{1-2}} \tag{7.74}$$

and, as for the case of momentum about the mass center, an absence of external angular impulses gave us an expression for conservation of angular momentum about O:

$$H_O(t_2) = H_O(t_1) \tag{7.75}$$

Finally, we saw that the kinetic energy of a rigid body took a far simpler form,

$$\mathcal{KE} = \frac{1}{2}m\bar{v}^2 + \frac{1}{2}\bar{I}\dot{\theta}^2 \tag{7.86}$$

than had the kinetic energy of a multiparticle system.

SYSTEM ANALYSIS (SA) EXERCISES

SA7.1 Evaluation of Head Rotation Effects—Take Two

Pilots of fighter aircraft are often required to perform high g maneuvers. During these maneuvers, the pilot's blood tends to collect in the lower extremities. The lack of cerebral blood flow will cause the average person to pass out at 4–5g's, which is known as GLOC (g-induced loss of consciousness). The maximum g level can be increased by performing an anti-g straining maneuver (muscular straining to help force the blood back up to the brain). Similarly, an anti-g suit consisting of pants (and sometimes a vest) can also provide protection by squeezing the blood back up to the head. Pulling g's can be very taxing, especially when a pilot has so many other things to worry about (such as enemy aircraft).

For training, human centrifuges (**Figure SA7.1.1*a***) such as those at Brooks and Wright-Patterson Air Force Bases are utilized, a simplified model of which is shown in **Figure SA7.1.1*b***. The gondola that carries the pilot will be approximated as a uniform cylinder having a weight of 900 kg, a radius of 1 m, and a length of 3 m. The 5 m long supporting arm rotates in a horizontal plane and weighs 450 kg. The pilot weighs 80 kg and his center of mass is located at the center of the gondola. His mass moment of inertia will be lumped in with that of the gondola. Treat the gondola as a uniform, solid cylinder with a weight of 990 kg.

Aerospace physiologists calculate the g level by looking at the normal force that acts along the longitudinal axis of the pilot. The number of g's is just the ratio of the normal force to the pilot's weight. As the centrifuge rotates, the gondola is free to roll outwards so that the resultant force acts down the long axis (z') of the pilot. **Figure SA7.1.1*c*** shows a cut-away that illustrates how the pilot's orientation changes. Note that the arm and box drawn are simply there to help visualize the motion—the actual gondola is a cylinder. **Figure SA7.1.1*d*** shows the needed coordinate axes and unit vectors.

Initial training involves a slow onset rate (1g/s) run up to a maximum of 9g's. After these initial runs, a rapid onset rate (ROR) of 6g/s is typically performed; this is more indicative of the g onset rate for high performance fighters. Finally, a Tactical Aircraft Combat Maneuver (TACM) may be performed. In a recent study, pilots were tasked with performing several of these difficult maneuvers. The sequence consisted of a Rapid Onset (6g/s onset), beginning at 2g. The centrifuge then goes to 9g for 5 seconds, 5g for 1 second, then 5 seconds at 8g, 2 seconds at 4g, and 1.5g for 3 seconds. The centrifuge was finally brought to a complete stop.

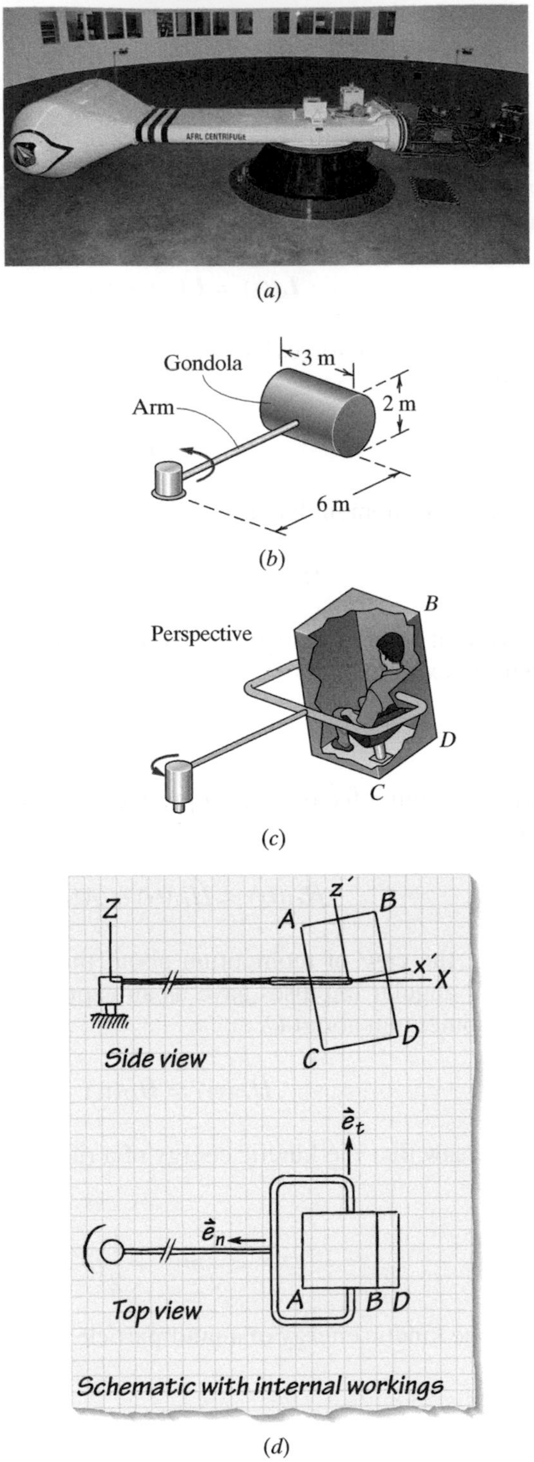

Figure SA7.1.1 Apparatus and g level versus time

Researchers at Brooks Air Force Base have asked you to determine what type of torque will be required to generate the TACM. They would also like to know how much force will be placed on the centrifuge's drive shaft.

Break this assignment into subtasks:

a. Find the angular velocity required to produce the required acceleration levels.

b. Determine the angular acceleration need to achieve a $6g$/s onset rate.

c. Determine the torque needed to produce this angular acceleration.

SA7.2 Inertias of Catapults

Read **SA6.4** as background information. In this exercise, we'll examine a catapult more closely (**Figure SA7.2.1**). This siege engine fires a missile using the energy gained from a dropping weight and the advantage of a lever arm. These were shown to excellent effect in *The Lord of the Rings* trilogy, hurling stones and even more gruesome loads.

a. Assume the counter-weight (CW) weighs 200 kg, the missile weighs 25 kg, and the throwing arm is uniform in density and shape with a weight of 50 kg, centered along the 5 m shown. Determine the center of mass of the composite system of (b).

b. Determine the composite system's (throwing arm plus counter-weight plus missile) mass moment of inertia for the data in **a** about the composite system's center of mass. Assume the throwing arm is a long slender rod and the counter-weight and missile are point loads.

c. Determine the mass moment of inertia of the composite system about the support pin A based on the data in **a**.

d. Compare the results for **a**, mass moment of inertia about composite center of mass, and **b**, mass moment of inertia about support pin A. Comment on the results relative to the efficiency of the catapult system.

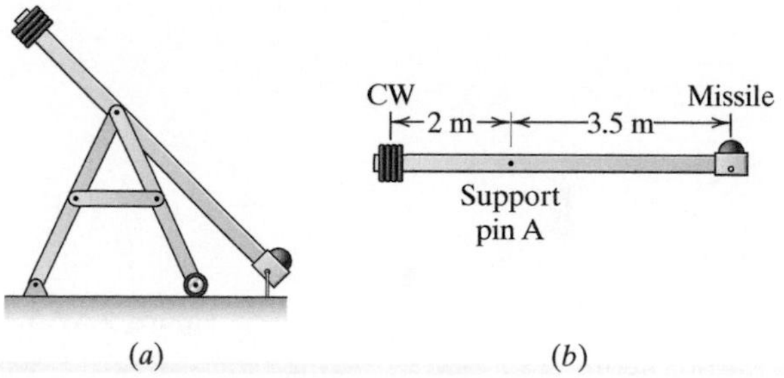

Figure SA7.2.1 Sideview and dimensions

SA7.3 Catapult Launches

Read **SA6.4** as background information. Ancient siege engines provided military commanders with the ability to engage an enemy from a distance; essentially it was the artillery of the armies past. Unfortunately, what is known of these medieval siege engines is limited to crude artist renditions and manuscript references. Hence, the hypothetical analysis of siege engines is still intriguing and challenging. Let us analyze one of the simplest forms of a siege engine, the catapult, as shown in **Figure SA7.3.1**. This siege engine fires a missile using the energy gained by a dropping weight and the advantage of a lever arm.

a. Assume the counter-weight (CW) is 200 kg, the missile weight is 25 kg, and the throwing arm is generally uniform in density and shape with a weight of 50 kg, centered along the 5 m shown. The composite system's mass moment of inertia about the support pin A has been previously calculated to be 990 kg·m^2. Assume the throwing arm starts from $\theta = 40°$ and rotates around the support pin A in a counterclockwise fashion. What is the initial acceleration of the missile?

b. Will the missile's acceleration be constant during the rotation of the throwing arm? Why or why not? Support your answer with calculations and discuss the results.

c. Recommend a physical change to the system to increase the missile's initial acceleration. Support your recommendation with calculations.

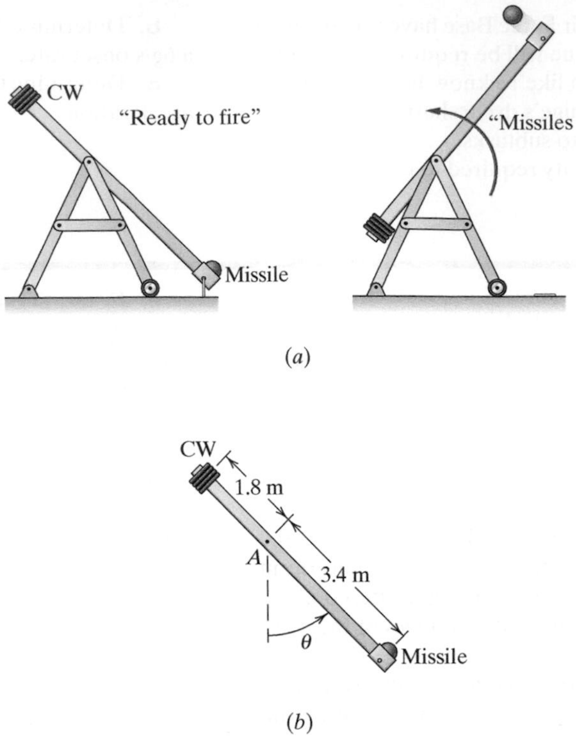

Figure SA7.3.1 Launch behavior and dimensions

SA7.4 More on Catapult Launches

Read **SA6.4** as background information and refer to **Figure SA7.3.1**.

a. Assume that the counter-weight (CW) is 181.4 kg, the missile weighs 22.7 kg, and the throwing arm is uniform in density and shape with a weight of 45.4 kg, centered along the 5.2 m shown. The composite system's mass moment of inertia about the support pin A has been previously calculated to be 980 kg·m^2. Assume the throwing arm starts from $\theta = 40°$, and rotates around the support pin A to the missile's release at $\theta = 135°$. Using the concepts of work and energy, determine the missile's release speed.

b. How might you redesign the machine to supply greater launch speeds?

OBJECTIVES

Upon completion of this chapter you will be able to:

- Determine the velocity and accelerations of three-dimensional bodies
- Calculate moments and products of inertia
- Utilize the angular momentum and energy of three-dimensional bodies
- Formulate the equations of motion for three-dimensional bodies

CHAPTER 8

KINEMATICS AND KINETICS OF RIGID BODIES IN THREE-DIMENSIONAL MOTION

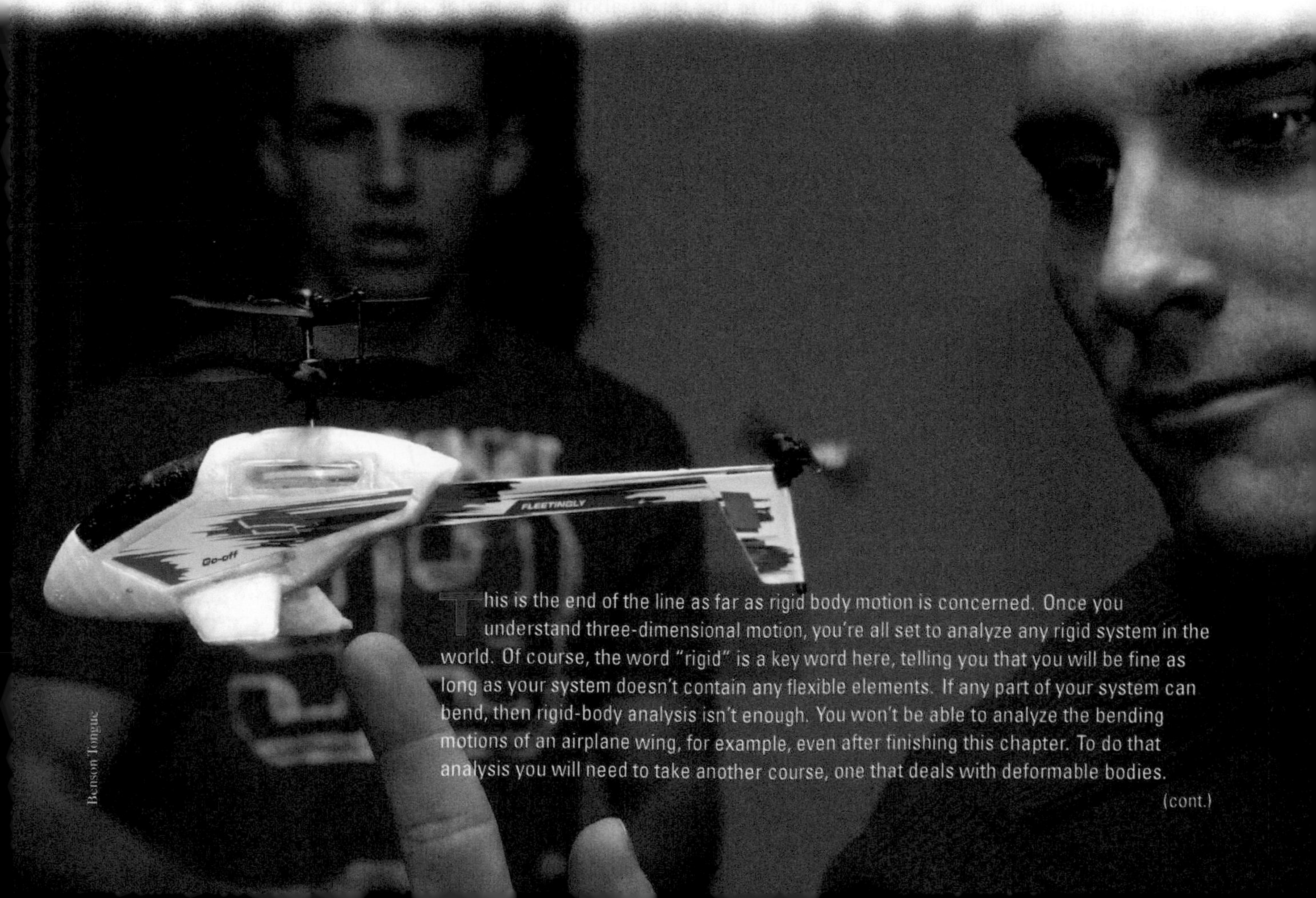

Benson Tongue

This is the end of the line as far as rigid body motion is concerned. Once you understand three-dimensional motion, you're all set to analyze any rigid system in the world. Of course, the word "rigid" is a key word here, telling you that you will be fine as long as your system doesn't contain any flexible elements. If any part of your system can bend, then rigid-body analysis isn't enough. You won't be able to analyze the bending motions of an airplane wing, for example, even after finishing this chapter. To do that analysis you will need to take another course, one that deals with deformable bodies.

(cont.)

(cont.)

Nonetheless, there's more than enough to occupy our time just dealing with rigid bodies.

The main complication we encounter when analyzing three-dimensional motion is that, rather than needing only three equations of motion the way we did in analyzing two-dimensional motion in Chapter 3, now we need six—three force balance equations and three moment balance equations. The physics changes a bit as well, as a consequence of the additional three equations. Because the direction of our angular velocity vector isn't fixed (as it was in two-dimensional motion), we have the possibility of gyroscopic motion. In addition, we will see that if we spin a body about an axis, it won't necessarily want to remain spinning about that axis. The possibility of stable and unstable rotations is another property that doesn't occur in the two-dimensional world—only in the three-dimensional.

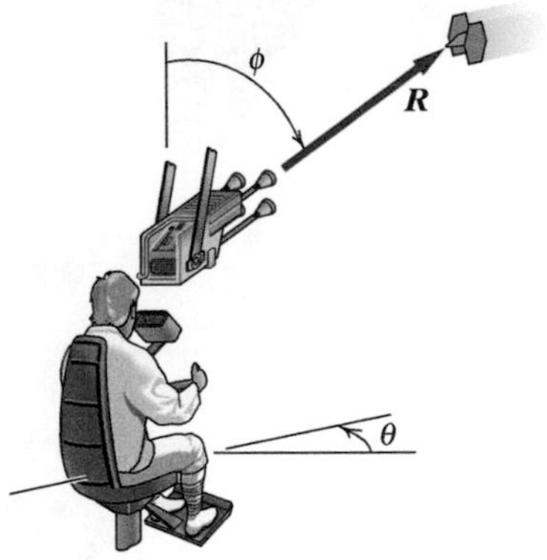

Figure 8.3 Luke shooting at TIE fighter

8.1 SPHERICAL COORDINATES

You can think of spherical coordinates as the three-dimensional analogue of polar coordinates. In polar coordinates, you need an angle (θ) to give you an orientation, and you also need to know how far out from the origin to go (r) to reach the end of your position vector. The complication in three dimensions is that you now need two angles (**Figure 8.1**). **Figure 8.2** shows the relationships between the spherical unit vectors $\boldsymbol{e}_R, \boldsymbol{e}_\phi, \boldsymbol{e}_\theta$ and the polar/Cartesian unit vectors.

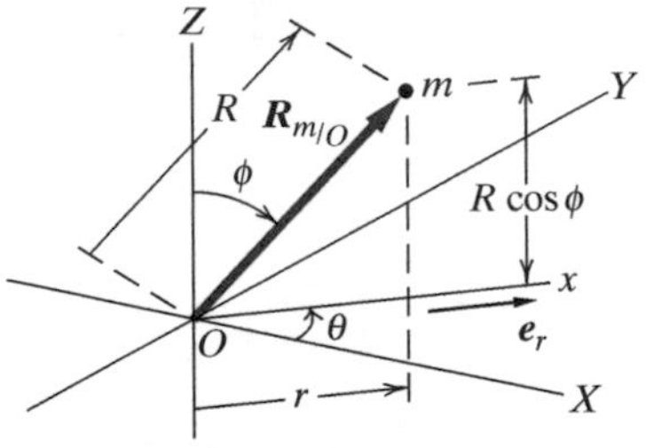

Figure 8.1 Spherical coordinates

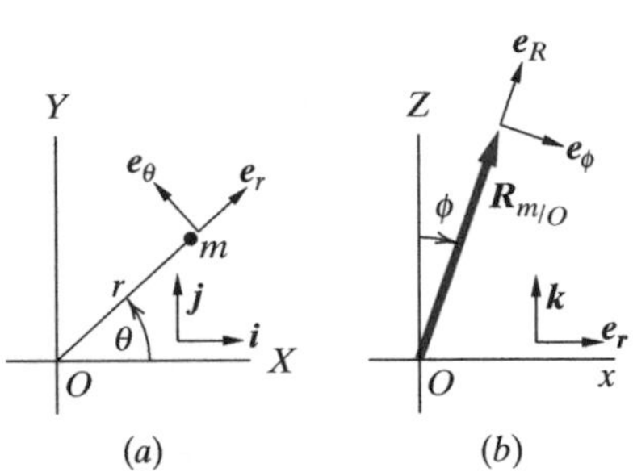

Figure 8.2 Alternative perspectives

Remember the scene in the original *Star Wars* movie when Luke and Han jumped into the Millennium Falcon's gun turrets to blast the pursuing TIE fighters? This scene provides a great way of visualizing spherical coordinates. **Figure 8.3** shows an idealized model. Luke had the ability to rotate the turret about a vertical axis, thereby changing θ. He could also rotate the gun up and down, thereby altering ϕ. Lastly, the distance from the gun to the TIE fighter defined R.

These are the same physical considerations that govern the spherical coordinate unit vectors. As you can see from **Figure 8.2*b***, the position vector $\boldsymbol{R}_{m/O}$ is given by

$$\boldsymbol{R}_{m/O} = R\boldsymbol{e}_R \tag{8.1}$$

We use R instead of r so that we can recover a polar representation when the three-dimensional representation is projected down onto the x–y plane. As **Figure 8.2** shows, the distance from the origin O to the projection of m on that plane is then given by r, and the angular orientation from the x axis is given by θ, just as it is in a normal polar coordinate set.

We can use **Figure 8.2** to define two sets of coordinate transformations:

	$\boldsymbol{i}$	$\boldsymbol{j}$	$\boldsymbol{k}$
$\boldsymbol{e}_r$	$\cos\theta$	$\sin\theta$	0
$\boldsymbol{e}_\theta$	$-\sin\theta$	$\cos\theta$	0
$\boldsymbol{k}$	0	0	1

(8.2)

and

	$\boldsymbol{e}_r$	$\boldsymbol{e}_\theta$	$\boldsymbol{k}$
$\boldsymbol{e}_\phi$	$\cos\phi$	0	$-\sin\phi$
$\boldsymbol{e}_\theta$	0	1	0
$\boldsymbol{e}_R$	$\sin\phi$	0	$\cos\phi$

(8.3)

that enable us to express a vector in terms of the ground-based unit vectors $\boldsymbol{i}, \boldsymbol{j}, \boldsymbol{k}$; the cylindrical coordinate unit vectors $\boldsymbol{e}_r, \boldsymbol{e}_\theta, \boldsymbol{k}$; or the spherical coordinate unit vectors $\boldsymbol{e}_R, \boldsymbol{e}_\theta, \boldsymbol{e}_\phi$.

Differentiating $\boldsymbol{R}_{m/O}$ yields

$$\dot{\boldsymbol{R}}_m = \dot{R}\boldsymbol{e}_R + R\dot{\phi}\boldsymbol{e}_\phi + R\dot{\theta}\sin\phi\boldsymbol{e}_\theta \tag{8.4}$$

and another differentiation produces

$$\begin{aligned}\ddot{\boldsymbol{R}}_m &= (\ddot{R} - R\dot{\phi}^2 - R\dot{\theta}^2\sin^2\phi)\boldsymbol{e}_R \\ &+ \left(\frac{1}{R}\frac{d}{dt}(R^2\dot{\phi}) - \frac{R\dot{\theta}^2\sin(2\phi)}{2}\right)\boldsymbol{e}_\phi \\ &+ \left(\frac{\sin\phi}{R}\frac{d}{dt}(R^2\dot{\theta}) + 2R\dot{\theta}\dot{\phi}\cos\phi\right)\boldsymbol{e}_\theta\end{aligned} \tag{8.5}$$

As you can see, this expression is a lot more complex than the expressions we saw in our two-dimensional analyses. To a computer program, however, it's still not a big deal to handle.

8.2 ANGULAR VELOCITY OF RIGID BODIES IN THREE-DIMENSIONAL MOTION

There's nothing conceptually different between a rigid body that's moving in three dimensions and one that's simply moving about in a plane. We'll still have to look at the forces and moments acting on the body and try to calculate the spatial response. The biggest difference is that, rather than having to concern ourselves with only one moment of inertia, now we'll have to deal with several. Just as we did in the two-dimensional work, we'll start by looking at velocity, move on to acceleration, and finally add the forces and moments to obtain our equations of motion.

Dealing with the translational motion of the body won't be much of a problem. Instead of tracking the center of mass's acceleration in two directions, such as $\boldsymbol{i}$ and $\boldsymbol{j}$, we'll track it in three: $\boldsymbol{i}, \boldsymbol{j}$, and $\boldsymbol{k}$. Rotation, on the other hand, is a little tricky. In the two-dimensional case we can rotate a body 10° and then rotate it an additional 20°, and end up with a body that's rotated 30°. Thus we might assume that rotations are additive and therefore vector-like. That's one of the important aspects of vectors—you can add them. The same is not true in three dimensions.

Here's an experiment that will illustrate this fact. Hold your right arm straight out in front of you, palm down, as shown in the leftmost part of **Figure 8.4*a*** and the leftmost part of **Figure 8.4*b***. We'll be considering your arm from elbow to fingers to be the rigid body. Now rotate this rigid body twice: first rotate it 90° clockwise so that your palm is facing the wall, with your pinky closest to the floor and your thumb closest to the ceiling (first rotation of **Figure 8.4*a***), then rotate your forearm 90° about the elbow so that your forearm is perpendicular to the floor and all your fingers point to the ceiling, as shown in the final drawing of **Figure 8.4*a***.

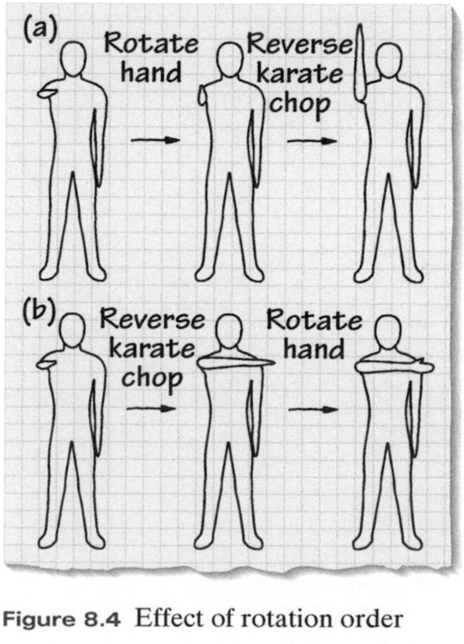

Figure 8.4 Effect of rotation order

Now do the same two rotations but in reverse order, as illustrated in **Figure 8.4*b***: with the order reversed, a 90° pivot about the elbow is followed by a 90° hand rotation. Note that the orientation of your hand is totally different in the two cases. What's the conclusion? That finite rotations don't add in three dimensions and therefore can't be viewed as vector quantities. That might seem bad, inasmuch as we've found vectors to be so useful in the past. But, luckily for us, even though finite rotations can't be added, infinitesimal rotations *can*. And because velocity is just an infinitesimal displacement divided by an infinitesimal time, this implies that velocities are vector quantities and can be added together with ease. The same can be said for accelerations.

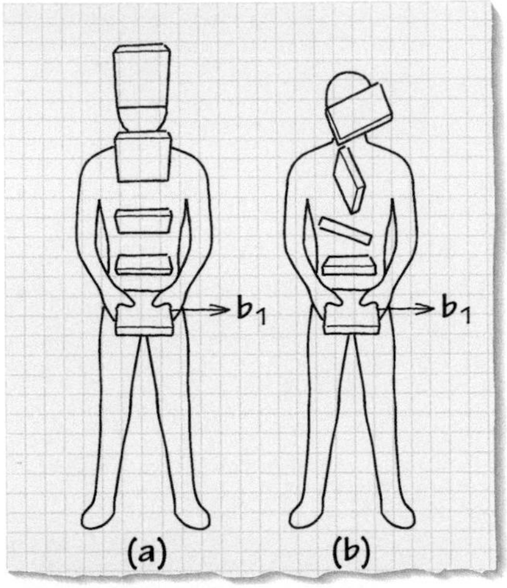

Figure 8.5 Stable and unstable rotations

This is excellent because rigid bodies behave in very strange ways when moving around in three-dimensional space, as **Figure 8.5** shows. This figure illustrates what happens when you toss a book upward (tied together so it doesn't fly open) and at the same time try to spin it smoothly about the $\boldsymbol{b}_1$ axis (an axis that runs laterally across the book). What you would expect is a motion that looks like **Figure 8.5*a***, a steady rotation about the $\boldsymbol{b}_1$ axis as the book moves upward. What actually happens, however, is shown in **Figure 8.5*b***. Almost immediately the book refuses to spin about the $\boldsymbol{b}_1$ axis and instead tumbles through the air in a somewhat random-looking manner. Try it yourself if you don't believe it; it's not hard to do. Just grab a convenient book (this one will do fine), secure the covers together with a rubber band or string, and try to flip it up as shown. Make sure you flip it in the orientation that's illustrated. If you try spinning it about the longitudinal axis, you will be able to spin it in a stable manner. It's only the lateral axis that causes problems.

So, our problem is to determine a reasonable way of characterizing how a rigid body is spinning so that we can then characterize its acceleration and solve its equations of motion. Because this book isn't meant as an in-depth exposition of three dimensional Newtonian dynamics, I am going to limit my discussion to the basics of three-dimensional motion. What this means for angular velocity and acceleration is that I will consider only rotations about body-fixed axes. Although other approaches can (and are) used at times, body-fixed rotations are by far the most common approach.

Figure 8.6 shows a rigid body (a rectangular box) that's being rotated in a prescribed way. In this and the two pictures to follow, the light blue box shows the original position and the dark blue box shows the new, rotated position of the body.

We start with the box aligned with the X, Y, Z ground-fixed axes. A body-fixed set of axes x', y', z' are initially coincident with X, Y, Z. We then rotate the body by an angle ψ about the z' axis. The angle ψ is called an **Euler angle**, and when the body is rotating about this axis the motion is often referred to as **precession**. In order to keep track of the body, we need to introduce a new set of unit vectors that align with the x', y', z' axes. We will use $\boldsymbol{b}_1, \boldsymbol{b}_2, \boldsymbol{b}_3$ for this purpose, as shown in **Figure 8.6**.

The next rotation results from rotating the box about the y' axis by an amount θ. A rotation about this axis is often referred to as **nutation**. Rotating the body in this way brings the box into the orientation shown in **Figure 8.7**, with x'', y'', z'' as the body-fixed axes and $\boldsymbol{c}_1, \boldsymbol{c}_2, \boldsymbol{c}_3$ as the new unit vector set. For our final rotation we will rotate the body about

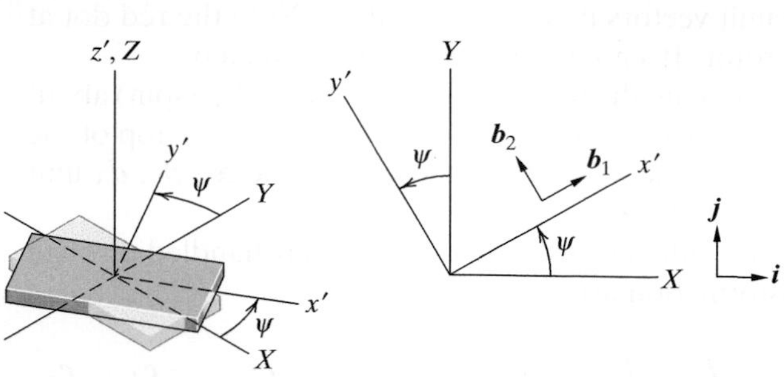

Figure 8.6 Precession

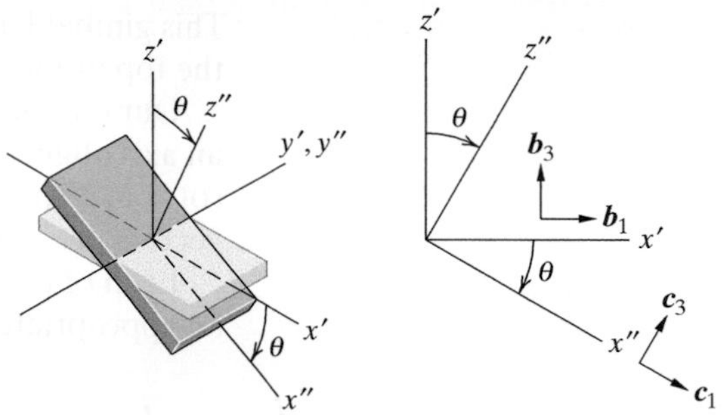

Figure 8.7 Nutation

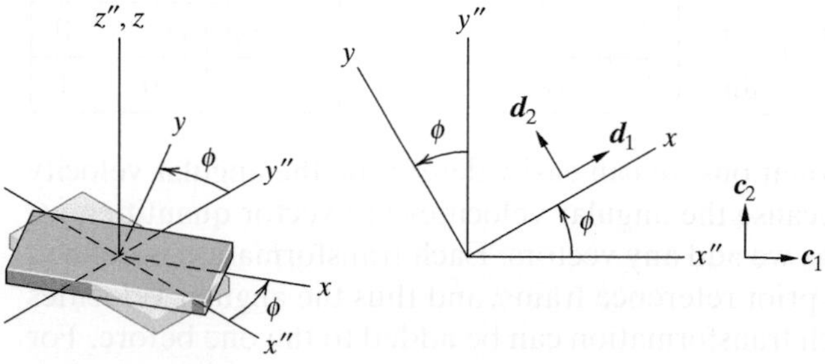

Figure 8.8 Spin

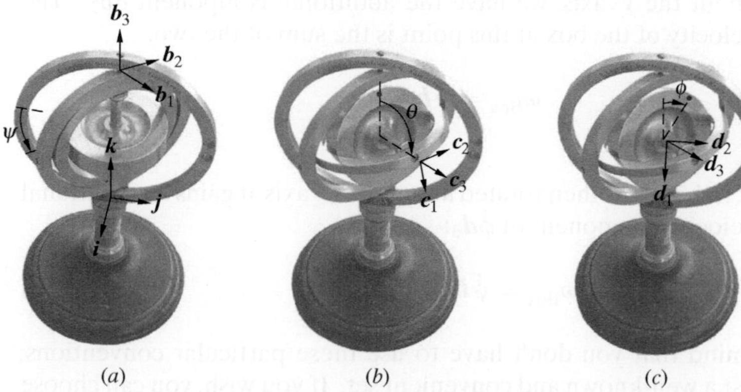

Figure 8.9 Euler angles illustrated on a gyroscope

the z'' axis by an amount ϕ. Rotations about this final axis are known as **spin**. This last rotation brings us to the final body-fixed axes, x, y, z, with associated unit vectors $\boldsymbol{d}_1, \boldsymbol{d}_2, \boldsymbol{d}_3$, as shown in **Figure 8.8**.

Figure 8.9 shows how a gyroscope will look if each of these rotations is imposed in turn. The inner rotor is the body that we're concerned with, and the supporting gimbal rings allow us to easily see how the different reference frames move with respect to one another. The outer gimbal remains fixed with respect to ground.

In **Figure 8.9*a*** the middle gimbal has been rotated by an angle ψ (precession), and the ground-fixed $(\boldsymbol{i}, \boldsymbol{j}, \boldsymbol{k})$ as well as gimbal-fixed unit vectors $(\boldsymbol{b}_1, \boldsymbol{b}_2, \boldsymbol{b}_3)$ are shown.

Figure 8.9*b* takes the system from its previous configuration and rotates the inner gimbal by an amount θ about a horizontal axis aligned with $\boldsymbol{b}_2$.

This gimbal has the unit vectors $\boldsymbol{c}_1, \boldsymbol{c}_2, \boldsymbol{c}_3$ fixed to it. Note the red dot at the top of the inner rotor. It's going to move in the next step.

Our final action is to rotate the inner rotor by an amount ϕ (spin) about an axis aligned with $\boldsymbol{c}_3$. As you can see, the dot that was at the top of the rotor in **Figure 8.9*b*** has now rotated around a bit. The $\boldsymbol{d}_1, \boldsymbol{d}_2, \boldsymbol{d}_3$ unit vectors are our final, body-fixed set.

Each of the three individual transformations is easily handled through an appropriate transformation array:

	$\boldsymbol{i}$	$\boldsymbol{j}$	$\boldsymbol{k}$
$\boldsymbol{b}_1$	$\cos\psi$	$\sin\psi$	0
$\boldsymbol{b}_2$	$-\sin\psi$	$\cos\psi$	0
$\boldsymbol{b}_3$	0	0	1

	$\boldsymbol{b}_1$	$\boldsymbol{b}_2$	$\boldsymbol{b}_3$
$\boldsymbol{c}_1$	$\cos\theta$	0	$-\sin\theta$
$\boldsymbol{c}_2$	0	1	0
$\boldsymbol{c}_3$	$\sin\theta$	0	$\cos\theta$

	$\boldsymbol{c}_1$	$\boldsymbol{c}_2$	$\boldsymbol{c}_3$
$\boldsymbol{d}_1$	$\cos\phi$	$\sin\phi$	0
$\boldsymbol{d}_2$	$-\sin\phi$	$\cos\phi$	0
$\boldsymbol{d}_3$	0	0	1

With these transformations we can easily determine the angular velocity of a rigid body. Because the angular velocities are vector quantities, we can add them just as we add any vectors. Each transformation is defined with respect to the prior reference frame, and thus the angular velocities associated with each transformation can be added to the one before. For the particular angles and transformation just defined, we see that the first angular velocity component of the box is just $\dot{\psi}\boldsymbol{b}_3$. If the box is then rotated about the y' axis, we have the additional component $\dot{\theta}\boldsymbol{c}_2$. The angular velocity of the box at this point is the sum of the two:

$$\boldsymbol{\omega}_{\text{Box}} = \dot{\psi}\boldsymbol{b}_3 + \dot{\theta}\boldsymbol{c}_2$$

Finally, if the box is then rotated about the z'' axis, it gains an additional angular velocity component of $\dot{\phi}\boldsymbol{d}_3$:

$$\boldsymbol{\omega}_{\text{Box}} = \dot{\psi}\boldsymbol{b}_3 + \dot{\theta}\boldsymbol{c}_2 + \dot{\phi}\boldsymbol{d}_3 \tag{8.6}$$

Keep in mind that you don't have to use these particular conventions; they're just a well-known and convenient set. If you wish, you can choose a different order and different axes to rotate around, such as x', then y'', and then z. And, of course, you can use whatever variables you wish for the angles.

>>> **Check out Example 8.1 (page 493) and Example 8.2 (page 494) for applications of this material.**

EXAMPLE 8.1 ANGULAR VELOCITY OF A SIMPLIFIED GYROSCOPE (Theory on page 492)

Figure 8.10 shows a simplified gyroscope. The outer gimbal rotates about a vertical Z axis with an angular speed ω_Z. As it does so, the horizontal rod $\overline{AB}$, which is initially aligned along the X axis, rotates about the Z axis as well and, consequently, so does the disk D, which is free to rotate around $\overline{AB}$ and does so with an angular speed ω_x with respect to the gimbal. Let's determine the angular velocity of the inner disk.

Goal Find the total angular velocity of the gyroscope's inner disk.

Given The configuration of the gyroscope and the angular speed of the outer gimbal and inner disk.

Draw **Figure 8.11** shows two sets of unit vectors—$\boldsymbol{i}, \boldsymbol{j}, \boldsymbol{k}$ fixed in space and $\boldsymbol{b}_1, \boldsymbol{b}_2, \boldsymbol{b}_3$ fixed to the rotating gimbal.

Formulate Equations Proceed in the same manner that led to (8.6).

Solve To find the disk's absolute angular velocity, we need to add the angular velocity of the gimbal (to which the disk is attached) and the angular velocity of the disk with respect to the gimbal. The outer gimbal has an angular velocity of $\omega_Z \boldsymbol{b}_3$, and the disk has an angular velocity of $\omega_x \boldsymbol{b}_1$ with respect to the gimbal. Thus the total angular velocity is given by

$$\boldsymbol{\omega}_D = \omega_Z \boldsymbol{b}_3 + \omega_x \boldsymbol{b}_1$$

Check Note that in this case the angular velocity is expressed in terms of the $\boldsymbol{b}_i$ unit vectors, which aren't actually body fixed. They're intermediate between unit vectors that rotate with the inner disk and ground-fixed unit vectors. When dealing with a symmetric spinning body like the circular disk of this example, this is all that's usually needed. There's nothing unique about any particular point of the disk, and as we will see, a unit-vector set that aligns with the disk (like the $\boldsymbol{b}_i$ set) is sufficient for solving kinetics problems involving the disk.

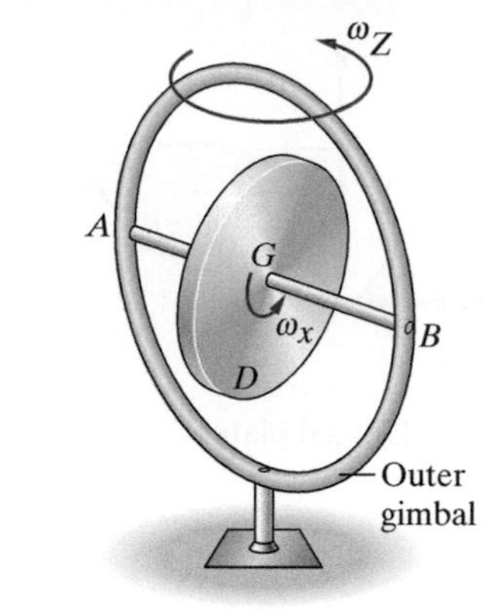

Figure 8.10 Simple gyroscope

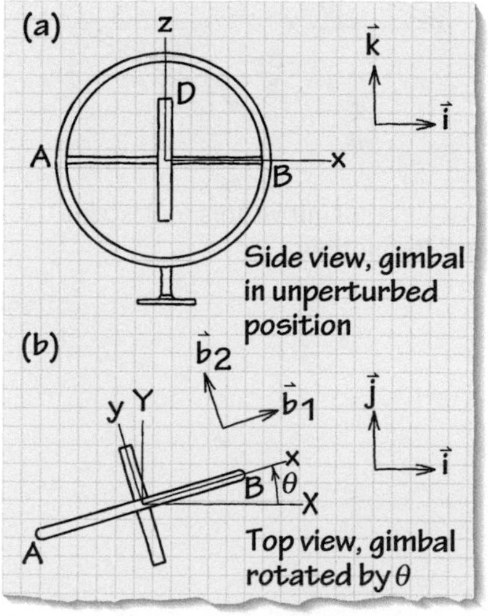

Figure 8.11 Labeling of gyroscope

EXAMPLE 8.2 ANGULAR VELOCITY OF A HINGED PLATE (Theory on page 492)

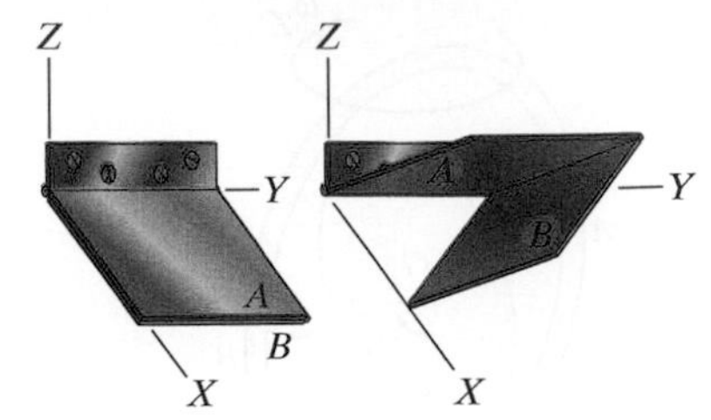

Figure 8.12 Hinged plates

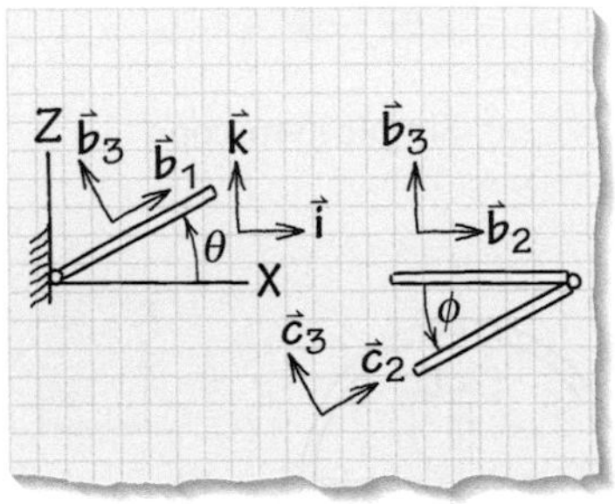

Figure 8.13 Unit vectors to describe plate B's motion

The gyroscope of Example 8.1 was nicely symmetric, and the bodies were all centered. There's no reason, however, that this has to be the case. **Figure 8.12** shows a system that clearly isn't symmetric or centered. What we have are two plates, A and B. Plate A is hinged to the wall along one of its edges, and plate B is hinged to plate A. The initial configuration has both plates lying in the X–Y plane. Find an expression for the angular velocity of plate B.

Goal Find the angular velocity of hinged plate B.

Given The initial orientation of the plates and the manner in which they're constrained to move.

Draw **Figure 8.13** shows the unit vectors we need to describe plate B's position and angular velocity.

Formulate Equations Proceeding as we did when deriving (8.6) will give us the required angular velocity components.

Solve **Figure 8.13** shows that if we define θ as the angle that plate A has rotated up, then its angular velocity is $-\dot{\theta}\boldsymbol{b}_2$. Plate B has this angular velocity as well and, in addition, can rotate with respect to plate A. If we define the angle between the two plates as ϕ, then the angular velocity of plate B with respect to plate A is $\dot{\phi}\boldsymbol{b}_1$. The total angular velocity for plate B is the sum of these two components:

$$\boldsymbol{\omega}_B = \dot{\phi}\boldsymbol{b}_1 - \dot{\theta}\boldsymbol{b}_2$$

Check In this example, as in Example 8.1, the angular velocity is expressed in terms of a unit-vector set that isn't fixed to plate B. Plate B isn't a centered, symmetric object like the circular disk of Example 8.1; therefore this isn't the form we would use if we were calculating the velocity of a point on plate B. If we were asked to find linear velocities of some point on B, we would express our angular velocity in terms of the $\boldsymbol{c}_i$ frame so that we could then take cross products with the position vector going to the particular point on plate B (which would be expressed in terms of $\boldsymbol{c}_1$ and $\boldsymbol{c}_2$).

8.3 ANGULAR ACCELERATION OF RIGID BODIES IN THREE-DIMENSIONAL MOTION

We can find angular acceleration by differentiating angular velocity. This is *not* necessarily the easiest way to find a body's angular acceleration because the components of the angular velocity vector may themselves be rotating. Remember back in Section 2.3 when we looked at polar coordinates? We differentiated $\boldsymbol{r} = r\boldsymbol{e}_r$ and we got $\boldsymbol{v} = \dot{r}\boldsymbol{e}_r + r\dot{\theta}\boldsymbol{e}_\theta$ (2.46). The second term came about because the unit vector $\boldsymbol{e}_r$ was rotating, giving rise to a term proportional to $\dot{\theta}\boldsymbol{e}_\theta$. The same thing can happen here. The angular velocity is oriented along different unit vectors, and if these vectors are themselves rotating, then we will have terms that reflect this, as you can see in Example 8.1. The disk had an angular velocity equal to $\omega_x\boldsymbol{b}_1$. Because $\boldsymbol{b}_1$ continuously changed its orientation as a result of the gimbal's spin, the angular velocity changed its orientation as well. When this type of situation occurs, you end up with angular acceleration because angular acceleration is just the time rate of change of angular velocity.

If you don't want to deal with the terms that show up because a vector is rotating, you can express the angular velocity in terms of the nonrotating $\boldsymbol{i}, \boldsymbol{j}, \boldsymbol{k}$ unit vectors. Of course, just because you *can* doesn't mean you *should*. This route is often a pain because it can involve quite a bit of coordinate transformations and the subsequent introduction of trigonometric terms, all of which then have to be differentiated with respect to time. Thus you might want to consider a less brute-force approach. Recall from Section 6.3 that the derivative of a vector $\boldsymbol{p}$ in a rotating reference frame is given by (6.24):

$$\left.\frac{d}{dt}\right|_N \boldsymbol{p} = \left.\frac{d}{dt}\right|_S \boldsymbol{p} + \boldsymbol{\omega}\times\boldsymbol{p} \tag{8.7}$$

where N refers to the fixed, inertial frame and S refers to the rotating frame. This enables us to simplify our angular acceleration procedure. All we need do is use the angular velocity as our vector $\boldsymbol{p}$:

$$\left.\frac{d}{dt}\right|_N \boldsymbol{\omega}_{\text{Body}} = \left.\frac{d}{dt}\right|_S \boldsymbol{\omega}_{\text{Body}} + \boldsymbol{\omega}_S\times\boldsymbol{\omega}_{\text{Body}} \tag{8.8}$$

where $\boldsymbol{\omega}_S$ is the rotational velocity of the rotating reference frame S and $\boldsymbol{\omega}_{\text{Body}}$ is the rotational velocity of the body.

>>> **Check out Example 8.3 (page 496) for an application of this material.**

EXAMPLE 8.3 ANGULAR ACCELERATION OF A SIMPLE GYROSCOPE (Theory on page 495)

We will take Example 8.1 a little further here. As before, **Figure 8.10** shows a simple gyroscope. The disk D spins about the horizontal rod $\overline{AB}$, part of a surrounding gimbal. The gimbal's angular velocity and acceleration are ω_Z and α_Z, respectively. The disk has an angular velocity ω_x and an angular acceleration α_x, both *with respect to the gimbal.* Find the disk's absolute angular acceleration.

Goal Given the absolute angular velocity and acceleration of the gimbal and the relative angular velocity and acceleration of the disk, find the absolute angular acceleration of the disk.

We will find the answers in two ways: one way is longer but straightforward, and the other way is shorter but more complex. For the long way we will express everything in the $\boldsymbol{i}, \boldsymbol{j}, \boldsymbol{k}$ coordinate system, and for the shorter way we will use the expression that lets us differentiate a vector in a rotating reference frame.

Draw **Figure 8.11** shows all the relevant quantities we will need.

Solve As you have already seen from Example 8.1, because angular velocities add, we find the absolute angular velocity of the disk D to be $\boldsymbol{\omega}_D = \omega_Z \boldsymbol{k} + \omega_x \boldsymbol{b}_1$. As **Figure 8.11** shows, the orientation of the gimbal with respect to the ground is governed by θ, and the coordinate transformation matrix is

	$\boldsymbol{i}$	$\boldsymbol{j}$	$\boldsymbol{k}$
$\boldsymbol{b}_1$	$\cos\theta$	$\sin\theta$	0
$\boldsymbol{b}_2$	$-\sin\theta$	$\cos\theta$	0
$\boldsymbol{b}_3$	0	0	1

The disk's angular velocity can therefore be rewritten as $\boldsymbol{\omega}_D = \omega_x \cos\theta \boldsymbol{i} + \omega_x \sin\theta \boldsymbol{j} + \omega_Z \boldsymbol{k}$.

Differentiating with respect to t gives us the disk's angular acceleration:

$$\boldsymbol{\alpha}_D = (\alpha_x \cos\theta - \omega_x\omega_Z \sin\theta)\boldsymbol{i} + (\alpha_x \sin\theta + \omega_x\omega_Z \cos\theta)\boldsymbol{j} + \alpha_Z \boldsymbol{k}$$

Now let's try the other route. The expression we need to evaluate is

$$\left.\frac{d}{dt}\right|_N \boldsymbol{\omega}_D = \left.\frac{d}{dt}\right|_S \boldsymbol{\omega}_D + \boldsymbol{\omega}_S \times \boldsymbol{\omega}_D$$

where the reference frame S is rotating with the outer gimbal. We've got

$$\left.\frac{d}{dt}\right|_S \boldsymbol{\omega}_D = \alpha_Z \boldsymbol{b}_3 + \alpha_x \boldsymbol{b}_1$$

$$\boldsymbol{\omega}_S \times \boldsymbol{\omega}_D = \omega_Z \boldsymbol{b}_3 \times (\omega_Z \boldsymbol{b}_3 + \omega_x \boldsymbol{b}_1) = \omega_Z\omega_x \boldsymbol{b}_2$$

Thus

$$\boldsymbol{\alpha}_D = \alpha_Z \boldsymbol{b}_3 + \alpha_x \boldsymbol{b}_1 + \omega_Z\omega_x \boldsymbol{b}_2$$

Check The simplest check is to ask whether the two expressions for $\boldsymbol{\alpha}_D$ give the same answer. They're expressed in terms of different unit-vector sets, and so all we need to do is use the coordinate transformation matrix to express them both in terms of the same set, and compare. Doing so (an exercise left to you) shows that they are indeed identical.

8.4 GENERAL MOTION OF AND ON THREE-DIMENSIONAL BODIES

We have now seen how to formulate the angular velocity and acceleration of a three-dimensional body. The next step is to use these two parameters as part of a general expression involving both linear motion and rotation. The formulations for linear velocity and linear acceleration are the familiar ones from Section 6.3:

$$\boldsymbol{v}_B = \boldsymbol{v}_A + \boldsymbol{\omega} \times \boldsymbol{r}_{B/A} \tag{8.9}$$

and

$$\boldsymbol{a}_B = \boldsymbol{a}_A + \boldsymbol{\alpha} \times \boldsymbol{r}_{B/A} + \boldsymbol{\omega} \times \left(\boldsymbol{\omega} \times \boldsymbol{r}_{B/A}\right) \tag{8.10}$$

The only difference is that now the angular velocity and acceleration terms ($\boldsymbol{\omega}$ and $\boldsymbol{\alpha}$) can include three components rather than the single component that a two-dimensional analysis imposed.

Similarly, the equations for motion on a three-dimensional body are identical in form to those derived in Section 6.4:

$$\boldsymbol{v}_B = \boldsymbol{v}_A + \boldsymbol{\omega} \times \boldsymbol{r}_{B/A} + \boldsymbol{v}_{rel} \tag{8.11}$$

$$\boldsymbol{a}_B = \boldsymbol{a}_A + \boldsymbol{\alpha} \times \boldsymbol{r}_{B/A} + \boldsymbol{\omega} \times \left(\boldsymbol{\omega} \times \boldsymbol{r}_{B/A}\right) + \boldsymbol{a}_{rel} + 2\boldsymbol{\omega} \times \boldsymbol{v}_{rel} \tag{8.12}$$

So you see that you already know what to do with three-dimensional kinematics. Of course, the actual calculations are more complicated because the angular velocity and angular acceleration have three components rather than one and the linear velocity and linear acceleration terms can have three components instead of their previous two. But aside from the added bookkeeping, there's no conceptual difficulty in dealing with three-dimensional motion.

Example 8.3 was fine, but because all the relevant rotation rates were given as part of the problem, it was pretty straightforward. More realistic, and more difficult, is the case where we have to impose constraints in order to determine all the needed rates. The next example looks at such a case and requires that we consider linear velocities.

>>> **Check out Example 8.4 (page 498) and Example 8.5 (page 500) for applications of this material.**

EXAMPLE 8.4 MOTION OF A DISK ATTACHED TO A BENT SHAFT (Theory on page 497)

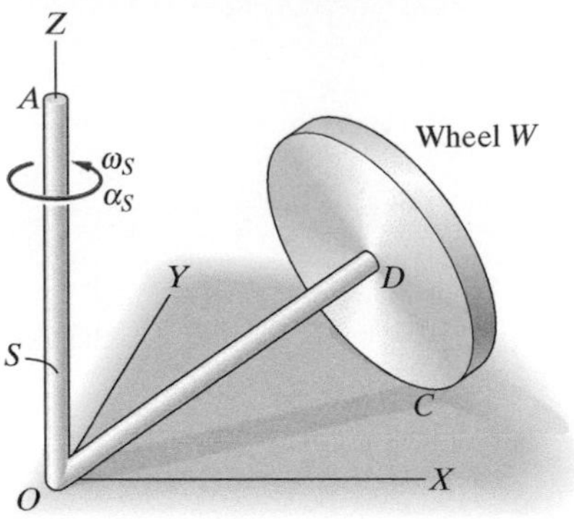

Figure 8.14 Rolling wheel on shaft

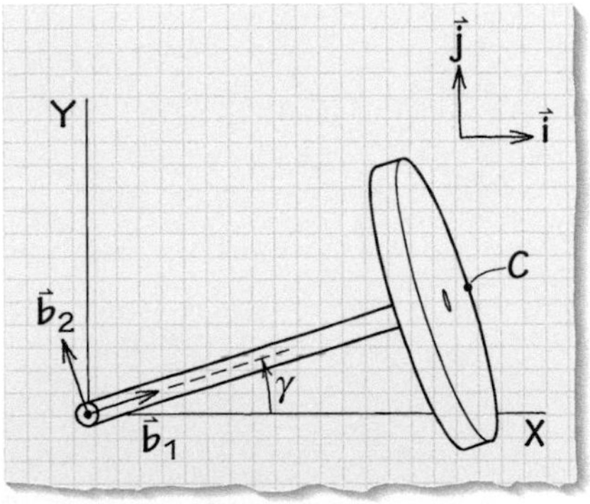

Figure 8.15 Rolling wheel on shaft—top view

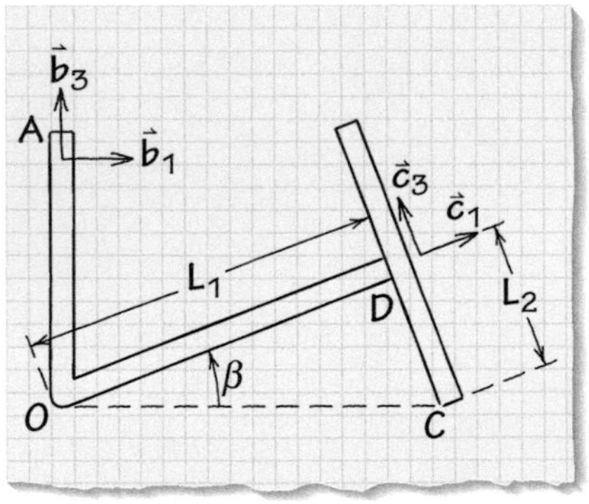

Figure 8.16 Rolling wheel on shaft—side view

Figure 8.14 shows the system under consideration. What we have is a bent shaft S that rotates about a vertical Z axis with angular velocity $\omega_S \boldsymbol{k}$ and angular acceleration $\alpha_S \boldsymbol{k}$. The arm $\overline{OD}$ has a length L_1. A circular wheel W with radius L_2 is attached at right angles to the end of arm $\overline{OD}$ and is free to rotate about it. The wheel rolls without slip and contacts the ground at C. Find the angular velocity and angular acceleration of the wheel.

Goal Use a no-slip rolling constraint to find the relationship between the rotation rates of the shaft and wheel.

Draw The key to simplifying our lives in this example is to introduce a variety of unit vectors that let us relate the wheel to the shaft to the ground. **Figure 8.15** and **Figure 8.16** show the three relevant sets. The $\boldsymbol{i}, \boldsymbol{j}, \boldsymbol{k}$ frame is fixed to the ground. The $\boldsymbol{b}_1, \boldsymbol{b}_2, \boldsymbol{b}_3$ unit vectors are attached to the bent shaft (with both $\boldsymbol{k}$ and $\boldsymbol{b}_3$ aligned with the vertical section $\overline{OA}$). Finally, $\boldsymbol{c}_1, \boldsymbol{c}_2, \boldsymbol{c}_3$, are also attached to the bent shaft, with $\boldsymbol{c}_1$ aligned with the bent section $\overline{OD}$. The two coordinate transformation matrices are

	$\boldsymbol{i}$	$\boldsymbol{j}$	$\boldsymbol{k}$
$\boldsymbol{b}_1$	$\cos\gamma$	$\sin\gamma$	0
$\boldsymbol{b}_2$	$-\sin\gamma$	$\cos\gamma$	0
$\boldsymbol{b}_3$	0	0	1

and

	$\boldsymbol{b}_1$	$\boldsymbol{b}_2$	$\boldsymbol{b}_3$
$\boldsymbol{c}_1$	$\cos\beta$	0	$\sin\beta$
$\boldsymbol{c}_2$	0	1.0	0
$\boldsymbol{c}_3$	$-\sin\beta$	0	$\cos\beta$

where β is the angle between $\overline{OD}$ and the X–Y plane.

Assume The constraints we will use will be the no-slip condition and the rotation about $\overline{OA}$. Because the bent shaft rotates about $\overline{OA}$, we know that the velocity at O is zero. And because we have a no-slip condition, we know that the velocity of the contact point C is also zero. That's useful because we're going to determine the velocity of D by going from O to D and then from C to D. Because we have to get the same velocity either way, we will be able to equate the expressions, find the wheel's rotational speed, and from there find the angular acceleration.

Solve Let's first determine D's linear velocity by starting at O. Because O is stationary, we have

$$
\begin{aligned}
\boldsymbol{v}_D &= \boldsymbol{\omega}_S \times \boldsymbol{r}_{D/O} \\
&= \omega_S \boldsymbol{b}_3 \times L_1(\cos\beta \boldsymbol{b}_1 + \sin\beta \boldsymbol{b}_3) \qquad (8.13) \\
&= \omega_S \cos\beta L_1 \boldsymbol{b}_2
\end{aligned}
$$

Now let's find D's linear velocity by starting at C. To do this, we will need an expression for the wheel's total angular velocity. As suggested earlier in the chapter, all we have to do is add up all the individual angular velocities. The wheel is attached to the bent shaft and so has angular velocity $\omega_S \boldsymbol{b}_3$. In addition, it spins about the end of arm $\overline{OD}$ as it rolls and thus has the additional component $\dot{\phi}\boldsymbol{c}_1$, where $\dot{\phi}$ is the wheel's (currently unknown) angular speed about the shaft. Thus we have

$$
\boldsymbol{\omega}_W = \dot{\phi}\boldsymbol{c}_1 + \omega_S \boldsymbol{b}_3
$$

Because C has zero velocity (for no-slip conditions), we can say

$$\begin{aligned} \boldsymbol{v}_D &= \boldsymbol{v}_C + \boldsymbol{\omega}_W \times \boldsymbol{r}_{D/C} = \boldsymbol{\omega}_W \times \boldsymbol{r}_{D/C} \\ &= (\dot{\phi}\boldsymbol{c}_1 + \omega_S \boldsymbol{b}_3) \times L_2 \boldsymbol{c}_3 \qquad (8.14) \\ &= [(\dot{\phi} + \omega_S \sin\beta)\boldsymbol{c}_1 + \omega_S \cos\beta \boldsymbol{c}_3] \times L_2 \boldsymbol{c}_3 \\ &= -(\dot{\phi} + \omega_S \sin\beta) L_2 \boldsymbol{c}_2 \end{aligned}$$

We know that (8.13) and (8.14) have to be equal, and because $\boldsymbol{b}_2 = \boldsymbol{c}_2$, we can equate the two equations and obtain

$$\begin{aligned} \omega_S \cos\beta L_1 &= -(\dot{\phi} + \omega_S \sin\beta) L_2 \\ \dot{\phi} &= -\frac{\omega_S (\cos\beta L_1 + \sin\beta L_2)}{L_2} \end{aligned} \qquad (8.15)$$

Knowing $\dot{\phi}$ allows us to solve for the wheel's angular velocity:

$$\boldsymbol{\omega}_W = \dot{\phi}\boldsymbol{c}_1 + \omega_S \boldsymbol{b}_3$$

$$\boldsymbol{\omega}_W = \omega_S \left[\sin\beta - \frac{(\cos\beta L_1 + \sin\beta L_2)}{L_2}\right] \boldsymbol{c}_1 + \omega_S \cos\beta \boldsymbol{c}_3 \qquad (8.16)$$

To find the wheel's angular acceleration, we need only differentiate (8.16) with respect to time:

$$\begin{aligned} \boldsymbol{\alpha}_W &= \alpha_S \left[\sin\beta - \frac{(\cos\beta L_1 + \sin\beta L_2)}{L_2}\right] \boldsymbol{c}_1 \\ &\quad + \omega_S \left[\sin\beta - \frac{(\cos\beta L_1 + \sin\beta L_2)}{L_2}\right] \dot{\boldsymbol{c}}_1 \qquad (8.17) \\ &\quad + \alpha_S \cos\beta \boldsymbol{c}_3 + \omega_S \cos\beta \dot{\boldsymbol{c}}_3 \end{aligned}$$

The needed time derivatives of the unit vectors are

$$\begin{aligned} \dot{\boldsymbol{c}}_1 &= \omega_S \boldsymbol{b}_3 \times \boldsymbol{c}_1 = \omega_S \cos\beta \boldsymbol{c}_2 \\ \dot{\boldsymbol{c}}_3 &= \omega_S \boldsymbol{b}_3 \times \boldsymbol{c}_3 = -\omega_S \sin\beta \boldsymbol{c}_2 \end{aligned}$$

and, using them, we will obtain

$$\begin{aligned} \boldsymbol{\alpha}_W &= \alpha_S \left[\sin\beta - \frac{(\cos\beta L_1 + \sin\beta L_2)}{L_2}\right] \boldsymbol{c}_1 \\ &\quad + \omega_S^2 \left[\cos\beta \left(\sin\beta - \frac{(\cos\beta L_1 + \sin\beta L_2)}{L_2}\right) - \cos\beta \sin\beta\right] \boldsymbol{c}_2 + \alpha_S \cos\beta \boldsymbol{c}_3 \end{aligned}$$

EXAMPLE 8.5 VELOCITY AND ACCELERATION OF A ROBOTIC MANIPULATOR (Theory on page 497)

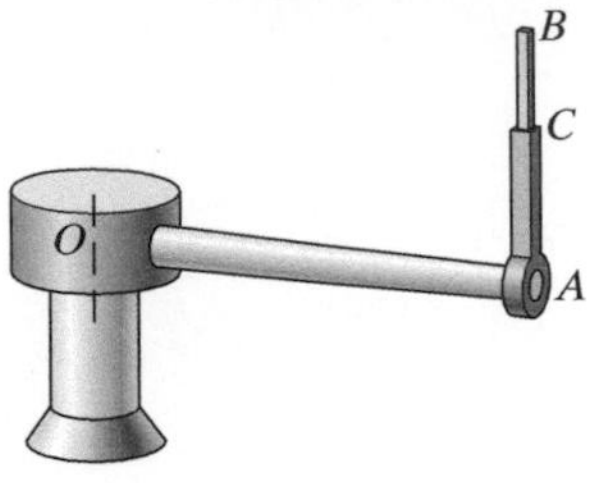

Figure 8.17 Robotic arm

In this example we will look at the velocity and acceleration of a robotic manipulator. **Figure 8.17** shows the system. Arm $\overline{OA}$ rotates in a horizontal plane, its angle of rotation given by θ. Arm $\overline{AB}$ is mounted at right angles to $\overline{OA}$ and is able to rotate about $\overline{OA}$, the angle of rotation being given by ϕ. When $\theta = \phi = 0$ the arm $\overline{OA}$ lies along the x axis and arm $\overline{AB}$ is vertical. Note that arm $\overline{AB}$ consists of two parts: a fixed link $\overline{AC}$ and an inner link that moves in and out of $\overline{AC}$, making $\overline{AB}$ what we call an **extensible** arm. Find general expressions for the linear velocity and linear acceleration of tip B in terms of the system's rotational rates and the rates of extension of B with respect to $\overline{AC}$.

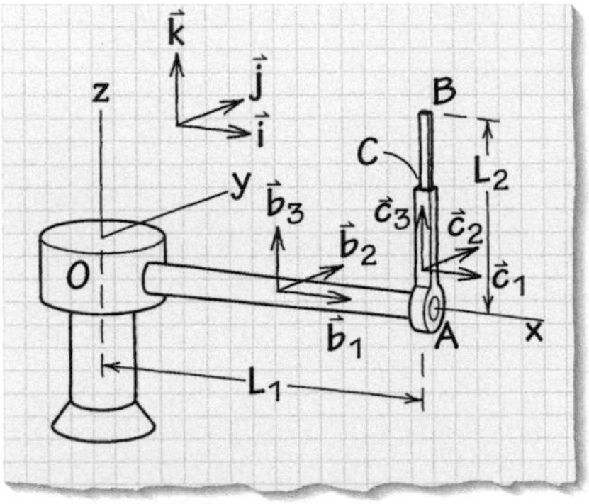

Figure 8.18 Unit vectors for robotic arm

Goal Find the linear velocity and linear acceleration of the robotic arm's tip B.

Draw **Figure 8.18** shows the unit vectors and dimensions we will need. The X, Y, Z axes are fixed to the ground, as are the unit vectors $\boldsymbol{i}, \boldsymbol{j}, \boldsymbol{k}$. $\boldsymbol{b}_1$, $\boldsymbol{b}_2$, $\boldsymbol{b}_3$ are fixed to $\overline{OA}$ and therefore rotate with it. $\boldsymbol{c}_1$, $\boldsymbol{c}_2$, $\boldsymbol{c}_3$ are fixed to the link $\overline{AC}$. For the instant shown, all the unit-vector sets align. The relationships between them are given by

	$\boldsymbol{i}$	$\boldsymbol{j}$	$\boldsymbol{k}$
$\boldsymbol{b}_1$	$\cos\theta$	$\sin\theta$	0
$\boldsymbol{b}_2$	$-\sin\theta$	$\cos\theta$	0
$\boldsymbol{b}_3$	0	0	1

and

	$\boldsymbol{b}_1$	$\boldsymbol{b}_2$	$\boldsymbol{b}_3$
$\boldsymbol{c}_1$	1	0	0
$\boldsymbol{c}_2$	0	$\cos\phi$	$\sin\phi$
$\boldsymbol{c}_3$	0	$-\sin\phi$	$\cos\phi$

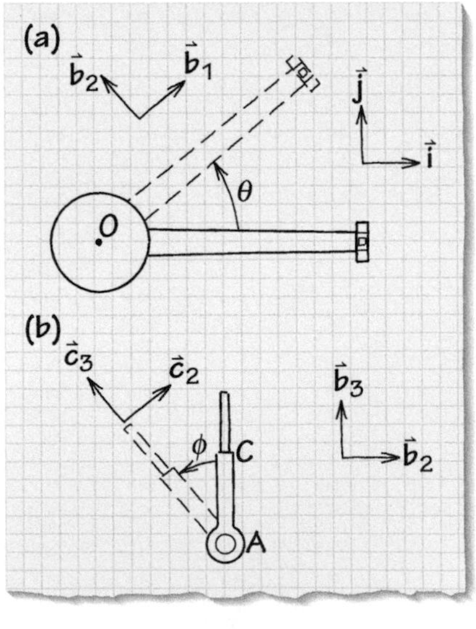

Figure 8.19 Individual rotations of robotic arm

The individual rotations are shown in **Figure 8.19**. **Figure 8.19*a*** shows how the primary arm $\overline{OA}$ rotates through an angle θ in the horizontal plane. Next, the secondary arm $\overline{AB}$ rotates with respect to the primary arm by an amount ϕ, as shown in **Figure 8.19*b***.

Assume We know that tip B moves in and out with respect to A because arm $\overline{AB}$ is extensible. We will view B as being a point that can move with respect to a rigid body, the rigid body being the fixed link $\overline{AC}$. Because $\boldsymbol{c}_1$, $\boldsymbol{c}_2$, $\boldsymbol{c}_3$ align with $\boldsymbol{b}_1$, $\boldsymbol{b}_2$, $\boldsymbol{b}_3$ at the instant we're analyzing the system, I will interchange the two at will to make the mathematics simpler. Keep in mind that this is okay only because the two sets are aligned; if they weren't, then we would have a harder time.

Solve The most straightforward way to approach this problem is to make our way out from the stationary point O, jumping first to A and then to B.

The linear velocity of A is

$$\begin{aligned} \boldsymbol{v}_A &= \boldsymbol{v}_O + \boldsymbol{\omega}_{\overline{OA}} \times \boldsymbol{r}_{A/O} = \dot{\theta}\boldsymbol{b}_3 \times L_1\boldsymbol{b}_1 \\ &= L_1\dot{\theta}\boldsymbol{b}_2 \end{aligned}$$

and the linear acceleration is

$$\begin{aligned}\boldsymbol{a}_A &= \boldsymbol{a}_O + \boldsymbol{\omega}_{\overline{OA}} \times \left(\boldsymbol{\omega}_{\overline{OA}} \times \boldsymbol{r}_{A/O}\right) + \boldsymbol{\alpha}_{\overline{OA}} \times \boldsymbol{r}_{A/O} \\ &= \dot{\theta}\boldsymbol{b}_3 \times (\dot{\theta}\boldsymbol{b}_3 \times L_1\boldsymbol{b}_1) + \ddot{\theta}\boldsymbol{b}_3 \times L_1\boldsymbol{b}_1 \\ &= -L_1\dot{\theta}^2\boldsymbol{b}_1 + L_1\ddot{\theta}\boldsymbol{b}_2\end{aligned}$$

That's halfway. Now we've got to get to B. Because B is moving with respect to $\overline{AC}$, we will need to use (8.11) and (8.12). And before that, we will need to find $\boldsymbol{\omega}_{\overline{AC}}$ and $\boldsymbol{\alpha}_{\overline{AC}}$. For $\boldsymbol{\omega}_{\overline{AC}}$ we just add the two given angular velocities:

$$\boldsymbol{\omega}_{\overline{AC}} = \dot{\phi}\boldsymbol{b}_1 + \dot{\theta}\boldsymbol{b}_3$$

The angular velocity $\boldsymbol{\alpha}_{\overline{AC}}$ is more involved. We apply

$$\left.\frac{d}{dt}\right|_N \boldsymbol{\omega}_{\overline{AC}} = \left.\frac{d}{dt}\right|_S \boldsymbol{\omega}_{\overline{AC}} + \boldsymbol{\omega}_{\overline{OA}} \times \boldsymbol{\omega}_{\overline{AC}}$$

where, as usual, N represents an inertial reference frame and S a rotating frame. The rotating frame of reference rotates with the same angular speed as arm $\overline{OA}$, and the vector of interest is the angular velocity vector $\boldsymbol{\omega}_{\overline{OA}}$. Evaluating the individual terms gives us

$$\left.\frac{d}{dt}\right|_S \boldsymbol{\omega}_{\overline{AC}} = \ddot{\phi}\boldsymbol{b}_1 + \ddot{\theta}\boldsymbol{b}_3 \qquad \text{and} \qquad \boldsymbol{\omega}_{\overline{OA}} \times \boldsymbol{\omega}_{\overline{AC}} = \dot{\theta}\boldsymbol{b}_3 \times (\dot{\phi}\boldsymbol{b}_1 + \dot{\theta}\boldsymbol{b}_3) = \dot{\theta}\dot{\phi}\boldsymbol{b}_2$$

which leads to $\boldsymbol{\alpha}_{\overline{AC}} = \ddot{\phi}\boldsymbol{b}_1 + \dot{\theta}\dot{\phi}\boldsymbol{b}_2 + \ddot{\theta}\boldsymbol{b}_3$.

Now that we have the angular rates under control, we can obtain the desired linear velocity and acceleration.

$$\begin{aligned}\boldsymbol{v}_B &= \boldsymbol{v}_A + \boldsymbol{\omega}_{\overline{AC}} \times \boldsymbol{r}_{B/A} + \boldsymbol{v}_{rel} \\ &= L_1\dot{\theta}\boldsymbol{b}_2 + (\dot{\phi}\boldsymbol{b}_1 + \dot{\theta}\boldsymbol{b}_3) \times L_2\boldsymbol{b}_3 + \dot{L}_2\boldsymbol{b}_3 \\ &= (L_1\dot{\theta} - L_2\dot{\phi})\boldsymbol{b}_2 + \dot{L}_2\boldsymbol{b}_3\end{aligned}$$

$$\begin{aligned}\boldsymbol{a}_B &= \boldsymbol{a}_A + \boldsymbol{\omega}_{\overline{AC}} \times \left(\boldsymbol{\omega}_{\overline{AC}} \times \boldsymbol{r}_{B/A}\right) + \boldsymbol{\alpha}_{\overline{AC}} \times \boldsymbol{r}_{B/A} + 2\boldsymbol{\omega}_{\overline{AC}} \times \boldsymbol{v}_{rel} + \boldsymbol{a}_{rel} \\ &= -L_1\dot{\theta}^2\boldsymbol{b}_1 + L_1\ddot{\theta}\boldsymbol{b}_2 + (\dot{\phi}\boldsymbol{b}_1 + \dot{\theta}\boldsymbol{b}_3) \times [(\dot{\phi}\boldsymbol{b}_1 + \dot{\theta}\boldsymbol{b}_3) \times L_2\boldsymbol{b}_3] \\ &\quad + (\ddot{\phi}\boldsymbol{b}_1 + \dot{\theta}\dot{\phi}\boldsymbol{b}_2 + \ddot{\theta}\boldsymbol{b}_3) \times L_2\boldsymbol{b}_3 + 2(\dot{\phi}\boldsymbol{b}_1 + \dot{\theta}\boldsymbol{b}_3) \times \dot{L}_2\boldsymbol{b}_3 + \ddot{L}_2\boldsymbol{b}_3 \\ &= (-L_1\dot{\theta}^2 + 2L_2\dot{\theta}\dot{\phi})\boldsymbol{b}_1 + (L_1\ddot{\theta} - L_2\ddot{\phi} - 2\dot{L}_2\dot{\phi})\boldsymbol{b}_2 + (-L_2\dot{\phi}^2 + \ddot{L}_2)\boldsymbol{b}_3\end{aligned}$$

Check We can do some quick checks to assure ourselves that the results are at least plausible. The first term of the acceleration, $-L_1\dot{\theta}^2$, is the centripetal acceleration generated at A as it rotates about O. The radial distance is L_1, and the angular speed is $\dot{\theta}$; hence the centripetal acceleration is $-L_1\dot{\theta}^2$. From **Figure 8.18** you can see that it should point in the negative $\boldsymbol{b}_1$ direction, in agreement with the mathematics. The second term, $2L_2\dot{\theta}\dot{\phi}$, looks like a coriolis term, judging from the factor of 2. If B is spinning about A at $\dot{\phi}$ and is a distance L_2 away from A, then its relative velocity is $-L_2\dot{\phi}\boldsymbol{b}_2$. Arm $\overline{OA}$ is spinning with speed $\dot{\theta}$, and so the coriolis term $2\boldsymbol{\omega}_{\overline{OA}} \times \boldsymbol{v}_{rel}$ gives us $2L_2\dot{\theta}\dot{\phi}\boldsymbol{b}_1$. The $L_1\ddot{\theta}$ term is the angular acceleration of A, and the $-L_2\ddot{\phi}$ term is the angular acceleration of B with respect to A. The $-2\dot{L}_2\dot{\phi}$ term is the coriolis acceleration that accounts for tip B's extension out of or retraction into $\overline{AC}$, while arm $\overline{AB}$ rotates about A. The $-L_2\dot{\phi}^2$ term is the centrifugal acceleration of B about A, and the last term, $\ddot{L}_2$, is just a_{rel}, the acceleration of B with respect to the rigid body $\overline{AC}$.

EXERCISES 8.4

8.4.1. **[Level 1]** A rotating disk D turns with angular speed ω_2 about an arm extending from the circular shaft $\overline{AB}$, which is itself rotating with an angular speed ω_1. What is the disk's angular velocity?

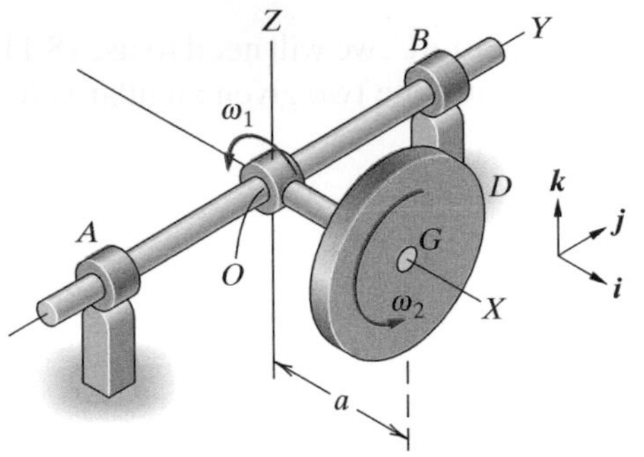

E8.4.1

8.4.2. **[Level 1]** The caster of a shopping cart is shown. The wheel frame F is rotating about the z axis with an angular velocity $\omega_1 \boldsymbol{b}_3$, and the wheel is rotating with respect to F at an angular velocity of $-\omega_2 \boldsymbol{b}_1$. Both ω_1 and ω_2 are constant. What is the wheel's angular acceleration? The $\boldsymbol{b}_i$ unit vectors are fixed with respect to the frame F.

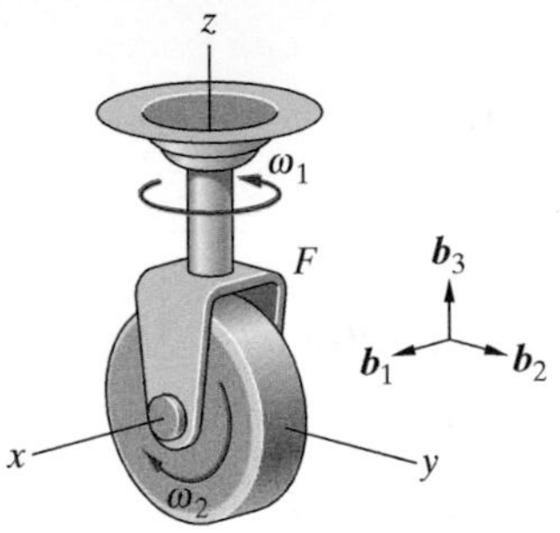

E8.4.2

8.4.3. **[Level 2]** Consider the illustrated robotic positioning device, where a motor at the fixed joint O controls the angular position θ of the inner arm $\overline{OA}$ about the vertical axis. Arm $\overline{OA}$ is extensible, and its length is denoted by L. The angular position ϕ of the outer arm $\overline{AB}$, which is attached to the end of arm $\overline{OA}$, is regulated by a motor at the joint A. Suppose that θ and ϕ are defined such that $\theta = \phi = 0$ when the robotic positioning device is in the depicted configuration. Find the velocity $\boldsymbol{v}_B$ and acceleration $\boldsymbol{a}_B$ of the robot's end effector at B if $\dot{L}$ and $\dot{\phi}$ are constant when $\theta = \phi = 90°$. Express your answer with respect to the rotating frame $\{\boldsymbol{b}_1, \boldsymbol{b}_2, \boldsymbol{b}_3\}$ fixed to the inner arm $\overline{OA}$.

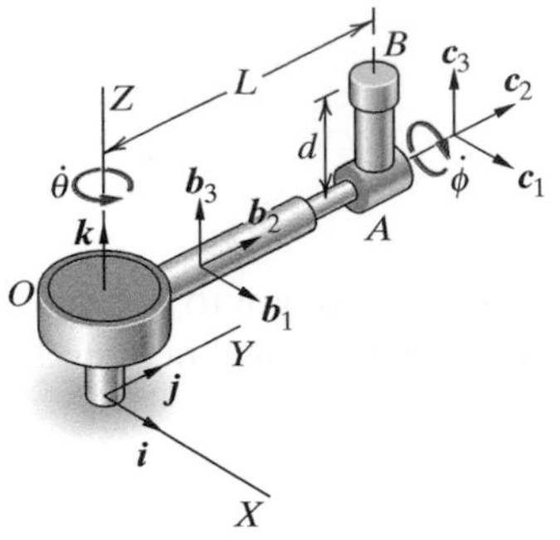

E8.4.3

8.4.4. **[Level 2]** The illustrated inner arm $\overline{OA}$ is rotating about the Z axis with a constant angular speed $\dot{\psi}$, and the outer arm $\overline{AB}$ is rotating with respect to the inner arm with a constant angular speed $\dot{\theta}$. What are the angular velocity of the outer arm $\overline{AB}$? What is its angular acceleration?

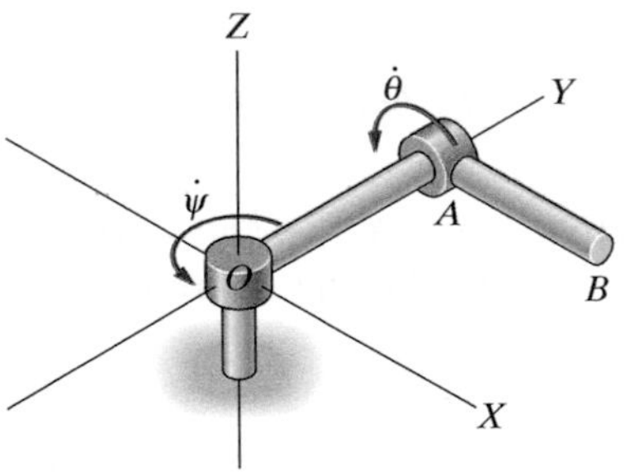

E8.4.4

8.4.5. **[Level 2]** Illustrated here is the device that supports the rotating plate of a microwave oven and allows it to rotate. It consists of a three-armed piece supported by three wheels of radius r. Assume that the device is rotating about O at a constant angular velocity of $2\boldsymbol{j}$ rad/s. What are the angular velocity and angular acceleration of the wheel A? $h = 0.1$ m and $r = 0.004$ m.

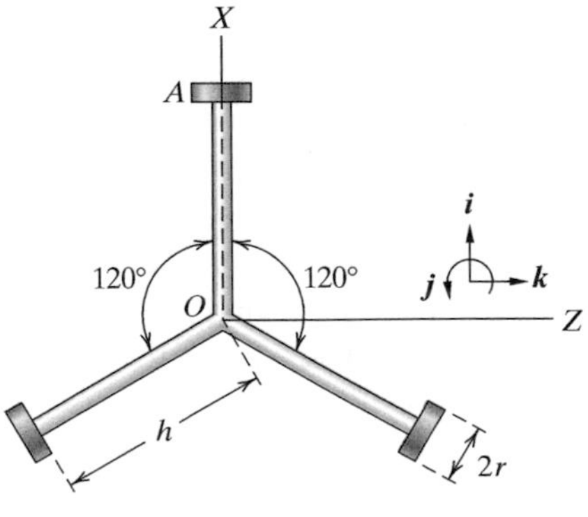

Top view

E8.4.5

8.4.6. **[Level 2]** The illustration shows a cyclist pedaling along a circular path, with (*b*) and (*c*) showing two views of the rear wheel (a side view and rear view, respectively). The wheel has a radius of 33 cm, and the cyclist is traveling at a constant 40 km/h. The cyclist's inclination angle ϕ is 30°. It takes him 6 s to complete one full trip around the circle. What is the wheel's angular acceleration?

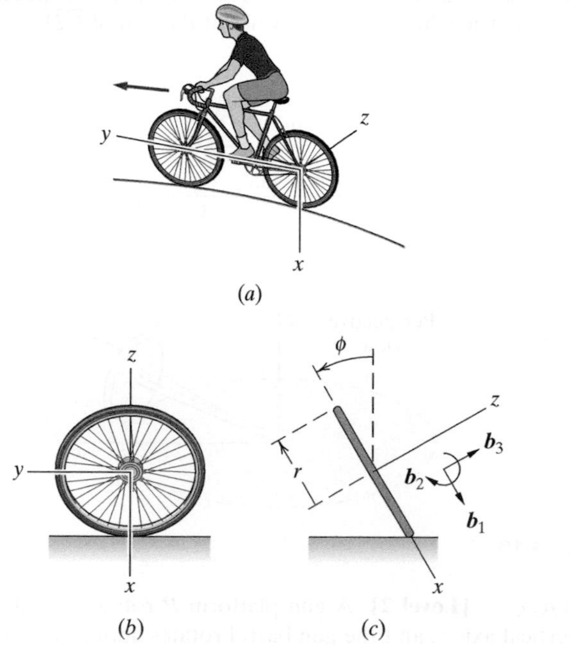

E8.4.6

8.4.7. **[Level 2]** The illustrated mechanism consists of two arms, where the inner arm $\overline{OA}$ is free to rotate about the vertical axis passing through the fixed joint O with a constant angular speed of $\dot{\theta}$. The outer arm $\overline{AB}$ simultaneously spins about joint A at a constant angular speed $\dot{\phi}$. If θ and ϕ are defined such that $\theta = \phi = 0$ in the depicted configuration, what are the velocity $\boldsymbol{v}_B$ and acceleration $\boldsymbol{a}_B$ of tip B at the illustrated instant? Express your answer with respect to the rotating frame $\{\boldsymbol{b}_1, \boldsymbol{b}_2, \boldsymbol{b}_3\}$ fixed to the inner arm $\overline{OA}$.

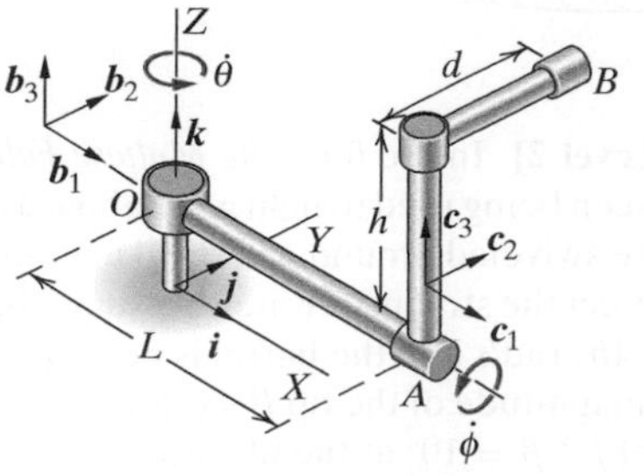

E8.4.7

8.4.8. **[Level 2]** Consider the illustrated biomechanical model of a person's left arm. The person rotates her upper arm $\overline{AB}$ with constant angular velocity $\omega_1\boldsymbol{b}_1$, so that it goes from pointing straight down to pointing directly ahead in 1 s. The $\boldsymbol{b}_i$ set of unit vectors is fixed to the upper arm $\overline{AB}$. At the same time that she's raising her upper arm, she rotates her forearm $\overline{BC}$ with angular velocity $\omega_2\boldsymbol{b}_2$ (ω_2 constant), so that there's a 90° bend between her upper arm and forearm at the end of 1 s. What is the forearm's angular acceleration during the motion?

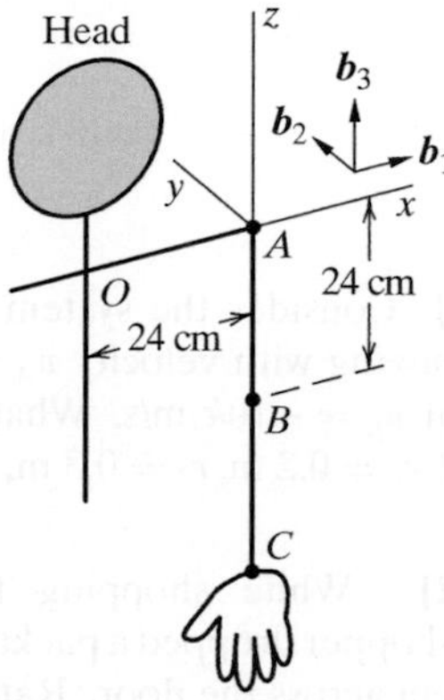

E8.4.8

8.4.9. **[Level 2]** Consider the system from **Exercise 8.4.8.** The difference is that now, instead of rotating at a constant speed ω_2, her forearm starts with zero angular speed and rotates with a constant angular acceleration relative to the upper arm such that it ends at the same place at the same time as in **Exercise 8.4.8**—that is, oriented at 90° to the upper arm. What is the forearm's angular acceleration for $0 \le t \le 1$ s?

8.4.10. **[Level 2]** Consider the system from **Exercise 8.4.8.** This time, the forearm $\overline{BC}$ rotates relative to the upper arm with constant angular velocity $\omega_2\boldsymbol{b}_3$. Physically, the upper arm and forearm ($\overline{AB}$ and $\overline{BC}$) remain in-line, but the forearm rotates along its longitudinal axis with respect to the upper arm. At $t = 1$ s the forearm has rotated by 45° with respect to its original orientation. (The thumb, originally pointing to the right of the person, ends up pointing upward, 45° beyond horizontal.) What are the angular velocity and angular acceleration of the forearm for $0 \le t \le 1$ s? Note that this isn't precisely how the forearm actually works, but it's not too bad an approximation. Look at your own arm and you will see that you can rotate your forearm quite a bit. In your case, however, it's not due to a rotational joint but to two arm bones that can move relative to each other.

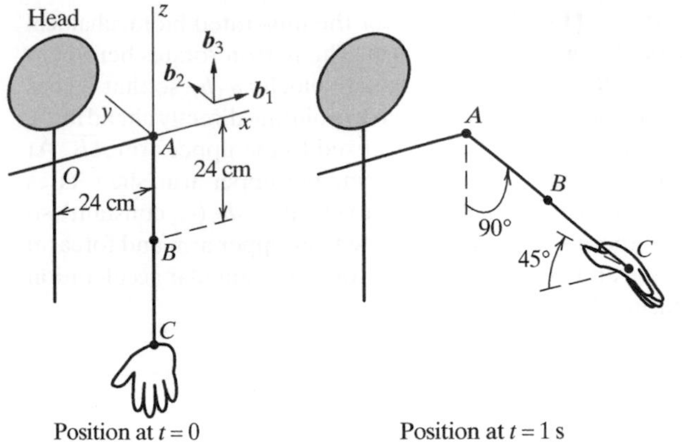

E8.4.10

8.4.11. **[Level 2]** Consider the system from **Exercise 8.4.23**. Point A is moving with velocity $\boldsymbol{v}_A = 10\boldsymbol{k}$ m/s, and point B is moving at $\boldsymbol{v}_B = -10\boldsymbol{k}$ m/s. What is the angular velocity of gear G_2? $r_1 = 0.2$ m, $r_2 = 0.3$ m, and $d = 0.5$ m.

8.4.12. **[Level 2]** While shopping for dessert, a dynamics-oriented shopper dropped a package of ice cream cones, scattering them across the floor. Rather than cleaning up the mess, the shopper decided to determine the maximum velocity of a point on the cone. Assume that the cone rolls without slip at a constant speed (completing a circle every 2 s). What is the maximum velocity seen by a point on the cone?

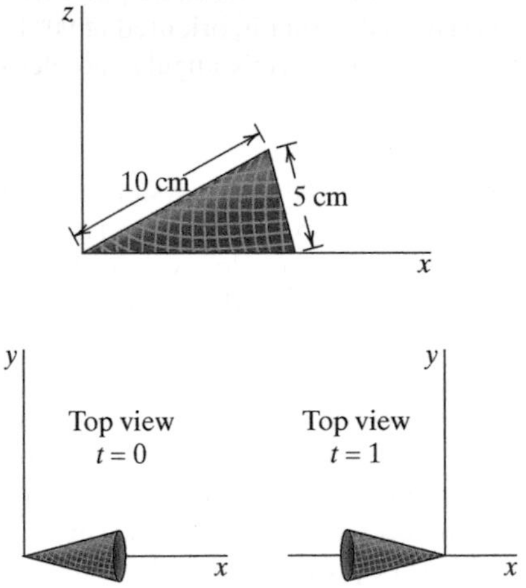

E8.4.12

8.4.13. **[Level 2]** Show that the angular velocity for the rotating wheel of Example 8.4 is always directed horizontally.

8.4.14. **[Level 2]** What is the acceleration of point D from Example 8.4? Express your answer in terms of ω_S, α_S, and $\boldsymbol{c}_1, \boldsymbol{c}_2$, and $\boldsymbol{c}_3$.

8.4.15. **[Level 2]** In Example 8.4 the angular acceleration has a component in the $\boldsymbol{b}_1$ direction and a component in the $\boldsymbol{b}_2$ direction. What physically is happening to the angular velocity vector that gives rise to both of these terms?

8.4.16. **[Level 2]** Find the acceleration of the contact point C for the illustrated rolling wheel. O is fixed and D revolves around the z axis with a constant angular speed of 3 rad/s. The wheel spins freely about the end of $\overline{OD}$.

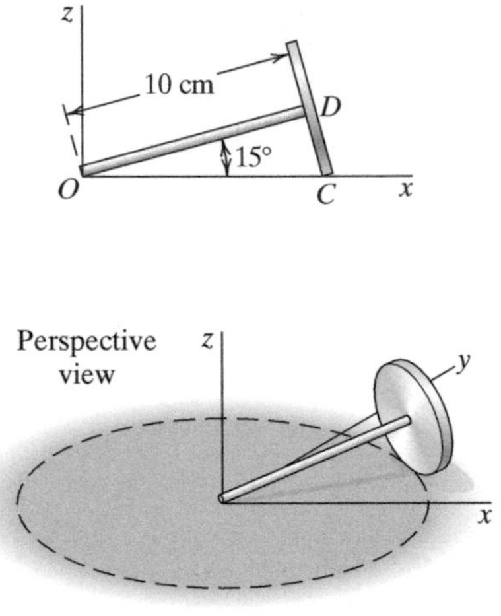

E8.4.16

8.4.17. **[Level 2]** A gun platform P rotates about the vertical axis z, and the gun barrel rotates about the x axis. The x, y, z frame rotates with the platform P and has associated unit vectors $\boldsymbol{b}_1, \boldsymbol{b}_2, \boldsymbol{b}_3$. The platform rotates at 2 rad/s, and its angular acceleration is 4 rad/s^2. The gun swivels up at a constant rate of 3 rad/s with respect to the platform. What are $\boldsymbol{v}_A$ and $\boldsymbol{a}_A$ at the illustrated position? The gun barrel lies in the y-z plane.

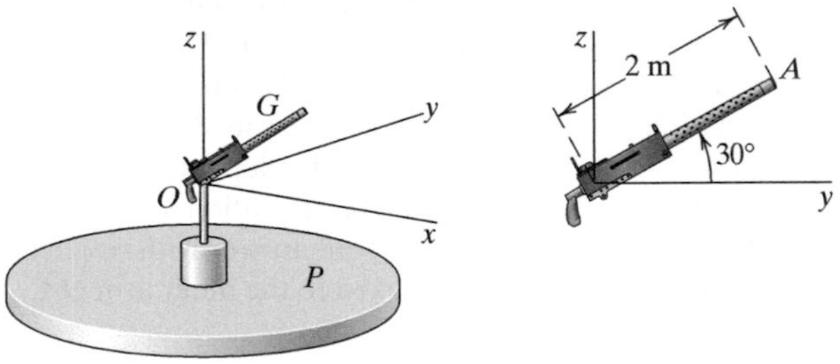

E8.4.17

8.4.18. **[Level 2]** In the film *The Maltese Falcon*, water cannons are seen being used to fight a fire aboard a ship. The cannon can be swiveled around a vertical axis and tilted up or down to direct the stream of water. Assume that the base is rotating at $4\boldsymbol{b}_3$ rad/s and the barrel is being pivoted up at $\dot{\beta}$ rad/s. The magnitude of the tip B's velocity is 6 m/s. What is the value of $\dot{\beta}$? $\beta = 10°$ at the illustrated instant?

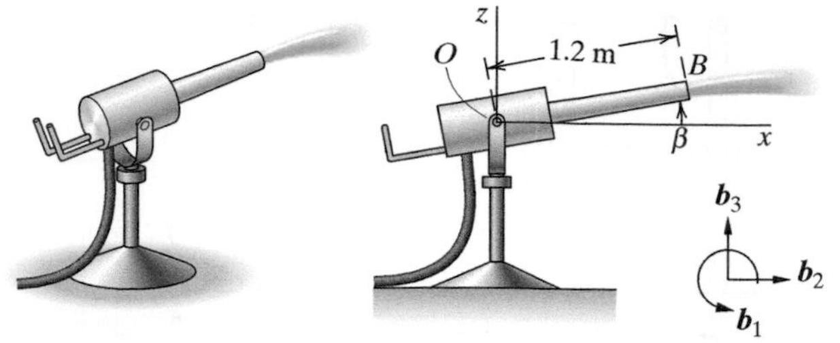

E8.4.18

8.4.19. **[Level 2]** A tethered model airplane is circling its attachment point O. The rate with which it circles is increasing at α_3 rad/s^2, and its current angular velocity is $\omega_3 \boldsymbol{b}_3$ rad/s. The propeller is turning with respect to the model at $\omega_2 \boldsymbol{b}_2$ rad/s, and its angular acceleration with respect to the model is $\alpha_2 \boldsymbol{b}_2$ rad/s^2. Determine the total angular acceleration of the propeller in two ways:

a. Express the propeller's angular velocity $\boldsymbol{\omega}_p$ in terms of $\boldsymbol{b}_1, \boldsymbol{b}_2, \boldsymbol{b}_3$ and differentiate each component with respect to time.

b. Apply the formula for differentiating a vector in a rotating frame,

$$\left.\frac{d}{dt}\right|_N \boldsymbol{p} = \left.\frac{d}{dt}\right|_S \boldsymbol{p} + \boldsymbol{\omega} \times \boldsymbol{p}$$

using $\boldsymbol{\omega}_p$ as the vector and the $\boldsymbol{b}_1, \boldsymbol{b}_2, \boldsymbol{b}_3$ frame as the rotating body frame.

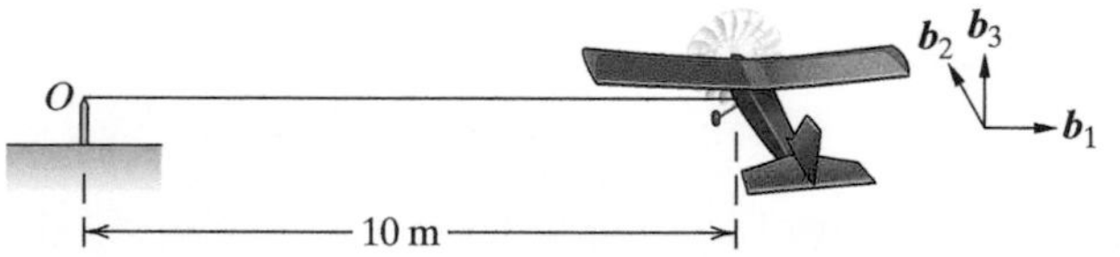

E8.4.19

8.4.20. **[Level 2]** Biomechanical studies often require the researcher to model the extremities of a person as interconnected joints and links. Consider the model in which a person turns and reaches up at the same time. The x, y, z axes are fixed to her torso. Assume that the person rotates her body about the z axis. A represents the shoulder joint, and B represents the elbow. $\overline{OA}$ rotates about the z axis at $\boldsymbol{\omega}_1 = 0.8\boldsymbol{b}_3$ rad/s and $\boldsymbol{\alpha}_1 = 2\boldsymbol{b}_3$ rad/s^2. $\overline{OA}$ remains parallel to the floor, while the shoulder-elbow link $\overline{AB}$ rotates around the x axis at $\boldsymbol{\omega}_2 = -1.6\boldsymbol{b}_1$ rad/s and $\boldsymbol{\alpha}_2 = -0.5\boldsymbol{b}_1$ rad/s^2. What are the velocity and acceleration of the elbow B at this instant? ($\overline{AB}$ is oriented vertically.)

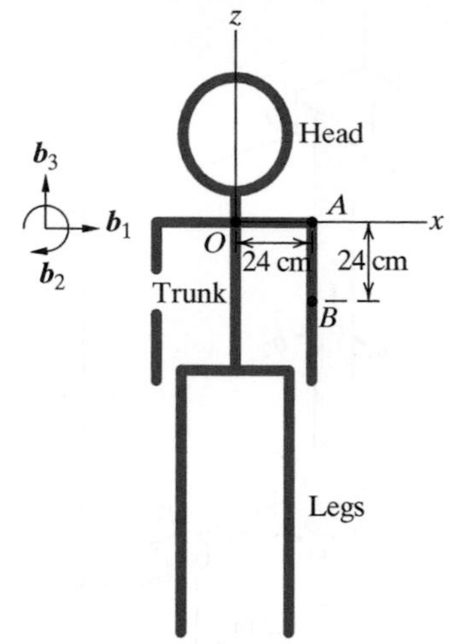

E8.4.20

8.4.21. **[Level 2]** A common type of fan is mounted on a rotating shaft so that the breeze can cover the whole room. The fan is oriented upward at an angle β, and its side-to-side motion is given by $\theta = \theta_0 \sin(\omega t)$. What is the velocity of a point T on the illustrated blade tip as a function of time?

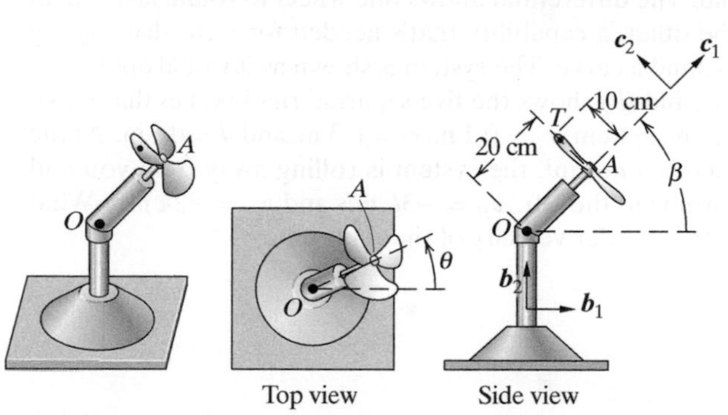

E8.4.21

8.4.22. **[Level 2]** Link $\overline{OB}$ rotates about O with angular velocity $\dot{\theta}\boldsymbol{b}_1$, link $\overline{BC}$ rotates about B with angular velocity $\dot{\phi}\boldsymbol{b}_3$, and a movable collar slides along the link $\overline{BC}$ with speed $5\boldsymbol{c}_2$ m/s. Find $\boldsymbol{a}_D$. $\dot{\theta} = 4$ rad/s and $\dot{\phi} = 5$ rad/s.

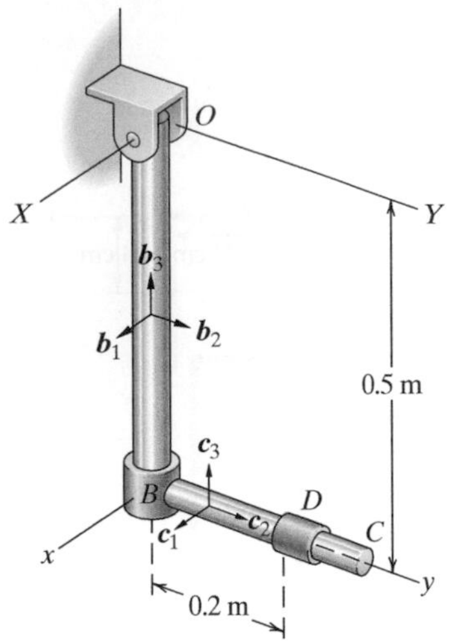

E8.4.22

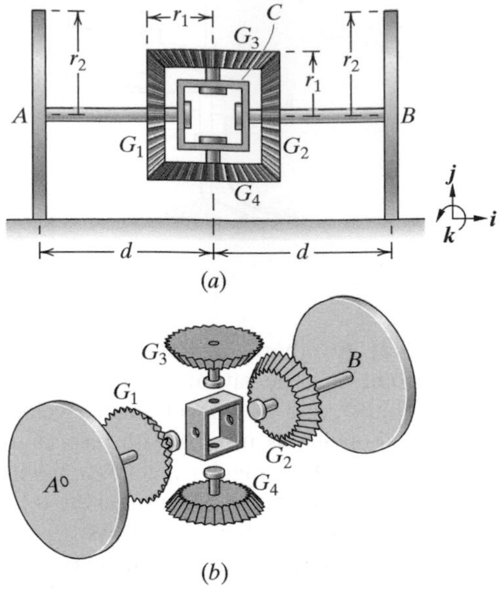

E8.4.23

8.4.23. **[Level 3]** The figure shows two wheels and four gears that are connected so as to form a simple differential. The differential allows one wheel to rotate faster than the other, a capability that's needed for a car that's going around a curve. The system is shown as it would operate in (*a*), and (*b*) shows the five separate rigid bodies that make up the system. $r_1 = 0.1$ m, $r_2 = 0.3$ m, and $d = 0.5$ m. At the pictured instant, the system is rolling away from you and turning to the left. $\boldsymbol{v}_B = -3\boldsymbol{k}$ m/s and $\boldsymbol{v}_A = -2\boldsymbol{k}$ m/s. What is the angular velocity of the central box C?

8.4.24. **[Level 3]** Consider the system from **Exercise 8.4.23**. Both points A and B are moving with velocity $10\boldsymbol{k}$ m/s. What is the angular velocity of gear G_3? $r_1 = 0.2$ m, $r_2 = 0.3$ m, and $d = 0.5$ m.

8.4.25. **[Level 3]** Consider the system from **Exercise 8.4.23**. Point A is stationary, and point B is moving at $\boldsymbol{v}_B = -6\boldsymbol{k}$ m/s. What is the angular velocity of gear G_4? $r_1 = 0.2$ m, $r_2 = 0.3$ m, and $d = 0.5$ m.

8.5 MOMENTS AND PRODUCTS OF INERTIA FOR A THREE-DIMENSIONAL BODY

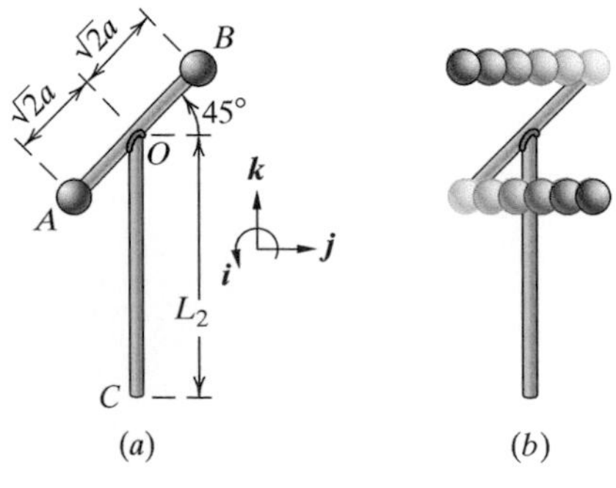

Figure 8.20 Space tool

Just as the angular velocity of a three-dimensional body is more complicated than the two-dimensional case, so too is the rotational inertia. When we were spinning rigid bodies in a plane, they could rotate only about one axis; therefore, the only rotational inertia we had to deal with was the rotational inertia about that axis. Here's a good way to see how things get complicated in three dimensions. Imagine you're an astronaut, floating in space and amusing yourself for a few minutes in observing Newtonian dynamics. One of your tools, a side view of which is shown in **Figure 8.20*a***, consists of a lightweight handle $\overline{CO}$ with a tilted rigid bar $\overline{AB}$ welded to it at O. At each end of the tilted bar is a solid sphere m. The spheres are so massive that we can neglect the mass of the handle and welded bar. What happens if you twirl it about $\overline{CO}$ and then release the tool? Extrapolating from what we've learned in our two-dimensional studies, you might expect the tool to rotate as shown in **Figure 8.20*b***, a situation in which both masses describe circular trajectories that lie in planes that are perpendicular to $\boldsymbol{k}$. This makes sense, but, interestingly, the reality is not so simple.

Take a look at **Figure 8.21*a***, which shows a side view of the tool as it's released, along with the velocity vectors of the spheres. You can verify from the picture that the two spheres have the same speed and are moving in opposite directions. What would such an initial motion imply in our old two-dimensional world? The answer is a simple rotation about O. And the angular momentum would be given by $-2mva\boldsymbol{j}$. That's right—$\boldsymbol{j}$. It's odd that you twirled it along the $\boldsymbol{k}$ direction, but the angular momentum has a term in the $\boldsymbol{j}$ direction. **Figure 8.21*b***, a top view, shows the situation that's more in line with our two-dimensional expectations. Here the angular momentum appears to be $2mva\boldsymbol{k}$. Because the tool has an angular momentum that has components in two directions, it makes sense that it will do more than simply rotate about the $\boldsymbol{k}$ axis and will, instead, tumble in a fairly complicated way.

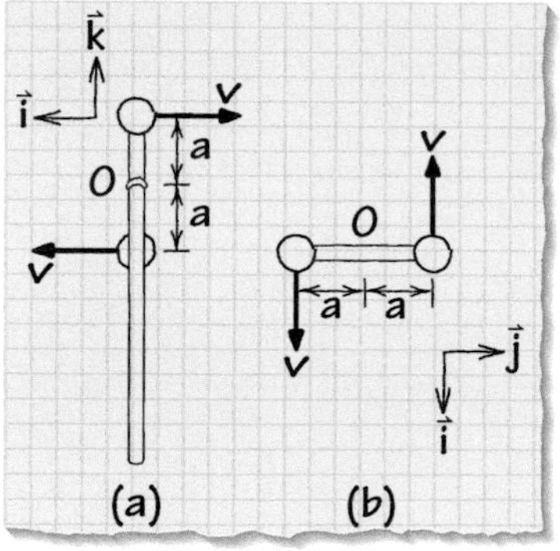

Figure 8.21 Spinning space tool

Given that the inertial characteristics are more complex in a three-dimensional body than in the two-dimensional case, let's try to characterize them. To begin, consider **Figure 8.22**. This shows an arbitrarily shaped rigid body, along with a set of body-fixed axes x, y, and z. The vector $\boldsymbol{r}$ points to a differential mass element dm. The **moment of inertia** I_{xx} is given by

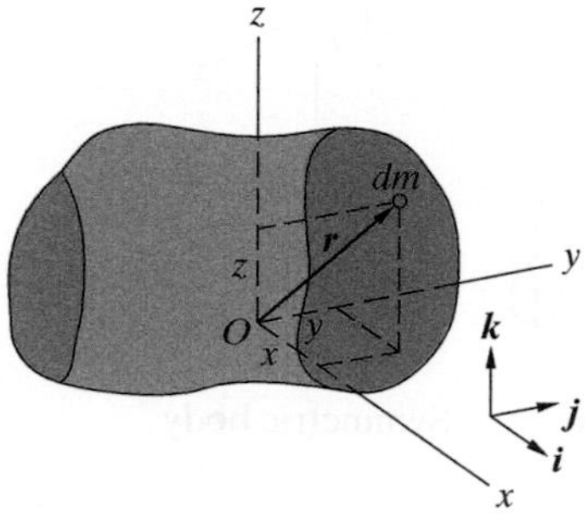

Figure 8.22 General three-dimensional body

$$I_{xx} = \int_{\text{Body}} (y^2 + z^2)\, dm \tag{8.18}$$

This moment of inertia is identical to what we would have found back in our two-dimensional work if we had been told that the body could rotate only about the x axis and if we were asked to find the corresponding mass moment of inertia about an axis passing through O.

As you might have guessed, there are two more moments of inertia:

$$I_{yy} = \int_{\text{Body}} (x^2 + z^2)\, dm \tag{8.19}$$

$$I_{zz} = \int_{\text{Body}} (x^2 + y^2)\, dm \tag{8.20}$$

These are the "expected from two-dimensional intuition" rotational inertias, the ones we would find simply by assuming that the body spins about a particular axis of rotation.

The new inertial quantity is the **product of inertia**, an inertia property that involves two axes simultaneously. The products of inertia are defined by

$$I_{xy} = \int_{\text{Body}} xy\, dm \tag{8.21}$$

$$I_{yz} = \int_{\text{Body}} yz\,dm \tag{8.22}$$

$$I_{zx} = \int_{\text{Body}} zx\,dm \tag{8.23}$$

As you can see from the definitions, $I_{xy} = I_{yx}, I_{xz} = I_{zx}, I_{yz} = I_{zy}$.

Moments of inertia are always nonzero for bodies that have finite dimensions and finite mass. Products of inertia, on the other hand, can be zero even if the mass and dimensions of the body are finite. For instance, consider the body shown in **Figure 8.23**. When you calculate I_{xz}, you will get zero because for each little bit of mass at (x, y, z), there will be an identical piece of mass at $(-x, y, z)$. During the integrations, all these (x, y, z) pairs will cancel each other out.

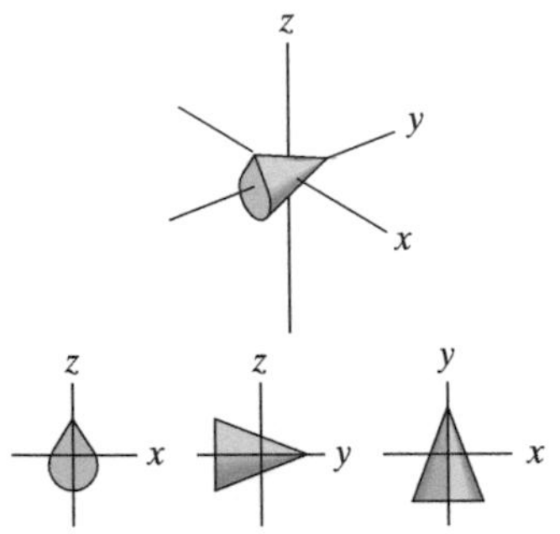

Figure 8.23 Symmetric body

This is a general finding. Whenever you have a plane of symmetry, the products of inertia that involve the coordinate axis normal to this plane are zero. In **Figure 8.23** the y–z plane is a plane of symmetry and the axis normal to this is the x axis. Thus I_{xy} and I_{xz} are zero. If you have two planes of symmetry, then all the products of inertia are zero. This condition also means that if you have a **body of revolution**, like that of **Figure 8.24**, then all the products of inertia will be zero. A body of revolution is one formed by spinning a curve about a fixed axis. All cross sections of the body normal to that axis will be circular. Bowling pins, AA batteries, CDs, and so forth are all bodies of revolution.

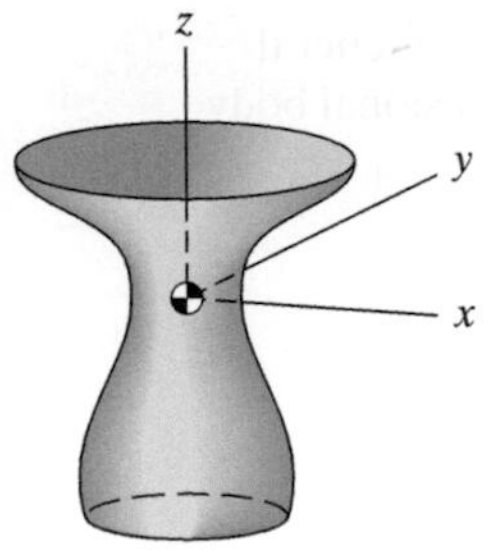

Figure 8.24 Body of revolution

8.6 PARALLEL AXIS EXPRESSIONS FOR INERTIAS

Remember the parallel axis theorem, equation (7.31)? It enabled us to calculate the rotational moment of inertia of a two-dimensional object about any point in the plane of the object as long as we knew the moment of inertia *about the object's center of mass* and the position of the new point with respect to the mass center. As you might expect, the result in the case of three-dimensional bodies is highly similar.

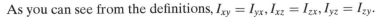

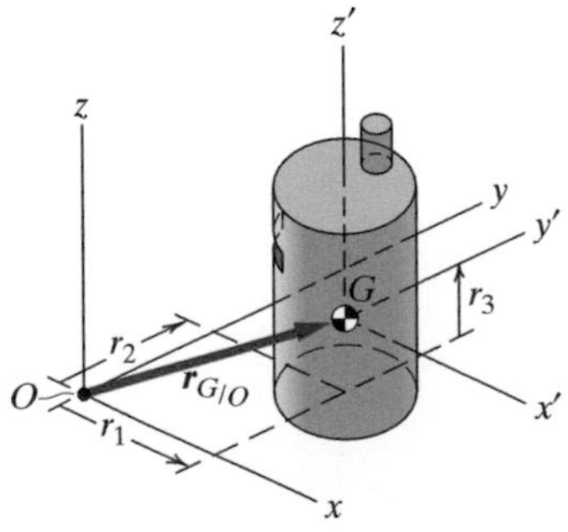

Figure 8.25 Solid body

Figure 8.25 shows a body of mass m, with a known center of mass G and known inertial properties $(\bar{I}_{x'x'}, \bar{I}_{y'y'}, \bar{I}_{z'z'}, \bar{I}_{x'y'}, \bar{I}_{y'z'}, \bar{I}_{z'x'})$ along three orthogonal axes (x', y', z'). What we want are the inertial properties as calculated about the axes x, y, z going through the point O. In an exact parallel with (7.31), the moments of inertia are given by

$$I_{xx} = \bar{I}_{x'x'} + m(r_2^2 + r_3^2) \tag{8.24}$$

$$I_{yy} = \bar{I}_{y'y'} + m(r_3^2 + r_1^2) \tag{8.25}$$

$$I_{zz} = \bar{I}_{z'z'} + m(r_1^2 + r_2^2) \tag{8.26}$$

and the products of inertia are given by

$$I_{xy} = \overline{I}_{x'y'} + mr_1r_2 \tag{8.27}$$

$$I_{yz} = \overline{I}_{y'z'} + mr_2r_3 \tag{8.28}$$

$$I_{zx} = \overline{I}_{z'x'} + mr_3r_1 \tag{8.29}$$

It's not too difficult to show this. **Figure 8.26** shows a projection onto the x–y plane of our body. I will calculate one moment of inertia and one product of inertia and leave the others to you to verify if you wish. The moment of inertia I_{zz}, from our two-dimensional work, is given by $I_{zz} = \overline{I}_{z'z'} + md^2$, where d is the distance from the mass center to the point O. From geometry you can see that $d^2 = r_1^2 + r_2^2$, giving us (8.26).

The product of inertia I_{xy} is given by

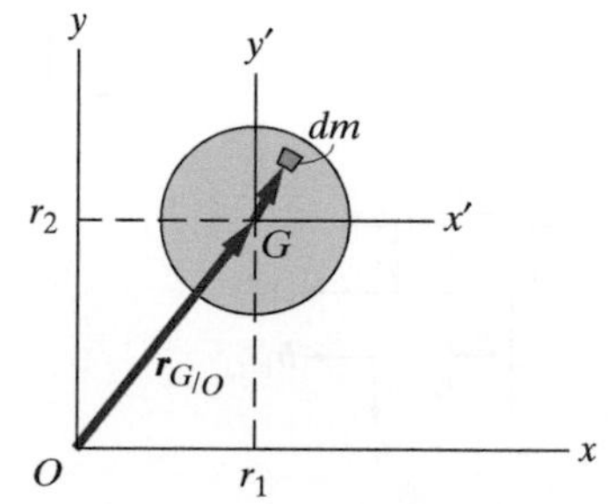

Figure 8.26 Projection of regular cylinder

$$\begin{aligned} I_{xy} &= \int_{\text{Body}} (r_1 + x')(r_2 + y')\, dm \\ &= \int_{\text{Body}} r_1 r_2\, dm + \int_{\text{Body}} x'y'\, dm \\ &= mr_1r_2 + \overline{I}_{x'y'} \end{aligned}$$

Armed with these formulas, we can determine a body's rotational inertial properties about any point in space, as long as the axes we're concerned with are parallel to the ones used to tabulate $\overline{I}_{x'x'}, \overline{I}_{y'y'}, \overline{I}_{z'z'}, \overline{I}_{x'y'}, \overline{I}_{y'z'}, \overline{I}_{z'x'}$.

>>> **Check out Example 8.6 (page 510) for an application of this material.**

EXAMPLE 8.6 INERTIAL PROPERTIES OF A FLAT PLATE (Theory on page 508)

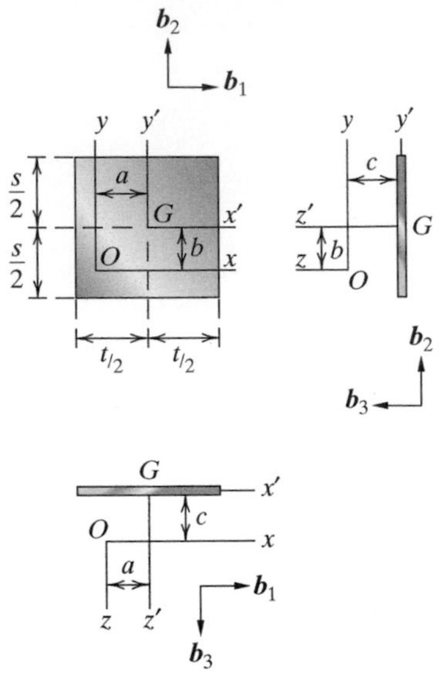

Figure 8.27 Flat, rectangular plate

Let's find the inertial properties of a flat plate (negligible thickness), both from the center of mass and from a corner of the plate. **Figure 8.27** shows three views of the system under consideration. The x', y', z' axes go through the body's mass center. The plate has an areal mass density ρ, and thus its total mass is ρst.

Goal Using the formulas just derived, determine the rotational moments of inertia and products of inertia about x, y, z and x', y', z'.

Draw **Figure 8.27** shows the relevant axes and dimensions.

Assume No additional assumptions are needed.

Formulate Equations We will be using (8.24)–(8.29).

Solve Because we can neglect any plate dimensions in the z' direction, the moments of inertia $\overline{I}_{x'x'}$ and $\overline{I}_{y'y'}$ are particularly easy to calculate. If you spin the plate about the x' axis, for example, it behaves like a rigid rod of length s and mass ρst. Thus the first two moments of inertia are

$$\overline{I}_{x'x'} = \frac{ms^2}{12}, \qquad \overline{I}_{y'y'} = \frac{mt^2}{12}$$

$$\overline{I}_{z'z'} = \int_{-\frac{s}{2}}^{\frac{s}{2}}\int_{-\frac{t}{2}}^{\frac{t}{2}} (x^2+y^2)\rho\,dx\,dy = \int_{-\frac{s}{2}}^{\frac{s}{2}}\left(\frac{t^3}{12}+ty^2\right)\rho\,dy = \rho\left(\frac{t^3 s}{12}+\frac{ts^3}{12}\right) = \frac{m}{12}(t^2+s^2)$$

$$\overline{I}_{x'y'} = \int_{-\frac{s}{2}}^{\frac{s}{2}}\int_{-\frac{t}{2}}^{\frac{t}{2}} xy\rho\,dx\,dy = \int_{-\frac{s}{2}}^{\frac{s}{2}} \rho\left(\frac{x^2}{2}\right)y\Big|_{-\frac{t}{2}}^{\frac{t}{2}} dy = 0$$

Both $\overline{I}_{z'x'}$ and $\overline{I}_{y'z'}$ are zero because $z' = 0$ for the all points on the flat plate. Note that we could have deduced these results from a symmetry argument as well, as discussed at the end of the previous section.

Now let's consider the moments of inertia and products of inertia about the x, y, z axes. From (8.24) through (8.26),

$$I_{xx} = \overline{I}_{x'x'} + m(b^2+c^2) = \frac{ms^2}{12} + m(b^2+c^2) = m\left(\frac{s^2}{12}+b^2+c^2\right)$$

$$I_{yy} = \overline{I}_{y'y'} + m(a^2+c^2) = \frac{mt^2}{12} + m(a^2+c^2) = m\left(\frac{t^2}{12}+a^2+c^2\right)$$

$$I_{zz} = \overline{I}_{z'z'} + m(a^2+b^2) = \frac{m}{12}(t^2+s^2) + m(a^2+b^2) = m\left(\frac{t^2+s^2}{12}+a^2+b^2\right)$$

and from (8.27) through (8.29),

$$I_{xy} = \overline{I}_{x'y'} + mab = mab, \qquad I_{zx} = \overline{I}_{z'x'} + ma(-c) = -mac, \qquad I_{yz} = \overline{I}_{y'z'} + mb(-c) = -mbc$$

EXERCISES 8.6

8.6.1. **[Level 1]** Determine the mass moments of inertia and products of inertia along the x, y, z axes for the illustrated rectangular body. Assume an areal density ρ. Express your answer in terms of the body's mass m.

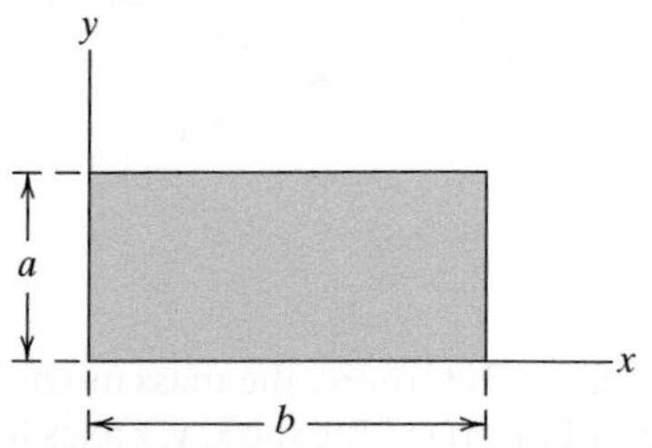

E8.6.1

8.6.2. **[Level 1]** Determine I_{xx}, I_{yy}, and I_{xy} for the illustrated solid cylinder. Assume a density ρ. Express your results first in terms of ρ and then in terms of the cylinder's mass m.

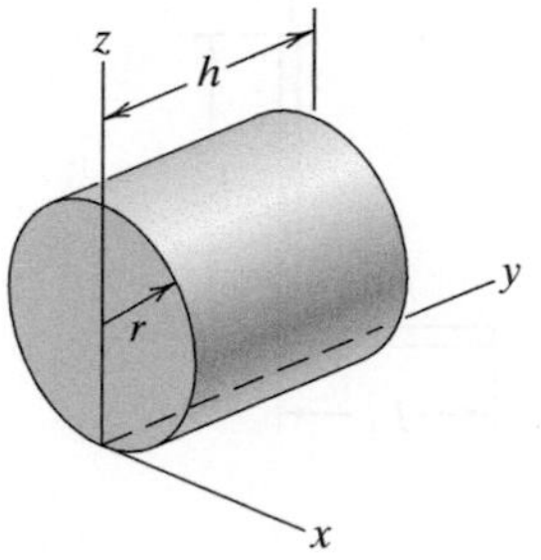

E8.6.2

8.6.3. **[Level 2]** Consider the illustrated body of areal density ρ and negligible thickness. Find the body's mass moments of inertia and products of inertia along the depicted x, y, z axes, and express your results in terms of the body's mass m.

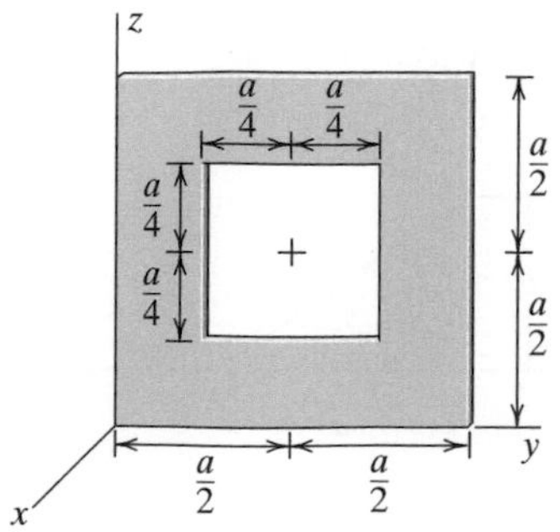

E8.6.3

8.6.4. **[Level 2]** Determine the mass moments of inertia and products of inertia along the x, y, z axes for the illustrated body. Assume a mass per unit length of ρ. Express your results in terms of the body's mass m.

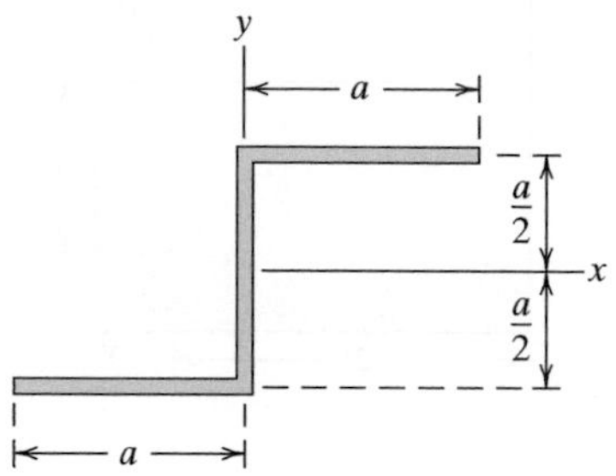

E8.6.4

8.6.5. **[Level 2]** Determine the mass moments of inertia and products of inertia along the x, y, z axes for the illustrated body. Express your results in terms of the body's areal density ρ.

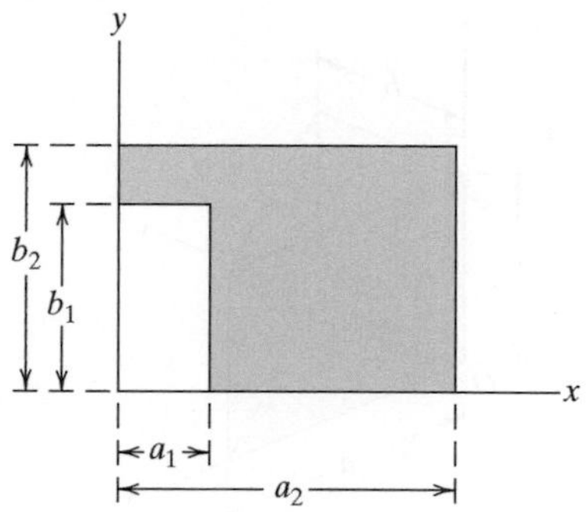

E8.6.5

8.6.6. **[Level 2]** Determine the mass moments of inertia and products of inertia along the x, y, z axes (which go through the body's mass center) for the illustrated washer. Express your results in terms of the body's density ρ. $h = 0.006$ m, the inner radius is 0.02 m, and the outer radius is 0.03 m.

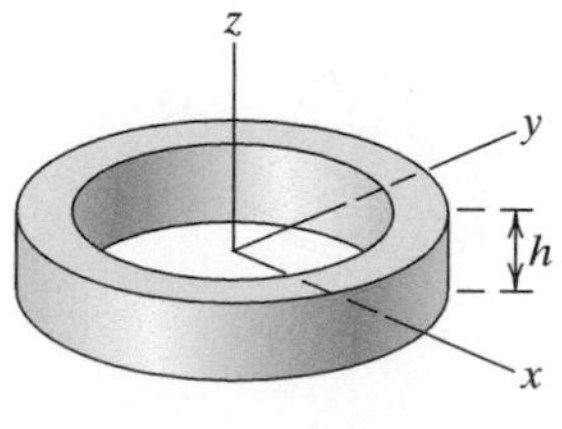

E8.6.6

8.6.7. **[Level 2]** What are the mass moments of inertia and products of inertia for the illustrated body about the depicted x, y, z axes? Express your answer in terms of the body's mass m.

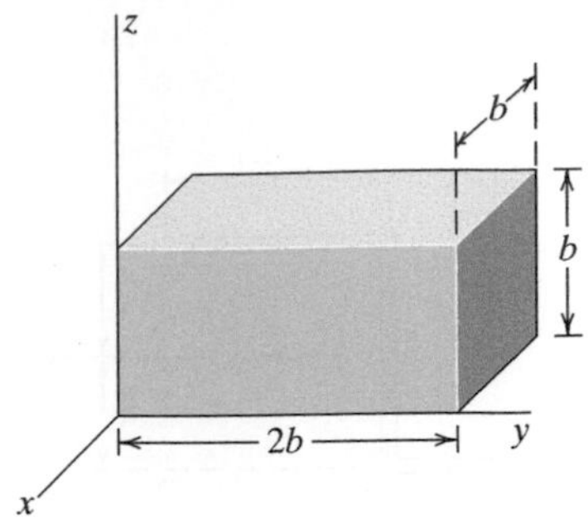

E8.6.7

8.6.8. **[Level 2]** Determine the mass moments of inertia and products of inertia along the x, y, z axes for the illustrated triangular prism. Assume a density ρ. Express your results first in terms of ρ and then in terms of the prism's mass m.

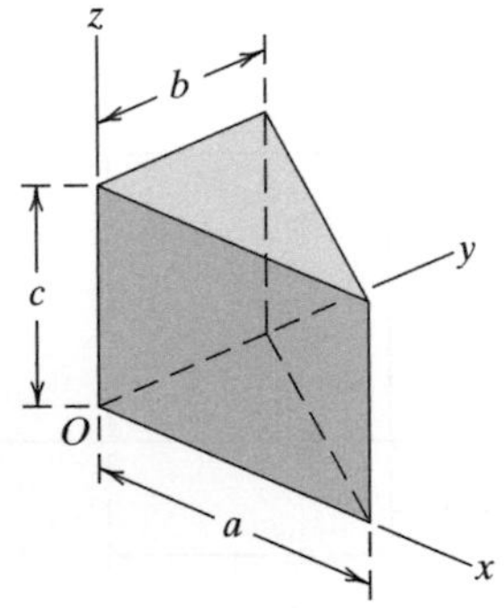

E8.6.8

8.6.9. **[Level 2]** Determine I_{xx}, I_{xy}, I_{yy}, and I_{zz} for the illustrated body, made up of two half-circles joined at 90° to each other. Assume an areal density ρ. Express your results first in terms of ρ and then in terms of the body's mass m.

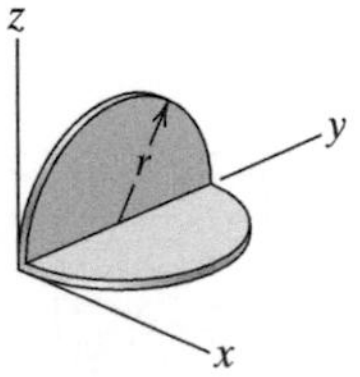

E8.6.9

8.6.10. **[Level 2]** Determine I_{xx}, I_{xy}, I_{yy}, and I_{zz} for the illustrated body consisting of two rectangular solids joined to each other as shown. The smaller piece has a density of 1000 kg/m^3, and the larger piece has a density of 800 kg/m^3. $a = 0.1$ m, $b = 0.02$ m, $c = 0.07$ m, and $d = 0.06$ m.

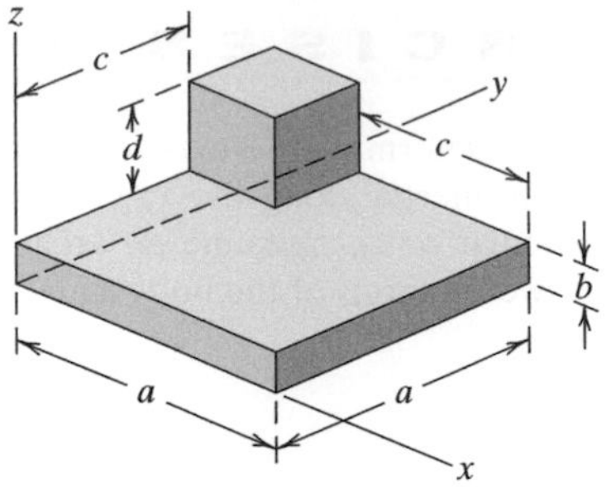

E8.6.10

8.6.11. **[Level 3]** Determine the mass moments of inertia and products of inertia along the x, y, z axes for the illustrated body. Then calculate the inertias about axes parallel to x, y, z that go through the body's mass center. Assume a mass per unit length of ρ. Express your results in terms of the body's mass m.

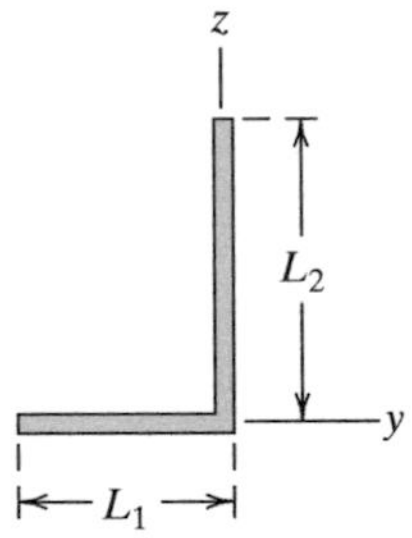

E8.6.11

8.6.12. **[Level 3]** Determine the mass moments of inertia and products of inertia along the x, y, z axes for the illustrated triangular body. Then calculate the inertias about axes parallel to x, y, z that go through the body's mass center. Assume an areal density ρ. Express your results in terms of ρ.

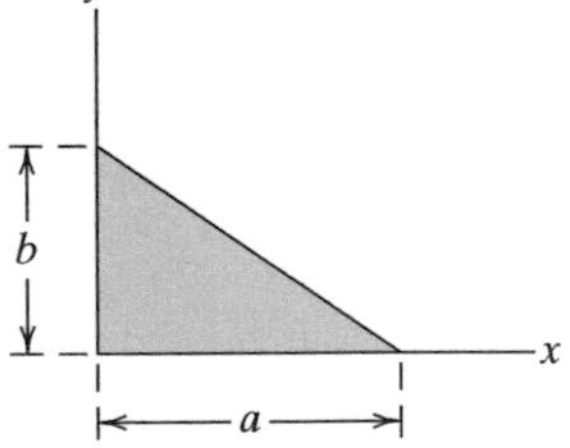

E8.6.12

8.6.13. **[Level 3]** Determine the mass moments of inertia and products of inertia along the x, y, z axes for the illustrated body. Express your results in terms of the body's areal density ρ.

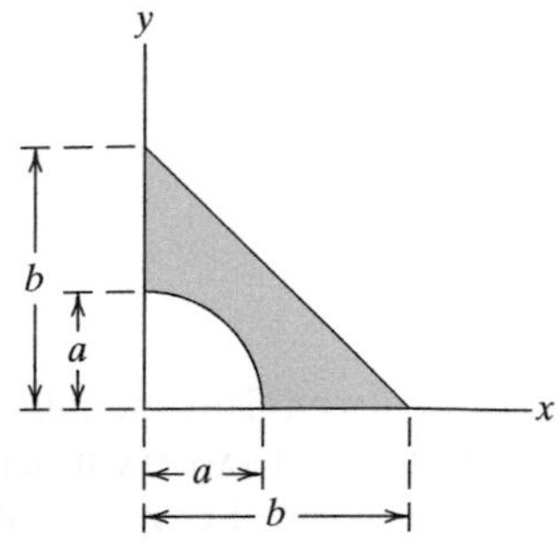

E8.6.13

8.6.14. **[Level 3]** Determine the mass moments of inertia and products of inertia along the x, y, z axes for the illustrated body. Assume a density ρ. $a = 0.1$ m, $b = 0.25$ m, $c = 0.15$ m, $d = 0.1$ m, $e = 0.1$ m, $f = 0.2$ m, and $g = 0.15$ m. Express your results in terms of the body's mass m.

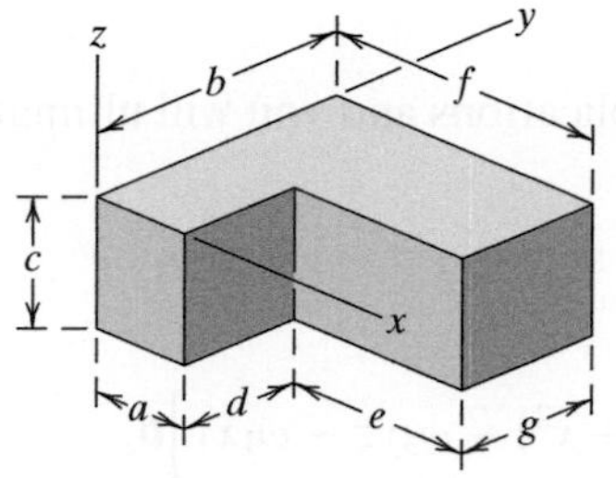

E8.6.14

8.6.15. **[Level 3]** Determine I_{xx}, I_{xy}, and I_{zz} for the illustrated body, made up of two half-circles joined to a rectangular plate. $a = 0.4$ m and $b = 0.2$ m. Assume an areal density ρ. Express your results first in terms of ρ and then in terms of the body's mass.

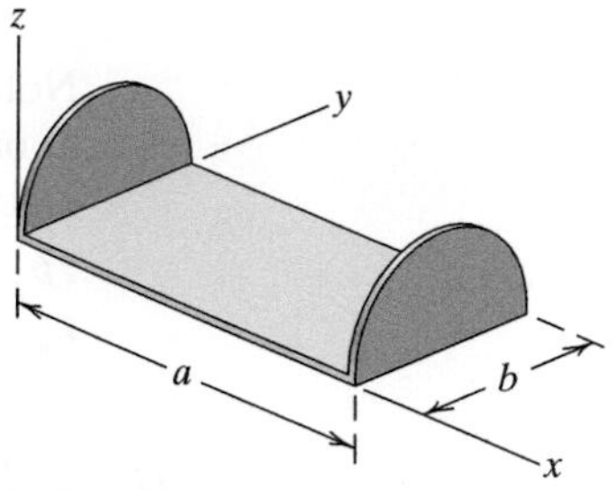

E8.6.15

8.7 ANGULAR MOMENTUM

Next we will calculate the angular momentum of a three-dimensional body, and then we will use this information to get our equations of motion. **Figure 8.28** shows a body plus a ground-fixed set of coordinate axes X, Y, Z and a body-fixed set x, y, z. The body's total angular momentum about G is found by integrating all the differential angular momenta associated with differential chunks of mass dm:

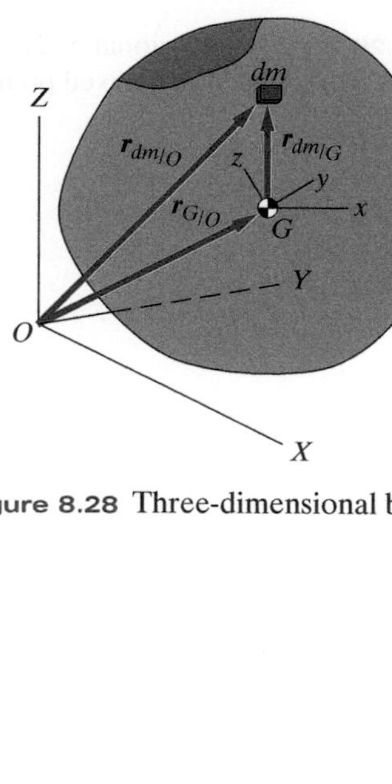

Figure 8.28 Three-dimensional body

$$
\begin{aligned}
\boldsymbol{H}_G &= \int_{\text{Body}} \boldsymbol{r}_{dm/G} \times \left(\boldsymbol{v}_G + \boldsymbol{\omega} \times \boldsymbol{r}_{dm/G}\right) dm \\
&= -\boldsymbol{v}_G \times \int_{\text{Body}} \boldsymbol{r}_{dm/G}\, dm + \int_{\text{Body}} \boldsymbol{r}_{dm/G} \times \left(\boldsymbol{\omega} \times \boldsymbol{r}_{dm/G}\right) dm \\
&= \int_{\text{Body}} \boldsymbol{r}_{dm/G} \times \left(\boldsymbol{\omega} \times \boldsymbol{r}_{dm/G}\right) dm
\end{aligned}
\tag{8.30}
$$

where the first term drops out because of the definition of the mass center.

It's even quicker to get the expression for angular momentum if the body is rotating about a fixed point O. In that case, from **Figure 8.29**, we have

$$\boldsymbol{H}_O = \int_{\text{Body}} \boldsymbol{r}_{dm/O} \times \left(\boldsymbol{\omega} \times \boldsymbol{r}_{dm/O}\right) dm \tag{8.31}$$

Note that (8.30) and (8.31) have the same form. This means we can look at one of the expressions in detail and not have to worry much about the second; its solution will have the exact same form. Let $\boldsymbol{r}_{dm/G} = x\boldsymbol{b}_1 + y\boldsymbol{b}_2 + z\boldsymbol{b}_3$ and $\boldsymbol{\omega} = \omega_1\boldsymbol{b}_1 + \omega_2\boldsymbol{b}_2 + \omega_3\boldsymbol{b}_3$, using body-fixed coordinates for both $\boldsymbol{r}_{dm/G}$ and $\boldsymbol{\omega}$. Then (8.30) becomes

$$\boldsymbol{H}_G = \int_{\text{Body}} (x\boldsymbol{b}_1 + y\boldsymbol{b}_2 + z\boldsymbol{b}_3) \times [(\omega_1\boldsymbol{b}_1 + \omega_2\boldsymbol{b}_2 + \omega_3\boldsymbol{b}_3) \times (x\boldsymbol{b}_1 + y\boldsymbol{b}_2 + z\boldsymbol{b}_3)]\, dm$$

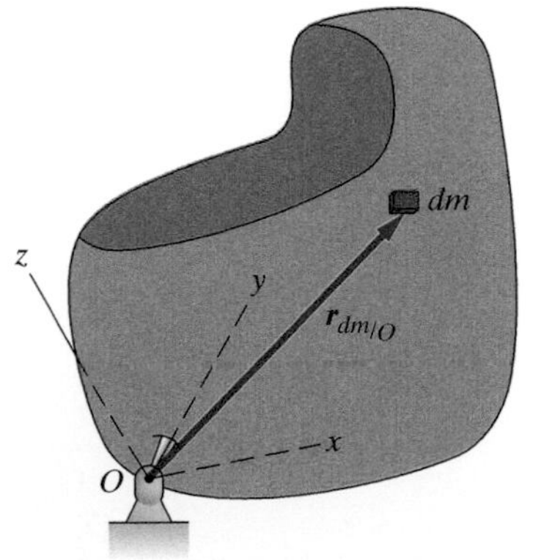

Figure 8.29 Three-dimensional body constrained to rotate about a fixed point

Carry through with all the multiplications and you will ultimately get

$$\begin{aligned}\boldsymbol{H}_G = \int_{\text{Body}} \Big(&\left[\omega_1(y^2 + z^2) - \omega_2 xy - \omega_3 zx\right]\boldsymbol{b}_1 \\ &+ \left[\omega_2(z^2 + x^2) - \omega_3 yz - \omega_1 xy\right]\boldsymbol{b}_2 \\ &+ \left[\omega_3(x^2 + y^2) - \omega_1 zx - \omega_2 yz\right]\boldsymbol{b}_3\Big)\, dm\end{aligned}$$

Using the definitions from (8.18–8.23) gives us

$$\begin{aligned}\boldsymbol{H}_G = &\left[\bar{I}_{xx}\omega_1 - \bar{I}_{xy}\omega_2 - \bar{I}_{xz}\omega_3\right]\boldsymbol{b}_1 \\ &+ \left[\bar{I}_{yy}\omega_2 - \bar{I}_{yz}\omega_3 - \bar{I}_{yx}\omega_1\right]\boldsymbol{b}_2 \\ &+ \left[\bar{I}_{zz}\omega_3 - \bar{I}_{zx}\omega_1 - \bar{I}_{zy}\omega_2\right]\boldsymbol{b}_3\end{aligned} \tag{8.32}$$

As already mentioned, we will have exactly the same form for a body that's rotated about a fixed point of rotation O. The equations for this case are

$$\begin{aligned}\boldsymbol{H}_O = &\left[I_{xx}\omega_1 - I_{xy}\omega_2 - I_{xz}\omega_3\right]\boldsymbol{b}_1 \\ &+ \left[I_{yy}\omega_2 - I_{yz}\omega_3 - I_{yx}\omega_1\right]\boldsymbol{b}_2 \\ &+ \left[I_{zz}\omega_3 - I_{zx}\omega_1 - I_{zy}\omega_2\right]\boldsymbol{b}_3\end{aligned} \tag{8.33}$$

The only difference between the two cases is whether the inertia components are calculated with respect to a fixed point of rotation or with respect to the body's mass center.

As you can see, both $\boldsymbol{H}_G$ and $\boldsymbol{H}_O$ have three orthogonal components. The angular momentum around the x axis (taking $\boldsymbol{H}_G$ as our example) involves $\overline{I}_{xx}\omega_1$ (expected from our two-dimensional work), and components involve rotation about the y and z axes ($-\overline{I}_{xy}\omega_2$ and $\overline{I}_{zx}\omega_3$, respectively). These last two are the terms that make life more interesting in three-dimensional analyses.

Pay close attention to the fact that the inertias are defined with respect to *body-fixed axes* x, y, z. The axes rotate with the body, and therefore, even as the body tumbles through space, the inertias remain *constant* because they're defined with respect to axes that are moving *along with the body*.

This isn't the only way to look at three-dimensional motions, for you could easily ask about the body's inertia with respect to the ground-fixed X, Y, Z axes. The problem with this approach is that the inertias would then *not* be constant in time but would change as the orientation of the body changed.

Can you think of any situations in which you could use axes that weren't fixed to the body, and yet, for a given motion, you still had constant inertial properties? How about something like the object shown in **Figure 8.30**? What if this object is spinning about the ground-fixed Z axis? Because it's a body of revolution about that axis, the inertial properties about the X axis or Y axis are the same. In fact, the inertial properties about any axis through O that lies in the X–Y plane will be the same as those through the X or Y axes. Thus, as long as the object is spinning only about the Z axis, the inertial properties about the ground-fixed X, Y, Z axes are constant. This is only a special case, however. For bodies that aren't rotationally symmetric, you will probably want to use body-fixed axes.

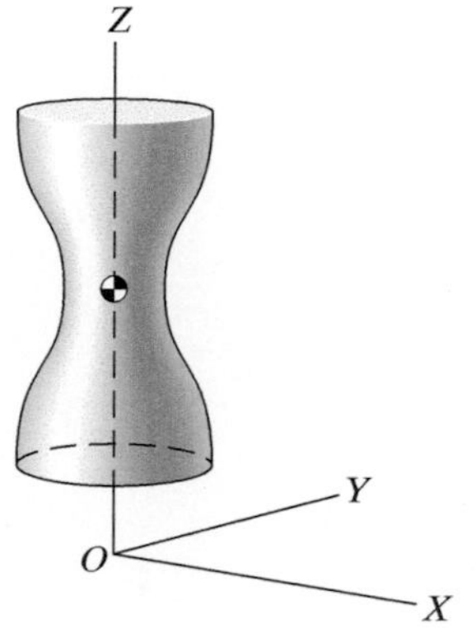

Figure 8.30 Rotationally symmetric body

A convenient way to express a body's inertial properties is in matrix form:

$$[I] \equiv \begin{bmatrix} I_{xx} & -I_{xy} & -I_{xz} \\ -I_{yx} & I_{yy} & -I_{yz} \\ -I_{zx} & -I_{zy} & I_{zz} \end{bmatrix} \tag{8.34}$$

where the matrix $[I]$ is referred to either as the **inertia matrix** or, occasionally, as the **inertia tensor**.

This form enables us to express our angular momenta very compactly. If we represent our total angular velocity as a column vector:

$$\boldsymbol{\omega} \equiv \begin{bmatrix} \omega_1 \\ \omega_2 \\ \omega_3 \end{bmatrix} \tag{8.35}$$

then the body's angular momentum is given by

$$\boldsymbol{H}_O = [I]\boldsymbol{\omega}$$

where the entries in the $\boldsymbol{H}_O$ vector are the three components given explicitly in (8.33).

It makes sense that as the axes x, y, z change, so do the elements of the inertia matrix. A different choice of body-fixed axes will produce a different inertia matrix. Although it isn't obvious, there always exists a

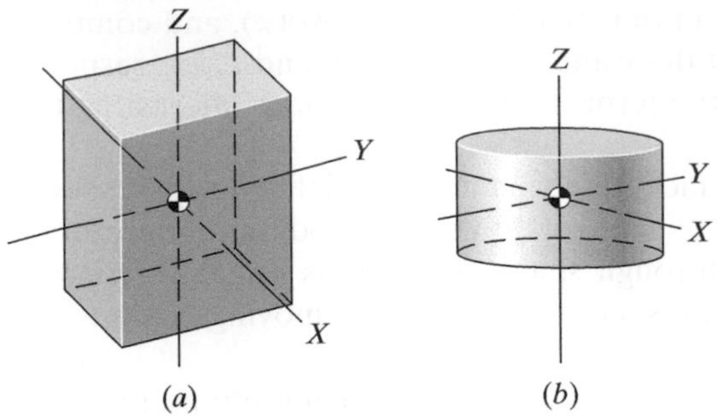

Figure 8.31 A couple of symmetric bodies

particular set of body-fixed axes for which all the products of inertia are zero. This results in the **principal inertia matrix**, which (again taking the inertias about the body's mass center) has the form

$$\left[\overline{I}\right] = \begin{bmatrix} \overline{I}_{xx} & 0 & 0 \\ 0 & \overline{I}_{yy} & 0 \\ 0 & 0 & \overline{I}_{zz} \end{bmatrix} \tag{8.36}$$

For symmetric bodies, it's pretty easy to pick out the principal axes, and **Figure 8.31** shows a couple of examples.

When the products of inertia are equal to zero, (8.32) takes on a particularly simple form. The axes corresponding to this diagonalized form are called **principal axes**, and I_{xx}, I_{yy}, and I_{zz} are known as the **principal moments of inertia**. If we use principal axes, our momentum expression (8.32) becomes

$$\boldsymbol{H}_G = \overline{I}_{xx}\omega_1\boldsymbol{b}_1 + \overline{I}_{yy}\omega_2\boldsymbol{b}_2 + \overline{I}_{zz}\omega_3\boldsymbol{b}_3 \tag{8.37}$$

Note that, even with this simplified form, we don't have the same relationship between the angular momentum and the angular velocity that we had with two-dimensional systems. In two-dimensional analyses, both the angular momentum and the angular velocity are constrained to be in the same direction (out of the plane of rotation) and thus are related by a single constant I_G:

$$\boldsymbol{H}_G = I_G\boldsymbol{\omega}$$

In the three-dimensional case (8.37), we have three components (such as $I_{xx}\omega_1\boldsymbol{b}_1$) that make up the final angular velocity.

The last momentum question I will address is how the angular momentum about some arbitrary point is related to H_G (**Figure 8.32**). The answer is: It's an exact parallel to the two-dimensional case. The angular momentum about a point O is just the angular momentum about the body's mass center G plus a term that accounts for the linear momentum

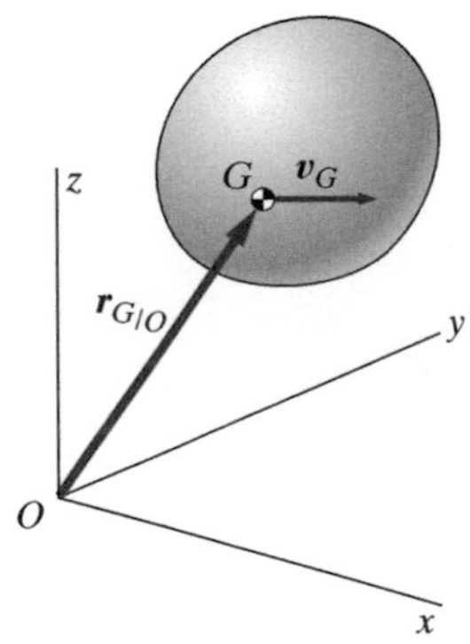

Figure 8.32 Momentum of a three-dimensional body

of the body about O:

$$\boldsymbol{H}_O = \boldsymbol{H}_G + \boldsymbol{r}_{G/O} \times m\boldsymbol{v}_G \tag{8.38}$$

>>> **Check out Example 8.7 (page 517) and Example 8.8 (page 518) for applications of this material.**

EXAMPLE 8.7 ANGULAR MOMENTUM OF A FLAT PLATE (Theory on page 517)

We calculated the rotational inertias of a rectangular flat plate back in Example 8.6, and so let's use this same body as we look at angular momentum. Find the angular momentum of this flat plate as it rotates about the y axis with a rotational speed ω.

Goal Use the known inertias and (8.32) to calculate the angular momentum.

Draw Refer to **Figure 8.27** for all the relevant axes and dimensions.

Solve We'll pick up only three terms from (8.32), the ones corresponding to ω_2. The total angular momentum is

$$H_G = -I_{xy}\omega\boldsymbol{b}_1 + I_{yy}\omega\boldsymbol{b}_2 - I_{zy}\omega\boldsymbol{b}_3 = -mab\omega\boldsymbol{b}_1 + \left(\frac{t^2}{12} + a^2 + c^2\right) m\omega\boldsymbol{b}_2 + mbc\omega\boldsymbol{b}_3$$

EXAMPLE 8.8 ANGULAR MOMENTUM OF A SIMPLE STRUCTURE (Theory on page 517)

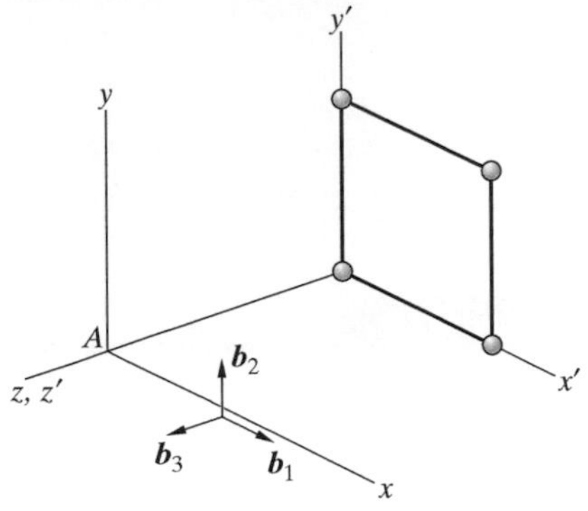

Figure 8.33 Simple four-mass structure

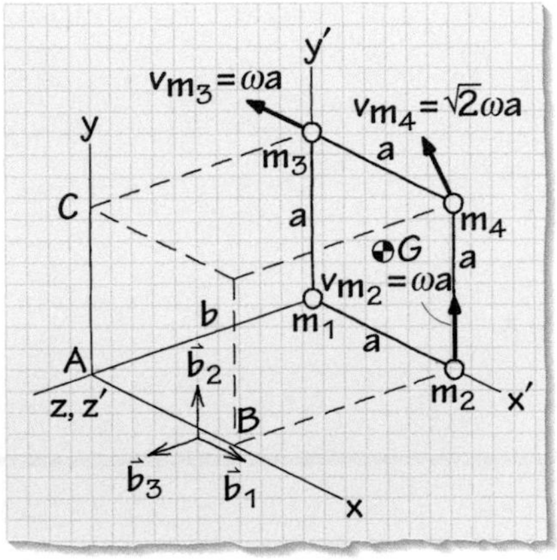

Figure 8.34 Velocities for four-mass structure

Okay. I don't know about you, but for me the last example is just a bit hard to believe. All we had was a simple plate spinning nicely about an axis, and yet we got angular momentum components in all three directions. The question is: "Where are these components coming from?" Let's look at a really simple body and pick out exactly how we get components in the $\boldsymbol{b}_1$ and $\boldsymbol{b}_2$ directions, even though the body has an angular velocity in the $\boldsymbol{b}_3$ direction. The body we will look at is shown in **Figure 8.33**, a massless square frame that has a mass m at each corner. The plane of the frame is parallel to the x–y plane and has angular velocity $\omega \boldsymbol{b}_3$.

Goal Show explicitly where the three components of the angular momentum originate.

Draw **Figure 8.34** identifies each mass (m_1–m_4), so that you can follow how each affects the overall angular momentum of the structure. For simplicity, all the masses are equal in magnitude (m), and I will refer to the total mass as m_T.

Solve From (8.33), when ω_1 and ω_2 are set equal to zero, we have

$$\boldsymbol{H}_A = -I_{xz}\omega \boldsymbol{b}_1 - I_{yz}\omega \boldsymbol{b}_2 + I_{zz}\omega \boldsymbol{b}_3$$

We know from symmetry that $\overline{I}_{z'x'}$ and $\overline{I}_{y'z'}$ are zero. The mass moment of inertia about the mass center $\overline{I}_{z'z'}$ is just

$$\overline{I}_{z'z'} = m_1\left(\frac{a}{\sqrt{2}}\right)^2 + m_2\left(\frac{a}{\sqrt{2}}\right)^2 + m_3\left(\frac{a}{\sqrt{2}}\right)^2 + m_4\left(\frac{a}{\sqrt{2}}\right)^2 = \frac{m_T a^2}{2}$$

The products of inertia I_{zx} and I_{yz} and the moment of inertia I_{zz} about the x, y, z axes are found from

$$(8.29) \Rightarrow \qquad I_{xz} = 0 + m_T(-b)\left(\frac{a}{2}\right) = -\frac{m_T ab}{2}$$

$$(8.28) \Rightarrow \qquad I_{yz} = 0 + m_T\left(\frac{a}{2}\right)(-b) = -\frac{m_T ab}{2}$$

$$(8.26) \Rightarrow \qquad I_{zz} = \frac{m_T a^2}{2} + m_T\left(\frac{a^2}{4} + \frac{a^2}{4}\right) = m_T a^2$$

The angular momentum about A is therefore

$$\boldsymbol{H}_A = \frac{m_T \omega ab}{2}\boldsymbol{b}_1 + \frac{m_T \omega ab}{2}\boldsymbol{b}_2 + m_T \omega a^2 \boldsymbol{b}_3 \qquad (8.39)$$

Now let's see where these terms *really* come from. **Figure 8.34** shows the velocity vector for each mass due to its rotation about the z axis. The easiest way to visualize what's going on is to imagine that the four masses are attached to a rigid, massless box, indicated by the dashed lines. If I am considering angular momentum about the x axis, then I imagine that the box is hinged along $\overline{AB}$, allowing the box to rotate about x but not about y or z. Next I look at the linear momentum of the individual masses and ask myself what angular momentum they produce about the x axis. It's an exact parallel to figuring out what moment (moment arm cross force) a force would produce, but in this case I'm asking what angular momentum (moment arm cross linear momentum) is produced.

Let's consider the box as being hinged along $\overline{AB}$, as suggested. In this case, m_3 will not contribute any angular momentum because its linear momentum vector is parallel to x. Mass m_4 has a linear velocity $\omega a\boldsymbol{b}_2 + \omega a\boldsymbol{b}_1$. The $\boldsymbol{b}_1$ component doesn't contribute any angular momentum (being parallel to x), but the $\boldsymbol{b}_2$ component contributes angular momentum $m_4\omega ab$. Finally, m_2 contributes angular momentum $m_2\omega ab$. The total angular momentum about x is therefore

$$(m_2 + m_4)\omega ab\boldsymbol{b}_1 = 2m\omega ab\boldsymbol{b}_1 = \frac{m_T\omega ab}{2}\boldsymbol{b}_1$$

To find the angular momentum about the y axis, I imagine that the box is now hinged along $\overline{AC}$. Mass m_3 produces an angular momentum $m_3\omega ab$. The $\boldsymbol{b}_1$ velocity component of m_4 plays a part here, contributing angular momentum $m_4\omega ab$, but m_2 doesn't contribute this time because its velocity is parallel to y. The total angular momentum about y is

$$(m_3 + m_4)\omega ab\boldsymbol{b}_2 = 2m\omega ab\boldsymbol{b}_2 = \frac{m_T\omega ab}{2}\boldsymbol{b}_2$$

Finally, I will hinge the box along $\overline{Am_1}$. In this configuration, m_3 contributes angular momentum $m_3\omega a^2$, m_4 contributes $2m_4\omega a^2$, and m_2 contributes $m_2\omega a^2$. This all combines to give

$$(m_3 + 2m_4 + m_2)\omega a^2\boldsymbol{b}_3 = 4m\omega a^2\boldsymbol{b}_3 = m_T\omega a^2\boldsymbol{b}_3$$

Our complete expression for the body's angular momentum about A is the sum of all these components:

$$\boldsymbol{H}_A = \frac{m_T\omega ab}{2}\boldsymbol{b}_1 + \frac{m_T\omega ab}{2}\boldsymbol{b}_2 + m_T\omega a^2\boldsymbol{b}_3 \qquad (8.40)$$

Equations (8.39) and (8.40) are, as would be hoped, identical. This example should indeed give you some insight into why a body rotating about a single axis can nonetheless have angular momenta around other axes as well.

EXERCISES 8.7

8.7.1. **[Level 2]** Find the angular momentum about O of the illustrated L-shaped structure that rotates with an angular velocity of $\omega\boldsymbol{j}$ about Y and has an areal density of ρ.

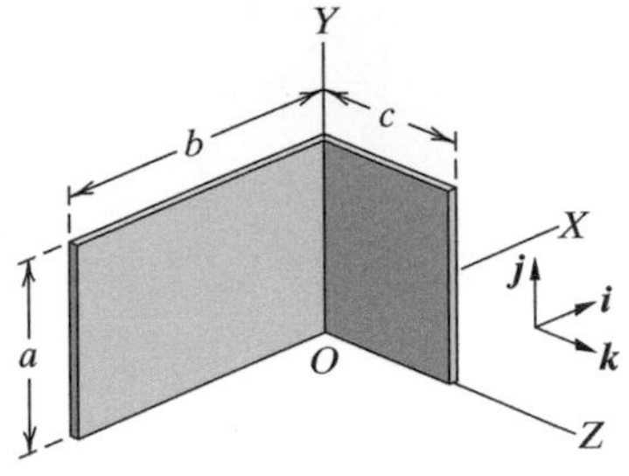

E8.7.1

8.7.2. **[Level 2]** As illustrated, a T-shaped rod of linear density $\rho = \frac{m}{L}$ is attached to a ball-and-socket joint at its fixed end O, and two balls with mass m are appended to the rod's free ends. Suppose the system's angular velocity with respect to its body-fixed x, y, z axes is $\boldsymbol{\omega} = \omega_1\boldsymbol{b}_1 + \omega_2\boldsymbol{b}_2 + \omega_3\boldsymbol{b}_3$ at the depicted instant. What is the system's instantaneous angular momentum $\boldsymbol{H}_O$ about O?

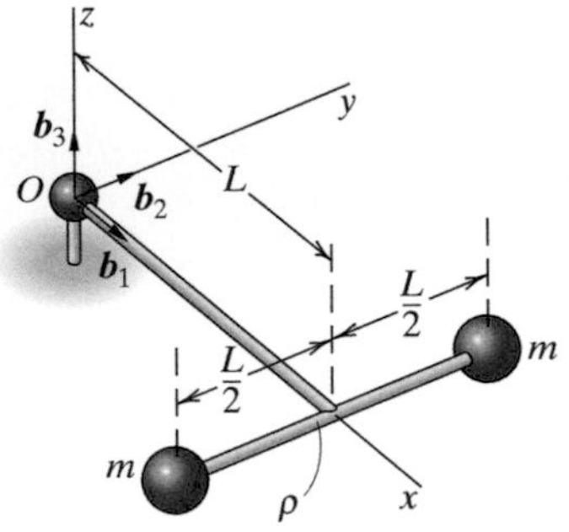

E8.7.2

8.7.3. **[Level 2]** What is the angular momentum about O of the illustrated solid rectangular body that is rotating about the Z axis with an angular velocity $\omega\boldsymbol{k}$? The body has mass m.

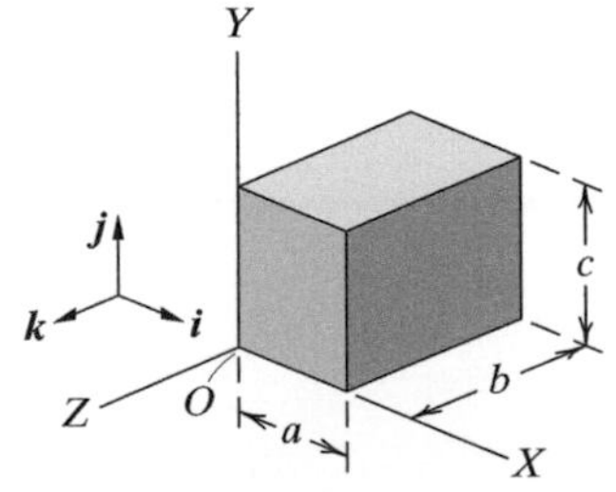

E8.7.3

8.7.4. **[Level 2]** What is the angular momentum about O of the illustrated solid washer that is rotating about the Z axis with an angular velocity $\omega\boldsymbol{k}$? The washer has mass m, inner radius R_1, outer radius R_2, and thickness h.

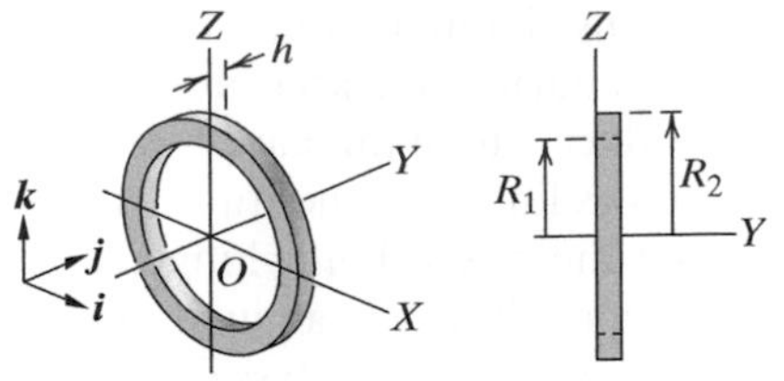

E8.7.4

8.7.5. **[Level 2]** The illustrated triangular plate of mass m is free to rotate about its fixed pivot O with an angular velocity of $\boldsymbol{\omega} = \omega_2\boldsymbol{b}_2 + \omega_3\boldsymbol{b}_3$ with respect to the plate's body-fixed x, y, z axes. Calculate the plate's angular momentum $\boldsymbol{H}_O$ about O.

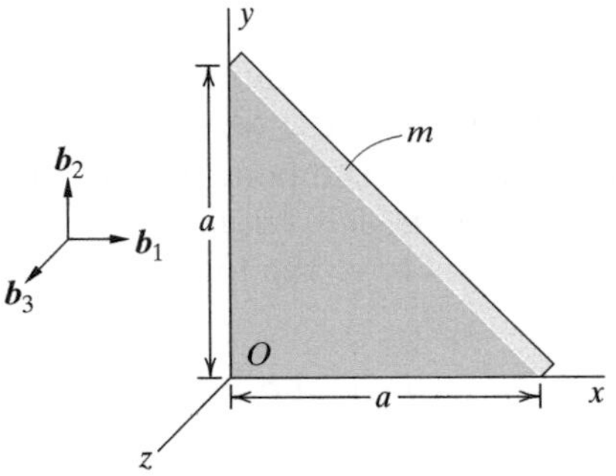

E8.7.5

8.7.6. **[Level 2]** The illustrated T-shaped structure (fixed at O) has a mass m and an angular velocity of $(-6\boldsymbol{i} + 10\boldsymbol{j})$ rad/s. What is the structure's angular momentum about the point O?

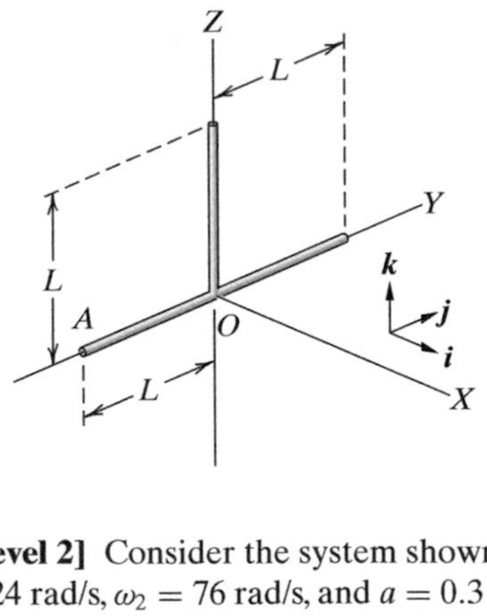

E8.7.6

8.7.7. **[Level 2]** Consider the system shown in **Exercise 8.4.1.** $\omega_1 = 24$ rad/s, $\omega_2 = 76$ rad/s, and $a = 0.3$ m. Treat the rotating body D as a flat disk with radius 0.25 m and a mass of 10 kg. What is the disk's angular momentum about O and about G?

8.7.8. **[Level 2]** Consider the system shown in **Exercise 8.4.5.** The framework has a mass of 0.005 kg, and each arm has a length of $h = 0.12$ m. The wheels have radii $r = 0.007$

m and negligible thickness. The framework rotates about the Y axis at 2.2 rad/s, and the wheels rotate without slip. Each wheel has a mass of 0.001 kg. What is the total angular momentum of the system about O?

8.7.9. **[Level 2]** The illustrated bent bar rotates about the z axis with angular velocity $\omega_3 \boldsymbol{b}_3$. Find $\boldsymbol{H}_O$. The bar has a mass per unit length of ρ.

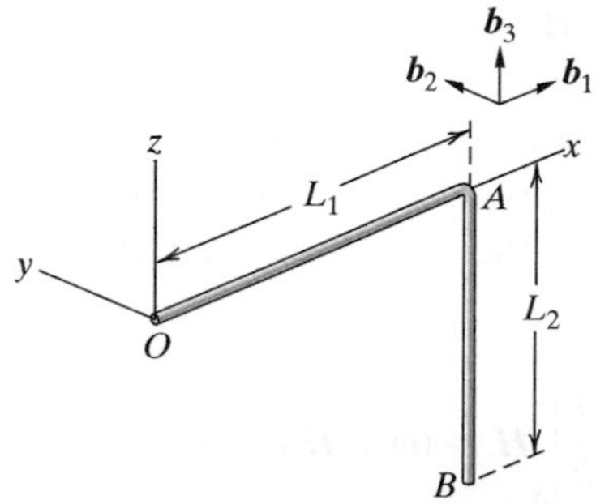

E8.7.9

8.7.10. **[Level 2]** The illustrated body is fixed at O, $\boldsymbol{\omega} = \omega \boldsymbol{b}_1$, and its areal density is ρ. Find $\boldsymbol{H}_O$.

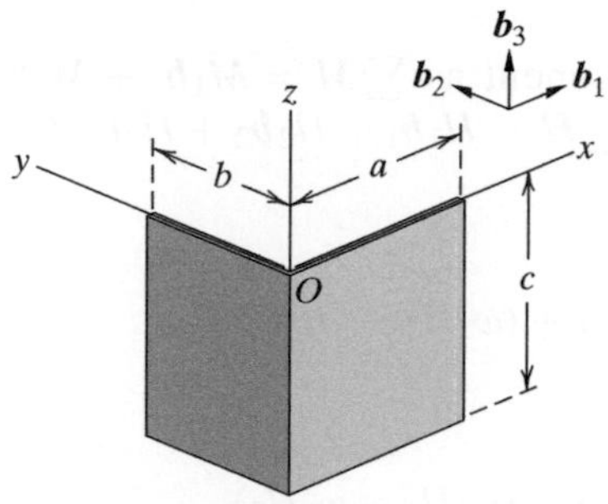

E8.7.10

8.7.11. **[Level 3]** The illustrated square framework, made up of four thin bars of length a, lying in the Y-Z plane, has an angular velocity of $(10\boldsymbol{j} + 18\boldsymbol{k})$ rad/s. What is the framework's angular momentum about the point O? Its mass center is fixed in space.

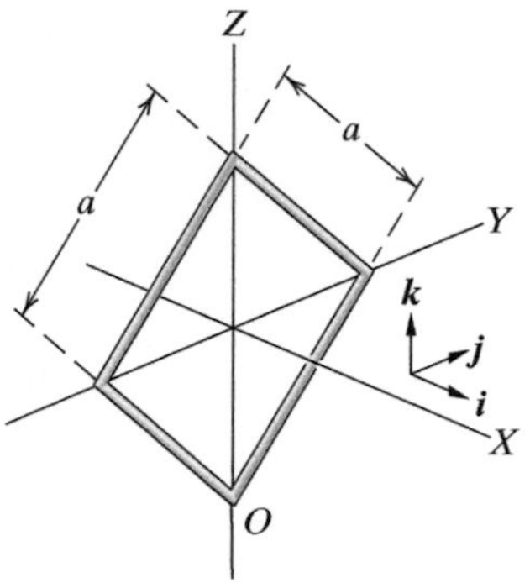

E8.7.11

8.8 EQUATIONS OF MOTION FOR A THREE-DIMENSIONAL BODY

You now have everything you need to determine the equations of motion for a three-dimensional rigid body. You know how to calculate the mass and rotational inertias, as well as the body's acceleration.

The translational equations of motion, assuming external forces and moments, are given by

$$\sum \boldsymbol{F} = \dot{\boldsymbol{L}}$$

which, for a constant mass, is just

$$\sum \boldsymbol{F} = \frac{d(m\boldsymbol{v})}{dt} = m\frac{d\boldsymbol{v}}{dt} = m\boldsymbol{a} \tag{8.41}$$

Our angular momentum equation of motion has the general form we saw in Section 5.2, namely,

$$\sum \boldsymbol{M}_G = \dot{\boldsymbol{H}}_G$$

for the case in which we are summing applied moments M_G about the center of mass G and

$$\sum \boldsymbol{M}_O = \dot{\boldsymbol{H}}_O$$

when we are summing applied moments M_O about a fixed point O.

Because we have expressed the body's momentum in terms of body-fixed axes, the time derivative $\dot{\boldsymbol{H}}$ of the angular momentum has two parts (6.24):

$$\dot{\boldsymbol{H}}\bigg|_N = \frac{d}{dt}\bigg|_S \boldsymbol{H} + (\boldsymbol{\omega} \times \boldsymbol{H})$$

where N represents the inertial reference frame and S represents the rotating frame of reference, which in this case coincides with our body's x, y, z axes.

If we express the applied moment as $\sum \boldsymbol{M} = M_1\boldsymbol{b}_1 + M_2\boldsymbol{b}_2 + M_3\boldsymbol{b}_3$ and the angular momentum as $\sum \boldsymbol{H} = H_1\boldsymbol{b}_1 + H_2\boldsymbol{b}_2 + H_3\boldsymbol{b}_3$, then we will have

$$\sum M_1 = \dot{H}_1 + (\omega_2 H_3 - \omega_3 H_2) \tag{8.42}$$

$$\sum M_2 = \dot{H}_2 + (\omega_3 H_1 - \omega_1 H_3) \tag{8.43}$$

$$\sum M_3 = \dot{H}_3 + (\omega_1 H_2 - \omega_2 H_1) \tag{8.44}$$

This expression simplifies if we use the principal axes for our x, y, z frame because all the products of inertia go to zero. In this case we will have what are termed **Euler's equations**:

$$\sum M_1 = \bar{I}_{xx}\alpha_1 + \omega_2\omega_3(\bar{I}_{zz} - \bar{I}_{yy}) \tag{8.45}$$

$$\sum M_2 = \bar{I}_{yy}\alpha_2 + \omega_3\omega_1(\bar{I}_{xx} - \bar{I}_{zz}) \tag{8.46}$$

$$\sum M_3 = \bar{I}_{zz}\alpha_3 + \omega_1\omega_2(\bar{I}_{yy} - \bar{I}_{xx}) \tag{8.47}$$

Notice that when we shift to principal axes we *don't* get equations of the form

$$\sum M_1 = \overline{I}_{xx}\alpha_1$$

$$\sum M_2 = \overline{I}_{yy}\alpha_2$$

$$\sum M_3 = \overline{I}_{zz}\alpha_3$$

as you might have expected. The only way we could avoid the angular velocity terms is if the two cross products of inertia for a particular equation were equal. For instance, if $\overline{I}_{yy} = \overline{I}_{xx}$, in (8.47) we would have

$$\sum M_3 = \overline{I}_{zz}\alpha_z - (0)\omega_y\omega_z = \overline{I}_{zz}\alpha_z$$

Is this possible? Yes. All we need is some rotational symmetry. Consider **Figure 8.35**, for example. Cutting the body in either the z–x or z–y plane produces identical areas, as shown in **Figure 8.36**. Therefore the moments of inertia $\overline{I}_{yy}$ and $\overline{I}_{xx}$ are the same.

This coupling between the equations is what makes three-dimensional motions so complicated and so interesting. For instance, consider the astronaut's tool we looked at in Section 8.5. We now have the mathematical equations needed to analyze its motion after we give it a quick twist, as we will do in the following example.

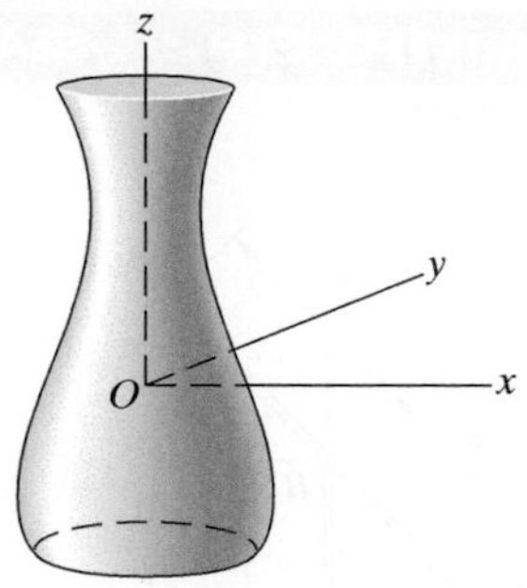

Figure 8.35 Rotationally symmetric body

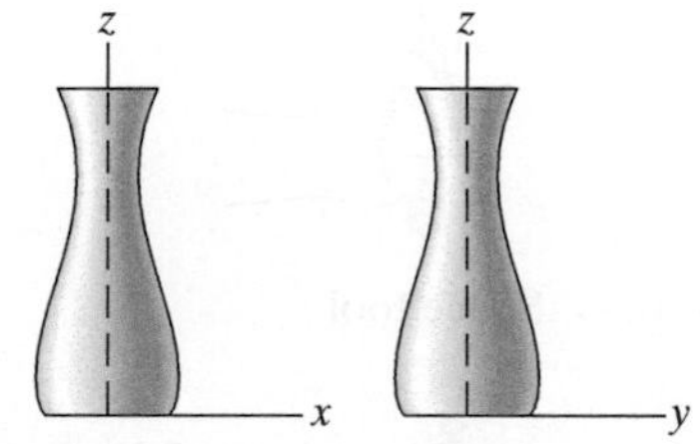

Figure 8.36 Cross-sectional view of rotationally symmetric body

>>> **Check out Example 8.9 (page 524) for an application of this material.**

EXAMPLE 8.9 REACTION FORCES OF A CONSTRAINED, ROTATING BODY (Theory on page 522)

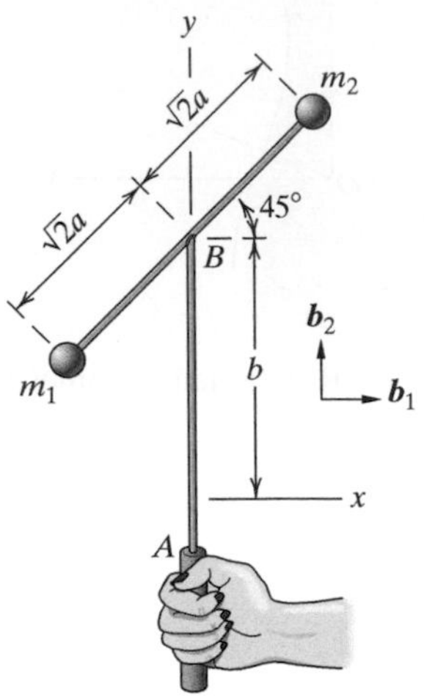

Figure 8.37 Space tool

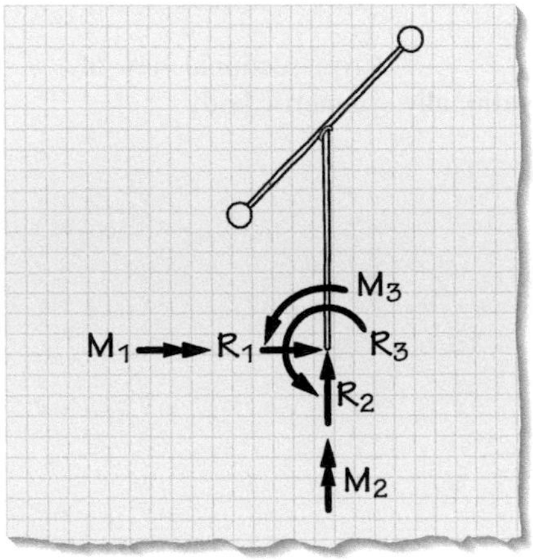

Figure 8.38 Space tool—FBD

Consider the astronaut's tool shown in **Figure 8.37**, where the only significant mass comes from the two tip spheres, m_1 and m_2, each of mass m. Here I have modified the tool of **Figure 8.20** so that now the lower end of the vertical shaft is encased in a sleeve. The sleeve contains a tiny motor (not shown) that can spin shaft $\overline{AB}$, while the astronaut holds the sleeve stationary. Determine the forces and moments that must be exerted on the shaft at A to allow the tool to rotate at a constant rate with angular velocity $\omega \boldsymbol{b}_2$. Assume that the experiment takes place in space and therefore gravitational forces are negligible.

Goal Given the rotation rate of the tool, find the constraining reaction forces $\boldsymbol{R}_i$ and moments $\boldsymbol{M}_i$.

Draw Body-fixed x, y, z axes are shown in **Figure 8.37**, and **Figure 8.38** shows the three reaction forces $\boldsymbol{R}_i$ and three applied moments $\boldsymbol{M}_i$.

Formulate Equations We will need the angular momentum $\boldsymbol{H}$ of the spinning tool as well as our force and moment balances. The angular momentum is found from (8.33):

$$\begin{aligned} \boldsymbol{H}_A = & \left[I_{xx}\omega_1 - I_{xy}\omega_2 - I_{xz}\omega_3\right]\boldsymbol{b}_1 \\ & + \left[I_{yy}\omega_2 - I_{yz}\omega_3 - I_{yx}\omega_1\right]\boldsymbol{b}_2 \\ & + \left[I_{zz}\omega_3 - I_{zx}\omega_1 - I_{zy}\omega_2\right]\boldsymbol{b}_3 \end{aligned} \tag{8.48}$$

The balance of linear momentum (8.44) is

$$\begin{aligned} M_1 &= \dot{H}_1 + (\omega_2 H_3 - \omega_3 H_2) \\ M_2 &= \dot{H}_2 + (\omega_3 H_1 - \omega_1 H_3) \\ M_3 &= \dot{H}_3 + (\omega_1 H_2 - \omega_2 H_1) \end{aligned} \tag{8.49}$$

and, of course, we have our old friend $\boldsymbol{F} = m\boldsymbol{a}$.

Assume The tool is free only to rotate about the y axis, and thus $\omega_1 = \omega_3 = 0$. This fact, along with the simple mass distribution, will lead to a highly simplified set of equations.

Solve Using $\omega_1 = \omega_3 = 0$ gives us an angular momentum of

$$\boldsymbol{H}_A = -I_{xy}\omega\boldsymbol{b}_1 + I_{yy}\omega\boldsymbol{b}_2 - I_{zy}\omega\boldsymbol{b}_3$$

With only the mass of the spheres to be concerned about, we can quickly solve for the needed inertias:

$$\begin{aligned} I_{xy} &= m(a)(b+a) + m(b-a)(-a) = 2ma^2 \\ I_{yy} &= ma^2 + ma^2 = 2ma^2 \\ I_{zy} &= m(b+a)(0) + m(b-a)(0) = 0 \end{aligned} \tag{8.50}$$

Thus

$$\boldsymbol{H}_A = -2ma^2\omega\boldsymbol{b}_1 + 2ma^2\omega\boldsymbol{b}_2$$

By observation we have $H_1 = -2ma^2\omega$, $H_2 = 2ma^2\omega$, and $H_3 = 0$. It's a given in the problem statement that $\dot{\omega} = 0$, which means $\dot{H}_1 = \dot{H}_2 = \dot{H}_3 = 0$. Thus our moment balance becomes

$$
\begin{aligned}
M_1 &= 0 \\
M_2 &= 0 \\
M_3 &= -\omega H_1 = 2ma^2\omega^2
\end{aligned}
$$

The center of mass is clearly stationary, and thus a force balance gives us, for our reaction forces $\boldsymbol{R}_i$,

$$
\begin{aligned}
R_1 &= 0 \\
R_2 &= 0 \\
R_3 &= 0
\end{aligned}
$$

This is perhaps a little surprising. Notice that the length of shaft $\overline{AB}$ doesn't enter into the results and that the only constraint is a moment constraint. Even though the two spheres are twirling around in different planes of rotation, no constraining forces are needed at A to support the motion.

EXERCISES 8.8

8.8.1. **[Level 1]** A popular tool for demonstrating gyroscopic principles is a bicycle wheel with a set of straight handlebars mounted to its axle. A student stands on a swiveling platform and holds the wheel away from his body so that the axle of the wheel lies in the horizontal plane. The wheel is given an initial spin ω about its axle in the direction indicated in the figure. There is no initial rotation of the platform ($\Omega_0 = 0$). The student then rotates the handlebars so that his right hand is directly above his left, and the wheel axle is oriented vertically in space. In what direction will the student rotate when this action is performed?

E8.8.1

8.8.2. **[Level 2]** Suppose the depicted system's angular velocity with respect to its body-fixed x, y, z axes is a constant $\boldsymbol{\omega} = \omega_2 \boldsymbol{b}_2 + \omega_3 \boldsymbol{b}_3$ at the illustrated instant. Find the forces and moments exerted on the system at its fixed pivot O for the given instant.

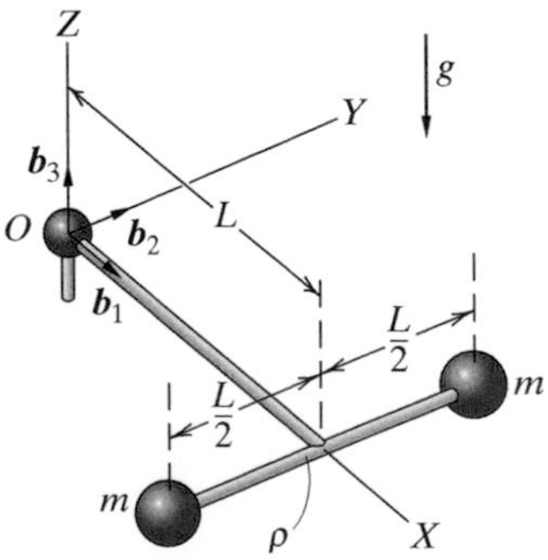

E8.8.2

8.8.3. **[Level 2]** The system shown is a simple model for a crankshaft. Each lumped mass is 2 kg. $a = 14$ cm and $b = 11$ cm. The crankshaft rotates at a constant rate of $\omega = 100$ rad/s. What are the forces acting at the two bearings? Neglect gravity.

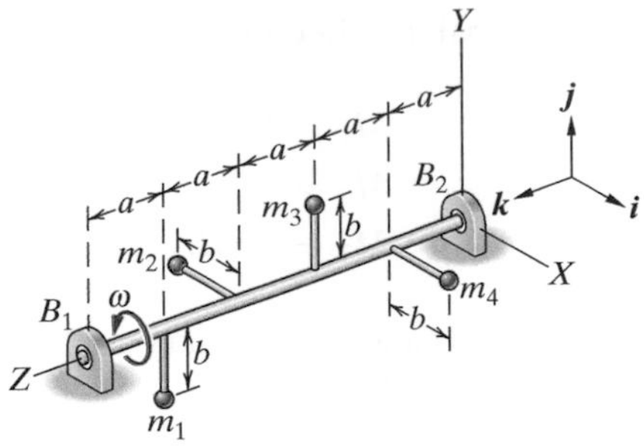

E8.8.3

8.8.4. **[Level 2]** The system shown in the figure is a circular sanding disk attached to a portable drill. The figure (b) shows a side view of the system. Because the disk isn't attached squarely to the shaft, it will produce unbalanced moments as it rotates. Find the moment that the shaft must exert on the disk in order for it to rotate at a constant angular speed ω. The disk has an areal density of ρ kg/m^2.

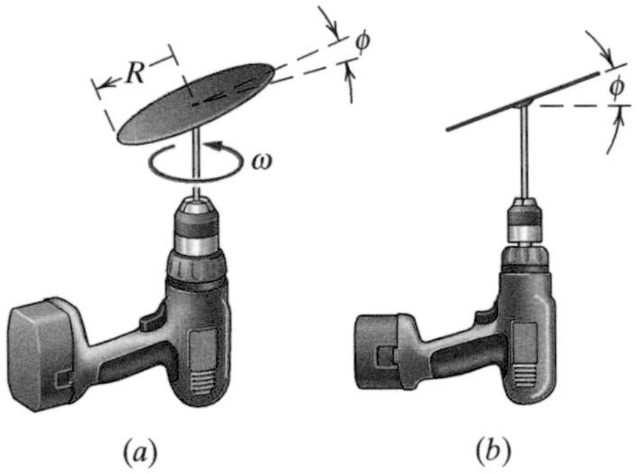

E8.8.4

8.8.5. **[Level 2]** A half-circular plate with radius $d/2$ is welded to a thin, circular shaft. Determine the forces acting at the bearing A if the shaft is rotating with a constant angular speed of ω. The shaft has mass m_1, and the plate has mass m_2.

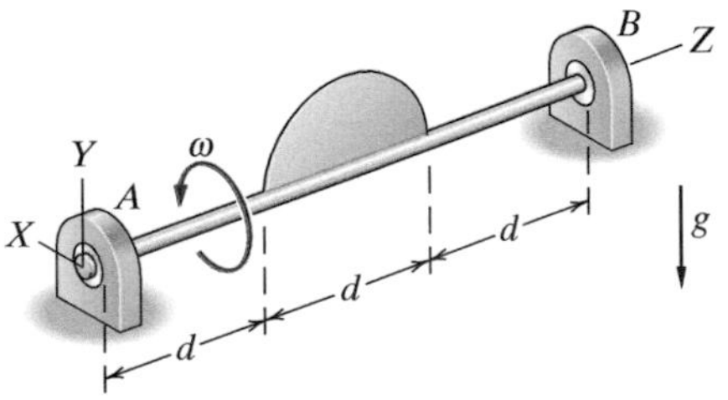

E8.8.5

8.8.6. **[Level 2]** The illustrated T-shaped object consists of a massless link $\overline{CO}$ attached to a horizontal bar $\overline{AB}$ of mass m. Calculate the constant rotation speed Ω that allows $\theta = 45°, \dot{\theta} = \ddot{\theta} = 0$. Does this differ from the result you would get if all the mass were concentrated at C? $L = 0.5$ m and $m = 2$ kg.

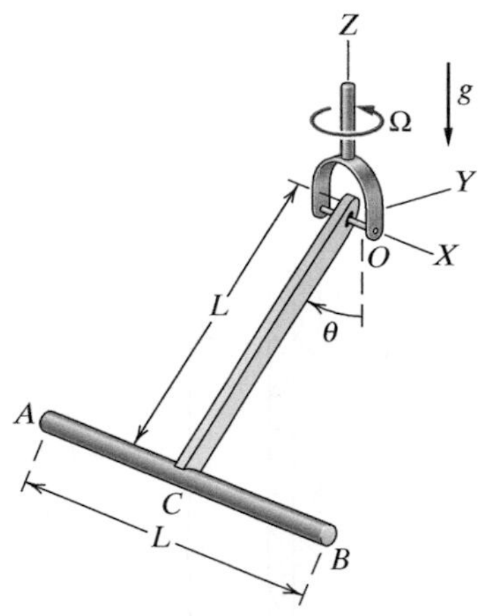

E8.8.6

8.8.7. **[Level 2]** Suppose the illustrated body is free to pivot about its fixed lower left corner O, and its angular velocity is a constant $\boldsymbol{\omega} = \omega_1 \boldsymbol{b}_1 + \omega_2 \boldsymbol{b}_2$ at the depicted instant. What are the forces and moments acting on the body at O for this instant? Express your answer in terms of the body's mass m.

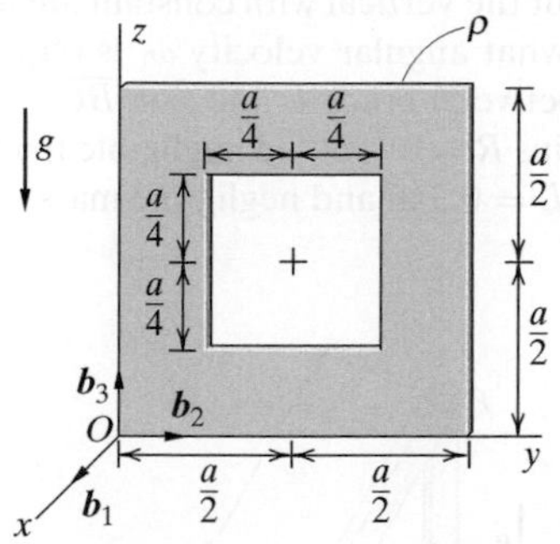

E8.8.7

8.8.8. **[Level 2]** A bicyclist coasting downhill figures he could impress a few friends by attempting to ride with no hands. Assume that, initially, the bike travels in a perfectly straight line with the mass centers of both the bike and the rider directly above the line connecting the contact points of both wheels. Naturally, it will be impossible for the rider to maintain this position, and there will be inevitable perturbations causing the bike to lean to one side or the other. When the bike does lean, describe what effect this will have on the motion of the front wheel, and why. Will the resulting motion of the front wheel tend to stabilize or destabilize the bike? Explain. Does it matter to which side, left or right, the lean occurs?

8.8.9. **[Level 2]** A 25 kg rigid body has the following moments and products of inertia about its mass center G in the x, y, z coordinate system shown: $I_{xx} = 0.05, I_{yy} = 0.03, I_{zz} = 0.1, I_{xy} = 0.02, I_{xz} = 0.04$, and $I_{yz} = 0.07$ kg·m^2. If $\boldsymbol{\omega} = (8\boldsymbol{i} - 12\boldsymbol{j} + 3\boldsymbol{k})$ rad/s, $\boldsymbol{\alpha} = (85\boldsymbol{i} - 100\boldsymbol{j} + 30\boldsymbol{k})$ rad/s^2, $\boldsymbol{v}_G = 2\boldsymbol{i}$ m/s, and $\boldsymbol{a}_G = 4\boldsymbol{i}$ m/s^2, what are the total force $\boldsymbol{F}$ and moment $\boldsymbol{M}_G$ acting on the body? Express the force and moment values in xyz components.

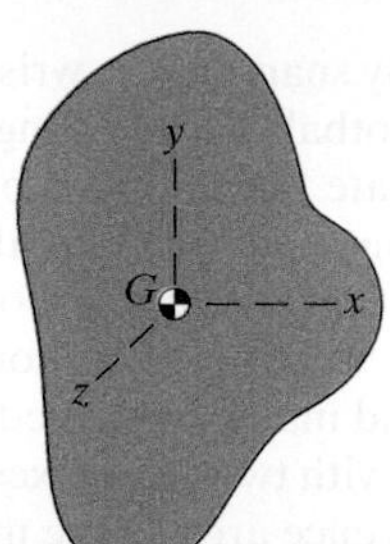

E8.8.9

8.8.10. **[Level 2]** The slender, uniform T-bar illustrated in the figure has mass $m = 3$ kg, and the length h is 30 cm. The body is initially at rest when a force $\boldsymbol{F} = (-50\boldsymbol{i} + 40\boldsymbol{j} + 20\boldsymbol{k})$ N is applied at the point $-\frac{h}{2}\boldsymbol{i}$. Determine the components of the angular acceleration $\boldsymbol{\alpha}$, as well as the linear accelerations $\boldsymbol{a}_G$ and $\boldsymbol{a}_P$ of the center of mass and point P at the instant the force is applied.

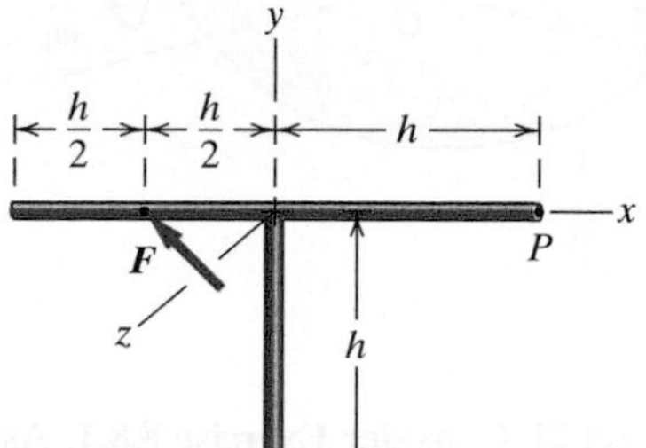

E8.8.10

8.8.11. **[Level 2]** The symmetric mechanism shown in the figure rotates about the vertical axis at constant angular speed $\Omega = 3$ rad/s in the direction shown. The vertical shaft is rigidly fixed to horizontal link $\overline{CD}$, which is pinned at one end to link $\overline{AC}$ and at the other to link $\overline{BD}$. Two identical disks of radius $R = 8$ cm and mass $m = 5$ kg are mounted at points A and B on the ends of links $\overline{AC}$ and $\overline{BD}$, respectively, and spin at equal rates ω about the axis of the respective link. A spring with an unstretched length of 30 cm and spring constant $k = 320$ N/m connects link $\overline{AC}$ to link $\overline{BD}$ at the illustrated points. Determine the angular speed ω, and its direction, that is required to maintain an angle $\theta = 30°$ between the vertical axis and links $\overline{AC}$ and $\overline{BD}$. The mass of the shafts and spring is negligible.

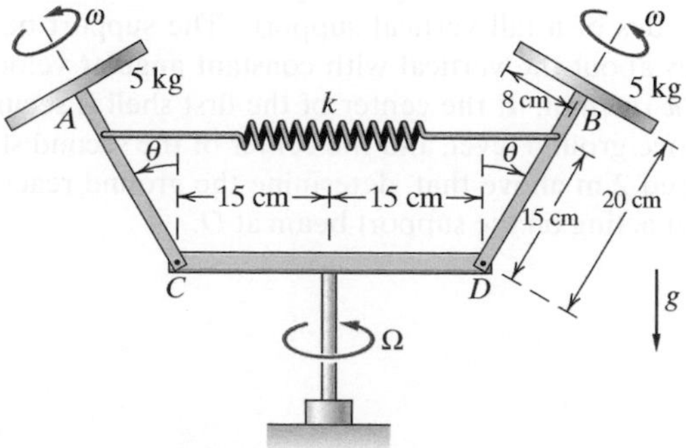

E8.8.11

8.8.12. **[Level 2]** By snapping his wrist, the quarterback has just released the football in a wobbling, "wounded duck" spiral (a prime candidate for an interception). A spin rate of 15 rad/s has been imparted to the football about its long axis, but the long axis itself is rotating about the flight direction at an angle of 15°. The football is composed of 0.26 kg of soft foam rubber and may be modeled as a solid prolate spheroid (an ellipsoid with two equal axes). It is 25 cm long with a 48 cm circumference around the middle. Determine the precession rate ω_1 of the long axis of the football, and calculate the rotational kinetic energy during flight. The principal moments of inertia for an ellipsoid are $I_1 = \frac{1}{5}m(a^2 + b^2)$, $I_2 = \frac{1}{5}m(a^2 + c^2)$, and $I_3 = \frac{1}{5}m(b^2 + c^2)$, where the semi-axes are of lengths a, b, and c.

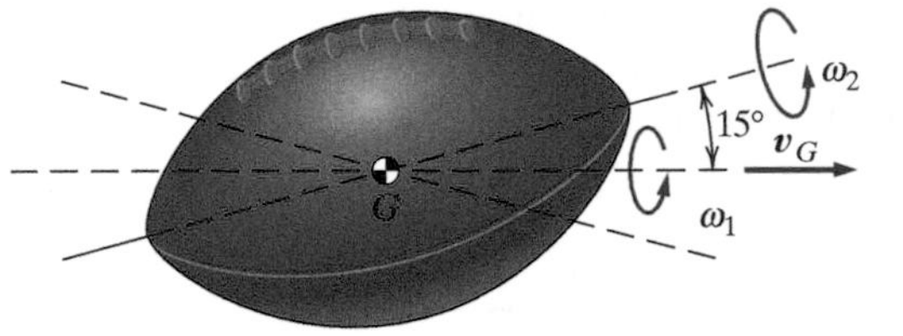

E8.8.12

8.8.13. **[Level 2]** Consider **Exercise 8.8.1**. Assume that torsional friction $T_f = 0.2$ N·m in the platform swivel slowly brings it to rest. Let $\omega = 18$ rad/s. If the student performed the described action *instantaneously*, without altering the position of his mass center or that of the wheel, determine the angle ϕ through which the platform rotates before coming to rest. The horizontal distance between the vertical z axis of the swivel and the wheel center is 55 cm. The mass of the wheel is 2 kg, and its principal moments of inertia are $I_{xx} = 0.2$ kg·m² and $I_{yy} = I_{zz} = 0.1$ kg·m². The student/swivel combination may be approximated as a vertical cylinder with a moment of inertia $I = 0.9$ kg·m² about the z axis.

8.8.14. **[Level 2]** Two massive hemispherical shells with 2 m diameters and weighing 115 kg each are secured to opposite sides of a tall vertical support. The support beam revolves about the vertical with constant angular velocity $\boldsymbol{\omega} = 30\boldsymbol{k}$ rev/min. If the center of the first shell is secured 4 m above ground level, and the center of the second shell is secured 2 m above that, determine the ground reaction moment acting on the support beam at O.

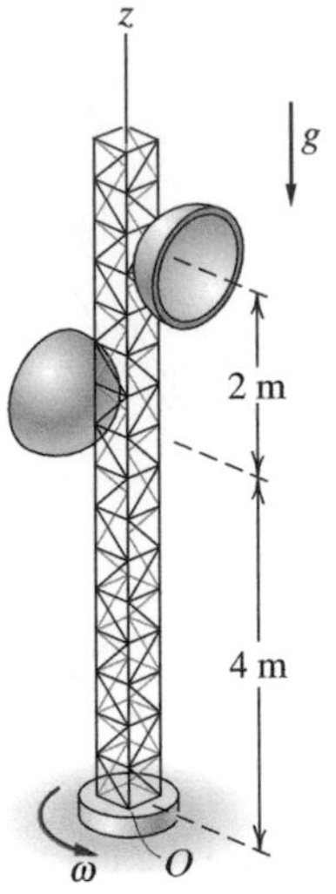

E8.8.14

8.8.15. **[Level 2]** Disk D spins with angular speed ω_1 about arm $\overline{BC}$, which is hinged to shaft $\overline{BA}$ at B. A brace E, extending from shaft $\overline{BA}$, supports arm $\overline{BC}$ at an angle $\theta = 30°$ from vertical, as shown in the figure. If shaft $\overline{BA}$ rotates about the vertical with constant angular velocity $\boldsymbol{\omega}_2 = 2\boldsymbol{k}$ rad/s, what angular velocity ω_1 is required for the support force between brace E and arm $\overline{BC}$ to go to zero? Disk D has radius $R = 0.1$ m and negligible thickness. Arm $\overline{BC}$ has length $L = 0.3$ m and negligible mass.

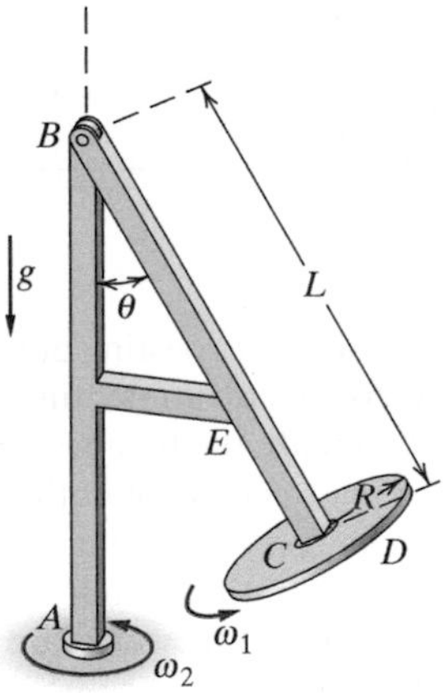

E8.8.15

8.8.16. **[Level 2]** Consider the system illustrated in the figure. Two thin, identical disks A and B spin freely about shaft $\overline{CD}$, which is suspended at point O by a vertical support. The shaft itself pivots freely at point O about a horizontal axis and the vertical. Let the spin rates of disks A and B be ω_1 and ω_2, respectively, in the directions indicated, and let the distances between point O and the centers of disks A and B be L_1 and L_2, respectively. Determine the ratio $(\omega_1 - \omega_2)/(L_1 - L_2)$ for which the system will rotate steadily (called precession) at a given rate Ω about the vertical axis, with shaft $\overline{CD}$ remaining in the horizontal position. The mass and radius of each disk are m and R, respectively, and the mass of the shaft may be neglected.

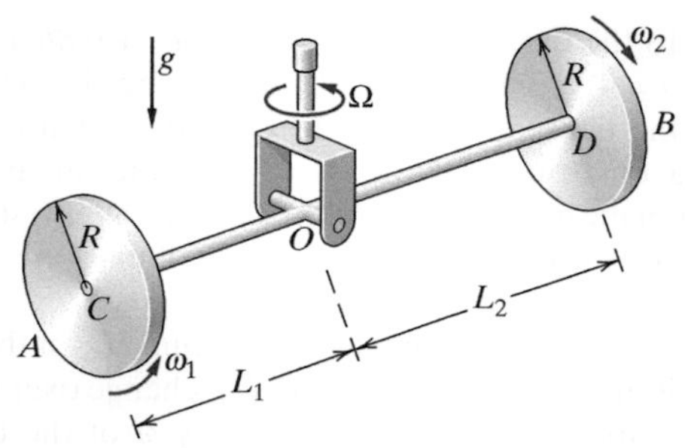

E8.8.16

8.8.17. **[Level 3]** A rod is attached by a horizontal pivot to a vertical shaft. Treat the rod as one-dimensional, with a linear density of ρ kg/m.

a. Find the full equations of motion for the system for $\ddot{\psi} = 0$.

b. Find the equilibrium conditions for which $\dot{\psi} = \dot{\psi}_0$ (constant), $\dot{\theta} = \ddot{\theta} = 0$, and $\theta = \theta_0 \neq 0$—that is, steady rotation with the rod tilted up from vertical at some constant angle.

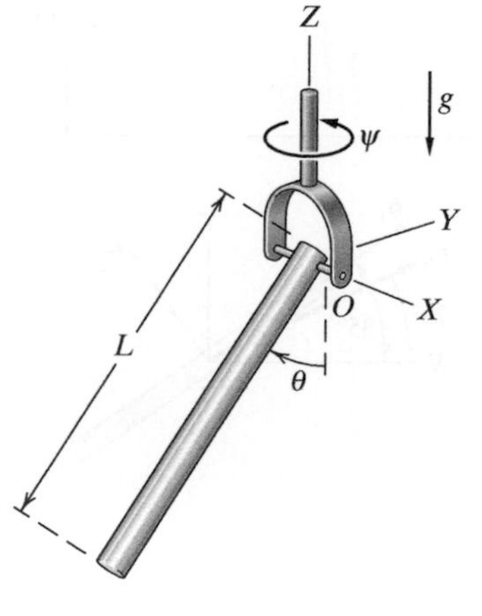

E8.8.17

8.8.18. **[Level 3]** The L-shaped bar has a total mass of 3 kg and is attached to a ball-and-socket support at end O. The bar may be modeled as a uniform slender rod with a right-angle bend and dimensions as given in the figure. Determine the components of the angular acceleration $\boldsymbol{\alpha}$ of the bar and support force $\boldsymbol{F}_O$ at the instant the bar is released from rest in the following two positions:

a. $\theta = 0$. The bar initially lies in the horizontal plane.

b. $\theta = 30°$. The 30 cm portion is initially elevated 30° above the horizontal, and the 40 cm portion lies in the horizontal X–Z plane.

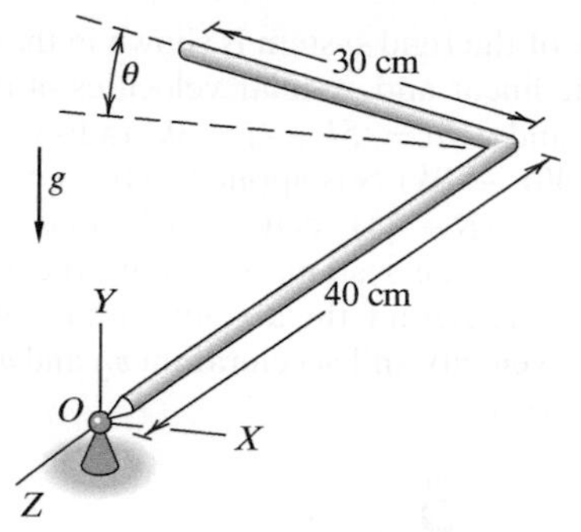

E8.8.18

8.8.19. **[Level 3]** The semicircular disk of negligible thickness has a total mass $m = 12$ kg and radius $R = 1.0$ m. The disk is connected to a ball-and-socket joint at point O, which is held fixed in space. If the initial angular velocity of the disk is $\boldsymbol{\omega} = (2\boldsymbol{j} + 5\boldsymbol{k})$ rad/s when the disk lies in the horizontal plane, determine the components of the angular acceleration $\boldsymbol{\alpha}$ and force $\boldsymbol{F}_O$ at point O at this instant.

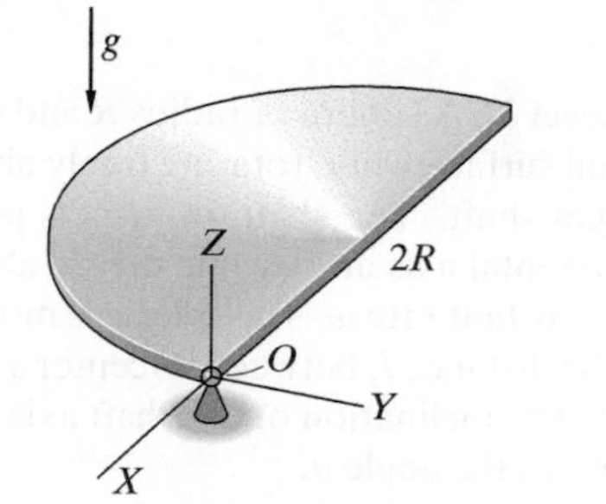

E8.8.19

8.8.20. **[Level 3]** The homogeneous rigid body depicted in the figure has a total weight of 18 kg and is acted on by a force $\boldsymbol{F}_Q = (89\boldsymbol{i} - 36\boldsymbol{j} + 44\boldsymbol{k})$ N at point Q. The body is also under the influence of gravity. If the angular velocity is $\boldsymbol{\omega} = (-5\boldsymbol{i} + 4\boldsymbol{j} - \boldsymbol{k})$ rad/s at the instant the force $\boldsymbol{F}_Q$ is applied, determine the acceleration of the mass center $\boldsymbol{a}_G$, the angular acceleration $\boldsymbol{\alpha}$, and the acceleration $\boldsymbol{a}_P$ of point P at this instant. The length h is 20 cm.

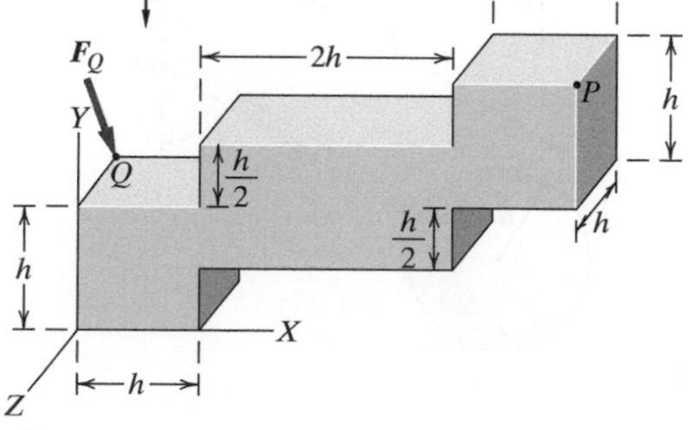

E8.8.20

8.8.21. **[Level 3]** A cylindrical shell of length $L = 80$ cm, radius $R = 5$ cm, and mass $M = 5$ kg is rigidly attached to two small $m = 3$ kg spherical masses via some massless rods.

The geometry of the rigid system is shown in the figure. At time $t = 0$, the linear and angular velocities of the system are $\boldsymbol{v}_G(0) = 0$ and $\boldsymbol{\omega}(0) = (5\boldsymbol{i} + 2\boldsymbol{j} - 3\boldsymbol{k})$ rad/s when a force $\boldsymbol{F} = (-60\boldsymbol{i} + 40\boldsymbol{j} - 50\boldsymbol{k})$ N is applied to the spherical mass at location $-\frac{L}{4}\boldsymbol{i} + (R + \frac{L}{4})\boldsymbol{j}$ relative to the mass center. No other forces act on the system. Compute the angular acceleration $\boldsymbol{\alpha}$ of the system, the acceleration $\boldsymbol{a}_G$ of the mass center, and the velocity and acceleration $\boldsymbol{v}_Q$ and $\boldsymbol{a}_Q$ of point Q, all at time $t = 0$.

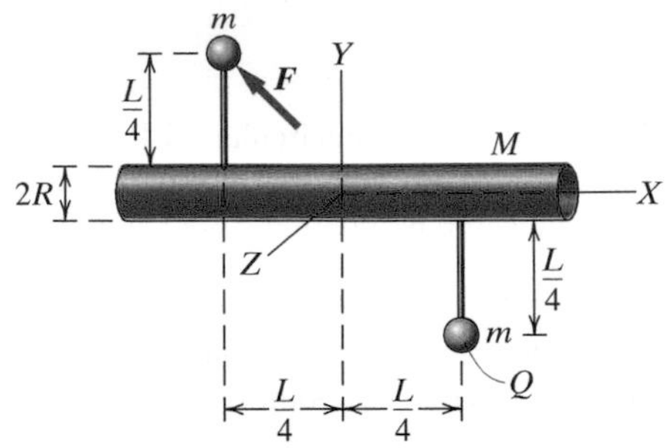

E8.8.21

8.8.22. **[Level 3]** A sphere of radius R and mass m rolls on a horizontal surface while rotating freely about the axis of a lightweight shaft. The shaft, in turn, is pinned freely about the horizontal axis at O, while driven about the vertical axis at a constant rate ω. The sphere is mounted to the shaft so that the distance L between its center and point O is held constant. The inclination of the shaft axis with respect to vertical defines the angle θ.

a. Find expressions for the vertical components of the reaction forces at points O and C (the contact point), in terms of m, g, L, θ, and ω.

b. Write an expression for the angle θ at which the vertical reaction at the contact point goes to zero, for a given rate ω. Assume that the sphere continues a "rolling" motion over the surface, although no frictional forces will be present. Note any restrictions on m, R, and L for which this will not be possible.

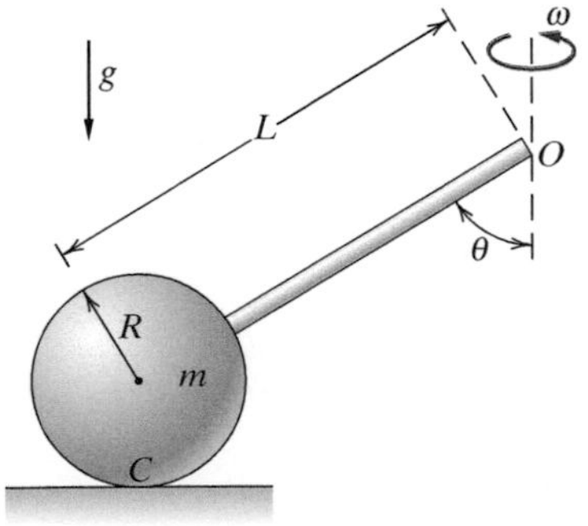

E8.8.22

8.8.23. **[Level 3]** In an Olympic discus event, a competitor has just completed his final throw, releasing the discus at a height of 1.5 m with a linear velocity of 25 m/s in a direction 25° above the horizontal. The angular velocity imparted to the disc is $\boldsymbol{\omega} = (-1.3\boldsymbol{b}_1 + 20\boldsymbol{b}_2 + 0.8\boldsymbol{b}_3)$ rad/s, where $\boldsymbol{b}_i : i = 1, 2, 3$ is the body-fixed reference frame illustrated in the figure, at the instant just after release. Initially, the geometric axis of the discus is oriented at an angle $\theta = 35°$ from the ground-fixed reference axes, as also shown. Approximate the discus as a circular disk of mass $m = 2$ kg, diameter $d = 22$ cm, and thickness $t = 4.5$ cm.

a. What is the initial angular momentum $\boldsymbol{H}_G$ of the discus about its mass center? How does $\boldsymbol{H}_G$ change over time?

b. Describe how the angular velocity $\boldsymbol{\omega}$ of the discus changes. Compute $\dot{\boldsymbol{\omega}}$ in the body-fixed reference frame. Under what release conditions would $\boldsymbol{\omega}$ remain constant?

c. Describe the trajectory of the discus mass center. What distance will the discus travel after release? Although realistic values have been given for all throw parameters, actual Olympic throws may travel 60 to 70 m under the same initial conditions. What do you think accounts for this difference? (*Hint:* Does a 25° release angle seem ideal?)

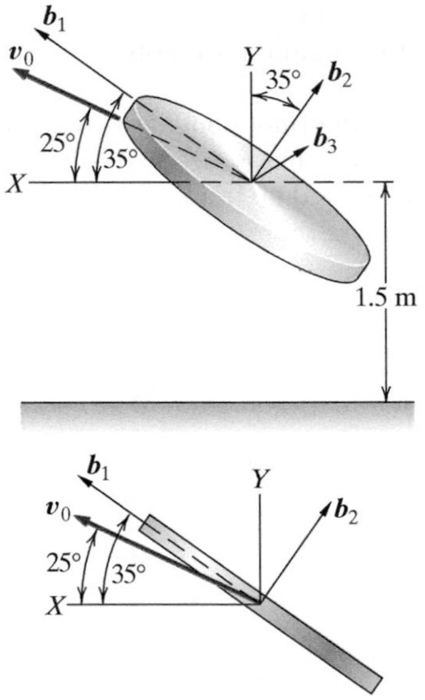

E8.8.23

8.8.24. **[Level 3]** A 5 kg solid cylinder has radius $R = 5$ cm, has thickness $t = 10$ cm, and spins freely about the revolving x axis at an initial rate $\omega_1 = 120$ rev/min. The massless 90°-bend shaft, to which the cylinder is attached, revolves freely about the vertical Y axis with an angular speed ω_2. $L = 15$ cm. If $\omega_2 = 40$ rev/min when a horizontal force $F = 40$ N is applied at point P on the cylinder, perpendicular to its axis of spin, determine

a. $\dot{\omega}_1, \dot{\omega}_2$, and the angular acceleration $\boldsymbol{\alpha}$ of the cylinder at the instant the force is applied. (Report in a convenient frame of reference.)

b. the reaction forces and moments at support A at this instant.

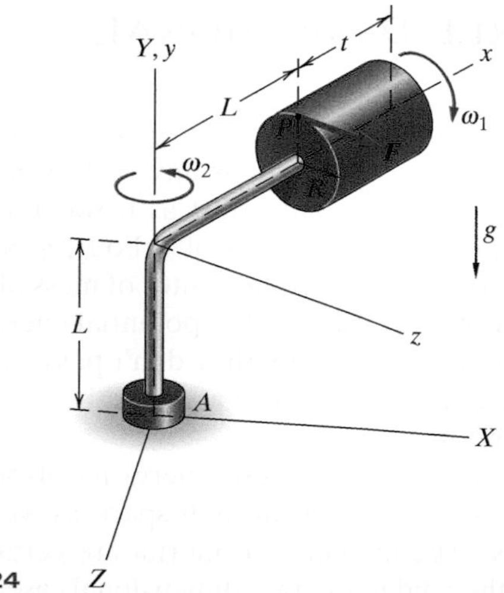

E8.8.24

8.8.25. **[Level 3]** The system in the figure is released from rest in the position shown. Disk m_1 and pulley m_2 are rigidly mounted to massless rod $\overline{AB}$, which is mounted horizontally in two frictionless ball-and-socket joints at points A and B. The disk is mounted off-center a distance d from its central axis, with the centroid initially at its vertical peak. A mass m_3 hangs from the pulley by a massless string. The radii of the disk and pulley, respectively, are r_1 and r_2. Determine the tension in the string, and the force reactions at supports A and B, when mass m_3 has descended a distance $s = \frac{9}{2}\pi r_2$ from its release position. Parameter values are $m_1 = 3$ kg, $m_2 = 0.5$ kg, $m_3 = 2$ kg, $r_1 = 10$ cm, $d = 6$ cm, $r_2 = 4$ cm, and $L_1 = L_2 = L_3 = 15$ cm.

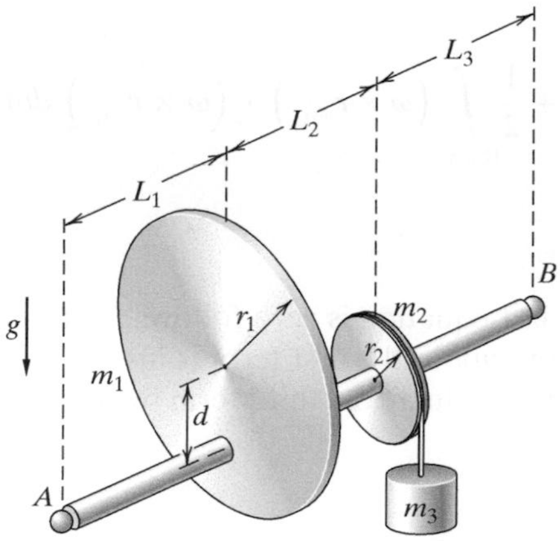

E8.8.25

8.8.26. **[Level 3]** A thin hemispherical shell of mass m and radius r is welded to a vertical shaft at O, as shown in the figure. The assembly rotates about the z axis of the shaft with constant angular velocity ω. Determine the bending moment, and its direction, at the spot of the weld in terms of m, ω, r, and g. Solve using two methods: (1) Equations of motion for the mass center G and (2) equations of motion for the fixed point O.

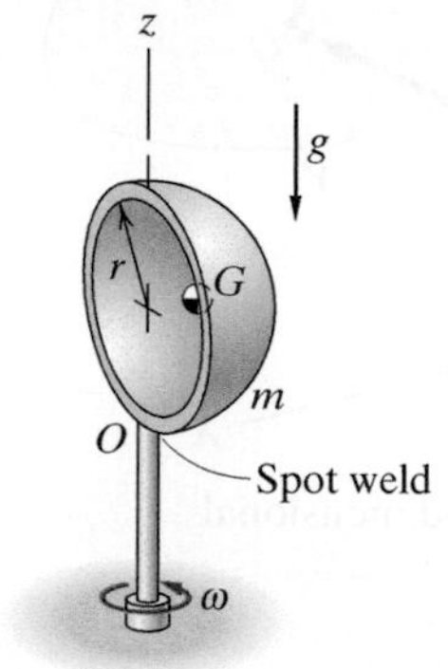

E8.8.26

8.8.27. **[Level 3]** The hemispherical shell has been welded to the vertical shaft at point O and is accelerated by an external motor torque $T = -0.05\boldsymbol{k}$ N·m acting on the shaft. The central axis of the shell is tilted at an angle $\theta = 40°$ from the vertical axis and intersects the shaft axis at point A, as depicted in the figure. The areal density of the shell is $\rho = 8$ kg/m^2, and its radius is $r = 6$ cm.

a. Determine the angular acceleration $\boldsymbol{\alpha}$ of the shell when its angular velocity is $\boldsymbol{\omega} = 10\boldsymbol{k}$ rad/s.

b. Determine the reaction moments at point O at the same instant. Report the values as components in the x, y, z coordinate frame that is aligned and rotating with the shaft.

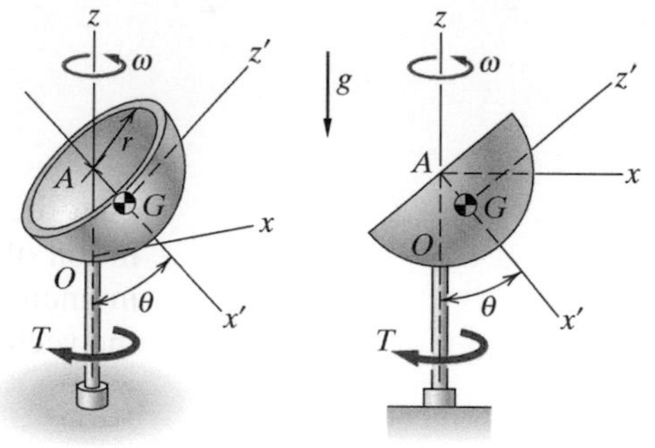

E8.8.27

8.9 ENERGY OF THREE-DIMENSIONAL BODIES

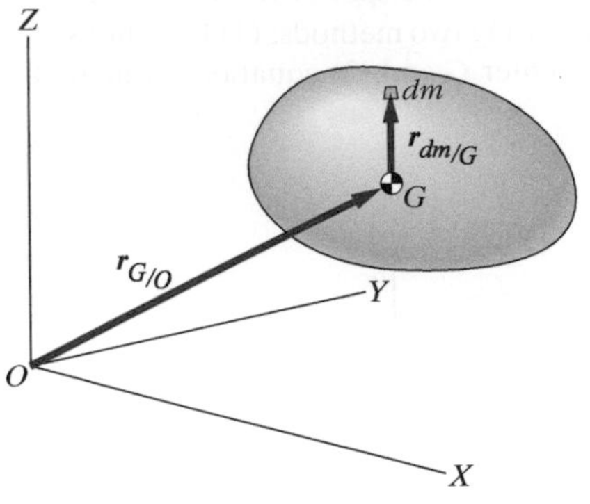

Figure 8.39 Three-dimensional dynamical potato

Although we can consider both potential and kinetic energy of three-dimensional bodies, only their kinetic energy needs any real examination. The gravitational potential energy remains just as it was for the two-dimensional case—mgh, where m is the mass of a body, g is the gravitational acceleration, and h is the height of the center of mass of the body above or below some reference level where the potential energy is defined to be zero. Because our bodies are rigid, they don't possess any internal energy due to structural deformation (such as would be the case for a squeezed rubber ball, for instance).

Although potential energy isn't a problem, kinetic energy merits some thought for any rigid body that is translating through space as well as rotating about up to three axes. The moments of inertia are certainly going to come into play, just as they did in the two-dimensional case. We can start with the basic definition of kinetic energy for a three-dimensional body (**Figure 8.39**) undergoing general motion:

$$
\begin{aligned}
\mathcal{KE} &= \frac{1}{2}\int_{\text{Body}} \left[\bar{\boldsymbol{v}} + \boldsymbol{\omega} \times \boldsymbol{r}_{dm/G}\right]^2 dm \\
&= \frac{1}{2}\int_{\text{Body}} \left[\bar{\boldsymbol{v}} + \boldsymbol{\omega} \times \boldsymbol{r}_{dm/G}\right] \cdot \left[\bar{\boldsymbol{v}} + \boldsymbol{\omega} \times \boldsymbol{r}_{dm/G}\right] dm \\
&= \frac{1}{2}\bar{\boldsymbol{v}} \cdot \bar{\boldsymbol{v}} \underbrace{\int_{\text{Body}} dm}_{=m} + \int_{\text{Body}} \bar{\boldsymbol{v}} \cdot \left(\boldsymbol{\omega} \times \boldsymbol{r}_{dm/G}\right) dm + \frac{1}{2}\int_{\text{Body}} \left(\boldsymbol{\omega} \times \boldsymbol{r}_{dm/G}\right) \cdot \left(\boldsymbol{\omega} \times \boldsymbol{r}_{dm/G}\right) dm \\
&= \frac{1}{2}\bar{\boldsymbol{v}} \cdot m\bar{\boldsymbol{v}} + \bar{\boldsymbol{v}} \cdot \left[\boldsymbol{\omega} \times \underbrace{\int_{\text{Body}} \boldsymbol{r}_{dm/G}\, dm}_{=0}\right] + \frac{1}{2}\int_{\text{Body}} \left(\boldsymbol{\omega} \times \boldsymbol{r}_{dm/G}\right) \cdot \left(\boldsymbol{\omega} \times \boldsymbol{r}_{dm/G}\right) dm
\end{aligned}
$$

That the second term in this last equation is zero follows from the definition of mass center. Also, the $m\bar{\boldsymbol{v}}$ in the first term is just L, the linear momentum of the body. We can evaluate the third term by remembering the vector relationship

$$\boldsymbol{A} \cdot (\boldsymbol{B} \times \boldsymbol{C}) = (\boldsymbol{A} \times \boldsymbol{B}) \cdot \boldsymbol{C}$$

Let $\boldsymbol{A} = \boldsymbol{\omega}$, $\boldsymbol{B} = \boldsymbol{r}_{dm/G}$, and $\boldsymbol{C} = \boldsymbol{\omega} \times \boldsymbol{r}_{dm/G}$. With this substitution we can see that the integrand of the third term in the final equation is in the form

$(\boldsymbol{A} \times \boldsymbol{B}) \cdot \boldsymbol{C}$. This lets us express $\mathcal{KE}$ as

$$\mathcal{KE} = \frac{1}{2}\overline{\boldsymbol{v}} \cdot \boldsymbol{L} + \frac{1}{2}\boldsymbol{\omega} \cdot \underbrace{\int_{\text{Body}} \boldsymbol{r}_{dm/G} \times \left(\boldsymbol{\omega} \times \boldsymbol{r}_{dm/G}\right) dm}_{=\boldsymbol{H}_G \text{ from (8.30)}} \tag{8.51}$$

$$= \frac{1}{2}\overline{\boldsymbol{v}} \cdot \boldsymbol{L} + \frac{1}{2}\boldsymbol{\omega} \cdot \boldsymbol{H}_G$$

This becomes even simpler if the rigid body is pinned at O. In this case we know that $\overline{\boldsymbol{v}} = \boldsymbol{\omega} \times \boldsymbol{r}_{G/O}$, which leads to

$$\mathcal{KE} = \frac{1}{2}\left(\boldsymbol{\omega} \times \boldsymbol{r}_{G/O}\right) \cdot \boldsymbol{L} + \frac{1}{2}\boldsymbol{\omega} \cdot \boldsymbol{H}_G$$

$$= \frac{1}{2}\boldsymbol{\omega} \cdot \left(\boldsymbol{r}_{G/O} \times \boldsymbol{L}\right) + \frac{1}{2}\boldsymbol{\omega} \cdot \boldsymbol{H}_G = \frac{1}{2}\boldsymbol{\omega} \cdot \left(\boldsymbol{r}_{G/O} \times \boldsymbol{L} + \boldsymbol{H}_G\right) \tag{8.52}$$

$$= \frac{1}{2}\boldsymbol{\omega} \cdot \boldsymbol{H}_O$$

>>>**Check out Example 8.10 (page 534) for an application of this material.**

EXAMPLE 8.10 KINETIC ENERGY OF A ROTATING DISK (Theory on page 533)

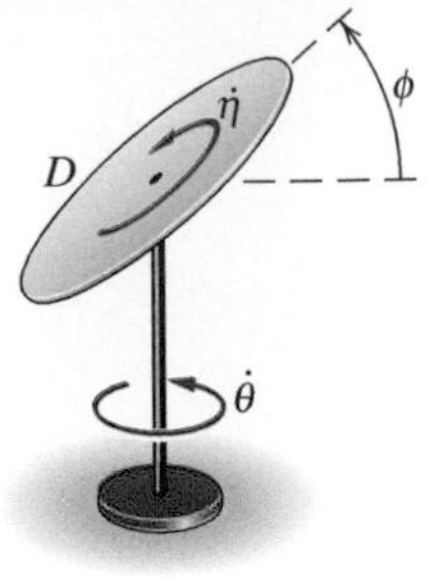

Figure 8.40 Spinning, rotating disk

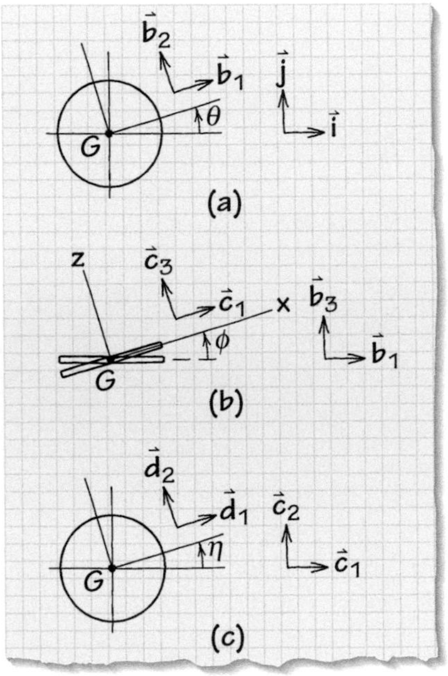

Figure 8.41 Unit-vector transformations

Consider the disk shown in **Figure 8.40**. The disk is supported by a vertical stand that rotates with angular speed $\dot{\theta}$. The plane of the disk is inclined by an angle ϕ with respect to the ground, and the disk spins along its central axis at an angular speed of $\dot{\eta}$. Determine the disk's kinetic energy. Assume that the disk's thickness is negligible. The disk's radius is R, and it has a mass m and areal density ρ.

Goal Determine the kinetic energy of a rotating disk.

Given Dimensions and mass of the disk, as well as its angular velocity components.

Draw The first step is to set up a convenient set of body-fixed unit vectors with which to calculate the angular momentum and angular velocities. **Figure 8.41** shows a sequence of unit-vector sets that lead from the ground set to the rotating disk. We start with the disk on the ground. When it rotates through an angle θ, it gets to the orientation shown in **Figure 8.41*a***. A set of axes originally aligned with the x, y, z axes is now rotated by the angle θ and defines the intermediate unit vectors $\boldsymbol{b}_1$, $\boldsymbol{b}_2$, $\boldsymbol{b}_3$. The disk is still flat on the floor. The coordinate transformation array from $\boldsymbol{i}, \boldsymbol{j}, \boldsymbol{k}$ to $\boldsymbol{b}_1, \boldsymbol{b}_2, \boldsymbol{b}_3$ is given by

	$\boldsymbol{i}$	$\boldsymbol{j}$	$\boldsymbol{k}$
$\boldsymbol{b}_1$	$\cos\theta$	$\sin\theta$	0
$\boldsymbol{b}_2$	$-\sin\theta$	$\cos\theta$	0
$\boldsymbol{b}_3$	0	0	1

Next, we tilt the disk up by an angle ϕ, as shown in **Figure 8.41*b***. This defines a new set of unit vectors, $\boldsymbol{c}_1$, $\boldsymbol{c}_2$, $\boldsymbol{c}_3$. The transformation array to go from $\boldsymbol{b}_1, \boldsymbol{b}_2, \boldsymbol{b}_3$ to $\boldsymbol{c}_1, \boldsymbol{c}_2, \boldsymbol{c}_3$ is

	$\boldsymbol{b}_1$	$\boldsymbol{b}_2$	$\boldsymbol{b}_3$
$\boldsymbol{c}_1$	$\cos\phi$	0	$\sin\phi$
$\boldsymbol{c}_2$	0	1	0
$\boldsymbol{c}_3$	$-\sin\phi$	0	$\cos\phi$

The disk is now in the correct basic orientation. Even though we could impose one more rotation, from $\boldsymbol{c}_1$, $\boldsymbol{c}_2$, $\boldsymbol{c}_3$ to $\boldsymbol{d}_1$, $\boldsymbol{d}_2$, $\boldsymbol{d}_3$ (**Figure 8.41*c***), there's no need to do so. The disk is circular, and therefore we can express our equations using the $\boldsymbol{c}_i$ unit vectors without any loss of accuracy. If the disk weren't circular, then we would have to go to the $\boldsymbol{d}_i$ vector set (which rotates with the body) in order to capture the differences due to the disk's varying orientation as it rotates. But, because the disk is circular, there's no change in the mass properties of the disk as it rotates, and sticking with the $\boldsymbol{c}_i$ vectors is both accurate and convenient.

Assume No additional assumptions are needed.

Formulate Equations We will apply (8.52), which means we need $\boldsymbol{\omega}$

and H_G. For this problem the point of rotation O is coincident with G, the mass center.

Solve The angular velocity of the disk is found by adding its angular velocity due to the rotation rate $\dot{\theta}$ about the stand and its rotation rate about $\dot{\eta}$ its central axis:

$$\begin{aligned}\boldsymbol{\omega}_D &= \dot{\theta}\boldsymbol{b}_3 + \dot{\eta}\boldsymbol{c}_3 \\ &= \dot{\theta}\sin\phi\boldsymbol{c}_1 + (\dot{\eta} + \dot{\theta}\cos\phi)\boldsymbol{c}_3\end{aligned}$$

The disk's products of inertia are all zero, a consequence of its symmetry. From Appendix B we see that the principal moments of inertia are

$$\overline{I}_{xx} = \frac{mR^2}{4}, \qquad \overline{I}_{yy} = \frac{mR^2}{4}, \qquad \overline{I}_{zz} = \frac{mR^2}{2}$$

and so $\boldsymbol{H}_G$ is given by

$$\begin{aligned}\boldsymbol{H}_G &= \overline{I}_{xx}\omega_1\boldsymbol{c}_1 + \overline{I}_{yy}\omega_2\boldsymbol{c}_2 + \overline{I}_{zz}\omega_3\boldsymbol{c}_3 \\ &= \frac{mR^2\dot{\theta}\sin\phi}{4}\boldsymbol{c}_1 + \frac{mR^2(\dot{\eta} + \dot{\theta}\cos\phi)}{2}\boldsymbol{c}_3\end{aligned}$$

Applying (8.52) gives us

$$\begin{aligned}\mathcal{KE} &= \frac{1}{2}\boldsymbol{\omega}_D \cdot \boldsymbol{H}_G \\ &= \frac{1}{2}[\dot{\theta}\sin\phi\boldsymbol{c}_1 + (\dot{\eta} + \dot{\theta}\cos\phi)\boldsymbol{c}_3] \cdot \left[\frac{mR^2\dot{\theta}\sin\phi}{4}\boldsymbol{c}_1 + \frac{mR^2(\dot{\eta} + \dot{\theta}\cos\phi)}{2}\boldsymbol{c}_3\right]\end{aligned}$$

$$\mathcal{KE} = \frac{mR^2\dot{\theta}^2\sin^2\phi}{8} + \frac{mR^2[\dot{\eta} + \dot{\theta}\cos\phi]^2}{4}$$

EXERCISES 8.9

8.9.1. **[Level 1]** The illustrated rectangular body (mass m) is both translating and rotating. $\boldsymbol{v}_G = (1.5\boldsymbol{b}_1 + 1.5\boldsymbol{b}_2 + 10\boldsymbol{b}_3)$ m/s, and its angular velocity is $(5\boldsymbol{b}_1 + 15\boldsymbol{b}_3)$ rad/s. What is the body's total kinetic energy?

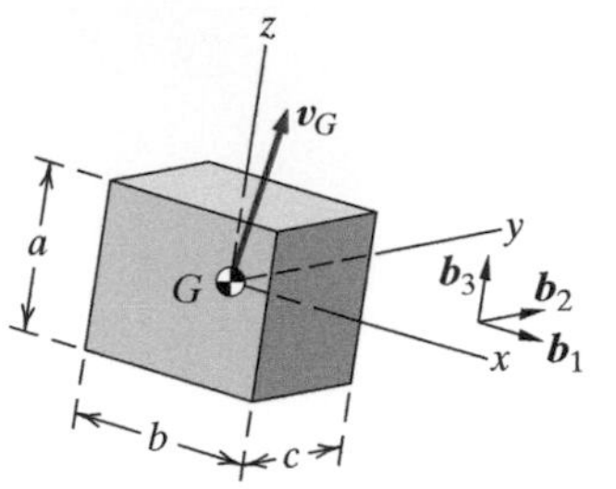

E8.9.1

8.9.2. **[Level 2]** Find the kinetic energy of the illustrated system if its angular velocity at the depicted instant is $\boldsymbol{\omega} = \omega_1\boldsymbol{b}_1 + \omega_2\boldsymbol{b}_2 + \omega_3\boldsymbol{b}_3$.

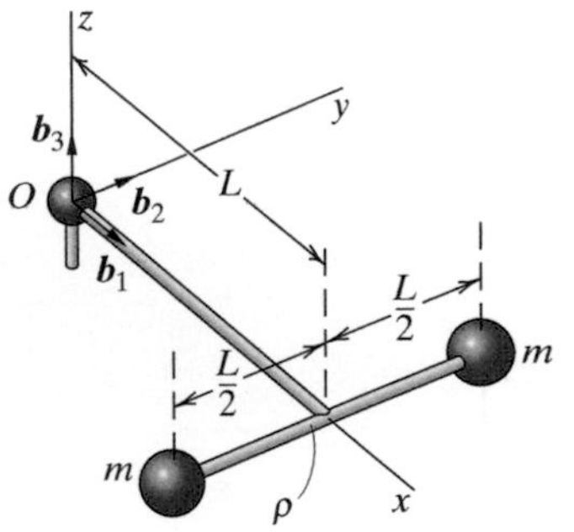

E8.9.2

8.9.3. **[Level 2]** The illustrated plate has an angular velocity of $(20\boldsymbol{i} - 10\boldsymbol{j})$ rad/s, $a = 0.5$ m, and $b = 0.4$ m and is pinned at O. By what percentage will the plate's kinetic energy change from its value with a finite width $h = 0.02$ m versus being approximated as a flat plate? Assume a mass m.

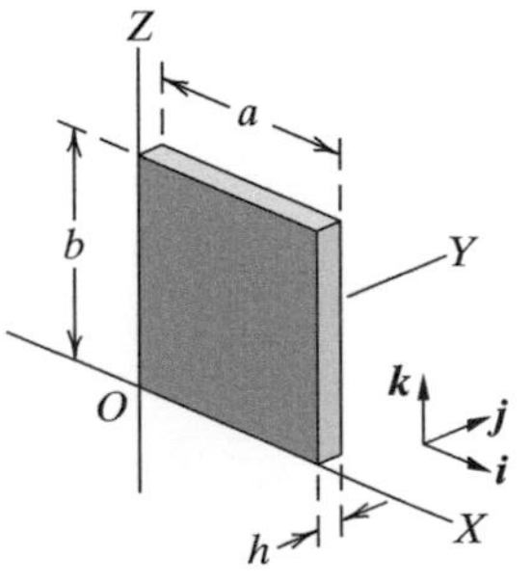

E8.9.3

8.9.4. **[Level 2]** The illustrated solid cone has an angular velocity of $(8\boldsymbol{i} + 6\boldsymbol{j})$ rad/s and is pinned at O. $r = 0.2$ m and $h = 0.6$ m. What is its kinetic energy? Assume a mass m.

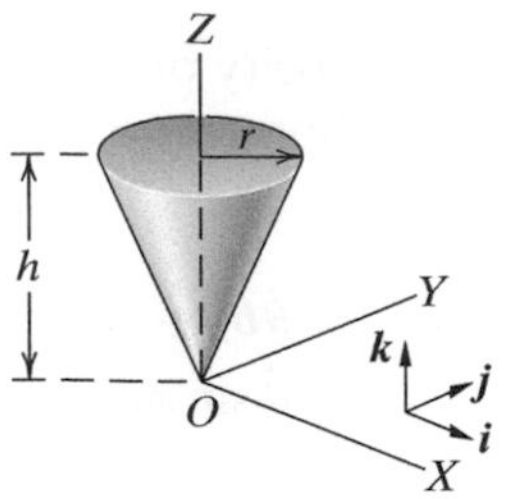

E8.9.4

8.9.5. **[Level 2]** Find the kinetic energy of the illustrated system if its angular velocity is $\boldsymbol{\omega} = \omega_2\boldsymbol{b}_2 + \omega_3\boldsymbol{b}_3$.

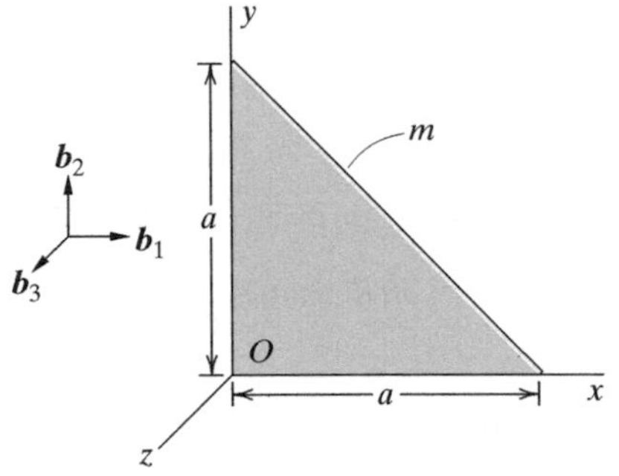

E8.9.5

8.9.6. **[Level 2]** The illustrated thin, flat triangular plate has an angular velocity of $(6\boldsymbol{i} + 6\boldsymbol{j})$ rad/s and is fixed in space at O. $a = 0.5$ m, $b = 0.4$ m, and the mass is m. What is the kinetic energy of the plate?

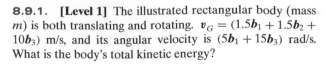

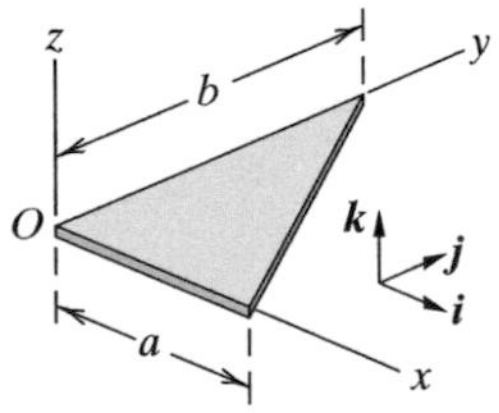

E8.9.6

8.9.7. **[Level 2]** Consider the system in **Exercise 8.4.4**. What is the kinetic energy of the arm $\overline{AB}$? Treat it as a thin rod (negligible diameter) of mass m.

8.9.8. **[Level 2]** Consider the system in **Exercise 8.4.1**. Assume that the disk has a radius of 0.23 m and thickness 0.01 m. $a = 0.3$ m, $\omega_1 = 10$ rad/s, $\omega_2 = 70$ rad/s, and the disk's mass is 1.2 kg. What is the disk's kinetic energy?

8.9.9. **[Level 2]** Consider the system in **Exercise 8.4.5**. Assume that the rotating framework has a mass of 0.005 kg and each thin, circular wheel has mass 0.002 kg. What is the system's total kinetic energy equal to?

8.9.10. **[Level 2]** Consider the system in **Exercise 8.4.12**. The rolling cone has a mass of 22g. What is its kinetic energy equal to?

8.9.11. **[Level 2]** Consider the system in **Exercise 8.7.1.** What is the body's kinetic energy?

8.9.12. **[Level 2]** Determine the kinetic energy of the system shown in **Exercise 8.7.3.**

8.9.13. **[Level 2]** Determine the kinetic energy of the system shown in **Exercise 8.7.4.**

8.10 JUST THE FACTS

This chapter dealt with three-dimensional dynamics and covered kinematics, kinetics, momentum, and energy. Although the only obvious difference between two- and three-dimensional motion is the addition of another coordinate, in reality we saw that the dynamics became substantially more complicated—far more so than our two-dimensional intuition would have predicted.

The chapter began with a look at spherical coordinates:

$$\boldsymbol{R}_{m/O} = R\boldsymbol{e}_R \tag{8.1}$$

$$\dot{\boldsymbol{R}}_m = \dot{R}\boldsymbol{e}_R + R\dot{\phi}\boldsymbol{e}_\phi + R\dot{\theta}\sin\phi\,\boldsymbol{e}_\theta \tag{8.4}$$

$$\begin{aligned}\ddot{\boldsymbol{R}}_m &= (\ddot{R} - R\dot{\phi}^2 - R\dot{\theta}^2\sin^2\phi)\boldsymbol{e}_R \\ &+ \left(\frac{1}{R}\frac{d}{dt}(R^2\dot{\phi}) + \frac{R\dot{\theta}^2\sin(2\phi)}{2}\right)\boldsymbol{e}_\phi \\ &+ \left(\frac{\sin\phi}{R}\frac{d}{dt}(R^2\dot{\theta}) - 2R\dot{\theta}\dot{\phi}\cos\phi\right)\boldsymbol{e}_\theta\end{aligned} \tag{8.5}$$

We then saw how to determine the orientation of a body in terms of **Euler angles** and from that how to obtain a body's angular velocity. Next on the list was differentiating our angular velocity to obtain a three-dimensional angular acceleration:

$$\left.\frac{d}{dt}\right|_N \boldsymbol{\omega}_{\text{Body}} = \left.\frac{d}{dt}\right|_S \boldsymbol{\omega}_{\text{Body}} + \boldsymbol{\omega}_S \times \boldsymbol{\omega}_{\text{Body}} \tag{8.8}$$

After handling angular velocity and acceleration, we were ready to look at linear velocities and accelerations. The velocity of a point fixed to a rigid body was given by

$$\boldsymbol{v}_B = \boldsymbol{v}_A + \boldsymbol{\omega} \times \boldsymbol{r}_{B/A} \tag{8.9}$$

and its acceleration was found from

$$\boldsymbol{a}_B = \boldsymbol{a}_A + \boldsymbol{\alpha} \times \boldsymbol{r}_{B/A} + \boldsymbol{\omega} \times \left(\boldsymbol{\omega} \times \boldsymbol{r}_{B/A}\right) \tag{8.10}$$

When we were concerned with the velocity of a point moving on a rigid body, we had

$$\boldsymbol{v}_B = \boldsymbol{v}_A + \boldsymbol{\omega} \times \boldsymbol{r}_{B/A} + \boldsymbol{v}_{rel} \tag{8.11}$$

for the point's velocity and

$$\boldsymbol{a}_B = \boldsymbol{a}_A + \boldsymbol{\alpha} \times \boldsymbol{r}_{B/A} + \boldsymbol{\omega} \times \left(\boldsymbol{\omega} \times \boldsymbol{r}_{B/A}\right) + \boldsymbol{a}_{rel} + 2\boldsymbol{\omega} \times \boldsymbol{v}_{rel} \tag{8.12}$$

for its acceleration.

The first really big difference between two- and three-dimensional dynamics was the complex form that a body's inertia exhibited. We defined three inertia properties, the **moments of inertia**:

$$I_{xx} = \int_{\text{Body}} (y^2 + z^2)\, dm \tag{8.18}$$

$$I_{yy} = \int_{\text{Body}} (x^2 + z^2)\, dm \tag{8.19}$$

$$I_{zz} = \int_{\text{Body}} (x^2 + y^2)\, dm \tag{8.20}$$

and the **products of inertia**:

$$I_{xy} = \int_{\text{Body}} xy\, dm \tag{8.21}$$

$$I_{yz} = \int_{\text{Body}} yz\, dm \tag{8.22}$$

$$I_{zx} = \int_{\text{Body}} zx\, dm \tag{8.23}$$

We then saw how to extend our parallel axis theorem from two to three dimensions:

$$I_{xx} = \overline{I}_{x'x'} + m(r_2^2 + r_3^2) \tag{8.24}$$

$$I_{yy} = \overline{I}_{y'y'} + m(r_3^2 + r_1^2) \tag{8.25}$$

$$I_{zz} = \overline{I}_{z'z'} + m(r_1^2 + r_2^2) \tag{8.26}$$

$$I_{xy} = \overline{I}_{x'y'} + mr_1r_2 \tag{8.27}$$

$$I_{yz} = \overline{I}_{y'z'} + mr_2r_3 \tag{8.28}$$

$$I_{zx} = \overline{I}_{z'x'} + mr_3r_1 \tag{8.29}$$

Angular momentum about the mass center was found from the general expression

$$\boldsymbol{H}_G = \int_{\text{Body}} \boldsymbol{r}_{dm/G} \times \left(\boldsymbol{\omega} \times \boldsymbol{r}_{dm/G}\right) dm \tag{8.30}$$

and about a point of rotation O from

$$\boldsymbol{H}_O = \int_{\text{Body}} \boldsymbol{r}_{dm/O} \times \left(\boldsymbol{\omega} \times \boldsymbol{r}_{dm/O}\right) dm \tag{8.31}$$

If the particular forms for the position and angular velocity were given by $\boldsymbol{r}_{dm/G} = x\boldsymbol{b}_1 + y\boldsymbol{b}_2 + z\boldsymbol{b}_3$ and $\boldsymbol{\omega} = \omega_1\boldsymbol{b}_1 + \omega_2\boldsymbol{b}_2 + \omega_3\boldsymbol{b}_3$, then these equations became

$$\begin{aligned}\boldsymbol{H}_G &= \left[\bar{I}_{xx}\omega_1 - \bar{I}_{xy}\omega_2 - \bar{I}_{zx}\omega_3\right]\boldsymbol{b}_1 \\ &+ \left[\bar{I}_{yy}\omega_2 - \bar{I}_{yz}\omega_3 - \bar{I}_{xy}\omega_1\right]\boldsymbol{b}_2 \\ &+ \left[\bar{I}_{zz}\omega_3 - \bar{I}_{zx}\omega_1 - \bar{I}_{zy}\omega_2\right]\boldsymbol{b}_3\end{aligned} \tag{8.32}$$

for the angular momentum about G and became

$$\begin{aligned}\boldsymbol{H}_O &= \left[I_{xx}\omega_1 - I_{xy}\omega_2 - I_{xz}\omega_3\right]\boldsymbol{b}_1 \\ &+ \left[I_{yy}\omega_2 - I_{yz}\omega_3 - I_{yx}\omega_1\right]\boldsymbol{b}_2 \\ &+ \left[I_{zz}\omega_3 - I_{zx}\omega_1 - I_{zy}\omega_2\right]\boldsymbol{b}_3\end{aligned} \tag{8.33}$$

when taken about O.

We found that a convenient way to display and use the various inertia properties was in an **inertia matrix**:

$$[I] \equiv \begin{bmatrix} I_{xx} & -I_{xy} & -I_{xz} \\ -I_{yx} & I_{yy} & -I_{yz} \\ -I_{zx} & -I_{zy} & I_{zz} \end{bmatrix} \tag{8.34}$$

and if the body was aligned along its **principal axes**, we obtained the **principal inertia matrix**:

$$\left[\bar{I}\right] = \begin{bmatrix} \bar{I}_{xx} & 0 & 0 \\ 0 & \bar{I}_{yy} & 0 \\ 0 & 0 & \bar{I}_{zz} \end{bmatrix} \tag{8.36}$$

Using the principal inertia matrix resulted in a simple form of the angular momentum about G:

$$\boldsymbol{H}_G = \bar{I}_{xx}\omega_1\boldsymbol{b}_1 + \bar{I}_{yy}\omega_2\boldsymbol{b}_2 + \bar{I}_{zz}\omega_3\boldsymbol{b}_3 \tag{8.37}$$

Just as in the two-dimensional case, the angular momentum about a point O was given by

$$\boldsymbol{H}_O = \boldsymbol{H}_G + \boldsymbol{r}_{G/O} \times m\boldsymbol{v}_G \tag{8.38}$$

We finally made it to the actual equation of motion for a body in three-dimensional motion. Our translational equations of motion were familiar:

$$\sum \boldsymbol{F} = m\boldsymbol{a} \tag{8.41}$$

The rotational equations of motion, for the general case, were given by

$$\sum M_1 = \dot{H}_1 + (\omega_2 H_3 - \omega_3 H_2) \tag{8.42}$$

$$\sum M_2 = \dot{H}_2 + (\omega_3 H_1 - \omega_1 H_3) \tag{8.43}$$

$$\sum M_3 = \dot{H}_3 + (\omega_1 H_2 - \omega_2 H_1) \tag{8.44}$$

and if we used principal axes, we obtained the simpler form known as **Euler's equations**:

$$\sum M_1 = \bar{I}_{xx}\alpha_1 + \omega_2\omega_3(\bar{I}_{zz} - \bar{I}_{yy}) \tag{8.45}$$

$$\sum M_2 = \bar{I}_{yy}\alpha_2 + \omega_3\omega_1(\bar{I}_{xx} - \bar{I}_{zz}) \tag{8.46}$$

$$\sum M_3 = \bar{I}_{zz}\alpha_3 + \omega_1\omega_2(\bar{I}_{yy} - \bar{I}_{xx}) \tag{8.47}$$

We then moved on to energy, specifically kinetic energy, and found that the kinetic energy of a body in general motion is given by

$$\mathcal{KE} = \frac{1}{2}\bar{\boldsymbol{v}} \cdot \boldsymbol{L} + \frac{1}{2}\boldsymbol{\omega} \cdot \boldsymbol{H}_G \tag{8.51}$$

If the body is rotating about a fixed point O, this simplifies to

$$\mathcal{KE} = \frac{1}{2}\boldsymbol{\omega} \cdot \boldsymbol{H}_O \tag{8.52}$$

SYSTEM ANALYSIS (SA) EXERCISES

SA8.1 Evaluation of Head Rotation Effects—Take Two

Spatial disorientation can obviously have severe consequences in the flight environment, and thus research is ongoing to assess and evaluate the causes of such disorientation. To help with both training and research, Brooks Air Force Base has used the Advanced Spatial Disorientation Demonstrator (ASDD), which enables researchers to create several different vestibular illusions. (See **Figure SA8.1.1*a*.**)

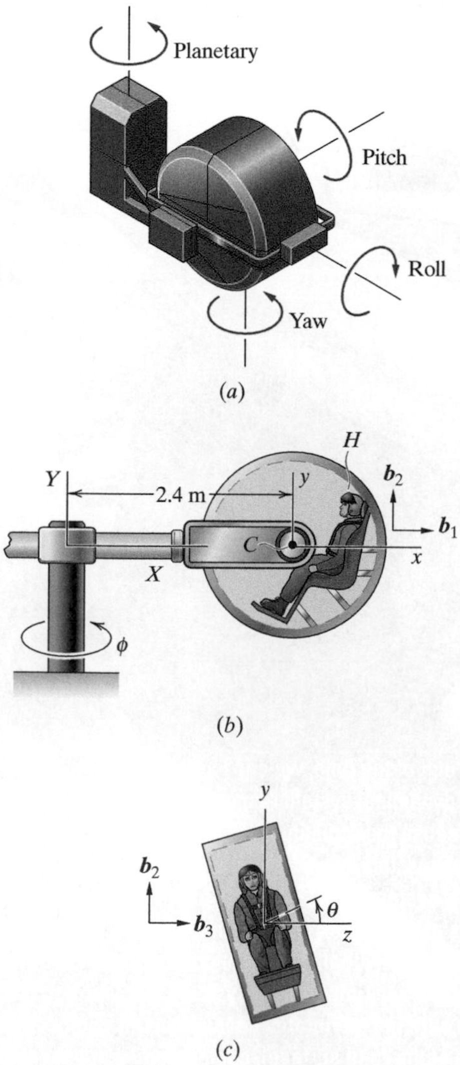

Figure SA8.1.1 ASDD degrees of freedom

To help test the capabilities of the device, researchers want to simulate a right barrel roll with an accompanying simulated forward acceleration. Planetary motion can be utilized to simulate forward acceleration (e.g., spinning the person about a fixed center and inducing centripetal acceleration), and motion about the roll axis can then be added. Specifications for the device are shown in **Table SA8.1.1**.

Table SA8.1.1 ASDD Performance Specifications

Spin Axis	Max Spin Rate	Max Rate of Change of Spin Rate
Planetary	±28 rpm	±15 deg/s^2
Pitch	±5 rpm	±50 deg/s^2
Roll	±10 rpm	±100 deg/s^2
Yaw	±25 rpm	±50 deg/s^2

The X, Y, Z axes are inertial directions with origin at point O. The x, y, z axes originate at the center of the cab (denoted by point C) and rotate with the 2.4 m long main arm of the centrifuge. Hence they rotate about a vertical axis as the arm rotates. The unit vectors $\boldsymbol{b}_1, \boldsymbol{b}_2, \boldsymbol{b}_3$ are aligned along the x, y, z axes. (See **Figure SA8.1.1*b*** and **Figure SA8.1.1c**.) At the instant shown, the x, y, z axes align with the X, Y, Z axes. With the cab in the configuration shown, the rider's head (point H) is located at $0.6\boldsymbol{b}_1 + 0.9\boldsymbol{b}_2$ m relative to point C.

Your Task: Consider the situation where just planetary and roll motion are occurring. **Figure SA8.1.1*c*** illustrates the orientation of the cab at a generic roll angle θ as viewed looking down the roll axis.

a. Produce an equation for the acceleration vector of the rider's head (point H) in terms of the variables $\theta, \dot{\theta}, \ddot{\theta}, \dot{\phi}$, and $\ddot{\phi}$, where θ is the roll angle and ϕ is the planetary angle.

b. Determine the maximum possible acceleration magnitude of the rider's head. What angular velocity and acceleration will occur at the vestibular system (i.e., at the head)? Note that in reality the maximum angular acceleration about a particular axis cannot occur simultaneously with the maximum angular velocity; however, for this worst-case analysis assume that both do occur simultaneously.

OBJECTIVES

Upon completion of this chapter you will be able to:

- Analyze the free vibrational response of simple, single-degree-of-freedom systems
- Know how to apply damping to modify the system's response
- Analyze system responses under the application of external sinusoidal forces

CHAPTER 9
VIBRATORY MOTIONS

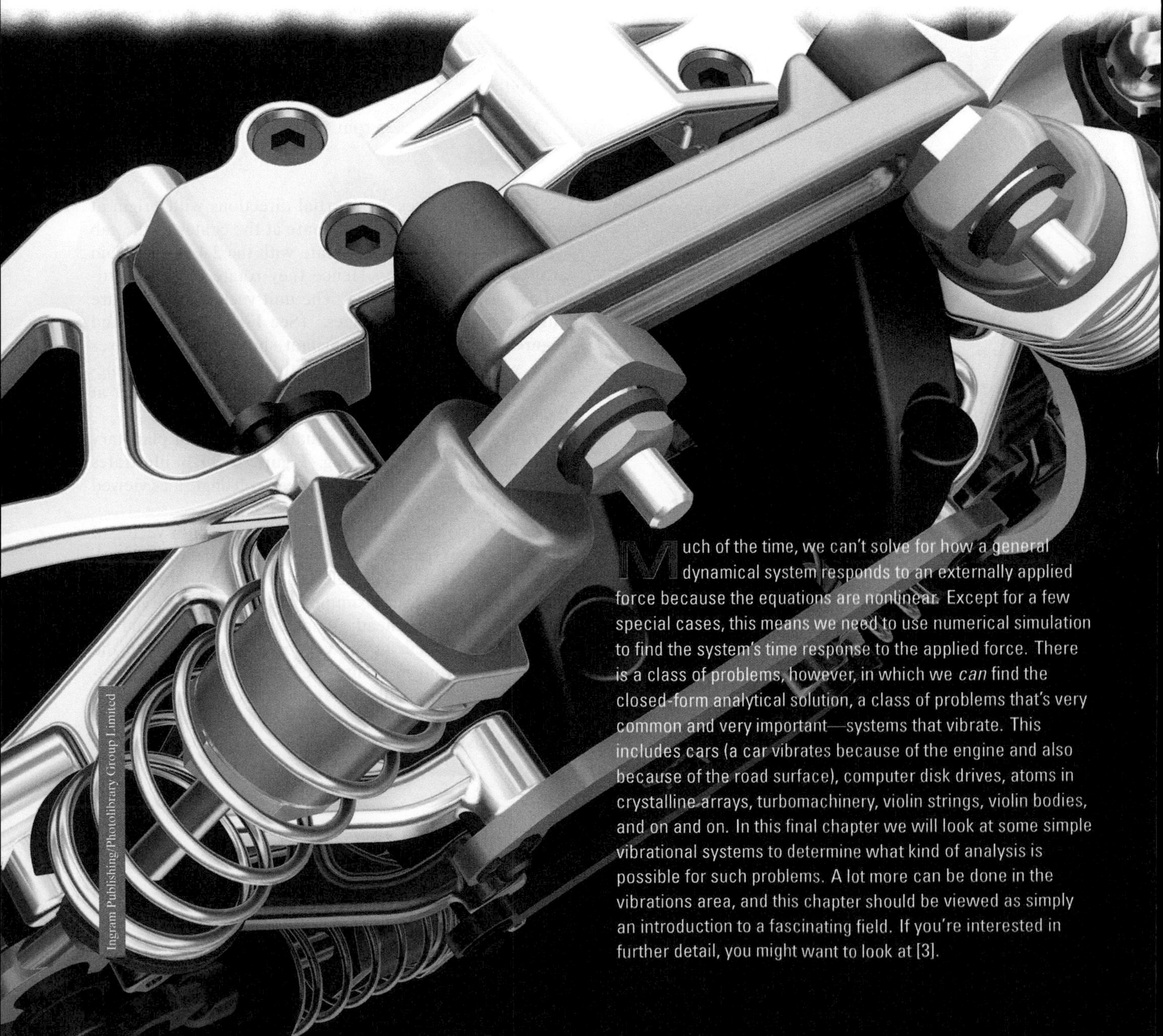

Much of the time, we can't solve for how a general dynamical system responds to an externally applied force because the equations are nonlinear. Except for a few special cases, this means we need to use numerical simulation to find the system's time response to the applied force. There is a class of problems, however, in which we *can* find the closed-form analytical solution, a class of problems that's very common and very important—systems that vibrate. This includes cars (a car vibrates because of the engine and also because of the road surface), computer disk drives, atoms in crystalline arrays, turbomachinery, violin strings, violin bodies, and on and on. In this final chapter we will look at some simple vibrational systems to determine what kind of analysis is possible for such problems. A lot more can be done in the vibrations area, and this chapter should be viewed as simply an introduction to a fascinating field. If you're interested in further detail, you might want to look at [3].

9.1 UNDAMPED, FREE RESPONSE FOR SINGLE-DEGREE-OF-FREEDOM SYSTEMS

The simplest vibratory model is a mass attached to a linear spring (**Figure 9.1**). A **linear spring** is one for which the deflection is proportional to the applied force, $f_{spring} = ky$, where f_{spring} is the force pulling on the spring, k is the spring constant, and y is the spring's deflection.

The spring has an unstretched length L, and in the absence of gravity (shown in **Figure 9.1*a***), the mass will remain motionless a distance L below the upper end of the spring. Imagine that gravity is now included. There exists an equilibrium position at which the force due to gravity (mg) is exactly countered by the force in the spring, as shown in **Figure 9.1*b***.

$$-ky_{eq} + mg = 0 \quad \Rightarrow \quad y_{eq} = \frac{mg}{k}$$

Left undisturbed, the mass will just sit there at the lower end of the stretched spring. If, however, we pull it down a little farther (as in **Figure 9.1*c***) and then release it, we will see some vibrations. Note in **Figure 9.1*c*** that a coordinate system is put into place with $y = 0$ at the mass's static equilibrium position and with positive y indicating that the mass is below that equilibrium.

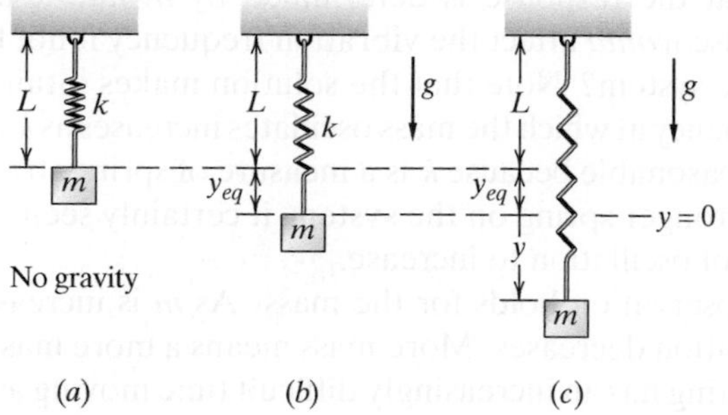

Figure 9.1 Spring-mass system

Constructing a FBD=IRD for the mass displaced a distance y (**Figure 9.2**) gives us $m\ddot{y} = mg - k(y_{eq} + y)$. Since $ky_{eq} = mg$, we have

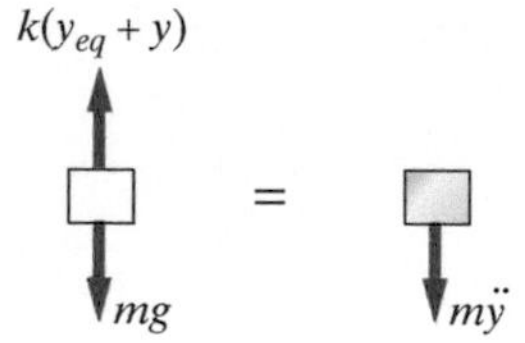

Figure 9.2 FBD=IRD for spring-mass system

$$m\ddot{y} + ky = 0 \tag{9.1}$$

Note that the coefficients of this equation are constant (m and k), it is second-order (two differentiations with respect to time), there is no forcing, and no nonlinearity (terms like x^2 or $\dot{x}^3x$, for example) is present. It is what we call a linear, constant-coefficient, homogeneous, second-order ordinary differential equation, and, unlike most other ODEs, it has an easily accessible solution in the form of sines and cosines:

$$y(t) = a_1 \cos \omega t + a_2 \sin \omega t \tag{9.2}$$

Initial conditions will take care of a_1 and a_2, but what about ω? To find out what that is equal to, just substitute the solution into (9.1). Doing so will give you

$$-m\omega^2(a_1 \cos \omega t + a_2 \sin \omega t) + k(a_1 \cos \omega t + a_2 \sin \omega t) = 0$$

This can be factored into

$$(a_1 \cos \omega t + a_2 \sin \omega t)\left(k - m\omega^2\right) = 0$$

For our assumed solution to hold, the left side of this equation must equal zero. The first term in parentheses cannot be zero for all times because $a_1 \cos \omega t + a_2 \sin \omega t$ is a sinusoidally varying function. Occasionally it is zero, but more often than not, it is nonzero. Therefore, $k - m\omega^2$ has to be zero, which means that because we know k and m, we can solve for ω:

$$\omega^2 = \frac{k}{m} \quad \Rightarrow \quad \omega = \sqrt{\frac{k}{m}}$$

Thus the frequency of undamped, free vibration is determined by the physical parameters m and k. Note that the word *free* refers to the lack of an external forcing. We'll consider the effect of external forcing later in the chapter.

The fact that the response is determined by m and k makes good sense—what else *would* affect the vibration frequency if not the physical elements of the system? Note that the solution makes intuitive sense as well. The frequency at which the mass oscillates increases as k is increased. This is pretty reasonable because k is a measure of spring strength. If you put a bigger, stronger spring on the system, it certainly seems normal for the frequency of oscillation to increase.

A similar observation holds for the mass. As m is increased, the frequency of vibration decreases. More mass means a more massive system, one that the spring has an increasingly difficult time moving as m gets bigger. Thus it's reasonable to expect the frequency of oscillation to become smaller. Because this ratio is so fundamental to the system's response, it is given its own name, the **natural frequency**, and its own symbol, ω_n:

$$\omega_n = \sqrt{\frac{k}{m}} \tag{9.3}$$

You may recall that a frequency can be specified in two ways: cycles per second (also known as hertz, Hz) and radians per second (rad/s). They're related by $\omega = 2\pi f$, where ω is in radians per second and f is in hertz. Reporting frequencies in hertz is useful for acoustic work, where the number of cycles per second has a direct and well-known correlation to the pitch of a note. For our work, however, we will usually be dealing with radians per second.

If we divide (9.1) by m, we obtain a **mass-normalized** form of the equation.

$$\ddot{y} + \omega_n^2 y = 0 \tag{9.4}$$

This form reflects the fact that, often, what we care primarily about is a system's frequency characteristics, not its particular m and k values.

If we are given the initial conditions $y(0) = y_0$ and $\dot{y}(0) = v_0$, then by using (9.2) we find

$$y(t) = y_0 \cos \omega_n t + \frac{v_0}{\omega_n} \sin \omega_n t \tag{9.5}$$

>>> **Check out Example 9.2 (page 548) and Example 9.1 (page 546) for applications of this material.**

If we use $\omega_n = 2$ rad/s, and initial conditions of $y_0 = 1.0$ cm and $v_0 = 0$, we will obtain a response that looks like the one shown in **Figure 9.3*a***—a pure cosine wave. Changing the initial conditions to $y_0 = 0$ and $v_0 = 2.0$ cm/s produces a pure sine wave, as shown in **Figure 9.3*b***. Both of these cases are somewhat special; most of the time we'll have not a pure sine wave (zero at the origin and positive slope) or a pure cosine wave (maximum at the origin and zero slope), but rather a wave that isn't zero at the origin or doesn't have a zero slope. **Figure 9.3*c*** shows just such a response, which is found by using the initial conditions $y_0 = 1.0$ cm and $v_0 = 2.0$ cm/s.

Figure 9.3*c* also shows T, the period of the signal. The frequency in cycles per second (or Hz) is $1/T$, and therefore $\omega = 2\pi/T$.

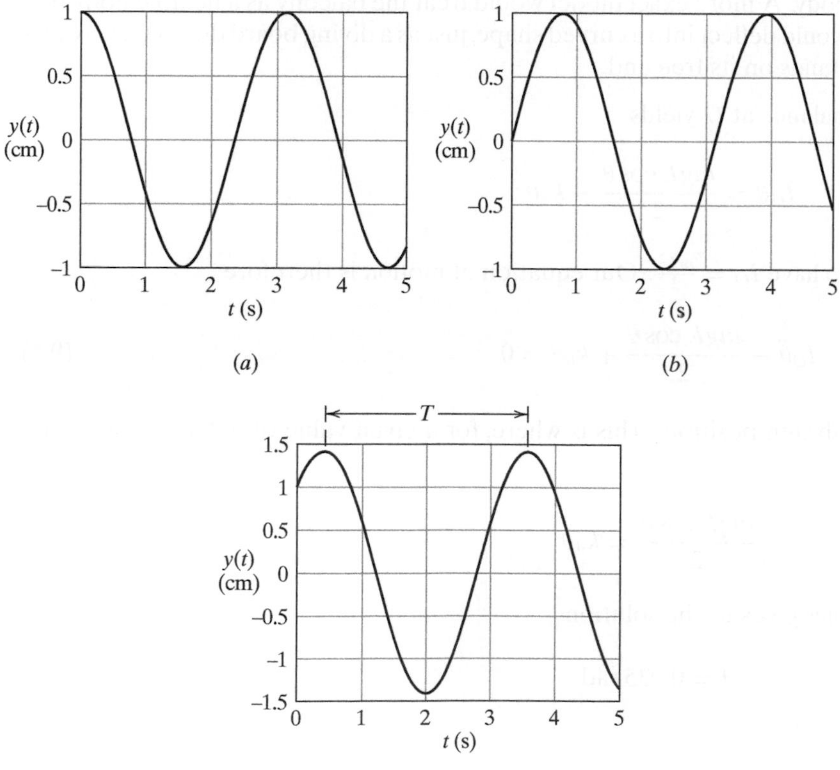

Figure 9.3 Sinusoidal time response

EXAMPLE 9.1 NATURAL FREQUENCY OF A CANTILEVERED BALCONY (Theory on page 545)

Figure 9.4 Cantilevered balcony

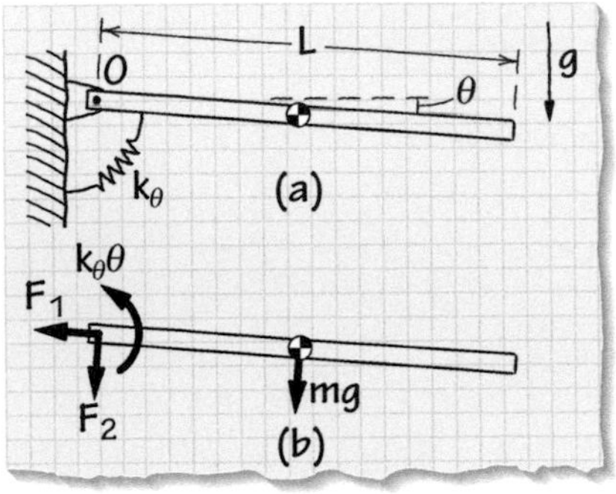

Figure 9.5 Single-degree-of-freedom model of cantilevered balcony

Figure 9.4 shows a cantilevered balcony. As a first cut at analyzing its response to an earthquake, we will model it as a rigid bar of mass m and model the stiffness of the structure as a single torsional spring (with stiffness k_θ) at the attachment point O. Include gravity, and solve for the equilibrium angle of the balcony and its natural frequency of oscillation. The mass is 1000 kg, the length of the balcony is 5 m, and $k_\theta = 7.00 \times 10^5$ N·m/rad.

Goal Find the equilibrium position under gravity that the balcony will exhibit, and determine the frequency of free vibrations about this equilibrium.

Given Physical parameters of the system.

Draw **Figure 9.5*a*** shows our simplified model. The torque at O from gravity will cause the bar to rotate down until there's a sufficient counter-torque in the torsional spring to hold it in equilibrium, both of which are indicated in the free-body diagram of **Figure 9.5*b***. Note that the reaction forces at the hinge produce no moments about it and are thus not part of our solution.

Assume We're assuming that the distributed flexibility of the balcony can be neglected and it can be viewed as a torsionally restrained, rigid body. A more exact model would treat the balcony as a flexible body that would deflect into a curved shape, just as a diving board does when a diver stands on its free end.

Formulate Equations A moment balance at O yields

$$I_O\ddot{\theta} = \frac{mgL\cos\theta}{2} - k_\theta\theta$$

where for a bar rotating about its end we have $I_O = \frac{mL^2}{3}$. Our equation of motion is therefore

$$I_O\ddot{\theta} - \frac{mgL\cos\theta}{2} + k_\theta\theta = 0 \tag{9.6}$$

Solve First we need to find the equilibrium position. This is where, for a given value of θ, $\dot{\theta}$ and $\ddot{\theta}$ are zero. From (9.6) we have

$$\frac{mgL\cos\theta}{2} = k_\theta\theta$$

Solving this for the given parameter values gives us the solution

$$\theta = 0.035 \text{ rad}$$

Thus

$$\theta_{eq} = 0.035 \text{ rad}$$

Now we need to find the equation for small oscillations about this equilibrium. To do so, redefine θ as

$$\theta = \theta_{eq} + \beta$$

In this formulation, β is the angular deviation from the equilibrium position $\theta_{eq} = 0.035\,\text{rad}$.
Substituting this into (9.6) gives us

$$I_O(\ddot{\theta}_{eq} + \ddot{\beta}) - \frac{mgL\cos(\theta_{eq} + \beta)}{2} + k_\theta(\theta_{eq} + \beta) = 0$$

Knowing that θ_{eq} is constant means $\ddot{\theta}_{eq} = 0$:

$$I_O\ddot{\beta} - \frac{mgL\cos(\theta_{eq} + \beta)}{2} + k_\theta\theta_{eq} + k_\theta\beta = 0$$

We now have to expand $\cos(\theta_{eq} + \beta)$. Then, because we're interested only in small oscillations about the equilibrium position, we will use the small-angle approximations $\sin\beta \approx \beta$ and $\cos\beta \approx 1$. The needed trigonometric expansion formula is

$$\cos(\theta_{eq} + \beta) = \cos\theta_{eq}\cos\beta - \sin\theta_{eq}\sin\beta \approx \cos\theta_{eq} - \beta\sin\theta_{eq}$$

giving us

$$I_O\ddot{\beta} + \left(k_\theta\theta_{eq} - \frac{mgL}{2}\cos\theta_{eq}\right) + \left(\frac{mgL}{2}\sin\theta_{eq} + k_\theta\right)\beta = 0$$

If you go ahead and evaluate the $k_\theta\theta_{eq} - \frac{mgL}{2}\cos\theta_{eq}$ term, you will see that it is precisely equal to zero. That is *always* going to be the case when expanding about an equilibrium position; the finite terms in the equation cancel, leaving us with the acceleration term and the modified stiffness term. This leaves us with

$$I_O\ddot{\beta} + \left(\frac{mgL}{2}\sin\theta_{eq} + k_\theta\right)\beta = 0$$

This is quite similar to what we would have gotten in the absence of gravity—namely, the mass moment of inertia $I_O\ddot{\beta}$ acceleration term and the stiffness due to the torsional spring $k_\theta\beta$. The additional term due to gravity adds a bit of stiffness to the problem.

Using the given parameter values gets us

$$\frac{(1000\,\text{kg})(5\,\text{m})^2}{3}\ddot{\beta} + \left(\frac{(1000\,\text{kg})(9.81\,\text{m/s}^2)(5\,\text{m})}{2}\sin 0.035 + 7.00\times 10^5\,\text{N}\cdot\text{m/rad}\right)\beta = 0$$

Evaluating the terms and mass-normalizing give us our final equation:

$$\ddot{\beta} + (84.1\,\text{s}^{-2})\beta = 0$$

$$\omega_n = \sqrt{84.1}\,\text{rad/s} = 9.17\,\text{rad/s}$$

EXAMPLE 9.2 DISPLACEMENT RESPONSE OF A SINGLE-STORY BUILDING (Theory on page 545)

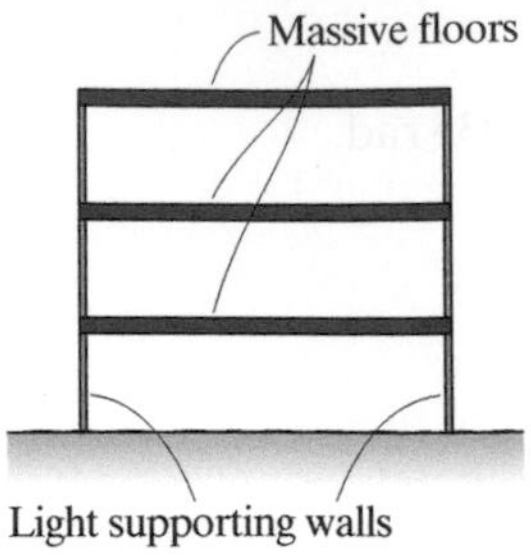

Figure 9.6 Schematic of a multistory building

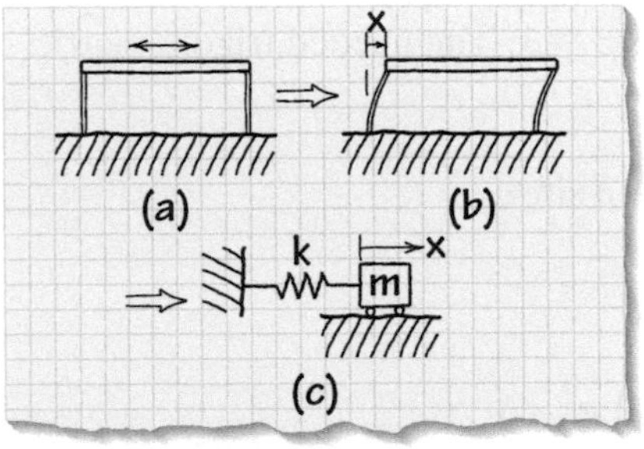

Figure 9.7 Approximating a single-story building as spring-mass system

You might not think it when looking at them but multistory buildings can quite reasonably be viewed as lumped masses (the individual floors) connected by springs (the flexible supporting walls). **Figure 9.6** shows such a structure, with several floors and connecting walls.

The basic building block is a single story, shown in **Figure 9.7*a***. **Figure 9.7*b*** shows what happens when the top moves to the right due to some excitation: the walls bend and, because of their elastic nature, try to pull the top back into its rest position. The system is well approximated by a lumped mass (the top of the structure) restrained by a linear spring (the walls), as shown in **Figure 9.7*c***.

You would want to model a building like this one whenever it experiences a loading sufficient to cause it to sway. Is this common? Absolutely. The top floors of skyscrapers routinely sway by several meters in the wind. And when you have an earthquake, the deflections get a lot worse. During an earthquake the ground moves laterally, and the building can respond strongly. As someone who's been on the top of a tall engineering building during a sizable earthquake, I can assure you it's no fun to feel the building, previously such a rock of stability, oscillate back and forth like a mass on a spring.

Take **Figure 9.7*c*** as our model and use the parameters $m = 1.20 \times 10^3$ kg and $k = 4.00 \times 10^4$ N/m. Apply initial conditions $x(0) = 1.20$ m, $\dot{x}(0) = 4.00$ m/s and determine the system's free vibrational response.

Goal Determine the free vibrational response of a spring-mass system.

Given $m = 1200\,\text{kg}, k = 4.00 \times 10^4\,\text{N/m}, x(0) = 1.20\,\text{m}, \dot{x}(0) = 4.00\,\text{m/s}$

Draw **Figure 9.7** shows our system model.

Assume No assumptions are needed.

Formulate Equations The equation of motion is $m\ddot{x} + kx = 0$, which, when we divide by m, becomes $\ddot{x} + \omega_n^2 x = 0$. The natural frequency is $\omega_n = \sqrt{\frac{k}{m}} = 5.77$ rad/s.

Solve

(9.5) $\Rightarrow$ $$x(t) = 1.20\cos 5.77t + 0.69\sin 5.77t$$

where $x(t)$ has the units of meters and the frequency is in radians per second.

Check We can differentiate our solution for $x(t)$ and obtain $\dot{x}(t)$:

$$\dot{x}(t) = [-6.92\sin 5.77t + 4.0\cos 5.77t]\,\text{m/s}$$

Evaluating the expressions for $x(t)$ and $\dot{x}(t)$ at $t = 0$ yields

$$x(0) = 1.20\,\text{m}, \qquad \dot{x}(0) = 4.00\,\text{m/s}$$

matching the original initial conditions.

EXERCISES 9.1

9.1.1. **[Level 1]** Derive expressions for the illustrated system's equation of motion and natural frequency. Neglect gravity, and take the pivot O to be frictionless. Consider small rotations of the bar.

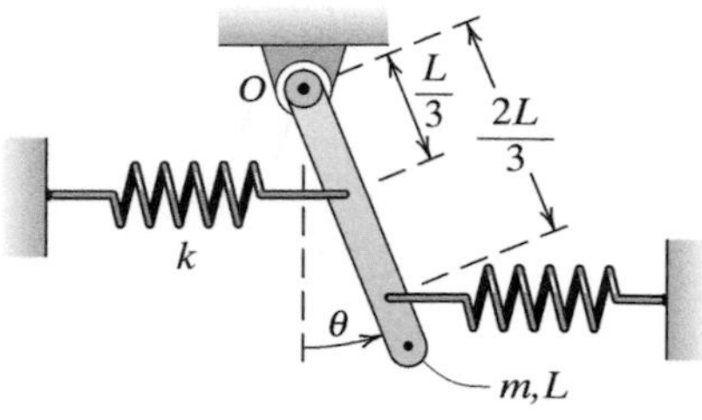

E9.1.1

9.1.2. **[Level 1]** The solution to a free-vibration problem can be represented as $a_1 \cos \omega t + a_2 \sin \omega t$ or, equally correctly, by $b_1 e^{i\omega t} + b_2 e^{-i\omega t}$. If the initial conditions are such that $a_1 = 2$ cm and $a_2 = -3$ cm, what are the values for b_1 and b_2?

9.1.3. **[Level 1]** What are the position, velocity, and acceleration at $t = 1.5$ s if $x(t) = 2 \cos 3t - \sin 3t$? $x(t)$ is given in meters.

9.1.4. **[Level 1]** What is the maximum value of $\dot{x}$ if $x(t) = -\cos 4t + 2 \sin 4t$? $x(t)$ is given in meters.

9.1.5. **[Level 1]** A block of mass $m = 5$ kg is elastically constrained as illustrated, where the stiffness of each spring is $k = 800$ N/m. Suppose the block is released from rest at a distance $x_0 = 0.25$ m from its equilibrium position. Find the block's equation of motion, and determine the maximum speed and acceleration that it experiences. Neglect gravity.

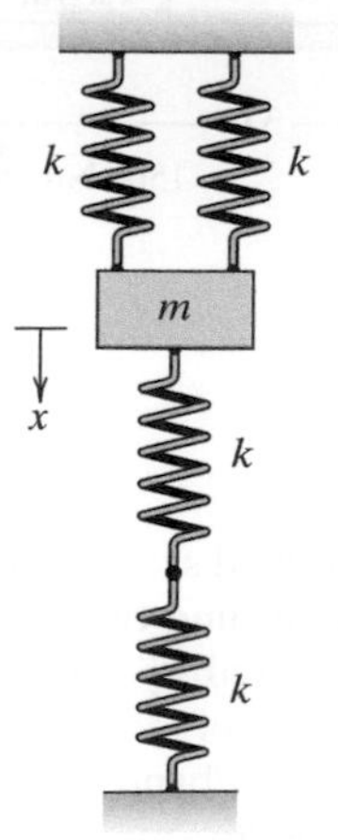

E9.1.5

9.1.6. **[Level 1]** Is there any difference in the natural frequency of the systems shown in (a) and (b)?

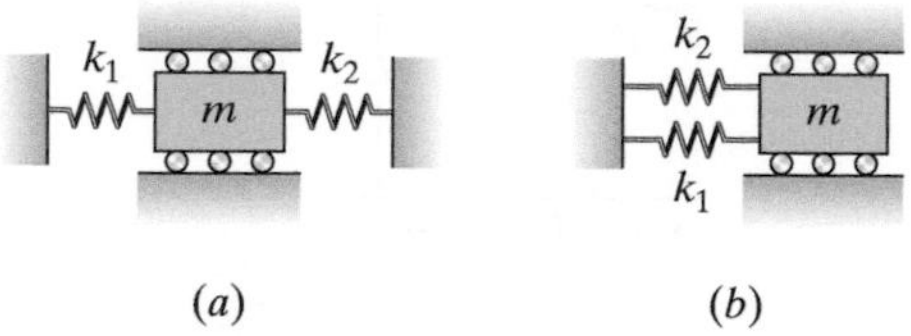

E9.1.6

9.1.7. **[Level 1]** The illustrated system has a natural frequency of 40 rad/s and a mass of 2 kg when measured on the earth's surface. By how much will the system's natural frequency change if it is brought to the surface of the moon?

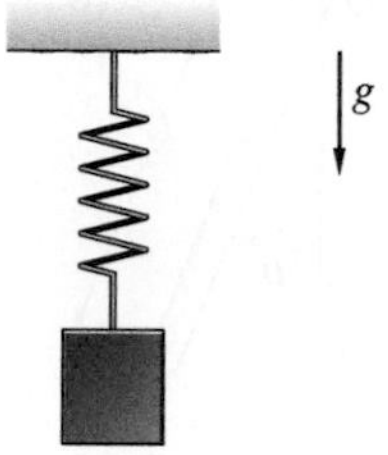

E9.1.7

9.1.8. **[Level 1]** If a 10 kg mass is placed on an unstretched spring, the spring deflects 0.1 m. What is the system's natural frequency of oscillation?

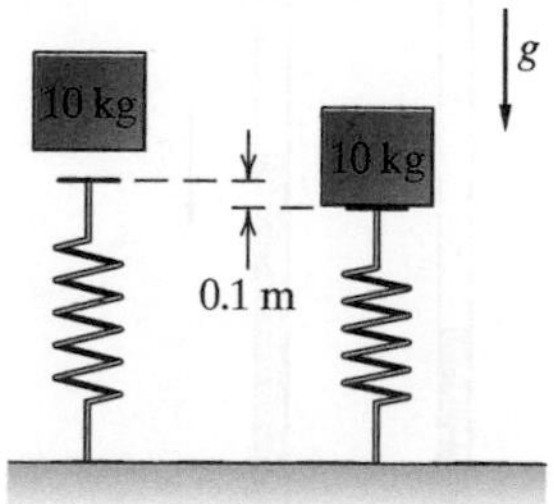

E9.1.8

9.1.9. **[Level 1]** The figure shows a uniform disk that can roll without slip and is attached by means of a frictionless axle at its center to a linear spring. Determine the natural frequency of oscillation for this system in terms of its given physical parameters.

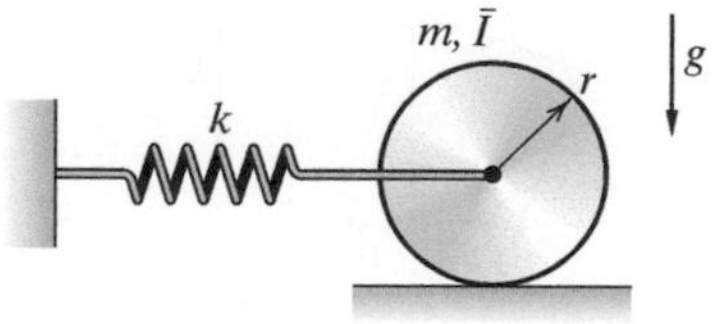

E9.1.9

9.1.10. **[Level 1]** A square plate is hinged in the middle of one side. Determine its natural frequency of oscillation.

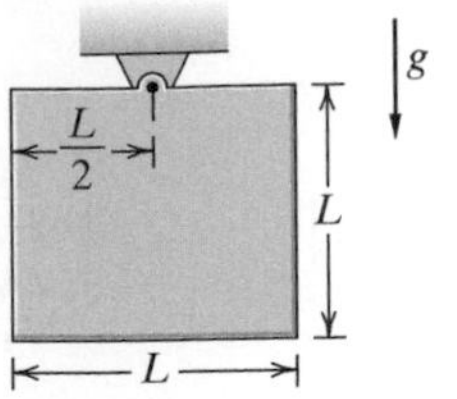

E9.1.10

9.1.11. [Level 1] Using a small-angle approximation ($\sin\theta \approx \theta$), show that the natural frequency of the illustrated pendulum is $\omega_n = \sqrt{\frac{3\,g}{2L}}$ rad/s. The pendulum is a rigid rod, pivoted at O, with mass m and length L.

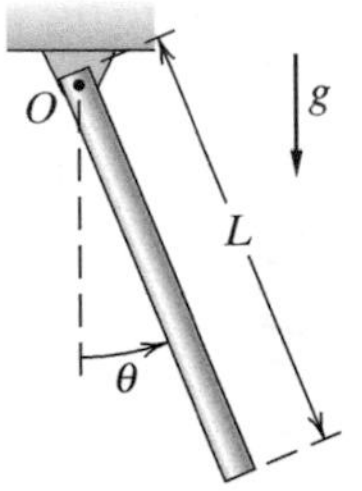

E9.1.11

9.1.12. [Level 1] Calculate the natural frequency of the water contained in a U-shaped tube that results from the water column being momentarily perturbed. The tube has an inner area of s, the water has density ρ, and the length of the water-filled section is l.

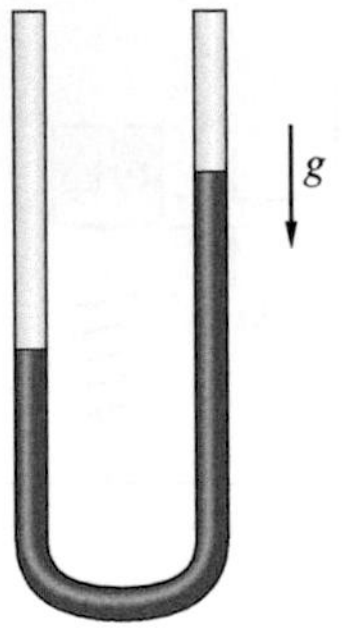

E9.1.12

9.1.13. [Level 1] Determine the natural frequency of oscillation about $\theta = 0$ for the illustrated system. $l = 1.5$ m and $m = 2$ kg.

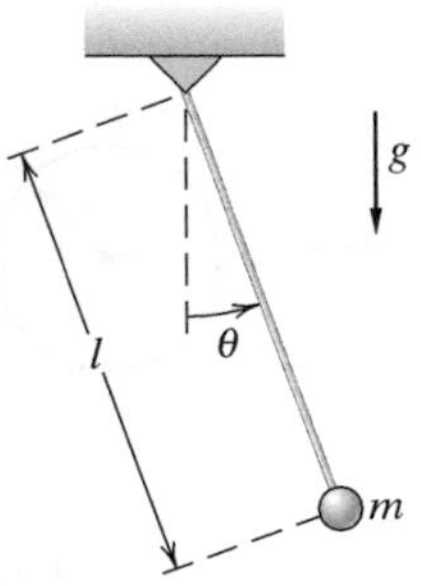

E9.1.13

9.1.14. [Level 1] Find the natural frequency about $\theta = 0$ for the system shown in the figure. $l_1 = 1$ m, $l_2 = 1.1$ m, and $m_1 = m_2 = 5$ kg.

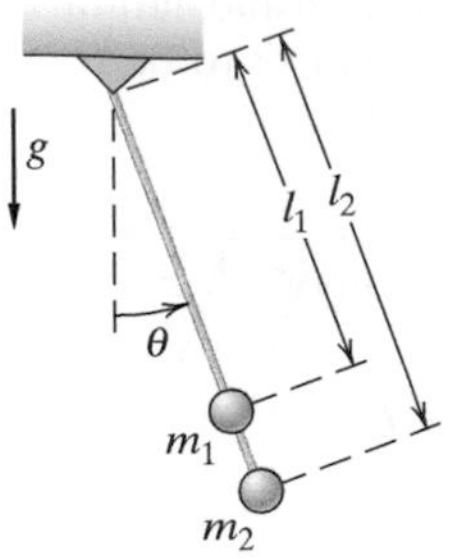

E9.1.14

9.1.15. [Level 1] What is the linearized natural frequency of oscillation of a point mass sliding without friction along a circular track?

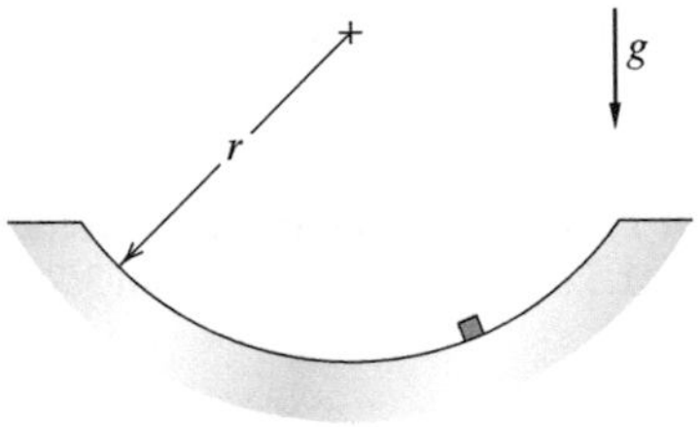

E9.1.15

9.1.16. [Level 1] What is the natural frequency of the illustrated system? Neglect gravity.

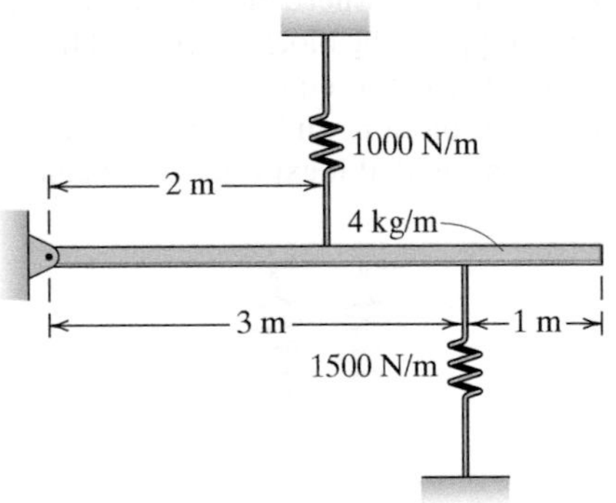

E9.1.16

9.1.17. [Level 1] A band saw can be modeled as a continuous loop that runs around two pulleys. The steel has some flexibility, represented by the springs in the figure. Neglect the mass of the steel loop, and determine the natural frequency of the pulley/band. Assume that the oscillation mode involves the pulleys moving in opposition to each other. The rotational inertia of each pulley about its shaft is equal to 0.001 kg·m^2, the spring stiffness is 10,000 N/m, and the radius of the pulleys is 4.0 cm.

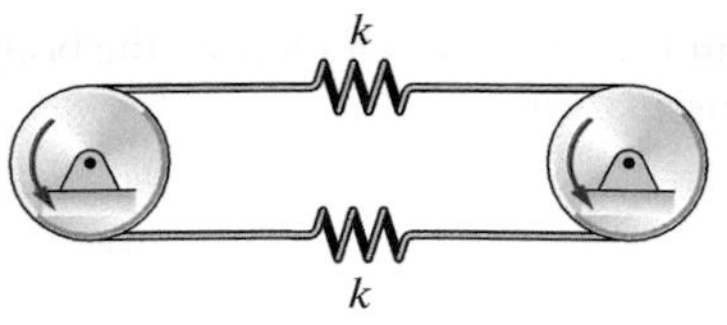

E9.1.17

9.1.18. **[Level 1]** Find the natural frequency for the illustrated system.

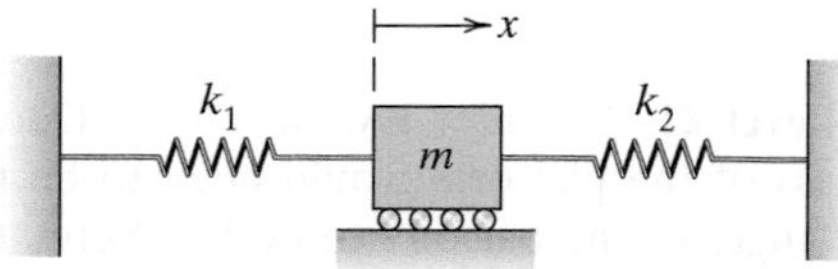

E9.1.18

9.1.19. **[Level 1]** Find the equation of motion for the illustrated system.

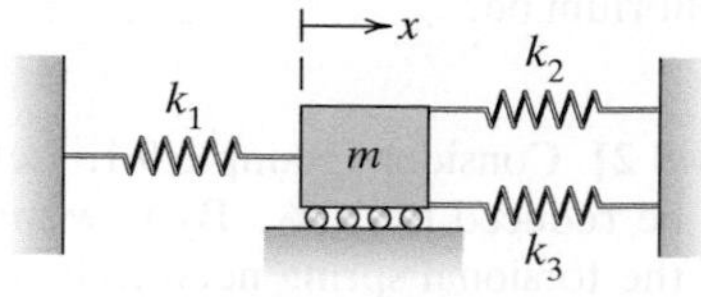

E9.1.19

9.1.20. **[Level 1]** Determine the equation of motion for the system shown here. Neglect the effects of gravity. What is the natural frequency for the system?

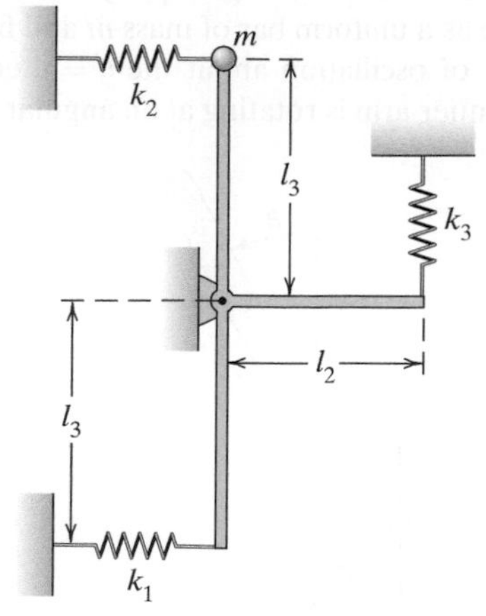

E9.1.20

9.1.21. **[Level 1]** Find the equations of motion and natural frequency for the illustrated system. Ignore gravity. The system is freely hinged at O.

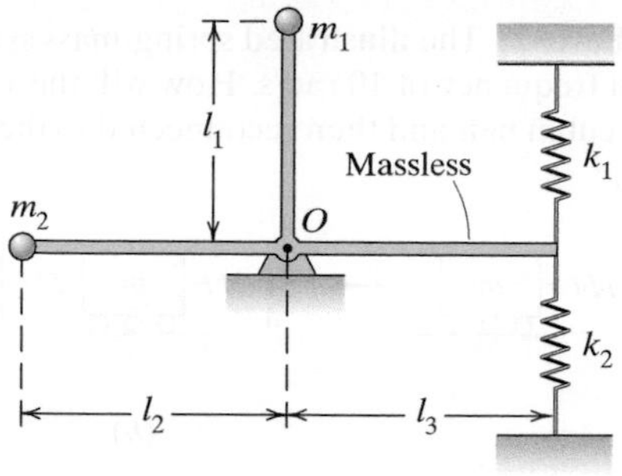

E9.1.21

9.1.22. **[Level 2]** What are the equation of motion and natural frequency for the depicted triangular plate? Friction at the fixed pivot O is negligible, and consider small rotations of the plate, which is an isosceles triangle.

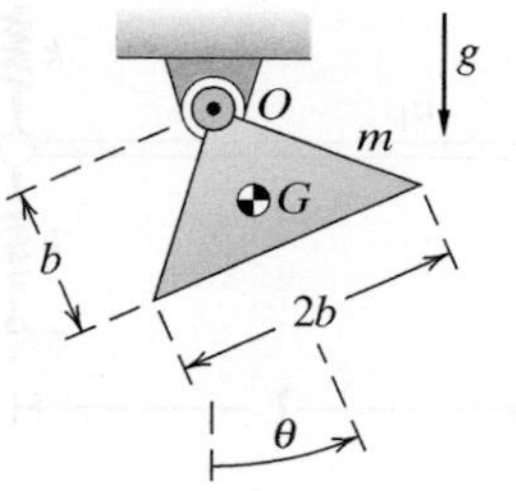

E9.1.22

9.1.23. **[Level 2]** Plot $x(t), \dot{x}(t)$, and $\ddot{x}(t)$, $(0 \leq t \leq 2\pi)$ for

$$x(t) = 2\cos t + 0.05 \sin 10t$$

What does this tell you about the advisability of differentiating a displacement signal in order to find the acceleration if the actual signal ($2\cos t$ in our case) is corrupted by high-frequency sensor noise (represented by $0.05 \sin 10t$)?

9.1.24. **[Level 2]** When block A with a mass of 8 kg is attached to block B, the natural frequency of the combined system is reduced by 10% as compared to its value with block B alone. What is the mass of block B?

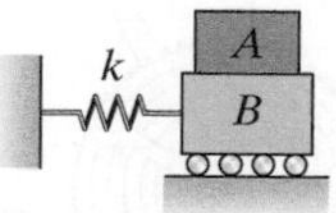

E9.1.24

9.1.25. **[Level 2]** When block A with a mass of 5 kg is attached to block B, the natural frequency of the combined system is reduced by 30% as compared to its value with block B. What is the mass of block B?

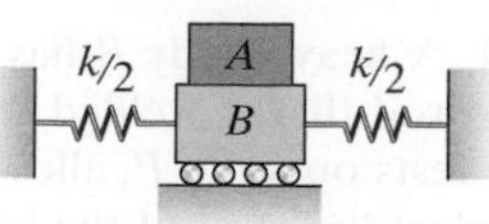

E9.1.25

9.1.26. **[Level 2]** The illustrated spring-mass system (*a*) has a natural frequency of 10 rad/s. How will this change if the spring is cut in half and then reconnected to the mass as shown in (*b*)?

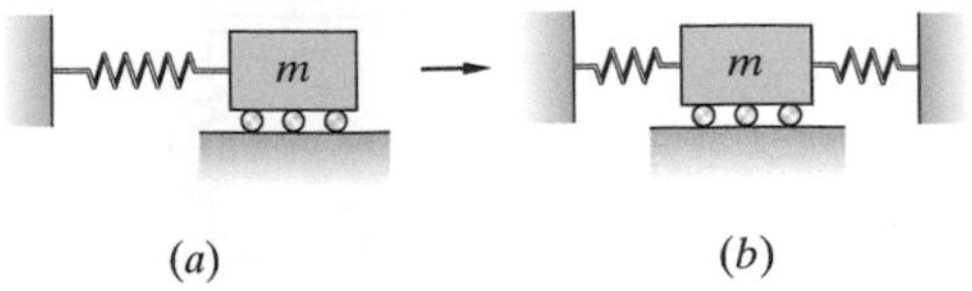

E9.1.26

9.1.27. **[Level 2]** What's the natural frequency of vibration for the illustrated system? The bar is rigid with mass m_1 and is pinned at O. Neglect gravity.

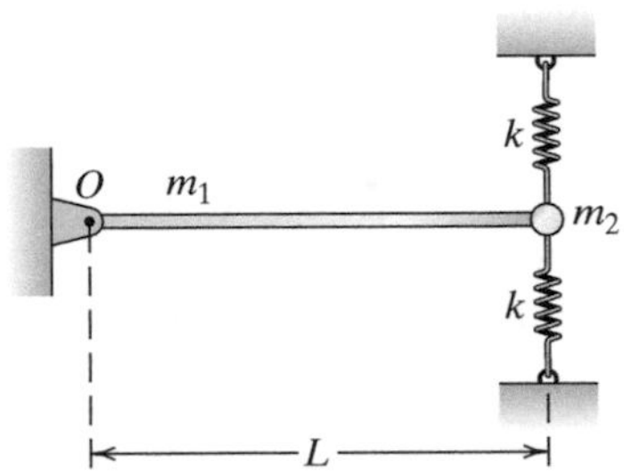

E9.1.27

9.1.28. **[Level 2]** In order to create **Exercise 7.5.23**, I had to calculate the rotational moment of inertia of a crankset (crank, bottom bracket, chain rings, and pedals). I didn't want to do it analytically because that would have taken a ridiculously long time. What I did was exploit what we have learned in this chapter—namely, the fact that a pendular system's natural frequency can be calculated from its period of oscillation. I took a 0.23 kg cylindrical piece of steel that had a diameter of 6.4 cm and taped it to the pedal, positioning it so that its center was 18 cm from the crank's center of rotation. I then pushed it and saw that the period of the oscillation was 1.5 s. What was I_G for the crankset (*not* including the test mass)?

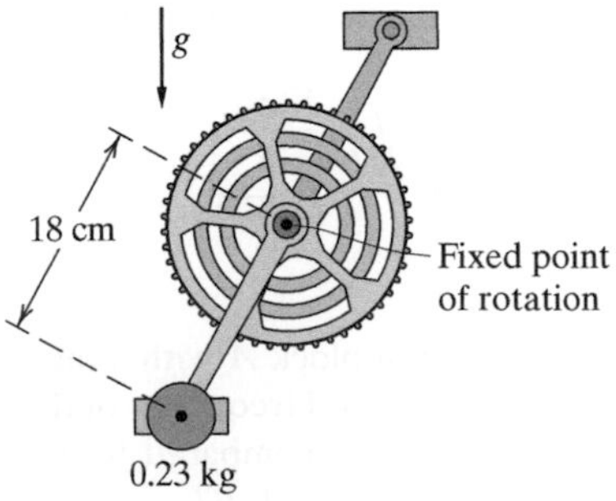

E9.1.28

9.1.29. **[Level 2]** A heavy body B has a handle in the form of a bent rigid rod that is welded to the body at A and C. The handle rests on a pin P, allowing the body to rotate about P. Neglect the mass of the handle. The body weighs 11 kg and $L = 0.42$ m. Determine the body's natural frequency of oscillation.

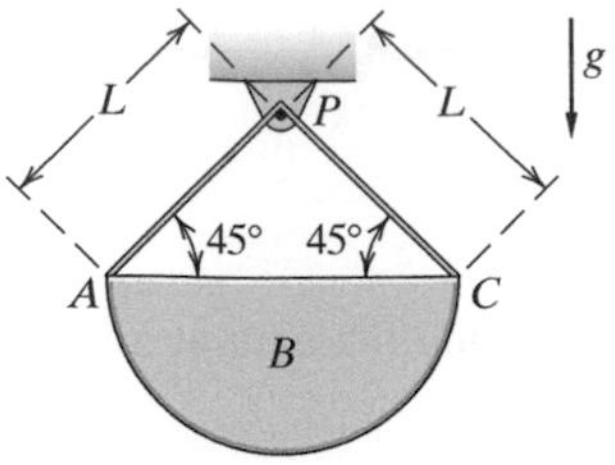

E9.1.29

9.1.30. **[Level 2]** Consider Example 9.1. Determine what the mass of the platform would need to be for the equilibrium angle of the balcony to be 5°. Solve for the associated natural frequency of the system.

9.1.31. **[Level 2]** Consider Example 9.1. Assume that four people, each with a mass of 75 kg, walk to the far edge of the balcony. What will the balcony's equilibrium angle be? What will the system's natural frequency of oscillation about this equilibrium be?

9.1.32. **[Level 2]** Consider Example 9.1. Let the mass of the balcony be reduced by 10%. By how much would the stiffness of the torsional spring need to be changed to yield the same equilibrium angle as in the example? Will the natural frequency for the system with these new values of mass and stiffness be the same?

9.1.33. **[Level 2]** The figure shows a simple model of a single helicopter blade. The central rotor spins at a constant speed ω and is connected through a pin joint to the blade. Treat the blade as a uniform bar of mass m and find its natural frequency of oscillation about the $\theta = 0$ equilibrium position. The inner arm is rotating at an angular speed ω.

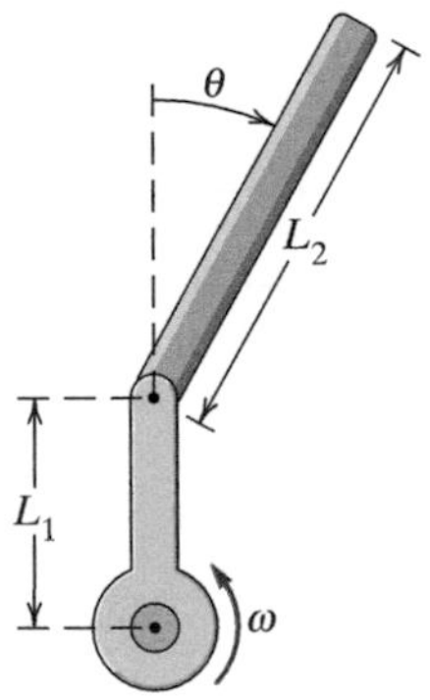

E9.1.33

9.1.34. **[Level 2]** Find the natural frequency of oscillation for the illustrated system. The mass per unit length of the thin L-shaped bar is 10 kg/m. $L_1 = 1.5$ m and $L_2 = 0.6$ m.

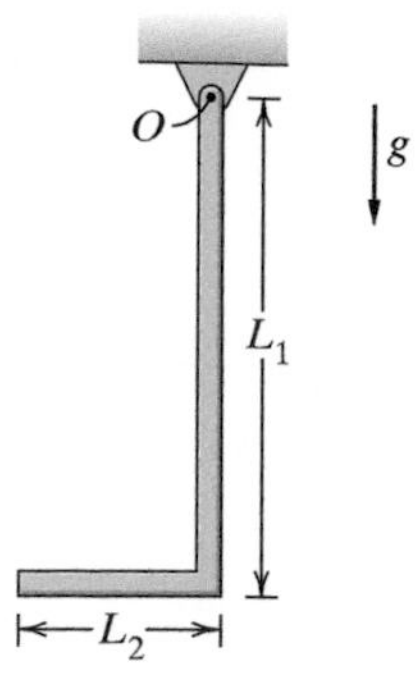

E9.1.34

9.1.35. **[Level 2]** Find the natural frequency of oscillation for the illustrated system. $m_A = 3$ kg, $m_B = 2$ kg, $m_C = 20$ kg, $r = 0.04$ m, and $k = 1000$ N/m.

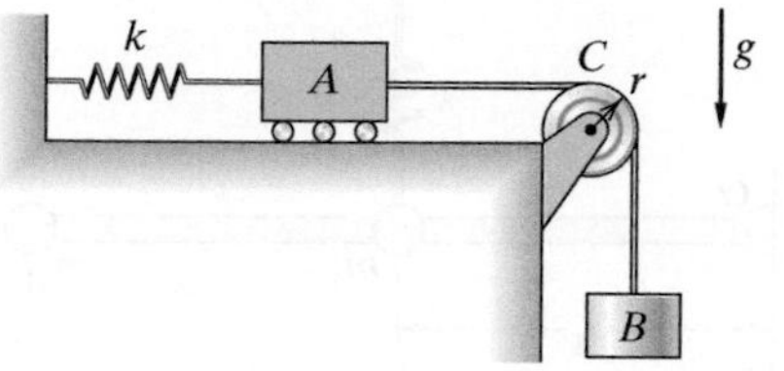

E9.1.35

9.1.36. **[Level 2]** Find the natural frequency of the illustrated system. $L = 1.4$ m. The rigid bar has a mass $m_1 = 5$ kg, $m_2 = 2$ kg, and the torsional spring, unstretched at $\theta = 0$, produces a moment given by $M = -k_\theta \theta$, where $k_\theta = 15$ N·m/rad.

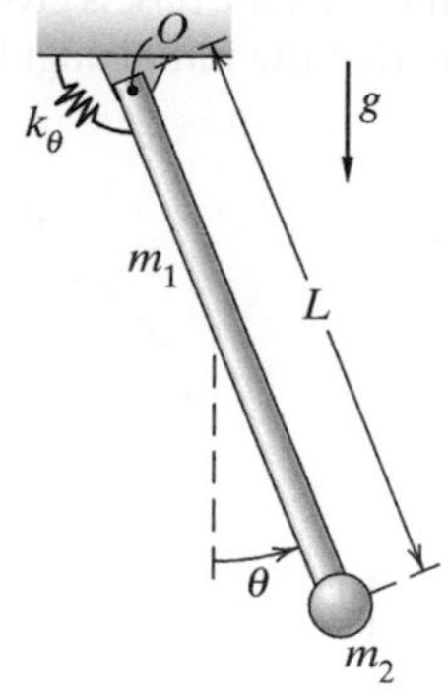

E9.1.36

9.1.37. **[Level 2]** A thin circular wooden disk is mounted on a central pivot and can spin freely. A passing woodpecker amused itself by pecking a perfectly circular hole in it, as shown. Because of the hole, the disk's mass center is now located off from the pivot, and if disturbed, the disk will oscillate like a pendulum. What is the natural frequency of the oscillation?

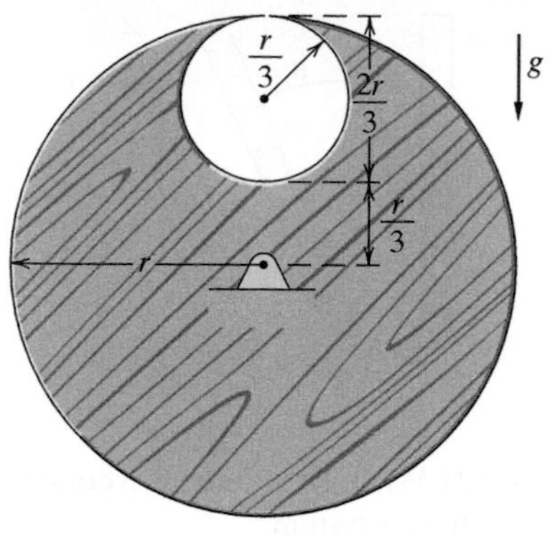

E9.1.37

9.1.38. **[Level 2]** Determine the natural frequency of oscillation for the system illustrated. Assume small angles of rotation.

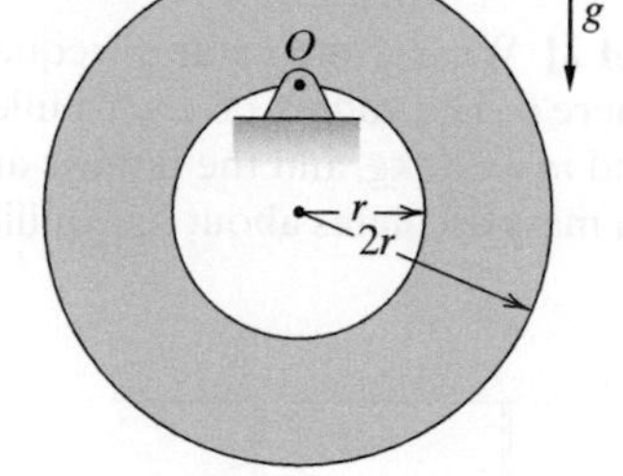

E9.1.38

9.1.39. **[Level 2]** **Exercise 7.3.9** asked for a numerical solution to the question of how fast a cylinder would be falling after 0.339 s. The relevant graphic is repeated here for convenience. Formulate the equation of motion in the form of an oscillator equation, find the system's natural frequency, and construct the complete solution.

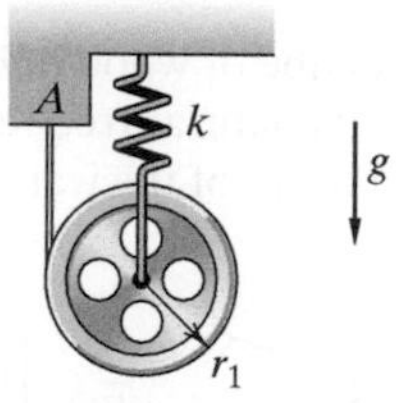

E9.1.39

9.1.40. **[Level 2]** The illustrated tube contains mercury (13,570 kg/m^3). The inner cross-sectional area of the tube is 16 mm^2, and the total mass of mercury is 22g. Determine the natural frequency of oscillation that results if the mercury level is momentarily perturbed. Approximate the system such that the mercury's free surface is always at right angles to the tube.

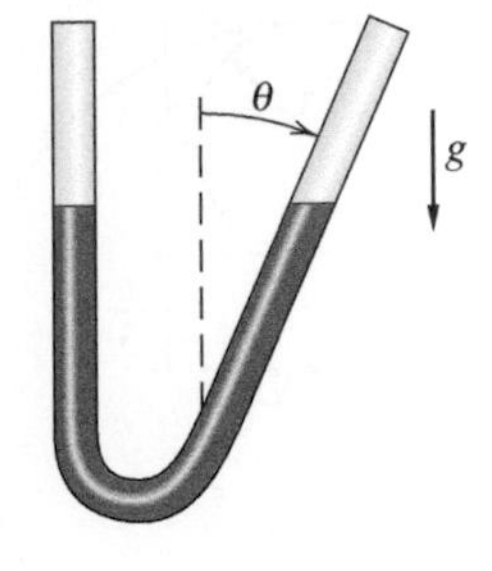

E9.1.40

9.1.41. **[Level 2]** What is the natural frequency of oscillation for the illustrated system?

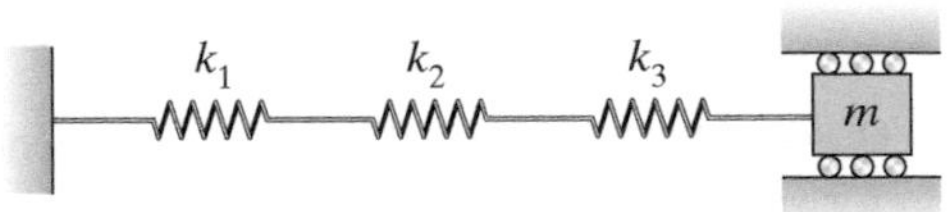

E9.1.41

9.1.42. **[Level 2]** What is the natural frequency for the system shown here? The radius of each pulley is 0.3 m, $k = 600$ N/m, and $m = 20$ kg, and the pulleys are massless. Assume that the mass oscillates about its equilibrium position.

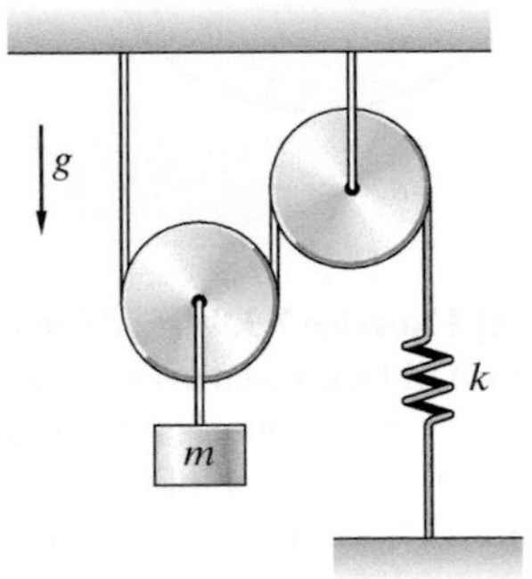

E9.1.42

9.1.43. **[Level 2]** A cube of wood bobs up and down in the water. Determine its natural frequency of oscillation, expressing the result in terms of the water's density and the wood's density, g and l.

E9.1.43

9.1.44. **[Level 2]** What is the natural frequency of oscillation for the uniform bar of mass m shown in the figure? Assume that it moves only within the x–z plane and that the supporting wires are massless and inextensible.

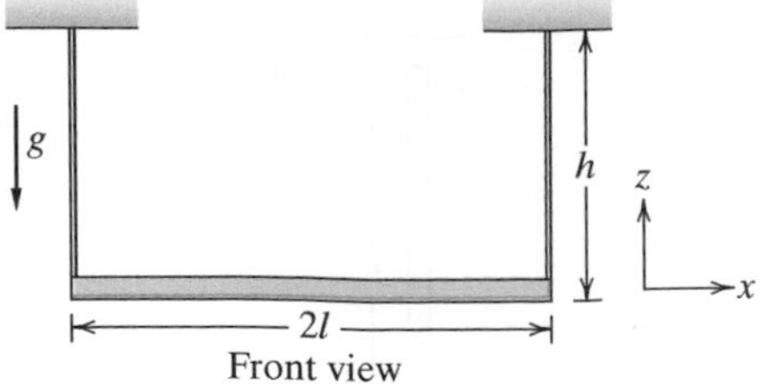

E9.1.44

9.1.45. **[Level 2]** What is the natural frequency for the illustrated system? The hinge at O is frictionless. $k = 10,000$ N/m, $m_1 = 5$ kg, $m_2 = 7$ kg, the mass per unit length of the rigid bar is 1 kg/m, $l_1 = 3$ m, and $l_2 = 5$ m. Neglect gravity.

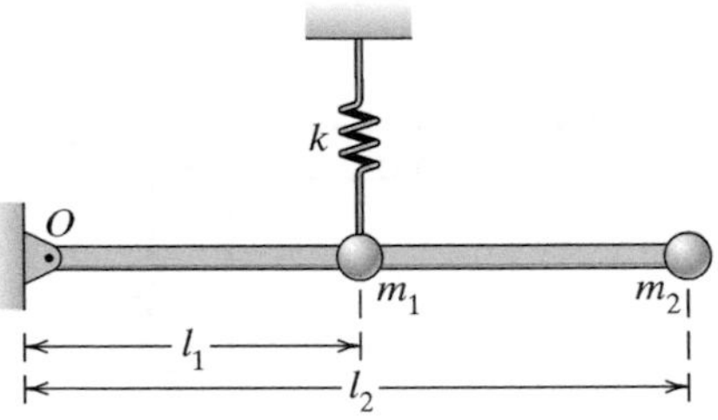

E9.1.45

9.1.46. **[Level 2]** A circular disk is supported by three massless wires. The disk has an areal density ρ. Determine the frequency of oscillation if the disk is given an initial twist and released. Assume that the initial angular displacement of the disk is small.

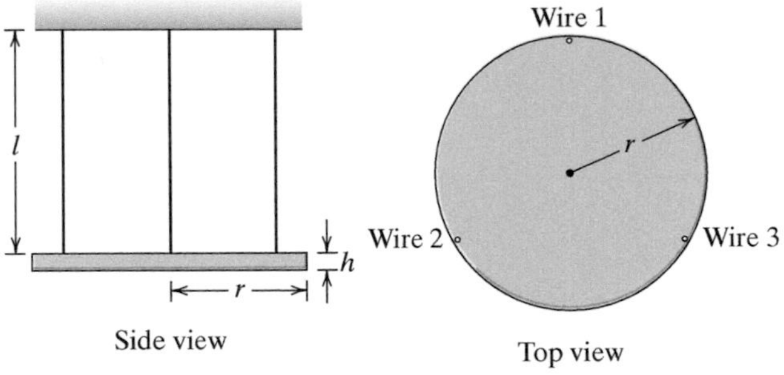

E9.1.46

9.1.47. **[Level 2]** Find ω_n for rotational motions of a uniform disk constrained by four wires. The tensioned restraining wires, all of length L, are massless with uniform tension T. Neglect gravity.

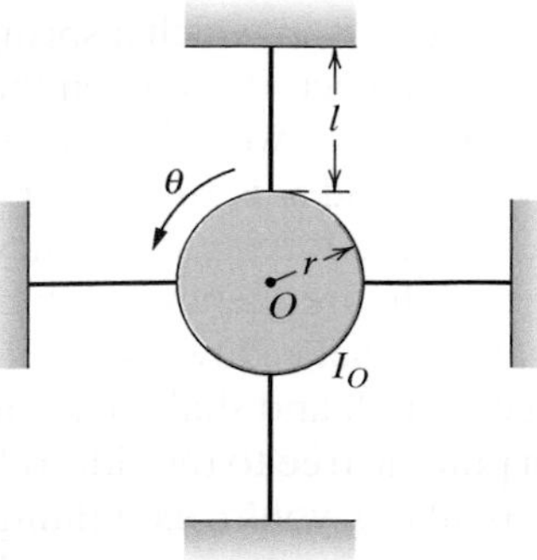

E9.1.47

9.1.48. **[Level 2]** Find the linearized equation of motion for the illustrated system. Assume that the cylinder has mass m and rolls without slipping on the illustrated circular path.

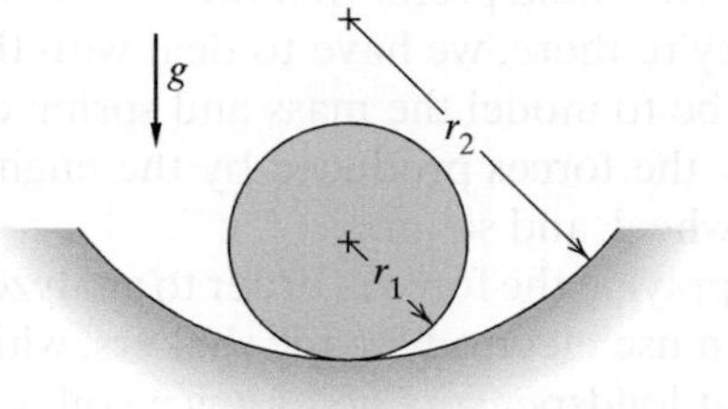

E9.1.48

9.1.49. **[Level 2]** A uniform disk A of mass m_A rolls without slip on block B (mass m_B). A spring with spring constant k connects to the center of disk A. The interface between the ground and block B is frictionless. Determine the system's natural frequency of oscillation.

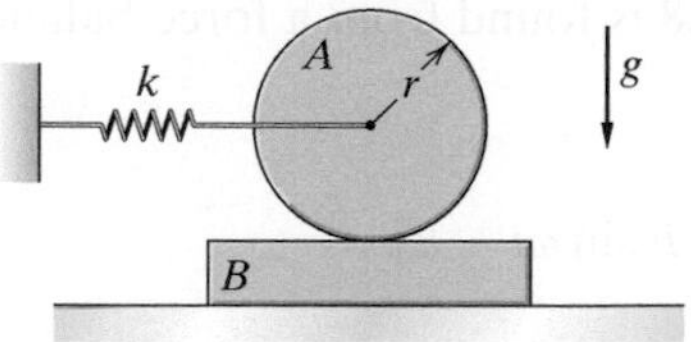

E9.1.49

9.1.50. **[Level 3]** Determine the natural frequency of oscillation for the illustrated half-disk under the assumption of rolling without slip.

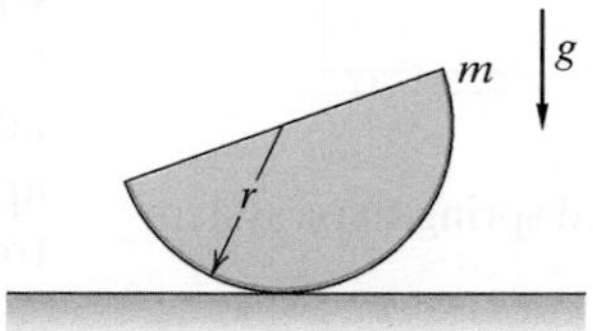

E9.1.50

9.1.51. **[Level 3]** The semicircular shell shown edge-on is rolled slightly to the right and then released. Determine the natural frequency of the ensuing oscillations.

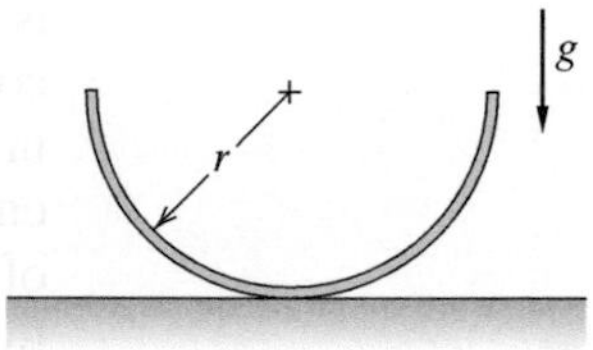

E9.1.51

9.1.52. **[Level 3]** A uniform disk A (mass $m_A = 10$ kg, radius $r = 0.25$ m) rolls without slip. A rigid link B (length 0.22 m, $m_B = 5$ kg) is attached via a frictionless pivot to the disk at its center O. Determine the linearized system's natural frequency of oscillation.

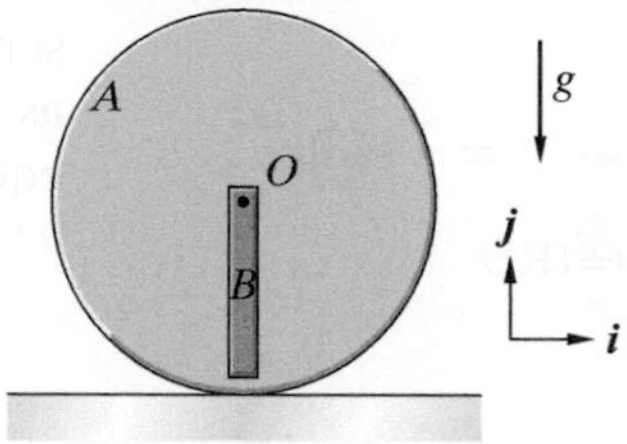

E9.1.52

9.2 UNDAMPED, SINUSOIDALLY FORCED RESPONSE FOR SINGLE-DEGREE-OF-FREEDOM SYSTEMS

Vibratory systems can experience a wide array of forcings—shock loadings, loadings that die away after some time (transient loadings), irregular loadings (random excitation), and so on. By far, the most common excitation with regard to vibratory systems, however, is a sinusoidal input. This is true for a few reasons. For one thing, we can solve for the response of a linear system with a sinusoidal input. Even better, we can approximate any periodic input as equivalent to a collection of sinusoidal inputs. You may recall learning about Fourier series back in a previous math class; this is where it becomes quite useful.

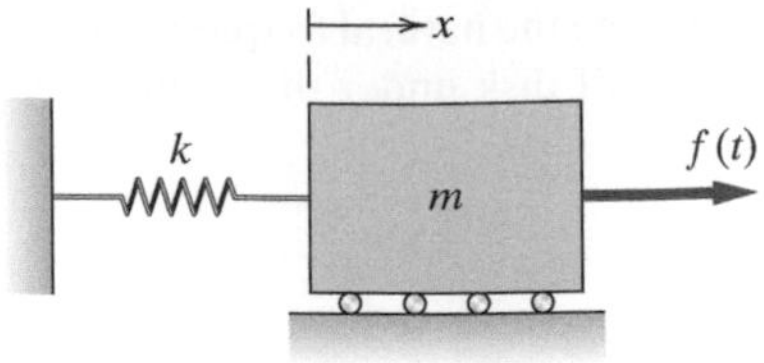

Figure 9.8 Forced spring-mass system

A simplified system is shown in **Figure 9.8**, in which a spring-mass system experiences a sinusoidal force $f(t)$ that acts directly on the mass. Because we're considering sinusoidal excitations, we will let $f(t) = F \sin \omega t$, where F is the magnitude of the forcing and ω is the associated forcing frequency. It's important to realize that the forcing frequency ω isn't related to the natural frequency ω_n. ω is the frequency at which a force is applied. For instance, you may want to shake down some fruit from a tree, and therefore you grab the tree's trunk and shake back and forth at some frequency ω. If, however, you pull the tree to the side, release it, and calculate the frequency at which it oscillates, you're examining its natural frequency ω_n. Each spring-mass system has only one natural frequency, but it can be forced at any frequency you choose.

Most of the time we're concerned with forces that we aren't controlling and aren't particularly happy to see. For instance, a four-cylinder engine is inherently imbalanced and will produce vibrations in the passenger compartment. Given a choice, we would prefer that they hadn't occurred in the first place, but since they're there, we have to deal with them. As engineers, our first job would be to model the mass and spring elements of the car and determine how the forces produced by the engine affect the driver's seat, the steering wheel, and so on.

At other times we will be supplying the force in order to analyze the system. Vibration engineers often use electrodynamic shakers, which work pretty much the same way that loudspeakers do: a voice coil is made to vibrate back and forth within a magnetic field. In a loudspeaker, the back-and-forth motion drives a paper or plastic cone that produces the sound you hear, whereas in a shaker, the moving element is attached directly to the test object being investigated. By applying a known force and observing the response of the object, we can deduce its natural frequency as well as its damping (which we will introduce in the next section). The equation of motion for **Figure 9.8** is found from a force balance (**Figure 9.9**):

Figure 9.9 FBD=IRD

$$m\ddot{x} = F \sin \omega t - kx$$

which gives us the equation of motion

$$m\ddot{x} + kx = F \sin \omega t \tag{9.7}$$

We are most concerned with the forced vibrational response of the mass, so we won't worry about matching initial conditions. In mathspeak this means that we want to look only at the particular (or forced) solution and not at the homogeneous (or unforced) solution. This is the solution that corresponds to the steady vibration you might feel coming from your seat while sitting at a stoplight. The engine is continually vibrating, and this vibration is transmitted to your seat with a particular amplitude that we can solve for. Assume a solution of the form

$$x(t) = X \sin \omega t \tag{9.8}$$

Differentiating this twice and plugging into our equation of motion yield

$$-\omega^2 mX \sin \omega t + kX \sin \omega t = F \sin \omega t$$

Grouping terms and canceling out the common $\sin \omega t$ factor give us

$$(k - m\omega^2)X = F$$

which means

$$X = \frac{F}{k - m\omega^2} \tag{9.9}$$

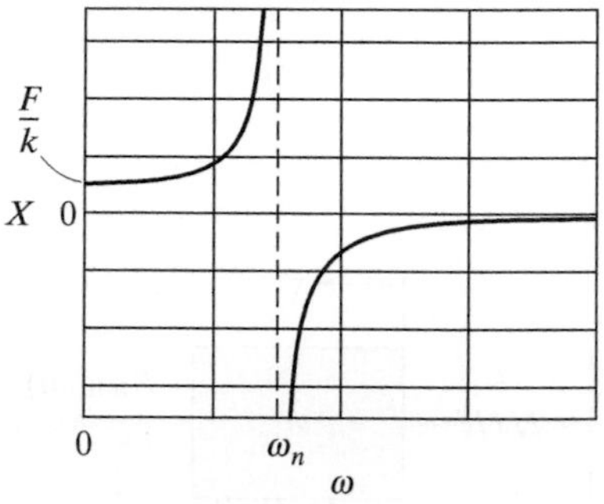

Figure 9.10 Forced response for spring-mass

Figure 9.10 shows what X looks like when plotted as a function of the frequency ω. The important points to notice are that the response is finite and positive at $\omega = 0$ (and equal to $\frac{F}{k}$), goes toward infinity as ω approaches ω_n, changes sign, and then approaches zero for high frequencies. Thus the responses for frequencies less than ω_n are **in-phase** with the forcing (that is, when the forcing is positive, so is the response) but **out-of-phase** for frequencies above ω_n.

>>> **Check out Example 9.3 (page 558) and Example 9.4 (page 559) for applications of this material.**

What does all this mean? The low-frequency limit of $\omega = 0$ is just the static response of the spring to a constant force with magnitude F. No dynamics is going on at all. As the frequency ω of the forcing increases toward ω_n, we get a *lot* of dynamic behavior. Most of us are familiar with the Tacoma Narrows bridge disaster, in which the bridge vibrated itself to destruction. Although not exactly accurate, it wouldn't be terribly wrong to say that the bridge encountered the same kind of problem that our mass would encounter if the forcing were near ω_n. Such a situation is called **resonance**, and it's almost always bad for a physical structure. It's a rare system that appreciates being forced into vibrations of immense amplitude, and most respond by breaking. For this reason, engineers usually try very hard to avoid the situation. One remedy is to add damping to the system, which we will look at in the next section.

As already mentioned, when the forcing frequency increases well beyond ω_n, the amplitude of the response drops toward zero. This is because the mass is becoming the dominant element in the system. Remember, the force associated with the mass is $m\ddot{x}$. If the response is $X \sin \omega t$, then the acceleration is $-\omega^2 X \sin \omega t$, which means the inertial force is $-\omega^2 mX \sin \omega t$. This term grows as the square of frequency and, as the frequency goes toward infinity, becomes the dominant term in the equation. Simply stated, the effective inertia of the mass becomes larger and larger as the forcing frequency continues to increase, meaning that a finite forcing produces a smaller and smaller response.

EXAMPLE 9.3 FORCED RESPONSE OF A SPRING-MASS SYSTEM (Theory on page 557)

In the Tacoma Narrows bridge disaster, the wind permitted a small periodic excitation to act on the bridge, which then began to oscillate. The oscillations eventually grew so large that the bridge collapsed. The designers of the bridge anticipated none of this catastrophic behavior. The design had been prepared from a static standpoint, which was the normal procedure: estimate the traffic on the bridge, multiply by some safety factor, and you're done. The fact that the bridge could act as a dynamical object and be driven to destruction by a relatively gentle breeze simply hadn't occurred to anyone. Contrary to what you might suppose, the periodic force didn't arise because the wind kept changing direction. Instead, a complex phenomenon known as **vortex shedding** allowed swirls of air (vortices) to come off the top of the bridge, then the bottom, back to the top, and so on. This shedding of vortices caused an oscillatory vertical pressure to act against the bridge, which is what actually caused the bridge to ultimately collapse.

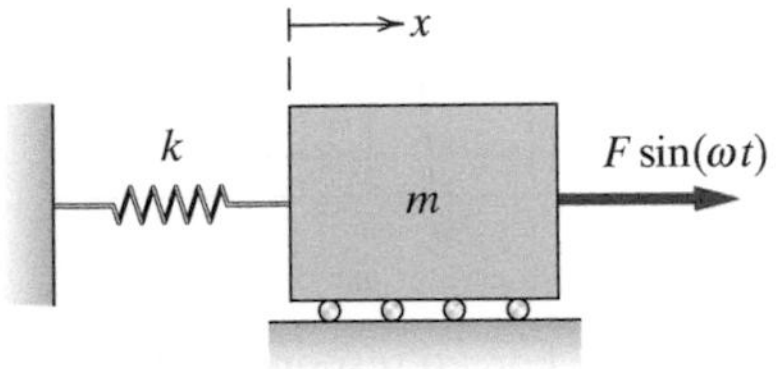

Figure 9.11 Simple model of Tacoma Narrows bridge

Let's look at a simple example of how a small forcing can cause a response big enough to destroy a bridge. **Figure 9.11** shows a spring-mass system with an applied external force, a model that will serve as a first approximation to the Tacoma Narrows bridge under wind loading. The mass-spring represents the bridge, and the force represents the aerodynamic excitation. For $m = 1.01 \times 10^5$ kg, $k = 3.64 \times 10^6$ N/m, and $f(t) = 1.00 \times 10^4 \sin 6t$ N, determine the steady-state response of the system. This is a fairly small loading—about 1% of the bridge's weight.

Goal Determine the steady-state response of the system.

Given System parameter values and magnitude of force exerted by wind.

Draw **Figure 9.11** shows our system of interest.

Formulate Equations The equation of motion for the given system is

$$m\ddot{x} + kx = F \sin \omega t$$

$$(1.01 \times 10^5\ \text{kg})\ddot{x} + (3.64 \times 10^6\ \text{N/m})x = (1.00 \times 10^4\ \text{N}) \sin 6t$$

Solve

$$(9.9) \Rightarrow \qquad X = \frac{1.00 \times 10^4\ \text{N}}{(3.64 \times 10^6\ \text{N/m}) - (1.01 \times 10^5\ \text{kg})(36\ \text{rad}^2/\text{s}^2)} = 2.5\ \text{m}$$

$$x(t) = 2.5 \sin 6t$$

with $x(t)$ given in meters.

The end result is a 2.5 m oscillation amplitude (5 m total excursion) from a forcing that was only 1% of the system's weight! So you can see that forcing an undamped system near its natural frequency leads to a very large response, even if the level of forcing is small.

EXAMPLE 9.4 TIME RESPONSE OF AN UNDAMPED SYSTEM (Theory on page 557)

In Example 9.3 we looked at the steady-state response of a system with a natural frequency of $\sqrt{\frac{3.64\times10^6\ \text{N/m}}{1.01\times10^5\ \text{m}}} = 6.003\ \text{rad/s}$ to an excitation of 6.00 rad/s. Now let's see what would happen if we forced the system at precisely its natural frequency. We will use MATLAB to quickly find the numerical result and change the mass to 1.00×10^5 kg and the stiffness to 3.60×10^6 N/m so that both the natural frequency and the forcing are equal to 6.00 rad/s. Let F again be 1.00×10^4 N. We will use initial conditions of zero for both deflection and speed.

Goal Find the dynamical response of a spring-mass system when forced at its natural frequency.

Given System parameter values and magnitude of the sinusoidal forcing.

Draw **Figure 9.11** shows our system of interest.

Formulate Equations Our governing equation is

$$(1.00 \times 10^5\ \text{kg})\ddot{x} + (3.60 \times 10^6\ \text{N/m})x = (1.00 \times 10^4\ \text{N})\sin 6t$$

and if we mass-normalize, it becomes

$$\ddot{x} + (36.0\ \text{N/m})x = (0.1\ \text{N})\sin 6t$$

Solve Integrating this equation with MATLAB for 10 s produces the output shown in **Figure 9.12**. Note that the oscillations build up linearly from zero, getting bigger and bigger as time goes on. Although the time scale for the Tacoma Narrows bridge buildup was quite a bit longer (hours rather than the seconds of this plot), the basic physics involved was the same.

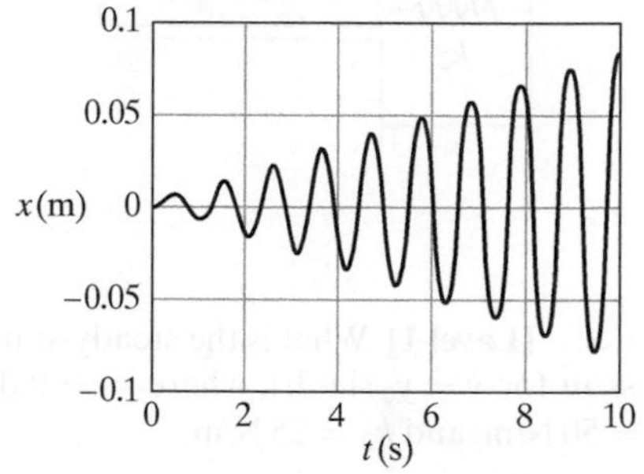

Figure 9.12 Increasing oscillations

Check That this example showed a linear increase in oscillation amplitude isn't a surprise. The solution to the unforced problem is a sinusoid at the system's natural frequency—in this case 6 rad/s. We then forced the system with a sinusoid of 6 rad/s. Specifically, we forced the differential equation with its own homogeneous solution. Whenever you do this, you will find solutions that grow with time. These are called **secular solutions**, and you probably saw them in your differential equations course. Because our equation was only second order, the highest secular term is t^1 (the highest power of t is one less than the order of the equation). If the equation had been third order, then we could have expected a t^2 term as well.

EXERCISES 9.2

9.2.1. [Level 1] The depicted mass-spring system experiences in-phase oscillation of amplitude $\bar{x} = 0.02$ m when its base oscillates at $\omega = 6$ rad/s with an amplitude of $\bar{y} = 0.01$ m. If $m = 10$ kg, what is the stiffness k of each spring? Neglect gravity.

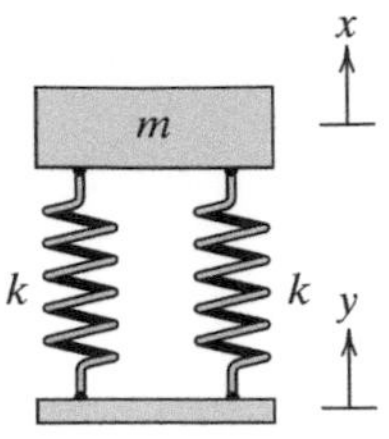

E9.2.1

9.2.2. [Level 1] What are the equations of motion for the illustrated system?

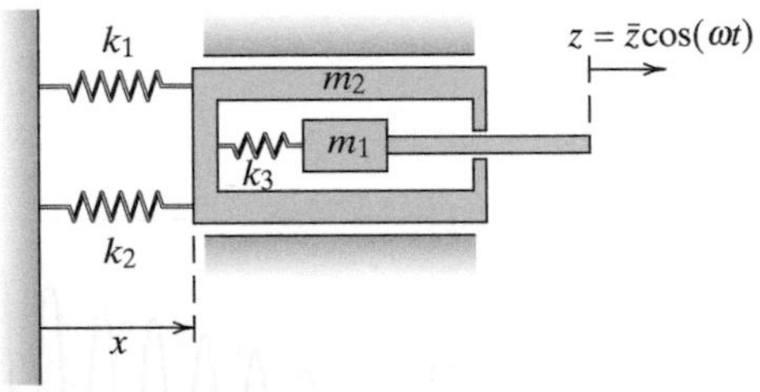

E9.2.2

9.2.3. [Level 1] What is the steady-state response of the mass m for $y = y_0 \sin(3t)$, where $y_0 = 0.02$ m? $m = 10$ kg, $k_1 = 50$ N/m, and $k_2 = 25$ N/m.

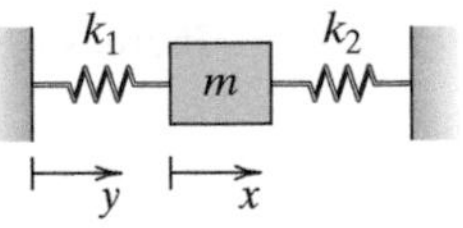

E9.2.3

9.2.4. [Level 1] The base of the illustrated mass-spring system oscillates at $\omega = 4$ rad/s with an amplitude of $\bar{y} = 0.03$ m. The system response is out-of-phase with amplitude $\bar{x} = 0.05$ m, and the stiffness of both springs is $k = 100$ N/m. Calculate the system's mass m.

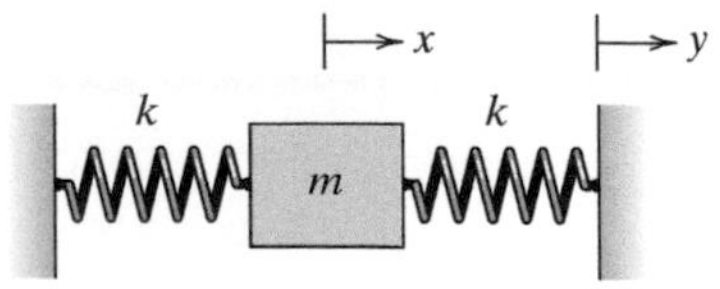

E9.2.4

9.2.5. [Level 1] What is the steady-state response of the mass m for $y = y_0 \cos(12t)$ where $y_0 = 0.02$ m? $m = 1.2$ kg, $k_1 = 100$ N/m, $k_2 = 100$ N/m, and $k_3 = 10$ N/m.

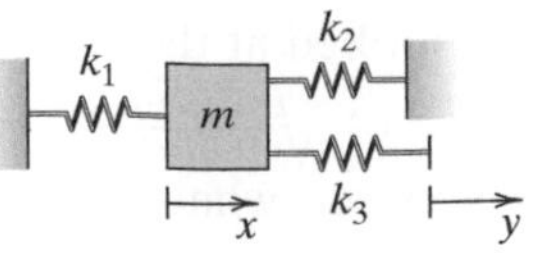

E9.2.5

9.2.6. [Level 1] The driver's seat of an automobile can be modeled as a single-degree-of-freedom system. The seat weighs 23 kg. If a sinusoidal displacement of $a\cos(\omega t)$ ($a = 0.64$ cm) is applied to the base of the supporting spring, what is the amplitude of the seat's steady-state vibration? Let $\omega = 60$ rad/s. $k = 1.23$ kN/cm.

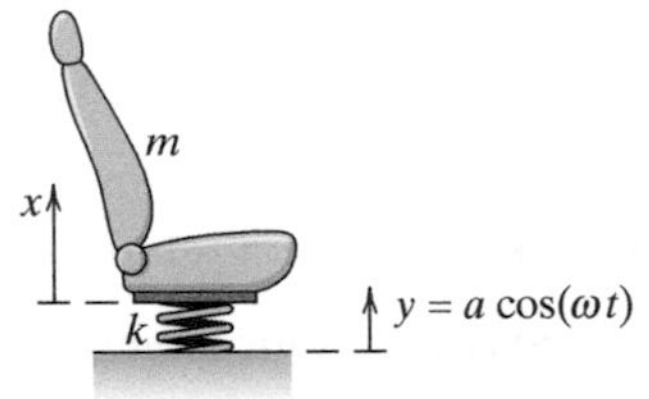

E9.2.6

9.2.7. [Level 1] The depicted mass-spring system is acted on by a periodic force $f(t) = F \sin \omega t$ with frequency $\omega = 5$ rad/s, and the resulting system response is in-phase with amplitude $\bar{x} = 0.04$ m. If $k_1 = 75$ N/m, $k_2 = 160$ N/m, and $m = 7$ kg, what is the magnitude F of the applied force $f(t)$?

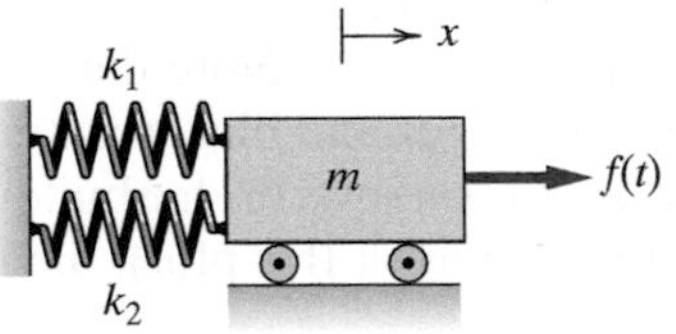

E9.2.7

9.2.8. [Level 1] A 17 kg wheel on a test stand has an imbalance eccentricity of 0.20 cm and is being spun at 20 Hz. The wheel is supported in the test stand by the illustrated spring combination. $k = 6.48$ kN/m. Assume vertical translation of the wheel/axle and neglect gravity.

a. Calculate the natural frequency of this arrangement.

b. Determine the steady-state vibration amplitude of the wheel.

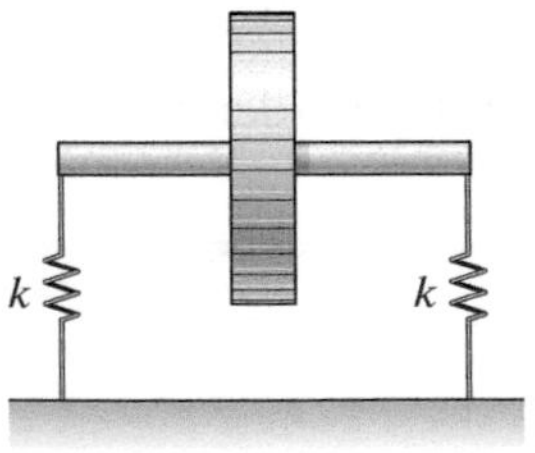

E9.2.8

9.2.9. [Level 1] An 80 kg motor is mounted on a 1000 kg block. The total stiffness of the supporting springs is 243,000 N/m. 60 kg of the 80 kg motor is rotating mass, and it has an eccentric imbalance of 3 mm. If the motor is rotating at 15.2 rad/s, what is the oscillation amplitude of the system, and what is the force transmitted to the floor? Assume purely vertical motion.

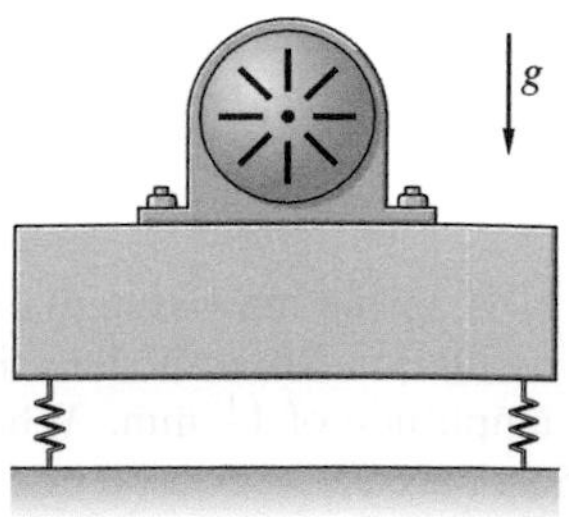

E9.2.9

9.2.10. **[Level 1]** A 4.5 kg rotor with a radius of 0.15 m is being spun at 50 rad/s, and the imbalance forces are recorded through force sensors in the support bearings. Each bearing sensor reacts only to vertical forces, and both record a maximum force of 1.5 N. To balance the rotor, how large a mass must be attached to the outside edge of the rotor?

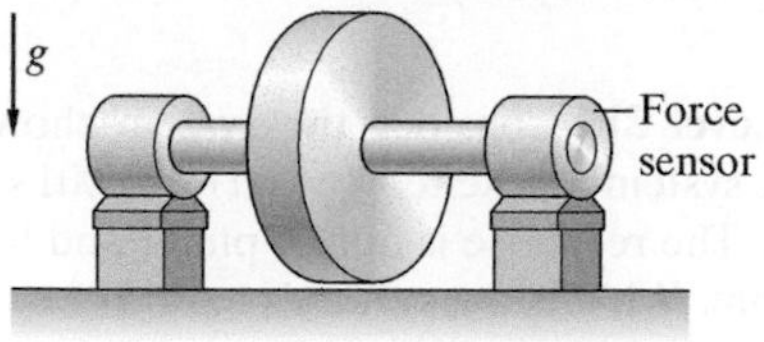

E9.2.10

9.2.11. **[Level 1]** Consider a spring-mass system acted on by a force $f(t)$. Because of the force, the mass has the response $x(t) = 0.003\sin(\omega t)$ (measured in meters). What is the force transmitted to the wall? $m = 10$ kg and $k = 5000$ N/m.

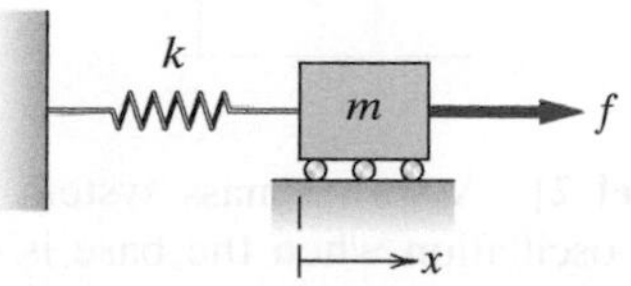

E9.2.11

9.2.12. **[Level 1]** A spring-mass system has a spring constant of 1600 N/m and a mass of 2 kg. What is the response amplitude to an excitation of

$$y(t) = (0.06\,\text{m})\sin(\omega t)$$

where $\omega = 10$ rad/s?

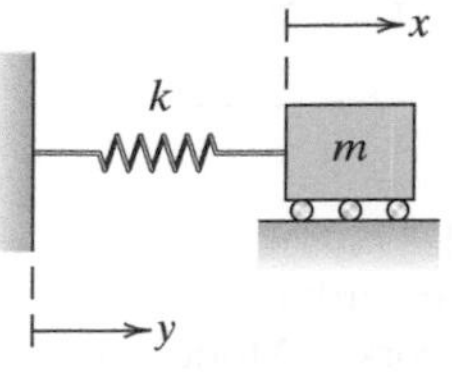

E9.2.12

9.2.13. **[Level 1]** Consider a car that's moving along an undulating road. The road's profile is given by

$$y(x) = 0.02\sin\left(\frac{2\pi x}{15\,\text{m}}\right)$$

where both x and y are in meters. Treat the car as a point mass m and assume that the orientation of the normal force between the car and road is vertical. $\dot{x}$ is constant and $m = 1300$ kg. What is the normal force developed between the car and the road as a function of the car's speed $\dot{x}$?

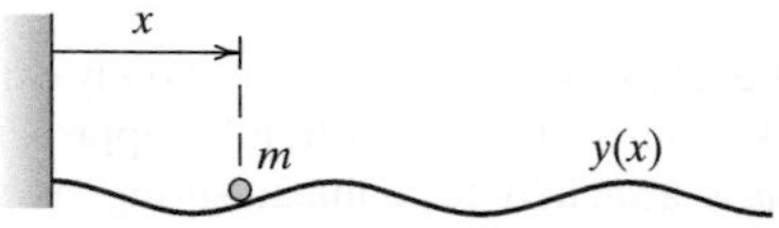

E9.2.13

9.2.14. **[Level 2]** Imagine that you're at the 2010 World's Fair and are planning to check out the ride RocketBlast. In this ride you're strapped into a chair in the RocketSled, and at $t = 0$ the whole thing rockets vertically at a constant acceleration of 1.5g. You doubt the validity of the 1.5g claim and want to verify it, so you build the device shown in the figure. $k = 4.6$ N/m, $m = 28$g, and the unstretched length of the spring is 2 cm. How large should h be so that the mass barely touches the bottom of the enclosure during the upward acceleration if you've got the device securely placed on the floor in front of you?

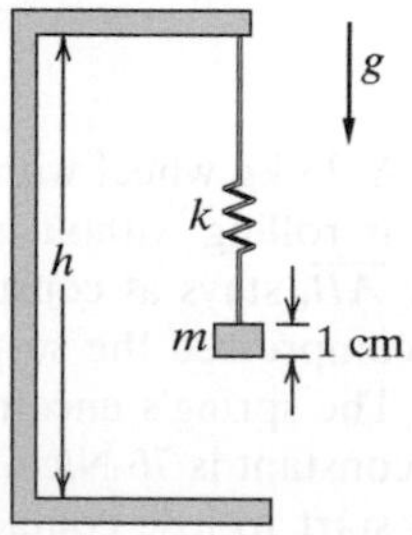

E9.2.14

9.2.15. **[Level 2]** A washing machine's support springs will compress 0.6 cm under the weight of the machine when it is placed on the floor. Knowing this, determine the ratio of the machine's oscillation amplitude to the floor's oscillation amplitude if the floor is vertically oscillating at 80π rad/s.

E9.2.15

9.2.16. **[Level 2]** When the top of the spring is oscillated at 300 rad/s and an amplitude of 0.002 m, the mass responds with an in-phase oscillation with amplitude 0.0011 m. How much will the spring stretch if statically loaded by the mass m?

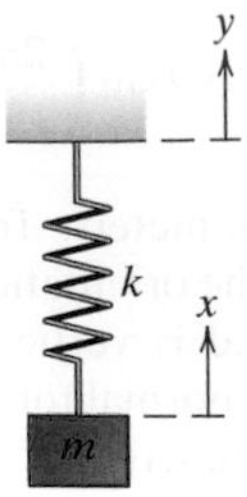

E9.2.16

9.2.17. [Level 2] When a tall, vertical body experiences an earthquake, the system can often be approximated as a lumped mass attached to a linear spring, as shown in the figure. Assume that the ground motion is given by $y = y_0 \sin(50t)$, $y_0 = 0.001$ m. The mass responds in an out-of-phase manner, with a magnitude of 0.004 m. Determine the system's natural frequency.

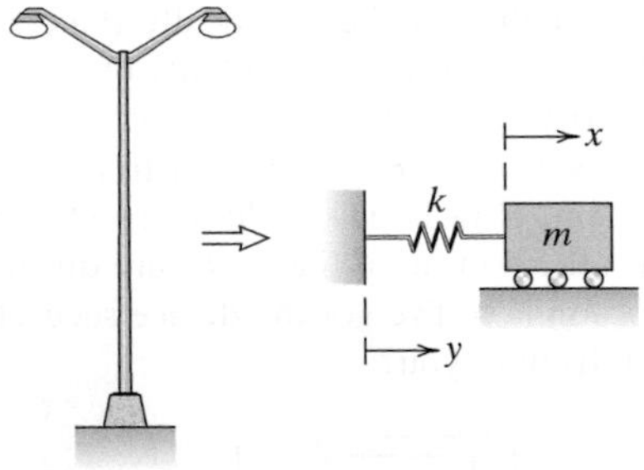

E9.2.17

9.2.18. [Level 2] A 23 kg wheel with an eccentric imbalance $e = 0.003$ m is rolling without slip. The vehicle body, represented by $\overline{AB}$, stays at constant height above the ground and has compressed the suspension spring to a length $h = 0.29$ m. The spring's uncompressed length is 0.3 m, and its spring constant is 76 N/cm. At what vehicle speed will the wheel start to lose contact with the road? $r = 0.3$ m.

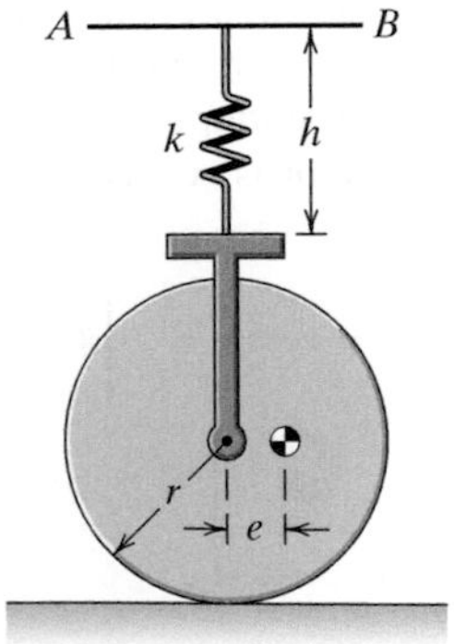

E9.2.18

9.2.19. [Level 2] Calculate the range of frequencies for which m's response amplitude will be less than or equal to 0.001 m. The forcing is given by $f(t) = F\cos(\omega t)$, with $F = 1.10$ N. $m = 10$ kg and $k = 1.00 \times 10^4$ N/m.

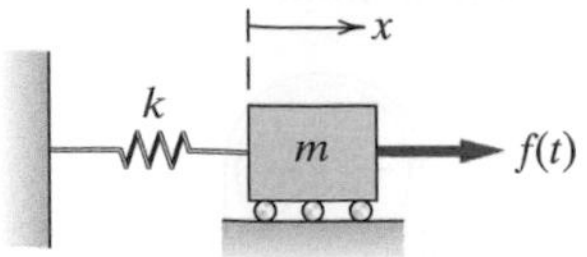

E9.2.19

9.2.20. [Level 2] A spring-mass system experiences a 4 mm out-of-phase oscillation when the base is oscillated at 200 rad/s with an amplitude of $1\frac{1}{3}$ mm. What is the static deflection of the 5 kg mass?

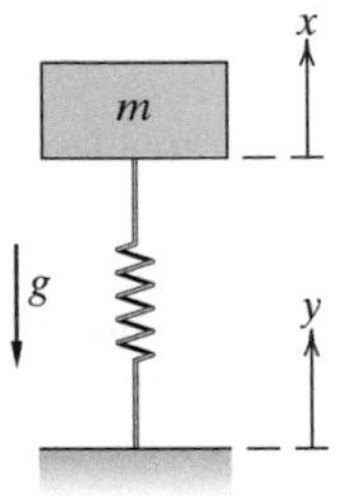

E9.2.20

9.2.21. [Level 2] Consider the system shown in the figure. This system is excited by $y(t) = 0.001\sin(\omega t)$ m. $\omega = 50$ rad/s. The response is out-of-phase and has a magnitude of 4 mm. What is the system's natural frequency?

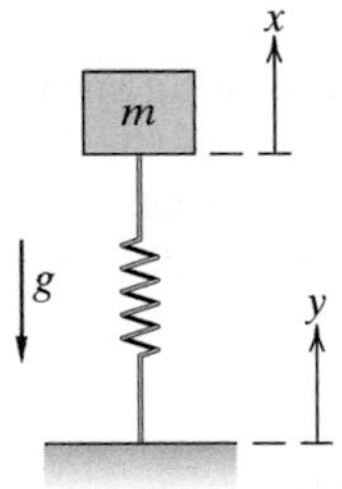

E9.2.21

9.2.22. [Level 2] A spring-mass system undergoes a 2 mm in-phase oscillation when the base is oscillating at 300 rad/s with an amplitude of 1.1 mm. What is the static deflection of the mass when it is placed on the uncompressed spring?

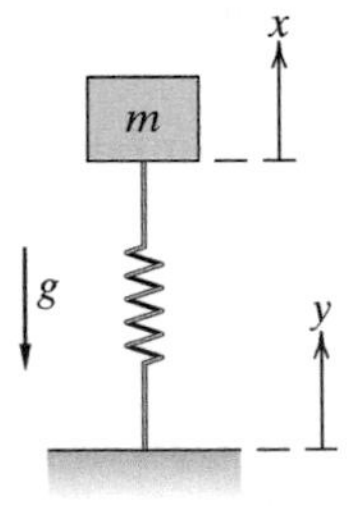

E9.2.22

9.2.23. [Level 2] One benefit of a suspension is that it reduces the level of disturbance generated by a rough road. Let's see how this works. Model the car as a single mass and the suspension as a spring. The road profile is given by $y = a\sin(\frac{2\pi x}{\lambda})$. The car travels to the right at v. What is the ratio of the car's acceleration with a suspension versus without one (that is, replacing the spring with a rigid, massless rod)? Does the suspension always help, or does it depend on the car's speed? Consider motions about the car's equilibrium position (that is, neglect gravity).

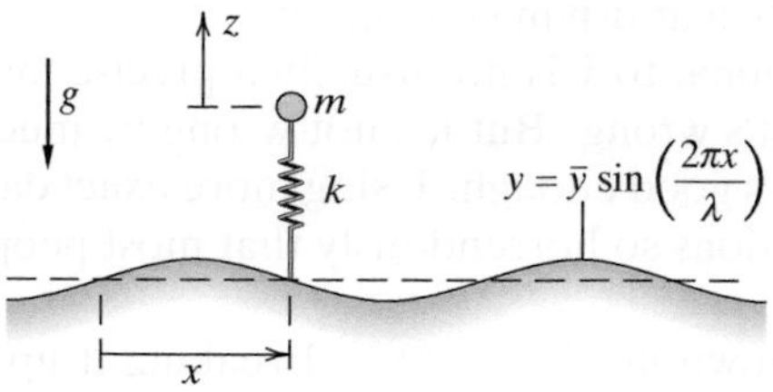

E9.2.23

9.2.24. **[Level 2]** Consider a car that's traveling along a road that has a sinusoidally varying profile. $\bar{y} = 3$ cm and $\lambda = 10$ m. When the car is stationary, the suspension springs are compressed from their uncompressed length of 0.5 m by 8 cm. Assume that the body of the car remains vertically fixed in position (0.42 m above the dashed line) as it travels to the right. At what speed will the tire (modeled as a lumped mass m) begin to lose contact with the road? $m = 15$ kg and $k = 4800$ N/m.

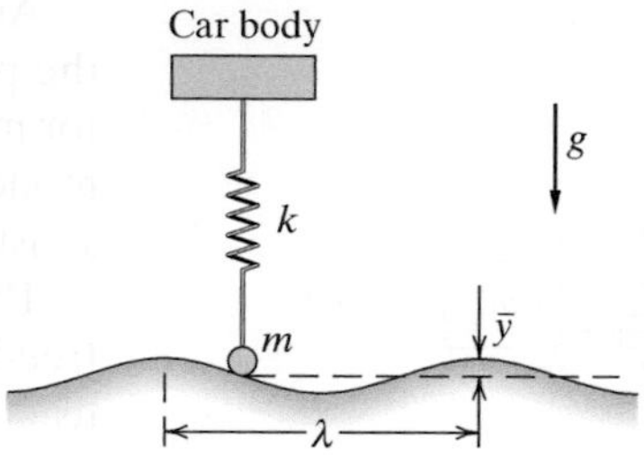

E9.2.24

9.3 DAMPED, FREE RESPONSE FOR SINGLE-DEGREE-OF-FREEDOM SYSTEMS

The only problem with the systems discussed in the preceding two sections is their absolute lack of damping—they had no energy dissipation. This isn't very realistic because all physical systems have damping. They may have just a little bit (a tuning fork, for example, is close to undamped), but there is always *some*. If there weren't, then any motion in the system would persist forever—perpetual motion, in other words.

All systems have some inherent damping, and we often deliberately engineer damping into the systems we are designing. For instance, **Figure 9.13** shows an automobile shock absorber and spring. If only the spring were present, the car would bounce up and down continually and make for a pretty unpleasant ride. The damper removes energy from the system whenever it's compressed or extended, eliminating the excess bounciness. In fact, one way you can tell that you need new shocks is to notice when your car starts to bounce a lot.

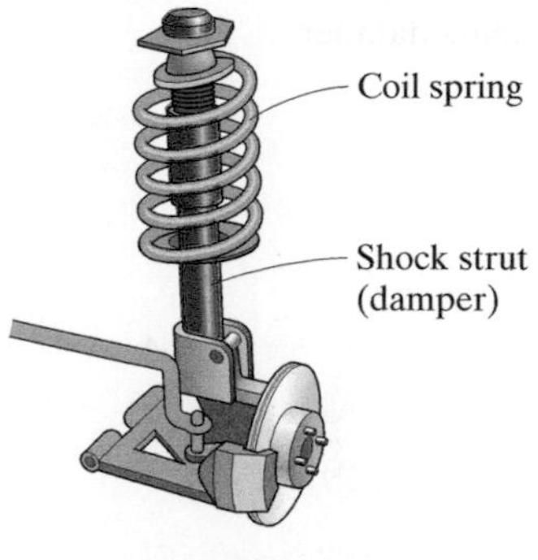

Figure 9.13 Damper and spring for an automobile suspension

In the next few pages we will introduce a mathematically tractable form of damping and see how it affects our spring-mass system's response for both free and forced cases.

(By the way, the correct word is *damping*. You'll very often hear people use *dampen* or *dampening*, as in "There's too much vibration and we'll have to dampen it." Well, they're all wet. (Wow, that was bad, wasn't it?) Anyway, leave dampening to gardeners; we'll work with damping.)

The simplest type of damping to add, and the only one discussed in this book, is called **viscous damping**, or **linear damping**. The viscous damper is attached to the mass and acts in parallel with the spring. The force generated by the damper is directly proportional to its speed ($c\dot{x}$), where c is the damping constant and acts to slow the mass down. This is what gives us the name *linear damper*; the damping force is linearly proportional to $\dot{x}$. This kind of damper is very much like that found in the shock absorber of a car's suspension. When the car body moves down, the shock absorber tube is compressed. This forces fluid through small orifices that produce an energy loss. Upward motions of the body extend the shock absorber tube, again inducing motion in the enclosed fluid and again taking energy out of the system. As you might guess, a thick fluid won't get through the orifices easily, leading to high damping levels, whereas thin fluids produce lower levels of damping. Some car manufacturers now have dampers that control the viscosity of the fluid in the damper electrically, allowing them to tailor the damping for particular driving conditions.

Assuming a force proportional to $\dot{x}$ is not usually a precise match to the physics. In other words, it's wrong. But it's not wrong by much, and for most practical purposes it's good enough. Using more exact damping models complicates the equations so horrendously that most people are satisfied using $c\dot{x}$.

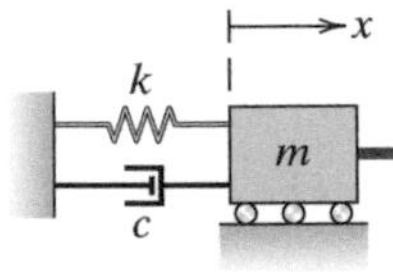

Figure 9.14 Forced spring-mass damper system

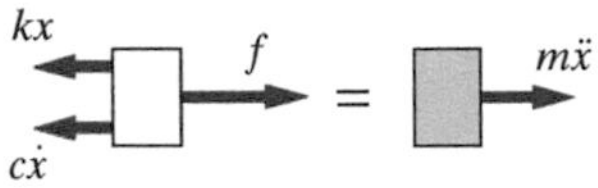

Figure 9.15 FBD=IRD for spring-mass damper

The relevant system is shown in **Figure 9.14**. Breaking it up into a free-body diagram and an inertial-response diagram (**Figure 9.15**) gives us

$$m\ddot{x} = -c\dot{x} - kx + f(t)$$

$$m\ddot{x} + c\dot{x} + kx = f(t) \tag{9.10}$$

As you can see, the only new item is the $c\dot{x}$ term, the force due to the damper. In the absence of forcing ($f = 0$), we have

$$m\ddot{x} + c\dot{x} + kx = 0 \tag{9.11}$$

and dividing by m gives us the mass-normalized form

$$\ddot{x} + \frac{c}{m}\dot{x} + \omega_n^2 x = 0 \tag{9.12}$$

Rewriting $\frac{c}{m}$ as $2\zeta\omega_n$ is a good idea because it makes life simpler later on and gives us a useful new parameter to consider—ζ, the **damping factor**. Notice that the dimensionless parameter ζ is related to the physical parameters c and m, which is somewhat akin to the way ω_n is related to the two physical parameters m and k. These two parameters ζ and ω_n allow us to characterize the damped response of our system, and because there are only two of them, it's simpler this way than having to deal with the three physical parameters m, c, and k. If you don't want to make this substitution, then just follow along and keep m, c, and k separate. Soon enough you will see that it's a pain doing it that way, and so you will happily switch over. From the definition we immediately find that

$$\zeta = \frac{c/m}{2\omega_n} = \frac{c\sqrt{m}}{2m\sqrt{k}} = \frac{c}{2\sqrt{mk}} \tag{9.13}$$

Rewriting (9.12) in terms of ζ and ω_n yields

$$\ddot{x} + 2\zeta\omega_n\dot{x} + \omega_n^2 x = 0 \tag{9.14}$$

All linear, homogeneous, constant-coefficient, ordinary differential equations have solutions that can be expressed as

$$x(t) = ae^{\lambda t} \tag{9.15}$$

that is, an exponential solution in time governed by a magnitude a and time-dependent behavior defined by the constant λ. There's no restriction on λ—it can be real, imaginary, or complex, depending on the particular equation of motion.

Differentiating (9.15) with respect to time, inserting the results into (9.14), and then factoring out the common terms give us

$$ae^{\lambda t}\left(\lambda^2 + 2\zeta\omega_n\lambda + \omega_n^2\right) = 0 \tag{9.16}$$

a can't be zero, or we would simply have the trivial solution $x(t) = 0$. The exponential function $e^{\lambda t}$ isn't identically zero. Thus, in order to have the left-hand side of (9.16) be zero so as to match the right-hand side, we need λ to satisfy

$$\lambda^2 + 2\zeta\omega_n\lambda + \omega_n^2 = 0$$

This is a quadratic in λ and has the solutions

$$\lambda_{1,2} = -\zeta\omega_n \pm \omega_n\sqrt{\zeta^2 - 1}$$

We have two roots, λ_1 and λ_2, which is as expected because the equation is a second-order differential equation.

Depending on whether ζ is greater than or less than unity, the argument of the square root will be positive or negative, respectively. If negative, we have a pair of imaginary roots. As it turns out, this is most often the case. The value of ζ for most vibrating structures is quite low—between 0.001 and 0.1. Therefore, let's pull the imaginary i out of the square root and express the solution as

$$\lambda_{1,2} = -\zeta\omega_n \pm i\omega_n\sqrt{1 - \zeta^2}$$

The total solution of (9.14) is therefore

$$x(t) = a_1 e^{(-\zeta\omega_n + i\omega_n\sqrt{1-\zeta^2})t} + a_2 e^{(-\zeta\omega_n - i\omega_n\sqrt{1-\zeta^2})t} \tag{9.17}$$

This can be written in a more familiar form by realizing that the real part of the exponential term represents an exponential decay of the response and that the imaginary part corresponds to the oscillations that are also present. The alternative form that captures both of these effects is

$$x(t) = e^{-\zeta\omega_n t}(b_1 \sin\omega_d t + b_2 \cos\omega_d t) \tag{9.18}$$

where

$$\omega_d \equiv \omega_n\sqrt{1 - \zeta^2} \tag{9.19}$$

with the subscript d indicating that ω_d is the **damped frequency of oscillation**, and b_1 and b_2 are constants determined by the problem's initial conditions.

>>> **Check out Example 9.5 (page 567) for an application of this material.**

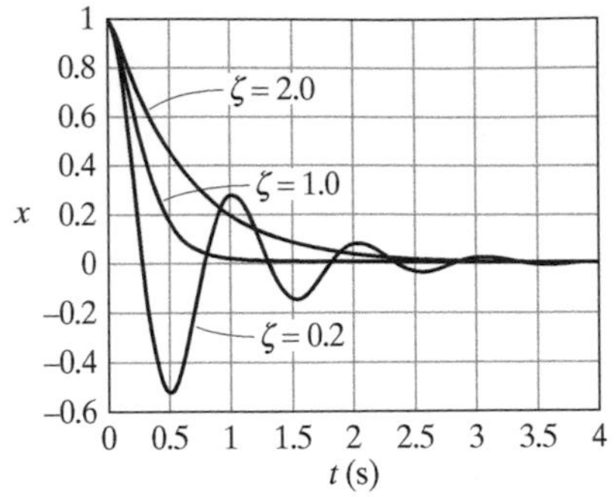

Figure 9.16 Underdamped, critically damped, and overdamped responses

As already noted, the response characteristics change for ζ below, equal to, or above 1.0. **Figure 9.16** shows typical responses for all three cases. The response of

$$\ddot{x} + 2\zeta\omega_n\dot{x} + \omega_n^2 x = 0$$

has been plotted for $\omega_n = 6.28$ and $\zeta = 0.2, 1.0$, and 2.0. As you can see, the response behavior for $\zeta = 0.2$ shows oscillations that die down in an exponential manner—the typical vibratory behavior we will be dealing with. This is referred to as **underdamped** behavior. The response associated with $\zeta = 2.0$ shows an **overdamped** case. In this situation there aren't any oscillations; the response simply decays in an exponential manner to zero. **Critically damped** systems ($\zeta = 1.0$) are the limiting case between oscillatory and nonoscillatory solutions. Notice also that the rate at which the response approaches zero is decreasing as the damping increases. As more damping is added to a system, it responds more sluggishly. What's shown is a basic engineering tradeoff. When we design a mechanism, we typically want it to react quickly, which occurs with low damping. At the same time, we don't want it to oscillate once the parts get to their desired positions. Rather, we want it to settle down quickly. Think of a welding robot, for instance. An automobile manufacturer would like it to swing quickly into position so that it can make a spot weld. If it's vibrating, then it can't make an accurate weld. And what kills vibrations? Damping. More damping is therefore good for this aspect of the design.

The problem is that we now have conflicting requirements. We want low damping for a quick response and high damping for low residual vibrations. Engineers face this kind of issue all the time. We can never optimize everything simultaneously—we always have to make tradeoffs in order to reach the best *overall* design.

EXAMPLE 9.5 VIBRATION RESPONSE OF A GOLF CLUB (Theory on page 565)

Figure 9.17 shows a golf club that's clamped at the grip end of the shaft. If the head is moved in the x direction and released, it will exhibit damped oscillations. One of the club designer's tasks is to optimize the vibrational characteristics to enhance a player's accuracy. One way to characterize the system's dynamic response is through its **settling time**. This term refers to the time it takes for the system's amplitude to decay by some fixed amount. For this example, we will define the settling time as the time necessary for the system's response amplitude to decrease by 90%.

We're told that the head of the club has a mass of $m = 0.2$ kg and that the damping and stiffness parameters are $c = 0.45$ N·s/m and $k = 400$ N/m, respectively. Assume initial conditions of $x(0) = 1.0$ cm, $\dot{x}(0) = 0$. How long will it take for the oscillations to die down to a 0.1 cm amplitude?

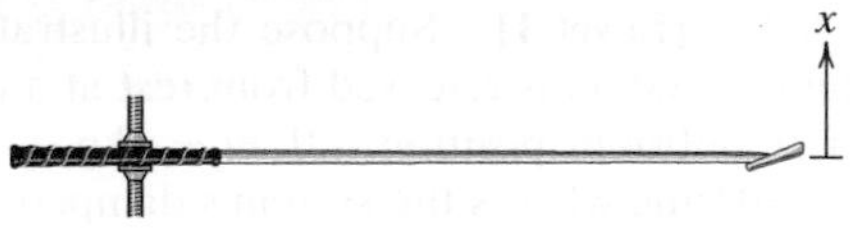

Figure 9.17 Clamped golf club

Goal Determine the length of time needed for the system's oscillation amplitude to drop from 1.0 cm to 0.1 cm.

Draw **Figure 9.18** shows our simple golf club model. m represents the mass of the club's head, and c, k represent the physical damping and stiffness properties, respectively.

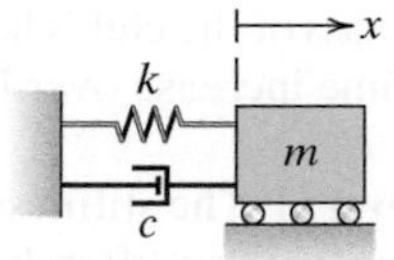

Figure 9.18 Simple spring-mass damper

Formulate Equations Our system equation is (9.11):

$$(0.2\,\text{kg})\ddot{x} + (0.45\,\text{N·s/m})\dot{x} + (400\,\text{N/m})x = 0$$

The natural frequency and damping factors are given in (9.3) and (9.13):

$$\omega_n = \sqrt{\frac{400\,\text{N/m}}{0.2\,\text{kg}}} = 44.7\,\text{rad/s}, \qquad \zeta = \frac{0.45\,\text{N·s/m}}{2(44.7\,\text{s}^{-1})(0.2\,\text{kg})} = 0.025$$

Solve From (9.18) you can see that the oscillating motion of the mass is reduced, as time goes on, by the exponentially decaying term $e^{-\zeta\omega_n t}$. Thus we need to determine when this function, which is 1.0 at $t = 0$, reaches 0.1:

$$e^{-\zeta\omega_n t} = e^{-(0.025)(44.7\,\text{s}^{-1})t} = 0.1 \quad \Rightarrow \quad t = 2.0\,\text{s}$$

EXERCISES 9.3

9.3.1. **[Level 1]** Suppose the illustrated spring-mass-damper system is released from rest at a distance x_0 from its equilibrium position. If $m = 4$ kg, $c = 12$ N·s/m, and $k = 30$ N/m, what is the system's damped frequency of oscillation ω_d, and how long does it take for the amplitude of its response to settle to 5% of its initial value?

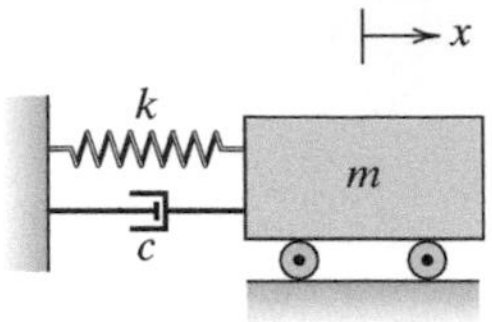

E9.3.1

9.3.2. **[Level 1]** Consider Example 9.5. If the clubmaker doubles the mass of the club's head, by what percentage will the settling time increase over its original time?

9.3.3. **[Level 1]** The stiffness and damping properties of a vibratory system can often be deduced by examining its impulse response, which is obtained experimentally by, say, very quickly striking the system with a hammer and measuring its oscillation. Consider the depicted mass-spring-damper system, whose mass is $m = 2$ kg but its stiffness k and damping coefficient c are unknown. If the system's impulse response is as shown, what are k and c?

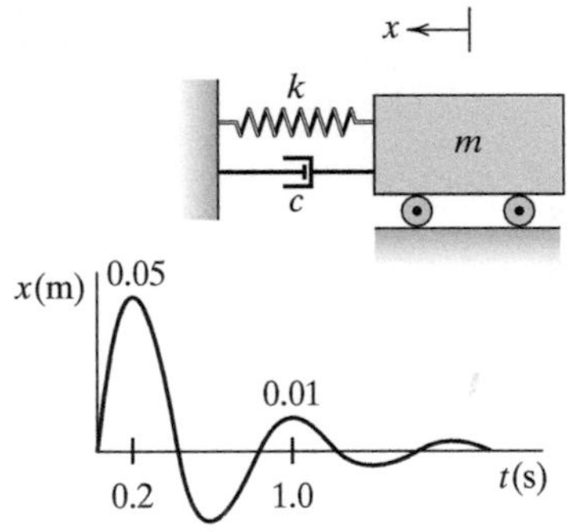

E9.3.3

9.3.4. **[Level 1]** Consider Example 9.5. By how much will the settling time change if the club's stiffness is doubled?

9.3.5. **[Level 1]** The illustrated spring-mass-damper system has mass $m = 5$ kg, stiffness $k = 100$ N/m, and damping coefficient $c = 55$ N·s/m. Find the system's response $x(t)$ if it is released from rest at a distance $x_0 = 0.1$ m from equilibrium.

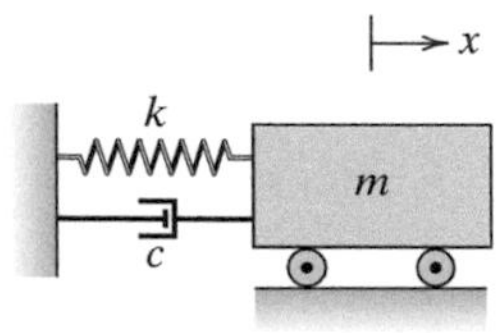

E9.3.5

9.3.6. **[Level 1]** A spring-mass damper has parameters $m = 1.4$ kg, $c = 40$ N·s/m, and $k = 6000$ N/m. By how much should c be changed for the system to have critical damping?

9.3.7. **[Level 1]** A spring-mass damper has parameters m, c, k and initial conditions $x(0) = x_0, \dot{x}(0) = v_0$. Derive an analytical expression for the response $x(t)$.

9.3.8. **[Level 2]** Consider Example 9.5. The design goal is to maintain the same settling time while at the same time increasing the club's mass by 20%. What must the club's damping be for this to be accomplished?

9.3.9. **[Level 2] Computational** A spring-mass damper with $m = 10$ kg and $k = 100$ N/m is given initial conditions $x(0) = 1.0$ m, $\dot{x}(0) = 0$. What minimum value of c is necessary to limit the first overshoot of the response to a value less than 0.25 m?

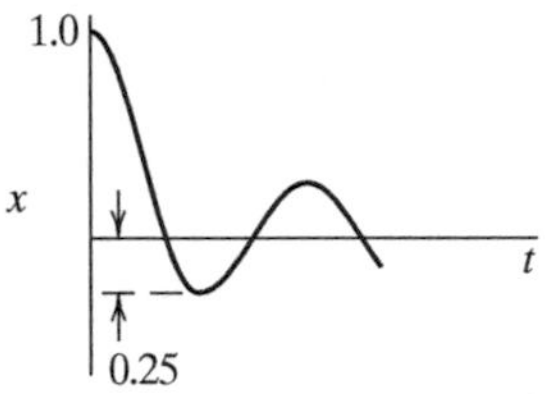

E9.3.9

9.3.10. **[Level 2] Computational** A spring-mass damper with $m = 10$ kg and $k = 1000$ N/m is given initial conditions $x(0) = 0, \dot{x}(0) = 11.0$ m/s. What minimum value of c is necessary to limit the first overshoot of the response to a value less than 0.60 m?

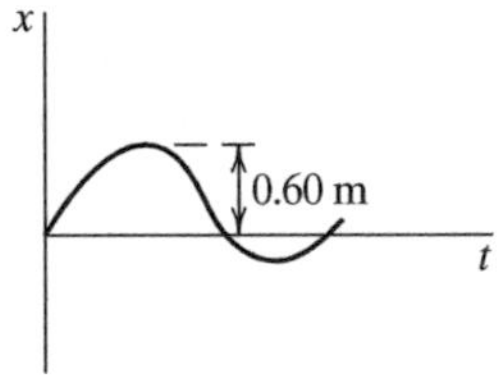

E9.3.10

9.3.11. **[Level 2] Computational** A spring-mass damper with $m = 10$ kg, $c = 40$ N·s/m, and $k = 1000$ N/m is given initial conditions $x(0) = 0, \dot{x}(0) = 6.0$ m/s. What is the time (t_m) needed for the response to reach its maximum value? If c is increased by 20%, by how much does the time needed to reach the maximum value change?

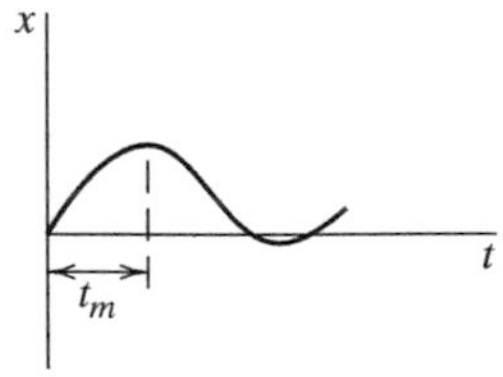

E9.3.11

9.4 DAMPED, SINUSOIDALLY FORCED RESPONSE FOR SINGLE-DEGREE-OF-FREEDOM SYSTEMS

The last system we will concern ourselves with is a forced system that includes damping. We derived the equation we need in the previous section—(9.10):

$$m\ddot{x} + c\dot{x} + kx = f(t) \tag{9.20}$$

Let's consider a sinusoidal force with amplitude $F, f(t) = F \sin \omega t$. Just as we saw in Section 9.3, using ζ and ω_n is preferable to using m, c, and k. Dividing (9.20) by m and using our sinusoidal forcing yield the final equation of motion:

$$\ddot{x} + 2\zeta\omega_n\dot{x} + \omega_n^2 x = \frac{F}{m}\sin \omega t \tag{9.21}$$

Solving this isn't difficult, but it's a bit tedious. We must assume a general solution of the form

$$x(t) = c_1 \cos \omega t + c_2 \sin \omega t \tag{9.22}$$

differentiate the appropriate number of times, plug the results into (9.21), and then pull two final equations out—one associated with the sine terms and the other with the cosine terms. Doing all this will get us

$\sin \omega t$: $$-2\zeta\omega\omega_n c_1 + (\omega_n^2 - \omega^2)c_2 = \frac{F}{m}$$

$\cos \omega t$: $$(\omega_n^2 - \omega^2)c_1 + 2\zeta\omega\omega_n c_2 = 0$$

We have two equations in two unknowns (c_1, c_2). Solving them yields

$$c_1 = \frac{-2\zeta\omega_n\omega\dfrac{F}{m}}{(\omega_n^2 - \omega^2)^2 + (2\zeta\omega\omega_n)^2} \tag{9.23}$$

$$c_2 = \frac{(\omega_n^2 - \omega^2)\dfrac{F}{m}}{(\omega_n^2 - \omega^2)^2 + (2\zeta\omega\omega_n)^2} \tag{9.24}$$

Note that in this formulation we acted on our equation with the input $F \sin \omega t$ and got an output of $c_1 \cos \omega t + c_2 \sin \omega t$. This isn't quite as useful a representation as it could be. What we normally want to know is the overall amplitude of the output as well as the **phase shift** between the input and output. The concept of a phase shift is easy to explain. Look at **Figure 9.19**, in which two functions are plotted: $x_1(t) = 2 \sin 0.5t$ and $x_2(t) = 2 \sin[0.5(t - 1.0)]$. (Note that units are not included so that the

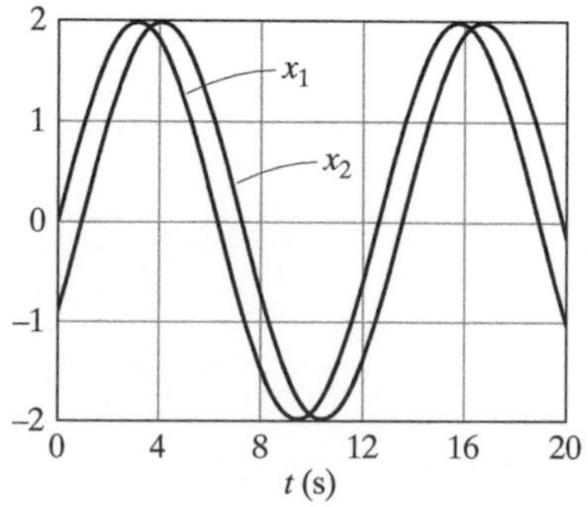

Figure 9.19 Sine and phase-shifted sine

expressions stay uncluttered. As usual, t is in units of seconds, the 0.5 is a frequency and therefore in units of rad/s, and the 1.0 in x_2 has units of seconds, to match t's units.)

You can see from the plot that x_2 is shifted by 1.0 s to the right (positive t) with respect to x_1. That's what the -1.0 does in the sine's argument of $x_2(t)$: it shifts the function to the right along the time axis. This is expressed as a **time shift** here, but it's easy enough to view it as a shift in phase by regrouping the terms in the sine's argument. Rewrite $x_2(t)$ as

$$x_2(t) = 2\sin[0.5(t - 1.0)] = 2\sin[0.5t - 0.5]$$

This shows that the sine wave has been shifted in phase by 0.5 rad.

Knowing the ratio of the output amplitude to the input tells us whether an input has a large or a small effect on the output, clearly a nice piece of information to have. We can determine this, and the phase shift, without much trouble. We start with the desired form of the response:

$$x(t) = C\sin(\omega t + \phi)$$

and then equate that with the general response (9.22):

$$x(t) = c_1\cos\omega t + c_2\sin\omega t = C\sin(\omega t + \phi) \tag{9.25}$$

We can expand $\sin(\omega t + \phi)$ using the trig identity:

$$\sin(\omega t + \phi) = \sin\omega t\cos\phi + \cos\omega t\sin\phi$$

Thus we have

$$c_1\cos\omega t + c_2\sin\omega t = C\cos\phi\sin\omega t + C\sin\phi\cos\omega t$$

Matching the sine and cosine components gives us

$$C\sin\phi = c_1 \tag{9.26}$$
$$C\cos\phi = c_2 \tag{9.27}$$

Dividing (9.26) by (9.27) gives us

$$\tan\phi = \frac{c_1}{c_2}$$

$$\phi = \tan^{-1}\left(\frac{c_1}{c_2}\right) \tag{9.28}$$

and squaring both sides and adding (9.26) and (9.27) let us derive

$$C = \sqrt{c_1^2 + c_2^2} \tag{9.29}$$

Substituting our known values of c_1 and c_2 in (9.23, 9.24) yields

$$C = \frac{F}{m\sqrt{(\omega_n^2 - \omega^2)^2 + (2\zeta\omega\omega_n)^2}} \tag{9.30}$$

$$\phi = -\tan^{-1}\left(\frac{2\zeta\omega\omega_n}{\omega_n^2 - \omega^2}\right) \tag{9.31}$$

>>> **Check out Example 9.7 (page 574) and Example 9.6 (page 572) for applications of this material.**

Figure 9.20 and **Figure 9.21** show what C and ϕ look like as a function of frequency for varying values of damping. The responses are plotted versus a normalized frequency (ω/ω_n), which makes them relevant to any particular parameter values you might be concerned with.

As **Figure 9.20** shows, as soon as we add damping we lose the infinite response at ω_n that we had for the undamped problem, but we still have a large response as long as the damping isn't too large. The phase shift ϕ is always negative and grows monotonically with frequency. Interestingly, it always is 90° at $\omega = \omega_n$.

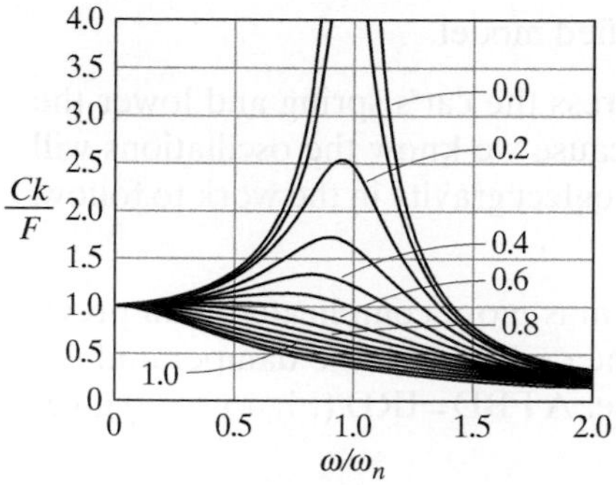

Figure 9.20 Amplitude response of spring-mass damper

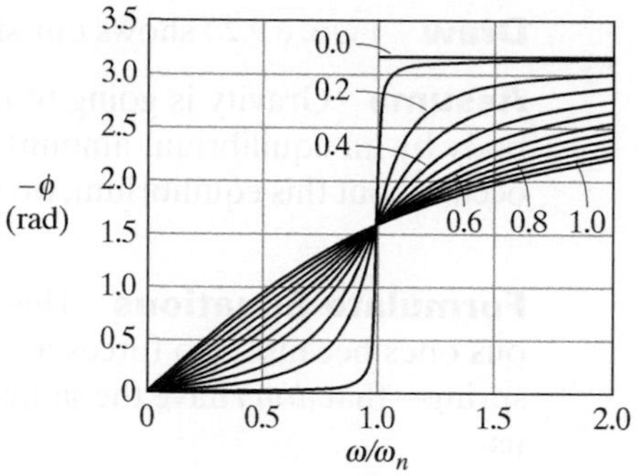

Figure 9.21 Phase shift in spring-mass damper

EXAMPLE 9.6 RESPONSE OF A SIMPLE CAR MODEL ON A WAVY ROAD (Theory on page 571)

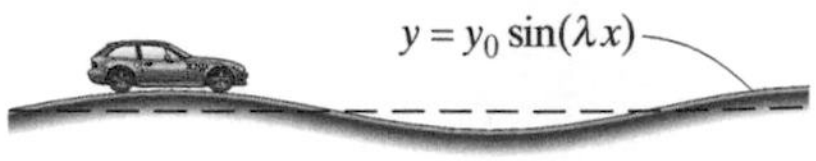

Figure 9.22 Car on a wavy road

In this example we will look at a slightly different kind of excitation, one known as **seismic forcing**. For seismic forcing, the excitation arises from motions of a supporting structure. The most obvious "seismic" example is the building that shakes because of ground motion due to an earthquake. Another example is the vibration of a driver's seat because of road irregularities. Consider the model of a car driving over a sinusoidally varying road profile shown in **Figure 9.22**. Our task is to determine the oscillation amplitude of the car as it travels along the road. We will approximate the car as a single spring-mass damper. The car is traveling at a steady 30 m/s, $\lambda = \frac{1}{15\text{ m}}$, $m = 2000$ kg, $c = 2.94 \times 10^4$ N·s/m, $k = 3.00 \times 10^5$ N/m, and $y_0 = 0.104$ m.

Goal Determine the amplitude of oscillation for a vehicle traveling along a wavy road.

Given Vehicle parameters and road profile.

Draw **Figure 9.23** shows our simplified model.

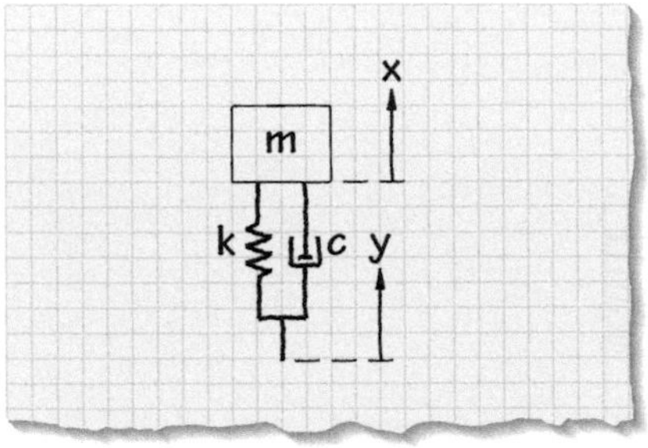

Figure 9.23 Simplified car model

Assume Gravity is going to compress the car's spring and lower the body by an equilibrium amount. Because we know the oscillations will occur about this equilibrium, we will neglect gravity in the work to follow.

Formulate Equations This system is more complicated than previous ones because two forces act on the car's body—the damper and the spring—that *don't* have the same phase. A FBD=IRD (**Figure 9.24**) gives us

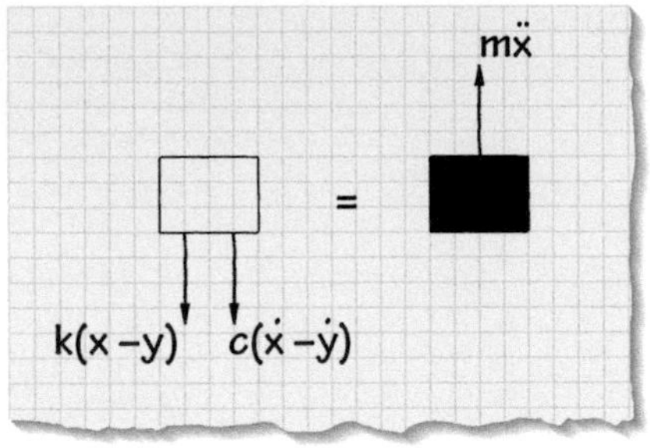

Figure 9.24 FBD=IRD for car and suspension

$$m\ddot{x} = -c(\dot{x} - \dot{y}) - k(x - y)$$

which means our equation of motion is

$$m\ddot{x} + c\dot{x} + kx = c\dot{y} + ky$$

To move farther, we have to re-express y in terms of t. For a car traveling at a speed v we have $x = vt$. Thus we have

$$y = y_0 \sin(\lambda x) = y_0 \sin(\lambda vt) = y_0 \sin(\omega t)$$

where $\omega = 2$ rad/s. Our equation of motion now looks like

$$m\ddot{x} + c\dot{x} + kx = c\omega y_0 \cos(\omega t) + ky_0 \sin(\omega t)$$

Because we're concerned only with the overall amplitude of the response, let's combine the sine and cosine inputs into a single sinusoidal function, using (9.29):

$$\text{force input} = F\cos(\omega t + \phi_1)$$

where

$$F = y_0\sqrt{k^2 + (c\omega)^2}$$

Mass normalizing gives us

$$\ddot{x} + 2\zeta\omega_n\dot{x} + \omega_n^2 x = \frac{F}{m}\cos(\omega t + \phi_1) \tag{9.32}$$

where $\omega_n = \sqrt{150}$ rad/s and $\zeta = \frac{c}{2m\omega_n} = 0.600$.

Solve Equation (9.30) gives us this solution of (9.32):

$$C = \frac{F}{m\sqrt{(\omega_n^2 - \omega^2)^2 + (2\zeta\omega\omega_n)^2}}$$
$$= \frac{3.18 \times 10^4\,\text{N}}{2000\,\text{kg}\sqrt{([150 - 2^2]\,\text{s}^{-2})^2 + (4(0.6)\sqrt{150}\,\text{s}^{-2})^2}} = 0.107\,\text{m}$$

Thus we have an amplitude response of

$$C = 0.107\,\text{m}$$

Check Does this result make physical sense? The car's natural frequency is $\sqrt{150}$ rad/s. The forcing frequency is merely 2 rad/s. This means that the system is being driven at a much lower frequency than its natural frequency. Look at **Figure 9.20**. For our system, ω/ω_n is $2/\sqrt{150} = 0.163$. At that low a frequency ratio, hardly any change takes place in the response owing to the system dynamics. For our problem, that means that the mass is essentially following the profile of the road and the springs aren't moving much. They are moving a little—hence the response amplitude (0.107 m) slightly larger than the input amplitude (0.104 m). If the car sped up such that ω/ω_n approached 1.0, then we would expect to see a much larger dynamic response.

EXAMPLE 9.7 **RESPONSE OF A SINUSOIDALLY FORCED, SPRING-MASS DAMPER** (Theory on page 570)

Consider a forced, damped oscillator with parameters $m = 4\,\text{kg}$, $c = 8\,\text{N·s/m}$, $k = 100\,\text{N/m}$, and $F = 2\,\text{N}$, which is being forced at 4 rad/s. Determine the steady-state response of the system, and put your solution into the form

$$x(t) = C\sin(4t + \phi)$$

Goal Solve the given oscillator problem, and put the solution into a form that makes the amplitude and phase shift of the output apparent.

Formulate Equations We can find ω_n and ζ from (9.3) and (9.13):

$$\omega_n = \sqrt{\frac{100\,\text{N/m}}{4\,\text{kg}}} = 5\,\text{rad/s}$$

$$\zeta = \frac{8\,\text{N·s/m}}{2\sqrt{(4\,\text{kg})(100\,\text{N/m})}} = 0.2$$

All that's needed is to use these parameters in (9.30) and (9.31).

Solve

$$C = \frac{2\,\text{N}}{(4\,\text{kg})\sqrt{[(5\,\text{rad/s})^2 - (4\,\text{rad/s})^2]^2 + [2(0.2)(4\,\text{rad/s})(5\,\text{rad/s})]^2}}$$
$$= 4.15 \times 10^{-2}\,\text{m}$$

$$\phi = -\tan^{-1}\left(\frac{2(0.2)(4\,\text{rad/s})(5\,\text{rad/s})}{(5\,\text{rad/s})^2 - (4\,\text{rad/s})^2}\right) = -0.727\,\text{rad}$$

$$x(t) = 4.15 \times 10^{-2}\sin(4t - 0.727)$$

Check Let's apply a logic check on the phase shift, which, as we see, is negative. A negative phase shift makes sense for the following reason. Recall that the phase shift shows the phase difference between the *output* and the *input*. A negative phase shift, or **phase lag**, reflects the physical reality that the system takes some time to react to an input. A positive phase shift would mean that the system somehow reacted to an input *before* it occurred, and that's not going to happen in the real world. Thus the negative phase shift can be taken as an indication that our solution makes physical sense.

EXERCISES 9.4

9.4.1. [Level 1] Suppose the depicted spring-mass-damper system has mass $m = 9$ kg, stiffness $k = 120$ N/m, and damping coefficient $c = 40$ N·s/m, and the attached platform oscillates according to $y(t) = \overline{y} \sin \omega t$, where $\overline{y} = 0.03$ m and $\omega = 6$ rad/s. Find the amplitude $\overline{x}$ of the system's response under the given excitation.

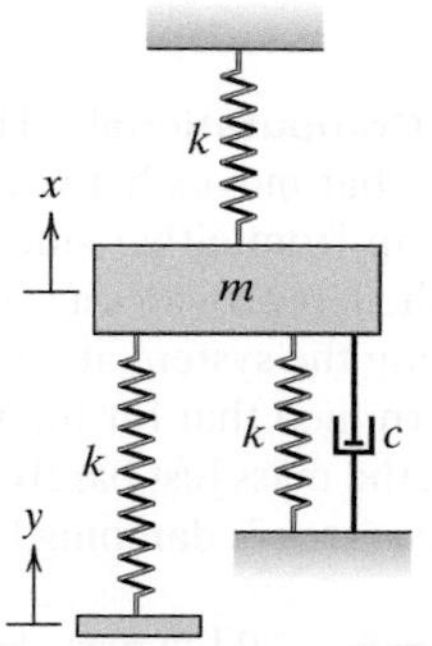

E9.4.1

9.4.2. [Level 1] Consider Example 9.7. What will the phase shift be if the forcing frequency is increased to 10 rad/s?

9.4.3. [Level 1] Obtaining an expression for the amplitude response $\overline{x}(\omega)$ of a forced spring-mass-damper system using trigonometric functions is quite tedious. An alternative approach involves expressing the steady-state system response $x(t)$ and forcing function $f(t)$ in terms of the complex exponential $e^{i\omega t}$, where ω is the frequency of the applied excitation. In particular, we can say that $x(t) = \overline{x}(i\omega)e^{i\omega t}$ and $f(t) = \overline{f}e^{i\omega t}$, where $\overline{x}(i\omega)$ is the system's *complex frequency response* and $\overline{f}$ is the amplitude of the excitation. The amplitude response is simply the magnitude of $\overline{x}(i\omega)$, or $\overline{x}(\omega) = |\overline{x}(i\omega)|$. Derive an expression for the amplitude response of the illustrated spring-mass-damper system by using complex exponentials. Neglect gravity.

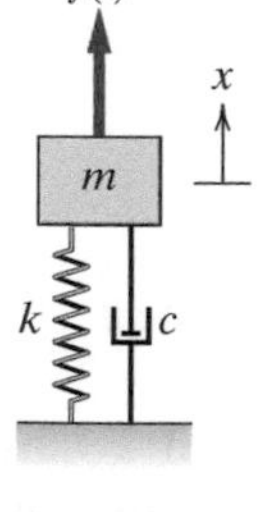

E9.4.3

9.4.4. [Level 2] Computational In the illustrated system, $k = 100$ N/m and $c = 10$ N·s/m. x is given by $x = x_0 \cos \omega t$, with $x_0 = 0.02$ m. Plot the magnitude of the force transmitted to the wall through the damper, the spring, and the combination of both spring and damper for $0 \leq \omega \leq 25$.

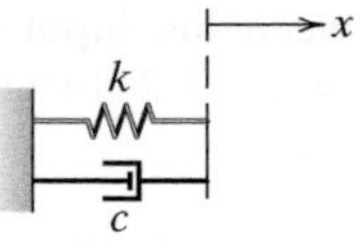

E9.4.4

9.4.5. [Level 2] Consider the depicted mass-spring-damper system, which is acted on by a periodic force $f(t) = \overline{f} \sin \omega t$. For what forcing frequency ω will the system's speed reach a maximum? Neglect gravity.

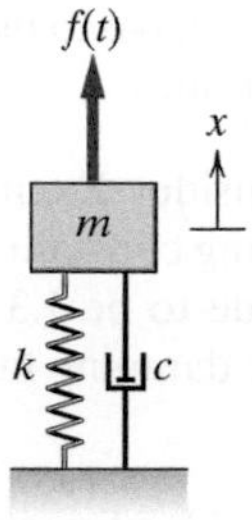

E9.4.5

9.4.6. [Level 2] Determine the total steady-state response for the illustrated system. $m = 2.5$ kg, $c = 1.0$ N·s/m, $k = 150$ N/m, $\omega = 8$ rad/s, and $f(t) = f_0 \sin \omega t$, $f_0 = 2$ N.

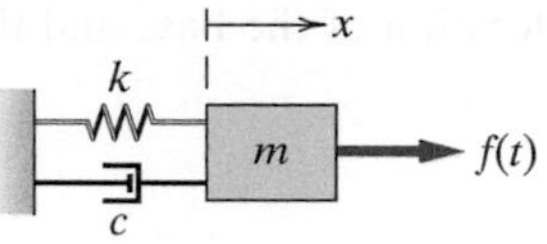

E9.4.6

9.4.7. [Level 2] Find $\dot{x}$ for the illustrated system. $m = 0.5$ kg, $c = 8.66$ N·s/m, $k_1 = 1.00 \times 10^4$ N/m, $k_2 = 5.00 \times 10^3$ N/m, $y(t) = y_0 \sin \omega t$, $y_0 = 0.02$ m, and $\omega = 150$ rad/s.

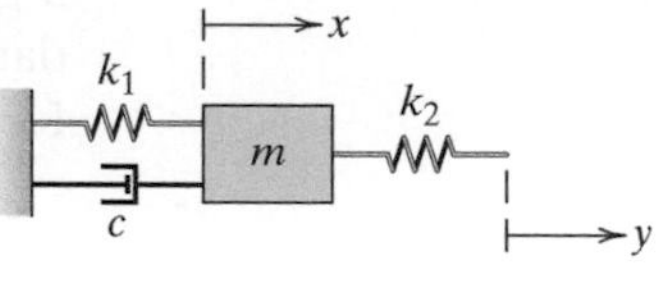

E9.4.7

9.4.8. [Level 2] Find $\dot{x}$ for the illustrated system. $m = 0.25$ kg, $c_1 = 4.00$ N·s/m, $c_2 = 2.00$ N·s/m, $k = 8.00 \times 10^3$ N/m, $y(t) = y_0 \sin \omega t$, $y_0 = 0.01$ m, and $\omega = 100$ rad/s.

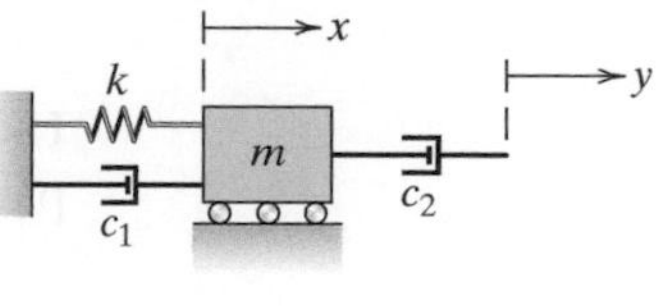

E9.4.8

9.4.9. [Level 2] Computational Consider Example 9.7. What is the peak response amplitude, and at what forcing frequency does it occur?

9.4.10. **[Level 2]** Determine the magnitude of the force transmitted to the wall for the illustrated system. What is the phase shift between the input force and the output force? $m = 0.01$ kg, $c = 1.2\,\text{N}\cdot\text{s/m}$, $k = 900$ N/m, and $f(t) = (0.02\,\text{N})\cos(250t)$.

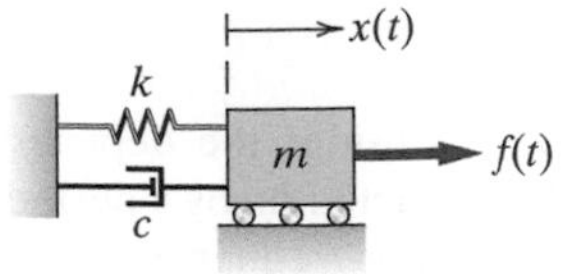

E9.4.10

9.4.11. **[Level 2]** **Computational** Consider Example 9.6. How fast would the car have to move to experience the maximum response amplitude?

9.4.12. **[Level 2]** Consider Example 9.6. How much would the supporting spring constant k have to be reduced for the response amplitude to be 1.3 times the road's amplitude? Assume that the damping factor stays constant at 0.6.

9.4.13. **[Level 2]** Consider Example 9.6. Assume that the car's speed has increased to 45 m/s. By what percentage will the car's response amplitude change compared to the response amplitude if it were traveling at 30 m/s?

9.4.14. **[Level 2]** What is the steady-state phase shift between the acceleration of the base and the acceleration response of the mass? $m = 4$ kg, $c = 4\,\text{N}\cdot\text{s/m}$, $k = 300$ N/m, and $y = (0.02\,\text{m})\sin(12t)$.

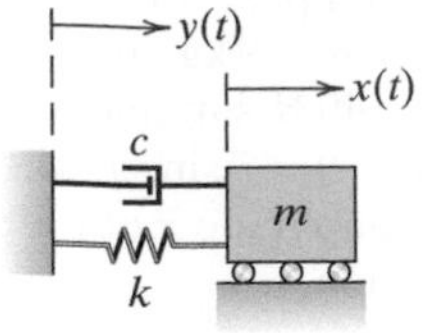

E9.4.14

9.4.15. **[Level 2]** **Computational** The illustrated system shows a mass m that moves between two fixed stops, the stops located 0.1 m from either side of the mass. The moving element on the left provides a seismic excitation to the system. By running the system at a variety of frequencies, it has been determined that for the worst-case (large-amplitude) response, the mass just barely contacts the fixed supports. What is the system's damping factor?

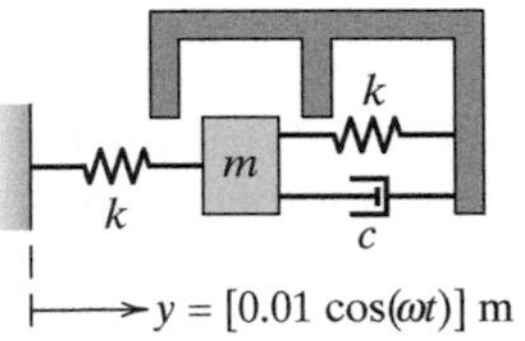

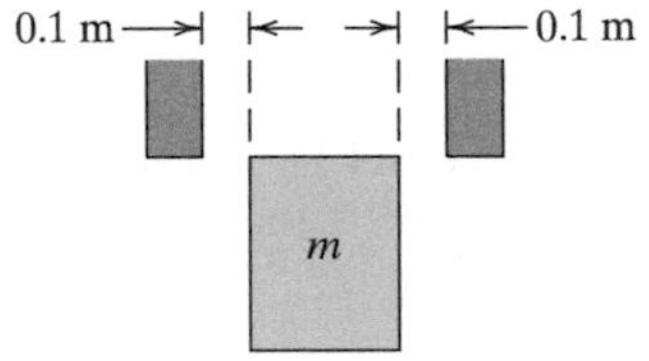

E9.4.15

9.5 JUST THE FACTS

In this chapter we learned how to deal with single-degree-of-freedom vibrating systems. Because these systems have analytical solutions, at least for free and sinusoidally forced vibrations, we were able to gain a great deal of physical insight into their behavior. We looked at both damped and undamped systems for the cases of no forcing and sinusoidal forcing.

The basic vibratory system involved only a single mass and spring:

$$m\ddot{y} + ky = 0 \tag{9.1}$$

and had the general solution

$$y(t) = a_1 \cos \omega t + a_2 \sin \omega t \tag{9.2}$$

Using this solution revealed the existence of a **natural frequency**, the particular frequency at which a spring-mass system will always vibrate:

$$\omega_n = \sqrt{\frac{k}{m}} \tag{9.3}$$

Dividing our equation of motion by the mass produced a **mass-normalized** form of the equation:

$$\ddot{y} + \omega_n^2 y = 0 \tag{9.4}$$

When solved for the initial conditions $y(0) = y_0, \dot{y}(0) = v_0$, this equation yielded the general solution

$$y(t) = y_0 \cos \omega_n t + \frac{v_0}{\omega_n} \sin \omega_n t \tag{9.5}$$

Adding an external forcing gave us the equation of motion

$$m\ddot{x} + kx = F \sin \omega t \tag{9.7}$$

and assuming a solution in the form

$$x(t) = X \sin \omega t \tag{9.8}$$

produced the result

$$X = \frac{F}{k - m\omega^2} \tag{9.9}$$

This shows that the amplitude of the response is frequency dependent. For frequencies below ω_n, the response is **in-phase** with the forcing; it is **out-of-phase** for frequencies above ω_n. The amplitude is finite at low frequencies, grows toward infinity as ω_n is approached, and then goes to zero as the frequency increases toward infinity. **Resonance** occurs when $\omega = \omega_n$, producing an infinite response for a finite input force.

The next step in complexity was to add a **viscous damper** to our system:

$$m\ddot{x} + c\dot{x} + kx = f(t) \tag{9.10}$$

The unforced case was simply

$$m\ddot{x} + c\dot{x} + kx = 0 \tag{9.11}$$

and the unforced, mass-normalized form was given by

$$\ddot{x} + 2\zeta\omega_n\dot{x} + \omega_n^2 x = 0 \tag{9.14}$$

where

$$\zeta = \frac{c/m}{2\omega_n} = \frac{c\sqrt{m}}{2m\sqrt{k}} = \frac{c}{2\sqrt{mk}} \tag{9.13}$$

The solution of this equation for underdamped systems was found to be

$$x(t) = e^{-\zeta\omega_n t}(b_1 \sin \omega_d t + b_2 \cos \omega_d t) \tag{9.18}$$

where

$$\omega_d \equiv \omega_n\sqrt{1 - \zeta^2} \tag{9.19}$$

ω_d being the **damped frequency of oscillation**.

Adding a sinusoidal forcing gave us the mass-normalized equation of motion

$$\ddot{x} + 2\zeta\omega_n\dot{x} + \omega_n^2 x = \frac{F}{m}\sin\omega t \tag{9.21}$$

which had a solution $x(t) = c_1\cos\omega t + c_2\sin\omega t$, where

$$c_1 = \frac{-2\zeta\omega_n\omega\dfrac{F}{m}}{(\omega_n^2 - \omega^2)^2 + (2\zeta\omega\omega_n)^2} \tag{9.23}$$

$$c_2 = \frac{(\omega_n^2 - \omega^2)\dfrac{F}{m}}{(\omega_n^2 - \omega^2)^2 + (2\zeta\omega\omega_n)^2} \tag{9.24}$$

An alternative solution form, $x(t) = C\sin(\omega_n t + \phi)$, was then derived, for which

$$C = \frac{F}{m\sqrt{(\omega_n^2 - \omega^2)^2 + (2\zeta\omega\omega_n)^2}} \tag{9.30}$$

$$\phi = -\tan^{-1}\left(\frac{2\zeta\omega\omega_n}{\omega_n^2 - \omega^2}\right) \tag{9.31}$$

SYSTEM ANALYSIS (SA) EXERCISES

SA9.1 Clothes Washer Vibrations

Figure SA9.1.1 shows a simplified model of a vertically loaded clothes washer. A situation can occur in which the clothes being washed gather together in a lump. When this lump is spun at high speed (spin cycle), the forces generated are sometimes sufficient to make the washer start to "walk" across the floor.

The simplified model consists of a washer body ($m_1 = 25$ kg), a drum that rotates within the washer body ($m_2 = 3$ kg), and two springs ($k = 500$ N/m), which model the supporting suspension of the washer. The drum has an internal radius of 0.3 m.

a. Assume that a 5 kg lump of clothes has formed (m_3). Assume that the spin rate of the washer can vary from 3 to 5 rad/s. Plot the force transmitted to the ground via the supporting springs as a function of spin frequency.

b. Will reducing the spring stiffness cause a reduction in the transmitted force levels?

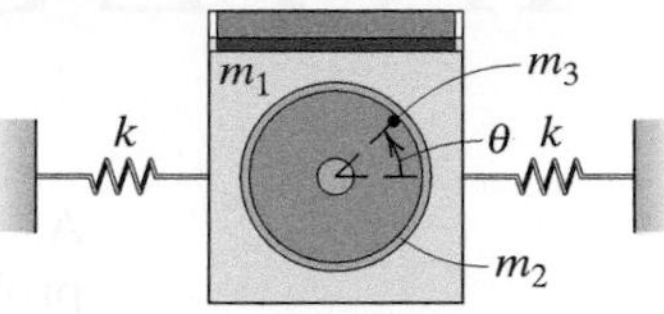

Figure SA9.1.1 Washer schematic

NUMERICAL INTEGRATION LIGHT

As mentioned earlier in the text, finding a solution to many dynamics problems requires the use of a computer and a numerical integration program. Although simple equations of motion can be integrated by hand, in the grand scheme of things, these are the exceptions rather than the rule. The point of this appendix is to show you one way to integrate an equation of motion that will be somewhat accurate and can be coded by hand in a short time. It's called *backward Euler integration* and is quite useful if you don't have a canned integration code handy. This approach is one of the simplest methods around and is only one of a huge body of techniques. As the title of this chapter suggests, your exposure to numerical methods will be very light. It's not reasonable for it to be otherwise because you're going to be spending most of the semester learning the theory of dynamics and an in-depth look at numerical methods requires at least an entire semester. An excellent text for those who want more detailed information is *Numerical Methods That Work* [4].

Although Euler integration is fine, *much* finer methods are available that are more accurate and allow users to specify what kind of error tolerance they wish. These methods come pre-packaged, along with convenient plotting routines, and are available to students at a low price. Probably the dominant package today is MATLAB, produced by The MathWorks, Inc. This software package began life as a set of tools for solving matrix problems and was initially targeted at people doing controls work. Its use has spread steadily, and it now contains a comprehensive set of toolboxes in many different areas.

MATLAB isn't the only program available, but in the interest of time (and because it's the one I use), I will limit my discussion to it. Just as for numerical integration, my coverage will be very light and will only involve those functions that are immediately useful to someone trying to integrate a system's equations of motion and to plot the resulting trajectories.

Earlier in the text, we said that a particle's velocity was found by taking the difference between its position *now* ($\boldsymbol{r}_m(t)$) and its position a small time in the future ($\boldsymbol{r}_m(t+\Delta)$), dividing by the time difference Δ and taking the limit as that time difference went to zero:

$$\boldsymbol{v}_m(t) = \lim_{\Delta \to 0} \left(\frac{\boldsymbol{r}_m(t+\Delta) - \boldsymbol{r}_m(t)}{\Delta} \right)$$

This showed us that the velocity of a particle always points along the particle's path. Let's now consider what we would get if we kept the same formula but didn't take the limit:

$$\tilde{\boldsymbol{v}}_m(t) = \frac{\boldsymbol{r}_m(t+\Delta) - \boldsymbol{r}_m(t)}{\Delta}$$

The tilde indicates that our $\boldsymbol{v}$ is an approximation to the actual velocity. Next, consider what you would do if I told you the particle's velocity as a function of time but didn't tell you where the particle was located. For instance, assume that

$$\boldsymbol{v}_m = \cos(\omega t)\boldsymbol{i}$$

You could combine these two equations (letting $\tilde{\boldsymbol{v}}_m = \boldsymbol{v}_m$) and get

$$\frac{\boldsymbol{r}_m(t+\Delta) - \boldsymbol{r}_m(t)}{\Delta} = \cos(\omega t)\boldsymbol{i}$$

Cross multiplying would then give you

$$\boldsymbol{r}_m(t+\Delta) = \boldsymbol{r}_m(t) + \Delta\cos(\omega t)\boldsymbol{i} \tag{A.1}$$

That's a numerical integration scheme. It tells us that the position of our particle at $t + \Delta$ is equal to its position at t plus a correction term that depends on the particle's velocity $\boldsymbol{v}$ and the time duration Δ. The approximation isn't particularly accurate, and nobody in his right mind would actually use it for any kind of serious analysis, but it shows what's fundamentally going on with numerical integration. All "good" integration codes do the same kind of thing, but they do it more accurately and, obviously, involve more complicated mathematics.

Let's go back to our method and see what you would get if you started at $t = 0$ and wanted to know the particle's position as time increases. You would start off by knowing where the particle is at $t = 0$: $\boldsymbol{r}_m(0)$. You would then ask where it is at some time Δ in the future. By using (A.1) you would get

$$\boldsymbol{r}_m(\Delta) = \boldsymbol{r}_m(0) + \Delta\cos(0)\boldsymbol{i}$$

To find the position at $t = 2\Delta$ you would just repeat the operation:

$$\boldsymbol{r}_m(2\Delta) = \boldsymbol{r}_m(\Delta) + \Delta\cos(\omega\Delta)\boldsymbol{i}$$

All you would need to do to automate this would be to create a looping program that sequentially ran this formula, getting you from $t = 0$ to $t = \Delta$ to $t = 2\Delta$ and on to however far you cared to calculate the motion.

What if you want to know where the particle is at some time other than $0, \Delta, 2\Delta, \ldots$? Perhaps at $t = 1.5\Delta$, for instance. You have two basic choices. You can try to estimate the answer by determining the position at $t = \Delta$ and $t = 2\Delta$ and then averaging the two. That's called linear interpolation. Linear interpolation is the simplest approach, but more complicated types of interpolation could be used as well.

Or, if you didn't want to interpolate, you could run a new integration with a smaller time step. Using a time step that's one-half as long as the old one would give you data at $t = 0, \frac{\Delta}{2}, \frac{2\Delta}{2}, \frac{3\Delta}{2}, \ldots$, which means that a data point now exists for the precise time that's of interest to you.

Interpolation is "free" in the sense that you have already gotten the data around the desired time, and the only computational expense is the need to implement the interpolation itself. Re-running the integration code with a new timestep is more costly from a computational point of view, for you need to calculate many more data points, most of which aren't of interest to you. This isn't always a particularly big deal, however, for computation is quite fast nowadays and unless your system of equations is very large, running a new integration isn't going to take more than a second or so.

As you might expect, dozens of different computational algorithms exist that do what you've just seen but in vastly more efficient and accurate ways. In the best of all possible worlds, you would become familiar with all the different methods, understand their strengths and weaknesses, and then use the ones that are most fitted to the problem you're analyzing. But even in this best of all possible worlds, you're not going to be wildly excited about having to code all these up yourself each time you want to attack a new problem. That's where canned codes come in. Several avenues exist, but the one we will look at here is MATLAB.

MATLAB is a sophisticated software package that encompasses its own programming language, a multitude of individual programs to address a wide range of mathematical problems, and a fairly sophisticated plotting capability. The word "MATLAB" is an abbreviation for "matrix laboratory" and reflects the prime purpose that guided its original development. At its core, MATLAB is all about matrix manipulation. I'm not even going to attempt to give an overview of everything that MATLAB can do—it would quickly become far too large. All I'm going to do is tell you the basics that you'll need to address the problems in this text. It certainly wouldn't hurt to learn more on your own but you can get by just with what you'll find in this appendix.

The name for a program in MATLAB is "M-file." MATLAB comes with a large number of M-files already written, and you can create whatever additional M-files you find helpful. Let's look at an example in which you would use both your own and a predefined MATLAB M-file—that of integrating a system's equation of motion. We will take the specific example of wanting a displacement versus time plot of

$$\ddot{x} + 2\dot{x} + 16x = 0.1\cos t$$

with initial conditions $x(0) = 0, \dot{x}(0) = 1$ and an integration time span of $0 \le t \le 2$. Note that no units are included because MATLAB works with numbers, not units. So the presumption is that I carried through all the units correctly and know how to interpret the numerical results in terms of the associated units.

The key to integrating our equation of motion is to transform it from its current form as a second-order, ordinary differential equation (ODE) into two first-order differential equations. This is called "putting the equation into state-variable form." To do it, we will create two new variables, y_1

and y_2, and define them as follows:

$$y_1 \equiv x$$
$$y_2 \equiv \dot{x}$$

As you might surmise, y_1 and y_2 are called **state variables.**

Substituting these variables into our equation of motion gives us

$$\dot{y_2} + 2y_2 + 16y_1 = 0.1 \cos t$$

$$\dot{y_2} = -2y_2 - 16y_1 + 0.1 \cos t$$

That's the first of our first-order ODEs. The second is found by simply observing that $\dot{x} = \dot{x}$, and therefore

$$\dot{y_1} = y_2$$

Our set of ODEs is thus

$$\dot{y_1} = y_2 \tag{A.2}$$
$$\dot{y_2} = -2y_2 - 16y_1 + 0.1 \cos t \tag{A.3}$$

MATLAB can now be invoked to solve this set of first-order differential equations. First, however, we have to decide what integrator to use. MATLAB has a few, with names like `ode45`, `ode23`, `ode113`, and so on. A well-rounded routine that can handle a wide range of problems is `ode45`, and that's the one we will be using. It employs an integration scheme called **Runge-Kutta**, a moderately sophisticated method.

Typing

```
[t,y]=ode45('AppAexample',tspan,y0);
```

in the MATLAB program's command line will solve for the system's trajectory *after* a few details have been taken care of. The first detail is that the integration code `ode45` needs to know what it is you want integrated. That means you need to write an M-file that contains the system equations (A.2)–(A.3). Open an editor and create a file named "AppAexample.m," which contains the following text:

```
function dy=AppAexample(t,y)
y1=y(1,1);
y2=y(2,1);
dy(1,1)=y2;
dy(2,1)=-2*y2-16*y1+0.1*cos(t);
```

The first line tells MATLAB that the file is a function that will output the state derivatives (`dy`) from knowing the time (`t`) and the states (`y`). Lines 2 and 3 aren't necessary—they just make it easier to type the equations. Notice the semicolons. A semicolon at the end of the line keeps MATLAB from printing out the numerical result of that line. If you wish, you can leave the semicolon off and see how you then get a bunch of numbers printed on the screen as the calculations progress. Lines 4 and

5 are the lines that tell MATLAB what it needs to integrate. Notice that MATLAB uses a “`*`” to indicate multiplication. It uses “`/`” for division (as in `4 = 8/2`), “+” for addition, and “`--`” for subtraction. If you need an exponent, you use “∧” (x squared is written as x∧2). Save this file and set the path so that MATLAB knows where on your computer to look in order to find the file.

Now we need to define `tspan` and `y0`. `tspan` provides the numerical integration algorithm with the time limits of integration. Say you want to integrate from $t = 0$ to $t = 2$. You need to define a vector that contains the start and end times, and you do so by typing

```
tspan=[0 2];
```

in the workspace. The square brackets indicate that you're defining a vector.

The initial conditions are handled similarly. In this case, you want a vector that contains the initial values for y_1 and y_2. Recalling that our initial conditions are $y_1 = x(0) = 0$ and $y_2 = \dot{x}(0) = 1$ means we need to type

```
y0=[0 1];
```

So now you can see that typing

```
[t,y]=ode45('AppAexample',tspan,y0);
```

in the Command Window tells MATLAB to grab the M-file “AppAexample,” apply initial conditions contained in “`y0`,” and use the integrator “`ode45`” for the timespan defined by “`tspan`.”

Typing this (and hitting “return”) will cause MATLAB to produce two column matrices. The first, “`t`,” will contain the time values that MATLAB used as it performed the integration. The second, “`y`,” will contain two columns of numbers. The first column shows the y_1 values, and the second shows the y_2 values corresponding to the times contained in “`t`.”

It's quite easy to plot this data. Simply type

```
plot(t,y);
```

and you will get a plot that looks like **Figure A.1**.

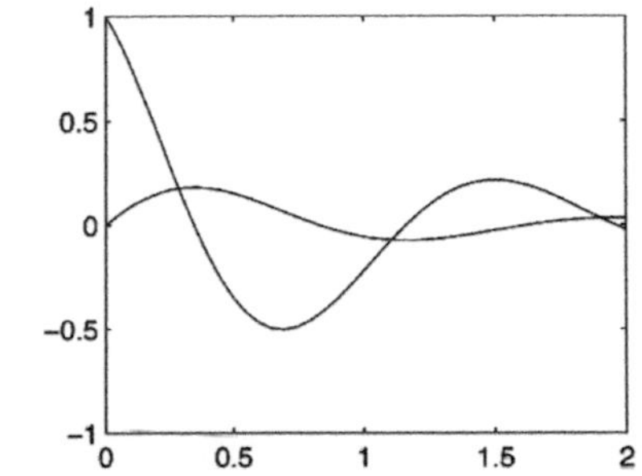

Figure A.1 Position and speed versus time

If you would like grid lines, just type

```
grid
```

after your plot command and MATLAB will add some nice grid lines, as shown in **Figure A.2**.

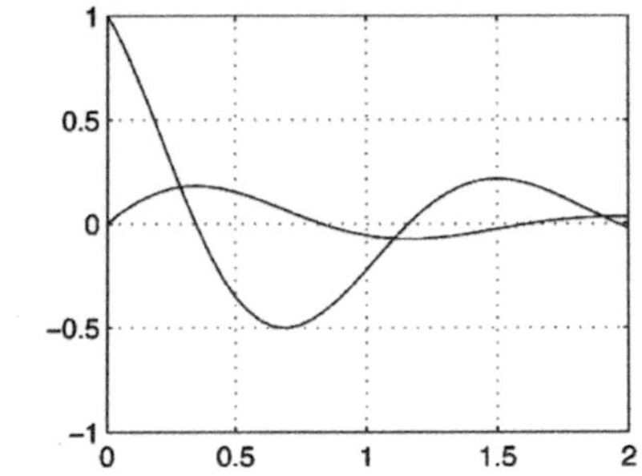

Figure A.2 Gridded plot

You will note that both y_1 and y_2 were plotted. If you just want to plot one of them, you can type

```
plot(t,y(:,1));
```

This tells MATLAB to plot everything in the first column of the matrix `y` versus `t` and will produce **Figure A.3**.

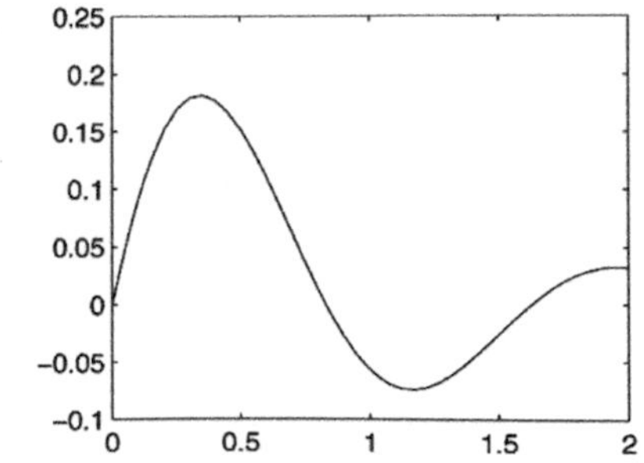

Figure A.3 Position versus time

Now let's see how to handle a more complicated system of equations. Assume we have two, coupled, second-order ODEs:

$$4\ddot{x} + 3\ddot{y} + 0.5\dot{y} + 10x + 12y = 0$$

$$2\ddot{x} - 3\ddot{y} + 1.2\dot{x} - 2x + 20y = 5\cos 4t$$

Here's how we would MATLABize them. Define state variables z_1–z_4:

$$\begin{aligned} z_1 &\equiv x \\ z_2 &\equiv \dot{x} \\ z_3 &\equiv y \\ z_4 &\equiv \dot{y} \end{aligned}$$

Our state equations can be written as

$$\dot{z}_1 = z_2$$

$$4\dot{z}_2 + 3\dot{z}_4 = -0.5z_4 - 10z_1 - 12z_3$$

$$\dot{z}_3 = z_4$$

$$2\dot{z}_2 - 3\dot{z}_4 = -1.2z_2 + 2z_1 - 20z_3 + 5\cos 4t$$

Putting this into matrix form gives us

$$\begin{bmatrix} 1 & 0 & 0 & 0 \\ 0 & 4 & 0 & 3 \\ 0 & 0 & 1 & 0 \\ 0 & 2 & 0 & -3 \end{bmatrix} \begin{Bmatrix} \dot{z}_1 \\ \dot{z}_2 \\ \dot{z}_3 \\ \dot{z}_4 \end{Bmatrix} = \begin{bmatrix} 0 & 1 & 0 & 0 \\ -10 & 0 & -12 & -0.5 \\ 0 & 0 & & 1 \\ 2 & -1.2 & -20 & 0 \end{bmatrix} \begin{Bmatrix} z_1 \\ z_2 \\ z_3 \\ z_4 \end{Bmatrix} + \begin{bmatrix} 0 \\ 0 \\ 0 \\ 5\cos 4t \end{bmatrix}$$

MATLAB makes it easy from here. This matrix equation is in the form

$$[A]\dot{Z} = [B]Z + [F]$$

where Z is a 4×1 vector containing the state variables z_1–z_4 and the matrices $[A]$, $[B]$, $[F]$ are given by

$$[A] = \begin{bmatrix} 1 & 0 & 0 & 0 \\ 0 & 4 & 0 & 3 \\ 0 & 0 & 1 & 0 \\ 0 & 2 & 0 & -3 \end{bmatrix}$$

$$[B] = \begin{bmatrix} 0 & 1 & 0 & 0 \\ -10 & 0 & -12 & -0.5 \\ 0 & 0 & & 1 \\ 2 & -1.2 & -20 & 0 \end{bmatrix}$$

$$[F] = \begin{bmatrix} 0 \\ 0 \\ 0 \\ 5\cos 4t \end{bmatrix}$$

Simply open up an M-file and enter the appropriate values for the matrices. Then, because MATLAB expects your equations to be in the form

$$\dot{Z} = f(Z)$$

(that is, with the explicit time derivative of the state variables on the left-hand side) you can use the inverse function "`inv`" to have MATLAB do the appropriate calculations for you. The last line in your M-file would thus be

```
dy=inv(A)*B*y+inv(A)*F
```

MATLAB has a huge number of predefined functions, some of which are in common use in dynamics problems. **Table A.1** lists some of the more useful ones. If you want more information, just type "`help`" followed by the function (i.e., **`help quad`**) in the command line. And with that, I'll end the tutorial. This is enough to let you integrate and plot the problems in this text. I do recommend that you spend a little time browsing through the MATLAB manual; you will find lots of nice features there that will make your life much easier in most of your classes. Those interested in a more in-depth look at simulation using MATLAB might also wish to consult *Simulations of Machines Using MATLAB® and SIMULINK®* [5], a compact and user-friendly book.

Table A.1 Table of MATLAB functions

Function	Property
`sqrt(y)`	calculates the square root of y
`sin(y)`	calculates the sine of y
`cos(y)`	calculates the cosine of y
`tan(y)`	calculates the tangent of y
`exp(y)`	calculates e^y
`linspace(a,b)`	creates a vector with 100 evenly spaced entries from a to b
`length(x)`	calculates the length of the vector x
`size(X)`	calculates the dimensions of the matrix X
`det(X)`	calculates the determinant of the matrix X
`X'`	calculates the transpose of the matrix X
`inv(X)`	calculates the inverse of the matrix X
`eig(X)`	calculates the eigenvalues of the matrix X
`abs(X)`	calculates the absolute values of the entries of the matrix X
`dot(a,b)`	calculates the dot product of the vectors a and b
`cross(a,b)`	calculates the cross product of the vectors a and b
`norm(x)`	calculates the Euclidean norm of the vector x
`fzero('func',x)`	calculates the zero closest to x of the function in the M-file func.m
`quad('func',a,b)`	calculates the integral of func.m from a to b
`plot(t,y)`	plots y versus t
`grid`	adds gridlines to a plot
`subplot(m,n,t)`	defines an m by n array of plots, with t the current one

PROPERTIES OF PLANE AND SOLID BODIES

Areas of Some Common Shapes

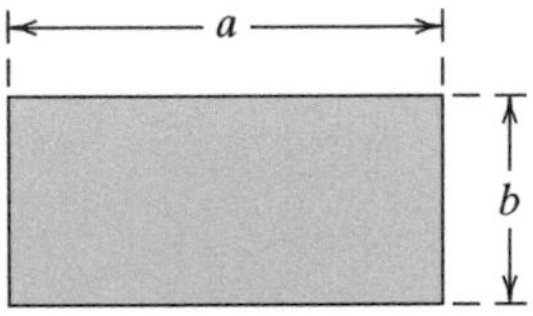

Rectangle

Area = ab

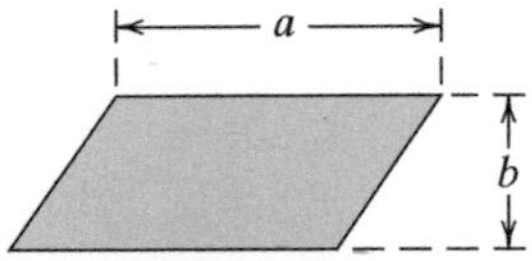

Parallelogram

Area = ab

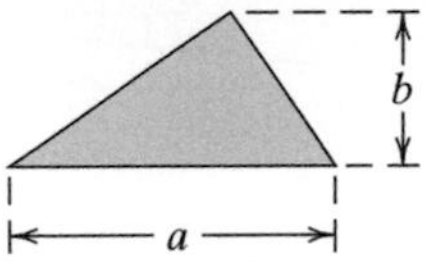

Triangle

Area = $\frac{ab}{2}$

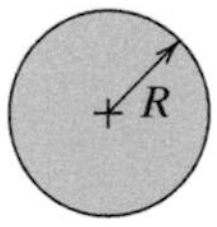

Circular Disk

Area = πR^2

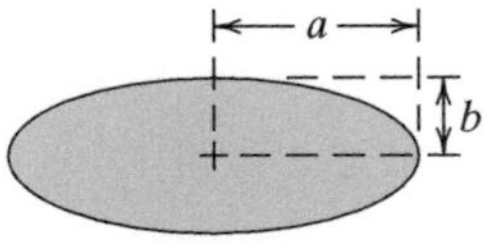

Ellipse

Area = πab

Volumes of Some Solid Bodies

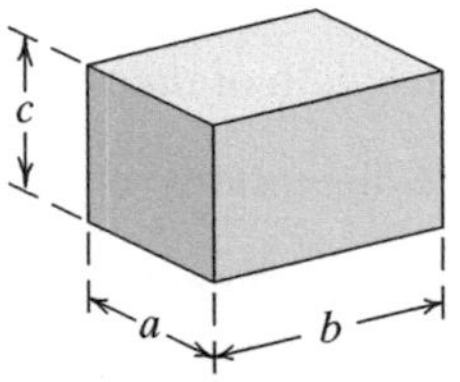

Rectangular Parallelepiped

Volume = abc

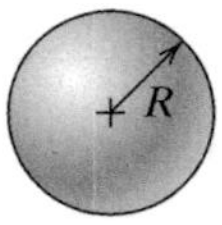

Solid Sphere

Volume = $\frac{4}{3}\pi R^3$

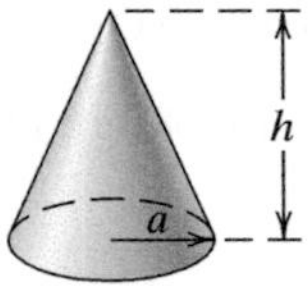

Circular Cone

Volume = $\frac{\pi a^2 h}{3}$

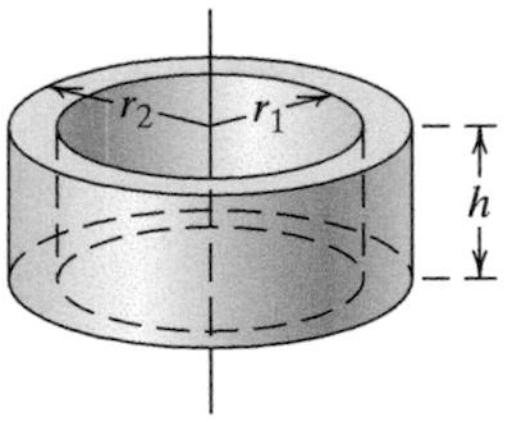

Annular Ring

Volume = $\pi(r_2^2 - r_1^2)\,h$

Inertia and Center of Mass Properties of Selected Bodies with Mass *m*

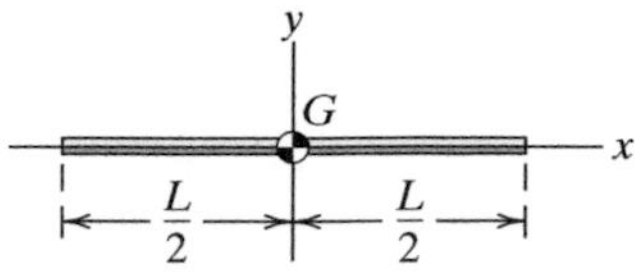

$$\bar{I}_{xx} = 0$$

$$\bar{I}_{yy} = \bar{I}_{zz} = \frac{mL^2}{12}$$

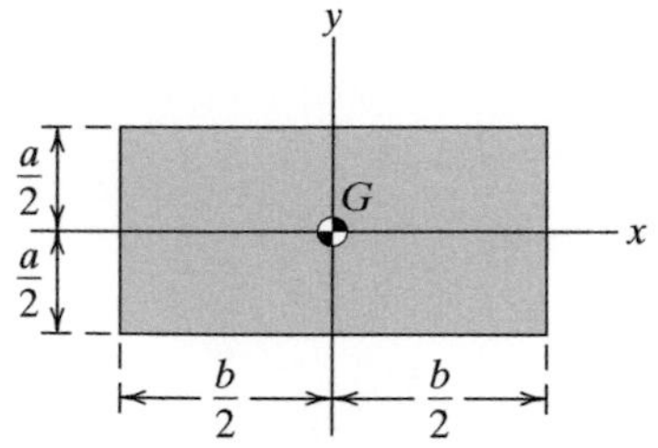

Flat Plate

$$\bar{I}_{xx} = \frac{ma^2}{12}$$

$$\bar{I}_{yy} = \frac{mb^2}{12}$$

$$\bar{I}_{zz} = \frac{m(a^2+b^2)}{12}$$

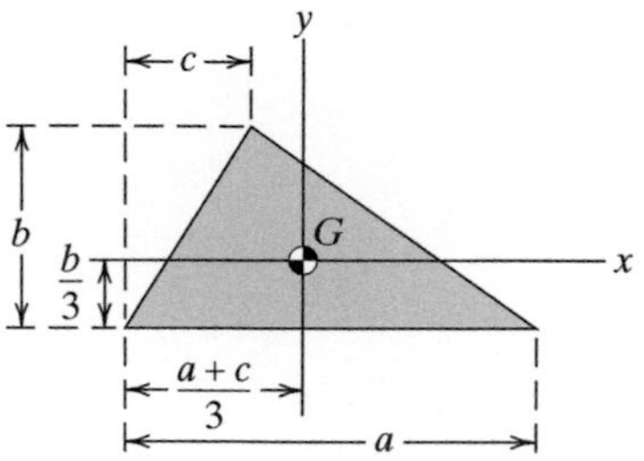

Flat Triangular Plate

$$\bar{I}_{xx} = \frac{mb^2}{18}$$

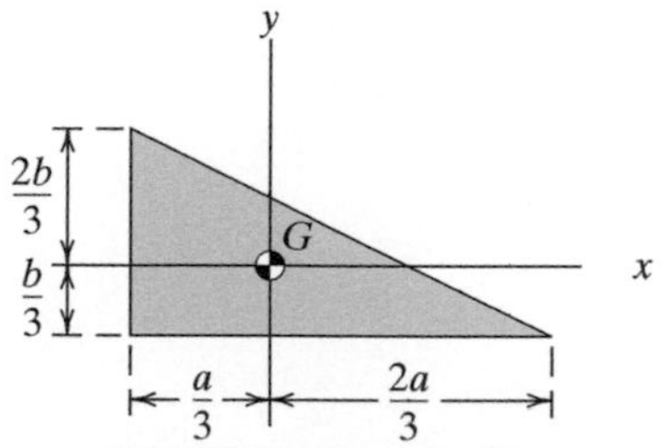

Flat, Right-triangular Plate

$$\bar{I}_{xx} = \frac{mb^2}{18}$$

$$\bar{I}_{yy} = \frac{ma^2}{18}$$

$$\bar{I}_{zz} = m\left(\frac{a^2}{18} + \frac{b^2}{18}\right)$$

$$\bar{I}_{xy} = -\frac{mab}{36}$$

$$\bar{I}_{xz} = \bar{I}_{yz} = 0$$

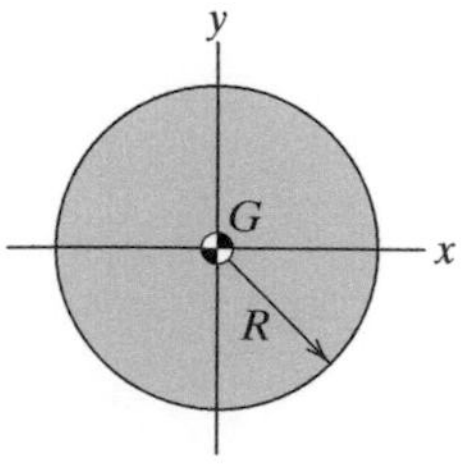

Circular Plate

$$\bar{I}_{xx} = \bar{I}_{yy} = \frac{mR^2}{4}$$

$$\bar{I}_{zz} = \frac{mR^2}{2}$$

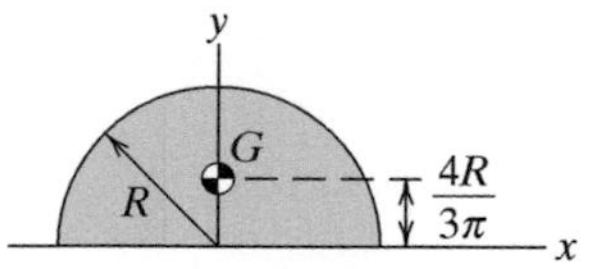

Semicircular Plate

$$I_{xx} = I_{yy} = \frac{mR^2}{4}$$

$$I_{zz} = \frac{mR^2}{2}$$

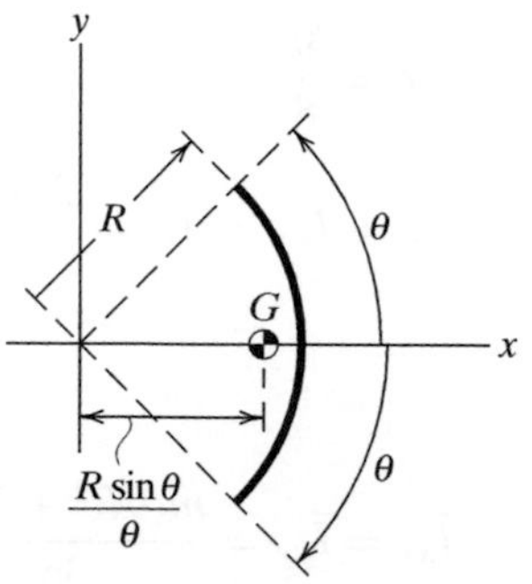

Circular Arc

$$I_{xx} = mR^2\left(\frac{1}{2} - \frac{\sin 2\theta}{4\theta}\right)$$

$$I_{yy} = mR^2\left(\frac{1}{2} + \frac{\sin 2\theta}{4\theta}\right)$$

$$I_{zz} = mR^2$$

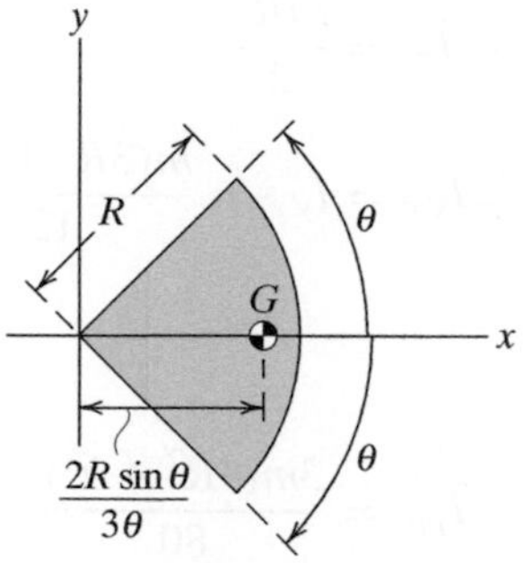

Circular Sector

$$I_{xx} = \frac{mR^2}{4}\left(1 - \frac{\sin 2\theta}{2\theta}\right)$$

$$I_{yy} = \frac{mR^2}{4}\left(1 + \frac{\sin 2\theta}{2\theta}\right)$$

$$I_{zz} = \frac{mR^2}{2}$$

$$I_{xy} = I_{yz} = I_{zx} = 0$$

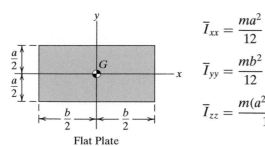

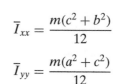

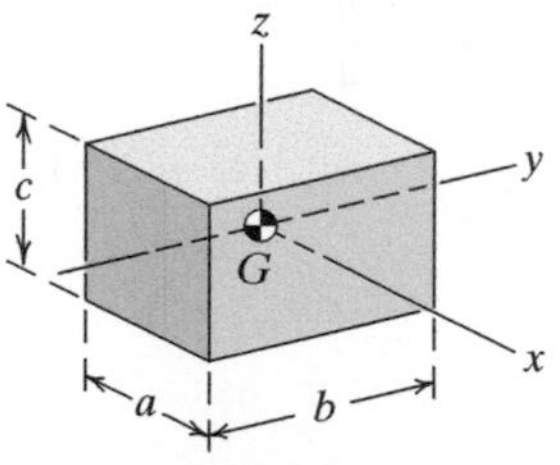

Rectangular Parallelepiped

$$\bar{I}_{xx} = \frac{m(c^2+b^2)}{12}$$

$$\bar{I}_{yy} = \frac{m(a^2+c^2)}{12}$$

$$\bar{I}_{zz} = \frac{m(a^2+b^2)}{12}$$

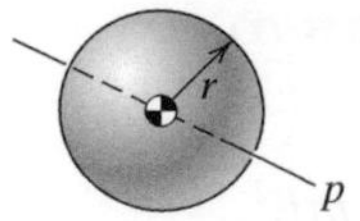

Solid Sphere

$$\overline{I}_{pp} = \frac{2mr^2}{5}$$

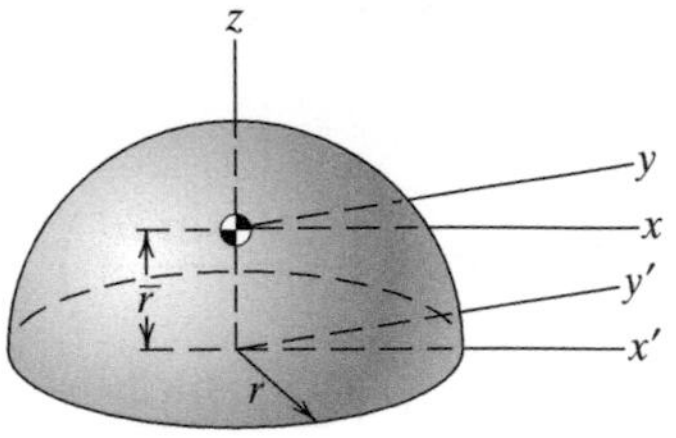

Hemisphere

$$\overline{I}_{xx} = \overline{I}_{yy} = \frac{83mr^2}{320}$$

$$\bar{r} = \frac{3r}{8}$$

$$I_{x'x'} = I_{y'y'} = I_{zz} = \frac{2mr^2}{5}$$

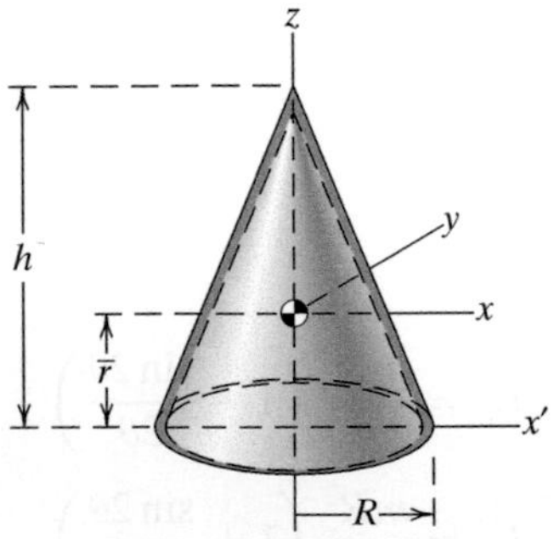

Hollow Circular Cone

$$\overline{I}_{xx} = \overline{I}_{yy} = \frac{m(9R^2 + 2h^2)}{36}$$

$$\bar{r} = \frac{h}{3}$$

$$\overline{I}_{zz} = \frac{mR^2}{2}$$

$$I_{x'x'} = I_{y'y'} = \frac{m(3R^2 + 2h^2)}{12}$$

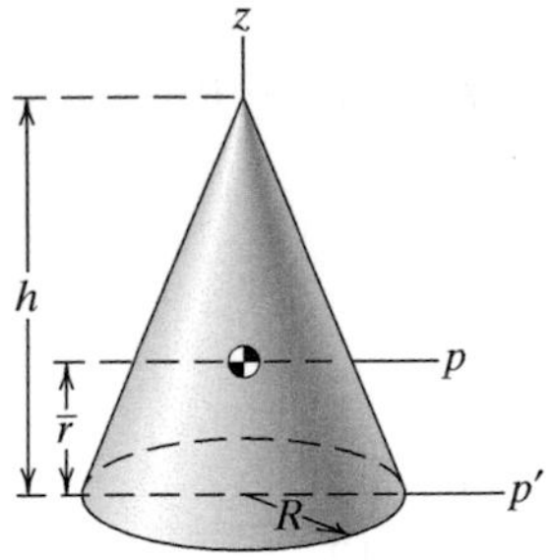

Solid Circular Cone

$$\overline{I}_{pp} = \frac{3m(4R^2 + h^2)}{80}$$

$$\overline{I}_{zz} = \frac{3mR^2}{10}$$

$$\bar{r} = \frac{h}{4}$$

$$I_{p'p'} = \frac{m(3R^2 + 2h^2)}{20}$$

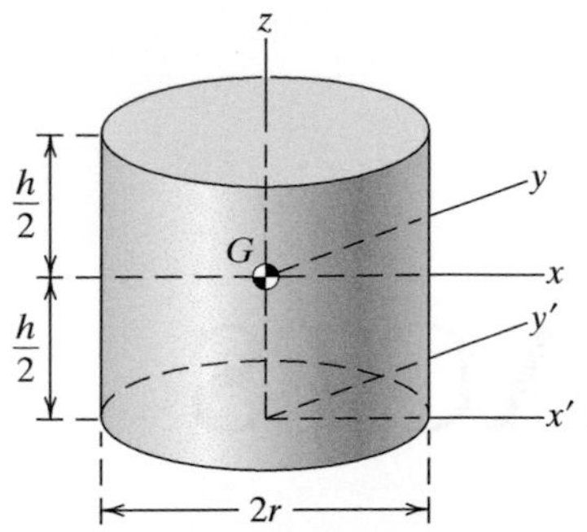

Circular Cylinder

$$\bar{I}_{xx} = \bar{I}_{yy} = \frac{m(3r^2 + h^2)}{12}$$

$$\bar{I}_{zz} = \frac{mr^2}{2}$$

$$I_{x'x'} = I_{y'y'} = \left(\frac{r^2}{4} + \frac{h^2}{3}\right)m$$

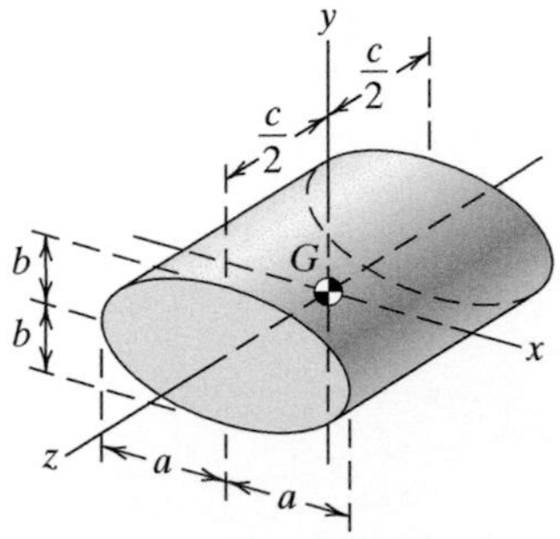

Solid Elliptical Cylinder

$$\bar{I}_{xx} = \frac{m(3b^2 + c^2)}{12}$$

$$\bar{I}_{yy} = \frac{m(3a^2 + c^2)}{12}$$

$$\bar{I}_{zz} = \frac{m(a^2 + b^2)}{4}$$

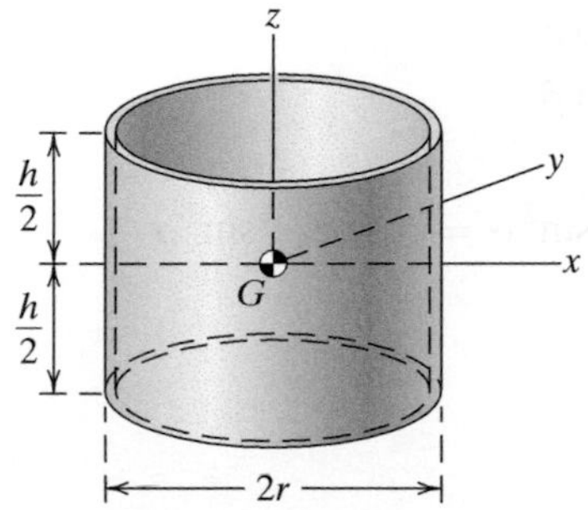

Thin Cylindrical Shell

$$\bar{I}_{xx} = \bar{I}_{yy} = \frac{mr^2}{2} + \frac{mh^2}{12}$$

$$\bar{I}_{zz} = mr^2$$

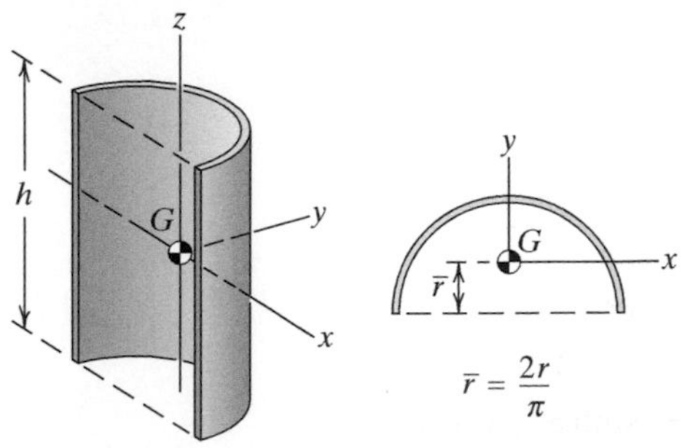

Thin, Semi-cylindrical Shell

$$\bar{I}_{xx} = mr^2\left(\frac{1}{2} - \frac{4}{\pi^2}\right) + \frac{mh^2}{12}$$

$$\bar{I}_{yy} = \frac{m(6r^2 + h^2)}{12}$$

$$\bar{I}_{zz} = mr^2\left(1 - \frac{4}{\pi^2}\right)$$

APPENDIX C

SOME USEFUL MATHEMATICAL FACTS

Trigonometric Identities (see Figure C.1)

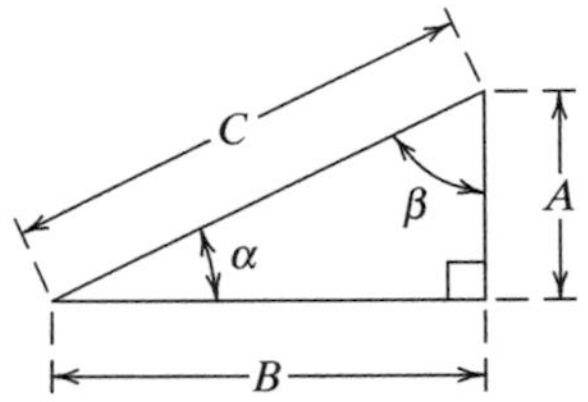

Figure C.1 Right triangle and included angles

$$
\begin{aligned}
\alpha + \beta &= 90^\circ \\
\sin\alpha &= \frac{A}{C} \\
\cos\alpha &= \frac{B}{C} \\
\tan\alpha &= \frac{A}{B} \\
A^2 + B^2 &= C^2 \\
\sin^2\alpha + \cos^2\alpha &= 1 \\
\sin(\alpha \pm \beta) &= \sin\alpha\cos\beta \pm \cos\alpha\sin\beta \\
\cos(\alpha \pm \beta) &= \cos\alpha\cos\beta \mp \sin\alpha\sin\beta \\
\sin 2\alpha &= 2\sin\alpha\cos\alpha \\
\cos 2\alpha &= 2\cos^2\alpha - 1 = 1 - 2\sin^2\alpha = \cos^2\alpha - \sin^2\alpha \\
\sin\alpha &= \frac{e^{i\alpha} - e^{-i\alpha}}{2i} \\
\cos\alpha &= \frac{e^{i\alpha} + e^{-i\alpha}}{2} \\
e^{i\alpha} &= \cos\alpha + i\sin\alpha \\
e^{-i\alpha} &= \cos\alpha - i\sin\alpha
\end{aligned}
$$

Hyberbolic Identities

$$
\begin{aligned}
\sinh\alpha &= \frac{e^{\alpha} - e^{-\alpha}}{2} \\
\cosh\alpha &= \frac{e^{\alpha} + e^{-\alpha}}{2} \\
e^{\alpha} &= \sinh\alpha + \cosh\alpha \\
e^{-\alpha} &= \cosh\alpha - \sinh\alpha
\end{aligned}
$$

Series Expansions

$$\sin x = x - \frac{x^3}{3!} + \frac{x^5}{5!} - \cdots$$

$$\cos x = 1 - \frac{x^2}{2!} + \frac{x^4}{4!} - \frac{x^5}{5!} + \cdots$$

$$\sinh x = x + \frac{x^3}{3!} + \frac{x^5}{5!} + \cdots$$

$$\cosh x = 1 + \frac{x^2}{2!} + \frac{x^4}{4!} + \cdots$$

$$e^x = 1 + x + \frac{x^2}{2!} + \frac{x^3}{3!} + \cdots$$

$$\frac{1}{1+\epsilon} = 1 - \epsilon + \epsilon^2 - \epsilon^3 + \cdots, (|\epsilon| < 1.0)$$

Derivative and Integral Relationships $\left(\frac{d(\cdot)}{dx} \equiv (\cdot)'\right)$

$$\frac{dx^n}{dx} = nx^{n-1}$$

$$\frac{dy^n}{dx} = ny^{n-1}y'$$

$$\frac{d(ab)}{dx} = ab' + ba'$$

$$\frac{d\left(\frac{a}{b}\right)}{dx} = \frac{ba' - ab'}{b^2}$$

$$\frac{d \sin x}{dx} = \cos x$$

$$\frac{d \cos x}{dx} = -\text{sinx}$$

$$\frac{d \tan x}{dx} = \frac{1}{\cos^2 x}$$

$$\frac{d \sinh x}{dx} = \cosh x$$

$$\frac{d \cosh x}{dx} = \sinh x$$

Vector Relationships

As you recall, a vector is comprised of a magnitude and a direction. **Figure C.2** shows an arbitrary vector $\boldsymbol{r}$ and its three component vectors $\boldsymbol{r}_x, \boldsymbol{r}_y, \boldsymbol{r}_z$ oriented along the $\boldsymbol{i}, \boldsymbol{j}, \boldsymbol{k}$ directions. This graphical representation can be expressed as

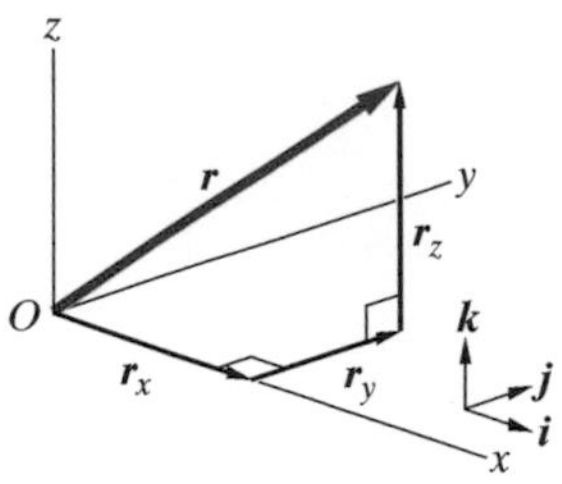

Figure C.2 Vector decomposition into components

$$\begin{aligned} \boldsymbol{r} &= \boldsymbol{r}_x + \boldsymbol{r}_y + \boldsymbol{r}_z \\ &= r_x \boldsymbol{i} + r_y \boldsymbol{j} + r_z \boldsymbol{k} \end{aligned}$$

and the relationship between the magnitudes of these vectors is given by

$$r = \|\boldsymbol{r}\| = \sqrt{r_x^2 + r_y^2 + r_z^2}$$

Constructing a unit vector $\boldsymbol{u}$ that points in the same direction as $\boldsymbol{r}$ is easily accomplished:

$$\boldsymbol{u} = \frac{\boldsymbol{r}}{\|\boldsymbol{r}\|}$$

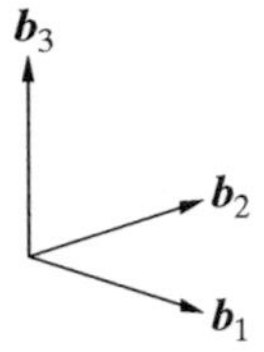

Figure C.3 Right-handed triad

Figure C.3 shows a set of mutually orthogonal unit vectors, $\boldsymbol{b}_1$, $\boldsymbol{b}_2$, $\boldsymbol{b}_3$, arranged in a **right-handed triad**. For such an arrangement the following relationships hold:

$$\begin{aligned} \boldsymbol{b}_1 \times \boldsymbol{b}_2 &= \boldsymbol{b}_3 \\ \boldsymbol{b}_2 \times \boldsymbol{b}_3 &= \boldsymbol{b}_1 \\ \boldsymbol{b}_3 \times \boldsymbol{b}_1 &= \boldsymbol{b}_2 \end{aligned}$$

Some useful cross-product relationships are:

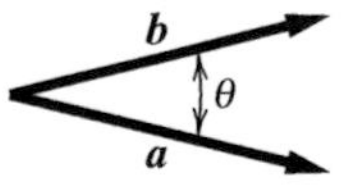

Figure C.4 Two vectors

$$\begin{aligned} \boldsymbol{a} \times \boldsymbol{a} &= 0 \\ \boldsymbol{a} \times \boldsymbol{b} &= -\boldsymbol{b} \times \boldsymbol{a} \\ \|\boldsymbol{a} \times \boldsymbol{b}\| &= ab \sin\theta \qquad \text{(see } \textbf{Figure C.4}\text{)} \\ \boldsymbol{a} \times (\boldsymbol{b} + \boldsymbol{c}) &= \boldsymbol{a} \times \boldsymbol{b} + \boldsymbol{a} \times \boldsymbol{c} \\ \alpha(\boldsymbol{a} \times \boldsymbol{b}) &= (\alpha \boldsymbol{a}) \times \boldsymbol{b} = \boldsymbol{a} \times (\alpha \boldsymbol{b}) \\ \boldsymbol{a} \times [\boldsymbol{b} \times (\boldsymbol{b} \times \boldsymbol{a})] &= 0 \end{aligned}$$

and some useful dot product relationships are:

$$\begin{aligned} \boldsymbol{a} \cdot \boldsymbol{b} &= ab \cos\theta \qquad \text{(see } \textbf{Figure C.4}\text{)} \\ \boldsymbol{a} \cdot (\boldsymbol{b} + \boldsymbol{c}) &= \boldsymbol{a} \cdot \boldsymbol{b} + \boldsymbol{a} \cdot \boldsymbol{c} \\ \alpha(\boldsymbol{a} \cdot \boldsymbol{b}) &= (\alpha \boldsymbol{a}) \cdot \boldsymbol{b} + \boldsymbol{a} \cdot (\alpha \boldsymbol{b}) \\ \boldsymbol{a} \cdot \boldsymbol{a} &= \|\boldsymbol{a}\|^2 = a^2 \end{aligned}$$

MATERIAL DENSITIES

Table D.1 Material densities

Material	Density (kg/m^3)
Air 20°C	1.29×10^0
Aluminum	2.71×10^3
Brick	2.10×10^3
Cork	2.00×10^2
Copper	8.94×10^3
Glass	2.60×10^3
Gold	1.93×10^4
Helium 0°C, 1 atm	1.79×10^{-1}
Hydrogen 0°C, 1 atm	8.99×10^{-2}
Iron, cast	7.20×10^3
Iron, wrought	7.60×10^3
Lead	1.13×10^4
Oil, olive	9.20×10^2
Rubber	1.10×10^3
Platinum	2.14×10^4
Silver	1.05×10^4
Steel	7.85×10^3
Titanium	4.54×10^3
Water, 20°C	9.98×10^2
Wood, balsa	1.20×10^2
Wood, oak	6.80×10^2
Wood, pine	5.90×10^2

Table D.2 Celestial data

Newton's gravitational constant, G	6.67×10^{-11} $m^3 \cdot kg^{-1} \cdot s^{-2}$
Distance, Earth to Sun	1.50×10^8 km
Radius, Earth	6.37×10^3 km
Mass, Earth	5.98×10^{24} kg
Distance, Earth to Moon	3.84×10^5 km
Radius, Moon	1.74×10^3 km
Mass, Moon	7.35×10^{22} kg
Radius, Mars	3.38×10^3 km
Mass, Mars	6.42×10^{23} kg
Radius, Sun	6.96×10^5 km
Mass, Sun	1.99×10^{30} kg

BIBLIOGRAPHY

1. *Auto Handbook*, Robert Bosch, GmbH, 1996.
2. Stratthern, P., *The Big Idea—Newton and Gravity*, Anchor Books, 1997.
3. Tongue, B. H., *Principles of Vibration*, Oxford University Press, 2nd ed., 2002.
4. Acton, F., *Numerical Methods That Work*, Mathematical Association of America, 1997.
5. Gardner, J. F., *Simulations of Machines Using MATLAB® and SIMULINK®*, Brooks/Cole, 2001.

Symbols
⇒, 8
→, 8

A
Absolute acceleration, 78
Acceleration
  absolute, 78, 82, 83
  angular, 62
  average, 25
  Cartesian coordinates, 19
  centripetal, 51
  constant, 20–21
  Coriolis, 50, 95
  cylindrical coordinates, 53
  forward, 207
  general, 20
  instantaneous, 19
  magnitude, 25
  normal, 65
  path coordinates, 65
  polar coordinates, 56
  position dependent, 22–26
  radial, 51
  relative, 78, 82, 83
  rigid body
    2D, 349–354
    3D, 489
  speed dependent, 22–28
  spherical coordinates, 489
  tangential, 65
  transverse, 52
Action/reaction, 2
Amplitude
  vibration, 556
Angular acceleration
  rigid body
    2D, 349
    3D, 495–496
Angular impulse, 155–160
Angular momentum
  2D, 457–462
  3D, 507, 513–519
  continuous flow, 290
  multiple particles, 273–278
  single particle, 155
Angular unit vector, 47
Angular velocity
  relative
    2D, 324
  rigid-body
    3D, 489–494
Anti-*g* straining maneuver, 484
Apogee, 176
Applied moment, 155
Areal density, 401, 402, 405
Array, 49
ASDD, 541
Atomic scale, 2

B
Ballistic trajectory, 56
Barrel roll, 541
Body of revolution, 508
Body-fixed rotation, 490
Brahe, Tycho, 165
Braking thruster, 300

C
Cantilevered balcony, 546
Cartesian coordinates, 33, 100
Cartesian unit vector, 47
Catapult, 381, 485
Center of curvature, 65
Center of mass, 259
Central body, 160
Centrifuge, 484
Centripetal acceleration, 51
Centroid, 531
Coefficient of restitution, 184, 187
Collision, 144, 183
Compression phase, 187
Conic section, 166
Conservation
  angular momentum, 164
  energy, 224, 471
  linear momentum, 147, 186, 263
  rope, 79, 80
Conservative force, 220
Constant acceleration, 20
Constant coefficient, 543
Constrained link, 321
Constraints, 76–86
Coordinate system
  body fixed, 34
  ground fixed, 34
Coordinate transformation
  2D, 34–39
  3D, 488
Coordinate transformation array, 43, 48, 353, 534
Coordinates
  Cartesian, 33–47
  cylindrical, 52–57
  normal/tangential, 64–71
  path, 64–76
  polar, 47–56
  spherical, 488
Coriolis acceleration, 50
Crankshaft, 400
Critically damped response, 566
Cross product, 274
Cross-product, 594
Curvilinear translation, 383–391
Cylindrical coordinates, 52

D
Damped frequency of oscillation, 566
Damping
  linear, 563
  viscous, 563
Damping factor, 564
Deformation
  elastic, 187
  plastic, 187
Deformation impulse, 185
Deformation phase, 185
Degree of freedom, 78
Degrees, 50
Derivative relationships, 593
Descartes, René, 33
Diagram
  Free Body, 102
  Inertial Response, 102
Differential, 13
Differential mass element, 397
Direct drive, 442
Directrix, 169
Displacement, 18
DOF, 78
Dot notation, 8
Dot product, 594
Dynamical potato, 259

E

Eccentricity, 170
Efficiency, 245–246
Ejection seat, 97
Elastic deformation, 187
Electrodynamic shaker, 556
Energy, 14, 209
  kinetic, 14, 210–214
  potential, 222–230
    gravity, 223, 226
    spring, 224, 226
Energy dissipation, 563
Equations of motion, 14
  3D, 540
  multiple particles, 261
  planar, 389
  single particle, 104, 163
Equilibrium angle, 546
Equilibrium position, 440, 543
Euclidean norm, 9
Euler angle, 490
Euler integration, 580
Euler's equations, 522
Exact differentials, 21
Expansion phase, 187
Extensible, 500
External forces, 260

F

FBD, 5, 102
FBD=IRD, 5, 386
Fixed reference frame, 34
Font
  italic, 8
  roman, 7
Force
  average, 148, 153
  conservative, 220
  Coriolis, 51
  external, 260
  internal, 186, 260
  magnitude, 556
  newton, 6
  radial, 164
Force balance, 258
  multiple bodies, 259
  multiple particles, 260, 261
  rigid body, 521
    planar motion, 383, 384
  single particle, 100
Forced vibration, 555–559, 569–573
Forcing
  natural, 556
Fourier series, 555
Free vibration, 544, 548
Free-body diagram, 5, 14, 102
Free-body diagram =
    Inertial-response diagram, 5
Frequency
  damped, 566
  forcing, 556
  natural, 544
Friction
  dynamic coefficient, 129, 140
  static coefficient, 108, 124, 125

G

Gear ratios, 443
Gearbox, 13
Geared drive, 442
General motion
  2D, 422–443
  3D, 497–501
Gimbal, 491
GLOC, 207, 484
Grade, 390
Gravitation, 11
Gravitational potential energy, 226
Gravity, 6, 11
Ground reaction, 436
Gyroscope, 491, 496

H

Half-car model, 440
Homogeneous equation, 543
Horsepower, 244
Human-scale, 2
Hyperbolic identities, 592

I

Imbalance eccentricity, 410
Impact
  direct, 183–190
  oblique, 193–199
Impulse
  angular, 155–160
  deformation, 185
  linear, 143–147
  restoration, 185
In-phase, 557
Independent variable, 27
Inertia
  matrix, 515
  moment of, 507
  parallel axis theorem, 508–510
  product of, 507
  tensor, 515
Inertia properties, 589–591
Inertial reference frame, 100
inertial response, 100
Inertial response diagram, 5, 14, 102
Inner rotor, 491
Input
  periodic, 555
  sinusoidal, 555
Instantaneous acceleration, 19
Integral relationships, 593
Integration
  Euler, 580
  numerical, 580
Internal forces, 186, 260
International System, 6
Interpolation, 581
IRD, 5, 102

J

Jerk, 381

K

Kepler, Johannes, 165
Kilogram, 6
Kilowatts, 244
Kinematics, 14, 316
  cylindrical coordinates, 52–57
  path coordinates, 64–76
  polar coordinates, 47–56
  straight-line motion, 17–28
  variable geometry pulleys, 97
Kinetic energy, 210–214
  mass center, 282
  multiple particles, 282
  rigid body
    general motion, 532
    planar motion, 469–475
  rotational, 284
  translational, 284
Knot, 208

L

Linear damper, 563
Linear damping, 563
Linear impulse, 143–147
Linear momentum, 143
  continuous flow, 288–304
  multiple particles, 262–267
  rigid body
    planar motion, 457–461
Linear spring, 543
Loudspeaker, 556
Lumped mass, 399
Lumped-mass momentum analysis,
    299

M

M-file, 389, 582
Magnitude, 9

Mass
continuous stream, 288
continuum, 397
kilogram, 6
lumped, 399
slug, 7
vibrational response, 556
Mass center, 14, 259
Mass inflow, 288–304
Mass moment of inertia, 399, 507
Mass outflow, 288–304
Mass unit, 6
Mass-normalized, 544
MATLAB, 389, 559, 582
Matrix formulation, 585
Meter, 6
Moment balance, 156, 258
continuous flow, 290
mass flow, 290
multiple particles, 273–276
rigid body, 522
general motion, 521–525
planar motion, 396–457
rigid body translation
planar motion, 384
Moment of inertia, 507
Momentum
angular, 155
linear, 143
Motion on a rigid body, 361–369
Moving reference frame, 34, 76–85
MSDD, 313
Multi-body systems, 258

N

Nano-engineering, 2
Natural frequency, 544
newton, 6
Newton's Law of gravitation, 12
Newton's laws
first, 2
second, 2, 6
third, 2
Newton, Isaac, 1
Nonconservative force, 224
Nonconstant mass systems, 299–304
Nonlinearity, 543
Nonzero net force, 2
Normal acceleration, 65
Normal unit vector, 65
Numerical integration, 580
Numerical methods, 580
Numerical precision convention, 12
Nutation, 490

O

Oblique impact, 193–199
Occipital condyles, 380
ODE, 543, 582
ode45, 431, 583
Off-axis collision, 193
One-body problem, 163
Orbital mechanics, 163
Orbiting body, 164
Ordinary differential equation, 543, 582
Orthogonal components, 317
Orthogonal vectors, 33
Oscillation amplitude, 558
Osculate, 168
Out-of-phase, 557
Out-of-plane axis, 95
Overdamped response, 566

P

Parabolic trajectory, 41
Parallel axis theorem, 404
Path coordinates, 64
Path independent work, 221
Pendulum, 224
Perigee, 176
Perpetual motion, 563
Phase lag, 574
Phase shift, 569
Pitch-up, 207
Planar motion, 382
curvilinear translation, 384–391
general, 422–443
instantaneous center of rotation, 333
linear/angular momentum, 457–461
motion on a rigid body, 361–369
relative velocity, 315–324
rotating reference frame, 346–354
rotation about a fixed point, 396–409
work/energy, 468–475
Plastic deformation, 187
Plot
acceleration-time, 18, 19
displacement-time, 18
speed-time, 18, 19
Polar coordinates, 47
Position, 18
Cartesian coordinates, 34
cylindrical coordinates, 52
polar coordinates, 47
spherical coordinates, 488
Position vector, 34, 48
Potential energy, 221–230
Potential function, 222
Pound, 7
Power, 243–248
constant, 245
Precession, 490
Principal inertia matrix, 516
Product of inertia, 507
Product rule, 274
Projectile motion, 41
Prolate spheroid, 528
Pulley
compound, 86
double, 86
simple, 85
variable geometry, 206
Push-pull maneuver, 207
Pythagorean theorem, 592

Q

Quad function, 216

R

Radial acceleration, 51
Radial unit vector, 47
Radial velocity, 49
Radians, 50
Radius of curvature, 65, 66
Radius of gyration, 399
Random excitation, 555
Reaction force, 2, 105, 524
Rectangular coordinates, 33
Rectilinear trajectory, 40
Reference frame
body-fixed, 34
fixed, 34
ground-fixed, 34
moving, 34
rotating, 346–354
Relative acceleration, 78
Relative angular velocity, 324
Relative motion, 76–86
Relative position vector, 77
Relative velocity, 318
Resonance, 557
Response
critically damped, 566
overdamped, 566
underdamped, 566
Restoration impulse, 185
Restoration phase, 185
Right-hand rule, 101
Right-handed triad, 594
Rigid-body, 314
Rigid-body constraints, 316
Rigid-body motion
general, 422–443

general planar, 316
rotation, 315
translation, 316
Robotic manipulator, 500
Roll
with slip, 433
without slip, 432
Rotating reference frame, 346–354
differentiation, 348
Rotation
about a fixed axis, 397
about a fixed point, 396
Rotation rate, 57
Rotational inertia, 399
Rotational motion, 383
Rotational symmetry, 523
Runge-Kutta, 583

S

Safety factor, 558
Scalar quantity, 8
Second, 6
Secular solutions, 559
Seismic forcing, 572
Semi-static, 2
Series expansions, 593
Settling time, 567
Shock absorber, 563
Shock loading, 555
SI, 6
Signed magnitude, 9, 18
Significant figures, 12
SIMULINK, 586
Skid pad, 69
Sliding constraint, 429
Slug, 7
Small-angle approximation, 547
Solution
homogeneous, 556
particular, 556
Solution procedure, 3
Somatogravic illusion, 207
Speed, 8, 18
angular, 49
Spherical coordinates, 488
Spin, 491
Spin response
stable, 490
unstable, 490
Spring potential energy, 226
Spring-mass system, 167, 556
State variables, 583
Static equilibrium, 441, 543, 547
Stationary enclosures with mass inflow/outflow, 288
Straight line motion, 383
Suspensions, 15
Sweet spot, 418
Swept area, 165

T

TACM, 484
Tangential acceleration, 65
Terminal speed, 28
Thrust, 102
Time shift, 570
Time step, 581
Torque, 155
Torque-driven wheel, 435
Torsional spring, 546
total applied forces, 100
Transformation array, 352, 492
Transient loading, 555
Translation rate
vertical, 57
Transverse acceleration, 52
Transverse velocity, 49
Trebuchet, 381
Trigonometric identities, 592

U

U.S. Customary System, 6
Underdamped response, 566
Unit vector
angular, 47
Cartesian, 33, 47
ground-fixed, 34
path, 65
polar, 47
radial, 47
Units, 6
USC, 6

V

Vasodilator response, 207
VDT, 380
Vector
direction, 594
magnitude, 594
momentum, 143
position, 34
quantity, 8
representation, 8
time rate of change, 48
Vector relationships, 594
Velocity, 8
absolute, 82
average, 18
cylindrical coordinates, 53
definition, 48
instantaneous, 18
path coordinates, 64–65
polar coordinates, 49
radial, 49, 51
relative, 82, 318
rigid body
2D, 349–354
3D, 489
spherical coordinates, 489
transverse, 49
Vertical drop tower, 380
Vestibular system, 207
Vibration
damped
forced response, 569–573
free response, 563–567
steady-state, 556
torsional, 546
undamped
forced response, 555–559
free response, 543–547
Vibrations, 15
Viscous damper, 563, 577
Volumetric flow rate, 289, 303
Vortex shedding, 558

W

Watt, 244
Weight, 6
Work, 14, 210
conservative, 222
multiple particles, 282–286
nonconservative, 222

Z

Zero gravitational potential, 226

Printed and bound by CPI Group (UK) Ltd, Croydon, CR0 4YY
20/10/2024
14576724-0001